Y0-CAL-404

Finite Fields

GIAN-CARLO ROTA, *Editor*

ENCYCLOPEDIA OF MATHEMATICS AND ITS APPLICATIONS

Volume		Section
1	LUIS A. SANTALÓ **Integral Geometry and Geometric Probability,** 1976 (2nd printing, with revisions, 1979)	Probability
2	GEORGE E. ANDREWS **The Theory of Partitions,** 1976 (2nd printing, 1981)	Number Theory
3	ROBERT J. McELIECE **The Theory of Information and Coding** A Mathematical Framework for Communication, 1977 (2nd printing, with revisions, 1979)	Probability
4	WILLARD MILLER, Jr. **Symmetry and Separation of Variables,** 1977	Special Functions
5	DAVID RUELLE **Thermodynamic Formalism** The Mathematical Structures of Classical Equilibrium Statistical Mechanics, 1978	Statistical Mechanics
6	HENRYK MINC **Permanents,** 1978	Linear Algebra
7	FRED S. ROBERTS **Measurement Theory** with Applications to Decisionmaking, Utility, and the Social Sciences, 1979	Mathematics and the Social Sciences
8	L. C. BIEDENHARN and J. D. LOUCK **Angular Momentum in Quantum Physics:** Theory and Application, 1981	Mathematics of Physics
9	L. C. BIEDENHARN and J. D. LOUCK **The Racah-Wigner Algebra in Quantum Theory**, 1981	Mathematics of Physics

GIAN-CARLO ROTA, *Editor*
ENCYCLOPEDIA OF MATHEMATICS AND ITS APPLICATIONS

Volume		Section
10	JOHN D. DOLLARD and CHARLES N. FRIEDMAN **Product Integration** with Application to Differential Equations, 1979	Analysis
11	WILLIAM B. JONES and W. J. THRON **Continued Fractions:** Analytic Theory and Applications, 1980	Analysis
12	NATHANIEL F. G. MARTIN and JAMES W. ENGLAND **Mathematical Theory of Entropy,** 1981	Real Variables
13	GEORGE A. BAKER, Jr. and PETER R. GRAVES-MORRIS **Padé Approximants, Part I** **Basic Theory,** 1981	Mathematics of Physics
14	GEORGE A. BAKER, Jr. and PETER R. GRAVES-MORRIS **Padé Approximants, Part II:** **Extensions and Applications,** 1981	Mathematics of Physics
15	E. C. BELTRAMETTI and G. CASSINELLI **The Logic of Quantum Mechanics,** 1981	Mathematics of Physics
16	G. D. JAMES and A. KERBER **The Representation Theory of the Symmetric Group,** 1981	Algebra
17	M. LOTHAIRE **Combinatorics on Words,** 1982	Algebra

GIAN-CARLO ROTA, *Editor*
ENCYCLOPEDIA OF MATHEMATICS AND ITS APPLICATIONS

Volume		Section
18	H. O. FATTORINI **The Abstract Cauchy Problem,** 1983	Analysis
19	G. G. LORENTZ, K. JETTER, and S. D. RIEMENSCHNEIDER **Birkhoff Interpolation,** 1983	Interpolation and Approximation
20	RUDOLF LIDL and HARALD NIEDERREITER **Finite Fields,** 1983	Algebra
21	WILLIAM T. TUTTE **Graph Theory,** 1984	Graph Theory and Combinatorics
22	JULIO R. BASTIDA **Field Extensions and** **Galois Theory,** 1984	Algebra

Other volumes in preparation

ENCYCLOPEDIA OF MATHEMATICS and Its Applications

GIAN-CARLO ROTA, Editor
Department of Mathematics
Massachusetts Institute of Technology
Cambridge, Massachusetts

GIAN-CARLO ROTA, *Editor*

ENCYCLOPEDIA OF MATHEMATICS AND ITS APPLICATIONS

Volume 20

Section: Algebra
P. M. Cohn and Roger Lyndon, *Section Editors*

Finite Fields

Rudolf Lidl
University of Tasmania
Hobart, Australia

Harald Niederreiter
Austrian Academy of Sciences
Vienna, Austria

Foreword by
P. M. Cohn
University of London
London, England

1983

Addison-Wesley Publishing Company
Advanced Book Program/World Science Division
Reading, Massachusetts

London • Amsterdam • Don Mills, Ontario • Sydney • Tokyo

Library of Congress Cataloging in Publication Data

Lidl, Rudolf
Finite fields.

(Encyclopedia of mathematics and its applications; v. 20. Section, Algebra)
Bibliography: p.
Includes indexes.
1. Finite fields (Algebra) I. Niederreiter, Harald, 1944– II. Title. III. Series: Encyclopedia of mathematics and its applications; v. 20. IV. Series: Encyclopedia of mathematics and its applications. Section, Algebra.
QA247.3.L53 1983 512′.32 83-2756
ISBN 0-201-13519-1

American Mathematical Society (MOS) Subject Classification Scheme (1980): 12CXX, 10G05, 05BXX, 51EXX, 62K10, 94BXX, 94CXX

Published simultaneously in Canada

Manufactured in the United States of America

ABCDEFGHIJ-HA-89876543

To Pamela and Gerlinde

Contents

Editor's Statement

A large body of mathematics consists of facts that can be presented and described much like any other natural phenomenon. These facts, at times explicitly brought out as theorems, at other times concealed within a proof, make up most of the applications of mathematics, and are the most likely to survive change of style and of interest.

This ENCYCLOPEDIA will attempt to present the factual body of all mathematics. Clarity of exposition, accessibility to the non-specialist, and a thorough bibliography are required of each author. Volumes will appear in no particular order, but will be organized into sections, each one comprising a recognizable branch of present-day mathematics. Numbers of volumes and sections will be reconsidered as times and needs change.

It is hoped that this enterprise will make mathematics more widely used where it is needed, and more accessible in fields in which it can be applied but where it has not yet penetrated because of insufficient information.

GIAN-CARLO ROTA

Foreword

Most modern algebra texts devote a few pages (but no more) to finite fields. So at first it may come as a surprise to see an entire book on the subject, and even more for it to appear in the *Encyclopedia of Mathematics and Its Applications*. But the reader of this book will find that the authors performed the very timely task of drawing together the different threads of development that have emanated from the subject. Foremost among these developments is the rapid growth of coding theory which already has been treated in R. J. McEliece's volume in this series. The present volume deals with coding theory in the wider context of polynomial theory over finite fields, and also establishes the connection with linear recurring series and shift registers.

On the pure side there is a good deal of number theory that is most naturally expressed in terms of finite fields. Much of this—for example, equations over finite fields and exponential sums—can serve as a paradigm for the more general case; and the authors have gone as far in their treatment as is reasonable, using elementary algebraic methods only. As a result the book can also serve as an introduction to these topics.

But finite fields also have properties that are not shared with other types of algebra; thus they (like finite Boolean algebras) are functionally complete. This means that every mapping of a finite field can be expressed as a polynomial. While the proof is not hard (it is an immediate consequence of the Lagrange interpolation formula), practical questions arise when we try to find polynomials effecting permutations. Such permutation polynomials are useful in several contexts, and methods of obtaining them are discussed here. True to its nature as a handbook of applications, this volume also gives various algorithms for factorizing polynomials (over both large and small finite fields).

The lengthy notes at the end of each chapter contain interesting historical perspectives, and the comprehensive bibliography helps to make this volume truly the handbook of finite fields.

P. M. COHN

Preface

The theory of finite fields is a branch of modern algebra that has come to the fore in the last 50 years because of its diverse applications in combinatorics, coding theory, and the mathematical study of switching circuits, among others. The origins of the subject reach back into the 17th and 18th century, with such eminent mathematicians as Pierre de Fermat (1601–1665), Leonhard Euler (1707–1783), Joseph-Louis Lagrange (1736–1813), and Adrien-Marie Legendre (1752–1833) contributing to the structure theory of special finite fields—namely, the so-called finite prime fields. The general theory of finite fields may be said to begin with the work of Carl Friedrich Gauss (1777–1855) and Evariste Galois (1811–1832), but it only became of interest for applied mathematicians in recent decades with the emergence of discrete mathematics as a serious discipline.

In this book, which is the first one devoted entirely to finite fields, we have aimed at presenting both the classical and the applications-oriented aspect of the subject. Thus, in addition to what has to be considered the essential core of the theory, the reader will find results and techniques that are of importance mainly because of their use in applications. Because of the vastness of the subject, limitations had to be imposed on the choice of material. In trying to make the book as self-contained as possible, we have refrained from discussing results or methods that belong properly to algebraic geometry or to the theory of algebraic function fields. Applications are described to the extent to which this can be done without too much digression. The only noteworthy prerequisite for the book is a background in linear algebra, on the level of a first course on this topic. A rudimentary knowledge of analysis is needed in a few passages. Prior exposure to abstract algebra is certainly helpful, although all the necessary information is summarized in Chapter 1.

Chapter 2 is basic for the rest of the book as it contains the general structure theory of finite fields as well as the discussion of concepts that are used throughout the book. Chapter 3 on the theory of polynomials and Chapter 4 on factorization algorithms for polynomials are closely linked and should best be studied together. A similar unit is formed by Chapters 5 and 6. Chapters 7 and 8 can be read independently of each other and depend mostly on Chapters 2 and 3. The applications presented in Chapter 9 draw on various material in the previous chapters. Chapter 10 supplements parts of Chapters 2 and 3.

Each chapter starts with a brief description of its contents, hence it should not be necessary to give a synopsis of the book here. As this volume is part of an encyclopedic series, we have attempted to provide as much information as possible in a limited space, which meant, in particular, the omission of a few cumbersome proofs. Bibliographical references have been relegated to the notes at the end of each chapter so as not to clutter the main text. These notes also provide the researcher in the field with a survey of the literature and a summary of further results. The bibliography at the end of the volume collects all the references given in the notes.

In order to enhance the attractiveness of this monograph as a textbook, we have inserted worked-out examples at appropriate points in the text and included lists of exercises for Chapters 1–9. These exercises range from routine problems to alternative proofs of key theorems, but contain also material going beyond what is covered in the text.

With regard to cross-references, we have numbered all items in the main text consecutively by chapters, regardless of whether they are definitions, theorems, examples, and so on. Thus, "Definition 2.41" refers to item 41 in Chapter 2 (which happens to be a definition) and "Remark 6.28" refers to item 28 in Chapter 6 (which happens to be a remark). In the same vein, "Exercise 5.31" refers to the list of exercises in Chapter 5.

It is with great pleasure that we express our gratitude to Professor Gian-Carlo Rota for inviting us to write this book and for his patience in waiting for the result of our effort. We gratefully acknowledge the help of Mrs. Melanie Barton, who typed the manuscript with great care and efficiency. The staff of Addison-Wesley deserves our thanks for its professionalism in the production of the book.

R. Lidl
H. Niederreiter

Chapter 1

Algebraic Foundations

This introductory chapter contains a survey of some basic algebraic concepts that will be employed throughout the book. Elementary algebra uses the operations of arithmetic such as addition and multiplication, but replaces particular numbers by symbols and thereby obtains formulas that, by substitution, provide solutions to specific numerical problems. In modern algebra the level of abstraction is raised further: instead of dealing with the familiar operations on real numbers, one treats general operations —processes of combining two or more elements to yield another element—in general sets. The aim is to study the common properties of all systems consisting of sets on which are defined a fixed number of operations interrelated in some definite way—for instance, sets with two binary operations behaving like $+$ and $\cdot$ for the real numbers.

Only the most fundamental definitions and properties of algebraic systems—that is, of sets together with one or more operations on the set—will be introduced, and the theory will be discussed only to the extent needed for our special purposes in the study of finite fields later on. We state some standard results without proof. With regard to sets we adopt the naive standpoint. We use the following sets of numbers: the set $\mathbb{N}$ of natural numbers, the set $\mathbb{Z}$ of integers, the set $\mathbb{Q}$ of rational numbers, the set $\mathbb{R}$ of real numbers, and the set $\mathbb{C}$ of complex numbers.

1. GROUPS

In the set of all integers the two operations addition and multiplication are well known. We can generalize the concept of operation to arbitrary sets. Let S be a set and let $S \times S$ denote the set of all ordered pairs (s, t) with $s \in S$, $t \in S$. Then a mapping from $S \times S$ into S will be called a *(binary) operation* on S. Under this definition we require that the image of $(s, t) \in S \times S$ must be in S; this is the *closure property* of an operation. By an *algebraic structure* or *algebraic system* we mean a set S together with one or more operations on S.

In elementary arithmetic we are provided with two operations, addition and multiplication, that have associativity as one of their most important properties. Of the various possible algebraic systems having a single associative operation, the type known as a group has been by far the most extensively studied and developed. The theory of groups is one of the oldest parts of abstract algebra as well as one particularly rich in applications.

1.1. Definition. A *group* is a set G together with a binary operation $*$ on G such that the following three properties hold:

1. $*$ is *associative*; that is, for any $a, b, c \in G$,
$$a * (b * c) = (a * b) * c.$$
2. There is an *identity* (or *unity*) *element* e in G such that for all $a \in G$,
$$a * e = e * a = a.$$
3. For each $a \in G$, there exists an *inverse element* $a^{-1} \in G$ such that
$$a * a^{-1} = a^{-1} * a = e.$$

If the group also satisfies

4. For all $a, b \in G$,
$$a * b = b * a,$$

then the group is called *abelian* (or *commutative*).

It is easily shown that the identity element e and the inverse element a^{-1} of a given element $a \in G$ are uniquely determined by the properties above. Furthermore, $(a * b)^{-1} = b^{-1} * a^{-1}$ for all $a, b \in G$. For simplicity, we shall frequently use the notation of ordinary multiplication to designate the operation in the group, writing simply ab instead of $a * b$. But it must be emphasized that by doing so we do not assume that the operation actually is ordinary multiplication. Sometimes it is also convenient to write $a + b$ instead of $a * b$ and $-a$ instead of a^{-1}, but this additive notation is usually reserved for abelian groups.

The associative law guarantees that expressions such as $a_1 a_2 \cdots a_n$ with $a_j \in G$, $1 \leqslant j \leqslant n$, are unambiguous, since no matter how we insert parentheses, the expression will always represent the same element of G. To indicate the n-fold composite of an element $a \in G$ with itself, where $n \in \mathbb{N}$, we shall write

$$a^n = aa \cdots a \qquad (n \text{ factors } a)$$

if using multiplicative notation, and we call a^n the nth power of a. If using additive notation for the operation $*$ on G, we write

$$na = a + a + \cdots + a \qquad (n \text{ summands } a).$$

Following customary notation, we have the following rules:

Multiplicative Notation	*Additive Notation*
$a^{-n} = (a^{-1})^n$	$(-n)a = n(-a)$
$a^n a^m = a^{n+m}$	$na + ma = (n+m)a$
$(a^n)^m = a^{nm}$	$m(na) = (mn)a$

For $n = 0 \in \mathbb{Z}$, one adopts the convention $a^0 = e$ in the multiplicative notation and $0a = 0$ in the additive notation, where the last "zero" represents the identity element of G.

1.2. Examples

(i) Let G be the set of integers with the operation of addition. The ordinary sum of two integers is a unique integer and the associativity is a familiar fact. The identity element is 0 (zero), and the inverse of an integer a is the integer $-a$. We denote this group by $\mathbb{Z}$.

(ii) The set consisting of a single element e, with the operation $*$ defined by $e * e = e$, forms a group.

(iii) Let G be the set of remainders of all the integers on division by 6—that is, $G = \{0, 1, 2, 3, 4, 5\}$—and let $a * b$ be the remainder on division by 6 of the ordinary sum of a and b. The existence of an identity element and of inverses is again obvious. In this case, it requires some computation to establish the associativity of $*$. This group can be readily generalized by replacing the integer 6 by any positive integer n. □

These examples lead to an interesting class of groups in which every element is a power of some fixed element of the group. If the group operation is written as addition, we refer to "multiple" instead of "power" of an element.

1.3. Definition. A multiplicative group G is said to be *cyclic* if there is an element $a \in G$ such that for any $b \in G$ there is some integer j with $b = a^j$.

Such an element a is called a *generator* of the cyclic group, and we write $G = \langle a \rangle$.

It follows at once from the definition that every cyclic group is commutative. We also note that a cyclic group may very well have more than one element that is a generator of the group. For instance, in the additive group $\mathbb{Z}$ both 1 and -1 are generators.

With regard to the "additive" group of remainders of the integers on division by n, the generalization of Example 1.2(iii), we find that the type of operation used there leads to an equivalence relation on the set of integers. In general, a subset R of $S \times S$ is called an *equivalence relation* on a set S if it has the following three properties:

(a) $(s, s) \in R$ for all $s \in S$ (*reflexivity*).
(b) If $(s, t) \in R$, then $(t, s) \in R$ (*symmetry*).
(c) If $(s, t), (t, u) \in R$, then $(s, u) \in R$ (*transitivity*).

The most obvious example of an equivalence relation is that of equality. It is an important fact that an equivalence relation R on a set S induces a partition of S—that is, a representation of S as the union of nonempty, mutually disjoint subsets of S. If we collect all elements of S equivalent to a fixed $s \in S$, we obtain the *equivalence class* of s, denoted by

$$[s] = \{t \in S : (s, t) \in R\}.$$

The collection of all distinct equivalence classes forms then the desired partition of S. We note that $[s] = [t]$ precisely if $(s, t) \in R$. Example 1.2(iii) suggests the following concept.

1.4. Definition. For arbitrary integers a, b and a positive integer n, we say that a is *congruent* to b modulo n, and write $a \equiv b \bmod n$, if the difference $a - b$ is a multiple of n—that is, if $a = b + kn$ for some integer k.

It is easily verified that "congruence modulo n" is an equivalence relation on the set $\mathbb{Z}$ of integers. The relation is obviously reflexive and symmetric. The transitivity also follows easily: if $a = b + kn$ and $b = c + ln$ for some integers k and l, then $a = c + (k + l)n$, so that $a \equiv b \bmod n$ and $b \equiv c \bmod n$ together imply $a \equiv c \bmod n$.

Consider now the equivalence classes into which the relation of congruence modulo n partitions the set $\mathbb{Z}$. These will be the sets

$$\begin{aligned}
[0] &= \{\ldots, -2n, -n, 0, n, 2n, \ldots\},\\
[1] &= \{\ldots, -2n+1, -n+1, 1, n+1, 2n+1, \ldots\},\\
&\vdots\\
[n-1] &= \{\ldots, -n-1, -1, n-1, 2n-1, 3n-1, \ldots\}.
\end{aligned}$$

We may define on the set $\{[0], [1], \ldots, [n-1]\}$ of equivalence classes a binary operation (which we shall again write as +, although it is certainly not ordinary addition) by

$$[a]+[b]=[a+b], \tag{1.1}$$

where a and b are any elements of the respective sets $[a]$ and $[b]$ and the sum $a+b$ on the right is the ordinary sum of a and b. In order to show that we have actually defined an operation—that is, that this operation is well defined—we must verify that the image element of the pair $([a],[b])$ is uniquely determined by $[a]$ and $[b]$ alone and does not depend in any way on the representatives a and b. We leave this proof as an exercise. Associativity of the operation in (1.1) follows from the associativity of ordinary addition. The identity element is [0] and the inverse of $[a]$ is $[-a]$. Thus the elements of the set $\{[0],[1],\ldots,[n-1]\}$ form a group.

1.5. Definition. The group formed by the set $\{[0],[1],\ldots,[n-1]\}$ of equivalence classes modulo n with the operation (1.1) is called the *group of integers modulo n* and denoted by $\mathbb{Z}_n$.

$\mathbb{Z}_n$ is actually a cyclic group with the equivalence class [1] as a generator, and it is a group of order n according to the following definition.

1.6. Definition. A group is called *finite* (resp. *infinite*) if it contains finitely (resp. infinitely) many elements. The number of elements in a finite group is called its *order*. We shall write $|G|$ for the order of the finite group G.

There is a convenient way of presenting a finite group. A table displaying the group operation, nowadays referred to as a *Cayley table*, is constructed by indexing the rows and the columns of the table by the group elements. The element appearing in the row indexed by a and the column indexed by b is then taken to be ab.

1.7. Example. The Cayley table for the group $\mathbb{Z}_6$ is:

+	[0]	[1]	[2]	[3]	[4]	[5]
[0]	[0]	[1]	[2]	[3]	[4]	[5]
[1]	[1]	[2]	[3]	[4]	[5]	[0]
[2]	[2]	[3]	[4]	[5]	[0]	[1]
[3]	[3]	[4]	[5]	[0]	[1]	[2]
[4]	[4]	[5]	[0]	[1]	[2]	[3]
[5]	[5]	[0]	[1]	[2]	[3]	[4]

□

A group G contains certain subsets that form groups in their own right under the operation of G. For instance, the subset $\{[0],[2],[4]\}$ of $\mathbb{Z}_6$ is easily seen to have this property.

1.8. Definition. A subset H of the group G is a *subgroup* of G if H is itself a group with respect to the operation of G. Subgroups of G other than the *trivial subgroups* $\{e\}$ and G itself are called *nontrivial subgroups* of G.

One verifies at once that for any fixed a in a group G, the set of all powers of a is a subgroup of G.

1.9. Definition. The subgroup of G consisting of all powers of the element a of G is called the subgroup *generated by* a and is denoted by $\langle a \rangle$. This subgroup is necessarily cyclic. If $\langle a \rangle$ is finite, then its order is called the *order* of the element a. Otherwise, a is called an element of *infinite order*.

Thus, a is of finite order k if k is the least positive integer such that $a^k = e$. Any other integer m with $a^m = e$ is then a multiple of k. If S is a nonempty subset of a group G, then the subgroup H of G consisting of all finite products of powers of elements of S is called the subgroup *generated by* S, denoted by $H = \langle S \rangle$. If $\langle S \rangle = G$, we say that S *generates* G, or that G is *generated by* S.

For a positive element n of the additive group $\mathbb{Z}$ of integers, the subgroup $\langle n \rangle$ is closely associated with the notion of congruence modulo n, since $a \equiv b \bmod n$ if and only if $a - b \in \langle n \rangle$. Thus the subgroup $\langle n \rangle$ defines an equivalence relation on $\mathbb{Z}$. This situation can be generalized as follows.

1.10. Theorem. *If H is a subgroup of G, then the relation R_H on G defined by $(a, b) \in R_H$ if and only if $a = bh$ for some $h \in H$, is an equivalence relation.*

The proof is immediate. The equivalence relation R_H is called *left congruence* modulo H. Like any equivalence relation, it induces a partition of G into nonempty, mutually disjoint subsets. These subsets (= equivalence classes) are called the *left cosets* of G modulo H and they are denoted by

$$aH = \{ah : h \in H\}$$

(or $a + H = \{a + h : h \in H\}$ if G is written additively), where a is a fixed element of G. Similarly, there is a decomposition of G into *right cosets* modulo H, which have the form $Ha = \{ha : h \in H\}$. If G is abelian, then the distinction between left and right cosets modulo H is unnecessary.

1.11. Example. Let $G = \mathbb{Z}_{12}$ and let H be the subgroup $\{[0], [3], [6], [9]\}$. Then the distinct (left) cosets of G modulo H are given by:

$$[0] + H = \{[0], [3], [6], [9]\},$$
$$[1] + H = \{[1], [4], [7], [10]\},$$
$$[2] + H = \{[2], [5], [8], [11]\}. \qquad \square$$

1.12. Theorem. *If H is a finite subgroup of G, then every (left or right) coset of G modulo H has the same number of elements as H.*

1.13. Definition. If the subgroup H of G only yields finitely many distinct left cosets of G modulo H, then the number of such cosets is called the *index* of H in G.

Since the left cosets of G modulo H form a partition of G, Theorem 1.12 implies the following important result.

1.14. Theorem. *The order of a finite group G is equal to the product of the order of any subgroup H and the index of H in G. In particular, the order of H divides the order of G and the order of any element $a \in G$ divides the order of G.*

The subgroups and the orders of elements are easy to describe for cyclic groups. We summarize the relevant facts in the subsequent theorem.

1.15. Theorem

(i) *Every subgroup of a cyclic group is cyclic.*
(ii) *In a finite cyclic group $\langle a \rangle$ of order m, the element a^k generates a subgroup of order $m/\gcd(k, m)$, where $\gcd(k, m)$ denotes the greatest common divisor of k and m.*
(iii) *If d is a positive divisor of the order m of a finite cyclic group $\langle a \rangle$, then $\langle a \rangle$ contains one and only one subgroup of index d. For any positive divisor f of m, $\langle a \rangle$ contains precisely one subgroup of order f.*
(iv) *Let f be a positive divisor of the order of a finite cyclic group $\langle a \rangle$. Then $\langle a \rangle$ contains $\phi(f)$ elements of order f. Here $\phi(f)$ is* Euler's function *and indicates the number of integers n with $1 \leqslant n \leqslant f$ that are relatively prime to f.*
(v) *A finite cyclic group $\langle a \rangle$ of order m contains $\phi(m)$ generators—that is, elements a^r such that $\langle a^r \rangle = \langle a \rangle$. The generators are the powers a^r with $\gcd(r, m) = 1$.*

Proof. (i) Let H be a subgroup of the cyclic group $\langle a \rangle$ with $H \neq \{e\}$. If $a^n \in H$, then $a^{-n} \in H$; hence H contains at least one power of a with a positive exponent. Let d be the least positive exponent such that $a^d \in H$, and let $a^s \in H$. Dividing s by d gives $s = qd + r$, $0 \leqslant r < d$, and $q, r \in \mathbb{Z}$. Thus $a^s(a^{-d})^q = a^r \in H$, which contradicts the minimality of d, unless $r = 0$. Therefore the exponents of all powers of a that belong to H are divisible by d, and so $H = \langle a^d \rangle$.

(ii) Put $d = \gcd(k, m)$. The order of $\langle a^k \rangle$ is the least positive integer n such that $a^{kn} = e$. The latter identity holds if and only if m divides kn, or equivalently, if and only if m/d divides n. The least positive n with this property is $n = m/d$.

(iii) If d is given, then $\langle a^d \rangle$ is a subgroup of order m/d, and so of index d, because of (ii). If $\langle a^k \rangle$ is another subgroup of index d, then its order is m/d, and so $d = \gcd(k, m)$ by (ii). In particular, d divides k, so that $a^k \in \langle a^d \rangle$ and $\langle a^k \rangle$ is a subgroup of $\langle a^d \rangle$. But since both groups have the same order, they are identical. The second part follows immediately because the subgroups of order f are precisely the subgroups of index m/f.

(iv) Let $|\langle a \rangle| = m$ and $m = df$. By (ii), an element a^k is of order f if and only if $\gcd(k, m) = d$. Hence, the number of elements of order f is equal to the number of integers k with $1 \leqslant k \leqslant m$ and $\gcd(k, m) = d$. We may write $k = dh$ with $1 \leqslant h \leqslant f$, the condition $\gcd(k, m) = d$ being now equivalent to $\gcd(h, f) = 1$. The number of these h is equal to $\phi(f)$.

(v) The generators of $\langle a \rangle$ are precisely the elements of order m, so that the first part is implied by (iv). The second part follows from (ii). □

When comparing the structures of two groups, mappings between the groups that preserve the operations play an important role.

1.16. Definition. A mapping $f: G \to H$ of the group G into the group H is called a *homomorphism* of G into H if f preserves the operation of G. That is, if $*$ and $\cdot$ are the operations of G and H, respectively, then f preserves the operation of G if for all $a, b \in G$ we have $f(a * b) = f(a) \cdot f(b)$. If, in addition, f is onto H, then f is called an *epimorphism* (or *homomorphism "onto"*) and H is a *homomorphic image* of G. A homomorphism of G into G is called an *endomorphism*. If f is a one-to-one homomorphism of G onto H, then f is called an *isomorphism* and we say that G and H are *isomorphic*. An isomorphism of G onto G is called an *automorphism*.

Consider, for instance, the mapping f of the additive group $\mathbb{Z}$ of the integers onto the group $\mathbb{Z}_n$ of the integers modulo n, defined by $f(a) = [a]$. Then

$$f(a+b) = [a+b] = [a] + [b] = f(a) + f(b) \quad \text{for } a, b \in \mathbb{Z},$$

and f is a homomorphism.

If $f: G \to H$ is a homomorphism and e is the identity element in G, then $ee = e$ implies $f(e)f(e) = f(e)$, so that $f(e) = e'$, the identity element in H. From $aa^{-1} = e$ we get $f(a^{-1}) = (f(a))^{-1}$ for all $a \in G$.

The automorphisms of a group G are often of particular interest, partly because they themselves form a group with respect to the usual composition of mappings, as can be easily verified. Important examples of automorphisms are the *inner automorphisms*. For fixed $a \in G$, define f_a by $f_a(b) = aba^{-1}$ for $b \in G$. Then f_a is an automorphism of G of the indicated type, and we get all inner automorphisms of G by letting a run through all elements of G. The elements b and aba^{-1} are said to be *conjugate*, and for a nonempty subset S of G the set $aSa^{-1} = \{asa^{-1} : s \in S\}$ is called a *conjugate of* S. Thus, the conjugates of S are just the images of S under the various inner automorphisms of G.

1.17. Definition. The *kernel* of the homomorphism $f: G \to H$ of the group G into the group H is the set

$$\ker f = \{a \in G: f(a) = e'\},$$

where e' is the identity element in H.

1.18. Example. For the homomorphism $f: \mathbb{Z} \to \mathbb{Z}_n$ given by $f(a) = [a]$, $\ker f$ consists of all $a \in \mathbb{Z}$ with $[a] = [0]$. Since this condition holds exactly for all multiples a of n, we have $\ker f = \langle n \rangle$, the subgroup of $\mathbb{Z}$ generated by n. □

It is easily checked that $\ker f$ is always a subgroup of G. Moreover, $\ker f$ has a special property: whenever $a \in G$ and $b \in \ker f$, then $aba^{-1} \in \ker f$. This leads to the following concept.

1.19. Definition. The subgroup H of the group G is called a *normal* subgroup of G if $aha^{-1} \in H$ for all $a \in G$ and all $h \in H$.

Every subgroup of an abelian group is normal since we then have $aha^{-1} = aa^{-1}h = eh = h$. We shall state some alternative characterizations of the property of normality of a subgroup.

1.20. Theorem

(i) *The subgroup H of G is normal if and only if H is equal to its conjugates, or equivalently, if and only if H is invariant under all the inner automorphisms of G.*
(ii) *The subgroup H of G is normal if and only if the left coset aH is equal to the right coset Ha for every $a \in G$.*

One important feature of a normal subgroup is the fact that the set of its (left) cosets can be endowed with a group structure.

1.21. Theorem. *If H is a normal subgroup of G, then the set of (left) cosets of G modulo H forms a group with respect to the operation $(aH)(bH) = (ab)H$.*

1.22. Definition. For a normal subgroup H of G, the group formed by the (left) cosets of G modulo H under the operation in Theorem 1.21 is called the *factor group* (or *quotient group*) of G modulo H and denoted by G/H.

If G/H is finite, then its order is equal to the index of H in G. Thus, by Theorem 1.14, we get for a finite group G,

$$|G/H| = \frac{|G|}{|H|}.$$

Each normal subgroup of a group G determines in a natural way a homomorphism of G and vice versa.

1.23. Theorem (Homomorphism Theorem). *Let $f: G \to f(G) = G_1$ be a homomorphism of a group G onto a group G_1. Then* $\ker f$ *is a normal subgroup of G, and the group G_1 is isomorphic to the factor group $G/\ker f$. Conversely, if H is any normal subgroup of G, then the mapping $\psi: G \to G/H$ defined by $\psi(a) = aH$ for $a \in G$ is a homomorphism of G onto G/H with* $\ker \psi = H$.

We shall now derive a relation known as the *class equation* for a finite group, which will be needed in Chapter 2, Section 6.

1.24. Definition. Let S be a nonempty subset of a group G. The *normalizer* of S in G is the set $N(S) = \{a \in G: aSa^{-1} = S\}$.

1.25. Theorem. *For any nonempty subset S of the group G, $N(S)$ is a subgroup of G and there is a one-to-one correspondence between the left cosets of G modulo $N(S)$ and the distinct conjugates aSa^{-1} of S.*

Proof. We have $e \in N(S)$, and if $a, b \in N(S)$, then a^{-1} and ab are also in $N(S)$, so that $N(S)$ is a subgroup of G. Now

$$aSa^{-1} = bSb^{-1} \Leftrightarrow S = a^{-1}bSb^{-1}a = (a^{-1}b)S(a^{-1}b)^{-1}$$
$$\Leftrightarrow a^{-1}b \in N(S) \Leftrightarrow b \in aN(S).$$

Thus, conjugates of S are equal if and only if they are defined by elements in the same left coset of G modulo $N(S)$, and so the second part of the theorem is shown. □

If we collect all elements conjugate to a fixed element a, we obtain a set called the *conjugacy class* of a. For certain elements the corresponding conjugacy class has only one member, and this will happen precisely for the elements of the center of the group.

1.26. Definition. For any group G, the *center* of G is defined as the set $C = \{c \in G: ac = ca \text{ for all } a \in G\}$.

It is straightforward to check that the center C is a normal subgroup of G. Clearly, G is abelian if and only if $C = G$. A counting argument leads to the following result.

1.27. Theorem (Class Equation). *Let G be a finite group with center C. Then*

$$|G| = |C| + \sum_{i=1}^{k} n_i,$$

where each n_i is $\geqslant 2$ and a divisor of $|G|$. In fact, $n_1, n_2, \ldots, n_k$ are the numbers of elements of the distinct conjugacy classes in G containing more than one member.

Proof. Since the relation "a is conjugate to b" is an equivalence relation on G, the distinct conjugacy classes in G form a partition of G. Thus, $|G|$ is equal to the sum of the numbers of elements of the distinct conjugacy classes. There are $|C|$ conjugacy classes (corresponding to the elements of C) containing only one member, whereas $n_1, n_2, \dots, n_k$ are the numbers of elements of the remaining conjugacy classes. This yields the class equation. To show that each n_i divides $|G|$, it suffices to note that n_i is the number of conjugates of some $a \in G$ and so equal to the number of left cosets of G modulo $N(\{a\})$ by Theorem 1.25. □

2. RINGS AND FIELDS

In most of the number systems used in elementary arithmetic there are two distinct binary operations: addition and multiplication. Examples are provided by the integers, the rational numbers, and the real numbers. We now define a type of algebraic structure known as a ring that shares some of the basic properties of these number systems.

1.28. Definition. A *ring* $(R, +, \cdot)$ is a set R, together with two binary operations, denoted by $+$ and $\cdot$, such that:

1. R is an abelian group with respect to $+$.
2. $\cdot$ is associative—that is, $(a \cdot b) \cdot c = a \cdot (b \cdot c)$ for all $a, b, c \in R$.
3. The *distributive laws* hold; that is, for all $a, b, c \in R$ we have $a \cdot (b + c) = a \cdot b + a \cdot c$ and $(b + c) \cdot a = b \cdot a + c \cdot a$.

We shall use R as a designation for the ring $(R, +, \cdot)$ and stress that the operations $+$ and $\cdot$ are not necessarily the ordinary operations with numbers. In following convention, we use 0 (called the *zero element*) to denote the identity element of the abelian group R with respect to addition, and the additive inverse of a is denoted by $-a$; also, $a + (-b)$ is abbreviated by $a - b$. Instead of $a \cdot b$ we will usually write ab. As a consequence of the definition of a ring one obtains the general property $a0 = 0a = 0$ for all $a \in R$. This, in turn, implies $(-a)b = a(-b) = -ab$ for all $a, b \in R$.

The most natural example of a ring is perhaps the ring of ordinary integers. If we examine the properties of this ring, we realize that it has properties not enjoyed by rings in general. Thus, rings can be further classified according to the following definitions.

1.29. Definition

(i) A ring is called a *ring with identity* if the ring has a multiplicative identity—that is, if there is an element e such that $ae = ea = a$ for all $a \in R$.

(ii) A ring is called *commutative* if $\cdot$ is commutative.

(iii) A ring is called an *integral domain* if it is a commutative ring with identity $e \neq 0$ in which $ab = 0$ implies $a = 0$ or $b = 0$.
(iv) A ring is called a *division ring* (or *skew field*) if the nonzero elements of R form a group under $\cdot$.
(v) A commutative division ring is called a *field*.

Since our study is devoted to fields, we emphasize again the definition of this concept. In the first place, a *field* is a set F on which two binary operations, called addition and multiplication, are defined and which contains two distinguished elements 0 and e with $0 \neq e$. Furthermore, F is an abelian group with respect to addition having 0 as the identity element, and the elements of F that are $\neq 0$ form an abelian group with respect to multiplication having e as the identity element. The two operations of addition and multiplication are linked by the distributive law $a(b + c) = ab + ac$. The second distributive law $(b + c)a = ba + ca$ follows automatically from the commutativity of multiplication. The element 0 is called the *zero element* and e is called the *multiplicative identity element* or simply the *identity*. Later on, the identity will usually be denoted by 1.

The property appearing in Definition 1.29(iii)—namely, that $ab = 0$ implies $a = 0$ or $b = 0$—is expressed by saying that there are *no zero divisors*. In particular, a field has no zero divisors, for if $ab = 0$ and $a \neq 0$, then multiplication by a^{-1} yields $b = a^{-1}0 = 0$.

In order to give an indication of the generality of the concept of ring, we present some examples.

1.30. Examples

(i) Let R be any abelian group with group operation $+$. Define $ab = 0$ for all $a, b \in R$; then R is a ring.
(ii) The integers form an integral domain, but not a field.
(iii) The even integers form a commutative ring without identity.
(iv) The functions from the real numbers into the real numbers form a commutative ring with identity under the definitions for $f + g$ and fg given by $(f + g)(x) = f(x) + g(x)$ and $(fg)(x) = f(x)g(x)$ for $x \in \mathbb{R}$.
(v) The set of all 2×2 matrices with real numbers as entries forms a noncommutative ring with identity with respect to matrix addition and multiplication. □

We have seen above that a field is, in particular, an integral domain. The converse is not true in general (see Example 1.30(ii)), but it will hold if the structures contain only finitely many elements.

1.31. Theorem. *Every finite integral domain is a field.*

Proof. Let the elements of the finite integral domain R be $a_1, a_2, \ldots, a_n$. For a fixed nonzero element $a \in R$, consider the products $aa_1, aa_2, \ldots, aa_n$. These are distinct, for if $aa_i = aa_j$, then $a(a_i - a_j) = 0$, and since $a \neq 0$ we must have $a_i - a_j = 0$, or $a_i = a_j$. Thus each element of R is of the form aa_i, in particular, $e = aa_i$ for some i with $1 \leqslant i \leqslant n$, where e is the identity of R. Since R is commutative, we have also $a_i a = e$, and so a_i is the multiplicative inverse of a. Thus the nonzero elements of R form a commutative group, and R is a field. □

1.32. Definition. A subset S of a ring R is called a *subring* of R provided S is closed under $+$ and $\cdot$ and forms a ring under these operations.

1.33. Definition. A subset J of a ring R is called an *ideal* provided J is a subring of R and for all $a \in J$ and $r \in R$ we have $ar \in J$ and $ra \in J$.

1.34. Examples

(i) Let R be the field $\mathbb{Q}$ of rational numbers. Then the set $\mathbb{Z}$ of integers is a subring of $\mathbb{Q}$, but not an ideal since, for example, $1 \in \mathbb{Z}$, $\frac{1}{2} \in \mathbb{Q}$, but $\frac{1}{2} \cdot 1 = \frac{1}{2} \notin \mathbb{Z}$.

(ii) Let R be a commutative ring, $a \in R$, and let $J = \{ra : r \in R\}$, then J is an ideal.

(iii) Let R be a commutative ring. Then the smallest ideal containing a given element $a \in R$ is the ideal $(a) = \{ra + na : r \in R, n \in \mathbb{Z}\}$. If R contains an identity, then $(a) = \{ra : r \in R\}$. □

1.35. Definition. Let R be a commutative ring. An ideal J of R is said to be *principal* if there is an $a \in R$ such that $J = (a)$. In this case, J is also called the principal ideal *generated by a*.

Since ideals are normal subgroups of the additive group of a ring, it follows immediately that an ideal J of the ring R defines a partition of R into disjoint cosets, called *residue classes* modulo J. The residue class of the element a of R modulo J will be denoted by $[a] = a + J$, since it consists of all elements of R that are of the form $a + c$ for some $c \in J$. Elements $a, b \in R$ are called *congruent* modulo J, written $a \equiv b \bmod J$, if they are in the same residue class modulo J, or equivalently, if $a - b \in J$ (compare with Definition 1.4). One can verify that $a \equiv b \bmod J$ implies $a + r \equiv b + r \bmod J$, $ar \equiv br \bmod J$, and $ra \equiv rb \bmod J$ for any $r \in R$ and $na \equiv nb \bmod J$ for any $n \in \mathbb{Z}$. If, in addition, $r \equiv s \bmod J$, then $a + r \equiv b + s \bmod J$ and $ar \equiv bs \bmod J$.

It is shown by a straightforward argument that the set of residue classes of a ring R modulo an ideal J forms a ring with respect to the operations

$$(a + J) + (b + J) = (a + b) + J, \tag{1.2}$$

$$(a + J)(b + J) = ab + J. \tag{1.3}$$

1.36. Definition. The ring of residue classes of the ring R modulo the ideal J under the operations (1.2) and (1.3) is called the *residue class ring* (or *factor ring*) of R modulo J and is denoted by R/J.

1.37. Example (The residue class ring $\mathbb{Z}/(n)$). As in the case of groups (compare with Definition 1.5), we denote the coset or residue class of the integer a modulo the positive integer n by $[a]$, as well as by $a+(n)$, where (n) is the principal ideal generated by n. The elements of $\mathbb{Z}/(n)$ are

$$[0]=0+(n),[1]=1+(n),\dots,[n-1]=n-1+(n). \qquad \square$$

1.38. Theorem. *$\mathbb{Z}/(p)$, the ring of residue classes of the integers modulo the principal ideal generated by a prime p, is a field.*

Proof. By Theorem 1.31 it suffices to show that $\mathbb{Z}/(p)$ is an integral domain. Now $[1]$ is an identity of $\mathbb{Z}/(p)$, and $[a][b]=[ab]=[0]$ if and only if $ab=kp$ for some integer k. But since p is prime, p divides ab if and only if p divides at least one of the factors. Therefore, either $[a]=[0]$ or $[b]=[0]$, so that $\mathbb{Z}/(p)$ contains no zero divisors. □

1.39. Example. Let $p=3$. Then $\mathbb{Z}/(p)$ consists of the elements [0], [1], and [2]. The operations in this field can be described by operation tables that are similar to Cayley tables for finite groups (see Example 1.7):

+	[0]	[1]	[2]
[0]	[0]	[1]	[2]
[1]	[1]	[2]	[0]
[2]	[2]	[0]	[1]

·	[0]	[1]	[2]
[0]	[0]	[0]	[0]
[1]	[0]	[1]	[2]
[2]	[0]	[2]	[1]

□

The residue class fields $\mathbb{Z}/(p)$ are our first examples of *finite fields*—that is, of fields that contain only finitely many elements. The general theory of such fields will be developed later on.

The reader is cautioned not to assume that in the formation of residue class rings all the properties of the original ring will be preserved in all cases. For example, the lack of zero divisors is not always preserved, as may be seen by considering the ring $\mathbb{Z}/(n)$, where n is a composite integer.

There is an obvious extension from groups to rings of the definition of a homomorphism. A mapping $\varphi: R \to S$ from a ring R into a ring S is called a *homomorphism* if for any $a, b \in R$ we have

$$\varphi(a+b)=\varphi(a)+\varphi(b) \quad \text{and} \quad \varphi(ab)=\varphi(a)\varphi(b).$$

Thus a homomorphism $\varphi: R \to S$ preserves both operations + and · of R and induces a homomorphism of the additive group of R into the additive group of S. The set

$$\ker \varphi = \{a \in R: \varphi(a)=0 \in S\}$$

is called the *kernel* of φ. Other concepts, such as that of an *isomorphism*, are analogous to those in Definition 1.16. The homomorphism theorem for rings, similar to Theorem 1.23 for groups, runs as follows.

1.40. Theorem (Homomorphism Theorem for Rings). *If φ is a homomorphism of a ring R onto a ring S, then* $\ker \varphi$ *is an ideal of R and S is* *isomorphic to the factor ring* $R/\ker\varphi$. *Conversely, if* J *is an ideal of the ring* R, *then the mapping* $\psi: R \to R/J$ *defined by* $\psi(a) = a + J$ *for* $a \in R$ *is a homomorphism of* R *onto* R/J *with kernel* J.

Mappings can be used to transfer a structure from an algebraic system to a set without structure. For instance, let R be a ring and let φ be a one-to-one and onto mapping from R to a set S; then by means of φ one can define a ring structure on S that converts φ into an isomorphism. In detail, let s_1 and s_2 be two elements of S and let r_1 and r_2 be the elements of R uniquely determined by $\varphi(r_1) = s_1$ and $\varphi(r_2) = s_2$. Then one defines $s_1 + s_2$ to be $\varphi(r_1 + r_2)$ and s_1s_2 to be $\varphi(r_1r_2)$, and all the desired properties are satisfied. This structure on S may be called the ring structure *induced by* φ. In case R has additional properties, such as being an integral domain or a field, then these properties are inherited by S. We use this principle in order to arrive at a more convenient representation for the finite fields $\mathbb{Z}/(p)$.

1.41. Definition. For a prime p, let $\mathbb{F}_p$ be the set $\{0, 1, \ldots, p-1\}$ of integers and let $\varphi: \mathbb{Z}/(p) \to \mathbb{F}_p$ be the mapping defined by $\varphi([a]) = a$ for $a = 0, 1, \ldots, p-1$. Then $\mathbb{F}_p$, endowed with the field structure induced by φ, is a finite field, called the *Galois field of order* p.

By what we have said before, the mapping $\varphi: \mathbb{Z}/(p) \to \mathbb{F}_p$ is then an isomorphism, so that $\varphi([a]+[b]) = \varphi([a]) + \varphi([b])$ and $\varphi([a][b]) = \varphi([a])\varphi([b])$. The finite field $\mathbb{F}_p$ has zero element 0, identity 1, and its structure is exactly the structure of $\mathbb{Z}/(p)$. Computing with elements of $\mathbb{F}_p$ therefore means ordinary arithmetic of integers with reduction modulo p.

1.42. Examples

(i) Consider $\mathbb{Z}/(5)$, isomorphic to $\mathbb{F}_5 = \{0, 1, 2, 3, 4\}$, with the isomorphism given by: $[0] \to 0$, $[1] \to 1$, $[2] \to 2$, $[3] \to 3$, $[4] \to 4$. The tables for the two operations $+$ and $\cdot$ for elements in $\mathbb{F}_5$ are as follows:

+	0	1	2	3	4
0	0	1	2	3	4
1	1	2	3	4	0
2	2	3	4	0	1
3	3	4	0	1	2
4	4	0	1	2	3

·	0	1	2	3	4
0	0	0	0	0	0
1	0	1	2	3	4
2	0	2	4	1	3
3	0	3	1	4	2
4	0	4	3	2	1

(ii) An even simpler and more important example is the finite field $\mathbb{F}_2$. The elements of this field of order two are 0 and 1, and the operation tables have the following form:

+	0	1
0	0	1
1	1	0

·	0	1
0	0	0
1	0	1

In this context, the elements 0 and 1 are called *binary elements*. □

If b is any nonzero element of the ring $\mathbb{Z}$ of integers, then the additive order of b is infinite; that is, $nb = 0$ implies $n = 0$. However, in the ring $\mathbb{Z}/(p)$, p prime, the additive order of every nonzero element b is p; that is, $pb = 0$, and p is the least positive integer for which this holds. It is of interest to formalize this property.

1.43. Definition. If R is an arbitrary ring and there exists a positive integer n such that $nr = 0$ for every $r \in R$, then the least such positive integer n is called the *characteristic* of R and R is said to have (positive) characteristic n. If no such positive integer n exists, R is said to have characteristic 0.

1.44. Theorem. *A ring $R \neq \{0\}$ of positive characteristic having an identity and no zero divisors must have prime characteristic.*

Proof. Since R contains nonzero elements, R has characteristic $n \geqslant 2$. If n were not prime, we could write $n = km$ with $k, m \in \mathbb{Z}$, $1 < k$, $m < n$. Then $0 = ne = (km)e = (ke)(me)$, and this implies that either $ke = 0$ or $me = 0$ since R has no zero divisors. It follows that either $kr = (ke)r = 0$ for all $r \in R$ or $mr = (me)r = 0$ for all $r \in R$, in contradiction to the definition of the characteristic n. □

1.45. Corollary. *A finite field has prime characteristic.*

Proof. By Theorem 1.44 it suffices to show that a finite field F has a positive characteristic. Consider the multiples $e, 2e, 3e, \dots$ of the identity. Since F contains only finitely many distinct elements, there exist integers k and m with $1 \leqslant k < m$ such that $ke = me$, or $(m-k)e = 0$, and so F has a positive characteristic. □

The finite field $\mathbb{Z}/(p)$ (or, equivalently, $\mathbb{F}_p$) obviously has characteristic p, whereas the ring $\mathbb{Z}$ of integers and the field $\mathbb{Q}$ of rational numbers have characteristic 0. We note that in a ring R of characteristic 2 we have $2a = a + a = 0$, hence $a = -a$ for all $a \in R$. A useful property of commutative rings of prime characteristic is the following.

1.46. Theorem. *Let R be a commutative ring of prime characteristic p. Then*

$$(a+b)^{p^n} = a^{p^n} + b^{p^n} \quad \textit{and} \quad (a-b)^{p^n} = a^{p^n} - b^{p^n}$$

for $a, b \in R$ and $n \in \mathbb{N}$.

Proof. We use the fact that

$$\binom{p}{i} = \frac{p(p-1)\cdots(p-i+1)}{1 \cdot 2 \cdot \cdots \cdot i} \equiv 0 \bmod p$$

for all $i \in \mathbb{Z}$ with $0 < i < p$, which follows from $\binom{p}{i}$ being an integer and the observation that the factor p in the numerator cannot be cancelled. Then by the binomial theorem (see Exercise 1.8),

$$(a+b)^p = a^p + \binom{p}{1} a^{p-1}b + \cdots + \binom{p}{p-1} ab^{p-1} + b^p = a^p + b^p,$$

and induction on n completes the proof of the first identity. By what we have shown, we get

$$a^{p^n} = ((a-b)+b)^{p^n} = (a-b)^{p^n} + b^{p^n},$$

and the second identity follows. □

Next we will show for the case of commutative rings with identity which ideals give rise to factor rings that are integral domains or fields. For this we need some definitions from ring theory.

Let R be a commutative ring with identity. An element $a \in R$ is called a *divisor* of $b \in R$ if there exists $c \in R$ such that $ac = b$. A *unit* of R is a divisor of the identity; two elements $a, b \in R$ are said to be *associates* if there is a unit ε of R such that $a = b\varepsilon$. An element $c \in R$ is called a *prime element* if it is no unit and if it has only the units of R and the associates of c as divisors. An ideal $P \neq R$ of the ring R is called a *prime ideal* if for $a, b \in R$ we have $ab \in P$ only if either $a \in P$ or $b \in P$. An ideal $M \neq R$ of R is called a *maximal ideal* of R if for any ideal J of R the property $M \subseteq J$ implies $J = R$ or $J = M$. Furthermore, R is said to be a *principal ideal domain* if R is an integral domain and if every ideal J of R is principal—that is, if there is a generating element a for J such that $J = (a) = \{ra : r \in R\}$.

1.47. Theorem. *Let R be a commutative ring with identity. Then:*

(i) *An ideal M of R is a maximal ideal if and only if R/M is a field.*
(ii) *An ideal P of R is a prime ideal if and only if R/P is an integral domain.*
(iii) *Every maximal ideal of R is a prime ideal.*
(iv) *If R is a principal ideal domain, then $R/(c)$ is a field if and only if c is a prime element of R.*

Proof.

(i) Let M be a maximal ideal of R. Then for $a \notin M$, $a \in R$, the set $J = \{ar + m : r \in R, m \in M\}$ is an ideal of R properly containing M, and therefore $J = R$. In particular, $ar + m = 1$ for some suitable $r \in R$, $m \in M$, where 1 denotes the multiplicative identity element of R. In other words, if $a + M \neq 0 + M$ is an element of R/M different from the zero element in R/M, then it possesses a multiplicative inverse, because $(a+M)(r+M) = ar + M = (1-m) + M = 1 + M$. Therefore, R/M is a field. Conversely, let R/M be a field and let $J \supseteq M$, $J \neq M$, be an ideal of R. Then for $a \in J$, $a \notin M$, the residue class $a + M$ has a multiplicative inverse, so that $(a+M)(r+M)=1+M$ for some $r \in R$. This implies $ar+m=1$ for some $m \in M$. Since J is an ideal, we have $1 \in J$ and therefore $(1)=R \subseteq J$, hence $J=R$. Thus M is a maximal ideal of R.

(ii) Let P be a prime ideal of R; then R/P is a commutative ring with identity $1+P \neq 0+P$. Let $(a+P)(b+P)=0+P$, hence $ab \in P$. Since P is a prime ideal, either $a \in P$ or $b \in P$; that is, either $a+P=0+P$ or $b+P=0+P$. Thus, R/P has no zero divisors and is therefore an integral domain. The converse follows immediately by reversing the steps of this proof.

(iii) This follows from (i) and (ii) since every field is an integral domain.

(iv) Let $c \in R$. If c is a unit, then $(c)=R$ and the ring $R/(c)$ consists only of one element and is no field. If c is neither a unit nor a prime element, then c has a divisor $a \in R$ that is neither a unit nor an associate of c. We note that $a \neq 0$, for if $a=0$, then $c=0$ and a would be an associate of c. We can write $c=ab$ with $b \in R$. Next we claim that $a \notin (c)$. For otherwise $a=cd=abd$ for some $d \in R$, or $a(1-bd)=0$. Since $a \neq 0$, this would imply $bd=1$, so that d would be a unit, which contradicts the fact that a is not an associate of c. It follows that $(c) \subseteq (a) \subseteq R$, where all containments are proper, and so $R/(c)$ cannot be a field because of (i). Finally, we are left with the case where c is a prime element. Then $(c) \neq R$ since c is no unit. Furthermore, if $J \supseteq (c)$ is an ideal of R, then $J=(a)$ for some $a \in R$ since R is a principal ideal domain. It follows that $c \in (a)$, and so a is a divisor of c. Consequently, a is either a unit or an associate of c, so that either $J=R$ or $J=(c)$. This shows that (c) is a maximal ideal of R. Hence $R/(c)$ is a field by (i). □

As an application of this theorem, let us consider the case $R=\mathbb{Z}$. We note that $\mathbb{Z}$ is a principal ideal domain since the additive subgroups of $\mathbb{Z}$ are already generated by a single element because of Theorem 1.15(i). A prime number p fits the definition of a prime element, and so Theorem 1.47(iv) yields another proof of the known result that $\mathbb{Z}/(p)$ is a field. Consequently, (p) is a maximal ideal and a prime ideal of $\mathbb{Z}$. For a composite integer n, the ideal (n) is not a prime ideal of $\mathbb{Z}$, and so $\mathbb{Z}/(n)$ is not even an integral domain. Other applications will follow in the next section when we consider residue class rings of polynomial rings over fields.

3. POLYNOMIALS

In elementary algebra one regards a polynomial as an expression of the form $a_0+a_1x+\cdots+a_nx^n$. The a_i's are called coefficients and are usually real or complex numbers; x is viewed as a variable: that is, substituting an arbitrary number α for x, a well-defined number $a_0 + a_1\alpha + \cdots + a_n\alpha^n$ is obtained. The arithmetic of polynomials is governed by familiar rules. The concept of polynomial and the associated operations can be generalized to a formal algebraic setting in a straightforward manner.

Let R be an arbitrary ring. A *polynomial* over R is an expression of the form

$$f(x) = \sum_{i=0}^{n} a_i x^i = a_0 + a_1 x + \cdots + a_n x^n,$$

where n is a nonnegative integer, the *coefficients* a_i, $0 \leqslant i \leqslant n$, are elements of R, and x is a symbol not belonging to R, called an *indeterminate* over R. Whenever it is clear which indeterminate is meant, we can use f as a designation for the polynomial $f(x)$. We adopt the convention that a term $a_i x^i$ with $a_i = 0$ need not be written down. In particular, the polynomial $f(x)$ above may then also be given in the equivalent form $f(x) = a_0 + a_1 x + \cdots + a_n x^n + 0x^{n+1} + \cdots + 0x^{n+h}$, where h is any positive integer. When comparing two polynomials $f(x)$ and $g(x)$ over R, it is therefore possible to assume that they both involve the same powers of x. The polynomials

$$f(x) = \sum_{i=0}^{n} a_i x^i \quad \text{and} \quad g(x) = \sum_{i=0}^{n} b_i x^i$$

over R are considered equal if and only if $a_i = b_i$ for $0 \leqslant i \leqslant n$. We define the *sum* of $f(x)$ and $g(x)$ by

$$f(x) + g(x) = \sum_{i=0}^{n} (a_i + b_i) x^i.$$

To define the *product* of two polynomials over R, let

$$f(x) = \sum_{i=0}^{n} a_i x^i \quad \text{and} \quad g(x) = \sum_{j=0}^{m} b_j x^j$$

and set

$$f(x)g(x) = \sum_{k=0}^{n+m} c_k x^k, \quad \text{where } c_k = \sum_{\substack{i+j=k \\ 0 \leqslant i \leqslant n,\, 0 \leqslant j \leqslant m}} a_i b_j.$$

It is easily seen that with these operations the set of polynomials over R forms a ring.

1.48. Definition. The ring formed by the polynomials over R with the above operations is called the *polynomial ring* over R and denoted by $R[x]$.

The zero element of $R[x]$ is the polynomial all of whose coefficients are 0. This polynomial is called the *zero polynomial* and denoted by 0. It should always be clear from the context whether 0 stands for the zero element of R or the zero polynomial.

1.49. Definition. Let $f(x)=\sum_{i=0}^{n} a_i x^i$ be a polynomial over R that is not the zero polynomial, so that we can suppose $a_n \neq 0$. Then a_n is called the *leading coefficient* of $f(x)$ and a_0 the *constant term*, while n is called the *degree* of $f(x)$, in symbols $n=\deg(f(x))=\deg(f)$. By convention, we set $\deg(0)=-\infty$. Polynomials of degree $\leqslant 0$ are called *constant polynomials*. If R has the identity 1 and if the leading coefficient of $f(x)$ is 1, then $f(x)$ is called a *monic polynomial*.

By computing the leading coefficient of the sum and the product of two polynomials, one finds the following result.

1.50. Theorem. *Let $f, g \in R[x]$. Then*

$$\deg(f+g) \leqslant \max(\deg(f), \deg(g)),$$
$$\deg(fg) \leqslant \deg(f)+\deg(g).$$

If R is an integral domain, we have

$$\deg(fg)=\deg(f)+\deg(g). \tag{1.4}$$

If one identifies constant polynomials with elements of R, then R can be viewed as a subring of $R[x]$. Certain properties of R are inherited by $R[x]$. The essential step in the proof of part (iii) of the subsequent theorem depends on (1.4).

1.51. Theorem. *Let R be a ring. Then:*

(i) *$R[x]$ is commutative if and only if R is commutative.*
(ii) *$R[x]$ is a ring with identity if and only if R has an identity.*
(iii) *$R[x]$ is an integral domain if and only if R is an integral domain.*

In the following chapters we will deal almost exclusively with polynomials over fields. Let F denote a field (not necessarily finite). The concept of divisibility, when specialized to the ring $F[x]$, leads to the following. The polynomial $g \in F[x]$ *divides* the polynomial $f \in F[x]$ if there exists a polynomial $h \in F[x]$ such that $f=gh$. We also say that g is a *divisor* of f, or that f is a *multiple* of g, or that f is *divisible* by g. The units of $F[x]$ are the divisors of the constant polynomial 1, which are precisely all nonzero constant polynomials.

As for the ring of integers, there is a division with remainder in polynomial rings over fields.

1.52. Theorem (Division Algorithm). *Let $g \neq 0$ be a polynomial in $F[x]$. Then for any $f \in F[x]$ there exist polynomials $q, r \in F[x]$ such that*

$$f=qg+r, \quad \textit{where } \deg(r)<\deg(g).$$

1.53. Example. Consider $f(x)=2x^5+x^4+4x+3 \in \mathbb{F}_5[x]$, $g(x)=3x^2+1 \in \mathbb{F}_5[x]$. We compute the polynomials $q, r \in \mathbb{F}_5[x]$ with $f=qg+r$ by using long division:

$$
\begin{array}{rllllll}
 & \multicolumn{6}{l}{4x^3 + 2x^2 + 2x + 1} \\ \cline{2-7}
3x^2+1 \,\big) & 2x^5 & +x^4 & & & +4x & +3 \\
 & -2x^5 & & -4x^3 & & & \\ \cline{2-4}
 & & x^4 & +x^3 & & & \\
 & & -x^4 & & -2x^2 & & \\ \cline{3-5}
 & & & x^3 & +3x^2 & +4x & \\
 & & & -x^3 & & -2x & \\ \cline{4-6}
 & & & & 3x^2 & +2x & +3 \\
 & & & & -3x^2 & & -1 \\ \cline{5-7}
 & & & & & 2x & +2
\end{array}
$$

Thus $q(x) = 4x^3 + 2x^2 + 2x + 1$, $r(x) = 2x + 2$, and obviously $\deg(r) < \deg(g)$. □

The fact that $F[x]$ permits a division algorithm implies by a standard argument that every ideal of $F[x]$ is principal.

1.54. Theorem. *$F[x]$ is a principal ideal domain. In fact, for every ideal $J \neq (0)$ of $F[x]$ there exists a uniquely determined monic polynomial $g \in F[x]$ with $J = (g)$.*

Proof. $F[x]$ is an integral domain by Theorem 1.51(iii). Suppose $J \neq (0)$ is an ideal of $F[x]$. Let $h(x)$ be a nonzero polynomial of least degree contained in J, let b be the leading coefficient of $h(x)$, and set $g(x) = b^{-1}h(x)$. Then $g \in J$ and g is monic. If $f \in J$ is arbitrary, the division algorithm yields $q, r \in F[x]$ with $f = qg + r$ and $\deg(r) < \deg(g) = \deg(h)$. Since J is an ideal, we get $f - qg = r \in J$, and by the definition of h we must have $r = 0$. Therefore, f is a multiple of g, and so $J = (g)$. If $g_1 \in F[x]$ is another monic polynomial with $J = (g_1)$, then $g = c_1 g_1$ and $g_1 = c_2 g$ with $c_1, c_2 \in F[x]$. This implies $g = c_1 c_2 g$, hence $c_1 c_2 = 1$, and c_1 and c_2 are constant polynomials. Since both g and g_1 are monic, it follows that $g = g_1$, and the uniqueness of g is established. □

1.55. Theorem. *Let $f_1, \ldots, f_n$ be polynomials in $F[x]$ not all of which are 0. Then there exists a uniquely determined monic polynomial $d \in F[x]$ with the following properties: (i) d divides each f_j, $1 \leqslant j \leqslant n$; (ii) any polynomial $c \in F[x]$ dividing each f_j, $1 \leqslant j \leqslant n$, divides d. Moreover, d can be expressed in the form*

$$d = b_1 f_1 + \cdots + b_n f_n \quad \text{with } b_1, \ldots, b_n \in F[x]. \tag{1.5}$$

Proof. The set J consisting of all polynomials of the form $c_1 f_1 + \cdots + c_n f_n$ with $c_1, \ldots, c_n \in F[x]$ is easily seen to be an ideal of $F[x]$. Since not all f_j are 0, we have $J \neq (0)$, and Theorem 1.54 implies that $J = (d)$

for some monic polynomial $d \in F[x]$. Property (i) and the representation (1.5) follow immediately from the construction of d. Property (ii) follows from (1.5). If d_1 is another monic polynomial in $F[x]$ satisfying (i) and (ii), then these properties imply that d and d_1 are divisible by each other, and so $(d) = (d_1)$. An application of the uniqueness part of Theorem 1.54 yields $d = d_1$. □

The monic polynomial d appearing in the theorem above is called the *greatest common divisor* of $f_1, \ldots, f_n$, in symbols $d = \gcd(f_1, \ldots, f_n)$. If $\gcd(f_1, \ldots, f_n) = 1$, then the polynomials $f_1, \ldots, f_n$ are said to be *relatively prime*. They are called *pairwise relatively prime* if $\gcd(f_i, f_j) = 1$ for $1 \leqslant i < j \leqslant n$.

The greatest common divisor of two polynomials $f, g \in F[x]$ can be computed by the *Euclidean algorithm*. Suppose, without loss of generality, that $g \neq 0$ and that g does not divide f. Then we repeatedly use the division algorithm in the following manner:

$$\begin{aligned}
f &= q_1 g + r_1 && 0 \leqslant \deg(r_1) < \deg(g) \\
g &= q_2 r_1 + r_2 && 0 \leqslant \deg(r_2) < \deg(r_1) \\
r_1 &= q_3 r_2 + r_3 && 0 \leqslant \deg(r_3) < \deg(r_2) \\
&\vdots && \vdots \\
r_{s-2} &= q_s r_{s-1} + r_s && 0 \leqslant \deg(r_s) < \deg(r_{s-1}) \\
r_{s-1} &= q_{s+1} r_s.
\end{aligned}$$

Here $q_1, \ldots, q_{s+1}$ and $r_1, \ldots, r_s$ are polynomials in $F[x]$. Since $\deg(g)$ is finite, the procedure must stop after finitely many steps. If the last nonzero remainder r_s has leading coefficient b, then $\gcd(f, g) = b^{-1} r_s$. In order to find $\gcd(f_1, \ldots, f_n)$ for $n > 2$ and nonzero polynomials f_i, one first computes $\gcd(f_1, f_2)$, then $\gcd(\gcd(f_1, f_2), f_3)$, and so on, by the Euclidean algorithm.

1.56. Example. The Euclidean algorithm applied to

$$f(x) = 2x^6 + x^3 + x^2 + 2 \in \mathbb{F}_3[x], \qquad g(x) = x^4 + x^2 + 2x \in \mathbb{F}_3[x]$$

yields:

$$\begin{aligned}
2x^6 + x^3 + x^2 + 2 &= (2x^2 + 1)(x^4 + x^2 + 2x) + x + 2 \\
x^4 + x^2 + 2x &= (x^3 + x^2 + 2x + 1)(x + 2) + 1 \\
x + 2 &= (x + 2)1.
\end{aligned}$$

Therefore $\gcd(f, g) = 1$ and f and g are relatively prime. □

A counterpart to the notion of greatest common divisor is that of least common multiple. Let $f_1, \ldots, f_n$ be nonzero polynomials in $F[x]$. Then one shows (see Exercise 1.25) that there exists a uniquely determined monic polynomial $m \in F[x]$ with the following properties: (i) m is a multiple of each f_j, $1 \leqslant j \leqslant n$; (ii) any polynomial $b \in F[x]$ that is a multiple of each f_j, $1 \leqslant j \leqslant n$, is a multiple of m. The polynomial m is called the *least common multiple* of $f_1, \ldots, f_n$ and denoted by $m = \text{lcm}(f_1, \ldots, f_n)$. For two nonzero polynomials $f, g \in F[x]$ we have

$$a^{-1}fg = \text{lcm}(f, g)\text{gcd}(f, g), \tag{1.6}$$

where a is the leading coefficient of fg. This relation conveniently reduces the calculation of $\text{lcm}(f, g)$ to that of $\text{gcd}(f, g)$. There is no direct analog of (1.6) for three or more polynomials. In this case, one uses the identity $\text{lcm}(f_1, \ldots, f_n) = \text{lcm}(\text{lcm}(f_1, \ldots, f_{n-1}), f_n)$ to compute the least common multiple.

The prime elements of the ring $F[x]$ are usually called irreducible polynomials. To emphasize this important concept, we give the definition again for the present context.

1.57. Definition. A polynomial $p \in F[x]$ is said to be *irreducible over* F (or *irreducible in* $F[x]$, or *prime in* $F[x]$) if p has positive degree and $p = bc$ with $b, c \in F[x]$ implies that either b or c is a constant polynomial.

Briefly stated, a polynomial of positive degree is irreducible over F if it allows only trivial factorizations. A polynomial in $F[x]$ of positive degree that is not irreducible over F is called *reducible over* F. The reducibility or irreducibility of a given polynomial depends heavily on the field under consideration. For instance, the polynomial $x^2 - 2 \in \mathbb{Q}[x]$ is irreducible over the field $\mathbb{Q}$ of rational numbers, but $x^2 - 2 = (x + \sqrt{2})(x - \sqrt{2})$ is reducible over the field of real numbers.

Irreducible polynomials are of fundamental importance for the structure of the ring $F[x]$ since the polynomials in $F[x]$ can be written as products of irreducible polynomials in an essentially unique manner. For the proof we need the following result.

1.58. Lemma. *If an irreducible polynomial p in $F[x]$ divides a product $f_1 \cdots f_m$ of polynomials in $F[x]$, then at least one of the factors f_j is divisible by p.*

Proof. Since p divides $f_1 \cdots f_m$, we get the identity $(f_1 + (p)) \cdots (f_m + (p)) = 0 + (p)$ in the factor ring $F[x]/(p)$. Now $F[x]/(p)$ is a field by Theorem 1.47(iv), and so $f_j + (p) = 0 + (p)$ for some j; that is, p divides f_j. □

1.59. Theorem (Unique Factorization in $F[x]$). *Any polynomial $f \in F[x]$ of positive degree can be written in the form*

$$f = ap_1^{e_1} \cdots p_k^{e_k}, \tag{1.7}$$

where $a \in F$, $p_1, \ldots, p_k$ are distinct monic irreducible polynomials in $F[x]$, and $e_1, \ldots, e_k$ are positive integers. Moreover, this factorization is unique apart from the order in which the factors occur.

Proof. The fact that any nonconstant $f \in F[x]$ can be represented in the form (1.7) is shown by induction on the degree of f. The case $\deg(f)=1$ is trivial since any polynomial in $F[x]$ of degree 1 is irreducible over F. Now suppose the desired factorization is established for all nonconstant polynomials in $F[x]$ of degree $< n$. If $\deg(f)=n$ and f is irreducible over F, then we are done since we can write $f = a(a^{-1}f)$, where a is the leading coefficient of f and $a^{-1}f$ is a monic irreducible polynomial in $F[x]$. Otherwise, f allows a factorization $f = gh$ with $1 \leqslant \deg(g) < n$, $1 \leqslant \deg(h) < n$, and $g, h \in F[x]$. By the induction hypothesis, g and h can be factored in the form (1.7), and so f can be factored in this form.

To prove uniqueness, suppose f has two factorizations of the form (1.7), say

$$f = ap_1^{e_1} \cdots p_k^{e_k} = bq_1^{d_1} \cdots q_r^{d_r}. \tag{1.8}$$

By comparing leading coefficients, we get $a = b$. Furthermore, the irreducible polynomial p_1 in $F[x]$ divides the right-hand side of (1.8), and so Lemma 1.58 shows that p_1 divides q_j for some $j, 1 \leqslant j \leqslant r$. But q_j is also irreducible in $F[x]$, so that we must have $q_j = cp_1$ with a constant polynomial c. Since q_j and p_1 are both monic, it follows that $q_j = p_1$. Thus we can cancel p_1 against q_j in (1.8) and continue in the same manner with the remaining identity. After finitely many steps of this type, we obtain that the two factorizations are identical apart from the order of the factors. □

We shall refer to (1.7) as the *canonical factorization* of the polynomial f in $F[x]$. If $F = \mathbb{Q}$, there is a method due to Kronecker for finding the canonical factorization of a polynomial in finitely many steps. This method is briefly described in Exercise 1.30. For polynomials over finite fields, factorization algorithms will be discussed in Chapter 4.

A central question about polynomials in $F[x]$ is to decide whether a given polynomial is irreducible or reducible over F. For our purposes, irreducible polynomials over $\mathbb{F}_p$ are of particular interest. To determine all monic irreducible polynomials over $\mathbb{F}_p$ of fixed degree n, one may first compute all monic reducible polynomials over $\mathbb{F}_p$ of degree n and then eliminate them from the set of monic polynomials in $\mathbb{F}_p[x]$ of degree n. If p or n is large, this method is not feasible, and we will develop more powerful methods in Chapter 3, Sections 2 and 3.

1.60. Example. Find all irreducible polynomials over $\mathbb{F}_2$ of degree 4 (note that a nonzero polynomial in $\mathbb{F}_2[x]$ is automatically monic). There are $2^4 = 16$ polynomials in $\mathbb{F}_2[x]$ of degree 4. Such a polynomial is reducible over $\mathbb{F}_2$ if and only if it has a divisor of degree 1 or 2. Therefore, we compute all products $(a_0 + a_1x + a_2x^2 + x^3)(b_0 + x)$ and $(a_0 + a_1x + x^2)(b_0 + b_1x + x^2)$ with $a_i, b_j \in \mathbb{F}_2$ and obtain all reducible polynomials over $\mathbb{F}_2$ of degree 4. Comparison with the 16 polynomials of degree 4 leaves us with the irreducible polynomials $f_1(x) = x^4 + x + 1$, $f_2(x) = x^4 + x^3 + 1$, $f_3(x) = x^4 + x^3 + x^2 + x + 1$ in $\mathbb{F}_2[x]$. □

Since the irreducible polynomials over a field F are exactly the prime elements of $F[x]$, the following result, one part of which was already used in Lemma 1.58, is an immediate consequence of Theorems 1.47(iv) and 1.54.

1.61. Theorem. *For $f \in F[x]$, the residue class ring $F[x]/(f)$ is a field if and only if f is irreducible over F.*

As a preparation for the next section, we shall take a closer look at the structure of the residue class ring $F[x]/(f)$, where f is an arbitrary nonzero polynomial in $F[x]$. We recall that as a residue class ring $F[x]/(f)$ consists of residue classes $g + (f)$ (also denoted by $[g]$) with $g \in F[x]$, where the operations are defined as in (1.2) and (1.3). Two residue classes $g + (f)$ and $h + (f)$ are identical precisely if $g \equiv h \bmod f$—that is, precisely if $g - h$ is divisible by f. This is equivalent to the requirement that g and h leave the same remainder after division by f. Each residue class $g + (f)$ contains a unique representative $r \in F[x]$ with $\deg(r) < \deg(f)$, which is simply the remainder in the division of g by f. The process of passing from g to r is called *reduction* mod f. The uniqueness of r follows from the observation that if $r_1 \in g + (f)$ with $\deg(r_1) < \deg(f)$, then $r - r_1$ is divisible by f and $\deg(r - r_1) < \deg(f)$, which is only possible if $r = r_1$. The distinct residue classes comprising $F[x]/(f)$ can now be described explicitly; namely, they are exactly the residue classes $r + (f)$, where r runs through all polynomials in $F[x]$ with $\deg(r) < \deg(f)$. Thus, if $F = \mathbb{F}_p$ and $\deg(f) = n \geqslant 0$, then the number of elements of $\mathbb{F}_p[x]/(f)$ is equal to the number of polynomials in $\mathbb{F}_p[x]$ of degree $< n$, which is p^n.

1.62. Examples

(i) Let $f(x) = x \in \mathbb{F}_2[x]$. The $p^n = 2^1$ polynomials in $\mathbb{F}_2[x]$ of degree < 1 determine all residue classes comprising $\mathbb{F}_2[x]/(x)$. Thus, $\mathbb{F}_2[x]/(x)$ consists of the residue classes [0] and [1] and is isomorphic to $\mathbb{F}_2$.

(ii) Let $f(x) = x^2 + x + 1 \in \mathbb{F}_2[x]$. Then $\mathbb{F}_2[x]/(f)$ has the $p^n = 2^2$ elements [0], [1], $[x]$, $[x+1]$. The operation tables for this residue class ring are obtained by performing the required operations with the polynomials determining the residue classes and by carrying out reduction mod f if necessary:

$+$	$[0]$	$[1]$	$[x]$	$[x+1]$
$[0]$	$[0]$	$[1]$	$[x]$	$[x+1]$
$[1]$	$[1]$	$[0]$	$[x+1]$	$[x]$
$[x]$	$[x]$	$[x+1]$	$[0]$	$[1]$
$[x+1]$	$[x+1]$	$[x]$	$[1]$	$[0]$

$\cdot$	[0]	[1]	[x]	[x+1]
[0]	[0]	[0]	[0]	[0]
[1]	[0]	[1]	[x]	[x+1]
[x]	[0]	[x]	[x+1]	[1]
[x+1]	[0]	[x+1]	[1]	[x]

By inspecting these tables, or from the irreducibility of f over $\mathbb{F}_2$ and Theorem 1.61, it follows that $\mathbb{F}_2[x]/(f)$ is a field. This is our first example of a finite field for which the number of elements is not a prime.

(iii) Let $f(x) = x^2 + 2 \in \mathbb{F}_3[x]$. Then $\mathbb{F}_3[x]/(f)$ consists of the $p^n = 3^2$ residue classes [0], [1], [2], [x], [x + 1], [x + 2], [2x], [2x + 1], [2x + 2]. The operation tables for $\mathbb{F}_3[x]/(f)$ are again produced by performing polynomial operations and using reduction mod f whenever necessary. Since $\mathbb{F}_3[x]/(f)$ is a commutative ring, we only have to compute the entries on and above the main diagonal.

+	[0]	[1]	[2]	[x]	[x+1]	[x+2]	[2x]	[2x+1]	[2x+2]
[0]	[0]	[1]	[2]	[x]	[x+1]	[x+2]	[2x]	[2x+1]	[2x+2]
[1]		[2]	[0]	[x+1]	[x+2]	[x]	[2x+1]	[2x+2]	[2x]
[2]			[1]	[x+2]	[x]	[x+1]	[2x+2]	[2x]	[2x+1]
[x]				[2x]	[2x+1]	[2x+2]	[0]	[1]	[2]
[x+1]					[2x+2]	[2x]	[1]	[2]	[0]
[x+2]						[2x+1]	[2]	[0]	[1]
[2x]							[x]	[x+1]	[x+2]
[2x+1]								[x+2]	[x]
[2x+2]									[x+1]

$\cdot$	[0]	[1]	[2]	[x]	[x+1]	[x+2]	[2x]	[2x+1]	[2x+2]
[0]	[0]	[0]	[0]	[0]	[0]	[0]	[0]	[0]	[0]
[1]		[1]	[2]	[x]	[x+1]	[x+2]	[2x]	[2x+1]	[2x+2]
[2]			[1]	[2x]	[2x+2]	[2x+1]	[x]	[x+2]	[x+1]
[x]				[1]	[x+1]	[2x+1]	[2]	[x+2]	[2x+2]
[x+1]					[2x+2]	[0]	[2x+2]	[0]	[x+1]
[x+2]						[x+2]	[x+2]	[2x+1]	[0]
[2x]							[1]	[2x+1]	[x+1]
[2x+1]								[x+2]	[0]
[2x+2]									[2x+2]

Note that $\mathbb{F}_3[x]/(f)$ is not a field (and not even an integral domain). This is in accordance with Theorem 1.61 since $x^2 + 2 = (x + 1)(x + 2)$ is reducible over $\mathbb{F}_3$. □

If F is again an arbitrary field and $f(x) \in F[x]$, then replacement of the indeterminate x in $f(x)$ by a fixed element of F yields a well-defined element of F. In detail, if $f(x)=a_0+a_1x+\cdots+a_nx^n \in F[x]$ and $b \in F$, then replacing x by b we get $f(b)=a_0+a_1b+\cdots+a_nb^n \in F$. In any polynomial identity in $F[x]$ we can substitute a fixed $b \in F$ for x and obtain a valid identity in F (*principle of substitution*).

1.63. Definition. An element $b \in F$ is called a *root* (or a *zero*) of the polynomial $f \in F[x]$ if $f(b)=0$.

An important connection between roots and divisibility is given by the following theorem.

1.64. Theorem. *An element $b \in F$ is a root of the polynomial $f \in F[x]$ if and only if $x-b$ divides $f(x)$.*

Proof. We use the division algorithm (see Theorem 1.52) to write $f(x)=q(x)(x-b)+c$ with $q \in F[x]$ and $c \in F$. Substituting b for x, we get $f(b)=c$, hence $f(x)=q(x)(x-b)+f(b)$. The theorem follows now from this identity. □

1.65. Definition. Let $b \in F$ be a root of the polynomial $f \in F[x]$. If k is a positive integer such that $f(x)$ is divisible by $(x-b)^k$, but not by $(x-b)^{k+1}$, then k is called the *multiplicity* of b. If $k=1$, then b is called a *simple root* (or a *simple zero*) of f, and if $k \geqslant 2$, then b is called a *multiple root* (or a *multiple zero*) of f.

1.66. Theorem. *Let $f \in F[x]$ with $\deg f = n \geqslant 0$. If $b_1,\ldots,b_m \in F$ are distinct roots of f with multiplicities $k_1,\ldots,k_m$, respectively, then $(x-b_1)^{k_1}\cdots(x-b_m)^{k_m}$ divides $f(x)$. Consequently, $k_1+\cdots+k_m \leqslant n$, and f can have at most n distinct roots in F.*

Proof. We note that each polynomial $x-b_j$, $1 \leqslant j \leqslant m$, is irreducible over F, and so $(x-b_j)^{k_j}$ occurs as a factor in the canonical factorization of f. Altogether, the factor $(x-b_1)^{k_1}\cdots(x-b_m)^{k_m}$ appears in the canonical factorization of f and is thus a divisor of f. By comparing degrees, we get $k_1+\cdots+k_m \leqslant n$, and $m \leqslant k_1+\cdots+k_m \leqslant n$ shows the last statement. □

1.67. Definition. If $f(x)=a_0+a_1x+a_2x^2+\cdots+a_nx^n \in F[x]$, then the *derivative* f' of f is defined by $f'=f'(x)=a_1+2a_2x+\cdots+na_nx^{n-1} \in F[x]$.

1.68. Theorem. *The element $b \in F$ is a multiple root of $f \in F[x]$ if and only if it is a root of both f and f'.*

There is a relation between the nonexistence of roots and irreducibility. If f is an irreducible polynomial in $F[x]$ of degree $\geqslant 2$, then Theorem 1.64 shows that f has no root in F. The converse holds for polynomials of degree 2 or 3, but not necessarily for polynomials of higher degree.

1.69. Theorem. *The polynomial $f \in F[x]$ of degree 2 or 3 is irreducible in $F[x]$ if and only if f has no root in F.*

Proof. The necessity of the condition was already noted. Conversely, if f has no root in F and were reducible in $F[x]$, we could write $f = gh$ with $g, h \in F[x]$ and $1 \leqslant \deg(g) \leqslant \deg(h)$. But $\deg(g)+\deg(h) = \deg(f) \leqslant 3$, hence $\deg(g) = 1$; that is, $g(x) = ax + b$ with $a, b \in F$, $a \neq 0$. Then $-ba^{-1}$ is a root of g, and so a root of f in F, a contradiction. □

1.70. Example. Because of Theorem 1.69, the irreducible polynomials in $\mathbb{F}_2[x]$ of degree 2 or 3 can be obtained by eliminating the polynomials with roots in $\mathbb{F}_2$ from the set of all polynomials in $\mathbb{F}_2[x]$ of degree 2 or 3. The only irreducible polynomial in $\mathbb{F}_2[x]$ of degree 2 is $f(x) = x^2 + x + 1$, and the irreducible polynomials in $\mathbb{F}_2[x]$ of degree 3 are $f_1(x) = x^3 + x + 1$ and $f_2(x) = x^3 + x^2 + 1$. □

In elementary analysis there is a well-known method for constructing a polynomial with real coefficients which assumes certain assigned values for given values of the indeterminate. The same method carries over to any field.

1.71. Theorem (Lagrange Interpolation Formula). *For $n \geqslant 0$, let $a_0, \ldots, a_n$ be $n+1$ distinct elements of F, and let $b_0, \ldots, b_n$ be $n+1$ arbitrary elements of F. Then there exists exactly one polynomial $f \in F[x]$ of degree $\leqslant n$ such that $f(a_i) = b_i$ for $i = 0, \ldots, n$. This polynomial is given by*

$$f(x) = \sum_{i=0}^{n} b_i \prod_{\substack{k=0 \\ k \neq i}}^{n} (a_i - a_k)^{-1}(x - a_k).$$

One can also consider polynomials in several indeterminates. Let R denote a commutative ring with identity and let $x_1, \ldots, x_n$ be symbols that will serve as indeterminates. We form the polynomial ring $R[x_1]$, then the polynomial ring $R[x_1, x_2] = R[x_1][x_2]$, and so on, until we arrive at $R[x_1, \ldots, x_n] = R[x_1, \ldots, x_{n-1}][x_n]$. The elements of $R[x_1, \ldots, x_n]$ are then expressions of the form

$$f = f(x_1, \ldots, x_n) = \sum a_{i_1 \cdots i_n} x_1^{i_1} \cdots x_n^{i_n}$$

with coefficients $a_{i_1 \cdots i_n} \in R$, where the summation is extended over finitely many n-tuples $(i_1, \ldots, i_n)$ of nonnegative integers and the convention $x_j^0 = 1$ $(1 \leqslant j \leqslant n)$ is observed. Such an expression is called a *polynomial in* $x_1, \ldots, x_n$ over R. Two polynomials $f, g \in R[x_1, \ldots, x_n]$ are equal if and only if all corresponding coefficients are equal. It is tacitly assumed that the indeterminates $x_1, \ldots, x_n$ commute with each other, so that, for instance, the expressions $x_1x_2x_3x_4$ and $x_4x_1x_3x_2$ are identified.

1.72. Definition. Let $f \in R[x_1, \ldots, x_n]$ be given by

$$f(x_1, \ldots, x_n) = \sum a_{i_1 \cdots i_n} x_1^{i_1} \cdots x_n^{i_n}.$$

If $a_{i_1\cdots i_n} \neq 0$, then $a_{i_1\cdots i_n}x_1^{i_1}\cdots x_n^{i_n}$ is called a *term* of f and $i_1+\cdots+i_n$ is the degree of the term. For $f \neq 0$ one defines the *degree* of f, denoted by $\deg(f)$, to be the maximum of the degrees of the terms of f. For $f=0$ one sets $\deg(f)=-\infty$. If $f=0$ or if all terms of f have the same degree, then f is called *homogeneous*.

Any $f \in R[x_1,\ldots,x_n]$ can be written as a finite sum of homogeneous polynomials. The degrees of polynomials in $R[x_1,\ldots,x_n]$ satisfy again the inequalities in Theorem 1.50, and if R is an integral domain, then (1.4) is valid and $R[x_1,\ldots,x_n]$ is an integral domain. If F is a field, then the polynomials in $F[x_1,\ldots,x_n]$ of positive degree can again be factored uniquely into a constant factor and a product of "monic" prime elements (using a suitable definition of "monic"), but for $n \geqslant 2$ there is no analog of the division algorithm (in the case of commuting indeterminates) and $F[x_1,\ldots,x_n]$ is not a principal ideal domain.

An important special class of polynomials in n indeterminates is that of symmetric polynomials.

1.73. Definition. A polynomial $f \in R[x_1,\ldots,x_n]$ is called *symmetric* if $f(x_{i_1},\ldots,x_{i_n}) = f(x_1,\ldots,x_n)$ for any permutation $i_1,\ldots,i_n$ of the integers $1,\ldots,n$.

1.74. Example. Let z be an indeterminate over $R[x_1,\ldots,x_n]$, and let $g(z)=(z-x_1)(z-x_2)\cdots(z-x_n)$. Then

$$g(z)=z^n-\sigma_1 z^{n-1}+\sigma_2 z^{n-2}+\cdots+(-1)^n\sigma_n$$

with

$$\sigma_k=\sigma_k(x_1,\ldots,x_n)=\sum_{1\leqslant i_1<\cdots<i_k\leqslant n} x_{i_1}\cdots x_{i_k} \quad (k=1,2,\ldots,n).$$

Thus:

$$\sigma_1 = x_1+x_2+\cdots+x_n,$$
$$\sigma_2 = x_1x_2+x_1x_3+\cdots+x_1x_n+x_2x_3+\cdots+x_2x_n+\cdots+x_{n-1}x_n,$$
$$\vdots$$
$$\sigma_n = x_1x_2\cdots x_n.$$

As g remains unaltered under any permutation of the x_i, all the σ_k are symmetric polynomials; they are also homogeneous. The polynomial $\sigma_k = \sigma_k(x_1,\ldots,x_n) \in R[x_1,\ldots,x_n]$ is called the kth *elementary symmetric polynomial* in the indeterminates $x_1,\ldots,x_n$ over R. The adjective "elementary" is used because of the so-called "fundamental theorem on symmetric polynomials," which states that for any symmetric polynomial $f \in R[x_1,\ldots,x_n]$ there exists a uniquely determined polynomial $h \in R[x_1,\ldots,x_n]$ such that $f(x_1,\ldots,x_n)=h(\sigma_1,\ldots,\sigma_n)$. □

1.75. Theorem (Newton's Formula). *Let $\sigma_1,\ldots,\sigma_n$ be the elementary symmetric polynomials in $x_1,\ldots,x_n$ over R, and let $s_0=n\in\mathbb{Z}$ and*

$s_k = s_k(x_1,\ldots,x_n) = x_1^k + \cdots + x_n^k \in R[x_1,\ldots,x_n]$ *for* $k \geqslant 1$. *Then the formula*

$$s_k - s_{k-1}\sigma_1 + s_{k-2}\sigma_2 + \cdots + (-1)^{m-1} s_{k-m+1}\sigma_{m-1} + (-1)^m \frac{m}{n} s_{k-m}\sigma_m = 0$$

holds for $k \geqslant 1$, *where* $m = \min(k, n)$.

1.76. Theorem (Waring's Formula). *With the same notation as in Theorem* 1.75, *we have*

$$s_k = \sum (-1)^{i_2 + i_4 + i_6 + \cdots} \frac{(i_1 + i_2 + \cdots + i_n - 1)!k}{i_1! i_2! \cdots i_n!} \sigma_1^{i_1} \sigma_2^{i_2} \cdots \sigma_n^{i_n}$$

for $k \geqslant 1$, *where the summation is extended over all n-tuples* $(i_1,\ldots,i_n)$ *of nonnegative integers with* $i_1 + 2i_2 + \cdots + ni_n = k$. *The coefficient of* $\sigma_1^{i_1}\sigma_2^{i_2} \cdots \sigma_n^{i_n}$ *is always an integer.*

4. FIELD EXTENSIONS

Let F be a field. A subset K of F that is itself a field under the operations of F will be called a *subfield* of F. In this context, F is called an *extension* (*field*) of K. If $K \neq F$, we say that K is a *proper subfield* of F.

If K is a subfield of the finite field $\mathbb{F}_p$, p prime, then K must contain the elements 0 and 1, and so all other elements of $\mathbb{F}_p$ by the closure of K under addition. It follows that $\mathbb{F}_p$ contains no proper subfields. We are thus led to the following concept.

1.77. Definition. A field containing no proper subfields is called a *prime field*.

By the above argument, any finite field of order p, p prime, is a prime field. Another example of a prime field is the field $\mathbb{Q}$ of rational numbers.

The intersection of any nonempty collection of subfields of a given field F is again a subfield of F. If we form the intersection of *all* subfields of F, we obtain the *prime subfield* of F. It is obviously a prime field.

1.78. Theorem. *The prime subfield of a field F is isomorphic to either* $\mathbb{F}_p$ *or* $\mathbb{Q}$, *according as the characteristic of F is a prime p or* 0.

1.79. Definition. Let K be a subfield of the field F and M any subset of F. Then the field $K(M)$ is defined as the intersection of all subfields of F containing both K and M and is called the extension (field) of K obtained by *adjoining* the elements in M. For finite $M = \{\theta_1,\ldots,\theta_n\}$ we write $K(M) = K(\theta_1,\ldots,\theta_n)$. If M consists of a single element $\theta \in F$, then $L = K(\theta)$ is said to be a *simple extension* of K and θ is called a *defining element* of L over K.

Obviously, $K(M)$ is the smallest subfield of F containing both K and M. We define now an important type of extension.

1.80. Definition. Let K be a subfield of F and $\theta \in F$. If θ satisfies a nontrivial polynomial equation with coefficients in K, that is, if $a_n\theta^n + \cdots + a_1\theta + a_0 = 0$ with $a_i \in K$ not all being 0, then θ is said to be *algebraic* over K. An extension L of K is called *algebraic* over K (or an *algebraic extension* of K) if every element of L is algebraic over K.

Suppose $\theta \in F$ is algebraic over K, and consider the set $J = \{f \in K[x]: f(\theta) = 0\}$. It is easily checked that J is an ideal of $K[x]$, and we have $J \neq (0)$ since θ is algebraic over K. It follows then from Theorem 1.54 that there exists a uniquely determined monic polynomial $g \in K[x]$ such that J is equal to the principal ideal (g). It is important to note that g is irreducible in $K[x]$. For, in the first place, g is of positive degree since it has the root θ; and if $g = h_1h_2$ in $K[x]$ with $1 \leqslant \deg(h_i) < \deg(g)$ $(i = 1, 2)$, then $0 = g(\theta) = h_1(\theta)h_2(\theta)$ implies that either h_1 or h_2 is in J and so divisible by g, which is impossible.

1.81. Definition. If $\theta \in F$ is algebraic over K, then the uniquely determined monic polynomial $g \in K[x]$ generating the ideal $J = \{f \in K[x]: f(\theta) = 0\}$ of $K[x]$ is called the *minimal polynomial* (or *defining polynomial*, or *irreducible polynomial*) of θ over K. By the *degree* of θ over K we mean the degree of g.

1.82. Theorem. *If $\theta \in F$ is algebraic over K, then its minimal polynomial g over K has the following properties*:

(i) *g is irreducible in $K[x]$.*
(ii) *For $f \in K[x]$ we have $f(\theta) = 0$ if and only if g divides f.*
(iii) *g is the monic polynomial in $K[x]$ of least degree having θ as a root.*

Proof. Property (i) was already noted and (ii) follows from the definition of g. As to (iii), it suffices to note that any monic polynomial in $K[x]$ having θ as a root must be a multiple of g, and so it is either equal to g or its degree is larger than that of g. □

We note that both the minimal polynomial and the degree of an algebraic element θ depend on the field K over which it is considered, so that one must be careful not to speak of the minimal polynomial or the degree of θ without specifying K, unless the latter is amply clear from the context.

If L is an extension field of K, then L may be viewed as a vector space over K. For the elements of L (= "vectors") form, first of all, an abelian group under addition. Moreover, each "vector" $\alpha \in L$ can be multiplied by a "scalar" $r \in K$ so that $r\alpha$ is again in L (here $r\alpha$ is simply the product of the field elements r and α of L) and the laws for multiplication by scalars are satisfied: $r(\alpha+\beta)=r\alpha+r\beta$, $(r+s)\alpha=r\alpha+s\alpha$, $(rs)\alpha=r(s\alpha)$, and $1\alpha=\alpha$, where $r, s \in K$ and $\alpha, \beta \in L$.

1.83. Definition. Let L be an extension field of K. If L, considered as a vector space over K, is finite-dimensional, then L is called a *finite extension* of K. The dimension of the vector space L over K is then called the *degree* of L over K, in symbols $[L: K]$.

1.84. Theorem. *If L is a finite extension of K and M is a finite extension of L, then M is a finite extension of K with*

$$[M: K]=[M: L][L: K].$$

Proof. Put $[M: L]=m$, $[L: K]=n$, and let $\{\alpha_1,\ldots,\alpha_m\}$ be a basis of M over L and $\{\beta_1,\ldots,\beta_n\}$ a basis of L over K. Then every $\alpha \in M$ is a linear combination $\alpha=\gamma_1\alpha_1+\cdots+\gamma_m\alpha_m$ with $\gamma_i \in L$ for $1 \leqslant i \leqslant m$, and writing each γ_i in terms of the basis elements β_j we get

$$\alpha=\sum_{i=1}^{m}\gamma_i\alpha_i=\sum_{i=1}^{m}\left(\sum_{j=1}^{n}r_{ij}\beta_j\right)\alpha_i=\sum_{i=1}^{m}\sum_{j=1}^{n}r_{ij}\beta_j\alpha_i$$

with coefficients $r_{ij} \in K$. If we can show that the mn elements $\beta_j\alpha_i$, $1 \leqslant i \leqslant m$, $1 \leqslant j \leqslant n$, are linearly independent over K, then we are done. So suppose we have

$$\sum_{i=1}^{m}\sum_{j=1}^{n}s_{ij}\beta_j\alpha_i=0$$

with coefficients $s_{ij} \in K$. Then

$$\sum_{i=1}^{m}\left(\sum_{j=1}^{n}s_{ij}\beta_j\right)\alpha_i=0,$$

and from the linear independence of the α_i over L we infer

$$\sum_{j=1}^{n}s_{ij}\beta_j=0 \quad \text{for } 1 \leqslant i \leqslant m.$$

But since the β_j are linearly independent over K, we conclude that all s_{ij} are 0. □

1.85. Theorem. *Every finite extension of K is algebraic over K.*

Proof. Let L be a finite extension of K and put $[L: K]=m$. For $\theta \in L$, the $m+1$ elements $1, \theta, \ldots, \theta^m$ must then be linearly dependent over K, and so we get a relation $a_0+a_1\theta+\cdots+a_m\theta^m=0$ with $a_i \in K$ not all being 0. This just says that θ is algebraic over K. □

For the study of the structure of a simple extension $K(\theta)$ of K obtained by adjoining an algebraic element, let F be an extension of K and let $\theta \in F$ be algebraic over K. It turns out that $K(\theta)$ is a finite (and therefore an algebraic) extension of K.

1.86. Theorem. *Let $\theta \in F$ be algebraic of degree n over K and let g be the minimal polynomial of θ over K. Then:*

(i) *$K(\theta)$ is isomorphic to $K[x]/(g)$.*
(ii) *$[K(\theta): K] = n$ and $\{1, \theta, \ldots, \theta^{n-1}\}$ is a basis of $K(\theta)$ over K.*
(iii) *Every $\alpha \in K(\theta)$ is algebraic over K and its degree over K is a divisor of n.*

Proof. (i) Consider the mapping $\tau: K[x] \to K(\theta)$, defined by $\tau(f) = f(\theta)$ for $f \in K[x]$, which is easily seen to be a ring homomorphism. We have $\ker \tau = \{f \in K[x]: f(\theta) = 0\} = (g)$ by the definition of the minimal polynomial. Let S be the image of τ; that is, S is the set of polynomial expressions in θ with coefficients in K. Then the homomorphism theorem for rings (see Theorem 1.40) yields that S is isomorphic to $K[x]/(g)$. But $K[x]/(g)$ is a field by Theorems 1.61 and 1.82(i), and so S is a field. Since $K \subseteq S \subseteq K(\theta)$ and $\theta \in S$, it follows from the definition of $K(\theta)$ that $S = K(\theta)$, and (i) is thus shown.

(ii) Since $S = K(\theta)$, any given $\alpha \in K(\theta)$ can be written in the form $\alpha = f(\theta)$ for some $f \in K[x]$. By the division algorithm, $f = qg + r$ with $q, r \in K[x]$ and $\deg(r) < \deg(g) = n$. Then $\alpha = f(\theta) = q(\theta)g(\theta) + r(\theta) = r(\theta)$, and so α is a linear combination of $1, \theta, \ldots, \theta^{n-1}$ with coefficients in K. On the other hand, if $a_0 + a_1\theta + \cdots + a_{n-1}\theta^{n-1} = 0$ for certain $a_i \in K$, then the polynomial $h(x) = a_0 + a_1x + \cdots + a_{n-1}x^{n-1} \in K[x]$ has θ as a root and is thus a multiple of g by Theorem 1.82(ii). Since $\deg(h) < n = \deg(g)$, this is only possible if $h = 0$—that is, if all $a_i = 0$. Therefore, the elements $1, \theta, \ldots, \theta^{n-1}$ are linearly independent over K and (ii) follows.

(iii) $K(\theta)$ is a finite extension of K by (ii), and so $\alpha \in K(\theta)$ is algebraic over K by Theorem 1.85. Furthermore, $K(\alpha)$ is a subfield of $K(\theta)$. If d is the degree of α over K, then (ii) and Theorem 1.84 imply that $n = [K(\theta): K] = [K(\theta): K(\alpha)][K(\alpha): K] = [K(\theta): K(\alpha)]d$, hence d divides n. □

The elements of the simple algebraic extension $K(\theta)$ of K are therefore polynomial expressions in θ. Any element of $K(\theta)$ can be uniquely represented in the form $a_0 + a_1\theta + \cdots + a_{n-1}\theta^{n-1}$ with $a_i \in K$ for $0 \leqslant i \leqslant n-1$.

It should be pointed out that Theorem 1.86 operates under the assumption that both K and θ are embedded in a larger field F. This is necessary in order that algebraic expressions involving θ make sense. We now want to construct a simple algebraic extension *ab ovo*—that is, without reference to a previously given larger field. The clue to this is contained in part (i) of Theorem 1.86.

1.87. Theorem. *Let $f \in K[x]$ be irreducible over the field K. Then there exists a simple algebraic extension of K with a root of f as a defining element.*

Proof. Consider the residue class ring $L = K[x]/(f)$, which is a field by Theorem 1.61. The elements of L are the residue classes $[h] = h + (f)$ with $h \in K[x]$. For any $a \in K$ we can form the residue class $[a]$ determined by the constant polynomial a, and if $a, b \in K$ are distinct, then $[a] \neq [b]$ since f has positive degree. The mapping $a \mapsto [a]$ gives an isomorphism from K onto a subfield K' of L, so that K' may be identified with K. In other words, we can view L as an extension of K. For every $h(x) = a_0 + a_1x + \cdots + a_mx^m \in K[x]$ we have $[h] = [a_0 + a_1x + \cdots + a_mx^m] = [a_0] + [a_1][x] + \cdots + [a_m][x]^m = a_0 + a_1[x] + \cdots + a_m[x]^m$ by the rules for operating with residue classes and the identification $[a_i] = a_i$. Thus, every element of L can be written as a polynomial expression in $[x]$ with coefficients in K. Since any field containing both K and $[x]$ must contain these polynomial expressions, L is a simple extension of K obtained by adjoining $[x]$. If $f(x) = b_0 + b_1x + \cdots + b_nx^n$, then $f([x]) = b_0 + b_1[x] + \cdots + b_n[x]^n = [b_0 + b_1x + \cdots + b_nx^n] = [f] = [0]$, so that $[x]$ is a root of f and L is a simple algebraic extension of K. □

1.88. Example. As an example of the formal process of root adjunction in Theorem 1.87, consider the prime field $\mathbb{F}_3$ and the polynomial $f(x) = x^2 + x + 2 \in \mathbb{F}_3[x]$, which is irreducible over $\mathbb{F}_3$. Let θ be a "root" of f; that is, θ is the residue class $x + (f)$ in $L = \mathbb{F}_3[x]/(f)$. The other root of f in L is then $2\theta + 2$, since $f(2\theta + 2) = (2\theta + 2)^2 + (2\theta + 2) + 2 = \theta^2 + \theta + 2 = 0$. By Theorem 1.86(ii), or by the known structure of a residue class field, the simple algebraic extension $L = \mathbb{F}_3(\theta)$ consists of the nine elements $0, 1, 2, \theta, \theta + 1, \theta + 2, 2\theta, 2\theta + 1, 2\theta + 2$. The operation tables for L can be constructed as in Example 1.62. □

We observe that in the above example we may adjoin either the root θ or the root $2\theta + 2$ of f and we would still obtain the same field. This situation is covered by the following result, which is easily established.

1.89. Theorem. *Let α and β be two roots of the polynomial $f \in K[x]$ that is irreducible over K. Then $K(\alpha)$ and $K(\beta)$ are isomorphic under an isomorphism mapping α to β and keeping the elements of K fixed.*

We are now asking for an extension field to which all roots of a given polynomial belong.

1.90. Definition. Let $f \in K[x]$ be of positive degree and F an extension field of K. Then f is said to *split in F* if f can be written as a product of linear factors in $F[x]$—that is, if there exist elements $\alpha_1, \alpha_2, \ldots, \alpha_n \in F$ such that

$$f(x) = a(x - \alpha_1)(x - \alpha_2) \cdots (x - \alpha_n),$$

where a is the leading coefficient of f. The field F is a *splitting field* of f over K if f splits in F and if, moreover, $F = K(\alpha_1, \alpha_2, \ldots, \alpha_n)$.

It is clear that a splitting field F of f over K is in the following sense the smallest field containing all the roots of f: no proper subfield of F that is an extension of K contains all the roots of f. By repeatedly applying the process used in Theorem 1.87, one obtains the first part of the subsequent result. The second part is an extension of Theorem 1.89.

1.91. Theorem (Existence and Uniqueness of Splitting Field). *If K is a field and f any polynomial of positive degree in $K[x]$, then there exists a splitting field of f over K. Any two splitting fields of f over K are isomorphic under an isomorphism which keeps the elements of K fixed and maps roots of f into each other.*

Since isomorphic fields may be identified, we can speak of *the* splitting field of f over K. It is obtained from K by adjoining finitely many algebraic elements over K, and therefore one can show on the basis of Theorems 1.84 and 1.86(ii) that the splitting field of f over K is a finite extension of K.

As an illustration of the usefulness of splitting fields, we consider the question of deciding whether a given polynomial has a multiple root (compare with Definition 1.65).

1.92. Definition. Let $f \in K[x]$ be a polynomial of degree $n \geqslant 2$ and suppose that $f(x) = a_0(x - \alpha_1) \cdots (x - \alpha_n)$ with $\alpha_1, \ldots, \alpha_n$ in the splitting field of f over K. Then the *discriminant* $D(f)$ of f is defined by

$$D(f) = a_0^{2n-2} \prod_{1 \leqslant i < j \leqslant n} (\alpha_i - \alpha_j)^2.$$

It is obvious from the definition of $D(f)$ that f has a multiple root if and only if $D(f) = 0$. Although $D(f)$ is defined in terms of elements of an extension of K, it is actually an element of K itself. For small n this can be seen by direct calculation. For instance, if $n = 2$ and $f(x) = ax^2 + bx + c = a(x - \alpha_1)(x - \alpha_2)$, then $D(f) = a^2(\alpha_1 - \alpha_2)^2 = a^2((\alpha_1 + \alpha_2)^2 - 4\alpha_1\alpha_2) = a^2(b^2a^{-2} - 4ca^{-1})$, hence

$$D(ax^2 + bx + c) = b^2 - 4ac,$$

a well-known expression from the theory of quadratic equations. If $n = 3$ and $f(x) = ax^3 + bx^2 + cx + d = a(x - \alpha_1)(x - \alpha_2)(x - \alpha_3)$, then $D(f) = a^4(\alpha_1 - \alpha_2)^2(\alpha_1 - \alpha_3)^3(\alpha_2 - \alpha_3)^3$, and a more involved computation yields

$$D(ax^3 + bx^2 + cx + d) = b^2c^2 - 4b^3d - 4ac^3 - 27a^2d^2 + 18abcd. \quad (1.9)$$

In the general case, consider first the polynomial $s \in K[x_1,\ldots,x_n]$ given by

$$s(x_1,\ldots,x_n) = a_0^{2n-2} \prod_{1 \leqslant i < j \leqslant n} (x_i - x_j)^2.$$

Then s is a symmetric polynomial, and by a result in Example 1.74 it can be written as a polynomial expression in the elementary symmetric polynomials $\sigma_1,\ldots,\sigma_n$—that is, $s = h(\sigma_1,\ldots,\sigma_n)$ for some $h \in K[x_1,\ldots,x_n]$. If $f(x) = a_0x^n + a_1x^{n-1} + \cdots + a_n = a_0(x-\alpha_1)\cdots(x-\alpha_n)$, then the definition of the elementary symmetric polynomials (see again Example 1.74) implies that $\sigma_k(\alpha_1,\ldots,\alpha_n) = (-1)^k a_k a_0^{-1} \in K$ for $1 \leqslant k \leqslant n$. Thus,

$$\begin{aligned} D(f) = s(\alpha_1,\ldots,\alpha_n) &= h(\sigma_1(\alpha_1,\ldots,\alpha_n),\ldots,\sigma_n(\alpha_1,\ldots,\alpha_n)) \\ &= h\left(-a_1a_0^{-1},\ldots,(-1)^n a_n a_0^{-1}\right) \in K. \end{aligned}$$

Since $D(f) \in K$, it should be possible to calculate $D(f)$ without having to pass to an extension field of K. This can be done via the notion of resultant. We note first that if a polynomial $f \in K[x]$ is given in the form $f(x) = a_0x^n + a_1x^{n-1} + \cdots + a_n$ and we accept the possibility that $a_0 = 0$, then n need not be the degree of f. We speak of n as the *formal degree* of f; it is always greater than or equal to $\deg(f)$.

1.93. Definition. Let $f(x) = a_0x^n + a_1x^{n-1} + \cdots + a_n \in K[x]$ and $g(x) = b_0x^m + b_1x^{m-1} + \cdots + b_m \in K[x]$ be two polynomials of formal degree n resp. m with $n, m \in \mathbb{N}$. Then the *resultant* $R(f,g)$ of the two polynomials is defined by the determinant

$$R(f,g) = \begin{vmatrix} a_0 & a_1 & \cdots & a_n & 0 & & \cdots & 0 \\ 0 & a_0 & a_1 & \cdots & a_n & 0 & \cdots & 0 \\ \vdots & & & & & & & \vdots \\ 0 & \cdots & 0 & a_0 & a_1 & & \cdots & a_n \\ b_0 & b_1 & \cdots & & b_m & 0 & \cdots & 0 \\ 0 & b_0 & b_1 & \cdots & & b_m & \cdots & 0 \\ \vdots & & & & & & & \vdots \\ 0 & \cdots & 0 & b_0 & b_1 & & \cdots & b_m \end{vmatrix} \begin{matrix} \left.\vphantom{\begin{matrix} a \\ a \\ a \\ a \end{matrix}}\right\} m \text{ rows} \\ \left.\vphantom{\begin{matrix} a \\ a \\ a \\ a \end{matrix}}\right\} n \text{ rows} \end{matrix}$$

of order $m + n$.

If $\deg(f) = n$ (i.e., if $a_0 \neq 0$) and $f(x) = a_0(x-\alpha_1)\cdots(x-\alpha_n)$ in the splitting field of f over K, then $R(f,g)$ is also given by the formula

$$R(f,g) = a_0^m \prod_{i=1}^{n} g(\alpha_i). \tag{1.10}$$

In this case, we obviously have $R(f,g) = 0$ if and only if f and g have a common root, which is the same as saying that f and g have a common divisor in $K[x]$ of positive degree.

Theorem 1.68 suggests a connection between the discriminant $D(f)$ and the resultant $R(f, f')$. Let $f \in K[x]$ with $\deg(f) = n \geqslant 2$ and leading coefficient a_0. Then we have, in fact, the identity

$$D(f) = (-1)^{n(n-1)/2} a_0^{-1} R(f, f'), \tag{1.11}$$

where f' is viewed as a polynomial of formal degree $n-1$. The last remark is needed since we may have $\deg(f') < n-1$ and even $f' = 0$ in case K has prime characteristic. At any rate, the identity (1.11) shows that we can obtain $D(f)$ by calculating a determinant of order $2n-1$ with entries in K.

NOTES

1. The definitions and theorems in this chapter can be found in nearly any of the introductory books on modern algebra. To mention a few: Birkhoff and MacLane [1], Fraleigh [1], Herstein [4], Kochendörffer [1], Lang [4], Rédei [10], van der Waerden [2].

There are various alternative definitions of a group; for example, a group may be defined as a nonempty set G together with an associative binary operation such that for all $a, b \in G$ the equations $ax = b$ and $ya = b$ have solutions in G. Apart from the examples already given, important illustrations of the group concept are furnished by matrix groups—that is, sets of matrices with entries in a field that form groups under matrix multiplication. Such groups will occur in Chapter 8. For many other examples of groups we refer to the textbooks mentioned above.

A square table such that in every row and in every column each element of a certain set occurs exactly once is called a *latin square*. Hence, the Cayley table for any finite group forms a latin square. However, not every latin square may be regarded as a Cayley table since the associative law need not hold. See Chapter 9, Section 4, and Dénes and Keedwell [1] for more information on latin squares.

In connection with cyclic groups one can prove easily that any infinite cyclic group is isomorphic to the additive group $\mathbb{Z}$ of the integers and any cyclic group of order n is isomorphic to $\mathbb{Z}_n$.

We mention the definitions of algebraic systems that are even simpler than groups, insofar as only a part of the group axioms is assumed. A set with a binary operation is called a *groupoid*; if, in addition, associativity is assumed, then we speak of a *semigroup*. A semigroup with an identity element is called a *monoid*.

2. There are various definitions of a ring. For instance, some authors drop the associativity of multiplication and call the structure introduced in Definition 1.28 an associative ring. The requirement of the

existence of a multiplicative identity in an integral domain is sometimes omitted.

The first abstract definition of a field was given by Weber [3]. The finite fields $\mathbb{F}_p$, p prime, were already studied extensively by Gauss [1] in the context of congruences in $\mathbb{Z}$ with respect to prime moduli.

The characteristic of a field is equal to the characteristic of its prime subfield. There are fields of prime characteristic that are not finite. To get examples, consider suitable extensions of $\mathbb{F}_p$, such as the field of rational functions over $\mathbb{F}_p$ or the algebraic closure of $\mathbb{F}_p$ (compare with the notes on Section 4).

Many properties of the integers can be translated into properties of the corresponding principal ideals in the ring $\mathbb{Z}$. This is based on the fact that the integer a divides the integer b if and only if the principal ideal (a) contains the principal ideal (b). Of particular interest are the prime numbers. According to the usual definition, a prime is an integer >1 that has no nontrivial divisors. Alternatively, one could define a prime as an integer >1 that divides a product of integers only if it divides at least one of the factors. Phrased in terms of ideals, these characterizations lead to the definition of maximal and prime ideals.

3. In the usual definition of a polynomial as an expression of the form $a_0 + a_1x + \cdots + a_nx^n$, the question of how the coefficients a_i and the indeterminate x are connected is glossed over or altogether avoided. There is, however, a way of giving a rigorous definition of a polynomial as an element of a polynomial ring.

For this definition of a polynomial ring, we consider the set S of all infinite sequences of the form

$$(a_0, a_1, \ldots, a_n, \ldots),$$

where the components a_i are elements of a commutative ring R with identity 1 and at most finitely many a_i are allowed to be different from 0. One can easily show that the set S forms a commutative ring with identity with respect to the following operations of addition and multiplication:

$$(a_0, a_1, \ldots) + (b_0, b_1, \ldots) = (a_0 + b_0, a_1 + b_1, \ldots),$$
$$(a_0, a_1, \ldots)(b_0, b_1, \ldots) = (a_0b_0, a_0b_1 + a_1b_0, \ldots),$$

the $(n+1)$st component in the product being $a_0b_n + a_1b_{n-1} + \cdots + a_{n-1}b_1 + a_nb_0$. The zero element of this ring S is obviously $(0,0,\ldots)$ and the identity is the sequence $(1,0,0,\ldots)$.

The set P of special sequences $(a_0,0,0,\ldots)$, where at most the first component is different from 0, forms a subring of S. This subring P and the given ring R are isomorphic via the mapping $(a_0,0,0,\ldots) \mapsto a_0$ from P onto R. Thus we identify these two rings and write $(a_0,0,0,\ldots) = a_0$. Hence R can be regarded as a subring of S, and S is an extension ring of R.

We introduce the notation $x=(0,1,0,\ldots)$ for this special sequence and verify that

$$x^n=(0,\ldots,0,1,0,\ldots) \text{ for } n\geqslant 1,$$

where 1 is the $(n+1)$st component. If we define $x^0=(1,0,0,\ldots)=1$, we have

$$\begin{aligned}(a_0,a_1,a_2,\ldots)&=(a_0,0,0,\ldots)+(0,a_1,0,\ldots)+(0,0,a_2,0,\ldots)+\cdots\\&=(a_0,0,0,\ldots)(1,0,0,\ldots)+(a_1,0,0,\ldots)(0,1,0,\ldots)\\&\quad+(a_2,0,0,\ldots)(0,0,1,0,\ldots)+\cdots\\&=(a_0,0,0,\ldots)1+(a_1,0,0,\ldots)x+(a_2,0,0,\ldots)x^2+\cdots\\&=a_0+a_1x+a_2x^2+\cdots+a_nx^n\\&=f(x)\end{aligned}$$

for any sequence belonging to S. Thus the elements of the ring S are the polynomials $f(x)\in R[x]$, defined as infinite sequences with only finitely many components $a_i\neq 0$.

We emphasize again that the reason for this kind of definition of polynomials $f(x)$ over R is to clarify the relation between the elements of R and the new element x. The process of passing from R to the ring S of polynomials in x is called ring adjunction of x to R. The polynomial ring $R[x]$ can also be regarded as a subring of the ring of formal power series over R, which will be introduced in Chapter 8.

By considering the properties of the ring of integers and of polynomial rings over fields, one soon notices similarities. Actually, both types of rings belong to the same special class of Euclidean rings. A *Euclidean ring* is a commutative ring R with at least two elements, that has no zero divisors, and for which there exists a mapping ν from the set of nonzero elements of R to the set of nonnegative integers such that: (i) if $a,b\in R$ with $ab\neq 0$, then $\nu(ab)\geqslant\nu(a)$; (ii) for $a,b\in R$ with $b\neq 0$, there exist elements $q,r\in R$ with $a=qb+r$ and either $r=0$ or $\nu(r)<\nu(b)$. The mapping ν is often called a (*Euclidean*) *valuation* on the ring R. We see at once that the integers form a Euclidean ring with "absolute value" as a valuation, and a polynomial ring over a field is a Euclidean ring with "degree" as a valuation. As a general result, one shows that any Euclidean ring is a principal ideal domain.

The property stated in Theorem 1.59 also holds in more general contexts and leads to the following definition. An integral domain in which a unique factorization theorem holds—that is, in which every nonunit $\neq 0$ can be expressed uniquely (up to units and the order of the factors) as a product of prime elements—is called a *unique factorization domain*. Thus, put succinctly, Theorem 1.59 says that $F[x]$ is a unique factorization domain. More generally, any principal ideal domain is also a unique factorization domain. The Chinese remainder theorem for $F[x]$ (see Exercise 1.37) is a special case of a general result of this type shown in Lang [4, Ch. 2].

Good sources for facts about polynomials in one and several indeterminates are Rédei [10] and van der Waerden [2]. A more advanced monograph on polynomials is Lausch and Nöbauer [1].

4. In this section, Theorems 1.86 and 1.87 are the key theorems. In fact, one could say that Theorem 1.87 constitutes one of the most fundamental results in the theory of fields. For this result, due to Kronecker [8], assures us that given any nonconstant polynomial over any field, there must be an extension field in which the polynomial has a root. Moreover, the proof of the theorem does more than merely prove existence, as it also provides a method for constructing the required field.

One can classify the elements in an extension F of a field K according to their relation to K. If $\theta \in F$, then either $K(\theta)$ is isomorphic to $K(x)$, the field of rational functions over K (also called the quotient field of $K[x]$), or θ is a root of an irreducible polynomial g in $K[x]$ and $K(\theta)$ is isomorphic to $K[x]/(g)$, as stated in Theorem 1.86. In the first case, θ is called *transcendental* over K, in the second case θ is algebraic over K, as we already know. Extensions F of K that are not algebraic extensions are called *transcendental extensions* of K. Examples of transcendental elements exist in abundance. For instance, most real numbers (such as e, π, $2^{\sqrt{2}},\ldots$) are transcendental over the field $\mathbb{Q}$ of rationals.

Splitting fields not only exist for a single nonconstant polynomial in $K[x]$, but for any collection of nonconstant polynomials over K. The splitting field over K of the collection of all nonconstant polynomials in $K[x]$ is called the *algebraic closure* $\overline{K}$ of K. It is an algebraic extension of K with the additional property that any nonconstant polynomial in $\overline{K}[x]$ splits in $\overline{K}$. For $K = \mathbb{Q}$ and $K = \mathbb{F}_p$, the algebraic closure $\overline{K}$ is an example of an algebraic extension that is not a finite extension of K.

The abstract theory of field extensions was developed in the fundamental paper of Steinitz [1]. Earlier investigations in this direction were carried out by Kneser [1], Kronecker [5], [8], and Weber [3].

EXERCISES

1.1. Prove that the identity element of a group is uniquely determined.

1.2. For a multiplicative group G, prove that a nonempty subset H of G is a subgroup of G if and only if $a, b \in H$ implies $ab^{-1} \in H$. If H is finite, then the condition can be replaced by: $a, b \in H$ implies $ab \in H$.

1.3. Let a be an element of finite order k in the multiplicative group G. Show that for $m \in \mathbb{Z}$ we have $a^m = e$ if and only if k divides m.

1.4. For $m \in \mathbb{N}$, Euler's function $\phi(m)$ is defined to be the number of integers k with $1 \leqslant k \leqslant m$ and $\gcd(k, m) = 1$. Show the following properties for $m, n, s \in \mathbb{N}$ and a prime p:

(a) $\phi(p^s) = p^s\left(1 - \dfrac{1}{p}\right)$;

(b) $\phi(mn) = \phi(m)\phi(n)$ if $\gcd(m, n) = 1$;

(c) $\phi(m) = m\left(1 - \dfrac{1}{p_1}\right)\cdots\left(1 - \dfrac{1}{p_r}\right)$, where $m = p_1^{e_1} \cdots p_r^{e_r}$ is the prime factor decomposition of m.

1.5. Calculate $\phi(490)$ and $\phi(768)$.

1.6. Use the class equation to show the following: if the order of a finite group is a prime power p^s, p prime, $s \geqslant 1$, then the order of its center is divisible by p.

1.7. Prove that in a ring R we have $(-a)(-b) = ab$ for all $a, b \in R$.

1.8. Prove that in a commutative ring R the formula

$$(a+b)^n = a^n + \binom{n}{1} a^{n-1}b + \cdots + \binom{n}{n-1} ab^{n-1} + b^n$$

holds for all $a, b \in R$ and $n \in \mathbb{N}$. (Binomial Theorem)

1.9. Let p be a prime number in $\mathbb{Z}$. For all integers a not divisible by p, show that p divides $a^{p-1} - 1$. (Fermat's Little Theorem)

1.10. Prove that for any prime p we have $(p-1)! \equiv -1 \bmod p$. (Wilson's Theorem)

1.11. Prove: if p is a prime, we have $\dbinom{p-1}{j} \equiv (-1)^j \bmod p$ for $0 \leqslant j \leqslant p-1$, $j \in \mathbb{Z}$.

1.12. A conjecture of Fermat stated that for all $n \geqslant 0$ the integer $2^{2^n} + 1$ is a prime. Euler found to the contrary that 641 divides $2^{32} + 1$. Confirm this by using congruences.

1.13. Prove: if $m_1, \ldots, m_k$ are positive integers that are pairwise relatively prime—that is, $\gcd(m_i, m_j) = 1$ for $1 \leqslant i < j \leqslant k$—then for any integers $a_1, \ldots, a_k$ the system of congruences $y \equiv a_i \bmod m_i$, $i = 1, 2, \ldots, k$, has a simultaneous solution y that is uniquely determined modulo $m = m_1 \cdots m_k$. (Chinese Remainder Theorem)

1.14. Solve the system of congruences $5x \equiv 20 \bmod 6$, $6x \equiv 6 \bmod 5$, $4x \equiv 5 \bmod 77$.

1.15. For a commutative ring R of prime characteristic p, show that

$$(a_1 + \cdots + a_s)^{p^n} = a_1^{p^n} + \cdots + a_s^{p^n}$$

for all $a_1, \ldots, a_s \in R$ and $n \in \mathbb{N}$.

1.16. Deduce from Exercise 1.11 that in a commutative ring R of prime characteristic p we have

$$(a-b)^{p-1} = \sum_{j=0}^{p-1} a^j b^{p-1-j} \quad \text{for all } a, b \in R.$$

1.17. Let F be a field and $f \in F[x]$. Prove that $\{g(f(x)) : g \in F[x]\}$ is equal to $F[x]$ if and only if $\deg(f) = 1$.

1.18. Show that $p^2(x) - xq^2(x) = xr^2(x)$ for $p, q, r \in \mathbb{R}[x]$ implies $p = q = r = 0$.

1.19. Show that if $f, g \in F[x]$, then the principal ideal (f) is contained in the principal ideal (g) if and only if g divides f.

1.20. Prove: if $f, g \in F[x]$ are relatively prime and not both constant, then there exist $a, b \in F[x]$ such that $af + bg = 1$ and $\deg(a) < \deg(g)$, $\deg(b) < \deg(f)$.

1.21. Let $f_1, \ldots, f_n \in F[x]$ with $\gcd(f_1, \ldots, f_n) = d$, so that $f_i = dg_i$ with $g_i \in F[x]$ for $1 \leqslant i \leqslant n$. Prove that $g_1, \ldots, g_n$ are relatively prime.

1.22. Prove that $\gcd(f_1, \ldots, f_n) = \gcd(\gcd(f_1, \ldots, f_{n-1}), f_n)$ for $n \geqslant 3$.

1.23. Prove: if $f, g, h \in F[x]$, f divides gh, and $\gcd(f, g) = 1$, then f divides h.

1.24. Use the Euclidean algorithm to compute $\gcd(f, g)$ for the polynomials f and g with coefficients in the indicated field F:

(a) $F = \mathbb{Q}$, $f(x) = x^7 + 2x^5 + 2x^2 - x + 2$, $g(x) = x^6 - 2x^5 - x^4 + x^2 + 2x + 3$

(b) $F = \mathbb{F}_2$, $f(x) = x^7 + 1$, $g(x) = x^5 + x^3 + x + 1$

(c) $F = \mathbb{F}_2$, $f(x) = x^5 + x + 1$, $g(x) = x^6 + x^5 + x^4 + 1$

(d) $F = \mathbb{F}_3$, $f(x) = x^8 + 2x^5 + x^3 + x^2 + 1$, $g(x) = 2x^6 + x^5 + 2x^3 + 2x^2 + 2$

1.25. Let $f_1, \ldots, f_n$ be nonzero polynomials in $F[x]$. By considering the intersection $(f_1) \cap \cdots \cap (f_n)$ of principal ideals, prove the existence and uniqueness of the monic polynomial $m \in F[x]$ with the properties attributed to the least common multiple of $f_1, \ldots, f_n$.

1.26. Prove (1.6).

1.27. If $f_1, \ldots, f_n \in F[x]$ are nonzero polynomials that are pairwise relatively prime, show that $\operatorname{lcm}(f_1, \ldots, f_n) = a^{-1} f_1 \cdots f_n$, where a is the leading coefficient of $f_1 \cdots f_n$.

1.28. Prove that $\operatorname{lcm}(f_1, \ldots, f_n) = \operatorname{lcm}(\operatorname{lcm}(f_1, \ldots, f_{n-1}), f_n)$ for $n \geqslant 3$.

1.29. Let $f_1, \ldots, f_n \in F[x]$ be nonzero polynomials. Write the canonical factorization of each f_i, $1 \leqslant i \leqslant n$, in the form

$$f_i = a_i \prod p^{e_i(p)},$$

where $a_i \in F$, the product is extended over all monic irreducible polynomials p in $F[x]$, the $e_i(p)$ are nonnegative integers, and for each i we have $e_i(p) > 0$ for only finitely many p. For each p set $m(p) = \min(e_1(p), \ldots, e_n(p))$ and $M(p) = \max(e_1(p), \ldots, e_n(p))$. Prove that

$$\gcd(f_1, \ldots, f_n) = \prod p^{m(p)},$$

$$\operatorname{lcm}(f_1, \ldots, f_n) = \prod p^{M(p)}.$$

1.30. Kronecker's method for finding divisors of degree $\leqslant s$ of a nonconstant polynomial $f \in \mathbb{Q}[x]$ proceeds as follows:

(1) By multiplying f by a constant, we can assume $f \in \mathbb{Z}[x]$.

(2) Choose distinct elements $a_0, \ldots, a_s \in \mathbb{Z}$ that are not roots of f and determine all divisors of $f(a_i)$ for each $i, 0 \leqslant i \leqslant s$.

(3) For each $(s+1)$-tuple $(b_0, \ldots, b_s)$ with b_i dividing $f(a_i)$ for $0 \leqslant i \leqslant s$, determine the polynomial $g \in \mathbb{Q}[x]$ with $\deg(g) \leqslant s$ and $g(a_i) = b_i$ for $0 \leqslant i \leqslant s$ (for instance, by the Lagrange interpolation formula).

(4) Decide which of these polynomials g in (3) are divisors of f.

If $\deg(f) = n \geqslant 1$ and s is taken to be the greatest integer $\leqslant n/2$, then f is irreducible in $\mathbb{Q}[x]$ in case the method only yields constant polynomials as divisors. Otherwise, Kronecker's method yields a nontrivial factorization. By applying the method again to the factors and repeating the process, one eventually gets the canonical factorization of f. Use this procedure to find the canonical factorization of

$$f(x) = \tfrac{1}{3}x^6 - \tfrac{5}{3}x^5 + 2x^4 - x^3 + 5x^2 - \tfrac{17}{3}x - 1 \in \mathbb{Q}[x].$$

1.31. Construct the addition and multiplication table for $\mathbb{F}_2[x]/(x^3 + x^2 + x)$. Determine whether or not this ring is a field.

1.32. Let $[x+1]$ be the residue class of $x+1$ in $\mathbb{F}_2[x]/(x^4+1)$. Find the residue classes comprising the principal ideal $([x+1])$ in $\mathbb{F}_2[x]/(x^4+1)$.

1.33. Let F be a field and $a, b, g \in F[x]$ with $g \neq 0$. Prove that the congruence $af \equiv b \bmod g$ has a solution $f \in F[x]$ if and only if $\gcd(a, g)$ divides b.

1.34. Solve the congruence $(x^2+1)f(x) \equiv 1 \bmod (x^3+1)$ in $\mathbb{F}_3[x]$, if possible.

1.35. Solve $(x^4 + x^3 + x^2 + 1)f(x) \equiv (x^2+1) \bmod (x^3+1)$ in $\mathbb{F}_2[x]$, if possible.

1.36. Prove that $R[x]/(x^4 + x^3 + x + 1)$ cannot be a field, no matter what the commutative ring R with identity is.

1.37. Prove: given a field F, nonzero polynomials $f_1, \ldots, f_k \in F[x]$ that are pairwise relatively prime, and arbitrary polynomials $g_1, \ldots, g_k \in F[x]$, then the simultaneous congruences $h \equiv g_i \bmod f_i, i = 1, 2, \ldots, k$, have a unique solution $h \in F[x]$ modulo $f = f_1 \cdots f_k$. (Chinese Remainder Theorem for $F[x]$)

1.38. Evaluate $f(3)$ for $f(x) = x^{214} + 3x^{152} + 2x^{47} + 2 \in \mathbb{F}_5[x]$.

1.39. Let p be a prime and $a_0, \ldots, a_n$ integers with p not dividing a_n. Show that $a_0 + a_1 y + \cdots + a_n y^n \equiv 0 \bmod p$ has at most n different solutions y modulo p.

1.40. If $p > 2$ is a prime, show that there are exactly two elements $a \in \mathbb{F}_p$ such that $a^2 = 1$.

1.41. Show: if $f \in \mathbb{Z}[x]$ and $f(0) \equiv f(1) \equiv 1 \bmod 2$, then f has no roots in $\mathbb{Z}$.

1.42. Let p be a prime and $f \in \mathbb{Z}[x]$. Show: $f(a) \equiv 0 \bmod p$ holds for all $a \in \mathbb{Z}$ if and only if $f(x) = (x^p - x)g(x) + ph(x)$ with $g, h \in \mathbb{Z}[x]$.

1.43. Let p be a prime integer and c an element of the field F. Show that $x^p - c$ is irreducible over F if and only if $x^p - c$ has no root in F.

1.44. Show that for a polynomial $f \in F[x]$ of positive degree the following conditions are equivalent:
(a) f is irreducible over F;
(b) the principal ideal (f) of $F[x]$ is a maximal ideal;
(c) the principal ideal (f) of $F[x]$ is a prime ideal.

1.45. Show the following properties of the derivative for polynomials in $F[x]$:
(a) $(f_1 + \cdots + f_m)' = f_1' + \cdots + f_m'$;
(b) $(fg)' = f'g + fg'$;
(c) $(f_1 \cdots f_m)' = \sum_{i=1}^{m} f_1 \cdots f_{i-1} f_i' f_{i+1} \cdots f_m$.

1.46. For $f \in F[x]$ and F of characteristic 0, prove that $f' = 0$ if and only if f is a constant polynomial. If F has prime characteristic p, prove that $f' = 0$ if and only if $f(x) = g(x^p)$ for some $g \in F[x]$.

1.47. Prove Theorem 1.68.

1.48. Prove that the nonzero polynomial $f \in F[x]$ has a multiple root (in some extension field of F) if and only if f and f' are not relatively prime.

1.49. Use the criterion in the previous exercise to determine whether the following polynomials have a multiple root:
(a) $f(x) = x^4 - 5x^3 + 6x^2 + 4x - 8 \in \mathbb{Q}[x]$
(b) $f(x) = x^6 + x^5 + x^4 + x^3 + 1 \in \mathbb{F}_2[x]$

1.50. The nth derivative $f^{(n)}$ of $f \in F[x]$ is defined recursively as follows: $f^{(0)} = f$, $f^{(n)} = (f^{(n-1)})'$ for $n \geqslant 1$. Prove that for $f, g \in F[x]$ we have

$$(fg)^{(n)} = \sum_{i=0}^{n} \binom{n}{i} f^{(n-i)} g^{(i)}.$$

1.51. Let F be a field and k a positive integer such that $k < p$ in case F has prime characteristic p. Prove: $b \in F$ is a root of $f \in F[x]$ of multiplicity k if and only if $f^{(i)}(b) = 0$ for $0 \leqslant i \leqslant k-1$ and $f^{(k)}(b) \neq 0$.

1.52. Show that the Lagrange interpolation formula can also be written in the form

$$f(x) = \sum_{i=0}^{n} b_i \big(g'(a_i)\big)^{-1} \frac{g(x)}{x - a_i} \quad \text{with } g(x) = \prod_{k=0}^{n} (x - a_k).$$

1.53. Determine a polynomial $f \in \mathbb{F}_5[x]$ with $f(0) = f(1) = f(4) = 1$ and $f(2) = f(3) = 3$.

1.54. Determine a polynomial $f \in \mathbb{Q}[x]$ of degree $\leqslant 3$ such that $f(-1) = -1$, $f(0) = 3$, $f(1) = 3$, and $f(2) = 5$.

1.55. Express $s_5(x_1, x_2, x_3, x_4) = x_1^5 + x_2^5 + x_3^5 + x_4^5 \in \mathbb{F}_3[x_1, x_2, x_3, x_4]$ in terms of the elementary symmetric polynomials $\sigma_1, \sigma_2, \sigma_3, \sigma_4$.

1.56. Prove that a subset K of a field F is a subfield if and only if the following conditions are satisfied:
(a) K contains at least two elements;
(b) if $a, b \in K$, then $a - b \in K$;
(c) if $a, b \in K$ and $b \neq 0$, then $ab^{-1} \in K$.

1.57. Prove that an extension L of the field K is a finite extension if and only if L can be obtained from K by adjoining finitely many algebraic elements over K.

1.58. Prove: if θ is algebraic over L and L is an algebraic extension of K, then θ is algebraic over K. Thus show that if F is an algebraic extension of L, then F is an algebraic extension of K.

1.59. Prove: if the degree $[L : K]$ is a prime, then the only fields F with $K \subseteq F \subseteq L$ are $F = K$ and $F = L$.

1.60. Construct the operation tables for the field $L = \mathbb{F}_3(\theta)$ in Example 1.88.

1.61. Show that $f(x) = x^4 + x + 1 \in \mathbb{F}_2[x]$ is irreducible over $\mathbb{F}_2$. Then construct the operation tables for the simple extension $\mathbb{F}_2(\theta)$, where θ is a root of f.

1.62. Calculate the discriminant $D(f)$ and decide whether or not f has a multiple root:
(a) $f(x) = 2x^3 - 3x^2 + x + 1 \in \mathbb{Q}[x]$
(b) $f(x) = 2x^4 + x^3 + x^2 + 2x + 2 \in \mathbb{F}_3[x]$

1.63. Deduce (1.9) from (1.11).

1.64. Prove that $f, g \in K[x]$ have a common root (in some extension field of K) if and only if f and g have a common divisor in $K[x]$ of positive degree.

1.65. Determine the common roots of the polynomials $x^7 - 2x^4 - x^3 + 2$ and $x^5 - 3x^4 - x + 3$ in $\mathbb{Q}[x]$.

1.66. Prove: if f and g are as in Definition 1.93, then $R(f, g) = (-1)^{mn} R(g, f)$.

1.67. Let $f, g \in K[x]$ be of positive degree and suppose that $f(x) = a_0(x - \alpha_1) \cdots (x - \alpha_n)$, $a_0 \neq 0$, and $g(x) = b_0(x - \beta_1) \cdots (x - \beta_m)$, $b_0 \neq 0$, in the splitting field of fg over K. Prove that

$$R(f, g) = (-1)^{mn} b_0^n \prod_{j=1}^{m} f(\beta_j) = a_0^m b_0^n \prod_{i=1}^{n} \prod_{j=1}^{m} (\alpha_i - \beta_j),$$

where n and m are also taken as the formal degrees of f and g, respectively.

1.68. Calculate the resultant $R(f, g)$ of the two given polynomials f and g (with the formal degree equal to the degree) and decide whether or not f and g have a common root:
(a) $f(x) = x^3 + x + 1$, $g(x) = 2x^5 + x^2 + 2 \in \mathbb{F}_3[x]$
(b) $f(x) = x^4 + x^3 + 1$, $g(x) = x^4 + x^2 + x + 1 \in \mathbb{F}_2[x]$

1.69. For $f \in K[x_1,\ldots,x_n]$, $n \geqslant 2$, an n-tuple $(\alpha_1,\ldots,\alpha_n)$ of elements α_i belonging to some extension L of K may be called a *zero* of f if $f(\alpha_1,\ldots,\alpha_n)=0$. Now let $f, g \in K[x_1,\ldots,x_n]$ with x_n actually appearing in f and g. Then f and g can be regarded as polynomials $\bar{f}(x_n)$ and $\bar{g}(x_n)$ in $K[x_1,\ldots,x_{n-1}][x_n]$ of positive degree. Their resultant with respect to x_n (with formal degree = degree) is $R(\bar{f}, \bar{g}) = R_{x_n}(f, g)$, which is a polynomial in $x_1,\ldots,x_{n-1}$. Show that f and g have a common zero $(\alpha_1,\ldots,\alpha_{n-1},\alpha_n)$ if and only if $(\alpha_1,\ldots,\alpha_{n-1})$ is a zero of $R(\bar{f}, \bar{g})$.

1.70. Using the result of the previous exercise, determine the common zeros of the polynomials $f(x, y) = x(y^2 - x)^2 + y^5$ and $g(x, y) = y^4 + y^3 - x^2$ in $\mathbb{Q}[x, y]$.

Chapter 2

Structure of Finite Fields

This chapter is of central importance since it contains various fundamental properties of finite fields and a description of methods for constructing finite fields.

The field of integers modulo a prime number is, of course, the most familiar example of a finite field, but many of its properties extend to arbitrary finite fields. The characterization of finite fields (see Section 1) shows that every finite field is of prime-power order and that, conversely, for every prime power there exists a finite field whose number of elements is exactly that prime power. Furthermore, finite fields with the same number of elements are isomorphic and may therefore be identified. The next two sections provide information on roots of irreducible polynomials, leading to an interpretation of finite fields as splitting fields of irreducible polynomials, and on traces, norms, and bases relative to field extensions.

Section 4 treats roots of unity from the viewpoint of general field theory, which will be needed occasionally in Section 6 as well as in Chapter 5. Section 5 presents different ways of representing the elements of a finite field. In Section 6 we give two proofs of the famous theorem of Wedderburn according to which every finite division ring is a field.

Many discussions in this chapter will be followed up, continued, and partly generalized in later chapters.

1. CHARACTERIZATION OF FINITE FIELDS

In the previous chapter we have already encountered a basic class of finite fields—that is, of fields with finitely many elements. For every prime p the residue class ring $\mathbb{Z}/(p)$ forms a finite field with p elements (see Theorem 1.38), which may be identified with the Galois field $\mathbb{F}_p$ of order p (see Definition 1.41). The fields $\mathbb{F}_p$ play an important role in general field theory since every field of characteristic p must contain an isomorphic copy of $\mathbb{F}_p$ by Theorem 1.78 and can thus be thought of as an extension of $\mathbb{F}_p$. This observation, together with the fact that every finite field has prime characteristic (see Corollary 1.45), is fundamental for the classification of finite fields. We first establish a simple necessary condition on the number of elements of a finite field.

2.1. Lemma. *Let F be a finite field containing a subfield K with q elements. Then F has q^m elements, where $m = [F: K]$.*

Proof. F is a vector space over K, and since F is finite, it is finite-dimensional as a vector space over K. If $[F: K] = m$, then F has a basis over K consisting of m elements, say $b_1, b_2, \ldots, b_m$. Thus every element of F can be uniquely represented in the form $a_1b_1 + a_2b_2 + \cdots + a_mb_m$, where $a_1, a_2, \ldots, a_m \in K$. Since each a_i can have q values, F has exactly q^m elements. □

2.2. Theorem. *Let F be a finite field. Then F has p^n elements, where the prime p is the characteristic of F and n is the degree of F over its prime subfield.*

Proof. Since F is finite, its characteristic is a prime p according to Corollary 1.45. Therefore the prime subfield K of F is isomorphic to $\mathbb{F}_p$ by Theorem 1.78 and thus contains p elements. The rest follows from Lemma 2.1. □

Starting from the prime fields $\mathbb{F}_p$, we can construct other finite fields by the process of root adjunction described in Chapter 1, Section 4. If $f \in \mathbb{F}_p[x]$ is an irreducible polynomial over $\mathbb{F}_p$ of degree n, then by adjoining a root of f to $\mathbb{F}_p$ we get a finite field with p^n elements. However, at this stage it is not clear whether for every positive integer n there exists an irreducible polynomial in $\mathbb{F}_p[x]$ of degree n. In order to establish that for every prime p and every $n \in \mathbb{N}$ there is a finite field with p^n elements, we use an approach suggested by the following results.

2.3. Lemma. *If F is a finite field with q elements, then every $a \in F$ satisfies $a^q = a$.*

Proof. The identity $a^q = a$ is trivial for $a = 0$. On the other hand, the nonzero elements of F form a group of order $q - 1$ under multiplication.

Thus $a^{q-1}=1$ for all $a \in F$ with $a \neq 0$, and multiplication by a yields the desired result. □

2.4. Lemma. *If F is a finite field with q elements and K is a subfield of F, then the polynomial $x^q - x$ in $K[x]$ factors in $F[x]$ as*

$$x^q - x = \prod_{a \in F} (x-a)$$

and F is a splitting field of $x^q - x$ over K.

Proof. The polynomial $x^q - x$ of degree q has at most q roots in F. By Lemma 2.3 we know q such roots—namely, all the elements of F. Thus the given polynomial splits in F in the indicated manner, and it cannot split in any smaller field. □

We are now able to prove the main characterization theorem for finite fields, the leading idea being contained in Lemma 2.4.

2.5. Theorem (Existence and Uniqueness of Finite Fields). *For every prime p and every positive integer n there exists a finite field with p^n elements. Any finite field with $q = p^n$ elements is isomorphic to the splitting field of $x^q - x$ over $\mathbb{F}_p$.*

Proof. (*Existence*) For $q = p^n$ consider $x^q - x$ in $\mathbb{F}_p[x]$, and let F be its splitting field over $\mathbb{F}_p$. This polynomial has q distinct roots in F since its derivative is $qx^{q-1} - 1 = -1$ in $\mathbb{F}_p[x]$ and so can have no common root with $x^q - x$ (compare with Theorem 1.68). Let $S = \{a \in F : a^q - a = 0\}$. Then S is a subfield of F since: (i) S contains 0 and 1; (ii) $a, b \in S$ implies by Theorem 1.46 that $(a-b)^q = a^q - b^q = a - b$, and so $a - b \in S$; (iii) for $a, b \in S$ and $b \neq 0$ we have $(ab^{-1})^q = a^q b^{-q} = ab^{-1}$, and so $ab^{-1} \in S$. But, on the other hand, $x^q - x$ must split in S since S contains all its roots. Thus $F = S$, and since S has q elements, F is a finite field with q elements.

(*Uniqueness*) Let F be a finite field with $q = p^n$ elements. Then F has characteristic p by Theorem 2.2 and so contains $\mathbb{F}_p$ as a subfield. It follows from Lemma 2.4 that F is a splitting field of $x^q - x$ over $\mathbb{F}_p$. Thus the desired result is a consequence of the uniqueness (up to isomorphisms) of splitting fields, which was noted in Theorem 1.91. □

The uniqueness part of Theorem 2.5 provides the justification for speaking of *the* finite field (or *the* Galois field) with q elements, or of *the* finite field (or *the* Galois field) of order q. We shall denote this field by $\mathbb{F}_q$, where it is of course understood that q is a power of the prime characteristic p of $\mathbb{F}_q$.

2.6. Theorem (Subfield Criterion). *Let $\mathbb{F}_q$ be the finite field with $q = p^n$ elements. Then every subfield of $\mathbb{F}_q$ has order p^m, where m is a positive divisor of n. Conversely, if m is a positive divisor of n, then there is exactly one subfield of $\mathbb{F}_q$ with p^m elements.*

Proof. It is clear that a subfield K of $\mathbb{F}_q$ has order p^m for some positive integer $m \leqslant n$. Lemma 2.1 shows that $q = p^n$ must be a power of p^m, and so m is necessarily a divisor of n.

Conversely, if m is a positive divisor of n, then $p^m - 1$ divides $p^n - 1$, and so $x^{p^m - 1} - 1$ divides $x^{p^n - 1} - 1$ in $\mathbb{F}_p[x]$. Consequently, $x^{p^m} - x$ divides $x^{p^n} - x = x^q - x$ in $\mathbb{F}_p[x]$. Thus, every root of $x^{p^m} - x$ is a root of $x^q - x$ and so belongs to $\mathbb{F}_q$. It follows that $\mathbb{F}_q$ must contain as a subfield a splitting field of $x^{p^m} - x$ over $\mathbb{F}_p$, and as we have seen in the proof of Theorem 2.5, such a splitting field has order p^m. If there were two distinct subfields of order p^m in $\mathbb{F}_q$, they would together contain more than p^m roots of $x^{p^m} - x$ in $\mathbb{F}_q$, an obvious contradiction. □

The proof of Theorem 2.6 shows that the unique subfield of $\mathbb{F}_{p^n}$ of order p^m, where m is a positive divisor of n, consists precisely of the roots of the polynomial $x^{p^m} - x \in \mathbb{F}_p[x]$ in $\mathbb{F}_{p^n}$.

2.7. Example. The subfields of the finite field $\mathbb{F}_{2^{30}}$ can be determined by listing all positive divisors of 30. The containment relations between these various subfields are displayed in the following diagram.

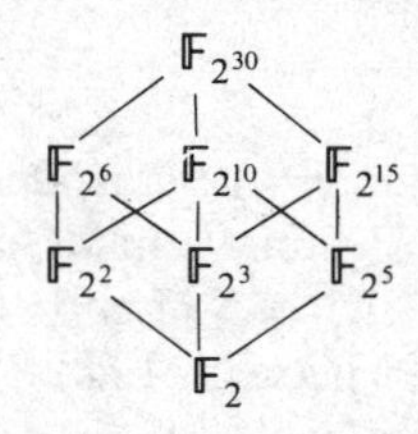

By Theorem 2.6, the containment relations are equivalent to divisibility relations among the positive divisors of 30. □

For a finite field $\mathbb{F}_q$ we denote by $\mathbb{F}_q^*$ the multiplicative group of nonzero elements of $\mathbb{F}_q$. The following result enunciates a useful property of this group.

2.8. *Theorem.* *For every finite field $\mathbb{F}_q$ the multiplicative group $\mathbb{F}_q^*$ of nonzero elements of $\mathbb{F}_q$ is cyclic.*

Proof. We may assume $q \geqslant 3$. Let $h = p_1^{r_1} p_2^{r_2} \cdots p_m^{r_m}$ be the prime factor decomposition of the order $h = q - 1$ of the group $\mathbb{F}_q^*$. For every i, $1 \leqslant i \leqslant m$, the polynomial $x^{h/p_i} - 1$ has at most h/p_i roots in $\mathbb{F}_q$. Since $h/p_i < h$, it follows that there are nonzero elements in $\mathbb{F}_q$ that are not roots of this polynomial. Let a_i be such an element and set $b_i = a_i^{h/p_i^{r_i}}$. We have $b_i^{p_i^{r_i}} = 1$, hence the order of b_i is a divisor of $p_i^{r_i}$ and is therefore of the form $p_i^{s_i}$ with $0 \leqslant s_i \leqslant r_i$. On the other hand,

$$b_i^{p_i^{r_i - 1}} = a_i^{h/p_i} \neq 1,$$

and so the order of b_i is $p_i^{r_i}$. We claim that the element $b = b_1 b_2 \cdots b_m$ has order h. Suppose, on the contrary, that the order of b is a proper divisor of h

and is therefore a divisor of at least one of the m integers h/p_i, $1 \leqslant i \leqslant m$, say of h/p_1. Then we have

$$1 = b^{h/p_1} = b_1^{h/p_1} b_2^{h/p_1} \cdots b_m^{h/p_1}.$$

Now if $2 \leqslant i \leqslant m$, then $p_i^{r_i}$ divides h/p_1, and hence $b_i^{h/p_1} = 1$. Therefore $b_1^{h/p_1} = 1$. This implies that the order of b_1 must divide h/p_1, which is impossible since the order of b_1 is $p_1^{r_1}$. Thus, $\mathbb{F}_q^*$ is a cyclic group with generator b. □

2.9. Definition. A generator of the cyclic group $\mathbb{F}_q^*$ is called a *primitive element* of $\mathbb{F}_q$.

It follows from Theorem 1.15(v) that $\mathbb{F}_q$ contains $\phi(q-1)$ primitive elements, where ϕ is Euler's function. The existence of primitive elements can be used to show a result that implies, in particular, that every finite field can be thought of as a simple algebraic extension of its prime subfield.

2.10. Theorem. *Let $\mathbb{F}_q$ be a finite field and $\mathbb{F}_r$ a finite extension field. Then $\mathbb{F}_r$ is a simple algebraic extension of $\mathbb{F}_q$ and every primitive element of $\mathbb{F}_r$ can serve as a defining element of $\mathbb{F}_r$ over $\mathbb{F}_q$.*

Proof. Let ζ be a primitive element of $\mathbb{F}_r$. We clearly have $\mathbb{F}_q(\zeta) \subseteq \mathbb{F}_r$. On the other hand, $\mathbb{F}_q(\zeta)$ contains 0 and all powers of ζ, and so all elements of $\mathbb{F}_r$. Therefore $\mathbb{F}_r = \mathbb{F}_q(\zeta)$. □

2.11. Corollary. *For every finite field $\mathbb{F}_q$ and every positive integer n, there exists an irreducible polynomial in $\mathbb{F}_q[x]$ of degree n.*

Proof. Let $\mathbb{F}_r$ be the extension field of $\mathbb{F}_q$ of order q^n, so that $[\mathbb{F}_r : \mathbb{F}_q] = n$. By Theorem 2.10 we have $\mathbb{F}_r = \mathbb{F}_q(\zeta)$ for some $\zeta \in \mathbb{F}_r$. Then the minimal polynomial of ζ over $\mathbb{F}_q$ is an irreducible polynomial in $\mathbb{F}_q[x]$ of degree n, according to Theorems 1.82(i) and 1.86(ii). □

2. ROOTS OF IRREDUCIBLE POLYNOMIALS

In this section we collect some information about the set of roots of an irreducible polynomial over a finite field.

2.12. Lemma. *Let $f \in \mathbb{F}_q[x]$ be an irreducible polynomial over a finite field $\mathbb{F}_q$ and let α be a root of f in an extension field of $\mathbb{F}_q$. Then for a polynomial $h \in \mathbb{F}_q[x]$ we have $h(\alpha) = 0$ if and only if f divides h.*

Proof. Let a be the leading coefficient of f and set $g(x) = a^{-1}f(x)$. Then g is a monic irreducible polynomial in $\mathbb{F}_q[x]$ with $g(\alpha) = 0$ and so it is the minimal polynomial of α over $\mathbb{F}_q$ in the sense of Definition 1.81. The rest follows from Theorem 1.82(ii). □

2.13. Lemma. *Let $f \in \mathbb{F}_q[x]$ be an irreducible polynomial over $\mathbb{F}_q$ of degree m. Then $f(x)$ divides $x^{q^n} - x$ if and only if m divides n.*

Proof. Suppose $f(x)$ divides $x^{q^n} - x$. Let α be a root of f in the splitting field of f over $\mathbb{F}_q$. Then $\alpha^{q^n} = \alpha$, so that $\alpha \in \mathbb{F}_{q^n}$. It follows that $\mathbb{F}_q(\alpha)$ is a subfield of $\mathbb{F}_{q^n}$. But since $[\mathbb{F}_q(\alpha):\mathbb{F}_q] = m$ and $[\mathbb{F}_{q^n}:\mathbb{F}_q] = n$, Theorem 1.84 shows that m divides n.

Conversely, if m divides n, then Theorem 2.6 implies that $\mathbb{F}_{q^n}$ contains $\mathbb{F}_{q^m}$ as a subfield. If α is a root of f in the splitting field of f over $\mathbb{F}_q$, then $[\mathbb{F}_q(\alpha):\mathbb{F}_q] = m$, and so $\mathbb{F}_q(\alpha) = \mathbb{F}_{q^m}$. Consequently, we have $\alpha \in \mathbb{F}_{q^n}$, hence $\alpha^{q^n} = \alpha$, and thus α is a root of $x^{q^n} - x \in \mathbb{F}_q[x]$. We infer then from Lemma 2.12 that $f(x)$ divides $x^{q^n} - x$. □

2.14. Theorem. *If f is an irreducible polynomial in $\mathbb{F}_q[x]$ of degree m, then f has a root α in $\mathbb{F}_{q^m}$. Furthermore, all the roots of f are simple and are given by the m distinct elements $\alpha, \alpha^q, \alpha^{q^2}, \ldots, \alpha^{q^{m-1}}$ of $\mathbb{F}_{q^m}$.*

Proof. Let α be a root of f in the splitting field of f over $\mathbb{F}_q$. Then $[\mathbb{F}_q(\alpha):\mathbb{F}_q] = m$, hence $\mathbb{F}_q(\alpha) = \mathbb{F}_{q^m}$, and in particular $\alpha \in \mathbb{F}_{q^m}$. Next we show that if $\beta \in \mathbb{F}_{q^m}$ is a root of f, then β^q is also a root of f. Write $f(x) = a_m x^m + \cdots + a_1 x + a_0$ with $a_i \in \mathbb{F}_q$ for $0 \leqslant i \leqslant m$. Then, using Lemma 2.3 and Theorem 1.46, we get

$$\begin{aligned} f(\beta^q) &= a_m \beta^{qm} + \cdots + a_1 \beta^q + a_0 = a_m^q \beta^{qm} + \cdots + a_1^q \beta^q + a_0^q \\ &= (a_m \beta^m + \cdots + a_1 \beta + a_0)^q = f(\beta)^q = 0. \end{aligned}$$

Therefore, the elements $\alpha, \alpha^q, \alpha^{q^2}, \ldots, \alpha^{q^{m-1}}$ are roots of f. It remains to prove that these elements are distinct. Suppose, on the contrary, that $\alpha^{q^j} = \alpha^{q^k}$ for some integers j and k with $0 \leqslant j < k \leqslant m-1$. By raising this identity to the power q^{m-k}, we get

$$\alpha^{q^{m-k+j}} = \alpha^{q^m} = \alpha.$$

It follows then from Lemma 2.12 that $f(x)$ divides $x^{q^{m-k+j}} - x$. By Lemma 2.13, this is only possible if m divides $m - k + j$. But we have $0 < m - k + j < m$, and so we arrive at a contradiction. □

2.15. Corollary. *Let f be an irreducible polynomial in $\mathbb{F}_q[x]$ of degree m. Then the splitting field of f over $\mathbb{F}_q$ is given by $\mathbb{F}_{q^m}$.*

Proof. Theorem 2.14 shows that f splits in $\mathbb{F}_{q^m}$. Furthermore, $\mathbb{F}_q(\alpha, \alpha^q, \alpha^{q^2}, \ldots, \alpha^{q^{m-1}}) = \mathbb{F}_q(\alpha) = \mathbb{F}_{q^m}$ for a root α of f in $\mathbb{F}_{q^m}$, where the second identity is taken from the proof of Theorem 2.14. □

2.16. Corollary. *Any two irreducible polynomials in $\mathbb{F}_q[x]$ of the same degree have isomorphic splitting fields.*

We introduce a convenient terminology for the elements appearing in Theorem 2.14, regardless of whether $\alpha \in \mathbb{F}_{q^m}$ is a root of an irreducible polynomial in $\mathbb{F}_q[x]$ of degree m or not.

2.17. Definition. Let $\mathbb{F}_{q^m}$ be an extension of $\mathbb{F}_q$ and let $\alpha \in \mathbb{F}_{q^m}$. Then the elements $\alpha, \alpha^q, \alpha^{q^2}, \ldots, \alpha^{q^{m-1}}$ are called the *conjugates* of α with respect to $\mathbb{F}_q$.

The conjugates of $\alpha \in \mathbb{F}_{q^m}$ with respect to $\mathbb{F}_q$ are distinct if and only if the minimal polynomial of α over $\mathbb{F}_q$ has degree m. Otherwise, the degree d of this minimal polynomial is a proper divisor of m, and then the conjugates of α with respect to $\mathbb{F}_q$ are the distinct elements $\alpha, \alpha^q, \ldots, \alpha^{q^{d-1}}$, each repeated m/d times.

2.18. Theorem. *The conjugates of $\alpha \in \mathbb{F}_q^*$ with respect to any subfield of $\mathbb{F}_q$ have the same order in the group $\mathbb{F}_q^*$.*

Proof. Since $\mathbb{F}_q^*$ is a cyclic group by Theorem 2.8, the result follows from Theorem 1.15(ii) and the fact that every power of the characteristic of $\mathbb{F}_q$ is relatively prime to the order $q-1$ of $\mathbb{F}_q^*$. □

2.19. Corollary. *If α is a primitive element of $\mathbb{F}_q$, then so are all its conjugates with respect to any subfield of $\mathbb{F}_q$.*

2.20. Example. Let $\alpha \in \mathbb{F}_{16}$ be a root of $f(x) = x^4 + x + 1 \in \mathbb{F}_2[x]$. Then the conjugates of α with respect to $\mathbb{F}_2$ are $\alpha, \alpha^2, \alpha^4 = \alpha + 1$, and $\alpha^8 = \alpha^2 + 1$, each of them being a primitive element of $\mathbb{F}_{16}$. The conjugates of α with respect to $\mathbb{F}_4$ are α and $\alpha^4 = \alpha + 1$. □

There is an intimate relationship between conjugate elements and certain automorphisms of a finite field. Let $\mathbb{F}_{q^m}$ be an extension of $\mathbb{F}_q$. By an *automorphism* σ of $\mathbb{F}_{q^m}$ over $\mathbb{F}_q$ we mean an automorphism of $\mathbb{F}_{q^m}$ that fixes the elements of $\mathbb{F}_q$. Thus, in detail, we require that σ be a one-to-one mapping from $\mathbb{F}_{q^m}$ onto itself with $\sigma(\alpha+\beta) = \sigma(\alpha) + \sigma(\beta)$ and $\sigma(\alpha\beta) = \sigma(\alpha)\sigma(\beta)$ for all $\alpha, \beta \in \mathbb{F}_{q^m}$ and $\sigma(a) = a$ for all $a \in \mathbb{F}_q$.

2.21. Theorem. *The distinct automorphisms of $\mathbb{F}_{q^m}$ over $\mathbb{F}_q$ are exactly the mappings $\sigma_0, \sigma_1, \ldots, \sigma_{m-1}$, defined by $\sigma_j(\alpha) = \alpha^{q^j}$ for $\alpha \in \mathbb{F}_{q^m}$ and $0 \leqslant j \leqslant m-1$.*

Proof. For each σ_j and all $\alpha, \beta \in \mathbb{F}_{q^m}$ we obviously have $\sigma_j(\alpha\beta) = \sigma_j(\alpha)\sigma_j(\beta)$, and also $\sigma_j(\alpha+\beta) = \sigma_j(\alpha) + \sigma_j(\beta)$ because of Theorem 1.46, so that σ_j is an endomorphism of $\mathbb{F}_{q^m}$. Furthermore, $\sigma_j(\alpha) = 0$ if and only if $\alpha = 0$, and so σ_j is one-to-one. Since $\mathbb{F}_{q^m}$ is a finite set, σ_j is an epimorphism and therefore an automorphism of $\mathbb{F}_{q^m}$. Moreover, we have $\sigma_j(a) = a$ for all $a \in \mathbb{F}_q$ by Lemma 2.3, and so each σ_j is an automorphism of $\mathbb{F}_{q^m}$ over $\mathbb{F}_q$.

The mappings $\sigma_0, \sigma_1, \ldots, \sigma_{m-1}$ are distinct since they attain distinct values for a primitive element of $\mathbb{F}_{q^m}$.

Now suppose that σ is an arbitrary automorphism of $\mathbb{F}_{q^m}$ over $\mathbb{F}_q$. Let β be a primitive element of $\mathbb{F}_{q^m}$ and let $f(x) = x^m + a_{m-1}x^{m-1} + \cdots + a_0 \in \mathbb{F}_q[x]$ be its minimal polynomial over $\mathbb{F}_q$. Then

$$0 = \sigma\left(\beta^m + a_{m-1}\beta^{m-1} + \cdots + a_0\right)$$
$$= \sigma(\beta)^m + a_{m-1}\sigma(\beta)^{m-1} + \cdots + a_0,$$

so that $\sigma(\beta)$ is a root of f in $\mathbb{F}_{q^m}$. It follows from Theorem 2.14 that $\sigma(\beta) = \beta^{q^j}$ for some j, $0 \leqslant j \leqslant m-1$. Since σ is a homomorphism, we get then $\sigma(\alpha) = \alpha^{q^j}$ for all $\alpha \in \mathbb{F}_{q^m}$. □

On the basis of Theorem 2.21 it is evident that the conjugates of $\alpha \in \mathbb{F}_{q^m}$ with respect to $\mathbb{F}_q$ are obtained by applying all automorphisms of $\mathbb{F}_{q^m}$ over $\mathbb{F}_q$ to the element α. The automorphisms of $\mathbb{F}_{q^m}$ over $\mathbb{F}_q$ form a group with the operation being the usual composition of mappings. The information provided in Theorem 2.21 shows that this group of automorphisms of $\mathbb{F}_{q^m}$ over $\mathbb{F}_q$ is a cyclic group of order m generated by σ_1.

3. TRACES, NORMS, AND BASES

In this section we adopt again the viewpoint of regarding a finite extension $F = \mathbb{F}_{q^m}$ of the finite field $K = \mathbb{F}_q$ as a vector space over K (compare with Chapter 1, Section 4). Then F has dimension m over K, and if $\{\alpha_1, \ldots, \alpha_m\}$ is a basis of F over K, each element $\alpha \in F$ can be uniquely represented in the form

$$\alpha = c_1\alpha_1 + \cdots + c_m\alpha_m \quad \text{with } c_j \in K \quad \text{for } 1 \leqslant j \leqslant m.$$

We introduce an important mapping from F to K which will turn out to be linear.

2.22. Definition. For $\alpha \in F = \mathbb{F}_{q^m}$ and $K = \mathbb{F}_q$, the *trace* $\mathrm{Tr}_{F/K}(\alpha)$ of α over K is defined by

$$\mathrm{Tr}_{F/K}(\alpha) = \alpha + \alpha^q + \cdots + \alpha^{q^{m-1}}.$$

If K is the prime subfield of F, then $\mathrm{Tr}_{F/K}(\alpha)$ is called the *absolute trace* of α and simply denoted by $\mathrm{Tr}_F(\alpha)$.

In other words, the trace of α over K is the sum of the conjugates of α with respect to K. Still another description of the trace may be obtained as follows. Let $f \in K[x]$ be the minimal polynomial of α over K; its degree d is a divisor of m. Then $g(x) = f(x)^{m/d} \in K[x]$ is called the *characteristic polynomial* of α over K. By Theorem 2.14, the roots of f in F are given by

$\alpha, \alpha^q, \ldots, \alpha^{q^{d-1}}$, and then a remark following Definition 2.17 implies that the roots of g in F are precisely the conjugates of α with respect to K. Hence

$$\begin{aligned} g(x) &= x^m + a_{m-1}x^{m-1} + \cdots + a_0 \\ &= (x-\alpha)(x-\alpha^q)\ldots(x-\alpha^{q^{m-1}}), \end{aligned} \tag{2.1}$$

and a comparison of coefficients shows that

$$\mathrm{Tr}_{F/K}(\alpha) = -a_{m-1}. \tag{2.2}$$

In particular, $\mathrm{Tr}_{F/K}(\alpha)$ is always an element of K.

2.23. Theorem. *Let $K = \mathbb{F}_q$ and $F = \mathbb{F}_{q^m}$. Then the trace function $\mathrm{Tr}_{F/K}$ satisfies the following properties:*

(i) $\mathrm{Tr}_{F/K}(\alpha+\beta) = \mathrm{Tr}_{F/K}(\alpha) + \mathrm{Tr}_{F/K}(\beta)$ *for all* $\alpha, \beta \in F$;
(ii) $\mathrm{Tr}_{F/K}(c\alpha) = c\,\mathrm{Tr}_{F/K}(\alpha)$ *for all* $c \in K$, $\alpha \in F$;
(iii) $\mathrm{Tr}_{F/K}$ *is a linear transformation from F onto K, where both F and K are viewed as vector spaces over K;*
(iv) $\mathrm{Tr}_{F/K}(a) = ma$ *for all* $a \in K$;
(v) $\mathrm{Tr}_{F/K}(\alpha^q) = \mathrm{Tr}_{F/K}(\alpha)$ *for all* $\alpha \in F$.

Proof.

(i) For $\alpha, \beta \in F$ we use Theorem 1.46 to get

$$\begin{aligned} \mathrm{Tr}_{F/K}(\alpha+\beta) &= \alpha+\beta+(\alpha+\beta)^q + \cdots + (\alpha+\beta)^{q^{m-1}} \\ &= \alpha+\beta+\alpha^q+\beta^q+\cdots+\alpha^{q^{m-1}}+\beta^{q^{m-1}} \\ &= \mathrm{Tr}_{F/K}(\alpha)+\mathrm{Tr}_{F/K}(\beta). \end{aligned}$$

(ii) For $c \in K$ we have $c^{q^j} = c$ for all $j \geqslant 0$ by Lemma 2.3. Therefore we obtain for $\alpha \in F$,

$$\begin{aligned} \mathrm{Tr}_{F/K}(c\alpha) &= c\alpha + c^q\alpha^q + \cdots + c^{q^{m-1}}\alpha^{q^{m-1}} \\ &= c\alpha + c\alpha^q + \cdots + c\alpha^{q^{m-1}} \\ &= c\,\mathrm{Tr}_{F/K}(\alpha). \end{aligned}$$

(iii) The properties (i) and (ii), together with the fact that $\mathrm{Tr}_{F/K}(\alpha) \in K$ for all $\alpha \in F$, show that $\mathrm{Tr}_{F/K}$ is a linear transformation from F into K. To prove that this mapping is onto, it suffices then to show the existence of an $\alpha \in F$ with $\mathrm{Tr}_{F/K}(\alpha) \neq 0$. Now $\mathrm{Tr}_{F/K}(\alpha) = 0$ if and only if α is a root of the polynomial $x^{q^{m-1}} + \cdots + x^q + x \in K[x]$ in F. But since this polynomial can have at most q^{m-1} roots in F and F has q^m elements, we are done.
(iv) This follows immediately from the definition of the trace function and Lemma 2.3.
(v) For $\alpha \in F$ we have $\alpha^{q^m} = \alpha$ by Lemma 2.3, and so $\mathrm{Tr}_{F/K}(\alpha^q) = \alpha^q + \alpha^{q^2} + \cdots + \alpha^{q^m} = \mathrm{Tr}_{F/K}(\alpha)$. □

The trace function $\mathrm{Tr}_{F/K}$ is not only in itself a linear transformation from F onto K, but serves for a description of all linear transformations from F into K (or, in an equivalent terminology, of all linear functionals on F) that has the advantage of being independent of a chosen basis.

2.24. *Theorem.* *Let F be a finite extension of the finite field K, both considered as vector spaces over K. Then the linear transformations from F into K are exactly the mappings L_β, $\beta \in F$, where $L_\beta(\alpha) = \mathrm{Tr}_{F/K}(\beta\alpha)$ for all $\alpha \in F$. Furthermore, we have $L_\beta \neq L_\gamma$ whenever β and γ are distinct elements of F.*

Proof. Each mapping L_β is a linear transformation from F into K by Theorem 2.23(iii). For $\beta, \gamma \in F$ with $\beta \neq \gamma$, we have $L_\beta(\alpha) - L_\gamma(\alpha) = \mathrm{Tr}_{F/K}(\beta\alpha) - \mathrm{Tr}_{F/K}(\gamma\alpha) = \mathrm{Tr}_{F/K}((\beta - \gamma)\alpha) \neq 0$ for suitable $\alpha \in F$ since $\mathrm{Tr}_{F/K}$ maps F onto K, and so the mappings L_β and L_γ are different. If $K = \mathbb{F}_q$ and $F = \mathbb{F}_{q^m}$, then the mappings L_β yield q^m different linear transformations from F into K. On the other hand, every linear transformation from F into K can be obtained by assigning arbitrary elements of K to the m elements of a given basis of F over K. Since this can be done in q^m different ways, the mappings L_β already exhaust all possible linear transformations from F into K. □

2.25. *Theorem.* *Let F be a finite extension of $K = \mathbb{F}_q$. Then for $\alpha \in F$ we have $\mathrm{Tr}_{F/K}(\alpha) = 0$ if and only if $\alpha = \beta^q - \beta$ for some $\beta \in F$.*

Proof. The sufficiency of the condition is obvious by Theorem 2.23(v). To prove the necessity, suppose $\alpha \in F = \mathbb{F}_{q^m}$ with $\mathrm{Tr}_{F/K}(\alpha) = 0$ and let β be a root of $x^q - x - \alpha$ in some extension field of F. Then $\beta^q - \beta = \alpha$ and

$$\begin{aligned}
0 = \mathrm{Tr}_{F/K}(\alpha) &= \alpha + \alpha^q + \cdots + \alpha^{q^{m-1}} \\
&= (\beta^q - \beta) + (\beta^q - \beta)^q + \cdots + (\beta^q - \beta)^{q^{m-1}} \\
&= (\beta^q - \beta) + \left(\beta^{q^2} - \beta^q\right) + \cdots + \left(\beta^{q^m} - \beta^{q^{m-1}}\right) \\
&= \beta^{q^m} - \beta,
\end{aligned}$$

so that $\beta \in F$. □

In case a chain of extension fields is considered, the composition of trace functions proceeds according to a very simple rule.

2.26. *Theorem* (Transitivity of Trace). *Let K be a finite field, let F be a finite extension of K and E a finite extension of F. Then*

$$\mathrm{Tr}_{E/K}(\alpha) = \mathrm{Tr}_{F/K}\left(\mathrm{Tr}_{E/F}(\alpha)\right) \quad \textit{for all } \alpha \in E.$$

Proof. Let $K=\mathbb{F}_q$, let $[F:K]=m$ and $[E:F]=n$, so that $[E:K]=mn$ by Theorem 1.84. Then for $\alpha \in E$ we have

$$\operatorname{Tr}_{F/K}(\operatorname{Tr}_{E/F}(\alpha)) = \sum_{i=0}^{m-1} \operatorname{Tr}_{E/F}(\alpha)^{q^i} = \sum_{i=0}^{m-1}\left(\sum_{j=0}^{n-1} \alpha^{q^{jm}}\right)^{q^i}$$

$$= \sum_{i=0}^{m-1}\sum_{j=0}^{n-1} \alpha^{q^{jm+i}} = \sum_{k=0}^{mn-1} \alpha^{q^k} = \operatorname{Tr}_{E/K}(\alpha). \qquad \square$$

Another interesting function from a finite field to a subfield is obtained by forming the product of the conjugates of an element of the field with respect to the subfield.

2.27. Definition. For $\alpha \in F=\mathbb{F}_{q^m}$ and $K=\mathbb{F}_q$, the *norm* $\mathrm{N}_{F/K}(\alpha)$ of α over K is defined by

$$\mathrm{N}_{F/K}(\alpha) = \alpha \cdot \alpha^q \cdot \cdots \cdot \alpha^{q^{m-1}} = \alpha^{(q^m-1)/(q-1)}.$$

By comparing the constant terms in (2.1), we see that $\mathrm{N}_{F/K}(\alpha)$ can be read off from the characteristic polynomial g of α over K—namely,

$$\mathrm{N}_{F/K}(\alpha) = (-1)^m a_0. \tag{2.3}$$

It follows, in particular, that $\mathrm{N}_{F/K}(\alpha)$ is always an element of K.

2.28. *Theorem.* *Let $K=\mathbb{F}_q$ and $F=\mathbb{F}_{q^m}$. Then the norm function $\mathrm{N}_{F/K}$ satisfies the following properties:*

(i) $\mathrm{N}_{F/K}(\alpha\beta) = \mathrm{N}_{F/K}(\alpha)\mathrm{N}_{F/K}(\beta)$ *for all* $\alpha, \beta \in F$;
(ii) $\mathrm{N}_{F/K}$ *maps* F *onto* K *and* F^* *onto* K^*;
(iii) $\mathrm{N}_{F/K}(a) = a^m$ *for all* $a \in K$;
(iv) $\mathrm{N}_{F/K}(\alpha^q) = \mathrm{N}_{F/K}(\alpha)$ *for all* $\alpha \in F$.

Proof. (i) follows immediately from the definition of the norm. We have already noted that $\mathrm{N}_{F/K}$ maps F into K. Since $\mathrm{N}_{F/K}(\alpha)=0$ if and only if $\alpha=0$, $\mathrm{N}_{F/K}$ maps F^* into K^*. Property (i) shows that $\mathrm{N}_{F/K}$ is a group homomorphism between these multiplicative groups. Since the elements of the kernel of $\mathrm{N}_{F/K}$ are exactly the roots of the polynomial $x^{(q^m-1)/(q-1)}-1 \in K[x]$ in F, the order d of the kernel satisfies $d \leqslant (q^m-1)/(q-1)$. By Theorem 1.23, the image of $\mathrm{N}_{F/K}$ has order $(q^m-1)/d$, which is $\geqslant q-1$. Therefore, $\mathrm{N}_{F/K}$ maps F^* onto K^* and so F onto K. Property (iii) follows from the definition of the norm and the fact that for $a \in K$ the conjugates of a with respect to K are all equal to a. Finally, we have $\mathrm{N}_{F/K}(\alpha^q) = \mathrm{N}_{F/K}(\alpha)^q = \mathrm{N}_{F/K}(\alpha)$ because of (i) and $\mathrm{N}_{F/K}(\alpha) \in K$, and so (iv) is shown. $\square$

2.29. Theorem (Transitivity of Norm). *Let K be a finite field, let F be a finite extension of K and E a finite extension of F. Then*

$$\mathrm{N}_{E/K}(\alpha) = \mathrm{N}_{F/K}(\mathrm{N}_{E/F}(\alpha)) \quad \textit{for all } \alpha \in E.$$

Proof. With the same notation as in the proof of Theorem 2.26, we have for $\alpha \in E$,

$$\begin{aligned}\mathrm{N}_{F/K}(\mathrm{N}_{E/F}(\alpha)) &= \mathrm{N}_{F/K}\left(\alpha^{(q^{mn}-1)/(q^m-1)}\right)\\ &= \left(\alpha^{(q^{mn}-1)/(q^m-1)}\right)^{(q^m-1)/(q-1)}\\ &= \alpha^{(q^{mn}-1)/(q-1)} = \mathrm{N}_{E/K}(\alpha).\end{aligned}$$ □

If $\{\alpha_1,\ldots,\alpha_m\}$ is a basis of the finite field F over a subfield K, the question arises as to the calculation of the coefficients $c_j(\alpha) \in K$, $1 \leqslant j \leqslant m$, in the unique representation

$$\alpha = c_1(\alpha)\alpha_1 + \cdots + c_m(\alpha)\alpha_m \tag{2.4}$$

of an element $\alpha \in F$. We note that $c_j : \alpha \mapsto c_j(\alpha)$ is a linear transformation from F into K, and thus, according to Theorem 2.24, there exists a $\beta_j \in F$ such that $c_j(\alpha) = \mathrm{Tr}_{F/K}(\beta_j\alpha)$ for all $\alpha \in F$. Putting $\alpha = \alpha_i$, $1 \leqslant i \leqslant m$, we see that $\mathrm{Tr}_{F/K}(\beta_j\alpha_i) = 0$ for $i \neq j$ and 1 for $i = j$. Furthermore, $\{\beta_1,\ldots,\beta_m\}$ is again a basis of F over K, for if

$$d_1\beta_1 + \cdots + d_m\beta_m = 0 \quad \text{with } d_i \in K \quad \text{for } 1 \leqslant i \leqslant m,$$

then by multiplying by a fixed α_i and applying the trace function $\mathrm{Tr}_{F/K}$, one shows that $d_i = 0$.

2.30. Definition. Let K be a finite field and F a finite extension of K. Then two bases $\{\alpha_1,\ldots,\alpha_m\}$ and $\{\beta_1,\ldots,\beta_m\}$ of F over K are said to be *dual* (or *complementary*) bases if for $1 \leqslant i, j \leqslant m$ we have

$$\mathrm{Tr}_{F/K}(\alpha_i\beta_j) = \begin{cases} 0 & \text{for } i \neq j,\\ 1 & \text{for } i = j.\end{cases}$$

In the discussion above we have shown that for any basis $\{\alpha_1,\ldots,\alpha_m\}$ of F over K there exists a dual basis $\{\beta_1,\ldots,\beta_m\}$. The dual basis is, in fact, uniquely determined since its definition implies that the coefficients $c_j(\alpha)$, $1 \leqslant j \leqslant m$, in (2.4) are given by $c_j(\alpha) = \mathrm{Tr}_{F/K}(\beta_j\alpha)$ for all $\alpha \in F$, and by Theorem 2.24 the element $\beta_j \in F$ is uniquely determined by the linear transformation c_j.

2.31. Example. Let $\alpha \in \mathbb{F}_8$ be a root of the irreducible polynomial $x^3 + x^2 + 1$ in $\mathbb{F}_2[x]$. Then $\{\alpha, \alpha^2, 1+\alpha+\alpha^2\}$ is a basis of $\mathbb{F}_8$ over $\mathbb{F}_2$. One checks easily that its uniquely determined dual basis is again $\{\alpha, \alpha^2, 1+\alpha+\alpha^2\}$. Such a basis that is its own dual basis is called a *self-dual basis*. The element $\alpha^5 \in \mathbb{F}_8$ can be uniquely represented in the form $\alpha^5 = c_1\alpha + c_2\alpha^2 +$

$c_3(1+\alpha+\alpha^2)$ with $c_1, c_2, c_3 \in \mathbb{F}_2$, and the coefficients are given by

$$c_1 = \mathrm{Tr}_{\mathbb{F}_8}(\alpha \cdot \alpha^5) = 0,$$

$$c_2 = \mathrm{Tr}_{\mathbb{F}_8}(\alpha^2 \cdot \alpha^5) = 1,$$

$$c_3 = \mathrm{Tr}_{\mathbb{F}_8}((1+\alpha+\alpha^2)\alpha^5) = 1,$$

so that $\alpha^5 = \alpha^2 + (1+\alpha+\alpha^2)$. □

The number of distinct bases of F over K is rather large (see Exercise 2.37), but there are two special types of bases of particular importance. The first is a *polynomial basis* $\{1, \alpha, \alpha^2, \ldots, \alpha^{m-1}\}$, made up of the powers of a defining element α of F over K. The element α is often taken to be a primitive element of F (compare with Theorem 2.10). Another type of basis is a normal basis defined by a suitable element of F.

2.32. Definition. Let $K = \mathbb{F}_q$ and $F = \mathbb{F}_{q^m}$. Then a basis of F over K of the form $\{\alpha, \alpha^q, \ldots, \alpha^{q^{m-1}}\}$, consisting of a suitable element $\alpha \in F$ and its conjugates with respect to K, is called a *normal basis* of F over K.

The basis $\{\alpha, \alpha^2, 1+\alpha+\alpha^2\}$ of $\mathbb{F}_8$ over $\mathbb{F}_2$ discussed in Example 2.31 is a normal basis of $\mathbb{F}_8$ over $\mathbb{F}_2$ since $1+\alpha+\alpha^2 = \alpha^4$. We shall show that a normal basis exists in the general case as well. The proof depends on two lemmas, one on a kind of linear independence property of certain group homomorphisms and one on linear operators.

2.33. Lemma (Artin Lemma). *Let $\psi_1, \ldots, \psi_m$ be distinct homomorphisms from a group G into the multiplicative group F^* of an arbitrary field F, and let $a_1, \ldots, a_m$ be elements of F that are not all 0. Then for some $g \in G$ we have*

$$a_1\psi_1(g) + \cdots + a_m\psi_m(g) \neq 0.$$

Proof. We proceed by induction on m. The case $m=1$ being trivial, we assume that $m>1$ and that the statement is shown for any $m-1$ distinct homomorphisms. Now take $\psi_1, \ldots, \psi_m$ and $a_1, \ldots, a_m$ as in the lemma. If $a_1 = 0$, the induction hypothesis immediately yields the desired result. Thus let $a_1 \neq 0$. Suppose we had

$$a_1\psi_1(g) + \cdots + a_m\psi_m(g) = 0 \quad \text{for all } g \in G. \tag{2.5}$$

Since $\psi_1 \neq \psi_m$, there exists $h \in G$ with $\psi_1(h) \neq \psi_m(h)$. Then, replacing g by hg in (2.5), we get

$$a_1\psi_1(h)\psi_1(g) + \cdots + a_m\psi_m(h)\psi_m(g) = 0 \quad \text{for all } g \in G.$$

After multiplication by $\psi_m(h)^{-1}$ we obtain

$$b_1\psi_1(g) + \cdots + b_{m-1}\psi_{m-1}(g) + a_m\psi_m(g) = 0 \quad \text{for all } g \in G,$$

where $b_i = a_i\psi_i(h)\psi_m(h)^{-1}$ for $1 \leqslant i \leqslant m-1$. By subtracting this identity from (2.5), we arrive at

$$c_1\psi_1(g)+\cdots+c_{m-1}\psi_{m-1}(g)=0 \quad \text{for all } g\in G,$$

where $c_i=a_i-b_i$ for $1\leqslant i\leqslant m-1$. But $c_1=a_1-a_1\psi_1(h)\psi_m(h)^{-1}\neq 0$, and we have a contradiction to the induction hypothesis. □

We recall a few concepts and facts from linear algebra. If T is a linear operator on the finite-dimensional vector space V over the (arbitrary) field K, then a polynomial $f(x)=a_nx^n+\cdots+a_1x+a_0\in K[x]$ is said to *annihilate* T if $a_nT^n+\cdots+a_1T+a_0I=0$, where I is the identity operator and 0 the zero operator on V. The uniquely determined monic polynomial of least positive degree with this property is called the *minimal polynomial* for T. It divides any other polynomial in $K[x]$ annihilating T. In particular, the minimal polynomial for T divides the *characteristic polynomial* $g(x)$ for T (Cayley-Hamilton theorem), which is given by $g(x)=\det(xI-T)$ and is a monic polynomial of degree equal to the dimension of V. A vector $\alpha\in V$ is called a *cyclic vector* for T if the vectors $T^k\alpha$, $k=0,1,\ldots,$ span V. The following is a standard result from linear algebra.

2.34. Lemma. *Let T be a linear operator on the finite-dimensional vector space V. Then T has a cyclic vector if and only if the characteristic and minimal polynomials for T are identical.*

2.35. Theorem (Normal Basis Theorem). *For any finite field K and any finite extension F of K, there exists a normal basis of F over K.*

Proof. Let $K=\mathbb{F}_q$ and $F=\mathbb{F}_{q^m}$ with $m\geqslant 2$. From Theorem 2.21 and the remarks following it, we know that the distinct automorphisms of F over K are given by $\varepsilon,\sigma,\sigma^2,\ldots,\sigma^{m-1}$, where ε is the identity mapping on F, $\sigma(\alpha)=\alpha^q$ for $\alpha\in F$, and a power σ^j refers to the j-fold composition of σ with itself. Because of $\sigma(\alpha+\beta)=\sigma(\alpha)+\sigma(\beta)$ and $\sigma(c\alpha)=\sigma(c)\sigma(\alpha)=c\sigma(\alpha)$ for $\alpha,\beta\in F$ and $c\in K$, the mapping σ may also be considered as a linear operator on the vector space F over K. Since $\sigma^m=\varepsilon$, the polynomial $x^m-1\in K[x]$ annihilates σ. Lemma 2.33, applied to $\varepsilon,\sigma,\sigma^2,\ldots,\sigma^{m-1}$ viewed as endomorphisms of F^*, shows that no nonzero polynomial in $K[x]$ of degree less than m annihilates σ. Consequently, x^m-1 is the minimal polynomial for the linear operator σ. Since the characteristic polynomial for σ is a monic polynomial of degree m that is divisible by the minimal polynomial for σ, it follows that the characteristic polynomial for σ is also given by x^m-1. Lemma 2.34 implies then the existence of an element $\alpha\in F$ such that $\alpha,\sigma(\alpha),\sigma^2(\alpha),\ldots$ span F. By dropping repeated elements, we see that $\alpha,\sigma(\alpha),\sigma^2(\alpha),\ldots,\sigma^{m-1}(\alpha)$ span F and thus form a basis of F over K. Since this basis consists of α and its conjugates with respect to K, it is a normal basis of F over K. □

An alternative proof of the normal basis theorem will be provided in Chapter 3, Section 4, by using so-called linearized polynomials.

We introduce an expression that allows us to decide whether a given set of elements forms a basis of an extension field.

2.36. Definition. Let K be a finite field and F an extension of K of degree m over K. Then the *discriminant* $\Delta_{F/K}(\alpha_1,\ldots,\alpha_m)$ of the elements $\alpha_1,\ldots,\alpha_m \in F$ is defined by the determinant of order m given by

$$\Delta_{F/K}(\alpha_1,\ldots,\alpha_m)=\begin{vmatrix} \mathrm{Tr}_{F/K}(\alpha_1\alpha_1) & \mathrm{Tr}_{F/K}(\alpha_1\alpha_2) & \cdots & \mathrm{Tr}_{F/K}(\alpha_1\alpha_m) \\ \mathrm{Tr}_{F/K}(\alpha_2\alpha_1) & \mathrm{Tr}_{F/K}(\alpha_2\alpha_2) & \cdots & \mathrm{Tr}_{F/K}(\alpha_2\alpha_m) \\ \vdots & \vdots & & \vdots \\ \mathrm{Tr}_{F/K}(\alpha_m\alpha_1) & \mathrm{Tr}_{F/K}(\alpha_m\alpha_2) & \cdots & \mathrm{Tr}_{F/K}(\alpha_m\alpha_m) \end{vmatrix}.$$

It follows from the definition that $\Delta_{F/K}(\alpha_1,\ldots,\alpha_m)$ is always an element of K. The following simple characterization of bases can now be given.

2.37. Theorem. *Let K be a finite field, F an extension of K of degree m over K, and $\alpha_1,\ldots,\alpha_m \in F$. Then $\{\alpha_1,\ldots,\alpha_m\}$ is a basis of F over K if and only if $\Delta_{F/K}(\alpha_1,\ldots,\alpha_m) \neq 0$.*

Proof. Let $\{\alpha_1,\ldots,\alpha_m\}$ be a basis of F over K. We prove that $\Delta_{F/K}(\alpha_1,\ldots,\alpha_m) \neq 0$ by showing that the row vectors of the determinant defining $\Delta_{F/K}(\alpha_1,\ldots,\alpha_m)$ are linearly independent. For suppose that

$$c_1\mathrm{Tr}_{F/K}(\alpha_1\alpha_j)+\cdots+c_m\mathrm{Tr}_{F/K}(\alpha_m\alpha_j)=0 \quad \text{for } 1 \leqslant j \leqslant m,$$

where $c_1,\ldots,c_m \in K$. Then with $\beta = c_1\alpha_1+\cdots+c_m\alpha_m$ we get $\mathrm{Tr}_{F/K}(\beta\alpha_j) = 0$ for $1 \leqslant j \leqslant m$, and since $\alpha_1,\ldots,\alpha_m$ span F, it follows that $\mathrm{Tr}_{F/K}(\beta\alpha)=0$ for all $\alpha \in F$. However, this is only possible if $\beta = 0$, and then $c_1\alpha_1+\cdots+c_m\alpha_m = 0$ implies $c_1=\cdots=c_m=0$.

Conversely, suppose that $\Delta_{F/K}(\alpha_1,\ldots,\alpha_m) \neq 0$ and $c_1\alpha_1+\cdots+c_m\alpha_m = 0$ for some $c_1,\ldots,c_m \in K$. Then

$$c_1\alpha_1\alpha_j+\cdots+c_m\alpha_m\alpha_j=0 \quad \text{for } 1 \leqslant j \leqslant m,$$

and by applying the trace function we get

$$c_1\mathrm{Tr}_{F/K}(\alpha_1\alpha_j)+\cdots+c_m\mathrm{Tr}_{F/K}(\alpha_m\alpha_j)=0 \quad \text{for } 1 \leqslant j \leqslant m.$$

But since the row vectors of the determinant defining $\Delta_{F/K}(\alpha_1,\ldots,\alpha_m)$ are linearly independent, it follows that $c_1=\cdots=c_m=0$. Therefore, $\alpha_1,\ldots,\alpha_m$ are linearly independent over K. □

There is another determinant of order m that serves the same purpose as the discriminant $\Delta_{F/K}(\alpha_1,\ldots,\alpha_m)$. The entries of this determinant are, however, elements of the extension field F. For $\alpha_1,\ldots,\alpha_m \in F$, let

A be the $m \times m$ matrix whose entry in the ith row and jth column is $\alpha_j^{q^{i-1}}$, where q is the number of elements of K. If A^{T} denotes the transpose of A, then a simple calculation shows that $A^{\mathrm{T}}A = B$, where B is the $m \times m$ matrix whose entry in the ith row and jth column is $\mathrm{Tr}_{F/K}(\alpha_i\alpha_j)$. By taking determinants, we obtain

$$\Delta_{F/K}(\alpha_1,\ldots,\alpha_m) = \det(A)^2.$$

The following result is now implied by Theorem 2.37.

2.38. Corollary. *Let $\alpha_1,\ldots,\alpha_m \in \mathbb{F}_{q^m}$. Then $\{\alpha_1,\ldots,\alpha_m\}$ is a basis of $\mathbb{F}_{q^m}$ over $\mathbb{F}_q$ if and only if*

$$\begin{vmatrix} \alpha_1 & \alpha_2 & \cdots & \alpha_m \\ \alpha_1^q & \alpha_2^q & \cdots & \alpha_m^q \\ \vdots & \vdots & & \vdots \\ \alpha_1^{q^{m-1}} & \alpha_2^{q^{m-1}} & \cdots & \alpha_m^{q^{m-1}} \end{vmatrix} \neq 0.$$

From the criterion above we are led to a relatively simple way of checking whether a given element gives rise to a normal basis.

2.39. Theorem. *For $\alpha \in \mathbb{F}_{q^m}$, $\{\alpha, \alpha^q, \alpha^{q^2},\ldots,\alpha^{q^{m-1}}\}$ is a normal basis of $\mathbb{F}_{q^m}$ over $\mathbb{F}_q$ if and only if the polynomials $x^m - 1$ and $\alpha x^{m-1} + \alpha^q x^{m-2} + \cdots + \alpha^{q^{m-2}}x + \alpha^{q^{m-1}}$ in $\mathbb{F}_{q^m}[x]$ are relatively prime.*

Proof. When $\alpha_1 = \alpha$, $\alpha_2 = \alpha^q,\ldots,\alpha_m = \alpha^{q^{m-1}}$, the determinant in Corollary 2.38 becomes

$$\pm\begin{vmatrix} \alpha & \alpha^q & \alpha^{q^2} & \cdots & \alpha^{q^{m-1}} \\ \alpha^{q^{m-1}} & \alpha & \alpha^q & \cdots & \alpha^{q^{m-2}} \\ \alpha^{q^{m-2}} & \alpha^{q^{m-1}} & \alpha & \cdots & \alpha^{q^{m-3}} \\ \vdots & \vdots & \vdots & & \vdots \\ \alpha^q & \alpha^{q^2} & \alpha^{q^3} & \cdots & \alpha \end{vmatrix} \tag{2.6}$$

after a suitable permutation of the rows. Now consider the resultant $R(f, g)$ of the polynomials $f(x) = x^m - 1$ and $g(x) = \alpha x^{m-1} + \alpha^q x^{m-2} + \cdots + \alpha^{q^{m-2}}x + \alpha^{q^{m-1}}$ of formal degree m resp. $m-1$, which is given by a determinant of order $2m-1$ in accordance with Definition 1.93. In this determinant, add the $(m+1)$st column to the first column, the $(m+2)$nd column to the second column, and so on, finally adding the $(2m-1)$st column to the $(m-1)$st column. The resulting determinant factorizes into the determinant of the diagonal matrix of order $m-1$ with entries -1 along the main diagonal and the determinant in (2.6). Therefore, $R(f, g)$ is, apart from the sign, equal to the determinant in (2.6). The statement of the theorem follows then from Corollary 2.38 and the fact that $R(f, g) \neq 0$ if and only if f and g are relatively prime. □

In connection with the preceding discussion, we mention without proof the following refinement of the normal basis theorem.

2.40. Theorem. *For any finite field F there exists a normal basis of F over its prime subfield that consists of primitive elements of F.*

4. ROOTS OF UNITY AND CYCLOTOMIC POLYNOMIALS

In this section we investigate the splitting field of the polynomial $x^n - 1$ over an arbitrary field K, where n is a positive integer. At the same time we obtain a generalization of the concept of a root of unity, well known for complex numbers.

2.41. Definition. Let n be a positive integer. The splitting field of $x^n - 1$ over a field K is called the *nth cyclotomic field* over K and denoted by $K^{(n)}$. The roots of $x^n - 1$ in $K^{(n)}$ are called the *nth roots of unity* over K and the set of all these roots is denoted by $E^{(n)}$.

A special case of this general definition is obtained if K is the field of rational numbers. Then $K^{(n)}$ is a subfield of the field of complex numbers and the nth roots of unity have their known geometric interpretation as the vertices of a regular polygon with n vertices on the unit circle in the complex plane.

For our purposes, the most important case is that of a finite field K. The basic properties of roots of unity can, however, be established without using this restriction. The structure of $E^{(n)}$ is determined by the relation of n to the characteristic of K, as the following theorem shows. When we refer to the characteristic p of K in this discussion, we permit the case $p = 0$ as well.

2.42. Theorem. *Let n be a positive integer and K a field of characteristic p. Then:*

(i) *If p does not divide n, then $E^{(n)}$ is a cyclic group of order n with respect to multiplication in $K^{(n)}$.*

(ii) *If p divides n, write $n = mp^e$ with positive integers m and e and m not divisible by p. Then $K^{(n)} = K^{(m)}$, $E^{(n)} = E^{(m)}$, and the roots of $x^n - 1$ in $K^{(n)}$ are the m elements of $E^{(m)}$, each attained with multiplicity p^e.*

Proof. (i) The case $n = 1$ is trivial. For $n \geqslant 2$, $x^n - 1$ and its derivative nx^{n-1} have no common roots, as nx^{n-1} only has the root 0 in $K^{(n)}$. Therefore, by Theorem 1.68, $x^n - 1$ cannot have multiple roots, and hence $E^{(n)}$ has n elements. Now if $\zeta, \eta \in E^{(n)}$, then $(\zeta\eta^{-1})^n = \zeta^n(\eta^n)^{-1} = 1$, thus

$\zeta\eta^{-1} \in E^{(n)}$. It follows that $E^{(n)}$ is a multiplicative group. Let $n = p_1^{e_1} p_2^{e_2} \cdots p_t^{e_t}$ be the prime factor decomposition of n. Then one shows by the same argument as in the proof of Theorem 2.8 that for each i, $1 \leqslant i \leqslant t$, there exists an element $\alpha_i \in E^{(n)}$ that is not a root of the polynomial $x^{n/p_i} - 1$, that $\beta_i = \alpha_i^{n/p_i^{e_i}}$ has order $p_i^{e_i}$, and that $E^{(n)}$ is a cyclic group with generator $\beta = \beta_1 \beta_2 \cdots \beta_t$.

(ii) This follows immediately from $x^n - 1 = x^{mp^e} - 1 = (x^m - 1)^{p^e}$ and part (i). □

2.43. Definition. Let K be a field of characteristic p and n a positive integer not divisible by p. Then a generator of the cyclic group $E^{(n)}$ is called a *primitive nth root of unity* over K.

By Theorem 1.15(v) we know that under the conditions of Definition 2.43 there are exactly $\phi(n)$ different primitive nth roots of unity over K. If ζ is one of them, then all primitive nth roots of unity over K are given by ζ^s, where $1 \leqslant s \leqslant n$ and $\gcd(s, n) = 1$. The polynomial whose roots are precisely the primitive nth roots of unity over K is of great interest.

2.44. Definition. Let K be a field of characteristic p, n a positive integer not divisible by p, and ζ a primitive nth root of unity over K. Then the polynomial

$$Q_n(x) = \prod_{\substack{s=1 \\ \gcd(s,n)=1}}^{n} (x - \zeta^s)$$

is called the *nth cyclotomic polynomial* over K.

The polynomial $Q_n(x)$ is clearly independent of the choice of ζ. The degree of $Q_n(x)$ is $\phi(n)$ and its coefficients obviously belong to the nth cyclotomic field over K. A simple argument will show that they are actually contained in the prime subfield of K. We use the product symbol $\prod_{d|n}$ to denote a product extended over all positive divisors d of a positive integer n.

2.45. Theorem. *Let K be a field of characteristic p and n a positive integer not divisible by p. Then:*

(i) $x^n - 1 = \prod_{d|n} Q_d(x)$;
(ii) *the coefficients of $Q_n(x)$ belong to the prime subfield of K, and to $\mathbb{Z}$ if the prime subfield of K is the field of rational numbers.*

Proof. (i) Each nth root of unity over K is a primitive dth root of unity over K for exactly one positive divisor d of n. In detail, if ζ is a primitive nth root of unity over K and ζ^s is an arbitrary nth root of unity over K, then $d = n/\gcd(s, n)$; that is, d is the order of ζ^s in $E^{(n)}$. Since

$$x^n - 1 = \prod_{s=1}^{n} (x - \zeta^s),$$

the formula in (i) is obtained by collecting those factors $(x-\zeta^s)$ for which ζ^s is a primitive dth root of unity over K.

(ii) This is proved by induction on n. Note that $Q_n(x)$ is a monic polynomial. For $n=1$ we have $Q_1(x)=x-1$, and the claim is obviously valid. Now let $n>1$ and suppose the proposition is true for all $Q_d(x)$ with $1\leqslant d<n$. Then we have by (i), $Q_n(x)=(x^n-1)/f(x)$, where $f(x)=\prod_{d\mid n,\, d<n}Q_d(x)$. The induction hypothesis implies that $f(x)$ is a polynomial with coefficients in the prime subfield of K or in $\mathbb{Z}$ in case the characteristic of K is 0. Using long division with x^n-1 and the monic polynomial $f(x)$, we see that the coefficients of $Q_n(x)$ belong to the prime subfield of K or to $\mathbb{Z}$, respectively. □

2.46. Example. Let r be a prime and $k\in\mathbb{N}$. Then

$$Q_{r^k}(x)=1+x^{r^{k-1}}+x^{2r^{k-1}}+\cdots+x^{(r-1)r^{k-1}}$$

since

$$Q_{r^k}(x)=\frac{x^{r^k}-1}{Q_1(x)Q_r(x)\cdots Q_{r^{k-1}}(x)}=\frac{x^{r^k}-1}{x^{r^{k-1}}-1}$$

by Theorem 2.45(i). For $k=1$ we simply have $Q_r(x)=1+x+x^2+\cdots+x^{r-1}$. □

An explicit expression for the nth cyclotomic polynomial generalizing the formula for $Q_{r^k}(x)$ in Example 2.46 will be given in Chapter 3, Section 2. For applications to finite fields it is useful to know some properties of cyclotomic fields.

2.47. ***Theorem.*** *The cyclotomic field $K^{(n)}$ is a simple algebraic extension of K. Moreover:*

(i) *If $K=\mathbb{Q}$, then the cyclotomic polynomial Q_n is irreducible over K and $[K^{(n)}:K]=\phi(n)$.*

(ii) *If $K=\mathbb{F}_q$ with $\gcd(q,n)=1$, then Q_n factors into $\phi(n)/d$ distinct monic irreducible polynomials in $K[x]$ of the same degree d, $K^{(n)}$ is the splitting field of any such irreducible factor over K, and $[K^{(n)}:K]=d$, where d is the least positive integer such that $q^d\equiv 1 \bmod n$.*

Proof. If there exists a primitive nth root of unity ζ over K, it is clear that $K^{(n)}=K(\zeta)$. Otherwise, we have the situation described in Theorem 2.42(ii), then $K^{(n)}=K^{(m)}$ and the result follows again. As to the remaining statements, we prove only (ii), the important case for our purposes. Let η be a primitive nth root of unity over $\mathbb{F}_q$. Then $\eta\in\mathbb{F}_{q^k}$ if and only if $\eta^{q^k}=\eta$, and the latter identity is equivalent to $q^k\equiv 1 \bmod n$. The smallest positive integer for which this holds is $k=d$, and so η is in $\mathbb{F}_{q^d}$, but in no proper subfield thereof. Thus the minimal polynomial of η over $\mathbb{F}_q$ has degree d, and since η is an arbitrary root of Q_n, the desired results follow. □

2.48. Example. Let $K = \mathbb{F}_{11}$ and $Q_{12}(x) = x^4 - x^2 + 1 \in \mathbb{F}_{11}[x]$. In the notation of Theorem 2.47(ii) we have $d = 2$. In detail, $Q_{12}(x)$ factors in the form $Q_{12}(x) = (x^2 + 5x + 1)(x^2 - 5x + 1)$, with both factors being irreducible in $\mathbb{F}_{11}[x]$. The cyclotomic field $K^{(12)}$ is equal to $\mathbb{F}_{121}$. □

A further connection between cyclotomic fields and finite fields is given by the following theorem.

2.49. *Theorem.* *The finite field $\mathbb{F}_q$ is the $(q-1)$st cyclotomic field over any one of its subfields.*

Proof. The polynomial $x^{q-1} - 1$ splits in $\mathbb{F}_q$ since its roots are exactly all nonzero elements of $\mathbb{F}_q$. Obviously, the polynomial cannot split in any proper subfield of $\mathbb{F}_q$, so that $\mathbb{F}_q$ is the splitting field of $x^{q-1} - 1$ over any one of its subfields. □

Since $\mathbb{F}_q^*$ is a cyclic group of order $q - 1$ by Theorem 2.8, there will exist, for any positive divisor n of $q - 1$, a cyclic subgroup $\{1, \alpha, \ldots, \alpha^{n-1}\}$ of $\mathbb{F}_q^*$ of order n (see Theorem 1.15(iii)). All elements of this subgroup are nth roots of unity over any subfield of $\mathbb{F}_q$ and the generating element α is a primitive nth root of unity over any subfield of $\mathbb{F}_q$.

We conclude this section with a lemma we shall need later on.

2.50. *Lemma.* *If d is a divisor of the positive integer n with $1 \leqslant d < n$, then $Q_n(x)$ divides $(x^n - 1)/(x^d - 1)$ whenever $Q_n(x)$ is defined.*

Proof. From Theorem 2.45(i) we know that $Q_n(x)$ divides

$$x^n - 1 = (x^d - 1) \cdot \frac{x^n - 1}{x^d - 1}.$$

Since d is a proper divisor of n, the polynomials $Q_n(x)$ and $x^d - 1$ have no common root, hence $\gcd(Q_n(x), x^d - 1) = 1$ and the proposition is true. □

5. REPRESENTATION OF ELEMENTS OF FINITE FIELDS

In this section we describe three different ways of representing the elements of a finite field $\mathbb{F}_q$ with $q = p^n$ elements, where p is the characteristic of $\mathbb{F}_q$.

The first method is based on principles expounded in Chapter 1, Section 4, and in the present chapter. We note that $\mathbb{F}_q$ is a simple algebraic extension of $\mathbb{F}_p$ by Theorem 2.10. In fact, if f is an irreducible polynomial in $\mathbb{F}_p[x]$ of degree n, then f has a root α in $\mathbb{F}_q$ according to Theorem 2.14, and so $\mathbb{F}_q = \mathbb{F}_p(\alpha)$. Then, by Theorem 1.86, every element of $\mathbb{F}_q$ can be uniquely expressed as a polynomial in α over $\mathbb{F}_p$ of degree less than n. We may also view $\mathbb{F}_q$ as the residue class ring $\mathbb{F}_p[x]/(f)$.

2.51. Example. To represent the elements of $\mathbb{F}_9$ in this way, we regard $\mathbb{F}_9$ as a simple algebraic extension of $\mathbb{F}_3$ of degree 2, which is obtained by adjunction of a root α of an irreducible quadratic polynomial over $\mathbb{F}_3$, say $f(x)=x^2+1\in\mathbb{F}_3[x]$. Thus $f(\alpha)=\alpha^2+1=0$ in $\mathbb{F}_9$, and the nine elements of $\mathbb{F}_9$ are given in the form $a_0+a_1\alpha$ with $a_0,\ a_1\in\mathbb{F}_3$. In detail, $\mathbb{F}_9=\{0,1,2,\alpha,1+\alpha,2+\alpha,2\alpha,1+2\alpha,2+2\alpha\}$. The operation tables for $\mathbb{F}_9$ may be constructed as in Example 1.62, with α playing the role of the residue class $[x]$. □

If we use Theorems 2.47 and 2.49, we get another possibility of expressing the elements of $\mathbb{F}_q$. Since $\mathbb{F}_q$ is the $(q-1)$st cyclotomic field over $\mathbb{F}_p$, we can construct it by finding the decomposition of the $(q-1)$st cyclotomic polynomial $Q_{q-1}\in\mathbb{F}_p[x]$ into irreducible factors in $\mathbb{F}_p[x]$, which are all of the same degree. A root of any one of these factors is then a primitive $(q-1)$st root of unity over $\mathbb{F}_p$ and therefore a primitive element of $\mathbb{F}_q$. Thus, $\mathbb{F}_q$ consists of 0 and appropriate powers of that primitive element.

2.52. Example. To apply this to the construction of $\mathbb{F}_9$, we note that $\mathbb{F}_9=\mathbb{F}_3^{(8)}$, the eighth cyclotomic field over $\mathbb{F}_3$. Now $Q_8(x)=x^4+1\in\mathbb{F}_3[x]$ by Example 2.46, and

$$Q_8(x)=(x^2+x+2)(x^2+2x+2)$$

is the decomposition of Q_8 into irreducible factors in $\mathbb{F}_3[x]$. Let ζ be a root of x^2+x+2; then ζ is a primitive eighth root of unity over $\mathbb{F}_3$. Thus, all nonzero elements of $\mathbb{F}_9$ can be expressed as powers of ζ, and so $\mathbb{F}_9=\{0,\zeta,\zeta^2,\zeta^3,\zeta^4,\zeta^5,\zeta^6,\zeta^7,\zeta^8\}$. We may arrange the nonzero elements of $\mathbb{F}_9$ in a so-called *index table*, where we list the elements ζ^i according to their exponents i. In order to establish the connection with the representation in Example 2.51, we observe that $x^2+x+2\in\mathbb{F}_3[x]$ has $\zeta=1+\alpha$ as a root, where $\alpha^2+1=0$ as in Example 2.51. Therefore, the index table for $\mathbb{F}_9$ may be written as follows:

i	ζ^i	i	ζ^i
1	$1+\alpha$	5	$2+2\alpha$
2	2α	6	α
3	$1+2\alpha$	7	$2+\alpha$
4	2	8	1

We see that we obtain, of course, the same elements as in Example 2.51, just in a different order. □

A third possibility of representing the elements of $\mathbb{F}_q$ is given by means of matrices. In general, the *companion matrix* of a monic polynomial

$f(x)=a_0+a_1x+\cdots+a_{n-1}x^{n-1}+x^n$ of positive degree n over a field is defined to be the $n \times n$ matrix

$$A=\begin{pmatrix} 0 & 0 & 0 & \cdots & 0 & -a_0 \\ 1 & 0 & 0 & \cdots & 0 & -a_1 \\ 0 & 1 & 0 & \cdots & 0 & -a_2 \\ \vdots & \vdots & \vdots & & \vdots & \vdots \\ 0 & 0 & 0 & \cdots & 1 & -a_{n-1} \end{pmatrix}.$$

It is well known in linear algebra that A satisfies the equation $f(A)=0$; that is, $a_0I+a_1A+\cdots+a_{n-1}A^{n-1}+A^n=0$, where I is the $n\times n$ identity matrix.

Thus, if A is the companion matrix of a monic irreducible polynomial f over $\mathbb{F}_p$ of degree n, then $f(A)=0$, and therefore A can play the role of a root of f. The polynomials in A over $\mathbb{F}_p$ of degree less than n yield a representation of the elements of $\mathbb{F}_q$.

2.53. Example. As in Example 2.51, let $f(x)=x^2+1\in\mathbb{F}_3[x]$. The companion matrix of f is

$$A=\begin{pmatrix} 0 & 2 \\ 1 & 0 \end{pmatrix}.$$

The field $\mathbb{F}_9$ can then be represented in the form $\mathbb{F}_9=\{0, I, 2I, A, I+A, 2I+A, 2A, I+2A, 2I+2A\}$. Explicitly:

$$0=\begin{pmatrix} 0 & 0 \\ 0 & 0 \end{pmatrix},\quad I=\begin{pmatrix} 1 & 0 \\ 0 & 1 \end{pmatrix},\quad 2I=\begin{pmatrix} 2 & 0 \\ 0 & 2 \end{pmatrix},\quad A=\begin{pmatrix} 0 & 2 \\ 1 & 0 \end{pmatrix},$$

$$I+A=\begin{pmatrix} 1 & 2 \\ 1 & 1 \end{pmatrix},\quad 2I+A=\begin{pmatrix} 2 & 2 \\ 1 & 2 \end{pmatrix},\quad 2A=\begin{pmatrix} 0 & 1 \\ 2 & 0 \end{pmatrix},$$

$$I+2A=\begin{pmatrix} 1 & 1 \\ 2 & 1 \end{pmatrix},\quad 2I+2A=\begin{pmatrix} 2 & 1 \\ 2 & 2 \end{pmatrix}.$$

With $\mathbb{F}_9$ given in this way, calculations in this finite field are then carried out by the usual rules of matrix algebra. For instance,

$$(2I+A)(I+2A)=\begin{pmatrix} 2 & 2 \\ 1 & 2 \end{pmatrix}\begin{pmatrix} 1 & 1 \\ 2 & 1 \end{pmatrix}=\begin{pmatrix} 0 & 1 \\ 2 & 0 \end{pmatrix}=2A. \qquad \square$$

In the same way, the method based on the factorization of the cyclotomic polynomial Q_{q-1} in $\mathbb{F}_p[x]$ can be adapted to yield a representation of the elements of $\mathbb{F}_q$ in terms of matrices.

2.54. Example. As in Example 2.52, let $h(x)=x^2+x+2\in\mathbb{F}_3[x]$ be an irreducible factor of the cyclotomic polynomial $Q_8\in\mathbb{F}_3[x]$. The companion matrix of h is

$$C=\begin{pmatrix} 0 & 1 \\ 1 & 2 \end{pmatrix}.$$

The field $\mathbb{F}_9$ can then be represented in the form

$$\mathbb{F}_9 = \{0, C, C^2, C^3, C^4, C^5, C^6, C^7, C^8\}.$$

Explicitly:

$$0 = \begin{pmatrix} 0 & 0 \\ 0 & 0 \end{pmatrix}, \quad C = \begin{pmatrix} 0 & 1 \\ 1 & 2 \end{pmatrix}, \quad C^2 = \begin{pmatrix} 1 & 2 \\ 2 & 2 \end{pmatrix},$$

$$C^3 = \begin{pmatrix} 2 & 2 \\ 2 & 0 \end{pmatrix}, \quad C^4 = \begin{pmatrix} 2 & 0 \\ 0 & 2 \end{pmatrix}, \quad C^5 = \begin{pmatrix} 0 & 2 \\ 2 & 1 \end{pmatrix},$$

$$C^6 = \begin{pmatrix} 2 & 1 \\ 1 & 1 \end{pmatrix}, \quad C^7 = \begin{pmatrix} 1 & 1 \\ 1 & 0 \end{pmatrix}, \quad C^8 = \begin{pmatrix} 1 & 0 \\ 0 & 1 \end{pmatrix}.$$

Calculations proceed by the rules of matrix algebra. For instance,

$$C^6 + C = \begin{pmatrix} 2 & 1 \\ 1 & 1 \end{pmatrix} + \begin{pmatrix} 0 & 1 \\ 1 & 2 \end{pmatrix} = \begin{pmatrix} 2 & 2 \\ 2 & 0 \end{pmatrix} = C^3. \qquad \square$$

6. WEDDERBURN'S THEOREM[1]

All results for finite fields are at the same time also true for all finite division rings by a famous theorem due to Wedderburn. This theorem states that in a finite ring in which all the field properties except commutativity of multiplication are assumed (i.e., in a finite division ring), the multiplication must also be commutative. Basically, the first proof we present of the theorem considers a subring of the finite divison ring that is a field and establishes a numerical relation between the multiplicative group of the field and the multiplicative group of the whole division ring. Using this relation and information about cyclotomic polynomials, one obtains a contradiction—unless the field is all of the division ring. Before we prove Wedderburn's theorem in detail, we mention some general principles that will be employed.

Let D be a division ring and F a subring that is a field (later on, we will express this more briefly by saying that F is a subfield of D). Then D can be viewed as a (left) vector space over F (compare with the discussion of the analogous situation for fields in Chapter 1, Section 4). If $F = \mathbb{F}_q$ and D is of finite dimension n over $\mathbb{F}_q$, then D has q^n elements. We shall write D^* for the multiplicative group of nonzero elements of D.

For a group G and a nonempty subset S of G, we defined the normalizer $N(S)$ of S in G in Definition 1.24. If S is a singleton $\{b\}$, we may also refer to $N(\{b\})$ as the normalizer of the element b in G. From Theorem

---

[1]This section can be omitted without losing necessary information for the following chapters.

1.25 we infer that if G is finite, then the number of elements in the conjugacy class of b is given by $|G|/|N(\langle b\rangle)|$.

2.55. ***Theorem*** (Wedderburn's Theorem). *Every finite division ring is a field.*

First Proof. Let D be a finite division ring and let $Z=\{z\in D: zd=dz \text{ for all } d\in D\}$ be the *center* of D. We omit the obvious verification that Z is a field. Thus $Z=\mathbb{F}_q$ for some prime power q. Now D is a vector space over Z of finite dimension n, and so D has q^n elements. We shall show that $D=Z$, or, equivalently, that $n=1$.

Let us suppose, on the contrary, that $n>1$. Now let $a\in D$ and define $N_a=\{b\in D: ab=ba\}$. Then N_a is a division ring and N_a contains Z. Thus N_a has q^r elements, where $1\leqslant r\leqslant n$. We wish to show that r divides n. Since N_a^* is a subgroup of D^*, we know that q^r-1 divides q^n-1. If $n=rm+t$ with $0\leqslant t<r$, then $q^n-1=q^{rm}q^t-1=q^t(q^{rm}-1)+(q^t-1)$. Now q^r-1 divides q^n-1 and also $q^{rm}-1$, thus it follows that q^r-1 divides q^t-1. But $q^t-1<q^r-1$, and so we must have $t=0$. This implies that r divides n.

We consider now the class equation for the group D^* (see Theorem 1.27). The center of D^* is Z^*, which has order $q-1$. For $a\in D^*$, the normalizer of a in D^* is exactly N_a^*. Therefore, a conjugacy class in D^* containing more than one member has $(q^n-1)/(q^r-1)$ elements, where r is a divisor of n with $1\leqslant r<n$. Hence the class equation becomes

$$q^n-1=q-1+\sum_{i=1}^{k}\frac{q^n-1}{q^{r_i}-1}, \tag{2.7}$$

where $r_1,\ldots,r_k$ are (not necessarily distinct) divisors of n with $1\leqslant r_i<n$ for $1\leqslant i\leqslant k$.

Now let Q_n be the nth cyclotomic polynomial over the field of rational numbers. Then $Q_n(q)$ is an integer by Theorem 2.45(ii). Furthermore, Lemma 2.50 implies that $Q_n(q)$ divides $(q^n-1)/(q^{r_i}-1)$ for $1\leqslant i\leqslant k$. We conclude then from (2.7) that $Q_n(q)$ divides $q-1$. However, this will lead to a contradiction. By definition, we have

$$Q_n(x)=\prod_{\substack{s=1\\ \gcd(s,n)=1}}^{n}(x-\zeta^s),$$

where the complex number ζ is a primitive nth root of unity over the field of rationals. Therefore, as complex numbers,

$$|Q_n(q)|=\prod_{\substack{s=1\\ \gcd(s,n)=1}}^{n}|q-\zeta^s|>\prod_{\substack{s=1\\ \gcd(s,n)=1}}^{n}(q-1)\geqslant q-1$$

since $n>1$ and $q\geqslant 2$. This inequality is incompatible with the statement that $Q_n(q)$ divides $q-1$. Hence we must have $n=1$ and $D=Z$, and the theorem is proved. □

Before we start with the second proof of Wedderburn's theorem, we establish some preparatory results. Let D be a finite division ring with center Z, and let F denote a *maximal subfield* of D; that is, F is a subfield of D such that the only subfield of D containing F is F itself. Then F is an extension of Z, for if there were an element $z \in Z$ with $z \notin F$, we could adjoin z to F and obtain a subfield of D properly containing F. From Theorem 2.10 we know that $F = Z(\xi)$, where $\xi \in F^*$ is a root of a monic irreducible polynomial $f \in Z[x]$.

If we view D as a vector space over F, then for each $a \in D$ the assignment $T_a(d) = da$ for $d \in D$ defines a linear operator T_a on this vector space. We consider now the linear operator T_ξ. If d is an eigenvector of T_ξ, then for some $\lambda \in F^*$ we have $d\xi = \lambda d$. This implies $d\xi d^{-1} = \lambda$ and hence $dF^*d^{-1} = F^*$, thus $d \in N(F^*)$, the normalizer of F^* in the group D^*. Conversely, if $d \in N(F^*)$, then $d\xi d^{-1} = \lambda$ for some $\lambda \in F^*$, and so d is an eigenvector of T_ξ. This proves the following result.

2.56. Lemma. *An element $d \in D^*$ is an eigenvector of T_ξ if and only if $d \in N(F^*)$.*

Let λ be an eigenvalue of T_ξ with eigenvector d, then $d\xi = \lambda d$. It follows that $0 = df(\xi) = f(\lambda)d$, hence λ must be a root of f. If d_0 is another eigenvector corresponding to the eigenvalue λ, then $d_0 d^{-1}\lambda d d_0^{-1} = \lambda$, and so the element $b = d_0 d^{-1}$ commutes with λ and, consequently, with every element of $F = Z(\lambda)$. Let P be the set of all polynomial expressions in b with coefficients in F. Then it is easily checked that P forms a finite integral domain, and so P is a finite field by Theorem 1.31. But P contains F, and thus $P = F$ by the maximality of F. In particular, we have $b \in F$, and since $d_0 = bd$, we conclude that every eigenspace of T_ξ has dimension 1. We use now the following result from linear algebra.

2.57. Lemma. *Let T be a linear operator on the finite-dimensional vector space V over the field K. Then V has a basis consisting of eigenvectors of T if and only if the minimal polynomial for T splits in K into distinct monic linear factors.*

Since $f(\xi) = 0$, the polynomial f annihilates the linear operator T_ξ. Furthermore, f splits in F into distinct monic linear factors by Theorem 2.14. The minimal polynomial for T_ξ divides f, and so it also splits in F into distinct monic linear factors. It follows then from Lemma 2.57 that D has a basis as a vector space over F consisting of eigenvectors of T_ξ. Since every eigenspace of T_ξ has dimension 1, the dimension m of D over F is equal to the number of distinct eigenvalues of T_ξ. Let $\xi = \xi_1, \xi_2, \ldots, \xi_m$ be the distinct eigenvalues of T_ξ and let $1 = d_1, d_2, \ldots, d_m$ be corresponding eigenvectors.

Because $N(F^*)$ is closed under multiplication, it follows from Lemma 2.56 that $d_i d_j$ must correspond to an eigenvalue ξ_k, say, and hence $d_i d_j \xi = \xi_k d_i d_j$. Using $d_j \xi = \xi_j d_j$, we obtain $d_i \xi_j = \xi_k d_i$, or $d_i \xi_j d_i^{-1} = \xi_k$. This shows that for each i, $1 \leqslant i \leqslant m$, the mapping that takes ξ_j to $d_i \xi_j d_i^{-1}$ permutes the eigenvalues among themselves. Consequently, the coefficients of $g(x) = (x - \xi_1) \cdots (x - \xi_m)$ commute with the eigenvectors $d_1, d_2, \ldots, d_m$ of T_ξ. Since the coefficients of g obviously belong to F and thus commute with all the elements of F, they commute with all the elements of D, since these can be written as linear combinations of $d_1, d_2, \ldots, d_m$ with coefficients in F. Thus the coefficients of g are elements of the center Z of D. Since $g(\xi) = 0$, Lemma 2.12 implies that f divides g. On the other hand, we have already observed that every eigenvalue of T_ξ must be a root of f, and so $f = g$. It follows that $[F: Z] = [Z(\xi): Z] = \deg(f) = m$. Now m is also the dimension of D over F, and so the argument in the proof of Theorem 1.84 shows that D is of dimension m^2 over Z. Since the latter dimension is independent of F, we conclude that every maximal subfield of D has the same degree over Z. We state this result in the following equivalent form.

2.58. Lemma. *All maximal subfields of D have the same order.*

Second Proof of Theorem 2.55. Let D be a finite division ring, and let Z, $F = Z(\xi)$, and $f \in Z[x]$ be as above. Let E be an arbitrary maximal subfield of D. Then, by Lemma 2.58, E and F have the same order, say q. In view of Lemma 2.4, both E and F are splitting fields of $x^q - x$ over Z. It follows then from Theorem 1.91 that there exists an isomorphism from F onto E that keeps the elements of Z fixed. The image $\eta \in E^*$ of ξ under this isomorphism is therefore a root of f in E, and so $E = Z(\eta)$. Consider the linear operator T_η on the vector space D over F. Since $f(\eta) = 0$, the polynomial f annihilates T_η. But f splits in F, and so there exists a root $\lambda \in F$ of f that is an eigenvalue of T_η. For a corresponding eigenvector d we have then $d\eta = \lambda d$, and this implies $E^* = d^{-1} F^* d$. Thus, E^* is a conjugate of the subgroup F^* of D^*.

For an arbitrary $c \in D^*$, the set of polynomial expressions in c with coefficients in Z forms a finite integral domain, and thus a finite field by Theorem 1.31. Hence, any element of D^* is contained in some subfield of D, and so in some maximal subfield of D. From what we have already shown, it follows that any element of D^* belongs to some conjugate of F^*. By Theorem 1.25, the number of distinct conjugates of F^* is given by $|D^*|/|N(F^*)|$, and so it is at most $|D^*|/|F^*|$. Since each conjugate of F^* contains the identity element of D^*, the union of the conjugates of F^* has at most

$$\frac{|D^*|}{|F^*|}(|F^*| - 1) + 1 = |D^*| - \frac{|D^*|}{|F^*|} + 1$$

elements. This number is less than $|D^*|$ except when $D^* = F^*$. Hence $D = F$, and D is a field. □

NOTES

1. This section marks the proper beginning of a theory of finite fields. Most textbooks on abstract algebra devote a few pages to finite fields. More extensive treatments can be found, for example, in Albert [3], Berlekamp [4], Birkhoff and Bartee [1], Carmichael [4], Dornhoff and Hohn [1], Herstein [4], Lüneburg [2], Rédei [10], [11], and von Ammon and Tröndle [1]. General field theory is discussed in detail in Browkin [2], Jacobson [2], Nagata [2], and Winter [1]. For finite rings see McDonald [1].

The concept of a finite field in its general meaning (i.e., not only prime fields $\mathbb{F}_p$) first occurred in a paper of Galois [1] in 1830, in the context of solving congruences mod p (or, equivalently, equations over $\mathbb{F}_p$) in suitable extension fields. Of course, by that time many properties of the special finite fields $\mathbb{F}_p$ had already been established, in particular by Fermat, Euler, Lagrange, Legendre, and Gauss (see Gauss [1]). After the seminal paper of Galois, further research on "higher congruences," as equations over finite fields were called then, was carried out by Schönemann [3], Serret [1], and Dedekind [1]. Rudiments of the theory are also contained in a posthumous manuscript of Gauss [4]. Surveys of this early work on finite fields can be found in the report of H. J. S. Smith [1] and in the books of Serret [2], C. Jordan [2], and Borel and Drach [1]. See also Niederreiter [14] for a discussion of the early history of the subject. For an account of the history of the theory of finite fields up to 1915 we refer to Dickson [40, Ch. 8]. The first modern treatment of the theory was presented by Dickson [7].

Theorems 2.5, 2.6, and 2.8 are the most important results of this section. There are various proofs for these theorems, which can be found in the textbooks mentioned above. It should be noted that many authors use the alternative notation $GF(q)$ for the finite field (or Galois field) of order q. The uniqueness part of Theorem 2.5 was first proved in this general form by Moore [1], [2]. Another classical proof of Theorem 2.5 is due to Dickson [6]. See also Szele [1]. In connection with Lemma 2.4 we note that a simple formula for $\prod_{a \in F}(x - a^n)$ is implicit in the work of Rados [5]; see also the related papers of Beeger [1], Lubelski [1], and Ore [3].

The method in the proof of Theorem 2.8 can be used to show more generally that every finite subgroup of the multiplicative group of a field is cyclic. The converse of Theorem 2.8 is also valid (see Exercise 2.10). Gilmer [1] determined all finite commutative rings with identity having a cyclic group of units. Čupona [1] characterized finite fields in terms of properties of orders of elements in the multiplicative group of the field.

In the case of a finite field $\mathbb{F}_p$ with p prime, primitive elements of $\mathbb{F}_p$ are also referred to as *primitive roots* modulo p in elementary number theory. The problem of finding primitive roots modulo p was already treated by Gauss [1]; see also Chebyshev [1], Desmarest [1], Frolov [1], Jacobi [3], Schönheim [1], and Stern [1]. Zassenhaus [4] presents an algorithm for determining primitive elements of any $\mathbb{F}_q$. See also Miller, Reed, and Truong [1], Reed and Truong [1], and Reed, Truong, and Miller [4] for the special case $\mathbb{F}_{p^2}$ with p being a Mersenne prime. The first extensive table of primitive roots was set up by Jacobi [3] in 1839. A more detailed bibliography of such tables will be given in Chapter 10. Primitive elements will be discussed again in the context of so-called primitive polynomials (see Chapter 3, Section 1).

Davenport [6] showed that if p is sufficiently large, say $p > p_0(n)$, and θ is a defining element of $\mathbb{F}_{p^n}$ over $\mathbb{F}_p$, then there exists $a \in \mathbb{F}_p$ such that $\theta - a$ is a primitive element of $\mathbb{F}_{p^n}$. On the other hand, for given $p > 2$ there exists an extension $\mathbb{F}_{p^n}$ and a defining element θ of $\mathbb{F}_{p^n}$ over $\mathbb{F}_p$ such that no element of the form $b\theta + c$, $b, c \in \mathbb{F}_p$, is a primitive element of $\mathbb{F}_{p^n}$. Quantitative refinements and generalizations were established by Carlitz [34], [41], Friedlander [1], and Schwarz [7]; see also Giudici and Margaglio [2] for the case of a quadratic extension of $\mathbb{F}_q$. Davenport and Lewis [3] established a result on the distribution of primitive elements in finite fields, which extended the work of Burgess [2] for prime fields $\mathbb{F}_p$. Later, Burgess [7] improved the result of Davenport and Lewis for fields $\mathbb{F}_{p^2}$ and Karacuba [6] improved it for general finite fields. H. Stevens [1] proved an elementary result on the distribution of primitive elements. Gerjets and Bergum [1] studied the distribution of primitive elements in $\mathbb{F}_{p^2}$ from an elementary standpoint. A simple proof of the existence of primitive elements of $\mathbb{F}_q$, q even, with absolute trace 1 was given by O. Moreno [2]. It follows from a result of I. M. Vinogradov [11] that for given integers a and b and all sufficiently large primes p there exists an integer c such that each of $c, c+a, c+b$ is a primitive root modulo p. Segal [1] showed that for given $r \geqslant 2$ and all sufficiently large primes p there exists an integer c such that each of $c+1, \ldots, c+r$ is a primitive root modulo p, and this was generalized by Carlitz [68]. See also Johnsen [1], Szalay [1], [2], Vegh [1], [2], [3], [4], [5], [6], and I. M. Vinogradov [8]. Madden [1], Segal [1], and I. M. Vinogradov [5] discuss the occurrence of primitive elements as values of a given polynomial. Erdös [1] lists problems and results on primitive roots modulo p. Jacobi's logarithm based on primitive elements (see Exercise 2.8) is tabulated in Jacobi [2] for prime fields $\mathbb{F}_p$, $p \leqslant 103$, and is proposed again in Conway [1] as a convenient computational device; see also Chapter 10, Section 1 and Table B. Gauss [1] gave a formula for the sum of all primitive elements of $\mathbb{F}_p$ and Stern [1] gave one for the sum of all elements of fixed order in $\mathbb{F}_p^*$. Forsyth [1] obtained a formula for the sum of kth powers of all primitive elements of $\mathbb{F}_p$; see also Czarnota [1]. Szymiczek [1] has a formula for the sum of kth powers of all primitive elements of $\mathbb{F}_q$ and, more generally, for the sum of kth powers of all elements of fixed order in $\mathbb{F}_q^*$.

In connection with Corollary 2.11 we note that a formula for the number of monic irreducible polynomials in $\mathbb{F}_q[x]$ of degree n will be given in Theorem 3.25. This leads to an alternative proof of Corollary 2.11 (see the remarks following Example 3.26).

2. The important Theorem 2.14 was established by Galois [1]. This theorem expresses the fact that every finite extension $\mathbb{F}_{q^m}$ of a finite field $\mathbb{F}_q$ is a *normal* extension; that is, it has the property that every irreducible polynomial in $\mathbb{F}_q[x]$ having a root in $\mathbb{F}_{q^m}$ must split in $\mathbb{F}_{q^m}$. More generally, an extension of an arbitrary field K is a finite normal extension if and only if it is the splitting field over K of a polynomial in $K[x]$. Theorem 2.14 shows also that every finite field is a *perfect* field, which means that every irreducible polynomial over the field has only simple roots. According to a well-known characterization, the perfect fields are exactly the fields of characteristic 0 and those fields of prime characteristic p for which any element of the field has a pth root in the field. The latter condition can be verified directly and easily for finite fields (compare with Exercise 2.12).

The automorphism σ_1 of $\mathbb{F}_{q^m}$ over $\mathbb{F}_q$, which generates all automorphisms of $\mathbb{F}_{q^m}$ over $\mathbb{F}_q$ by Theorem 2.21, is called the *Frobenius automorphism* of $\mathbb{F}_{q^m}$ over $\mathbb{F}_q$. The group of automorphisms of $\mathbb{F}_{q^m}$ over $\mathbb{F}_q$ is also referred to as the *Galois group* of $\mathbb{F}_{q^m}$ over $\mathbb{F}_q$. This group plays a fundamental role in Galois theory. See Artin [8], Gaal [1], Jacobson [2], Lang [4, Ch. 8], and van der Waerden [2, Ch. 8] for a general account of Galois theory and Dickson [10] and Scarpis [3] for the special case considered here. The Galois group of $\mathbb{F}_{q^m}$ over $\mathbb{F}_q$ is cyclic according to Theorem 2.21, and thus $\mathbb{F}_{q^m}$ is a *cyclic* extension of $\mathbb{F}_q$.

Irreducible polynomials over finite fields will be considered further in Chapter 3, especially Sections 2, 3, and 5.

3. The results in Theorem 2.25 and Exercise 2.33 are special cases of Hilbert's Theorem 90 for traces and norms, valid for any finite cyclic extension of a field (Hilbert [2]). See also Exercises 2.30 and 2.31 for alternative proofs of Theorem 2.25. We refer to Albert [3, Ch. 4], Bourbaki [1, Sec. 11], Jacobson [2], and Winter [1] for proofs of general versions of Hilbert's Theorem 90.

In connection with the concept of a self-dual basis (see Example 2.31) we note that Seroussi and Lempel [1] have shown that $F = \mathbb{F}_{q^m}$ has a self-dual basis over $K = \mathbb{F}_q$ if and only if either q is even or both q and m are odd. In the same paper it is proved that F always has a *trace-orthogonal* basis over K—that is, a basis $\{\alpha_1, \ldots, \alpha_m\}$ with $\mathrm{Tr}_{F/K}(\alpha_i \alpha_j) = 0$ for $i \neq j$. For $q = 2$ these results were shown earlier by Lempel [2]. In MacWilliams and Sloane [2, Ch. 4] it is shown that $\mathbb{F}_{2^m}$ has a self-dual normal basis over $\mathbb{F}_2$ provided that m is odd. This need not be true for even m (see Exercise 2.41).

Lemma 2.34 is shown, for instance, in Hoffman and Kunze [1, Ch. 7], and this textbook may be used as a good reference for other facts from linear algebra. Theorem 2.35 is a special case of a general normal basis theorem for finite Galois extensions (Albert [3, Ch. 4], Berger and Reiner [1], Deuring [1], Jacobson [2], Rédei [10, Ch. 11], Waterhouse [3]). An alternative proof of Theorem 2.35, together with a formula for the number of different normal bases of $\mathbb{F}_{q^m}$ over $\mathbb{F}_q$, will be presented in Chapter 3 (see Theorem 3.73 and the remarks following it). The normal basis theorem for finite fields was conjectured by Eisenstein [6] and partly proved by Schönemann [4]. The first complete proof was given in Hensel [1]. See also Krasner [2] for a different type of proof. Tables of normal bases and dual bases can be found in Conway [1]; see also Chapter 10, Section 1 and Table B. An application of the normal basis theorem to coding theory occurs in Camion [1].

Moore [3] gave the value of the determinant in Corollary 2.38 as $\prod(\sum_{j=1}^{m} b_j \alpha_j)$, where the product is extended over all nonzero m-tuples $(b_1, \ldots, b_m) \in \mathbb{F}_q^m$ for which the nonzero b_j with largest index j is equal to 1. See also Lemma 3.51 for a simpler proof of this formula. Carlitz [85] proves some related identities for determinants. A direct proof of Corollary 2.38 not using Moore's formula was given by Dickson [2], [7, Part I, Ch. 4]. Theorem 2.39 was shown in an equivalent form by Davenport [9]. The same paper also contains a proof of Theorem 2.40. For F of sufficiently large order, this result was already established earlier by Carlitz [35]. H. W. Lenstra [1] showed that a normal basis consisting of primitive elements exists relative to any subfield of F. If m in Theorem 2.39 is a power of the characteristic of $\mathbb{F}_q$, then one arrives at the simpler condition that the trace of α over $\mathbb{F}_q$ be $\neq 0$; see also Perlis [1], Burde [5], and Childs and Orzech [1] for this case.

An efficient algorithm for generating basis vectors for all subspaces of fixed dimension of a vector space over $\mathbb{F}_q$ was developed by Calabi and Wilf [1]. Combinatorial problems for vector spaces over $\mathbb{F}_q$ were considered by Baum and Neuwirth [1], Bu [1], Constantin and Courteau [1], Jamison [1], A. Lee [1], Luh [1], and Wolfmann [1]. Brawley and Hankins [1] enumerated bases of the vector space of $m \times n$ matrices over $\mathbb{F}_q$ according to the ranks of the basis matrices.

4. An explicit formula for cyclotomic polynomials will be given in Theorem 3.27. The result of Theorem 2.47(i) was first shown by Kronecker [1]. Other classical proofs are due to Arndt [1], Dedekind [2], and Lebesgue [3]. Proofs can also be found in Lang [4, Ch. 8], Rédei [10, Ch. 8], and van der Waerden [2, Ch. 8]. The factorization of cyclotomic polynomials

over prime fields $\mathbb{F}_p$ was already discussed by Gauss [4], Schönemann [3], and Pellet [5] in the 19th century. See also Ballieu [1], Chowla and Vijayaraghavan [1], Golomb [7], Guerrier [1], Lubelski [2], McLain and Edgar [1], Rédei [10, Ch. 8], and van de Vooren-van Veen [1]. The case of arbitrary $\mathbb{F}_q$ was treated by Rauter [2]. It follows from Theorem 2.47(ii) and elementary number theory that Q_n is irreducible over $\mathbb{F}_q$ if and only if $n = r^k$, $2r^k$, or 4, where r is an odd prime and $k \geqslant 0$, and q is a primitive root modulo n. See Chapter 4, especially Example 4.6, for methods of finding the factorization of cyclotomic polynomials over finite fields explicitly. Exercise 2.57 contains a list of further properties of cyclotomic polynomials.

Reed, Truong, and Miller [1], [2] develop efficient algorithms for calculating certain roots of unity in special finite fields; see also Liu, Reed, and Truong [1]. Althaus and Leake [1] give a formula for the inverse of a Vandermonde matrix whose entries are roots of unity; see also Knuth [2, Ch. 1].

5. In addition to the methods mentioned already, we note that finite fields can also be represented as residue class rings modulo prime ideals in rings of algebraic integers, a viewpoint emphasized by Dedekind [3]. See also Burde [7] and Niederreiter [14]. Zassenhaus [4] presents an algorithm for constructing finite extensions of $\mathbb{F}_q$, and Yu. P. Vasil'ev [1] discusses this problem from the viewpoint of computer implementation; see also Chor [1]. See Scognamiglio [1] for the representation of elements of finite fields by matrices. In this connection the work of Beard [1], [2], [3], [4] and Beard and McConnel [1] is also of interest. We refer to Hoffman and Kunze [1, Ch. 7] for the result on companion matrices used in this section.

Hohler [1] constructs $\mathbb{F}_{p^2}$ from the prime field $\mathbb{F}_p$ in a way similar to the construction of the complex numbers from the real numbers. Further representations of elements of finite fields can be found in Bartee and Schneider [1], Fadini [1], Mönnig [1], and Neikirk [1]. Finite fields that can be viewed as subfields of $\mathbb{Z}/(m)$ were characterized by Nymann [1]. Raktoe [1] shows how rings such as polynomial rings over $\mathbb{Z}/(m)$ may be built from finite fields and polynomial rings over finite fields.

6. In 1905 Wedderburn determined all finite division rings. Since the original paper of Wedderburn [1] many proofs of this result have been given, and they may be broadly classified as depending mainly on number theory, group theory, linear algebra, the theory of finite-dimensional algebras, and cohomology theory. Wedderburn's paper contains three proofs of the result. The first is based on linear algebra and the theory of minimal polynomials, it is, however, false as noted by Artin [2]; see Hinz [1] for a rectification of this proof. The other two proofs depend on a lemma from number theory which states that if n and b are integers $\geqslant 2$ such that every

prime factor of $b^n - 1$ divides $b^m - 1$ for some $1 \leqslant m < n$, then either $n = 2$ and $b + 1$ is a power of 2, or $n = 6$ and $b = 2$ (Zsigmondy [1], Birkhoff and Vandiver [1]; see also Artin [5]). Dickson [8] gives a proof of Wedderburn's theorem using this result, and he notes that Wedderburn arrived at his last two proofs after he had seen Dickson's.

Our first proof of Theorem 2.55 is due to Witt [1]. This is the shortest and neatest proof along these lines. In the last part of the argument, the use of complex numbers can be avoided by methods of elementary number theory (see Klobe [1] and K. Rogers [1]). Our second proof follows D. E. Taylor [1]. A proof of Lemma 2.57 can be found, for instance, in Hoffman and Kunze [1, Ch. 6].

There are various proofs of Wedderburn's theorem involving group theory. The argument of Zassenhaus [2] depends on the lemma that a finite group in which the normalizer and the centralizer of every abelian subgroup are identical is abelian. For other group-theoretic proofs see Brandis [1], Kaczynski [1], and Scott [1, Ch. 14]. A proof based on cohomology arguments is presented in Blanchard [1, Ch. 4]. Herstein [2] uses a combination of ring-theoretic and group-theoretic methods. A proof employing polynomials over division rings was given by Artin [2].

Proofs of Wedderburn's theorem based on the theory of finite-dimensional algebras and on results such as Lemma 2.58 can be found in Blanchard [1, Ch. 3], Bourbaki [2, Sec. 11], Nagahara and Tominaga [1], and van der Waerden [3, Ch. 14]. An interesting variant occurs in Joly [5] where Chevalley's theorem on equations over finite fields (see Corollary 6.6) is used in a crucial step of the argument.

For further proofs and comments on the history of Wedderburn's theorem see Artin [4], Herstein [3, Ch. 3], [4], and Rédei [10, Ch. 8].

A well-known generalization of Wedderburn's theorem is provided by a result of Jacobson [1] to the effect that if R is a ring such that for every $a \in R$ there exists an integer $n(a) > 1$, depending on a, with $a^{n(a)} = a$, then R is a field. See also Herstein [1], [2], [3, Ch. 3], [4], Laffey [1], Nagahara and Tominaga [1], K. Rogers [2], and Wamsley [1] for proofs of Jacobson's theorem. In another direction, Wedderburn's theorem has been generalized by relaxing the condition of associativity of multiplication in the finite division ring (Albert [2], McCrimmon [1]). A characterization of finite prime fields in the class of near-rings with identity was given by Clay and Malone [1] and Maxson [1].

EXERCISES

2.1. Prove that $x^2 + 1$ is irreducible over $\mathbb{F}_{11}$ and show directly that $\mathbb{F}_{11}[x]/(x^2 + 1)$ has 121 elements. Prove also that $x^2 + x + 4$ is irreducible over $\mathbb{F}_{11}$ and show that $\mathbb{F}_{11}[x]/(x^2 + 1)$ is isomorphic to $\mathbb{F}_{11}[x]/(x^2 + x + 4)$.

2.2. Show that the sum of all elements of a finite field is 0, except for $\mathbb{F}_2$.
2.3. Let a, b be elements of $\mathbb{F}_{2^n}$, n odd. Show that $a^2 + ab + b^2 = 0$ implies $a = b = 0$.
2.4. Determine all primitive elements of $\mathbb{F}_7$.
2.5. Determine all primitive elements of $\mathbb{F}_{17}$.
2.6. Determine all primitive elements of $\mathbb{F}_9$.
2.7. Write all elements of $\mathbb{F}_{25}$ as linear combinations of basis elements over $\mathbb{F}_5$. Then find a primitive element β of $\mathbb{F}_{25}$ and determine for each $\alpha \in \mathbb{F}_{25}^*$ the least nonnegative integer n such that $\alpha = \beta^n$.
2.8. If the elements of $\mathbb{F}_q^*$ are represented as powers of a fixed primitive element $b \in \mathbb{F}_q$, then addition in $\mathbb{F}_q$ is facilitated by the introduction of *Jacobi's logarithm* $L(n)$ defined by the equation $1 + b^n = b^{L(n)}$, where the case $b^n = -1$ is excluded. Show that we have then $b^m + b^n = b^{m+L(n-m)}$ whenever L is defined. Construct a table of Jacobi's logarithm for $\mathbb{F}_9$ and $\mathbb{F}_{17}$.
2.9. Prove: for any field F, every finite subgroup of the multiplicative group F^* is cyclic.
2.10. Let F be any field. If F^* is cyclic, show that F is finite.
2.11. Prove: if F is a finite field, then $H \cup \{0\}$ is a subfield of F for every subgroup H of the multiplicative group F^* if and only if the order of F^* is either 1 or a prime number of the form $2^p - 1$ with a prime p.
2.12. For every finite field $\mathbb{F}_q$ of characteristic p, show that there exists exactly one pth root for each element of $\mathbb{F}_q$.
2.13. For a finite field $\mathbb{F}_q$ with q odd, show that an element $a \in \mathbb{F}_q^*$ has a square root in $\mathbb{F}_q$ if and only if $a^{(q-1)/2} = 1$.
2.14. Prove that for given $k \in \mathbb{N}$ the element $a \in \mathbb{F}_q^*$ is the kth power of some element of $\mathbb{F}_q$ if and only if $a^{(q-1)/d} = 1$, where $d = \gcd(q-1, k)$.
2.15. Prove: every element of $\mathbb{F}_q$ is the kth power of some element of $\mathbb{F}_q$ if and only if $\gcd(q-1, k) = 1$.
2.16. Let k be a positive divisor of $q-1$ and $a \in \mathbb{F}_q$ be such that the equation $x^k = a$ has no solution in $\mathbb{F}_q$. Prove that the same equation has a solution in $\mathbb{F}_{q^m}$ if m is divisible by k, and that the converse holds for a prime number k.
2.17. Prove that $f(x)^q = f(x^q)$ for $f \in \mathbb{F}_q[x]$.
2.18. Show that any quadratic polynomial in $\mathbb{F}_q[x]$ splits over $\mathbb{F}_{q^2}$ into linear factors.
2.19. Show that for $a \in \mathbb{F}_q$ and $n \in \mathbb{N}$ the polynomial $x^{q^n} - x + na$ is divisible by $x^q - x + a$ over $\mathbb{F}_q$.
2.20. Find all automorphisms of a finite field.
2.21. If F is a field and $\Psi: F \to F$ is the mapping defined by $\Psi(a) = a^{-1}$ if $a \neq 0$, $\Psi(a) = 0$ if $a = 0$, show that Ψ is an automorphism of F if and only if F has at most four elements.
2.22. Prove: if p is a prime and n a positive integer, then n divides $\phi(p^n - 1)$. (*Hint:* Use Corollary 2.19.)

2.23. Let $\mathbb{F}_q$ be a finite field of characteristic p. Prove that $f \in \mathbb{F}_q[x]$ satisfies $f'(x)=0$ if and only if f is the pth power of some polynomial in $\mathbb{F}_q[x]$.

2.24. Let F be a finite extension of the finite field K with $[F:K]=m$ and let $f(x)=x^d+b_{d-1}x^{d-1}+\cdots+b_0 \in K[x]$ be the minimal polynomial of $\alpha \in F$ over K. Prove that $\mathrm{Tr}_{F/K}(\alpha)=-(m/d)b_{d-1}$ and $\mathrm{N}_{F/K}(\alpha)=(-1)^m b_0^{m/d}$.

2.25. Let F be a finite extension of the finite field K and $\alpha \in F$. The mapping $L: \beta \in F \mapsto \alpha\beta \in F$ is a linear transformation of F, considered as a vector space over K. Prove that the characteristic polynomial $g(x)$ of α over K is equal to the characteristic polynomial of the linear transformation L; that is, $g(x)=\det(xI-L)$, where I is the identity transformation.

2.26. Consider the same situation as in Exercise 2.25. Prove that $\mathrm{Tr}_{F/K}(\alpha)$ is equal to the trace of the linear transformation L and that $\mathrm{N}_{F/K}(\alpha) = \det(L)$.

2.27. Prove properties (i) and (ii) of Theorem 2.23 by using the interpretation of $\mathrm{Tr}_{F/K}(\alpha)$ obtained in Exercise 2.26.

2.28. Prove properties (i) and (iii) of Theorem 2.28 by using the interpretation of $\mathrm{N}_{F/K}(\alpha)$ obtained in Exercise 2.26.

2.29. Let F be a finite extension of the finite field K of characteristic p. Prove that $\mathrm{Tr}_{F/K}(\alpha^{p^n})=(\mathrm{Tr}_{F/K}(\alpha))^{p^n}$ for all $\alpha \in F$ and $n \in \mathbb{N}$.

2.30. Give an alternative proof of Theorem 2.25 by viewing F as a vector space over K and showing by dimension arguments that the kernel of the linear transformation $\mathrm{Tr}_{F/K}$ is equal to the range of the linear operator L on F defined by $L(\beta)=\beta^q-\beta$ for $\beta \in F$.

2.31. Give an alternative proof of the necessity of the condition in Theorem 2.25 by showing that if $\alpha \in F$ with $\mathrm{Tr}_{F/K}(\alpha)=0$, $\gamma \in F$ with $\mathrm{Tr}_{F/K}(\gamma)=-1$, and $\delta_j=\alpha+\alpha^q+\cdots+\alpha^{q^{j-1}}$, then

$$\beta=\sum_{j=1}^{[F:K]} \delta_j \gamma^{q^{j-1}}$$

satisfies $\beta^q-\beta=\alpha$.

2.32. Let F be a finite extension of $K=\mathbb{F}_q$ and $\alpha=\beta^q-\beta$ for some $\beta \in F$. Prove that $\alpha=\gamma^q-\gamma$ with $\gamma \in F$ if and only if $\beta-\gamma \in K$.

2.33. Let F be a finite extension of $K=\mathbb{F}_q$. Prove that for $\alpha \in F$ we have $\mathrm{N}_{F/K}(\alpha)=1$ if and only if $\alpha=\beta^{q-1}$ for some $\beta \in F^*$.

2.34. Prove $\sum_{j=0}^{m-1} x^{q^j}-c=\prod(x-\alpha)$ for all $c \in K=\mathbb{F}_q$, where the product is extended over all $\alpha \in F=\mathbb{F}_{q^m}$ with $\mathrm{Tr}_{F/K}(\alpha)=c$.

2.35. Prove

$$x^{q^m}-x=\prod_{c \in \mathbb{F}_q}\left(\sum_{j=0}^{m-1} x^{q^j}-c\right)$$

for any $m \in \mathbb{N}$.

2.36. Consider $\mathbb{F}_{q^m}$ as a vector space over $\mathbb{F}_q$ and prove that for every linear operator L on $\mathbb{F}_{q^m}$ there exists a uniquely determined m-tuple $(\alpha_0, \alpha_1, \ldots, \alpha_{m-1})$ of elements of $\mathbb{F}_{q^m}$ such that

$$L(\beta) = \alpha_0\beta + \alpha_1\beta^q + \cdots + \alpha_{m-1}\beta^{q^{m-1}} \text{ for all } \beta \in \mathbb{F}_{q^m}.$$

2.37. Prove that if the order of basis elements is taken into account, then the number of different bases of $\mathbb{F}_{q^m}$ over $\mathbb{F}_q$ is

$$(q^m - 1)(q^m - q)(q^m - q^2)\cdots(q^m - q^{m-1}).$$

2.38. Prove: if $\{\alpha_1, \ldots, \alpha_m\}$ is a basis of $F = \mathbb{F}_{q^m}$ over $K = \mathbb{F}_q$, then $\mathrm{Tr}_{F/K}(\alpha_i) \neq 0$ for at least one i, $1 \leqslant i \leqslant m$.

2.39. Prove that there exists a normal basis $\{\alpha, \alpha^q, \ldots, \alpha^{q^{m-1}}\}$ of $F = \mathbb{F}_{q^m}$ over $K = \mathbb{F}_q$ with $\mathrm{Tr}_{F/K}(\alpha) = 1$.

2.40. Let K be a finite field, $F = K(\alpha)$ a finite simple extension of degree n, and $f \in K[x]$ the minimal polynomial of α over K. Let

$$\frac{f(x)}{x - \alpha} = \beta_0 + \beta_1 x + \cdots + \beta_{n-1}x^{n-1} \in F[x] \quad \text{and} \quad \gamma = f'(\alpha).$$

Prove that the dual basis of $\{1, \alpha, \ldots, \alpha^{n-1}\}$ is $\{\beta_0\gamma^{-1}, \beta_1\gamma^{-1}, \ldots, \beta_{n-1}\gamma^{-1}\}$.

2.41. Show that there is a self-dual normal basis of $\mathbb{F}_4$ over $\mathbb{F}_2$, but no self-dual normal basis of $\mathbb{F}_{16}$ over $\mathbb{F}_2$ (see Example 2.31 for the definition of a self-dual basis).

2.42. Construct a self-dual basis of $\mathbb{F}_{16}$ over $\mathbb{F}_2$ (see Example 2.31 for the definition of a self-dual basis).

2.43. Prove that the dual basis of a normal basis of $\mathbb{F}_{q^m}$ over $\mathbb{F}_q$ is again a normal basis of $\mathbb{F}_{q^m}$ over $\mathbb{F}_q$.

2.44. Let F be an extension of the finite field K with basis $\{\alpha_1, \ldots, \alpha_m\}$ over K. Let $\beta_1, \ldots, \beta_m \in F$ with $\beta_i = \sum_{j=1}^m b_{ij}\alpha_j$ for $1 \leqslant i \leqslant m$ and $b_{ij} \in K$. Let B be the $m \times m$ matrix whose (i, j) entry is b_{ij}. Prove that $\Delta_{F/K}(\beta_1, \ldots, \beta_m) = \det(B)^2 \Delta_{F/K}(\alpha_1, \ldots, \alpha_m)$.

2.45. Let $K = \mathbb{F}_q$ and $F = \mathbb{F}_{q^m}$. Prove that for $\alpha \in F$ we have

$$\Delta_{F/K}(1, \alpha, \ldots, \alpha^{m-1}) = \prod_{0 \leqslant i < j \leqslant m-1} (\alpha^{q^i} - \alpha^{q^j})^2.$$

2.46. Prove that for $\alpha \in F = \mathbb{F}_{q^m}$ with $m \geqslant 2$ and $K = \mathbb{F}_q$ the discriminant $\Delta_{F/K}(1, \alpha, \ldots, \alpha^{m-1})$ is equal to the discriminant of the characteristic polynomial of α over K.

2.47. Determine the primitive 4th and 8th roots of unity in $\mathbb{F}_9$.

2.48. Determine the primitive 9th roots of unity in $\mathbb{F}_{19}$.

2.49. Let ζ be an nth root of unity over a field K. Prove that $1 + \zeta + \zeta^2 + \cdots + \zeta^{n-1} = 0$ or n according as $\zeta \neq 1$ or $\zeta = 1$.

2.50. For $n \geqslant 2$ let $\zeta_1, \ldots, \zeta_n$ be all the (not necessarily distinct) nth roots of unity over an arbitrary field K. Prove that $\zeta_1^k + \cdots + \zeta_n^k = n$ for $k = 0$ and $\zeta_1^k + \cdots + \zeta_n^k = 0$ for $k = 1, 2, \ldots, n-1$.

2.51. For an arbitrary field K and an odd positive integer n, show that $K^{(2n)} = K^{(n)}$.

2.52. Let K be an arbitrary field. Prove that the cyclotomic field $K^{(d)}$ is a subfield of $K^{(n)}$ for any positive divisor d of $n \in \mathbb{N}$. Determine the minimal polynomial over $K^{(4)}$ of a root of unity that can serve as a defining element of $K^{(12)}$ over $K^{(4)}$.

2.53. Prove that for p prime the $p-1$ primitive pth roots of unity over $\mathbb{Q}$ are linearly independent over $\mathbb{Q}$ and therefore form a basis of $\mathbb{Q}^{(p)}$ over $\mathbb{Q}$.

2.54. Let K be an arbitrary field and $n \geqslant 2$. Prove that the polynomial $x^{n-1} + x^{n-2} + \cdots + x + 1$ is irreducible over K only if n is a prime number.

2.55. Find the least prime p such that $x^{22} + x^{21} + \cdots + x + 1$ is irreducible over $\mathbb{F}_p$.

2.56. Find the ten least primes p such that $x^{p-1} + x^{p-2} + \cdots + x + 1$ is irreducible over $\mathbb{F}_2$.

2.57. Prove the following properties of cyclotomic polynomials over a field for which the polynomials exist:

(a) $Q_{mp}(x) = Q_m(x^p)/Q_m(x)$ if p is prime and $m \in \mathbb{N}$ is not divisible by p;

(b) $Q_{mp}(x) = Q_m(x^p)$ for all $m \in \mathbb{N}$ divisible by the prime p;

(c) $Q_{mp^k}(x) = Q_{mp}(x^{p^{k-1}})$ if p is a prime and $m, k \in \mathbb{N}$ are arbitrary;

(d) $Q_{2n}(x) = Q_n(-x)$ if $n \geqslant 3$ and n odd;

(e) $Q_n(0) = 1$ if $n \geqslant 2$;

(f) $Q_n(x^{-1})x^{\phi(n)} = Q_n(x)$ if $n \geqslant 2$;

(g)

$$Q_n(1) = \begin{cases} 0 & \text{if } n = 1, \\ p & \text{if } n \text{ is a power of the prime } p, \\ 1 & \text{if } n \text{ has at least two distinct prime factors;} \end{cases}$$

(h)

$$Q_n(-1) = \begin{cases} 0 & \text{if } n = 2, \\ -2 & \text{if } n = 1, \\ p & \text{if } n \text{ is 2 times a power of the prime } p, \\ 1 & \text{otherwise.} \end{cases}$$

2.58. Give the matrix representation for the elements of $\mathbb{F}_8$ using the irreducible polynomial $x^3 + x + 1$ over $\mathbb{F}_2$.

2.59. Let ζ be a primitive element of $F = \mathbb{F}_{16}$ with $\zeta^4 + \zeta + 1 = 0$. For $k \geqslant 0$ write $\zeta^k = \sum_{m=0}^{3} a_{km}\zeta^m$ with $a_{km} \in \mathbb{F}_2$, and let M_k be the 4×4 matrix whose (i, j) entry is $a_{k+i-1, j-1}$. Show that the 15 matrices M_k, $0 \leqslant k \leqslant 14$, and the 4×4 zero matrix form a field (with respect to addition and matrix multiplication over $\mathbb{F}_2$) which is isomorphic to F. For $0 \leqslant k \leqslant 14$ prove that $\mathrm{Tr}_F(\zeta^k) = \text{trace of } M_k = a_{k3}$.

Chapter 3

Polynomials over Finite Fields

The theory of polynomials over finite fields is important for investigating the algebraic structure of finite fields as well as for many applications. Above all, irreducible polynomials—the prime elements of the polynomial ring over a finite field—are indispensable for constructing finite fields and computing with the elements of a finite field.

Section 1 introduces the notion of the order of a polynomial. An important fact is the connection between minimal polynomials of primitive elements (so-called primitive polynomials) and polynomials of the highest possible order for a given degree. Results about irreducible polynomials going beyond those discussed in the previous chapters are presented in Section 2. The next section is devoted to constructive aspects of irreducibility and deals also with the problem of calculating the minimal polynomial of an element in an extension field.

Certain special types of polynomials are discussed in the last two sections. Linearized polynomials are singled out by the property that all the exponents occurring in them are powers of the characteristic. The remarkable theory of these polynomials enables us, in particular, to give an alternative proof of the normal basis theorem. Binomials and trinomials—that is, two-term and three-term polynomials—form another class of polynomials for which special results of considerable interest can be established. We remark that another useful collection of polynomials—namely, that of cyclotomic polynomials—was already considered in Chapter

2, Section 4, and that some additional information on cyclotomic polynomials is contained in Section 2 of the present chapter.

1. ORDER OF POLYNOMIALS AND PRIMITIVE POLYNOMIALS

Besides the degree, there is another important integer attached to a nonzero polynomial over a finite field, namely its order. The definition of the order of a polynomial is based on the following result.

3.1. **Lemma.** *Let $f \in \mathbb{F}_q[x]$ be a polynomial of degree $m \geqslant 1$ with $f(0) \neq 0$. Then there exists a positive integer $e \leqslant q^m - 1$ such that $f(x)$ divides $x^e - 1$.*

Proof. The residue class ring $\mathbb{F}_q[x]/(f)$ contains $q^m - 1$ nonzero residue classes. The q^m residue classes $x^j + (f)$, $j = 0, 1, \ldots, q^m - 1$, are all nonzero, and so there exist integers r and s with $0 \leqslant r < s \leqslant q^m - 1$ such that $x^s \equiv x^r \bmod f(x)$. Since x and $f(x)$ are relatively prime, it follows that $x^{s-r} \equiv 1 \bmod f(x)$; that is, $f(x)$ divides $x^{s-r} - 1$ and $0 < s - r \leqslant q^m - 1$. □

Since a nonzero constant polynomial divides $x - 1$, these polynomials can be included in the following definition.

3.2. Definition. Let $f \in \mathbb{F}_q[x]$ be a nonzero polynomial. If $f(0) \neq 0$, then the least positive integer e for which $f(x)$ divides $x^e - 1$ is called the *order* of f and denoted by $\mathrm{ord}(f) = \mathrm{ord}(f(x))$. If $f(0) = 0$, then $f(x) = x^h g(x)$, where $h \in \mathbb{N}$ and $g \in \mathbb{F}_q[x]$ with $g(0) \neq 0$ are uniquely determined; $\mathrm{ord}(f)$ is then defined to be $\mathrm{ord}(g)$.

The order of the polynomial f is sometimes also called the *period* of f or the *exponent* of f. The order of an irreducible polynomial f can be characterized in the following alternative fashion.

3.3. **Theorem.** *Let $f \in \mathbb{F}_q[x]$ be an irreducible polynomial over $\mathbb{F}_q$ of degree m and with $f(0) \neq 0$. Then $\mathrm{ord}(f)$ is equal to the order of any root of f in the multiplicative group $\mathbb{F}_{q^m}^*$.*

Proof. According to Corollary 2.15, $\mathbb{F}_{q^m}$ is the splitting field of f over $\mathbb{F}_q$. The roots of f have the same order in the group $\mathbb{F}_{q^m}^*$ by Theorem 2.18. Let $\alpha \in \mathbb{F}_{q^m}^*$ be any root of f. Then we obtain from Lemma 2.12 that we have $\alpha^e = 1$ if and only if $f(x)$ divides $x^e - 1$. The result follows now from the definitions of $\mathrm{ord}(f)$ and the order of α in the group $\mathbb{F}_{q^m}^*$. □

3.4. **Corollary.** *If $f \in \mathbb{F}_q[x]$ is an irreducible polynomial over $\mathbb{F}_q$ of degree m, then $\mathrm{ord}(f)$ divides $q^m - 1$.*

Proof. If $f(x)=cx$ with $c\in\mathbb{F}_q^*$, then $\operatorname{ord}(f)=1$ and the result is trivial. Otherwise, the result follows from Theorem 3.3 and the fact that $\mathbb{F}_{q^m}^*$ is a group of order q^m-1. □

For reducible polynomials the result of Corollary 3.4 need not be valid (see Example 3.10). There is another interpretation of $\operatorname{ord}(f)$ based on associating a square matrix to f in a canonical fashion and considering the order of this matrix in a certain group of matrices (see Lemma 8.26).

Theorem 3.3 leads to a formula for the number of monic irreducible polynomials of given degree and given order. We use again ϕ to denote Euler's function introduced in Theorem 1.15(iv). The following terminology will be convenient: if n is a positive integer and the integer b is relatively prime to n, then the least positive integer k for which $b^k\equiv 1 \bmod n$ is called the *multiplicative order* of b modulo n.

3.5. Theorem. *The number of monic irreducible polynomials in $\mathbb{F}_q[x]$ of degree m and order e is equal to $\phi(e)/m$ if $e\geqslant 2$ and m is the multiplicative order of q modulo e, equal to 2 if $m=e=1$, and equal to 0 in all other cases. In particular, the degree of an irreducible polynomial in $\mathbb{F}_q[x]$ of order e must be equal to the multiplicative order of q modulo e.*

Proof. Let f be an irreducible polynomial in $\mathbb{F}_q[x]$ with $f(0)\neq 0$. Then, according to Theorem 3.3, we have $\operatorname{ord}(f)=e$ if and only if all roots of f are primitive eth roots of unity over $\mathbb{F}_q$. In other words, we have $\operatorname{ord}(f)=e$ if and only if f divides the cyclotomic polynomial Q_e. By Theorem 2.47(ii), any monic irreducible factor of Q_e has the same degree m, the least positive integer such that $q^m\equiv 1 \bmod e$, and the number of such factors is given by $\phi(e)/m$. For $m=e=1$, we also have to take into account the monic irreducible polynomial $f(x)=x$. □

Values of $\operatorname{ord}(f)$ are available in tabulated form, at least for irreducible polynomials f (see Chapter 10, Section 2). Since any polynomial of positive degree can be written as a product of irreducible polynomials, the computation of orders of polynomials can be achieved if one knows how to determine the order of a power of an irreducible polynomial and the order of the product of pairwise relatively prime polynomials. The subsequent discussion is devoted to these questions.

3.6. Lemma. *Let c be a positive integer. Then the polynomial $f\in\mathbb{F}_q[x]$ with $f(0)\neq 0$ divides x^c-1 if and only if $\operatorname{ord}(f)$ divides c.*

Proof. If $e=\operatorname{ord}(f)$ divides c, then $f(x)$ divides x^e-1 and x^e-1 divides x^c-1, so that $f(x)$ divides x^c-1. Conversely, if $f(x)$ divides x^c-1, we have $c\geqslant e$, so that we can write $c=me+r$ with $m\in\mathbb{N}$ and $0\leqslant r<e$. Since $x^c-1=(x^{me}-1)x^r+(x^r-1)$, it follows that $f(x)$ divides x^r-1, which is only possible for $r=0$. Therefore, e divides c. □

3.7. **Corollary.** *If e_1 and e_2 are positive integers, then the greatest common divisor of $x^{e_1}-1$ and $x^{e_2}-1$ in $\mathbb{F}_q[x]$ is x^d-1, where d is the greatest common divisor of e_1 and e_2.*

Proof. Let $f(x)$ be the (monic) greatest common divisor of $x^{e_1}-1$ and $x^{e_2}-1$. Since x^d-1 is a common divisor of $x^{e_i}-1$, $i=1,2$, it follows that x^d-1 divides $f(x)$. On the other hand, $f(x)$ is a common divisor of $x^{e_i}-1$, $i=1,2$, and so Lemma 3.6 implies that $\operatorname{ord}(f)$ divides e_1 and e_2. Consequently, $\operatorname{ord}(f)$ divides d, and hence $f(x)$ divides x^d-1 by Lemma 3.6. Altogether, we have shown that $f(x)=x^d-1$. □

Since powers of x are factored out in advance when determining the order of a polynomial, we need not consider powers of the irreducible polynomials $g(x)$ with $g(0)=0$.

3.8. **Theorem.** *Let $g\in\mathbb{F}_q[x]$ be irreducible over $\mathbb{F}_q$ with $g(0)\neq 0$ and $\operatorname{ord}(g)=e$, and let $f=g^b$ with a positive integer b. Let t be the smallest integer with $p^t\geqslant b$, where p is the characteristic of $\mathbb{F}_q$. Then $\operatorname{ord}(f)=ep^t$.*

Proof. Setting $c=\operatorname{ord}(f)$ and noting that the divisibility of x^c-1 by $f(x)$ implies the divisibility of x^c-1 by $g(x)$, we obtain that e divides c by Lemma 3.6. Furthermore, $g(x)$ divides x^e-1; therefore, $f(x)$ divides $(x^e-1)^b$ and, *a fortiori*, it divides $(x^e-1)^{p^t}=x^{ep^t}-1$. Thus according to Lemma 3.6, c divides ep^t. It follows from what we have shown so far that c is of the form $c=ep^u$ with $0\leqslant u\leqslant t$. We note now that x^e-1 has only simple roots, since e is not a multiple of p because of Corollary 3.4. Therefore, all the roots of $x^{ep^u}-1=(x^e-1)^{p^u}$ have multiplicity p^u. But $g(x)^b$ divides $x^{ep^u}-1$, whence $p^u\geqslant b$ by comparing multiplicities of roots, and so $u\geqslant t$. Thus we get $u=t$ and $c=ep^t$. □

3.9. **Theorem.** *Let $g_1,\ldots,g_k$ be pairwise relatively prime nonzero polynomials over $\mathbb{F}_q$, and let $f=g_1\cdots g_k$. Then $\operatorname{ord}(f)$ is equal to the least common multiple of $\operatorname{ord}(g_1),\ldots,\operatorname{ord}(g_k)$.*

Proof. It is easily seen that it suffices to consider the case where $g_i(0)\neq 0$ for $1\leqslant i\leqslant k$. Set $e=\operatorname{ord}(f)$ and $e_i=\operatorname{ord}(g_i)$ for $1\leqslant i\leqslant k$, and let $c=\operatorname{lcm}(e_1,\ldots,e_k)$. Then each $g_i(x), 1\leqslant i\leqslant k$, divides $x^{e_i}-1$, and so $g_i(x)$ divides x^c-1. Because of the pairwise relative primality of the polynomials $g_1,\ldots,g_k$, we obtain that $f(x)$ divides x^c-1. An application of Lemma 3.6 shows that e divides c. On the other hand, $f(x)$ divides x^e-1, and so each $g_i(x)$, $1\leqslant i\leqslant k$, divides x^e-1. Again by Lemma 3.6, it follows that each e_i, $1\leqslant i\leqslant k$, divides e, and therefore c divides e. Thus we conclude that $e=c$. □

By using the same argument as above, one may, in fact, show that the order of the least common multiple of finitely many nonzero polynomials is equal to the least common multiple of the orders of the polynomials.

3.10. Example. Let us compute the order of $f(x)=x^{10}+x^9+x^3+x^2+1\in\mathbb{F}_2[x]$. The canonical factorization of $f(x)$ over $\mathbb{F}_2$ is given by $f(x)=(x^2+x+1)^3(x^4+x+1)$. Since $\operatorname{ord}(x^2+x+1)=3$, we get $\operatorname{ord}((x^2+x+1)^3)=12$ by Theorem 3.8. Furthermore, $\operatorname{ord}(x^4+x+1)=15$, and so Theorem 3.9 implies that $\operatorname{ord}(f)$ is equal to the least common multiple of 12 and 15; that is, $\operatorname{ord}(f)=60$. Note that $\operatorname{ord}(f)$ does not divide $2^{10}-1$, which shows that Corollary 3.4 need not hold for reducible polynomials. □

On the basis of the information provided above, one arrives then at the following general formula for the order of a polynomial. It suffices to consider polynomials of positive degree and with nonzero constant term.

3.11. Theorem. *Let $\mathbb{F}_q$ be a finite field of characteristic p, and let $f\in\mathbb{F}_q[x]$ be a polynomial of positive degree and with $f(0)\neq 0$. Let $f=af_1^{b_1}\cdots f_k^{b_k}$, where $a\in\mathbb{F}_q$, $b_1,\ldots,b_k\in\mathbb{N}$, and $f_1,\ldots,f_k$ are distinct monic irreducible polynomials in $\mathbb{F}_q[x]$, be the canonical factorization of f in $\mathbb{F}_q[x]$. Then $\operatorname{ord}(f)=ep^t$, where e is the least common multiple of $\operatorname{ord}(f_1),\ldots,\operatorname{ord}(f_k)$ and t is the smallest integer with $p^t\geqslant\max(b_1,\ldots,b_k)$.*

A *method of determining the order* of an irreducible polynomial f in $\mathbb{F}_q[x]$ with $f(0)\neq 0$ is based on the observation that the order e of f is the least positive integer such that $x^e\equiv 1 \bmod f(x)$. Furthermore, by Corollary 3.4, e divides q^m-1, where $m=\deg(f)$. Assuming $q^m>2$, we start from the prime factor decomposition

$$q^m-1=\prod_{j=1}^{s}p_j^{r_j}.$$

For $1\leqslant j\leqslant s$ we calculate the residues of $x^{(q^m-1)/p_j} \bmod f(x)$. This is accomplished by multiplying together a suitable combination of the residues of $x, x^q, x^{q^2},\ldots,x^{q^{m-1}} \bmod f(x)$. If $x^{(q^m-1)/p_j}\not\equiv 1 \bmod f(x)$, then e is a multiple of $p_j^{r_j}$. If $x^{(q^m-1)/p_j}\equiv 1 \bmod f(x)$, then e is not a multiple of $p_j^{r_j}$. In the latter case we check to see whether e is a multiple of $p_j^{r_j-1}, p_j^{r_j-2},\ldots,p_j$ by calculating the residues of

$$x^{(q^m-1)/p_j^2}, x^{(q^m-1)/p_j^3},\ldots,x^{(q^m-1)/p_j^{r_j}} \bmod f(x).$$

This computation is repeated for each prime factor of q^m-1.

A key step in the method above is the factorization of the integer q^m-1. There exist extensive tables for the complete factorization of numbers of this form, especially for the case $q=2$.

We compare now the orders of polynomials obtained from each other by simple algebraic transformations. The following is a typical example.

3.12. Definition. Let

$$f(x)=a_nx^n+a_{n-1}x^{n-1}+\cdots+a_1x+a_0\in\mathbb{F}_q[x]$$

with $a_n \neq 0$. Then the *reciprocal polynomial* f^* of f is defined by

$$f^*(x) = x^n f\left(\frac{1}{x}\right) = a_0 x^n + a_1 x^{n-1} + \cdots + a_{n-1}x + a_n.$$

3.13. Theorem. *Let f be a nonzero polynomial in $\mathbb{F}_q[x]$ and f^* its reciprocal polynomial. Then* $\operatorname{ord}(f) = \operatorname{ord}(f^*)$.

Proof. First consider the case $f(0) \neq 0$. Then the result follows from the fact that $f(x)$ divides $x^e - 1$ if and only if $f^*(x)$ does. If $f(0) = 0$, write $f(x) = x^h g(x)$ with $h \in \mathbb{N}$ and $g \in \mathbb{F}_q[x]$ satisfying $g(0) \neq 0$. Then from what we have already shown it follows that $\operatorname{ord}(f) = \operatorname{ord}(g) = \operatorname{ord}(g^*) = \operatorname{ord}(f^*)$, where the last identity is valid since $g^* = f^*$. □

There is also a close relationship between the orders of $f(x)$ and $f(-x)$. Since $f(x) = f(-x)$ for a field of characteristic 2, it suffices to consider finite fields of odd characteristic.

3.14. Theorem. *For odd q, let $f \in \mathbb{F}_q[x]$ be a polynomial of positive degree with $f(0) \neq 0$. Let e and E be the orders of $f(x)$ and $f(-x)$, respectively. Then $E = e$ if e is a multiple of 4 and $E = 2e$ if e is odd. If e is twice an odd number, then $E = e/2$ if all irreducible factors of f have even order and $E = e$ otherwise.*

Proof. Since $\operatorname{ord}(f(x)) = e$, $f(x)$ divides $x^{2e} - 1$, and so $f(-x)$ divides $(-x)^{2e} - 1 = x^{2e} - 1$. Thus E divides $2e$ by Lemma 3.6. By the same argument, e divides $2E$, and so E can only be $2e$, e, or $e/2$. If e is a multiple of 4, then both e and E are even. Since $f(x)$ divides $x^e - 1$, $f(-x)$ divides $(-x)^e - 1 = x^e - 1$, and so E divides e. Similarly, e divides E, and thus it follows that $E = e$. If e is odd, then $f(-x)$ divides $(-x)^e - 1 = -x^e - 1$ and so $x^e + 1$. But then $f(-x)$ cannot divide $x^e - 1$, and so we must have $E = 2e$.

In the remaining case we have $e = 2h$ with an odd integer h. Let f be a power of an irreducible polynomial in $\mathbb{F}_q[x]$. Then $f(x)$ divides $(x^h - 1)(x^h + 1)$ and $f(x)$ does not divide $x^h - 1$ since $\operatorname{ord}(f) = 2h$. But $x^h - 1$ and $x^h + 1$ are relatively prime, and this implies that $f(x)$ divides $x^h + 1$. Consequently, $f(-x)$ divides $(-x)^h + 1 = -x^h + 1$ and so $x^h - 1$. It follows that $E = e/2$. Note that by Theorem 3.8 the power of an irreducible polynomial has even order if and only if the irreducible polynomial itself has even order.

For general f we have a factorization $f = g_1 \cdots g_k$, where each g_i is a power of an irreducible polynomial and $g_1, \ldots, g_k$ are pairwise relatively prime. Furthermore, $2h = \operatorname{lcm}(\operatorname{ord}(g_1), \ldots, \operatorname{ord}(g_k))$ according to Theorem 3.9. We arrange the g_i in such a way that $\operatorname{ord}(g_i) = 2h_i$ for $1 \leqslant i \leqslant m$ and $\operatorname{ord}(g_i) = h_i$ for $m+1 \leqslant i \leqslant k$, where the h_i are odd integers with $\operatorname{lcm}(h_1, \ldots, h_k) = h$. By what we have already shown, we get $\operatorname{ord}(g_i(-x)) = h_i$ for $1 \leqslant i \leqslant m$ and $\operatorname{ord}(g_i(-x)) = 2h_i$ for $m+1 \leqslant i \leqslant k$. Then Theorem 3.9 yields

$$E = \operatorname{lcm}(h_1, \ldots, h_m, 2h_{m+1}, \ldots, 2h_k),$$

and so $E = h = e/2$ if $m = k$ and $E = 2h = e$ if $m < k$. These formulas are equivalent to those given in the last part of the theorem. □

It follows from Lemma 3.1 and Definition 3.2 that the order of a polynomial of degree $m \geqslant 1$ over $\mathbb{F}_q$ is at most $q^m - 1$. This bound is attained for an important class of polynomials—namely, so-called primitive polynomials. The definition of a primitive polynomial is based on the notion of primitive element introduced in Definition 2.9.

3.15. Definition. A polynomial $f \in \mathbb{F}_q[x]$ of degree $m \geqslant 1$ is called a *primitive* polynomial over $\mathbb{F}_q$ if it is the minimal polynomial over $\mathbb{F}_q$ of a primitive element of $\mathbb{F}_{q^m}$.

Thus, a primitive polynomial over $\mathbb{F}_q$ of degree m may be described as a monic polynomial that is irreducible over $\mathbb{F}_q$ and has a root $\alpha \in \mathbb{F}_{q^m}$ that generates the multiplicative group of $\mathbb{F}_{q^m}$. Primitive polynomials can also be characterized as follows.

3.16. Theorem. *A polynomial $f \in \mathbb{F}_q[x]$ of degree m is a primitive polynomial over $\mathbb{F}_q$ if and only if f is monic, $f(0) \neq 0$, and* $\operatorname{ord}(f) = q^m - 1$.

Proof. If f is primitive over $\mathbb{F}_q$, then f is monic and $f(0) \neq 0$. Since f is irreducible over $\mathbb{F}_q$, we get $\operatorname{ord}(f) = q^m - 1$ from Theorem 3.3 and the fact that f has a primitive element of $\mathbb{F}_{q^m}$ as a root.

Conversely, the property $\operatorname{ord}(f) = q^m - 1$ implies that $m \geqslant 1$. Next, we claim that f is irreducible over $\mathbb{F}_q$. Suppose f were reducible over $\mathbb{F}_q$. Then f is either a power of an irreducible polynomial or it can be written as a product of two relatively prime polynomials of positive degree. In the first case, we have $f = g^b$ with $g \in \mathbb{F}_q[x]$ irreducible over $\mathbb{F}_q$, $g(0) \neq 0$, and $b \geqslant 2$. Then, according to Theorem 3.8, $\operatorname{ord}(f)$ is divisible by the characteristic of $\mathbb{F}_q$, but $q^m - 1$ is not, a contradiction. In the second case, we have $f = g_1 g_2$ with relatively prime monic polynomials $g_1, g_2 \in \mathbb{F}_q[x]$ of positive degree m_1 and m_2, respectively. If $e_i = \operatorname{ord}(g_i)$ for $i = 1, 2$, then $\operatorname{ord}(f) \leqslant e_1 e_2$ by Theorem 3.9. Furthermore, $e_i \leqslant q^{m_i} - 1$ for $i = 1, 2$ by Lemma 3.1, hence

$$\operatorname{ord}(f) \leqslant (q^{m_1} - 1)(q^{m_2} - 1) < q^{m_1 + m_2} - 1 = q^m - 1,$$

a contradiction. Therefore, f is irreducible over $\mathbb{F}_q$, and it follows then from Theorem 3.3 that f is a primitive polynomial over $\mathbb{F}_q$. □

We remark that the condition $f(0) \neq 0$ in the theorem above is only needed to rule out the non-primitive polynomial $f(x) = x$ in case $q = 2$ and $m = 1$. Still another characterization of primitive polynomials is based on the following auxiliary result.

3.17. Lemma. *Let $f \in \mathbb{F}_q[x]$ be a polynomial of positive degree with $f(0) \neq 0$. Let r be the least positive integer for which x^r is congruent* mod $f(x)$ *to some element of $\mathbb{F}_q$, so that $x^r \equiv a \bmod f(x)$ with a uniquely determined*

$a \in \mathbb{F}_q^*$. Then $\operatorname{ord}(f) = hr$, *where h is the order of a in the multiplicative group* $\mathbb{F}_q^*$.

Proof. Put $e = \operatorname{ord}(f)$. Since $x^e \equiv 1 \bmod f(x)$, we must have $e \geqslant r$. Thus we can write $e = sr + t$ with $s \in \mathbb{N}$ and $0 \leqslant t < r$. Now

$$1 \equiv x^e \equiv x^{sr+t} \equiv a^s x^t \bmod f(x), \tag{3.1}$$

thus $x^t \equiv a^{-s} \bmod f(x)$, and because of the definition of r this is only possible if $t = 0$. The congruence (3.1) yields then $a^s \equiv 1 \bmod f(x)$, thus $a^s = 1$, and so $s \geqslant h$ and $e \geqslant hr$. On the other hand, $x^{hr} \equiv a^h \equiv 1 \bmod f(x)$, and so $e = hr$. □

3.18. Theorem. *The monic polynomial $f \in \mathbb{F}_q[x]$ of degree $m \geqslant 1$ is a primitive polynomial over $\mathbb{F}_q$ if and only if $(-1)^m f(0)$ is a primitive element of $\mathbb{F}_q$ and the least positive integer r for which x^r is congruent* $\bmod f(x)$ *to some element of $\mathbb{F}_q$ is $r = (q^m - 1)/(q-1)$. In case f is primitive over $\mathbb{F}_q$, we have $x^r \equiv (-1)^m f(0) \bmod f(x)$.*

Proof. If f is primitive over $\mathbb{F}_q$, then f has a root $\alpha \in \mathbb{F}_{q^m}$, which is a primitive element of $\mathbb{F}_{q^m}$. By calculating the norm $\mathrm{N}_{\mathbb{F}_{q^m}/\mathbb{F}_q}(\alpha)$ both by Definition 2.27 and by (2.3) and observing that f is the characteristic polynomial of α over $\mathbb{F}_q$, we arrive at the identity

$$(-1)^m f(0) = \alpha^{(q^m - 1)/(q-1)}. \tag{3.2}$$

It follows that the order of $(-1)^m f(0)$ in $\mathbb{F}_q^*$ is $q - 1$; that is, $(-1)^m f(0)$ is a primitive element of $\mathbb{F}_q$. Since f is the minimal polynomial of α over $\mathbb{F}_q$, the identity (3.2) implies that

$$x^{(q^m-1)/(q-1)} \equiv (-1)^m f(0) \bmod f(x),$$

and so $r \leqslant (q^m - 1)/(q-1)$. But Theorem 3.16 and Lemma 3.17 yield $q^m - 1 = \operatorname{ord}(f) \leqslant (q-1)r$, thus $r = (q^m - 1)/(q-1)$.

Conversely, suppose the conditions of the theorem are satisfied. It follows from $r = (q^m - 1)/(q-1)$ and Lemma 3.17 that $\operatorname{ord}(f)$ is relatively prime to q. Then Theorem 3.11 shows that f has a factorization of the form $f = f_1 \cdots f_k$, where the f_i are distinct monic irreducible polynomials over $\mathbb{F}_q$. If $m_i = \deg(f_i)$, then $\operatorname{ord}(f_i)$ divides $q^{m_i} - 1$ for $1 \leqslant i \leqslant k$ according to Corollary 3.4. Now $q^{m_i} - 1$ divides

$$d = (q^{m_1} - 1) \cdots (q^{m_k} - 1)/(q-1)^{k-1},$$

thus $\operatorname{ord}(f_i)$ divides d for $1 \leqslant i \leqslant k$. It follows from Lemma 3.6 that $f_i(x)$ divides $x^d - 1$ for $1 \leqslant i \leqslant k$, and so $f(x)$ divides $x^d - 1$. If $k \geqslant 2$, then

$$d < (q^{m_1 + \cdots + m_k} - 1)/(q-1) = (q^m - 1)/(q-1) = r,$$

a contradiction to the definition of r. Thus $k = 1$ and f is irreducible over $\mathbb{F}_q$.

If $\beta \in \mathbb{F}_{q^m}$ is a root of f, then the argument leading to (3.2) shows that $\beta^r = (-1)^m f(0)$, and so $x^r \equiv (-1)^m f(0) \bmod f(x)$. Since the order of $(-1)^m f(0)$ in $\mathbb{F}_q^*$ is $q-1$, it follows from Lemma 3.17 that $\operatorname{ord}(f) = q^m - 1$, so that f is primitive over $\mathbb{F}_q$ by Theorem 3.16. □

3.19. Example. Consider the polynomial $f(x) = x^4 + x^3 + x^2 + 2x + 2 \in \mathbb{F}_3[x]$. Since f is irreducible over $\mathbb{F}_3$, one can use the method outlined after Theorem 3.11 to show that $\operatorname{ord}(f) = 80 = 3^4 - 1$. Consequently, f is primitive over $\mathbb{F}_3$ by Theorem 3.16. We have $x^{40} \equiv 2 \bmod f(x)$ in accordance with Theorem 3.18. □

2. IRREDUCIBLE POLYNOMIALS

We recall that a polynomial $f \in \mathbb{F}_q[x]$ is *irreducible* over $\mathbb{F}_q$ if f has positive degree and every factorization of f in $\mathbb{F}_q[x]$ must involve a constant polynomial (see Definition 1.57). Elementary properties of irreducible polynomials over $\mathbb{F}_q$ were discussed in Chapter 2, Section 2.

3.20. Theorem. *For every finite field $\mathbb{F}_q$ and every $n \in \mathbb{N}$, the product of all monic irreducible polynomials over $\mathbb{F}_q$ whose degrees divide n is equal to $x^{q^n} - x$.*

Proof. According to Lemma 2.13, the monic irreducible polynomials over $\mathbb{F}_q$ occurring in the canonical factorization of $g(x) = x^{q^n} - x$ in $\mathbb{F}_q[x]$ are precisely those whose degrees divide n. Since $g'(x) = -1$, Theorem 1.68 implies that g has no multiple roots in its splitting field over $\mathbb{F}_q$, and so each monic irreducible polynomial over $\mathbb{F}_q$ whose degree divides n occurs exactly once in the canonical factorization of g in $\mathbb{F}_q[x]$. □

3.21. Corollary. *If $N_q(d)$ is the number of monic irreducible polynomials in $\mathbb{F}_q[x]$ of degree d, then*

$$q^n = \sum_{d \mid n} d N_q(d) \quad \text{for all } n \in \mathbb{N}, \tag{3.3}$$

where the sum is extended over all positive divisors d of n.

Proof. The identity (3.3) follows from Theorem 3.20 by comparing the degree of $g(x) = x^{q^n} - x$ with the total degree of the canonical factorization of $g(x)$. □

With a little elementary number theory we can derive from (3.3) an explicit formula for the number of monic irreducible polynomials in $\mathbb{F}_q[x]$ of fixed degree. We need an arithmetic function, called the Moebius function, which is defined as follows.

3.22. Definition. The *Moebius function* μ is the function on $\mathbb{N}$ defined by

$$\mu(n)=\begin{cases}1 & \text{if } n=1,\\ (-1)^k & \text{if } n \text{ is the product of } k \text{ distinct primes},\\ 0 & \text{if } n \text{ is divisible by the square of a prime.}\end{cases}$$

As in (3.3), we use the summation symbol $\Sigma_{d\mid n}$ to denote a sum extended over all positive divisors d of $n\in\mathbb{N}$. A similar convention applies to the product symbol $\Pi_{d\mid n}$.

3.23. Lemma. *For $n\in\mathbb{N}$ the Moebius function μ satisfies*

$$\sum_{d\mid n}\mu(d)=\begin{cases}1 & \text{if } n=1,\\ 0 & \text{if } n>1.\end{cases}$$

Proof. For $n>1$ we have to take into account only those positive divisors d of n for which $\mu(d)\neq 0$—that is, for which $d=1$ or d is a product of distinct primes. Thus, if $p_1, p_2,\ldots,p_k$ are the distinct prime divisors of n, we get

$$\begin{aligned}\sum_{d\mid n}\mu(d)&=\mu(1)+\sum_{i=1}^{k}\mu(p_i)+\sum_{1\leqslant i_1<i_2\leqslant k}\mu(p_{i_1}p_{i_2})+\cdots+\mu(p_1p_2\cdots p_k)\\ &=1+\binom{k}{1}(-1)+\binom{k}{2}(-1)^2+\cdots+\binom{k}{k}(-1)^k\\ &=(1+(-1))^k=0.\end{aligned}$$

The case $n=1$ is trivial. □

3.24. Theorem (Moebius Inversion Formula)

(i) Additive case: *Let h and H be two functions from $\mathbb{N}$ into an additively written abelian group G. Then*

$$H(n)=\sum_{d\mid n}h(d)\quad \textit{for all } n\in\mathbb{N} \tag{3.4}$$

if and only if

$$h(n)=\sum_{d\mid n}\mu\left(\frac{n}{d}\right)H(d)=\sum_{d\mid n}\mu(d)H\left(\frac{n}{d}\right)\quad \textit{for all } n\in\mathbb{N}. \tag{3.5}$$

(ii) Multiplicative case: *Let h and H be two functions from $\mathbb{N}$ into a multiplicatively written abelian group G. Then*

$$H(n)=\prod_{d\mid n}h(d)\quad \textit{for all } n\in\mathbb{N} \tag{3.6}$$

if and only if

$$h(n)=\prod_{d\mid n}H(d)^{\mu(n/d)}=\prod_{d\mid n}H\left(\frac{n}{d}\right)^{\mu(d)}\quad \textit{for all } n\in\mathbb{N}. \tag{3.7}$$

Proof. Assuming (3.4) and using Lemma 3.23, we get

$$\sum_{d|n} \mu\left(\frac{n}{d}\right) H(d) = \sum_{d|n} \mu(d) H\left(\frac{n}{d}\right) = \sum_{d|n} \mu(d) \sum_{c|n/d} h(c)$$

$$= \sum_{c|n} \sum_{d|n/c} \mu(d) h(c) = \sum_{c|n} h(c) \sum_{d|n/c} \mu(d) = h(n)$$

for all $n \in \mathbb{N}$. The converse is derived by a similar calculation. The proof of part (ii) follows immediately from the proof of part (i) if we replace the sums by products and the multiples by powers. □

3.25. Theorem. *The number $N_q(n)$ of monic irreducible polynomials in $\mathbb{F}_q[x]$ of degree n is given by*

$$N_q(n) = \frac{1}{n} \sum_{d|n} \mu\left(\frac{n}{d}\right) q^d = \frac{1}{n} \sum_{d|n} \mu(d) q^{n/d}.$$

Proof. We apply the additive case of the Moebius inversion formula to the group $G = \mathbb{Z}$, the additive group of integers. Let $h(n) = nN_q(n)$ and $H(n) = q^n$ for all $n \in \mathbb{N}$. Then (3.4) is satisfied because of the identity (3.3), and so (3.5) already gives the desired formula. □

3.26. Example. The number of monic irreducible polynomials in $\mathbb{F}_q[x]$ of degree 20 is given by

$$N_q(20) = \tfrac{1}{20}\left(\mu(1) q^{20} + \mu(2) q^{10} + \mu(4) q^5 + \mu(5) q^4 + \mu(10) q^2 + \mu(20) q\right)$$

$$= \tfrac{1}{20}\left(q^{20} - q^{10} - q^4 + q^2\right).$$ □

It should be noted that the formula in Theorem 3.25 shows again that for every finite field $\mathbb{F}_q$ and every $n \in \mathbb{N}$ there exists an irreducible polynomial in $\mathbb{F}_q[x]$ of degree n (compare with Corollary 2.11). Namely, using $\mu(1) = 1$ and $\mu(d) \geqslant -1$ for all $d \in \mathbb{N}$, a crude estimate yields

$$N_q(n) \geqslant \frac{1}{n}\left(q^n - q^{n-1} - q^{n-2} - \cdots - q\right) = \frac{1}{n}\left(q^n - \frac{q^n - q}{q - 1}\right) > 0.$$

As another application of the Moebius inversion formula, we establish an explicit formula for the nth cyclotomic polynomial Q_n.

3.27. Theorem. *For a field K of characteristic p and $n \in \mathbb{N}$ not divisible by p, the nth cyclotomic polynomial Q_n over K satisfies*

$$Q_n(x) = \prod_{d|n} (x^d - 1)^{\mu(n/d)} = \prod_{d|n} (x^{n/d} - 1)^{\mu(d)}.$$

Proof. We apply the multiplicative case of the Moebius inversion formula to the multiplicative group G of nonzero rational functions over K. Let $h(n) = Q_n(x)$ and $H(n) = x^n - 1$ for all $n \in \mathbb{N}$. Then Theorem 2.45(i) shows that (3.6) is satisfied, and so (3.7) yields the desired result. □

3.28. Example. For fields K over which Q_{12} is defined, we have

$$\begin{aligned} Q_{12}(x) &= \prod_{d \mid 12} (x^{12/d} - 1)^{\mu(d)} \\ &= (x^{12} - 1)^{\mu(1)}(x^6 - 1)^{\mu(2)}(x^4 - 1)^{\mu(3)}(x^3 - 1)^{\mu(4)} \\ &\quad \cdot (x^2 - 1)^{\mu(6)}(x - 1)^{\mu(12)} \\ &= \frac{(x^{12} - 1)(x^2 - 1)}{(x^6 - 1)(x^4 - 1)} = x^4 - x^2 + 1. \end{aligned}$$

□

The explicit formula in Theorem 3.27 can be used to establish the basic properties of cyclotomic polynomials (compare with Exercise 3.35).

In Theorem 3.25 we determined the *number* of monic irreducible polynomials in $\mathbb{F}_q[x]$ of fixed degree. We present now a formula for the *product* of all monic irreducible polynomials in $\mathbb{F}_q[x]$ of fixed degree.

3.29. Theorem. *The product $I(q, n; x)$ of all monic irreducible polynomials in $\mathbb{F}_q[x]$ of degree n is given by*

$$I(q, n; x) = \prod_{d \mid n} (x^{q^d} - x)^{\mu(n/d)} = \prod_{d \mid n} (x^{q^{n/d}} - x)^{\mu(d)}.$$

Proof. It follows from Theorem 3.20 that

$$x^{q^n} - x = \prod_{d \mid n} I(q, d; x).$$

We apply the multiplicative case of the Moebius inversion formula to the multiplicative group G of nonzero rational functions over $\mathbb{F}_q$, putting $h(n) = I(q, n; x)$ and $H(n) = x^{q^n} - x$ for all $n \in \mathbb{N}$, and we obtain the desired formula. □

3.30. Example. For $q = 2$, $n = 4$ we get

$$\begin{aligned} I(2,4; x) &= (x^{16} - x)^{\mu(1)}(x^4 - x)^{\mu(2)}(x^2 - x)^{\mu(4)} \\ &= \frac{x^{16} - x}{x^4 - x} = \frac{x^{15} - 1}{x^3 - 1} \\ &= x^{12} + x^9 + x^6 + x^3 + 1. \end{aligned}$$

□

All monic irreducible polynomials in $\mathbb{F}_q[x]$ of degree n can be determined by factoring $I(q, n; x)$. For this purpose it is advantageous to have $I(q, n; x)$ available in a partially factored form. This is achieved by the following result.

3.31. Theorem. *Let $I(q, n; x)$ be as in Theorem 3.29. Then for $n > 1$ we have*

$$I(q, n; x) = \prod_{m} Q_m(x), \tag{3.8}$$

where the product is extended over all positive divisors m of $q^n - 1$ *for which n is the multiplicative order of q modulo m, and where* $Q_m(x)$ *is the mth cyclotomic polynomial over* $\mathbb{F}_q$.

Proof. For $n > 1$ let S be the set of elements of $\mathbb{F}_{q^n}$ that are of degree n over $\mathbb{F}_q$. Then every $\alpha \in S$ has a minimal polynomial over $\mathbb{F}_q$ of degree n and is thus a root of $I(q, n; x)$. On the other hand, if β is a root of $I(q, n; x)$, then β is a root of some monic irreducible polynomial in $\mathbb{F}_q[x]$ of degree n, which implies that $\beta \in S$. Therefore,

$$I(q, n; x) = \prod_{\alpha \in S} (x - \alpha).$$

If $\alpha \in S$, then $\alpha \in \mathbb{F}_{q^n}^*$, and so the order of α in that multiplicative group is a divisor of $q^n - 1$. We note that $\gamma \in \mathbb{F}_{q^n}^*$ is an element of a proper subfield $\mathbb{F}_{q^d}$ of $\mathbb{F}_{q^n}$ if and only if $\gamma^{q^d} = \gamma$—that is, if and only if the order of γ divides $q^d - 1$. Thus, the order m of an element α of S must be such that n is the least positive integer with $q^n \equiv 1 \bmod m$—that is, such that n is the multiplicative order of q modulo m. For a positive divisor m of $q^n - 1$ with this property, let S_m be the set of elements of S of order m. Then S is the disjoint union of the subsets S_m, so that we can write

$$I(q, n; x) = \prod_{m} \prod_{\alpha \in S_m} (x - \alpha).$$

Now S_m contains exactly all elements of $\mathbb{F}_{q^n}^*$ of order m. In other words, S_m is the set of primitive mth roots of unity over $\mathbb{F}_q$. From the definition of cyclotomic polynomials (see Definition 2.44), it follows that

$$\prod_{\alpha \in S_m} (x - \alpha) = Q_m(x),$$

and so (3.8) is established. □

3.32. Example. We determine all (monic) irreducible polynomials in $\mathbb{F}_2[x]$ of degree 4. The identity (3.8) yields $I(2,4; x) = Q_5(x)Q_{15}(x)$. By Theorem 2.47(ii), $Q_5(x) = x^4 + x^3 + x^2 + x + 1$ is irreducible in $\mathbb{F}_2[x]$. By the same theorem, $Q_{15}(x)$ factors into two irreducible polynomials in $\mathbb{F}_2[x]$ of degree 4. Since $Q_5(x+1) = x^4 + x^3 + 1$ is irreducible in $\mathbb{F}_2[x]$, this polynomial must divide $Q_{15}(x)$, and so

$$Q_{15}(x) = x^8 + x^7 + x^5 + x^4 + x^3 + x + 1 = (x^4 + x^3 + 1)(x^4 + x + 1).$$

Therefore, the irreducible polynomials in $\mathbb{F}_2[x]$ of degree 4 are $x^4 + x^3 + x^2 + x + 1$, $x^4 + x^3 + 1$, and $x^4 + x + 1$. □

Irreducible polynomials often arise as minimal polynomials of elements of an extension field. Minimal polynomials were introduced in Definition 1.81 and their fundamental properties established in Theorem 1.82. With special reference to finite fields, we summarize now the most useful facts about minimal polynomials.

3.33. Theorem. *Let α be an element of the extension field $\mathbb{F}_{q^m}$ of $\mathbb{F}_q$. Suppose that the degree of α over $\mathbb{F}_q$ is d and that $g \in \mathbb{F}_q[x]$ is the minimal polynomial of α over $\mathbb{F}_q$. Then:*

(i) *g is irreducible over $\mathbb{F}_q$ and its degree d divides m.*
(ii) *A polynomial $f \in \mathbb{F}_q[x]$ satisfies $f(\alpha)=0$ if and only if g divides f.*
(iii) *If f is a monic irreducible polynomial in $\mathbb{F}_q[x]$ with $f(\alpha)=0$, then $f=g$.*
(iv) *$g(x)$ divides $x^{q^d}-x$ and $x^{q^m}-x$.*
(v) *The roots of g are $\alpha, \alpha^q, \ldots, \alpha^{q^{d-1}}$, and g is the minimal polynomial over $\mathbb{F}_q$ of all these elements.*
(vi) *If $\alpha \neq 0$, then $\operatorname{ord}(g)$ is equal to the order of α in the multiplicative group $\mathbb{F}_{q^m}^*$.*
(vii) *g is a primitive polynomial over $\mathbb{F}_q$ if and only if α is of order q^d-1 in $\mathbb{F}_{q^m}^*$.*

Proof. (i) The first part follows from Theorem 1.82(i) and the second part from Theorem 1.86.

(ii) This follows from Theorem 1.82(ii).

(iii) This is an immediate consequence of (ii).

(iv) This follows from (i) and Lemma 2.13.

(v) The first part follows from (i) and Theorem 2.14 and the second part from (iii).

(vi) Since $\alpha \in \mathbb{F}_{q^d}^*$ and $\mathbb{F}_{q^d}^*$ is a subgroup of $\mathbb{F}_{q^m}^*$, the result is contained in Theorem 3.3.

(vii) If g is primitive over $\mathbb{F}_q$, then $\operatorname{ord}(g)=q^d-1$, and so α is of order q^d-1 in $\mathbb{F}_{q^m}^*$ because of (vi). Conversely, if α is of order q^d-1 in $\mathbb{F}_{q^m}^*$ and so in $\mathbb{F}_{q^d}^*$, then α is a primitive element of $\mathbb{F}_{q^d}$, and therefore g is primitive over $\mathbb{F}_q$ by Definition 3.15. □

3. CONSTRUCTION OF IRREDUCIBLE POLYNOMIALS

We first describe a general principle of obtaining new irreducible polynomials from known ones. It depends on an auxiliary result from number theory. We recall that if n is a positive integer and the integer b is relatively prime to n, then the least positive integer k for which $b^k \equiv 1 \bmod n$ is called the *multiplicative order* of b modulo n. We note that this multiplicative order divides any other positive integer h for which $b^h \equiv 1 \bmod n$.

3.34. Lemma. *Let $s \geqslant 2$ and $e \geqslant 2$ be relatively prime integers and let m be the multiplicative order of s modulo e. Let $t \geqslant 2$ be an integer whose prime factors divide e but not $(s^m-1)/e$. Assume also that $s^m \equiv 1 \bmod 4$ if $t \equiv 0 \bmod 4$. Then the multiplicative order of s modulo et is equal to mt.*

Proof. We proceed by induction on the number of prime factors of t, each counted with its multiplicity. First, let t be a prime number. Writing $d = (s^m - 1)/e$, we have $s^m = 1 + de$, and so

$$\begin{aligned} s^{mt} &= (1+de)^t \\ &= 1 + \binom{t}{1} de + \binom{t}{2} d^2e^2 + \cdots + \binom{t}{t-1} d^{t-1}e^{t-1} + d^te^t. \end{aligned}$$

In the last expression, each term except the first and the last is divisible by et because of a property of binomial coefficients noted in the proof of Theorem 1.46. Furthermore, the last term is divisible by et since t divides e. Therefore, $s^{mt} \equiv 1 \bmod et$, and so the multiplicative order k of s modulo et divides mt. Also, $s^k \equiv 1 \bmod et$ implies $s^k \equiv 1 \bmod e$, and so k is divisible by m. Since t is a prime number, k can only be m or mt. If $k = m$, then $s^m \equiv 1 \bmod et$, hence $de \equiv 0 \bmod et$ and t divides d, a contradiction. Thus we must have $k = mt$.

Now suppose that t has at least two prime factors and write $t = rt_0$, where r is a prime factor of t. By what we have already shown, the multiplicative order of s modulo er is equal to mr. If we can prove that each prime factor of t_0 divides er but not $d_0 = (s^{mr} - 1)/er$, then the induction hypothesis applied to t_0 yields that the multiplicative order of s modulo $ert_0 = et$ is equal to $mrt_0 = mt$. Let r_0 be a prime factor of t_0. Since every prime factor of t divides e, it is trivial that r_0 divides er. We write again $d = (s^m - 1)/e$. We have $s^{mr} - 1 = c(s^m - 1)$ with $c = s^{m(r-1)} + \cdots + s^m + 1$, thus $d_0 = c(s^m - 1)/er = cd/r$. Furthermore, since $s^m \equiv 1 \bmod e$ and r divides e, we get $s^m \equiv 1 \bmod r$, and so $c \equiv r \equiv 0 \bmod r$. Thus c/r is an integer. Since r_0 does not divide d, it suffices to demonstrate that r_0 does not divide c/r in order to prove that r_0 does not divide $d_0 = cd/r$. We note that $s^m \equiv 1 \bmod r_0$, and so $c \equiv r \bmod r_0$. If $r_0 \neq r$, then $c/r \equiv 1 \bmod r_0$, thus r_0 does not divide c/r. Now let $r_0 = r$. Then $s^m \equiv 1 + br \bmod r^2$ for some $b \in \mathbb{Z}$, hence $s^{mj} \equiv (1+br)^j \equiv 1 + jbr \bmod r^2$ for all $j \geqslant 0$, and thus

$$c = r + br \sum_{j=0}^{r-1} j \equiv r + br\frac{r(r-1)}{2} \bmod r^2.$$

It follows that

$$\frac{c}{r} \equiv 1 + b\frac{r(r-1)}{2} \bmod r.$$

If r is odd, then $c/r \equiv 1 \bmod r$, so that $r_0 = r$ does not divide c/r. In the remaining case we have $r_0 = r = 2$. Then $t \equiv 0 \bmod 4$, and so $s^m \equiv 1 \bmod 4$ by hypothesis. Since $c = s^m + 1$ in this case, we get $c \equiv 2 \bmod 4$, and thus $c/r = c/2 \equiv 1 \bmod 2$. It follows again that r_0 does not divide c/r. □

3.35. Theorem. *Let $f_1(x), f_2(x), \ldots, f_N(x)$ be all the distinct monic irreducible polynomials in $\mathbb{F}_q[x]$ of degree m and order e, and let $t \geqslant 2$ be an* *integer whose prime factors divide e but not* $(q^m - 1)/e$. *Assume also that* $q^m \equiv 1 \bmod 4$ *if* $t \equiv 0 \bmod 4$. *Then* $f_1(x^t), f_2(x^t), \ldots, f_N(x^t)$ *are all the distinct monic irreducible polynomials in* $\mathbb{F}_q[x]$ *of degree mt and order et.*

Proof. The condition on e implies $e \geqslant 2$. According to Theorem 3.5, monic irreducible polynomials in $\mathbb{F}_q[x]$ of degree m and order $e \geqslant 2$ exist only if m is the multiplicative order of q modulo e, and then $N = \phi(e)/m$. By Lemma 3.34, the multiplicative order of q modulo et is equal to mt, and since $\phi(et)/mt = \phi(e)/m$ by the formula in Exercise 1.4, part (c), it follows that the number of monic irreducible polynomials in $\mathbb{F}_q[x]$ of degree mt and order et is also equal to N. Therefore, it remains to show that each of the polynomials $f_j(x^t)$, $1 \leqslant j \leqslant N$, is irreducible in $\mathbb{F}_q[x]$ and of order et. Since the roots of each $f_j(x)$ are primitive eth roots of unity over $\mathbb{F}_q$ by Theorem 3.3, it follows that $f_j(x)$ divides the cyclotomic polynomial $Q_e(x)$ over $\mathbb{F}_q$. Then $f_j(x^t)$ divides $Q_e(x^t)$, and repeated use of the property enunciated in Exercise 2.57, part (b), shows that $Q_e(x^t) = Q_{et}(x)$. Thus $f_j(x^t)$ divides $Q_{et}(x)$. According to Theorem 2.47(ii), the degree of each irreducible factor of $Q_{et}(x)$ in $\mathbb{F}_q[x]$ is equal to the multiplicative order of q modulo et, which is mt. Since $f_j(x^t)$ has degree mt, it follows that $f_j(x^t)$ is irreducible in $\mathbb{F}_q[x]$. Furthermore, since $f_j(x^t)$ divides $Q_{et}(x)$, the order of $f_j(x^t)$ is et. □

3.36. Example. The irreducible polynomials in $\mathbb{F}_2[x]$ of degree 4 and order 15 are $x^4 + x + 1$ and $x^4 + x^3 + 1$. Then the irreducible polynomials in $\mathbb{F}_2[x]$ of degree 12 and order 45 are $x^{12} + x^3 + 1$ and $x^{12} + x^9 + 1$. The irreducible polynomials in $\mathbb{F}_2[x]$ of degree 60 and order 225 are $x^{60} + x^{15} + 1$ and $x^{60} + x^{45} + 1$. The irreducible polynomials in $\mathbb{F}_2[x]$ of degree 100 and order 375 are $x^{100} + x^{25} + 1$ and $x^{100} + x^{75} + 1$. □

The case in which $t \equiv 0 \bmod 4$ and $q^m \equiv -1 \bmod 4$ is not covered in Theorem 3.35. Here we must have $q \equiv -1 \bmod 4$ and m odd. The result referring to this case is somewhat more complicated than Theorem 3.35.

3.37. Theorem. *Let* $f_1(x), f_2(x), \ldots, f_N(x)$ *be all the distinct monic irreducible polynomials in* $\mathbb{F}_q[x]$ *of odd degree m and of order e. Let* $q = 2^a u - 1$, $t = 2^b v$ *with* $a, b \geqslant 2$, *where u and v are odd and all prime factors of t divide e but not* $(q^m - 1)/e$. *Let k be the smaller of a and b. Then each of the polynomials* $f_j(x^t)$ *factors as a product of* 2^{k-1} *monic irreducible polynomials* $g_{ij}(x)$ *in* $\mathbb{F}_q[x]$ *of degree* $mt2^{1-k}$. *The* $2^{k-1}N$ *polynomials* $g_{ij}(x)$ *are all the distinct monic irreducible polynomials in* $\mathbb{F}_q[x]$ *of degree* $mt2^{1-k}$ *and order et.*

Proof. If $v \geqslant 3$, then Theorem 3.35 implies that $f_1(x^v), f_2(x^v), \ldots, f_N(x^v)$ are all the distinct monic irreducible polynomials in $\mathbb{F}_q[x]$ of odd degree mv and of order ev. Thus we will be done once the special case $t = 2^b$ is settled.

Let now $t = 2^b$, and note that as in the proof of Theorem 3.35 we obtain that m is the multiplicative order of q modulo e, $N = \phi(e)/m$, and each $f_j(x^t)$ divides $Q_{et}(x)$. By Theorem 2.47(ii), $Q_{et}(x)$ factors into distinct monic irreducible polynomials in $\mathbb{F}_q[x]$ of degree d, where d is the multiplicative order of q modulo et. Since $q^d \equiv 1 \bmod et$, we have $q^d \equiv 1 \bmod e$, and so m divides d. Consider first the case $a \geqslant b$. Then $q^{2m}-1 = (q^m-1)(q^m+1)$, and the first factor is divisible by e, whereas the second factor is divisible by t since $q \equiv -1 \bmod 2^a$ implies $q \equiv -1 \bmod t$, and thus $q^m \equiv (-1)^m \equiv -1 \bmod t$. Altogether, we get $q^{2m} \equiv 1 \bmod et$, and so d can only be m or $2m$. If $d = m$, then $q^m \equiv 1 \bmod et$, hence $q^m \equiv 1 \bmod t$, a contradiction. Thus $d = 2m = m2^{b-k+1}$ since $k = b$ in this case.

Now consider the case $a < b$. We prove by induction on h that

$$q^{m2^h} \equiv 1 + w2^{a+h} \bmod 2^{a+h+1} \quad \text{for all } h \in \mathbb{N}, \tag{3.9}$$

where w is odd. For $h = 1$ we get

$$\begin{aligned} q^{2m} &= (2^a u - 1)^{2m} \\ &= 1 - 2^{a+1}um + \sum_{n=2}^{2m} \binom{2m}{n}(-1)^{2m-n}2^{na}u^n \equiv 1 + w2^{a+1} \bmod 2^{a+2} \end{aligned}$$

with $w = -um$. If (3.9) is shown for some $h \in \mathbb{N}$, then

$$q^{m2^h} = 1 + w2^{a+h} + c2^{a+h+1} \quad \text{for some } c \in \mathbb{Z}.$$

It follows that

$$q^{m2^{h+1}} = (1 + w2^{a+h} + c2^{a+h+1})^2 \equiv 1 + w2^{a+h+1} \bmod 2^{a+h+2},$$

and so the proof of (3.9) is complete. Applying (3.9) with $h = b - a + 1$, we get $q^{m2^{b-a+1}} \equiv 1 \bmod 2^{b+1}$. Furthermore, $q^m \equiv 1 \bmod e$ implies $q^{m2^{b-a+1}} \equiv 1 \bmod e$, and so $q^{m2^{b-a+1}} \equiv 1 \bmod L$, where L is the least common multiple of 2^{b+1} and e. Now e is even since all prime factors of t divide e, but also $e \not\equiv 0 \bmod 4$ since $q^m \equiv 1 \bmod e$ and $q^m \equiv -1 \bmod 4$. Therefore, $L = e2^b = et$, and thus $q^{m2^{b-a+1}} \equiv 1 \bmod et$. On the other hand, using (3.9) with $h = b - a$ we get

$$q^{m2^{b-a}} \equiv 1 + w2^b \not\equiv 1 \bmod 2^{b+1},$$

which implies $q^{m2^{b-a}} \not\equiv 1 \bmod et$. Consequently, we must have $d = m2^{b-a+1} = m2^{b-k+1}$ since $k = a$ in this case. Therefore, the formula $d = m2^{b-k+1} = mt2^{1-k}$ is valid in both cases.

Since $Q_{et}(x)$ factors into distinct monic irreducible polynomials in $\mathbb{F}_q[x]$ of degree $mt2^{1-k}$, each $f_j(x^t)$ factors into such polynomials. By comparing degrees, the number of factors is found to be 2^{k-1}. Since each irreducible factor $g_{ij}(x)$ of $f_j(x^t)$ divides $Q_{et}(x)$, each $g_{ij}(x)$ is of order et. The various polynomials $g_{ij}(x)$, $1 \leqslant i \leqslant 2^{k-1}$, $1 \leqslant j \leqslant N$, are distinct, for otherwise one such polynomial, say $g(x)$, would divide $f_{j_1}(x^t)$ and $f_{j_2}(x^t)$ for $j_1 \neq j_2$, and then any root β of $g(x)$ would lead to a common root β^t of $f_{j_1}(x)$ and $f_{j_2}(x)$, a contradiction. By Theorem 3.5, the number of monic irreducible polynomials in $\mathbb{F}_q[x]$ of degree $mt2^{1-k}$ and order et is $\phi(et)/mt2^{1-k} = 2^{k-1}\phi(et)/mt = 2^{k-1}\phi(e)/m = 2^{k-1}N$, and so the $g_{ij}(x)$ yield all such polynomials. □

We will show how, from a given irreducible polynomial of order e, all the irreducible polynomials whose orders divide e may be obtained. Since in all cases $g(x) = x$ will be among the latter polynomials, we only consider polynomials g with $g(0) \neq 0$. Let f be a monic irreducible polynomial in $\mathbb{F}_q[x]$ of degree m and order e and with $f(0) \neq 0$. Let $\alpha \in \mathbb{F}_{q^m}$ be a root of f, and for every $t \in \mathbb{N}$ let $g_t \in \mathbb{F}_q[x]$ be the minimal polynomial of α^t over $\mathbb{F}_q$. Let $T = \{t_1, t_2, \ldots, t_n\}$ be a set of positive integers such that for each $t \in \mathbb{N}$ there exists a uniquely determined i, $1 \leqslant i \leqslant n$, with $t \equiv t_i q^b \bmod e$ for some integer $b \geqslant 0$. Such a set T can, for instance, be constructed as follows. Put $t_1 = 1$ and, when $t_1, t_2, \ldots, t_{j-1}$ have been constructed, let t_j be the least positive integer such that $t_j \not\equiv t_i q^b \bmod e$ for $1 \leqslant i < j$ and all integers $b \geqslant 0$. This procedure stops after finitely many steps.

With the notation introduced above, we have then the following general result.

3.38. Theorem. *The polynomials $g_{t_1}, g_{t_2}, \ldots, g_{t_n}$ are all the distinct monic irreducible polynomials in $\mathbb{F}_q[x]$ whose orders divide e and whose constant terms are nonzero.*

Proof. Each g_{t_i} is monic and irreducible in $\mathbb{F}_q[x]$ by definition and satisfies $g_{t_i}(0) \neq 0$. Furthermore, since g_{t_i} has the root α^{t_i} whose order in the group $\mathbb{F}_{q^m}^*$ divides the order of α, it follows from Theorem 3.3 that $\operatorname{ord}(g_{t_i})$ divides e.

Let g be an arbitrary monic irreducible polynomial in $\mathbb{F}_q[x]$ of order d dividing e and with $g(0) \neq 0$. If β is a root of g, then $\beta^d = 1$ implies $\beta^e = 1$, and so β is an eth root of unity over $\mathbb{F}_q$. Since α is a primitive eth root of unity over $\mathbb{F}_q$, it follows from Theorem 2.42(i) that $\beta = \alpha^t$ for some $t \in \mathbb{N}$. Then the definition of the set T implies that $t \equiv t_i q^b \bmod e$ for some i, $1 \leqslant i \leqslant n$, and some $b \geqslant 0$. Hence $\beta = \alpha^t = (\alpha^{t_i})^{q^b}$, and so β is a root of g_{t_i} because of Theorem 2.14. Since g is the minimal polynomial of β over $\mathbb{F}_q$, it follows from Theorem 3.33(iii) that $g = g_{t_i}$.

It remains to show that the polynomials g_{t_i}, $1 \leqslant i \leqslant n$, are distinct. Suppose $g_{t_i} = g_{t_j}$ for $i \neq j$. Then α^{t_i} and α^{t_j} are roots of g_{t_i}, and so $\alpha^{t_j} = (\alpha^{t_i})^{q^b}$ for some $b \geqslant 0$. This implies $t_j \equiv t_i q^b \bmod e$, but since we also have $t_j \equiv t_j q^0 \bmod e$, we obtain a contradiction to the definition of the set T. □

The minimal polynomial g_t of $\alpha^t \in \mathbb{F}_{q^m}$ over $\mathbb{F}_q$ is usually calculated by means of the characteristic polynomial f_t of $\alpha^t \in \mathbb{F}_{q^m}$ over $\mathbb{F}_q$. From the discussion following Definition 2.22 we know that $f_t = g_t^r$, where $r = m/k$ and k is the degree of g_t. Since g_t is irreducible in $\mathbb{F}_q[x]$, k is the multiplicative order of q modulo $d = \operatorname{ord}(g_t)$, and d is equal to the order of

α^t in the group $\mathbb{F}_{q^m}^*$, which is $e/\gcd(t,e)$ by Theorem 1.15(ii). Therefore d, and so k and r, can be determined easily.

Several methods are known for calculating f_t. One of them is based on a useful relationship between f_t and the given polynomial f.

3.39. Theorem. *Let f be a monic irreducible polynomial in $\mathbb{F}_q[x]$ of degree m. Let $\alpha \in \mathbb{F}_{q^m}$ be a root of f, and for $t \in \mathbb{N}$ let f_t be the characteristic polynomial of $\alpha^t \in \mathbb{F}_{q^m}$ over $\mathbb{F}_q$. Then*

$$f_t(x^t) = (-1)^{m(t+1)} \prod_{j=1}^{t} f(\omega_j x),$$

where $\omega_1, \ldots, \omega_t$ are the tth roots of unity over $\mathbb{F}_q$ counted according to multiplicity.

Proof. Let $\alpha = \alpha_1, \alpha_2, \ldots, \alpha_m$ be all the roots of f. Then $\alpha_1^t, \alpha_2^t, \ldots, \alpha_m^t$ are the roots of f_t counted according to multiplicity. Thus

$$\begin{aligned} f_t(x^t) &= \prod_{i=1}^{m} \left(x^t - \alpha_i^t\right) \\ &= \prod_{i=1}^{m} \prod_{j=1}^{t} (x - \alpha_i \omega_j) \\ &= \prod_{i=1}^{m} \prod_{j=1}^{t} \omega_j \left(\omega_j^{-1} x - \alpha_i\right). \end{aligned}$$

A comparison of coefficients in the identity

$$x^t - 1 = \prod_{j=1}^{t} (x - \omega_j)$$

shows that

$$\prod_{j=1}^{t} \omega_j = (-1)^{t+1},$$

and so

$$\begin{aligned} f_t(x^t) &= (-1)^{m(t+1)} \prod_{j=1}^{t} \prod_{i=1}^{m} \left(\omega_j^{-1} x - \alpha_i\right) \\ &= (-1)^{m(t+1)} \prod_{j=1}^{t} f\left(\omega_j^{-1} x\right) = (-1)^{m(t+1)} \prod_{j=1}^{t} f(\omega_j x) \end{aligned}$$

since $\omega_1^{-1}, \ldots, \omega_t^{-1}$ run exactly through all tth roots of unity over $\mathbb{F}_q$. □

3.40. Example. Consider the irreducible polynomial $f(x) = x^4 + x + 1$ in $\mathbb{F}_2[x]$. To calculate f_3, we note that the third roots of unity over $\mathbb{F}_2$ are 1, ω, and ω^2, where ω is a root of x^2+x+1 in $\mathbb{F}_4$. Then

$$\begin{aligned} f_3(x^3) &= (-1)^{16} f(x) f(\omega x) f(\omega^2 x) \\ &= (x^4+x+1)(\omega x^4+\omega x+1)(\omega^2 x^4+\omega^2 x+1) \\ &= x^{12}+x^9+x^6+x^3+1, \end{aligned}$$

so that $f_3(x)=x^4+x^3+x^2+x+1$. □

Another method of calculating f_t is based on matrix theory. Let $f(x)=x^m-a_{m-1}x^{m-1}-\cdots-a_1x-a_0$ and let A be the *companion matrix* of f, which is defined to be the $m\times m$ matrix

$$A=\begin{pmatrix} 0 & 0 & \cdots & 0 & a_0 \\ 1 & 0 & \cdots & 0 & a_1 \\ 0 & 1 & \cdots & 0 & a_2 \\ \vdots & \vdots & & \vdots & \vdots \\ 0 & 0 & \cdots & 1 & a_{m-1} \end{pmatrix}.$$

Then f is the *characteristic polynomial* of A in the sense of linear algebra; that is, $f(x)=\det(xI-A)$ with I being the $m\times m$ identity matrix over $\mathbb{F}_q$. For each $t\in\mathbb{N}$, f_t is the characteristic polynomial of A^t, the tth power of A. Thus, by calculating the powers of A one obtains the polynomials f_t.

3.41. Example. It is of interest to determine which polynomials f_t are irreducible in $\mathbb{F}_q[x]$. From the discussion prior to Theorem 3.39 it follows immediately that f_t is irreducible in $\mathbb{F}_q[x]$ if and only if $k=m$, that is, if and only if m is the multiplicative order of q modulo $d=e/\gcd(t,e)$. Consider, for instance, the case $q=2$, $m=6$, $e=63$. Since the multiplicative order of q modulo a divisor of e must be a divisor of m, the only possibilities for the multiplicative order apart from m are $k=1,2,3$. Then $q^k-1=1$, 3, 7, and $q^k\equiv 1 \bmod d$ is only possible when $d=1,3,7$. Thus f_t is reducible in $\mathbb{F}_2[x]$ precisely if $\gcd(t,63)=9,21,63$. Since it suffices to consider values of t with $1\leqslant t\leqslant 63$, it follows that f_t is irreducible in $\mathbb{F}_2[x]$ except when $t=9,18,21,27,36,42,45,54,63$. □

In practice, irreducible polynomials often arise as minimal polynomials of elements in an extension field. If in the discussion above we let f be a primitive polynomial over $\mathbb{F}_q$, so that $e=q^m-1$, then the powers of α run through all nonzero elements of $\mathbb{F}_{q^m}$. Therefore, the methods outlined above can be used to calculate the minimal polynomial over $\mathbb{F}_q$ of each element of $\mathbb{F}_{q^m}^*$.

A straightforward *method of determining minimal polynomials* is the following one. Let θ be a defining element of $\mathbb{F}_{q^m}$ over $\mathbb{F}_q$, so that $\{1,\theta,\ldots,\theta^{m-1}\}$ is a basis of $\mathbb{F}_{q^m}$ over $\mathbb{F}_q$. In order to find the minimal polynomial g of $\beta\in\mathbb{F}_{q^m}^*$ over $\mathbb{F}_q$, we express the powers $\beta^0,\beta^1,\ldots,\beta^m$ in terms of the basis elements. Let

$$\beta^{i-1} = \sum_{j=1}^{m} b_{ij}\theta^{j-1} \quad \text{for } 1 \leqslant i \leqslant m+1.$$

We write g in the form $g(x) = c_m x^m + \cdots + c_1 x + c_0$. We want g to be the monic polynomial of least positive degree with $g(\beta) = 0$. The condition $g(\beta) = c_m \beta^m + \cdots + c_1 \beta + c_0 = 0$ leads to the homogeneous system of linear equations

$$\sum_{i=1}^{m+1} c_{i-1} b_{ij} = 0 \quad \text{for } 1 \leqslant j \leqslant m \tag{3.10}$$

with unknowns $c_0, c_1, \ldots, c_m$. Let B be the matrix of coefficients of the system—that is, B is the $(m+1) \times m$ matrix whose (i, j) entry is b_{ij}—and let r be the rank of B. Then the dimension of the space of solutions of the system is $s = m + 1 - r$, and since $1 \leqslant r \leqslant m$, we have $1 \leqslant s \leqslant m$. Therefore, we can prescribe values for s of the unknowns $c_0, c_1, \ldots, c_m$, and then the remaining ones are uniquely determined. If $s = 1$, we set $c_m = 1$, and if $s > 1$, we set $c_m = c_{m-1} = \cdots = c_{m-s+2} = 0$ and $c_{m-s+1} = 1$.

3.42. Example. Let $\theta \in \mathbb{F}_{64}$ be a root of the irreducible polynomial $x^6 + x + 1$ in $\mathbb{F}_2[x]$. For $\beta = \theta^3 + \theta^4$ we have

$$\begin{aligned}
\beta^0 &= 1 \\
\beta^1 &= \theta^3 + \theta^4 \\
\beta^2 &= 1 + \theta + \theta^2 + \theta^3 \\
\beta^3 &= \theta + \theta^2 + \theta^3 \\
\beta^4 &= \theta + \theta^2 + \theta^4 \\
\beta^5 &= 1 + \theta^3 + \theta^4 \\
\beta^6 &= 1 + \theta + \theta^2 + \theta^4
\end{aligned}$$

Therefore, the matrix B is given by

$$B = \begin{pmatrix} 1 & 0 & 0 & 0 & 0 & 0 \\ 0 & 0 & 0 & 1 & 1 & 0 \\ 1 & 1 & 1 & 1 & 0 & 0 \\ 0 & 1 & 1 & 1 & 0 & 0 \\ 0 & 1 & 1 & 0 & 1 & 0 \\ 1 & 0 & 0 & 1 & 1 & 0 \\ 1 & 1 & 1 & 0 & 1 & 0 \end{pmatrix}$$

and its rank is $r = 3$. Hence $s = m + 1 - r = 4$, so that we set $c_6 = c_5 = c_4 = 0$, $c_3 = 1$. The remaining coefficients are determined from (3.10), and this yields $c_2 = 1$, $c_1 = 0$, $c_0 = 1$. Consequently, the minimal polynomial of β over $\mathbb{F}_2$ is $g(x) = x^3 + x^2 + 1$. □

Still another *method of determining minimal polynomials* is based on Theorem 3.33(v). If we wish to find the minimal polynomial g of $\beta \in \mathbb{F}_{q^m}$

over $\mathbb{F}_q$, we calculate the powers $\beta, \beta^q, \beta^{q^2}, \ldots$ until we find the least positive integer d for which $\beta^{q^d} = \beta$. This integer d is the degree of g, and g itself is given by

$$g(x) = (x - \beta)(x - \beta^q) \cdots (x - \beta^{q^{d-1}}).$$

The elements $\beta, \beta^q, \ldots, \beta^{q^{d-1}}$ are the distinct conjugates of β with respect to $\mathbb{F}_q$, and g is the minimal polynomial over $\mathbb{F}_q$ of all these elements.

3.43. Example. We compute the minimal polynomials over $\mathbb{F}_2$ of all elements of $\mathbb{F}_{16}$. Let $\theta \in \mathbb{F}_{16}$ be a root of the primitive polynomial $x^4 + x + 1$ over $\mathbb{F}_2$, so that every nonzero element of $\mathbb{F}_{16}$ can be written as a power of θ. We have the following index table for $\mathbb{F}_{16}$:

i	θ^i	i	θ^i
0	1	8	$1+\theta^2$
1	θ	9	$\theta+\theta^3$
2	θ^2	10	$1+\theta+\theta^2$
3	θ^3	11	$\theta+\theta^2+\theta^3$
4	$1+\theta$	12	$1+\theta+\theta^2+\theta^3$
5	$\theta+\theta^2$	13	$1+\theta^2+\theta^3$
6	$\theta^2+\theta^3$	14	$1+\theta^3$
7	$1+\theta+\theta^3$		

The minimal polynomials of the elements β of $\mathbb{F}_{16}$ over $\mathbb{F}_2$ are:

$\beta = 0$: $g_1(x) = x$.

$\beta = 1$: $g_2(x) = x + 1$.

$\beta = \theta$: The distinct conjugates of θ with respect to $\mathbb{F}_2$ are $\theta, \theta^2, \theta^4, \theta^8$, and the minimal polynomial is

$$\begin{aligned} g_3(x) &= (x-\theta)(x-\theta^2)(x-\theta^4)(x-\theta^8) \\ &= x^4 + x + 1. \end{aligned}$$

$\beta = \theta^3$: The distinct conjugates of θ^3 with respect to $\mathbb{F}_2$ are $\theta^3, \theta^6, \theta^{12}, \theta^{24} = \theta^9$, and the minimal polynomial is

$$\begin{aligned} g_4(x) &= (x-\theta^3)(x-\theta^6)(x-\theta^9)(x-\theta^{12}) \\ &= x^4 + x^3 + x^2 + x + 1. \end{aligned}$$

$\beta = \theta^5$: Since $\beta^4 = \beta$, the distinct conjugates of this element with respect to $\mathbb{F}_2$ are θ^5, θ^{10}, and the minimal polynomial is

$$g_5(x) = (x-\theta^5)(x-\theta^{10}) = x^2 + x + 1.$$

$\beta = \theta^7$: The distinct conjugates of θ^7 with respect to $\mathbb{F}_2$ are $\theta^7, \theta^{14}, \theta^{28} = \theta^{13}, \theta^{56} = \theta^{11}$, and the minimal polynomial is

$$\begin{aligned} g_6(x) &= (x-\theta^7)(x-\theta^{11})(x-\theta^{13})(x-\theta^{14}) \\ &= x^4 + x^3 + 1. \end{aligned}$$

These elements, together with their conjugates with respect to $\mathbb{F}_2$, exhaust $\mathbb{F}_{16}$. □

An important problem is that of the *determination of primitive polynomials*. One approach is based on the fact that the product of all primitive polynomials over $\mathbb{F}_q$ of degree m is equal to the cyclotomic polynomial Q_e with $e = q^m - 1$ (see Theorem 2.47(ii) and Exercise 3.42). Therefore, all primitive polynomials over $\mathbb{F}_q$ of degree m can be determined by applying one of the factorization algorithms in Chapter 4 to the cyclotomic polynomial Q_e.

Another method depends on constructing a primitive element of $\mathbb{F}_{q^m}$ and then determining the minimal polynomial of this element over $\mathbb{F}_q$ by the methods described above. To find a primitive element of $\mathbb{F}_{q^m}$, one starts from the order $q^m - 1$ of such an element in the group $\mathbb{F}_{q^m}^*$ and factors it in the form $q^m - 1 = h_1 \cdots h_k$, where the positive integers $h_1, \ldots, h_k$ are pairwise relatively prime. If for each $i, 1 \leqslant i \leqslant k$, one can find an element $\alpha_i \in \mathbb{F}_{q^m}^*$ of order h_i, then the product $\alpha_1 \cdots \alpha_k$ has order $q^m - 1$ and is thus a primitive element of $\mathbb{F}_{q^m}$.

3.44. Example. We determine a primitive polynomial over $\mathbb{F}_3$ of degree 4. Since $3^4 - 1 = 16 \cdot 5$, we first construct two elements of $\mathbb{F}_{81}^*$ of order 16 and 5, respectively. The elements of order 16 are the roots of the cyclotomic polynomial $Q_{16}(x) = x^8 + 1 \in \mathbb{F}_3[x]$. Since the multiplicative order of 3 modulo 16 is 4, Q_{16} factors into two monic irreducible polynomials in $\mathbb{F}_3[x]$ of degree 4. Now

$$\begin{aligned} x^8 + 1 &= (x^4 - 1)^2 - x^4 \\ &= (x^4 - 1 + x^2)(x^4 - 1 - x^2), \end{aligned}$$

and so $f(x) = x^4 - x^2 - 1$ is irreducible over $\mathbb{F}_3$ and with a root θ of f we have $\mathbb{F}_{81} = \mathbb{F}_3(\theta)$. Furthermore, θ is an element of $\mathbb{F}_{81}^*$ of order 16. To find an element α of order 5, we write $\alpha = a + b\theta + c\theta^2 + d\theta^3$ with $a, b, c, d \in \mathbb{F}_3$, and since we must have $\alpha^{10} = 1$, we get

$$\begin{aligned} 1 = \alpha^9\alpha &= (a + b\theta^9 + c\theta^{18} + d\theta^{27})(a + b\theta + c\theta^2 + d\theta^3) \\ &= (a - b\theta + c\theta^2 - d\theta^3)(a + b\theta + c\theta^2 + d\theta^3) \\ &= (a + c\theta^2)^2 - (b\theta + d\theta^3)^2 = a^2 + (2ac - b^2)\theta^2 + (c^2 - 2bd)\theta^4 - d^2\theta^6 \\ &= a^2 + c^2 - d^2 + bd + (c^2 + d^2 - b^2 - ac + bd)\theta^2. \end{aligned}$$

A comparison of coefficients yields

$$a^2 + c^2 - d^2 + bd = 1, \qquad c^2 + d^2 - b^2 - ac + bd = 0.$$

Setting $a = d = 0$, we get $b^2 = c^2 = 1$. Take $b = c = 1$, and then it is easily checked that $\alpha = \theta + \theta^2$ has order 5. Therefore, $\zeta = \theta\alpha = \theta^2 + \theta^3$ has order 80 and is thus a primitive element of $\mathbb{F}_{81}$. The minimal polynomial g of ζ

over $\mathbb{F}_3$ is

$$\begin{aligned} g(x) &= (x-\zeta)(x-\zeta^3)(x-\zeta^9)(x-\zeta^{27}) \\ &= (x-\theta^2-\theta^3)(x-1+\theta+\theta^2)(x-\theta^2+\theta^3)(x-1-\theta+\theta^2) \\ &= x^4+x^3+x^2-x-1, \end{aligned}$$

and we have thus obtained a primitive polynomial over $\mathbb{F}_3$ of degree 4. □

3.45. Example. We determine a primitive polynomial over $\mathbb{F}_2$ of degree 6. Since $2^6-1=9\cdot 7$, we first construct two elements of $\mathbb{F}_{64}^*$ of order 9 and 7, respectively. The multiplicative order of 2 modulo 9 is 6, and so the cyclotomic polynomial $Q_9(x)=x^6+x^3+1$ is irreducible over $\mathbb{F}_2$. A root θ of Q_9 has order 9 and $\mathbb{F}_{64}=\mathbb{F}_2(\theta)$. An element $\alpha\in\mathbb{F}_{64}^*$ of order 7 satisfies $\alpha^8=\alpha$, thus writing $\alpha=\sum_{i=0}^{5}a_i\theta^i$ with $a_i\in\mathbb{F}_2$, $0\leqslant i\leqslant 5$, we get

$$\begin{aligned} \sum_{i=0}^{5} a_i\theta^i &= \left(\sum_{i=0}^{5} a_i\theta^i\right)^8 \\ &= \sum_{i=0}^{5} a_i\theta^{8i} \\ &= a_0+a_1\theta^8+a_2\theta^7+a_3\theta^6+a_4\theta^5+a_5\theta^4 \\ &= a_0+a_3+a_2\theta+a_1\theta^2+a_3\theta^3+(a_2+a_5)\theta^4+(a_1+a_4)\theta^5, \end{aligned}$$

and a comparison of coefficients yields $a_3=0$, $a_1=a_2$, $a_4=a_2+a_5$. Choose $a_0=a_3=a_4=0$, $a_1=a_2=a_5=1$, so that $\alpha=\theta+\theta^2+\theta^5$ is an element of order 7. Thus, $\zeta=\theta\alpha=1+\theta^2$ is a primitive element of $\mathbb{F}_{64}$. Then $\zeta^2=1+\theta^4$, $\zeta^3=\theta^2+\theta^3+\theta^4$, $\zeta^4=1+\theta^2+\theta^5$, $\zeta^5=1+\theta+\theta^5$, $\zeta^6=1+\theta^2+\theta^3+\theta^4+\theta^5$. An application of the method in Example 3.42 yields the minimal polynomial $g(x)=x^6+x^4+x^3+x+1$ of ζ over $\mathbb{F}_2$ and thus a primitive polynomial over $\mathbb{F}_2$ of degree 6. □

If a primitive polynomial g over $\mathbb{F}_q$ of degree m is known, all other such primitive polynomials can be obtained by considering a root θ of g in $\mathbb{F}_{q^m}$ and determining the minimal polynomials over $\mathbb{F}_q$ of all elements θ^t, where t runs through all positive integers $\leqslant q^m-1$ that are relatively prime to q^m-1. The calculation of these minimal polynomials is carried out by the methods described earlier in this section.

It is useful to be able to decide whether an irreducible polynomial over a finite field remains irreducible over a certain finite extension field. The following results address themselves to this question.

3.46. Theorem. *Let f be an irreducible polynomial over $\mathbb{F}_q$ of degree n and let $k\in\mathbb{N}$. Then f factors into d irreducible polynomials in $\mathbb{F}_{q^k}[x]$ of the same degree n/d, where $d=\gcd(k,n)$.*

Proof. Since the case $f(0)=0$ is trivial, we can assume $f(0)\neq 0$. Let g be an irreducible factor of f in $\mathbb{F}_{q^k}[x]$. If $\text{ord}(f)=e$, then also $\text{ord}(g)=e$ by Theorem 3.3 since the roots of g are also roots of f. By Theorem 3.5 the multiplicative order of q modulo e is n and the degree of g is equal to the multiplicative order of q^k modulo e. The powers q^j, $j=0,1,\ldots,$ considered modulo e, form a cyclic group of order n. Thus it follows from Theorem 1.15(ii) that the multiplicative order of q^k modulo e is n/d, and so the degree of g is n/d. □

3.47. Corollary. *An irreducible polynomial over $\mathbb{F}_q$ of degree n remains irreducible over $\mathbb{F}_{q^k}$ if and only if k and n are relatively prime.*

Proof. This is an immediate consequence of Theorem 3.46. □

3.48. Example. Consider the primitive polynomial $g(x)=x^6+x^4+x^3+x+1$ over $\mathbb{F}_2$ from Example 3.45 as a polynomial over $\mathbb{F}_{16}$. Then, in the notation of Theorem 3.46, we have $n=6$, $k=4$, and thus $d=2$. Therefore, g factors in $\mathbb{F}_{16}[x]$ into two irreducible cubic polynomials. Using the notation of Example 3.45, let g_1 be the factor that has $\zeta=1+\theta^2$ as a root. The other roots of g_1 must be the conjugates ζ^{16} and $\zeta^{256}=\zeta^4$ with respect to $\mathbb{F}_{16}$. Since these elements are also conjugates with respect to $\mathbb{F}_4$, it follows that g_1 is actually in $\mathbb{F}_4[x]$. Now $\beta=\zeta^{21}$ is a primitive third root of unity over $\mathbb{F}_2$, and so $\mathbb{F}_4=\{0,1,\beta,\beta^2\}$. Furthermore,

$$\begin{aligned} g_1(x) &= (x-\zeta)(x-\zeta^4)(x-\zeta^{16}) \\ &= x^3+(\zeta+\zeta^4+\zeta^{16})x^2+(\zeta^5+\zeta^{17}+\zeta^{20})x+\zeta^{21}. \end{aligned}$$

We have $\zeta^4=1+\theta^2+\theta^5$, $\zeta^{16}=1+\theta^5$, and so $\zeta+\zeta^4+\zeta^{16}=1$. Similarly, we obtain $\zeta^5+\zeta^{17}+\zeta^{20}=1$, so that $g_1(x)=x^3+x^2+x+\beta$. By dividing g by g_1 we get the second factor and thus the factorization

$$g(x)=(x^3+x^2+x+\beta)(x^3+x^2+x+\beta^2)$$

in $\mathbb{F}_4[x]$, and hence in $\mathbb{F}_{16}[x]$. The two factors of g are primitive polynomials over $\mathbb{F}_4$, but not over $\mathbb{F}_{16}$. By Corollary 3.47, the polynomial g remains irreducible over certain other extension fields of $\mathbb{F}_2$, such as $\mathbb{F}_{32}$ and $\mathbb{F}_{128}$. □

4. LINEARIZED POLYNOMIALS

Both in theory and in applications the special class of polynomials to be introduced below is of importance. A useful feature of these polynomials is the structure of the set of roots that facilitates the determination of the roots. Let q, as usual, denote a prime power.

3.49. Definition. A polynomial of the form

$$L(x) = \sum_{i=0}^{n} \alpha_i x^{q^i}$$

with coefficients in an extension field $\mathbb{F}_{q^m}$ of $\mathbb{F}_q$ is called a *q-polynomial* over $\mathbb{F}_{q^m}$.

If the value of q is fixed once and for all or is clear from the context, it is also customary to speak of a *linearized polynomial*. This terminology stems from the following property of linearized polynomials. If F is an arbitrary extension field of $\mathbb{F}_{q^m}$ and $L(x)$ is a linearized polynomial (i.e., a q-polynomial) over $\mathbb{F}_{q^m}$, then

$$L(\beta+\gamma) = L(\beta) + L(\gamma) \quad \text{for all } \beta, \gamma \in F, \tag{3.11}$$

$$L(c\beta) = cL(\beta) \quad \text{for all } c \in \mathbb{F}_q \text{ and all } \beta \in F. \tag{3.12}$$

The identity (3.11) follows immediately from Theorem 1.46 and (3.12) follows from the fact that $c^{q^i} = c$ for $c \in \mathbb{F}_q$ and $i \geqslant 0$. Thus, if F is considered as a vector space over $\mathbb{F}_q$, then the linearized polynomial $L(x)$ induces a linear operator on F.

The special character of the set of roots of a linearized polynomial is shown by the following result.

3.50. Theorem. *Let $L(x)$ be a nonzero q-polynomial over $\mathbb{F}_{q^m}$ and let the extension field $\mathbb{F}_{q^s}$ of $\mathbb{F}_{q^m}$ contain all the roots of $L(x)$. Then each root of $L(x)$ has the same multiplicity, which is either* 1 *or a power of q, and the roots form a linear subspace of $\mathbb{F}_{q^s}$, where $\mathbb{F}_{q^s}$ is regarded as a vector space over $\mathbb{F}_q$.*

Proof. It follows from (3.11) and (3.12) that any linear combination of roots with coefficients in $\mathbb{F}_q$ is again a root, and so the roots of $L(x)$ form a linear subspace of $\mathbb{F}_{q^s}$. If

$$L(x) = \sum_{i=0}^{n} \alpha_i x^{q^i},$$

then $L'(x) = \alpha_0$, so that $L(x)$ has only simple roots in case $\alpha_0 \neq 0$. Otherwise, we have $\alpha_0 = \alpha_1 = \cdots = \alpha_{k-1} = 0$, but $\alpha_k \neq 0$ for some $k \geqslant 1$, and then

$$L(x) = \sum_{i=k}^{n} \alpha_i x^{q^i} = \sum_{i=k}^{n} \alpha_i^{q^{mk}} x^{q^i} = \left(\sum_{i=k}^{n} \alpha_i^{q^{(m-1)k}} x^{q^{i-k}} \right)^{q^k},$$

which is the q^kth power of a linearized polynomial having only simple roots. In this case, each root of $L(x)$ has multiplicity q^k. □

There is also a partial converse of Theorem 3.50, which is given by Theorem 3.52. It depends on a result about certain determinants which extends Corollary 2.38.

3.51. Lemma. *Let $\beta_1, \beta_2, \ldots, \beta_n$ be elements of $\mathbb{F}_{q^m}$. Then*

$$\begin{vmatrix} \beta_1 & \beta_1^q & \beta_1^{q^2} & \cdots & \beta_1^{q^{n-1}} \\ \beta_2 & \beta_2^q & \beta_2^{q^2} & \cdots & \beta_2^{q^{n-1}} \\ \vdots & \vdots & \vdots & & \vdots \\ \beta_n & \beta_n^q & \beta_n^{q^2} & \cdots & \beta_n^{q^{n-1}} \end{vmatrix} = \beta_1 \prod_{j=1}^{n-1} \prod_{c_1,\ldots,c_j \in \mathbb{F}_q} \left(\beta_{j+1} - \sum_{k=1}^{j} c_k \beta_k \right), \tag{3.13}$$

and so the determinant is $\neq 0$ if and only if $\beta_1, \beta_2, \ldots, \beta_n$ are linearly independent over $\mathbb{F}_q$.

Proof. Let D_n be the determinant on the left-hand side of (3.13). We prove (3.13) by induction on n and note that the formula is trivial for $n = 1$ if the empty product on the right-hand side is interpreted as 1. Suppose the formula is shown for some $n \geqslant 1$. Consider the polynomial

$$D(x) = \begin{vmatrix} \beta_1 & \beta_1^q & \cdots & \beta_1^{q^{n-1}} & \beta_1^{q^n} \\ \beta_2 & \beta_2^q & \cdots & \beta_2^{q^{n-1}} & \beta_2^{q^n} \\ \vdots & \vdots & & \vdots & \vdots \\ \beta_n & \beta_n^q & \cdots & \beta_n^{q^{n-1}} & \beta_n^{q^n} \\ x & x^q & \cdots & x^{q^{n-1}} & x^{q^n} \end{vmatrix}.$$

By expansion along the last row we get

$$D(x) = D_n x^{q^n} + \sum_{i=0}^{n-1} \alpha_i x^{q^i}$$

with $\alpha_i \in \mathbb{F}_{q^m}$ for $0 \leqslant i \leqslant n-1$. Assume first that $\beta_1, \ldots, \beta_n$ are linearly independent over $\mathbb{F}_q$. We have $D(\beta_k) = 0$ for $1 \leqslant k \leqslant n$, and since $D(x)$ is a q-polynomial over $\mathbb{F}_{q^m}$, all linear combinations $c_1\beta_1 + \cdots + c_n\beta_n$ with $c_k \in \mathbb{F}_q$ for $1 \leqslant k \leqslant n$ are roots of $D(x)$. Thus $D(x)$ has q^n distinct roots, so that we obtain a factorization

$$D(x) = D_n \prod_{c_1,\ldots,c_n \in \mathbb{F}_q} \left(x - \sum_{k=1}^{n} c_k \beta_k \right). \tag{3.14}$$

If $\beta_1, \ldots, \beta_n$ are linearly dependent over $\mathbb{F}_q$, then $D_n = 0$ and $\sum_{k=1}^{n} b_k \beta_k = 0$ for some $b_1, \ldots, b_n \in \mathbb{F}_q$, not all of which are 0. It follows that

$$\sum_{k=1}^{n} b_k \beta_k^{q^j} = \left(\sum_{k=1}^{n} b_k \beta_k \right)^{q^j} = 0 \quad \text{for } j = 0, 1, \ldots, n,$$

and so the first n row vectors in the determinant defining $D(x)$ are linearly dependent over $\mathbb{F}_q$. Thus $D(x)=0$, and the identity (3.14) is satisfied in all cases. Consequently,

$$D_{n+1}=D(\beta_{n+1})=D_n \prod_{c_1,\dots,c_n\in\mathbb{F}_q}\left(\beta_{n+1}-\sum_{k=1}^{n}c_k\beta_k\right),$$

and (3.13) is established. □

3.52. Theorem. *Let U be a linear subspace of $\mathbb{F}_{q^m}$, considered as a vector space over $\mathbb{F}_q$. Then for any nonnegative integer k the polynomial*

$$L(x)=\prod_{\beta\in U}(x-\beta)^{q^k}$$

is a q-polynomial over $\mathbb{F}_{q^m}$.

Proof. Since the q^kth power of a q-polynomial over $\mathbb{F}_{q^m}$ is again such a polynomial, it suffices to consider the case $k=0$. Let $\{\beta_1,\dots,\beta_n\}$ be a basis of U over $\mathbb{F}_q$. Then the determinant D_n on the left-hand side of (3.13) is $\neq 0$ by Lemma 3.51, and so

$$L(x)=\prod_{\beta\in U}(x-\beta)$$

$$=\prod_{c_1,\dots,c_n\in\mathbb{F}_q}\left(x-\sum_{k=1}^{n}c_k\beta_k\right)=D_n^{-1}D(x)$$

by (3.14), which shows already that $L(x)$ is a q-polynomial over $\mathbb{F}_{q^m}$. □

The properties of linearized polynomials lead to the following *method of determining roots* of such polynomials. Let

$$L(x)=\sum_{i=0}^{n}\alpha_i x^{q^i}$$

be a q-polynomial over $\mathbb{F}_{q^m}$, and suppose we want to find all roots of $L(x)$ in the finite extension F of $\mathbb{F}_{q^m}$. As we noted above, the mapping L: $\beta\in F\mapsto L(\beta)\in F$ is a linear operator on the vector space F over $\mathbb{F}_q$. Therefore, L can be represented by a matrix over $\mathbb{F}_q$. Specifically, let $\{\beta_1,\dots,\beta_s\}$ be a basis of F over $\mathbb{F}_q$, so that every $\beta\in F$ can be written in the form

$$\beta=\sum_{j=1}^{s}c_j\beta_j \quad \text{with } c_j\in\mathbb{F}_q \quad \text{for } 1\leqslant j\leqslant s;$$

then

$$L(\beta)=\sum_{j=1}^{s}c_jL(\beta_j).$$

Now let

$$L(\beta_j) = \sum_{k=1}^{s} b_{jk}\beta_k \quad \text{for } 1 \leqslant j \leqslant s,$$

where $b_{jk} \in \mathbb{F}_q$ for $1 \leqslant j, k \leqslant s$, and let B be the $s \times s$ matrix over $\mathbb{F}_q$ whose (j, k) entry is b_{jk}. Then, if

$$(c_1, \ldots, c_s)B = (d_1, \ldots, d_s),$$

we have

$$L(\beta) = \sum_{k=1}^{s} d_k \beta_k.$$

Therefore, the equation $L(\beta) = 0$ is equivalent to

$$(c_1, \ldots, c_s)B = (0, \ldots, 0). \tag{3.15}$$

This is a homogeneous system of s linear equations for $c_1, \ldots, c_s$. If r is the rank of the matrix B, then (3.15) has q^{s-r} solution vectors $(c_1, \ldots, c_s)$. Each solution vector $(c_1, \ldots, c_s)$ yields a root $\beta = \sum_{j=1}^{s} c_j \beta_j$ of $L(x)$ in F. Thus, the problem of finding the roots of $L(x)$ in F is reduced to the easier problem of solving a homogeneous system of linear equations.

3.53. Example. Consider the linearized polynomial $L(x) = x^9 - x^3 - \alpha x \in \mathbb{F}_9[x]$, where α is a root of the primitive polynomial $x^2 + x - 1$ over $\mathbb{F}_3$. In order to find the roots of $L(x)$ in $\mathbb{F}_{81}$, we choose the basis $\{1, \zeta, \zeta^2, \zeta^3\}$ of $\mathbb{F}_{81}$ over $\mathbb{F}_3$, where ζ is a root of the primitive polynomial $x^4 + x^3 + x^2 - x - 1$ over $\mathbb{F}_3$ (compare with Example 3.44). Because of the orders involved, we must have $\alpha = \zeta^{10j}$ with $j = 1$, 3, 5, or 7, and since $\zeta^{20} + \zeta^{10} - 1 = 0$, we can take $\alpha = \zeta^{10} = -1 + \zeta + \zeta^2 - \zeta^3$. Next, we calculate

$$\begin{aligned}
L(1) &= -\alpha = 1 - \zeta - \zeta^2 + \zeta^3, \\
L(\zeta) &= \zeta^9 - \zeta^3 - \alpha\zeta = -\zeta - \zeta^2 - \zeta^3, \\
L(\zeta^2) &= \zeta^{18} - \zeta^6 - \alpha\zeta^2 = -1 + \zeta^3, \\
L(\zeta^3) &= \zeta^{27} - \zeta^9 - \alpha\zeta^3 = 1 - \zeta^3,
\end{aligned}$$

and so we get

$$B = \begin{pmatrix} 1 & -1 & -1 & 1 \\ 0 & -1 & -1 & -1 \\ -1 & 0 & 0 & 1 \\ 1 & 0 & 0 & -1 \end{pmatrix}.$$

The system (3.15) has two linearly independent solutions, such as $(0, 0, 1, 1)$ and $(-1, 1, 0, 1)$. All solutions of (3.15) are obtained by forming all linear combinations of these two vectors with coefficients in $\mathbb{F}_3$. The roots of $L(x)$ in $\mathbb{F}_{81}$ are then $\theta_1 = 0$, $\theta_2 = \zeta^2 + \zeta^3$, $\theta_3 = -\zeta^2 - \zeta^3$, $\theta_4 = -1 + \zeta + \zeta^3$,

$\theta_5 = 1 - \zeta - \zeta^3$, $\theta_6 = -1 + \zeta + \zeta^2 - \zeta^3$, $\theta_7 = 1 - \zeta - \zeta^2 + \zeta^3$, $\theta_8 = 1 - \zeta + \zeta^2$, $\theta_9 = -1 + \zeta - \zeta^2$. □

This method of finding roots can also be applied to a somewhat more general class of polynomials—namely, affine polynomials.

3.54. Definition. A polynomial of the form $A(x) = L(x) - \alpha$, where $L(x)$ is a q-polynomial over $\mathbb{F}_{q^m}$ and $\alpha \in \mathbb{F}_{q^m}$, is called an *affine q-polynomial* over $\mathbb{F}_{q^m}$.

An element $\beta \in F$ is a root of $A(x)$ if and only if $L(\beta) = \alpha$. In the notation of (3.15), the equation $L(\beta) = \alpha$ is equivalent to

$$(c_1, \ldots, c_s)B = (d_1, \ldots, d_s), \tag{3.16}$$

where $\alpha = \sum_{k=1}^{s} d_k \beta_k$. The system (3.16) of linear equations is solved for $c_1, \ldots, c_s$, and each solution vector $(c_1, \ldots, c_s)$ yields a root $\beta = \sum_{j=1}^{s} c_j \beta_j$ of $A(x)$ in F.

The fact that roots are easier to determine for affine polynomials suggests the following *method of finding the roots* of an arbitrary polynomial $f(x)$ over $\mathbb{F}_{q^m}$ of positive degree in an extension field F of $\mathbb{F}_{q^m}$. First determine a nonzero affine q-polynomial $A(x)$ over $\mathbb{F}_{q^m}$ that is divisible by $f(x)$—that is, a so-called *affine multiple* of $f(x)$. Next, obtain all the roots of $A(x)$ in F by the method described above. Since the roots of $f(x)$ in F must be among the roots of $A(x)$ in F, it suffices then to calculate $f(\beta)$ for all roots β of $A(x)$ in F in order to locate the roots of $f(x)$ in F.

The only point that remains to be settled is how to determine an affine multiple $A(x)$ of $f(x)$. This can be achieved as follows. Let $n \geqslant 1$ be the degree of $f(x)$. For $i = 0, 1, \ldots, n-1$, calculate the unique polynomial $r_i(x)$ of degree $\leqslant n-1$ with $x^{q^i} \equiv r_i(x) \bmod f(x)$. Then determine elements $\alpha_i \in \mathbb{F}_{q^m}$, not all 0, such that $\sum_{i=0}^{n-1} \alpha_i r_i(x)$ is a constant polynomial. This involves $n-1$ conditions concerning the vanishing of the coefficients of x^j, $1 \leqslant j \leqslant n-1$, and thus leads to a homogeneous system of $n-1$ linear equations for the n unknowns $\alpha_0, \alpha_1, \ldots, \alpha_{n-1}$. Such a system always has a nontrivial solution. Once a nontrivial solution has been fixed, we have $\sum_{i=0}^{n-1} \alpha_i r_i(x) = \alpha$ for some $\alpha \in \mathbb{F}_{q^m}$. It follows that

$$\sum_{i=0}^{n-1} \alpha_i x^{q^i} \equiv \sum_{i=0}^{n-1} \alpha_i r_i(x) \equiv \alpha \bmod f(x),$$

and so

$$A(x) = \sum_{i=0}^{n-1} \alpha_i x^{q^i} - \alpha$$

is a nonzero affine q-polynomial over $\mathbb{F}_{q^m}$ divisible by $f(x)$. It is clear that we may take $A(x)$ to be a monic polynomial.

3.55. Example. Let $f(x)=x^4+\theta^2x^3+\theta x^2+x+\theta\in\mathbb{F}_4[x]$, where θ is a root of $x^2+x+1\in\mathbb{F}_2[x]$. We want to find the roots of $f(x)$ in $\mathbb{F}_{64}$. We first determine an affine multiple $A(x)$ of $f(x)$ by using the method described above with $q=2$. Modulo $f(x)$ we have $x\equiv x=r_0(x)$, $x^2\equiv x^2=r_1(x)$, $x^4\equiv\theta^2x^3+\theta x^2+x+\theta=r_2(x)$, $x^8\equiv\theta x^3+\theta x^2+x+\theta=r_3(x)$. The condition that $\alpha_0r_0(x)+\alpha_1r_1(x)+\alpha_2r_2(x)+\alpha_3r_3(x)$ should be a constant polynomial leads to the system

$$\begin{aligned}\alpha_0 \quad + \quad \alpha_2+\ \alpha_3 &= 0\\ \alpha_1+\ \theta\alpha_2+\theta\alpha_3 &= 0\\ \theta^2\alpha_2+\theta\alpha_3 &= 0.\end{aligned}$$

We choose $\alpha_3=1$ and then obtain $\alpha_2=\theta^2, \alpha_1=\theta^2, \alpha_0=\theta$. Furthermore,

$$\alpha=\alpha_0r_0(x)+\alpha_1r_1(x)+\alpha_2r_2(x)+\alpha_3r_3(x)=\theta^2,$$

and so

$$A(x)=\alpha_3x^8+\alpha_2x^4+\alpha_1x^2+\alpha_0x-\alpha=x^8+\theta^2x^4+\theta^2x^2+\theta x+\theta^2.$$

Next, we calculate the roots of $A(x)$ in $\mathbb{F}_{64}$. We have to solve the equation $L(x)=\theta^2$ with the 2-polynomial $L(x)=x^8+\theta^2x^4+\theta^2x^2+\theta x$ over $\mathbb{F}_4$. Let ζ be a root of the primitive polynomial x^6+x+1 over $\mathbb{F}_2$. Then $\{1,\zeta,\zeta^2,\zeta^3,\zeta^4,\zeta^5\}$ is a basis of $\mathbb{F}_{64}$ over $\mathbb{F}_2$. Since θ is a primitive third root of unity over $\mathbb{F}_2$, we can take $\theta=\zeta^{21}=1+\zeta+\zeta^3+\zeta^4+\zeta^5$. Using $\theta^2=\theta+1=\zeta+\zeta^3+\zeta^4+\zeta^5$, we obtain

$$\begin{aligned}L(1) &= \zeta+\zeta^3+\zeta^4+\zeta^5\\ L(\zeta) &= \zeta+\zeta^2+\zeta^5\\ L(\zeta^2) &= \zeta^2+\zeta^3+\zeta^4+\zeta^5\\ L(\zeta^3) &= \zeta+\zeta^3+\zeta^4\\ L(\zeta^4) &= \zeta^5\\ L(\zeta^5) &= \zeta^2+\zeta^3+\zeta^4\end{aligned}$$

Thus the matrix B in (3.16) is given by

$$B=\begin{pmatrix}0&1&0&1&1&1\\0&1&1&0&0&1\\0&0&1&1&1&1\\0&1&0&1&1&0\\0&0&0&0&0&1\\0&0&1&1&1&0\end{pmatrix}.$$

From the representation for θ^2 given above it follows that the vector $(d_1,\ldots,d_s)$ in (3.16) is equal to $(0,1,0,1,1,1)$. The general solution of the system (3.16) is then

$$(1,0,0,0,0,0)+a_1(0,1,1,1,0,0)+a_2(1,1,1,0,1,0)+a_3(1,1,0,0,0,1)$$

with $a_1, a_2, a_3 \in \mathbb{F}_2$. Thus the roots of $A(x)$ in $\mathbb{F}_{64}$ are $\eta_1 = 1$, $\eta_2 = \zeta + \zeta^5$, $\eta_3 = \zeta + \zeta^2 + \zeta^4$, $\eta_4 = 1 + \zeta^2 + \zeta^4 + \zeta^5$, $\eta_5 = 1 + \zeta + \zeta^2 + \zeta^3$, $\eta_6 = \zeta^2 + \zeta^3 + \zeta^5$, $\eta_7 = \zeta^3 + \zeta^4$, $\eta_8 = 1 + \zeta + \zeta^3 + \zeta^4 + \zeta^5 = \theta$. By calculating $f(\eta_j)$ for $j = 1, 2, \ldots, 8$, we find that the roots of $f(x)$ in $\mathbb{F}_{64}$ are $\eta_3, \eta_5, \eta_7, \eta_8$. □

The method of determining the roots of an affine polynomial shows, in particular, that these roots form an affine subspace—that is, a translate of a linear subspace. This can also be deduced from abstract principles, together with a statement concerning multiplicities.

3.56. Theorem. *Let $A(x)$ be an affine q-polynomial over $\mathbb{F}_{q^m}$ of positive degree and let the extension field $\mathbb{F}_{q^s}$ of $\mathbb{F}_{q^m}$ contain all the roots of $A(x)$. Then each root of $A(x)$ has the same multiplicity, which is either* 1 *or a power of q, and the roots form an affine subspace of $\mathbb{F}_{q^s}$, where $\mathbb{F}_{q^s}$ is regarded as a vector space over $\mathbb{F}_q$.*

Proof. The result about the multiplicities is shown in the same way as in the proof of Theorem 3.50. Now let $A(x) = L(x) - \alpha$, where $L(x)$ is a q-polynomial over $\mathbb{F}_{q^m}$, and let β be a fixed root of $A(x)$. Then $\gamma \in \mathbb{F}_{q^s}$ is a root of $A(x)$ if and only if $L(\gamma) = \alpha = L(\beta)$ if and only if $L(\gamma - \beta) = 0$ if and only if $\gamma \in \beta + U$, where U is the linear subspace of $\mathbb{F}_{q^s}$ consisting of the roots of $L(x)$. Thus the roots of $A(x)$ form an affine subspace of $\mathbb{F}_{q^s}$. □

3.57. Theorem. *Let T be an affine subspace of $\mathbb{F}_{q^m}$, considered as a vector space over $\mathbb{F}_q$. Then for any nonnegative integer k the polynomial*

$$A(x) = \prod_{\gamma \in T} (x - \gamma)^{q^k}$$

is an affine q-polynomial over $\mathbb{F}_{q^m}$.

Proof. Let $T = \eta + U$, where U is a linear subspace of $\mathbb{F}_{q^m}$. Then

$$L(x) = \prod_{\beta \in U} (x - \beta)^{q^k}$$

is a q-polynomial over $\mathbb{F}_{q^m}$ according to Theorem 3.52. Furthermore,

$$A(x) = \prod_{\gamma \in T} (x - \gamma)^{q^k} = \prod_{\beta \in U} (x - \eta - \beta)^{q^k} = L(x - \eta),$$

and $L(x - \eta)$ is easily seen to be an affine q-polynomial over $\mathbb{F}_{q^m}$. □

The ordinary product of linearized polynomials need not be a linearized polynomial. However, the composition $L_1(L_2(x))$ of two q-polynomials $L_1(x)$, $L_2(x)$ over $\mathbb{F}_{q^m}$ is again a q-polynomial. Instead of the word composition (or substitution) we use the phrase "symbolic multiplication." Thus, we define *symbolic multiplication* by

$$L_1(x) \otimes L_2(x) = L_1(L_2(x)).$$

If we consider only q-polynomials over $\mathbb{F}_q$, then a simple investigation shows that symbolic multiplication is commutative, associative, and distributive (with respect to ordinary addition). In fact, the set of q-polynomials over $\mathbb{F}_q$ forms an integral domain under the operations of symbolic multiplication and ordinary addition. The operation of symbolic multiplication can be related to the conventional arithmetic of polynomials by means of the following notion.

3.58. Definition. The polynomials

$$l(x)=\sum_{i=0}^{n}\alpha_i x^i \quad\text{and}\quad L(x)=\sum_{i=0}^{n}\alpha_i x^{q^i}$$

over $\mathbb{F}_{q^m}$ are called *q-associates* of each other. More specifically, $l(x)$ is the *conventional q-associate* of $L(x)$ and $L(x)$ is the *linearized q-associate* of $l(x)$.

3.59. Lemma. *Let $L_1(x)$ and $L_2(x)$ be q-polynomials over $\mathbb{F}_q$ with conventional q-associates $l_1(x)$ and $l_2(x)$. Then $l(x)=l_1(x)l_2(x)$ and $L(x)=L_1(x)\otimes L_2(x)$ are q-associates of each other.*

Proof. The equations

$$l(x)=\sum_i a_i x^i=\sum_j b_j x^j\sum_k c_k x^k=l_1(x)l_2(x)$$

and

$$L(x)=\sum_i a_i x^{q^i}=\sum_j b_j\left(\sum_k c_k x^{q^k}\right)^{q^j}=\sum_j b_j\sum_k c_k x^{q^{j+k}}=L_1(x)\otimes L_2(x)$$

are each true if and only if

$$a_i=\sum_{j+k=i} b_j c_k \quad\text{for every } i. \qquad \square$$

If $L_1(x)$ and $L(x)$ are q-polynomials over $\mathbb{F}_q$, we say that $L_1(x)$ *symbolically divides* $L(x)$ (or that $L(x)$ is *symbolically divisible* by $L_1(x)$) if $L(x)=L_1(x)\otimes L_2(x)$ for some q-polynomial $L_2(x)$ over $\mathbb{F}_q$. The following criterion is then an immediate consequence of Lemma 3.59.

3.60. Corollary. *Let $L_1(x)$ and $L(x)$ be q-polynomials over $\mathbb{F}_q$ with conventional q-associates $l_1(x)$ and $l(x)$. Then $L_1(x)$ symbolically divides $L(x)$ if and only if $l_1(x)$ divides $l(x)$.*

3.61. Example. Let $L(x)$ be a q-polynomial over $\mathbb{F}_q$ that symbolically divides $x^{q^m}-x$ for some $m\in\mathbb{N}$. Then there exists a q-polynomial $L_1(x)$ over $\mathbb{F}_q$ such that

$$x^{q^m}-x=L(x)\otimes L_1(x)=L_1(x)\otimes L(x)=L_1(L(x)). \qquad (3.17)$$

This can be applied as follows. Let α be a fixed element of $\mathbb{F}_{q^m}$. Then the affine polynomial $L(x)-\alpha$ has at least one root in $\mathbb{F}_{q^m}$ if and only if $L_1(\alpha)=0$, and if $L_1(\alpha)=0$, then actually all the roots of $L(x)-\alpha$ are in $\mathbb{F}_{q^m}$. For if $\beta \in \mathbb{F}_{q^m}$ is a root of $L(x)-\alpha$, then $L(\beta)=\alpha$, and substituting x by β in (3.17) yields $L_1(\alpha)=\beta^{q^m}-\beta=0$. Conversely, suppose $L_1(\alpha)=0$ and let γ be a root of $L(x)-\alpha$ in some extension field of $\mathbb{F}_{q^m}$; then $L(\gamma)=\alpha$, and substituting x by γ in (3.17) yields $\gamma^{q^m}-\gamma=L_1(\alpha)=0$, so that $\gamma \in \mathbb{F}_{q^m}$. The polynomial $L_1(x)$ can be calculated by letting $l(x)$ be the conventional q-associate of $L(x)$, determining $l_1(x)=(x^m-1)/l(x)$, and then taking $L_1(x)$ to be the linearized q-associate of $l_1(x)$. This application contains Theorem 2.25 as a special case, as one sees easily by choosing $L(x)=x^q-x$. □

It is an important fact that although symbolic multiplication and ordinary multiplication are quite different operations, the divisibility concepts for linearized polynomials based on these operations are equivalent.

3.62. Theorem. *Let $L_1(x)$ and $L(x)$ be q-polynomials over $\mathbb{F}_q$ with conventional q-associates $l_1(x)$ and $l(x)$. Then the following properties are equivalent: (i) $L_1(x)$ symbolically divides $L(x)$; (ii) $L_1(x)$ divides $L(x)$ in the ordinary sense; (iii) $l_1(x)$ divides $l(x)$.*

Proof. Since the equivalence of (i) and (iii) has been established in Corollary 3.60, it suffices to show the equivalence of (i) and (ii). If $L_1(x)$ symbolically divides $L(x)$, then

$$L(x)=L_1(x)\otimes L_2(x)=L_2(x)\otimes L_1(x)=L_2(L_1(x))$$

for some q-polynomial $L_2(x)$ over $\mathbb{F}_q$. Let

$$L_2(x)=\sum_{i=0}^{n} a_i x^{q^i},$$

then

$$L(x)=a_0L_1(x)+a_1L_1(x)^q+\cdots+a_nL_1(x)^{q^n},$$

and so $L_1(x)$ divides $L(x)$ in the ordinary sense. Conversely, suppose $L_1(x)$ divides $L(x)$ in the ordinary sense, where we can assume that $L_1(x)$ is nonzero. Using the division algorithm, we write $l(x)=k(x)l_1(x)+r(x)$, where $\deg(r(x))<\deg(l_1(x))$, and turning to linearized q-associates we get in an obvious notation $L(x)=K(x)\otimes L_1(x)+R(x)$. By what we have already shown, $L_1(x)$ divides $K(x)\otimes L_1(x)$ in the ordinary sense, and so $L_1(x)$ divides $R(x)$ in the ordinary sense. But since $\deg(R(x))<\deg(L_1(x))$, $R(x)$ must be the zero polynomial, and this proves that $L_1(x)$ symbolically divides $L(x)$. □

This result can be used to establish an interesting relationship between an irreducible polynomial and the irreducible factors of its linearized q-associate.

3.63. *Theorem.* *Let $f(x)$ be irreducible in $\mathbb{F}_q[x]$ and let $F(x)$ be its linearized q-associate. Then the degree of every irreducible factor of $F(x)/x$ in $\mathbb{F}_q[x]$ is equal to* $\operatorname{ord}(f(x))$.

Proof. Since the case $f(0)=0$ is trivial, we can assume $f(0)\neq 0$. Put $e=\operatorname{ord}(f(x))$ and let $h(x)\in\mathbb{F}_q[x]$ be an irreducible factor of $F(x)/x$ of degree d. Then $f(x)$ divides x^e-1, and so by Theorem 3.62 $F(x)$ divides $x^{q^e}-x$. It follows that $h(x)$ divides $x^{q^e}-x$, hence d divides e by Theorem 3.20. By the division algorithm, we can write $x^d-1=g(x)f(x)+r(x)$ with $g(x), r(x)\in\mathbb{F}_q[x]$ and $\deg(r(x))<\deg(f(x))$. Turning to linearized q-associates, we get

$$x^{q^d}-x=G(x)\otimes F(x)+R(x),$$

and since $h(x)$ divides $x^{q^d}-x$ and $G(x)\otimes F(x)$, it follows that $h(x)$ divides $R(x)$. If $r(x)$ is not the zero polynomial, then $r(x)$ and $f(x)$ are relatively prime, and so by Theorem 1.55 there exist polynomials $s(x)$, $k(x)\in\mathbb{F}_q[x]$ with

$$s(x)r(x)+k(x)f(x)=1.$$

Turning to linearized q-associates, we get

$$S(x)\otimes R(x)+K(x)\otimes F(x)=x.$$

Since $h(x)$ divides $R(x)$ and $F(x)$, it follows that $h(x)$ divides x, which is impossible. Thus $r(x)$ is the zero polynomial, so that $f(x)$ divides x^d-1, and therefore e divides d by Lemma 3.6. Altogether, we have shown $d=e$. □

We say that a q-polynomial $L(x)$ over $\mathbb{F}_q$ of degree >1 is *symbolically irreducible* over $\mathbb{F}_q$ if the only symbolic decompositions $L(x)=L_1(x)\otimes L_2(x)$ with q-polynomials $L_1(x)$, $L_2(x)$ over $\mathbb{F}_q$ are those for which one of the factors has degree 1. A symbolically irreducible polynomial is always reducible in the ordinary sense since any linearized polynomial of degree >1 has the nontrivial factor x. By using Lemma 3.59, one shows immediately that the q-polynomial $L(x)$ is symbolically irreducible over $\mathbb{F}_q$ if and only if its conventional q-associate $l(x)$ is irreducible over $\mathbb{F}_q$.

Every q-polynomial $L(x)$ over $\mathbb{F}_q$ of degree >1 has a *symbolic factorization* into symbolically irreducible polynomials over $\mathbb{F}_q$ and this factorization is essentially unique, in the sense that all other symbolic factorizations are obtained by rearranging factors and by multiplying factors by nonzero elements of $\mathbb{F}_q$. Using the correspondence between linearized polynomials and their conventional q-associates, one sees that the symbolic factorization of $L(x)$ is obtained by writing down the canonical factorization in $\mathbb{F}_q[x]$ of its conventional q-associate $l(x)$ and then turning to linearized q-associates.

3.64. Example. Consider the 2-polynomial $L(x)=x^{16}+x^8+x^2+x$ over $\mathbb{F}_2$. Its conventional 2-associate $l(x)=x^4+x^3+x+1$ has the canonical factorization $l(x) = (x^2+x+1)(x+1)^2$ in $\mathbb{F}_2[x]$. Thus,

$$L(x) = (x^4+x^2+x) \otimes (x^2+x) \otimes (x^2+x)$$

is the symbolic factorization of $L(x)$ into symbolically irreducible polynomials over $\mathbb{F}_2$. □

For two or more q-polynomials over $\mathbb{F}_q$, not all of them 0, we may define their *greatest common symbolic divisor* to be the monic q-polynomial over $\mathbb{F}_q$ of highest degree that symbolically divides all of them. In order to compare this notion with that of the ordinary greatest common divisor, we note first that the roots of the greatest common divisor are exactly the common roots of the given q-polynomials. Since the intersection of linear subspaces is another linear subspace, it follows that the roots of the greatest common divisor form a linear subspace of some extension field $\mathbb{F}_{q^m}$, considered as a vector space over $\mathbb{F}_q$. Furthermore, by applying the first part of Theorem 3.50 to the given q-polynomials, we conclude that each root of the greatest common divisor has the same multiplicity, which is either 1 or a power of q. Therefore, Theorem 3.52 implies that the greatest common divisor is a q-polynomial. It follows then from Theorem 3.62 that *the greatest common divisor and the greatest common symbolic divisor are identical*. An efficient way of calculating the greatest common (symbolic) divisor of q-polynomials over $\mathbb{F}_q$ is to consider the conventional q-associates and determine their greatest common divisor; then the linearized q-associate of this greatest common divisor is the greatest common (symbolic) divisor of the given q-polynomials.

By Theorem 3.50 the roots of a nonzero q-polynomial over $\mathbb{F}_q$ form a vector space over $\mathbb{F}_q$. The roots have the additional property that the qth power of a root is again a root. A finite-dimensional vector space M over $\mathbb{F}_q$ that is contained in some extension field of $\mathbb{F}_q$ and has the property that the qth power of every element of M is again in M is called a *q-modulus*. On the basis of this concept we can establish the following criterion.

3.65. Theorem. *The monic polynomial $L(x)$ is a q-polynomial over $\mathbb{F}_q$ if and only if each root of $L(x)$ has the same multiplicity, which is either* 1 *or a power of q, and the roots form a q-modulus.*

Proof. The necessity of the conditions follows from Theorem 3.50 and the remarks above. Conversely, the given conditions and Theorem 3.52 imply that $L(x)$ is a q-polynomial over some extension field of $\mathbb{F}_q$. If M is the q-modulus consisting of the roots of $L(x)$, then

$$L(x) = \prod_{\beta \in M} (x-\beta)^{q^k}$$

for some nonnegative integer k. Since $M = \{\beta^q : \beta \in M\}$, we obtain

$$L(x)^q = \prod_{\beta \in M} (x^q - \beta^q)^{q^k} = \prod_{\beta \in M} (x^q - \beta)^{q^k} = L(x^q).$$

If

$$L(x)=\sum_{i=0}^{n}\alpha_i x^{q^i},$$

then

$$\sum_{i=0}^{n}\alpha_i^q x^{q^{i+1}}=L(x)^q=L(x^q)=\sum_{i=0}^{n}\alpha_i x^{q^{i+1}},$$

so that for $0\leqslant i\leqslant n$ we have $\alpha_i^q=\alpha_i$ and thus $\alpha_i\in\mathbb{F}_q$. Therefore, $L(x)$ is a q-polynomial over $\mathbb{F}_q$. □

Any q-polynomial over $\mathbb{F}_q$ of degree q is symbolically irreducible over $\mathbb{F}_q$. For q-polynomials of degree $>q$, the notion of q-modulus can be used to characterize symbolically irreducible polynomials.

3.66. Theorem. *The q-polynomial $L(x)$ over $\mathbb{F}_q$ of degree $>q$ is symbolically irreducible over $\mathbb{F}_q$ if and only if $L(x)$ has simple roots and the q-modulus M consisting of the roots of $L(x)$ contains no q-modulus other than $\{0\}$ and M itself.*

Proof. Suppose $L(x)$ is symbolically irreducible over $\mathbb{F}_q$. If $L(x)$ had multiple roots, then Theorem 3.65 would imply that we could write $L(x)=L_1(x)^q$ with a q-polynomial $L_1(x)$ over $\mathbb{F}_q$ of degree >1. But then $L(x)=x^q\otimes L_1(x)$, a contradiction to the symbolic irreducibility of $L(x)$. Thus $L(x)$ has only simple roots. Furthermore, if N is a q-modulus contained in M, then Theorem 3.65 shows that $L_2(x)=\prod_{\beta\in N}(x-\beta)$ is a q-polynomial over $\mathbb{F}_q$. Since $L_2(x)$ divides $L(x)$ in the ordinary sense, it symbolically divides $L(x)$ by Theorem 3.62. But $L(x)$ is symbolically irreducible over $\mathbb{F}_q$, and so $\deg(L_2(x))$ must be either 1 or $\deg(L(x))$; that is, N is either $\{0\}$ or M.

To prove the sufficiency of the condition, suppose that $L(x)=L_1(x)\otimes L_2(x)$ is a symbolic decomposition with q-polynomials $L_1(x)$, $L_2(x)$ over $\mathbb{F}_q$. Then $L_1(x)$ symbolically divides $L(x)$, and so it divides $L(x)$ in the ordinary sense by Theorem 3.62. It follows that $L_1(x)$ has simple roots and that the q-modulus N consisting of the roots of $L_1(x)$ is contained in M. Consequently, N is either $\{0\}$ or M, and so $\deg(L_1(x))$ is either 1 or $\deg(L(x))$. Thus, either $L_1(x)$ or $L_2(x)$ is of degree 1, which means that $L(x)$ is symbolically irreducible over $\mathbb{F}_q$. □

3.67. Definition. Let $L(x)$ be a nonzero q-polynomial over $\mathbb{F}_{q^m}$. A root ζ of $L(x)$ is called a *q-primitive root* over $\mathbb{F}_{q^m}$ if it is not a root of any nonzero q-polynomial over $\mathbb{F}_{q^m}$ of lower degree.

This concept may also be viewed as follows. Let $g(x)$ be the minimal polynomial of ζ over $\mathbb{F}_{q^m}$. Then ζ is a q-primitive root of $L(x)$ over $\mathbb{F}_{q^m}$ if and only if $g(x)$ divides $L(x)$ and $g(x)$ does not divide any nonzero q-polynomial over $\mathbb{F}_{q^m}$ of lower degree.

Given an element ζ of a finite extension field of $\mathbb{F}_{q^m}$, one can always find a nonzero q-polynomial over $\mathbb{F}_{q^m}$ for which ζ is a q-primitive root over $\mathbb{F}_{q^m}$. To see this, we proceed as in the construction of an affine multiple. Let $g(x)$ be the minimal polynomial of ζ over $\mathbb{F}_{q^m}$, let n be the degree of $g(x)$, and calculate for $i = 0, 1, \ldots, n$ the unique polynomial $r_i(x)$ of degree $\leqslant n-1$ with $x^{q^i} \equiv r_i(x) \bmod g(x)$. Then determine elements $\alpha_i \in \mathbb{F}_{q^m}$, not all 0, such that $\sum_{i=0}^{n} \alpha_i r_i(x) = 0$. This involves n conditions concerning the vanishing of the coefficients of x^j, $0 \leqslant j \leqslant n-1$, and thus leads to a homogeneous system of n linear equations for the $n+1$ unknowns $\alpha_0, \alpha_1, \ldots, \alpha_n$. Such a system always has a nontrivial solution, and with such a solution we get

$$L(x) = \sum_{i=0}^{n} \alpha_i x^{q^i} \equiv \sum_{i=0}^{n} \alpha_i r_i(x) \equiv 0 \bmod g(x),$$

so that $L(x)$ is a nonzero q-polynomial over $\mathbb{F}_{q^m}$ divisible by $g(x)$. By choosing the α_i in such a way that $L(x)$ is monic and of the lowest possible degree, one finds that ζ is a q-primitive root of $L(x)$ over $\mathbb{F}_{q^m}$. It is easily seen that this monic q-polynomial $L(x)$ over $\mathbb{F}_{q^m}$ of least positive degree that is divisible by $g(x)$ is uniquely determined; it is called the *minimal q-polynomial* of ζ over $\mathbb{F}_{q^m}$.

3.68. Theorem. *Let ζ be an element of a finite extension field of $\mathbb{F}_{q^m}$ and let $M(x)$ be its minimal q-polynomial over $\mathbb{F}_{q^m}$. Then a q-polynomial $K(x)$ over $\mathbb{F}_{q^m}$ has ζ as a root if and only if $K(x) = L(x) \otimes M(x)$ for some q-polynomial $L(x)$ over $\mathbb{F}_{q^m}$. In particular, for the case $m = 1$ this means that $K(x)$ has ζ as a root if and only if $K(x)$ is symbolically divisible by $M(x)$.*

Proof. If $K(x) = L(x) \otimes M(x) = L(M(x))$, it follows immediately that $K(\zeta) = 0$. Conversely, let

$$M(x) = \sum_{j=0}^{t} \gamma_j x^{q^j} \quad \text{with } \gamma_t = 1$$

and suppose

$$K(x) = \sum_{h=0}^{r} \alpha_h x^{q^h} \quad \text{with } r \geqslant t$$

has ζ as a root. Put $s = r - t$ and $\gamma_j = 0$ for $j < 0$, and consider the following system of $s+1$ linear equations in the $s+1$ unknowns $\beta_0, \beta_1, \ldots, \beta_s$:

$$\begin{aligned}
\beta_0 + \gamma_{t-1}^{q}\beta_1 + \gamma_{t-2}^{q^2}\beta_2 + \cdots + \gamma_{t-s}^{q^s}\beta_s &= \alpha_t \\
\beta_1 + \gamma_{t-1}^{q^2}\beta_2 + \cdots + \gamma_{t-s+1}^{q^s}\beta_s &= \alpha_{t+1} \\
\ddots \qquad \vdots \qquad &\quad \vdots \\
\beta_{s-1} + \gamma_{t-1}^{q^s}\beta_s &= \alpha_{r-1} \\
\beta_s &= \alpha_r.
\end{aligned}$$

It is clear that this system has a unique solution involving elements $\beta_0, \beta_1, \ldots, \beta_s$ of $\mathbb{F}_{q^m}$. With

$$L(x) = \sum_{i=0}^{s} \beta_i x^{q^i} \quad \text{and} \quad R(x) = K(x) - L(M(x))$$

we get

$$\begin{aligned}
R(x) &= \sum_{h=0}^{r} \alpha_h x^{q^h} - \sum_{i=0}^{s} \beta_i \left(\sum_{j=0}^{t} \gamma_j x^{q^j} \right)^{q^i} \\
&= \sum_{h=0}^{r} \alpha_h x^{q^h} - \sum_{i=0}^{s} \beta_i \sum_{j=0}^{t} \gamma_j^{q^i} x^{q^{i+j}} \\
&= \sum_{h=0}^{r} \alpha_h x^{q^h} - \sum_{h=0}^{r} \left(\sum_{i=0}^{s} \gamma_{h-i}^{q^i} \beta_i \right) x^{q^h} \\
&= \sum_{h=0}^{r} \left(\alpha_h - \sum_{i=0}^{s} \gamma_{h-i}^{q^i} \beta_i \right) x^{q^h}.
\end{aligned}$$

It follows from the system above that $R(x)$ has degree $< q^t$. But since $R(\zeta) = K(\zeta) - L(M(\zeta)) = 0$, the definition of $M(x)$ implies that $R(x)$ is the zero polynomial. Therefore, we have $K(x) = L(M(x)) = L(x) \otimes M(x)$. □

We consider now the problem of determining the number N_L of q-primitive roots over $\mathbb{F}_q$ of a nonzero q-polynomial $L(x)$ over $\mathbb{F}_q$. If $L(x)$ has multiple roots, then by Theorem 3.65 we can write $L(x) = L_1(x)^q$ with a q-polynomial $L_1(x)$ over $\mathbb{F}_q$. Since every root of $L(x)$ is then also a root of $L_1(x)$, we have $N_L = 0$. Thus we can assume that $L(x)$ has only simple roots. If $L(x)$ has degree 1, it is obvious that $N_L = 1$. If $L(x)$ has degree $q^n > 1$ and is monic (without loss of generality), let

$$L(x) = \underbrace{L_1(x) \otimes \cdots \otimes L_1(x)}_{e_1} \otimes \cdots \otimes \underbrace{L_r(x) \otimes \cdots \otimes L_r(x)}_{e_r}$$

be the symbolic factorization of $L(x)$ with distinct monic symbolically irreducible polynomials $L_i(x)$ over $\mathbb{F}_q$. We obtain N_L by subtracting from the total number q^n of roots the number of roots of $L(x)$ that are already roots of some nonzero q-polynomial over $\mathbb{F}_q$ of degree $< q^n$. If ζ is a root of $L(x)$ of the latter kind and $M(x)$ is the minimal q-polynomial of ζ over $\mathbb{F}_q$, then $\deg(M(x)) < q^n$ and $M(x)$ symbolically divides $L(x)$ by Theorem 3.68. It follows that $M(x)$ symbolically divides one of the polynomials $K_i(x)$, $1 \leqslant i \leqslant r$, obtained from the symbolic factorization of $L(x)$ by omitting the symbolic factor $L_i(x)$, in which case $K_i(\zeta) = 0$ by Theorem 3.68. Since every root of $K_i(x)$ is automatically a root of $L(x)$, it follows that N_L is q^n minus the number of ζ that are roots of some $K_i(x)$. If q^{n_i} is the degree of $L_i(x)$, then the degree, and thus the number of roots, of $K_i(x)$ is q^{n-n_i}. If $i_1, \ldots, i_s$ are distinct subscripts, then the number of common roots of $K_{i_1}(x), \ldots, K_{i_s}(x)$ is equal to the degree of the greatest common divisor, which is the same as the degree of the greatest common symbolic divisor (see the discussion following Example 3.64). Using symbolic factorizations, one finds that this degree is equal to

$$q^{n-n_{i_1}-\cdots-n_{i_s}}.$$

Altogether, the inclusion-exclusion principle of combinatorics yields

$$\begin{aligned} N_L &= q^n - \sum_{i=1}^{r} q^{n-n_i} + \sum_{1 \leqslant i < j \leqslant r} q^{n-n_i-n_j} \mp \cdots + (-1)^r q^{n-n_1-\cdots-n_r} \\ &= q^n(1-q^{-n_1}) \cdots (1-q^{-n_r}). \end{aligned}$$

This expression can also be interpreted in a different way. Let $l(x)$ be the conventional q-associate of $L(x)$. Then

$$l(x) = l_1(x)^{e_1} \cdots l_r(x)^{e_r}$$

is the canonical factorization of $l(x)$ in $\mathbb{F}_q[x]$, where $l_i(x)$ is the conventional q-associate of $L_i(x)$. We define an analog of Euler's ϕ-function (see Exercise 1.4) for nonzero $f \in \mathbb{F}_q[x]$ by letting $\Phi_q(f(x)) = \Phi_q(f)$ denote the number of polynomials in $\mathbb{F}_q[x]$ that are of smaller degree than f as well as relatively prime to f. The following result will then imply the identity $N_L = \Phi_q(l(x))$ for the case under consideration.

3.69. Lemma. *The function Φ_q defined for nonzero polynomials in $\mathbb{F}_q[x]$ has the following properties:*

(i) $\Phi_q(f) = 1$ *if* $\deg(f) = 0$;
(ii) $\Phi_q(fg) = \Phi_q(f)\Phi_q(g)$ *whenever f and g are relatively prime*;
(iii) *if* $\deg(f) = n \geqslant 1$, *then*

$$\Phi_q(f) = q^n(1-q^{-n_1}) \cdots (1-q^{-n_r}),$$

where the n_i are the degrees of the distinct monic irreducible polynomials appearing in the canonical factorization of f in $\mathbb{F}_q[x]$.

Proof. Property (i) is trivial. For property (ii), let $\Phi_q(f)=s$ and $\Phi_q(g)=t$, and let $f_1,\ldots,f_s$ resp. $g_1,\ldots,g_t$ be the polynomials counted by $\Phi_q(f)$ resp. $\Phi_q(g)$. If $h\in\mathbb{F}_q[x]$ is a polynomial with $\deg(h)<\deg(fg)$ and $\gcd(fg,h)=1$, then $\gcd(f,h)=\gcd(g,h)=1$, and so $h\equiv f_i \bmod f$, $h\equiv g_j \bmod g$ for a unique ordered pair (i,j) with $1\leqslant i\leqslant s$, $1\leqslant j\leqslant t$. On the other hand, given an ordered pair (i,j), the Chinese remainder theorem for $\mathbb{F}_q[x]$ (see Exercise 1.37) shows that there exists a unique $h\in\mathbb{F}_q[x]$ with $h\equiv f_i \bmod f$, $h\equiv g_j \bmod g$, and $\deg(h)<\deg(fg)$. This h satisfies $\gcd(f,h)=\gcd(g,h)=1$, and so $\gcd(fg,h)=1$. Therefore, there is a one-to-one correspondence between the st ordered pairs (i,j) and the polynomials $h\in\mathbb{F}_q[x]$ with $\deg(h)<\deg(fg)$ and $\gcd(fg,h)=1$. Consequently, $\Phi_q(fg)=st=\Phi_q(f)\Phi_q(g)$.

For an irreducible polynomial b in $\mathbb{F}_q[x]$ of degree m and a positive integer e, we can calculate $\Phi_q(b^e)$ directly. The polynomials $h\in\mathbb{F}_q[x]$ with $\deg(h)<\deg(b^e)=em$ that are not relatively prime to b^e are exactly those divisible by b, and they are thus of the form $h=gb$ with $\deg(g)<em-m$. Since there are q^{em-m} different choices for g, we get $\Phi_q(b^e)=q^{em}-q^{em-m}=q^{em}(1-q^{-m})$. Property (iii) follows now from property (ii). □

3.70. Theorem. *Let $L(x)$ be a nonzero q-polynomial over $\mathbb{F}_q$ with conventional q-associate $l(x)$. Then the number N_L of q-primitive roots of $L(x)$ over $\mathbb{F}_q$ is given by $N_L=0$ if $L(x)$ has multiple roots and by $N_L=\Phi_q(l(x))$ if $L(x)$ has simple roots.*

Proof. This follows from Lemma 3.69 and the discussion preceding it. □

3.71. Corollary. *Every nonzero q-polynomial over $\mathbb{F}_q$ with simple roots has at least one q-primitive root over $\mathbb{F}_q$.*

Earlier in this section we introduced the notion of a q-modulus. The results about q-primitive roots can be used to construct a special type of basis for a q-modulus.

3.72. Theorem. *Let M be a q-modulus of dimension $m\geqslant 1$ over $\mathbb{F}_q$. Then there exists an element $\zeta\in M$ such that $\{\zeta,\zeta^q,\zeta^{q^2},\ldots,\zeta^{q^{m-1}}\}$ is a basis of M over $\mathbb{F}_q$.*

Proof. According to Theorem 3.65, $L(x)=\prod_{\beta\in M}(x-\beta)$ is a q-polynomial over $\mathbb{F}_q$. By Corollary 3.71, $L(x)$ has a q-primitive root ζ over $\mathbb{F}_q$. Then $\zeta,\zeta^q,\zeta^{q^2},\ldots,\zeta^{q^{m-1}}$ are elements of M. If these elements were linearly dependent over $\mathbb{F}_q$, then ζ would be a root of a nonzero q-polynomial over $\mathbb{F}_q$ of degree less than $q^m=\deg(L(x))$, a contradiction to the definition of a q-primitive root of $L(x)$ over $\mathbb{F}_q$. Therefore, these m elements are linearly independent over $\mathbb{F}_q$, and so they form a basis of M over $\mathbb{F}_q$. □

3.73. **Theorem.** *In $\mathbb{F}_{q^m}$ there exist exactly $\Phi_q(x^m-1)$ elements ζ such that $\{\zeta, \zeta^q, \zeta^{q^2}, \ldots, \zeta^{q^{m-1}}\}$ is a basis of $\mathbb{F}_{q^m}$ over $\mathbb{F}_q$.*

Proof. Since $\mathbb{F}_{q^m}$ can be viewed as a q-modulus, the argument in the proof of Theorem 3.72 applies. Here

$$L(x) = \prod_{\beta \in \mathbb{F}_{q^m}} (x-\beta) = x^{q^m} - x$$

by Lemma 2.4, and every q-primitive root of $L(x)$ over $\mathbb{F}_q$ yields a basis of the desired type. On the other hand, if $\zeta \in \mathbb{F}_{q^m}$ is not a q-primitive root of $L(x)$ over $\mathbb{F}_q$, then $\zeta, \zeta^q, \zeta^{q^2}, \ldots, \zeta^{q^{m-1}}$ are linearly dependent over $\mathbb{F}_q$, and so they do not form a basis of $\mathbb{F}_{q^m}$ over $\mathbb{F}_q$. Consequently, the number of $\zeta \in \mathbb{F}_{q^m}$ such that $\{\zeta, \zeta^q, \zeta^{q^2}, \ldots, \zeta^{q^{m-1}}\}$ is a basis of $\mathbb{F}_{q^m}$ over $\mathbb{F}_q$ is equal to the number of q-primitive roots of $L(x)$ over $\mathbb{F}_q$, which is given by $\Phi_q(x^m-1)$ according to Theorem 3.70. □

This result provides a refinement of the normal basis theorem (compare with Definition 2.32 and Theorem 2.35). Since each of the elements $\zeta, \zeta^q, \zeta^{q^2}, \ldots, \zeta^{q^{m-1}}$ generates the same normal basis of $\mathbb{F}_{q^m}$ over $\mathbb{F}_q$, the number of different normal bases of $\mathbb{F}_{q^m}$ over $\mathbb{F}_q$ is given by $(1/m)\Phi_q(x^m-1)$.

3.74. Example. We calculate the number of different normal bases of $\mathbb{F}_{64}$ over $\mathbb{F}_2$. Since $64 = 2^6$, this number is given by $\frac{1}{6}\Phi_2(x^6-1)$. From the canonical factorization

$$x^6 - 1 = (x+1)^2(x^2+x+1)^2$$

in $\mathbb{F}_2[x]$ and Lemma 3.69(iii) it follows that

$$\Phi_2(x^6-1) = 2^6\left(1-\tfrac{1}{2}\right)\left(1-\tfrac{1}{4}\right) = 24,$$

and so there are four different normal bases of $\mathbb{F}_{64}$ over $\mathbb{F}_2$. □

5. BINOMIALS AND TRINOMIALS

A *binomial* is a polynomial with two nonzero terms, one of them being the constant term. Irreducible binomials can be characterized explicitly. For this purpose it suffices to consider nonlinear, monic binomials.

3.75. **Theorem.** *Let $t \geqslant 2$ be an integer and $a \in \mathbb{F}_q^*$. Then the binomial $x^t - a$ is irreducible in $\mathbb{F}_q[x]$ if and only if the following two conditions are satisfied: (i) each prime factor of t divides the order e of a in $\mathbb{F}_q^*$, but not $(q-1)/e$; (ii) $q \equiv 1 \bmod 4$ if $t \equiv 0 \bmod 4$.*

Proof. Suppose (i) and (ii) are satisfied. Then we note that $f(x)=x-a$ is an irreducible polynomial in $\mathbb{F}_q[x]$ of order e, and so $f(x^t)=x^t-a$ is irreducible in $\mathbb{F}_q[x]$ by Theorem 3.35.

Suppose (i) is violated. Then there exists a prime factor r of t that either divides $(q-1)/e$ or does not divide e. In the first case, we have $rs=(q-1)/e$ for some $s\in\mathbb{N}$. The subgroup of $\mathbb{F}_q^*$ consisting of rth powers has order $(q-1)/r=es$ and thus contains the subgroup of order e of $\mathbb{F}_q^*$ generated by a. In particular, $a=b^r$ for some $b\in\mathbb{F}_q^*$, and so $x^t-a=x^{t_1r}-b^r$ has the factor $x^{t_1}-b$. In the remaining case, r divides neither $(q-1)/e$ nor e, and so r does not divide $q-1$. Then $r_1r\equiv 1 \bmod (q-1)$ for some $r_1\in\mathbb{N}$, and thus $x^t-a=x^{t_1r}-a^{r_1r}$ has the factor $x^{t_1}-a^{r_1}$.

Suppose (i) is satisfied and (ii) is violated. Then $t=4t_2$ for some $t_2\in\mathbb{N}$ and $q\not\equiv 1 \bmod 4$. But (i) implies that e is even, and since e divides $q-1$, q must be odd. Hence $q\equiv 3 \bmod 4$. The fact that x^t-a is reducible in $\mathbb{F}_q[x]$ is then a consequence of Theorem 3.37. This can also be seen directly as follows. First we note that the information on e and q yields $e\equiv 2 \bmod 4$. Moreover, $a^{e/2}=-1$, and so $x^t-a=x^t+a^{(e/2)+1}=x^t+a^d$, where $d=(e/2)+1$ is even. Now

$$\begin{aligned} a^d &= 4(2^{-1}a^{d/2})^2 \\ &= 4(2^{-1}a^{d/2})^{q+1}=4c^4 \quad \text{with } c=(2^{-1}a^{d/2})^{(q+1)/4}, \end{aligned}$$

and this leads to the decomposition

$$\begin{aligned} x^t-a &= x^{4t_2}+4c^4 \\ &= (x^{2t_2}+2cx^{t_2}+2c^2)(x^{2t_2}-2cx^{t_2}+2c^2). \qquad \square \end{aligned}$$

If $q\equiv 3 \bmod 4$, we can write q in the form $q=2^Au-1$ with $A\geqslant 2$ and u odd. Suppose condition (i) in Theorem 3.75 is satisfied and t is divisible by 2^A. We write $t=Bv$ with $B=2^{A-1}$ and v even. Then $k=A$ in Theorem 3.37, so that with $f(x)=x-a$ the polynomial $f(x^t)=x^t-a$ factors as a product of B monic irreducible polynomials in $\mathbb{F}_q[x]$ of degree $t/B=v$. These irreducible factors can be determined explicitly. We note that as in the last part of the proof of Theorem 3.75, $d=(e/2)+1$ is even. Since $\gcd(2B,q-1)=2$, there exists $r\in\mathbb{N}$ with $2Br\equiv d \bmod (q-1)$. Setting $b=a^r\in\mathbb{F}_q$, we get then the following canonical factorization.

3.76. Theorem. *With the conditions and the notation introduced above, let*

$$F(x)=\sum_{i=0}^{B/2}\frac{(B-i-1)!B}{i!(B-2i)!}x^{B-2i}\in\mathbb{F}_q[x].$$

Then the roots $c_1,\ldots,c_B$ of $F(x)$ are all in $\mathbb{F}_q$, and in $\mathbb{F}_q[x]$ we have the *canonical factorization*

$$x^t - a = \prod_{j=1}^{B} \left(x^v - bc_j x^{v/2} - b^2 \right).$$

Proof. For a nonzero element γ in an extension field of $\mathbb{F}_q$ we have

$$(x-\gamma)(x+\gamma^{-1}) = x^2 - \beta x - 1 \text{ with } \beta = \gamma - \gamma^{-1}.$$

Using the statement and the notation of Waring's formula (see Theorem 1.76), we get

$$\begin{aligned} s_B(x_1, x_2) &= x_1^B + x_2^B \\ &= \sum_{\substack{i_1 + 2i_2 = B \\ i_1, i_2 \geqslant 0}} (-1)^{i_2} \frac{(i_1 + i_2 - 1)!B}{i_1! i_2!} \sigma_1(x_1, x_2)^{i_1} \sigma_2(x_1, x_2)^{i_2} \\ &= \sum_{i_2 = 0}^{B/2} (-1)^{i_2} \frac{(B - i_2 - 1)!B}{(B - 2i_2)! i_2!} (x_1 + x_2)^{B - 2i_2} (x_1 x_2)^{i_2}. \end{aligned}$$

Putting $x_1 = \gamma$, $x_2 = -\gamma^{-1}$, we obtain

$$\gamma^B + \gamma^{-B} = \sum_{i=0}^{B/2} (-1)^i \frac{(B-i-1)!B}{i!(B-2i)!} \beta^{B-2i} (-1)^i = F(\beta).$$

If c_j is a root of $F(x)$ in some extension field of $\mathbb{F}_q$ and γ_j is such that $\gamma_j - \gamma_j^{-1} = c_j$, then $\gamma_j^B + \gamma_j^{-B} = F(c_j) = 0$, and so $\gamma_j^{2B} = -1$. Since $q+1 = 2Bu$ with u odd, we get $\gamma_j^{q+1} = -1$, hence $\gamma_j^q = -\gamma_j^{-1}$. Then

$$c_j^q = \left(\gamma_j - \gamma_j^{-1} \right)^q = \gamma_j^q - \gamma_j^{-q} = -\gamma_j^{-1} + \gamma_j = c_j,$$

and so $c_j \in \mathbb{F}_q$. Since $F(x)$ is monic, we have

$$F(x) = \prod_{j=1}^{B} (x - c_j),$$

hence

$$\gamma^B + \gamma^{-B} = F(\beta) = \prod_{j=1}^{B} (\beta - c_j) = \prod_{j=1}^{B} \left(\gamma - \gamma^{-1} - c_j \right).$$

It follows that

$$\gamma^{2B} + 1 = \prod_{j=1}^{B} \left(\gamma^2 - c_j \gamma - 1 \right).$$

Since this identity holds for any element γ of any extension field of $\mathbb{F}_q$ (also for $\gamma = 0$), we get the polynomial identity

$$x^{2B} + 1 = \prod_{j=1}^{B} \left(x^2 - c_j x - 1 \right).$$

By substituting $b^{-1}x^{v/2}$ for x and multiplying by b^{2B}, we get a factorization of $x^{Bv}+b^{2B}=x^t+a^{2Br}=x^t+a^d=x^t-a$ (compare with the final portion of the proof of Theorem 3.75 for the last step). The resulting factors are irreducible in $\mathbb{F}_q[x]$ because we know already that the canonical factorization of x^t-a involves B irreducible polynomials in $\mathbb{F}_q[x]$ of degree v (see the discussion preceding Theorem 3.76). □

3.77. Example. We factor the binomial $x^{24}-3$ in $\mathbb{F}_7[x]$. Here $q=2^3-1$, so that $A=3$, $B=4$, and $v=6$. Furthermore, the element $a=3$ is of order $e=6$ in $\mathbb{F}_7^*$, and so condition (i) in Theorem 3.75 is satisfied and Theorem 3.76 can be applied. We have $d=4$, and a solution of the congruence $8r\equiv 4 \bmod 6$ is given by $r=2$. Therefore, $b=a^2=2$. Furthermore, $F(x)=x^4+4x^2+2$ has the roots ± 1 and ± 3 in $\mathbb{F}_7$. Thus $x^{24}-3=(x^6-2x^3-4)(x^6+2x^3-4)(x^6+x^3-4)(x^6-x^3-4)$ is the canonical factorization in $\mathbb{F}_7[x]$. □

A *trinomial* is a polynomial with three nonzero terms, one of them being the constant term. We first consider trinomials that are also affine polynomials.

3.78. Theorem. *Let $a\in\mathbb{F}_q$ and let p be the characteristic of $\mathbb{F}_q$. Then the trinomial x^p-x-a is irreducible in $\mathbb{F}_q[x]$ if and only if it has no root in $\mathbb{F}_q$.*

Proof. If β is a root of x^p-x-a in some extension field of $\mathbb{F}_q$, then by the proof of Theorem 3.56 the set of roots of x^p-x-a is $\beta+U$, where U is the set of roots of the linearized polynomial x^p-x. But $U=\mathbb{F}_p$, and so

$$x^p-x-a=\prod_{b\in\mathbb{F}_p}(x-\beta-b).$$

Suppose now that x^p-x-a has a factor $g\in\mathbb{F}_q[x]$ with $1\leqslant r=\deg(g)<p$ and g monic. Then

$$g(x)=\prod_{i=1}^{r}(x-\beta-b_i)$$

for certain $b_i\in\mathbb{F}_p$. A comparison of the coefficients of x^{r-1} shows that $r\beta+b_1+\cdots+b_r$ is an element of $\mathbb{F}_q$. Since r has a multiplicative inverse in $\mathbb{F}_q$, it follows that $\beta\in\mathbb{F}_q$. Thus we have shown that if x^p-x-a factors nontrivially in $\mathbb{F}_q[x]$, then it has a root in $\mathbb{F}_q$. The converse is trivial. □

3.79. Corollary. *With the notation of Theorem 3.78, the trinomial x^p-x-a is irreducible in $\mathbb{F}_q[x]$ if and only if $\mathrm{Tr}_{\mathbb{F}_q}(a)\neq 0$.*

Proof. By Theorem 2.25, x^p-x-a has a root in $\mathbb{F}_q$ if and only if the absolute trace $\mathrm{Tr}_{\mathbb{F}_q}(a)$ is 0. The rest follows from Theorem 3.78. □

Since for $b \in \mathbb{F}_q^*$ the polynomial $f(x)$ is irreducible over $\mathbb{F}_q$ if and only if $f(bx)$ is irreducible over $\mathbb{F}_q$, the criteria above hold also for trinomials of the form $b^p x^p - bx - a$.

If we consider more general trinomials of the above type for which the degree is a higher power of the characteristic, then these criteria need not be valid any longer. In fact, the following decomposition formula can be established.

3.80. Theorem. *For $x^q - x - a$ with a being an element of the subfield $K = \mathbb{F}_r$ of $F = \mathbb{F}_q$, we have the decomposition*

$$x^q - x - a = \prod_{j=1}^{q/r} \left(x^r - x - \beta_j\right) \tag{3.18}$$

in $\mathbb{F}_q[x]$, where the β_j are the distinct elements of $\mathbb{F}_q$ with $\mathrm{Tr}_{F/K}(\beta_j) = a$.

Proof. For a given β_j, let γ be a root of $x^r - x - \beta_j$ in some extension field of $\mathbb{F}_q$. Then $\gamma^r - \gamma = \beta_j$, and also

$$\begin{aligned} a &= \mathrm{Tr}_{F/K}(\beta_j) \\ &= \mathrm{Tr}_{F/K}(\gamma^r - \gamma) \\ &= (\gamma^r - \gamma) + (\gamma^r - \gamma)^r + (\gamma^r - \gamma)^{r^2} + \cdots + (\gamma^r - \gamma)^{q/r} = \gamma^q - \gamma, \end{aligned}$$

so that γ is a root of $x^q - x - a$. Since $x^r - x - \beta_j$ has only simple roots, $x^r - x - \beta_j$ divides $x^q - x - a$. Now the polynomials $x^r - x - \beta_j$, $1 \leqslant j \leqslant q/r$, are pairwise relatively prime, and so the polynomial on the right-hand side of (3.18) divides $x^q - x - a$. A comparison of degrees and of leading coefficients shows that the two sides of (3.18) are identical. □

3.81. Example. Consider $x^9 - x - 1$ in $\mathbb{F}_9[x]$. Viewing $\mathbb{F}_9$ as $\mathbb{F}_3(\alpha)$, where α is a root of the irreducible polynomial $x^2 - x - 1$ in $\mathbb{F}_3[x]$, we find that the elements of $\mathbb{F}_9$ with absolute trace equal to 1 are $-1, \alpha, 1 - \alpha$. Thus (3.18) yields the decomposition

$$x^9 - x - 1 = (x^3 - x + 1)(x^3 - x - \alpha)(x^3 - x - 1 + \alpha).$$

Since all three factors are irreducible in $\mathbb{F}_9[x]$, we have also obtained the canonical factorization of $x^9 - x - 1$ in $\mathbb{F}_9[x]$. □

The information about irreducible trinomials can be applied to the construction of new irreducible polynomials from given ones.

3.82. Theorem. *Let $f(x) = x^m + a_{m-1}x^{m-1} + \cdots + a_0$ be an irreducible polynomial over the finite field $\mathbb{F}_q$ of characteristic p and let $b \in \mathbb{F}_q$. Then the polynomial $f(x^p - x - b)$ is irreducible over $\mathbb{F}_q$ if and only if the absolute trace $\mathrm{Tr}_{\mathbb{F}_q}(mb - a_{m-1})$ is $\neq 0$.*

Proof. Suppose $\mathrm{Tr}_{\mathbb{F}_q}(mb - a_{m-1}) \neq 0$. Put $K = \mathbb{F}_q$ and let F be the splitting field of f over K. If $\alpha \in F$ is a root of f, then, according to Theorem 2.14, all the roots of f are given by $\alpha, \alpha^q, \ldots, \alpha^{q^{m-1}}$ and $F = K(\alpha)$. Furthermore, $\mathrm{Tr}_{F/K}(\alpha) = -a_{m-1}$ by (2.2), and using Theorem 2.26 we get

$$\mathrm{Tr}_F(\alpha + b) = \mathrm{Tr}_K\big(\mathrm{Tr}_{F/K}(\alpha + b)\big) = \mathrm{Tr}_K(-a_{m-1} + mb) \neq 0.$$

By Corollary 3.79, the trinomial $x^p - x - (\alpha + b)$ is irreducible over F. Thus $[F(\beta): F] = p$, where β is a root of $x^p - x - (\alpha + b)$. It follows from Theorem 1.84 that

$$[F(\beta): K] = [F(\beta): F][F: K] = pm.$$

Now $\alpha = \beta^p - \beta - b$, so that $\alpha \in K(\beta)$ and $K(\beta) = K(\alpha, \beta) = F(\beta)$. Hence $[K(\beta): K] = pm$ and the minimal polynomial of β over K has degree pm. But $f(\beta^p - \beta - b) = f(\alpha) = 0$, and so β is a root of the monic polynomial $f(x^p - x - b) \in K[x]$ of degree pm. Theorem 3.33(ii) shows that $f(x^p - x - b)$ is the minimal polynomial of β over K. By Theorem 3.33(i), $f(x^p - x - b)$ is irreducible over $K = \mathbb{F}_q$.

If $\mathrm{Tr}_{\mathbb{F}_q}(mb - a_{m-1}) = 0$, then $x^p - x - (\alpha + b)$ is reducible over F, and so $[F(\beta): F] < p$ for any root β of $x^p - x - (\alpha + b)$. The same arguments as above show that β is a root of $f(x^p - x - b)$ and that $[F(\beta): K] < pm$, hence $f(x^p - x - b)$ is reducible over $K = \mathbb{F}_q$. □

For certain types of reducible trinomials we can establish the form of the canonical factorization. The hypothesis for this result involves the irreducibility of a binomial, which can be checked by Theorem 3.75.

3.83. Theorem. *Let $f(x) = x^r - ax - b \in \mathbb{F}_q[x]$, where $r > 2$ is a power of the characteristic of $\mathbb{F}_q$, and suppose that the binomial $x^{r-1} - a$ is irreducible over $\mathbb{F}_q$. Then $f(x)$ is the product of a linear polynomial and an irreducible polynomial over $\mathbb{F}_q$ of degree $r - 1$.*

Proof. Since $f'(x) = -a \neq 0$, $f(x)$ has only simple roots. If p is the characteristic of $\mathbb{F}_q$, then $f(x)$ is an affine p-polynomial over $\mathbb{F}_q$. Hence, Theorem 3.56 shows that the difference γ of two distinct roots of $f(x)$ is a root of the p-polynomial $x^r - ax$, and so a root of $x^{r-1} - a$. From $r - 1 > 1$ and the hypothesis about this binomial, it follows that γ is not an element of $\mathbb{F}_q$, and so there exists a root α of $f(x)$ that is not an element of $\mathbb{F}_q$. Then $\alpha^q \neq \alpha$ is also a root of $f(x)$ and, by what we have already shown, $\alpha^q - \alpha$ is a root of the irreducible polynomial $x^{r-1} - a$ over $\mathbb{F}_q$, so that $[\mathbb{F}_q(\alpha^q - \alpha): \mathbb{F}_q] = r - 1$. Since $\mathbb{F}_q(\alpha^q - \alpha) \subseteq \mathbb{F}_q(\alpha)$, it follows that $m = [\mathbb{F}_q(\alpha): \mathbb{F}_q]$ is a multiple of $r - 1$. On the other hand, α is a root of the polynomial $f(x)$ of degree r, so that $m \leqslant r$. Because of $r > 2$, this is only possible if $m = r - 1$. Thus the minimal polynomial of α over $\mathbb{F}_q$ is an irreducible polynomial over $\mathbb{F}_q$ of degree $r - 1$ that divides $f(x)$. The result follows now immediately. □

In the special case of prime fields, one can characterize the primitive polynomials among trinomials of a certain kind.

3.84. Theorem. *For a prime p, the trinomial $x^p - x - a \in \mathbb{F}_p[x]$ is a primitive polynomial over $\mathbb{F}_p$ if and only if a is a primitive element of $\mathbb{F}_p$ and* $\operatorname{ord}(x^p - x - 1) = (p^p - 1)/(p-1)$.

Proof. Suppose first that $f(x) = x^p - x - a$ is a primitive polynomial over $\mathbb{F}_p$. Then a must be a primitive element of $\mathbb{F}_p$ because of Theorem 3.18. If β is a root of $g(x) = x^p - x - 1$ in some extension field of $\mathbb{F}_p$, then

$$0 = ag(\beta) = a(\beta^p - \beta - 1) = a^p\beta^p - a\beta - a = f(a\beta),$$

and so $\alpha = a\beta$ is a root of $f(x)$. Consequently, we have $\beta^r \neq 1$ for $0 < r < (p^p - 1)/(p-1)$, for otherwise $\alpha^{r(p-1)} = 1$ with $0 < r(p-1) < p^p - 1$, a contradiction to α being a primitive element of $\mathbb{F}_{p^p}$. On the other hand, $g(x)$ is irreducible over $\mathbb{F}_p$ by Corollary 3.79, and so

$$g(x) = x^p - x - 1 = (x-\beta)(x-\beta^p)\cdots\left(x-\beta^{p^{p-1}}\right).$$

A comparison of the constant terms leads to $\beta^{(p^p-1)/(p-1)} = 1$, hence $\operatorname{ord}(x^p - x - 1) = (p^p-1)/(p-1)$ on account of Theorem 3.3.

Conversely, if the conditions of the theorem are satisfied, then a and β have orders $p-1$ and $(p^p-1)/(p-1)$, respectively, in the multiplicative group $\mathbb{F}_{p^p}^*$. Now

$$\begin{aligned}(p^p-1)/(p-1) &= 1 + p + p^2 + \cdots + p^{p-1} \equiv 1 + 1 + 1 + \cdots + 1 \\ &\equiv p \equiv 1 \bmod (p-1),\end{aligned}$$

so that $p-1$ and $(p^p-1)/(p-1)$ are relatively prime. Therefore, $\alpha = a\beta$ has order $(p-1)\cdot(p^p-1)/(p-1) = p^p - 1$ in $\mathbb{F}_{p^p}^*$. Hence α is a primitive element of $\mathbb{F}_{p^p}$ and $f(x)$ is a primitive polynomial over $\mathbb{F}_p$. □

3.85. Example. For $p = 5$ we have $(p^p - 1)/(p-1) = 781 = 11 \cdot 71$. From the proof of Theorem 3.84 it follows that $x^{781} \equiv 1 \bmod (x^5 - x - 1)$, and since $x^{11} \not\equiv 1 \bmod (x^5 - x - 1)$ and $x^{71} \not\equiv 1 \bmod (x^5 - x - 1)$, we obtain $\operatorname{ord}(x^5 - x - 1) = 781$. Now 2 and 3 are primitive elements of $\mathbb{F}_5$, and so $x^5 - x - 2$ and $x^5 - x - 3$ are primitive polynomials over $\mathbb{F}_5$ by Theorem 3.84. □

For a trinomial $x^2 + x + a$ over a finite field $\mathbb{F}_q$ of odd characteristic, it is easily seen that it is irreducible over $\mathbb{F}_q$ if and only if a is not of the form $a = 4^{-1} - b^2$, $b \in \mathbb{F}_q$. Thus, there are exactly $(q-1)/2$ choices for $a \in \mathbb{F}_q$ that make $x^2 + x + a$ irreducible over $\mathbb{F}_q$. More generally, the number of $a \in \mathbb{F}_q$ that make $x^n + x + a$ irreducible over $\mathbb{F}_q$ is usually asymptotic to q/n, according to the following result.

3.86. Theorem. *Let $\mathbb{F}_q$ be a finite field of characteristic p. For an integer $n \geqslant 2$ such that $2n(n-1)$ is not divisible by p, let $T_n(q)$ denote the number of $a \in \mathbb{F}_q$ for which the trinomial $x^n + x + a$ is irreducible over $\mathbb{F}_q$. Then there is a constant B_n, depending only on n, such that*

$$\left| T_n(q) - \frac{q}{n} \right| \leqslant B_n q^{1/2}.$$

We omit the proof, as it depends on an elaborate investigation of certain Galois groups.

In Definition 1.92 we defined the discriminant of a polynomial. The following result gives an explicit formula for the discriminant of a trinomial.

3.87. Theorem. *The discriminant of the trinomial $x^n + ax^k + b \in \mathbb{F}_q[x]$ with $n > k \geqslant 1$ is given by*

$$D(x^n + ax^k + b) = (-1)^{n(n-1)/2} b^{k-1} \cdot \left(n^N b^{N-K} - (-1)^N (n-k)^{N-K} k^K a^N\right)^d,$$

where $d = \gcd(n, k)$, $N = n/d$, $K = k/d$.

NOTES

1. Presentations of material on polynomials over finite fields can also be found in Albert [3, Ch. 5] and Berlekamp [4]; see also Blake and Mullin [1], MacWilliams and Sloane [2], and McDonald [1], to mention a few more recently published books. These books contain results relevant to all the sections of this chapter. Further results on polynomials and some additional references to material not covered in the text will be added at the end of the notes to Section 5.

The basic results enunciated in Lemma 3.1 and Corollary 3.4 were already shown by Gauss [4]. The study of $\text{ord}(f)$ was continued by Serret [3] and Pellet [1], where one can find Theorem 3.5; see also Bachmann [4, Ch. 7] and Dickson [7, Part I, Ch. 3]. For Theorems 3.8 and 3.9 see, for example, Ward [5]. A simple method for the determination of $\text{ord}(f)$ for irreducible f over $\mathbb{F}_p$ of degree m with $(p^m - 1)/(p-1)$ prime is given in Garakov [2]. As was mentioned in the text, the order of a polynomial (Definition 3.2) is also called the period (Berlekamp [4]) or the exponent (Albert [3], Dickson [7]) of the polynomial. The concept of order of polynomials in several indeterminates was introduced by Long [1], [2]. If $f(x)$ is a polynomial over $\mathbb{F}_q$ with $f(0) \neq 0$, then the least positive integer e such that $f(x)$ divides $x^e - c$ for some $c \in \mathbb{F}_q$ is called the *integral order* of $f(x)$ by Bose, Chowla, and Rao [1] or the *subexponent* of $f(x)$ by Hirschfeld [4], [5]. A connection between the order and the integral order of an irreducible polynomial is implicit in the work of Ward [6]; compare also with Lemma 3.17.

Tables of irreducible polynomials and their orders are given, for example, by Chang and Godwin [1], Golomb [4, Ch. 3], Marsh [1], and McEliece [3] (see also Chapter 10). As the key step in finding the order of an irreducible polynomial is the factorization of $q^m - 1$, we list some references on such factorizations. The classical tables of such factorizations are those of Cunningham and Woodall [1], Cunningham [1], and Kraïtchik [1]. D. H. Lehmer [4], [5] developed an effective sieving machine for factors and in [1], [2], [3], [7], [10] methods for primality testing and factoring of integers $q^m - 1$; see also Kraïtchik [2] and J. C. P. Miller [1]. Brillhart and Selfridge [1] give a collection of complete factorizations of numbers of the form $2^m - 1$, and the paper of Brillhart, Lehmer, and Selfridge [1] contains further information on the factorization of such numbers. Knuth [3, Ch. 4] and H. C. Williams [2] give surveys of various factorization methods.

In connection with Definition 3.12 we note that an interesting class of polynomials is that of *self-reciprocal polynomials*—that is, polynomials f with $f^* = f$ (see also Exercises 3.13–3.15, 3.24, and 3.93). Levine and Brawley [1] determined the number of monic irreducible self-reciprocal polynomials of given degree for degrees 2 and 4 and Carlitz [105] for any even degree. See Fredman [1], Golomb [5], Hong and Bossen [1], R. L. Miller [1], and Varshamov and Garakov [1] for further results on self-reciprocal polynomials. Knee and Goldman [1] and Levine and Brawley [1] studied *quasi-self-reciprocal polynomials*—that is, polynomials f with $f^* = \pm f$. Albert [5] describes the property that two irreducible polynomials over $\mathbb{F}_2$ of the same degree are reciprocals of each other in terms of the polynomials whose roots are the products of roots of the original polynomials. For Theorem 3.14 see Chang and Godwin [1]. MacWilliams and Odlyzko [1] used the connection between the factorization of a polynomial and that of its reciprocal polynomial to disprove a conjecture on the correlation of finite sequences in $\mathbb{F}_2$.

Primitive polynomials are sometimes also called *indexing polynomials* (see Alanen and Knuth [2] and Sugimoto [1]). They are, of course, closely related to primitive elements of finite fields. E. J. Watson [1] lists one primitive polynomial over $\mathbb{F}_2$ for each degree $n \leqslant 100$ and Stahnke [1] for $n \leqslant 168$. Sugimoto [1] gives a table of primitive polynomials over prime fields $\mathbb{F}_p$ for $3 \leqslant p \leqslant 47$. See also Alanen and Knuth [1], [2], Bussey [1], [2], Marsh [1], Peterson and Weldon [1], and the tables in Chapter 10 for lists of primitive polynomials. Bose, Chowla, and Rao [1], [2] characterize primitive polynomials of degree 2 and 3. Carlitz [96] shows that there is exactly one primitive polynomial over $\mathbb{F}_q$ of degree n if and only if $q = 2$, $n \leqslant 2$, or $q = 3$, $n = 1$. Beard [5] and Beard and West [1] follow Carlitz [35] and study primitive elements and primitive polynomials of various types. Bilharz [1] proved an analog for primitive polynomials of Artin's conjecture on primitive roots, under an assumption on the location of zeros of congruence zeta-functions (see the notes to Chapter 6, Section 4, for these functions). This assumption was proved by Davenport [7] and in an even stronger form by Weil [1], [2], [3]. Brown and Zassenhaus [1] study the frequency of primes p for which a given irreducible polynomial over $\mathbb{Q}$ with integer coefficients is a primitive polynomial modulo p. Mironchikov [1] shows a result related to Theorem 3.16; namely, that if $f \in \mathbb{F}_2[x]$ with $\deg(f) = 2m$, m even, and $\operatorname{ord}(f) = 2^m + 1$, then f is irreducible over $\mathbb{F}_2$. Agou [12] characterizes irreducible polynomials of given order and so, in particular, primitive polynomials. Theorem 3.18 is due to Alanen and Knuth [2] for finite prime fields. Hirschfeld [4], [5] studies *subprimitive polynomials*—that is, polynomials over $\mathbb{F}_q$ of degree m and having subexponent $(q^m - 1)/(q - 1)$. According to Theorem 3.18, every primitive polynomial over $\mathbb{F}_q$ is also subprimitive.

2. The first extensive treatment of irreducible polynomials and of irreducible factors of polynomials in one indeterminate over $\mathbb{F}_q$ is given by Dickson [7], who builds on earlier work of C. Jordan [2], Pellet [7], and Serret [2], [3]. Theorem 3.25 was already shown by Gauss [4] for finite prime fields; see also Dedekind [1] and Schönemann [3]. Proofs can also be found in Albert [3, Ch. 5], Bachmann [4, Ch. 7], Berlekamp [4, Ch. 3], Dickson [7, Part I, Ch. 2], Jeger [1], Serret [3], Simmons [1], and van de Vooren-van Veen [1]. A general theory of Moebius functions and Moebius inversion was developed by Rota [1]. Some general principles underlying the enumeration of irreducible polynomials were pointed out by Fredman [1]. Lenskoĭ [1] and K. S. Williams [11] obtain asymptotic formulas for the number of monic irreducible polynomials over $\mathbb{F}_q$ of degree $\leqslant n$ and degree n, respectively. For an abstract viewpoint on such formulas see Knopfmacher [1], [2]. Carlitz [40] generalizes a result of Dickson [29] to give an asymptotic formula for the number of irreducible polynomials over $\mathbb{F}_q$ with certain preassigned coefficients; see also S. D. Cohen [7], Dress [1], Hayes [2], Uchiyama [1], [4], Varshamov [3], [5], and K. S. Williams [15]. Levine and Brawley [1], Carlitz [105], and Fredman [1] study the number of monic irreducible self-reciprocal polynomials of fixed degree. Fredman [1] discusses the enumeration of other special types of irreducible polynomials. Golomb [6] establishes a bijective correspondence between irreducible polynomials of degree n over $\mathbb{F}_q$ and aperiodic necklaces with n beads in q colors and thus determines the number of such necklaces. See also R. L. Miller [1] for related enumeration problems.

In connection with Theorem 3.27 we mention that references for cyclotomic polynomials have been collected in the notes to Chapter 2, Section 4.

The formula for $I(q, n; x)$ in Theorem 3.29 is essentially due to Dedekind [1]; see also Serret [2], [4], [5], Dickson [3], [7, Part I, Ch. 2],

Bachmann [4, Ch. 7], and Albert [3, Ch. 5]. Carlitz [3] proves formulas for the least common multiple (see also Walker [1]) and the product of all monic polynomials over $\mathbb{F}_q$ of fixed degree (see also K. S. Williams [26]). The latter formula is also of use in the calculation of products of characteristic polynomials extended over certain classes of matrices over $\mathbb{F}_q$ (see Carlitz [104]).

For tables of irreducible polynomials see Alanen and Knuth [1], [2], Bussey [1], [2], Chang and Godwin [1], Church [1], Garakov [3], Golomb [4, Ch. 3], Marsh [1], Mossige [1], Peterson and Weldon [1], and Chapter 10. Conway [1] gives excerpts of more extensive tables for finite fields of order $p^n \leqslant 1024$, p prime, $n \geqslant 2$, which list for each element of such a field the characteristic and minimal polynomial over each proper subfield; see also Chapter 10, Section 1 and Table B.

Irreducible cubic polynomials have been studied in detail by Cailler [1], Carlitz [103], Dickson [9], [29], Gel'fand [1], [2], Mirimanoff [1], and K. S. Williams [32]. For irreducible quartic polynomials see Albert [3, Ch. 5], Carlitz [73], [103], Dickson [9], Leonard [3], Leonard and Williams [1], and Skolem [4]. Serret [4], [5] and Dickson [3], [7, Part I, Ch. 3] characterized irreducible polynomials whose degrees are powers of the characteristic of the underlying finite field. Golomb and Lempel [1] study the irreducibility of polynomials that are given by second-order recursions. For irreducible binomials and trinomials see Section 5 and the notes to it. Agou [1], [3] shows that a monic polynomial $f(x)$ over $\mathbb{F}_q$ of degree $n \geqslant 1$ is irreducible over $\mathbb{F}_q$ if and only if $f(x)$ divides the coefficients of $(y-x)(y-x^q)\cdots(y-x^{q^{n-1}})-f(y)$ considered as a polynomial in y, and in Agou [3] a similar criterion is given for a monic polynomial to be a power of an irreducible polynomial. Agou [9] establishes a criterion for $f(g(x))$ to be irreducible over $\mathbb{F}_q$, where $f, g \in \mathbb{F}_q[x]$ are monic and f is irreducible over $\mathbb{F}_q$. This criterion was used in Agou [9], [13], [15], [16] to characterize irreducible polynomials of special types such as $f(x^{p^r}-ax)$, $f(x^p-x-b)$, and others. Such irreducible compositions of polynomials are also studied in S. D. Cohen [2], [10], Long [3], [4], Ore [6], and in Section 5. For compositions of the form $f(x^t)$ see Section 3.

The action of a group of affine transformations on the set of all irreducible polynomials over $\mathbb{F}_q$ is considered by Dodunekov [1], whereas the action of the group of linear fractional transformations with coefficients in the prime field $\mathbb{F}_p$ on the set of irreducible polynomials over $\mathbb{F}_p$ of a given degree is studied by Brahana [1], [2] and Hanneken [1], [2], [3], [4], [5]. The action of this group for $p=2$ on the roots of irreducible polynomials over $\mathbb{F}_2$ is discussed in Golomb [5].

Kornblum [1] shows an analog for $\mathbb{F}_q[x]$ of Dirichlet's theorem on primes in arithmetic progressions, namely that if $g, h \in \mathbb{F}_q[x]$ are both nonzero and relatively prime, then there exist infinitely many monic irreducible polynomials f over $\mathbb{F}_q$ with $f \equiv g \bmod h$, even if one requires that deg(f) is in a prescribed arithmetic progression. Artin [1], S. D. Cohen [7], Hayes [2], Johnsen [2], and Rhin [3] give quantitative refinements of this result. Davenport [6] applies Kornblum's result in the context of studying primitive elements.

Carlitz [15], [95], [99] considered the enumeration of irreducible polynomials in several indeterminates, and this work was extended by S. D. Cohen [1], [3] and Fredman [2]. A procedure for evaluating the number of irreducible polynomials in $\mathbb{F}_q[x_1,\ldots,x_n]$ with prescribed degree in each x_i was presented by Prabhu and Bose [1]. Carlitz [12] calls a polynomial in several indeterminates over $\mathbb{F}_q$ *factorable* if it can be factored into a product of linear polynomials over some finite extension of $\mathbb{F}_q$. In the same paper, the number of factorable polynomials and of irreducible factorable polynomials over $\mathbb{F}_q$ of given degree is determined. Factorable polynomials are also studied in Agou [7], Carlitz [15], Long [1], [2], [5], and K. S. Williams [14]. For absolutely irreducible polynomials in several indeterminates see Chapter 6, Section 4.

3. Theorems 3.35 and 3.37, at least for finite prime fields, are due to Serret [2]; see also Albert [3, Ch. 5] and Dickson [7, Part I, Ch. 3]. A direct method of obtaining irreducible polynomials of given degree and order was discussed by Golomb [8]. Theorems 3.38 and 3.39 were shown by Daykin [6]. This paper also contains other results on the characteristic polynomial f_t of α^t. For the case where α is a primitive element, an algorithm for the calculation of f_t was described by Alanen and Knuth [2]. Golomb [6] presents an algorithm for determining the minimal polynomial of α^t from that of α over $\mathbb{F}_2$; see also Gordon [1]. For the case where the minimal polynomial of α over $\mathbb{F}_2$ is a trinomial, see also Bajoga [1] and Bajoga and Walbesser [1]. Berlekamp [4, Ch. 4] and MacWilliams and Sloane [2, Ch. 4] also discuss algorithms for the calculation of minimal polynomials. See Conway [1] for tables of characteristic and minimal polynomials; compare also with Chapter 10, Section 1.

Irreducibility criteria for polynomials of the form $f(x^t)$ were established by Agou [9], [10], [11], Butler [2], S. D. Cohen [2], Pellet [7], Petterson [3], and Serret [2], [3]. Berlekamp [4, Ch. 6] and Varshamov and Ananiashvili [1] discuss the relationship between the order of $f(x^t)$ and that of $f(x)$.

Other classical methods for constructing irreducible and primitive polynomials can be found in Albert [3, Ch. 5], Dickson [3], [7, Part I, Ch. 3], Pellet [4], [5], and Serret [2]. An algorithm for constructing all irreducible polynomials over a finite prime field was developed by Popovici [1], [2]. Rabin [1] and Calmet and Loos [1] describe probabilistic algorithms for generating irreducible polynomials. Varshamov and Antonjan [1] describe a method of constructing further irreducible polynomials over $\mathbb{F}_2$ from a given irreducible polynomial; see also Varshamov [2]. Swift [1] and Varshamov [4] show how to construct irreducible polynomials over $\mathbb{F}_2$ from primitive polynomials. In Varshamov [4] there is also a matrix-theoretic method for constructing from an irreducible polynomial of degree n all irreducible polynomials of degrees dividing n. Lempel [1] and Swift [1] describe methods of obtaining primitive polynomials over $\mathbb{F}_2$, and Varshamov and Gamkrelidze [1] have construction methods for $\mathbb{F}_p$. Alanen and Knuth [2] have algorithms for calculating further primitive polynomials over $\mathbb{F}_p$ from one such polynomial.

Theorem 3.46 was shown by Pellet [1] for finite prime fields and by Dickson [7, Part I, Ch. 3] for general finite fields; see also Agou [4].

4. Many results of this section go back to the fundamental papers of Ore [4], [5], [6], [7]. Some of the results of Ore were anticipated by Rella [1]. In Ore [4] p-polynomials over arbitrary fields of characteristic p are studied. The theory of linearized polynomials over finite fields is developed extensively in Ore [5], where also several results of Dickson [3] are generalized and the operation of symbolic multiplication is investigated. In Ore [6] one finds a continuation of these studies as well as the important refinement of the normal basis theorem (see Theorems 3.72 and 3.73). This technique was also used by Artin [3] to establish a normal basis theorem for certain infinite fields of characteristic p. Further references on the normal basis theorem are given in the notes to Chapter 2, Section 3. The formula in Lemma 3.51 is due to Moore [3]. A short proof was given by Dickson [30].

Linearized polynomials are also studied in Carcanague [1], [2], Carlitz [21], and T. P. Vaughan [1]. Special classes of linearized polynomials appear in Carlitz [7], [17], [20], [34], [35]. Daykin [5] considers affine polynomials and determines the degrees and numbers of their irreducible factors. The method of finding roots of polynomials by considering affine multiples is due to Berlekamp, Rumsey, and Solomon [1]. Connections between circulant matrices and linearized polynomials are explored in Carlitz [21], [91], Ore [7], Silva [1], and T. P. Vaughan [1]. For relations to vector spaces and linear transformations over finite fields see Brawley, Carlitz, and Vaughan [1], Burde [2], [5], Pele [1], Rella [1], Ulbrich [1], and T. P. Vaughan [1], [2]. Peterson [2] uses the results of Pele [1] in coding theory. Jamison [1] applies linearized polynomials to a problem of covering a vector space over $\mathbb{F}_q$ by cosets of subspaces. Segre and Bartocci [1] apply linearized polynomials over $\mathbb{F}_2$ to finite projective geometries. Surveys of the theory of linearized and affine polynomials can be found in Berlekamp [4, Ch. 11], MacWilliams and Sloane [2, Ch. 4], McDonald [1, Ch. 2], and Rédei [10, Ch. 8]. Linearized polynomials over more general fields of characteristic p were studied further in Artin [3], Crampton and Whaples [1], Krasner [1], and Whaples [1].

Theorem 3.63 is due to Ore [6]; see also Zierler [2]. In the special case where $f(x)$ is a primitive polynomial over $\mathbb{F}_q$ (in which $F(x)/x$ is then irreducible over $\mathbb{F}_q$) one obtains a result noted by Marsh and Gleason [1]. For further results in this direction see Dickson [30], Mills [3], and Varshamov [2], [3], [5].

The formula for $\Phi_q(f)$ in Lemma 3.69(iii) was shown by Dedekind [1] for q prime and by O. H. Mitchell [1] in the general case. For further results on Φ_q we refer to Carlitz [26], [28]. The group structure of the set of polynomials counted by Φ_q was investigated by Claasen [1], [2]; see also Smits [1]. An analog of Φ_q for polynomials in several indeterminates was considered by S. D. Cohen [4]. Kühne [3] gives a formula for the number of monic polynomials of given degree $< \deg(f)$ that are relatively prime to $f \in \mathbb{F}_p[x]$.

5. Theorem 3.75 was essentially shown by Serret [2] for finite prime fields. Further characterizations of irreducible binomials can be found in Albert [3, Ch. 5], Capelli [1], [2], [3], Dickson [7, Part I, Ch. 3]. Lowe and Zelinsky [1], Rédei [10, Ch. 11], and Schwarz [4]. The factorization in Theorem 3.76 is again due to Serret [2]; see also Albert [3, Ch. 5] and Dickson [7, Part I, Ch. 3]. Shiva and Allard [1] discuss a method for factoring $x^{2^k-1}+1$ over $\mathbb{F}_2$. The factorization of $x^{q-1}-a$ over $\mathbb{F}_q$ is considered in Dickson [30]; see also Agou [14]. Schwarz [7] has a formula for the number of monic irreducible factors of fixed degree for a given binomial and Rédei [9] gives a short proof of it. See also Agou [10], Butler [2], and Schwarz [4]. Gay and Vélez [1] prove a formula for the degree of the splitting field of an irreducible binomial over an arbitrary field that was shown by Darbi [1] for fields of characteristic 0. Agou [4] studies the factorization of an irreducible binomial over $\mathbb{F}_q$ in an extension field of $\mathbb{F}_q$. Beard and West [2] and McEliece [3] tabulate factorizations of the binomials x^n-1. The factorization of more general polynomials $g(x)^t - a$ over finite prime fields is considered in Ore [2] and Petterson [3]. Applications of factorizations of binomials are contained in Agou [10], Berlekamp [2], and T. P. Vaughan [1].

Theorem 3.78 and Corollary 3.79 were first shown by Pellet [1]. The fact that $x^p - x - a$ is irreducible over $\mathbb{F}_p$ if $a \in \mathbb{F}_p^*$ was already established by Serret [2], [3]. See also Dickson [3], [7, Part I, Ch. 3] and Albert [3, Ch. 5] for these results. Theorem 3.80 is due to Dickson [3], [7, Part I, Ch. 3], but in the special case $a = 0$ it was already noted by Mathieu [1]. Theorem 3.82 was shown in this general form by Varshamov [3], [5]; see also Agou [9]. The case $b = 0$ received considerable attention much earlier. The corresponding result for $b = 0$ and finite prime fields was stated by Pellet [1] and proved in Pellet [9]. Polynomials $f(x^p - x)$ over $\mathbb{F}_p$ with $\deg(f)$ a power of p were treated by Serret [4], [5]. The case $b = 0$ for arbitrary finite fields was considered in Dickson [7, Part I, Ch. 3] and Albert [3, Ch. 5]. More general types of polynomials such as $f(x^{p^r} - ax)$, $f(x^{p^{2r}} - ax^{p^r} - bx)$ and others have also been studied; see Agou [9], [13], [14], [15], [16], [17], [18], [19], [20], S. D. Cohen [10], Long [2], [3], [4], [5], Long and Vaughan [1], [2], and Ore [6]. Theorem 3.83 generalizes results of Dickson [30] and Albert [3, Ch. 5]. See Schwarz [12] for further results in this direction. The number of roots of trinomials $x^r - ax - b \in \mathbb{F}_q[x]$ with r a power of the characteristic of $\mathbb{F}_q$ was studied by Liang [1], Segre [10], and Vilanova [1]. Theorem 3.84 can be found in Dickson [7, Part I, Ch. 3].

S. Chowla [17] conjectured that for fixed n the number of polynomials $x^n + x + a \in \mathbb{F}_p[x]$ that are irreducible over $\mathbb{F}_p$ is asymptotic to p/n as $p \to \infty$, and he showed this for $n = 3$. This case is also considered in Carlitz [103]. Leonard [1] extended the proof to $n = 4$ (see also Leonard [3]), and Leonard [2] proved a weakened form of Chowla's conjecture for $n = 5$. K. S. Williams [13] settled the case $n \leqslant 5$. The general proof of Chowla's conjecture (and our Theorem 3.86) is due to S. D. Cohen [5] and Ree [1]. S. D. Cohen [5] shows more generally an asymptotic result on the number of irreducible polynomials of the form $f(x) + ag(x)$ with given $f, g \in \mathbb{F}_q[x]$ and varying $a \in \mathbb{F}_q$. See also S. D. Cohen [6], Hayes [6], and Leonard [5]. Further work on trinomials $x^n + x + a$ over $\mathbb{F}_p$ is contained in Mortimer and Williams [1], Sato and Yorinaga [1], Uchiyama [8], and K. S. Williams [13]. Cazacu and Simovici [1] present an algorithm for determining irreducible trinomials $ax^n + x + 1$ over finite fields of characteristic 2. Proofs of Theorem 3.87 are given in Berlekamp [4, Ch. 6] and Swan [1].

Tables of irreducible trinomials over $\mathbb{F}_2$ can be found in Golomb [4, Ch. 5], Golomb, Welch, and Hales [1], Zierler [7], and Zierler and Brillhart [1], [2]. Fredricksen and Wisniewski [1] considered special classes of irreducible trinomials over $\mathbb{F}_2$. Primitive trinomials over $\mathbb{F}_2$ are listed in Rodemich and Rumsey [1], Zierler [6], and Zierler and Brillhart [1], [2]. Results and tables on orders of trinomials over $\mathbb{F}_2$ are contained in Arakelov and Tenengol'c [1], Golomb [4, Ch. 5], Golomb, Welch, and Hales [1], Young [1], and Zierler [8]. Bajoga [1] and Bajoga and Walbesser [1] consider elements for which the minimal polynomial over $\mathbb{F}_2$ is a trinomial. Some irreducible trinomials over $\mathbb{F}_p$ of the form $x^n + x + a$ are tabulated in Mortimer and Williams [1].

Beard and West [3], Golomb [4, Ch. 5], Golomb, Welch, and Hales [1], Zierler [7], and Zierler and Brillhart [1], [2] tabulate factorizations of trinomials. Golomb [4, Ch. 5] conjectured that the degree of every irreducible factor of the trinomial $x^{2r+1} + x^{r-1} + 1$, $r = 2^n$, over $\mathbb{F}_2$ divides $6n$. This was shown by Mills and Zierler [1] in the even stronger form that every such degree divides either $2n$ or $3n$ but not n, and a generalization for any finite field was obtained by Carlitz [112]. Mills [3] proved that the degree of every irreducible factor of $x^{r+1} + x + 1$, $r = q^n$, over $\mathbb{F}_q$ divides $3n$. Many other results of this type can be found in Golomb [4, Ch. 5] and Marsh, Mills, Ward, Rumsey, and Welch [1]. The factorization behavior of the trinomials $x^4 + ax^2 + b$ over finite prime fields was considered by Carlitz [73], [103].

In what follows we give references for some topics on polynomials not mentioned in this chapter. There is a considerable amount of literature on generalizations of classical topics in number theory to $\mathbb{F}_q[x]$. The problem of representing a polynomial as a sum of irreducible polynomials (general form of *Goldbach's problem* for $\mathbb{F}_q[x]$) is treated in Car [2], Cherly [2], Hayes [1], [4], and Webb [1]. The representation of polynomials as a sum of two irreducible polynomials and a square is studied in Car [6]; see also Webb [1]. The problem of representing a polynomial as a sum of powers of polynomials (*Waring's problem* for $\mathbb{F}_q[x]$) was considered by Car [1], [3], [4], Joly [1], R. M. Kubota [1], K. R. Matthews [1], Paley [2], and Webb [2], [3]. The special case of sums of squares has been studied extensively; see Carlitz [5], [8], [13], [19], [26], [27], [116], Carlitz and Cohen [2], E. Cohen [1], [2], Joly [2], Leahey [1], Stevens and Kuty [1], Verner [1], [2], and Webb [2]. For representations by quadratic forms see Carlitz [54] and Carlitz and Cohen [3], and for simultaneous representations by quadratic and linear forms see Carlitz [94]. Sums of powers with polynomial coefficients are considered in Carlitz and Cohen [1] and E. Cohen [3], [4]. For more general forms see Carlitz [32] and E. Cohen [4], [6]. Problems of additive number theory in $\mathbb{F}_q[x]$ are studied by Carlitz [48] and Cherly [1], [2], [3].

The functions Φ_q (see Lemma 3.69) and μ_q (see Exercise 3.75) are special cases of arithmetic functions on $\mathbb{F}_q[x]$. Such functions have been studied systematically by Carlitz [1], [2], [3], [26], [28], [30], [36], but the special function Φ_q appears already in Dedekind [1]. An analog of Φ_q for polynomials in several indeterminates was studied by S. D. Cohen [4]. Further results on arithmetic functions can be found in Carlitz [6], [14], [19], [20], Carlitz and Cohen [1], [2], E. Cohen [3], [6], S. D. Cohen [3], [4], Dress [1], Rhin [3], Shader [1], [4], and Silva [2]. See Knopfmacher [1], [2] for an abstract approach. A concept for $\mathbb{F}_q[x]$ analogous to that of a perfect number was studied by Beard [6], Beard, Bullock, and Harbin [1], Beard, Doyle, and Mandelberg [1], Beard and Harbin [1], and Beard, O'Connell, and West [1]. For the sum of divisors of a polynomial over $\mathbb{F}_2$ see Canaday [1]. Johnsen [2] considers sieve methods in $\mathbb{F}_q[x]$. Relations between arithmetic in $\mathbb{F}_q[x]$ and modern algebraic geometry appear in Goss [3]. A law of quadratic reciprocity for monic irreducible polynomials over $\mathbb{F}_p$ was shown by Dedekind [1] and a different proof was given by Artin [1]; see also Vaidyanathaswamy [1]. A higher reciprocity law for monic irreducible polynomials over $\mathbb{F}_q$ was established by Kühne [1] and proved again by F. K. Schmidt [2] and Carlitz [1], [2], [4]; see also Ore [6], Pocklington [2], Schwarz [2], and Whiteman [1]. Polynomial congruences have been considered by Carlitz [7], [8], [9], [10], [11], [17], E. Cohen [5], Lenskoĭ [2], K. N. Rao [1], [2], [3], and Shader [2], [3].

Carlitz [7] began to study the functions $\psi_m(t) = \prod(t - f(x))$, where the product is taken over all polynomials $f(x)$ over $\mathbb{F}_q$ of degree less than m. This work was continued in Carlitz [10], [19], [20], [31] and Wagner [1]. See Bundschuh [2], Carlitz [119], Geijsel [1], Wade [1], [2], [3], and Wagner [2], [3] for applications. Carlitz [18] introduced polynomials analogous to cyclotomic polynomials. An analog of the Bernoulli polynomials for finite fields was defined in Carlitz [23]. A theorem of the von Staudt type for such polynomials was proved in Carlitz [24], extending results in Carlitz [16], [22].

See also Carlitz [25], Dickey, Kairies, and Shank [1], Goss [1], [2], and Herget [1], [2].

Carlitz [31] showed that the polynomials $f \in \mathbb{F}_q[x]$ with $f(x+a) = f(x)$ for all $a \in \mathbb{F}_q$ are exactly those of the form $f(x) = \sum_k c_k (x^q - x)^k$ with $c_k \in \mathbb{F}_q$; see also Dodunekov [2]. Generalizations to several indeterminates are presented in Carlitz [119]. Mullen [13] characterizes for fixed $a \in \mathbb{F}_q^*$, $b \in \mathbb{F}_q$ the polynomials $f \in \mathbb{F}_q[x]$ with $f(bx+a) = bf(x) + a$; the case $b = 1$ was treated earlier by Wells [6]. Wagner [2] determines the linear operators L on $\mathbb{F}_q[x]$ with $L(f+g) \equiv L(f) \bmod g$ for all $f, g \in \mathbb{F}_q[x]$.

Shehadeh [1] determines the maximum possible number of consecutive 0's or 1's and their distribution among the coefficients of certain polynomials over $\mathbb{F}_2$, such as $(x^n - 1)/f(x)$ with $f(x)$ primitive over $\mathbb{F}_2$ of degree k and $n = 2^k - 1$.

A concept of uniform distribution for sequences of polynomials over finite fields was introduced by Hodges [23]. Further references are Dijksma [1], [2], [3], [4], Hodges [26], [27], Kuipers [2], Kuipers and Scheelbeek [1], Meijer and Dijksma [1], and Webb [4]. Uniform distribution in the field of Laurent series $\sum_{i=-\infty}^{m} c_i x^i$ over $\mathbb{F}_q$ was defined by Carlitz [33] and studied further by de Mathan [1], [2], [3], Dijksma [2], [3], [4], Hodges [27], Kuipers [1], Long and Webb [1], Meijer and Dijksma [1], Rhin [1], [2], [3], and Webb [4]. A survey of these theories of uniform distribution can be found in Kuipers and Niederreiter [1, Ch. 5]. Related results on diophantine approximations in the field of Laurent series over $\mathbb{F}_q$ are contained in Bateman and Duquette [1], Carlitz [32], [33], Deshouillers [1], [2], [3], Dubois and Paysan-Le Roux [1], Grandet-Hugot [1], [2], and Houndonougbo [2].

Kustaanheimo and Qvist [1] develop a sort of complex analysis for $\mathbb{F}_q[x]$ by working with a suitable notion of derivative and showing, for example, an analog of the Cauchy-Riemann differential equations.

EXERCISES

3.1. Determine the order of the polynomial $(x^2+x+1)^5(x^3+x+1)$ over $\mathbb{F}_2$.

3.2. Determine the order of the polynomial $x^7 - x^6 + x^4 - x^2 + x$ over $\mathbb{F}_3$.

3.3. Determine $\operatorname{ord}(f)$ for all monic irreducible polynomials f in $\mathbb{F}_3[x]$ of degree 3.

3.4. Prove that the polynomial $x^8 + x^7 + x^3 + x + 1$ is irreducible over $\mathbb{F}_2$ and determine its order.

3.5. Let $f \in \mathbb{F}_q[x]$ be a polynomial of degree $m \geqslant 1$ with $f(0) \neq 0$ and suppose that the roots $\alpha_1, \ldots, \alpha_m$ of f in the splitting field of f over $\mathbb{F}_q$ are all simple. Prove that $\operatorname{ord}(f)$ is equal to the least positive integer e such that $\alpha_i^e = 1$ for $1 \leqslant i \leqslant m$.

3.6. Prove that $\operatorname{ord}(Q_e) = e$ for all e for which the cyclotomic polynomial $Q_e \in \mathbb{F}_q[x]$ is defined.

3.7. Let f be irreducible over $\mathbb{F}_q$ with $f(0) \neq 0$. For $e \in \mathbb{N}$ relatively prime to q, prove that $\operatorname{ord}(f) = e$ if and only if f divides the cyclotomic polynomial Q_e.

3.8. Let $f \in \mathbb{F}_q[x]$ be as in Exercise 3.5 and let $b \in \mathbb{N}$. Find a general formula showing the relationship between $\operatorname{ord}(f^b)$ and $\operatorname{ord}(f)$.

3.9. Let $\mathbb{F}_q$ be a finite field of characteristic p, and let $f \in \mathbb{F}_q[x]$ be a polynomial of positive degree with $f(0) \neq 0$. Prove that $\operatorname{ord}(f(x^p)) = p \operatorname{ord}(f(x))$.

3.10. Let f be an irreducible polynomial in $\mathbb{F}_q[x]$ with $f(0) \neq 0$ and $\operatorname{ord}(f) = e$, and let r be a prime not dividing q. Prove: (i) if r divides e, then every irreducible factor of $f(x^r)$ in $\mathbb{F}_q[x]$ has order er; (ii) if r does not divide e, then one irreducible factor of $f(x^r)$ in $\mathbb{F}_q[x]$ has order e and the other factors have order er.

3.11. Deduce from Exercise 3.10 that if $f \in \mathbb{F}_q[x]$ is a polynomial of positive degree with $f(0) \neq 0$, and if r is a prime not dividing q, then $\operatorname{ord}(f(x^r)) = r \operatorname{ord}(f(x))$.

3.12. Prove that the reciprocal polynomial of an irreducible polynomial f over $\mathbb{F}_q$ with $f(0) \neq 0$ is again irreducible over $\mathbb{F}_q$.

3.13. A nonzero polynomial $f \in \mathbb{F}_q[x]$ is called *self-reciprocal* if $f = f^*$. Prove that if $f = gh$, where g and h are irreducible in $\mathbb{F}_q[x]$ and f is self-reciprocal, then either (i) $h^* = ag$ with $a \in \mathbb{F}_q^*$; or (ii) $g^* = bg$, $h^* = bh$ with $b = \pm 1$.

3.14. Prove: if f is a self-reciprocal irreducible polynomial in $\mathbb{F}_q[x]$ of degree $m > 1$, then m must be even.

3.15. Prove: if f is a self-reciprocal irreducible polynomial in $\mathbb{F}_q[x]$ of degree > 1 and of order e, then every irreducible polynomial in $\mathbb{F}_q[x]$ of degree > 1 whose order divides e is self-reciprocal.

3.16. Show that $x^6 + x^5 + x^2 + x + 1$ is a primitive polynomial over $\mathbb{F}_2$.

3.17. Show that $x^8 + x^6 + x^5 + x + 1$ is a primitive polynomial over $\mathbb{F}_2$.

3.18. Show that $x^5 - x + 1$ is a primitive polynomial over $\mathbb{F}_3$.

3.19. Let $f \in \mathbb{F}_q[x]$ be monic of degree $m \geqslant 1$. Prove that f is primitive over $\mathbb{F}_q$ if and only if f is an irreducible factor over $\mathbb{F}_q$ of the cyclotomic polynomial $Q_d \in \mathbb{F}_q[x]$ with $d = q^m - 1$.

3.20. Determine the number of primitive polynomials over $\mathbb{F}_q$ of degree m.

3.21. If $m \in \mathbb{N}$ is not a prime, prove that not every monic irreducible polynomial over $\mathbb{F}_q$ of degree m can be a primitive polynomial over $\mathbb{F}_q$.

3.22. If m is a prime, prove that all monic irreducible polynomials over $\mathbb{F}_q$ of degree m are primitive over $\mathbb{F}_q$ if and only if $q = 2$ and $2^m - 1$ is a prime.

3.23. If f is a primitive polynomial over $\mathbb{F}_q$, prove that $f(0)^{-1} f^*$ is again primitive over $\mathbb{F}_q$.

3.24. Prove that the only self-reciprocal primitive polynomials are $x+1$ and x^2+x+1 over $\mathbb{F}_2$ and $x+1$ over $\mathbb{F}_3$ (see Exercise 3.13 for the definition of a self-reciprocal polynomial).

3.25. Prove: if $f(x)$ is irreducible in $\mathbb{F}_q[x]$, then $f(ax+b)$ is irreducible in $\mathbb{F}_q[x]$ for any $a, b \in \mathbb{F}_q$ with $a \neq 0$.

3.26. Prove that $N_q(n) \leqslant (1/n)(q^n - q)$ with equality if and only if n is prime.

3.27. Prove that

$$N_q(n) \geqslant \frac{1}{n}q^n - \frac{q}{n(q-1)}(q^{n/2}-1).$$

3.28. Give a detailed proof of the fact that (3.5) implies (3.4).

3.29. Prove that the Moebius function μ satisfies $\mu(mn) = \mu(m)\mu(n)$ for all $m, n \in \mathbb{N}$ with $\gcd(m, n) = 1$.

3.30. Prove the identity

$$\sum_{d \mid n} \frac{\mu(d)}{d} = \frac{\phi(n)}{n} \quad \text{for all } n \in \mathbb{N}.$$

3.31. Prove that $\sum_{d \mid n} \mu(d)\phi(d) = 0$ for every even integer $n \geqslant 2$.

3.32. Prove the identity $\sum_{d \mid n} |\mu(d)| = 2^k$, where k is the number of distinct prime factors of $n \in \mathbb{N}$.

3.33. Prove that $N_q(n)$ is divisible by eq provided that $n \geqslant 2$, e is a divisor of $q-1$, and $\gcd(eq, n) = 1$.

3.34. Calculate the cyclotomic polynomials Q_{12} and Q_{30} from the explicit formula in Theorem 3.27.

3.35. Establish the properties of cyclotomic polynomials listed in Exercise 2.57, Parts (a)–(f), by using the explicit formula in Theorem 3.27.

3.36. Prove that the cyclotomic polynomial Q_n with $\gcd(n, q) = 1$ is irreducible over $\mathbb{F}_q$ if and only if the multiplicative order of q modulo n is $\phi(n)$.

3.37. If Q_n is irreducible over $\mathbb{F}_2$, prove that n must be a prime $\equiv \pm 3 \bmod 8$ or a power of such a prime. Show also that this condition is not sufficient.

3.38. Prove that Q_{15} is reducible over any finite field over which it is defined.

3.39. Prove that for $n \in \mathbb{N}$ there exists an integer b relatively prime to n whose multiplicative order modulo n is $\phi(n)$ if and only if $n = 1, 2, 4, p^r$, or $2p^r$, where p is an odd prime and $r \in \mathbb{N}$.

3.40. Dirichlet's theorem on primes in arithmetic progressions states that any arithmetic progression of integers $b, b+n, \ldots, b+kn, \ldots$ with $n \in \mathbb{N}$ and $\gcd(b, n) = 1$ contains infinitely many primes. Use this theorem to prove the following: the integers $n \in \mathbb{N}$ for which there exists a finite field $\mathbb{F}_q$ with $\gcd(n, q) = 1$ over which the cyclotomic polynomial Q_n is irreducible are exactly given by $n = 1, 2, 4, p^r$, or $2p^r$, where p is an odd prime and $r \in \mathbb{N}$.

3.41. Prove that Q_{19} and Q_{27} are two cyclotomic polynomials over $\mathbb{F}_2$ of the same degree that are both irreducible over $\mathbb{F}_2$.

3.42. If $e \geqslant 2$, $\gcd(e, q) = 1$, and m is the multiplicative order of q modulo e, prove that the product of all monic irreducible polynomials in $\mathbb{F}_q[x]$ of degree m and order e is equal to the cyclotomic polynomial Q_e over $\mathbb{F}_q$.

3.43. Find the factorization of $x^{32} - x$ into irreducible polynomials over $\mathbb{F}_2$.

3.44. Calculate $I(2,6; x)$ from the formula in Theorem 3.29.

3.45. Calculate $I(2,6; x)$ from the formula in Theorem 3.31.

3.46. Prove that

$$I(q, n; x) = \prod_{d|n} \left(x^{q^d - 1} - 1\right)^{\mu(n/d)} \quad \text{for } n > 1.$$

3.47. Prove that over a finite field of odd order q the polynomial $\frac{1}{2}(1 + x^{(q+1)/2} + (1 - x)^{(q+1)/2})$ is the square of a polynomial.

3.48. Determine all irreducible polynomials in $\mathbb{F}_2[x]$ of degree 6 and order 21 and then all irreducible polynomials in $\mathbb{F}_2[x]$ of degree 294 and order 1029.

3.49. Determine all monic irreducible polynomials in $\mathbb{F}_3[x]$ of degree 3 and order 26 and then all monic irreducible polynomials in $\mathbb{F}_3[x]$ of degree 6 and order 104.

3.50. Proceed as in Example 3.41 to determine which polynomials f_t are irreducible in $\mathbb{F}_q[x]$ in the case $q = 5$, $m = 4$, $e = 78$.

3.51. In the notation of Example 3.41, prove that if t is a prime with $t - 1$ dividing $m - 1$, then f_t is irreducible in $\mathbb{F}_2[x]$.

3.52. Given the irreducible polynomial $f(x) = x^3 - x^2 + x + 1$ over $\mathbb{F}_3$, calculate f_2 and f_5 by the matrix-theoretic method.

3.53. Calculate f_2 and f_5 in the previous exercise by using the result of Theorem 3.39.

3.54. Use a root of the primitive polynomial $x^3 - x + 1$ over $\mathbb{F}_3$ to represent all elements of $\mathbb{F}_{27}^*$ and compute the minimal polynomials over $\mathbb{F}_3$ of all elements of $\mathbb{F}_{27}$.

3.55. Let $\theta \in \mathbb{F}_{64}$ be a root of the irreducible polynomial $x^6 + x + 1$ in $\mathbb{F}_2[x]$. Find the minimal polynomial of $\beta = 1 + \theta^2 + \theta^3$ over $\mathbb{F}_2$.

3.56. Let $\theta \in \mathbb{F}_{64}$ be a root of the irreducible polynomial $x^6 + x^4 + x^3 + x + 1$ in $\mathbb{F}_2[x]$. Find the minimal polynomial of $\beta = 1 + \theta + \theta^5$ over $\mathbb{F}_2$.

3.57. Determine all primitive polynomials over $\mathbb{F}_3$ of degree 2.

3.58. Determine all primitive polynomials over $\mathbb{F}_4$ of degree 2.

3.59. Determine a primitive polynomial over $\mathbb{F}_5$ of degree 3.

3.60. Factor the polynomial $g \in \mathbb{F}_3[x]$ from Example 3.44 in $\mathbb{F}_9[x]$ to obtain primitive polynomials over $\mathbb{F}_9$.

3.61. Factor the polynomial $g \in \mathbb{F}_2[x]$ from Example 3.45 in $\mathbb{F}_8[x]$ to obtain primitive polynomials over $\mathbb{F}_8$.

3.62. Find the roots of the following linearized polynomials in their splitting fields:
(a) $L(x) = x^8 + x^4 + x^2 + x \in \mathbb{F}_2[x]$;
(b) $L(x) = x^9 + x \in \mathbb{F}_3[x]$.

3.63. Find the roots of the following polynomials in the indicated fields by first determining an affine multiple:
(a) $f(x) = x^7 + x^6 + x^3 + x^2 + 1 \in \mathbb{F}_2[x]$ in $\mathbb{F}_{32}$;
(b) $f(x) = x^4 + \theta x^3 - x^2 - (\theta + 1)x + 1 - \theta \in \mathbb{F}_9[x]$ in $\mathbb{F}_{729}$, where θ is a root of $x^2 - x - 1 \in \mathbb{F}_3[x]$.

3.64. Prove that for every polynomial f over $\mathbb{F}_{q^m}$ of positive degree there exists a nonzero q-polynomial over $\mathbb{F}_{q^m}$ that is divisible by f.

3.65. Prove that the greatest common divisor of two or more nonzero q-polynomials over $\mathbb{F}_{q^m}$ is again a q-polynomial, but that their least common multiple need not necessarily be a q-polynomial.

3.66. Determine the greatest common divisor of the following linearized polynomials:
(a) $L_1(x) = x^{64} + x^{16} + x^8 + x^4 + x^2 + x \in \mathbb{F}_2[x]$,
$L_2(x) = x^{32} + x^8 + x^2 + x \in \mathbb{F}_2[x]$;
(b) $L_1(x) = x^{243} - x^{81} - x^9 + x^3 + x \in \mathbb{F}_3[x]$,
$L_2(x) = x^{81} + x \in \mathbb{F}_3[x]$.

3.67. Determine the symbolic factorization of the following linearized polynomials into symbolically irreducible polynomials over the given prime fields:
(a) $L(x) = x^{32} + x^{16} + x^8 + x^4 + x^2 + x \in \mathbb{F}_2[x]$;
(b) $L(x) = x^{81} - x^9 - x^3 - x \in \mathbb{F}_3[x]$.

3.68. Prove that the q-polynomial $L_1(x)$ over $\mathbb{F}_{q^m}$ divides the q-polynomial $L(x)$ over $\mathbb{F}_{q^m}$ if and only if $L(x) = L_2(x) \otimes L_1(x)$ for some q-polynomial $L_2(x)$ over $\mathbb{F}_{q^m}$.

3.69. Prove that the greatest common divisor of two or more affine q-polynomials over $\mathbb{F}_{q^m}$, not all of them 0, is again an affine q-polynomial.

3.70. If $A_1(x) = L_1(x) - \alpha_1$ and $A_2(x) = L_2(x) - \alpha_2$ are affine q-polynomials over $\mathbb{F}_{q^m}$ and $A_1(x)$ divides $A_2(x)$, prove that the q-polynomial $L_1(x)$ divides the q-polynomial $L_2(x)$.

3.71. Let $f(x)$ be irreducible in $\mathbb{F}_q[x]$ with $f(0) \neq 0$ and let $F(x)$ be its linearized q-associate. Prove that $F(x)/x$ is irreducible in $\mathbb{F}_q[x]$ if and only if $f(x)$ is a primitive polynomial over $\mathbb{F}_q$ or a nonzero constant multiple of such a polynomial.

3.72. Let ζ be an element of a finite extension field of $\mathbb{F}_{q^m}$. Prove that a q-polynomial $K(x)$ over $\mathbb{F}_{q^m}$ has ζ as a root if and only if $K(x)$ is divisible by the minimal q-polynomial of ζ over $\mathbb{F}_{q^m}$.

3.73. For a nonzero polynomial $f \in \mathbb{F}_q[x]$, prove that $\Sigma \Phi_q(g) = q^{\deg(f)}$, where the sum is extended over all monic divisors $g \in \mathbb{F}_q[x]$ of f.

3.74. For a nonzero polynomial $f \in \mathbb{F}_q[x]$ and $g \in \mathbb{F}_q[x]$ with $\gcd(f, g) = 1$, prove that $g^k \equiv 1 \bmod f$, where $k = \Phi_q(f)$.

3.75. The function μ_q is defined on the set S of nonzero polynomials f over $\mathbb{F}_q$ by $\mu_q(f)=1$ if $\deg(f)=0$, $\mu_q(f)=0$ if f has at least one multiple root, and $\mu_q(f)=(-1)^k$ if $\deg(f)\geqslant 1$ and f has only simple roots, where k is the number of irreducible factors in the canonical factorization of f in $\mathbb{F}_q[x]$. Let Σ denote a sum extended over all monic divisors $g\in\mathbb{F}_q[x]$ of f. Prove the following properties:

(a) $\sum\mu_q(g)=\begin{cases}1 & \text{if } \deg(f)=0,\\ 0 & \text{if } \deg(f)\geqslant 1;\end{cases}$

(b) $\mu_q(fg)=\mu_q(f)\mu_q(g)$ for all $f, g\in S$ with $\gcd(f,g)=1$;

(c) $\Sigma q^{\deg(g)}\mu_q(f/g)=\Phi_q(f)$ for all $f\in S$;

(d) if ψ is a mapping from S into an additively written abelian group G with $\psi(cf)=\psi(f)$ for all $c\in\mathbb{F}_q^*$ and $f\in S$, and if $\Psi(f)=\Sigma\psi(g)$ for all $f\in S$, then $\psi(f)=\Sigma\mu_q(f/g)\Psi(g)=\Sigma\mu_q(g)\Psi(f/g)$ for all $f\in S$.

3.76. Prove that the number of different normal bases of $\mathbb{F}_{q^m}$ over $\mathbb{F}_q$ is

$$\frac{1}{m}\prod_{d\mid m}\left(q^{\phi(d)}-1\right)$$

provided that $\gcd(m,q)=1$ and the multiplicative order of q modulo m is $\phi(m)$.

3.77. Refer to Example 2.31 for the definition of a self-dual basis and show that there exists a self-dual normal basis of $\mathbb{F}_{2^m}$ over $\mathbb{F}_2$ whenever m is odd. (*Hint*: Show first that the number of different normal bases of $\mathbb{F}_{2^m}$ over $\mathbb{F}_2$ is odd whenever m is odd.)

3.78. For a prime r and $a\in\mathbb{F}_q$, prove that x^r-a is either irreducible in $\mathbb{F}_q[x]$ or has a root in $\mathbb{F}_q$.

3.79. For an odd prime r, an integer $n\geqslant 1$, and $a\in\mathbb{F}_q$, prove that $x^{r^n}-a$ is irreducible in $\mathbb{F}_q[x]$ if and only if a is not an rth power of an element of $\mathbb{F}_q$.

3.80. Find the canonical factorization of the following binomials over the given prime fields:

(a) $f(x)=x^8+1\in\mathbb{F}_3[x]$;

(b) $f(x)=x^{27}-4\in\mathbb{F}_{19}[x]$;

(c) $f(x)=x^{88}-10\in\mathbb{F}_{23}[x]$.

3.81. Prove that under the conditions of Theorem 3.76 the roots of the polynomial $F(x)$ introduced there are simple.

3.82. Prove that the resultant of two binomials x^n-a and x^m-b in $\mathbb{F}_q[x]$ is given by $(-1)^n(b^{n/d}-a^{m/d})^d$ with $d=\gcd(n,m)$, where n and m are considered to be the formal degrees of the binomials (compare with Definition 1.93).

3.83. For a nonzero element b of a prime field $\mathbb{F}_p$, prove that the trinomial x^p-x-b is irreducible in $\mathbb{F}_{p^n}[x]$ if and only if n is not divisible by p.

3.84. Prove that any polynomial of the form $x^q - ax - b \in \mathbb{F}_q[x]$ with $a \neq 1$ has a root in $\mathbb{F}_q$.

3.85. Prove: if $x^p - x - a$ is irreducible over the field $\mathbb{F}_q$ of characteristic p and β is a root of this trinomial in an extension field of $\mathbb{F}_q$, then $x^p - x - a\beta^{p-1}$ is irreducible over $\mathbb{F}_q(\beta)$.

3.86. Prove: if $f(x) = x^m + a_{m-1}x^{m-1} + \cdots + a_0$ is irreducible over the field $\mathbb{F}_q$ of characteristic p and $b \in \mathbb{F}_q$ is such that $\mathrm{Tr}_{\mathbb{F}_q}(mb - a_{m-1}) = 0$, then $f(x^p - x - b)$ is the product of p irreducible polynomials over $\mathbb{F}_q$ of degree m.

3.87. If m and p are distinct primes and the multiplicative order of p modulo m is $m-1$, prove that $\sum_{i=0}^{m-1}(x^p - x)^i$ is irreducible over $\mathbb{F}_p$.

3.88. Find the canonical factorization of the given polynomial over the indicated field:
(a) $f(x) = x^8 - \alpha x - 1 \in \mathbb{F}_{64}[x]$, where α satisfies $\alpha^3 = \alpha + 1$;
(b) $f(x) = x^9 - \alpha x + \alpha \in \mathbb{F}_9[x]$, where α satisfies $\alpha^2 = \alpha + 1$.

3.89. Let $A(x) = L(x) - a \in \mathbb{F}_q[x]$ be an affine p-polynomial of degree $r > 2$, and suppose the p-polynomial $L(x)$ is such that $L(x)/x$ is irreducible over $\mathbb{F}_q$. Prove that $A(x)$ is the product of a linear polynomial and an irreducible polynomial over $\mathbb{F}_q$ of degree $r-1$.

3.90. Prove: the trinomial $x^n + ax^k + b \in \mathbb{F}_q[x]$, $n > k \geqslant 1$, q even, has multiple roots if and only if n and k are both even.

3.91. Prove that the degree of every irreducible factor of $x^{2^n} + x + 1$ in $\mathbb{F}_2[x]$ divides $2n$.

3.92. Prove that the degree of every irreducible factor of $x^{2^n+1} + x + 1$ in $\mathbb{F}_2[x]$ divides $3n$.

3.93. Recall the notion of a self-reciprocal polynomial defined in Exercise 3.13. Prove that if $f \in \mathbb{F}_2[x]$ is a self-reciprocal polynomial of positive degree, then f divides a trinomial in $\mathbb{F}_2[x]$ only if $\mathrm{ord}(f)$ is a multiple of 3. Prove also that the converse holds if f is irreducible over $\mathbb{F}_2$.

3.94. Prove that for odd $d \in \mathbb{N}$ the cyclotomic polynomial $Q_d \in \mathbb{F}_2[x]$ divides a trinomial in $\mathbb{F}_2[x]$ if and only if d is a multiple of 3.

3.95. Let $f(x) = x^n + ax^k + b \in \mathbb{F}_q[x]$, $n > k \geqslant 1$, be a trinomial and let $m \in \mathbb{N}$ be a multiple of $\mathrm{ord}(f)$. Prove that $f(x)$ divides the trinomial $g(x) = x^{m-k} + b^{-1}x^{n-k} + ab^{-1}$.

3.96. Prove that the trinomial $x^{2n} + x^n + 1$ is irreducible over $\mathbb{F}_2$ if and only if $n = 3^k$ for some nonnegative integer k.

3.97. Prove that the trinomial $x^{4n} + x^n + 1$ is irreducible over $\mathbb{F}_2$ if and only if $n = 3^k 5^m$ for some nonnegative integers k and m.

Chapter 4

Factorization of Polynomials

Any nonconstant polynomial over a field can be expressed as a product of irreducible polynomials. In the case of finite fields, some reasonably efficient algorithms can be devised for the actual calculation of the irreducible factors of a given polynomial of positive degree.

The availability of feasible factorization algorithms for polynomials over finite fields is important for coding theory and for the study of linear recurrence relations in finite fields. Beyond the realm of finite fields, there are various computational problems in algebra and number theory that depend in one way or another on the factorization of polynomials over finite fields. We mention the factorization of polynomials over the ring of integers, the determination of the decomposition of rational primes in algebraic number fields, the calculation of the Galois group of an equation over the rationals, and the construction of field extensions.

We shall present several algorithms for the factorization of polynomials over finite fields. The decision on the choice of algorithm for a specific factorization problem usually depends on whether the underlying finite field is "small" or "large." In Section 1 we describe those algorithms that are better adapted to "small" finite fields and in the next section those that work better for "large" finite fields. Some of these algorithms reduce the problem of factoring polynomials to that of finding the roots of certain other polynomials. Therefore, Section 3 is devoted to the discussion of the latter problem from the computational viewpoint.

1. FACTORIZATION OVER SMALL FINITE FIELDS

Any polynomial $f \in \mathbb{F}_q[x]$ of positive degree has a canonical factorization in $\mathbb{F}_q[x]$ by Theorem 1.59. For the discussion of factorization algorithms it will suffice to consider only monic polynomials. Our goal is thus to express a monic polynomial $f \in \mathbb{F}_q[x]$ of positive degree in the form

$$f = f_1^{e_1} \cdots f_k^{e_k}, \tag{4.1}$$

where $f_1, \ldots, f_k$ are distinct monic irreducible polynomials in $\mathbb{F}_q[x]$ and $e_1, \ldots, e_k$ are positive integers.

First we simplify our task by showing that the problem can be reduced to that of factoring a polynomial *with no repeated factors*, which means that the exponents $e_1, \ldots, e_k$ in (4.1) are all equal to 1 (or, equivalently, that the polynomial has no multiple roots). To this end, we calculate

$$d(x) = \gcd(f(x), f'(x)),$$

the greatest common divisor of $f(x)$ and its derivative, by the Euclidean algorithm.

If $d(x) = 1$, then we know that $f(x)$ has no repeated factors because of Theorem 1.68. If $d(x) = f(x)$, we must have $f'(x) = 0$. Hence $f(x) = g(x)^p$, where $g(x)$ is a suitable polynomial in $\mathbb{F}_q[x]$ and p is the characteristic of $\mathbb{F}_q$. If necessary, the reduction process can be continued by applying the method to $g(x)$.

If $d(x) \neq 1$ and $d(x) \neq f(x)$, then $d(x)$ is a nontrivial factor of $f(x)$ and $f(x)/d(x)$ has no repeated factors. The factorization of $f(x)$ is achieved by factoring $d(x)$ and $f(x)/d(x)$ separately. In case $d(x)$ still has repeated factors, further applications of the reduction process will have to be carried out.

By applying this process sufficiently often, the original problem is reduced to that of factoring a certain number of polynomials with no repeated factors. The canonical factorizations of these polynomials lead directly to the canonical factorization of the original polynomial. Therefore, we may restrict the attention to polynomials with no repeated factors. The following theorem is crucial.

4.1. Theorem. *If $f \in \mathbb{F}_q[x]$ is monic and $h \in \mathbb{F}_q[x]$ is such that $h^q \equiv h \bmod f$, then*

$$f(x) = \prod_{c \in \mathbb{F}_q} \gcd(f(x), h(x) - c). \tag{4.2}$$

Proof. Each greatest common divisor on the right-hand side of (4.2) divides $f(x)$. Since the polynomials $h(x) - c$, $c \in \mathbb{F}_q$, are pairwise relatively prime, so are the greatest common divisors with $f(x)$, and thus the product of these greatest common divisors divides $f(x)$. On the other hand, $f(x)$

divides

$$h(x)^q - h(x) = \prod_{c \in \mathbb{F}_q} (h(x) - c),$$

and so $f(x)$ divides the right-hand side of (4.2). Thus, the two sides of (4.2) are monic polynomials that divide each other, and therefore they must be equal. □

In general, (4.2) does not yield the complete factorization of f since $\gcd(f(x), h(x) - c)$ may be reducible in $\mathbb{F}_q[x]$. If $h(x) \equiv c \bmod f(x)$ for some $c \in \mathbb{F}_q$, then Theorem 4.1 gives a trivial factorization of f and therefore is of no use. However, if h is such that Theorem 4.1 yields a nontrivial factorization of f, we say that h is an *f-reducing polynomial*. Any h with $h^q \equiv h \bmod f$ and $0 < \deg(h) < \deg(f)$ is obviously f-reducing. In order to obtain factorization algorithms on the basis of Theorem 4.1, we have to find methods of constructing f-reducing polynomials. It should be clear at this stage already that since the factorization provided by (4.2) depends on the calculation of q greatest common divisors, a direct application of this formula will only be feasible for small finite fields $\mathbb{F}_q$.

The first method of constructing f-reducing polynomials makes use of the Chinese remainder theorem for polynomials (see Exercise 1.37). Let us assume that f has no repeated factors, so that $f = f_1 \cdots f_k$ is a product of distinct monic irreducible polynomials over $\mathbb{F}_q$. If $(c_1, \ldots, c_k)$ is any k-tuple of elements of $\mathbb{F}_q$, the Chinese remainder theorem implies that there is a unique $h \in \mathbb{F}_q[x]$ with $h(x) \equiv c_i \bmod f_i(x)$ for $1 \leqslant i \leqslant k$ and $\deg(h) < \deg(f)$. The polynomial $h(x)$ satisfies the condition

$$h(x)^q \equiv c_i^q = c_i \equiv h(x) \bmod f_i(x) \quad \text{for } 1 \leqslant i \leqslant k,$$

and therefore

$$h^q \equiv h \bmod f, \qquad \deg(h) < \deg(f). \tag{4.3}$$

On the other hand, if h is a solution of (4.3), then the identity

$$h(x)^q - h(x) = \prod_{c \in \mathbb{F}_q} (h(x) - c)$$

implies that every irreducible factor of f divides one of the polynomials $h(x) - c$. Thus, all solutions of (4.3) satisfy $h(x) \equiv c_i \bmod f_i(x)$, $1 \leqslant i \leqslant k$, for some k-tuple $(c_1, \ldots, c_k)$ of elements of $\mathbb{F}_q$. Consequently, there are exactly q^k solutions of (4.3).

We find these solutions by reducing (4.3) to a system of linear equations. With $n = \deg(f)$ we construct the $n \times n$ matrix $B = (b_{ij})$, $0 \leqslant i, j \leqslant n - 1$, by calculating the powers $x^{iq} \bmod f(x)$. Specifically, let

$$x^{iq} \equiv \sum_{j=0}^{n-1} b_{ij} x^j \bmod f(x) \quad \text{for } 0 \leqslant i \leqslant n-1. \tag{4.4}$$

Then $h(x)=a_0+a_1x+\cdots+a_{n-1}x^{n-1}\in\mathbb{F}_q[x]$ is a solution of (4.3) if and only if

$$(a_0,a_1,\ldots,a_{n-1})B=(a_0,a_1,\ldots,a_{n-1}). \tag{4.5}$$

This follows from the fact that (4.5) holds if and only if

$$\begin{aligned} h(x) &= \sum_{j=0}^{n-1} a_j x^j \\ &= \sum_{j=0}^{n-1}\sum_{i=0}^{n-1} a_i b_{ij} x^j \\ &\equiv \sum_{i=0}^{n-1} a_i x^{iq} = h(x)^q \bmod f(x). \end{aligned}$$

The system (4.5) may be written in the equivalent form

$$(a_0,a_1,\ldots,a_{n-1})(B-I)=(0,0,\ldots,0), \tag{4.6}$$

where I is the $n\times n$ identity matrix over $\mathbb{F}_q$. By the considerations above, the system (4.6) has q^k solutions. Thus, *the dimension of the null space of the matrix* $B-I$ *is* k, the number of distinct monic irreducible factors of f, and *the rank of* $B-I$ *is* $n-k$.

Since the constant polynomial $h_1(x)=1$ is always a solution of (4.3), the vector $(1,0,\ldots,0)$ is always a solution of (4.6), as can also be checked directly. There will exist polynomials $h_2(x),\ldots,h_k(x)$ of degree $\leqslant n-1$ such that the vectors corresponding to $h_1(x), h_2(x),\ldots,h_k(x)$ form a basis for the null space of $B-I$. The polynomials $h_2(x),\ldots,h_k(x)$ have positive degree and are thus f-reducing.

In this approach, an important role is played by the determination of the rank r of the matrix $B-I$. We have $r=n-k$ as noted above, so that once the rank r is found, we know that *the number of distinct monic irreducible factors of* f *is given by* $n-r$. On the basis of this information we can then decide when the factorization procedure can be stopped. The rank of $B-I$ can be determined by using row and column operations to reduce the matrix to echelon form. However, since we also want to solve the system (4.6), it is advisable to use only column operations because they leave the null space invariant. Thus, we are allowed to multiply any column of the matrix $B-I$ by a nonzero element of $\mathbb{F}_q$ and to add any multiple of one of its columns to a different column. The rank r is the number of nonzero columns in the column echelon form.

Having found r, we form $k=n-r$. If $k=1$, we know that f is irreducible over $\mathbb{F}_q$ and the procedure terminates. In this case, the only solutions of (4.3) are the constant polynomials and the null space of $B-I$ contains only the vectors of the form $(c,0,\ldots,0)$ with $c\in\mathbb{F}_q$. If $k\geqslant 2$, we take the f-reducing basis polynomial $h_2(x)$ and calculate

$\gcd(f(x), h_2(x)-c)$ for all $c \in \mathbb{F}_q$. The result will be a nontrivial factorization of $f(x)$ afforded by (4.2). If the use of $h_2(x)$ does not succeed in splitting $f(x)$ into k factors, we calculate $\gcd(g(x), h_3(x)-c)$ for all $c \in \mathbb{F}_q$ and all nontrivial factors $g(x)$ found so far. This procedure is continued until k factors of $f(x)$ are obtained.

The process described above must eventually yield all the factors. For if we consider two distinct monic irreducible factors of $f(x)$, say $f_1(x)$ and $f_2(x)$, then by the argument following (4.3) there exist elements c_{j1}, $c_{j2} \in \mathbb{F}_q$ such that $h_j(x) \equiv c_{j1} \bmod f_1(x)$, $h_j(x) \equiv c_{j2} \bmod f_2(x)$ for $1 \leqslant j \leqslant k$. Suppose we had $c_{j1} = c_{j2}$ for $1 \leqslant j \leqslant k$. Then, since any solution $h(x)$ of (4.3) is a linear combination of $h_1(x), \ldots, h_k(x)$ with coefficients in $\mathbb{F}_q$, there would exist for any such $h(x)$ an element $c \in \mathbb{F}_q$ with $h(x) \equiv c \bmod f_1(x)$, $h(x) \equiv c \bmod f_2(x)$. But the argument leading to (4.3) shows, in particular, that there is a solution $h(x)$ of (4.3) with $h(x) \equiv 0 \bmod f_1(x)$, $h(x) \equiv 1 \bmod f_2(x)$. This contradiction proves that $c_{j1} \neq c_{j2}$ for some j with $1 \leqslant j \leqslant k$ (in fact, since $h_1(x) = 1$, we will have $j \geqslant 2$). Therefore, $h_j(x) - c_{j1}$ will be divisible by $f_1(x)$, but not by $f_2(x)$. Hence any two distinct monic irreducible factors of $f(x)$ will be separated by some $h_j(x)$.

This factorization algorithm based on determining f-reducing polynomials by solving the system (4.6) is called *Berlekamp's algorithm*.

4.2. Example. Factor $f(x) = x^8 + x^6 + x^4 + x^3 + 1$ over $\mathbb{F}_2$ by Berlekamp's algorithm. Since $\gcd(f(x), f'(x)) = 1$, $f(x)$ has no repeated factors. We have to compute $x^{iq} \bmod f(x)$ for $q = 2$ and $0 \leqslant i \leqslant 7$. This yields the following congruences mod $f(x)$:

$$
\begin{aligned}
x^0 &\equiv 1 \\
x^2 &\equiv x^2 \\
x^4 &\equiv x^4 \\
x^6 &\equiv x^6 \\
x^8 &\equiv 1 + x^3 + x^4 + x^6 \\
x^{10} &\equiv 1 + x^2 + x^3 + x^4 + x^5 \\
x^{12} &\equiv x^2 + x^4 + x^5 + x^6 + x^7 \\
x^{14} &\equiv 1 + x + x^3 + x^4 + x^5
\end{aligned}
$$

Therefore, the 8×8 matrix B is given by

$$
B = \begin{pmatrix}
1 & 0 & 0 & 0 & 0 & 0 & 0 & 0 \\
0 & 0 & 1 & 0 & 0 & 0 & 0 & 0 \\
0 & 0 & 0 & 0 & 1 & 0 & 0 & 0 \\
0 & 0 & 0 & 0 & 0 & 0 & 1 & 0 \\
1 & 0 & 0 & 1 & 1 & 0 & 1 & 0 \\
1 & 0 & 1 & 1 & 1 & 1 & 0 & 0 \\
0 & 0 & 1 & 0 & 1 & 1 & 1 & 1 \\
1 & 1 & 0 & 1 & 1 & 1 & 0 & 0
\end{pmatrix}
$$

and $B - I$ is given by

$$B - I = \begin{pmatrix} 0 & 0 & 0 & 0 & 0 & 0 & 0 & 0 \\ 0 & 1 & 1 & 0 & 0 & 0 & 0 & 0 \\ 0 & 0 & 1 & 0 & 1 & 0 & 0 & 0 \\ 0 & 0 & 0 & 1 & 0 & 0 & 1 & 0 \\ 1 & 0 & 0 & 1 & 0 & 0 & 1 & 0 \\ 1 & 0 & 1 & 1 & 1 & 0 & 0 & 0 \\ 0 & 0 & 1 & 0 & 1 & 1 & 0 & 1 \\ 1 & 1 & 0 & 1 & 1 & 1 & 0 & 1 \end{pmatrix}.$$

The matrix $B - I$ has rank 6, and the two vectors $(1,0,0,0,0,0,0,0)$ and $(0,1,1,0,0,1,1,1)$ form a basis of the null space of $B - I$. The corresponding polynomials are $h_1(x) = 1$ and $h_2(x) = x + x^2 + x^5 + x^6 + x^7$. We calculate $\gcd(f(x), h_2(x) - c)$ for $c \in \mathbb{F}_2$ by the Euclidean algorithm and obtain $\gcd(f(x), h_2(x)) = x^6 + x^5 + x^4 + x + 1$, $\gcd(f(x), h_2(x) - 1) = x^2 + x + 1$. The desired canonical factorization is therefore

$$f(x) = (x^6 + x^5 + x^4 + x + 1)(x^2 + x + 1). \qquad \square$$

A second method of obtaining f-reducing polynomials is based on the explicit construction of a family of polynomials among which at least one f-reducing polynomial can be found. Let f be again a monic polynomial of degree n with no repeated factors. Let $f = f_1 \cdots f_k$ be its canonical factorization in $\mathbb{F}_q[x]$ with $\deg(f_j) = n_j$ for $1 \leqslant j \leqslant k$. If N is the least positive integer with $x^{q^N} \equiv x \bmod f(x)$, then it follows from Theorem 3.20 that $N = \operatorname{lcm}(n_1, \ldots, n_k)$, and it is also easily seen that N is the degree of the splitting field F of f over $\mathbb{F}_q$. Let the polynomial $T \in \mathbb{F}_q[x]$ be given by $T(x) = x + x^q + x^{q^2} + \cdots + x^{q^{N-1}}$ and define $T_i(x) = T(x^i)$ for $i = 0, 1, \ldots$. The following result guarantees that in the case of interest, namely, when f is reducible, there are f-reducing polynomials among the T_i.

4.3. Theorem. *If f is reducible in $\mathbb{F}_q[x]$, then at least one of the polynomials T_i, $1 \leqslant i \leqslant n - 1$, is f-reducing.*

Proof. It is immediate that any polynomial T_i satisfies $T_i^q \equiv T_i \bmod f$. Suppose now that for all T_i, $1 \leqslant i \leqslant n - 1$, the factorization of f afforded by (4.2) were trivial. This means that there exist elements $c_1, \ldots, c_{n-1} \in \mathbb{F}_q$ such that $T_i(x) \equiv c_i \bmod f(x)$ for $1 \leqslant i \leqslant n - 1$. With $c_0 = N$, viewed as an element of $\mathbb{F}_q$, we get $T(x^i) \equiv c_i \bmod f(x)$ for $0 \leqslant i \leqslant n - 1$. For any

$$g(x) = \sum_{i=1}^{n-1} a_i x^i \in \mathbb{F}_q[x]$$

of degree less than n we have then

$$T(g(x)) = T\left(\sum_{i=0}^{n-1} a_i x^i\right) = \sum_{i=0}^{n-1} a_i T(x^i) \equiv \sum_{i=0}^{n-1} a_i c_i \bmod f(x).$$

Putting

$$c(g)=\sum_{i=0}^{n-1} a_i c_i \in \mathbb{F}_q,$$

we obtain

$$T(g(x)) \equiv c(g) \bmod f_j(x) \quad \text{for } 1 \leqslant j \leqslant k. \tag{4.7}$$

Since $N=\operatorname{lcm}(n_1,\ldots,n_k)$, at least one of the integers N/n_j, say N/n_1, is not divisible by the characteristic of $\mathbb{F}_q$. Let θ_1 be a root of f_1 in the splitting field F_1 of f_1 over $\mathbb{F}_q$. Because of Theorem 2.23(iii) there exists $g_1 \in \mathbb{F}_q[x]$ with

$$\operatorname{Tr}_{F_1/\mathbb{F}_q}(g_1(\theta_1))=1. \tag{4.8}$$

Since $k \geqslant 2$ by assumption, we can apply the Chinese remainder theorem to obtain a polynomial $g \in \mathbb{F}_q[x]$ of degree $<n$ with

$$g \equiv g_1 \bmod f_1,\ g \equiv 0 \bmod f_2. \tag{4.9}$$

From (4.8) and (4.9) we deduce that

$$\operatorname{Tr}_{F_1/\mathbb{F}_q}(g(\theta_1))=1,$$

and Theorems 2.23(iv) and 2.26 imply that

$$\operatorname{Tr}_{F/\mathbb{F}_q}(g(\theta_1))=N/n_1.$$

Because of the definitions of the trace and of the element θ_1, it follows that

$$T(g(x)) \equiv N/n_1 \bmod f_1(x).$$

However, the second congruence in (4.9) leads to $T(g(x)) \equiv 0 \bmod f_2(x)$, and since $N/n_1 \neq 0$ as an element of $\mathbb{F}_q$, we get a contradiction to (4.7). Therefore, at least one of the T_i, $1 \leqslant i \leqslant n-1$, is f-reducing. □

4.4. Example. Factor $f(x)=x^{17}+x^{14}+x^{13}+x^{12}+x^{11}+x^{10}+x^9+x^8+x^7+x^5+x^4+x+1$ over $\mathbb{F}_2$. We have $\gcd(f(x), f'(x))=x^{10}+x^8+1$, and so $f_0(x)=f(x)/\gcd(f(x),f'(x))=x^7+x^5+x^4+x+1$ has no repeated factors. We factor f_0 by finding an f_0-reducing polynomial of the type described above. To this end, we calculate the powers $x, x^2, x^4, \ldots \bmod f_0(x)$ until we obtain the least positive integer N with $x^{2^N} \equiv x \bmod f_0(x)$. We simplify the notation by identifying a polynomial $\sum_{i=0}^{n-1} a_i x^i$ with the n-tuple $a_0 a_1 \cdots a_{n-1}$ of its coefficients, so that, for instance, $f_0(x)=11001101$. The calculation of the required powers of $x \bmod f_0(x)$ is facilitated by the observation that squaring a polynomial $a_0 a_1 \cdots a_6 \bmod f_0(x)$ is the same as multiplying the vector $a_0 a_1 \cdots a_6$ by the 7×7 matrix of even powers

$x^0, x^2, \ldots, x^{12} \bmod f_0(x)$. This matrix is obtained from

$$\begin{array}{rcccccc}
x^0 \equiv 1 & 0 & 0 & 0 & 0 & 0 & 0 \\
x^2 \equiv 0 & 0 & 1 & 0 & 0 & 0 & 0 \\
x^4 \equiv 0 & 0 & 0 & 0 & 1 & 0 & 0 \\
x^6 \equiv 0 & 0 & 0 & 0 & 0 & 0 & 1 \\
x^8 \equiv 0 & 1 & 1 & 0 & 0 & 1 & 1 \\
x^{10} \equiv 1 & 0 & 1 & 1 & 0 & 0 & 1 \\
x^{12} \equiv 0 & 1 & 0 & 0 & 1 & 0 & 1
\end{array}$$

where all the congruences are mod $f_0(x)$. Therefore we get mod $f_0(x)$:

$$\begin{array}{rcccccc}
x \equiv 0 & 1 & 0 & 0 & 0 & 0 & 0 \\
x^2 \equiv 0 & 0 & 1 & 0 & 0 & 0 & 0 \\
x^4 \equiv 0 & 0 & 0 & 0 & 1 & 0 & 0 \\
x^8 \equiv 0 & 1 & 1 & 0 & 0 & 1 & 1 \\
x^{16} \equiv 1 & 1 & 0 & 1 & 0 & 0 & 0 \\
x^{32} \equiv 1 & 0 & 1 & 0 & 0 & 0 & 1 \\
x^{64} \equiv 1 & 1 & 0 & 0 & 0 & 0 & 1 \\
x^{128} \equiv 1 & 1 & 1 & 0 & 1 & 0 & 1 \\
x^{256} \equiv 1 & 0 & 0 & 0 & 0 & 1 & 0 \\
x^{512} \equiv 0 & 0 & 1 & 1 & 0 & 0 & 1 \\
x^{1024} \equiv 0 & 1 & 0 & 0 & 0 & 0 & 0
\end{array}$$

Thus $N = 10$ and

$$T_1(x) = \sum_{j=0}^{9} x^{2^j} \equiv 1 \quad 1 \quad 1 \quad 0 \quad 0 \quad 0 \quad 1 \bmod f_0(x).$$

Since $T_1(x)$ is not congruent to a constant mod $f_0(x)$, $T_1(x)$ is f_0-reducing. We have

$$\begin{aligned}
\gcd(f_0(x), T_1(x)) &= \gcd(1\ 1\ 0\ 0\ 1\ 1\ 0\ 1, 1\ 1\ 1\ 0\ 0\ 0\ 1) \\
&= x^5 + x^4 + x^3 + x^2 + 1, \\
\gcd(f_0(x), T_1(x) - 1) &= \gcd(1\ 1\ 0\ 0\ 1\ 1\ 0\ 1, 0\ 1\ 1\ 0\ 0\ 0\ 1) \\
&= x^2 + x + 1,
\end{aligned}$$

and so

$$f_0(x) = (x^5 + x^4 + x^3 + x^2 + 1)(x^2 + x + 1).$$

The second factor is obviously irreducible in $\mathbb{F}_2[x]$. Since $N = 10$ is the least common multiple of the degrees of the irreducible factors of $f_0(x)$, any nontrivial factorization of the first factor would lead to a value of N different from 10, so that the first factor is also irreducible in $\mathbb{F}_2[x]$.

It remains to factor $\gcd(f(x), f'(x)) = x^{10} + x^8 + 1$. We have $x^{10} + x^8 + 1 = (x^5 + x^4 + 1)^2$, and by checking whether $x^5 + x^4 + 1$ is divisible by one of the irreducible factors of $f_0(x)$, we find that $x^5 + x^4 + 1 = (x^3 + x + 1)(x^2 + x + 1)$, with $x^3 + x + 1$ irreducible in $\mathbb{F}_2[x]$. Hence

$$f(x) = (x^5 + x^4 + x^3 + x^2 + 1)(x^3 + x + 1)^2(x^2 + x + 1)^3$$

is the canonical factorization of $f(x)$ in $\mathbb{F}_2[x]$. □

It should be noted that, in general, the f-reducing polynomials T_i do not yield the complete factorization of f since the T_i are not able to separate those irreducible factors f_j for which N/n_j is divisible by the characteristic of $\mathbb{F}_q$. In practice, however, one calculates the first f-reducing T_i and then calculates new T_i for each of the resulting factors. In this way, one eventually obtains the complete factorization of f.

It is, however, possible to construct a related set of polynomials R_i that are capable of separating all the irreducible factors of f at once. We assume, without loss of generality, that $f(0) \neq 0$. Let $\text{ord}(f(x)) = e$, so that $f(x)$ divides $x^e - 1$. Since f has no repeated factors, e and q are relatively prime by Corollary 3.4 and Theorem 3.9. For each $i \geqslant 0$ let m_i be the least positive integer with

$$x^{iq^{m_i}} \equiv x^i \bmod f(x). \tag{4.10}$$

Then we define

$$R_i(x) = x^i + x^{iq} + x^{iq^2} + \cdots + x^{iq^{m_i - 1}}.$$

Since (4.10) is equivalent to

$$iq^{m_i} \equiv i \bmod e, \tag{4.11}$$

which is in turn equivalent to $q^{m_i} \equiv 1 \bmod (e/\gcd(e, i))$, it follows that m_i can also be described as the multiplicative order of q modulo $e/\gcd(e, i)$. A comparison with the definition of $T_i(x)$ shows that

$$T_i(x) \equiv \frac{N}{m_i} R_i(x) \bmod f(x).$$

It is clear that $R_i^q \equiv R_i \bmod f$ for all i, so that the R_i can be used in (4.2) in place of h. We prove now the claim about the R_i made above.

4.5. Theorem. *Let f be monic and reducible in $\mathbb{F}_q[x]$ with no repeated factors, and suppose that $f(0) \neq 0$ and $\text{ord}(f) = e$. Then, if all the polynomials R_i, $1 \leqslant i \leqslant e - 1$, are used in* (4.2), *they will separate all irreducible factors of f.*

Proof. Let $h(x) = \sum_{i=0}^{e-1} a_i x^i \in \mathbb{F}_q[x]$ be a solution of $h(x)^q \equiv h(x) \bmod (x^e - 1)$. If we interpret subscripts mod e, then $h(x) \equiv \sum_{i=0}^{e-1} a_{iq} x^{iq} \bmod (x^e - 1)$ since iq, $i = 0, 1, \ldots, e - 1$, runs through all residues mod e as q and e are relatively prime. Since $h(x)^q = \sum_{i=0}^{e-1} a_i x^{iq}$, we get

$$\sum_{i=0}^{e-1} a_i x^{iq} \equiv \sum_{i=0}^{e-1} a_{iq} x^{iq} \operatorname{mod}(x^e - 1).$$

By considering the exponents mod e, it follows that corresponding coefficients are identical. Thus $a_i = a_{iq}$ for all i, and so $a_i = a_{iq} = a_{iq^2} = \cdots$ for all i. Since m_i is the least positive integer for which (4.11) holds, we obtain

$$h(x) \equiv \sum_{i \in J} a_i R_i(x) \operatorname{mod}(x^e - 1),$$

where the set J contains exactly one representative from each equivalence class of residues mod e determined by the equivalence relation $\sim$ which is defined by $i_1 \sim i_2$ if and only if $i_1 \equiv i_2 q^t \bmod e$ for some $t \geqslant 0$. Thus, for suitable $b_i \in \mathbb{F}_q$ we have

$$h(x) \equiv \sum_{i=0}^{e-1} b_i R_i(x) \operatorname{mod}(x^e - 1). \tag{4.12}$$

Let now $f_1(x)$ and $f_2(x)$ be two distinct monic irreducible factors of $f(x)$, and so of $x^e - 1$. By the argument leading to (4.3), there is a solution $h(x) \in \mathbb{F}_q[x]$ of $h(x)^q \equiv h(x) \operatorname{mod}(x^e - 1)$, $\deg(h(x)) < e$, with

$$h(x) \equiv 0 \bmod f_1(x), \qquad h(x) \equiv 1 \bmod f_2(x). \tag{4.13}$$

Since $R_i^q \equiv R_i \bmod f$, the argument subsequent to (4.3) shows that there exist elements $c_{i1}, c_{i2} \in \mathbb{F}_q$ with $R_i(x) \equiv c_{i1} \bmod f_1(x)$, $R_i(x) \equiv c_{i2} \bmod f_2(x)$ for $0 \leqslant i \leqslant e-1$. If we had $c_{i1} = c_{i2}$ for $0 \leqslant i \leqslant e-1$, then it would follow from (4.12) that $h(x) \equiv c \bmod f_1(x)$, $h(x) \equiv c \bmod f_2(x)$ for some $c \in \mathbb{F}_q$, a contradiction to (4.13). Thus $c_{i1} \neq c_{i2}$ for some i with $0 \leqslant i \leqslant e-1$, and since $R_0(x) = 1$, we must have $i \geqslant 1$. Then $R_i(x) - c_{i1}$ will be divisible by $f_1(x)$, but not by $f_2(x)$. Hence the use of this $R_i(x)$ in (4.2) will separate $f_1(x)$ from $f_2(x)$. □

The argument in the proof of Theorem 4.5 shows, of course, that the polynomials R_i, with i running through the nonzero elements of the set J, are already separating all irreducible factors of f. However, the determination of the set J depends on knowing the order e, and a direct calculation of e (i.e., one that does not have recourse to the canonical factorization of f) will be lengthy in most cases.

This problem does not arise in the special cases $f(x) = x^e - 1$ and $f(x) = Q_e(x)$, the eth cyclotomic polynomial, since it is trivial that $\operatorname{ord}(x^e - 1) = \operatorname{ord}(Q_e(x)) = e$. The polynomials R_i are, in fact, well suited for factoring these binomials and cyclotomic polynomials.

4.6. Example. We determine the canonical factorization of the cyclotomic polynomial $Q_{52}(x)$ in $\mathbb{F}_3[x]$. According to Theorem 3.27 we have

$$\begin{aligned} Q_{52}(x) &= \frac{(x^{52}-1)(x^2-1)}{(x^{26}-1)(x^4-1)} \\ &= x^{24}-x^{22}+x^{20}-x^{18}+x^{16}-x^{14}+x^{12} \\ &\quad -x^{10}+x^8-x^6+x^4-x^2+1. \end{aligned}$$

Now $R_1(x) = x + x^3 + x^9 + x^{27} + x^{81} + x^{243}$, and since $x^{26} \equiv -1 \bmod Q_{52}(x)$, we get $R_1(x) \equiv 0 \bmod Q_{52}(x)$, so that R_1 is not Q_{52}-reducing. With $R_2(x) = x^2 + x^6 + x^{18}$ we get

$$\begin{aligned} \gcd(Q_{52}(x), R_2(x)) &= x^6 - x^2 + 1, \\ \gcd(Q_{52}(x), R_2(x)+1) &= x^6 + x^4 - x^2 + 1, \\ \gcd(Q_{52}(x), R_2(x)-1) &= x^{12} + x^{10} - x^8 + x^6 + x^4 + x^2 + 1 = g(x), \end{aligned}$$

say, so that (4.2) yields

$$Q_{52}(x) = (x^6 - x^2 + 1)(x^6 + x^4 - x^2 + 1)g(x).$$

By Theorem 2.47(ii), $Q_{52}(x)$ is the product of four irreducible factors in $\mathbb{F}_3[x]$ of degree 6. Thus, it remains to factor $g(x)$. Since $R_3(x) = x^3 + x^9 + x^{27} + x^{81} + x^{243} + x^{729} \equiv 0 \bmod Q_{52}(x)$, we next use $R_4(x) = x^4 + x^{12} + x^{36}$. We note that $x^{12} \equiv -x^{10} + x^8 - x^6 - x^4 - x^2 - 1 \bmod g(x)$, $x^{36} \equiv -x^{10} \bmod g(x)$, and so

$$R_4(x) \equiv x^{10} + x^8 - x^6 - x^2 - 1 \bmod g(x).$$

Therefore,

$$\begin{aligned} \gcd(g(x), R_4(x)) &= \gcd(g(x), x^{10} + x^8 - x^6 - x^2 - 1) = 1, \\ \gcd(g(x), R_4(x)+1) &= \gcd(g(x), x^{10} + x^8 - x^6 - x^2) = x^6 - x^4 + x^2 + 1, \\ \gcd(g(x), R_4(x)-1) &= \gcd(g(x), x^{10} + x^8 - x^6 - x^2 + 1) = x^6 - x^4 + 1. \end{aligned}$$

Thus,

$$Q_{52}(x) = (x^6 - x^2 + 1)(x^6 + x^4 - x^2 + 1)(x^6 - x^4 + x^2 + 1)(x^6 - x^4 + 1)$$

is the desired canonical factorization. □

2. FACTORIZATION OVER LARGE FINITE FIELDS

If $\mathbb{F}_q$ is a finite field with a large number q of elements, the practical implementation of the methods in the previous section will become more difficult. We may still be able to find an f-reducing polynomial with a reasonable effort, but a direct application of the basic formula (4.2) will be problematic since it requires the calculation of q greatest common divisors. Thus, to make the use of f-reducing polynomials feasible for large finite fields, it is imperative that we devise ways of reducing the number of elements $c \in \mathbb{F}_q$ for which the greatest common divisor in (4.2) needs to be calculated. We note that in the context of factorization we consider q to be "large" if q is (substantially) bigger than the degree of the polynomial to be factored.

Let f again be a monic polynomial in $\mathbb{F}_q[x]$ with no repeated factors, let $\deg(f) = n$, and let k be the number of distinct monic irreducible factors of f. Suppose that $h \in \mathbb{F}_q[x]$ satisfies $h^q \equiv h \bmod f$ and $0 < \deg(h) < n$, so that h is f-reducing. Since the various greatest common divisors in (4.2) are pairwise relatively prime, it is clear that at most k of these greatest common divisors will be $\neq 1$. The problem is to find an *a priori* characterization of those $c \in \mathbb{F}_q$ for which $\gcd(f(x), h(x) - c) \neq 1$.

One such characterization can be obtained by using the theory of resultants (see Definition 1.93 and the remarks following it). Let $R(f(x), h(x) - c)$ be the resultant of $f(x)$ and $h(x) - c$, where the degrees of the two polynomials are taken as the formal degrees in the definition of the resultant. Then $\gcd(f(x), h(x) - c) \neq 1$ if and only if $R(f(x), h(x) - c) = 0$. We are thus led to consider

$$F(y) = R(f(x), h(x) - y),$$

which, from the representation of the resultant as a determinant, is seen to be a polynomial in y of degree $\leqslant n$. Then we have $\gcd(f(x), h(x) - c) \neq 1$ if and only if c is a root of $F(y)$ in $\mathbb{F}_q$.

The polynomial $F(y)$ may be calculated from the definition, which involves the evaluation of a determinant of order $\leqslant 2n - 1$ whose entries are either elements of $\mathbb{F}_q$ or linear polynomials in y. In many cases it will, however, be preferable to use the following method. Choose $n + 1$ distinct elements $c_0, c_1, \ldots, c_n \in \mathbb{F}_q$ and calculate the resultants $r_i = R(f(x), h(x) - c_i)$ for $0 \leqslant i \leqslant n$. Then the unique polynomial $F(y)$ of degree $\leqslant n$ with $F(c_i) = r_i$ for $0 \leqslant i \leqslant n$ is obtained from the Lagrange interpolation formula (see Theorem 1.71). This method has the advantage that if any of the r_i are 0, we automatically get roots of the polynomial $F(y)$ in $\mathbb{F}_q$. At any rate, the question of isolating the elements $c \in \mathbb{F}_q$ with $\gcd(f(x), h(x) - c) \neq 1$ is now reduced to that of finding the roots of a polynomial in $\mathbb{F}_q$. Computational methods for dealing with this problem will be discussed in the next section.

4.7. Example. Factor $f(x) = x^6 - 3x^5 + 5x^4 - 9x^3 - 5x^2 + 6x + 7$ over $\mathbb{F}_{23}$. Since $\gcd(f(x), f'(x)) = 1$, $f(x)$ has no repeated factors. We proceed by Berlekamp's algorithm and calculate $x^{23i} \bmod f(x)$ for $0 \leqslant i \leqslant 5$. This yields the 6×6 matrix

$$B = \begin{pmatrix} 1 & 0 & 0 & 0 & 0 & 0 \\ 5 & 0 & -1 & 8 & -3 & -10 \\ -10 & 10 & 10 & 0 & 1 & -9 \\ 0 & 7 & 9 & -8 & 10 & -11 \\ 11 & 0 & -4 & 7 & 7 & 2 \\ -3 & 0 & -10 & 9 & 2 & -9 \end{pmatrix},$$

and thus $B - I$ is given by

$$B - I = \begin{pmatrix} 0 & 0 & 0 & 0 & 0 & 0 \\ 5 & -1 & -1 & 8 & -3 & -10 \\ -10 & 10 & 9 & 0 & 1 & -9 \\ 0 & 7 & 9 & -9 & 10 & -11 \\ 11 & 0 & -4 & 7 & 6 & 2 \\ -3 & 0 & -10 & 9 & 2 & -10 \end{pmatrix}.$$

Reduction to column echelon form shows that $B - I$ has rank $r = 3$, so that f has $k = 6 - r = 3$ distinct monic irreducible factors in $\mathbb{F}_{23}[x]$. A basis for the null space of $B - I$ is given by the vectors $h_1 = (1,0,0,0,0,0)$, $h_2 = (0,4,2,1,0,0)$, $h_3 = (0,-2,9,0,1,1)$, which correspond to the polynomials $h_1(x) = 1$, $h_2(x) = x^3 + 2x^2 + 4x$, $h_3(x) = x^5 + x^4 + 9x^2 - 2x$. We take the f-reducing polynomial $h_2(x)$ and consider

$$F(y) = R(f(x), h_2(x) - y)$$

$$= \begin{vmatrix} 1 & -3 & 5 & -9 & -5 & 6 & 7 & 0 & 0 \\ 0 & 1 & -3 & 5 & -9 & -5 & 6 & 7 & 0 \\ 0 & 0 & 1 & -3 & 5 & -9 & -5 & 6 & 7 \\ 1 & 2 & 4 & -y & 0 & 0 & 0 & 0 & 0 \\ 0 & 1 & 2 & 4 & -y & 0 & 0 & 0 & 0 \\ 0 & 0 & 1 & 2 & 4 & -y & 0 & 0 & 0 \\ 0 & 0 & 0 & 1 & 2 & 4 & -y & 0 & 0 \\ 0 & 0 & 0 & 0 & 1 & 2 & 4 & -y & 0 \\ 0 & 0 & 0 & 0 & 0 & 1 & 2 & 4 & -y \end{vmatrix}.$$

In this case a direct computation of $F(y)$ is feasible, and we obtain $F(y) = y^6 + 4y^5 + 3y^4 - 7y^3 + 10y^2 + 11y + 7$. Since f has three distinct monic irreducible factors in $\mathbb{F}_{23}[x]$, the polynomial F can have at most three roots in $\mathbb{F}_{23}$. By using either the methods to be discussed in the next section or trial and error, one determines the roots of F in $\mathbb{F}_{23}$ to be -3, 2, and 6. Furthermore,

$$\gcd(f(x), h_2(x) + 3) = x - 4,$$

$$\gcd(f(x), h_2(x) - 2) = x^2 - x + 7,$$

$$\gcd(f(x), h_2(x) - 6) = x^3 + 2x^2 + 4x - 6,$$

so that

$$f(x)=(x-4)(x^2-x+7)(x^3+2x^2+4x-6)$$

is the canonical factorization of $f(x)$ in $\mathbb{F}_{23}[x]$. □

Another method of characterizing the elements $c \in \mathbb{F}_q$ for which the greatest common divisors in (4.2) need to be calculated is based on the following considerations. With the notation as above, let C be the set of all $c \in \mathbb{F}_q$ such that $\gcd(f(x), h(x)-c) \neq 1$. Then (4.2) implies

$$f(x)=\prod_{c \in C} \gcd(f(x), h(x)-c), \tag{4.14}$$

and so $f(x)$ divides $\prod_{c \in C}(h(x)-c)$. We introduce the polynomial

$$G(y)=\prod_{c \in C}(y-c).$$

Then $f(x)$ divides $G(h(x))$ and the polynomial $G(y)$ may be characterized as follows.

4.8. Theorem. *Among all the polynomials $g \in \mathbb{F}_q[y]$ such that $f(x)$ divides $g(h(x))$, the polynomial $G(y)$ is the unique monic polynomial of least degree.*

Proof. We have already shown that the monic polynomial $G(y)$ is such that $f(x)$ divides $G(h(x))$. It is easily seen that the polynomials $g \in \mathbb{F}_q[y]$ with $f(x)$ dividing $g(h(x))$ form a nonzero ideal of $\mathbb{F}_q[y]$. By Theorem 1.54, this ideal is a principal ideal generated by a uniquely determined monic polynomial $G_0 \in \mathbb{F}_q[y]$. It follows that $G_0(y)$ divides $G(y)$, and so

$$G_0(y)=\prod_{c \in C_1}(y-c)$$

for some subset C_1 of C. Furthermore, $f(x)$ divides $G_0(h(x))= \prod_{c \in C_1}(h(x)-c)$, and hence

$$f(x)=\prod_{c \in C_1} \gcd(f(x), h(x)-c).$$

A comparison with (4.14) shows that $C_1=C$. Therefore $G_0(y)=G(y)$, and the theorem follows. □

This result is applied in the following manner. Let m be the number of elements of the set C. Then we write

$$G(y)=\prod_{c \in C}(y-c)=\sum_{j=0}^{m} b_j y^j$$

with coefficients $b_j \in \mathbb{F}_q$. Now $f(x)$ divides $G(h(x))$, so that we have

$$\sum_{j=0}^{m} b_j h(x)^j \equiv 0 \bmod f(x).$$

Since $b_m = 1$, this may be viewed as a nontrivial linear dependence relation over $\mathbb{F}_q$ of the residues of $1, h(x), h(x)^2, \ldots, h(x)^m \bmod f(x)$. Theorem 4.8 says that with the normalization $b_m = 1$ this linear dependence relation is unique, and that the residues of $1, h(x), h(x)^2, \ldots, h(x)^{m-1} \bmod f(x)$ are linearly independent over $\mathbb{F}_q$. The bound $m \leqslant k$ follows from (4.14).

The polynomial G can thus be determined by calculating the residues mod $f(x)$ of $1, h(x), h(x)^2, \ldots$ until we find the smallest power of $h(x)$ that is linearly dependent (over $\mathbb{F}_q$) on its predecessors. The coefficients of this first linear dependence relation, in the normalized form, are the coefficients of G. We know that we need not go beyond $h(x)^k$ to find this linear dependence relation, and k can be obtained from Berlekamp's algorithm. The elements of C are now precisely the roots of the polynomial G. This method of reducing the problem of finding the elements of C to that of calculating the roots of a polynomial in $\mathbb{F}_q$ is called the *Zassenhaus algorithm*.

4.9. Example. Consider again the polynomial $f \in \mathbb{F}_{23}[x]$ from Example 4.7. From Berlekamp's algorithm we obtained $k = 3$ and the f-reducing polynomial $h(x) = x^3 + 2x^2 + 4x \in \mathbb{F}_{23}[x]$. We apply the Zassenhaus algorithm in order to determine the elements $c \in \mathbb{F}_{23}$ for which $\gcd(f(x), h(x) - c) \neq 1$. We have

$$h(x) \equiv x^3 + 2x^2 + 4x \bmod f(x),$$

$$h(x)^2 \equiv 7x^5 + 7x^4 + 2x^3 - 2x^2 - 6x - 7 \bmod f(x),$$

and so it is clear that $h(x)^2$ is not linearly dependent on 1 and $h(x)$. Therefore, $h(x)^3$ must be the smallest power of $h(x)$ that is linearly dependent on its predecessors. We have

$$h(x)^3 \equiv -11x^5 - 11x^4 - x^3 - 9x^2 - 5x - 2 \bmod f(x),$$

and the linear dependence relation is

$$h(x)^3 - 5h(x)^2 + 11h(x) - 10 \equiv 0 \bmod f(x),$$

so that $G(y) = y^3 - 5y^2 + 11y - 10$. By using either the methods to be discussed in the next section or trial and error, one determines the roots of G to be -3, 2, and 6. The canonical factorization of f in $\mathbb{F}_{23}[x]$ is then obtained as in the last part of Example 4.7. □

A method that is conceptually more complicated, but of great theoretical interest, is based on the use of matrices of polynomials. By a *matrix of polynomials* we mean here a matrix whose entries are elements of $\mathbb{F}_q[x]$.

4.10. Definition. A square matrix of polynomials is called *nonsingular* if its determinant is a nonzero polynomial, and it is called *unimodular* if its determinant is a nonzero element of $\mathbb{F}_q$.

4.11. Definition. Two square matrices P and Q of polynomials are said to be *equivalent* if there exists a unimodular matrix U of polynomials and a nonsingular matrix E with entries in $\mathbb{F}_q$ such that $P = UQE$.

It is easily verified that this notion of equivalence is an equivalence relation, in the sense that it is reflexive, symmetric, and transitive.

We have seen in Section 1 that there are polynomials $h_2, \ldots, h_k \in \mathbb{F}_q[x]$ with $0 < \deg(h_i) < \deg(f)$ for $2 \leqslant i \leqslant k$, which together with $h_1 = 1$ are solutions of $h^q \equiv h \bmod f$ that are linearly independent over $\mathbb{F}_q$. Clearly, the polynomials h_i may be taken to be monic. The following theorem is fundamental.

4.12. Theorem. *Let $f = f_1 \cdots f_k$, where $f_1, \ldots, f_k$ are distinct monic irreducible polynomials in $\mathbb{F}_q[x]$, and let $h_2, \ldots, h_k \in \mathbb{F}_q[x]$ be monic polynomials with $0 < \deg(h_i) < \deg(f)$ for $2 \leqslant i \leqslant k$, which together with $h_1 = 1$ are solutions of $h^q \equiv h \bmod f$ that are linearly independent over $\mathbb{F}_q$. Then the diagonal matrix of polynomials*

$$D = \begin{pmatrix} f_1 & 0 & 0 & \cdots & 0 \\ 0 & f_2 & 0 & \cdots & 0 \\ 0 & 0 & f_3 & \cdots & 0 \\ \vdots & \vdots & \vdots & \ddots & \vdots \\ 0 & 0 & 0 & \cdots & f_k \end{pmatrix}$$

is equivalent to the matrix of polynomials

$$A = \begin{pmatrix} f & 0 & 0 & \cdots & 0 \\ h_2 & -1 & 0 & \cdots & 0 \\ h_3 & 0 & -1 & \cdots & 0 \\ \vdots & \vdots & \vdots & \ddots & \vdots \\ h_k & 0 & 0 & \cdots & -1 \end{pmatrix}.$$

Proof. By the argument following (4.3) we have $h_i(x) \equiv e_{ij} \bmod f_j(x)$ with $e_{ij} \in \mathbb{F}_q$ for $1 \leqslant i, j \leqslant k$. Let E be the $k \times k$ matrix whose (i, j) entry is e_{ij}. We show first that E is nonsingular. Otherwise, there would exist elements $d_1, \ldots, d_k \in \mathbb{F}_q$, not all zero, such that

$$\sum_{i=1}^{k} d_i e_{ij} = 0 \quad \text{for } 1 \leqslant j \leqslant k.$$

This implies that

$$\sum_{i=1}^{k} d_i h_i \equiv 0 \bmod f_j \quad \text{for } 1 \leqslant j \leqslant k,$$

and so $\sum_{i=1}^{k} d_i h_i \equiv 0 \bmod f$. Since $\deg(h_i) < \deg(f)$ for $1 \leqslant i \leqslant k$, it follows that $\sum_{i=1}^{k} d_i h_i = 0$, a contradiction to the linear independence of $h_1, \ldots, h_k$.

Next we note that AE is a nonsingular matrix of polynomials. Thus we can write $D = (D(AE)^{-1})AE$, so that the theorem is established once we have shown that $U = D(AE)^{-1}$ is a unimodular matrix of polynomials.

Let $b_{ij} \in \mathbb{F}_q[x]$ be the (i, j) entry of AE. Then $b_{1j} = fe_{1j} = f \equiv 0 \bmod f_j$ for $1 \leqslant j \leqslant k$, and for $2 \leqslant i \leqslant k$ we have $b_{ij} = h_i e_{1j} - e_{ij} = h_i - e_{ij} \equiv 0 \bmod f_j$ for $1 \leqslant j \leqslant k$, so that

$$b_{ij} \equiv 0 \bmod f_j \quad \text{for} \quad 1 \leqslant i, j \leqslant k. \tag{4.15}$$

Now

$$(AE)^{-1} = \frac{1}{\det(AE)}(B_{ij})_{1 \leqslant i,j \leqslant k} = \frac{(-1)^{k-1}}{\det(E)f}(B_{ij})_{1 \leqslant i,j \leqslant k},$$

where B_{ij} is the cofactor of the (j, i) entry in AE, and

$$U = D(AE)^{-1} = \frac{(-1)^{k-1}}{\det(E)f}(f_i B_{ij})_{1 \leqslant i,j \leqslant k}.$$

Since (4.15) implies that $B_{ij} \equiv 0 \bmod (f/f_i)$, it follows that each entry of U is a polynomial over $\mathbb{F}_q$. Furthermore,

$$\det(U) = \frac{\det(D)}{\det(AE)} = \frac{(-1)^{k-1}}{\det(E)},$$

which is a nonzero element of $\mathbb{F}_q$. Thus, U is a unimodular matrix of polynomials. □

Theorem 4.12 leads to the theoretical possibility of determining the irreducible factors of f by diagonalizing the matrix A. The number k as well as the entries $h_2, \ldots, h_k$ in the first column of A can be obtained with relative ease by Berlekamp's algorithm. The algorithm that achieves the diagonalization of A is, however, quite complicated.

The *diagonalization algorithm* is based on the use of the following *elementary operations*: (i) permute any pair of rows (columns); (ii) multiply any row (column) by an element of $\mathbb{F}_q^*$; (iii) multiply some row (column) by a monomial (element of $\mathbb{F}_q$) and add the result to any other row (column).

The elementary row operations may be performed by multiplying the original matrix from the left by an appropriate unimodular matrix of polynomials, whereas the elementary column operations may be performed by multiplying the original matrix from the right by an appropriate nonsingular matrix with entries in $\mathbb{F}_q$. Therefore, the new matrix obtained by any of these elementary operations is equivalent to the original matrix.

One can show that A is equivalent to a matrix R of polynomials with the property that for each row of R the degree of the diagonal entry is greater than the degrees of the other entries in the row. The matrix R can be computed from A by performing at most $(2\Delta + k - 1)(k - 1)$ elementary operations, where $\Delta = \deg(h_2) + \cdots + \deg(h_k)$.

We note that the diagonal entries of R can be permuted by carrying out suitable row and column permutations. We can thus obtain a matrix S that, in addition to the property of R stated above, satisfies $\deg(s_{ii}) \geqslant \deg(s_{jj})$ for $1 \leqslant i \leqslant j \leqslant k$, where the s_{ii} are the diagonal entries of S. By multiplying the rows of S by appropriate elements of $\mathbb{F}_q^*$, if necessary, we may assume that the s_{ii} are monic polynomials. A matrix S of polynomials with all these properties is called a *normalized matrix*.

The diagonal entries of the matrix D in Theorem 4.12 may also be arranged in such a way that $\deg(f_i) \geqslant \deg(f_j)$ for $1 \leqslant i \leqslant j \leqslant k$. The resulting equivalent matrix, which we again call D, is then diagonal and normalized. Using the fact that the normalized matrix S is equivalent to D, one can then show that $\deg(s_{ii}) = \deg(f_i)$ for $1 \leqslant i \leqslant k$. Thus, one can read off the degrees of the various irreducible factors of f from the diagonal entries of S. Furthermore, if d is a positive integer which occurs as the degree of some s_{ii}, and if $S^{(d)}$ is the square submatrix of S whose main diagonal contains exactly all s_{ii} of degree d, then one can prove that the determinant of $S^{(d)}$ is equal to the determinant of the corresponding submatrix of D. Thus $\det(S^{(d)}) = g_d$, where g_d is the product of all f_i of degree d. In this way we are led to the partial factorization

$$f = \prod_d g_d, \tag{4.16}$$

where the product is over all positive integers d that occur as the degree of some f_i.

In summary, we see that the matrix S can be used to obtain the following information about the distinct monic irreducible factors of f: the degrees of these factors, the number of these factors of given degree, and the product of all these factors of given degree. If the f_i have distinct degrees, or, equivalently, if the s_{ii} have distinct degrees, then (4.16) represents already the canonical factorization of f in $\mathbb{F}_q[x]$.

If (4.16) is not yet the canonical factorization, then one can proceed in various ways. An obvious option is the application of one of the methods discussed earlier to factor the polynomials g_d. One can also continue with the diagonalization algorithm in order to obtain the diagonal matrix D equivalent to the normalized matrix S.

For the latter purpose, we assume as above that D is put in normalized form. In addition to the properties mentioned above, it is then also true that each of the submatrices $S^{(d)}$ is equivalent to the corresponding submatrix $D^{(d)}$ of D. It is therefore sufficient to diagonalize each of the submatrices $S^{(d)}$ separately. By the equivalence of $S^{(d)}$ and $D^{(d)}$ we have $S^{(d)} = UD^{(d)}E$ for some unimodular matrix U of polynomials and some nonsingular matrix E with entries in $\mathbb{F}_q$. We may then write

$$S^{(d)} = S_0^{(d)} + S_1^{(d)}x + \cdots + S_d^{(d)}x^d,$$

$$U = U_0 + U_1 x + \cdots + U_m x^m,$$

$$D^{(d)} = D_0^{(d)} + D_1^{(d)}x + \cdots + D_d^{(d)}x^d,$$

where the $S_r^{(d)}$, $D_r^{(d)}$, and U_t, $0 \leqslant r \leqslant d$, $0 \leqslant t \leqslant m$, are matrices with entries in $\mathbb{F}_q$, $U_m \neq 0$, and $S_d^{(d)} = D_d^{(d)} = I$, the identity matrix of appropriate order. A comparison of the matrix coefficients of the highest powers of x on both sides of the equation $S^{(d)} = UD^{(d)}E$ yields $I = U_m IE$ and $m = 0$. Thus, $U = U_0 = E^{-1}$ and hence $S^{(d)} = E^{-1}D^{(d)}E$.

Comparing the matrix coefficients of like powers of x in the last identity gives $S_r^{(d)} = E^{-1}D_r^{(d)}E$ for $0 \leqslant r \leqslant d$. Consequently, $S_r^{(d)}$ and $D_r^{(d)}$ have the same characteristic polynomial and eigenvalues, and since $D_r^{(d)}$ is diagonal, its eigenvalues are exactly its diagonal entries. Therefore, the latter can be determined by finding the roots of the characteristic polynomial of $S_r^{(d)}$, which must all be in $\mathbb{F}_q$. As in the earlier methods, we have thus again reduced the factorization problem to that of finding the roots of certain polynomials in $\mathbb{F}_q$.

The partial factorization (4.16) can also be obtained by an entirely different method. To this end, we extend the definition of g_d by letting g_i, $i \geqslant 1$, be the product of all monic irreducible polynomials in $\mathbb{F}_q[x]$ of degree i that divide f. In particular, $g_i(x) = 1$ in case f has no irreducible factor in $\mathbb{F}_q[x]$ of degree i. We can thus write

$$f = \prod_{i \geqslant 1} g_i.$$

It is trivial that only those i with $i \leqslant \deg(f)$ need to be considered. We calculate now recursively the polynomials $r_0(x)$, $r_1(x), \ldots$ and $F_0(x)$, $F_1(x), \ldots$ as well as $d_1(x), d_2(x), \ldots$. We start with

$$r_0(x) = x,\ F_0(x) = f(x),$$

and for $i \geqslant 1$ we use the formulas

$$r_i(x) \equiv r_{i-1}(x)^q \bmod F_{i-1}(x), \deg(r_i) < \deg(F_{i-1}),$$
$$d_i(x) = \gcd(F_{i-1}(x), r_i(x) - x),$$
$$F_i(x) = F_{i-1}(x)/d_i(x).$$

The algorithm can be stopped when $d_i(x) = F_{i-1}(x)$.

4.13. Theorem. *With the notation above, we have $d_i(x) = g_i(x)$ for all $i \geqslant 1$.*

Proof. Using the fact that F_i divides F_{i-1}, a straightforward induction shows that

$$r_i(x) \equiv x^{q^i} \bmod F_{i-1}(x) \quad \text{for all } i \geqslant 1. \tag{4.17}$$

We prove now by induction that

$$F_{i-1} = \prod_{j \geqslant i} g_j \quad \text{and} \quad d_i = g_i \quad \text{for all } i \geqslant 1. \tag{4.18}$$

For $i = 1$ the first identity holds since $F_0 = f$. As to the second identity, we have

$$d_1(x) = \gcd(F_0(x), r_1(x) - x) = \gcd(f(x), x^q - x)$$

by (4.17), and since $x^q - x$ is the product of all monic linear polynomials in $\mathbb{F}_q[x]$, it follows that d_1 is the product of all monic linear polynomials in $\mathbb{F}_q[x]$ dividing f, and hence $d_1 = g_1$. Now assume that (4.18) is shown for some $i \geqslant 1$. Then

$$F_i = F_{i-1}/d_i = F_{i-1}/g_i = \prod_{j \geqslant i+1} g_j, \tag{4.19}$$

which proves the first identity in (4.18) for $i+1$. Furthermore,

$$d_{i+1}(x) = \gcd(F_i(x), r_{i+1}(x) - x) = \gcd\left(F_i(x), x^{q^{i+1}} - x\right)$$

by (4.17). According to Theorem 3.20, $x^{q^{i+1}} - x$ is the product of all monic irreducible polynomials in $\mathbb{F}_q[x]$ whose degrees divide $i+1$. Consequently, d_{i+1} is the product of all monic irreducible polynomials in $\mathbb{F}_q[x]$ that divide F_i and whose degrees divide $i+1$. It follows then from (4.19) that $d_{i+1} = g_{i+1}$. □

In the algorithm above, the most complicated step from the viewpoint of calculation is that of obtaining r_i by computing the qth power of $r_{i-1} \bmod F_{i-1}$. A common technique of cutting down the amount of calculation somewhat is based on computing first the residues $\bmod F_{i-1}$ of $r_{i-1}, r_{i-1}^2, r_{i-1}^4, \ldots, r_{i-1}^{2^e}$ by repeated squaring and reduction mod F_{i-1}, where 2^e is the largest power of 2 that is $\leqslant q$, and then multiplying together an appropriate combination of these residues mod F_{i-1} to obtain the residue of

$r_{i-1}^q \bmod F_{i-1}$. For instance, to get the residue of $r_{i-1}^{23} \bmod F_{i-1}$, one would multiply together the residues of r_{i-1}^{16}, r_{i-1}^4, r_{i-1}^2, and $r_{i-1} \bmod F_{i-1}$.

Instead of working with the repeated squaring technique, we could employ the matrix B from Berlekamp's algorithm in Section 1 to calculate r_i from r_{i-1}. We write $n = \deg(f)$ and

$$r_{i-1}(x) = \sum_{j=0}^{n-1} r_{i-1}^{(j)} x^j,$$

and define $(s_i^{(0)}, s_i^{(1)}, \ldots, s_i^{(n-1)}) \in \mathbb{F}_q^n$ by the matrix identity

$$\left(s_i^{(0)}, s_i^{(1)}, \ldots, s_i^{(n-1)}\right) = \left(r_{i-1}^{(0)}, r_{i-1}^{(1)}, \ldots, r_{i-1}^{(n-1)}\right) B, \tag{4.20}$$

where B is the $n \times n$ matrix in (4.5). With

$$s_i(x) = \sum_{j=0}^{n-1} s_i^{(j)} x^j \tag{4.21}$$

we get then $r_{i-1}(x)^q \equiv s_i(x) \bmod f(x)$, hence $r_{i-1}(x)^q \equiv s_i(x) \bmod F_{i-1}(x)$, and thus

$$r_i(x) \equiv s_i(x) \bmod F_{i-1}(x).$$

Therefore, once the matrix B has been calculated, we compute r_i from r_{i-1} in each step by reduction mod F_{i-1} of the polynomial s_i obtained from (4.20) and (4.21).

4.14. Example. We consider $f(x) = x^6 - 3x^5 + 5x^4 - 9x^3 - 5x^2 + 6x + 7 \in \mathbb{F}_{23}[x]$ as in Example 4.7. Then

$$B = \begin{pmatrix} 1 & 0 & 0 & 0 & 0 & 0 \\ 5 & 0 & -1 & 8 & -3 & -10 \\ -10 & 10 & 10 & 0 & 1 & -9 \\ 0 & 7 & 9 & -8 & 10 & -11 \\ 11 & 0 & -4 & 7 & 7 & 2 \\ -3 & 0 & -10 & 9 & 2 & -9 \end{pmatrix}.$$

We start the algorithm with $r_0(x) = x$, $F_0(x) = f(x)$. From (4.20) and (4.21) we get $s_1(x) = -10x^5 - 3x^4 + 8x^3 - x^2 + 5$, and reduction mod $F_0(x)$ yields $r_1(x) = s_1(x)$. By Theorem 4.13 we have $g_1(x) = d_1(x) = \gcd(F_0(x), r_1(x) - x) = x - 4$. Furthermore, $F_1(x) = F_0(x)/d_1(x) = x^5 + x^4 + 9x^3 + 4x^2 + 11x + 4$.

In the second iteration, we use again (4.20) and (4.21) to obtain $s_2(x) = 5x^5 - 8x^4 + 9x^3 - 10x^2 - 11$, and reduction mod $F_1(x)$ leads to $r_2(x) = 10x^4 + 10x^3 - 7x^2 - 9x - 8$. By Theorem 4.13 we have $g_2(x) = d_2(x) = \gcd(F_1(x), r_2(x) - x) = x^2 - x + 7$. Furthermore, $F_2(x) = F_1(x)/d_2(x) = x^3 + 2x^2 + 4x - 6$. But, according to the first part of (4.18), all irreducible factors of $F_2(x)$ have degree $\geqslant 3$, so that $F_2(x)$ itself must be irreducible in $\mathbb{F}_{23}[x]$ and $g_3(x) = F_2(x)$. Thus, we arrive at the partial factorization

$$f(x) = (x-4)(x^2 - x + 7)(x^3 + 2x^2 + 4x - 6),$$

which, in this case, is already the canonical factorization of $f(x)$ in $\mathbb{F}_{23}[x]$. □

3. CALCULATION OF ROOTS OF POLYNOMIALS

We have seen in the preceding section that the problem of determining the canonical factorization of a polynomial can often be reduced to that of finding the roots of an auxiliary polynomial in a finite field. The calculation of roots of a polynomial is, of course, a matter of independent interest as well.

In general, one will be interested in determining the roots of a polynomial in an extension of the field from which the coefficients are taken. However, it suffices to consider the situation in which we are asked to find the roots of a polynomial $f \in \mathbb{F}_q[x]$ of positive degree in $\mathbb{F}_q$, since a polynomial over a subfield can always be viewed as a polynomial over $\mathbb{F}_q$.

It is clear that every factorization algorithm is, in particular, a root-finding algorithm since the roots of f in $\mathbb{F}_q$ can be read off from the linear factors that occur in the canonical factorization of f in $\mathbb{F}_q[x]$. Thus, the algorithms presented in the earlier sections of this chapter can also be used for the determination of roots. However, these algorithms will often not be the most efficient procedures for the more specialized task of calculating roots. Therefore, we shall discuss methods that are better suited to this particular purpose.

As a first step, one may isolate that part of f which contains the roots of f in $\mathbb{F}_q$. This is achieved by calculating $\gcd(f(x), x^q - x)$. Since $x^q - x$ is the product of all monic linear polynomials in $\mathbb{F}_q[x]$, this greatest common divisor is the product of all monic linear polynomials over $\mathbb{F}_q$ dividing f, and so its roots are precisely the roots of f in $\mathbb{F}_q$. Therefore, we may assume, without loss of generality, that the polynomial for which we want to find the roots in $\mathbb{F}_q$ is a product of distinct monic linear polynomials over $\mathbb{F}_q$.

A useful method of finding roots of polynomials was already discussed in Chapter 3, Section 4. It is based on the determination of an affine multiple of the given polynomial. See Example 3.55 for an illustration of this method.

In order to arrive at other methods, we consider first the case of a prime field $\mathbb{F}_p$. As we have seen above, it suffices to deal with polynomials of the form

$$f(x) = \prod_{i=1}^{n} (x - c_i),$$

where $c_1,\ldots,c_n$ are distinct elements of $\mathbb{F}_p$. If p is small, then it is feasible to determine the roots of f by trial and error, that is, by simply calculating $f(0), f(1),\ldots,f(p-1)$.

For large p the following method may be employed. For $b \in \mathbb{F}_p$, p odd, we consider

$$f(x-b)=\prod_{i=1}^{n}(x-(b+c_i)).$$

We note that $f(x-b)$ divides $x^p - x = x(x^{(p-1)/2}+1)(x^{(p-1)/2}-1)$. If x is a factor of $f(x-b)$, then $f(-b)=0$ and a root of f has been found. If x is not a factor of $f(x-b)$, then we have

$$f(x-b)=\gcd\big(f(x-b),x^{(p-1)/2}+1\big)\gcd\big(f(x-b),x^{(p-1)/2}-1\big). \tag{4.22}$$

The identity (4.22) is now used as follows. We calculate the residue mod $f(x-b)$ of $x^{(p-1)/2}$—for example, by the repeated squaring technique discussed after Theorem 4.13. If $x^{(p-1)/2} \not\equiv \pm 1 \bmod f(x-b)$, then (4.22) yields a nontrivial partial factorization of $f(x-b)$. Replacing x by $x+b$, we get then a nontrivial partial factorization of $f(x)$. In the rather unlikely case where $x^{(p-1)/2} \equiv \pm 1 \bmod f(x-b)$, we try another value of b. Thus, by using, if necessary, several choices for b, we will find either a root of f or a nontrivial partial factorization of f. Continuing this process, we will eventually obtain all the roots of f. It should be noted that, strictly speaking, this is not a deterministic, but a probabilistic root-finding algorithm, as it depends on the random selection of several elements $b \in \mathbb{F}_p$.

4.15. Example. Find the roots of $f(x)=x^6-7x^5+3x^4-7x^3+4x^2-x-2 \in \mathbb{F}_{17}[x]$ contained in $\mathbb{F}_{17}$. The roots of $f(x)$ in $\mathbb{F}_{17}$ are precisely the roots of $g(x)=\gcd(f(x), x^{17}-x)$ in $\mathbb{F}_{17}$. By the Euclidean algorithm we obtain $g(x)=x^4+6x^3-5x^2+7x-2$. To find the roots of $g(x)$, we use the algorithm above and first select $b=0$. A straightforward calculation yields $x^{(p-1)/2}=x^8 \equiv 1 \bmod g(x)$, and so this value of b does not afford a nontrivial partial factorization of $g(x)$. Next we choose $b=1$. Then $g(x-1)=x^4+2x^3-3x-2$ and $x^8 \equiv -4x^3-7x^2+8x-5 \bmod g(x-1)$, so that $b=1$ yields a nontrivial partial factorization of $g(x-1)$. We have

$$\begin{aligned}\gcd\big(g(x-1),x^8+1\big) &= \gcd(x^4+2x^3-3x-2,\,-4x^3-7x^2+8x-4)\\ &= x^2-7x+4\end{aligned}$$

and

$$\begin{aligned}\gcd\big(g(x-1),x^8-1\big) &= \gcd(x^4+2x^3-3x-2,\,-4x^3-7x^2+8x-6)\\ &= x^2-8x+8,\end{aligned}$$

hence (4.22) implies

$$g(x-1) = (x^2 - 7x + 4)(x^2 - 8x + 8),$$

which leads to the partial factorization

$$g(x) = (x^2 - 5x - 2)(x^2 - 6x + 1) = g_1(x) g_2(x),$$

say. In order to factor $g_1(x)$ and $g_2(x)$, we try $b = 2$. We have $g_1(x-2) = x^2 + 8x - 5$ and $x^8 \equiv -8x + 2 \bmod g_1(x-2)$. Furthermore,

$$\gcd\bigl(g_1(x-2), x^8 + 1\bigr) = \gcd(x^2 + 8x - 5, -8x + 3) = x + 6,$$

and long division yields $g_1(x-2) = (x+6)(x+2)$, so that

$$g_1(x) = (x+8)(x+4).$$

Turning to $g_2(x)$, we have $g_2(x-2) = x^2 + 7x = x(x+7)$, thus -2 is a root of $g_2(x)$ and

$$g_2(x) = (x+2)(x-8).$$

Combining these factorizations, we get

$$g(x) = (x+8)(x+4)(x+2)(x-8).$$

Therefore, the roots of $g(x)$, and thus of $f(x)$, in $\mathbb{F}_{17}$ are $-8, -4, -2, 8$. □

Next we discuss a root-finding algorithm for large finite fields $\mathbb{F}_q$ with small characteristic p. As before, it suffices to consider the case where

$$f(x) = \prod_{i=1}^{n} (x - \gamma_i)$$

with distinct elements $\gamma_1, \dots, \gamma_n \in \mathbb{F}_q$. Let $q = p^m$ and define the polynomial

$$S(x) = \sum_{j=0}^{m-1} x^{p^j}.$$

We note that for $\gamma \in \mathbb{F}_q$ we have $S(\gamma) = \mathrm{Tr}_{\mathbb{F}_q}(\gamma) \in \mathbb{F}_p$, where $\mathrm{Tr}_{\mathbb{F}_q}$ is the absolute trace function (see Definition 2.22). Because of Theorem 2.23(iii), the equation $S(\gamma) = c$ has p^{m-1} solutions $\gamma \in \mathbb{F}_q$ for every $c \in \mathbb{F}_p$, and this observation leads to the identity

$$x^q - x = \prod_{c \in \mathbb{F}_p} (S(x) - c). \tag{4.23}$$

Since $f(x)$ divides $x^q - x$, we get

$$\prod_{c \in \mathbb{F}_p} (S(x) - c) \equiv 0 \bmod f(x),$$

and so

$$f(x) = \prod_{c \in \mathbb{F}_p} \gcd(f(x), S(x) - c). \tag{4.24}$$

This yields a partial factorization of $f(x)$ that calls for the calculation of p greatest common divisors. If p is small, this is certainly a feasible method.

It can, however, happen that the factorization in (4.24) is trivial—namely, precisely when $S(x) \equiv c \bmod f(x)$ for some $c \in \mathbb{F}_p$. In this case, other auxiliary polynomials related to $S(x)$ have to be used. Let β be a defining element of $\mathbb{F}_q$ over $\mathbb{F}_p$, so that $\{1, \beta, \beta^2, \ldots, \beta^{m-1}\}$ is a basis of $\mathbb{F}_q$ over $\mathbb{F}_p$. For $j = 0, 1, \ldots, m-1$ we substitute $\beta^j x$ for x in (4.23) and we get

$$(\beta^j)^q x^q - \beta^j x = \prod_{c \in \mathbb{F}_p} (S(\beta^j x) - c).$$

Since $(\beta^j)^q = \beta^j$, we obtain

$$x^q - x = \beta^{-j} \prod_{c \in \mathbb{F}_p} (S(\beta^j x) - c).$$

This yields the following generalization of (4.24):

$$f(x) = \prod_{c \in \mathbb{F}_p} \gcd(f(x), S(\beta^j x) - c) \quad \text{for } 0 \leqslant j \leqslant m-1. \tag{4.25}$$

We show now that if $n = \deg(f) \geqslant 2$, then there exists at least one j, $0 \leqslant j \leqslant m-1$, for which the partial factorization in (4.25) is nontrivial. For suppose, on the contrary, that all the partial factorizations in (4.25) are trivial. Then for each j, $0 \leqslant j \leqslant m-1$, there exists a $c_j \in \mathbb{F}_p$ with

$$S(\beta^j x) \equiv c_j \bmod f(x).$$

In particular, we get

$$S(\beta^j \gamma_1) = S(\beta^j \gamma_2) = c_j \quad \text{for } 0 \leqslant j \leqslant m-1.$$

By the linearity of the trace it follows that

$$\mathrm{Tr}_{\mathbb{F}_q}((\gamma_1 - \gamma_2)\beta^j) = 0 \quad \text{for } 0 \leqslant j \leqslant m-1$$

and

$$\mathrm{Tr}_{\mathbb{F}_q}((\gamma_1 - \gamma_2)\alpha) = 0 \quad \text{for all } \alpha \in \mathbb{F}_q.$$

Using the second part of Theorem 2.24, we conclude that $\gamma_1 - \gamma_2 = 0$, which is a contradiction. Thus, for at least one j the partial factorization in (4.25) is nontrivial.

The defining element β of $\mathbb{F}_q$ over $\mathbb{F}_p$ used in (4.25) is chosen as a root of a known irreducible polynomial in $\mathbb{F}_p[x]$ of degree m. Once a nontrivial factorization of the form (4.25) has been found, the method is applied to the nontrivial factors by employing other values of j. The argument above shows also that all distinct roots of f can eventually be separated by using all the values of j in (4.25).

4.16. Example. Consider $\mathbb{F}_{64} = \mathbb{F}_2(\beta)$, where β is a root of the irreducible polynomial $x^6 + x + 1$ in $\mathbb{F}_2[x]$, and let

$$f(x) = x^4 + (\beta^5 + \beta^4 + \beta^3 + \beta^2)x^3 + (\beta^5 + \beta^4 + \beta^2 + \beta + 1)x^2 + (\beta^4 + \beta^3 + \beta)x + \beta^3 + \beta \in \mathbb{F}_{64}[x].$$

Using

$$x^6 \equiv (\beta^5 + \beta + 1)x^3 + (\beta^4 + \beta^3 + \beta^2)x^2 + (\beta^5 + \beta^3 + \beta^2 + 1)x + \beta^5 + \beta^4 + \beta^2 + 1 \bmod f(x),$$

we get the following congruences mod $f(x)$ by repeated squaring:

$$\begin{aligned}
x &\equiv x\\
x^2 &\equiv x^2\\
x^4 &\equiv (\beta^5+\beta^4+\beta^3+\beta^2)x^3 + (\beta^5+\beta^4+\beta^2+\beta+1)x^2 + (\beta^4+\beta^3+\beta)x + \beta^3+\beta\\
x^8 &\equiv (\beta^4+\beta^3+\beta^2)x^3 + (\beta^5+\beta+1)x^2 + (\beta^5+\beta+1)x + \beta^5+\beta^4\\
x^{16} &\equiv (\beta^5+\beta^3+\beta)x^3 + (\beta^3+\beta)x^2 + \beta^5 x + \beta^4+\beta^3+\beta^2+\beta+1\\
x^{32} &\equiv (\beta^5+\beta^2+1)x^3 + (\beta^5+\beta^4+\beta^3+\beta^2+\beta+1)x^2 + (\beta^4+\beta^2)x + \beta^5+\beta^3\\
x^{64} &\equiv x
\end{aligned}$$

Thus, $f(x)$ divides $x^{64} - x$ and so has four distinct roots in $\mathbb{F}_{64}$. We consider now $S(x) = x + x^2 + x^4 + x^8 + x^{16} + x^{32}$. From the congruences above we obtain

$$S(x) \equiv (\beta^5 + \beta^3 + \beta^2 + \beta + 1)x^3 + \beta^5 x^2 + (\beta^3 + \beta^2)x + \beta^3 + \beta^2 + 1 \bmod f(x),$$

and therefore

$$\begin{aligned}
\gcd(f(x), S(x)) &= \gcd(f(x), (\beta^5 + \beta^3 + \beta^2 + \beta + 1)x^3 + \beta^5 x^2 + (\beta^3 + \beta^2)x + \beta^3 + \beta^2 + 1)\\
&= x^3 + (\beta^4 + \beta^3 + \beta^2)x^2 + (\beta^5 + \beta^2 + 1)x + \beta^3 + \beta^2 = g(x)
\end{aligned}$$

say, and

$$\gcd(f(x), S(x) - 1) = \gcd(f(x), (\beta^5 + \beta^3 + \beta^2 + \beta + 1)x^3 + \beta^5 x^2 + (\beta^3 + \beta^2)x + \beta^3 + \beta^2) = x + \beta^5.$$

Then (4.24) yields

$$f(x) = g(x)(x + \beta^5). \tag{4.26}$$

To find the roots of $g(x)$, we next use (4.25) with $j = 1$. We have

$$\begin{aligned}
S(\beta x) &= \beta x + \beta^2 x^2 + \beta^4 x^4 + \beta^8 x^8 + \beta^{16} x^{16} + \beta^{32} x^{32}\\
&= \beta x + \beta^2 x^2 + \beta^4 x^4 + (\beta^3 + \beta^2)x^8 + (\beta^4 + \beta + 1)x^{16} + (\beta^3 + 1)x^{32},
\end{aligned}$$

and the congruences above yield

$$S(\beta x) \equiv (\beta^2+1)x^3 + (\beta^3+\beta+1)x^2 + (\beta^5+\beta^4+\beta^3+\beta^2+\beta+1)x$$
$$+\beta^4+\beta^2+\beta \bmod f(x).$$

Since $g(x)$ divides $f(x)$, this congruence holds also mod $g(x)$, and so

$$\begin{aligned} S(\beta x) &\equiv (\beta^2+1)x^3 + (\beta^3+\beta+1)x^2 + (\beta^5+\beta^4+\beta^3+\beta^2+\beta+1)x \\ &\quad +\beta^4+\beta^2+\beta \\ &\equiv (\beta^5+\beta^2)x^2 + \beta^3 x + \beta^5+\beta^3+\beta \bmod g(x). \end{aligned}$$

Thus,

$$\begin{aligned} \gcd(g(x), S(\beta x)) &= \gcd(g(x), (\beta^5+\beta^2)x^2+\beta^3x+\beta^5+\beta^3+\beta) \\ &= x^2+(\beta^3+1)x+\beta^4+\beta^3+\beta^2+\beta = h(x), \end{aligned}$$

say, and

$$\begin{aligned} \gcd(g(x), S(\beta x)-1) &= \gcd(g(x), (\beta^5+\beta^2)x^2+\beta^3x+\beta^5+\beta^3+\beta+1) \\ &= x+\beta^4+\beta^2+1. \end{aligned}$$

Then (4.25) with $j=1$ yields

$$g(x) = h(x)(x+\beta^4+\beta^2+1). \tag{4.27}$$

To find the roots of $h(x)$, we use (4.25) with $j=2$. We have

$$\begin{aligned} S(\beta^2 x) &= \beta^2 x + \beta^4 x^2 + \beta^8 x^4 + \beta^{16} x^8 + \beta^{32} x^{16} + \beta^{64} x^{32} \\ &= \beta^2 x + \beta^4 x^2 + (\beta^3+\beta^2)x^4 + (\beta^4+\beta+1)x^8 \\ &\quad + (\beta^3+1)x^{16} + \beta x^{32}, \end{aligned}$$

and a similar calculation as for $S(\beta x)$ yields

$$S(\beta^2 x) \equiv (\beta^5+\beta^2+1)x+\beta^5+\beta^3+\beta^2 \bmod h(x).$$

Therefore,

$$\begin{aligned} \gcd(h(x), S(\beta^2 x)) &= \gcd(h(x), (\beta^5+\beta^2+1)x+\beta^5+\beta^3+\beta^2) \\ &= x+\beta+1 \end{aligned}$$

and

$$\begin{aligned} \gcd(h(x), S(\beta^2 x)-1) &= \gcd(h(x), (\beta^5+\beta^2+1)x+\beta^5+\beta^3+\beta^2+1) \\ &= x+\beta^3+\beta, \end{aligned}$$

so that from (4.25) with $j=2$ we get

$$h(x) = (x+\beta+1)(x+\beta^3+\beta). \tag{4.28}$$

Combining (4.26), (4.27), and (4.28), we arrive at the factorization

$$f(x)=(x+\beta+1)(x+\beta^3+\beta)(x+\beta^4+\beta^2+1)(x+\beta^5),$$

and so the roots of $f(x)$ are $\beta+1$, $\beta^3+\beta$, $\beta^4+\beta^2+1$, and β^5. □

Finally we consider the root-finding problem for large finite fields $\mathbb{F}_q$ with large characteristic p. As we have seen before, it suffices to know how to treat polynomials of the form

$$f(x)=\prod_{i=1}^{n}(x-\gamma_i)$$

with distinct elements $\gamma_1,\ldots,\gamma_n\in\mathbb{F}_q$. To check whether $f(x)$ has this form, we need only verify the congruence $x^q\equiv x \bmod f(x)$ (compare with the first part of Example 4.16). We can assume that q is the least power of p for which this holds. The polynomial $f(x)$ will, of course, be given by its standard representation

$$f(x)=\sum_{j=0}^{n}\alpha_j x^j,$$

where $\alpha_j\in\mathbb{F}_q$ for $0\leqslant j\leqslant n$ and $\alpha_n=1$.

It will be our first aim to find a nontrivial factor of $f(x)$. To exclude a trivial case, we can assume $n\geqslant 2$. Let $q=p^m$ and define the polynomials

$$f_k(x)=\sum_{j=0}^{n}\alpha_j^{p^k}x^j \quad \text{for } 0\leqslant k\leqslant m-1, \tag{4.29}$$

so that $f_0(x)=f(x)$ and each $f_k(x)$ is a monic polynomial over $\mathbb{F}_q$. Furthermore,

$$f_k\left(\gamma_i^{p^k}\right)=\sum_{j=0}^{n}\alpha_j^{p^k}\gamma_i^{jp^k}=\left(\sum_{j=0}^{n}\alpha_j\gamma_i^j\right)^{p^k}=0$$

for $1\leqslant i\leqslant n$, $0\leqslant k\leqslant m-1$, and so

$$f_k(x)=\prod_{i=1}^{n}\left(x-\gamma_i^{p^k}\right) \quad \text{for } 0\leqslant k\leqslant m-1.$$

We calculate now the polynomial

$$F(x)=\prod_{k=0}^{m-1}f_k(x). \tag{4.30}$$

This is a polynomial over $\mathbb{F}_p$ since

$$F(x)=\prod_{k=0}^{m-1}\prod_{i=1}^{n}\left(x-\gamma_i^{p^k}\right)=\prod_{i=1}^{n}\prod_{k=0}^{m-1}\left(x-\gamma_i^{p^k}\right)=\prod_{i=1}^{n}F_i(x)^{m/d_i},$$

where $F_i(x)$ is the minimal polynomial of γ_i over $\mathbb{F}_p$ and d_i is its degree (compare with the discussion following Definition 2.22). The $F_i(x)$ are therefore the irreducible factors of $F(x)$ in $\mathbb{F}_p[x]$, but certain $F_i(x)$ could be identical. Thus, the canonical factorization of $F(x)$ in $\mathbb{F}_p[x]$ has the form

$$F(x) = G_1(x) \cdots G_r(x), \tag{4.31}$$

where the $G_t(x)$, $1 \leqslant t \leqslant r$, are powers of the distinct $F_i(x)$. This canonical factorization can be obtained by one of the factorization algorithms in Section 2 of this chapter. Since $f(x) = f_0(x)$ divides $F(x)$, it follows from (4.31) that

$$f(x) = \prod_{t=1}^{r} \gcd(f(x), G_t(x)). \tag{4.32}$$

In most cases, (4.32) will provide a nontrivial partial factorization of $f(x)$. The factorization will be trivial precisely if $\gcd(f(x), G_t(x)) = f(x)$ for some t, $1 \leqslant t \leqslant r$, which is equivalent to $r = 1$ and $f(x)$ dividing $F_1(x)$. A comparison of degrees shows then $n \leqslant d_1 = m$. Furthermore, the roots of $f(x)$ are then all conjugate with respect to $\mathbb{F}_p$. Thus, by labelling the roots of $f(x)$ suitably, we can write

$$\gamma_i = \gamma_1^{p^{b_i}} \quad \text{for} \quad 1 \leqslant i \leqslant n, \text{ with } 0 = b_1 < b_2 < \cdots < b_n < m.$$

We set $b_{n+1} = m$ and

$$d = \min_{1 \leqslant i \leqslant n} (b_{i+1} - b_i).$$

It is clear that $d \leqslant m/n$. The following two possibilities can occur:

(A) $b_{i+1} - b_i > d$ for some i, $1 \leqslant i \leqslant n$;
(B) $b_{i+1} - b_i = d$ for all i, $1 \leqslant i \leqslant n$.

In case (A) we note that the set of roots of $f(x)$ is

$$\left\{\gamma_1^{p^{b_1}}, \gamma_1^{p^{b_2}}, \ldots, \gamma_1^{p^{b_n}}\right\}$$

and the set of roots of $f_d(x)$ is

$$\left\{\gamma_1^{p^{b_1+d}}, \gamma_1^{p^{b_2+d}}, \ldots, \gamma_1^{p^{b_n+d}}\right\}.$$

The condition in (A) implies that these two sets of roots are not identical. On the other hand, since $b_{i+1} - b_i = d$ for some i, $1 \leqslant i \leqslant n$, the two sets of roots have a common element. Thus, $\gcd(f(x), f_d(x)) \neq f(x)$ and $\neq 1$; that is, $\gcd(f(x), f_d(x))$ is a nontrivial factor of $f(x)$. We observe also that in this case we have $d < m/n$.

In case (B) a comparison of the sets of roots of $f(x)$ and $f_d(x)$ shows that $f(x) = f_d(x)$, whereas $\gcd(f(x), f_k(x)) = 1$ for $1 \leqslant k < d$. Moreover, we have $d = m/n$, so that n divides m, and also $b_i = d(i-1)$ for $1 \leqslant i \leqslant n$. It follows that

$$\gamma_i = \gamma_1^{p^{d(i-1)}} \quad \text{for } 1 \leqslant i \leqslant n,$$

hence the γ_i are exactly all the conjugates of γ_1 with respect to $\mathbb{F}_{p^d}$. Consequently, $f(x)$ is the minimal polynomial of γ_1 over $\mathbb{F}_{p^d}$ and thus irreducible over $\mathbb{F}_{p^d}$.

Therefore, corresponding to the cases (A) and (B) above we have the following alternatives:

(A) $\gcd(f(x), f_k(x))$ is a nontrivial factor of $f(x)$ for some k, $1 \leqslant k < m/n$;

(B) $\gcd(f(x), f_k(x)) = 1$ for $1 \leqslant k < d = m/n \in \mathbb{N}$ and $f(x) = f_d(x)$ is the minimal polynomial of γ_1 over $\mathbb{F}_{p^d}$.

In alternative (A) our aim of finding a nontrivial factor of $f(x)$ has been achieved.

Further work is needed in alternative (B). Let β again denote a defining element of $\mathbb{F}_q$ over $\mathbb{F}_p$. Then $\mathbb{F}_{p^d}(\beta) = \mathbb{F}_q = \mathbb{F}_{p^m}$, and so β is of degree $m/d = n$ over $\mathbb{F}_{p^d}$. In particular, we have $\beta^j \notin \mathbb{F}_{p^d}$ for $1 \leqslant j \leqslant n-1$. Now let the coefficients α_j of $f(x)$ be such that $\alpha_{j_0} \neq 0$ for some j_0 with $1 \leqslant j_0 \leqslant n-1$. Consider

$$\bar{f}(x) = \beta^{-n} f(\beta x), \tag{4.33}$$

which is a monic polynomial of degree n over $\mathbb{F}_q$. Since $\beta^{n-j_0} \notin \mathbb{F}_{p^d}$ and $\alpha_{j_0} \in \mathbb{F}_{p^d}^*$, it follows that the coefficient of x^{j_0} in $\bar{f}(x)$ is not an element of $\mathbb{F}_{p^d}$. Thus $\bar{f}(x)$ is not a polynomial over $\mathbb{F}_{p^d}$, and so the alternative (B) cannot occur if the procedure above is applied to $\bar{f}(x)$. Since $f(x) = \beta^n \bar{f}(\beta^{-1}x)$, any nontrivial factor of $\bar{f}(x)$ yields immediately a nontrivial factor of $f(x)$.

It remains to consider the case where alternative (B) is valid and $\alpha_j = 0$ for $1 \leqslant j \leqslant n-1$. Then $f(x)$ is the binomial $x^n + \alpha_0 \in \mathbb{F}_{p^d}[x]$. Now n is not a multiple of p, for otherwise we would have $f(x) = (x^{n/p} + \alpha_0^{p^{d-1}})^p$, which would contradict the irreducibility of $f(x)$ over $\mathbb{F}_{p^d}$. We set

$$\bar{f}(x) = \beta^{-n} f(\beta x + 1), \tag{4.34}$$

and then it is easily seen from $\beta^{-1} \notin \mathbb{F}_{p^d}$ that the coefficient of x^{n-1} in $\bar{f}(x)$ is not in $\mathbb{F}_{p^d}$. Thus, the alternative (B) cannot occur if the procedure described above is applied to $\bar{f}(x)$. Since $f(x) = \beta^n \bar{f}(\beta^{-1}(x-1))$, any nontrivial factor of $\bar{f}(x)$ yields immediately a nontrivial factor of $f(x)$.

This root-finding algorithm is thus carried out as follows. We first form the polynomials $f_k(x)$ according to (4.29) and then the polynomial $F(x) \in \mathbb{F}_p[x]$ according to (4.30). Next, we apply a factorization algorithm to obtain the canonical factorization (4.31) of $F(x)$ in $\mathbb{F}_p[x]$. This leads to the partial factorization of $f(x)$ given by (4.32). Should this factorization be trivial, we calculate $\gcd(f(x), f_k(x))$ for $1 \leqslant k < m/n$. If this also does not produce a nontrivial factor of $f(x)$, we transform $f(x)$ into $\bar{f}(x)$ by either (4.34) or (4.33), depending on whether $f(x)$ is a binomial or not. As we have shown above, an application of the algorithm to $\bar{f}(x)$ is bound to yield a nontrivial factor of $\bar{f}(x)$ and thus of $f(x)$. Once a nontrivial factor of $f(x)$ has been found, the procedure is continued with the resulting factors in place of $f(x)$, until $f(x)$ is split up completely into linear factors.

NOTES

1. The factorization algorithm based on the matrix B (Berlekamp's algorithm) was first developed in Berlekamp [3] and republished in Berlekamp [4, Ch. 6]. The fact that the matrix B can be used to determine the number of irreducible factors of f had been noted earlier. Petr [1] showed in the case where $e_1 = \cdots = e_k = 1$ in (4.1) that the characteristic polynomial $\det(xI - B)$ of B is given by $(x^{n_1} - 1) \cdots (x^{n_k} - 1)$, where $n_i = \deg(f_i)$ for $1 \leqslant i \leqslant k$. The general case was dealt with by Schwarz [1] who proved that $\det(xI - B) = x^m(x^{n_1} - 1) \cdots (x^{n_k} - 1)$ with $m = \sum_{i=1}^{k} n_i(e_i - 1)$; see also Schwarz [11]. The result that the rank of $B - I$ is equal to $n - k$ was established by Butler [1] and a new proof was given by Schwarz [11]. An interpretation of the number k as the dimension of a vector space of linear recurring sequences (see Chapter 8) was found by Willett [5]. Berlekamp's algorithm occurs in the following surveys: Childs [1, Part II, Ch. 12], Collins [3], Knuth [3, Ch. 4], Lidl and Pilz [1, Ch. 7], Mignotte [3], and Zimmer [2, Ch. 2].

The algorithms based on the polynomials T_i (see Theorem 4.3) and R_i (see Theorem 4.5) are due to McEliece [2]. Factorizations of the binomials $x^n - 1$ obtained by these algorithms are tabulated in McEliece [3]. Surveys of these and other factorization algorithms can be found in Collins [3], Knuth [3, Ch. 4], Lidl and Wiesenbauer [1, Ch. 2], Mignotte [3], and Zimmer [2, Ch. 2].

The matrix B is also of use in the determination of the numbers σ_d, $1 \leqslant d \leqslant n$, where σ_d is the number of distinct monic irreducible factors of f of degree d. Schwarz [11] showed that the σ_d satisfy the following system of linear equations:

$$\sum_{d=1}^{n} \gcd(j, d)\sigma_d = n - h_j \quad \text{for } 1 \leqslant j \leqslant n,$$

where h_j is the rank of the matrix $B^j - I$. Special cases were considered earlier in Rédei [9] and Schwarz [4], [7]. This system was used by Gunji and Arnon [1] in an algorithm for determining the σ_d. A congruence modulo the characteristic p of $\mathbb{F}_q$ for the σ_d was established by Schwarz [13]—namely,

$$d\sigma_d \equiv \sum_{t \mid d} \mu(d/t)\operatorname{Tr}(B^t) \bmod p,$$

where μ is the Moebius function, $\operatorname{Tr}(B^t)$ is the trace of the matrix B^t, and the sum is over all positive divisors t of d. The special case $d = 1$ of this congruence appears already in Schwarz [3]. The latter paper contains also congruences modulo p for the σ_d in terms of the $n \times n$ matrix whose (i, j) entry is s_{i+j}, where s_r denotes the sum of rth powers of the roots of f. Formulas for the σ_d in terms of the ranks of related matrices were obtained in Schwarz [14]. Horáková and Schwarz [1] established formulas for the σ_d in terms of the ranks of circulant matrices involving the coefficients of f. See also Ward [10] for related work. Connections between properties of the polynomial f and those of the corresponding matrix B were also investigated by Chen and Li [1] and T'u [1]. The matrix B is used in a general factorization algorithm of Kempfert [1] and in a criterion of Sims [1] for a commutative finite-dimensional algebra over $\mathbb{F}_q$ to be a field.

A useful result on the number of irreducible factors of a polynomial is the following. Let $f \in \mathbb{F}_q[x]$, q odd, with $\deg(f) = n \geqslant 2$ and discriminant $D \neq 0$ (see Definition 1.92), and let k be the number of monic irreducible factors of f; then $\eta(D) = (-1)^{n-k}$, where η is the quadratic character of $\mathbb{F}_q$ (see Example 5.10). This was first shown for prime fields by Pellet [2] and then proved in a more general form by Stickelberger [2], and is known in the literature as Stickelberger's parity theorem. Proofs of Stickelberger's parity theorem can also be found in Berlekamp [4, Ch. 6], Carlitz [67], Childs [1, Part III, Ch. 15], Dalen [1], Hensel [2], Lubelski [2], Rédei [6], [10, Ch. 11], Schwarz [3], Segre [10], Skolem [5], Swan [1], and Voronoï [2], and a special case was considered in Dickson [9]. See Stickelberger [2], Carlitz [44], Segre [10], and Berlekamp [4, Ch. 6], [10] for versions of Stickelberger's parity theorem in the case where q is even. The fact that Stickelberger's parity theorem can be used to prove the law of quadratic reciprocity (see Theorem 5.17) was already noted by Pellet [2]; see also Berlekamp [4, Ch. 6], Childs [1, Part III, Ch. 16], Mirimanoff and Hensel [1], Rédei [10, Ch. 11], and Swan [1].

Factorizations of some special classes of polynomials have been treated elsewhere. For cyclotomic polynomials see Chapter 2, Section 4, for linearized and affine polynomials see Chapter 3, Section 4, and for binomials and trinomials see Chapter 3, Section 5. For results on the number of roots, and thus the number of linear factors, of a polynomial we refer to Chapter 6, Section 1. For factorization tables and tables of irreducible polynomials see Chapter 10.

The factorization behavior of quartic polynomials was studied by Carlitz [70], [73], Giudici and Margaglio [1], Grebenjuk [1], Leonard [3], Leonard and Williams [1], and Skolem [4]. The factorization of polynomials $f(x^t)$, f irreducible, was considered in Agou [10], [11], Butler [2], and Pellet [1]. Results on the degrees of factors of polynomials $g(x)^t - a$ were obtained by Ore [2], and this work was generalized by Petterson [1], [2], [3] who treated $f(g(x)^t)$ with f irreducible and similar extensions. Agou [14], [19], [20], Long [3], [4], Long and Vaughan [1], [2], and Varshamov [3], [5] studied the factorization of polynomials of the form $f(L(x))$ with f irreducible and

L linearized, and analogous questions for several indeterminates were dealt with in Carlitz and Long [1] and Long [5]. The degrees and the number of irreducible factors of $(cx+d)x^{q^m}-(ax+b)$ over $\mathbb{F}_q$ were determined by Daykin [3]. K. S. Williams [25] gives a factorization of the so-called Dickson polynomials (see Chapter 7, Section 2) and Sergeev [1] factors a closely related class of polynomials. Brillhart [1] has results on the factorization of Euler and Bernoulli polynomials over $\mathbb{F}_p$. Golomb and Lempel [1] show results on the factorization of polynomials given by second-order recursions. Feit and Rees [1] have a criterion for a polynomial $f \in \mathbb{F}_q[x]$ to be a product of linear factors over $\mathbb{F}_q$ in terms of sums of powers of the roots of f. See Šatunovskiĭ [1], Schönemann [2], and Thouvenot and Châtelet [1] for earlier results in this direction for finite prime fields. Craven and Csordas [1] characterize the sequences (c_n), $n=0,1,\ldots,$ such that for all polynomials $\Sigma a_n x^n$ that split in $\mathbb{F}_q$ the polynomial $\Sigma c_n a_n x^n$ also splits in $\mathbb{F}_q$.

S. D. Cohen [5], [6] studies the distribution of various factorization patterns among polynomials of the form $f(x)+ag(x)$ with given $f, g \in \mathbb{F}_q[x]$ and varying $a \in \mathbb{F}_q$; see also Leonard [5]. The distribution of factorization patterns in residue classes modulo a given polynomial or in sets of polynomials of fixed degree with preassigned coefficients is considered in S. D. Cohen [7]. K. S. Williams [11] gives an asymptotic formula for the number $N(q,d,s,e)$ of monic polynomials of fixed degree d over $\mathbb{F}_q$ having exactly s distinct monic irreducible factors over $\mathbb{F}_q$ of degree e. Car [5], [7] and S. D. Cohen [3] show asymptotic formulas for the number of monic polynomials over $\mathbb{F}_q$ of fixed degree and having a certain factorization pattern; see Gogia and Luthar [1] for the case where the degree is bounded by a given positive integer. Zsigmondy [2] determines the number of monic polynomials of fixed degree n over $\mathbb{F}_p$ that have a given number of distinct roots in $\mathbb{F}_p$, and for $1 \leqslant d \leqslant n$ the number of monic polynomials of degree n over $\mathbb{F}_p$ having no irreducible factor of degree d is also calculated.

In the actual implementation of a factorization algorithm various field operations need to be performed. For instance, in order to make the given polynomial f monic, the multiplicative inverse of the leading coefficient of f needs to be calculated. Various methods have been devised for carrying out the arithmetic in a finite field efficiently. For the calculation of multiplicative inverses see Collins [1], Davida [1], Schönhage [1], and Willett [6]. Efficient methods for the multiplication of field elements are discussed in Fiduccia and Zalcstein [1] and Winograd [1], [2]; compare also with Bini and Capovani [1]. Liu, Reed, and Truong [1] and Reed, Truong, and Miller [1], [2] present algorithms for calculating certain primitive roots of unity. The paper of Pohlig and Hellman [1] contains a fast algorithm for computing the exponent r in $b^r = a$ for given $a, b \in \mathbb{F}_q^*$, where b is a primitive element of $\mathbb{F}_q$; see also Herlestam and Johannesson [1], Pollard [3], and Zierler [9], as well as Chapter 10, Section 1 and Table A. A survey of arithmetic algorithms in $\mathbb{F}_q$ is given in Mignotte [3]. The implementation of finite field arithmetic on switching circuits is discussed in Bartee and Schneider [1], Berlekamp [4, Ch. 2], Gill [2, Ch. 6], MacWilliams and Sloane [2, Ch. 3], Peterson and Weldon [1, Ch. 7], Redinbo [1], Tanaka, Kasahara, Tezuka, and Kasahara [1], and Willett [6]; for a computer implementation see Calmet and Loos [2].

A useful computational device is the *discrete Fourier transform* for $\mathbb{F}_q$ introduced by Pollard [1]. Let b be an element of order d in the multiplicative group $\mathbb{F}_q^*$. Then a finite sequence $a_0, a_1, \ldots, a_{d-1}$ of elements of $\mathbb{F}_q$ is transformed into the finite sequence

$$A_i = \sum_{j=0}^{d-1} a_j b^{ij}, \quad i = 0, 1, \ldots, d-1.$$

The inverse transform is given by

$$a_j = d^{-1} \sum_{i=0}^{d-1} A_i b^{-ij}, \quad j = 0, 1, \ldots, d-1.$$

Both transforms can be calculated quickly by a method analogous to the fast Fourier transform for complex numbers. The discrete Fourier transform occurs under the name of Mattson-Solomon polynomial in coding theory (see Mattson and Solomon [1] and MacWilliams and Sloane [2, Ch. 8]). Pollard [1] shows how to use the discrete Fourier transform for the arithmetic in $\mathbb{F}_q$. Further work on this transform and its applications can be found in Agarwal and Burrus [1], Blahut [1], Fateman [1], Golomb, Reed, and Truong [1], Liu, Reed, and Truong [1], [2], McClellan and Rader [1], Nussbaumer [1, Ch. 8], Pollard [2], Preparata and Sarwate [1], Redinbo [1], Reed, Scholtz, Truong, and Welch [1], Reed and Truong [1], [2], [3], Reed, Truong, and Welch [1], Rice [1], and Sarwate [1], [2].

The implementation of a factorization algorithm also depends on operations with polynomials. Efficient methods for multiplying two polynomials over $\mathbb{F}_q$ are described in Bassalygo [1], Borodin and Munro [1], Rice [2], and Schönhage [2]; see also Fateman [1] and Pollard [1]. For the calculation of greatest common divisors and the Euclidean algorithm for polynomials in one or several indeterminates see Agou [2], [5], Barnett [1], Blankinship [1], N. K. Bose [1], W. S. Brown [1], Collins [3], Dickson [33], Emre and Hüseyin [1], Knuth [3, Ch. 4], Maroulas and Barnett [1], McEliece and Shearer [1], Moses [1], and Vogt and Bose [1]. Further work on polynomial arithmetic can be found in Bhanu Murthy and Sampath [1], Lal [1], Mihaĭljuk [1], [2], and Murzaev [1]. For surveys of polynomial operations over finite fields see Aho, Hopcroft, and Ullman [1], Berlekamp [4, Ch. 2], Birkhoff and Bartee [1, Ch. 11], Borodin and Munro [1], Gill [2, Ch. 6], Knuth [3, Ch. 4], and Peterson and Weldon [1, Ch. 7].

2. The factorization algorithm based on the consideration of resultants goes back to a proposal in the first edition of Knuth [3, Ch. 4]. The calculation of polynomial resultants by interpolation was suggested by

Collins [2]. The Zassenhaus algorithm was introduced in Zassenhaus [5]. The factorization algorithm involving the diagonalization of matrices of polynomials is due to Berlekamp [6]. The algorithm dealt with in Theorem 4.13 is from Golomb, Welch, and Hales [1]; see also Agou [8] and Knuth [3, Ch. 4]. We refer to Cantor and Zassenhaus [1] for a probabilistic algorithm in this connection.

Zassenhaus [4], Kempfert [1], and Knuth [3, Ch. 4] considered an algorithm in which, for given $f \in \mathbb{F}_q[x]$ with no repeated factors and $\deg(f) = n \geqslant 2$, one computes successively $d_i = \gcd(f(x), x^{q^i} - x)$ for $i = 1, 2, \ldots, \lfloor n/2 \rfloor$. If $d_i = 1$ for all i, then f is irreducible over $\mathbb{F}_q$. Otherwise, let j be the smallest index with $d_j \neq 1$; then f is reducible over $\mathbb{F}_q$ and d_j is the product of all irreducible factors of f of degree j. Further factorization algorithms can be found in Ananiashvili, Varshamov, Gorovoĭ, and Parhomenko [1], Arwin [2], Beard [5], Camion [2], [3], Dyn'kin and Agaronov [1], Rabin [1], and Varshamov and Ostianu [1]. A comparative study of the efficiency of various factorization algorithms was carried out by Moenck [1]. Methods for the calculation of the degree of the splitting field of a polynomial over $\mathbb{F}_p$ were given by Speiser [1] and Wegner [4]; see Mignotte [1], [2], [4] for general $\mathbb{F}_q$.

Effective factorization algorithms for finite prime fields are important tools for the factorization of polynomials over the integers. The basic idea is to start from the factorization of a given $f \in \mathbb{Z}[x]$ modulo a suitably chosen prime p and then refine it to a factorization over $\mathbb{Z}$. Berlekamp [6], [7] and Knuth [3, Ch. 4] take an *a priori* bound B_0 on the absolute value of any coefficient of any possible factor of f over $\mathbb{Z}$, then select a prime $p > 2B_0$ and note that the factors of f over $\mathbb{Z}$ must occur among the known factors modulo p. Values for B_0 can be found in Childs [1, Part II, Ch. 13], Knuth [3, Ch. 4], Mignotte [4], Zassenhaus [5], and Zimmer [2, Ch. 2]. Zassenhaus [5], [6], [7] considers a p-adic routine in which one starts from the factorization of f modulo a smaller prime p and then refines it by a constructive version of Hensel's lemma to a factorization modulo a power $p^k > 2B_0$. For further work see Kempfert [1], Lenstra, Lenstra, and Lovász [1], and Lloyd [1], [2]; see also Childs [1, Part II, Ch. 13]. Surveys of factorization methods over $\mathbb{Z}$ are presented in Collins [3], Knuth [3, Ch. 4], and Zimmer [2, Ch. 2]. It should be noted that a polynomial $f \in \mathbb{Z}[x]$ may be irreducible over the rationals, but reducible modulo p for all primes p. An example of such a polynomial was already given by Hilbert [1]. A particularly simple example is due to Schwarz [4]—namely, $f(x) = x^4 + 1$. For other examples see M. A. Lee [1] and Pólya and Szegö [1, Sec. VIII, Problem 129]. Similarly, there are polynomials over $\mathbb{Z}$ that have no linear factor over the rationals, but have a linear factor modulo p for all primes p (see, e.g., Hasse [1], Skolem [3], and van der Waerden [1]).

Factorization algorithms for polynomials over algebraic number fields were developed by A. K. Lenstra [1] and Weinberger and Rothschild

[1]. Methods of factoring polynomials in several indeterminates over the rationals or over algebraic number fields are discussed in Collins [3], Moses [1], Musser [1], Viry [1], [2], P. S. Wang [1], [2], Wang and Rothschild [1], and Weinberger and Rothschild [1].

Results on matrices of polynomials can be found in Albert [3, Ch. 3], Gantmacher [1, Ch. 6], Hoffman and Kunze [1, Ch. 7], Krishnamurthy [1], and Maroulas and Barnett [1].

3. The methods described in this section are from Berlekamp [6]. Some improvements for prime fields $\mathbb{F}_p$ with $p-1$ divisible by a large power of 2 have been found by Moenck [1]. The root-finding algorithm based on the consideration of affine multiples and described in Chapter 3, Section 4, was developed by Berlekamp, Rumsey, and Solomon [1]; see also Berlekamp [4, Ch. 11]. Another method of finding roots for finite fields of large characteristic was proposed by Rabin [1]; see also Cantor and Zassenhaus [1]. Calmet and Loos [2] discuss the computer implementation of these algorithms. The theoretical question of when a polynomial equation over $\mathbb{F}_q$ is solvable by radicals was studied by Fray and Gilmer [1] and Mann [6]. The latter paper contains also expressions for the roots of f in terms of roots of unity over $\mathbb{F}_q$ and polynomials in the coefficients of f, in the case where f is irreducible over $\mathbb{F}_q$ and of degree not divisible by the characteristic of $\mathbb{F}_q$. If f has roots in $\mathbb{F}_q$, a complicated expression for these roots depending on a primitive element of $\mathbb{F}_q$ was given by Prešić [1]. An explicit expression for $\gcd(f(x), x^{q-1}-1)$ in case q is prime appears in Rados [4].

Special procedures have been developed for finding the roots of polynomials of small degree. Even the problem of determining the roots of $x^2 - a \in \mathbb{F}_p[x]$, p prime, is nontrivial when p is large. Some methods occur already in Gauss [1, Ch. 6]. See also Adleman, Manders, and Miller [1], Chang [1], Cipolla [1], [2], D. H. Lehmer [10], Pocklington [1], Schönheim [1], Shanks [2], H. J. S. Smith [1], Tamarkine and Friedmann [1], Tonelli [1], Uspensky and Heaslet [1, Ch. 10], and Vandiver [2] for this case. Methods of finding the roots of quadratic polynomials over finite fields of characteristic 2 are discussed in Berlekamp [4, Ch. 6] and Berlekamp, Rumsey, and Solomon [1]. For cubic and quartic polynomials see Arnoux [1, Ch. 9], Arwin [1], Cailler [1], Cauchy [3], Cordone [1], Dickson [31], Escott [1], Grebenjuk [2], Hirschfeld [5, Ch. 1], Ivanov [1], Matveeva [1], Mignosi [1], Mirimanoff [1], Oltramare [1], Sansone [1], [2], [3], [4], [5], Scarpis [1], Segre [10], and Williams and Zarnke [1]. For quintic polynomials see Arwin [2] and Gorbov and Šmidt [1]. Special methods of calculating roots for polynomials of small degree arise also in the context of decoding algorithms in algebraic coding theory; see, for example, Blokh [1], Chien and Cunningham [1], and Polkinghorn [1]. Explicit formulas for the roots of binomials over $\mathbb{F}_p$ can be found in Cipolla [4], Dickson [40, Ch. 7], Furquim de Almeida [1], Lindgren [1], and Scorza [1]. An efficient algorithm for finding the roots of binomials over $\mathbb{F}_p$ was developed by Adleman, Manders, and Miller [1]; see Mignotte [5] for arbitrary $\mathbb{F}_q$. The case of binomials over $\mathbb{F}_p$ of prime degree was treated by H. C. Williams [1].

Conditions for a polynomial over $\mathbb{F}_q$ to have all its roots in $\mathbb{F}_q$ can be found in Feit and Rees [1], and in Šatunovskiĭ [1], Schönemann [2], and Thouvenot and Châtelet [1] for the case where q is prime. Mignotte [4] presents a fast algorithm for testing whether all roots of $f \in \mathbb{F}_q[x]$ are in $\mathbb{F}_q$. Cipolla [5] and Mignosi [2] have formulas for the roots of f in case they are all in $\mathbb{F}_q$. Rédei [11, Ch. 5] determines the polynomials $f(x) = x^q + a_k x^k + a_{k-1}x^{k-1} + \cdots + a_0 \in \mathbb{F}_q[x]$ with $k \leqslant (q+1)/2$ that have all their roots in $\mathbb{F}_q$. See also Rédei [7] for an earlier result in this direction. Gerst and Brillhart [1] study conditions under which $f \in \mathbb{Z}[x]$ splits into distinct linear factors modulo p for many primes p; see Lubelski [3] for a related example.

For results on the number of roots of polynomials in a given finite field we refer to Chapter 6, Section 1, and the notes to that section.

EXERCISES

4.1. Factor $x^{12} + x^7 + x^5 + x^4 + x^3 + x^2 + 1$ over $\mathbb{F}_2$ by Berlekamp's algorithm.

4.2. Factor $x^7 + x^6 + x^5 - x^3 + x^2 - x - 1$ over $\mathbb{F}_3$ by Berlekamp's algorithm.

4.3. Let $\mathbb{F}_4 = \mathbb{F}_2(\theta)$ and factor $x^5 + \theta x^4 + x^3 + (1+\theta)x + \theta$ over $\mathbb{F}_4$ by Berlekamp's algorithm.

4.4. Use Berlekamp's algorithm to prove that $x^6 - x^3 - x - 1$ is irreducible in $\mathbb{F}_3[x]$.

4.5. Use Berlekamp's algorithm to determine the number of distinct monic irreducible factors of $x^4 + 1$ in $\mathbb{F}_p[x]$ for all odd primes p.

4.6. Use the polynomials T_i in Section 1 to factor $x^5 + x^4 + 1$ over $\mathbb{F}_2$.

4.7. Determine the splitting field of $x^8 + x^6 + x^5 + x^4 + x^3 + x^2 + 1$ over $\mathbb{F}_2$.

4.8. Determine the splitting field of $x^6 - x^4 - x^2 - x + 1$ over $\mathbb{F}_3$.

4.9. Use the polynomials R_i in Section 1 to factor the polynomial of Exercise 4.1 over $\mathbb{F}_2$.

4.10. Find the canonical factorization of $x^8 + x^6 + x^4 + x^3 + 1$ in $\mathbb{F}_2[x]$ by using the polynomials R_i in Section 1.

4.11. Determine the canonical factorization of the cyclotomic polynomial $Q_{31}(x)$ in $\mathbb{F}_2[x]$.

4.12. Factor $f(x) = x^8 + x^3 + 1$ over $\mathbb{F}_2$ and determine $\operatorname{ord}(f(x))$.

4.13. Factor $f(x) = x^9 + x^8 + x^7 + x^4 + x^3 + x + 1$ over $\mathbb{F}_2$ and determine $\operatorname{ord}(f(x))$.

4.14. Prove in detail that if f is a nonzero polynomial over a field and $d = \gcd(f, f')$, then f/d has no repeated factors. (*Note*: Count nonzero constant polynomials among the polynomials with no repeated factors.)

4.15. Let f be a monic polynomial of positive degree with integer coefficients. Prove that if f has no repeated factors, then there are only finitely many primes p such that f, considered as a polynomial over $\mathbb{F}_p$, has repeated factors.

4.16. Determine the number of monic polynomials in $\mathbb{F}_q[x]$ of degree $n \geqslant 1$ with no repeated factors.

4.17. Let f be a monic polynomial over $\mathbb{F}_q$ and let $g_1, \ldots, g_r$ be nonzero polynomials over $\mathbb{F}_q$ that are pairwise relatively prime. Prove that if f divides $g_1 \cdots g_r$, then $f = \prod_{i=1}^r \gcd(f, g_i)$.

4.18. Use Berlekamp's algorithm to prove the following special case of Theorem 3.75: the binomial $x^t - a$, where t is a prime divisor of $q-1$ and $a \in \mathbb{F}_q^*$, is irreducible in $\mathbb{F}_q[x]$ if and only if $a^{(q-1)/t} \neq 1$.

4.19. Let f be an irreducible polynomial in $\mathbb{F}_q[x]$ of degree n and define the $n \times n$ matrix $B = (b_{ij})$ by (4.4). Prove that the characteristic polynomial $\det(xI - B)$ of B is equal to $x^n - 1$.

4.20. Let $f = f_1 \cdots f_k$ be a product of k distinct monic irreducible polynomials $f_1, \ldots, f_k$ in $\mathbb{F}_q[x]$ of degree $n_1, \ldots, n_k$, respectively. Put $\deg(f) = n = n_1 + \cdots + n_k$ and define the $n \times n$ matrix $B = (b_{ij})$ by (4.4). Prove that the characteristic polynomial $\det(xI - B)$ of B is equal to $(x^{n_1} - 1) \cdots (x^{n_k} - 1)$.

4.21. In the notation of Section 1, prove that the polynomials T_i do not separate those irreducible factors f_j of f for which N/n_j is divisible by the characteristic of $\mathbb{F}_q$.

4.22. Let $f \in \mathbb{F}_q[x]$ be monic of degree $n \geqslant 1$. Define $h \in \mathbb{F}_q[x, y]$ by

$$h(x, y) = (y - x)(y - x^q)\left(y - x^{q^2}\right) \cdots \left(y - x^{q^{n-1}}\right) - f(y)$$

and write

$$h(x, y) = s_{n-1}(x) y^{n-1} + \cdots + s_1(x) y + s_0(x).$$

Prove that f is irreducible over $\mathbb{F}_q$ if and only if f divides s_j for $0 \leqslant j \leqslant n-1$.

4.23. Use the criterion in the preceding exercise to prove that $x^7 + x^6 + x^3 + x^2 + 1$ is reducible over $\mathbb{F}_2$.

4.24. Prove that the quadratic polynomial $f(x) = x^2 + bx + c$ is irreducible over $\mathbb{F}_q$ if and only if $f(x)$ divides $x^q + x + b$.

4.25. Let f be an irreducible polynomial in $\mathbb{F}_q[x]$ of degree m and let λ be a root of f in $\mathbb{F}_{q^m}$. Let g and h be nonzero polynomials in $\mathbb{F}_q[x]$. Prove that $h(x)^m f(g(x)/h(x))$ is irreducible in $\mathbb{F}_q[x]$ if and only if $g(x) - \lambda h(x)$ is irreducible in $\mathbb{F}_{q^m}[x]$.

4.26. Use the method in Example 4.7 to factor $x^4 + 3x^3 + 4x^2 + 2x - 1$ over $\mathbb{F}_{13}$.

4.27. Use the method in Example 4.7 to factor $x^3 - 6x^2 - 8x - 8$ over $\mathbb{F}_{19}$.

4.28. Use the Zassenhaus algorithm to factor $x^4+3x^3+4x^2+2x-1$ over $\mathbb{F}_{13}$.

4.29. Use the Zassenhaus algorithm to factor x^3-6x^2-8x-8 over $\mathbb{F}_{19}$.

4.30. Use the Zassenhaus algorithm to factor $x^5+3x^4+2x^3-6x^2+5$ over $\mathbb{F}_{17}$.

4.31. Factor $x^4-7x^3+4x^2+2x+4$ over $\mathbb{F}_{17}$.

4.32. Factor $x^4-3x^3+4x^2-6x-8$ over $\mathbb{F}_{19}$.

4.33. Prove in detail that equivalence of square matrices of polynomials as defined by Definition 4.11 is reflexive, symmetric, and transitive.

4.34. Use the method in Example 4.14 to factor x^3-6x^2-8x-8 over $\mathbb{F}_{19}$.

4.35. Use the method in Example 4.14 to factor $x^5+3x^4+2x^3-6x^2+5$ over $\mathbb{F}_{17}$.

4.36. Use the method in Example 4.14 to obtain a partial factorization of $x^7-2x^6-4x^4+3x^3-5x^2+3x+5$ over $\mathbb{F}_{11}$ and complete the factorization by another method.

4.37. Find the roots of $f(x)=x^5-x^4+2x^3+x^2-x-2\in\mathbb{F}_5[x]$ contained in $\mathbb{F}_5$.

4.38. Find the roots of $f(x)=x^5+6x^4+2x^3-6x^2-5x+5\in\mathbb{F}_{13}[x]$ contained in $\mathbb{F}_{13}$.

4.39. Prove that all the roots of $f(x)=x^3+8x^2+6x-7\in\mathbb{F}_{19}[x]$ are contained in $\mathbb{F}_{19}$ and find them.

4.40. Let $\mathbb{F}_{32}=\mathbb{F}_2(\beta)$, where β is a root of the irreducible polynomial x^5+x^2+1 over $\mathbb{F}_2$. Prove that all the roots of $f(x)=x^3+(\beta^4+\beta^3+1)x^2+\beta^2x+\beta^4+\beta^3+\beta+1\in\mathbb{F}_{32}[x]$ are contained in $\mathbb{F}_{32}$ and find them.

4.41. Let $\mathbb{F}_{27}=\mathbb{F}_3(\beta)$, where β is a root of the irreducible polynomial x^3-x+1 over $\mathbb{F}_3$. Prove that all the roots of $f(x)=x^3+x^2-(\beta^2-\beta+1)x+\beta^2-1\in\mathbb{F}_{27}[x]$ are contained in $\mathbb{F}_{27}$ and find them.

4.42. Let $\mathbb{F}_{169}=\mathbb{F}_{13}(\beta)$, where β is a root of the irreducible polynomial x^2-x-1 over $\mathbb{F}_{13}$. Find the roots of $f(x)=x^2+(3\beta+1)x+\beta+5\in\mathbb{F}_{169}[x]$ contained in $\mathbb{F}_{169}$.

4.43. If the polynomial $f(x-b)$ in (4.22) is quadratic with constant term $c\neq 0$, prove that the factorization in (4.22) is nontrivial if and only if c is not the square of an element of $\mathbb{F}_p$.

4.44. Let β be a defining element of $F=\mathbb{F}_{2^m}$ over $\mathbb{F}_2$. Prove:

(a) There exists k, $0\leqslant k\leqslant m-1$, with $\mathrm{Tr}_F(\beta^k)=1$.

(b) For each $i=0,1,\ldots,m-1$ there exists an $\alpha_i\in F$ such that

$$\alpha_i^2+\alpha_i=\begin{cases}\beta^i & \text{if } \mathrm{Tr}_F(\beta^i)=0,\\ \beta^i+\beta^k & \text{if } \mathrm{Tr}_F(\beta^i)=1.\end{cases}$$

(c) If $\gamma=\sum_{i=0}^{m-1}c_i\beta^i$, $c_i\in\mathbb{F}_2$, and $\mathrm{Tr}_F(\gamma)=0$, then the roots of $x^2+x+\gamma$ are $\sum_{i=0}^{m-1}c_i\alpha_i$ and $1+\sum_{i=0}^{m-1}c_i\alpha_i$.

Chapter 5

Exponential Sums

Exponential sums are important tools in number theory for solving problems involving integers—and real numbers in general—that are often intractable by other means. Analogous sums can be considered in the framework of finite fields and turn out to be useful in studying the number of solutions of equations over finite fields (see Chapter 6) and in various applications of finite fields.

A basic role in setting up exponential sums for finite fields is played by special group homomorphisms called characters. It is necessary to distinguish between two types of characters—namely, additive and multiplicative characters—depending on whether reference is made to the additive or the multiplicative group of the finite field. Exponential sums are formed by using the values of one or more characters and possibly combining them with weights or with other function values. If we only sum the values of a single character, we speak of a character sum.

In Section 1 we lay the foundation by first discussing characters of finite abelian groups and then specializing to finite fields. Section 2 is devoted to Gaussian sums, which are arguably the most important types of exponential sums for finite fields as they govern the transition from the additive to the multiplicative structure and vice versa. They also appear in many other contexts in algebra and number theory. The closely related Jacobi sums are studied in the next section. These sums are of particular importance for the applications to equations over finite fields presented in Chapter 6.

Character sums with polynomial arguments (sometimes called Weil sums) are treated in Section 4 as far as this is feasible with fairly elementary means—that is, without the use of algebraic geometry. The "elementary" approach has now been developed to such an extent that it yields all the basic estimates for Weil sums. The method goes hand in hand with a detailed study of certain types of equations over finite fields. Therefore, several loose ends can only be tied up in Chapter 6 where these equations will be discussed.

In Section 5 we consider special character sums of number-theoretic significance—namely, Kloosterman sums and Jacobsthal sums. A result of theoretical interest for quadratic character sums offers a pretext for introducing continued fraction expansions of rational functions over finite fields.

1. CHARACTERS

Let G be a finite abelian group (written multiplicatively) of order $|G|$ with identity element 1_G. A *character* χ of G is a homomorphism from G into the multiplicative group U of complex numbers of absolute value 1—that is, a mapping from G into U with $\chi(g_1g_2)=\chi(g_1)\chi(g_2)$ for all $g_1, g_2 \in G$. Since $\chi(1_G)=\chi(1_G)\chi(1_G)$, we must have $\chi(1_G)=1$. Furthermore,

$$(\chi(g))^{|G|}=\chi(g^{|G|})=\chi(1_G)=1$$

for every $g \in G$, so that the values of χ are $|G|$th roots of unity. We note also that $\chi(g)\chi(g^{-1})=\chi(gg^{-1})=\chi(1_G)=1$, and so $\chi(g^{-1})=(\chi(g))^{-1}=\overline{\chi(g)}$ for every $g \in G$, where the bar denotes complex conjugation.

Among the characters of G we have the *trivial* character χ_0 defined by $\chi_0(g)=1$ for all $g \in G$; all other characters of G are called *nontrivial.* With each character χ of G there is associated the *conjugate* character $\bar{\chi}$ defined by $\bar{\chi}(g)=\overline{\chi(g)}$ for all $g \in G$. Given finitely many characters $\chi_1,\ldots,\chi_n$ of G, one can form the product character $\chi_1 \cdots \chi_n$ by setting $(\chi_1 \cdots \chi_n)(g)=\chi_1(g)\cdots\chi_n(g)$ for all $g \in G$. If $\chi_1=\cdots=\chi_n=\chi$, we write χ^n for $\chi_1 \cdots \chi_n$. It is obvious that the set $G^\wedge$ of characters of G forms an abelian group under this multiplication of characters. Since the values of characters of G can only be $|G|$th roots of unity, $G^\wedge$ is finite.

After briefly considering the special case of a finite cyclic group, we establish some basic facts about characters.

5.1. Example. Let G be a finite cyclic group of order n, and let g be a generator of G. For a fixed integer j, $0 \leqslant j \leqslant n-1$, the function

$$\chi_j(g^k)=e^{2\pi ijk/n}, \quad k=0,1,\ldots,n-1,$$

defines a character of G. On the other hand, if χ is any character of G, then $\chi(g)$ must be an nth root of unity, say $\chi(g)=e^{2\pi ij/n}$ for some j, $0 \leqslant j \leqslant$

$n-1$, and it follows that $\chi = \chi_j$. Therefore, $G^\wedge$ consists exactly of the characters $\chi_0, \chi_1, \ldots, \chi_{n-1}$. □

5.2. Theorem. *Let H be a subgroup of the finite abelian group G and let ψ be a character of H. Then ψ can be extended to a character of G; that is, there exists a character χ of G with $\chi(h) = \psi(h)$ for all $h \in H$.*

Proof. We may suppose that H is a proper subgroup of G. Choose $a \in G$ with $a \notin H$, and let H_1 be the subgroup of G generated by H and a. Let m be the least positive integer for which $a^m \in H$. Then every element $g \in H_1$ can be written uniquely in the form $g = a^j h$ with $0 \leqslant j < m$ and $h \in H$. Define a function ψ_1 on H_1 by $\psi_1(g) = \omega^j \psi(h)$, where ω is a fixed complex number satisfying $\omega^m = \psi(a^m)$. To check that ψ_1 is indeed a character of H_1, let $g_1 = a^k h_1$, $0 \leqslant k < m$, $h_1 \in H$, be another element of H_1. If $j + k < m$, then $\psi_1(gg_1) = \omega^{j+k}\psi(hh_1) = \psi_1(g)\psi_1(g_1)$. If $j + k \geqslant m$, then $gg_1 = a^{j+k-m}(a^m hh_1)$, and so

$$\begin{aligned} \psi_1(gg_1) &= \omega^{j+k-m}\psi(a^m hh_1) \\ &= \omega^{j+k-m}\psi(a^m)\psi(hh_1) = \omega^{j+k}\psi(hh_1) = \psi_1(g)\psi_1(g_1). \end{aligned}$$

It is obvious that $\psi_1(h) = \psi(h)$ for $h \in H$. If $H_1 = G$, then we are done. Otherwise, we can continue the process above until, after finitely many steps, we obtain an extension of ψ to G. □

5.3. Corollary. *For any two distinct elements $g_1, g_2 \in G$ there exists a character χ of G with $\chi(g_1) \neq \chi(g_2)$.*

Proof. It suffices to show that for $h = g_1 g_2^{-1} \neq 1_G$ there exists a character χ of G with $\chi(h) \neq 1$. This follows, however, from Example 5.1 and Theorem 5.2 by letting H be the cyclic subgroup of G generated by h. □

5.4. Theorem. *If χ is a nontrivial character of the finite abelian group G, then*

$$\sum_{g \in G} \chi(g) = 0. \tag{5.1}$$

If $g \in G$ with $g \neq 1_G$, then

$$\sum_{\chi \in G^\wedge} \chi(g) = 0. \tag{5.2}$$

Proof. Since χ is nontrivial, there exists $h \in G$ with $\chi(h) \neq 1$. Then

$$\chi(h) \sum_{g \in G} \chi(g) = \sum_{g \in G} \chi(hg) = \sum_{g \in G} \chi(g),$$

because if g runs through G, so does hg. Thus we have

$$(\chi(h) - 1) \sum_{g \in G} \chi(g) = 0,$$

which already implies (5.1). For the second part, we note that the function $\hat{g}$ defined by $\hat{g}(\chi)=\chi(g)$ for $\chi \in G^\wedge$ is a character of the finite abelian group $G^\wedge$. This character is nontrivial since, by Corollary 5.3, there exists $\chi \in G^\wedge$ with $\chi(g) \neq \chi(1_G)=1$. Therefore from (5.1) applied to the group $G^\wedge$,

$$\sum_{\chi \in G^\wedge} \chi(g)=\sum_{\chi \in G^\wedge} \hat{g}(\chi)=0. \qquad \square$$

5.5. Theorem. *The number of characters of a finite abelian group G is equal to $|G|$.*

Proof. This follows from

$$|G^\wedge|=\sum_{g \in G} \sum_{\chi \in G^\wedge} \chi(g)=\sum_{\chi \in G^\wedge} \sum_{g \in G} \chi(g)=|G|,$$

where we used (5.2) in the first identity and (5.1) in the last identity. $\square$

The statements of Theorems 5.4 and 5.5 can be combined into the *orthogonality relations for characters*. Let χ and ψ be characters of G. Then

$$\frac{1}{|G|} \sum_{g \in G} \chi(g)\overline{\psi(g)}=\begin{cases}0 & \text{for } \chi \neq \psi, \\ 1 & \text{for } \chi=\psi.\end{cases} \tag{5.3}$$

The first part follows, of course, by applying (5.1) to the character $\chi\bar{\psi}$; the second part is trivial.

Furthermore, if g and h are elements of G, then

$$\frac{1}{|G|} \sum_{\chi \in G^\wedge} \chi(g)\overline{\chi(h)}=\begin{cases}0 & \text{for } g \neq h, \\ 1 & \text{for } g=h.\end{cases} \tag{5.4}$$

Here, the first part is obtained from (5.2) applied to the element gh^{-1}, whereas the second part follows from Theorem 5.5.

Character theory is often used to obtain expressions for the number of solutions of equations in a finite abelian group G. Let f be an arbitrary map from the cartesian product $G^n=G \times \cdots \times G$ (n factors) into G. Then, for fixed $h \in G$, the number $N(h)$ of n-tuples $(g_1, \ldots, g_n) \in G^n$ with $f(g_1, \ldots, g_n)=h$ is given by

$$N(h)=\frac{1}{|G|} \sum_{g_1 \in G} \cdots \sum_{g_n \in G} \sum_{\chi \in G^\wedge} \chi(f(g_1, \ldots, g_n))\overline{\chi(h)}, \tag{5.5}$$

on account of (5.4).

A character χ of G may be nontrivial on G, but still annihilate a whole subgroup H of G, in the sense that $\chi(h)=1$ for all $h \in H$. The set of all characters of G annihilating a given subgroup H is called the *annihilator* of H in $G^\wedge$.

5.6. Theorem. *Let H be a subgroup of the finite abelian group G. Then the annihilator of H in $G^\wedge$ is a subgroup of $G^\wedge$ of order $|G|/|H|$.*

Proof. Let A be the annihilator in question. Then it is obvious from the definition that A is a subgroup of $G^\wedge$. Let $\chi \in A$; then $\mu(gH) = \chi(g)$, $g \in G$, is a well-defined character of the factor group G/H. Conversely, if μ is a character of G/H, then $\chi(g) = \mu(gH)$, $g \in G$, defines a character of G annihilating H. Distinct elements of A correspond to distinct characters of G/H. Therefore, A is in one-to-one correspondence with the character group $(G/H)^\wedge$, and so the order of A is equal to the order of $(G/H)^\wedge$, which is $|G/H| = |G|/|H|$ according to Theorem 5.5. □

In a finite field $\mathbb{F}_q$ there are two finite abelian groups that are of significance—namely, the additive group and the multiplicative group of the field. Therefore, we will have to make an important distinction between the characters pertaining to these two group structures. In both cases, explicit formulas for the characters can be given.

Consider first the *additive group* of $\mathbb{F}_q$. Let p be the characteristic of $\mathbb{F}_q$; then the prime field contained in $\mathbb{F}_q$ is $\mathbb{F}_p$, which we identify with $\mathbb{Z}/(p)$. Let $\mathrm{Tr}: \mathbb{F}_q \to \mathbb{F}_p$ be the absolute trace function from $\mathbb{F}_q$ to $\mathbb{F}_p$ (see Definition 2.22). Then the function χ_1 defined by

$$\chi_1(c) = e^{2\pi i \,\mathrm{Tr}(c)/p} \quad \text{for all } c \in \mathbb{F}_q \tag{5.6}$$

is a character of the additive group of $\mathbb{F}_q$, since for $c_1, c_2 \in \mathbb{F}_q$ we have $\mathrm{Tr}(c_1 + c_2) = \mathrm{Tr}(c_1) + \mathrm{Tr}(c_2)$, and so $\chi_1(c_1 + c_2) = \chi_1(c_1)\chi_1(c_2)$. Instead of "character of the additive group of $\mathbb{F}_q$," we shall henceforth use the term *additive character* of $\mathbb{F}_q$. The character χ_1 in (5.6) will be called the *canonical additive character* of $\mathbb{F}_q$. All additive characters of $\mathbb{F}_q$ can be expressed in terms of χ_1.

5.7. Theorem. *For $b \in \mathbb{F}_q$, the function χ_b with $\chi_b(c) = \chi_1(bc)$ for all $c \in \mathbb{F}_q$ is an additive character of $\mathbb{F}_q$, and every additive character of $\mathbb{F}_q$ is obtained in this way.*

Proof. For $c_1, c_2 \in \mathbb{F}_q$ we have

$$\begin{aligned}\chi_b(c_1 + c_2) &= \chi_1(bc_1 + bc_2)\\ &= \chi_1(bc_1)\chi_1(bc_2) = \chi_b(c_1)\chi_b(c_2),\end{aligned}$$

and the first part is established. Since Tr maps $\mathbb{F}_q$ onto $\mathbb{F}_p$ by Theorem 2.23(iii), χ_1 is a nontrivial character. Therefore, if $a, b \in \mathbb{F}_q$ with $a \neq b$, then

$$\frac{\chi_a(c)}{\chi_b(c)} = \frac{\chi_1(ac)}{\chi_1(bc)} = \chi_1((a-b)c) \neq 1$$

for suitable $c \in \mathbb{F}_q$, and so χ_a and χ_b are distinct characters. Hence, if b runs through $\mathbb{F}_q$, we get q distinct additive characters χ_b. On the other hand, $\mathbb{F}_q$ has exactly q additive characters by Theorem 5.5, and so the list of additive characters of $\mathbb{F}_q$ is already complete. □

By setting $b=0$ in Theorem 5.7, we obtain the trivial additive character χ_0, for which $\chi_0(c)=1$ for all $c \in \mathbb{F}_q$.

Let E be a finite extension field of $\mathbb{F}_q$, let χ_1 be the canonical additive character of $\mathbb{F}_q$, and let μ_1 be the canonical additive character of E defined in analogy with (5.6), where Tr is of course replaced by the absolute trace function Tr_E from E to $\mathbb{F}_p$. Then χ_1 and μ_1 are connected by the identity

$$\chi_1\left(\mathrm{Tr}_{E/\mathbb{F}_q}(\beta)\right)=\mu_1(\beta) \quad \text{for all } \beta \in E, \tag{5.7}$$

where $\mathrm{Tr}_{E/\mathbb{F}_q}$ is the trace function from E to $\mathbb{F}_q$. This follows from the transitivity relation

$$\mathrm{Tr}_E(\beta)=\mathrm{Tr}\left(\mathrm{Tr}_{E/\mathbb{F}_q}(\beta)\right) \quad \text{for all } \beta \in E,$$

which was shown in Theorem 2.26.

Characters of the *multiplicative group* $\mathbb{F}_q^*$ of $\mathbb{F}_q$ are called *multiplicative characters* of $\mathbb{F}_q$. Since $\mathbb{F}_q^*$ is a cyclic group of order $q-1$ by Theorem 2.8, its characters can be easily determined.

5.8. Theorem. *Let g be a fixed primitive element of $\mathbb{F}_q$. For each $j=0,1,\ldots,q-2$, the function ψ_j with*

$$\psi_j(g^k)=e^{2\pi ijk/(q-1)} \quad \textit{for } k=0,1,\ldots,q-2$$

defines a multiplicative character of $\mathbb{F}_q$, and every multiplicative character of $\mathbb{F}_q$ is obtained in this way.

Proof. This follows immediately from Example 5.1. □

No matter what g is, the character ψ_0 will always represent the trivial multiplicative character, which satisfies $\psi_0(c)=1$ for all $c \in \mathbb{F}_q^*$.

5.9. Corollary. *The group of multiplicative characters of $\mathbb{F}_q$ is cyclic of order $q-1$ with identity element ψ_0.*

Proof. Every character ψ_j in Theorem 5.8 with j relatively prime to $q-1$ is a generator of the group in question. □

5.10. Example. Let q be odd and let η be the real-valued function on $\mathbb{F}_q^*$ with $\eta(c)=1$ if c is the square of an element of $\mathbb{F}_q^*$ and $\eta(c)=-1$ otherwise. Then η is a multiplicative character of $\mathbb{F}_q$. It can also be obtained from the characters in Theorem 5.8 by setting $j=(q-1)/2$. The character η annihilates the subgroup of $\mathbb{F}_q^*$ consisting of the squares of elements of $\mathbb{F}_q^*$, and by Theorem 5.6 it is the only nontrivial character of $\mathbb{F}_q^*$ with this property. This uniquely determined character η is called the *quadratic character* of $\mathbb{F}_q$. If q is an odd prime, then for $c \in \mathbb{F}_q^*$ we have $\eta(c)=\left(\frac{c}{q}\right)$, the Legendre symbol from elementary number theory. □

The orthogonality relations (5.3) and (5.4), when applied to additive or multiplicative characters of $\mathbb{F}_q$, yield several fundamental identities. We consider first the case of additive characters, in which we use the notation from Theorem 5.7. Then, for additive characters χ_a and χ_b we have

$$\sum_{c \in \mathbb{F}_q} \chi_a(c)\overline{\chi_b(c)} = \begin{cases} 0 & \text{for } a \neq b, \\ q & \text{for } a = b. \end{cases} \tag{5.8}$$

In particular,

$$\sum_{c \in \mathbb{F}_q} \chi_a(c) = 0 \quad \text{for } a \neq 0. \tag{5.9}$$

Furthermore, for elements $c, d \in \mathbb{F}_q$ we obtain

$$\sum_{b \in \mathbb{F}_q} \chi_b(c)\overline{\chi_b(d)} = \begin{cases} 0 & \text{for } c \neq d, \\ q & \text{for } c = d. \end{cases} \tag{5.10}$$

For multiplicative characters ψ and τ of $\mathbb{F}_q$ we have

$$\sum_{c \in \mathbb{F}_q^*} \psi(c)\overline{\tau(c)} = \begin{cases} 0 & \text{for } \psi \neq \tau, \\ q-1 & \text{for } \psi = \tau. \end{cases} \tag{5.11}$$

In particular,

$$\sum_{c \in \mathbb{F}_q^*} \psi(c) = 0 \text{ for } \psi \neq \psi_0. \tag{5.12}$$

If $c, d \in \mathbb{F}_q^*$, then

$$\sum_{\psi} \psi(c)\overline{\psi(d)} = \begin{cases} 0 & \text{for } c \neq d, \\ q-1 & \text{for } c = d, \end{cases} \tag{5.13}$$

where the sum is extended over all multiplicative characters ψ of $\mathbb{F}_q$.

2. GAUSSIAN SUMS

Let ψ be a multiplicative and χ an additive character of $\mathbb{F}_q$. Then the *Gaussian sum* $G(\psi, \chi)$ is defined by

$$G(\psi, \chi) = \sum_{c \in \mathbb{F}_q^*} \psi(c)\chi(c).$$

The absolute value of $G(\psi, \chi)$ can obviously be at most $q-1$, but is in general much smaller, as the following theorem shows. We recall that ψ_0 denotes the trivial multiplicative character and χ_0 the trivial additive character of $\mathbb{F}_q$.

5.11. Theorem. *Let ψ be a multiplicative and χ an additive character of $\mathbb{F}_q$. Then the Gaussian sum $G(\psi, \chi)$ satisfies*

$$G(\psi,\chi)=\begin{cases} q-1 & \text{for } \psi=\psi_0, \chi=\chi_0, \\ -1 & \text{for } \psi=\psi_0, \chi\neq\chi_0, \\ 0 & \text{for } \psi\neq\psi_0, \chi=\chi_0. \end{cases} \tag{5.14}$$

If $\psi \neq \psi_0$ and $\chi \neq \chi_0$, then

$$|G(\psi,\chi)| = q^{1/2}. \tag{5.15}$$

Proof. The first case in (5.14) is trivial, the third case follows from (5.12), and in the second case we have

$$G(\psi_0,\chi)=\sum_{c\in\mathbb{F}_q^*}\chi(c)=\sum_{c\in\mathbb{F}_q}\chi(c)-\chi(0)=-1$$

by (5.9). For $\psi \neq \psi_0$ and $\chi \neq \chi_0$ we get

$$\begin{aligned}|G(\psi,\chi)|^2 &= \overline{G(\psi,\chi)}\;G(\psi,\chi)\\ &= \sum_{c\in\mathbb{F}_q^*}\sum_{c_1\in\mathbb{F}_q^*}\overline{\psi(c)}\;\overline{\chi(c)}\;\psi(c_1)\chi(c_1)\\ &= \sum_{c\in\mathbb{F}_q^*}\sum_{c_1\in\mathbb{F}_q^*}\psi(c^{-1}c_1)\chi(c_1-c).\end{aligned}$$

In the inner sum we substitute $c^{-1}c_1 = d$. Then,

$$\begin{aligned}|G(\psi,\chi)|^2 &= \sum_{c\in\mathbb{F}_q^*}\sum_{d\in\mathbb{F}_q^*}\psi(d)\chi(c(d-1))\\ &= \sum_{d\in\mathbb{F}_q^*}\psi(d)\Big(\sum_{c\in\mathbb{F}_q}\chi(c(d-1))-\chi(0)\Big)\\ &= \sum_{d\in\mathbb{F}_q^*}\psi(d)\sum_{c\in\mathbb{F}_q}\chi(c(d-1))\end{aligned}$$

by (5.12). The inner sum has the value q if $d=1$ and the value 0 if $d \neq 1$, according to (5.9). Therefore, $|G(\psi,\chi)|^2 = \psi(1)q = q$, and (5.15) is established. □

The study of the behavior of Gaussian sums under various transformations of the additive or multiplicative character leads to a number of useful identities.

5.12. Theorem. *Gaussian sums for the finite field $\mathbb{F}_q$ satisfy the following properties*:

(i) $G(\psi,\chi_{ab})=\overline{\psi(a)}\;G(\psi,\chi_b)$ *for* $a\in\mathbb{F}_q^*, b\in\mathbb{F}_q$;
(ii) $G(\psi,\bar{\chi})=\psi(-1)\underline{G(\psi,\chi)}$;
(iii) $G(\bar{\psi},\chi)=\psi(-1)\overline{G(\psi,\chi)}$;

(iv) $G(\psi, \chi)G(\bar{\psi}, \chi) = \psi(-1)q$ *for* $\psi \neq \psi_0, \chi \neq \chi_0$;
(v) $G(\psi^p, \chi_b) = G(\psi, \chi_{\sigma(b)})$ *for* $b \in \mathbb{F}_q$, *where* p *is the characteristic of* $\mathbb{F}_q$ *and* $\sigma(b) = b^p$.

Proof. (i) For $c \in \mathbb{F}_q$ we have $\chi_{ab}(c) = \chi_1(abc) = \chi_b(ac)$ by the definition in Theorem 5.7. Therefore,

$$G(\psi, \chi_{ab}) = \sum_{c \in \mathbb{F}_q^*} \psi(c)\chi_{ab}(c) = \sum_{c \in \mathbb{F}_q^*} \psi(c)\chi_b(ac).$$

Now set $ac = d$. Then

$$\begin{aligned} G(\psi, \chi_{ab}) &= \sum_{d \in \mathbb{F}_q^*} \psi(a^{-1}d)\chi_b(d) \\ &= \psi(a^{-1}) \sum_{d \in \mathbb{F}_q^*} \psi(d)\chi_b(d) \\ &= \overline{\psi(a)}\; G(\psi, \chi_b). \end{aligned}$$

(ii) We have $\chi = \chi_b$ for a suitable $b \in \mathbb{F}_q$ and $\bar{\chi}(c) = \chi_b(-c) = \chi_{-b}(c)$ for $c \in \mathbb{F}_q$. Therefore, by using (i) with $a = -1$ and noting that $\psi(-1) = \pm 1$, we get

$$G(\psi, \bar{\chi}) = G(\psi, \chi_{-b}) = \overline{\psi(-1)}\; G(\psi, \chi_b) = \psi(-1)G(\psi, \chi).$$

(iii) It follows from (ii) that $G(\bar{\psi}, \chi) = \bar{\psi}(-1)G(\bar{\psi}, \bar{\chi}) = \psi(-1)\overline{G(\psi, \chi)}$.

(iv) By combining (iii) and (5.15), we obtain $G(\psi, \chi)G(\bar{\psi}, \chi) = \psi(-1)G(\psi, \chi)\overline{G(\psi, \chi)} = \psi(-1)|G(\psi, \chi)|^2 = \psi(-1)q$.

(v) Since $\mathrm{Tr}(a) = \mathrm{Tr}(a^p)$ for $a \in \mathbb{F}_q$ by Theorem 2.23(v), we have $\chi_1(a) = \chi_1(a^p)$ according to (5.6). Thus, for $c \in \mathbb{F}_q$ we get $\chi_b(c) = \chi_1(bc) = \chi_1(b^p c^p) = \chi_{\sigma(b)}(c^p)$, and so

$$G(\psi^p, \chi_b) = \sum_{c \in \mathbb{F}_q^*} \psi^p(c)\chi_b(c) = \sum_{c \in \mathbb{F}_q^*} \psi(c^p)\chi_{\sigma(b)}(c^p).$$

But c^p runs through $\mathbb{F}_q^*$ as c runs through $\mathbb{F}_q^*$, and the desired result follows. □

5.13. Remark. In connection with the properties above, the value $\psi(-1)$ is of interest. We obviously have $\psi(-1) = \pm 1$. Let m be the *order* of ψ; that is, m is the least positive integer such that $\psi^m = \psi_0$. Then m divides $q-1$ since $\psi^{q-1} = \psi_0$. The values of ψ are mth roots of unity; in particular, -1 can only appear as a value of ψ if m is even. If g is a primitive element of $\mathbb{F}_q$, then $\psi(g) = \zeta$, a primitive mth root of unity. If m is even (and so q odd), then $\psi(-1) = \psi(g^{(q-1)/2}) = \zeta^{(q-1)/2}$, which is -1 precisely if $(q-1)/2 \equiv m/2 \bmod m$, or, equivalently, $(q-1)/m \equiv 1 \bmod 2$. Therefore, $\psi(-1) = -1$ if and only if m is even and $(q-1)/m$ is odd. In all other cases we have $\psi(-1) = 1$. □

Gaussian sums occur in a variety of contexts, for example in the following. Let ψ be a multiplicative character of $\mathbb{F}_q$; then, using (5.10), we may write

$$\psi(c)=\frac{1}{q}\sum_{d\in\mathbb{F}_q^*}\psi(d)\sum_{b\in\mathbb{F}_q}\chi_b(c)\overline{\chi_b(d)}$$

$$=\frac{1}{q}\sum_{b\in\mathbb{F}_q}\chi_b(c)\sum_{d\in\mathbb{F}_q^*}\psi(d)\bar{\chi}_b(d)$$

for any $c\in\mathbb{F}_q^*$. Therefore,

$$\psi(c)=\frac{1}{q}\sum_{\chi}G(\psi,\bar{\chi})\chi(c)\quad\text{for } c\in\mathbb{F}_q^*, \tag{5.16}$$

where the sum is extended over all additive characters χ of $\mathbb{F}_q$. This may be thought of as the Fourier expansion of ψ in terms of the additive characters of $\mathbb{F}_q$, with Gaussian sums appearing as Fourier coefficients.

Similarly, if χ is an additive character of $\mathbb{F}_q$, then, using (5.13), we may write

$$\chi(c)=\frac{1}{q-1}\sum_{d\in\mathbb{F}_q^*}\chi(d)\sum_{\psi}\psi(c)\overline{\psi(d)}$$

$$=\frac{1}{q-1}\sum_{\psi}\psi(c)\sum_{d\in\mathbb{F}_q^*}\bar{\psi}(d)\chi(d)\quad\text{for } c\in\mathbb{F}_q^*.$$

Thus we obtain

$$\chi(c)=\frac{1}{q-1}\sum_{\psi}G(\bar{\psi},\chi)\psi(c)\quad\text{for } c\in\mathbb{F}_q^*, \tag{5.17}$$

where the sum is extended over all multiplicative characters ψ of $\mathbb{F}_q$. This can be interpreted as the Fourier expansion of the restriction of χ to $\mathbb{F}_q^*$ in terms of the multiplicative characters of $\mathbb{F}_q$, again with Gaussian sums as Fourier coefficients. Therefore, Gaussian sums are instrumental in the transition from the additive to the multiplicative structure (or vice versa) of a finite field.

Before we establish further properties of Gaussian sums, we develop a useful general principle. Let Φ be the set of monic polynomials over $\mathbb{F}_q$, and let λ be a complex-valued function on Φ which is multiplicative in the sense that

$$\lambda(gh)=\lambda(g)\lambda(h)\quad\text{for all } g,h\in\Phi, \tag{5.18}$$

and which satisfies $|\lambda(g)|\leqslant 1$ for all $g\in\Phi$ and $\lambda(1)=1$. With Φ_k denoting the subset of Φ containing the polynomials of degree k, consider the power series

$$L(z) = \sum_{k=0}^{\infty} \left(\sum_{g \in \Phi_k} \lambda(g) \right) z^k. \tag{5.19}$$

Since there are q^k polynomials in Φ_k, the coefficient of z^k is in absolute value $\leqslant q^k$, and so the power series converges absolutely for $|z| < q^{-1}$. Because of (5.18) and unique factorization in $\mathbb{F}_q[x]$, we may write

$$\begin{aligned} L(z) &= \sum_{g \in \Phi} \lambda(g) z^{\deg(g)} \\ &= \prod_f \left(1 + \lambda(f) z^{\deg(f)} + \lambda(f^2) z^{\deg(f^2)} + \cdots \right) \\ &= \prod_f \left(1 + \lambda(f) z^{\deg(f)} + \lambda(f)^2 z^{2\deg(f)} + \cdots \right), \end{aligned}$$

where the product is taken over all monic irreducible polynomials f in $\mathbb{F}_q[x]$. It follows that

$$L(z) = \prod_f \left(1 - \lambda(f) z^{\deg(f)}\right)^{-1}.$$

Now apply logarithmic differentiation and multiply the result by z to get

$$z \frac{d \log L(z)}{dz} = \sum_f \frac{\lambda(f) \deg(f) z^{\deg(f)}}{1 - \lambda(f) z^{\deg(f)}}.$$

Expansion of $(1 - \lambda(f) z^{\deg(f)})^{-1}$ into a geometric series leads to

$$\begin{aligned} z \frac{d \log L(z)}{dz} &= \sum_f \lambda(f) \deg(f) z^{\deg(f)} \\ &\quad \cdot \left(1 + \lambda(f) z^{\deg(f)} + \lambda(f)^2 z^{2\deg(f)} + \cdots \right) \\ &= \sum_f \deg(f) \left(\lambda(f) z^{\deg(f)} + \lambda(f)^2 z^{2\deg(f)} \right. \\ &\quad \left. + \lambda(f)^3 z^{3\deg(f)} + \cdots \right), \end{aligned}$$

and collecting equal powers of z we obtain

$$z \frac{d \log L(z)}{dz} = \sum_{s=1}^{\infty} L_s z^s \tag{5.20}$$

with

$$L_s = \sum_f \deg(f) \lambda(f)^{s/\deg(f)}, \tag{5.21}$$

where the sum is extended over all monic irreducible polynomials f in $\mathbb{F}_q[x]$ with $\deg(f)$ dividing s.

Now suppose there exists a positive integer t such that

$$\sum_{g \in \Phi_k} \lambda(g) = 0 \quad \text{for all } k > t. \tag{5.22}$$

Then $L(z)$ is a complex polynomial of degree $\leqslant t$ with constant term 1, so that we can write

$$L(z) = (1 - \omega_1 z)(1 - \omega_2 z) \cdots (1 - \omega_t z) \tag{5.23}$$

with complex numbers $\omega_1, \omega_2, \ldots, \omega_t$. It follows that

$$\begin{aligned} z\frac{d \log L(z)}{dz} &= -\sum_{m=1}^{t} \frac{\omega_m z}{1 - \omega_m z} \\ &= -\sum_{m=1}^{t} \omega_m z \sum_{j=0}^{\infty} \omega_m^j z^j \\ &= -\sum_{j=0}^{\infty} \left(\sum_{m=1}^{t} \omega_m^{j+1} \right) z^{j+1} = -\sum_{s=1}^{\infty} \left(\sum_{m=1}^{t} \omega_m^s \right) z^s, \end{aligned}$$

and comparison with (5.20) yields

$$L_s = -\omega_1^s - \omega_2^s - \cdots - \omega_t^s \quad \text{for all } s \geqslant 1. \tag{5.24}$$

As an application of the principle expressed in (5.24), we consider the following situation. Let χ be an additive and ψ a multiplicative character of $\mathbb{F}_q$, and let E be a finite extension field of $\mathbb{F}_q$. Then χ and ψ can be "lifted" to E by setting $\chi'(\beta) = \chi(\mathrm{Tr}_{E/\mathbb{F}_q}(\beta))$ for $\beta \in E$ and $\psi'(\beta) = \psi(\mathrm{N}_{E/\mathbb{F}_q}(\beta))$ for $\beta \in E^*$. From the additivity of the trace and the multiplicativity of the norm it follows that χ' is an additive and ψ' a multiplicative character of E. The following theorem establishes an important relationship between the Gaussian sum $G(\psi, \chi)$ in $\mathbb{F}_q$ and the Gaussian sum $G(\psi', \chi')$ in E.

5.14. Theorem (Davenport-Hasse Theorem). *Let χ be an additive and ψ a multiplicative character of $\mathbb{F}_q$, not both of them trivial. Suppose χ and ψ are lifted to characters χ' and ψ', respectively, of the finite extension field E of $\mathbb{F}_q$ with $[E:\mathbb{F}_q] = s$. Then*

$$G(\psi', \chi') = (-1)^{s-1} G(\psi, \chi)^s.$$

Proof. It is convenient to extend the definition of ψ by setting $\psi(0) = 0$. We use the notation of the discussion leading to (5.24); in particular, Φ denotes again the set of monic polynomials over $\mathbb{F}_q$. We define λ by setting $\lambda(1) = 1$ as required, and for $g \in \Phi$ of positive degree, say $g(x) = x^k - c_1 x^{k-1} + \cdots + (-1)^k c_k$, we set $\lambda(g) = \psi(c_k)\chi(c_1)$. The multiplicative property (5.18) is then easily checked. For $k > 1$ we split up Φ_k according to the values of c_1 and c_k. Each given pair (c_1, c_k) occurs q^{k-2}

times in Φ_k, and so

$$\sum_{g \in \Phi_k} \lambda(g) = q^{k-2} \sum_{c_1, c_k \in \mathbb{F}_q} \psi(c_k)\chi(c_1)$$
$$= q^{k-2}\left(\sum_{c \in \mathbb{F}_q^*} \psi(c)\right)\left(\sum_{c \in \mathbb{F}_q} \chi(c)\right).$$

Since one of χ and ψ is nontrivial, it follows from either (5.9) or (5.12) that

$$\sum_{g \in \Phi_k} \lambda(g) = 0 \text{ for } k > 1.$$

Therefore, (5.22) is satisfied with $t = 1$. Furthermore, Φ_1 comprises the linear polynomials $x - c$ with $c \in \mathbb{F}_q$, and so

$$\sum_{g \in \Phi_1} \lambda(g) = \sum_{c \in \mathbb{F}_q} \psi(c)\chi(c) = \sum_{c \in \mathbb{F}_q^*} \psi(c)\chi(c) = G(\psi, \chi).$$

Thus, $L(z) = 1 + G(\psi, \chi)z$ from (5.19), hence $\omega_1 = -G(\psi, \chi)$ by (5.23). Now we consider L_s, which, by (5.21) and the multiplicativity of λ, is given by

$$L_s = \sum_f \deg(f)\lambda(f)^{s/\deg(f)}$$
$$= \sum_f{}^* \deg(f)\lambda\left(f^{s/\deg(f)}\right),$$

where the sum is extended over all monic irreducible polynomials f in $\mathbb{F}_q[x]$ with $\deg(f)$ dividing s, and where the asterisk indicates that $f(x) = x$ is excluded. Each such f has $\deg(f)$ distinct nonzero roots in E, and each root β of f has as its characteristic polynomial over $\mathbb{F}_q$ the polynomial

$$f(x)^{s/\deg(f)} = x^s - c_1 x^{s-1} + \cdots + (-1)^s c_s,$$

say, where $c_1 = \mathrm{Tr}_{E/\mathbb{F}_q}(\beta)$ and $c_s = \mathrm{N}_{E/\mathbb{F}_q}(\beta)$ by (2.2) and (2.3). Therefore,

$$\lambda\left(f^{s/\deg(f)}\right) = \psi(c_s)\chi(c_1) = \psi\left(\mathrm{N}_{E/\mathbb{F}_q}(\beta)\right)\chi\left(\mathrm{Tr}_{E/\mathbb{F}_q}(\beta)\right)$$
$$= \psi'(\beta)\chi'(\beta),$$

and so

$$L_s = \sum_f{}^* \deg(f)\lambda\left(f^{s/\deg(f)}\right) = \sum_f{}^* \sum_{\substack{\beta \in E \\ f(\beta) = 0}} \psi'(\beta)\chi'(\beta).$$

If f runs through the range of summation above, then β runs exactly through all elements of E^*. Consequently,

$$L_s = \sum_{\beta \in E^*} \psi'(\beta)\chi'(\beta) = G(\psi', \chi'),$$

and an application of (5.24) yields

$$G(\psi', \chi') = -(-G(\psi, \chi))^s,$$

which completes the proof. □

For certain special characters, the associated Gaussian sums can be evaluated explicitly. We thereby obtain formulas that go beyond the trivial cases listed in (5.14). A celebrated formula of this kind holds for the quadratic character η considered in Example 5.10.

5.15. Theorem. *Let $\mathbb{F}_q$ be a finite field with $q = p^s$, where p is an odd prime and $s \in \mathbb{N}$. Let η be the quadratic character of $\mathbb{F}_q$ and let χ_1 be the canonical additive character of $\mathbb{F}_q$. Then*

$$G(\eta, \chi_1) = \begin{cases} (-1)^{s-1} q^{1/2} & \text{if } p \equiv 1 \bmod 4, \\ (-1)^{s-1} i^s q^{1/2} & \text{if } p \equiv 3 \bmod 4. \end{cases}$$

Proof. Using Theorem 5.12(iv) and $\bar{\eta} = \eta$, we obtain $G(\eta, \chi_1)^2 = \eta(-1)q$, and since $\eta(-1) = 1$ for $q \equiv 1 \bmod 4$ and $\eta(-1) = -1$ for $q \equiv 3 \bmod 4$ by Remark 5.13, it follows that

$$G(\eta, \chi_1) = \begin{cases} \pm q^{1/2} & \text{if } q \equiv 1 \bmod 4, \\ \pm i q^{1/2} & \text{if } q \equiv 3 \bmod 4. \end{cases} \tag{5.25}$$

The difficulty of the proof lies in the determination of the correct signs.

We first consider the case $s = 1$. Let V be the set of all complex-valued functions on $\mathbb{F}_p^*$; it is a $(p-1)$-dimensional vector space over the complex numbers. A basis for V is formed by the characteristic functions $f_1, f_2, \ldots, f_{p-1}$ of elements of $\mathbb{F}_p^*$; that is, $f_j(c) = 1$ if $c = j$ and 0 otherwise, where $j = 1, 2, \ldots, p-1$. From the orthogonality relation (5.11) it follows easily that the multiplicative characters $\psi_0, \psi_1, \ldots, \psi_{p-2}$ of $\mathbb{F}_p$ described in Theorem 5.8 also form a basis for V. Let $\zeta = e^{2\pi i/p}$, and define a linear operator T on V by letting Th for $h \in V$ be given by

$$(Th)(c) = \sum_{k=1}^{p-1} \zeta^{ck} h(k) \quad \text{for } c = 1, 2, \ldots, p-1. \tag{5.26}$$

Then Theorem 5.12(i) implies that $T\psi = G(\psi, \chi_1)\bar{\psi}$ for every multiplicative character ψ of $\mathbb{F}_p$. Since $\psi = \bar{\psi}$ precisely for the trivial character and the quadratic character, the matrix T in the basis $\psi_0, \psi_1, \ldots, \psi_{p-2}$ contains two diagonal entries—namely, $G(\psi_0, \chi_1) = -1$ and $G(\eta, \chi_1)$—and a collection of blocks

$$\begin{pmatrix} 0 & G(\bar{\psi}, \chi_1) \\ G(\psi, \chi_1) & 0 \end{pmatrix}$$

corresponding to pairs $\psi, \bar{\psi}$ of conjugate characters that are nontrivial and nonquadratic. If we compute the determinant of T, then each block contributes

$$-G(\psi,\chi_1)G(\bar{\psi},\chi_1)=-\psi(-1)p$$

by Theorem 5.12(iv). Thus we obtain

$$\det(T)=-G(\eta,\chi_1)(-p)^{(p-3)/2}\prod_{j=1}^{(p-3)/2}\psi_j(-1). \tag{5.27}$$

Now $\psi_j(-1)=\psi_1^j(-1)=(-1)^j$, and so

$$\prod_{j=1}^{(p-3)/2}\psi_j(-1)=(-1)^{1+2+\cdots+(p-3)/2}=(-1)^{(p-1)(p-3)/8}. \tag{5.28}$$

Furthermore, since

$$i^{(p-1)^2/4}=\begin{cases}1 & \text{if } p\equiv 1 \bmod 4,\\ i & \text{if } p\equiv 3 \bmod 4,\end{cases}$$

it follows from (5.25) that

$$G(\eta,\chi_1)=\pm i^{(p-1)^2/4}p^{1/2}. \tag{5.29}$$

Combining (5.27), (5.28), and (5.29), we get

$$\begin{aligned}\det(T)&=\pm(-1)^{(p-1)/2}i^{(p-1)^2/4}(-1)^{(p-1)(p-3)/8}p^{(p-2)/2}\\&=\pm(-1)^{(p-1)/2}i^{(p-1)^2/4+(p-1)(p-3)/4}p^{(p-2)/2},\end{aligned}$$

hence

$$\det(T)=\pm(-1)^{(p-1)/2}i^{(p-1)(p-2)/2}p^{(p-2)/2}. \tag{5.30}$$

Now we compute $\det(T)$ utilizing the matrix of T in the basis $f_1, f_2,\ldots,f_{p-1}$. From (5.26) we find

$$\begin{aligned}\det(T)&=\det\left((\zeta^{jk})_{1\leqslant j,k\leqslant p-1}\right)=\det\left((\zeta^j\zeta^{j(k-1)})_{1\leqslant j,k\leqslant p-1}\right)\\&=\zeta^{1+2+\cdots+(p-1)}\det\left((\zeta^{j(k-1)})_{1\leqslant j,k\leqslant p-1}\right)\\&=\det\left((\zeta^{j(k-1)})_{1\leqslant j,k\leqslant p-1}\right),\end{aligned}$$

which is a Vandermonde determinant. Therefore,

$$\det(T)=\prod_{1\leqslant m<n\leqslant p-1}(\zeta^n-\zeta^m).$$

With $\delta = e^{\pi i/p}$ we get

$$\det(T) = \prod_{1 \leqslant m < n \leqslant p-1} (\delta^{2n} - \delta^{2m})$$

$$= \prod_{1 \leqslant m < n \leqslant p-1} \delta^{n+m}(\delta^{n-m} - \delta^{-(n-m)})$$

$$= \prod_{1 \leqslant m < n \leqslant p-1} \delta^{n+m} \prod_{1 \leqslant m < n \leqslant p-1} \left(2i \sin \frac{\pi(n-m)}{p}\right).$$

Since

$$\sum_{1 \leqslant m < n \leqslant p-1} (n+m) = \sum_{n=2}^{p-1} \sum_{m=1}^{n-1} (n+m)$$

$$= \frac{3}{2} \sum_{n=2}^{p-1} n(n-1) = \frac{3}{2} \sum_{n=1}^{p-2} (n^2+n)$$

$$= \frac{3}{2}\left(\frac{(p-2)(p-1)(2p-3)}{6} + \frac{(p-2)(p-1)}{2}\right)$$

$$= \frac{p(p-1)(p-2)}{2},$$

the first product is equal to

$$\delta^{p(p-1)(p-2)/2} = (-1)^{(p-1)(p-2)/2} = ((-1)^{p-2})^{(p-1)/2} = (-1)^{(p-1)/2}.$$

Furthermore,

$$A = \prod_{1 \leqslant m < n \leqslant p-1} \left(2 \sin \frac{\pi(n-m)}{p}\right) > 0,$$

and so

$$\det(T) = (-1)^{(p-1)/2} i^{(p-1)(p-2)/2} A \quad \text{with } A > 0.$$

Comparison with (5.30) shows that the plus sign always applies in (5.29), and the theorem is established for $s = 1$.

The general case follows from Theorem 5.14 since the canonical additive character of $\mathbb{F}_p$ is lifted to the canonical additive character of $\mathbb{F}_q$ by (5.7) and the quadratic character of $\mathbb{F}_p$ is lifted to the quadratic character of $\mathbb{F}_q$. □

Because of (5.14) and Theorem 5.12(i), a formula for $G(\eta, \chi)$ can also be established for any additive character χ of $\mathbb{F}_q$.

We turn to another special formula for Gaussian sums which applies to a wider range of multiplicative characters but needs a restriction on the underlying field. We shall have to use the notion of order of a multiplicative character as introduced in Remark 5.13.

5.16. Theorem (Stickelberger's Theorem). *Let q be a prime power, let ψ be a nontrivial multiplicative character of $\mathbb{F}_{q^2}$ of order m dividing $q+1$, and let χ_1 be the canonical additive character of $\mathbb{F}_{q^2}$. Then,*

$$G(\psi,\chi_1)=\begin{cases} q & \text{if } m \text{ odd or } \dfrac{q+1}{m} \text{ even},\\[2mm] -q & \text{if } m \text{ even and } \dfrac{q+1}{m} \text{ odd}.\end{cases}$$

Proof. We write $E=\mathbb{F}_{q^2}$ and $F=\mathbb{F}_q$. Let γ be a primitive element of E and set $g=\gamma^{q+1}$. Then $g^{q-1}=1$, so that $g\in F$; furthermore, g is a primitive element of F. Every $\alpha\in E^*$ can be written in the form $\alpha=g^j\gamma^k$ with $0\leqslant j<q-1$ and $0\leqslant k<q+1$. Since $\psi(g)=\psi^{q+1}(\gamma)=1$, we have

$$\begin{aligned} G(\psi,\chi_1)&=\sum_{j=0}^{q-2}\sum_{k=0}^{q}\psi(g^j\gamma^k)\chi_1(g^j\gamma^k)\\ &=\sum_{k=0}^{q}\psi^k(\gamma)\sum_{j=0}^{q-2}\chi_1(g^j\gamma^k)\\ &=\sum_{k=0}^{q}\psi^k(\gamma)\sum_{b\in F^*}\chi_1(b\gamma^k). \end{aligned}\tag{5.31}$$

If τ_1 is the canonical additive character of F, then $\chi_1(b\gamma^k)=\tau_1(\mathrm{Tr}_{E/F}(b\gamma^k))$ by (5.7). Therefore,

$$\begin{aligned}\sum_{b\in F^*}\chi_1(b\gamma^k)&=\sum_{b\in F^*}\tau_1\big(b\,\mathrm{Tr}_{E/F}(\gamma^k)\big)\\ &=\begin{cases}-1 & \text{for } \mathrm{Tr}_{E/F}(\gamma^k)\neq 0,\\ q-1 & \text{for } \mathrm{Tr}_{E/F}(\gamma^k)=0,\end{cases}\end{aligned}\tag{5.32}$$

because of (5.9). Now $\mathrm{Tr}_{E/F}(\gamma^k)=\gamma^k+\gamma^{kq}$, and so

$$\mathrm{Tr}_{E/F}(\gamma^k)=0 \quad \text{if and only if } \gamma^{k(q-1)}=-1. \tag{5.33}$$

If q is odd, the last condition is equivalent to $k=(q+1)/2$, and then by (5.32),

$$\sum_{b\in F^*}\chi_1(b\gamma^k)=\begin{cases}-1 & \text{for } 0\leqslant k<q+1,\ k\neq\dfrac{q+1}{2},\\[2mm] q-1 & \text{for } k=\dfrac{q+1}{2}.\end{cases}$$

Together with (5.31) we get

$$\begin{aligned} G(\psi, \chi_1) &= -\sum_{\substack{k=0 \\ k \neq (q+1)/2}}^{q} \psi^k(\gamma) + (q-1)\psi^{(q+1)/2}(\gamma) \\ &= -\sum_{k=0}^{q} \psi^k(\gamma) + q\psi^{(q+1)/2}(\gamma) \\ &= q\psi^{(q+1)/2}(\gamma) \end{aligned}$$

since $\psi(\gamma) \neq 1$ and $\psi^{q+1}(\gamma) = 1$. Now $\psi^{(q+1)/2}(\gamma) = 1$ if $(q+1)/m$ is even and -1 if $(q+1)/m$ is odd, and thus for q odd we have

$$G(\psi, \chi_1) = \begin{cases} q & \text{if } \dfrac{q+1}{m} \text{ even,} \\[2ex] -q & \text{if } \dfrac{q+1}{m} \text{ odd.} \end{cases} \tag{5.34}$$

If q is even, then the condition in (5.33) is equivalent to $\gamma^{k(q-1)} = 1$, and the only k with $0 \leqslant k < q+1$ satisfying this property is $k = 0$. Then by (5.32),

$$\sum_{b \in F^*} \chi_1(b\gamma^k) = \begin{cases} -1 & \text{for } 1 \leqslant k \leqslant q, \\ q-1 & \text{for } k = 0, \end{cases}$$

and (5.31) yields

$$G(\psi, \chi_1) = -\sum_{k=1}^{q} \psi^k(\gamma) + q - 1 = -\sum_{k=0}^{q} \psi^k(\gamma) + q = q.$$

Combined with (5.34), this implies the theorem. □

We conclude this section by showing how to use Gaussian sums to establish a classical result of number theory, namely the law of quadratic reciprocity. We recall from Example 5.10 that if p is an odd prime and η is the quadratic character of $\mathbb{F}_p$, then for $c \not\equiv 0 \bmod p$ the Legendre symbol $\left(\frac{c}{p}\right)$ is defined by $\left(\frac{c}{p}\right) = \eta(c)$.

5.17. Theorem (Law of Quadratic Reciprocity). *For any distinct odd primes p and r we have*

$$\left(\frac{p}{r}\right)\left(\frac{r}{p}\right) = (-1)^{(p-1)(r-1)/4}$$

Proof. Let η be the quadratic character of $\mathbb{F}_p$, let χ_1 be the canonical additive character of $\mathbb{F}_p$, and put $G = G(\eta, \chi_1)$. Then it follows from (5.25) that $G^2 = (-1)^{(p-1)/2}p = \tilde{p}$, and so

$$G^r = (G^2)^{(r-1)/2} G = \tilde{p}^{(r-1)/2} G. \tag{5.35}$$

Let R be the ring of algebraic integers; that is, R consists of all complex numbers that are roots of monic polynomials with integer coefficients. Since the values of (additive and multiplicative) characters of finite fields are complex roots of unity, and since every complex root of unity is an algebraic integer, the values of Gaussian sums are algebraic integers. In particular, $G \in R$. Let (r) be the principal ideal of R generated by r. Then the residue class ring $R/(r)$ has characteristic r, and thus an application of Theorem 1.46 yields

$$G^r = \left(\sum_{c \in \mathbb{F}_p^*} \eta(c)\chi_1(c) \right)^r \equiv \sum_{c \in \mathbb{F}_p^*} \eta^r(c)\chi_1^r(c) \bmod (r).$$

Now

$$\sum_{c \in \mathbb{F}_p^*} \eta^r(c)\chi_1^r(c) = \sum_{c \in \mathbb{F}_p^*} \eta(c)\chi_r(c) = G(\eta, \chi_r) = \eta(r)G$$

by Theorem 5.12(i), and so

$$G^r \equiv \eta(r)G \bmod (r).$$

Together with (5.35) we get

$$\tilde{p}^{(r-1)/2}G \equiv \eta(r)G \bmod (r),$$

and multiplication by G leads to

$$\tilde{p}^{(r-1)/2}\tilde{p} \equiv \eta(r)\tilde{p} \bmod (r)$$

because of $G^2 = \tilde{p}$. Since the numbers on both sides of the congruence above are, in fact, elements of $\mathbb{Z}$, it follows that

$$\tilde{p}^{(r-1)/2}\tilde{p} \equiv \eta(r)\tilde{p} \bmod r$$

as a congruence in $\mathbb{Z}$. But $\tilde{p}$ and r are relatively prime, hence

$$\tilde{p}^{(r-1)/2} \equiv \eta(r) \bmod r.$$

Now $\tilde{p} = (-1)^{(p-1)/2}p$ and $p^{r-1} \equiv 1 \bmod r$, thus multiplication by $p^{(r-1)/2}$ yields

$$(-1)^{(p-1)(r-1)/4} \equiv p^{(r-1)/2}\eta(r) \bmod r. \tag{5.36}$$

We have $p^{(r-1)/2} \equiv \pm 1 \bmod r$, and the plus sign applies if and only if p is congruent to a square mod r. Thus,

$$p^{(r-1)/2} \equiv \left(\frac{p}{r}\right) \bmod r.$$

Since $\eta(r) = \left(\frac{r}{p}\right)$, we get from (5.36)

$$(-1)^{(p-1)(r-1)/4} \equiv \left(\frac{p}{r}\right)\left(\frac{r}{p}\right) \bmod r.$$

But the integers on both sides of this congruence can only be ± 1, and since $r \geqslant 3$, the congruence holds only if the two sides are identical. □

3. JACOBI SUMS

If λ is a multiplicative character of $\mathbb{F}_q$, then λ is defined for all nonzero elements of $\mathbb{F}_q$. It is now convenient to extend the definition of λ by setting $\lambda(0)=1$ if λ is the trivial character and $\lambda(0)=0$ if λ is a nontrivial character. With this definition, we have then

$$\sum_{c\in\mathbb{F}_q}\lambda(c)=\begin{cases} q & \text{if } \lambda \text{ is trivial,}\\ 0 & \text{if } \lambda \text{ is nontrivial.}\end{cases} \tag{5.37}$$

Furthermore, the property $\lambda(a_1a_2)=\lambda(a_1)\lambda(a_2)$ holds then for all $a_1, a_2 \in \mathbb{F}_q$.

Let $\lambda_1,\dots,\lambda_k$ be k multiplicative characters of $\mathbb{F}_q$ and let $a\in\mathbb{F}_q$ be fixed. We define the sum

$$J_a(\lambda_1,\dots,\lambda_k)=\sum_{c_1+\cdots+c_k=a}\lambda_1(c_1)\cdots\lambda_k(c_k),$$

where the summation is extended over all k-tuples $(c_1,\dots,c_k)$ of elements of $\mathbb{F}_q$ with $c_1+\cdots+c_k=a$. Thus the sum contains q^{k-1} terms.

If $a\neq 0$, we can put $c_1=ab_1,\dots,c_k=ab_k$. Then $b_1+\cdots+b_k=1$ and

$$\begin{aligned} J_a(\lambda_1,\dots,\lambda_k) &= \sum_{b_1+\cdots+b_k=1}\lambda_1(ab_1)\cdots\lambda_k(ab_k)\\ &=\lambda_1(a)\cdots\lambda_k(a)\sum_{b_1+\cdots+b_k=1}\lambda_1(b_1)\cdots\lambda_k(b_k)\\ &=(\lambda_1\cdots\lambda_k)(a)J_1(\lambda_1,\dots,\lambda_k). \end{aligned}\tag{5.38}$$

Because of this simple relationship, it suffices to consider the sums $J_0(\lambda_1,\dots,\lambda_k)$ and $J_1(\lambda_1,\dots,\lambda_k)$. The second sum is more important for applications, and so we use a slightly simpler notation for it.

5.18. Definition. Let $\lambda_1,\dots,\lambda_k$ be k multiplicative characters of $\mathbb{F}_q$. Then the sum

$$J(\lambda_1,\dots,\lambda_k)=\sum_{c_1+\cdots+c_k=1}\lambda_1(c_1)\cdots\lambda_k(c_k),$$

with the summation extended over all k-tuples $(c_1,\dots,c_k)$ of elements of $\mathbb{F}_q$ satisfying $c_1+\cdots+c_k=1$, is called a *Jacobi sum* in $\mathbb{F}_q$.

If $k=1$, then $J(\lambda_1)=\lambda_1(1)=1$ for any multiplicative character λ_1 of $\mathbb{F}_q$. Thus, Jacobi sums are only of interest for $k\geqslant 2$. For such k, it is immediate from the definition that the value of $J(\lambda_1,\dots,\lambda_k)$ is independent of the order in which the characters λ_i are listed. The same is true of $J_0(\lambda_1,\dots,\lambda_k)$.

The Jacobi sums $J(\lambda_1,\ldots,\lambda_k)$ as well as the sums $J_0(\lambda_1,\ldots,\lambda_k)$ can be evaluated easily in case some of the characters λ_i are trivial.

5.19. Theorem. *If the multiplicative characters $\lambda_1,\ldots,\lambda_k$ of $\mathbb{F}_q$ are trivial, then*

$$J(\lambda_1,\ldots,\lambda_k)=J_0(\lambda_1,\ldots,\lambda_k)=q^{k-1}. \tag{5.39}$$

If some, but not all, of the λ_i are trivial, then

$$J(\lambda_1,\ldots,\lambda_k)=J_0(\lambda_1,\ldots,\lambda_k)=0. \tag{5.40}$$

Proof. The identities (5.39) are obvious, since in both cases we have a sum over q^{k-1} terms, with each term being equal to 1. For the proof of (5.40) we can assume that the characters are listed in such a way that $\lambda_1,\ldots,\lambda_h$ are nontrivial and $\lambda_{h+1},\ldots,\lambda_k$ are trivial, where $1\leqslant h\leqslant k-1$. Then

$$\begin{aligned}J(\lambda_1,\ldots,\lambda_k)&=\sum_{c_1+\cdots+c_k=1}\lambda_1(c_1)\cdots\lambda_k(c_k)\\&=\sum_{c_1+\cdots+c_k=1}\lambda_1(c_1)\cdots\lambda_h(c_h).\end{aligned}$$

For fixed $c_1,\ldots,c_h\in\mathbb{F}_q$, there are q^{k-h-1} solutions $(c_{h+1},\ldots,c_k)$ of the equation $c_{h+1}+\cdots+c_k=1-c_1-\cdots-c_h$. Therefore,

$$\begin{aligned}J(\lambda_1,\ldots,\lambda_k)&=q^{k-h-1}\sum_{c_1,\ldots,c_h\in\mathbb{F}_q}\lambda_1(c_1)\cdots\lambda_h(c_h)\\&=q^{k-h-1}\left(\sum_{c_1\in\mathbb{F}_q}\lambda_1(c_1)\right)\cdots\left(\sum_{c_h\in\mathbb{F}_q}\lambda_h(c_h)\right)=0,\end{aligned}$$

where the last identity follows from (5.37). A similar argument shows that $J_0(\lambda_1,\ldots,\lambda_k)=0$. □

In order to treat the case where all λ_i are nontrivial, we first establish a result that exhibits a further relationship between the sum $J_0(\lambda_1,\ldots,\lambda_k)$ and Jacobi sums.

5.20. Theorem. *If $\lambda_1,\ldots,\lambda_k$ are multiplicative characters of $\mathbb{F}_q$ with λ_k nontrivial, then*

$$J_0(\lambda_1,\ldots,\lambda_k)=0 \tag{5.41}$$

if $\lambda_1\cdots\lambda_k$ is nontrivial and

$$J_0(\lambda_1,\ldots,\lambda_k)=\lambda_k(-1)(q-1)J(\lambda_1,\ldots,\lambda_{k-1}) \tag{5.42}$$

if $\lambda_1\cdots\lambda_k$ is trivial.

Proof. Since the case $k=1$ is obvious, we may assume $k \geqslant 2$. Then

$$J_0(\lambda_1,\ldots,\lambda_k) = \sum_{a \in \mathbb{F}_q} \left(\sum_{c_1 + \cdots + c_{k-1} = -a} \lambda_1(c_1)\cdots\lambda_{k-1}(c_{k-1}) \right) \lambda_k(a)$$

$$= \sum_{a \in \mathbb{F}_q} J_{-a}(\lambda_1,\ldots,\lambda_{k-1})\lambda_k(a).$$

Now $\lambda_k(0)=0$ since λ_k is nontrivial, hence an application of (5.38) yields

$$J_0(\lambda_1,\ldots,\lambda_k) = \sum_{a \in \mathbb{F}_q^*} J_{-a}(\lambda_1,\ldots,\lambda_{k-1})\lambda_k(a)$$

$$= J(\lambda_1,\ldots,\lambda_{k-1}) \sum_{a \in \mathbb{F}_q^*} (\lambda_1\cdots\lambda_{k-1})(-a)\lambda_k(a)$$

$$= (\lambda_1\cdots\lambda_{k-1})(-1) J(\lambda_1,\ldots,\lambda_{k-1}) \sum_{a \in \mathbb{F}_q^*} (\lambda_1\cdots\lambda_k)(a).$$

If $\lambda_1\cdots\lambda_k$ is nontrivial, the last sum is 0 by (5.12), and so (5.41) is shown. If $\lambda_1\cdots\lambda_k$ is trivial, the last sum is $q-1$, and (5.42) follows from $(\lambda_1\cdots\lambda_{k-1})(-1) = \bar{\lambda}_k(-1) = \lambda_k(-1)$. □

If all λ_i are nontrivial, there exists an important connection between Jacobi sums and Gaussian sums that will allow us to determine the absolute value of Jacobi sums.

5.21. Theorem. *If $\lambda_1,\ldots,\lambda_k$ are nontrivial multiplicative characters of $\mathbb{F}_q$ and χ is a nontrivial additive character of $\mathbb{F}_q$, then*

$$J(\lambda_1,\ldots,\lambda_k) = \frac{G(\lambda_1,\chi)\cdots G(\lambda_k,\chi)}{G(\lambda_1\cdots\lambda_k,\chi)} \tag{5.43}$$

if $\lambda_1\cdots\lambda_k$ is nontrivial and

$$J(\lambda_1,\ldots,\lambda_k) = -\lambda_k(-1)J(\lambda_1,\ldots,\lambda_{k-1})$$

$$= -\frac{1}{q}G(\lambda_1,\chi)\cdots G(\lambda_k,\chi) \tag{5.44}$$

if $\lambda_1\cdots\lambda_k$ is trivial.

Proof. Since each λ_i is nontrivial, we have $\lambda_i(0)=0$ and

$$G(\lambda_i,\chi) = \sum_{c_i \in \mathbb{F}_q} \lambda_i(c_i)\chi(c_i).$$

Therefore,

$$\begin{aligned}G(\lambda_1,\chi)\cdots G(\lambda_k,\chi) &= \left(\sum_{c_1\in\mathbb{F}_q}\lambda_1(c_1)\chi(c_1)\right)\cdots\left(\sum_{c_k\in\mathbb{F}_q}\lambda_k(c_k)\chi(c_k)\right)\\ &= \sum_{c_1,\ldots,c_k\in\mathbb{F}_q}\lambda_1(c_1)\cdots\lambda_k(c_k)\chi(c_1+\cdots+c_k)\\ &= \sum_{a\in\mathbb{F}_q}\chi(a)\sum_{c_1+\cdots+c_k=a}\lambda_1(c_1)\cdots\lambda_k(c_k)\\ &= \sum_{a\in\mathbb{F}_q}\chi(a)J_a(\lambda_1,\ldots,\lambda_k).\end{aligned}$$

If $\lambda_1\cdots\lambda_k$ is nontrivial, then $J_0(\lambda_1,\ldots,\lambda_k)=0$ by (5.41). Together with (5.38) we get

$$\begin{aligned}G(\lambda_1,\chi)\cdots G(\lambda_k,\chi) &= J(\lambda_1,\ldots,\lambda_k)\sum_{a\in\mathbb{F}_q^*}(\lambda_1\cdots\lambda_k)(a)\chi(a)\\ &= J(\lambda_1,\ldots,\lambda_k)G(\lambda_1\cdots\lambda_k,\chi).\end{aligned}$$

Now $G(\lambda_1\cdots\lambda_k,\chi)\neq 0$ by (5.15) for $\lambda_1\cdots\lambda_k$ nontrivial, and (5.43) follows.

If $\lambda_1\cdots\lambda_k$ is trivial, we have $J_a(\lambda_1,\ldots,\lambda_k)=J(\lambda_1,\ldots,\lambda_k)$ for all $a\in\mathbb{F}_q^*$ by (5.38), and so

$$\begin{aligned}J_0(\lambda_1,\ldots,\lambda_k)+(q-1)J(\lambda_1,\ldots,\lambda_k) &= \sum_{a\in\mathbb{F}_q}J_a(\lambda_1,\ldots,\lambda_k)\\ &= \sum_{c_1,\ldots,c_k\in\mathbb{F}_q}\lambda_1(c_1)\cdots\lambda_k(c_k)\\ &= \left(\sum_{c_1\in\mathbb{F}_q}\lambda_1(c_1)\right)\cdots\left(\sum_{c_k\in\mathbb{F}_q}\lambda_k(c_k)\right)\\ &= 0\end{aligned}$$

by (5.37), so that the first identity in (5.44) follows from (5.42). Furthermore, since $\lambda_1\cdots\lambda_{k-1}$ is nontrivial, we can apply (5.43) and obtain

$$\begin{aligned}\lambda_k(-1)J(\lambda_1,\ldots,\lambda_{k-1}) &= \frac{\lambda_k(-1)G(\lambda_1,\chi)\cdots G(\lambda_{k-1},\chi)}{G(\lambda_1\cdots\lambda_{k-1},\chi)}\\ &= \frac{\lambda_k(-1)G(\lambda_1,\chi)\cdots G(\lambda_{k-1},\chi)G(\lambda_k,\chi)}{G(\bar{\lambda}_k,\chi)G(\lambda_k,\chi)}\\ &= \frac{1}{q}G(\lambda_1,\chi)\cdots G(\lambda_k,\chi),\end{aligned}$$

using Theorem 5.12(iv) in the last step. The second identity in (5.44) is thus shown. □

5.22. Theorem. *Let $\lambda_1,\dots,\lambda_k$ be nontrivial multiplicative characters of $\mathbb{F}_q$. Then*

$$|J(\lambda_1,\dots,\lambda_k)| = q^{(k-1)/2} \tag{5.45}$$

if $\lambda_1 \cdots \lambda_k$ is nontrivial and

$$|J(\lambda_1,\dots,\lambda_k)| = q^{(k-2)/2} \tag{5.46}$$

if $\lambda_1 \cdots \lambda_k$ is trivial.

Proof. The identity (5.45) follows from (5.15) and (5.43), and (5.46) follows from (5.15) and (5.44). □

5.23. Corollary. *If $\lambda_1,\dots,\lambda_k$ are nontrivial multiplicative characters of $\mathbb{F}_q$ and $\lambda_1 \cdots \lambda_k$ is trivial, then*

$$|J_0(\lambda_1,\dots,\lambda_k)| = (q-1)q^{(k-2)/2}.$$

Proof. This follows from (5.42) and (5.45). □

5.24. Example. We present another proof of the law of quadratic reciprocity (see Theorem 5.17) by using properties of Jacobi sums. Let p and r be distinct odd primes, and as in the proof of Theorem 5.17 let η be the quadratic character of $\mathbb{F}_p$, χ_1 the canonical additive character of $\mathbb{F}_p$, and $G = G(\eta, \chi_1)$. Let J be the Jacobi sum in $\mathbb{F}_p$ defined by

$$J = \sum_{c_1 + \cdots + c_r = 1} \eta(c_1)\cdots\eta(c_r).$$

Since η^{r+1} is trivial, the second identity in (5.44) yields

$$G^{r+1} = \eta(-1)pJ = \tilde{p}J \text{ with } \tilde{p} = (-1)^{(p-1)/2}p.$$

On the other hand, $G^2 = \tilde{p}$ from the proof of Theorem 5.17, and so

$$G^{r+1} = (G^2)^{(r+1)/2} = \tilde{p}^{(r+1)/2}.$$

A comparison yields

$$J = \tilde{p}^{(r-1)/2}. \tag{5.47}$$

Now we inspect the terms of the sum J. Since η only attains the values $0, \pm 1$, each term of J is an integer. If $c_1 = \cdots = c_r$, then the common value must be $r^{-1} \in \mathbb{F}_p$ and the corresponding term of J has the value $\eta^r(r^{-1}) = \eta(r^{-1}) = \eta(r)$. If the c_i are not all equal, then there are r different r-tuples obtained from $(c_1,\dots,c_r)$ by cyclic permutation. The corresponding terms of J all have the same value, and so the sum over those r terms is $\equiv 0 \bmod r$. By splitting up J in this manner, we thus obtain $J \equiv \eta(r) \bmod r$. Together with (5.47) we get

$$\tilde{p}^{(r-1)/2} \equiv \eta(r) \bmod r.$$

The proof is then completed as in Theorem 5.17. □

5.25. Example. We use Jacobi sums to show that every prime $p \equiv 1 \bmod 4$ can be written as a sum of two squares of integers. Since 4 divides $p-1$, it follows from Corollary 5.9 that there exists a multiplicative character λ of $\mathbb{F}_p$ of order 4 (see Remark 5.13 for the definition of order). Then λ assumes only the values $0, \pm 1$, and $\pm i$, thus it is clear that with $\eta = \lambda^2$ being the quadratic character of $\mathbb{F}_p$ we get

$$J(\lambda, \eta) = \sum_{c_1 + c_2 = 1} \lambda(c_1)\eta(c_2) = A + Bi$$

for some integers A and B. Now (5.45) implies that

$$p = |J(\lambda, \eta)|^2 = A^2 + B^2,$$

and the claim is established. We note that a prime $p \equiv 3 \bmod 4$ cannot be written in this form since the square of an integer is congruent to 0 or $1 \bmod 4$, and so $A^2 + B^2$ is never congruent to $3 \bmod 4$. The only remaining prime—namely $p = 2$—is obviously the sum of two squares of integers since $2 = 1^2 + 1^2$. □

There is an analog of Theorem 5.14 for Jacobi sums. We use again the concept of "lifted" characters introduced before Theorem 5.14.

5.26. *Theorem.* *Let $\lambda_1, \ldots, \lambda_k$ be multiplicative characters of $\mathbb{F}_q$, not all of which are trivial. Suppose $\lambda_1, \ldots, \lambda_k$ are lifted to characters $\lambda_1', \ldots, \lambda_k'$, respectively, of the finite extension field E of $\mathbb{F}_q$ with $[E : \mathbb{F}_q] = s$. Then*

$$J(\lambda_1', \ldots, \lambda_k') = (-1)^{(s-1)(k-1)} J(\lambda_1, \ldots, \lambda_k)^s. \tag{5.48}$$

Proof. We note that trivial characters are lifted to trivial characters and nontrivial characters are lifted to nontrivial characters. Thus, if some λ_i are trivial, then both sides of (5.48) are 0 by (5.40). If all λ_i are nontrivial and $\lambda_1 \cdots \lambda_k$ is also nontrivial, then with a nontrivial additive character χ of $\mathbb{F}_q$ we obtain from (5.43) and Theorem 5.14 that

$$\begin{aligned} J(\lambda_1', \ldots, \lambda_k') &= \frac{G(\lambda_1', \chi') \cdots G(\lambda_k', \chi')}{G(\lambda_1' \cdots \lambda_k', \chi')} \\ &= \frac{(-1)^{s-1} G(\lambda_1, \chi)^s \cdots (-1)^{s-1} G(\lambda_k, \chi)^s}{(-1)^{s-1} G(\lambda_1 \cdots \lambda_k, \chi)^s} \\ &= (-1)^{(s-1)(k-1)} J(\lambda_1, \ldots, \lambda_k)^s. \end{aligned}$$

If all λ_i are nontrivial and $\lambda_1 \cdots \lambda_k$ is trivial, then (5.44) and Theorem 5.14

yield

$$\begin{aligned} J(\lambda_1',\dots,\lambda_k') &= -\frac{1}{q^s}G(\lambda_1',\chi')\cdots G(\lambda_k',\chi') \\ &= -\frac{1}{q^s}(-1)^{(s-1)k}G(\lambda_1,\chi)^s\cdots G(\lambda_k,\chi)^s \\ &= (-1)^{(s-1)k}(-1)^{-(s-1)}\left(-\frac{1}{q}G(\lambda_1,\chi)\cdots G(\lambda_k,\chi)\right)^s \\ &= (-1)^{(s-1)(k-1)}J(\lambda_1,\dots,\lambda_k)^s. \end{aligned}$$

□

For the case $k=2$, which often appears in applications of Jacobi sums, some results of special interest can be established. We use again the concept of the order of a multiplicative character as defined in Remark 5.13.

5.27. Theorem. *Let λ be a multiplicative character of $\mathbb{F}_q$ of order $m \geqslant 2$ and let χ be a nontrivial additive character of $\mathbb{F}_q$. Then*

$$G(\lambda,\chi)^m = \lambda(-1)qJ(\lambda,\lambda)J(\lambda,\lambda^2)\cdots J(\lambda,\lambda^{m-2}). \tag{5.49}$$

Proof. First suppose $m \geqslant 3$. Then from (5.43) we have

$$\frac{G(\lambda,\chi)G(\lambda^j,\chi)}{G(\lambda^{j+1},\chi)} = J(\lambda,\lambda^j) \text{ for } 1 \leqslant j \leqslant m-2.$$

Multiplying together these $m-2$ identities, we get

$$\frac{G(\lambda,\chi)^{m-1}}{G(\lambda^{m-1},\chi)} = J(\lambda,\lambda)J(\lambda,\lambda^2)\cdots J(\lambda,\lambda^{m-2}). \tag{5.50}$$

Since λ^m is trivial, we have $\lambda^{m-1}=\bar{\lambda}$, hence

$$G(\lambda,\chi)G(\lambda^{m-1},\chi) = \lambda(-1)q \tag{5.51}$$

by Theorem 5.12(iv). Multiplying together (5.50) and (5.51), we obtain the desired result. If $m=2$, the empty product of Jacobi sums in (5.49) is interpreted to be 1, and then the result is contained in (5.51). □

Another result for $k=2$ leads to a remarkable relationship between Gaussian sums. We use the notation for additive characters introduced in Theorem 5.7.

5.28. Theorem (Davenport-Hasse Relation). *Let λ and ψ be multiplicative characters of $\mathbb{F}_q$ such that λ has order $m \geqslant 2$ and ψ^m is nontrivial, and let χ_b be a nontrivial additive character of $\mathbb{F}_q$. Then*

$$\frac{G(\psi,\chi_b)^m}{G(\psi^m,\chi_{mb})} = \prod_{j=1}^{m-1} J(\psi,\lambda^j).$$

5.29. Corollary. *Under the same conditions as in Theorem* 5.28, *we have*

$$\prod_{j=0}^{m-1} G(\psi\lambda^j, \chi_b) = q^{(m-1)/2} G(\psi^m, \chi_{mb}) \tag{5.52}$$

if m *is odd and*

$$\prod_{j=0}^{m-1} G(\psi\lambda^j, \chi_b) = (-1)^{(q-1)(m-2)/8} q^{(m-2)/2} G(\eta, \chi_b) G(\psi^m, \chi_{mb}) \tag{5.53}$$

if m *is even, where* η *is the quadratic character of* $\mathbb{F}_q$.

The corollary is deduced from the theorem as follows. We note first that each character $\psi\lambda^j$ is nontrivial, for otherwise $(\psi\lambda^j)^m = \psi^m\lambda^{jm} = \psi^m$ would be trivial, a contradiction. Thus we can apply (5.43) to the identity in Theorem 5.28, and this yields

$$\frac{G(\psi, \chi_b)^m}{G(\psi^m, \chi_{mb})} = G(\psi, \chi_b)^{m-1} \prod_{j=1}^{m-1} \frac{G(\lambda^j, \chi_b)}{G(\psi\lambda^j, \chi_b)}.$$

A rearrangement of terms leads to the identity

$$\prod_{j=0}^{m-1} G(\psi\lambda^j, \chi_b) = G(\psi^m, \chi_{mb}) \prod_{j=1}^{m-1} G(\lambda^j, \chi_b). \tag{5.54}$$

If m is *odd*, then

$$\prod_{j=1}^{m-1} G(\lambda^j, \chi_b) = \prod_{j=1}^{(m-1)/2} G(\lambda^j, \chi_b) G(\lambda^{m-j}, \chi_b),$$

and since λ^m is trivial, we have $\lambda^{m-j} = \overline{\lambda^j}$, so that an application of Theorem 5.12(iv) yields

$$\prod_{j=1}^{m-1} G(\lambda^j, \chi_b) = q^{(m-1)/2} \prod_{j=1}^{(m-1)/2} \lambda^j(-1).$$

Now $\lambda(-1) = 1$ by Remark 5.13, hence

$$\prod_{j=1}^{m-1} G(\lambda^j, \chi_b) = q^{(m-1)/2}.$$

Together with (5.54) we get (5.52).

If m is *even*, we use again Theorem 5.12(iv) to obtain

$$\prod_{j=1}^{m-1} G(\lambda^j,\chi_b)=G(\lambda^{m/2},\chi_b)\prod_{j=1}^{(m-2)/2} G(\lambda^j,\chi_b)G(\lambda^{m-j},\chi_b)$$

$$=G(\eta,\chi_b)\prod_{j=1}^{(m-2)/2} G(\lambda^j,\chi_b)G(\overline{\lambda^j},\chi_b)$$

$$=q^{(m-2)/2}G(\eta,\chi_b)\prod_{j=1}^{(m-2)/2} \lambda^j(-1).$$

By Remark 5.13 we have $\lambda(-1)=1$ if $(q-1)/m$ is even and $\lambda(-1)=-1$ if $(q-1)/m$ is odd; hence $\lambda(-1)=(-1)^{(q-1)/m}$. Thus

$$\prod_{j=1}^{m-1} G(\lambda^j,\chi_b)=q^{(m-2)/2}G(\eta,\chi_b)\prod_{j=1}^{(m-2)/2} (-1)^{j(q-1)/m}$$

$$=q^{(m-2)/2}G(\eta,\chi_b)(-1)^{(q-1)(m-2)/8},$$

and together with (5.54) we get (5.53).

The known proofs of Theorem 5.28 depend either on algebraic number theory or on the theory of finite extension fields of the field of rational functions over $\mathbb{F}_q$ (*algebraic function fields* over $\mathbb{F}_q$). There are, however, certain cases in which an elementary proof can be given—for example, when ψ is the quadratic character or when m is a power of 2.

We consider first the case where ψ is the quadratic character η of $\mathbb{F}_q$. The m and q must be odd. We prove the identity in Theorem 5.28 by verifying the equivalent identity (5.52). For the left-hand side of (5.52) we get

$$\prod_{j=0}^{m-1} G(\eta\lambda^j,\chi_b)=G(\eta,\chi_b)\prod_{j=1}^{(m-1)/2} G(\eta\lambda^j,\chi_b)G(\eta\lambda^{m-j},\chi_b)$$

$$=G(\eta,\chi_b)\prod_{j=1}^{(m-1)/2} G(\eta\lambda^j,\chi_b)G(\overline{\eta\lambda^j},\chi_b)$$

$$=G(\eta,\chi_b)q^{(m-1)/2}\prod_{j=1}^{(m-1)/2} (\eta\lambda^j)(-1)$$

by Theorem 5.12(iv). Now $\lambda(-1)=1$ and $\eta(-1)=(-1)^{(q-1)/2}$ by Remark 5.13, and so

$$\prod_{j=0}^{m-1} G(\eta\lambda^j,\chi_b)=q^{(m-1)/2}(-1)^{(q-1)(m-1)/4}G(\eta,\chi_b).$$

For the right-hand side of (5.52) we get

$$q^{(m-1)/2}G(\eta^m,\chi_{mb})=q^{(m-1)/2}G(\eta,\chi_{mb})=q^{(m-1)/2}\eta(m)G(\eta,\chi_b)$$

by Theorem 5.12(i), so that it remains to show that

$$\eta(m) = (-1)^{(q-1)(m-1)/4}. \tag{5.55}$$

Let p be the characteristic of $\mathbb{F}_q$ and $q = p^s$. Since the quadratic character η of $\mathbb{F}_q$ can be obtained by lifting (in the sense of Theorem 5.14) the quadratic character η_p of $\mathbb{F}_p$, we have

$$\eta(m) = \eta_p\left(\mathrm{N}_{\mathbb{F}_q/\mathbb{F}_p}(m)\right) = \eta_p(m^s).$$

Let $m = r_1 \cdots r_t$, where the r_i are (not necessarily distinct) odd primes that are different from p since m divides $q-1$. Then

$$\eta(m) = \left[\eta_p(r_1)\cdots\eta_p(r_t)\right]^s = \left[\left(\frac{r_1}{p}\right)\cdots\left(\frac{r_t}{p}\right)\right]^s$$

because η_p is given by the Legendre symbol. The law of quadratic reciprocity (see Theorem 5.17) yields

$$\begin{aligned}\eta(m) &= \left[\left(\frac{p}{r_1}\right)(-1)^{(p-1)(r_1-1)/4}\cdots\left(\frac{p}{r_t}\right)(-1)^{(p-1)(r_t-1)/4}\right]^s \\ &= \left(\frac{q}{r_1}\right)\cdots\left(\frac{q}{r_t}\right)\left[(-1)^{us}\right]^{(p-1)/2},\end{aligned}$$

where

$$u = \frac{r_1 - 1}{2} + \cdots + \frac{r_t - 1}{2}.$$

We note that

$$\frac{q-1}{p-1} = p^{s-1} + p^{s-2} + \cdots + 1 \equiv s \bmod 2.$$

Furthermore, for any two odd integers v and w we have

$$\frac{vw-1}{2} - \frac{v-1}{2} - \frac{w-1}{2} = \frac{(v-1)(w-1)}{2} \equiv 0 \bmod 2,$$

and so

$$\frac{v-1}{2} + \frac{w-1}{2} \equiv \frac{vw-1}{2} \bmod 2.$$

By applying this repeatedly, we obtain

$$u = \frac{r_1 - 1}{2} + \cdots + \frac{r_t - 1}{2} \equiv \frac{r_1\cdots r_t - 1}{2} \equiv \frac{m-1}{2} \bmod 2.$$

Altogether, we have

$$us \equiv \frac{m-1}{2}\cdot\frac{q-1}{p-1} \bmod 2,$$

and hence,

$$\eta(m)=\left(\frac{q}{r_1}\right)\cdots\left(\frac{q}{r_t}\right)(-1)^{(q-1)(m-1)/4}.$$

But

$$\left(\frac{q}{r_i}\right)=1 \quad \text{for } 1\leqslant i\leqslant t$$

since $q\equiv 1 \bmod m$ implies $q\equiv 1 \bmod r_i$. Thus (5.55) and with it the Davenport-Hasse relation is established in this special case.

Next we consider the Davenport-Hasse relation for the case where m is a power of 2. If $m=2$, then λ is the quadratic character η of $\mathbb{F}_q$ and q is odd. By Theorem 5.12(i) and (5.43) we have

$$\begin{aligned}\frac{G(\psi,\chi_b)^2}{G(\psi^2,\chi_{2b})}&=\frac{G(\psi,\chi_b)^2}{\bar{\psi}(4)G(\psi^2,\chi_b)}\\&=\psi(4)J(\psi,\psi)\\&=\psi(4)\sum_{c_1+c_2=1}\psi(c_1)\psi(c_2)\\&=\psi(4)\sum_{c\in\mathbb{F}_q}\psi(c-c^2).\end{aligned}$$

For fixed $d\in\mathbb{F}_q$, the equation $x-x^2=d$ has two solutions in $\mathbb{F}_q$ if $1-4d$ is the square of an element of $\mathbb{F}_q^*$, one solution in $\mathbb{F}_q$ if $1-4d=0$, and no solution in $\mathbb{F}_q$ if $1-4d$ is not the square of an element of $\mathbb{F}_q$. Thus, the number of solutions in $\mathbb{F}_q$ is given by $1+\eta(1-4d)$. This yields

$$\begin{aligned}\frac{G(\psi,\chi_b)^2}{G(\psi^2,\chi_{2b})}&=\psi(4)\sum_{d\in\mathbb{F}_q}(1+\eta(1-4d))\psi(d)\\&=\psi(4)\sum_{d\in\mathbb{F}_q}\psi(d)+\sum_{d\in\mathbb{F}_q}\psi(4d)\eta(1-4d)\\&=J(\psi,\eta),\end{aligned}$$

where we used (5.37) in the last step. The Davenport-Hasse relation is thus shown for $m=2$.

Now let $m\geqslant 4$ be a power of 2, and suppose that the Davenport-Hasse relation, in the form (5.53), has already been shown for all smaller powers of

2. Applying the relation for $m/2$, we get

$$\prod_{j=0}^{m-1} G(\psi\lambda^j, \chi_b) = \prod_{j=0}^{(m/2)-1} G(\psi\lambda^{2j}, \chi_b) \prod_{j=0}^{(m/2)-1} G(\psi\lambda\lambda^{2j}, \chi_b)$$

$$= (-1)^{(q-1)(m-4)/16} q^{(m-4)/4} G(\eta, \chi_b) G(\psi^{m/2}, \chi_{(m/2)b})$$

$$\cdot (-1)^{(q-1)(m-4)/16} q^{(m-4)/4} G(\eta, \chi_b) G(\psi^{m/2}\eta, \chi_{(m/2)b})$$

$$= q^{(m-4)/2} G(\eta, \chi_b)^2 G(\psi^{m/2}, \chi_{(m/2)b}) G(\psi^{m/2}\eta, \chi_{(m/2)b}).$$

Since $G(\eta, \chi_b)^2 = \eta(-1)q$ by Theorem 5.12(iv) and $\eta(-1) = (-1)^{(q-1)/2}$ by Remark 5.13, we obtain

$$\prod_{j=0}^{m-1} G(\psi\lambda^j, \chi_b) = (-1)^{(q-1)/2} q^{(m-2)/2}$$

$$\cdot G(\psi^{m/2}, \chi_{(m/2)b}) G(\psi^{m/2}\eta, \chi_{(m/2)b}).$$

Using (5.53) with $m = 2$, we get

$$G(\psi^{m/2}, \chi_{(m/2)b}) G(\psi^{m/2}\eta, \chi_{(m/2)b}) = G(\eta, \chi_{(m/2)b}) G(\psi^m, \chi_{mb}),$$

and an application of Theorem 5.12(i) yields

$$\prod_{j=0}^{m-1} G(\psi\lambda^j, \chi_b) = (-1)^{(q-1)/2} q^{(m-2)/2} \eta\left(\frac{m}{2}\right) G(\eta, \chi_b) G(\psi^m, \chi_{mb}). \tag{5.56}$$

We now determine $\eta(2)$. Since $q^2 \equiv 1 \bmod 8$, there exists an element $\gamma \in \mathbb{F}_{q^2}^*$ of order 8. Then $\gamma^4 = -1$, hence $(\gamma + \gamma^{-1})^2 = \gamma^{-2}(\gamma^4 + 1) + 2 = 2$. Thus 2 is the square of an element of $\mathbb{F}_q$ if and only if $\gamma + \gamma^{-1} \in \mathbb{F}_q$—that is, if and only if $(\gamma + \gamma^{-1})^q = \gamma + \gamma^{-1}$. The last condition is equivalent to $\gamma^q + \gamma^{-q} = \gamma + \gamma^{-1}$, and so equivalent to $(\gamma^{q+1} - 1)(\gamma^{q-1} - 1) = 0$. This means that $\gamma^{q+1} = 1$ or $\gamma^{q-1} = 1$, and since γ has order 8, we obtain that $\eta(2) = 1$ if and only if $q \equiv \pm 1 \bmod 8$.

To determine $\eta(m/2)$, we note that if $m \geqslant 8$, we must have $q \equiv 1 \bmod 8$, and so $\eta(m/2) = 1$. If $m = 4$, we have $q \equiv 1 \bmod 4$, and then $\eta(2) = 1$ if $q \equiv 1 \bmod 8$ and $\eta(2) = -1$ if $q \equiv 5 \bmod 8$. In all cases we can write

$$\eta\left(\frac{m}{2}\right) = (-1)^{(q-1)(m-6)/8}.$$

Together with (5.56) we then obtain (5.53). Thus the Davenport-Hasse relation is established for the case where m is a power of 2.

4. CHARACTER SUMS WITH POLYNOMIAL ARGUMENTS

Let χ be a nontrivial additive character of $\mathbb{F}_q$ and let the polynomial $f \in \mathbb{F}_q[x]$ be of positive degree. We consider sums of the form

$$\sum_{c \in \mathbb{F}_q} \chi(f(c)),$$

which are sometimes referred to as *Weil sums*. The problem of evaluating such character sums explicitly is difficult. One usually has to be satisfied with estimates for the absolute value of the sum.

In certain special cases, these character sums can be treated elementarily. For instance, if f is linear, then it follows easily from (5.9) that the sum is 0. The treatment of the case where f is a binomial is based on the following result, which establishes an interesting relationship with Gaussian sums.

5.30. Theorem. *Let χ be a nontrivial additive character of $\mathbb{F}_q$, $n \in \mathbb{N}$, and λ a multiplicative character of $\mathbb{F}_q$ of order $d = \gcd(n, q-1)$. Then*

$$\sum_{c \in \mathbb{F}_q} \chi(ac^n + b) = \chi(b) \sum_{j=1}^{d-1} \bar{\lambda}^j(a) G(\lambda^j, \chi)$$

for any $a, b \in \mathbb{F}_q$ with $a \neq 0$.

Proof. Let τ be the nontrivial additive character of $\mathbb{F}_q$ defined by $\tau(c) = \chi(ac)$ for $c \in \mathbb{F}_q$. Then

$$\sum_{c \in \mathbb{F}_q} \chi(ac^n + b) = \chi(b) \sum_{c \in \mathbb{F}_q} \chi(ac^n) = \chi(b) \sum_{c \in \mathbb{F}_q} \tau(c^n). \tag{5.57}$$

By (5.17) we have

$$\tau(c^n) = \frac{1}{q-1} \sum_{\psi} G(\bar{\psi}, \tau) \psi(c^n) \quad \text{for } c \in \mathbb{F}_q^*,$$

where the sum is extended over all multiplicative characters ψ of $\mathbb{F}_q$. Thus,

$$\sum_{c \in \mathbb{F}_q} \tau(c^n) = \tau(0) + \sum_{c \in \mathbb{F}_q^*} \tau(c^n) = 1 + \frac{1}{q-1} \sum_{\psi} G(\bar{\psi}, \tau) \sum_{c \in \mathbb{F}_q^*} \psi^n(c).$$

The inner sum in the last expression is $q-1$ if ψ^n is trivial and 0 if ψ^n is nontrivial, because of (5.12). Now ψ^n is trivial if and only if the order of ψ divides d. Since $\bar{\lambda}$ is of order d, the characters ψ with order dividing d are exactly given by $\psi = \bar{\lambda}^j$ with $j = 0, 1, \ldots, d-1$. Therefore,

$$\sum_{c \in \mathbb{F}_q} \tau(c^n) = 1 + \sum_{j=0}^{d-1} G(\lambda^j, \tau) = \sum_{j=1}^{d-1} G(\lambda^j, \tau),$$

where we used (5.14). Together with Theorem 5.12(i) and (5.57) we get the desired result. □

5.31. Corollary. *If χ is a nontrivial additive character of $\mathbb{F}_q$ and* $\gcd(n, q-1) = 1$, *then*

$$\sum_{c \in \mathbb{F}_q} \chi(ac^n + b) = 0$$

for any $a, b \in \mathbb{F}_q$ with $a \neq 0$.

5.32. Theorem. *Let χ be a nontrivial additive character of $\mathbb{F}_q$, $n \in \mathbb{N}$, and $d = \gcd(n, q-1)$. Then*

$$\left| \sum_{c \in \mathbb{F}_q} \chi(ac^n + b) \right| \leqslant (d-1) q^{1/2}$$

for any $a, b \in \mathbb{F}_q$ with $a \neq 0$.

Proof. This follows immediately from Theorem 5.30 and (5.15). □

For $n = 2$ and q odd, Theorem 5.30 attains a particularly simple form, which can be used to evaluate character sums for any quadratic polynomial.

5.33. Theorem. *Let χ be a nontrivial additive character of $\mathbb{F}_q$ with q odd, and let $f(x) = a_2x^2 + a_1x + a_0 \in \mathbb{F}_q[x]$ with $a_2 \neq 0$. Then*

$$\sum_{c \in \mathbb{F}_q} \chi(f(c)) = \chi\left(a_0 - a_1^2(4a_2)^{-1}\right) \eta(a_2) G(\eta, \chi),$$

where η is the quadratic character of $\mathbb{F}_q$.

Proof. For $c \in \mathbb{F}_q$ we have

$$f(c) = a_2c^2 + a_1c + a_0 = a_2\left[c + a_1(2a_2)^{-1}\right]^2 + a_0 - a_1^2(4a_2)^{-1}.$$

Thus, putting $c_1 = c + a_1(2a_2)^{-1}$ and $b = a_0 - a_1^2(4a_2)^{-1}$, we get

$$\sum_{c \in \mathbb{F}_q} \chi(f(c)) = \sum_{c_1 \in \mathbb{F}_q} \chi\left(a_2c_1^2 + b\right) = \chi(b)\eta(a_2)G(\eta, \chi)$$

by Theorem 5.30. □

The character sums in question can also be evaluated explicitly in case f is an affine p-polynomial over $\mathbb{F}_q$ (see Definition 3.54).

5.34. Theorem. *Let $\mathbb{F}_q$ be of characteristic p and let*

$$f(x) = a_r x^{p^r} + a_{r-1} x^{p^{r-1}} + \cdots + a_1 x^p + a_0 x + a$$

be an affine p-polynomial over $\mathbb{F}_q$. Let χ_b, $b \in \mathbb{F}_q^*$, be a nontrivial additive character of $\mathbb{F}_q$ in the notation of Theorem 5.7. Then

$$\sum_{c \in \mathbb{F}_q} \chi_b(f(c)) = \begin{cases} \chi_b(a)q & \text{if } ba_r + b^p a_{r-1}^p + \cdots + b^{p^{r-1}} a_1^{p^{r-1}} + b^{p^r} a_0^{p^r} = 0, \\ 0 & \text{otherwise}. \end{cases}$$

Proof. We have

$$\sum_{c\in\mathbb{F}_q}\chi_b(f(c))=\chi_b(a)\sum_{c\in\mathbb{F}_q}\chi_1(L(c)),$$

where

$$L(x)=ba_rx^{p^r}+ba_{r-1}x^{p^{r-1}}+\cdots+ba_1x^p+ba_0x$$

is a p-polynomial over $\mathbb{F}_q$. If we put $\tau(c)=\chi_1(L(c))$ for all $c\in\mathbb{F}_q$, then (3.11) implies that τ is an additive character of $\mathbb{F}_q$. Thus

$$\sum_{c\in\mathbb{F}_q}\chi_1(L(c))=\sum_{c\in\mathbb{F}_q}\tau(c)=\begin{cases} q & \text{if } \tau \text{ is trivial,}\\ 0 & \text{otherwise.}\end{cases}$$

It remains to characterize those p-polynomials $L(x)$ for which τ is trivial. Let $q=p^s$ and Tr the absolute trace function from $\mathbb{F}_q$ to $\mathbb{F}_p$. Then according to (5.6), τ is trivial if and only if

$$\operatorname{Tr}(L(c))=\sum_{j=0}^{s-1}L(c)^{p^j}=0 \quad \text{for all } c\in\mathbb{F}_q.$$

This will be true if and only if we have the polynomial congruence

$$\sum_{j=0}^{s-1}L(x)^{p^j}\equiv 0 \bmod (x^q-x). \tag{5.58}$$

Now

$$\sum_{j=0}^{s-1}L(x)^{p^j}=\sum_{j=0}^{s-1}\left(\sum_{i=0}^{r}ba_ix^{p^i}\right)^{p^j}=\sum_{j=0}^{s-1}\sum_{i=0}^{r}b^{p^j}a_i^{p^j}x^{p^{i+j}},$$

and since $c^{p^m}=c^{p^n}$ for all $c\in\mathbb{F}_q$ and $x^{p^m}\equiv x^{p^n} \bmod (x^q-x)$ whenever $m\equiv n \bmod s$, we get

$$\sum_{j=0}^{s-1}L(x)^{p^j}\equiv\sum_{k=0}^{s-1}\left(\sum_{i=0}^{r}b^{p^{k-i}}a_i^{p^{k-i}}\right)x^{p^k} \bmod (x^q-x).$$

Thus (5.58) is satisfied if and only if

$$\sum_{i=0}^{r}b^{p^{k-i}}a_i^{p^{k-i}}=0 \quad \text{for } k=0,1,\ldots,s-1.$$

This holds if and only if

$$\sum_{i=0}^{r}b^{p^{r-i}}a_i^{p^{r-i}}=\left(\sum_{i=0}^{r}b^{p^{k-i}}a_i^{p^{k-i}}\right)^{p^{r-k}}=0,$$

and the proof is complete. □

Theorem 5.34 contains, in particular, a formula for the character sums in Theorem 5.33 for the case not considered there—namely when q is even.

5.35. Corollary. *Let $f(x)=a_2x^2+a_1x+a_0\in\mathbb{F}_q[x]$ with q even, and let χ_b, $b\in\mathbb{F}_q^*$, be as in Theorem 5.34. Then*

$$\sum_{c\in\mathbb{F}_q}\chi_b(f(c))=\begin{cases}\chi_b(a_0)q & \text{if } a_2=ba_1^2,\\ 0 & \text{otherwise.}\end{cases}$$

We turn now to a general method of obtaining estimates for character sums with polynomial arguments. Since the case of a linear polynomial is trivial, we may assume that the polynomial is of degree $\geqslant 2$. The following result is fundamental. We recall that an additive character χ of $\mathbb{F}_q$ can be lifted to an additive character $\chi^{(s)}$ of the extension field $E=\mathbb{F}_{q^s}$ by setting $\chi^{(s)}(\beta)=\chi(\mathrm{Tr}_{E/\mathbb{F}_q}(\beta))$ for $\beta\in E$.

5.36. Theorem. *Let $f\in\mathbb{F}_q[x]$ be of degree $n\geqslant 2$ with $\gcd(n,q)=1$ and let χ be a nontrivial additive character of $\mathbb{F}_q$. Then there exist complex numbers $\omega_1,\ldots,\omega_{n-1}$, only depending on f and χ, such that for any positive integer s we have*

$$\sum_{\gamma\in\mathbb{F}_{q^s}}\chi^{(s)}(f(\gamma))=-\omega_1^s-\cdots-\omega_{n-1}^s.$$

Proof. Let $f(x)=b_nx^n+\cdots+b_1x+b_0$ with $b_n\neq 0$. For fixed $k\geqslant 1$ we have then

$$f(x_1)+\cdots+f(x_k)=b_ns_n(x_1,\ldots,x_k)+\cdots+b_1s_1(x_1,\ldots,x_k)+kb_0 \tag{5.59}$$

with

$$s_j(x_1,\ldots,x_k)=x_1^j+\cdots+x_k^j \quad \text{for } j\geqslant 1.$$

For $1\leqslant r\leqslant k$, let $\sigma_r=\sigma_r(x_1,\ldots,x_k)$ be the rth elementary symmetric polynomial in the indeterminates $x_1,\ldots,x_k$ over $\mathbb{F}_q$ (see Example 1.74). Then for $j\geqslant 1$ we get by Waring's formula (see Theorem 1.76),

$$s_j(x_1,\ldots,x_k)=\sum(-1)^{i_2+i_4+i_6+\cdots}\frac{(i_1+i_2+\cdots+i_k-1)!\,j}{i_1!i_2!\cdots i_k!}\sigma_1^{i_1}\sigma_2^{i_2}\cdots\sigma_k^{i_k},$$

where the summation is extended over all k-tuples $(i_1,\ldots,i_k)$ of nonnegative integers with $i_1+2i_2+\cdots+ki_k=j$. For $j=1$ we have $s_1(x_1,\ldots,x_k)=\sigma_1$. For $2\leqslant j\leqslant k$, there is one solution of $i_1+2i_2+\cdots+ki_k=j$ with $i_j=1$ and with all other $i_r=0$, and the term corresponding to this solution is $(-1)^{j-1}j\sigma_j$. All other solutions of $i_1+2i_2+\cdots+ki_k=j$ have $i_j=i_{j+1}=\cdots=i_k=0$, and so the corresponding terms involve only $\sigma_1,\ldots,\sigma_{j-1}$. Thus,

$$s_1(x_1,\ldots,x_k)=\sigma_1,$$

$$s_j(x_1,\ldots,x_k)=(-1)^{j-1}j\sigma_j+G_j(\sigma_1,\ldots,\sigma_{j-1}) \quad \text{for } 2\leqslant j\leqslant k,$$

$$s_j(x_1,\ldots,x_k)=H_j(\sigma_1,\ldots,\sigma_k) \quad \text{for } j>k,$$

where G_j is a polynomial in $j-1$ and H_j a polynomial in k indeterminates over $\mathbb{F}_q$. It follows then from (5.59) that

$$f(x_1)+\cdots+f(x_k)=\begin{cases}(-1)^{n-1}nb_n\sigma_n+G(\sigma_1,\ldots,\sigma_{n-1}) & \text{for } k\geqslant n,\\ H(\sigma_1,\ldots,\sigma_k) & \text{for } 1\leqslant k<n,\end{cases} \tag{5.60}$$

where G is a polynomial in $n-1$ and H a polynomial in k indeterminates over $\mathbb{F}_q$.

We define now a function λ from the set Φ of monic polynomials over $\mathbb{F}_q$ into the set of complex numbers of absolute value 1 as follows. We put $\lambda(1)=1$. Furthermore, if g belongs to the subset Φ_k of Φ containing the polynomials of degree $k\geqslant 1$, then let $g(x)=(x-\alpha_1)\cdots(x-\alpha_k)$ be the factorization of g in its splitting field over $\mathbb{F}_q$. Since $\sigma_r(\alpha_1,\ldots,\alpha_k)\in\mathbb{F}_q$ for $1\leqslant r\leqslant k$, it follows from (5.60) that $f(\alpha_1)+\cdots+f(\alpha_k)\in\mathbb{F}_q$. We put

$$\lambda(g)=\chi(f(\alpha_1)+\cdots+f(\alpha_k)).$$

If $h(x)=(x-\beta_1)\cdots(x-\beta_m)\in\Phi$, then

$$\begin{aligned}\lambda(gh)&=\chi(f(\alpha_1)+\cdots+f(\alpha_k)+f(\beta_1)+\cdots+f(\beta_m))\\ &=\chi(f(\alpha_1)+\cdots+f(\alpha_k))\chi(f(\beta_1)+\cdots+f(\beta_m))\\ &=\lambda(g)\lambda(h),\end{aligned}$$

and so (5.18) is satisfied. We consider now the sum

$$\sum_{g\in\Phi_k}\lambda(g)$$

for fixed $k\geqslant n$. For

$$g(x)=x^k+\sum_{r=1}^{k}(-1)^r a_r x^{k-r}=(x-\alpha_1)\cdots(x-\alpha_k)\in\Phi_k$$

we have $\sigma_r(\alpha_1,\ldots,\alpha_k)=a_r$ for $1\leqslant r\leqslant k$, and so (5.60) implies that

$$f(\alpha_1)+\cdots+f(\alpha_k)=(-1)^{n-1}nb_na_n+G(a_1,\ldots,a_{n-1}).$$

Since $\gcd(n,q)=1$, we have $b=(-1)^{n-1}nb_n\neq 0$, hence

$$\begin{aligned}\sum_{g\in\Phi_k}\lambda(g)&=\sum_{a_1,\ldots,a_k\in\mathbb{F}_q}\chi(ba_n+G(a_1,\ldots,a_{n-1}))\\ &=q^{k-n}\sum_{a_1,\ldots,a_n\in\mathbb{F}_q}\chi(ba_n)\chi(G(a_1,\ldots,a_{n-1}))\\ &=q^{k-n}\left(\sum_{a_n\in\mathbb{F}_q}\chi(ba_n)\right)\left(\sum_{a_1,\ldots,a_{n-1}\in\mathbb{F}_q}\chi(G(a_1,\ldots,a_{n-1}))\right)=0\end{aligned}$$

by (5.9). Thus (5.22) is satisfied with $t=n-1$. It follows then from (5.24) that there exist complex numbers $\omega_1,\ldots,\omega_{n-1}$ such that

$$L_s = -\sum_{j=1}^{n-1} \omega_j^s \quad \text{for all } s \geqslant 1. \tag{5.61}$$

We calculate now L_s from (5.21). According to this formula, we have

$$L_s = \sum_g \deg(g)\lambda\left(g^{s/\deg(g)}\right),$$

where the sum is extended over all monic irreducible polynomials g in $\mathbb{F}_q[x]$ with $\deg(g)$ dividing s. For such g, let $\gamma \in E = \mathbb{F}_{q^s}$ be a root of g. Then $g^{s/\deg(g)}$ is the characteristic polynomial of γ over $\mathbb{F}_q$; that is,

$$g(x)^{s/\deg(g)} = (x-\gamma)(x-\gamma^q)\cdots\left(x-\gamma^{q^{s-1}}\right),$$

and so

$$\lambda\left(g^{s/\deg(g)}\right) = \chi\left(f(\gamma)+f(\gamma^q)+\cdots+f\left(\gamma^{q^{s-1}}\right)\right).$$

The last expression remains the same if γ is replaced by any of its distinct conjugates $\gamma^q, \gamma^{q^2},\ldots,\gamma^{q^{\deg(g)-1}}$, hence we can write

$$\deg(g)\lambda\left(g^{s/\deg(g)}\right) = \sum_{\substack{\gamma\in E\\ g(\gamma)=0}} \chi\left(f(\gamma)+f(\gamma^q)+\cdots+f\left(\gamma^{q^{s-1}}\right)\right)$$

and

$$\begin{aligned} L_s &= \sum_g \sum_{\substack{\gamma\in E\\ g(\gamma)=0}} \chi\left(f(\gamma)+f(\gamma^q)+\cdots+f\left(\gamma^{q^{s-1}}\right)\right)\\ &= \sum_{\gamma\in E} \chi\left(f(\gamma)+f(\gamma^q)+\cdots+f\left(\gamma^{q^{s-1}}\right)\right). \end{aligned}$$

Now

$$\begin{aligned} \chi\left(f(\gamma)+f(\gamma^q)+\cdots+f\left(\gamma^{q^{s-1}}\right)\right) &= \chi\left(f(\gamma)+f(\gamma)^q+\cdots+f(\gamma)^{q^{s-1}}\right)\\ &= \chi\left(\mathrm{Tr}_{E/\mathbb{F}_q}(f(\gamma))\right) = \chi^{(s)}(f(\gamma)), \end{aligned}$$

thus

$$L_s = \sum_{\gamma\in E} \chi^{(s)}(f(\gamma)),$$

and the theorem follows from (5.61). □

5.37. Theorem. *The complex numbers $\omega_1,\ldots,\omega_{n-1}$ in Theorem* 5.36 *are all of absolute value $q^{1/2}$.*

Theorem 5.37 was originally shown by deep methods of algebraic geometry. A more elementary proof can be given by using the theory of equations over finite fields. In the next chapter we will prove the weaker statement that $|\omega_j| \leqslant q^{1/2}$ for $1 \leqslant j \leqslant n-1$ (see Theorem 6.60), which is sufficient for the application to the estimation of character sums.

5.38. *Theorem* (Weil's Theorem). *Let $f \in \mathbb{F}_q[x]$ be of degree $n \geqslant 1$ with* $\gcd(n, q)=1$ *and let χ be a nontrivial additive character of $\mathbb{F}_q$. Then*

$$\left|\sum_{c \in \mathbb{F}_q} \chi(f(c))\right| \leqslant (n-1)q^{1/2}.$$

Proof. The case $n=1$ is trivial. For $n \geqslant 2$ we can apply Theorem 5.36 to get

$$\sum_{c \in \mathbb{F}_q} \chi(f(c)) = -\omega_1 - \cdots - \omega_{n-1}.$$

By using either Theorem 5.37 or the weaker statement that $|\omega_j| \leqslant q^{1/2}$ for $1 \leqslant j \leqslant n-1$, we arrive at the desired inequality. □

Some restriction on f is needed in order to guarantee the validity of Weil's theorem. We note that if $\gcd(n, q)>1$—that is, if $\deg(f)$ is divisible by the characteristic p of $\mathbb{F}_q$—then problems may arise. Consider, for instance, the case where $f(x)=x^p-x$ and $\chi=\chi_1$, the canonical additive character of $\mathbb{F}_q$ defined by (5.6). Then by a property of the absolute trace function contained in Theorem 2.23(v) we have $\chi_1(f(c))=1$ for all $c \in \mathbb{F}_q$, and so the estimate in Weil's theorem is false for $q \geqslant p^2$. More generally, a similar phenomenon occurs whenever $f = g^p - g + b$ for some $g \in \mathbb{F}_q[x]$ and $b \in \mathbb{F}_q$. However, if f is not of this form, then the estimate in Weil's theorem is still valid (see the notes).

By using similar techniques we can treat character sums with multiplicative characters. We use again the convention that $\psi(0)=0$ for a nontrivial multiplicative character ψ of $\mathbb{F}_q$. Furthermore, we note that any multiplicative character ψ of $\mathbb{F}_q$ can be lifted to a multiplicative character $\psi^{(s)}$ of the extension field $E=\mathbb{F}_{q^s}$ by setting $\psi^{(s)}(\beta)=\psi(\mathrm{N}_{E/\mathbb{F}_q}(\beta))$ for $\beta \in E$.

5.39. Theorem. *Let ψ be a multiplicative character of $\mathbb{F}_q$ of order $m>1$ and let $f \in \mathbb{F}_q[x]$ be a monic polynomial of positive degree that is not an mth power of a polynomial. Let d be the number of distinct roots of f in its splitting field over $\mathbb{F}_q$ and suppose that $d \geqslant 2$. Then there exist complex numbers $\omega_1, \ldots, \omega_{d-1}$, only depending on f and ψ, such that for any positive integer s we have*

$$\sum_{\gamma \in \mathbb{F}_{q^s}} \psi^{(s)}(f(\gamma)) = -\omega_1^s - \cdots - \omega_{d-1}^s.$$

Proof. We proceed as in the proof of Theorem 5.36. We define a function λ from Φ into the set of complex number of absolute value $\leqslant 1$ by first putting $\lambda(1)=1$. For $g \in \Phi_k$, $k \geqslant 1$, consider the resultant $R(g, f)$ with the formal degree equal to the degree for both g and f (see Definition 1.93). Then $R(g, f) \in \mathbb{F}_q$, and we put $\lambda(g)=\psi(R(g, f))$. If $g(x)=(x-\alpha_1)\cdots(x-\alpha_k)$ is the factorization of g in its splitting field over $\mathbb{F}_q$, then by (1.10) we have $R(g, f)=f(\alpha_1)\cdots f(\alpha_k)$, so that we can also write

$$\lambda(g)=\psi(f(\alpha_1)\cdots f(\alpha_k)).$$

Then (5.18) is clearly satisfied. Let

$$f=f_1^{e_1}\cdots f_r^{e_r}$$

be the canonical factorization of f in $\mathbb{F}_q[x]$, where $f_1,\ldots,f_r$ are distinct monic irreducible polynomials in $\mathbb{F}_q[x]$. By Exercise 1.66 we have $R(g, f)=(-1)^{kn}R(f, g)$, where $n=\deg(f)$. Using again the formula (1.10), we get

$$R(g, f)=(-1)^{kn}R(f_1, g)^{e_1}\cdots R(f_r, g)^{e_r}. \tag{5.62}$$

For $1 \leqslant i \leqslant r$, let $d_i=\deg(f_i)$, let E_i be the extension field of $\mathbb{F}_q$ with $[E_i:\mathbb{F}_q]=d_i$, and let β_i be a fixed root of f_i in E_i. Then all the roots of f_i are given by the conjugates of β_i with respect to $\mathbb{F}_q$, and so

$$R(f_i, g)=\mathrm{N}_{E_i/\mathbb{F}_q}(g(\beta_i)) \quad \text{for } 1 \leqslant i \leqslant r.$$

Together with (5.62) we get

$$\begin{aligned}\lambda(g)&=\psi((-1)^{kn})\psi^{e_1}\left(\mathrm{N}_{E_1/\mathbb{F}_q}(g(\beta_1))\right)\cdots\psi^{e_r}\left(\mathrm{N}_{E_r/\mathbb{F}_q}(g(\beta_r))\right)\\&=\varepsilon_k\tau_1(g(\beta_1))\cdots\tau_r(g(\beta_r)),\end{aligned} \tag{5.63}$$

where $\varepsilon_k=\psi((-1)^{kn})$ and τ_i, $1 \leqslant i \leqslant r$, is the multiplicative character obtained by lifting ψ^{e_i} to E_i. Since, by hypothesis, f is not an mth power, at least one of the e_i is not a multiple of m, hence at least one of the ψ^{e_i}, and so at least one of the τ_i, is nontrivial. We consider now the sum

$$\sum_{g \in \Phi_k} \lambda(g)$$

for $k \geqslant d$. We note that $d=d_1+\cdots+d_r$. Let the map $S:\Phi_k \to E_1 \times \cdots \times E_r$ be defined by

$$S(g)=(g(\beta_1),\ldots,g(\beta_r)) \quad \text{for } g \in \Phi_k.$$

Let $(\nu_1,\ldots,\nu_r) \in E_1 \times \cdots \times E_r$ be given. Each ν_i, $1 \leqslant i \leqslant r$, can be represented in the form $\nu_i=h_i(\beta_i)$ with $h_i \in \mathbb{F}_q[x]$. Then $S(g)=(\nu_1,\ldots,\nu_r)$ if and only if g is a solution of the system of congruences

$$g \equiv h_i \bmod f_i \quad \text{for } 1 \leqslant i \leqslant r.$$

By the Chinese remainder theorem for $\mathbb{F}_q[x]$ (see Exercise 1.37), this system of congruences has a unique solution $G \in \mathbb{F}_q[x]$ with $\deg(G)<d_1+\cdots+d_r=d$. Then all solutions $g \in \Phi_k$ of the system are given by $g=Ff_1\cdots f_r+G$, where F is an arbitrary monic polynomial over $\mathbb{F}_q$ of degree $k-d$. Since

there are exactly q^{k-d} choices for F, there are exactly q^{k-d} polynomials $g \in \Phi_k$ with $S(g) = (g(\beta_1), \ldots, g(\beta_r)) = (\nu_1, \ldots, \nu_r)$. Using this fact and (5.63), we get

$$\sum_{g \in \Phi_k} \lambda(g) = \varepsilon_k q^{k-d} \sum_{\substack{\nu_1, \ldots, \nu_r \\ \nu_i \in E_i}} \tau_1(\nu_1) \cdots \tau_r(\nu_r)$$

$$= \varepsilon_k q^{k-d} \left(\sum_{\nu_1 \in E_1} \tau_1(\nu_1) \right) \cdots \left(\sum_{\nu_r \in E_r} \tau_r(\nu_r) \right) = 0,$$

since at least one of the τ_i is nontrivial, as we have noted before. Thus (5.22) is satisfied with $t = d-1$. It follows then from (5.24) that there exist complex numbers $\omega_1, \ldots, \omega_{d-1}$ such that

$$L_s = -\sum_{j=1}^{d-1} \omega_j^s \quad \text{for all } s \geqslant 1.$$

We calculate now L_s from (5.21) by using arguments analogous to those in the proof of Theorem 5.36. With $E = \mathbb{F}_{q^s}$ this yields

$$L_s = \sum_{\gamma \in E} \psi\left(f(\gamma) f(\gamma^q) \cdots f\left(\gamma^{q^{s-1}}\right)\right),$$

and since

$$\psi\left(f(\gamma) f(\gamma^q) \cdots f\left(\gamma^{q^{s-1}}\right)\right) = \psi\left(f(\gamma) f(\gamma)^q \cdots f(\gamma)^{q^{s-1}}\right)$$

$$= \psi\left(\mathrm{N}_{E/\mathbb{F}_q}(f(\gamma))\right) = \psi^{(s)}(f(\gamma)),$$

it follows that

$$L_s = \sum_{\gamma \in E} \psi^{(s)}(f(\gamma)),$$

and this completes the proof. □

5.40. Theorem. *The complex numbers $\omega_1, \ldots, \omega_{d-1}$ in Theorem* 5.39 *are all of absolute value $q^{1/2}$.*

The remarks following Theorem 5.37 apply also to Theorem 5.40. In particular, an elementary proof of the weaker statement that $|\omega_j| \leqslant q^{1/2}$ for $1 \leqslant j \leqslant d-1$ will be given in Theorem 6.56.

5.41. Theorem. *Let ψ be a multiplicative character of $\mathbb{F}_q$ of order $m > 1$ and let $f \in \mathbb{F}_q[x]$ be a monic polynomial of positive degree that is not an mth power of a polynomial. Let d be the number of distinct roots of f in its splitting field over $\mathbb{F}_q$. Then for every $a \in \mathbb{F}_q$ we have*

$$\left| \sum_{c \in \mathbb{F}_q} \psi(af(c)) \right| \leqslant (d-1) q^{1/2}.$$

Proof. The case $d=1$ is easily checked, so that we may assume $d \geqslant 2$. Then an application of Theorem 5.39 yields

$$\sum_{c \in \mathbb{F}_q} \psi(af(c)) = \psi(a) \sum_{c \in \mathbb{F}_q} \psi(f(c)) = -\psi(a)(\omega_1 + \cdots + \omega_{d-1}).$$

By using either Theorem 5.40 or the weaker statement that $|\omega_j| \leqslant q^{1/2}$ for $1 \leqslant j \leqslant d-1$, we arrive at the desired inequality. □

In the case not covered by Theorem 5.41, namely when f is an mth power of a polynomial, the estimate given there need not necessarily hold. For instance, if $f = g^m$ with $g \in \mathbb{F}_q[x]$ not having a root in $\mathbb{F}_q$, then $\psi(f(c)) = \psi^m(g(c)) = 1$ for all $c \in \mathbb{F}_q$ since ψ^m is the trivial character, and the estimate in Theorem 5.41 need not be valid.

The case where the character ψ is the quadratic character will be considered in somewhat greater detail in the next section.

5. FURTHER RESULTS ON CHARACTER SUMS

We consider first a type of character sum that can be treated by the methods of Section 4 and that is of interest in number theory.

5.42. Definition. Let χ be a nontrivial additive character of $\mathbb{F}_q$ and let $a, b \in \mathbb{F}_q$. Then the sum

$$K(\chi; a, b) = \sum_{c \in \mathbb{F}_q^*} \chi(ac + bc^{-1})$$

is called a *Kloosterman sum*.

The cases where $ab = 0$ are trivial. If $a = b = 0$, then $K(\chi; ab) = q-1$, and if exactly one of a and b is 0, then $K(\chi; a, b) = -1$. We note in general that the value of a Kloosterman sum is always real, for if $K = K(\chi; a, b)$ and $\bar{K}$ is its complex conjugate, then

$$\bar{K} = \sum_{c \in \mathbb{F}_q^*} \chi(-ac - bc^{-1}) = \sum_{c \in \mathbb{F}_q^*} \chi\big(a(-c) + b(-c)^{-1}\big) = K,$$

since $-c$ runs again through $\mathbb{F}_q^*$.

As in Section 4, we use $\chi^{(s)}$ to denote the additive character obtained by lifting χ to the extension field $\mathbb{F}_{q^s}$.

5.43. Theorem. *Let χ be a nontrivial additive character of $\mathbb{F}_q$ and let $a, b \in \mathbb{F}_q$ with $ab \neq 0$. Then there exist numbers ω_1 and ω_2 (only depending on χ, a, and b) that are either complex conjugates or both real, such that for any positive integer s we have*

$$K(\chi^{(s)}; a, b) = \sum_{\gamma \in \mathbb{F}_{q^s}^*} \chi^{(s)}(a\gamma + b\gamma^{-1}) = -\omega_1^s - \omega_2^s.$$

Proof. We proceed as in the proofs of Theorems 5.36 and 5.39 by defining a suitable function λ from Φ into the set of complex numbers of absolute value $\leqslant 1$. We put $\lambda(1)=1$. Furthermore, if $g \in \Phi_k$, $k \geqslant 1$, say

$$g(x) = \sum_{r=0}^{k} (-1)^r c_r x^{k-r} \quad \text{with } c_0 = 1,$$

we set $\lambda(g)=0$ if $c_k = 0$ and

$$\lambda(g) = \chi(ac_1 + bc_{k-1}c_k^{-1}) \quad \text{if } c_k \neq 0.$$

It is easily checked that $\lambda(gh)=\lambda(g)\lambda(h)$ for all $g, h \in \Phi$. For $k \geqslant 3$ we have

$$\begin{aligned}\sum_{g \in \Phi_k} \lambda(g) &= \sum_{c_1, \ldots, c_{k-1} \in \mathbb{F}_q} \sum_{c_k \in \mathbb{F}_q^*} \chi(ac_1 + bc_{k-1}c_k^{-1}) \\ &= q^{k-3}\Big(\sum_{c_1 \in \mathbb{F}_q} \chi(ac_1)\Big)\Big(\sum_{c_{k-1} \in \mathbb{F}_q} \sum_{c_k \in \mathbb{F}_q^*} \chi(bc_{k-1}c_k^{-1})\Big) = 0,\end{aligned}$$

so that (5.22) holds with $t = 2$. From (5.19) we obtain

$$L(z) = 1 + \Big(\sum_{g \in \Phi_1} \lambda(g)\Big)z + \Big(\sum_{g \in \Phi_2} \lambda(g)\Big)z^2.$$

With $K = K(\chi; a, b)$ we get

$$\sum_{g \in \Phi_1} \lambda(g) = \sum_{c \in \mathbb{F}_q^*} \chi(ac + bc^{-1}) = K.$$

Furthermore,

$$\sum_{g \in \Phi_2} \lambda(g) = \sum_{c_2 \in \mathbb{F}_q^*} \sum_{c_1 \in \mathbb{F}_q} \chi(c_1(a + bc_2^{-1})) = q$$

since the inner sum is equal to q if $c_2 = -a^{-1}b$ and equal to 0 otherwise. Thus $L(z) = 1 + Kz + qz^2 = (1-\omega_1 z)(1-\omega_2 z)$, where ω_1 and ω_2 are either complex conjugates or both real because $L(z)$ has real coefficients. By (5.24) we have

$$L_s = -\omega_1^s - \omega_2^s \quad \text{for all } s \geqslant 1. \tag{5.64}$$

It remains to evaluate L_s. From (5.21) we get

$$L_s = \sum_g \deg(g)\lambda(g^{s/\deg(g)}) = \sum_g{}^{*} \deg(g)\lambda(g^{s/\deg(g)}),$$

with the sum being over all monic irreducible polynomials g in $\mathbb{F}_q[x]$ with $\deg(g)$ dividing s, and where the asterisk indicates that $g(x)=x$ is excluded. Each such g has $\deg(g)$ distinct nonzero roots in $E = \mathbb{F}_{q^s}$, and each root γ of g has as its characteristic polynomial over $\mathbb{F}_q$ the polynomial

$$g(x)^{s/\deg(g)} = (x-\gamma)(x-\gamma^q)\cdots\left(x-\gamma^{q^{s-1}}\right)$$
$$= x^s - c_1x^{s-1} + \cdots + (-1)^{s-1}c_{s-1}x + (-1)^s c_s,$$

say. Then $c_1 = \mathrm{Tr}_{E/\mathbb{F}_q}(\gamma)$, $c_s = \gamma\gamma^q\cdots\gamma^{q^{s-1}}$, and

$$c_{s-1}c_s^{-1} = \gamma^{-1} + \gamma^{-q} + \cdots + \gamma^{-q^{s-1}} = \mathrm{Tr}_{E/\mathbb{F}_q}(\gamma^{-1}).$$

Therefore,

$$\lambda\left(g^{s/\deg(g)}\right) = \chi\left(a\,\mathrm{Tr}_{E/\mathbb{F}_q}(\gamma) + b\,\mathrm{Tr}_{E/\mathbb{F}_q}(\gamma^{-1})\right) = \chi^{(s)}(a\gamma + b\gamma^{-1}),$$

and so

$$L_s = \sum_g{}^{*} \deg(g)\lambda\left(g^{s/\deg(g)}\right) = \sum_g{}^{*} \sum_{\substack{\gamma\in E\\ g(\gamma)=0}} \chi^{(s)}(a\gamma + b\gamma^{-1}).$$

If g runs through the range of summation above, then γ runs exactly through all elements of E^*. Thus,

$$L_s = \sum_{\gamma\in E^*} \chi^{(s)}(a\gamma + b\gamma^{-1}) = K(\chi^{(s)}; a, b),$$

and the desired result follows from (5.64). □

5.44. Theorem. *The numbers* ω_1 *and* ω_2 *in Theorem* 5.43 *satisfy* $|\omega_1| = |\omega_2| = q^{1/2}$.

An elementary proof of Theorem 5.44, using the theory of equations over finite fields, will be given in Example 6.63 for odd q.

5.45. Theorem. *If* χ *is a nontrivial additive character of* $\mathbb{F}_q$ *and* $a, b \in \mathbb{F}_q$ *are not both* 0, *then the Kloosterman sum* $K(\chi; a, b)$ *satisfies*

$$|K(\chi; a, b)| \leqslant 2q^{1/2}.$$

Proof. Because of a remark following Definition 5.42, the result is trivial if one of a and b is 0. If $ab \neq 0$, then Theorem 5.43 yields $K(\chi; a, b) = -\omega_1 - \omega_2$, and the desired inequality follows from Theorem 5.44. □

The result of Theorem 5.43 can be used to prove a reduction formula linking the "lifted" Kloosterman sum $K(\chi^{(s)}; a, b)$ as defined in that theorem to the Kloosterman sum $K(\chi; a, b)$ in the ground field $\mathbb{F}_q$.

5.46. Theorem. *Let* χ *be a nontrivial additive character of* $\mathbb{F}_q$, *let* $a, b \in \mathbb{F}_q$ *with* $ab \neq 0$, *and put* $K = K(\chi; a, b)$. *Then for any positive integer* s *we have*

$$K(\chi^{(s)}; a, b) = \sum_{j=0}^{\lfloor s/2\rfloor} (-1)^{s-j-1}\frac{s}{s-j}\binom{s-j}{j} q^j K^{s-2j},$$

where $\lfloor s/2\rfloor$ *denotes the greatest integer* $\leqslant s/2$.

Proof. The symmetric polynomial $x_1^s + x_2^s$ can be expressed in terms of the elementary symmetric polynomials $x_1 + x_2$ and x_1x_2 by means of Waring's formula (see Theorem 1.76). This yields

$$x_1^s + x_2^s = \sum_{i_1 + 2i_2 = s} (-1)^{i_2} \frac{(i_1 + i_2 - 1)!s}{i_1!i_2!} (x_1 + x_2)^{i_1} (x_1x_2)^{i_2},$$

where i_1 and i_2 assume nonnegative integral values. Putting $i_1 = s - 2j$, $i_2 = j$ we get

$$x_1^s + x_2^s = \sum_{j=0}^{\lfloor s/2 \rfloor} (-1)^j \frac{s}{s-j} \binom{s-j}{j} (x_1 + x_2)^{s-2j} (x_1x_2)^j.$$

Now apply Theorem 5.43 and substitute $x_1 = \omega_1$, $x_2 = \omega_2$ in the identity above. We note that $\omega_1^s + \omega_2^s = -K(\chi^{(s)}; a, b)$ and $\omega_1 + \omega_2 = -K$, and also $\omega_1\omega_2 = q$ on account of the identity $1 + Kz + qz^2 = (1 - \omega_1 z)(1 - \omega_2 z)$ shown in the proof of Theorem 5.43. The desired formula follows immediately. □

The Kloosterman sums $K^{(s)} = K(\chi^{(s)}; a, b)$ can also be calculated recursively in a convenient manner. The identity

$$\omega_1^s + \omega_2^s = \left(\omega_1^{s-1} + \omega_2^{s-1}\right)(\omega_1 + \omega_2) - \left(\omega_1^{s-2} + \omega_2^{s-2}\right)\omega_1\omega_2$$

together with Theorem 5.43 shows that

$$K^{(s)} = -K^{(s-1)}K - K^{(s-2)}q \quad \text{for } s \geqslant 2,$$

where we put $K^{(0)} = -2$ and $K^{(1)} = K = K(\chi; a, b)$.

If q is odd, then there is a simple way of linking nontrivial Kloosterman sums with the quadratic character η of $\mathbb{F}_q$. We employ again the standard convention $\eta(0) = 0$.

5.47. Theorem. *If χ is a nontrivial additive character of $\mathbb{F}_q$, q odd, and $a, b \in \mathbb{F}_q$ are not both 0, then the Kloosterman sum $K(\chi; a, b)$ can be represented in the form*

$$K(\chi; a, b) = \sum_{c \in \mathbb{F}_q} \chi(c)\eta(c^2 - 4ab).$$

Proof. If one of a and b is 0, then $K(\chi; a, b) = -1$, which is easily seen to be the value of the right-hand side as well. For $ab \neq 0$ we write

$$K(\chi; a, b) = \sum_{c \in \mathbb{F}_q^*} \chi(ac + bc^{-1}) = \sum_{d \in \mathbb{F}_q} \chi(d)N(d),$$

where $N(d)$ is the number of $c \in \mathbb{F}_q^*$ with $ac + bc^{-1} = d$. This equation is equivalent to the quadratic equation $ac^2 - dc + b = 0$. Thus $N(d) = 2$, 1, or 0 depending on whether $\eta(d^2 - 4ab) = 1$, 0, or -1. In other words, $N(d) =$

$1+\eta(d^2-4ab)$. It follows that

$$\begin{aligned}K(\chi;a,b) &= \sum_{d\in\mathbb{F}_q}\chi(d)\left(1+\eta(d^2-4ab)\right)\\ &= \sum_{d\in\mathbb{F}_q}\chi(d)+\sum_{d\in\mathbb{F}_q}\chi(d)\eta(d^2-4ab)\\ &= \sum_{d\in\mathbb{F}_q}\chi(d)\eta(d^2-4ab),\end{aligned}$$

where we used (5.9) in the last step. □

We consider now sums involving only the quadratic character η of $\mathbb{F}_q$, q odd, and having polynomial arguments—that is, sums of the form

$$\sum_{c\in\mathbb{F}_q}\eta(f(c)) \tag{5.65}$$

with $f\in\mathbb{F}_q[x]$. The case of linear f is trivial, and for quadratic f one can still establish an explicit formula.

5.48. Theorem. *Let $f(x)=a_2x^2+a_1x+a_0\in\mathbb{F}_q[x]$ with q odd and $a_2\neq 0$. Put $d=a_1^2-4a_0a_2$ and let η be the quadratic character of $\mathbb{F}_q$. Then*

$$\sum_{c\in\mathbb{F}_q}\eta(f(c))=\begin{cases}-\eta(a_2) & \text{if } d\neq 0,\\ (q-1)\eta(a_2) & \text{if } d=0.\end{cases}$$

Proof. Multiplying the sum by $\eta(4a_2^2)=1$, we get

$$\begin{aligned}\sum_{c\in\mathbb{F}_q}\eta(f(c)) &= \eta(a_2)\sum_{c\in\mathbb{F}_q}\eta\left(4a_2^2c^2+4a_1a_2c+4a_0a_2\right)\\ &= \eta(a_2)\sum_{c\in\mathbb{F}_q}\eta\left((2a_2c+a_1)^2-d\right)=\eta(a_2)\sum_{b\in\mathbb{F}_q}\eta(b^2-d).\end{aligned} \tag{5.66}$$

The result for the case $d=0$ follows now immediately. For $d\neq 0$ we write

$$\sum_{b\in\mathbb{F}_q}\eta(b^2-d)=-q+\sum_{b\in\mathbb{F}_q}\left(1+\eta(b^2-d)\right),$$

and since $1+\eta(b^2-d)$ is the number of $c\in\mathbb{F}_q$ with $c^2=b^2-d$, we obtain

$$\sum_{b\in\mathbb{F}_q}\eta(b^2-d)=-q+S(d), \tag{5.67}$$

where $S(d)$ is the number of ordered pairs (b,c) with $b,c\in\mathbb{F}_q$ and $b^2-c^2=d$. To solve this equation, we put $b+c=u$, $b-c=v$ and note that the ordered pairs (b,c) and (u,v) are in one-to-one correspondence since q is odd. Thus $S(d)$ is equal to the number of ordered pairs (u,v) with $u,v\in\mathbb{F}_q$ and $uv=d$, hence $S(d)=q-1$. Together with (5.66) and (5.67), this implies the desired formula. □

5.49. Definition. For $a \in \mathbb{F}_q^*$, q odd, $n \in \mathbb{N}$, and η the quadratic character of $\mathbb{F}_q$, the sum

$$H_n(a) = \sum_{c \in \mathbb{F}_q} \eta(c^{n+1} + ac) = \sum_{c \in \mathbb{F}_q} \eta(c)\eta(c^n + a)$$

is called a *Jacobsthal sum*.

It follows from Theorem 5.48 that $H_1(a) = -1$ for all $a \in \mathbb{F}_q^*$. There is a companion sum

$$I_n(a) = \sum_{c \in \mathbb{F}_q} \eta(c^n + a) \tag{5.68}$$

which is related to Jacobsthal sums in the following way.

5.50. ***Theorem.*** *The identity*

$$I_{2n}(a) - I_n(a) + H_n(a)$$

holds for all $n \in \mathbb{N}$ *and* $a \in \mathbb{F}_q^*$.

Proof. We have

$$I_{2n}(a) = \sum_{c \in \mathbb{F}_q} \eta(c^{2n} + a) = \sum_{d \in \mathbb{F}_q} N(d)\eta(d^n + a),$$

where $N(d)$ is the number of $c \in \mathbb{F}_q$ with $c^2 = d$. But $N(d) = 1 + \eta(d)$, and so

$$I_{2n}(a) = \sum_{d \in \mathbb{F}_q} (1 + \eta(d))\eta(d^n + a) = I_n(a) + H_n(a). \qquad \square$$

The sums in (5.68) can be evaluated easily for $n = 1$ and $n = 2$. We have $I_1(a) = 0$ and $I_2(a) = -1$ for all $a \in \mathbb{F}_q^*$, where the second result follows from either Theorem 5.48 or Theorem 5.50. In general, the sums $I_n(a)$ can be expressed in terms of Jacobi sums.

5.51. ***Theorem.*** *For all* $n \in \mathbb{N}$ *and* $a \in \mathbb{F}_q^*$ *we have*

$$I_n(a) = \eta(a) \sum_{j=1}^{d-1} \lambda^j(-a) J(\lambda^j, \eta),$$

where λ *is a multiplicative character of* $\mathbb{F}_q$ *of order* $d = \gcd(n, q-1)$.

Proof. We write

$$I_n(a) = \sum_{c \in \mathbb{F}_q} \eta(c^n + a) = \sum_{b \in \mathbb{F}_q} \eta(b + a)M(b), \tag{5.69}$$

where $M(b)$ is the number of $c \in \mathbb{F}_q$ with $c^n = b$. For $b \neq 0$ we can use (5.13) to get

$$M(b) = \frac{1}{q-1} \sum_{c \in \mathbb{F}_q^*} \sum_{\psi} \psi(c^n)\bar{\psi}(b) = \frac{1}{q-1} \sum_{\psi} \bar{\psi}(b) \sum_{c \in \mathbb{F}_q^*} \psi^n(c).$$

By (5.12) the inner sum in the last expression is $q-1$ if ψ^n is trivial and 0 if ψ^n is nontrivial. By an argument in the proof of Theorem 5.30, ψ^n is trivial if and only if $\psi = \bar{\lambda}^j$, $j = 0, 1, \ldots, d-1$. Therefore,

$$M(b) = \sum_{j=0}^{d-1} \lambda^j(b), \tag{5.70}$$

and since $M(0) = 1$, (5.70) holds also for $b = 0$. Combining (5.69) and (5.70) and using (5.38), we get

$$\begin{aligned} I_n(a) &= \sum_{b \in \mathbb{F}_q} \eta(b+a) \sum_{j=0}^{d-1} \lambda^j(b) = \eta(-1) \sum_{j=0}^{d-1} \sum_{b \in \mathbb{F}_q} \lambda^j(b)\eta(-b-a) \\ &= \eta(-1) \sum_{j=0}^{d-1} J_{-a}(\lambda^j, \eta) = \eta(-1) \sum_{j=0}^{d-1} (\lambda^j \eta)(-a) J(\lambda^j, \eta) \\ &= \eta(a) \sum_{j=0}^{d-1} \lambda^j(-a) J(\lambda^j, \eta). \end{aligned}$$

On account of (5.40), the term corresponding to $j = 0$ can be omitted. □

5.52. *Theorem.* *For $n \in \mathbb{N}$ and $a \in \mathbb{F}_q^*$, we have $H_n(a) = 0$ if the largest power of 2 dividing $q-1$ also divides n. Otherwise we have*

$$H_n(a) = \eta(a)\lambda(-1) \sum_{j=0}^{d-1} \lambda^{2j+1}(a) J(\lambda^{2j+1}, \eta),$$

where $d = \gcd(n, q-1)$ and λ is a multiplicative character of $\mathbb{F}_q$ of order $2d$.

Proof. We have $H_n(a) = I_{2n}(a) - I_n(a)$ by Theorem 5.50. If the largest power of 2 dividing $q-1$ also divides n, then $\gcd(2n, q-1) = \gcd(n, q-1)$, and Theorem 5.51 implies $H_n(a) = 0$. Otherwise we have $\gcd(2n, q-1) = 2d$, so that Theorem 5.51 yields

$$I_{2n}(a) = \eta(a) \sum_{j=1}^{2d-1} \lambda^j(-a) J(\lambda^j, \eta).$$

Since λ^2 is of order d, the same theorem shows that

$$I_n(a) = \eta(a) \sum_{j=1}^{d-1} \lambda^{2j}(-a) J(\lambda^{2j}, \eta).$$

It follows that

$$\begin{aligned} H_n(a) &= \eta(a) \sum_{j=0}^{d-1} \lambda^{2j+1}(-a) J(\lambda^{2j+1}, \eta) \\ &= \eta(a)\lambda(-1) \sum_{j=0}^{d-1} \lambda^{2j+1}(a) J(\lambda^{2j+1}, \eta). \end{aligned}$$

□

These results and Theorem 5.22 yield the estimates $|I_n(a)| \leqslant (d-1)q^{1/2}$ and $|H_n(a)| \leqslant dq^{1/2}$, which are at least as good as those implied by Theorem 5.41.

5.53. Example. We have shown in Example 5.25 that every prime $p \equiv 1 \bmod 4$ can be written in the form $p = A^2 + B^2$ with integers A and B. Using congruences mod 4, it is easily seen that one of the integers, say A, must be odd and the other must be even. Since the sign of A is at our disposal, we can take $A \equiv -1 \bmod 4$. We prove now that such an A can be calculated directly by means of a suitable Jacobsthal sum.

If λ is a multiplicative character of $\mathbb{F}_p$ of order 4, then it follows from Theorem 5.52 and $\lambda^3 = \bar{\lambda}$ that

$$\begin{aligned} H_2(1) &= \lambda(-1)\big(J(\lambda,\eta)+J(\lambda^3,\eta)\big) = \lambda(-1)\big(J(\lambda,\eta)+\overline{J(\lambda,\eta)}\big) \\ &= 2\lambda(-1)\operatorname{Re} J(\lambda,\eta), \end{aligned}$$

hence $\operatorname{Re} J(\lambda,\eta) = \pm\frac{1}{2}H_2(1)$. On the other hand, we have seen in Example 5.25 that $p = (\operatorname{Re} J(\lambda,\eta))^2 + (\operatorname{Im} J(\lambda,\eta))^2$. We will now show that $\frac{1}{2}H_2(1) \equiv -1 \bmod 4$. Since $\eta(-1) = 1$ for $p \equiv 1 \bmod 4$ by Remark 5.13, we can write

$$\begin{aligned} H_2(1) &= \sum_{c=1}^{p-1} \eta(c)\eta(c^2+1) \\ &= \sum_{c=1}^{(p-1)/2} \eta(c)\eta(c^2+1) + \sum_{c=1}^{(p-1)/2} \eta(-c)\eta(c^2+1) \\ &= 2\sum_{c=1}^{(p-1)/2} \eta(c)\eta(c^2+1), \end{aligned}$$

so that

$$\tfrac{1}{2}H_2(1) = \sum_{c=1}^{(p-1)/2} \eta(c)\eta(c^2+1). \tag{5.71}$$

From Theorem 5.48 we get

$$-1 = \sum_{c=0}^{p-1} \eta(c^2+1) = 1 + 2\sum_{c=1}^{(p-1)/2} \eta(c^2+1),$$

hence

$$-1 = \sum_{c=1}^{(p-1)/2} \eta(c^2+1). \tag{5.72}$$

Subtracting (5.72) from (5.71), we obtain

$$\tfrac{1}{2}H_2(1) + 1 = \sum_{c=1}^{(p-1)/2} (\eta(c)-1)\eta(c^2+1).$$

For c in this range of summation, we have

$$(\eta(c)-1)(\eta(c^2+1)-1) \equiv 0 \bmod 4 \text{ whenever } \eta(c^2+1) \neq 0,$$

since both factors on the left-hand side are even. Thus,

$$(\eta(c)-1)\eta(c^2+1) \equiv \eta(c)-1 \bmod 4 \text{ whenever } \eta(c^2+1) \neq 0.$$

Now $\eta(c^2+1)=0$ if and only if $c^2 \equiv -1 \bmod p$, and there exists a unique integer c_1, $1 \leqslant c_1 \leqslant (p-1)/2$, solving this congruence. Consequently,

$$\tfrac{1}{2}H_2(1)+1 \equiv \sum_{\substack{c=1 \\ c \neq c_1}}^{(p-1)/2} (\eta(c)-1) \equiv \sum_{c=1}^{(p-1)/2} (\eta(c)-1)-(\eta(c_1)-1)$$

$$\equiv \sum_{c=1}^{(p-1)/2} \eta(c)+\frac{3-p}{2}-\eta(c_1) \bmod 4.$$

Furthermore,

$$0=\sum_{c=1}^{p-1} \eta(c)=2\sum_{c=1}^{(p-1)/2} \eta(c)$$

and $\eta(c_1)=\lambda^2(c_1)=\lambda(c_1^2)=\lambda(-1)$, so that

$$\tfrac{1}{2}H_2(1)+1 \equiv \frac{3-p}{2}-\lambda(-1) \bmod 4.$$

Now Remark 5.13 yields

$$\lambda(-1)=\begin{cases} 1 & \text{if } p \equiv 1 \bmod 8, \\ -1 & \text{if } p \equiv 5 \bmod 8, \end{cases}$$

and $(3-p)/2 \equiv 1 \bmod 4$ if $p \equiv 1 \bmod 8$, $(3-p)/2 \equiv -1 \bmod 4$ if $p \equiv 5 \bmod 8$, hence $\frac{1}{2}H_2(1)+1 \equiv 0 \bmod 4$ and $\frac{1}{2}H_2(1) \equiv -1 \bmod 4$, as claimed.

One can show that with the normalization $A \equiv -1 \bmod 4$ the integer A is uniquely determined. Suppose $p=A^2+B^2=C^2+D^2$ with A, C odd and B, D even. If $h, k \in \mathbb{Z}$ are such that $A \equiv hB \bmod p, C \equiv kD \bmod p$, then $A^2+B^2 \equiv C^2+D^2 \equiv 0 \bmod p$ implies $h^2+1 \equiv k^2+1 \equiv 0 \bmod p$, hence $C \equiv \pm hD \bmod p$. Thus we can write $C=\pm C_1$ with $C_1 \equiv hD \bmod p$. Then in

$$p^2=(A^2+B^2)(C_1^2+D^2)=(AC_1+BD)^2+(AD-BC_1)^2$$

the numbers in parentheses on the right are divisible by p. Dividing by p^2 throughout, we get an expression for 1 as a sum of two squares of integers, with the first one being odd. The only possibility for this is $1=(\pm 1)^2+0^2$. Thus $AD-BC_1=0$, and since $\gcd(A, B)=\gcd(C_1, D)=1$, it follows that $A=\pm C_1$, hence $A=\pm C$. If now $A \equiv C \equiv -1 \bmod 4$, then $A=C$; that is, A is uniquely determined.

Thus, $\frac{1}{2}H_2(1)$ gives this unique integer A, and the integer B is then uniquely determined up to sign. In fact, by using the argument in the beginning of the example, one sees easily that one can take $B=\frac{1}{2}H_2(a)$, where $a\in\mathbb{F}_p$ is such that $\eta(a)=-1$. □

There is a remarkable connection between character sums with the quadratic character η and the *continued fraction algorithm* for rational functions over $\mathbb{F}_q$ (i.e., for fractions of the form f/g with $f, g\in\mathbb{F}_q[x]$, $g\neq 0$). We only need the rudiments of the theory of such continued fractions, which is quite analogous to the classical theory of continued fractions for rational numbers. The continued fraction algorithm is just another way of looking at the Euclidean algorithm discussed in Chapter 1, Section 3. We change the notation slightly to suit our present needs.

Let r_0 and r_1 be two polynomials over $\mathbb{F}_q$ with $r_1\neq 0$. By applying the Euclidean algorithm, we can write $r_0=A_0r_1+r_2$, $r_1=A_1r_2+r_3$, and in general

$$r_i=A_ir_{i+1}+r_{i+2}\quad \text{for } i=0,1,\ldots,s, \tag{5.73}$$

where $0\leqslant \deg(r_{i+1})<\deg(r_i)$ for $i=1,\ldots,s$ and $r_{s+2}=0$. Furthermore, $A_0, A_1,\ldots,A_s$ are polynomials over $\mathbb{F}_q$ with $A_1,\ldots,A_s$ having positive degree. From (5.73) we get

$$\frac{r_i}{r_{i+1}}=A_i+\frac{1}{r_{i+1}/r_{i+2}}\qquad \text{for } i=0,1,\ldots,s-1$$

and $r_s/r_{s+1}=A_s$. This yields

$$\frac{r_0}{r_1}=A_0+\frac{1}{r_1/r_2},$$

$$\frac{r_0}{r_1}=A_0+\cfrac{1}{A_1+\cfrac{1}{r_2/r_3}},$$

and continuing in this manner,

$$\frac{r_0}{r_1}=A_0+\cfrac{1}{A_1+\cfrac{1}{A_2+\cfrac{\ddots}{+\cfrac{1}{A_s}}}}=[A_0,A_1,A_2,\ldots,A_s],$$

where the symbol on the extreme right abbreviates the continued fraction expansion given by the middle term. A convenient way of calculating the rational function represented by a continued fraction is based on the following recursive procedure. Define polynomials P_i, Q_i, $i=-1,0,\ldots,s$, by

$$P_{-1}=1, \quad P_0=A_0, \quad P_i=A_iP_{i-1}+P_{i-2} \quad \text{for } i=1,\ldots,s, \tag{5.74}$$

$$Q_{-1}=0, \quad Q_0=1, \quad Q_i=A_iQ_{i-1}+Q_{i-2} \quad \text{for } i=1,\ldots,s. \tag{5.75}$$

It is clear that $\deg(P_{i-1})<\deg(P_i)$ for $i=1,\ldots,s$ and $\deg(Q_{i-1})<\deg(Q_i)$ for $i=0,1,\ldots,s$. Corollaries 5.55 and 5.58 below show that P_i and Q_i are the numerator and the denominator, respectively, of the rational function in reduced form represented by $[A_0, A_1,\ldots,A_i]$. It will be useful to extend the definition of degree by setting $\deg(\rho)=\deg(f)-\deg(g)$ for a rational function $\rho=f/g$. The standard conventions $\deg(0)=-\infty$ and $-\infty-n=-\infty$ for $n\in\mathbb{Z}$ are again in force.

5.54. Lemma. *For any rational function ρ of nonnegative degree we have*

$$[A_0, A_1,\ldots,A_{i-1},\rho]=\frac{\rho P_{i-1}+P_{i-2}}{\rho Q_{i-1}+Q_{i-2}} \quad \text{for } i=1,\ldots,s+1.$$

Proof. We proceed by induction. For $i=1$ both sides are equal to $A_0+\rho^{-1}$. If the statement is shown for some i, $1\leqslant i<s+1$, then since $A_i+\rho^{-1}$ is of positive degree, we get

$$[A_0, A_1,\ldots,A_i,\rho]=[A_0, A_1,\ldots,A_{i-1}, A_i+\rho^{-1}]=\frac{(A_i+\rho^{-1})P_{i-1}+P_{i-2}}{(A_i+\rho^{-1})Q_{i-1}+Q_{i-2}}$$

$$=\frac{P_i+\rho^{-1}P_{i-1}}{Q_i+\rho^{-1}Q_{i-1}}=\frac{\rho P_i+P_{i-1}}{\rho Q_i+Q_{i-1}}$$

by using (5.74) and (5.75). □

5.55. Corollary. *For $i=0,1,\ldots,s$ we have*

$$[A_0, A_1,\ldots,A_i]=\frac{P_i}{Q_i}.$$

Proof. This is trivial for $i=0$. For $1\leqslant i\leqslant s$ put $\rho=A_i$ in Lemma 5.54 and use (5.74) and (5.75). □

5.56. Lemma. *For $i=0,1,\ldots,s$ we have*

$$\frac{r_0}{r_1}=\frac{P_i+\beta_iP_{i-1}}{Q_i+\beta_iQ_{i-1}},$$

where $\beta_i=r_{i+2}/r_{i+1}$ is a rational function of negative degree.

Proof. We proceed by induction. For $i=0$ we have

$$\frac{P_0+\beta_0P_{-1}}{Q_0+\beta_0Q_{-1}}=A_0+\beta_0=A_0+\frac{r_2}{r_1}=\frac{A_0r_1+r_2}{r_1}=\frac{r_0}{r_1}$$

by (5.73). Suppose the statement is shown for some $i, 0 \leqslant i < s$. Then

$$\frac{P_{i+1}+\beta_{i+1}P_i}{Q_{i+1}+\beta_{i+1}Q_i}=\frac{(A_{i+1}+\beta_{i+1})P_i+P_{i-1}}{(A_{i+1}+\beta_{i+1})Q_i+Q_{i-1}}$$

by (5.74) and (5.75), and from (5.73) we obtain $\beta_i^{-1}=A_{i+1}+\beta_{i+1}$. Thus

$$\frac{P_{i+1}+\beta_{i+1}P_i}{Q_{i+1}+\beta_{i+1}Q_i}=\frac{\beta_i^{-1}P_i+P_{i-1}}{\beta_i^{-1}Q_i+Q_{i-1}}=\frac{P_i+\beta_iP_{i-1}}{Q_i+\beta_iQ_{i-1}}=\frac{r_0}{r_1}$$

by induction hypothesis. □

5.57. Lemma. *For $i=0,1,\ldots,s$ we have*

$$P_iQ_{i-1}-P_{i-1}Q_i=(-1)^{i-1}.$$

Proof. We proceed by induction. For $i=0$ we have $P_0Q_{-1}-P_{-1}Q_0=-1$. Suppose the identity is shown for some i, $0 \leqslant i < s$. Then

$$\begin{aligned} P_{i+1}Q_i-P_iQ_{i+1}&=(A_{i+1}P_i+P_{i-1})Q_i-P_i(A_{i+1}Q_i+Q_{i-1})\\ &=-(P_iQ_{i-1}-P_{i-1}Q_i)=(-1)^i \end{aligned}$$

by (5.74), (5.75), and the induction hypothesis. □

5.58. Corollary. *For $i=0,1,\ldots,s$ we have* $\gcd(P_i,Q_i)=1$.

Proof. If $d_i=\gcd(P_i,Q_i)$, then Lemma 5.57 shows that d_i divides $(-1)^{i-1}$, and so $d_i=1$. □

We are now in a position to discuss the application to character sums. For q odd, put $G(x)=x^q-x$, let $f\in\mathbb{F}_q[x]$ be a polynomial of positive degree with no roots in $\mathbb{F}_q$, and set $F(x)=f(x)^{(q-1)/2}$. Consider the continued fraction expansions

$$\frac{F(x)-1}{G(x)}=[A_0,A_1,\ldots,A_s] \tag{5.76}$$

and

$$\frac{F(x)+1}{G(x)}=[a_0,a_1,\ldots,a_t]. \tag{5.77}$$

It is clear that $A_0=a_0$. Define n_f to be the largest integer m such that $A_i=a_i$ for $i=0,1,\ldots,m$. We cannot have $n_f=s=t$, for otherwise the two continued fractions would be identical. The following result shows, however, that the two continued fractions resemble each other to a great extent.

5.59. Lemma. *Under the conditions above, we have either $n_f=s=t-1$ or $n_f=t=s-1$.*

Proof. Define P_i and Q_i by (5.74) and (5.75) and define p_i and q_i analogously using the a_i. By Corollaries 5.55 and 5.58, P_s/Q_s and p_t/q_t are reduced forms of $(F(x)-1)/G(x)$ and $(F(x)+1)/G(x)$, respectively, so that

$$Q_s(x) = \frac{b_1 G(x)}{\gcd(F(x)-1, G(x))}, \qquad q_t(x) = \frac{b_2 G(x)}{\gcd(F(x)+1, G(x))} \tag{5.78}$$

with $b_1, b_2 \in \mathbb{F}_q^*$. For every $c \in \mathbb{F}_q$ we have $f(c) \neq 0$ by hypothesis, hence $f(c)^{q-1} = 1$, and so $G(x)$ divides $f(x)^{q-1} - 1 = (F(x)-1)(F(x)+1)$. Thus,

$$\gcd(F(x)-1, G(x))\gcd(F(x)+1, G(x)) = G(x),$$

and it follows from (5.78) that

$$Q_s(x) q_t(x) = bG(x) \quad \text{with } b \in \mathbb{F}_q^*. \tag{5.79}$$

Write $n = n_f$ and suppose that $n < s$ and $n < t$. By Lemma 5.56 we have

$$\frac{F(x)-1}{G(x)} = \frac{P_{n+1} + \beta_{n+1} P_n}{Q_{n+1} + \beta_{n+1} Q_n}, \qquad \frac{F(x)+1}{G(x)} = \frac{p_{n+1} + \gamma_{n+1} p_n}{q_{n+1} + \gamma_{n+1} q_n}$$

with rational functions $\beta_{n+1}, \gamma_{n+1}$ of negative degree. Subtracting the first identity from the second, we get

$$\frac{2}{G(x)} = \frac{N}{(Q_{n+1} + \beta_{n+1} Q_n)(q_{n+1} + \gamma_{n+1} q_n)}, \tag{5.80}$$

where the numerator N is given by

$$\begin{aligned} N = {} & p_{n+1} Q_{n+1} - P_{n+1} q_{n+1} + \beta_{n+1}(p_{n+1} Q_n - P_n q_{n+1}) \\ & + \gamma_{n+1}(p_n Q_{n+1} - P_{n+1} q_n) + \beta_{n+1}\gamma_{n+1}(p_n Q_n - P_n q_n). \end{aligned}$$

The definition of n implies that $P_i = p_i$ and $Q_i = q_i$ for $i = -1, 0, \ldots, n$. Thus, $p_{n+1} Q_n - P_n q_{n+1} = (-1)^n$ and $p_n Q_{n+1} - P_{n+1} q_n = (-1)^{n+1}$ by Lemma 5.57, and $p_n Q_n - P_n q_n = 0$. Furthermore, a simple calculation using (5.74), (5.75), and Lemma 5.57 shows that

$$p_{n+1} Q_{n+1} - P_{n+1} q_{n+1} = (-1)^n (A_{n+1} - a_{n+1}).$$

Altogether, we get

$$N = (-1)^n (A_{n+1} - a_{n+1} + \beta_{n+1} - \gamma_{n+1}).$$

Now $A_{n+1} \neq a_{n+1}$ by the definition of n, and so $\deg(N) \geqslant 0$. A comparison of degrees in (5.80) yields

$$\begin{aligned} q & \leqslant \deg(G) + \deg(N) = \deg(Q_{n+1}) + \deg(q_{n+1}) \\ & \leqslant \deg(Q_s) + \deg(q_t) = q \end{aligned}$$

by (5.79). Thus we must have equality throughout. Hence $\deg(N) = 0$, and so $\deg(A_{n+1}) = \deg(a_{n+1})$. Then $\deg(Q_{n+1}) = \deg(q_{n+1}) = q/2$, a contradiction since q is odd.

Thus we have either $n=s$ or $n=t$. Suppose that $n=s$. Then $t>s$, and we can write

$$\frac{F(x)-1}{G(x)}=\frac{P_s}{Q_s}, \qquad \frac{F(x)+1}{G(x)}=\frac{p_{s+1}+\gamma_{s+1}p_s}{q_{s+1}+\gamma_{s+1}q_s}.$$

Proceeding as above, we get

$$\frac{2}{G(x)}=\frac{(-1)^s}{Q_s(q_{s+1}+\gamma_{s+1}q_s)},$$

and a comparison of degrees yields

$$q=\deg(Q_s)+\deg(q_{s+1})\leqslant\deg(Q_s)+\deg(q_t)=q$$

by (5.79). Thus $\deg(q_{s+1})=\deg(q_t)$, hence $s+1=t$ since $\deg(q_i)$ increases with i. Similarly, $n=t$ leads to the conclusion $t+1=s$. □

After these preparations, we can now derive in a simple way a formula for character sums involving the quadratic character and the polynomial f. The exact nature of the formula depends on which alternative in Lemma 5.59 is satisfied.

5.60. Theorem. *Let η be the quadratic character of $\mathbb{F}_q$, q odd, and let $f\in\mathbb{F}_q[x]$ be a polynomial of positive degree with no roots in $\mathbb{F}_q$. Then*

$$\sum_{c\in\mathbb{F}_q}\eta(f(c))=\begin{cases}\deg(a_t) & \text{if } n_f=s,\\ -\deg(A_s) & \text{if } n_f=t,\end{cases}$$

where A_s and a_t are obtained from (5.76) *and* (5.77), *respectively.*

Proof. Let $N(1)$ be the number of $c\in\mathbb{F}_q$ with $\eta(f(c))=1$, and define $N(-1)$ analogously. Then

$$\sum_{c\in\mathbb{F}_q}\eta(f(c))=N(1)-N(-1). \tag{5.81}$$

Since $\eta(f(c))=1$ if and only if $F(c)=f(c)^{(q-1)/2}=1$, $N(1)$ is equal to the degree of $\gcd(F(x)-1,\ G(x))$, and so $N(1)=q-\deg(Q_s)$ by the first identity in (5.78). Similarly, $N(-1)=q-\deg(q_t)$. If we have the alternative $n_f=t=s-1$ in Lemma 5.59, then $q_t=Q_t=Q_{s-1}$ and $N(1)-N(-1)=-\deg(Q_s)+\deg(Q_{s-1})=-\deg(A_s)$ by (5.75). In case $n_f=s=t-1$, we get $N(1)-N(-1)=\deg(q_t)-\deg(q_{t-1})=\deg(a_t)$. The result follows then from (5.81). □

The results of Lemma 5.59 and Theorem 5.60 break down if f has roots in $\mathbb{F}_q$. Consider, for instance, the case $f(x)=x$. Then the character sum has the value 0. On the other hand, $(F(x)-1)/G(x)=[0,x^{(q+1)/2}+x]$ and $(F(x)+1)/G(x)=[0,x^{(q+1)/2}-x]$, so that $s=t=1$ and $\deg(A_s)=\deg(a_t)=(q+1)/2$.

NOTES

1. A detailed discussion of characters of finite abelian groups can be found in Hall [6, Ch. 13]. The fact that a finite abelian group G has as many characters as elements (see Theorem 5.5) was first shown by Weber [2]. It is not hard to prove that the group $G^\wedge$ of characters of G is isomorphic to G (see Exercise 5.5). Special properties of the quadratic character of $\mathbb{F}_{p^2}$, p an odd prime, were noted by Giudici [1], [2] and Hardman and Jordan [1]. Cartier [1] studies quadratic characters of $\mathbb{F}_q$ and of groups of nonsingular matrices over $\mathbb{F}_q$. Pellegrino [2] considers the behavior of the quadratic character of elements of $\mathbb{F}_q$ under linear fractional transformations of $\mathbb{F}_q$. The additive characters of the residue class ring $\mathbb{F}_q[x]/(f)$ were determined explicitly by Carlitz [27]. For general accounts of exponential sums see Hua [12] and Katz [4].

2. Gaussian sums for finite prime fields appear already in the work of Lagrange [4] on solving algebraic equations, and they are often called "resolvents" or "cyclotomic resolvents" in the older literature. Gauss mentions them and shows (5.15) in his *Disquisitiones Arithmeticae* (Gauss [1, Sec. VII]) and establishes some of the properties in Theorem 5.12 (Gauss [2], [5]). Proofs of elementary results on Gaussian sums are also contained in the classical papers of Cauchy [2], [4], Eisenstein [1], Jacobi [1], [2], Kummer [3], [4], [5], [6], and Lebesgue [2]. Accounts of this early work can be found in Bachmann [1], [2], Dickson, Mitchell, Vandiver, and Wahlin [1, Sec. 19], and H. J. S. Smith [1]. Gaussian sums for general finite fields were first considered by Stickelberger [1]. For interesting historical remarks we refer to Berndt and Evans [4] and Weil [11]. Modern expositions of various parts of the theory are given in Apostol [2], Gras [1], Hasse [15], Ireland and Rosen [1], Joly [5], Lang [3], [5], and W. M. Schmidt [3].

Theorem 5.14 is due to Davenport and Hasse [1]. Our proof is essentially that of Weil [6], with the simplifications in Ireland and Rosen [1, Ch. 11] and McEliece and Rumsey [1]. Proofs can also be found in Lang [5, Ch. 1] and W. M. Schmidt [3, Ch. 2], and for particularly elementary arguments we refer to Schmid [1] and Stepanov [14]. A more general result was obtained by Deligne [4] via cohomology methods. See Hayes [3] for a version of the Davenport-Hasse theorem which applies to Gaussian sums for residue class rings $\mathbb{F}_q[x]/(f)$.

The evaluation of the quadratic Gaussian sum in Theorem 5.15 was first achieved by Gauss [2] in the case $s=1$. A variety of different proofs have since been discovered. The proof in the text is due to Waterhouse [2] and based on ideas of Schur [3]. It has the advantage of using as much as possible only algebraic principles. The matrix $(e^{2\pi ijk/m})_{1\leqslant j,k\leqslant m}$ appearing in Schur's proof was further studied by Carlitz [80] who determined its eigenvalues, and by McClellan and Parks [1] and Morton [1] who determined its eigenvectors. Another matrix-theoretic evaluation of quadratic

Gaussian sums is due to Carlitz [74]. See also Bressoud [1], Carlitz [106], Cauchy [5], Kronecker [2], Mordell [12], and Shanks [1] for mostly algebraic proofs. Analytic proofs can be found in Bambah and Chowla [1], Dirichlet [1], Estermann [2], Karamata and Tomić [1], Kronecker [9], Landau [3], Mordell [1], and Weber [6], among others. In some of these papers the sum

$$\sum_{n=0}^{p-1} e^{2\pi i n^2/p}$$

is evaluated, which is easily seen to be identical with the quadratic Gaussian sum for $\mathbb{F}_p$. The books of Apostol [2, Ch. 8], Borevich and Shafarevich [1, Ch. 5], S. Chowla [16, Ch. 2], Davenport [8, Ch. 2], Landau [5, Ch. 4], and Lang [3, Ch. 4] contain various proofs. We refer to Berndt and Evans [4] for an excellent comprehensive account of the techniques that have been used in evaluating quadratic Gaussian sums. In connection with Theorem 5.15 the following result of S. Chowla [13], [14] and Mordell [13] (see also Narkiewicz [1, Ch. 6]) is of interest: if p is an odd prime and ψ a multiplicative character of $\mathbb{F}_p$, then $G(\psi, \chi_1)p^{-1/2}$ is a root of unity only if ψ is the quadratic character. Generalizations of this result to arbitrary finite fields have been obtained by Evans [1] and Yokoyama [1]. See also Evans [8] and Stickelberger [1] for related results. Carlitz [81] shows that a sum of the form $B = \sum_{n=1}^{p-1} c_n e^{2\pi i n/p}$ with p an odd prime and $c_n = \pm 1$ satisfies $|B| = p^{1/2}$ only if B is a quadratic Gaussian sum for $\mathbb{F}_p$. Another characterization of quadratic Gaussian sums among sums of the form B is due to Rédei [7], [11, Ch. 6]. If the c_n are arbitrary integers and still $|B| = p^{1/2}$, then B is closely related to a quadratic Gaussian sum (see Cavior [3]).

Theorem 5.16 is due to Stickelberger [1]. A proof was also given in Carlitz [71] and Baumert and McEliece [1], and the latter paper contains evaluations of some other special Gaussian sums as well. Related evaluations can be found in Berndt and Evans [1], [2], [4], Evans [1], Ishimura [1], McEliece [5], and Myerson [5], but some of these formulas contain ambiguities.

The law of quadratic reciprocity (Theorem 5.17) was established by Gauss [1] who gave several proofs (see also Gauss [2], [4]). One of his proofs is based on Gaussian sums (Gauss [2]); see also Cauchy [2], Eisenstein [3], Hasse [15, Ch. 8], and Ireland and Rosen [1, Ch. 6]. Some arguments in the proof of Theorem 5.17 become simpler if Gaussian sums with values in finite fields are used; see Exercises 5.26 and 5.27 as well as Burde [6], Hausner [1], Holzer [1, §§18, 19], Ireland and Rosen [1, Ch. 7], Kloosterman [6], Serre [1, Ch. 1], and Zassenhaus [3]. Another method of proving the law of quadratic reciprocity via finite fields depends on considering the factorization of the polynomial $(x^r - 1)/(x - 1)$ over $\mathbb{F}_p$. This idea appeared first in a posthumous paper of Gauss [4] and was later used in essentially equivalent forms by Pellet [2], Mirimanoff and Hensel [1], and Swan [1]. This type of proof is also reproduced in the books of Bachmann [4, Ch. 7],

Berlekamp [4, Ch. 6], Childs [1, Part III, Ch. 16], and Rédei [10, Ch. 11]. Other proofs using finite fields are due to Agou [6], Brewer [1], S. Chowla [18], Furquim de Almeida [2], Lebesgue [4], Pellet [9], Rešetuha [2], and Skolem [6]. See also Example 5.24 in Section 3. A systematic exposition of the various techniques for proving the law of quadratic reciprocity can be found in Pieper [1]. Bachmann [4, Ch. 6] presents a useful history of various proofs. The recent proofs of E. Brown [1] and Frame [1] are also of interest.

Gaussian sums are also instrumental in proving higher reciprocity laws; see Eisenstein [2] for the cubic case and Eisenstein [4] for the biquadratic case. Eisenstein's proof of the cubic reciprocity law is reproduced in Bachmann [1, Ch. 14] and Ireland and Rosen [1, Ch. 9]; see also Bachmann [1, Ch. 13] for the biquadratic case. The proof of the cubic reciprocity law given by Joly [4] depends also on Gaussian sums. If Gaussian sums with values in finite fields are used, then one gets again some simplifications; see Burde [6], [9] for the cubic and biquadratic case, respectively. A proof of the cubic reciprocity law based on finite fields was also given by Skolem [6]. Gaussian sums are used in the reciprocity law of Western [1] and in the so-called rational reciprocity laws of Evans [9], Leonard and Williams [6], and K. S. Williams [34]. See Weil [11] for an account of the role of Gaussian sums in reciprocity laws.

The book of Hasse [16] and the survey article of Wyman [1] present interesting discussions of general reciprocity laws. A different type of reciprocity law connected with finite fields was considered by Dedekind [1] who established a law of quadratic reciprocity for monic irreducible polynomials over finite prime fields; see also Artin [1] and Vaidyanathaswamy [1]. A higher reciprocity law of this type for general finite fields was shown by Kühne [1] and was rediscovered by F. K. Schmidt [2] and Carlitz [1], [2]; see also Carlitz [4], Ore [6], Pocklington [2], Schwarz [2], and Whiteman [1].

A considerable amount of work has been done on Gaussian sums with cubic characters because of a long-standing conjecture of Kummer. If ψ is a multiplicative character of $\mathbb{F}_p$, $p \equiv 1 \bmod 3$, of order 3, then Kummer [2] conjectured on the basis of calculations in Kummer [1] that the ratio of occurrences of $G(\psi, \chi_1)p^{-1/2}$ in the subsets

$$I_1 = \{e^{\pi i t} : |t| \leqslant \tfrac{1}{3}\},\ I_2 = \{e^{\pi i t} : \tfrac{1}{3} < |t| \leqslant \tfrac{2}{3}\},$$

and

$$I_3 = \{e^{\pi i t} : \tfrac{2}{3} < |t| \leqslant 1\}$$

of the unit circle approaches $3:2:1$ as $p \to \infty$. More extensive numerical work carried out by von Neumann and Goldstine [1], Beyer [1], E. Lehmer [5], Cassels [2], and Fröberg [1], among others, tended to suggest that the correct asymptotic ratio is closer to $1:1:1$. Theoretical results of C. J. Moreno [1] and S. J. Patterson [3] also pointed in that direction. Finally, Heath-Brown and Patterson [1] succeeded in proving much more, namely that the values of $G(\psi, \chi_1)p^{-1/2}$ are uniformly distributed on the unit circle as p runs through the primes $\equiv 1 \bmod 3$. This proof builds on earlier work of T. Kubota [5], [6] and S. J. Patterson [1], [2]. See also Deligne [5] for another look at the work of S. J. Patterson. An earlier attempt by A. I. Vinogradov [1] to disprove Kummer's conjecture was fallacious. C. R. Matthews [1] proved a conjectured formula of Cassels [3], [4], [5] to the effect that the value of a cubic Gaussian sum over $\mathbb{F}_p$ is given by

$$G(\psi, \chi_1) = J(\psi, \psi)H(\psi)p^{1/3},$$

where $J(\psi, \psi)$ denotes a Jacobi sum (see Section 3) and $H(\psi)$ is a product of values of the Weierstrass $\wp$-function. A simple algorithm for computing this expression is due to McGettrick [1]. For further work on cubic Gaussian sums we refer to Hasse [15, Ch. 20], Krätzel [1], Loxton [1], [2], [3], and Rešetuha [1] as well as the survey article of Berndt and Evans [4].

C. R. Matthews [2] proved conjectured formulas of Loxton [2], [3] and McGettrick [2] giving the values of biquadratic Gaussian sums $G(\psi, \chi_1)$ over $\mathbb{F}_p$. The distribution of $G(\psi, \chi_1)p^{-1/2}$ on the unit circle was investigated by Hasse [15, Ch. 20], T. Kubota [4], E. Lehmer [5], and Yamamoto [1], before S. J. Patterson [4] showed that these values are uniformly distributed on the unit circle as p runs through the primes $\equiv 1 \bmod 4$ and that, in fact, a corresponding theorem holds for Gaussian sums of any higher order. Further results on biquadratic Gaussian sums can be found in Berndt and Evans [4], Hasse [15, Ch. 20], Krätzel [1], and T. Kubota [4]. Another type of equidistribution result for Gaussian sums is due to R. A. Smith [3], [4] (see also Katz [4, Ch. 1]) who showed that if ψ runs through the nontrivial multiplicative characters of $\mathbb{F}_p$, then the distribution of the $p-2$ values $G(\psi, \chi_1)p^{-1/2}$ on the unit circle approaches the uniform distribution as $p \to \infty$.

We have noted in the proof of Theorem 5.17 that the values of Gaussian sums are algebraic integers. It is then of interest to find the factorization of such a value (or, more exactly, of the principal ideal generated by such a value) in the ring of integers of the appropriate algebraic number field. This was achieved in the important paper of Stickelberger [1]. Expositions of this work are presented in Gras [1], Joly [5], and Lang [3, Ch. 4], [5, Ch. 1]. See Fröhlich [2] for a different method of proof. Stickelberger [1] also gives congruences for the values of Gaussian sums; see also Dwork [2] and Gras [1].

A result of great theoretical interest is the formula of Gross and Koblitz [1] expressing the value of a Gaussian sum as a product of values of a p-adic gamma-function. See also Boyarsky [1], Koblitz [3, Ch. 3], and Lang [6, Ch. 15].

An interesting multiplicative relation between Gaussian sums first stated by Jacobi [2] can be found in Corollary 5.29. A conjecture of Hasse [15, p. 465] that the relations generated from Theorem 5.12(iv) and Corollary

5.29 yield all the multiplicative relations between Gaussian sums was disproved by Yamamoto [2]; see also Yamamoto [4]. Other identities involving products of Gaussian sums are given in Boyarsky [1], Evans [7], Grant [1], and Helversen-Pasotto [1], [2], [3], [4].

Gaussian sums have also been considered in various other settings. In number theory Gaussian sums for residue class rings $\mathbb{Z}/(m)$, $m \geqslant 2$ not necessarily prime, are very useful. They are defined by means of an additive character of $\mathbb{Z}/(m)$ and a character of the group of units of $\mathbb{Z}/(m)$. An extensive theory of such Gaussian sums is presented in Hasse [15, Ch. 20]. These sums are also discussed in the books of Apostol [2, Ch. 8], Ayoub [1, Ch. 5], Lang [3, Ch. 4], [5, Ch. 3], and Narkiewicz [1, Ch. 6], among others. Quadratic Gaussian sums of this type satisfy a reciprocity law which was suggested by Cauchy [5] and proved by Schaar [1] and Kronecker [3]; see also Bochner [1], Landsberg [2], and Lerch [1]. Generalizations of this reciprocity law are given in Berndt [1], Berndt and Evans [4], Guinand [1], and Siegel [3]. Other relations between quadratic Gaussian sums for $\mathbb{Z}/(m)$ and closely related sums can be found, for example, in Carlitz [107], S. Chowla [1], [2], and Menon [1]. Gaussian sums for $\mathbb{Z}/(m)$ with a restricted range of summation are treated in Berndt and Evans [3] and D. H. Lehmer [11].

Gaussian sums for algebraic number fields were introduced by Hecke [1]. See Hasse [10], Hecke [4, Ch. 8], and Narkiewicz [1, Ch. 6] for accounts of this theory. For further work, in particular reciprocity laws for quadratic Gaussian sums of this type, we refer to Barner [2], Hecke [2], [3], Kloosterman [3], T. Kubota [1], Kunert [1], Mordell [3], Shiratani [1], and Siegel [3]. Hecke's Gaussian sums were generalized by Hasse [12], [13] who considered so-called Galois Gaussian sums. See also T. Kubota [1], [2], Lakkis [1], [2], [3], and the more recent work of Fröhlich and Taylor [1], Martinet [1], and M. J. Taylor [1], [2]. Still another type of Gaussian sum for algebraic number fields was considered by T. Kubota [3].

Carlitz [27] and Hayes [3] worked with Gaussian sums for the residue class rings $\mathbb{F}_q[x]/(f)$ and Schmid [2] introduced Gaussian sums for rings of Witt vectors over finite fields. Schmid and Teichmüller [1] considered Gaussian sums for a certain class of rings built from finite fields. Kondo [1] studied Gaussian sums for matrix rings over finite fields, and quadratic Gaussian sums for such rings were discussed by Porter [16], [17]. A general theory of Gaussian sums for finite rings was developed by Lamprecht [1], [2], [3]; see also Kutzko [1]. Gaussian sums for so-called quadratic characters of finite abelian groups were studied by Springer [3]. See Berndt and Evans [4] for further references on generalizations of Gaussian sums.

There are many applications of Gaussian sums in the theory of finite fields, in number theory, and in combinatorics. We refer to Chapter 6 for the use of Gaussian sums in determining the number of solutions of certain equations over finite fields. Gaussian sums are instrumental for obtaining unitary representations of degree p of the character group of $\mathbb{F}_p^*$ (Burde [1]). An application to the curves $y^p - y = x^{p^a} - 1$ over $\mathbb{F}_p$ occurs in Yamada [1]. O. Moreno [1] uses Gaussian sums in determining the number of elements of $\mathbb{F}_{2^m}$ having absolute trace 0 and a prescribed power character. The role of Gaussian sums in proving reciprocity laws has already been noted above. Among other applications in number theory we mention the following: criteria for residuacity (Ankeny [3], Evans [6], Hasse [14], Hayashi [2], Muskat [1], [3], Whiteman [8], K. S. Williams [31]); Waring's problem (Ayoub [1, Ch. 4], Barrucand [1], Hardy and Littlewood [2], [3], Landau [2, Ch. 6], R. C. Vaughan [1, Ch. 4]); primality testing (H. W. Lenstra [2]); Dirichlet L-functions and their functional equations (Apostol [1], Ayoub [1, App. B], S. Chowla [16, Ch. 1], Hasse [11, Ch. 1], Lang [5, Ch. 3]); functional equations for Dirichlet series associated to modular forms (Shimura [1, Ch. 3]); abelian number fields (Gras [1], Leopoldt [1]); and class number formulas (Bergström [1], Hasse [9]). Gaussian sums arise in the theory of difference sets (Berndt and Chowla [1], Berndt and Evans [1], Evans [4], [10], Menon [2], Muskat and Whiteman [1], Yamamoto [3]) and in connection with the weight distribution of cyclic codes (Baumert and McEliece [1], McEliece [5], McEliece and Rumsey [1], Niederreiter [8]). An interesting relation with finite Fourier transforms appears in Auslander and Tolimieri [1]. Gaussian sums occur also in the functional equation of zeta-functions associated with certain representations of $GL(n, \mathbb{F}_q)$, the group of nonsingular $n \times n$ matrices over $\mathbb{F}_q$ (see Springer [2]).

3. Jacobi sums for finite prime fields and with $k = 2$ were mentioned by Jacobi [1] in a letter to Gauss. The earliest publications discussing Jacobi sums are Cauchy [2] and Jacobi [2], and to complicate the matter of priority they also appear in a posthumous paper of Gauss [5]. These sources as well as Cauchy [4], Eisenstein [1], and Lebesgue [2] contain already all the basic properties of such sums. See also Bachmann [1], Dickson, Mitchell, Vandiver, and Wahlin [1, Sec. 19], H. J. S. Smith [1], and Weil [11] for accounts of this early work. Exponential sums that are equivalent to Jacobi sums for general finite fields and with $k = 2$ were introduced by Kummer [6]. The study of these sums was taken up again by Stickelberger [1]. Jacobi sums with general $k \geqslant 2$ seem to appear first in Weil [6] and Vandiver [16]. Proofs of the elementary properties of general Jacobi sums can be found in these papers and in Faircloth and Vandiver [1]. Expository accounts of these properties of Jacobi sums are presented in Hasse [15], Ireland and Rosen [1], and Joly [5].

Although the analog of Theorem 5.26 for Gaussian sums—namely, Theorem 5.14—was shown by Davenport and Hasse [1], Theorem 5.26 itself was already shown earlier by H. H. Mitchell [1]. Theorem 5.28 is due to Davenport and Hasse [1], whereas the statement of Corollary 5.29 appears already in Jacobi [2] for the case of finite prime fields. See also Gras [1] and

Lang [5, Ch. 2] for proofs of Theorem 5.28. The elementary proofs of Theorem 5.28 are from Berndt and Evans [1] and Hasse [15, Ch. 20] for the case $\psi = \eta$ and from Berndt and Evans [2] for the case where m is a power of 2. In Berndt and Evans [2] and Gras [1] it is also shown that the problem of providing an elementary proof of Theorem 5.28 can be reduced to the case where m is prime.

Evaluations of Jacobi sums with $k = 2$ and characters of small order have been attempted by many authors, often in connection with the theory of cyclotomy. In many cases there remain ambiguities. We mention Berndt and Evans [2], Dickson [46], Evans [2], [3], Ireland and Rosen [1], Ishimura [1], E. Lehmer [7], [8], Muskat and Zee [1], Tanner [1], [3], and Zee [1], [2], and for particularly detailed treatments we refer to Berndt and Evans [1] and Muskat [6]. The survey article of Berndt and Evans [4] also contains information on this topic. A result of Evans [1] shows that if λ is a multiplicative character of $\mathbb{F}_p$ of order > 2, then no power $J(\lambda, \ldots, \lambda)^n$ is real, where the Jacobi sum has at least two arguments and n is a nonzero integer. A considerably less explicit result in this direction was proved earlier by Yokoyama [1].

If λ is a multiplicative character of $\mathbb{F}_q$ of order m, then $J(\lambda^r, \lambda^s)$ is obviously an algebraic integer in the mth cyclotomic field over $\mathbb{Q}$. The problem of finding the factorization of the principal ideal generated by this algebraic integer in the ring of integers of $\mathbb{Q}^{(m)}$ is easier than in the case of Gaussian sums. It was solved in Kummer [3], [6] for m prime and in Kummer [7] for m composite. See Lang [5, Ch. 1] for a modern exposition of these results. Congruences for Jacobi sums were obtained by Kronecker [6] and Schwering [1]; see also Parnami, Agrawal, and Rajwade [2].

Like Gaussian sums, Jacobi sums can also be considered in other settings. Jacobi sums for algebraic number fields have proved to be important since Weil [7], [10] could show that they yield so-called Hecke characters (or "grössencharacters") of abelian extensions of $\mathbb{Q}$. See also Deligne [4] and Lang [5, Ch. 1]. Deligne [4] gives also a cohomological interpretation of Jacobi sums; see also Katz [5]. So-called Galois Jacobi sums and their factorization are discussed in Fröhlich [1]. Hall [7] defines Jacobi sums for group rings over cyclotomic fields. Lamprecht [3] develops a general theory of Jacobi sums for finite rings; see also Kutzko [1]. Ono [8] introduced Jacobi sums for finite abelian groups.

Apart from the instances provided by Examples 5.24 and 5.25, Jacobi sums can be applied in many other problems of number theory. Proofs of the cubic reciprocity law can be based on Jacobi sums (Ireland and Rosen [1, Ch. 9], Joly [4], Weil [11]), and this applies also to the biquadratic case (Bachmann [1, Ch. 13]) and to higher reciprocity laws (Evans [9], Leonard and Williams [6], Western [1], K. S. Williams [34]). Jacobi sums are used in establishing criteria for residuacity (see Berndt and Evans [1], Evans [6], Hasse [14], Leonard, Mortimer, and Williams [1], Muskat [2], [3], [5], Western [2], Whiteman [8]). Other results like Example

5.25 can be obtained by means of Jacobi sums; the papers of Berndt and Evans [1], [2] are a rich source for such results, see also Ireland and Rosen [1, Ch. 8] and Leonard and Williams [2]. An application of Jacobi sums to primality testing appears in Adleman [1], Adleman, Pomerance, and Rumely [1], and Cohen and Lenstra [1]. Carlitz [76] and D. H. Lehmer [9] use Jacobi sums in the study of the matrix $(\psi(i-j))_{1 \leqslant i, j \leqslant p-1}$ with ψ being a multiplicative character of $\mathbb{F}_p$. Iwasawa [1] relates Jacobi sums to class numbers of cyclotomic fields.

With regard to applications in the theory of finite fields, we note that Jacobi sums are not only intimately connected with Gaussian sums, but appear also in the study of other exponential sums; see, for example, Theorems 5.51 and 5.52 as well as Berndt and Evans [1], [2], Leonard and Williams [4], Singh and Rajwade [1], and Whiteman [14]. Applications to equations over finite fields will be discussed in Chapter 6. We note here only the close relationship between Jacobi sums and so-called cyclotomic numbers. If b is a primitive element of $\mathbb{F}_q$ and e a given positive divisor of $q-1$, then the *cyclotomic number* $(h, k)_e$ of order e is defined as the number of ordered pairs (s, t) with

$$b^{es+h}+1=b^{et+k}, \quad 0 \leqslant s, t<(q-1)/e.$$

The connection between cyclotomic numbers and Jacobi sums was already observed by Kummer [4], [6]. Most investigations have been restricted to the case where q is a prime, but see, for example, Hall [7], H. H. Mitchell [1], Myerson [5], Parnami, Agrawal, and Rajwade [3], Storer [2], [4], and Vandiver [14], [17] for the general case. The following additional references on connections between cyclotomic numbers and Jacobi sums should provide a guide to the literature: Bachmann [1, Ch. 15], Baumert and Fredricksen [1], Berndt and Evans [1], Bruck [2], Dickson [26], [44], [45], [46], Evans and Hill [1], Leonard and Williams [5], Muskat [4], [6], [7], Muskat and Whiteman [1], Schwering [1], Storer [1], Whiteman [5], [9], [10], [11], [14] (see the notes to Chapter 6, Section 3, for evaluations of cyclotomic numbers and relations to equations over finite fields). Some of this work also has a bearing on the theory of difference sets. For applications of Jacobi sums in this theory see Baumert and Fredricksen [1], Berndt and Evans [1], Menon [2], Muskat and Whiteman [1], Storer [1], Whiteman [10], [11], and Yamamoto [3].

4. The character sums in Theorem 5.30, which on account of this theorem are closely related to Gaussian sums, are sometimes also called Gaussian sums. The elementary estimate in Theorem 5.32 is due to Hardy and Littlewood [3] for finite prime fields and appears in essentially equivalent form in Hua and Vandiver [1] for arbitrary finite fields; see also W. M. Schmidt [3, Ch. 2]. A slight improvement can be obtained if n is in a certain range (see Mit'kin [5]). The sums can be evaluated for small n; see Berndt and Evans [1], [2] as well as the survey article of Berndt and Evans [4]. The

evaluations for $n=2$ are given in Theorem 5.33 and Corollary 5.35. The character sums with general n occur in analytic number theory in connection with Waring's problem of representing positive integers as sums of nth powers; see Ayoub [1, Ch. 4], Barrucand [1], Hardy and Littlewood [2], [3], Kloosterman [4], Landau [2, Ch. 6], and R. C. Vaughan [1, Chs. 2, 4]. Theorem 5.34 is due to Carlitz [120].

Theorems 5.38 and 5.41 were shown by Weil [5] on the basis of his proof of the Riemann hypothesis for curves over finite fields. For a discussion of the latter topic see the notes to Chapter 6, Section 4. As we mentioned in the discussion following Theorem 5.38, the condition on f in that theorem can be relaxed. In fact, Carlitz and Uchiyama [1] show that it suffices to take f not of the form $g^p - g + b$ with $g \in \mathbb{F}_q[x]$, $b \in \mathbb{F}_q$. For f of this form the estimate need not be valid (see again the discussion following Theorem 5.38). The completion of our proof of Theorem 5.38 in Chapter 6, Section 4, will show that it depends on finding good estimates for the number of solutions of the equation $y^q - y = f(x)$ in extension fields of $\mathbb{F}_q$. An elementary method for establishing such estimates is due to Stepanov [3], [5], with simplifications obtained by Mit'kin [1] and W. M. Schmidt [3, Ch. 2]. The transition from such estimates to Theorem 5.38 is also explained in Postnikov [1] and W. M. Schmidt [3, Ch. 2]. An estimate like Theorem 5.38 was conjectured by Hasse [5] and Mordell [6]. Earlier estimates for finite prime fields had instead of the exponent $\frac{1}{2}$ the exponent $1-2^{1-n}+\varepsilon$ for any $\varepsilon > 0$ (Hardy and Littlewood [1] with a slight improvement by Kamke [1]) and then $1-1/n$ (Mordell [4] with a slight improvement by Davenport [4]). As a general estimate, Theorem 5.38 is best possible according to a result of W. M. Schmidt [3, Ch. 2]. Other lower bounds for the absolute value of these character sums have been given by Anderson and Stiffler [1], Karacuba [5], Knižnerman and Sokolinskiĭ [1], Korobov and Mit'kin [1], Mit'kin [3], and Tietäväinen [2]. Cavior [3] determines the number of polynomials over $\mathbb{F}_p$ such that the corresponding character sum has absolute value $p^{1/2}$. A result on the statistical distribution of $|\sum_{c \in \mathbb{F}_q} \chi(f(c))| q^{-1/2}$ was shown by Odoni [1].

I. M. Vinogradov [5], Davenport and Heilbronn [1], Akuliničev [1], and Karacuba [5] treat the case $f(x) = ax^n + bx$, $a, b \in \mathbb{F}_p^*$, $2 \leqslant n \leqslant p-1$, with the latter showing that

$$\left| \sum_{c \in \mathbb{F}_p} \chi(f(c)) \right| \leqslant (n-1)^{1/4} p^{3/4}$$

for a nontrivial additive character χ of $\mathbb{F}_p$. This is better than Theorem 5.38 for $n > 1 + p^{1/3}$. The special cases $n=3$ and $n=4$ were considered by Carlitz [122], [124] and Mordell [28], [29], respectively. Birch [2] studied the average behavior of such sums as a, b vary over $\mathbb{F}_p$ in the case $n=3$. In the case $f(x) = ax^{p+1} + bx \in \mathbb{F}_q[x]$ with $a \neq 0$ and p the characteristic of $\mathbb{F}_q$, the character sums were evaluated by Carlitz [124], [125]. An elementary proof of the result of Carlitz and Uchiyama noted above was given by K. S. Williams [27] for the case where q is even and $\deg(f) \leqslant 6$.

Analogs of the character sums in Theorem 5.38 for residue class rings $\mathbb{Z}/(m)$ have also been studied; see, for example, J. R. Chen [1], Hua [1], [3], [6], [7], [9, Ch. 1], [12], and Karacuba [4], with recent improvements in J. R. Chen [2], Körner and Stähle [1], Loxton and Smith [1], Nečaev [7], Nečaev and Topunov [1], R. A. Smith [6], and Stečkin [1].

The completion of our proof of Theorem 5.41 in Chapter 6, Section 4, will show that it depends on finding good estimates for the number of solutions of the equation $y^m = f(x)$ in extension fields of $\mathbb{F}_q$. An elementary method for establishing such estimates is due to Stepanov [2], [7] and W. M. Schmidt [1], [3]; compare with the notes to Chapter 6, Section 4. Weaker bounds for the character sums in Theorem 5.41 were given earlier in Davenport [7] and in some special cases in Davenport [1], [2], [3] and I. M. Vinogradov [11]. Theorem 5.41 is best possible as a general estimate (see W. M. Schmidt [3, Ch. 2]). The case where f is a trinomial was studied by Ono [7]. For results on the sums in Theorem 5.41 with ψ being the quadratic character see also Section 5 and the notes to it. Some authors have studied so-called *hybrid sums*—that is, sums of the form

$$\sum_{c \in \mathbb{F}_q} \psi(f(c))\chi(g(c))$$

with $f, g \in \mathbb{F}_q[x]$, ψ a nontrivial multiplicative and χ a nontrivial additive character of $\mathbb{F}_q$. If trivial exceptions are ruled out, such sums are again at most of the order of magnitude $q^{1/2}$; see Perel'muter [1], [2] and W. M. Schmidt [3, Ch. 2]. Related sums were considered by K. S. Williams [20].

Character sums with polynomials in several indeterminates have been studied extensively. For the case of two indeterminates, let $f \in \mathbb{F}_q[x, y]$ be of degree $n \geqslant 2$ and nondegenerate in the sense that f is not expressible as a polynomial in one indeterminate after a nonsingular linear transformation of indeterminates. If χ is a nontrivial additive character of $\mathbb{F}_q$, then Hua and Min [2], [3] established a bound for

$$\sum_{c, d \in \mathbb{F}_q} \chi(f(c, d))$$

of the order $q^{2-(2/n)}$, thus improving substantially an earlier estimate of Kamke [1]. In the case $n = 3$ Hua and Min [2], [3] obtained a better bound of the order $q^{5/4}$; see also Davenport and Lewis [1] for the same result. The best possible order q in the case $n = 3$ was obtained by Bombieri and Davenport [1], but for some special cubic polynomials this was shown earlier by Mordell [11], [18]. Character sums with cubic polynomials in two indeterminates were also considered by Carlitz [122]. Some simple hybrid double sums—that is, double sums involving both additive and multiplicative characters—were studied by S. Chowla [21] and Chowla and Smith [1]. If χ is as above and f is a cubic polynomial over $\mathbb{F}_q$ in three indeterminates which is nondegenerate—that is, f is not expressible as a polynomial in fewer than three indeterminates after a nonsingular linear transformation of indeterminates—then Davenport and Lewis [1] have given a bound for

$$\sum_{c_1, c_2, c_3 \in \mathbb{F}_q} \chi(f(c_1, c_2, c_3))$$

of the best possible order q^2; see also Mordell [15] for some special cases.

For general sums of the form

$$S(f) = \sum_{c_1, \ldots, c_r \in \mathbb{F}_q} \chi(f(c_1, \ldots, c_r))$$

with $f \in \mathbb{F}_q[x_1, \ldots, x_r]$, an analog of Theorem 5.36 was shown by Bombieri [3], [4]. If χ is replaced by a multiplicative character, such a result was given by Perel'muter [3]. Nontrivial estimates for the sums $S(f)$ were obtained by Min [2] and Uchiyama [7], but the real breakthrough came with the fundamental paper of Deligne [3] verifying the Weil conjectures for algebraic varieties over finite fields (compare with the notes for Chapter 6, Section 4). It follows from Deligne's work that if f is of a degree n not divisible by the characteristic p of $\mathbb{F}_q$ and if the homogeneous part of f of degree n is nonsingular in a certain sense, then

$$|S(f)| \leqslant (n-1)^r q^{r/2}$$

(see Deligne [3], [6], Katz [4, Ch. 5], Serre [3]), a result conjectured by Bombieri [4]. Deligne [4] generalizes the result of Carlitz and Uchiyama mentioned above in the following way: if f is of arbitrary degree n and not of the form $g^p - g + b$ with $g \in \mathbb{F}_q[x_1, \ldots, x_r]$, $b \in \mathbb{F}_q$, then

$$|S(f)| \leqslant (n-1) q^{r-(1/2)}.$$

A detailed investigation of the sums $S(f)$ can be found in Katz [4]. It is easy to show that the average order of magnitude of $|S(f)|$ is $q^{r/2}$ (see Carlitz [47]). Estimates for special polynomials f were given before the work of Deligne; see Davenport and Lewis [1] for cubic f, Mordell [14] for

$$f(x_1, \ldots, x_r) = a_1 x_1^{h_1} + \cdots + a_r x_r^{h_r} + b x_1^{k_1} \cdots x_r^{k_r},$$

Mordell [24] for the cases

$$f(x_1, \ldots, x_r) = (a_1 x_1 + \cdots + a_r x_r) x_1^{m_1} \cdots x_r^{m_r}$$

and

$$f(x_1, \ldots, x_r) = g(x_1^2, \ldots, x_r^2)$$

with a quadratic polynomial g, and Ono [2] for the case where f is a semi-invariant of a connected algebraic group. If f is a quadratic form, then the sums $S(f)$ can be evaluated explicitly (see Exercises 6.27–6.30). For cohomological interpretations of $S(f)$ see Katz [4, Ch. 3] and Springer [4]. These sums have also been discussed from a general standpoint by Ono [1]. For sums with multiplicative characters and hybrid sums involving polynomials in several indeterminates see Davenport and Lewis [1], Katz [4], Perel'muter [11], [12], and Serre [3]. Sums $S(f)$ for $\mathbb{Z}/(m)$ are treated in Arhipov, Karacuba, and Čubarikov [1], Čubarikov [1], and Loxton and Smith [2].

Sums like $S(f)$ above, but with the r-tuples $(c_1,\ldots,c_r) \in \mathbb{F}_q^r$ restricted to a curve or variety in $\mathbb{F}_q^r$, were first considered by Bombieri [4]. See also Adolphson and Sperber [1], Bombieri [7], Chalk and Smith [1], Hooley [4], [5], [6], Laumon [1], Milne [2], Perel'muter [9], Serre [3], R. A. Smith [2], and K. S. Williams [6], as well as the detailed treatment in Katz [4].

So-called *incomplete sums* arise also from a restriction of the range of summation, namely by summing only over "intervals" or "boxes." Incomplete sums have been considered mostly for prime fields $\mathbb{F}_p$. For incomplete sums of the form $\sum_{c=N+1}^{N+H} \chi(f(c))$, where χ is a nontrivial additive character of $\mathbb{F}_p$ or $\mathbb{Z}/(m)$ and f is a polynomial, see Davenport and Heilbronn [1], Hua [11], [12, Sec. 14], Karacuba [3], [7], Korobov [2], [3], [4], [5], and Lebedev [1]. If f is a polynomial in r indeterminates and the summation is, correspondingly, over all $(c_1,\ldots,c_r) \in \mathbb{Z}^r$ with $a_i \leqslant c_i \leqslant b_i$ for $1 \leqslant i \leqslant r$, then see Mordell [22] and Serre [3].

For incomplete sums with nontrivial multiplicative characters ψ of $\mathbb{F}_p$ there is a classical inequality of Pólya [1] and I. M. Vinogradov [1] (see also Schur [2]) to the effect that

$$\left| \sum_{c=1}^{H} \psi(c) \right| < p^{1/2} \log p.$$

Slight improvements of the constant were obtained by Landau [1] and I. M. Vinogradov [12]; see also Whyburn [2]. The fact that the left-hand side can be of the order of magnitude $p^{1/2} \log\log p$ was shown by S. Chowla [3] (under a generalized Riemann hypothesis) and Bateman, Chowla, and Erdös [1] (with no unproved hypothesis) who thus improved a result of Paley [1] for $\mathbb{Z}/(m)$. An upper bound of this order of magnitude was established by Montgomery and Vaughan [1] under the assumption of the Riemann hypothesis for Dirichlet L-functions. For applications of the Pólya-Vinogradov inequality to number theory see Hua [12, Sec. 14]. Sokolovskiĭ [1] showed that for any ψ and $\mathbb{F}_p$, $p > 2$, there exists N with

$$\left| \sum_{c=N+1}^{N+(p-1)/2} \psi(c) \right| \geqslant \tfrac{1}{2}\left(p - \frac{1}{p} \right)^{1/2},$$

thus improving a result of Sárközy [1]. An extension of the Pólya-Vinogradov inequality to general finite fields $\mathbb{F}_q$, $q = p^n$, was established by

Davenport and Lewis [3]: if ψ is a nontrivial multiplicative character of $\mathbb{F}_q$, then

$$\left|\sum_{c\in B}\psi(c)\right|<q^{1/2}(1+\log p)^n,$$

with B being the "box" consisting of the elements $c=c_1\alpha_1+\cdots+c_n\alpha_n$ with $0\leqslant N_j<c_j\leqslant N_j+H_j<p$ for $1\leqslant j\leqslant n$, where $\{\alpha_1,\ldots,\alpha_n\}$ is a basis of $\mathbb{F}_q$ over $\mathbb{F}_p$ and N_j, H_j are given integers. Estimates for incomplete sums with multiplicative characters in terms of the number H of summands were first given by Burgess [1] for the quadratic character of $\mathbb{F}_p$ and by Y. Wang [1], [2] and Burgess [2], [3], [5] for arbitrary nontrivial multiplicative characters of $\mathbb{F}_p$. The results are of the following type: for any $\varepsilon>0$ there exists $\delta>0$ such that for $p>p_0(\varepsilon)$ and $H>p^{(1/4)+\varepsilon}$ the estimate

$$\left|\sum_{c=N+1}^{N+H}\psi(c)\right|<Hp^{-\delta}$$

holds for all nontrivial multiplicative characters ψ of $\mathbb{F}_p$ and all N (see Burgess [2]). An analog of a Burgess-Wang type of inequality for general $\mathbb{F}_q$ was first shown by Davenport and Lewis [3], with further refinements being due to Burgess [7], Friedlander [3], and J. H. Jordan [3] in the special case $q=p^2$ and to Burgess [10], Friedlander [2], and Karacuba [6], [8] in the general case. The distribution of the values of $\sum_{c=N+1}^{N+H}\eta(c)$ with η being the quadratic character of $\mathbb{F}_p$ was investigated by Davenport and Erdös [1], Montgomery [2], Usol'cev [1], and Wolke [1]. Incomplete sums for $\mathbb{F}_p$ of the form $\sum_{c=N+1}^{N+H}\psi(f(c))$ with $f\in\mathbb{F}_p[x]$ were considered by I. M. Vinogradov [4], Segal [1], and Burgess [6]. Lower bounds for the absolute value of such sums with $\psi=\eta$ were shown by Karacuba [9], Mit'kin [3], and Stepanov [11]. Incomplete multiple sums for $\mathbb{F}_p$ with multiplicative characters were studied by Burgess [8], [9] for binary quadratic forms and by Gillett [1] for general polynomials in several indeterminates.

5. Kloosterman sums appeared first in Kloosterman [1] in connection with a problem of representing integers by a quadratic form. This paper contains also an estimate for nontrivial Kloosterman sums of the order $p^{3/4}$ for finite prime fields $\mathbb{F}_p$ (see also Estermann [1], Salié [1]). A bound of the order $p^{2/3}$ was obtained by Davenport [4] and Salié [2]. Hasse [5] noted that the estimate in Theorem 5.45 would follow from the Riemann hypothesis for curves over finite fields. The truth of this hypothesis was established by Weil [1], [2], [3] who then showed in Weil [5] the estimate for Kloosterman sums. Weil's proof of Theorem 5.45 was simplified by Carlitz and Uchiyama [1] who also considered the case of even q not treated by Weil. Two proofs of Theorem 5.45 can be found in W. M. Schmidt [3, Ch. 2]; they are based essentially on the method of Stepanov [4], [5].

Theorem 5.46 is due to Carlitz [111]. Using Dickson polynomials defined in (7.6), the formula in Theorem 5.46 can also be written in the form $K(\chi^{(s)}; a, b) = -g_s(-K, q) = (-1)^{s-1}g_s(K, q)$. The identity in Theorem 5.47 is a special case of a transformation formula of Jacobsthal [1]. Proofs of Theorem 5.47 can also be found in Davenport [4], Salié [1], W. M. Schmidt [3, Ch. 2], and K. S. Williams [16]. Various identities for sums of Kloosterman sums or sums of products of Kloosterman sums are given in Davenport [4], Lehmer and Lehmer [1], [3], Salié [1], and Whiteman [5]; see Kutzko [1] for another approach. Some of these identities involve connections with Jacobsthal sums. Carlitz [109] discusses some elementary properties of Kloosterman sums for even q.

Because of the origin of Kloosterman sums in a problem on quadratic forms, it is not surprising that these sums can be used in the study of exponential sums involving quadratic forms (see Carlitz [45], [46], [109], S. Chowla [20], Malyšev [2], I. M. Vinogradov [4]). For generalizations of Carlitz [46] see Carlitz [72] and Hodges [16]. Dwork [11] looked at Kloosterman sums from the viewpoint of p-adic cohomology.

Kloosterman sums have also been considered for the residue class rings $\mathbb{Z}/(m)$. An estimate for such sums on the basis of Weil's estimate was established by Hooley [1]; see also Estermann [3], [4]. In the case $m = p^k$, p prime, $k \geqslant 2$, the Kloosterman sums were evaluated explicitly by Salié [1]; see also Malyšev [3, Ch. 2], Whiteman [2], and K. S. Williams [19], [28]. The observation that Kloosterman sums for $\mathbb{Z}/(m)$ appear in connection with Fourier coefficients of modular forms was made in Kloosterman [2]. This remarkable relationship has been exploited further; see, for example, Bruggeman [1], Deshouillers and Iwaniec [1], Iwaniec [1], N. V. Kuznecov [1], Linnik [2], Malyšev [4], Parson [1], Petersson [1], Proskurin [1], [2], Rademacher [1], [2], and Selberg [2]. For further work on Kloosterman sums for $\mathbb{Z}/(m)$ we refer to Hooley [2], Kloosterman [1], [2], Malyšev [3, Ch. 2], Salié [1], Selberg [1], and R. A. Smith [5]. Andruhaev [2] studied Kloosterman sums for residue class rings of the ring of Gaussian integers. Incomplete Kloosterman sums for $\mathbb{Z}/(m)$ were considered by Kloosterman [1], [2] and Rademacher [1]; see also Hooley [3, Ch. 2].

The *generalized Kloosterman sums* defined in Exercise 5.83 were introduced by Davenport [4]. If ψ is the quadratic character, such a sum is also called *Salié's sum* since Salié [1] first proved the formulas for it in Exercises 5.84 and 5.85 for finite prime fields. Proofs of formulas for Salié's sum were also given by D. H. Lehmer [6], Malyšev [3, Ch. 2], Mordell [26], [28], [30], and K. S. Williams [16], [21], [22]. Salié's sum plays a role in the study of the partition function (see D. H. Lehmer [6]). A Weil-type estimate for generalized Kloosterman sums was shown by S. Chowla [22] in the case where ψ is not the quadratic character. Generalized Kloosterman sums appear also in Carlitz [46], Hodges [16], Kloosterman [5], and I. M. Vinogradov [5], and in a related form in Andruhaev [1], Knopp [1],

Malyšev [1], [3, Ch. 2], [4], and Rohrbach [1]. For further generalizations of Kloosterman sums in connection with the theory of modular forms see Bruggeman [1], Deshouillers and Iwaniec [1], Proskurin [1], [3], Rankin [1, Ch. 5], and Selberg [2]. In a different direction, a significant generalization of Kloosterman sums for finite fields arises when sums of the form $\Sigma\chi(R(c))$ are considered, where R is a rational function over $\mathbb{F}_q$ and the summation is over all $c \in \mathbb{F}_q$ for which $R(c)$ is defined; see Bombieri [4], Elistratov [7], Perel'muter [1], [2], [9], and Stepanov [5]. Hybrid sums with rational functions are studied in Mordell [27] and Perel'muter [1], [2], [9], and an even more general sum appears in K. S. Williams [20].

Multiple Kloosterman sums (or *hyper-Kloosterman sums*) were introduced by Mordell [14]. In the nontrivial case they are of the form

$$K_r = \sum_{c_1,\ldots,c_r \in \mathbb{F}_q^*} \chi\left(a_1c_1 + \cdots + a_rc_r + bc_1^{-1}\cdots c_r^{-1}\right)$$

with χ a nontrivial additive character of $\mathbb{F}_q$ and $a_1,\ldots,a_r,\ b \in \mathbb{F}_q^*$. Mordell [14] proved the estimate $|K_r| \leqslant q^{(r+1)/2}$ for q prime; see also Carlitz [100] and R. A. Smith [4], with improvements in special cases by Carlitz [100], [108]. The definitive estimate

$$|K_r| \leqslant (r+1)q^{r/2}$$

was established by Deligne [4]. It is of course the direct analog of Theorem 5.45. See Katz [4, Chs. 2, 5] and Serre [3] for background on Deligne's proof. Bombieri [7] obtains a bound of the same order of magnitude (relative to q) by using preparatory work in Deligne [4]. For further results on multiple Kloosterman sums see Carlitz [100], [113], Kutzko [1], Lehmer and Lehmer [4], McEliece and Rumsey [1], and Sperber [2]. Multiple Kloosterman sums for $\mathbb{Z}/(m)$ have been considered by Carlitz [98], R. A. Smith [3], [4], [5], and Weinstein [1]. For hybrid multiple Kloosterman sums we refer to Deligne [4], Katz [4, Ch. 5], and Mordell [26], [31]. Multiple exponential sums with other rational functions are treated in Carlitz [113], Katz [4, Ch. 5], Lehmer and Lehmer [1], Mordell [10], [27], [28], [31], and Sperber [1], [3]. Kloosterman sums for nonsingular matrices over finite fields are considered in Hodges [6], for skew matrices in Hodges [8], [25], and for symmetric matrices and Hermitian matrices in Hodges [3], [10], [17].

Theorem 5.48 was shown by Jacobsthal [1], [2] for finite prime fields. See Hall [8, Ch. 14] and D. H. Lehmer [8] for applications of this identity. The sums $H_n(a)$ and $I_n(a)$ were first studied by Jacobsthal [1], [2], mainly for $n=1$ and $n=2$. Various identities linking these sums have been obtained by Berndt and Evans [1], E. Lehmer [4], Postnikov and Stepanov [1], von Schrutka [1], and Whiteman [6]. Jacobsthal sums have been evaluated for small n, sometimes in the essentially equivalent form of determining the number of solutions of equations $y^2 = x^n + a$ in $\mathbb{F}_q$; for the nontrivial case $n=4$, $a=1$, and q prime this appears already in Gauss [3]. For further evaluations see Andrianov [1], Brewer [2], S. Chowla [6], Davenport and Hasse [1], Evans [2], [3], Hasse [15, Ch. 10], Hudson and Williams [1], Ireland and Rosen [1, Ch. 11], E. Lehmer [1], [4], [7], Leonard and Williams [7], Morlaye [2], Rajwade [3], [8], Singh and Rajwade [1], Whiteman [3], [6], K. S. Williams [30], and particularly the papers of Berndt and Evans [1], [2] which provide detailed information on this topic. Congruences modulo p for the values of Jacobsthal sums for $\mathbb{F}_p$ were established by E. Lehmer [6], Nashier and Rajwade [1], and Whiteman [6]. Jacobsthal [1] already proved the inequality $|H_2(a)| < 2p^{1/2}$ in the prime case and S. Chowla [11] showed that both the constant factor and the exponent are best possible in general. A lower bound for $\max_{a \in \mathbb{F}_q^*} |H_n(a)|$ was given by Postnikov and Stepanov [1]. Lower bounds for the absolute value of Jacobsthal sums can also be obtained from those for more general character sums with polynomial arguments (see the notes to Section 4). For prime fields $\mathbb{F}_p$, Karacuba [9] showed that $I_n(a) = p$ is possible for $a \in \mathbb{F}_p^*$ and Postnikov and Stepanov [1] showed that $H_n(a) = p - 1$ is possible for $a \in \mathbb{F}_p^*$, in both cases under the provision that n is of the order of magnitude $p/\log p$.

As Theorems 5.51 and 5.52 demonstrate, Jacobsthal sums are closely related to other classical exponential sums such as Jacobi sums; see also Berndt and Evans [1], E. Lehmer [7], and Singh and Rajwade [1]. Connections between Jacobsthal sums and so-called Brewer sums are explored in Berndt and Evans [2], Giudici, Muskat, and Robinson [1], S. F. Robinson [1], and Whiteman [13], [14], [15]. Generalized Jacobsthal sums, i.e. sums of the form

$$\sum_{c \in \mathbb{F}_p} \lambda(c)\psi(c^n + a)$$

with nontrivial multiplicative characters λ, ψ of $\mathbb{F}_p$ are considered in Walum [1]. Double Jacobsthal sums occur in Lehmer and Lehmer [1].

The application of Jacobsthal sums in Example 5.53 is due to Jacobsthal [2]. Proofs of this result are also given in Berndt and Evans [1], Burde [4], S. Chowla [16, Ch. 4], Hasse [15, Ch. 10], and Whiteman [3], [6]. Other formulas for the integers A and B in Example 5.53 can be found in Bachmann [1, Ch. 10], S. Chowla [16, Ch. 5], and Whiteman [6]. Further applications of Jacobsthal sums to quadratic partitions of primes occur in Berndt and Evans [1], [2], S. Chowla [5], [6], Hasse [15, Ch. 10], E. Lehmer [4], Nashier and Rajwade [1], Rajwade [3], Rosenberg [1], von Schrutka [1], Whiteman [3], [6], and K. S. Williams [29], [30]. Related sums are also useful in connection with quadratic partitions of primes; see Berndt and Evans [2], Brewer [2], [3], and Whiteman [4], [13]. Relations between Jacobsthal sums and cyclotomy arise in Giudici, Muskat, and Robinson [1], Rajwade [3], and Whiteman [3], [6], [14], and an application of these sums to a residuacity criterion appears in Leonard, Mortimer, and Williams [1]. Carlitz [50] uses Jacobsthal sums in evaluating the number of solutions of certain equations in more than two indeterminates. Andrianov [1] establishes a connection between $I_3(1)$ for $\mathbb{F}_p$ and the number of representations of p by a certain quadratic form in four indeterminates; see also Fomenko [1] for a related result.

Theorem 5.60 is due to Davenport [10]. The underlying theory of continued fractions was already developed by Artin [1]; see also de Mathan [3] for a further study of such continued fractions. Baum and Sweet [1], [2] proved important results on continued fractions over $\mathbb{F}_2$ with partial quotients of small degree. Houndonougbo [1] studied the length of continued fraction expansions. Applications of the continued fraction algorithm to coding theory occur in Goppa [1], Mills [4], Reed, Scholtz, Truong, and Welch [1], Reed and Truong [4], Reed, Truong, and Miller [3], and Welch and Scholtz [1]. Special continued fractions for elements of $\mathbb{F}_q$ are studied in Borho [1].

Sums involving the quadratic character η which have received considerable attention are the so-called *Brewer sums*. Here one takes a Dickson polynomial $g_k(x, a)$ over $\mathbb{F}_q$, q odd, with $a \in \mathbb{F}_q^*$ as in (7.6) and forms the sum

$$\Lambda_k(a) = \sum_{c \in \mathbb{F}_q} \eta(g_k(c, a)).$$

In the special case $a = 1$ these sums were introduced in Brewer [2] and in the general case in Brewer [3]. We have $\Lambda_k(a) = 0$ whenever $\gcd(k, q^2 - 1) = 1$ (see P. Chowla [2] and Corollary 7.17). Further evaluations of Brewer sums for small values of k can be found in Berndt and Evans [2], Brewer [2], [3], Giudici, Muskat, and Robinson [1], Leonard and Williams [4], Rajwade [7], S. F. Robinson [1], Whiteman [13], [14], [15], and K. S. Williams [35], with Berndt and Evans [2] being a particularly good source for such results. In these papers one also finds relations to quadratic partitions of primes and to cyclotomy. Brewer sums are related to *Eisenstein sums* defined in Exercise 5.68 and first introduced by Eisenstein [5]; see Berndt and Evans [2], Giudici, Muskat, and Robinson [1], and Whiteman [15] for connections between these two types of character sums.

For sums with the quadratic character and general polynomial arguments, Korobov [6] and Mit'kin [2] have established bounds which in some cases are better than the Weil estimate in Theorem 5.41. Lower bounds for the absolute value of such sums were shown by Knižnerman and Sokolinskiĭ [1] and Mit'kin [3], and for the corresponding incomplete sums by Stepanov [11]. The case where the polynomial is a product of distinct monic linear factors has received some attention in connection with the distribution of quadratic residues; see, for example, Burde [4], Davenport [1], Hasse [15, Ch. 10], Hopf [1], I. M. Vinogradov [10], and Yamauchi [1]. For quadratic character sums with other special polynomials see Abdullaev [1], [2], Abdullaev and Kogan [1], Birch [2], S. Chowla [19], Davenport [3], Fomenko [1], Olson [1], Ono [3], Rajwade and Parnami [1], Rosenberg [1], Salié [3], Tuškina [1], and K. S. Williams [36], [37], [38]. Perel'muter [10], [11], [12] estimated multiple sums with the quadratic character.

The connection between quadratic character sums and the distribution of quadratic residues and nonresidues modulo p (or, more generally, of squares and nonsquares in $\mathbb{F}_q$) can be seen from Exercises 5.63 and 5.64. We refer also to Aladov [1], Bergum and Jordan [1], Burde [4], Davenport [1], [3], Giudici [1], Hardman and Jordan [1], Hasse [4], [15, Ch. 10], Hopf [1], Johnsen [1], Katz [4, Ch. 1], Koutský [1], Pellegrino [1], Rocci [1], W. M. Schmidt [3, Ch. 2], I. M. Vinogradov [4], and von Grosschmid [1] for this topic. Other types of distribution problems for quadratic residues and nonresidues have been studied extensively in number theory; see, for example, Ankeny [1], A. Brauer [2], Burgess [1], Davenport and Erdös [1], Dörge [1], Elliott [2], Gel'fond and Linnik [1, Ch. 9], Hua [12, Sec. 14], R. H. Hudson [2], [3], [5], Perron [1], Salié [3], Usol'cev [1], and I. M. Vinogradov [3]. See Raber [1] and Ralston [1] for a geometric approach. References pertaining to the topic of Exercises 5.65 and 5.66 are Burde [1], [3], [7], Davenport [2], [7], Jänichen [1], Koutský [2], Moroz [1], Segal [1], and I. M. Vinogradov [6], [11]. A. Brauer [1], [3] proved that if integers $m, h \geqslant 2$ and a complex mth root of unity ε are given, then for all sufficiently large primes $p \equiv 1 \bmod m$ and all multiplicative characters ψ of $\mathbb{F}_p$ of order m there exists $c \in \mathbb{F}_p$ with $\psi(c+i) = \varepsilon$ for $i = 0, 1, \ldots, h-1$. For related work on the distribution of values of multiplicative characters and the distribution of powers modulo p and in finite fields see Bierstedt and Mills [1], A. Brauer [4], [5], Brillhart, Lehmer, and Lehmer [1], Buhštab [1], Burgess [4], [6], [10], [11], Chowla and Chowla [2], Dunton [2], Elliott [1], [3], [4], [5], Gel'fond and Linnik [1, Ch. 9], Graham [1], Hua [12, Sec. 14], R. H. Hudson [1], [4], [5], [6], J. H. Jordan [1], [2], [4], [5], Lehmer and Lehmer [2], Lehmer, Lehmer, and Mills [1], Lehmer, Lehmer, Mills, and Selfridge [1], Metsänkylä [1], Mills [2], Montgomery [1, Ch. 13], Norton [2], [3], [4], [5], [6], Rabung and Jordan [1], Singh [1], [5], Stephens [2], H. Stevens [1], Stevens and Kuty [1], I. M. Vinogradov [2], [3], [7], [9], Y. Wang [3], and Whyburn [1]. Combinatorial properties of the subgroup of $\mathbb{F}_q^*$ consisting of the mth powers are discussed in Cameron, Hall, van Lint, Springer , and van Tilborg [1], Evans [5], Muskat and Street [1], and Street and Whitehead [1].

Exponential sums with linear recurring sequences as arguments will be treated in Chapter 8, Section 7.

EXERCISES

5.1. Let G be a finite abelian group, H a proper subgroup of G, and $g \in G$, $g \notin H$. Prove that there exists a character χ of G that annihilates H, but for which $\chi(g) \neq 1$.

5.2. Let H be a subgroup of the finite abelian group G. Prove that the annihilator A of H in $G^\wedge$ is isomorphic to G/H and that $G^\wedge/A$ is isomorphic to H.

5.3. Let G be a finite abelian group and $m \in \mathbb{N}$. Prove that $g \in G$ is an mth power of an element of G if and only if $\chi(g)=1$ for all characters χ of G for which χ^m is trivial.

5.4. Let $G_1,\ldots,G_k$ be finite abelian groups. Define multiplication of k-tuples $(g_1,\ldots,g_k)$, $(h_1,\ldots,h_k)$ with $g_i, h_i \in G_i$ for $1 \leqslant i \leqslant k$ by

$$(g_1,\ldots,g_k)(h_1,\ldots,h_k) = (g_1h_1,\ldots,g_kh_k).$$

Show that with this operation the set of all such k-tuples forms again a finite abelian group, the so-called *direct product* $G_1 \otimes \cdots \otimes G_k$. Then prove that $(G_1 \otimes \cdots \otimes G_k)^\wedge$ is isomorphic to $G_1^\wedge \otimes \cdots \otimes G_k^\wedge$.

5.5. Use the structure theorem for finite abelian groups, which says in its simplest form that every such group is isomorphic to a direct product of finite cyclic groups, to prove that $G^\wedge$ is isomorphic to G whenever G is a finite abelian group.

5.6. For additive characters of $\mathbb{F}_q$ in the notation of Theorem 5.7, show that $\chi_a\chi_b = \chi_{a+b}$ for all $a, b \in \mathbb{F}_q$. Thus prove without reference to Exercise 5.5 that the group of additive characters of $\mathbb{F}_q$ is isomorphic to the additive group of $\mathbb{F}_q$.

5.7. If χ_1 is the canonical additive character of the finite field $\mathbb{F}_q$ of characteristic p, prove that $\chi_1(c^{p^j}) = \chi_1(c)$ for all $c \in \mathbb{F}_q$ and $j \in \mathbb{N}$.

5.8. If ψ is a multiplicative character of $\mathbb{F}_{q^s}$ of order m, prove that the restriction of ψ to $\mathbb{F}_q$ is a multiplicative character of order $m/\gcd(m,(q^s-1)/(q-1))$.

5.9. With the notation of Exercise 5.8, prove that the restriction of ψ to $\mathbb{F}_q$ is the trivial character if and only if m divides $(q^s-1)/(q-1)$.

5.10. Let ψ be a multiplicative character of $\mathbb{F}_q$ and let ψ' be the lifted character of the extension field $\mathbb{F}_{q^s}$. Prove that $\psi'(c) = \psi^s(c)$ for $c \in \mathbb{F}_q^*$.

5.11. Prove that a multiplicative character τ of $\mathbb{F}_{q^s}$ is equal to a character ψ' lifted from $\mathbb{F}_q$ if and only if τ^{q-1} is trivial.

5.12. If $q \equiv 1 \bmod m$ and ψ varies over all multiplicative characters of $\mathbb{F}_q$ of order dividing m, prove that the lifted character ψ' of $\mathbb{F}_{q^s}$ varies over all multiplicative characters of $\mathbb{F}_{q^s}$ of order dividing m.

5.13. Prove that an additive character χ of the finite extension field E of $\mathbb{F}_q$ is equal to a character lifted from $\mathbb{F}_q$ if and only if $\chi = \mu_b$ with $b \in \mathbb{F}_q$, where μ_1 is the canonical additive character of E.

5.14. Prove for $c \in \mathbb{F}_q^*$ that

$$\sum_{d|(q-1)} \frac{\mu(d)}{\phi(d)} \sum_{\psi^{(d)}} \psi^{(d)}(c) = \begin{cases} \dfrac{q-1}{\phi(q-1)} & \text{if } c \text{ is a primitive element of } \mathbb{F}_q, \\ 0 & \text{otherwise,} \end{cases}$$

where in the outer sum d runs through all positive divisors of $q-1$

and in the inner sum $\psi^{(d)}$ runs through the $\phi(d)$ multiplicative characters of $\mathbb{F}_q$ of order d. Here μ denotes the Moebius function (see Definition 3.22) and ϕ Euler's function (see Theorem 1.15 (iv)).

5.15. Show that $\eta(2) = (-1)^{(q^2-1)/8}$, where η is the quadratic character of $\mathbb{F}_q$, q odd.

5.16. For $r \in \mathbb{N}$ prove $G(\psi^{p^r}, \chi_b) = G(\psi, \chi_{\rho(b)})$, where $\rho(b) = b^{p^r}$ for $b \in \mathbb{F}_q$ and p is the characteristic of $\mathbb{F}_q$.

5.17. Prove $\Sigma_\chi G(\psi, \chi) = 0$ for all multiplicative characters ψ of $\mathbb{F}_q$, where the sum is extended over all additive characters χ of $\mathbb{F}_q$.

5.18. Prove $\Sigma_\psi G(\psi, \chi) = (q-1)\chi(1)$ for all additive characters χ of $\mathbb{F}_q$, where the sum is extended over all multiplicative characters ψ of $\mathbb{F}_q$.

5.19. For the quadratic character η of $\mathbb{F}_q$, $q = p^s$, p an odd prime, $s \in \mathbb{N}$, and an additive character χ_b, $b \in \mathbb{F}_q$, in the notation of Theorem 5.7, prove that

$$G(\eta, \chi_b) = \eta(b)(-1)^{(q+1)/2} i^{s(p^2+2p+5)/4} q^{1/2}.$$

5.20. If q is odd and η is the quadratic character of $\mathbb{F}_q$, prove that $G(\eta, \chi_a) G(\eta, \chi_b) = \eta(-ab) q$ for $a, b \in \mathbb{F}_q^*$.

5.21. Use the law of quadratic reciprocity to evaluate the Legendre symbols $\left(\frac{13}{59}\right)$ and $\left(\frac{7}{61}\right)$.

5.22. Determine all primes p such that $\left(\dfrac{-3}{p}\right) = 1$.

5.23. Determine all odd prime powers q such that the quadratic character η of $\mathbb{F}_q$ satisfies $\eta(3) = 1$.

5.24. Prove that the polynomial $x^2 + ax + b \in \mathbb{F}_q[x]$, q odd, is irreducible in $\mathbb{F}_q[x]$ if and only if $\eta(a^2 - 4b) = -1$.

5.25. Determine whether the polynomial $x^2 + 12x + 41$ is irreducible in $\mathbb{F}_{227}[x]$.

5.26. Let p and r be distinct odd primes, let $s \in \mathbb{N}$ be such that $r^s \equiv 1 \bmod p$, and let ζ be an element of order p in $\mathbb{F}_{r^s}^*$. For $k \in \mathbb{Z}$ define

$$G_k = \sum_{c=1}^{p-1} \left(\frac{c}{p}\right) \zeta^{kc} \in \mathbb{F}_{r^s}.$$

Prove the following properties: (i) $G_k = \left(\dfrac{k}{p}\right) G_1$; (ii) $G_1^2 = (-1)^{(p-1)/2} p$, where the last expression is viewed as an element of $\mathbb{F}_r$.

5.27. Use the results of Exercise 5.26 to prove the law of quadratic reciprocity.

5.28. Prove $J(\lambda_1, \ldots, \lambda_k) = -\lambda_k(-1) J(\lambda_1, \ldots, \lambda_{k-1})$ if $\lambda_1 \cdots \lambda_k$ is trivial and λ_k is nontrivial.

5.29. Prove $\Sigma_{a \in \mathbb{F}_q} J_a(\lambda_1, \ldots, \lambda_k) = 0$ if at least one of the λ_i is nontrivial.

5.30. Prove that

$$\sum_\lambda J(\lambda, \lambda_1, \ldots, \lambda_k) = (q-1) J_0(\lambda_1, \ldots, \lambda_k) + J(\lambda_1, \ldots, \lambda_k),$$

where the sum is extended over all multiplicative characters λ of $\mathbb{F}_q$.

5.31. For a nontrivial additive character χ of $\mathbb{F}_q$ and $a, b_1, \ldots, b_k \in \mathbb{F}_q$, prove that

$$\sum_{c_1 + \cdots + c_k = a} \chi(b_1 c_1 + \cdots + b_k c_k) = \begin{cases} q^{k-1}\chi(ab_1) & \text{if } b_1 = b_2 = \cdots = b_k, \\ 0 & \text{otherwise.} \end{cases}$$

5.32. Use (5.16) and the result of Exercise 5.31 to give an alternative proof of the formulas for $J(\lambda_1, \ldots, \lambda_k)$ in terms of Gaussian sums given in Theorem 5.21. (*Note*: Show first that with the standard convention for $\psi(0)$, (5.16) holds also for $c = 0$ and nontrivial ψ.)

5.33. Use (5.16) and the result of Exercise 5.31 to prove that if $\lambda_1, \ldots, \lambda_k$ are nontrivial multiplicative characters of $\mathbb{F}_q$ with $\lambda_1 \cdots \lambda_k$ trivial, then

$$J_0(\lambda_1, \ldots, \lambda_k) = \left(1 - \frac{1}{q}\right) G(\lambda_1, \chi) \cdots G(\lambda_k, \chi),$$

where χ is a nontrivial additive character of $\mathbb{F}_q$.

5.34. Prove that

$$J(\lambda_1, \ldots, \lambda_k) = \frac{1}{q} G(\lambda_1, \chi) \cdots G(\lambda_k, \chi) G(\overline{\lambda_1 \cdots \lambda_k}, \bar{\chi})$$

for $\lambda_1, \ldots, \lambda_k$, and χ nontrivial.

5.35. Prove

$$J(\lambda_1, \lambda_2) J(\lambda_1\lambda_2, \lambda_3) = J(\lambda_1, \lambda_3) J(\lambda_1\lambda_3, \lambda_2)$$

if all λ_i and $\lambda_1\lambda_2$, $\lambda_1\lambda_3$ are nontrivial.

5.36. Let $\lambda_1, \ldots, \lambda_k$ be nontrivial multiplicative characters of $\mathbb{F}_q$ and $1 \leqslant h < k$. Prove that

$$J(\lambda_1, \ldots, \lambda_k) = J(\lambda_1, \ldots, \lambda_h) J(\lambda_1 \cdots \lambda_h, \lambda_{h+1}, \ldots, \lambda_k)$$

provided that $\lambda_1 \cdots \lambda_h$ is nontrivial.

5.37. Let $\lambda_1, \ldots, \lambda_k$ be $k \geqslant 3$ multiplicative characters of $\mathbb{F}_q$ with λ_1 and λ_k nontrivial and $\lambda_1 \cdots \lambda_k$ trivial. Prove that

$$J(\lambda_1, \ldots, \lambda_{k-1}) = (\lambda_2 \cdots \lambda_{k-1})(-1) J(\lambda_2, \ldots, \lambda_k).$$

5.38. Let ψ be a multiplicative character of $\mathbb{F}_q$ of order $m \geqslant 2$ and let χ be a nontrivial additive character of $\mathbb{F}_q$. Prove that

$$G(\psi, \chi)^k = \begin{cases} -qJ(\psi, \ldots, \psi) & \text{if } m \text{ divides } k, \\ G(\psi^k, \chi) J(\psi, \ldots, \psi) & \text{otherwise,} \end{cases}$$

where the Jacobi sums depend on k characters ψ.

5.39. Prove that the Jacobi sum $J(\eta,\ldots,\eta)$ depending on k quadratic characters η of $\mathbb{F}_q$, q odd, is given by

$$J(\eta,\ldots,\eta)=\begin{cases}(-1)^{(k(q-1)+4)/4}q^{(k-2)/2} & \text{for even } k,\\ (-1)^{(k-1)(q-1)/4}q^{(k-1)/2} & \text{for odd } k.\end{cases}$$

5.40. If λ is a multiplicative character of $\mathbb{F}_q$ of odd order $m \geqslant 3$, prove that

$$J(\lambda,\lambda^2,\ldots,\lambda^{m-1})=-q^{(m-3)/2}.$$

5.41. For a multiplicative character λ of $\mathbb{F}_q$ of order 3 and a nontrivial additive character χ of $\mathbb{F}_q$, prove that $G(\lambda,\chi)^3=qJ(\lambda,\lambda)$.

5.42. If η and ψ are multiplicative characters of $\mathbb{F}_{q^2}$ of orders 2 and 3, respectively, and if $q\equiv -1 \bmod 6$, prove that $J(\eta,\psi)=q$. (*Hint*: Use Stickelberger's theorem.)

5.43. If ψ is a nontrivial multiplicative character of $\mathbb{F}_q$, q odd, and η is the quadratic character of $\mathbb{F}_q$, prove that $J(\psi,\eta)=\psi(4)J(\psi,\psi)$.

5.44. Use the Davenport-Hasse relation to prove that $J(\psi^2,\eta)=\psi(4)J(\psi,\psi\eta)$, where η is the quadratic character of $\mathbb{F}_q$, q odd, and the multiplicative character ψ of $\mathbb{F}_q$ is such that ψ^4 is nontrivial.

5.45. If ψ and η are as in Exercise 5.44, prove that $J(\psi,\psi\eta)=\psi(4)J(\psi^2,\psi^2)$.

5.46. Prove that $\psi(16)J(\psi,\psi)=\eta(-1)J(\bar{\psi}\eta,\bar{\psi}\eta)$, where ψ is of order $\geqslant 3$ and η is the quadratic character of $\mathbb{F}_q$, q odd.

5.47. Prove that $J(\psi,\bar{\psi}\eta)=\psi(-4)J(\psi,\psi)$, where ψ and η are as in Exercise 5.46.

5.48. Let $q\equiv 1 \bmod 4$ and let ψ be a multiplicative character of $\mathbb{F}_q$ of order 4. Prove that $J(\psi,\psi)=\psi(-1)J(\psi,\eta)$.

5.49. Let $q\equiv 1 \bmod 3$ and let λ be a multiplicative character of $\mathbb{F}_q$ of order 3. Prove that $J(\lambda,\lambda)=A+B\omega$, where A, B are integers with $A\equiv -1 \bmod 3$ and $B\equiv 0 \bmod 3$ and where ω is a complex primitive third root of unity. (*Hint*: Use Theorem 5.27 and congruences in the ring of algebraic integers.)

5.50. For a prime $p\equiv 1 \bmod 4$ let λ be a multiplicative character of $\mathbb{F}_p$ of order 4. Prove that $J(\lambda,\lambda)=A+Bi$, where A and B are integers such that $p=A^2+B^2$ and $A\equiv -1 \bmod 4$.

5.51. Prove that every prime $p\equiv 1 \bmod 3$ can be written in the form $p=A^2-AB+B^2$ with integers A and B.

5.52. If λ and ψ are as in Theorem 5.28 and m is odd, prove that the Davenport-Hasse relation is equivalent to the statement that

$$J(\psi,\psi\lambda,\ldots,\psi\lambda^{m-1})=\bar{\psi}(m^m)q^{(m-1)/2}.$$

5.53. Let ψ be a multiplicative character of $\mathbb{F}_q$ of order m and χ an additive character of $\mathbb{F}_q$. Prove that

$$G(\psi,\chi)^m=A_0+A_1\zeta+\cdots+A_{m-1}\zeta^{m-1}$$

with integers $A_0, A_1, \ldots, A_{m-1}$ and a complex primitive mth root of unity ζ.

5.54. Prove that

$$\sum_{c \in \mathbb{F}_q} \psi(c+a)\bar{\psi}(c+b) = -1$$

for $a, b \in \mathbb{F}_q$ with $a \neq b$, where ψ is a nontrivial multiplicative character of $\mathbb{F}_q$.

5.55. Let ψ be a nontrivial multiplicative character of $\mathbb{F}_q$ and let S be a subset of $\mathbb{F}_q$ with h elements. Prove that

$$\sum_{c \in \mathbb{F}_q} \left| \sum_{a \in S} \psi(c+a) \right|^2 = h(q-h).$$

5.56. Let $\lambda_1, \lambda_2, \lambda_3$ be nontrivial multiplicative characters of $\mathbb{F}_q$ and let $a_1, a_2 \in \mathbb{F}_q$ with $a_1 \neq a_2$. Prove that

$$\sum_{b \in \mathbb{F}_q} \left| \sum_{c \in \mathbb{F}_q} \lambda_1(c+a_1)\lambda_2(c+a_2)\lambda_3(c+b) \right|^2$$

$$= \begin{cases} q^2 - 3q & \text{if } \lambda_1\lambda_2 \text{ nontrivial,} \\ q^2 - 2q - 1 & \text{if } \lambda_1\lambda_2 \text{ trivial.} \end{cases}$$

5.57. Let ψ be a multiplicative character of $\mathbb{F}_q$ of order $m > 1$. For $a \in \mathbb{F}_q$ prove

$$\sum_{c \in \mathbb{F}_q} \psi(ac^n) = \begin{cases} (q-1)\psi(a) & \text{if } m \text{ divides } n, \\ 0 & \text{otherwise.} \end{cases}$$

5.58. For $n \in \mathbb{N}$, $a, b \in \mathbb{F}_q^*$, and ψ a nontrivial multiplicative character of $\mathbb{F}_q$, prove that

$$\sum_{c \in \mathbb{F}_q} \psi(ac^n + b) = \psi(b) \sum_{j=1}^{d-1} \bar{\lambda}^j(a)\lambda^j(-b)J(\lambda^j, \psi),$$

where λ is a multiplicative character of $\mathbb{F}_q$ of order $d = \gcd(n, q-1)$.

5.59. Prove that $\sum_{c \in \mathbb{F}_q} \eta(f(c)) = 0$ if $q \equiv 3 \bmod 4$, η is the quadratic character of $\mathbb{F}_q$, and $f \in \mathbb{F}_q[x]$ is an odd polynomial—that is, a polynomial with $f(-x) = -f(x)$.

5.60. Let $f(x) = a_2x^2 + a_1x + a_0 \in \mathbb{F}_q[x]$ with q odd and $a_2 \neq 0$. Put $d = a_1^2 - 4a_0a_2$, let ψ be a multiplicative character of $\mathbb{F}_q$ of order $\geqslant 3$, and let η be the quadratic character of $\mathbb{F}_q$. Prove that

$$\sum_{c \in \mathbb{F}_q} \psi(f(c)) = \bar{\psi}(4a_2)\psi(-d)\eta(d)J(\psi, \eta).$$

5.61. Prove that for a p-polynomial

$$L(x)=a_r x^{p^r}+a_{r-1}x^{p^{r-1}}+\cdots+a_1x^p+a_0x$$

over $\mathbb{F}_q$ the condition $a_r+a_{r-1}^p+\cdots+a_1^{p^{r-1}}+a_0^{p^r}=0$ (see Theorem 5.34) is satisfied if and only if $L(x)=M(x)^p-M(x)$ for some p-polynomial $M(x)$ over $\mathbb{F}_q$.

5.62. In connection with Theorem 5.39 prove the following result. Let $m\in\mathbb{N}$ and $f\in\mathbb{F}_q[x]$ a monic polynomial of positive degree which is the mth power of a polynomial (over some extension field of $\mathbb{F}_q$); then there is a monic polynomial $g\in\mathbb{F}_q[x]$ with $f=g^m$.

5.63. Let $a_1,\ldots,a_k$ be k distinct elements of $\mathbb{F}_q$, q odd, and let $\varepsilon_1,\ldots,\varepsilon_k$ be k given integers, each of which is 1 or -1. Let $N(\varepsilon_1,\ldots,\varepsilon_k)$ denote the number of $c\in\mathbb{F}_q$ with $\eta(c+a_j)=\varepsilon_j$ for $1\leqslant j\leqslant k$, where η is the quadratic character of $\mathbb{F}_q$. Prove that

$$N(\varepsilon_1,\ldots,\varepsilon_k)=\frac{1}{2^k}\sum_{c\in\mathbb{F}_q}[1+\varepsilon_1\eta(c+a_1)]\cdots[1+\varepsilon_k\eta(c+a_k)]-A,$$

where $0\leqslant A\leqslant k/2$.

5.64. Use Exercise 5.63 and estimates for character sums to prove that

$$\left|N(\varepsilon_1,\ldots,\varepsilon_k)-\frac{q}{2^k}\right|\leqslant\left(\frac{k-2}{2}+\frac{1}{2^k}\right)q^{1/2}+\frac{k}{2}.$$

5.65. Let ψ be a multiplicative character of $\mathbb{F}_q$ of order $m\geqslant 2$, let $a_1,\ldots,a_k$ be k distinct elements of $\mathbb{F}_q$, and let $\varepsilon_1,\ldots,\varepsilon_k$ be k given complex mth roots of unity. Let $N(\varepsilon_1,\ldots,\varepsilon_k)$ denote the number of $c\in\mathbb{F}_q$ with $\psi(c+a_j)=\varepsilon_j$ for $1\leqslant j\leqslant k$. Prove that

$$N(\varepsilon_1,\ldots,\varepsilon_k)=\frac{1}{m^k}\sum_{c\in\mathbb{F}_q}\prod_{j=1}^{k}\{1+\varepsilon_j^{-1}\psi(c+a_j)+\varepsilon_j^{-2}\psi^2(c+a_j)$$

$$+\cdots+\varepsilon_j^{-m+1}\psi^{m-1}(c+a_j)\}-A,$$

where $0\leqslant A\leqslant k/m$.

5.66. Use Exercise 5.65 and estimates for character sums to prove that

$$\left|N(\varepsilon_1,\ldots,\varepsilon_k)-\frac{q}{m^k}\right|\leqslant\left(k-1-\frac{k}{m}+\frac{1}{m^k}\right)q^{1/2}+\frac{k}{m}.$$

5.67. Let m and k be given positive integers. Prove that there exists $q_0(m,k)$ such that for any prime power $q\geqslant q_0(m,k)$ and for any $a_1,\ldots,a_k\in\mathbb{F}_q$ there is at least one $c\in\mathbb{F}_q$ such that each of $c+a_1,\ldots,$ $c+a_k$ is an mth power of some element of $\mathbb{F}_q^*$.

5.68. For an odd prime p, let $\gamma\in\mathbb{F}_{p^2}^*$ be an element of order $2(p-1)$ and let ψ be a multiplicative character of $\mathbb{F}_{p^2}$. Define the *Eisenstein sum*

$E(\psi)$ by

$$E(\psi) = \sum_{c \in \mathbb{F}_p} \psi(1 + c\gamma).$$

Prove that $E(\psi)$ is independent of the specific choice of γ.

5.69. Prove that the Eisenstein sum $E(\psi)$ is also given by

$$E(\psi) = \frac{p\psi(2)G(\bar{\psi}^*, \chi_1)}{G(\bar{\psi}, \mu_1)},$$

where χ_1 and μ_1 are the canonical additive characters of $\mathbb{F}_p$ and $\mathbb{F}_{p^2}$, respectively, and where $\bar{\psi}^*$ is the restriction of $\bar{\psi}$ to $\mathbb{F}_p^*$. (*Hint*: Use (5.16) and properties of Gaussian sums.) If ψ^* is nontrivial, prove that this can also be written in the form

$$E(\psi) = \frac{\psi(2)G(\psi, \mu_1)}{G(\psi^*, \chi_1)}.$$

5.70. For $a \in \mathbb{F}_q$ and a multiplicative character ψ of $\mathbb{F}_{q^s}$ define the *generalized Eisenstein sum* $E_s(\psi; a)$ by

$$E_s(\psi; a) = \sum_{\mathrm{Tr}(\alpha) = a} \psi(\alpha),$$

where the sum is extended over all $\alpha \in \mathbb{F}_{q^s}$ for which the trace of α over $\mathbb{F}_q$ is equal to a. Prove that the Eisenstein sum $E(\psi)$ is also a generalized Eisenstein sum since $E(\psi) = E_2(\psi; 2)$.

5.71. Let T_0 be the set of all $\alpha \in \mathbb{F}_{q^s}$ for which the trace of α over $\mathbb{F}_q$ is equal to 0. Let B be the set of all $\beta \in \mathbb{F}_{q^s}$ with the property that $\alpha\beta \in T_0$ whenever $\alpha \in T_0$. Prove that $B = \mathbb{F}_q$.

5.72. Prove that the generalized Eisenstein sum $E_s(\psi; a)$ is also given by

$$E_s(\psi; a) = \frac{q^{s-1}\psi(a)G(\bar{\psi}^*, \chi_1)}{G(\bar{\psi}, \mu_1)},$$

where χ_1 and μ_1 are the canonical additive characters of $\mathbb{F}_q$ and $\mathbb{F}_{q^s}$, respectively, and where $\bar{\psi}^*$ is the restriction of $\bar{\psi}$ to $\mathbb{F}_q^*$. (*Hint*: Use (5.16), Exercise 5.71, and properties of Gaussian sums.) If ψ^* is nontrivial, prove that this can also be written in the form

$$E_s(\psi; a) = \frac{\psi(a)G(\psi, \mu_1)}{G(\psi^*, \chi_1)}.$$

5.73. Prove that $E_s(\psi^p; a) = E_s(\psi; a^p)$ and $E_s(\psi^q; a) = E_s(\psi; a)$, where p is the characteristic of the underlying finite fields $\mathbb{F}_q$ and $\mathbb{F}_{q^s}$.

5.74. For $a \in \mathbb{F}_q^*$ and a nontrivial multiplicative character ψ of $\mathbb{F}_{q^2}$ of order m dividing $q+1$, prove that

$$E_2(\psi; a) = (-1)^{(mq-q-1)/m}.$$

(*Hint*: Use Stickelberger's theorem.)

5.75. Suppose that $a \in \mathbb{F}_q^*$ and that, in the notation of Exercise 5.72, ψ^* is nontrivial. Prove $|E_s(\psi; a)| = q^{(s-1)/2}$. (*Note*: According to Exercise 5.9, ψ^* is nontrivial if and only if the order of ψ does not divide $(q^s-1)/(q-1)$.)

5.76. Prove the following property of Eisenstein sums: if the restriction ψ^* of ψ to $\mathbb{F}_p^*$ is nontrivial, then

$$\frac{E(\psi)^2}{E(\psi^2)} = \frac{J(\psi, \psi)}{J(\psi^*, \psi^*)}.$$

5.77. Prove the following property of generalized Eisenstein sums: if the restrictions $\psi_1^*, \ldots, \psi_k^*$ of $\psi_1, \ldots, \psi_k$ to $\mathbb{F}_q^*$ are nontrivial, if $\psi_1^* \cdots \psi_k^*$ is nontrivial, and if $a \in \mathbb{F}_q^*$, then

$$\frac{E_s(\psi_1; a) \cdots E_s(\psi_k; a)}{E_s(\psi_1 \cdots \psi_k; a)} = \frac{J(\psi_1, \ldots, \psi_k)}{J(\psi_1^*, \ldots, \psi_k^*)}.$$

5.78. Prove that if ψ^* is nontrivial, then $E(\psi^{p+1}) = -J(\psi^*, \eta_p)$, where η_p is the quadratic character of $\mathbb{F}_p$.

5.79. Prove that Kloosterman sums have the property $K(\chi; a, b) = K(\chi; b, a)$ for all $a, b \in \mathbb{F}_q$.

5.80. Prove that $K(\chi; a, b) = K(\chi; ab, 1) = K(\chi; 1, ab)$ if $a, b \in \mathbb{F}_q$ are not both 0.

5.81. In the notation of Theorem 5.43, prove that

$$K(\chi; a, b)^4 + K(\chi^{(4)}; a, b) + 4qK(\chi^{(2)}; a, b) = 6q^2 \quad \text{for } a, b \in \mathbb{F}_q^*.$$

5.82. Let $f, g \in \mathbb{F}_q[x]$, q odd, with $\deg(f) = \deg(g) = 2$, and let χ be a nontrivial additive character of $\mathbb{F}_q$. Express $\sum_c \chi(f(c)g(c)^{-1})$ in terms of Kloosterman sums, where c runs through all elements of $\mathbb{F}_q$ with $g(c) \neq 0$.

5.83. For a multiplicative character ψ of $\mathbb{F}_q$, an additive character χ of $\mathbb{F}_q$, and $a, b \in \mathbb{F}_q$ define a *generalized Kloosterman sum* by

$$K(\psi, \chi; a, b) = \sum_{c \in \mathbb{F}_q^*} \psi(c)\chi(ac + bc^{-1}).$$

Prove that such a sum reduces to a Gaussian sum whenever $ab = 0$, in the sense that

$$K(\psi,\chi;a,b)=\begin{cases}\psi(b)G(\bar{\psi},\chi) & \text{if } a=0, b\neq 0,\\ \bar{\psi}(a)G(\psi,\chi) & \text{if } a\neq 0, b=0,\\ G(\psi,\chi_0) & \text{if } a=b=0.\end{cases}$$

5.84. Prove that if η is the quadratic character of $\mathbb{F}_q$, q odd, and $a,b\in\mathbb{F}_q$ with $\eta(ab)=-1$, then $K(\eta,\chi;a,b)=0$ for any additive character χ of $\mathbb{F}_q$.

5.85. Prove that if η is the quadratic character of $\mathbb{F}_q$, q odd, and $a,b\in\mathbb{F}_q$ with $ab=d^2$ for some $d\in\mathbb{F}_q^*$, then

$$K(\eta,\chi;a,b)=\eta(b)G(\eta,\chi)(\chi(2d)+\chi(-2d))$$

for any additive character χ of $\mathbb{F}_q$. (*Hint*: Use (5.16) and Theorem 5.47.)

5.86. Prove that if χ is a nontrivial additive character of $\mathbb{F}_q$, q odd, η is the quadratic character of $\mathbb{F}_q$, ψ is a multiplicative character of $\mathbb{F}_q$ of order $\geqslant 3$, and $a,b\in\mathbb{F}_q$ with $b\neq 0$, then

$$K(\psi,\chi;a,b)=\frac{\psi(4b)G(\eta,\chi)}{G(\bar{\psi}\eta,\chi)}\sum_{c\in\mathbb{F}_q}\chi(c)(\bar{\psi}\eta)(c^2-4ab).$$

5.87. Show directly that $H_1(a)=H_1(1)$ for $a\in\mathbb{F}_q^*$. Then evaluate $\sum_{a\in\mathbb{F}_q^*}H_1(a)$ and thus obtain an alternative proof of the fact that $H_1(a)=-1$ for $a\in\mathbb{F}_q^*$.

5.88. Prove that $H_n(a)=\eta(a)I_n(a^{-1})-1$ for $a\in\mathbb{F}_q^*$ and odd $n\in\mathbb{N}$. (*Note*: This shows again that $H_1(a)=-1$ for $a\in\mathbb{F}_q^*$.)

5.89. Prove that

$$I_{2n}(a)=\eta(a)H_n(a^{-1})+H_n(a)+\eta(a)$$

for $a\in\mathbb{F}_q^*$ and odd $n\in\mathbb{N}$.

5.90. Prove that $I_4(a)=-1$ whenever $q\equiv 3\bmod 4$ and $a\in\mathbb{F}_q^*$.

5.91. Prove that for a prime $p\equiv 1\bmod 4$,

$$I_4(-1)=\begin{cases}-1+2A & \text{if } p\equiv 1\bmod 8,\\ -1-2A & \text{if } p\equiv 5\bmod 8,\end{cases}$$

where $p=A^2+B^2$ with integers $A\equiv -1\bmod 4$ and B.

5.92. Determine the continued fraction expansion of the rational function $(x^5+x^3+x^2+x)/(x^6+x^5+x^4+x^3+1)$ over $\mathbb{F}_2$ and of the rational function $(x^6-x^5-x^4+x^3-x^2-1)/(x^5-x^3-x^2+1)$ over $\mathbb{F}_3$.

5.93. If P_i and Q_i are defined by (5.74) and (5.75), respectively, prove that

$$P_iQ_{i-2}-P_{i-2}Q_i=(-1)^iA_i \quad \text{for } i=1,\ldots,s.$$

5.94. Prove that

$$\deg\left(\frac{r_0}{r_1} - \frac{P_i}{Q_i}\right) = -\deg(Q_i Q_{i+1}) \quad \text{for } i = 0, 1, \ldots, s-1.$$

5.95. Prove that the rational functions P_i/Q_i are best approximations to r_0/r_1 in the sense that if f/g is a rational function such that for some i, $0 \leqslant i \leqslant s-1$, we have

$$\deg\left(\frac{r_0}{r_1} - \frac{f}{g}\right) < \deg\left(\frac{r_0}{r_1} - \frac{P_i}{Q_i}\right),$$

then $\deg(g) \geqslant \deg(Q_{i+1})$.

Chapter 6

Equations over Finite Fields

We consider polynomial equations of the form $f_1(x_1,\ldots,x_n)=f_2(x_1,\ldots,x_n)$ with $f_1, f_2 \in \mathbb{F}_q[x_1,\ldots,x_n]$. By the number of solutions of this equation in $\mathbb{F}_q^n$ we mean the number of n-tuples $(c_1,\ldots,c_n)\in\mathbb{F}_q^n$ for which $f_1(c_1,\ldots,c_n)=f_2(c_1,\ldots,c_n)$. The equation is often put in the equivalent form $f(x_1,\ldots,x_n)=0$ with $f=f_1-f_2$. In special cases one can give explicit formulas for the number of solutions, but in general one will have to be satisfied with estimates. Instances of results of each type can be found throughout this chapter.

In Section 1 we present some classical theorems such as those of König-Rados, Chevalley, and Warning. We also establish elementary upper bounds for the number of solutions and results on the expected order of magnitude.

Sections 2 and 3 are devoted to special classes of equations—namely, quadratic equations and diagonal equations, respectively. Exponential sums turn out to be very useful tools in the study of these equations. In the case of quadratic equations, an essential role in simplifying the determination of the number of solutions is played by the reduction theory for quadratic forms.

The impact of information about the number of solutions of equations on the estimation of character sums is demonstrated in Section 4. By studying equations of the form $y^m=f(x)$ and $y^q-y=f(x)$, the proofs of important inequalities for character sums stated in Chapter 5, Section 4, can be completed. The methods employed in investigating these equations are rather intricate, but elementary in the sense that they depend neither on algebraic geometry nor on a detailed knowledge of algebraic function fields.

1. ELEMENTARY RESULTS ON THE NUMBER OF SOLUTIONS

We start with the simplest case, namely that of a polynomial equation in one indeterminate. Let $f \in \mathbb{F}_q[x]$ be of positive degree and consider the equation $f(x)=0$. The solutions of this equation in $\mathbb{F}_q$ are just the distinct roots of f in $\mathbb{F}_q$. As we have seen in Chapter 4, Section 3, $\gcd(f(x), x^q - x)$ is that part of f which contains exactly all the roots of f in $\mathbb{F}_q$, each with multiplicity 1. Therefore, the number of solutions of $f(x)=0$ in $\mathbb{F}_q$ is equal to the degree of $\gcd(f(x), x^q - x)$. For the actual calculation of the solutions, we refer again to Chapter 4, Section 3.

The number of solutions of $f(x)=0$ in $\mathbb{F}_q$ can also be determined by the use of matrix theory. Since it is trivial to decide whether 0 is a solution, it suffices to consider only the nonzero solutions of the equation. These solutions are exactly the distinct roots of $\gcd(f(x), x^{q-1}-1)$ in $\mathbb{F}_q$. Therefore, we may assume without loss of generality that $\deg(f) \leqslant q-1$. Furthermore, since $b^{q-1}=1$ for $b \in \mathbb{F}_q^*$, the nonzero solutions of

$$f(x) = a_0 + a_1 x + \cdots + a_{q-2}x^{q-2} + a_{q-1}x^{q-1} = 0$$

in $\mathbb{F}_q$ are the same as the nonzero solutions of

$$(a_0 + a_{q-1}) + a_1 x + \cdots + a_{q-2}x^{q-2} = 0$$

in $\mathbb{F}_q$. Thus we can even assume $\deg(f) \leqslant q-2$.

Let now

$$f(x) = a_0 + a_1 x + \cdots + a_{q-2}x^{q-2} \in \mathbb{F}_q[x].$$

We associate with f the $(q-1)\times(q-1)$ matrix A given by

$$A = \begin{pmatrix} a_0 & a_1 & \cdots & a_{q-3} & a_{q-2} \\ a_1 & a_2 & \cdots & a_{q-2} & a_0 \\ \vdots & \vdots & & \vdots & \vdots \\ a_{q-2} & a_0 & \cdots & a_{q-4} & a_{q-3} \end{pmatrix}. \tag{6.1}$$

This is a *left circulant* matrix, in which each row is obtained from the preceding row by a cyclic shift of the entries to the left.

6.1. Theorem (König-Rados Theorem). *Let*

$$f(x) = a_0 + a_1 x + \cdots + a_{q-2}x^{q-2} \in \mathbb{F}_q[x].$$

Then the number of nonzero solutions of the equation $f(x)=0$ *in* $\mathbb{F}_q$ *is equal to* $q-1-r$, *where* r *is the rank of the matrix* A *in* (6.1).

Proof. Let $b_1,\ldots,b_{q-1}$ be the distinct elements of $\mathbb{F}_q^*$ and use them to set up the $(q-1)\times(q-1)$ Vandermonde matrix

$$B=\begin{pmatrix} 1 & 1 & \cdots & 1 \\ b_1 & b_2 & \cdots & b_{q-1} \\ b_1^2 & b_2^2 & \cdots & b_{q-1}^2 \\ \vdots & \vdots & & \vdots \\ b_1^{q-2} & b_2^{q-2} & \cdots & b_{q-1}^{q-2} \end{pmatrix}.$$

Using $b^{q-1}=1$ for $b\in\mathbb{F}_q^*$, we obtain

$$AB=\begin{pmatrix} f(b_1) & f(b_2) & \cdots & f(b_{q-1}) \\ b_1^{-1}f(b_1) & b_2^{-1}f(b_2) & \cdots & b_{q-1}^{-1}f(b_{q-1}) \\ b_1^{-2}f(b_1) & b_2^{-2}f(b_2) & \cdots & b_{q-1}^{-2}f(b_{q-1}) \\ \vdots & \vdots & & \vdots \\ b_1^{-(q-2)}f(b_1) & b_2^{-(q-2)}f(b_2) & \cdots & b_{q-1}^{-(q-2)}f(b_{q-1}) \end{pmatrix}.$$

If N is the number of nonzero solutions of $f(x)=0$ in $\mathbb{F}_q$, we can assume that $N\leqslant q-2$ (the case $N=q-1$ occurs only if A is the zero matrix having $r=0$) and that the b_i have been ordered in such a way that $f(b_i)\neq 0$ for $1\leqslant i\leqslant q-1-N$ and $f(b_i)=0$ for $q-N\leqslant i\leqslant q-1$. Then the entries in the last N columns of AB are all 0, and so the rank of AB is at most $q-1-N$. On the other hand, the principal minor of AB of order $q-1-N$ is equal to

$$f(b_1)\cdots f(b_{q-1-N})\begin{vmatrix} 1 & 1 & \cdots & 1 \\ b_1^{-1} & b_2^{-1} & \cdots & b_{q-1-N}^{-1} \\ b_1^{-2} & b_2^{-2} & \cdots & b_{q-1-N}^{-2} \\ \vdots & \vdots & & \vdots \\ b_1^{-(q-2-N)} & b_2^{-(q-2-N)} & \cdots & b_{q-1-N}^{-(q-2-N)} \end{vmatrix},$$

which is $\neq 0$ since the Vandermonde determinant involves the distinct elements $b_1^{-1},\ldots,b_{q-1-N}^{-1}$. Therefore the rank of AB is $q-1-N$. But B is nonsingular as $b_1,\ldots,b_{q-1}$ are distinct, hence AB and A have the same rank. It follows that $r=q-1-N$, or $N=q-1-r$. □

6.2. Example. Let $f(x) = 3 + x - 3x^2 + 2x^3 \in \mathbb{F}_7[x]$. Then the associated matrix A is

$$\begin{pmatrix} 3 & 1 & -3 & 2 & 0 & 0 \\ 1 & -3 & 2 & 0 & 0 & 3 \\ -3 & 2 & 0 & 0 & 3 & 1 \\ 2 & 0 & 0 & 3 & 1 & -3 \\ 0 & 0 & 3 & 1 & -3 & 2 \\ 0 & 3 & 1 & -3 & 2 & 0 \end{pmatrix}$$

and has rank $r = 6$. Thus by the König-Rados theorem, $f(x) = 0$ has no nonzero solution in $\mathbb{F}_7$. Since 0 is obviously not a solution of $f(x) = 0$, the equation has no solution at all in $\mathbb{F}_7$. In other words, f is irreducible over $\mathbb{F}_7$. □

We turn now to polynomials in several indeterminates. Elementary results on the number of solutions of $f(x_1, \ldots, x_n) = 0$ can be established for the case where the number n of indeterminates is greater than the degree of f (for the definition of the degree of a polynomial in several indeterminates see Definition 1.72).

6.3. Lemma. *Let k be a nonnegative integer. Then*

$$\sum_{c \in \mathbb{F}_q} c^k = \begin{cases} 0 \text{ if } k = 0 \quad \text{or} \quad k \text{ is not divisible by } q-1, \\ -1 \text{ if } k > 0 \quad \text{and} \quad k \text{ is divisible by } q-1. \end{cases}$$

Proof. For $k = 0$ we use the convention $0^0 = 1$; then the statement is trivial. If $k > 0$, choose a primitive element b of $\mathbb{F}_q$ and write

$$\sum_{c \in \mathbb{F}_q} c^k = \sum_{c \in \mathbb{F}_q^*} c^k = \sum_{j=0}^{q-2} b^{jk} = \sum_{j=0}^{q-2} (b^k)^j.$$

Summing the geometric series, we obtain the result. □

6.4. Lemma. *Let $f \in \mathbb{F}_q[x_1, \ldots, x_n]$ with* $\deg(f) < n(q-1)$. *Then*

$$\sum_{c_1, \ldots, c_n \in \mathbb{F}_q} f(c_1, \ldots, c_n) = 0.$$

Proof. By linearity it suffices to prove the identity for monomials $x_1^{k_1} \cdots x_n^{k_n}$ with $k_1 + \cdots + k_n < n(q-1)$. It follows from this inequality that there is a k_j with $0 \leqslant k_j < q - 1$. Then

$$\sum_{c_1, \ldots, c_n \in \mathbb{F}_q} c_1^{k_1} \cdots c_n^{k_n} = \left(\sum_{c_1 \in \mathbb{F}_q} c_1^{k_1} \right) \cdots \left(\sum_{c_n \in \mathbb{F}_q} c_n^{k_n} \right) = 0$$

because of Lemma 6.3. □

6.5. ***Theorem*** (Warning's Theorem). *Let* $f \in \mathbb{F}_q[x_1,\dots,x_n]$ *with* $\deg(f) < n$. *Then the number of solutions of the equation* $f(x_1,\dots,x_n) = 0$ *in* $\mathbb{F}_q^n$ *is divisible by the characteristic* p *of* $\mathbb{F}_q$.

Proof. Consider the polynomial $F = 1 - f^{q-1}$, which has the property that $F(c_1,\dots,c_n) = 1$ whenever $f(c_1,\dots,c_n) = 0$ and $F(c_1,\dots,c_n) = 0$ whenever $f(c_1,\dots,c_n) \neq 0$. Therefore,

$$\sum_{c_1,\dots,c_n \in \mathbb{F}_q} F(c_1,\dots,c_n) = N, \tag{6.2}$$

the number of solutions of $f(x_1,\dots,x_n) = 0$ in $\mathbb{F}_q^n$. On the other hand, the condition $\deg(f) < n$ implies $\deg(F) < n(q-1)$, and so Lemma 6.4 shows that the sum in (6.2) is 0. Hence N, viewed as an element of $\mathbb{F}_q$, is equal to 0, which means that N is divisible by p. □

6.6. ***Corollary*** (Chevalley's Theorem). *Let* $f \in \mathbb{F}_q[x_1,\dots,x_n]$ *with* $f(0,\dots,0) = 0$ *and* $\deg(f) < n$. *Then the equation* $f(x_1,\dots,x_n) = 0$ *has a nontrivial solution in* $\mathbb{F}_q^n$, *that is, there exists* $(c_1,\dots,c_n) \in \mathbb{F}_q^n$ *with* $(c_1,\dots,c_n) \neq (0,\dots,0)$ *and* $f(c_1,\dots,c_n) = 0$.

Proof. The condition $f(0,\dots,0) = 0$ implies that the number N of solutions of the equation in question satisfies $N \geqslant 1$. An application of Theorem 6.5 yields then $N \geqslant p \geqslant 2$. □

The condition $\deg(f) < n$ in Theorem 6.5 and Corollary 6.6 is best possible. In the following, we construct an example of a polynomial f in n indeterminates with $\deg(f) = n$ for which the conclusions of Warning's theorem and Chevalley's theorem fail to hold.

6.7. Example. Let $n \in \mathbb{N}$ and let $\{\alpha_1,\dots,\alpha_n\}$ be a basis of $E = \mathbb{F}_{q^n}$ over $\mathbb{F}_q$. Put

$$f(x_1,\dots,x_n) = \prod_{j=0}^{n-1} \left(\alpha_1^{q^j} x_1 + \cdots + \alpha_n^{q^j} x_n\right).$$

Since the $\alpha_i^{q^j}$, $j = 0,1,\dots,n-1$, are the conjugates of α_i with respect to $\mathbb{F}_q$, the coefficients of f are in $\mathbb{F}_q$. It is clear that $\deg(f) = n$. Now let $(c_1,\dots,c_n) \in \mathbb{F}_q^n$ and put $\gamma = c_1\alpha_1 + \cdots + c_n\alpha_n \in E$. Then

$$\begin{aligned} f(c_1,\dots,c_n) &= \prod_{j=0}^{n-1} \left(\alpha_1^{q^j} c_1 + \cdots + \alpha_n^{q^j} c_n\right) \\ &= \prod_{j=0}^{n-1} (c_1\alpha_1 + \cdots + c_n\alpha_n)^{q^j} \\ &= \mathrm{N}_{E/\mathbb{F}_q}(\gamma). \end{aligned}$$

Thus $f(c_1,\dots,c_n) = 0$ is equivalent to $\mathrm{N}_{E/\mathbb{F}_q}(\gamma) = 0$, which holds only for

$\gamma = 0$—that is, only for $c_1 = \cdots = c_n = 0$. Hence the equation $f(x_1,\ldots,x_n) = 0$ has only the solution $(0,\ldots,0)$ in $\mathbb{F}_q^n$. This shows that the conclusions of Theorem 6.5 and Corollary 6.6 are not valid for f. □

Warning's theorem and Chevalley's theorem can easily be extended to systems of equations. In this case, one is interested in the number of common solutions of the equations.

6.8. Theorem. *Let $f_1,\ldots,f_m \in \mathbb{F}_q[x_1,\ldots,x_n]$ with* $\deg(f_1) + \cdots + \deg(f_m) < n$. *Then the number of $(c_1,\ldots,c_n) \in \mathbb{F}_q^n$ with $f_i(c_1,\ldots,c_n) = 0$ for $1 \leqslant i \leqslant m$ is divisible by the characteristic p of $\mathbb{F}_q$.*

Proof. Put $F = (1 - f_1^{q-1}) \cdots (1 - f_m^{q-1})$ and proceed as in the proof of Theorem 6.5. □

6.9. Corollary. *Let $f_1,\ldots,f_m \in \mathbb{F}_q[x_1,\ldots,x_n]$ with $f_i(0,\ldots,0) = 0$ for $1 \leqslant i \leqslant m$ and* $\deg(f_1) + \cdots + \deg(f_m) < n$. *Then there exists $(c_1,\ldots,c_n) \in \mathbb{F}_q^n$ with $(c_1,\ldots,c_n) \neq (0,\ldots,0)$ and $f_i(c_1,\ldots,c_n) = 0$ for $1 \leqslant i \leqslant m$.*

In the next theorem we show that if under the hypotheses of Theorem 6.8 the system of equations has at least one common solution, it must have many more. We first introduce some notation and terminology. Let S be the set of $(c_1,\ldots,c_n) \in \mathbb{F}_q^n$ with $f_i(c_1,\ldots,c_n) = 0$ for $1 \leqslant i \leqslant m$. We write $|T|$ for the cardinality (= number of elements) of a finite set T, so that, for instance, $|S|$ denotes the number of common solutions in $\mathbb{F}_q^n$ of the system of equations $f_i(x_1,\ldots,x_n) = 0$, $1 \leqslant i \leqslant m$. We recall from Chapter 3, Section 4, the notion of an *affine subspace* of a vector space, by which we mean a translate of a linear subspace. The *dimension* of an affine subspace is by definition the dimension of the corresponding linear subspace. Two affine subspaces are said to be *parallel* if they are obtained by translation from the same linear subspace.

6.10. Lemma. *Suppose the hypotheses of Theorem* 6.8 *are satisfied. If W_1 and W_2 are two parallel affine subspaces of $\mathbb{F}_q^n$ of dimension $d =$* $\deg(f_1) + \cdots + \deg(f_m)$, *then*

$$|W_1 \cap S| \equiv |W_2 \cap S| \bmod p.$$

Proof. The case $W_1 = W_2$ is trivial, and so we may assume $W_1 \neq W_2$. After an appropriate invertible linear change of coordinates in $\mathbb{F}_q^n$ (which does not affect the degrees of the f_i), we may suppose that W_1 and W_2 are defined by

$$W_1 = \{(c_1,\ldots,c_n) \in \mathbb{F}_q^n : c_1 = c_2 = \cdots = c_{n-d} = 0\},$$

$$W_2 = \{(c_1,\ldots,c_n) \in \mathbb{F}_q^n : c_1 = 1,\ c_2 = \cdots = c_{n-d} = 0\}.$$

We introduce the polynomial

$$G(x_1,\dots,x_n) = (-1)^{n-d}\left(x_1^{q-2} + \cdots + x_1 + 1\right)\left(x_2^{q-1} - 1\right)\cdots\left(x_{n-d}^{q-1} - 1\right)$$

with $\deg(G) = (n-d)(q-1) - 1$; furthermore, G is -1 on W_1, 1 on W_2 and 0 elsewhere. Then we put

$$H = \left(1 - f_1^{q-1}\right)\cdots\left(1 - f_m^{q-1}\right)G.$$

We have

$$\deg(H) \leqslant d(q-1) + (n-d)(q-1) - 1 = n(q-1) - 1 < n(q-1),$$

and H is -1 on $W_1 \cap S$, 1 on $W_2 \cap S$ and 0 elsewhere. Thus

$$\sum_{c_1,\dots,c_n \in \mathbb{F}_q} H(c_1,\dots,c_n) = |W_2 \cap S| - |W_1 \cap S|.$$

On the other hand, Lemma 6.4 shows that this sum is 0, and the desired result follows. □

6.11. Theorem. *Let $f_1,\dots,f_m \in \mathbb{F}_q[x_1,\dots,x_n]$ with $d = \deg(f_1) + \cdots + \deg(f_m) < n$. If the number of N of $(c_1,\dots,c_n) \in \mathbb{F}_q^n$ with $f_i(c_1,\dots,c_n) = 0$ for $1 \leqslant i \leqslant m$ satisfies $N \geqslant 1$, then $N \geqslant q^{n-d}$.*

Proof. We distinguish two cases. In the first case, suppose there exists at least one affine subspace W_1 of $\mathbb{F}_q^n$ of dimension d with $|W_1 \cap S| \not\equiv 0 \bmod p$. Then by Lemma 6.10 we have $|W_2 \cap S| \not\equiv 0 \bmod p$ for any affine subspace W_2 parallel to W_1, so in particular $|W_2 \cap S| \geqslant 1$. Now S can be written as the disjoint union $S = \cup_W (W \cap S)$, where W runs through the q^{n-d} distinct affine subspaces parallel to W_1. Therefore,

$$N = |S| = \sum_W |W \cap S| \geqslant q^{n-d}.$$

In the remaining case we have $|W \cap S| \equiv 0 \bmod p$ for all affine subspaces W of $\mathbb{F}_q^n$ of dimension d. Since $N = |S| \geqslant 1$ by hypothesis, there exists an integer k, $1 \leqslant k \leqslant d$, such that for any affine subspace V of dimension k we have $|V \cap S| \equiv 0 \bmod p$, but there is an affine subspace U of dimension $k-1$ such that $|U \cap S| \not\equiv 0 \bmod p$. Fix one such affine subspace U. Now consider all affine subspaces V of dimension k containing U, of which there are exactly

$$\frac{q^{n-k+1} - 1}{q-1} = q^{n-k} + \cdots + q + 1.$$

For each such V consider the set-theoretic difference $V \backslash U$. Then

$$|(V \backslash U) \cap S| = |V \cap S| - |U \cap S| \not\equiv 0 \bmod p,$$

hence $|(V \backslash U) \cap S| \geqslant 1$. Since U and the differences $V \backslash U$ form a partition of $\mathbb{F}_q^n$, it follows that

$$N = |S| = |U \cap S| + \sum_V |(V \setminus U) \cap S| \geqslant q^{n-k} + \cdots + q + 2 > q^{n-d}. \quad \square$$

6.12. Example. The inequality $N \geqslant q^{n-d}$ in Theorem 6.11 is best possible, even for $m = 1$, in the sense that for any positive integers d and n with $d < n$ there is a polynomial $f_1 \in \mathbb{F}_q[x_1, \ldots, x_n]$ of degree d such that the equation $f_1(x_1, \ldots, x_n) = 0$ has exactly q^{n-d} solutions in $\mathbb{F}_q^n$. Let $g \in \mathbb{F}_q[x_1, \ldots, x_d]$ be defined like the polynomial f in Example 6.7, but with n replaced by d. Then set $f_1(x_1, \ldots, x_n) = g(x_1, \ldots, x_d)$, so that the indeterminates $x_{d+1}, \ldots, x_n$ do not appear in f_1. By what we have shown in Example 6.7, we have $f_1(c_1, \ldots, c_n) = 0$ if and only if $c_1 = \cdots = c_d = 0$. Since $c_{d+1}, \ldots, c_n$ can be arbitrary elements of $\mathbb{F}_q$, the equation $f_1(x_1, \ldots, x_n) = 0$ has exactly q^{n-d} solutions in $\mathbb{F}_q^n$. $\square$

Elementary *upper bounds* on the number of solutions of equations over a finite field can also be given. The following result may be viewed as a generalization of the fact that a polynomial in one indeterminate of degree $d \geqslant 0$ can have at most d roots.

6.13. Theorem. *Let $f \in \mathbb{F}_q[x_1, \ldots, x_n]$ with $\deg(f) = d \geqslant 0$. Then the equation $f(x_1, \ldots, x_n) = 0$ has at most dq^{n-1} solutions in $\mathbb{F}_q^n$.*

Proof. If $d = 0$, then f is a nonzero constant and the result is trivial. If $d = 1$, then

$$f(x_1, \ldots, x_n) = a_1 x_1 + \cdots + a_n x_n + b = 0$$

has q^{n-1} solutions in $\mathbb{F}_q^n$ since at least one a_i is $\neq 0$, so that the corresponding x_i is uniquely determined once values have been assigned to the other indeterminates in an arbitrary manner. The result is clearly true for $n = 1$.

We have thus shown the theorem if either $d \leqslant 1$ or $n = 1$. We proceed now by double induction. Suppose $n > 1$, $d > 1$, and that the result is true for nonzero polynomials in at most n indeterminates of degree less than d and for nonzero polynomials in less than n indeterminates of degree at most d. We must prove the result for a polynomial $f(x_1, \ldots, x_n)$ in n indeterminates of degree d. We distinguish two cases.

Case 1: $f(x_1, \ldots, x_n)$ is divisible by $x_1 - c$ for some $c \in \mathbb{F}_q$. Then

$$f(x_1, \ldots, x_n) = (x_1 - c) g(x_1, \ldots, x_n),$$

where g is a nonzero polynomial of degree less than d. Using the induction hypothesis and elementary counting arguments, we find that the number of solutions of $f(x_1, \ldots, x_n) = 0$ in $\mathbb{F}_q^n$ is at most $q^{n-1} + (d-1)q^{n-1} = dq^{n-1}$.

Case 2: $f(x_1, \ldots, x_n)$ is not divisible by $x_1 - c$ for any $c \in \mathbb{F}_q$. Then for any $c \in \mathbb{F}_q$, $f(c, x_2, \ldots, x_n)$ is a nonzero polynomial in $n-1$ indeterminates of degree at most d. By the induction hypothesis, the equation $f(c, x_2, \ldots, x_n) = 0$ has at most dq^{n-2} solutions in $\mathbb{F}_q^{n-1}$. Since we have q

choices for $c \in \mathbb{F}_q$, the number of solutions of $f(x_1,\ldots,x_n)=0$ in $\mathbb{F}_q^n$ is at most $q \cdot dq^{n-2} = dq^{n-1}$. □

6.14. Example. The upper bound dq^{n-1} in Theorem 6.13 is only of interest when $d \leqslant q$. In this case the bound can actually be attained. Consider, for instance, the polynomial

$$f(x_1,\ldots,x_n)=(x_1-c_1)(x_1-c_2)\cdots(x_1-c_d)$$

with distinct elements $c_1, c_2,\ldots,c_d \in \mathbb{F}_q$. We have $\deg(f)=d$, and it is seen immediately that the equation $f(x_1,\ldots,x_n)=0$ has exactly dq^{n-1} solutions in $\mathbb{F}_q^n$. □

According to Definition 1.72, the polynomial f is called *homogeneous* if all its terms have the same degree. For such an f of positive degree, the equation $f(x_1,\ldots,x_n)=0$ always has the trivial solution $(0,\ldots,0)$. By considering the nontrivial solutions (if any), one can obtain a slight improvement of the upper bound in Theorem 6.13 for homogeneous polynomials.

*6.15. **Theorem.*** *Let $f \in \mathbb{F}_q[x_1,\ldots,x_n]$ be homogeneous with $\deg(f) = d \geqslant 1$. Then the equation $f(x_1,\ldots,x_n)=0$ has at most $d(q^{n-1}-1)$ nontrivial solutions in $\mathbb{F}_q^n$.*

Proof. If either $d=1$ or $n=1$, the result is seen easily. We proceed now by double induction as in the proof of Theorem 6.13. Suppose $n>1$, $d>1$, and that the result is true for nonconstant homogeneous polynomials in at most n indeterminates of degree less than d and for nonconstant homogeneous polynomials in less than n indeterminates of degree at most d. Take a homogeneous polynomial $f(x_1,\ldots,x_n)$ of degree d and distinguish two cases.

Case 1: $f(x_1,\ldots,x_n)$ is divisible by x_1. Then

$$f(x_1,\ldots,x_n)=x_1 \cdot g(x_1,\ldots,x_n),$$

where g is a nonconstant homogeneous polynomial of degree less than d. Using the induction hypothesis and elementary counting arguments, we find that the number of nontrivial solutions of $f(x_1,\ldots,x_n)=0$ in $\mathbb{F}_q^n$ is at most

$$(q^{n-1}-1)+(d-1)(q^{n-1}-1)=d(q^{n-1}-1).$$

Case 2: $f(x_1,\ldots,x_n)$ is not divisible by x_1. Then for $c \in \mathbb{F}_q^*$, $f(c, x_2,\ldots,x_n)$ is a polynomial in $n-1$ indeterminates of degree d, and so Theorem 6.13 implies that the equation $f(x_1,\ldots,x_n)=0$ has at most $(q-1)dq^{n-2}$ solutions $(c_1,\ldots,c_n) \in \mathbb{F}_q^n$ with $c_1 \neq 0$. Furthermore, $f(0, x_2,\ldots,x_n)$ is a homogeneous polynomial in $n-1$ indeterminates of degree d, and so the induction hypothesis implies that $f(x_1,\ldots,x_n)=0$ has at most $d(q^{n-2}-1)$ nontrivial solutions of the form $(0, c_2,\ldots,c_n) \in \mathbb{F}_q^n$. Altogether, the number of nontrivial solutions of $f(x_1,\ldots,x_n)=0$ in $\mathbb{F}_q^n$ is at most

$$(q-1)dq^{n-2}+d(q^{n-2}-1)=d(q^{n-1}-1). \qquad \square$$

We investigate now the *average number of solutions* of a polynomial equation. For a positive integer d, let Ω_d be the set of all $f \in \mathbb{F}_q[x_1,\ldots,x_n]$ with $\deg(f) \leqslant d$. Let $\omega(d)$ be the number of n-tuples $(i_1,\ldots,i_n)$ of nonnegative integers with $i_1+\cdots+i_n \leqslant d$. Then the cardinality $|\Omega_d|$ of Ω_d is $q^{\omega(d)}$. For $f \in \Omega_d$ let $N(f)$ be the number of solutions of $f(x_1,\ldots,x_n)=0$ in $\mathbb{F}_q^n$.

6.16. Theorem. *With the notation above, we have*

$$\frac{1}{|\Omega_d|}\sum_{f\in\Omega_d} N(f)=q^{n-1}.$$

Proof. We can write

$$\sum_{f\in\Omega_d} N(f)=\sum_{f\in\Omega_d}\sum_{\substack{(c_1,\ldots,c_n)\in\mathbb{F}_q^n\\ f(c_1,\ldots,c_n)=0}} 1=\sum_{(c_1,\ldots,c_n)\in\mathbb{F}_q^n}\sum_{\substack{f\in\Omega_d\\ f(c_1,\ldots,c_n)=0}} 1.$$

For fixed $(c_1,\ldots,c_n)\in\mathbb{F}_q^n$, we get all polynomials

$$f(x_1,\ldots,x_n)=\sum a_{i_1\cdots i_n}x_1^{i_1}\cdots x_n^{i_n}\in\Omega_d$$

with $f(c_1,\ldots,c_n)=0$ by choosing the coefficients $a_{i_1\cdots i_n}$ with $0<i_1+\cdots+i_n\leqslant d$ arbitrarily and then determining the constant term $a_{0\cdots 0}$ to make $f(c_1,\ldots,c_n)=0$. Thus, the number of $f\in\Omega_d$ with $f(c_1,\ldots,c_n)=0$ is equal to $q^{\omega(d)-1}$. Hence

$$\sum_{f\in\Omega_d} N(f)=q^n q^{\omega(d)-1}=|\Omega_d|q^{n-1},$$

and the result follows. $\square$

A polynomial equation in n indeterminates has thus on the average q^{n-1} solutions in $\mathbb{F}_q^n$. We consider next the average deviation from the expected value.

6.17. Theorem. *With the notation above, we have*

$$\frac{1}{|\Omega_d|}\sum_{f\in\Omega_d}\left(N(f)-q^{n-1}\right)^2=q^{n-1}-q^{n-2}.$$

Proof. With $\mathbf{b}=(b_1,\ldots,b_n)\in\mathbb{F}_q^n$ and $\mathbf{c}=(c_1,\ldots,c_n)\in\mathbb{F}_q^n$ we get

$$\sum_{f\in\Omega_d} N(f)^2=\sum_{f\in\Omega_d}\left(\sum_{\substack{\mathbf{c}\in\mathbb{F}_q^n\\ f(\mathbf{c})=0}} 1\right)^2=\sum_{f\in\Omega_d}\sum_{\substack{\mathbf{b}\in\mathbb{F}_q^n\\ f(\mathbf{b})=0}}\sum_{\substack{\mathbf{c}\in\mathbb{F}_q^n\\ f(\mathbf{c})=0}} 1$$

$$=\sum_{\mathbf{b},\mathbf{c}\in\mathbb{F}_q^n}\sum_{\substack{f\in\Omega_d\\ f(\mathbf{b})=f(\mathbf{c})=0}} 1.$$

If $\mathbf{b}=\mathbf{c}$, then we have seen in the proof of Theorem 6.16 that the value of the inner sum is $q^{\omega(d)-1}$. If $\mathbf{b}\neq\mathbf{c}$, then $f(\mathbf{b})=f(\mathbf{c})=0$ gives a system of two linear equations for the coefficients of f of rank 2, which therefore has $q^{\omega(d)-2}$ solutions. It follows that

$$\begin{aligned}\sum_{f\in\Omega_d} N(f)^2 &= \sum_{\mathbf{c}\in\mathbb{F}_q^n}|\Omega_d|q^{-1}+\sum_{\substack{\mathbf{b},\mathbf{c}\in\mathbb{F}_q^n\\ \mathbf{b}\neq\mathbf{c}}}|\Omega_d|q^{-2}\\ &= q^n|\Omega_d|q^{-1}+q^n(q^n-1)|\Omega_d|q^{-2}\\ &= |\Omega_d|\left(q^{2n-2}+q^{n-1}-q^{n-2}\right).\end{aligned}$$

Using this identity and Theorem 6.16, we obtain

$$\begin{aligned}\sum_{f\in\Omega_d}\left(N(f)-q^{n-1}\right)^2 &= \sum_{f\in\Omega_d}N(f)^2-2q^{n-1}\sum_{f\in\Omega_d}N(f)+q^{2n-2}\sum_{f\in\Omega_d}1\\ &= |\Omega_d|\left(q^{2n-2}+q^{n-1}-q^{n-2}\right)\\ &\quad -2q^{n-1}|\Omega_d|q^{n-1}+q^{2n-2}|\Omega_d|\\ &= |\Omega_d|\left(q^{n-1}-q^{n-2}\right),\end{aligned}$$

and the result follows. □

The average value of $(N(f)-q^{n-1})^2$ is thus $q^{n-1}-q^{n-2}$. One may therefore expect that $|N(f)-q^{n-1}|$ is often of the order of magnitude $q^{(n-1)/2}$. We will see various instances of this expected behavior in the following sections.

2. QUADRATIC FORMS

A *quadratic form* (in n indeterminates) over $\mathbb{F}_q$ is a homogeneous polynomial in $\mathbb{F}_q[x_1,\ldots,x_n]$ of degree 2, or the zero polynomial. If q is odd, which is the case of principal interest, we can write the mixed terms $b_{ij}x_ix_j$ $(1\leqslant i<j\leqslant n)$ as $\frac{1}{2}b_{ij}x_ix_j+\frac{1}{2}b_{ij}x_jx_i$, and this leads to the representation

$$f(x_1,\ldots,x_n)=\sum_{i,j=1}^{n}a_{ij}x_ix_j \quad \text{with } a_{ij}=a_{ji}$$

for any quadratic form f over $\mathbb{F}_q$. We then associate with f the $n\times n$ matrix A whose (i,j) entry is a_{ij}. The matrix A is called the *coefficient matrix* of f. Let M^{T} denote the transpose of a matrix M. Then $A^{\mathrm{T}}=A$; that is, A is symmetric. If $\mathbf{x}$ is the column vector of indeterminates $x_1,\ldots,x_n$, then f is given by $\mathbf{x}^{\mathrm{T}}A\mathbf{x}$.

6.18. Example. Consider the quadratic form $f(x_1,x_2)=2x_1^2+x_1x_2+x_2^2$ in two indeterminates (a so-called *binary quadratic form*) over $\mathbb{F}_5$. The coefficient matrix of f is

$$A = \begin{pmatrix} 2 & 3 \\ 3 & 1 \end{pmatrix},$$

and we have

$$\mathbf{x}^{\mathrm{T}}A\mathbf{x} = (x_1 \quad x_2)\begin{pmatrix} 2 & 3 \\ 3 & 1 \end{pmatrix}\begin{pmatrix} x_1 \\ x_2 \end{pmatrix} = 2x_1^2 + x_1x_2 + x_2^2 = f(x_1, x_2). \qquad \square$$

If f is a quadratic form over $\mathbb{F}_q$ and $b \in \mathbb{F}_q$, then an explicit formula for the number of solutions of the equation $f(x_1,\ldots,x_n) = b$ in $\mathbb{F}_q^n$ can be given. To arrive at this formula, we first transform f by a linear substitution of indeterminates into a simpler form. We note, in general, that a linear substitution can be expressed by the matrix identity $\mathbf{x} = C\mathbf{y}$, where C is an $n \times n$ matrix over $\mathbb{F}_q$ and $\mathbf{y}$ is the column vector of new indeterminates $y_1,\ldots,y_n$. If C is nonsingular, we speak of a *nonsingular linear substitution*.

6.19. Definition. For any finite field $\mathbb{F}_q$, two quadratic forms f and g over $\mathbb{F}_q$ are called *equivalent* if f can be transformed into g by means of a nonsingular linear substitution of indeterminates.

Equivalence of quadratic forms is easily seen to be an equivalence relation. Furthermore, if f and g are equivalent, then for any $b \in \mathbb{F}_q$ the equations $f(x_1,\ldots,x_n) = b$ and $g(x_1,\ldots,x_n) = b$ have the same number of solutions in $\mathbb{F}_q^n$, since the matrix C can be used to establish a one-to-one correspondence between the solution vectors. For odd q, the coefficient matrices A, B of two equivalent quadratic forms over $\mathbb{F}_q$ are related by $B = C^{\mathrm{T}}AC$, since $(C\mathbf{y})^{\mathrm{T}}A(C\mathbf{y}) = \mathbf{y}^{\mathrm{T}}(C^{\mathrm{T}}AC)\mathbf{y}$.

We study now in detail the case where q is odd. We shall show that every quadratic form over $\mathbb{F}_q$ is equivalent to a *diagonal quadratic form* $a_1x_1^2 + \cdots + a_nx_n^2$ over $\mathbb{F}_q$. We use the following terminology: the quadratic form f over $\mathbb{F}_q$ *represents* $a \in \mathbb{F}_q$ if the equation $f(x_1,\ldots,x_n) = a$ has a solution in $\mathbb{F}_q^n$.

6.20. Lemma. *If q is odd and the quadratic form $f \in \mathbb{F}_q[x_1,\ldots,x_n]$, $n \geqslant 2$, represents $a \in \mathbb{F}_q^*$, then f is equivalent to $ax_1^2 + g(x_2,\ldots,x_n)$, where g is a quadratic form over $\mathbb{F}_q$ in $n-1$ indeterminates.*

Proof. By hypothesis there exists $(c_1,\ldots,c_n) \in \mathbb{F}_q^n$ with $f(c_1,\ldots,c_n) = a$. Since $a \neq 0$, not all c_i are 0, and so we can find a nonsingular $n \times n$ matrix C over $\mathbb{F}_q$ for which the entries in the first column are $c_1,\ldots,c_n$. If we apply to f the linear substitution determined by C, we obtain a quadratic form in $y_1,\ldots,y_n$ for which the coefficient of y_1^2 is $f(c_1,\ldots,c_n) = a$. Thus f is equivalent to a quadratic form of the type

$$\begin{aligned} &ay_1^2 + 2b_2y_1y_2 + \cdots + 2b_ny_1y_n + h(y_2,\ldots,y_n) \\ &\quad = a\left(y_1 + b_2a^{-1}y_2 + \cdots + b_na^{-1}y_n\right)^2 + g(y_2,\ldots,y_n) \end{aligned}$$

with suitable $b_2, \ldots, b_n \in \mathbb{F}_q$ and quadratic forms h, g over $\mathbb{F}_q$. The nonsingular linear substitution $x_1 = y_1 + b_2 a^{-1} y_2 + \cdots + b_n a^{-1} y_n$, $x_2 = y_2, \ldots, x_n = y_n$ yields then a quadratic form of the desired type. □

6.21. Theorem. *Every quadratic form over $\mathbb{F}_q$, q odd, is equivalent to a diagonal quadratic form.*

Proof. We proceed by induction on the number n of indeterminates. If $n = 1$, then $f(x_1) = a_{11}x_1^2$ is already diagonal. Now let $n \geqslant 2$ and suppose the result holds for quadratic forms in $n-1$ indeterminates. Let $f(x_1, \ldots, x_n)$ be a quadratic form in n indeterminates. The theorem is true if f is the zero polynomial. If f is nonzero, either some $a_{ii} \neq 0$, in which case f represents $a_{ii} \neq 0$, or all $a_{ii} = 0$, but $a_{ij} = a_{ji} \neq 0$ for some $i \neq j$, in which case f represents $2a_{ij} \neq 0$ since $f(c_1, \ldots, c_n) = 2a_{ij}$ when $c_i = c_j = 1$ and $c_k = 0$ for $k \neq i, j$. At any rate, f represents then some element $a_1 \in \mathbb{F}_q^*$, so that f is equivalent to $a_1 x_1^2 + g(x_2, \ldots, x_n)$ by Lemma 6.20. By induction hypothesis, g is equivalent to a diagonal quadratic form $a_2 x_2^2 + \cdots + a_n x_n^2$, and so f is equivalent to $a_1 x_1^2 + a_2 x_2^2 + \cdots + a_n x_n^2$. □

If the quadratic form $f \in \mathbb{F}_q[x_1, \ldots, x_n]$ is equivalent to $a_1 x_1^2 + \cdots + a_n x_n^2$, then some of the a_i may be 0. Since multiplication of matrices by nonsingular matrices preserves ranks, equivalent quadratic forms have coefficient matrices of the same rank. In particular, the number of nonzero a_i in the diagonal quadratic form above is equal to the rank of the coefficient matrix A of f. If A is of rank n, we say that f is *nondegenerate*. Equivalently, we may define the *determinant* $\det(f)$ of f to be the determinant of A and call f nondegenerate if $\det(f) \neq 0$.

We note for later use that if f and g are equivalent, say by means of the linear substitution determined by the nonsingular matrix C, then

$$\det(g) = \det(f)\det(C)^2. \tag{6.3}$$

This follows, of course, from the identity $B = C^{\mathrm{T}} A C$ for the coefficient matrices A, B of f, g.

For a nonzero quadratic form $f \in \mathbb{F}_q[x_1, \ldots, x_n]$, a diagonal quadratic form equivalent to it can, without loss of generality, be written as $a_1 x_1^2 + \cdots + a_k x_k^2$, where $1 \leqslant k \leqslant n$ and all $a_i \neq 0$. Since for any $b \in \mathbb{F}_q$ the number of solutions of $a_1 x_1^2 + \cdots + a_k x_k^2 = b$ in $\mathbb{F}_q^n$ is q^{n-k} times the number of solutions of the same equation in $\mathbb{F}_q^k$, it suffices to consider the case where $k = n$—that is, where f is nondegenerate.

In the sequel, the function introduced below and some of its simple properties will turn out to be useful.

6.22. Definition. For any finite field $\mathbb{F}_q$ the integer-valued function v on $\mathbb{F}_q$ is defined by $v(b) = -1$ for $b \in \mathbb{F}_q^*$ and $v(0) = q - 1$.

6.23. Lemma. *For any finite field $\mathbb{F}_q$ we have*

$$\sum_{c \in \mathbb{F}_q} v(c) = 0, \tag{6.4}$$

and for any $b \in \mathbb{F}_q$,

$$\sum_{c_1 + \cdots + c_m = b} v(c_1) \cdots v(c_k) = \begin{cases} 0 & \text{if } 1 \leqslant k < m, \\ v(b) q^{m-1} & \text{if } k = m, \end{cases} \tag{6.5}$$

where the sum is over all $c_1, \ldots, c_m \in \mathbb{F}_q$ with $c_1 + \cdots + c_m = b$.

Proof. The identity (6.4) is trivial. Furthermore, for $1 \leqslant k < m$ we have

$$\begin{aligned} &\sum_{c_1 + \cdots + c_m = b} v(c_1) \cdots v(c_k) \\ &= \sum_{c_1, \ldots, c_k \in \mathbb{F}_q} v(c_1) \cdots v(c_k) \sum_{c_{k+1} + \cdots + c_m = b - c_1 - \cdots - c_k} 1 \\ &= q^{m-k-1} \sum_{c_1, \ldots, c_k \in \mathbb{F}_q} v(c_1) \cdots v(c_k) \\ &= q^{m-k-1} \left(\sum_{c_1 \in \mathbb{F}_q} v(c_1) \right) \cdots \left(\sum_{c_k \in \mathbb{F}_q} v(c_k) \right) = 0 \end{aligned}$$

by (6.4). If $k = m$ in (6.5), we use induction on m. The case $m = 1$ being trivial, suppose the formula is shown for some $m \geqslant 1$. Then by the first part of (6.5),

$$\begin{aligned} &\sum_{c_1 + \cdots + c_m + c_{m+1} = b} v(c_1) \cdots v(c_m) v(c_{m+1}) \\ &= \sum_{c_1 + \cdots + c_m + c_{m+1} = b} v(c_1) \cdots v(c_m) [v(c_{m+1}) + 1] \\ &= \sum_{c_1, \ldots, c_m \in \mathbb{F}_q} v(c_1) \cdots v(c_m) [v(b - c_1 - \cdots - c_m) + 1] \\ &= q \sum_{c_1 + \cdots + c_m = b} v(c_1) \cdots v(c_m). \end{aligned}$$

The last step is valid since the expression in square brackets is 0 unless $c_1 + \cdots + c_m = b$, when it has the value q. The rest follows from the induction hypothesis. □

From now on, we will often use $N(\cdots)$ to denote the number of solutions of the equation between parentheses in the underlying finite field, considering only the indeterminates actually written down. For instance, $N(a_1 x_1^2 + a_2 x_2^2 = b)$ refers to the number of solutions of the indicated equation in $\mathbb{F}_q^2$.

We resume the study of quadratic forms f over $\mathbb{F}_q$, q odd. As we have seen, it suffices to determine $N(f(x_1,\ldots,x_n)=b)$ for nondegenerate f. It is convenient to distinguish the cases of even and odd n. We consider first a special equation.

6.24. Lemma. *For odd q, let $b\in\mathbb{F}_q$, $a_1, a_2\in\mathbb{F}_q^*$, and η be the quadratic character of $\mathbb{F}_q$. Then*

$$N(a_1x_1^2+a_2x_2^2=b)=q+v(b)\eta(-a_1a_2).$$

Proof. With $c_1, c_2\in\mathbb{F}_q$ we obtain using (5.37),

$$\begin{aligned}
N(a_1x_1^2+a_2x_2^2=b) &= \sum_{c_1+c_2=b} N(a_1x_1^2=c_1)N(a_2x_2^2=c_2)\\
&= \sum_{c_1+c_2=b}\left[1+\eta(c_1a_1^{-1})\right]\left[1+\eta(c_2a_2^{-1})\right]\\
&= q+\eta(a_1)\sum_{c_1\in\mathbb{F}_q}\eta(c_1)+\eta(a_2)\sum_{c_2\in\mathbb{F}_q}\eta(c_2)\\
&\quad +\eta(a_1a_2)\sum_{c_1+c_2=b}\eta(c_1c_2)\\
&= q+\eta(a_1a_2)\sum_{c\in\mathbb{F}_q}\eta(bc-c^2).
\end{aligned}$$

Now the last sum is equal to $v(b)\eta(-1)$ by Theorem 5.48, and the result follows. □

6.25. Remark. The result above shows, in particular, that the equation $a_1x_1^2+a_2x_2^2=b$ always has a solution in $\mathbb{F}_q^2$. This can also be established by an easy counting argument. Let $S=\{a_1c_1^2: c_1\in\mathbb{F}_q\}$ with cardinality $|S|=(q+1)/2$ and $T=\{b-a_2c_2^2: c_2\in\mathbb{F}_q\}$ with $|T|=(q+1)/2$. Since $|S|+|T|>q$, S and T must have a common element c. Then $c=a_1c_1^2=b-a_2c_2^2$ for some $c_1, c_2\in\mathbb{F}_q$, hence $a_1c_1^2+a_2c_2^2=b$. Since for even q every element of $\mathbb{F}_q$ is a square, the remark is trivially valid for this case as well. □

6.26. Theorem. *Let f be a nondegenerate quadratic form over $\mathbb{F}_q$, q odd, in an even number n of indeterminates. Then for $b\in\mathbb{F}_q$ the number of solutions of the equation $f(x_1,\ldots,x_n)=b$ in $\mathbb{F}_q^n$ is*

$$q^{n-1}+v(b)q^{(n-2)/2}\eta\left((-1)^{n/2}\Delta\right),$$

where η is the quadratic character of $\mathbb{F}_q$ and $\Delta=\det(f)$.

Proof. Let $a_1x_1^2+\cdots+a_nx_n^2$ be a diagonal quadratic form equivalent to f. Since equivalence preserves the number of solutions, as well as the value of $\eta(\Delta)$ by (6.3), it suffices to establish the result for the equation $a_1x_1^2+\cdots+a_nx_n^2=b$, where all $a_i\neq 0$. With $m=n/2$ and $c_1,\ldots,c_m\in\mathbb{F}_q$

we get from Lemma 6.24 and (6.5),

$$
\begin{aligned}
N(a_1x_1^2 + &\cdots + a_nx_n^2 = b) \\
&= \sum_{c_1+\cdots+c_m=b} N(a_1x_1^2 + a_2x_2^2 = c_1)\cdots N(a_{n-1}x_{n-1}^2 + a_nx_n^2 = c_m) \\
&= \sum_{c_1+\cdots+c_m=b} [q + v(c_1)\eta(-a_1a_2)]\cdots[q + v(c_m)\eta(-a_{n-1}a_n)] \\
&= q^{m-1}q^m + \eta((-1)^m a_1\cdots a_n) \sum_{c_1+\cdots+c_m=b} v(c_1)\cdots v(c_m) \\
&= q^{n-1} + v(b)q^{(n-2)/2}\eta((-1)^{n/2}a_1\cdots a_n).
\end{aligned}
$$

□

6.27. Theorem. *Let f be a nondegenerate quadratic form over $\mathbb{F}_q$, q odd, in an odd number n of indeterminates. Then for $b \in \mathbb{F}_q$ the number of solutions of the equation $f(x_1,\ldots,x_n) = b$ in $\mathbb{F}_q^n$ is*

$$q^{n-1} + q^{(n-1)/2}\eta((-1)^{(n-1)/2}b\Delta),$$

where η is the quadratic character of $\mathbb{F}_q$ and $\Delta = \det(f)$.

Proof. As in Theorem 6.26, it suffices to establish the formula for the equation $a_1x_1^2 + \cdots + a_nx_n^2 = b$, where all $a_i \neq 0$. The formula is valid for $n = 1$. For $n \geqslant 3$ we apply Theorem 6.26 and obtain with $c_1, c_2 \in \mathbb{F}_q$,

$$
\begin{aligned}
N(a_1x_1^2 + &\cdots + a_nx_n^2 = b) \\
&= \sum_{c_1+c_2=b} N(a_1x_1^2 = c_1)N(a_2x_2^2 + \cdots + a_nx_n^2 = c_2) \\
&= \sum_{c_1+c_2=b} [1 + \eta(c_1a_1)] \\
&\quad \cdot [q^{n-2} + v(c_2)q^{(n-3)/2}\eta((-1)^{(n-1)/2}a_2\cdots a_n)] \\
&= q^{n-1} + q^{n-2}\eta(a_1)\sum_{c_1\in\mathbb{F}_q}\eta(c_1) \\
&\quad + q^{(n-3)/2}\eta((-1)^{(n-1)/2}a_2\cdots a_n)\sum_{c_2\in\mathbb{F}_q}v(c_2) \\
&\quad + q^{(n-3)/2}\eta((-1)^{(n-1)/2}a_1\cdots a_n)\sum_{c_1+c_2=b}\eta(c_1)v(c_2) \\
&= q^{n-1} + q^{(n-3)/2}\eta((-1)^{(n-1)/2}a_1\cdots a_n)\sum_{c\in\mathbb{F}_q}\eta(c)v(b-c),
\end{aligned}
$$

where we used (5.37) and (6.4) in the last step. Now

$$\sum_{c\in\mathbb{F}_q}\eta(c)v(b-c) = \sum_{c\in\mathbb{F}_q}\eta(c)[v(b-c)+1] = q\eta(b),$$

and the result follows □

We present now an *alternative proof* of Theorems 6.26 and 6.27 based on the method in the proof of Lemma 6.24 and properties of Jacobi sums. It suffices again to consider the equation $a_1x_1^2 + \cdots + a_nx_n^2 = b$, where all $a_i \neq 0$. Write ψ_0 for the trivial multiplicative character of $\mathbb{F}_q$, $\psi_1 = \eta$, and $N = N(a_1x_1^2 + \cdots + a_nx_n^2 = b)$. Then with $c_1,\ldots,c_n \in \mathbb{F}_q$,

$$\begin{aligned} N &= \sum_{c_1 + \cdots + c_n = b} N(a_1x_1^2 = c_1)\cdots N(a_nx_n^2 = c_n) \\ &= \sum_{c_1 + \cdots + c_n = b} [1 + \psi_1(c_1a_1^{-1})]\cdots[1 + \psi_1(c_na_n^{-1})] \\ &= \sum_{c_1 + \cdots + c_n = b} [\psi_0(c_1a_1) + \psi_1(c_1a_1)]\cdots[\psi_0(c_na_n) + \psi_1(c_na_n)] \\ &= \sum_{c_1 + \cdots + c_n = b}\ \sum_{i_1,\ldots,i_n = 0}^{1} \psi_{i_1}(c_1a_1)\cdots\psi_{i_n}(c_na_n) \\ &= \sum_{i_1,\ldots,i_n = 0}^{1} \psi_{i_1}(a_1)\cdots\psi_{i_n}(a_n) \sum_{c_1 + \cdots + c_n = b} \psi_{i_1}(c_1)\cdots\psi_{i_n}(c_n) \\ &= \sum_{i_1,\ldots,i_n = 0}^{1} \psi_{i_1}(a_1)\cdots\psi_{i_n}(a_n) J_b(\psi_{i_1},\ldots,\psi_{i_n}). \end{aligned}$$

By (5.38), (5.39), and (5.40), only the two terms with $(i_1,\ldots,i_n) = (0,\ldots,0)$ and $(1,\ldots,1)$ remain, yielding the formula

$$N = q^{n-1} + \eta(a_1\cdots a_n)J_b(\eta,\ldots,\eta), \tag{6.6}$$

where there are n copies of η in the Jacobi sum.

For $b \neq 0$ we use (5.38) to get

$$N = q^{n-1} + \eta(a_1\cdots a_n)\eta^n(b)J(\eta,\ldots,\eta). \tag{6.7}$$

If n is even, we can apply (5.44) and Theorem 5.12(iv) and obtain with a nontrivial additive character χ of $\mathbb{F}_q$,

$$\begin{aligned} J(\eta,\ldots,\eta) &= -\frac{1}{q}G(\eta,\chi)^n = -\frac{1}{q}\left[G(\eta,\chi)^2\right]^{n/2} = -\frac{1}{q}[\eta(-1)q]^{n/2} \\ &= -q^{(n-2)/2}\eta\left((-1)^{n/2}\right), \end{aligned}$$

hence from (6.7),

$$N = q^{n-1} - q^{(n-2)/2}\eta\left((-1)^{n/2}a_1\cdots a_n\right),$$

which agrees with Theorem 6.26. If n is odd, then (5.43) and Theorem 5.12(iv) yield

$$\begin{aligned} J(\eta,\ldots,\eta) &= G(\eta,\chi)^{n-1} = \left[G(\eta,\chi)^2\right]^{(n-1)/2} = [\eta(-1)q]^{(n-1)/2} \\ &= q^{(n-1)/2}\eta\left((-1)^{(n-1)/2}\right), \end{aligned} \tag{6.8}$$

hence from (6.7),

$$N=q^{n-1}+q^{(n-1)/2}\eta\left((-1)^{(n-1)/2}ba_1\cdots a_n\right),$$

which agrees with Theorem 6.27.

Now consider $b=0$. If n is even, we can apply (5.42) to get

$$J_0(\eta,\ldots,\eta)=(q-1)\eta(-1)J(\eta,\ldots,\eta),$$

where there are $n-1$ copies of η in the last Jacobi sum. Thus by (6.8),

$$J_0(\eta,\ldots,\eta)=(q-1)q^{(n-2)/2}\eta\left((-1)^{n/2}\right),$$

and so (6.6) implies that

$$N=q^{n-1}+(q-1)q^{(n-2)/2}\eta\left((-1)^{n/2}a_1\cdots a_n\right),$$

which agrees with Theorem 6.26. If n is odd, then $J_0(\eta,\ldots,\eta)=0$ by (5.41), hence (6.6) yields $N=q^{n-1}$, which agrees with Theorem 6.27.

6.28. Remark. For odd q we can also determine the number of solutions in $\mathbb{F}_q^n$ of the equation $h(x_1,\ldots,x_n)=b$, $b\in\mathbb{F}_q$, where h is a polynomial over $\mathbb{F}_q$ of degree 2 (not necessarily a quadratic form). We have $h=f+g$ with a quadratic form f and $\deg(g)\leqslant 1$. By carrying out a nonsingular linear substitution that transforms f into an equivalent diagonal quadratic form, we obtain an equation

$$a_1x_1^2+\cdots+a_kx_k^2+b_1x_1+\cdots+b_nx_n=b\quad (1\leqslant k\leqslant n,\text{ all } a_i\neq 0)$$

which has the same number of solutions as the original one. If $k<n$, we can assume without loss of generality that $b_n\neq 0$. Then the number of solutions is q^{n-1}, since we can substitute arbitrary elements of $\mathbb{F}_q$ for $x_1,\ldots,x_{n-1}$ and the value of x_n is then uniquely determined. If $k=n$, the substitution $x_i=y_i-b_i(2a_i)^{-1}$, $1\leqslant i\leqslant n$, yields $a_1y_1^2+\cdots+a_ny_n^2=c$ for some $c\in\mathbb{F}_q$ and preserves the number of solutions. Theorems 6.26 and 6.27 now give the desired information. □

We consider now $N(f(x_1,\ldots,x_n)=b)$ for a quadratic form f over a finite field of characteristic 2. We use the same strategy as before—namely, to reduce f to an equivalent quadratic form of simpler type. The quadratic form f in n indeterminates is called *nondegenerate* if f is not equivalent to a quadratic form in fewer than n indeterminates (if this definition were applied to the case of odd q, it would result in the same concept as that introduced earlier). If suffices again to discuss the case where f is nondegenerate.

6.29. Lemma. *A nondegenerate quadratic form $f\in\mathbb{F}_q[x_1,\ldots,x_n]$, q even, $n\geqslant 3$, is equivalent to $x_1x_2+g(x_3,\ldots,x_n)$, where g is a nondegenerate quadratic form over $\mathbb{F}_q$ in $n-2$ indeterminates.*

Proof. We show first that f is equivalent to a quadratic form in which the coefficient of x_1^2 is 0. We write

$$f(x_1,\ldots,x_n)=\sum_{1\leqslant i\leqslant j\leqslant n} a_{ij}x_ix_j. \tag{6.9}$$

If some $a_{ii}=0$, then by renaming the indeterminates we get $a_{11}=0$. Thus we can assume that all $a_{ii}\neq 0$. If we had $a_{ij}=0$ for all $i<j$, then

$$f(x_1,\ldots,x_n)=a_{11}x_1^2+\cdots+a_{nn}x_n^2=\left(a_{11}^{q/2}x_1+\cdots+a_{nn}^{q/2}x_n\right)^2,$$

which is equivalent to a quadratic form in one indeterminate, a contradiction. Hence, by naming the indeterminates suitably, we can assume $a_{23}\neq 0$. Separating the terms of f involving x_2, we write

$$\begin{aligned} f(x_1,\ldots,x_n)&=a_{22}x_2^2+x_2(a_{12}x_1+a_{23}x_3+\cdots+a_{2n}x_n)\\ &\quad+g_1(x_1,x_3,\ldots,x_n)\end{aligned}$$

and then we carry out the nonsingular linear substitution

$$\begin{aligned} x_3&=a_{23}^{-1}(a_{12}y_1+y_3+a_{24}y_4+\cdots+a_{2n}y_n),\\ x_i&=y_i \text{ for } i\neq 3,\end{aligned}$$

which yields

$$a_{22}y_2^2+y_2y_3+g_2(y_1,y_3,\ldots,y_n).$$

Now with b_{11} being the coefficient of y_1^2 in g_2, we apply the nonsingular linear substitution

$$y_2=\left(a_{22}^{-1}b_{11}\right)^{q/2}z_1+z_2,\qquad y_i=z_i\quad \text{for } i\neq 2.$$

The coefficient of z_1^2 in the resulting quadratic form is then 0.

Now let f be as in (6.9) with $a_{11}=0$. Since f is nondegenerate, not all a_{1j} can be 0, and so we may assume that $a_{12}\neq 0$. The nonsingular linear substitution

$$\begin{aligned} x_2&=a_{12}^{-1}(y_2+a_{13}y_3+\cdots+a_{1n}y_n),\\ x_i&=y_i \text{ for } i\neq 2,\end{aligned}$$

transforms f into a quadratic form of the type

$$y_1y_2+\sum_{2\leqslant i\leqslant j\leqslant n} c_{ij}y_iy_j.$$

The nonsingular linear substitution

$$\begin{aligned} y_1&=z_1+c_{22}z_2+\cdots+c_{2n}z_n,\\ y_i&=z_i\quad \text{for } i\neq 1,\end{aligned}$$

then yields an equivalent quadratic form $z_1z_2+g(z_3,\ldots,z_n)$, where g must clearly be nondegenerate. □

6.30. Theorem. *Let $f \in \mathbb{F}_q[x_1,\ldots,x_n]$, q even, be a nondegenerate quadratic form. If n is odd, then f is equivalent to*

$$x_1x_2 + x_3x_4 + \cdots + x_{n-2}x_{n-1} + x_n^2.$$

If n is even, then f is either equivalent to

$$x_1x_2 + x_3x_4 + \cdots + x_{n-1}x_n$$

or to a quadratic form of the type

$$x_1x_2 + x_3x_4 + \cdots + x_{n-1}x_n + x_{n-1}^2 + ax_n^2,$$

where $a \in \mathbb{F}_q$ satisfies $\mathrm{Tr}_{\mathbb{F}_q}(a) = 1$.

Proof. If n is odd, then using induction on n and Lemma 6.29 one shows that f is equivalent to a quadratic form of the type $x_1x_2 + x_3x_4 + \cdots + x_{n-2}x_{n-1} + ax_n^2$ with $a \in \mathbb{F}_q^*$. Replacing x_n by $a^{-q/2}x_n$, one obtains the desired quadratic form.

If n is even, then using induction on n and Lemma 6.29 one shows that f is equivalent to a quadratic form of the type

$$x_1x_2 + x_3x_4 + \cdots + x_{n-3}x_{n-2} + bx_{n-1}^2 + cx_{n-1}x_n + dx_n^2$$

with $b, c, d \in \mathbb{F}_q$. Since f is nondegenerate, we must have $c \neq 0$, for otherwise the identity

$$bx_{n-1}^2 + dx_n^2 = \left(b^{q/2}x_{n-1} + d^{q/2}x_n\right)^2$$

would enable us to find an equivalent quadratic form in fewer than n indeterminates. If $b = 0$, then

$$cx_{n-1}x_n + dx_n^2 = (cx_{n-1} + dx_n)x_n$$

is equivalent to $x_{n-1}x_n$, and we are done. If $b \neq 0$, then replacing x_{n-1} by $b^{-q/2}x_{n-1}$ and x_n by $b^{q/2}c^{-1}x_n$, we see that $bx_{n-1}^2 + cx_{n-1}x_n + dx_n^2$ is equivalent to $x_{n-1}^2 + x_{n-1}x_n + ax_n^2$ for some $a \in \mathbb{F}_q$. In case the polynomial $x^2 + x + a$ is reducible in $\mathbb{F}_q[x]$, we have

$$x^2 + x + a = (x + c_1)(x + c_2)$$

for some $c_1, c_2 \in \mathbb{F}_q$, hence

$$x_{n-1}^2 + x_{n-1}x_n + ax_n^2 = (x_{n-1} + c_1x_n)(x_{n-1} + c_2x_n)$$

is equivalent to $x_{n-1}x_n$. If $x^2 + x + a$ is irreducible in $\mathbb{F}_q[x]$, then we have $\mathrm{Tr}_{\mathbb{F}_q}(a) = 1$ by Corollary 3.79, and the result is established in all cases. □

Because of the invariance of $N(f(x_1,\ldots,x_n) = b)$ under equivalence, one may restrict the attention to the special quadratic forms in Theorem 6.30. The discussion of a particular case is needed first. We use again the function v introduced in Definition 6.22.

6.31. Lemma. *For even q, let $a \in \mathbb{F}_q$ with $\operatorname{Tr}_{\mathbb{F}_q}(a) = 1$ and $b \in \mathbb{F}_q$. Then*

$$N\left(x_1^2 + x_1x_2 + ax_2^2 = b\right) = q - v(b).$$

Proof. Since $x^2 + x + a$ is irreducible in $\mathbb{F}_q[x]$ by Corollary 3.79, we have

$$x^2 + x + a = (x + \alpha)(x + \alpha^q)$$

with $\alpha \in \mathbb{F}_{q^2}$, $\alpha \notin \mathbb{F}_q$, and so

$$f(x_1, x_2) = x_1^2 + x_1x_2 + ax_2^2 = (x_1 + \alpha x_2)(x_1 + \alpha^q x_2).$$

For $(c_1, c_2) \in \mathbb{F}_q^2$ we get then

$$f(c_1, c_2) = (c_1 + \alpha c_2)(c_1 + \alpha^q c_2) = (c_1 + \alpha c_2)(c_1 + \alpha c_2)^q = (c_1 + \alpha c_2)^{q+1}.$$

Now $\{1, \alpha\}$ is a basis of $\mathbb{F}_{q^2}$ over $\mathbb{F}_q$ and thus the ordered pairs (c_1, c_2) are in one-to-one correspondence with the elements $\gamma = c_1 + \alpha c_2 \in \mathbb{F}_{q^2}$. Hence $N(f(x_1, x_2) = b)$ is equal to the number of $\gamma \in \mathbb{F}_{q^2}$ with $\gamma^{q+1} = b$. Therefore,

$$N(f(x_1, x_2) = 0) = 1 = q - v(0).$$

If $b \neq 0$, then since $\mathbb{F}_{q^2}^*$ is cyclic and $b^{(q^2-1)/(q+1)} = b^{q-1} = 1$, there are $q + 1$ elements $\gamma \in \mathbb{F}_{q^2}$ with $\gamma^{q+1} = b$. Hence, $N(f(x_1, x_2) = b) = q + 1 = q - v(b)$. □

6.32. Theorem. *Let $\mathbb{F}_q$ be a finite field with q even and let $b \in \mathbb{F}_q$. Then for odd n, the number of solutions of the equation*

$$x_1x_2 + x_3x_4 + \cdots + x_{n-2}x_{n-1} + x_n^2 = b$$

in $\mathbb{F}_q^n$ is q^{n-1}. For even n, the number of solutions of the equation

$$x_1x_2 + x_3x_4 + \cdots + x_{n-1}x_n = b$$

in $\mathbb{F}_q^n$ is $q^{n-1} + v(b)q^{(n-2)/2}$. For even n and $a \in \mathbb{F}_q$ with $\operatorname{Tr}_{\mathbb{F}_q}(a) = 1$, the number of solutions of the equation

$$x_1x_2 + x_3x_4 + \cdots + x_{n-1}x_n + x_{n-1}^2 + ax_n^2 = b$$

in $\mathbb{F}_q^n$ is $q^{n-1} - v(b)q^{(n-2)/2}$.

Proof. Since the equation $x^2 = c$ has a unique solution in $\mathbb{F}_q$ for any $c \in \mathbb{F}_q$, we have

$$N\left(x_1x_2 + x_3x_4 + \cdots + x_{n-2}x_{n-1} + x_n^2 = b\right) = q^{n-1}$$

for odd n because we can assign arbitrary values to $x_1, \ldots, x_{n-1}$ and the value of x_n is then uniquely determined.

Next, we note that $N(x_1x_2 = b)$ is $q - 1$ if $b \neq 0$ and $2q - 1$ if $b = 0$, and so $N(x_1x_2 = b) = q + v(b)$ in both cases. Now for even n, say $n = 2m$, we have with $c_1,\ldots,c_m \in \mathbb{F}_q$,

$$\begin{aligned} N(x_1x_2 + x_3x_4 + \cdots + x_{n-1}x_n = b) \\ &= \sum_{c_1+\cdots+c_m=b} N(x_1x_2 = c_1)\cdots N(x_{n-1}x_n = c_m) \\ &= \sum_{c_1+\cdots+c_m=b} [q+v(c_1)]\cdots[q+v(c_m)] \\ &= q^{m-1}q^m + \sum_{c_1+\cdots+c_m=b} v(c_1)\cdots v(c_m) = q^{n-1} + v(b)q^{(n-2)/2}, \end{aligned}$$

where we used (6.5).

In the remaining case, the formula for the number of solutions is valid for $n = 2$ by Lemma 6.31. For $n \geqslant 4$ we use the result of the previous case and Lemma 6.31 to get with $c_1, c_2 \in \mathbb{F}_q$,

$$\begin{aligned} N\big(x_1x_2 + x_3x_4 + \cdots + x_{n-1}x_n + x_{n-1}^2 + ax_n^2 = b\big) \\ &= \sum_{c_1+c_2=b} N(x_1x_2 + \cdots + x_{n-3}x_{n-2} = c_1) \\ &\quad \cdot N\big(x_{n-1}x_n + x_{n-1}^2 + ax_n^2 = c_2\big) \\ &= \sum_{c_1+c_2=b} \big[q^{n-3} + v(c_1)q^{(n-4)/2}\big]\big[q - v(c_2)\big] \\ &= q^{n-1} + q^{(n-2)/2} \sum_{c_1\in\mathbb{F}_q} v(c_1) - q^{n-3} \sum_{c_2\in\mathbb{F}_q} v(c_2) \\ &\quad - q^{(n-4)/2} \sum_{c_1+c_2=b} v(c_1)v(c_2) \\ &= q^{n-1} - v(b)q^{(n-2)/2}, \end{aligned}$$

where we applied (6.4) and (6.5) in the last step. □

3. DIAGONAL EQUATIONS

In the preceding section we were led to the study of equations involving diagonal quadratic forms. These equations are a special case of the general family of diagonal equations. A *diagonal equation* (over $\mathbb{F}_q$) is an equation of the type

$$a_1x_1^{k_1} + \cdots + a_nx_n^{k_n} = b \tag{6.10}$$

with positive integers $k_1,\ldots,k_n$, coefficients $a_1,\ldots,a_n \in \mathbb{F}_q^*$, and $b \in \mathbb{F}_q$.

The number of solutions

$$N = N\big(a_1x_1^{k_1} + \cdots + a_nx_n^{k_n} = b\big)$$

of (6.10) in $\mathbb{F}_q^n$ can be expressed in terms of Jacobi sums. With $c_1,\ldots,c_n \in \mathbb{F}_q$ we have

$$\begin{aligned} N &= \sum_{c_1+\cdots+c_n=b} N(a_1x_1^{k_1}=c_1)\cdots N(a_nx_n^{k_n}=c_n) \\ &= \sum_{c_1+\cdots+c_n=b} N(x_1^{k_1}=a_1^{-1}c_1)\cdots N(x_n^{k_n}=a_n^{-1}c_n). \end{aligned}$$

From (5.70) we get

$$N(x^k=c)=\sum_{j=0}^{d-1}\lambda^j(c),$$

where λ is a multiplicative character of $\mathbb{F}_q$ of order $d=\gcd(k,q-1)$. For $i=1,\ldots,n$ let $d_i=\gcd(k_i,q-1)$ and λ_i a multiplicative character of $\mathbb{F}_q$ of order d_i. Then

$$\begin{aligned} N &= \sum_{c_1+\cdots+c_n=b}\left(\sum_{j_1=0}^{d_1-1}\lambda_1^{j_1}(a_1^{-1}c_1)\right)\cdots\left(\sum_{j_n=0}^{d_n-1}\lambda_n^{j_n}(a_n^{-1}c_n)\right) \\ &= \sum_{j_1=0}^{d_1-1}\cdots\sum_{j_n=0}^{d_n-1}\lambda_1^{j_1}(a_1^{-1})\cdots\lambda_n^{j_n}(a_n^{-1})\sum_{c_1+\cdots+c_n=b}\lambda_1^{j_1}(c_1)\cdots\lambda_n^{j_n}(c_n) \\ &= \sum_{j_1=0}^{d_1-1}\cdots\sum_{j_n=0}^{d_n-1}\bar{\lambda}_1^{j_1}(a_1)\cdots\bar{\lambda}_n^{j_n}(a_n)J_b(\lambda_1^{j_1},\ldots,\lambda_n^{j_n}). \end{aligned}$$

If $(j_1,\ldots,j_n)=(0,\ldots,0)$, then $J_b(\lambda_1^{j_1},\ldots,\lambda_n^{j_n})=q^{n-1}$ by (5.38) and (5.39). If some, but not all of the j_i are 0, then $J_b(\lambda_1^{j_1},\ldots,\lambda_n^{j_n})=0$ by (5.38) and (5.40). Therefore,

$$N=q^{n-1}+\sum_{j_1=1}^{d_1-1}\cdots\sum_{j_n=1}^{d_n-1}\bar{\lambda}_1^{j_1}(a_1)\cdots\bar{\lambda}_n^{j_n}(a_n)J_b(\lambda_1^{j_1},\ldots,\lambda_n^{j_n}). \tag{6.11}$$

We distinguish now the cases $b=0$ and $b\neq 0$. If $b=0$, then (5.41) shows that $J_0(\lambda_1^{j_1},\ldots,\lambda_n^{j_n})=0$ whenever $\lambda_1^{j_1}\cdots\lambda_n^{j_n}$ is nontrivial. We therefore arrive at the following result.

6.33. Theorem. *The number N of solutions of the diagonal equation $a_1x_1^{k_1}+\cdots+a_nx_n^{k_n}=0$ in $\mathbb{F}_q^n$ is given by*

$$N=q^{n-1}+\sum_{(j_1,\ldots,j_n)\in T}\bar{\lambda}_1^{j_1}(a_1)\cdots\bar{\lambda}_n^{j_n}(a_n)J_0(\lambda_1^{j_1},\ldots,\lambda_n^{j_n}),$$

where T is the set of all $(j_1,\ldots,j_n)\in\mathbb{Z}^n$ such that $1\leqslant j_i\leqslant d_i-1$ for $1\leqslant i\leqslant n$ and $\lambda_1^{j_1}\cdots\lambda_n^{j_n}$ is trivial, and where λ_i is a multiplicative character of $\mathbb{F}_q$ of order $d_i=\gcd(k_i,q-1)$.

For $b\neq 0$ we apply (5.38) to get $J_b(\lambda_1^{j_1},\ldots,\lambda_n^{j_n})=(\lambda_1^{j_1}\cdots\lambda_n^{j_n})(b)$ $J(\lambda_1^{j_1},\ldots,\lambda_n^{j_n})$, and this leads to the following result.

6.34. Theorem. *For $b \in \mathbb{F}_q^*$ the number N of solutions of the diagonal equation $a_1x_1^{k_1} + \cdots + a_nx_n^{k_n} = b$ in $\mathbb{F}_q^n$ is given by*

$$N = q^{n-1} + \sum_{j_1=1}^{d_1-1} \cdots \sum_{j_n=1}^{d_n-1} \lambda_1^{j_1}(ba_1^{-1}) \cdots \lambda_n^{j_n}(ba_n^{-1}) J(\lambda_1^{j_1}, \ldots, \lambda_n^{j_n}),$$

where λ_i is a multiplicative character of $\mathbb{F}_q$ of order $d_i = \gcd(k_i, q-1)$.

6.35. Remark. The formulas above show that N does not depend directly on the exponents k_i, but only on the greatest common divisors d_i. One may therefore assume, without loss of generality, that the exponents in (6.10) are divisors of $q-1$. □

Estimates for N can be deduced easily from Theorems 6.33 and 6.34. For positive integers $d_1, \ldots, d_n$ let $M(d_1, \ldots, d_n)$ denote the number of n-tuples $(j_1, \ldots, j_n) \in \mathbb{Z}^n$ such that $1 \leqslant j_i \leqslant d_i - 1$ for $1 \leqslant i \leqslant n$ and $(j_1/d_1) + \cdots + (j_n/d_n) \in \mathbb{Z}$.

6.36. Theorem. *The number N of solutions of the diagonal equation $a_1x_1^{k_1} + \cdots + a_nx_n^{k_n} = 0$ in $\mathbb{F}_q^n$ satisfies*

$$|N - q^{n-1}| \leqslant M(d_1, \ldots, d_n)(q-1)q^{(n-2)/2},$$

where $d_i = \gcd(k_i, q-1)$ for $1 \leqslant i \leqslant n$.

Proof. Corollary 5.23 and the formula in Theorem 6.33 yield

$$|N - q^{n-1}| \leqslant |T|(q-1)q^{(n-2)/2}.$$

The group of multiplicative characters of $\mathbb{F}_q$ is cyclic by Corollary 5.9. Let λ be a generator of this group. Then we can take $\lambda_i = \lambda^{(q-1)/d_i}$, $1 \leqslant i \leqslant n$, in Theorem 6.33. Furthermore,

$$\lambda_1^{j_1} \cdots \lambda_n^{j_n} = \lambda^{(j_1(q-1)/d_1) + \cdots + (j_n(q-1)/d_n)}$$

is trivial if and only if $(j_1/d_1) + \cdots + (j_n/d_n) \in \mathbb{Z}$, and so $|T| = M(d_1, \ldots, d_n)$. □

6.37. Theorem. *For $b \in \mathbb{F}_q^*$ the number N of solutions of the diagonal equation $a_1x_1^{k_1} + \cdots + a_nx_n^{k_n} = b$ in $\mathbb{F}_q^n$ satisfies*

$$|N - q^{n-1}| \leqslant \left[(d_1-1)\cdots(d_n-1) - (1 - q^{-1/2})M(d_1, \ldots, d_n)\right]q^{(n-1)/2},$$

where $d_i = \gcd(k_i, q-1)$ for $1 \leqslant i \leqslant n$.

Proof. Theorem 5.22 and the formula in Theorem 6.34 yield

$$\begin{aligned}|N - q^{n-1}| &\leqslant \left[(d_1-1)\cdots(d_n-1) - |T|\right]q^{(n-1)/2} + |T|q^{(n-2)/2}\\ &= \left[(d_1-1)\cdots(d_n-1) - (1-q^{-1/2})|T|\right]q^{(n-1)/2}.\end{aligned}$$

It remains to note that $|T| = M(d_1, \ldots, d_n)$ from the proof of Theorem 6.36. □

It is trivial that $M(d_1, \ldots, d_n) \leqslant (d_1-1)\cdots(d_n-1)$. We can have $M(d_1, \ldots, d_n) = 0$, for instance if one of the d_i is relatively prime to all the others. To prove this, suppose for the sake of concreteness that $\gcd(d_1, d_i) = 1$ for $2 \leqslant i \leqslant n$. If for some $(j_1,\ldots,j_n) \in \mathbb{Z}^n$ with $1 \leqslant j_i \leqslant d_i - 1$ for $1 \leqslant i \leqslant n$ we had $(j_1/d_1) + \cdots + (j_n/d_n) = m \in \mathbb{Z}$, then $j_1 d_2 \cdots d_n + j d_1 = m d_1 d_2 \cdots d_n$ for some $j \in \mathbb{Z}$, hence $j_1 d_2 \cdots d_n \equiv 0 \bmod d_1$. But $\gcd(d_1, d_2 \cdots d_n) = 1$, thus $j_1 \equiv 0 \bmod d_1$, a contradiction. If $M(d_1,\ldots,d_n) = 0$, then we have of course $N = q^{n-1}$ in Theorem 6.36.

A general formula for $M(d_1,\ldots,d_n)$ may be obtained as follows. Put $D = \operatorname{lcm}(d_1,\ldots,d_n)$ and $e(t) = e^{2\pi i t}$ for $t \in \mathbb{R}$, and observe that

$$\frac{1}{D}\sum_{h=0}^{D-1} e\left(h\left(\frac{j_1}{d_1} + \cdots + \frac{j_n}{d_n}\right)\right) = \begin{cases} 1 & \text{if } \dfrac{j_1}{d_1} + \cdots + \dfrac{j_n}{d_n} \in \mathbb{Z}, \\ 0 & \text{otherwise.} \end{cases}$$

Then

$$\begin{aligned} M(d_1,\ldots,d_n) &= \sum_{j_1=1}^{d_1-1} \cdots \sum_{j_n=1}^{d_n-1} \frac{1}{D} \sum_{h=0}^{D-1} e\left(h\left(\frac{j_1}{d_1} + \cdots + \frac{j_n}{d_n}\right)\right) \\ &= \frac{1}{D}\sum_{h=0}^{D-1}\left(\sum_{j_1=1}^{d_1-1} e\left(j_1\frac{h}{d_1}\right)\right)\cdots\left(\sum_{j_n=1}^{d_n-1} e\left(j_n\frac{h}{d_n}\right)\right) \\ &= \frac{1}{D}\sum_{h=0}^{D-1}\left(\sum_{j_1=0}^{d_1-1} e\left(j_1\frac{h}{d_1}\right)-1\right)\cdots\left(\sum_{j_n=0}^{d_n-1} e\left(j_n\frac{h}{d_n}\right)-1\right) \\ &= \frac{1}{D}\sum_{h=0}^{D-1}\left[(-1)^n + \sum_{r=1}^{n}(-1)^{n-r}\sum_{1\leqslant i_1<i_2<\cdots<i_r\leqslant n}\right. \\ &\qquad \left.\prod_{s=1}^{r}\left(\sum_{j_{i_s}=0}^{d_{i_s}-1} e\left(j_{i_s}\frac{h}{d_{i_s}}\right)\right)\right]. \end{aligned}$$

Now for $d \in \mathbb{N}$,

$$\sum_{j=0}^{d-1} e\left(j\frac{h}{d}\right) = \begin{cases} d & \text{if } h \equiv 0 \bmod d, \\ 0 & \text{otherwise,} \end{cases}$$

and so the product in the last expression for $M(d_1,\ldots,d_n)$ is equal to $d_{i_1}\cdots d_{i_r}$ if h is divisible by $\operatorname{lcm}(d_{i_1},\ldots,d_{i_r})$ and equal to 0 otherwise. It follows that

$$\begin{aligned} M(d_1,\ldots,d_n) &= (-1)^n + \frac{1}{D}\sum_{h=0}^{D-1}\sum_{r=1}^{n}(-1)^{n-r}\sum_{\substack{1\leqslant i_1<i_2<\cdots<i_r\leqslant n \\ h\equiv 0 \bmod \operatorname{lcm}(d_{i_1},\ldots,d_{i_r})}} d_{i_1}\cdots d_{i_r} \\ &= (-1)^n + \frac{1}{D}\sum_{r=1}^{n}(-1)^{n-r}\sum_{1\leqslant i_1<i_2<\cdots<i_r\leqslant n} d_{i_1}\cdots d_{i_r} \\ &\qquad \sum_{\substack{h=0 \\ h\equiv 0 \bmod \operatorname{lcm}(d_{i_1},\ldots,d_{i_r})}}^{D-1} 1. \end{aligned}$$

The innermost sum is equal to $D/\operatorname{lcm}(d_{i_1},\ldots,d_{i_r})$, and so we obtain

$$M(d_1,\ldots,d_n)=(-1)^n+\sum_{r=1}^{n}(-1)^{n-r}\sum_{1\leqslant i_1<i_2<\cdots<i_r\leqslant n}\frac{d_{i_1}\cdots d_{i_r}}{\operatorname{lcm}(d_{i_1},\ldots,d_{i_r})}. \tag{6.12}$$

We present now an *alternative approach* to the evaluation of the number N of solutions of (6.10) in $\mathbb{F}_q^n$, based on Gaussian sums. By (5.5) we have

$$N=\frac{1}{q}\sum_{c_1,\ldots,c_n\in\mathbb{F}_q}\sum_{\chi}\chi\left(a_1c_1^{k_1}+\cdots+a_nc_n^{k_n}\right)\bar{\chi}(b),$$

where the inner sum is over all additive characters χ of $\mathbb{F}_q$. Rearranging and separating the contribution from the trivial character χ_0, we get

$$\begin{aligned}N&=\frac{1}{q}\sum_{\chi}\bar{\chi}(b)\sum_{c_1,\ldots,c_n\in\mathbb{F}_q}\chi\left(a_1c_1^{k_1}\right)\cdots\chi\left(a_nc_n^{k_n}\right)\\&=q^{n-1}+\frac{1}{q}\sum_{\chi\neq\chi_0}\bar{\chi}(b)\left(\sum_{c_1\in\mathbb{F}_q}\chi\left(a_1c_1^{k_1}\right)\right)\cdots\left(\sum_{c_n\in\mathbb{F}_q}\chi\left(a_nc_n^{k_n}\right)\right).\end{aligned}$$

An application of Theorem 5.30 yields

$$\begin{aligned}N&=q^{n-1}+\frac{1}{q}\sum_{\chi\neq\chi_0}\bar{\chi}(b)\left(\sum_{j_1=1}^{d_1-1}\bar{\lambda}_1^{j_1}(a_1)G\left(\lambda_1^{j_1},\chi\right)\right)\cdots\left(\sum_{j_n=1}^{d_n-1}\bar{\lambda}_n^{j_n}(a_n)G\left(\lambda_n^{j_n},\chi\right)\right)\\&=q^{n-1}+\frac{1}{q}\sum_{j_1=1}^{d_1-1}\cdots\sum_{j_n=1}^{d_n-1}\bar{\lambda}_1^{j_1}(a_1)\cdots\bar{\lambda}_n^{j_n}(a_n)\sum_{\chi\neq\chi_0}\bar{\chi}(b)G\left(\lambda_1^{j_1},\chi\right)\cdots G\left(\lambda_n^{j_n},\chi\right),\end{aligned}$$

where λ_i is, as before, a multiplicative character of $\mathbb{F}_q$ of order $d_i=\gcd(k_i,q-1)$. For the inner sum we use Theorem 5.12(i) to obtain

$$\begin{aligned}&\sum_{\chi\neq\chi_0}\bar{\chi}(b)G\left(\lambda_1^{j_1},\chi\right)\cdots G\left(\lambda_n^{j_n},\chi\right)\\&\quad=\sum_{a\in\mathbb{F}_q^*}\bar{\chi}_a(b)G\left(\lambda_1^{j_1},\chi_a\right)\cdots G\left(\lambda_n^{j_n},\chi_a\right)\\&\quad=G\left(\lambda_1^{j_1},\chi_1\right)\cdots G\left(\lambda_n^{j_n},\chi_1\right)\sum_{a\in\mathbb{F}_q^*}\bar{\chi}_b(a)\bar{\lambda}_1^{j_1}(a)\cdots\bar{\lambda}_n^{j_n}(a)\\&\quad=G\left(\lambda_1^{j_1},\chi_1\right)\cdots G\left(\lambda_n^{j_n},\chi_1\right)G\left(\bar{\lambda}_1^{j_1}\cdots\bar{\lambda}_n^{j_n},\bar{\chi}_b\right),\end{aligned}$$

and so

$$N=q^{n-1}+\frac{1}{q}\sum_{j_1=1}^{d_1-1}\cdots\sum_{j_n=1}^{d_n-1}G\left(\lambda_1^{j_1},\chi_{a_1}\right)\cdots G\left(\lambda_n^{j_n},\chi_{a_n}\right)G\left(\bar{\lambda}_1^{j_1}\cdots\bar{\lambda}_n^{j_n},\bar{\chi}_b\right). \tag{6.13}$$

For $b=0$ we can apply (5.14) to get

$$N=q^{n-1}+\left(1-\frac{1}{q}\right)\sum_{(j_1,\ldots,j_n)\in T} G\left(\lambda_1^{j_1},\chi_{a_1}\right)\cdots G\left(\lambda_n^{j_n},\chi_{a_n}\right), \quad (6.14)$$

where T is as in Theorem 6.33. The formula in that theorem can be recovered by using Theorem 5.12(i), (5.42), and (5.44). For $b\neq 0$, (6.13) can be transformed into the formula in Theorem 6.34 by using Theorem 5.12(i) and the result of Exercise 5.34.

The diagonal equation (6.10), with exponents and coefficients as indicated there, can also be considered as an equation over any finite extension of $\mathbb{F}_q$. For a positive integer s, let $E=\mathbb{F}_{q^s}$ and let N_s be the number of solutions of (6.10) in E^n. We study the dependence of N_s on s under the assumption that all exponents k_i are divisors of $q-1$.

For $b=0$ we obtain N_s from (6.14). If μ_1 is the canonical additive character of E, then each χ_{a_i} has to be replaced by μ_{a_i}. But for $a\in\mathbb{F}_q$ we have $\mu_a(\beta)=\chi_a(\mathrm{Tr}_{E/\mathbb{F}_q}(\beta))$ for all $\beta\in E$ by (5.7), and so $\mu_a=(\chi_a)'$, the character obtained by lifting χ_a to E (compare with the discussion preceding Theorem 5.14). Furthermore, each λ_i of order k_i in (6.14) has to be replaced by a multiplicative character of E of order k_i. However, since the norm function maps E^* onto $\mathbb{F}_q^*$ by Theorem 2.28(ii), the lifted character λ_i' has the same order as λ_i and thus suits our purpose. Therefore, applying (6.14) and Theorem 5.14 we get

$$\begin{aligned}
N_s &= q^{s(n-1)}+\left(1-\frac{1}{q^s}\right)\sum_{(j_1,\ldots,j_n)\in T} G\left((\lambda_1^{j_1})',(\chi_{a_1})'\right)\cdots G\left((\lambda_n^{j_n})',(\chi_{a_n})'\right)\\
&= q^{s(n-1)}+(-1)^{n(s-1)}\left(1-\frac{1}{q^s}\right)\sum_{(j_1,\ldots,j_n)\in T} G\left(\lambda_1^{j_1},\chi_{a_1}\right)^s\cdots G\left(\lambda_n^{j_n},\chi_{a_n}\right)^s\\
&= \left(q^{n-1}\right)^s+(-1)^n\sum_{(j_1,\ldots,j_n)\in T}\left((-1)^n G\left(\lambda_1^{j_1},\chi_{a_1}\right)\cdots G\left(\lambda_n^{j_n},\chi_{a_n}\right)\right)^s\\
&\quad -(-1)^n\sum_{(j_1,\ldots,j_n)\in T}\left(\frac{1}{q}(-1)^n G\left(\lambda_1^{j_1},\chi_{a_1}\right)\cdots G\left(\lambda_n^{j_n},\chi_{a_n}\right)\right)^s,
\end{aligned}$$

which is of the form

$$N_s=\nu_1^s+\cdots+\nu_t^s-\omega_1^s-\cdots-\omega_u^s \quad (6.15)$$

with algebraic numbers $\nu_1,\ldots,\nu_t$, $\omega_1,\ldots,\omega_u$ independent of s that satisfy $|\nu_h|=q^{m_h/2}$ and $|\omega_j|=q^{n_j/2}$ with $m_h, n_j\in\mathbb{Z}$.

For $b \neq 0$ we obtain N_s from (6.13). The same arguments as above lead to the formula

$$N_s = q^{s(n-1)} + \frac{1}{q^s} \sum_{j_1=1}^{k_1-1} \cdots \sum_{j_n=1}^{k_n-1} G\big((\lambda_1^{j_1})', (\chi_{a_1})'\big) \cdots G\big((\lambda_n^{j_n})', (\chi_{a_n})'\big)$$

$$\cdot G\big((\bar{\lambda}_1^{j_1} \cdots \bar{\lambda}_n^{j_n})', (\bar{\chi}_b)'\big)$$

$$= q^{s(n-1)} + \frac{1}{q^s}(-1)^{(n+1)(s-1)} \sum_{j_1=1}^{k_1-1} \cdots \sum_{j_n=1}^{k_n-1} G\big(\lambda_1^{j_1}, \chi_{a_1}\big)^s \cdots G\big(\lambda_n^{j_n}, \chi_{a_n}\big)^s$$

$$\cdot G\big(\bar{\lambda}_1^{j_1} \cdots \bar{\lambda}_n^{j_n}, \bar{\chi}_b\big)^s$$

$$= (q^{n-1})^s + (-1)^{n+1} \sum_{j_1=1}^{k_1-1} \cdots \sum_{j_n=1}^{k_n-1}$$

$$\left(\frac{1}{q}(-1)^{n+1} G\big(\lambda_1^{j_1}, \chi_{a_1}\big) \cdots G\big(\lambda_n^{j_n}, \chi_{a_n}\big) G\big(\bar{\lambda}_1^{j_1} \cdots \bar{\lambda}_n^{j_n}, \bar{\chi}_b\big)\right)^s,$$

which is again of the form (6.15).

The estimates in Theorems 6.36 and 6.37 can be employed to establish the existence of solutions for diagonal equations, at least for sufficiently large q.

6.38. Example. Let k be a given positive integer. We show that for every finite field $\mathbb{F}_q$ with q sufficiently large, say

$$q > \tfrac{1}{4}\Big[(k-1)(k-2) + \sqrt{k(k-1)(k^2-5k+8)}\,\Big]^2,$$

every element $b \in \mathbb{F}_q$ can be written as a sum of two kth powers. Since the case $b = 0$ is trivial, we consider $b \neq 0$ and let N be the number of solutions of the equation $x_1^k + x_2^k = b$ in $\mathbb{F}_q^2$. With $d = \gcd(k, q-1)$ we get from Theorem 6.37,

$$|N - q| \leqslant \big[(d-1)^2 - (1 - q^{-1/2}) M(d, d)\big] q^{1/2}.$$

Now $M(d, d) = d - 1$ by (6.12) or by the definition, thus using $d \leqslant k$ we obtain

$$|N - q| \leqslant \big[(d-1)^2 - (1 - q^{-1/2})(d-1)\big] q^{1/2} = (d-1)(d-2) q^{1/2} + d - 1$$

$$\leqslant (k-1)(k-2) q^{1/2} + k - 1,$$

and in particular

$$N \geqslant q - (k-1)(k-2) q^{1/2} - k + 1.$$

It suffices then to find a $u_0 \in \mathbb{R}$ such that for $q > u_0$ the right-hand side is positive. We take $u_0 = t_0^2$, where t_0 is the maximum of the roots of the quadratic polynomial $t^2-(k-1)(k-2)t-k+1$, and arrive at the desired result. For $k=2$ see also Remark 6.25. □

Some information on the number of solutions of diagonal equations can also be obtained without the use of exponential sums. We present several examples in the sequel. Arithmetic properties of multinomial coefficients are often needed in this approach, and they can be deduced from Lemma 6.39 below. As a byproduct of this auxiliary result we get the fact that multinomial coefficients are integers, which may be shown by combinatorial arguments as well. For a prime p, let $E_p(r)$ be the largest exponent j such that p^j divides $r \in \mathbb{N}$. Furthermore, we denote by $\lfloor t \rfloor$ the greatest integer $\leqslant t \in \mathbb{R}$.

6.39. Lemma. *For any nonnegative integer m and any prime p we have*

$$E_p(m!) = \sum_{i=1}^{\infty} \left\lfloor \frac{m}{p^i} \right\rfloor = \frac{m-s}{p-1},$$

where s is the sum of digits in the representation of m to the base p.

Proof. If $p^i > m$, then $\lfloor m/p^i \rfloor = 0$, and so all but finitely many terms in the infinite series vanish. The first identity is valid for $0! = 1$. For $h \in \mathbb{N}$ we have

$$\left\lfloor \frac{h}{p^i} \right\rfloor - \left\lfloor \frac{h-1}{p^i} \right\rfloor = \begin{cases} 1 & \text{if } p^i \text{ divides } h, \\ 0 & \text{otherwise,} \end{cases}$$

hence for $m \geqslant 1$,

$$\begin{aligned} E_p(m!) = E_p(1\cdot 2\cdot \cdots \cdot m) &= \sum_{h=1}^{m} E_p(h) = \sum_{h=1}^{m} \sum_{i=1}^{\infty} \left(\left\lfloor \frac{h}{p^i} \right\rfloor - \left\lfloor \frac{h-1}{p^i} \right\rfloor \right) \\ &= \sum_{i=1}^{\infty} \sum_{h=1}^{m} \left(\left\lfloor \frac{h}{p^i} \right\rfloor - \left\lfloor \frac{h-1}{p^i} \right\rfloor \right) = \sum_{i=1}^{\infty} \left\lfloor \frac{m}{p^i} \right\rfloor. \end{aligned}$$

To prove the second identity, let $m = b_u p^u + b_{u-1} p^{u-1} + \cdots + b_0$ with $0 \leqslant b_i < p$ for $0 \leqslant i \leqslant u$ be the representation of m to the base p. Then

$$\begin{aligned} \left\lfloor \frac{m}{p} \right\rfloor &= b_u p^{u-1} + b_{u-1} p^{u-2} + \cdots + b_2 p + b_1 \\ \left\lfloor \frac{m}{p^2} \right\rfloor &= b_u p^{u-2} + b_{u-1} p^{u-3} + \cdots + b_2 \\ &\vdots \\ \left\lfloor \frac{m}{p^u} \right\rfloor &= b_u \end{aligned}$$

and $\lfloor m/p^i \rfloor = 0$ for $i > u$. Adding up these identities, we get

$$\sum_{i=1}^{\infty}\left\lfloor\frac{m}{p^i}\right\rfloor = b_u\frac{p^u-1}{p-1}+b_{u-1}\frac{p^{u-1}-1}{p-1}+\cdots+b_1$$

$$=\frac{1}{p-1}\left(b_u p^u + b_{u-1}p^{u-1}+\cdots+b_1 p + b_0 - b_u - b_{u-1} - \cdots - b_1 - b_0\right)$$

$$=\frac{m-s}{p-1}. \qquad \square$$

6.40. Corollary. *If $m_0, m_1, \ldots, m_n$ are nonnegative integers and $m = m_0 + m_1 + \cdots + m_n$, then the multinomial coefficient*

$$\frac{m!}{m_0! m_1! \cdots m_n!}$$

is an integer.

Proof. It suffices to show that every prime p divides the numerator to at least as high a power as it divides the denominator. Using

$$\lfloor t_0 + t_1 + \cdots + t_n \rfloor \geqslant \lfloor t_0 \rfloor + \lfloor t_1 \rfloor + \cdots + \lfloor t_n \rfloor \text{ for } t_0, t_1, \ldots, t_n \in \mathbb{R}$$

and Lemma 6.39, we get indeed

$$E_p(m!) = \sum_{i=1}^{\infty}\left\lfloor\frac{m}{p^i}\right\rfloor \geqslant \sum_{i=1}^{\infty}\left(\left\lfloor\frac{m_0}{p^i}\right\rfloor + \left\lfloor\frac{m_1}{p^i}\right\rfloor + \cdots + \left\lfloor\frac{m_n}{p^i}\right\rfloor\right)$$
$$= E_p(m_0!) + E_p(m_1!) + \cdots + E_p(m_n!) = E_p(m_0! m_1! \cdots m_n!). \square$$

6.41. Theorem. *The number of solutions in $\mathbb{F}_q^n$ of the diagonal equation* (6.10) *is divisible by the characteristic p of $\mathbb{F}_q$ provided that*

$$\frac{1}{d_1} + \cdots + \frac{1}{d_n} > 1, \tag{6.16}$$

where $d_i = \gcd(k_i, q-1)$ for $i = 1, \ldots, n$.

Proof. Let N be the number of solutions and let $\bar{N}$ be N considered as an element of $\mathbb{F}_q$. Put

$$G(x_1, \ldots, x_n) = a_0 + a_1 x_1^{k_1} + \cdots + a_n x_n^{k_n} \quad \text{with } a_0 = -b.$$

Since N is also the number of solutions of $G(x_1,\ldots,x_n)=0$ in $\mathbb{F}_q^n$, we get

$$\begin{aligned}\bar{N} &= \sum_{c_1,\ldots,c_n\in\mathbb{F}_q}\left(1-G(c_1,\ldots,c_n)^{q-1}\right) = -\sum_{c_1,\ldots,c_n\in\mathbb{F}_q} G(c_1,\ldots,c_n)^{q-1}\\ &= -\sum_{c_1,\ldots,c_n\in\mathbb{F}_q}\left(a_0+a_1c_1^{k_1}+\cdots+a_nc_n^{k_n}\right)^{q-1}\\ &= -\sum_{c_1,\ldots,c_n\in\mathbb{F}_q}\ \sum_{\substack{h_0,h_1,\ldots,h_n\geqslant 0\\ h_0+h_1+\cdots+h_n=q-1}}\frac{(q-1)!}{h_0!h_1!\cdots h_n!}a_0^{h_0}a_1^{h_1}\cdots\\ &\qquad a_n^{h_n}c_1^{k_1h_1}\cdots c_n^{k_nh_n}\\ &= -\sum_{\substack{h_0,h_1,\ldots,h_n\geqslant 0\\ h_0+h_1+\cdots+h_n=q-1}}\frac{(q-1)!}{h_0!h_1!\cdots h_n!}a_0^{h_0}a_1^{h_1}\cdots\\ &\qquad a_n^{h_n}\left(\sum_{c_1\in\mathbb{F}_q}c_1^{k_1h_1}\right)\cdots\left(\sum_{c_n\in\mathbb{F}_q}c_n^{k_nh_n}\right).\end{aligned}$$

Because of Lemma 6.3, we only have to take into account those terms with $h_i>0$ and k_ih_i divisible by $q-1$ for $1\leqslant i\leqslant n$. The latter condition is equivalent to h_i being divisible by $(q-1)/d_i$. Thus we obtain

$$\bar{N}=(-1)^{n+1}\sum_{(h_0,h_1,\ldots,h_n)\in H}\frac{(q-1)!}{h_0!h_1!\cdots h_n!}a_0^{h_0}a_1^{h_1}\cdots a_n^{h_n}, \qquad (6.17)$$

where H is the set of $(n+1)$-tuples $(h_0,h_1,\ldots,h_n)$ of integers with $h_0+h_1+\cdots+h_n=q-1$, $h_0\geqslant 0$, and h_i a positive multiple of $(q-1)/d_i$ for $1\leqslant i\leqslant n$.

Now suppose (6.16) is satisfied. If $(h_0,h_1,\ldots,h_n)\in H$, then $h_i\geqslant(q-1)/d_i$ for $1\leqslant i\leqslant n$, hence

$$h_0+h_1+\cdots+h_n\geqslant\frac{q-1}{d_1}+\cdots+\frac{q-1}{d_n}>q-1,$$

a contradiction. Thus the set H is empty, so $\bar{N}=0$ by (6.17), which means that N is divisible by p. □

If $k_1=\cdots=k_n=d$, a divisor of $q-1$, then (6.16) reduces to the inequality $n>d$ and we obtain a special case of Theorem 6.5.

6.42. Theorem. *Let p be the characteristic of $\mathbb{F}_q$. Suppose $d_i=$* gcd$(k_i,q-1)$ *divides $p-1$ for $1\leqslant i\leqslant n$ and*

$$\frac{1}{d_1}+\cdots+\frac{1}{d_n}=1.$$

Then the diagonal equation (6.10) *has at least one solution in $\mathbb{F}_q^n$.*

Proof. We use (6.17) and observe that under the condition $(1/d_1) + \cdots + (1/d_n) = 1$ the set H consists only of the $(n+1)$-tuple $(0, h_1, \ldots, h_n)$ with $h_i = (q-1)/d_i$ for $1 \leqslant i \leqslant n$, so that

$$\bar{N} = (-1)^{n+1} \frac{(q-1)!}{h_1! \cdots h_n!} a_1^{h_1} \cdots a_n^{h_n}.$$

Now $q = p^r$ for some $r \in \mathbb{N}$, hence

$$q-1 = (p-1)p^{r-1} + (p-1)p^{r-2} + \cdots + (p-1)$$

is the representation of $q-1$ to the base p and

$$E_p((q-1)!) = \frac{q-1-r(p-1)}{p-1} = \frac{q-1}{p-1} - r$$

by Lemma 6.39. Since d_i divides $p-1$, the representation of $h_i = (q-1)/d_i$ to the base p is given by

$$h_i = \frac{p-1}{d_i} p^{r-1} + \frac{p-1}{d_i} p^{r-2} + \cdots + \frac{p-1}{d_i},$$

and so

$$E_p(h_i!) = \frac{h_i - r(p-1)/d_i}{p-1} = \frac{h_i}{p-1} - \frac{r}{d_i}.$$

Using $(1/d_1) + \cdots + (1/d_n) = 1$, it follows that

$$E_p(h_1!) + \cdots + E_p(h_n!) = E_p((q-1)!),$$

which means that $(q-1)!/h_1! \cdots h_n!$ is not divisible by p. Since all $a_i \neq 0$, we therefore have $\bar{N} \neq 0$, hence $N \neq 0$. □

In some cases this method can also be applied to equations that are not diagonal. If the polynomial g in the subsequent theorem is taken to be constant, the result could also be deduced from Theorem 6.42.

6.43. Theorem. *Let p be the characteristic of $\mathbb{F}_q$, let k be a positive divisor of $p-1$, and let $a_1, \ldots, a_k \in \mathbb{F}_q^*$. If $g \in \mathbb{F}_q[x_1, \ldots, x_k]$ with* $\deg(g) < k$, *then the equation*

$$a_1 x_1^k + \cdots + a_k x_k^k = g(x_1, \ldots, x_k)$$

has at least one solution in $\mathbb{F}_q^k$.

Proof. We proceed as in the proof of Theorem 6.41, using

$$G(x_1, \ldots, x_k) = a_1 x_1^k + \cdots + a_k x_k^k - g(x_1, \ldots, x_k).$$

Then

$$\bar{N} = -\sum_{c_1,\ldots,c_k \in \mathbb{F}_q} \left(a_1c_1^k + \cdots + a_kc_k^k - g(c_1,\ldots,c_k)\right)^{q-1}$$

$$= -\sum_{c_1,\ldots,c_k \in \mathbb{F}_q} \sum_{\substack{h_0,h_1,\ldots,h_k \geqslant 0 \\ h_0+h_1+\cdots+h_k=q-1}} \frac{(q-1)!}{h_0!h_1!\cdots h_k!}\left(-g(c_1,\ldots,c_k)\right)^{h_0}$$

$$\cdot a_1^{h_1}\cdots a_k^{h_k}c_1^{kh_1}\cdots c_k^{kh_k}$$

$$= -\sum_{\substack{h_0,h_1,\ldots,h_k \geqslant 0 \\ h_0+h_1+\cdots+h_k=q-1}} \frac{(q-1)!}{h_0!h_1!\cdots h_k!}(-1)^{h_0}a_1^{h_1}\cdots a_k^{h_k}$$

$$\sum_{c_1,\ldots,c_k \in \mathbb{F}_q} g(c_1,\ldots,c_k)^{h_0}c_1^{kh_1}\cdots c_k^{kh_k}.$$

If $h_0 > 0$, then

$$h_0\deg(g) + kh_1 + \cdots + kh_k < k(h_0 + h_1 + \cdots + h_k) = k(q-1),$$

and Lemma 6.4 implies that the inner sum vanishes. Therefore,

$$\bar{N} = -\sum_{\substack{h_1,\ldots,h_k \geqslant 0 \\ h_1+\cdots+h_k=q-1}} \frac{(q-1)!}{h_1!\cdots h_k!}a_1^{h_1}\cdots a_k^{h_k}\left(\sum_{c_1 \in \mathbb{F}_q} c_1^{kh_1}\right)\cdots\left(\sum_{c_k \in \mathbb{F}_q} c_k^{kh_k}\right).$$

By Lemma 6.3 we can only get a nonzero term if h_i is a positive multiple of $h = (q-1)/k$ for all $i = 1,\ldots,k$. Because of $h_1 + \cdots + h_k = q-1$, this leaves only the choice $h_1 = \cdots = h_k = h$. Thus

$$\bar{N} = (-1)^{k+1}\frac{(q-1)!}{h!^k}(a_1\cdots a_k)^h.$$

As in the proof of Theorem 6.42, one shows that $(q-1)!/h!^k$ is not divisible by p. It follows that $\bar{N} \neq 0$, and so $N \neq 0$. □

6.44. Corollary. *If k is a positive divisor of $p-1$ and $f_1(x_1),\ldots,f_k(x_k)$ are polynomials over $\mathbb{F}_q$ of degree k, then the equation $f_1(x_1) + \cdots + f_k(x_k) = 0$ has at least one solution in $\mathbb{F}_q^k$.*

4. THE STEPANOV-SCHMIDT METHOD

We consider special types of equations of the form $F(x, y) = 0$ with $F \in \mathbb{F}_q[x, y]$ for which nontrivial results on the number of solutions can be established by "elementary" methods—that is, methods not depending on algebraic geometry or on the extensive use of algebraic function fields. Our main aim will be to prove the results that are needed for the estimation of character sums in Chapter 5, Section 4 (see Theorems 5.38 and 5.41).

The first type of equation that we study is $y^m = f(x)$. The following result provides a starting point for analyzing the number of solutions of such an equation.

6.45. Lemma. *Let m be a positive divisor of $q-1$, $f \in \mathbb{F}_q[x]$, and $g = f^{(q-1)/m}$. Then the number N of solutions of $y^m = f(x)$ in $\mathbb{F}_q^2$ is given by $N = |T_0| + m|T_1|$, where T_0 is the set of $c \in \mathbb{F}_q$ with $f(c) = 0$ and T_1 is the set of $c \in \mathbb{F}_q$ with $g(c) = 1$. Furthermore, $|T_0| + |T_1| + |T_2| = q$, where T_2 is the set of $c \in \mathbb{F}_q$ with*

$$g(c)^{m-1} + g(c)^{m-2} + \cdots + g(c) + 1 = 0.$$

Proof. We distinguish the solutions $x = c_1$, $y = c_2$ with $c_2 = 0$ and $c_2 \neq 0$. The number of solutions $(c_1, 0)$ is just N_0. If $c_2 \neq 0$, then $f(c_1) \neq 0$, and since $\mathbb{F}_q^*$ is cyclic, $f(c_1)$ is an mth power of an element of $\mathbb{F}_q^*$ if and only if $g(c_1) = f(c_1)^{(q-1)/m} = 1$. If $g(c_1) = 1$, then there are m elements $c_2 \in \mathbb{F}_q^*$ with $c_2^m = f(c_1)$. Hence $N = |T_0| + m|T_1|$.

Clearly, the sets T_0, T_1, T_2 are pairwise disjoint. On the other hand, the identity

$$0 = f(c)^q - f(c) = f(c)(g(c)-1)\left(g(c)^{m-1} + g(c)^{m-2} + \cdots + g(c) + 1\right),$$

valid for every $c \in \mathbb{F}_q$, shows that each element of $\mathbb{F}_q$ belongs to one of the three sets. Hence $|T_0| + |T_1| + |T_2| = q$. □

The key step in the elementary method of Stepanov and Schmidt is the construction of an auxiliary polynomial that has well-chosen elements of $\mathbb{F}_q$ as multiple roots. The next lemma is instrumental in showing that the polynomial to be constructed is nonzero. The following terminology and notation is used. For an arbitrary field K, a polynomial $F \in K[x, y]$ of positive degree is called *absolutely irreducible* (over K) if it is irreducible (i.e., does not allow a proper factorization) over any algebraic extension of K. Furthermore, let $K(x)$ denote the field of rational functions over K—that is, the field consisting of the fractions of the form f/g with $f, g \in K[x]$, $g \neq 0$.

6.46. Lemma. *Let $m \geqslant 2$ be a divisor of $q-1$ and let $f \in \mathbb{F}_q[x]$ with $\deg(f) = k \geqslant 1$ be such that $y^m - f(x)$ is absolutely irreducible. Put $g = f^{(q-1)/m}$ and let $h_0, h_1, \ldots, h_{m-1}$ be polynomials of the form*

$$h_i(x) = \sum_{j=0}^{u} e_{ij}(x) x^{qj} \quad \text{for } 0 \leqslant i \leqslant m-1,$$

where $e_{ij} \in \mathbb{F}_q[x]$ with $\deg(e_{ij}) \leqslant q/m - k$. If

$$h_0 + h_1 g + \cdots + h_{m-1} g^{m-1} = 0, \tag{6.18}$$

then all $e_{ij} = 0$.

Proof. We prove the lemma first under the additional condition $f(0) \neq 0$. Put

$$A(y; h_0, \ldots, h_{m-1}) = h_0 + h_1 y + \cdots + h_{m-1} y^{m-1};$$

then for indeterminates $y_1, \ldots, y_m$,

$$B(y_1, \ldots, y_m) = \prod_{i=1}^{m} A(y_i; h_0, \ldots, h_{m-1})$$

is a symmetric polynomial in $y_1, \ldots, y_m$. By the fundamental theorem on symmetric polynomials (see Example 1.74), $B(y_1, \ldots, y_m)$ can be expressed as a polynomial in the elementary symmetric polynomials $\sigma_1(y_1, \ldots, y_m), \ldots, \sigma_m(y_1, \ldots, y_m)$. Let $\zeta_1 = 1, \zeta_2, \ldots, \zeta_m$ be the mth roots of unity in $\mathbb{F}_q$ and substitute $y_i = \zeta_i y$, $1 \leqslant i \leqslant m$. From $x^m - 1 = (x - \zeta_1) \cdots (x - \zeta_m)$ we get $\sigma_i(\zeta_1 y, \ldots, \zeta_m y) = 0$ for $1 \leqslant i \leqslant m-1$ and $\sigma_m(\zeta_1 y, \ldots, \zeta_m y) = -y^m$. Therefore, $B(\zeta_1 y, \ldots, \zeta_m y)$ is a polynomial in y^m, say $G(y^m)$, and comparing the degrees in y we obtain

$$m \deg(G) = \deg(B(\zeta_1 y, \ldots, \zeta_m y)) \leqslant m(m-1),$$

hence $\deg(G) \leqslant m-1$. We can thus write

$$\prod_{i=1}^{m} A(\zeta_i y; h_0, \ldots, h_{m-1}) = \sum_{i=0}^{m-1} C_i(h_0, \ldots, h_{m-1}) y^{mi}, \qquad (6.19)$$

where each C_i is a polynomial of degree at most m. Since $A(\zeta_1 g; h_0, \ldots, h_{m-1}) = 0$ by (6.18) and $g^m = f^{q-1}$, it follows that

$$\sum_{i=0}^{m-1} C_i(h_0, \ldots, h_{m-1}) f^{(q-1)i} = 0,$$

and multiplying by f^{m-1},

$$\sum_{i=0}^{m-1} C_i(h_0, \ldots, h_{m-1}) f^{qi} f^{m-1-i} = 0.$$

Considering this identity mod x^q and noting that $h_i(x) \equiv e_{i0}(x) \bmod x^q$, $f(x)^q \equiv f(0) \bmod x^q$, we obtain

$$\sum_{i=0}^{m-1} C_i(e_{00}(x), \ldots, e_{m-1,0}(x)) f(0)^i f(x)^{m-1-i} \equiv 0 \bmod x^q.$$

The degree in x of the left-hand side is $\leqslant m(q/m - k) + (m-1)k < q$, and so we must have

$$\sum_{i=0}^{m-1} C_i(e_{00}, \ldots, e_{m-1,0}) f(0)^i f^{m-1-i} = 0,$$

hence

$$\sum_{i=0}^{m-1} C_i(e_{00}, \ldots, e_{m-1,0}) \left(\frac{f(0)}{f} \right)^i = 0. \qquad (6.20)$$

In a suitable algebraic extension of $\mathbb{F}_q$ there is an element α with $\alpha^m = f(0)$. Then $(\alpha y)^m - f(x) = f(0)(y^m - f(0)^{-1}f(x))$, and so $y^m - f(0)^{-1}f(x)$, is absolutely irreducible. Now $y^m - f(0)^{-1}f(x)$, considered as a polynomial in y over $\mathbb{F}_q(x)$, has a root Y in an extension of $\mathbb{F}_q(x)$. From (6.20) we get

$$\sum_{i=0}^{m-1} C_i(e_{00},\ldots,e_{m-1,0})Y^{-mi} = 0,$$

and so (6.19) yields

$$\prod_{i=1}^{m} A(\zeta_i Y^{-1}; e_{00},\ldots,e_{m-1,0}) = 0.$$

Thus $A(\zeta_i Y^{-1}; e_{00},\ldots,e_{m-1,0}) = 0$ for some i, and after multiplication by Y^{m-1} we obtain

$$e_{00}Y^{m-1} + \zeta_i e_{10}Y^{m-2} + \cdots + \zeta_i^{m-1}e_{m-1,0} = 0.$$

But Y is of degree m over $\mathbb{F}_q(x)$ since $y^m - f(0)^{-1}f(x)$ is absolutely irreducible, and so $e_{00} = e_{10} = \cdots = e_{m-1,0} = 0$. Then we can divide (6.18) by x^q, and the same procedure yields $e_{01} = e_{11} = \cdots = e_{m-1,1} = 0$. Continuing in this way, we get that all $e_{ij} = 0$.

It remains to show that the general case can be reduced to the case where $f(0) \neq 0$. If $k \geqslant q$, then $\deg(e_{ij}) < 0$ by hypothesis, hence $e_{ij} = 0$. Thus we may assume $k < q$, so that there is a $c \in \mathbb{F}_q$ with $f(c) \neq 0$. Put $f_1(x) = f(x+c)$ and $g_1(x) = g(x+c)$; then $y^m - f_1(x)$ is absolutely irreducible,

$$h_0(x+c) + h_1(x+c)g_1(x) + \cdots + h_{m-1}(x+c)g_1(x)^{m-1} = 0,$$

and the polynomials $h_i(x+c)$ are again of the required form. Since $f_1(0) \neq 0$, the result already established shows that the corresponding e_{ij} are all 0, hence all $h_i(x+c) = 0$, thus all $h_i = 0$, and so the original e_{ij} are all 0. □

In our construction of an auxiliary polynomial with certain multiple roots, it will be necessary to have a convenient method for detecting multiplicities. For a field of characteristic 0, derivatives provide such a method, but for a field of prime characteristic p, the method of derivatives can only be applied in a limited manner (compare with Exercise 1.51). For instance, for the polynomial x^p all derivatives vanish at 0, but 0 is a root of multiplicity p only. We therefore introduce modified derivatives, so-called hyperderivatives, that will prove more useful.

Let K be an arbitrary field. For $n = 0, 1, \ldots$ and $\sum_{j=0}^{d} a_j x^j \in K[x]$ we define the *nth hyperderivative* by

$$E^{(n)}\left(\sum_{j=0}^{d} a_j x^j\right) = \sum_{j=0}^{d} \binom{j}{n} a_j x^{j-n}.$$

Here we use a standard convention for binomial coefficients—namely, $\binom{j}{n} = 0$ for $n > j$—which guarantees that the nth hyperderivative is again a polynomial over K. If K is of characteristic 0, then

$$E^{(n)}(f) = \frac{1}{n!} f^{(n)} \quad \text{for all } f \in K[x].$$

It is clear that $E^{(n)}$ is linear, in the sense that $E^{(n)}(cf) = cE^{(n)}(f)$ and $E^{(n)}(f_1 + f_2) = E^{(n)}(f_1) + E^{(n)}(f_2)$ for all $c \in K$ and $f, f_1, f_2 \in K[x]$. The following formulas for hyperderivatives will be needed.

6.47. Lemma. *For $f_1, \ldots, f_t \in K[x]$ we have*

$$E^{(n)}(f_1 \cdots f_t) = \sum_{\substack{n_1, \ldots, n_t \geqslant 0 \\ n_1 + \cdots + n_t = n}} E^{(n_1)}(f_1) \cdots E^{(n_t)}(f_t). \tag{6.21}$$

Proof. By the linearity of $E^{(n)}$, it suffices to show the formula for monomials, say $f_j(x) = x^{k_j}$ for $1 \leqslant j \leqslant t$. Then (6.21) is equivalent to

$$\binom{k_1 + \cdots + k_t}{n} = \sum_{\substack{n_1, \ldots, n_t \geqslant 0 \\ n_1 + \cdots + n_t = n}} \binom{k_1}{n_1} \cdots \binom{k_t}{n_t},$$

and this identity can be verified by comparing the coefficients of x^n on both sides of

$$(x+1)^{k_1 + \cdots + k_t} = (x+1)^{k_1} \cdots (x+1)^{k_t}. \qquad \square$$

6.48. Corollary. *For $c \in K$, $E^{(n)}((x-c)^t) = \binom{t}{n}(x-c)^{t-n}$.*

Proof. Apply Lemma 6.47 with $f_i(x) = x - c$ for $1 \leqslant i \leqslant t$. Since $E^{(1)}(x-c) = 1$ and $E^{(n)}(x-c) = 0$ for $n \geqslant 2$, only those terms in the sum in (6.21) will remain for which each n_i is either 0 or 1. The number of such terms is $\binom{t}{n}$, and each term is $(x-c)^{t-n}$. $\square$

6.49. Corollary. *For $0 \leqslant n \leqslant t$ and $f, w \in K[x]$ we have $E^{(n)}(wf^t) = w_1 f^{t-n}$, where $w_1 \in K[x]$ with $\deg(w_1) \leqslant \deg(w) + n(\deg(f) - 1)$.*

Proof. In

$$E^{(n)}(wf^t) = \sum_{\substack{n_0, n_1, \ldots, n_t \geqslant 0 \\ n_0 + n_1 + \cdots + n_t = n}} E^{(n_0)}(w) E^{(n_1)}(f) \cdots E^{(n_t)}(f)$$

every term is divisible by f^{t-n}, so that $E^{(n)}(wf^t) = w_1 f^{t-n}$ for some $w_1 \in K[x]$. Furthermore,

$$\begin{aligned} \deg(w_1) &= \deg(E^{(n)}(wf^t)) - (t-n)\deg(f) \\ &\leqslant \deg(w) + t\deg(f) - n - (t-n)\deg(f) \\ &= \deg(w) + n(\deg(f) - 1). \end{aligned} \qquad \square$$

6.50. Corollary. *Suppose K is of prime characteristic p. Let $h(x)=v(x, x^{p^s})$ for some polynomial $v(x, y)$ over K and $s \in \mathbb{N}$. Then for $0 \leqslant n < p^s$, $E^{(n)}(h)$ is given by the nth partial hyperderivative of $v(x, y)$ with respect to x, with a subsequent substitution $y = x^{p^s}$.*

Proof. By linearity it suffices to consider the case where $v(x, y)= x^j y^k$. We note first that for $0 < n < p^s$ the binomial coefficient

$$\binom{p^s}{n} = \frac{p^s}{n}\binom{p^s-1}{n-1}$$

is divisible by p. Thus

$$E^{(n)}(x^{p^s}) = \binom{p^s}{n} x^{p^s-n} = 0$$

in a field of characteristic p. Therefore, if we apply Lemma 6.47 with $f_1(x)=x^j$, $f_2(x)=\cdots=f_{k+1}(x)=x^{p^s}$, we get

$$E^{(n)}(h) = \binom{j}{n} x^{j-n} x^{kp^s}.$$

The same result is obtained by calculating the nth partial hyperderivative of $v(x, y)$ with respect to x and substituting $y = x^{p^s}$ afterwards. □

The following result shows that hyperderivatives are an appropriate device for detecting multiplicities of roots.

6.51. Lemma. *Let f be a polynomial over an arbitrary field K. Suppose $c \in K$ is a root of $E^{(n)}(f)$ for $n = 0, 1, \ldots, M-1$. Then c is a root of f of multiplicity at least M.*

Proof. Write $f(x) = a_0 + a_1(x-c) + \cdots + a_d(x-c)^d$. From Corollary 6.48 we get

$$E^{(n)}(f(x)) = a_n + \binom{n+1}{n} a_{n+1}(x-c) + \cdots + \binom{d}{n} a_d (x-c)^{d-n}.$$

Substituting $x = c$, we obtain $a_n = 0$ for $n = 0, 1, \ldots, M-1$, and so $(x-c)^M$ divides $f(x)$. The conclusion follows now from Definition 1.65. □

We are now in a position to prove the fundamental lemma in which a suitable auxiliary polynomial is constructed.

6.52. Lemma. *Let $m \geqslant 2$ be a divisor of $q-1$ and let $f \in \mathbb{F}_q[x]$ with $\deg(f) = k \geqslant 1$ be such that $y^m - f(x)$ is absolutely irreducible. Let $B \in \mathbb{F}_q[x]$ with $1 \leqslant \deg(B) = r < m$, and let T be the set of $c \in \mathbb{F}_q$ with either $B(g(c)) = 0$ or $f(c) = 0$, where $g = f^{(q-1)/m}$. Let $M \geqslant k+1$ be an integer with $(M+3)^2 \leqslant 2q/m$. Then there exists a nonzero polynomial $h \in \mathbb{F}_q[x]$ such that every $c \in T$ is a root of h of multiplicity at least M and*

$$\deg(h) < \frac{r}{m} qM + 4kq.$$

Proof. We try a polynomial of the form

$$h(x) = f(x)^M \sum_{i=0}^{m-1} \sum_{j=0}^{u} e_{ij}(x) g(x)^i x^{qj}, \tag{6.22}$$

where the e_{ij} are polynomials with coefficients in $\mathbb{F}_q$ to be determined and $\deg(e_{ij}) \leqslant q/m - k$, and where

$$u = \left\lfloor \frac{r}{m}(M + k + 1) \right\rfloor. \tag{6.23}$$

We need to calculate $E^{(n)}(h)$ for $n = 0, 1, \ldots, M-1$. We note first that since g is a power of f,

$$E^{(n)}\left(f^M e_{ij} g^i\right) = f^{M-n} e_{ijn} g^i$$

by Corollary 6.49, where

$$\deg(e_{ijn}) \leqslant \deg(e_{ij}) + n(k-1) \leqslant \frac{q}{m} - k + n(k-1) \leqslant \frac{q}{m} + n(k-1) - 1. \tag{6.24}$$

Now $h(x) = v(x, x^q)$ with

$$v(x, y) = \sum_{i=0}^{m-1} \sum_{j=0}^{u} f(x)^M e_{ij}(x) g(x)^i y^j,$$

and since $M \leqslant q$, we can apply Corollary 6.50 for $0 \leqslant n \leqslant M-1$ to get

$$E^{(n)}(h(x)) = f(x)^{M-n} \sum_{i=0}^{m-1} \sum_{j=0}^{u} e_{ijn}(x) g(x)^i x^{qj}. \tag{6.25}$$

If $c \in \mathbb{F}_q$ satisfies $B(c) = 0$, we have

$$c^r = b_0 + b_1 c + \cdots + b_{r-1} c^{r-1} \quad \text{with } b_0, b_1, \ldots, b_{r-1} \in \mathbb{F}_q,$$

hence for $i \geqslant 0$,

$$c^i = \sum_{t=0}^{r-1} b_{ti} c^t \quad \text{with } b_{ti} \in \mathbb{F}_q.$$

For $c \in \mathbb{F}_q$ with $B(g(c)) = 0$ we thus get

$$g(c)^i = \sum_{t=0}^{r-1} b_{ti} g(c)^t \quad \text{for } i \geqslant 0.$$

For such c we can then use (6.25) and $c^q = c$ to obtain

$$\begin{aligned}(E^{(n)}(h))(c) &= f(c)^{M-n} \sum_{i=0}^{m-1} \sum_{j=0}^{u} e_{ijn}(c) g(c)^i c^j \\ &= f(c)^{M-n} \sum_{t=0}^{r-1} s_{tn}(c) g(c)^t,\end{aligned}$$

where

$$s_{tn}(x) = \sum_{i=0}^{m-1} \sum_{j=0}^{u} b_{ti} e_{ijn}(x) x^j.$$

We will get $(E^{(n)}(h))(c) = 0$, $0 \leqslant n \leqslant M-1$, for all $c \in \mathbb{F}_q$ with $B(g(c)) = 0$ provided the polynomials s_{tn}, $0 \leqslant t \leqslant r-1$, $0 \leqslant n \leqslant M-1$, are all 0. From (6.24) we have

$$\deg(s_{tn}) \leqslant \frac{q}{m} + n(k-1) - 1 + u.$$

Thus, if S denotes the total number of coefficients of the s_{tn}, $0 \leqslant t \leqslant r-1$, $0 \leqslant n \leqslant M-1$, then

$$\begin{aligned} S &\leqslant r \sum_{n=0}^{M-1} \left(\frac{q}{m} + n(k-1) + u \right) = rM\left(\frac{q}{m} + u \right) + \tfrac{1}{2} r(k-1) M(M-1) \\ &\leqslant \frac{rq}{m} M + rM \frac{r}{m}(M+k+1) + \tfrac{1}{2} r(k-1) M^2 \end{aligned}$$

by (6.23). Using $r < m$ we obtain

$$S < \frac{rq}{m} M + \tfrac{1}{2} rM^2 (k+1) + rM(k+1). \tag{6.26}$$

Let A be the number of possible coefficients of all e_{ij}, $0 \leqslant i \leqslant m-1$, $0 \leqslant j \leqslant u$. Then on account of (6.23),

$$\begin{aligned} A &\geqslant \left(\frac{q}{m} - k \right) m(u+1) \geqslant (q - km) \frac{r}{m}(M+k+1) \\ &= \frac{rq}{m} M + \frac{rq}{m}(k+1) - rk(M+k+1). \end{aligned}$$

Since $M \geqslant k+1$, we get

$$A \geqslant \frac{rq}{m} M + \frac{rq}{m}(k+1) - 2rkM. \tag{6.27}$$

The condition that all $s_{tn} = 0$ yields S homogeneous linear equations in the A coefficients of the e_{ij}. If $S < A$, then we can obtain a nontrivial solution for these coefficients, and so the resulting e_{ij} will not all be 0. Because of (6.26) and (6.27), the inequality $S < A$ is guaranteed if

$$\tfrac{1}{2} rM^2 (k+1) + rM(k+1) \leqslant \frac{rq}{m}(k+1) - 2rkM,$$

and for this it suffices to have

$$\tfrac{1}{2} M^2 (k+1) + 3(k+1) M \leqslant \frac{q}{m}(k+1),$$

or, equivalently,

$$M^2 + 6M \leqslant \frac{2q}{m},$$

which is valid because of the hypothesis that $(M+3)^2 \leqslant 2q/m$.

With a choice of e_{ij} as above, we then set up the polynomial h in (6.22). We have $h \neq 0$, for otherwise Lemma 6.46 would imply that all $e_{ij} = 0$, a contradiction. By the construction of h and Lemma 6.51, every $c \in \mathbb{F}_q$ with $B(g(c)) = 0$ is a root of h of multiplicity at least M. Since h has a factor f^M, every $c \in T$ is a root of h of multiplicity at least M. Furthermore, (6.22) and (6.23) yield

$$\begin{aligned}
\deg(h) &\leqslant kM + \frac{q}{m} - k + (m-1)\frac{q-1}{m}k + qu \\
&< kM + \frac{q}{m} + qk + \frac{qr}{m}(M + k + 1) \\
&< kq^{1/2} + \frac{r}{m}qM + q\left(\frac{1}{m} + 2k + 1\right) < \frac{r}{m}qM + 4kq. \qquad \square
\end{aligned}$$

We establish a preliminary estimate for the number of solutions of $y^m = f(x)$, valid for sufficiently large q. Later on, this result will be improved in two ways—namely, by weakening the hypothesis and by sharpening the estimate (see Theorem 6.57).

6.53. Theorem. *Let $m \geqslant 2$ be a divisor of $q-1$ and let $f \in \mathbb{F}_q[x]$ with $\deg(f) = k \geqslant 1$ be such that $y^m - f(x)$ is absolutely irreducible. Then for $q \geqslant 100mk^2$, the number N of solutions of the equation $y^m = f(x)$ in $\mathbb{F}_q^2$ satisfies*

$$|N - q| < 4km^{3/2}q^{1/2}.$$

Proof. Let h be the polynomial constructed in Lemma 6.52. Since $h \neq 0$, Theorem 1.66 shows that $M|T| \leqslant \deg(h)$, and the estimate for $\deg(h)$ in Lemma 6.52 yields

$$|T| < \frac{r}{m}q + \frac{4k}{M}q.$$

Now choose $M = \lfloor (2q/m)^{1/2} \rfloor - 3$. Since $q \geqslant 100mk^2$, we have

$$M \geqslant (2q/m)^{1/2} - 4 \geqslant (q/m)^{1/2} \geqslant k + 1,$$

and so M satisfies all the conditions of Lemma 6.52. Using $M \geqslant (q/m)^{1/2}$ we get

$$|T| < \frac{r}{m}q + 4km^{1/2}q^{1/2}. \tag{6.28}$$

First choose $B(x) = x - 1$ in Lemma 6.52. Then $r = 1$, and in the notation of Lemma 6.45 we have $|T| = |T_0| + |T_1|$, hence from (6.28),

$$|T_0| + |T_1| < \frac{q}{m} + 4km^{1/2}q^{1/2}.$$

Thus

$$N = |T_0| + m|T_1| \leqslant m(|T_0| + |T_1|) < q + 4km^{3/2}q^{1/2}. \tag{6.29}$$

Next choose $B(x)=x^{m-1}+x^{m-2}+\cdots+x+1$. Then $r=m-1$ and $|T|=|T_0|+|T_2|$, so that (6.28) yields

$$|T_0|+|T_2|<\frac{m-1}{m}q+4km^{1/2}q^{1/2}.$$

Applying again Lemma 6.45, we get

$$|T_1|=q-|T_0|-|T_2|>\frac{q}{m}-4km^{1/2}q^{1/2},$$

thus

$$N\geqslant m|T_1|>q-4km^{3/2}q^{1/2},$$

and together with (6.29) we have established the desired result. □

The condition of absolute irreducibility of $y^m-f(x)$ can be put in a more tractable form, according to the following general criterion.

6.54. Lemma. *Let K be an arbitrary field, let $f\in K[x]$ with* $\deg(f)\geqslant 1$, *and let $m\in\mathbb{N}$. Suppose*

$$f(x)=a(x-\alpha_1)^{e_1}\cdots(x-\alpha_d)^{e_d}$$

is the factorization of f in its splitting field over K, where $a\in K$ and $\alpha_1,\ldots,\alpha_d$ are the distinct roots of f. Then $y^m-f(x)$ is absolutely irreducible if and only if $\gcd(m,e_1,\ldots,e_d)=1$.

Proof. We show that $y^m-f(x)$ is not absolutely irreducible if and only if $\gcd(m,e_1,\ldots,e_d)>1$. Suppose $y^m-f(x)$ is reducible over an algebraic extension L of K, where we can assume, without loss of generality, that the polynomial y^m-1 splits in L, say $y^m-1=(y-\zeta_1)\cdots(y-\zeta_m)$. We have

$$y^m-f(x)=F(x,y)G(x,y)$$

with $F,G\in L[x,y]$, $\deg(F)>0$, $\deg(G)>0$. Now consider $y^m-f(x)$ as a polynomial in y over $L(x)$. If Y is a root of $y^m-f(x)$ in the splitting field over $L(x)$, then

$$y^m-f(x)=(y-\zeta_1Y)\cdots(y-\zeta_mY).$$

Because of unique factorization, we have with a nonzero α in L,

$$F(x,y)=\alpha(y-\zeta_{j_1}Y)\cdots(y-\zeta_{j_n}Y)$$

for some $j_1,\ldots,j_n\in\{1,\ldots,m\}$ with $1\leqslant n<m$. Viewing both sides as polynomials in y and comparing constant terms, we get

$$(-1)^n\alpha\zeta_{j_1}\cdots\zeta_{j_n}Y^n\in L[x],$$

hence $Y^n\in L[x]$. Let w be the least positive integer for which $Y^w\in L(x)$. Then $w\leqslant n<m$, and every $u\in\mathbb{N}$ with $Y^u\in L(x)$ is a multiple of w. In particular, m is a multiple of w since $Y^m=f\in L(x)$. With $t=m/w>1$

and $Y^w = g/h$, $g, h \in L[x]$, $h \neq 0$, we get then $f = (g/h)^t$, hence $fh^t = g^t$. By comparing the multiplicities of the roots α_i, $1 \leqslant i \leqslant d$, on both sides of this identity, we see that t divides each e_i. Therefore t divides $\gcd(m, e_1, \ldots, e_d)$, and so $\gcd(m, e_1, \ldots, e_d) > 1$.

Conversely, let $e = \gcd(m, e_1, \ldots, e_d) > 1$ and let K_1 be the splitting field of f over K. In a suitable finite extension of K_1 there is an element β with $\beta^e = a$. Put $s = m/e$ and

$$f_1(x) = \beta(x - \alpha_1)^{e_1/e} \cdots (x - \alpha_d)^{e_d/e}.$$

Then

$$\begin{aligned} y^m - f(x) &= (y^s)^e - f_1(x)^e \\ &= (y^s - f_1(x))\left(y^{s(e-1)} + y^{s(e-2)} f_1(x) + \cdots + f_1(x)^{e-1}\right), \end{aligned}$$

and so $y^m - f(x)$ is not absolutely irreducible. □

6.55. Lemma. *Let $\omega_1, \ldots, \omega_n$ be complex numbers, and let $B > 0$, $C > 0$ be constants such that*

$$|\omega_1^s + \cdots + \omega_n^s| \leqslant CB^s \quad \textit{for } s = 1, 2, \ldots. \tag{6.30}$$

Then $|\omega_j| \leqslant B$ for $1 \leqslant j \leqslant n$.

Proof. Let z be a complex variable. For sufficiently small values of $|z|$, we have

$$\log(1 - \omega_j z) = -\sum_{s=1}^{\infty} \frac{1}{s} \omega_j^s z^s \quad \text{for } 1 \leqslant j \leqslant n,$$

thus

$$\log((1 - \omega_1 z) \cdots (1 - \omega_n z)) = -\sum_{s=1}^{\infty} \frac{1}{s} (\omega_1^s + \cdots + \omega_n^s) z^s. \tag{6.31}$$

By (6.30), the series on the right-hand side of (6.31) converges for $|z| < B^{-1}$. Hence the function on the left-hand side of (6.31) is analytic for $|z| < B^{-1}$. Thus $1 - \omega_j z \neq 0$ if $|z| < B^{-1}$, and therefore $|\omega_j| \leqslant B$ for $1 \leqslant j \leqslant n$. □

We can now prove the first main result, which settles a claim made in Chapter 5, Section 4.

6.56. Theorem. *The complex numbers $\omega_1, \ldots, \omega_{d-1}$ in Theorem 5.39 satisfy $|\omega_j| \leqslant q^{1/2}$ for $1 \leqslant j \leqslant d - 1$.*

Proof. We use the notation and the hypotheses of Theorem 5.39 and put $k = \deg(f)$. Choose $r \in \mathbb{N}$ such that $q^r \geqslant 100mk^2$ and f splits in $\mathbb{F}_{q^r}$, say

$$f(x) = (x - \alpha_1)^{e_1} \cdots (x - \alpha_d)^{e_d},$$

where $\alpha_1,\ldots,\alpha_d$ are the distinct roots of f. Since f is not an mth power, $e=\gcd(m,e_1,\ldots,e_d)$ is a proper divisor of m. Let

$$g(x)=(x-\alpha_1)^{e_1/e}\cdots(x-\alpha_d)^{e_d/e}\in\mathbb{F}_{q^r}[x],$$

so that $f=g^e$. Fix $s\in\mathbb{N}$ and let $E=\mathbb{F}_{q^{rs}}$, $\lambda=\psi^{(rs)}$, $\tau=\lambda^e$. Then λ has order m, τ has order $n=m/e>1$, and

$$\sum_{\gamma\in E}\lambda(f(\gamma))=\sum_{\gamma\in E}\lambda(g(\gamma)^e)=\sum_{\gamma\in E}\tau(g(\gamma)). \tag{6.32}$$

Let the complex number ρ be a primitive nth root of unity and for $i=0,1,\ldots,n-1$ let U_i be the set of $\alpha\in E$ with $\tau(\alpha)=\rho^i$. For fixed $\zeta\in U_1$ we have $\alpha\in U_i$ if and only if $\alpha\zeta^{-i}\in U_0$, which is in turn equivalent to having $\alpha\zeta^{-i}=\beta^n$ for some $\beta\in E^*$. Let A_i be the number of $\gamma\in E$ with $g(\gamma)\in U_i$—that is, with $\zeta^{-i}g(\gamma)=\beta^n$ for some $\beta\in E^*$. Let B_i be the number of solutions of $y^n=\zeta^{-i}g(x)$ in E^2 with a nonzero value for y. Then $A_i=B_i/n$. Let N_i be the total number of solutions of $y^n=\zeta^{-i}g(x)$ in E^2. Since

$$\gcd(n,e_1/e,\ldots,e_d/e)=\gcd(m/e,e_1/e,\ldots,e_d/e)=1,$$

Lemma 6.54 shows that $y^n-\zeta^{-i}g(x)$ is absolutely irreducible. Furthermore, m divides $q-1$, hence n divides $q^{rs}-1$, and so Theorem 6.53 can be applied and yields

$$|N_i-q^{rs}|<4(k/e)n^{3/2}q^{rs/2}\quad\text{for } 0\leqslant i\leqslant n-1.$$

Now $|B_i-N_i|\leqslant k/e$, thus

$$|B_i-q^{rs}|\leqslant 5(k/e)n^{3/2}q^{rs/2}\quad\text{for } 0\leqslant i\leqslant n-1.$$

Write

$$A_i=\frac{1}{n}q^{rs}+R_i,$$

then

$$|R_i|\leqslant 5(k/e)n^{1/2}q^{rs/2}\quad\text{for } 0\leqslant i\leqslant n-1.$$

From (6.32) we get

$$\begin{aligned}\left|\sum_{\gamma\in E}\lambda(f(\gamma))\right|&=\left|\sum_{\gamma\in E}\tau(g(\gamma))\right|=\left|\sum_{i=0}^{n-1}A_i\rho^i\right|=\left|\sum_{i=0}^{n-1}\left(\frac{1}{n}q^{rs}+R_i\right)\rho^i\right|\\&=\left|\sum_{i=0}^{n-1}R_i\rho^i\right|\leqslant\sum_{i=0}^{n-1}|R_i|\leqslant 5(k/e)n^{3/2}q^{rs/2}.\end{aligned}$$

It follows from Theorem 5.39 that

$$|\omega_1^{rs}+\cdots+\omega_{d-1}^{rs}|\leqslant 5(k/e)n^{3/2}q^{rs/2}.$$

As this holds for $s=1,2,\ldots$, Lemma 6.55 implies that $|\omega_j^r|\leqslant q^{r/2}$, and so $|\omega_j|\leqslant q^{1/2}$ for $1\leqslant j\leqslant d-1$. □

The estimate for character sums in Theorem 5.41 is now proved completely. This estimate can then be used to improve upon Theorem 6.53.

6.57. Theorem. *Let $m \in \mathbb{N}$ and let $f \in \mathbb{F}_q[x]$ with $\deg(f) \geqslant 1$ be such that $y^t - f(x)$ is absolutely irreducible, where $t = \gcd(m, q-1)$. Then the number N of solutions of the equation $y^m = f(x)$ in $\mathbb{F}_q^2$ satisfies*

$$|N - q| \leqslant (t-1)(d-1)q^{1/2},$$

where d is the number of distinct roots of f in its splitting field over $\mathbb{F}_q$.

Proof. By (5.70) we get

$$N = \sum_{c \in \mathbb{F}_q} N(y^m = f(c)) = \sum_{c \in \mathbb{F}_q} \sum_{j=0}^{t-1} \lambda^j(f(c)) = q + \sum_{j=1}^{t-1} \sum_{c \in \mathbb{F}_q} \lambda^j(f(c)),$$

where λ is a multiplicative character of $\mathbb{F}_q$ of order t. If $t = 1$, the result follows, so we assume $t > 1$. Let

$$f(x) = af_1(x) = a(x - \alpha_1)^{e_1} \cdots (x - \alpha_d)^{e_d},$$

where $a \in \mathbb{F}_q$ and $\alpha_1, \ldots, \alpha_d$ are the distinct roots of f. Since $y^t - f(x)$ is absolutely irreducible, we have $\gcd(t, e_1, \ldots, e_d) = 1$ by Lemma 6.54. For $1 \leqslant j \leqslant t-1$, λ^j is of order $r_j > 1$ with r_j dividing t, and f_1 cannot be an r_jth power of a polynomial. Thus, Theorem 5.41 yields

$$|N - q| = \left| \sum_{j=1}^{t-1} \sum_{c \in \mathbb{F}_q} \lambda^j(af_1(c)) \right| \leqslant (t-1)(d-1)q^{1/2}. \qquad \square$$

In order to complete the proof of Theorem 5.38, we need to study equations of the type $y^q - y = f(x)$ over finite extensions of $\mathbb{F}_q$ and, more generally, the distribution of the traces of values of f. The key step is again the construction of a suitable auxiliary polynomial. We use the following notation. For $f \in \mathbb{F}_q[x]$, $b \in \mathbb{F}_q$, and a given finite extension E of $\mathbb{F}_q$, let $T(b)$ be the set of $\gamma \in E$ with $\mathrm{Tr}_{E/\mathbb{F}_q}(f(\gamma)) = b$.

6.58. Lemma. *Let $f \in \mathbb{F}_q[x]$ with $\deg(f) \geqslant 1$ and $\gcd(\deg(f), q) = 1$. Let $E = \mathbb{F}_{q^s}$ with $s \geqslant 3$ and fix $b \in \mathbb{F}_q$. Let M be an integer divisible by q and with $0 < M\deg(f) \leqslant q^{s-k-1}$, where k is the greatest integer $\leqslant s/2$. Then there exists a nonzero polynomial $h \in \mathbb{F}_q[x]$ such that every $\gamma \in T(b)$ is a root of h of multiplicity at least M and*

$$\deg(h) < Mq^{s-1} + q^s \deg(f).$$

Proof. Put $g = f^{q^k} + f^{q^{k+1}} + \cdots + f^{q^{s-1}}$, $m = \deg(f)$, and let m_0 be the smaller of the numbers m and q. We try a polynomial of the form

$$h(x) = \sum_{i=0}^{q-1} \sum_{j=0}^{u} e_{ij}(x) g(x)^i x^{jq^s}, \tag{6.33}$$

where the e_{ij} are polynomials with coefficients in $\mathbb{F}_q$ to be determined and $\deg(e_{ij}) < m_0 q^{s-2}$, and where $u = M/q$. Since

$$g(x) = f(x^{q^k}) + f(x^{q^{k+1}}) + \cdots + f(x^{q^{s-1}}),$$

we can write $h(x) = v(x, x^{q^k})$ with

$$v(x, y) = \sum_{i=0}^{q-1} \sum_{j=0}^{u} e_{ij}(x)\left(f(y) + f(y^q) + \cdots + f\left(y^{q^{s-k-1}}\right)\right)^i y^{jq^{s-k}},$$

From $s \leqslant 2k+1$ we get $M \leqslant q^{s-k-1} \leqslant q^k$, and thus we can apply Corollary 6.50 for $0 \leqslant n \leqslant M-1$ to get

$$E^{(n)}(h(x)) = \sum_{i=0}^{q-1} \sum_{j=0}^{u} e_{ijn}(x) g(x)^i x^{jq^s}$$

with $e_{ijn}(x) = E^{(n)}(e_{ij}(x))$. If $\gamma \in T(b)$, then $b = \mathrm{Tr}_{E/\mathbb{F}_q}(f(\gamma)) = G(\gamma) + g(\gamma)$, where $G = f + f^q + \cdots + f^{q^{k-1}}$. Using also $\gamma^{q^s} = \gamma$, we obtain

$$(E^{(n)}(h))(\gamma) = \sum_{i=0}^{q-1} \sum_{j=0}^{u} e_{ijn}(\gamma)(b - G(\gamma))^i \gamma^j = r_n(\gamma)$$

with

$$r_n(x) = \sum_{i=0}^{q-1} \sum_{j=0}^{u} e_{ijn}(x)(b - G(x))^i x^j.$$

In order to guarantee that every $\gamma \in T(b)$ is a root of h of multiplicity at least M, it suffices in view of Lemma 6.51 that the polynomials r_n, $0 \leqslant n \leqslant M-1$, are all 0. Because of $u = M/q \leqslant q^{k-1}$, we have

$$\begin{aligned} \deg(r_n) &< m_0 q^{s-2} + m(q-1)q^{k-1} + u \leqslant m_0 q^{s-2} + m(q-1)q^{k-1} + q^{k-1} \\ &\leqslant m_0 q^{s-2} + mq^k, \end{aligned}$$

and $\deg(r_n) < m_0 q^{s-2} + mq^k - 1$ if $n \geqslant 1$. Thus, if S denotes the total number of coefficients of the r_n, $0 \leqslant n \leqslant M-1$, then

$$S < M\left(m_0 q^{s-2} + mq^k\right) \leqslant Mm_0 q^{s-2} + q^{s-1}. \tag{6.34}$$

The number A of possible coefficients of all e_{ij}, $0 \leqslant i \leqslant q-1$, $0 \leqslant j \leqslant u$, satisfies

$$A = q(u+1)m_0 q^{s-2} = Mm_0 q^{s-2} + m_0 q^{s-1} \geqslant Mm_0 q^{s-2} + q^{s-1}. \tag{6.35}$$

The condition that all $r_n = 0$ yields S homogeneous linear equations in the A coefficients of the e_{ij}. Since $S < A$ by (6.34) and (6.35), there is a nontrivial solution for these coefficients, and then the resulting e_{ij} will not all be 0.

With such a choice of e_{ij} we define h according to (6.33). Then, using $u = M/q$ and $m_0 \leqslant m$, we have

$$\deg(h) < m_0 q^{s-2} + m(q-1)q^{s-1} + uq^s < Mq^{s-1} + mq^s.$$

It remains to verify that $h \neq 0$. This is achieved by showing that the nonzero summands

$$d_{ij}(x) = e_{ij}(x)g(x)^i x^{jq^s}$$

of h have distinct degrees. Now

$$\deg(d_{ij}) = \deg(e_{ij}) + imq^{s-1} + jq^s,$$

and so for $d_{ij} \neq 0$,

$$q^{s-1}(im + jq) \leqslant \deg(d_{ij}) < q^{s-1} + q^{s-1}(im + jq),$$

since $\deg(e_{ij}) < m_0 q^{s-2} \leqslant q^{s-1}$. To prove that the degrees of the nonzero d_{ij} are distinct, it suffices then to show that for pairs $(i, j) \neq (i', j')$ with $0 \leqslant i, i' \leqslant q-1$, $0 \leqslant j, j' \leqslant u$ we have $im + jq \neq i'm + j'q$. Suppose $im + jq = i'm + j'q$. Then $im \equiv i'm \bmod q$, and since $\gcd(m, q) = 1$ by hypothesis, we get $i \equiv i' \bmod q$, hence $i = i'$ and $j = j'$, a contradiction. □

On the basis of Lemma 6.58, we can now establish a preliminary estimate for the number $N(b) = |T(b)|$—that is, for the number of $\gamma \in E$ with $\mathrm{Tr}_{E/\mathbb{F}_q}(f(\gamma)) = b$. This result will be improved later on (see Theorem 6.61). The present estimate suffices for the proof of our second main result, given in Theorem 6.60 below.

6.59. Theorem. *Let $f \in \mathbb{F}_q[x]$ with $\deg(f) = n \geqslant 1$ and $\gcd(n, q) = 1$. Then for any finite extension $E = \mathbb{F}_{q^s}$ of $\mathbb{F}_q$ we have*

$$|N(b) - q^{s-1}| < 2n^2 q^{(s/2)+4} \quad \textit{for all } b \in \mathbb{F}_q. \tag{6.36}$$

Proof. If $q^s < n^2 q^4$, then the trivial estimate $0 \leqslant N(b) \leqslant q^s$ shows that (6.36) is satisfied. Thus we may assume $q^s \geqslant n^2 q^4$. If $k = \lfloor s/2 \rfloor$ as in Lemma 6.58 (where $\lfloor t \rfloor$ is the greatest integer $\leqslant t$), then

$$q^{s-k-2} \geqslant q^{(s-4)/2} \geqslant n,$$

and so

$$M = \left\lfloor \frac{1}{n} q^{s-k-2} \right\rfloor q$$

is a positive multiple of q. Clearly, $Mn \leqslant q^{s-k-1}$, and so all the hypotheses of Lemma 6.58 hold. With the polynomial $h \neq 0$ constructed there, we get $N(b)M \leqslant \deg(h)$ by Theorem 1.66, and the estimate for $\deg(h)$ in Lemma 6.58 yields

$$N(b) < q^{s-1} + \frac{nq^s}{M}.$$

Since

$$M \geqslant \frac{1}{2n} q^{s-k-1},$$

we obtain

$$N(b) < q^{s-1} + 2n^2 q^{k+1} \quad \text{for all } b \in \mathbb{F}_q.$$

Consequently,

$$N(b) = q^s - \sum_{\substack{c \in \mathbb{F}_q \\ c \neq b}} N(c) > q^s - (q-1)q^{s-1} - 2(q-1)n^2 q^{k+1}$$

$$> q^{s-1} - 2n^2 q^{k+2},$$

so altogether

$$|N(b) - q^{s-1}| < 2n^2 q^{k+2} \leqslant 2n^2 q^{(s/2)+2},$$

and (6.36) follows again. □

6.60. Theorem. *The complex numbers* $\omega_1, \ldots, \omega_{n-1}$ *in Theorem* 5.36 *satisfy* $|\omega_j| \leqslant q^{1/2}$ *for* $1 \leqslant j \leqslant n-1$.

Proof. We use the notation and the hypotheses of Theorem 5.36. With $E = \mathbb{F}_{q^s}$ we have

$$\sum_{\gamma \in E} \chi^{(s)}(f(\gamma)) = \sum_{\gamma \in E} \chi\left(\mathrm{Tr}_{E/\mathbb{F}_q}(f(\gamma))\right) = \sum_{b \in \mathbb{F}_q} N(b)\chi(b).$$

If we put $N(b) = q^{s-1} + R(b)$, then $|R(b)| < 2n^2 q^{(s/2)+4}$ by (6.36), thus using (5.9),

$$\left|\sum_{\gamma \in E} \chi^{(s)}(f(\gamma))\right| = \left|\sum_{b \in \mathbb{F}_q} (q^{s-1} + R(b))\chi(b)\right| = \left|\sum_{b \in \mathbb{F}_q} R(b)\chi(b)\right|$$

$$\leqslant \sum_{b \in \mathbb{F}_q} |R(b)| \leqslant 2n^2 q^{(s/2)+5}.$$

It follows from Theorem 5.36 that

$$|\omega_1^s + \cdots + \omega_{n-1}^s| \leqslant 2n^2 q^5 \cdot q^{s/2} \quad \text{for } s = 1, 2, \ldots,$$

hence Lemma 6.55 implies $|\omega_j| \leqslant q^{1/2}$ for $1 \leqslant j \leqslant n-1$. □

The estimate for character sums in Theorem 5.38 is now proved completely. This estimate can then be used to improve upon Theorem 6.59.

6.61. Theorem. *Let* $f \in \mathbb{F}_q[x]$ *with* $\deg(f) = n \geqslant 1$ *and* $\gcd(n, q) = 1$. *Then for* $b \in \mathbb{F}_q$, *the number* $N(b)$ *of* $\gamma \in E = \mathbb{F}_{q^s}$ *with* $\mathrm{Tr}_{E/\mathbb{F}_q}(f(\gamma)) = b$

satisfies

$$|N(b)-q^{s-1}| \leqslant \left(1-\frac{1}{q}\right)(n-1)q^{s/2}.$$

Proof. By (5.10) we have

$$N(b)=\frac{1}{q}\sum_{\gamma\in E}\sum_{\chi}\chi\left(\mathrm{Tr}_{E/\mathbb{F}_q}(f(\gamma))\right)\bar{\chi}(b),$$

where the inner sum is over all additive characters χ of $\mathbb{F}_q$. Changing the order of summation and separating the contribution from the trivial character χ_0, we get

$$N(b)=q^{s-1}+\frac{1}{q}\sum_{\chi\neq\chi_0}\bar{\chi}(b)\sum_{\gamma\in E}\chi^{(s)}(f(\gamma)),$$

hence

$$|N(b)-q^{s-1}| \leqslant \frac{1}{q}\sum_{\chi\neq\chi_0}\left|\sum_{\gamma\in E}\chi^{(s)}(f(\gamma))\right| \leqslant \left(1-\frac{1}{q}\right)(n-1)q^{s/2}$$

by Theorem 5.38, where we note that the lifted character $\chi^{(s)}$ is nontrivial whenever $\chi\neq\chi_0$. □

6.62. Corollary. *Let* $f\in\mathbb{F}_q[x]$ *with* $\deg(f)=n\geqslant 1$ *and* $\gcd(n,q)=1$, *and let* $E=\mathbb{F}_{q^s}$. *Then the number* N *of solutions of the equation* $y^q-y=f(x)$ *in* E^2 *satisfies*

$$|N-q^s| \leqslant (q-1)(n-1)q^{s/2}.$$

Proof. For $\gamma\in E$ we have $\beta^q-\beta=f(\gamma)$ for some $\beta\in E$ if and only if $\mathrm{Tr}_{E/\mathbb{F}_q}(f(\gamma))=0$, according to Theorem 2.25. Furthermore, for each $\gamma\in E$ with $\beta_0^q-\beta_0=f(\gamma)$ for some $\beta_0\in E$, there are altogether q elements $\beta\in E$ satisfying $\beta^q-\beta=f(\gamma)$—namely, the elements $\beta=\beta_0+c$ with $c\in\mathbb{F}_q$. Thus $N=qN(0)$, and the rest follows from Theorem 6.61. □

6.63. Example. As a further application of the method introduced in this section, we present a proof of Theorem 5.44 for odd q. We use the notation of Theorem 5.43 and write $E=\mathbb{F}_{q^s}$. Then

$$K(\chi^{(s)};a,b)=\sum_{\gamma\in E^*}\chi\left(\mathrm{Tr}_{E/\mathbb{F}_q}(a\gamma+b\gamma^{-1})\right)=\sum_{c\in\mathbb{F}_q}M(c)\chi(c),\tag{6.37}$$

where $M(c)$ is the number of $\gamma\in E^*$ with $\mathrm{Tr}_{E/\mathbb{F}_q}(a\gamma+b\gamma^{-1})=c$. If β_0 is a fixed element of E with $\mathrm{Tr}_{E/\mathbb{F}_q}(\beta_0)=c$, then Theorem 2.25 implies that $\mathrm{Tr}_{E/\mathbb{F}_q}(a\gamma+b\gamma^{-1})=c$ if and only if $a\gamma+b\gamma^{-1}=\beta^q-\beta+\beta_0$ for some $\beta\in E$, or equivalently, $a\gamma^2-(\beta^q-\beta+\beta_0)\gamma+b=0$. Let N be the number of solutions of $ay^2-(x^q-x+\beta_0)y+b=0$ in E^2. Then one sees as in the proof of Corollary 6.62 that $N=qM(c)$. For odd q an obvious manipulation shows that N is also the number of solutions of $y^2=(x^q-x+\beta_0)^2-4ab=$

$f(x)$ in E^2. Since $f'(x) = -2(x^q - x + \beta_0)$, f has only simple roots in view of Theorem 1.68, hence $y^2 - f(x)$ is absolutely irreducible by Lemma 6.54. Then Theorem 6.57 yields

$$|N - q^s| \leqslant (2q-1)q^{s/2}.$$

Thus, if we write $M(c) = q^{s-1} + R(c)$ and recall that $M(c) = N/q$, we obtain $|R(c)| \leqslant 2q^{s/2}$. Together with (6.37) and (5.9) we get

$$|K(\chi^{(s)}; a, b)| = \left| \sum_{c \in \mathbb{F}_q} (q^{s-1} + R(c))\chi(c) \right| = \left| \sum_{c \in \mathbb{F}_q} R(c)\chi(c) \right|$$
$$\leqslant \sum_{c \in \mathbb{F}_q} |R(c)| \leqslant 2q \cdot q^{s/2},$$

and so by Theorem 5.43,

$$|\omega_1^s + \omega_2^s| \leqslant 2q \cdot q^{s/2}.$$

As this holds for $s = 1, 2, \ldots$, Lemma 6.55 implies that $|\omega_j| \leqslant q^{1/2}$ for $j = 1, 2$. Moreover, in the proof of Theorem 5.43 we have established the polynomial identity

$$L(z) = 1 + Kz + qz^2 = (1 - \omega_1 z)(1 - \omega_2 z),$$

which shows, in particular, that $\omega_1 \omega_2 = q$. Thus we must have $|\omega_1| = |\omega_2| = q^{1/2}$. □

NOTES

1. The first important result on the number of solutions of equations over finite fields is that of Lagrange [2] to the effect that a polynomial in one indeterminate of degree $n \geqslant 0$ over $\mathbb{F}_p$, p prime, has at most n roots. This is of course true for any field (see Theorem 1.66). Conditions for all n roots to be in the ground field $\mathbb{F}_q$ are due to Feit and Rees [1], and to Šatunovskiĭ [1], Schönemann [2], and Thouvenot and Châtelet [1] in case q is prime. See also the notes to Chapter 4, Section 3. The values $b \in \mathbb{F}_q$ such that $f(x) + b \in \mathbb{F}_q[x]$ has deg(f) distinct nonzero roots in $\mathbb{F}_q$ are studied in Mignosi [5], [6].

Theorem 6.1 was shown by J. König and the first published account appears in Raussnitz [1]. Shortly after, proofs were also given by Gegenbauer [1], Kronecker [7], and Rados [1]. Further proofs can be found in Gegenbauer [6], [7] and Rédei [10, Ch. 8], and in Tazawa [1] for a special case. A related result, using the sum of all principal minors of fixed order rather than the rank of a circulant matrix, was established by Rédei and Turán [1]. The formulas of Horáková and Schwarz [1] and Schwarz [14] expressing the number of distinct monic irreducible factors of given degree for a polynomial in terms of the ranks of associated matrices can be viewed as extensions of the König-Rados theorem. The number of solutions of $f(x)=0$ that are mth powers in $\mathbb{F}_q^*$ can also be given in terms of the rank of a circulant matrix (see Raussnitz [1], Segre [3], [4], and T. P. Vaughan [1]). Gegenbauer [4] expresses the number of common nonzero roots of two polynomials in similar terms. The decision of whether there is a common root at all can be based on the theory of resultants (see Chapter 1, Section 4); see also Rados [3] and Vogt and Bose [1] for this question. Gegenbauer [2] uses resultants in formulas for the number of common nonzero roots of two polynomials and for the number of distinct nonzero roots of a single polynomial over $\mathbb{F}_p$; see also Ore [1]. Other types of formulas for the number of solutions of $f(x)=0$ in finite prime fields can be found in Bellman [1] and Cazacu [1]. Ore [7] shows an estimate for the number of solutions of an equation $a_0+a_1x+\cdots+a_{p-2}x^{p-2}=0$ in $\mathbb{F}_p$ in which the coefficients satisfy a linear recurrence relation.

The number N of solutions of $f(x)=0$ in $\mathbb{F}_q$ can be determined modulo the characteristic p of $\mathbb{F}_q$ by means of a simple principle due to Lebesgue [1]—namely, that

$$N \equiv \sum_{c\in\mathbb{F}_q}\left(1-f(c)^{q-1}\right) \bmod p.$$

This was developed further by Cipolla [3], Dickson [19], and Hurwitz [1]. A different principle was employed by Schwarz [3] who showed that if f is monic and has no repeated factors, then $N\equiv \operatorname{Tr}(B) \bmod p$, where $\operatorname{Tr}(B)$ denotes the trace of the matrix $B=(b_{ij})$ determined by (4.4). This method was extended in Schwarz [13]; see also the notes to Chapter 4, Section 1. A congruence for N modulo p is also given in Mignosi [7]. A congruence modulo p for the number of common roots of finitely many polynomials that are not roots of any member of another finite set of polynomials can be found in Mignosi [3].

There are many results on the number of solutions of $f(x)=0$ for special classes of polynomials f. For cubic and quartic f see Bose, Chowla, and Rao [3], Carlitz [103], Cazacu and Simovici [1], Cordone [1], Dickson [40, Ch. 8], Grebenjuk [2], Leonard [3], [4], Mirimanoff [1], Oltramare [1], Rédei [5], [6], [10, Ch. 11], Schwarz [3], Segre [10], Skolem [1], [2], [4], Thouvenot [1], Thouvenot and Châtelet [1], and Voronoï [1, Ch. 1], [2]. The case where f is a binomial is particularly easy; for general $\mathbb{F}_q$ it was treated by Dedekind [1], but for finite prime fields it was known much earlier (see the extensive literature in Dickson [40, Ch. 7]). For trinomials see Leonard [2], Liang [1], Segre [10], and Vilanova [1]. For linearized and affine polynomials f see Berlekamp [4, Ch. 11], Liang [1], Segre [10], and Vilanova [1]. Carlitz [118] proved that for $q>k_1>k_2>\cdots>k_s\geqslant 1$ there exist $a_1,\ldots,a_s\in\mathbb{F}_q$ such that $a_1x^{k_1}+\cdots+a_sx^{k_s}+1$ has at least s distinct roots in $\mathbb{F}_q$. Leonard [2] derived from an argument of Birch and Swinnerton-Dyer

[1] an asymptotic formula for the number of $b \in \mathbb{F}_q$ such that $f(x)+b$ has a prescribed number of roots in $\mathbb{F}_q$, under the assumption that f is of a certain general type. Relations between quadratic partitions of a prime p and the number of solutions of quadratic and cubic equations over $\mathbb{F}_p$ were established by Whiteman [4]. Methods of determining the roots of polynomials are discussed in Chapter 4, Section 3.

Corollaries 6.6 and 6.9 are due to Chevalley [1], whereas the refinements contained in Theorems 6.5, 6.8, and 6.11 were shown immediately afterwards by Warning [1]. The result of Corollary 6.6 was conjectured by Dickson [28] and shown there for homogeneous f of degree $\leqslant 3$ and in Dickson [32] for homogeneous f over $\mathbb{F}_2$. Proofs of the Chevalley-Warning theorems can also be found in Ax [1], Borevich and Shafarevich [1, Ch. 1], Greenberg [1, Ch. 2], Ireland and Rosen [1, Ch. 10], Joly [5], W. M. Schmidt [3, Ch. 4], and Serre [1, Ch. 1]. For a different proof of Lemma 6.3 see Dickson [2], [7, Part I, Ch. 4], for the analogous multiple sum see W. L. G. Williams [5], and for an analog for square matrices see Brawley, Carlitz, and Levine [1]. Chevalley's theorem can be used in the proof of Wedderburn's theorem (Theorem 2.55); see Joly [5] and McCrimmon [1]. For other applications we refer to Ax [2], Ax and Kochen [1], Carlitz [32], and L'vov [1]. Results of Chevalley-Warning type for systems of equations are also contained in Segre [7]. In the case $\deg(f)=n$ Carlitz [65] shows that if $f \in \mathbb{F}_q[x_1,\ldots,x_n]$ is homogeneous of degree n and $N(f=0)$ is not divisible by the characteristic p of $\mathbb{F}_q$, while $g \in \mathbb{F}_q[x_1,\ldots,x_n]$ is arbitrary of degree $<n$, then $f(x_1,\ldots,x_n)=g(x_1,\ldots,x_n)$ has at least one solution in $\mathbb{F}_q^n$. If, however, $N(f=0)$ is divisible by p and also $g(0,\ldots,0)=0$, then the number of solutions of $f(x_1,\ldots,x_n)=g(x_1,\ldots,x_n)$ in $\mathbb{F}_q^n$ is divisible by p, and this result can be extended to systems of such equations (Carlitz [87]). See also Terjanian [1] and Joly [5] for related results in the case $\deg(f)=n$. Frattini [2] gives a condition such that a homogeneous equation is solvable by values of the indeterminates distinct from each other and from 0. The result of R. Brauer [1] on systems of homogeneous equations over arbitrary fields can be viewed as a general version of Corollary 6.9. A different extension of Corollary 6.9 for homogeneous polynomials is contained in the work of W. Fulton [1] on varieties over finite fields. Chevalley-Warning theorems for $\mathbb{Z}/(p^k)$ were established by Browkin [1] and Schanuel [1].

A considerable refinement of Theorem 6.5 is the theorem of Ax [1] to the effect that if $f \in \mathbb{F}_q[x_1,\ldots,x_n]$ with $\deg(f)=d<n$ and b is the largest integer $<n/d$, then $N(f=0)$ is divisible by q^b. See also Joly [5] for a proof of this result. A generalization to systems of equations was shown by Katz [1]. Related work is contained in Delsarte and McEliece [1]. See Carlitz [110] for an application of the theorem of Ax.

The homogeneous polynomials $f(x_1,\ldots,x_n)$ constructed in Example 6.7 are called *norm forms* and were introduced by Dickson [16], [28]. Carlitz [78] corrects a proof of Dickson [16] and shows that if for odd $q \geqslant 13$ we have a homogeneous cubic $f \in \mathbb{F}_q[x_1, x_2, x_3]$ with the property that f vanishes on $\mathbb{F}_q^3$ only at $(0,0,0)$, then f must be a norm form. Systems of equations involving norm forms were studied in Carlitz [87]. Norm forms are special cases of so-called factorable polynomials (see Carlitz [12] and the notes to Chapter 3, Section 2). Terjanian [1] proved that if $f \in \mathbb{F}_q[x_1,\ldots,x_n]$ of degree n vanishes on $\mathbb{F}_q^n$ only at $(0,\ldots,0)$, then for every $g \in \mathbb{F}_q[x_1,\ldots,x_n]$ with $\deg(g) < n$ there is a solution of $f(x_1,\ldots,x_n) = g(x_1,\ldots,x_n)$ in $\mathbb{F}_q^n$; see also Joly [5].

Theorem 6.13 is due to Ore [1]; see also Borevich and Shafarevich [1, Ch. 1] and W. M. Schmidt [3, Ch. 4]. The last reference contains also the Theorems 6.15, 6.16, and 6.17. The fact that for nonzero $f \in \mathbb{F}_q[x_1,\ldots,x_n]$ the number of solutions of $f(x_1,\ldots,x_n) = 0$ in $\mathbb{F}_q^n$ is at most of the order of magnitude q^{n-1} was also noted by Min [2] and Lang and Weil [1], and by Hua [9, Ch. 2] for finite prime fields. Extensions of Theorem 6.13 to systems of equations can be found in Chalk and Williams [1] and W. M. Schmidt [3, Ch. 4].

Theorems of König-Rados type for equations in several indeterminates and for systems of such equations were shown by Gegenbauer [5], Rados [2], and Segre [4], [5], [7]. Congruences modulo the characteristic of $\mathbb{F}_q$ for the number of solutions of $f(x_1,\ldots,x_n) = 0$ in $\mathbb{F}_q^n$ can be obtained by an obvious extension of the principle of Lebesgue mentioned above; see Lebesgue [1], Hurwitz [1], Dickson [19], and Segre [3], [4]. An elementary approach to the number of solutions of this equation was used by Cazacu [2]. Vandiver [3] has an expression for the number of solutions of $f(x_1,\ldots,x_n) = 0$ in values of x_i that are m_ith powers in $\mathbb{F}_q^*$ for $1 \leqslant i \leqslant n$. Rédei [3] shows that $f(x_1,\ldots,x_n) = 0$ has no solutions in $\mathbb{F}_q^n$ if and only if $f^{q-1} - 1$ is a linear combination of $x_i^q - x_i$, $1 \leqslant i \leqslant n$, with coefficients in $\mathbb{F}_q[x_1,\ldots,x_n]$. A. Robinson [1, Ch. 2] proved that if a system of polynomial equations in several indeterminates over $\mathbb{Z}$ is such that it has at most m solutions in any extension field of $\mathbb{Q}$, then for all sufficiently large primes p the system obtained by considering the coefficients modulo p has at most m solutions in any extension field of $\mathbb{F}_p$; see also Gilmer and Mott [1].

Various elementary results have been proved for special types of equations in several indeterminates. For equations involving general homogeneous polynomials in nonoverlapping sets of indeterminates see Carlitz [62] and Segre [10], and for the special case of factorable polynomials see Carlitz [58] and K. S. Williams [14]. The number of solutions of $x_1^{m_1} \cdots x_n^{m_n} = f(y_1,\ldots,y_r)$, where $m_1,\ldots,m_n \in \mathbb{N}$ are relatively prime, can be expressed in terms of $N(f = b)$, $b \in \mathbb{F}_q$ (Carlitz [39]); see Porter [9] and van Meter [2], [3] for extensions. Equations and systems of equations involving elementary symmetric polynomials are considered in Aberth [1], Avanesov [1], Carlitz [64] , Fine [1], Mordell [7], Rédei [11, Ch. 6], Schwarz [15], and Sedláček [1]. The equation $f(x)(y-z) + f(y)(z-x) + f(z)(x-y) = 0$ with a polynomial f over $\mathbb{F}_q$ in one indeterminate is treated in Ceccherini and

Hirschfeld [1]. Some attention has been given to multilinear equations. A *k-linear equation* is an equation of the form

$$a_1x_{11}\cdots x_{1k}+a_2x_{21}\cdots x_{2k}+\cdots+a_nx_{n1}\cdots x_{nk}=a$$

in the kn indeterminates x_{ij} with $a_1, a_2,\ldots,a_n$, $a\in\mathbb{F}_q$. See Carlitz [56], E. Cohen [8], Hodges [2], Joly [5], Porter [4], [5], [6], and van Meter [2] for results on multilinear equations and systems thereof. Extensions to the case where higher powers of the x_{ij} are allowed can be found in Carlitz [39], Porter [7], [9], [11], and van Meter [1], [2], [3]. For further elementary results on special systems of equations see Babaev and Ismoilov [1], Corson [1], Klein [1], [2], Mignosi [4], and Segre and Bartocci [1]. The number of solutions of $\det(x_{ij})_{1\leqslant i,j\leqslant n}=a$ was determined by C. Jordan [7] and Fine and Niven [1]; see Carlitz [58], [67] for related equations.

The papers of Ax [2], [3] and Fried and Sacerdote [1] discuss algorithms for deciding the solvability of equations for whole families of finite fields. For matrix equations over finite fields see the notes to Section 2, and for equations with polynomials over finite fields as unknowns see the notes to Chapter 3. Some functional equations over finite fields are discussed in Dickey, Kairies, and Shank [1], Dunn and Lidl [1], Herget [1], and Lüneburg and Plaumann [1].

2. The history of the theory of equations over finite fields involving quadratic forms goes back to Lagrange [3] who showed that $x^2+by^2=c$ with $b,c\in\mathbb{F}_p^*$, p prime, is always solvable in $\mathbb{F}_p^2$. This was needed as a step in the proof of his famous theorem that every positive integer can be written as the sum of four squares of integers. The simple argument in Remark 6.25 is due to Cauchy [1] who deduced it from a general combinatorial principle which was later rediscovered by Davenport [5]. The same argument was used by Dickson [43]. Weber [5, Sec. 64] showed that every element of a finite field can be written as a sum of two squares. Formulas for the number of solutions of $x^2+by^2=c$ in $\mathbb{F}_p^2$ were given by Libri [1], Lebesgue [1], Schönemann [1], and Hermite [1]. For more recent work on this equation see Singh [2], [3] and Somer [1]. Solutions of $ax^2+by^2=c$ in $\mathbb{F}_p^2$ with small values of x and y were investigated by Mordell [9], R. A. Smith [1], and K. S. Williams [12]. The quadratic residue character of the solutions of $x^2+y^2=z^2$ was studied by Burde [8].

Lebesgue [1] established the formula for the number of solutions of $x_1^2+\cdots+x_n^2=b$ in $\mathbb{F}_p^n$, and for arbitrary diagonal quadratic forms over $\mathbb{F}_p$ it was first shown by C. Jordan [1]; see also C. Jordan [2, Secs. 197–200] and Lebesgue [5]. Theorem 6.21 for $\mathbb{F}_p$, p odd, was obtained in C. Jordan [5], [6], and with it the number of solutions for arbitrary quadratic forms over these fields. The extension to any finite field of odd characteristic was carried out in Dickson [4], where one can also find Theorem 6.30. For other accounts of these results and alternative proofs see Bachmann [3, Part II, Ch. 7], Berlekamp [4, Ch. 16], E. Cohen [13], Dickson [7, Part I, Ch. 4; Part II, Chs. 7, 8], Hull [1], Ireland and Rosen [1, Ch. 8], Joly [5], Nagata [1], and W. M. Schmidt [3, Ch. 4]. The method in Remark 6.28 for arbitrary polynomials over $\mathbb{F}_q$ of degree 2 does not work for even q, but the number of solutions can be determined for this case as well (see Carlitz [109]). Kantor [1] determined the number of different values attained by quadratic forms over $\mathbb{F}_p$, p odd.

Systems of equations $f_i(x_1,\ldots,x_n)=b_i\in\mathbb{F}_q$, $1\leqslant i\leqslant k$, with quadratic forms $f_1,\ldots,f_k$ over $\mathbb{F}_q$ have been studied by Birch and Lewis [2], Birch, Lewis, and Murphy [1], Carlitz [56], Dem'janov [2], Lewis and Schuur [1], Mordell [8], Nordon [1], [2], [3], and Weil [8]. For systems involving quadratic and linear equations see Carlitz [101], [102], E. Cohen [10], [11], [12], J. D. Fulton [9], Jung [1], [2], O'Connor [1], O'Connor and Pall [1], and Tietäväinen [3]. Carlitz [101] and Tietäväinen [3] proved a conjecture of E. Cohen [11] to the effect that a system over $\mathbb{F}_q$, q odd, consisting of the equation $f(x_1,\ldots,x_n)=b$ with a nondegenerate quadratic form f and a system of t linear equations in the x_i of full rank is always solvable in $\mathbb{F}_q^n$ if $n\geqslant 2t+2$, but that for $n=2t+1$ there exist such systems that are not solvable in $\mathbb{F}_q^n$. Carlitz [101] also has a similar result for even q. Systems involving quadratic and bilinear forms were considered by Carlitz [56] and Porter [1], [2], [3], [6], [15].

Carlitz [39], [43], [52], [63] studied equations of the form $f(x_1,\ldots,x_n)=g(x_1,\ldots,x_n)$ with f a quadratic polynomial and g of a special type. Special attention was given to the case where $n=3$ or 4 and $g(x_1,\ldots,x_n)=ax_1\cdots x_n+b$ with $a,b\in\mathbb{F}_q$ (see Carlitz [50], [52], [75] and Rosati [1]). Carlitz [43] treated also the equation $f_1f_2+\cdots+f_{2n-1}f_{2n}=b$, where the quadratic forms f_i have no indeterminates in common. Another special type of equation involving quadratic forms is considered in Carlitz [47]. For cubic and quartic forms see Campbell [1], [3], [8], [10], Carlitz [64], Cicchese [1], [2], [3], [4], Davenport and Lewis [1], [4], [6], de Groote [1], [2], Lewis [1], Lewis and Schuur [1], Manin [4, Ch. 4], Mordell [5], [21], and Segre [10], as well as the notes to Sections 3 and 4.

The character sums with quadratic forms in Exercises 6.27–6.30 can be viewed as generalizations of quadratic Gaussian sums (compare with Chapter 5, Section 2). They were first evaluated for finite prime fields, and even for $\mathbb{Z}/(m)$, by C. Jordan [4] and Weber [1]. For further work see Bachmann [3, Part II, Ch. 7], Braun [1], Callahan and Smith [1], Carlitz [59], [109], J. D. Fulton [9], Linnik [2], Malyšev [2], [3, Ch. 1], and R. A. Smith [4]. For related character sums see Carlitz [45], [46], [109], [113], Ono [4], Porter [16], [17], Springer [3], and the references in Berndt and Evans [4] on generalizations of Gaussian sums.

An extensive reduction theory and a theory of invariants for quadratic forms over finite fields was developed by Dickson [11], [12], [13], [18], [22],

[24]. Analogous results for systems of quadratic forms were shown in Dickson [20], [22], [27]. The theory was extended to forms of higher degree in Dickson [15], [24], [29], [30], [34], [36], [37], [38], [39]. Dickson's work is summarized in Dickson [35], [42, Ch. 19]. For a continuation of this work see, for example, Glenn [1], Hazlett [1], [2], [3], [4], Wiley [1], W. L. G. Williams [1], [2], [3], [4], and the expository account in Rutherford [1]. Later investigations of invariants of forms over finite fields were carried out by Almkvist [1], Campbell [9], [11], and Carlitz [47], [53], [59]. It should be noted that the concept of an invariant for such forms occurs already in Hurwitz [1]. The basic Theorem 6.21 is of course valid for any field of characteristic $\neq 2$ since the proof does not use the finiteness of the underlying field; see also Borevich and Shafarevich [1, Suppl.]. The theory of quadratic forms over arbitrary fields of characteristic 2 is more involved. Albert [1] and Arf [1] are basic references, with the latter containing the discovery of an important invariant. For a simplified approach to the Arf invariant see Dieudonné [1], Dye [1], Klingenberg and Witt [1], Springer [1], Tietze [1], and Witt [2]. A relation between generalized Gaussian sums and the Arf invariant appears in Žmud' [1]. Meyer [1] presents an account of the work of Dickson and Arf as it applies to finite fields. We refer also to the work of Campbell [2], [4], [5], [6], [7] on quadratic forms over finite fields of characteristic 2 and to T'u [2] for cubic forms over $\mathbb{F}_2$. Bilinear forms over fields of characteristic 2 are discussed in Pless [2]. For quadratic forms over $\mathbb{F}_q[x]$ see Byers [1] and Carlitz [69].

Applications of quadratic and bilinear forms over finite fields to coding theory appear in Cameron and Seidel [1], P. Delsarte [3], Delsarte and Goethals [2], Goethals [1], Lempel and Winograd [1], and Snapper [1]. For applications to finite geometries see e.g. Artin [7, Ch. 3], Cordes [1], [2], Dai and Feng [1], [2], Feng and Dai [1], [2], Kozel and Šakleina [1], and Snapper [1]. For applications to algebras over finite fields see e.g. Ono [5], [6]. Kaplan [1] proves the law of quadratic reciprocity (see Theorem 5.17) by considering the number of solutions of $x_1^2 + \cdots + x_r^2 \equiv r \bmod p$.

Theorem 6.21 may also be interpreted as saying that every symmetric matrix A over $\mathbb{F}_q$, q odd, is congruent to a diagonal matrix D in the sense that $D = C^{\mathrm{T}}AC$ for some nonsingular matrix C over $\mathbb{F}_q$ (compare with Newman [1, Ch. 4] and Mateos Mateos [1]). For related results on similarity, orthogonal similarity, etc. of matrices over finite fields see Albert [1], Porter [8], [13], Porter and Adams [1], and Porter and Hanson [1], [2]. Because of the connection between quadratic forms and their coefficient matrices for odd q, the number of linear substitutions transforming a quadratic form with coefficient matrix A into an equivalent quadratic form with coefficient matrix B is the same as the number of nonsingular matrices X with $X^{\mathrm{T}}AX = B$. This number was determined by Siegel [1] and Carlitz [54], and for even q the number of solutions X of this matrix equation with A, B symmetric was obtained by Feng and Dai [1], [2] and J. D. Fulton [5], [7];

see Buckhiester [2] for the case $B = 0$. For the case where X is of fixed but not necessarily full rank see Hodges [21]. With $A^{\mathrm{T}} = -A$ defining a *skew-symmetric matrix*, the matrix equation $X^{\mathrm{T}}AX = B$ with given skew-symmetric matrices A, B was discussed by Carlitz [51] and Hodges [22]. The same matrix equation with other types of matrices A, B was considered by Buckhiester [2], [3], [4], [5], [6]. For $X^{\mathrm{T}}X = 0$ see also Carlitz [114] and Perkins [1], [2]. Hodges [9] discusses $X^{\mathrm{T}}A + A^{\mathrm{T}}X = B$ for B symmetric and $X^{\mathrm{T}}A - A^{\mathrm{T}}X = B$ for B skew-symmetric.

The matrix equation

$$X_n^{\mathrm{T}} \cdots X_1^{\mathrm{T}} A X_1 \cdots X_n = B$$

with several unknown matrices $X_1, \ldots, X_n$, and given matrices A, B that are either both symmetric or both skew-symmetric, was studied by Mousouris and Porter [1], Porter and Riveland [1], and Riveland and Porter [1], [2]. See Porter [12], [22] for the extension of the matrix equations of Hodges [9] to several unknowns. Hodges [1], [4] determined the number of solutions of $X_1 + \cdots + X_n = A$, where the X_i are either symmetric or skew-symmetric; see also Porter and Mousouris [5]. Systems of matrix equations involving symmetric or skew-symmetric matrices were studied in Hodges [17], [25].

For a matrix A over $\mathbb{F}_{q^2}$ define A^* by applying the Frobenius automorphism of $\mathbb{F}_{q^2}$ over $\mathbb{F}_q$ to the entries and transposing, and call A *Hermitian* if $A^* = A$. The matrix equation $X^*AX = B$ with A, B Hermitian was studied by Carlitz and Hodges [1], J. D. Fulton [6], Hodges [24], and Wan and Yang [1], [2] and is related to Hermitian forms (see Dickson [11], Albert [1]). The more general equation

$$X_n^* \cdots X_1^* A X_1 \cdots X_n = B$$

with unknown matrices $X_1, \ldots, X_n$ and given Hermitian matrices A, B was discussed by Mousouris and Porter [2]. Hodges [9] considered $X^*A + A^*X = B$ with B Hermitian and Porter [20] the extension to several unknown matrices. Systems of matrix equations involving Hermitian matrices appear in Hodges [25]. Applications of Hermitian forms to finite geometries were studied in detail by Segre [11].

The matrix equation $AX = B$ was considered by Hodges [5], Porter [21], and Porter and Mousouris [2]. See Hodges [19] and Porter and Mousouris [2] for the slightly more general equation $AXC = B$. The papers Hodges [2], [18] are devoted to $XAY = B$ with X, Y unknown. Further extensions appear in Dalla and Porter [1], Porter [10], and Porter and Mousouris [1], where $AX_1 \cdots X_n = B$ is studied, and in Dalla and Porter [2], Porter [14], and Porter and Mousouris [1], where $X_1 \cdots X_n A Y_1 \cdots Y_m = B$ is studied. The last paper contains also results on $AX_1 \cdots X_n C = B$. The matrix equation $AX_1 \cdots X_n = BY_1 \cdots Y_m$ is considered in Porter and Mousouris [3]. For $XA + CY = B$ and extensions to more unknowns see Hodges [28], [29], Plesken [1], Porter [19], and Porter and Mousouris [4]. For $A_1 X_1$

$+ \cdots + A_n X_n = B$ and a slightly more general equation see Porter [18] and Hodges [30], respectively.

Some attention has been given to *involutory matrices*—that is, matrices X satisfying $X^2 = I$ with I an identity matrix. The number of $n \times n$ involutory matrices over $\mathbb{F}_q$ was determined by Hodges [12]; see also Brawley [2]. Enumeration problems for special classes of involutory matrices were solved by Brawley and Levine [2], J. D. Fulton [1], [2], and Perkins and Fulton [1]. Numbers of solutions of some equations in involutory matrices were discussed by J. D. Fulton [3], [4] and Levine and Brawley [1]. For involutory matrices over $\mathbb{Z}/(m)$ see Brawley and Gamble [1]. The matrix equation $f(X) = 0$ for a given polynomial f over $\mathbb{F}_q$ was studied by Hodges [11]; see also Brawley and Mullen [1]. Hodges [14] determined the number of solutions of the system $a_1 X + b_1 Y = c_1 I$, $a_2 X^2 + b_2 Y^2 = c_2 I$, where $a_1, a_2, b_1, b_2 \in \mathbb{F}_q^*$, $c_1, c_2 \in \mathbb{F}_q$.

The number of $n \times n$ matrices over $\mathbb{F}_p$, p prime, with a prescribed value of the determinant was calculated by C. Jordan [7] and Fine and Niven [1]. Determinantal equations involving unknown matrices were studied by Carlitz [51], [55], [60] and Hodges [13], [20]. The notes to Chapter 8, Section 1, will contain results on the number of matrices of certain types, such as matrices of given size and rank.

3. The interest in diagonal equations beyond the linear and quadratic case arose from the theory of cyclotomy where equations of the form $ax^k + by^k = 1$ have to be considered. This theory originated with Gauss; see Gauss [1] for the case $k = 3$ and Gauss [3] for the case $k = 4$. See also Libri [2] and Lebesgue [1] for other early work on these cases. For general k the relations to cyclotomic numbers (see the notes to Chapter 5, Section 3) and to Jacobi and Gaussian sums were fully developed by Kummer [4], [5], [6]. The classical account of this work is Bachmann [1]; for a modern treatment see Storer [1]. Authors of the 19th century (see also Carey [1], Pellet [8], Pepin [1], and Schwering [1]) considered only prime fields $\mathbb{F}_p$. The extension to arbitrary finite fields is due to H. H. Mitchell [1], [2]. The interest in cyclotomy was renewed through the important work of Dickson [44], [45], [46], [47]. The evaluations of cyclotomic numbers of low order in Dickson's papers were carried further in more recent times; see Baumert and Fredricksen [1], Berndt and Evans [1], Bruck [2], Evans and Hill [1], Leonard and Williams [3], [5], Muskat [4], [6], [7], Muskat and Whiteman [1], Parnami, Agrawal, and Rajwade [3], Storer [2], Wells and Muskat [1], and Whiteman [9], [10], [11]. For further work on cyclotomic numbers, such as relations among them, we refer to Baumert, Mills, and Ward [1], Hall [7], Hull [1], Kutzko [1], Myerson [5], Rédei [8], Storer [3], [4], Vandiver [9], [11], [12], [14], [15], [17], [19], [20], [21], [22], Venkatarayudu [1], and Whiteman [3], [5], [6], [14].

The equation $ax^k + by^k = c$ has also been considered independently of the theory of cyclotomy. Skolem [3] showed that for any prime $p \neq 7$ the equation $ax^3 + by^3 = c$ with $a, b, c \in \mathbb{F}_p^*$ has a solution in $\mathbb{F}_p^2$; see also Dunton [1] and Nagell [1]. For further work on the case $k = 3$, in addition to that of Gauss, Libri, and Lebesgue quoted above, we refer to E. Cohen [9], Ireland and Rosen [1, Ch. 8], Pepin [1], Singh [4], and Vaidyanathaswamy [2]. For $k = 4$ see also Parnami, Agrawal, and Rajwade [1]. Results for general k can be found in I. Chowla [1], [3], S. Chowla [12], Ireland and Rosen [1, Ch. 8], Small [1], [2], [3], and Vandiver [3], [4], [13]. A relation between the equation $x^k + y^k = 1$ over $\mathbb{F}_p$ and a graph coloring problem occurs in Greenwood and Gleason [1].

The more general equation $ax^{k_1} + by^{k_2} = c$ was first considered by Pellet [3] and Piuma [1]. See I. Chowla [2], Davenport and Hasse [1], Hua and Vandiver [3], Mordell [5], Vandiver [5], [6], [9], [12], [14], [15], [17], [18], [19], [20], [21], [22], [23], [25], Whiteman [5], and K. S. Williams [23] for further results on this equation. Numbers of solutions of such equations have been tabulated by Lehmer and Vandiver [1], Pearson and Vandiver [1], Selfridge, Nicol, and Vandiver [1], and Vandiver [24]. If one of the $k_i = 2$, we arrive at elliptic and hyperelliptic equations (compare with the notes to Section 4). For a related trinomial equation see Albert [4].

The equation $ax^k + by^k + cz^k = d$, mostly with $a = b = 1$, $c = \pm 1$, $d = 0$, has been studied extensively in connection with Fermat's last theorem. Early results on the existence of nontrivial solutions (i.e., solutions with $x, y, z \neq 0$) of this equation over finite prime fields for $d = 0$ are due to Cornacchia [1], Dickson [25], [26], Hurwitz [2], Mantel [2], Pellet [8], Pepin [1], Schur [1], and Wendt [1]. Cases where only trivial solutions exist were listed in Cornacchia [1] and Dickson [14], [17], [21]. Surveys of this work are given in Bachmann [6], Dickson [41, Ch. 26], Mordell [2], and Ribenboim [2, Ch. 12]. See also Klösgen [1] for a detailed study. Ankeny and Erdös [1] gave a condition so that $x^k + y^k + z^k = 0$ has only trivial solutions in $\mathbb{F}_p^3$ and Vandiver [6], [10] studied $ax^k + by^k + cz^k = 0$ over arbitrary finite fields. Dickson [48] used cyclotomy to derive results on the number of solutions of $ax^k + by^k + cz^k = d$ over $\mathbb{F}_p$. The equation $ax^3 + by^3 + cz^3 = d$ was treated by Chowla, Cowles, and Cowles [1], [2], [3], [4], E. Cohen [7], [9], Lewis [2], Schupfer [1], and Selmer [1]. For $x^4 + y^4 - z^4 = 0$ over $\mathbb{F}_p$ see Dickson [26]. The equation $x^k + y^k + z^k = 0$ over $\mathbb{F}_{q^6}$ with $k = q + 1$ or $q^5 + 1$ was considered by Ennola [1]. Segre [11] and Hirschfeld [1] determined the number of solutions of $x^{q+1} + y^{q+1} + z^{q+1} = 0$ over $\mathbb{F}_{q^2}$ and $\mathbb{F}_{q^4}$. Relations between diagonal equations over finite fields and diophantine equations such as the one in Fermat's last theorem were further investigated by Vandiver [1], [3], [4], [13], [18], [21], [23]. See Khadzhiivanov and Nenov [1] for a recent contribution.

An approach to the diagonal equation (6.10) based on exponential sums was already employed by Pellet [6] in the case where $b = 0$ and all the k_i are equal. The method described in Section 3 for the general case was developed at about the same time by Furtado Gomide [1], Hua and

Vandiver [1], [2], and Weil [6]. For further work on (6.10) using exponential sums see Ankeny [2], Borevich and Shafarevich [1, Ch. 1], Faircloth [1], [2], Faircloth and Vandiver [2], Hua and Vandiver [4], Ireland and Rosen [1, Chs. 8,10], Joly [3], [5], Mordell [23, Ch. 6], Pearson and Vandiver [1], W. M. Schmidt [3, Ch. 4], and Whiteman [7]. The identity (6.15) expresses the fact that the zeta-function of the hypersurface defined by (6.10) satisfies the Weil conjectures (see Weil [6]); compare also with Ireland and Rosen [1, Ch. 11] and the notes to Section 4. For results analogous to Theorems 6.16 and 6.17, but with averages over the equations (6.10) with fixed $k_1,\ldots,k_n$, b and variable $a_1,\ldots,a_n$ see Carlitz and Corson [1], [2] and W. M. Schmidt [3, Ch. 4]. Estimates for the number of solutions of (6.10) can also be obtained without the use of exponential sums; see Mordell [5] and W. M. Schmidt [3, Ch. 4]. An asymptotic formula for the number of solutions was given by Carlitz [34]. Vandiver [7] has a systematic procedure for finding the number of solutions. Gegenbauer [5] considered (6.10) over $\mathbb{F}_p$, p an odd prime, with the k_i being $1, 2$, or $(p-1)/2$. Results on the distribution of the solutions of (6.10) over $\mathbb{F}_p$ were obtained by Chalk [1], Tietäväinen [7], and K. S. Williams [23]. The number of solutions of (6.10) with some of the x_i having prescribed order in $\mathbb{F}_q^*$ was estimated by Carlitz [37], [57].

The special case of (6.10) where all the k_i are equal has received some attention, especially in connection with Waring's problem. Recursion formulas for the number of solutions were found by Lebesgue [1] and Hull [1]. A geometric viewpoint in the study of such equations was used by Segre [10], [11]. For further work on this case see Dem'janov [1], Dickson [46], [47], [48], Hull [1], Myerson [2], [3], and K. S. Williams [17]. The case where all $k_i = 3$ is treated in Chowla, Cowles, and Cowles [1], Myerson [1], Segre [10], and K. S. Williams [1], for all $k_i = 4$ see Myerson [1] and Segre [10], for all $k_i = 5$ see Hayashi [1] and Segre [10], and for all $k_i = 7$ see Segre [10]. The case where the equation is considered over $\mathbb{F}_{q^2}$ and the common value of the k_i is a divisor of $q+1$ was dealt with by Carlitz [71]. For the connections with Waring's problem see Hardy and Littlewood [3], [4] as well as Ayoub [1, Ch. 4], Ellison [1], Hua [9, Ch. 8], Huston [1], Kloosterman [4], Landau [2, Part VI], [4], Siegel [2], and R. C. Vaughan [1, Ch. 2].

The result of Example 6.38 is essentially that of Small [2]. By Remark 6.25, every element of $\mathbb{F}_q$ can be written as a sum of two squares. Except for $q = 4$ and 7, every element of $\mathbb{F}_q$ can be written as a sum of two cubes; see Skolem [3] and Nagell [1] for q prime and Singh [4] for the general case. There are instances in which certain elements of $\mathbb{F}_q$ cannot be written as a sum of any number of kth powers, since the sums of kth powers may all be contained in a proper subfield of $\mathbb{F}_q$. In fact, if $q = p^e$, p prime, then the necessary and sufficient condition for every element of $\mathbb{F}_q$ to be a sum of kth powers is that k has no factor of the form $(q-1)/(p^d-1)$ with d being a divisor of e satisfying $1 \leqslant d < e$. This was shown by Tornheim [1] for k prime and by Bhaskaran [1] for arbitrary k; see also

Anderson [1] and Joly [1], [5]. *Waring's problem* for $\mathbb{F}_p$ asks for the determination of the number $g(k, p)$, defined as the least positive integer n such that any element of $\mathbb{F}_p$ can be written as a sum of n kth powers. The problem is trivial if $d = \gcd(k, p-1) \geqslant (p-1)/2$. For $d < (p-1)/2$ it was shown by Hardy and Littlewood [4] that $g(k, p) \leqslant k$; see also Chowla, Mann, and Straus [1], Dickson [47], Joly [5], Landau [2, Part VI], and Tornheim [1] for comparable results. The first estimate of the type $g(k, p) = O(k^c)$ with an exponent $c < 1$ was established by I. Chowla [4]. Dodson [2] improved the exponent to $c = \frac{7}{8}$, Tietäväinen [13] obtained $c = \frac{3}{5} + \varepsilon$ for any $\varepsilon > 0$, and Dodson and Tietäväinen [1] showed $g(k, p) = O(k^{1/2}(\log k)^2)$, always under the assumption that $d < (p-1)/2$. The last estimate is best possible in the sense that in the same paper infinitely many k are constructed for which $g(k, p) \geqslant (\sqrt{3k} - 1)/2$ for some p with $d < (p-1)/2$. For further work on $g(k, p)$ see Bovey [4] and Small [1], [3]. An analogous question for general $\mathbb{F}_q$ was considered in Schwarz [5], Stemmler [1], Tietäväinen [9], and Tornheim [1]. For more general rings see Chinburg [1] and Joly [1]. Odlyzko and Stanley [1] estimated the number of subsets of $\mathbb{F}_p^*$ for which the sum of kth powers of the elements is equal to a given element of $\mathbb{F}_p$.

Conditions under which the diagonal equation $a_1x_1^k + \cdots + a_nx_n^k = b$ with given $a_i \in \mathbb{F}_q^*$ has a solution for every $b \in \mathbb{F}_q$ were given by Landau [2, Part VI], Dickson [47], Rédei [3], and Chowla, Mann, and Straus [1] for q prime and by Schwarz [6], [8], [9], [10] and Tietäväinen [9] in the general case. For $b = 0$ the question of the nontrivial solvability of this equation arises. It follows from Corollary 6.6 that there is a nontrivial solution (i.e., one for which not all $x_i = 0$) whenever $n > k$. Slight improvements on this result were obtained by Dickson [47] (see also Davenport and Lewis [5], Dočev and Dimitrov [1], and Joly [5]), Gray [1], Lewis [3], and Tietäväinen [9]. Let $G(k, p)$ be the least positive integer n such that the equation $a_1x_1^k + \cdots + a_nx_n^k = 0$ has a nontrivial solution in $\mathbb{F}_p^n$ for any $a_1, \ldots, a_n \in \mathbb{F}_p$. Significant progress was achieved by S. Chowla [9], [15] who showed for any $\varepsilon > 0$ that $G(k, p) < (2 + \varepsilon)\log_2 k$ for all sufficiently large primes k, where $\log_2$ denotes the logarithm to the base 2. A similar result was shown by Chowla and Shimura [1] and Tietäväinen [1], [2] for all sufficiently large odd k. An upper bound for $G(k, p)$ of the order $\log_2 k$ was established by S. Chowla [8], [10] under the condition that -1 is a kth power in $\mathbb{F}_p$, and this result was extended to arbitrary $\mathbb{F}_q$ by Tietäväinen [1], [2] and Heisler [2]. Tietäväinen [11] proved the best possible result $G(k, p) < (1 + \varepsilon)\log_2 k$ for all sufficiently large odd k. Dodson [2] showed $G(k, p) < k^{(2/3)+\varepsilon}$ for all sufficiently large even k not divisible by $p - 1$, and Tietäväinen [16] improved the exponent to $\frac{1}{2} + \varepsilon$. Similar investigations for $\mathbb{Z}/(p^r)$ were carried out by Bovey [1], [2], [3], S. Chowla [15], Chowla and Shimura [1], Davenport and Lewis [5], Dodson [1], [3], [4], Hardy and Littlewood [4], Hua [9, Ch. 8], Huston [1], Norton [1], and Tietäväinen [12]. Instances where the equation $a_1x_1^k + \cdots + a_nx_n^k = 0$ over $\mathbb{F}_p$ has no solution $(c_1, \ldots, c_n) \in \mathbb{F}_p^n$

with all $c_i \neq 0$ can be found in Gegenbauer [3] and Vandiver [8]. Lower bounds for the number of elements of $\mathbb{F}_q$ represented by $a_1x_1^k + \cdots + a_nx_n^k$ were obtained by Chowla, Mann, and Straus [1] (see also Mann [3, Ch. 2]) for q prime and by Diderrich and Mann [1] for arbitrary $\mathbb{F}_q$. E. Lehmer [2] and Kaplan [1] used results on the number of solutions of $x_1^k + \cdots + x_n^k = b$ in $\mathbb{F}_p^n$ in the study of residuacity and reciprocity laws; see also K. S. Williams [33]. Hodges [15] considered the determinantal equation $a_1\det(X_1)^k + \cdots + a_n\det(X_n)^k = b$ in unknown square matrices $X_1,\ldots,X_n$ over $\mathbb{F}_q$ of the same size.

For Theorems 6.41 and 6.42 see Morlaye [1] and Joly [5]. The method of proof goes back to Lebesgue [1], who established congruences modulo p for the number of solutions of diagonal equations over $\mathbb{F}_p$. Further results of this type were shown by Joly [3], [5] and Morlaye [1]. Theorem 6.43 is due to Carlitz [65] who extended a result of Schwarz [6] for constant g. Corollary 6.44 was also noted by Carlitz [65]. Equations of the form $f(x_1)+ \cdots + f(x_n) = b \in \mathbb{F}_p$ with an arbitrary polynomial f over $\mathbb{F}_p$ were already considered by Dickson [47]; see also Hua [2] and Hua and Min [1]. The case of cubic f was studied in greater detail in Ghent [1] and Hua [4], [5], [8]. General equations of the form $f_1(x_1)+ \cdots + f_n(x_n) = b \in \mathbb{F}_q$ with polynomials $f_1,\ldots,f_n$ over $\mathbb{F}_q$ were treated by Carlitz [43], Carlitz, Lewis, Mills, and Straus [1], and Tietäväinen [1], [2], [6], [7], and the even more general case where the f_i are polynomials in nonoverlapping sets of indeterminates was considered by Carlitz and Corson [1], [2]. The case where the f_i are special rational functions appears in Carlitz [43] and Brenner and Carlitz [1].

Systems of two diagonal equations were investigated in Corson [1], Davenport and Lewis [6], [7], Markovič [1], Spackman [1], and Tietäväinen [8]. More general systems of diagonal equations were considered in Akhtar [1], Carlitz and Wells [1], Davenport and Lewis [8], Hua [9, Ch. 11], [10], Karacuba [1], [2], Korobov [3], [7], Linnik [1], Min [1], Rédei [7], Spackman [1], [2], Tietäväinen [1], [2], [6], and Wells [2]. In Tietäväinen [6] it is shown that the system $\sum_{j=1}^{n} a_{ij}x_j^k = 0$, $1 \leqslant i \leqslant t$, over $\mathbb{F}_q$ has a nontrivial solution in $\mathbb{F}_q^n$ for odd $d = \gcd(k, q-1) > 1$ provided that $n \geqslant 2t(1+\log_2(d-1))$. Spackman [2] has results on the distribution of solutions and the existence of small solutions for systems of diagonal equations. Tietäväinen [1], [2], [6] considered systems $\sum_{j=1}^{n} f_{ij}(x_j) = 0$, $1 \leqslant i \leqslant t$, where the f_{ij} are polynomials over $\mathbb{F}_q$ with $f_{ij}(0) = 0$, and gives conditions for the existence of a nontrivial solution. See van der Corput [1] for such systems over the residue class rings $\mathbb{Z}/(p^r)$. For the result of Exercise 6.72 and generalizations thereof see Carlitz and Wells [1] and Wells [2].

4. The method described in this section was developed by S. A. Stepanov and refined and simplified by W. M. Schmidt. It represents a successful attempt to prove by elementary means the results of A. Weil who used sophisticated techniques of algebraic geometry. The method was first

introduced in Stepanov [1] where a result of the type of Theorem 6.53 was shown for the equation $y^2 = f(x)$ over $\mathbb{F}_p$ with f of odd degree (see also Elistratov [6] for a later result of this type). The result of Theorem 6.57 for this equation was then proved in Stepanov [6]. In Stepanov [2] the more general equation $y^m = f(x)$ over $\mathbb{F}_p$ with $\deg(f)$ coprime to m was treated by the method. Equations of the form $y^q - y = f(x)$ were first handled by the method in Stepanov [3], [5]. Another elementary method for these equations was used by Mit'kin [1]. Both types of equations are special cases of the equation

$$f(x, y) = y^m + a_1(x)y^{m-1} + \cdots + a_m(x) = 0.$$

Under the assumptions that $f(x, y)$ is irreducible over the ground field $\mathbb{F}_p$, that $\deg(a_m) = k$ is coprime to m, and that $\deg(a_i) < ik/m$ for $1 \leqslant i \leqslant m-1$, Stepanov [7], [8] showed an estimate of the form $|N - p| \leqslant Cp^{1/2}$ for the number N of solutions of $f(x, y) = 0$ in $\mathbb{F}_p^2$, where the constant C depends only on k and m; see also Stepanov [10]. W. M. Schmidt [1] realized that these conditions on $f(x, y)$ imply that $f(x, y)$ is absolutely irreducible and generalized Stepanov's result by showing that for absolutely irreducible $f(x, y)$ the number N of solutions of $f(x, y) = 0$ in $\mathbb{F}_q^2$ satisfies $|N - q| \leqslant Cq^{1/2}$ with a constant C depending only on $\deg(f)$; see also W. M. Schmidt [3, Ch. 3]. Surveys of Stepanov's method can be found in Stepanov [9], [10], [12], [13]. An excellent comprehensive account of the method is given in the monograph of W. M. Schmidt [3].

Prior to the work of S. A. Stepanov, nontrivial estimates for the number of solutions of $y^m = f(x)$ were obtained by Mordell [5] through elementary means. The character sum estimates of Davenport [1], [2], [3], [7] would also yield such results. See also the remarks below on elliptic and hyperelliptic equations. The idea of using the hyperderivatives $E^{(n)}$ in the study of functions over fields of nonzero characteristic is due to Hasse [7] and Teichmüller [1], who also proved the basic properties of $E^{(n)}$. For an alternative proof of Lemma 6.55 based on diophantine approximations see Lang [1, Ch. 5]; compare also with Exercises 6.68 and 6.69.

The equation $f(x, y) = 0$ is approached via algebraic geometry by thinking of this equation as defining a curve in an affine or projective space over $\mathbb{F}_q$. If $f \in \mathbb{F}_q[x, y]$ is absolutely irreducible and N_1 denotes the number of $\mathbb{F}_q$-rational points (i.e., points in homogeneous coordinates whose coordinate ratios are in $\mathbb{F}_q$) on the projective curve, then

$$|N_1 - q - 1| \leqslant 2gq^{1/2},$$

where g is the genus of the curve. This result was announced in Weil [1], [2] and proved in detail in Weil [3]. It was noted by Lang and Weil [1] that the constant $2g$ cannot in general be replaced by a smaller one. Since $2g \leqslant (d-1)(d-2)$ with $d = \deg(f)$ (see, e.g., Artin [9, Ch. 16]), one obtains for the number N of solutions of $f(x, y) = 0$ in $\mathbb{F}_q^2$ (i.e., for the number of "finite" points on the projective curve) the estimate

$$|N-q| \leqslant (d-1)(d-2)q^{1/2}+d^2.$$

Such an estimate cannot be expected if f is not absolutely irreducible (see Exercises 6.64 and 6.65). These estimates are connected with the so-called *Riemann hypothesis* for curves over finite fields in the following way. For $s \in \mathbb{N}$ let N_s denote the number of $\mathbb{F}_{q^s}$-rational points on the projective curve, and define the *zeta-function* of the curve by

$$Z(t) = \exp\left(\sum_{s=1}^{\infty} (N_s/s)t^s \right),$$

the series being convergent for $|t| < q^{-1}$ by trivial estimates for N_s. Then Weil has shown that $Z(t)$ is in fact a rational function of the form

$$Z(t) = \frac{L(t)}{(1-t)(1-qt)},$$

where $L(t)$ is a polynomial of degree $2g$ with integer coefficients and constant term 1. If one writes

$$L(t) = \prod_{j=1}^{2g} (1-\omega_j t),$$

then the ω_j are algebraic integers. The Riemann hypothesis (proved by Weil) expresses the fact that $|\omega_j| = q^{1/2}$ for $1 \leqslant j \leqslant 2g$. An argument quite similar to that leading to (5.24) shows

$$N_s = q^s + 1 - \sum_{j=1}^{2g} \omega_j^s \quad \text{for all } s,$$

and together with the Riemann hypothesis one obtains in particular the estimate for N_1 given above. Furthermore, the zeta-function satisfies a functional equation, which amounts to the fact that the ω_j can be paired off into complex conjugates.

Other proofs of the Riemann hypothesis for curves over finite fields were given after the work of Weil. Mattuck and Tate [1] deduced the crucial part of Weil's proof from the Riemann-Roch theorem; see Grothendieck [1] for a related proof. The Riemann-Roch theorem was also used in the proof of Bombieri [5], which is otherwise elementary; see also Bombieri [6]. For further proofs or surveys of Weil's results see Deuring [3], Joly [5], Lang [1, Ch. 6], Monsky [1], and Swinnerton-Dyer [3]. The fact that N_s is asymptotically equal to q^s as $s \to \infty$ was shown by Bombieri [1] with much less technical apparatus; see also Andrews [1] and Chowla and Hasse [1]. Pimenov [1] studied the distribution of $N_s - (q^s+1)$ for curves of genus $g > 1$. An investigation of the zeros of $Z(t)$ was carried out by Elistratov [4], [5]. A connection between the Riemann hypothesis for curves and the geometry of numbers over fields of Laurent series of characteristic p was noted by Armitage [2]. Improvements of the upper bound on N_1 were given by Ihara [1] and Manin [5].

A formally different approach to the equation $f(x, y) = 0$ is based on the study of the extension of the field $\mathbb{F}_q(x)$ of rational functions over $\mathbb{F}_q$ defined by the equation—that is, on the study of the corresponding algebraic function field. This viewpoint enabled Artin [1], Davenport and Hasse [1], and Hasse [2], [4], [8] to prove the Riemann hypothesis in some special cases before Weil. In this context one defines a function ζ (called "congruence zeta-function") by analogy with the Dedekind zeta-function for algebraic number fields. The congruence zeta-function is, however, intimately linked with the zeta-function Z for the curve in that $\zeta(u) = Z(q^{-u})$. The Riemann hypothesis amounts to the fact that all zeros of ζ have real part equal to $\frac{1}{2}$. This is quite analogous to the still unproved Riemann hypothesis in classical analytic number theory, one minor difference being that the congruence zeta-function has no trivial zeros. The congruence zeta-function was introduced by Artin [1] in the case of a quadratic extension of $\mathbb{F}_p(x)$ and by F. K. Schmidt [2], [3] in the general case. F. K. Schmidt [3] proved the functional equation for ζ, which can be written as $F(u) = F(1-u)$ with $F(u) = q^{(g-1)u}\zeta(u)$. Hasse [3] showed that if θ is the maximum real part of the zeros of ζ, then with N_1 defined as above,

$$|N_1 - q - 1| \leqslant 2gq^{\theta},$$

and also $\frac{1}{2} \leqslant \theta < 1$ if $g > 0$. The upper bound for θ was improved by Davenport [7] before Weil could show $\theta = \frac{1}{2}$. The connection between the Riemann hypothesis for ζ and sharp estimates for character sums was noted in Hasse [5]. Proofs of the Riemann hypothesis using the theory of algebraic function fields were given after Weil by Igusa [1] and Roquette [1], [2]; see also Eichler [1, Ch. 5] and Hasse [18]. For the general theory of congruence zeta-functions (or zeta-functions for algebraic function fields over $\mathbb{F}_q$) see also Deuring [4, Ch. 3]. A lower bound for the genus g in terms of the degree of the algebraic function field over $\mathbb{F}_q(x)$ was established by Armitage [1]. As far as the theory of algebraic function fields with finite fields of constants is concerned, we remark that the rudiments of such a theory already appear in Kühne [2]. The study of this theory was taken up again by Artin [1], F. K. Schmidt [1], Sengenhorst [1], and Rauter [1], [2] and led to the work discussed above.

An important special case is that of *elliptic equations* (*elliptic curves*, *elliptic function fields*) over $\mathbb{F}_q$. They are singled out by the property $g = 1$ and can be put in the form $y^2 = f(x)$ with q odd, $\deg(f) = 3$ or 4, and f of discriminant $\neq 0$. Hasse [4], [6] announced the proof of the Riemann hypothesis for this case and gave the details in Hasse [8]. In particular, he showed that the number N of solutions of the equation in $\mathbb{F}_q^2$ satisfies

$$|N - q + \eta(a)| < 2q^{1/2},$$

where η is the quadratic character of $\mathbb{F}_q$ and a the coefficient of x^4 in $f(x)$. In Artin [1] and Hasse [2] the Riemann hypothesis was proved for some special elliptic function fields; see also Hasse [17]. Weaker estimates for N were obtained by Mordell [5]. An elementary proof of the estimate for N stated above was given by Manin [1] in the case $\deg(f) = 3$ and q prime; see also Elistratov [3], Gel'fond and Linnik [1, Ch. 10], and Zimmer [1]. The rationality of the zeta-function of an elliptic curve can be proved quite easily (see, e.g., Robert [1, Ch. 4]).

The history of elliptic equations goes back to Gauss [3] who discussed the equation $y^2 = ax^4 + b$ over $\mathbb{F}_p$. In the last entry of his diary Gauss posed the question of determining the number of solutions of $x^2y^2 + x^2 + y^2 = 1$ in $\mathbb{F}_p^2$. This problem was first solved by Herglotz [1], but it was noticed later that the equation can easily be transformed into $y^2 = 1 - x^4$; see, for example, Ireland and Rosen [1, Ch. 11]. The elliptic equations $y^2 = ax^3 + b$ and $y^2 = ax^4 + b$ have received special attention; see Davenport and Hasse [1], Frattini [1], Hasse [15, Ch. 10], Joly [5], E. Lehmer [4], O. Neumann [1], Rajwade [2], [3], and Singh and Rajwade [1]. For the case $y^2 = ax^3 + bx$ we refer to Davenport and Hasse [1], Hasse [15, Ch. 10], Joly [5], E. Lehmer [4], Morlaye [2], Rajwade [4], and Singh and Rajwade [1]. Some other special elliptic equations with cubic f were considered in Abdullaev [1], [2], Baldisserri [1], Olson [1], Rajwade [1], [9], Rajwade and Parnami [1], Tuškina [1], and K. S. Williams [38]. Because of the obvious identity

$$N = \sum_{c \in \mathbb{F}_q} (1 + \eta(f(c)))$$

for the number N of solutions of $y^2 = f(x)$ in $\mathbb{F}_q^2$, the calculation of N is equivalent to the calculation of character sums involving η. Thus, we refer also to Chapter 5, Section 5, and to the notes for that section. The question of whether $N > 0$ was considered by Châtelet [1], [2], Frattini [1], and F. K. Schmidt [3]; see also Joly [5]. The possible values of N for elliptic cubic curves over $\mathbb{F}_q$, $q \leqslant 13$, were determined by de Groote and Hirschfeld [1]. Birch [2] studied the behavior of N for $y^2 = x^3 - ax - b$ and $q = p$ as a, b vary over $\mathbb{F}_p$. Yoshida [1] considered all elliptic curves over $\mathbb{F}_{p^m}$ and determined the asymptotic distribution of N as $m \to \infty$ for fixed p; see Deligne [6] for a generalization. Results on N as the elliptic curve varies over a family can also be found in Milne [2]. The behavior of N for $y^2 \equiv x^3 - Ax - B \bmod p$ with fixed $A, B \in \mathbb{Z}$ and variable p is the subject of important conjectures of Birch and Swinnerton-Dyer [2] and Tate [1]; see also Swinnerton-Dyer [1] and Tate [2] as well as the numerical work of Yamamoto, Naganuma, and Doi [1] on Tate's conjecture. The set of $\mathbb{F}_q$-rational points on an elliptic curve over $\mathbb{F}_q$ can be endowed with a group structure; see, for example, Bedocchi [1] and Borosh, Moreno, and Porta [1], [2] for results on this group. Bedocchi [2] studied isomorphism classes of elliptic curves over $\mathbb{F}_q$. Significant results on the general theory of elliptic function fields, resp. elliptic curves, are contained in Deuring [2] and Waterhouse [1]. Surveys of elliptic equations are presented in Cassels [1] and Zimmer [2, Ch. 11].

An equation of the form $y^2 = f(x)$ over $\mathbb{F}_q$, q odd, with an arbitrary polynomial f is called a *hyperelliptic equation*. We exclude now the work on elliptic equations quoted in the previous paragraph. A general theory of hyperelliptic equations was first developed by Artin [1] who also stated the Riemann hypothesis for this case and showed $\frac{1}{2} \leqslant \theta < 1$ for q prime and $\deg(f) \geqslant 3$, where θ is the maximum real part of the zeros of the congruence zeta-function. Mordell [5] gave nontrivial estimates for the number N of solutions in various cases. In view of the connection between N and character sums involving η, the work of Davenport [1], [3] on such character sums also implies estimates for N. The Riemann hypothesis for $y^2 = ax^m + b$ with m dividing $q-1$ was shown shortly afterwards by Davenport and Hasse [1]. The elementary method of Stepanov [1], [6] for hyperelliptic equations was refined somewhat by Korobov [6], Mit'kin [2], and Stark [1], thus yielding slight improvements on Theorem 6.57 in certain cases. Ono [7] notes a possible improvement for the case $y^2 = x^m + ax + b$. The number of solutions of some special hyperelliptic equations was calculated by E. Lehmer [4] and Rajwade [3], [5], [6], [8]. See also the work on character sums $\sum_{c \in \mathbb{F}_q} \eta(f(c))$ in Chapter 5, Section 5, and the notes for that section. Mit'kin [4] considered the solvability of hyperelliptic equations over $\mathbb{F}_p$. Davenport and Lewis [2] studied the "defect" $N(b) - p$ as b varies over $\mathbb{F}_p$, where $N(b)$ is the number of solutions of $y^2 = f(x) + b$ in $\mathbb{F}_p^2$ and $f \in \mathbb{F}_p[x]$ is fixed. Stephens [1] proved a conjecture of Chowla and Chowla [1] on the existence of small solutions (in terms of x) of $y^2 = (x + a_1) \cdots (x + a_n)$ over $\mathbb{F}_p$ with distinct $a_1, \ldots, a_n \in \mathbb{F}_p$. A special case was treated elementarily by Singh [6]. Segre [4] considered a system of two hyperelliptic equations.

As to other special cases of $f(x, y) = 0$ not yet covered under other categories, we note various instances of $y^m = f(x)$ treated before Weil by Davenport [2], [7], Davenport and Hasse [1], and Mordell [5]; see also Ono [8] for a recent elementary discussion of another case. In the paper of Davenport and Lewis [2] mentioned in the preceding paragraph, the "defect" $N(b) - p$ is also considered for the equation $y^3 = f(x) + b$ over $\mathbb{F}_p$ with $\deg(f) = 3$ or 4. An elementary approach to the Riemann hypothesis for $y^3 = f(x)$ with $\deg(f) = 3$ was developed by Elistratov [1], [2]. Special cases of $y^p - y = f(x)$ appear in Davenport and Hasse [1] and Yamada [1]. Carlitz [125] determined the number of solutions of $y^p - y = ax^{p+1} + bx$ in $\mathbb{F}_q^2$. Stepanov [4] proved Theorem 5.45 on Kloosterman sums by treating the equation $y^p - y = ax + 1/x$ by his elementary method; compare also with Example 6.63 and W. M. Schmidt [3, Ch. 2]. Mordell [25] characterized the quadratic and cubic polynomials $f \in \mathbb{F}_p[x, y]$ with the property that $f(x, y)$

$=0$ has at least one solution in $\mathbb{F}_p$ whenever a value of one of the indeterminates in $\mathbb{F}_p$ is prescribed. Goppa [2] used the theory of curves over $\mathbb{F}_q$ in the construction of codes; see also Manin [5].

For equations $f(x_1,\ldots,x_n)=0$ with $n \geqslant 3$ and systems of such equations, the first general result of importance was that of Lang and Weil [1] which can be conveniently stated in terms of projective varieties. If V is an absolutely irreducible variety in the n-dimensional projective space over $\mathbb{F}_q$ and V is of dimension r and degree d, then the number N_1 of $\mathbb{F}_q$-rational points of V satisfies

$$|N_1 - q^r| \leqslant (d-1)(d-2)q^{r-(1/2)} + A(n,r,d)q^{r-1},$$

where the constant $A(n,r,d)$ depends only on the indicated parameters. A slightly weaker result was shown at the same time by Nisnevich [1]. For the case of a curve $(r=1)$ one obtains, of course, the estimate of Weil stated earlier. Further remarks on the result of Lang and Weil can be found in Segre [9]. For the case of a hypersurface $(r=n-1)$ the following lower bound was established by W. M. Schmidt [2] by an adaptation of Stepanov's method: if $f \in \mathbb{F}_q[x_1,\ldots,x_n]$ is absolutely irreducible and $\deg(f)=d$, then the number N of solutions of $f(x_1,\ldots,x_n)=0$ in $\mathbb{F}_q^n$ satisfies

$$N > q^{n-1} - (d-1)(d-2)q^{n-(3/2)} - 6d^2q^{n-2}$$

for sufficiently large q. For a detailed treatment of the Lang-Weil estimate for hypersurfaces see W. M. Schmidt [3, Ch. 5]. The case of arbitrary dimension is discussed in Joly [5] and W. M. Schmidt [3, Ch. 6]. A Lang-Weil estimate for special classes of varieties was obtained by Carlitz and Wells [1] and Wells [2].

Birch and Lewis [1] used the Lang-Weil estimate to show that if $f \in \mathbb{F}_q[x_1,\ldots,x_n]$ is absolutely irreducible and homogeneous, then $f(x_1,\ldots,x_n)=0$ has more than $\frac{1}{2}q^{n-1}$ nonsingular solutions in $\mathbb{F}_q^n$ for sufficiently large q (a *singular* point in $\mathbb{F}_q^n$ is one at which all partial derivatives of f are 0). See Lewis and Schuur [1] for further results on nonsingular solutions of such equations, in particular for cubic f; for the latter case see also Birch and Lewis [1] and G. L. Watson [1]. Finite sets of homogeneous polynomials of identical degree >1 for which any nontrivial linear combination only has the trivial singular point at $(0,\ldots,0)$ were constructed by Carlitz [66]. Homogeneous polynomials f over $\mathbb{F}_q$ of small degree for which $f(x_1,\ldots,x_n)=0$ only has singular solutions in $\mathbb{F}_q^n$ were characterized by Lewis [1]. Formulas for the number of points on special varieties were obtained by Corson [1], Manin [2], and Swinnerton-Dyer [2]; see also Manin [4, Ch. 4]. The number of solutions of $f(x_1)=f(x_2)=\cdots=f(x_r)$, $f \in \mathbb{F}_q[x]$, in different values of the x_i was estimated by Birch and Swinnerton-Dyer [1] and K. S. Williams [24].

The theory of zeta-functions with its important consequences for the number of solutions of equations can be generalized to varieties. Let V be a variety defined in an n-dimensional affine or projective space over $\mathbb{F}_q$, let N_s be the number of points in $\mathbb{F}_{q^s}^n$ (resp. $\mathbb{F}_{q^s}$-rational points in the projective case) on V, and define the *zeta-function* of V by

$$Z(V;t)=\exp\left(\sum_{s=1}^{\infty}(N_s/s)t^s\right),$$

the series being convergent for $|t|<q^{-n}$. Weil [6] stated several conjectures on the zeta-function, in particular that $Z(V;t)$ is a rational function in t, that it satisfies a functional equation, and that its zeros and poles have certain prescribed absolute values. The rationality of $Z(V;t)$ was proved by Dwork [2] using p-adic techniques; see also Dwork [1] for preliminary results. Later, Grothendieck [2] proved the same result by methods of algebraic geometry. For further discussions or other proofs of the rationality of zeta-functions see Bruhat [1], Dwork [9], Joly [5], Koblitz [2, Ch. 5], [4], Milne [1, Ch. 6], Monsky [1], and Reich [1]. The method of Dwork was extended by Kiefe [1], who showed that the logarithmic derivative of the zeta-function of more general sets V is a rational function. The fact that the zeta-function of a hypersurface defined by a diagonal equation is a rational function was already noted by Weil [6]; compare also with (6.15) and Exercise 6.49 as well as with the discussion of this case in Ireland and Rosen [1, Ch. 11]. Other special cases of Weil's conjecture on the rationality of $Z(V;t)$ were shown before Dwork by Carlitz [42], [56], J. Delsarte [1], Furtado Gomide [1], Sampson and Washnitzer [1], and Weil [8].

If V is a nonsingular absolutely irreducible projective variety of dimension r, then Weil [6] conjectured further that the rational function $Z(V;t)$ has the form

$$Z(V;t)=\frac{P_1(t)P_3(t)\cdots P_{2r-1}(t)}{P_0(t)P_2(t)\cdots P_{2r}(t)}$$

with each P_h being a polynomial with integer coefficients and constant term 1. Moreover, $P_0(t)=1-t$, $P_{2r}(t)=1-q^rt$, and

$$P_h(t)=\prod_{i=1}^{B_h}(1-\alpha_{hi}t)\quad\text{for } 0\leqslant h\leqslant 2r,$$

where α_{hi} is an algebraic integer with $|\alpha_{hi}|=q^{h/2}$. An interpretation of the degree B_h of P_h as a Betti number is also suggested. The form of $Z(V;t)$ leads to the representation

$$N_s=\sum_{h=0}^{2r}(-1)^h\sum_{i=1}^{B_h}\alpha_{hi}^s$$

for the number N_s of $\mathbb{F}_{q^s}$-rational points on V, valid for all $s\in\mathbb{N}$. Discussions of the Weil conjectures from various viewpoints can be found in Deuring [3], Joly [5], Kleiman [1], B. Mazur [1], Milne [1, Ch. 6], Monsky [1],

Swinnerton-Dyer [3], and Tate [1]. The first breakthrough was achieved by Dwork [3], [5], [8] who established for the case of hypersurfaces the factorization of $Z(V;t)$, the interpretation of B_h, and the functional equation

$$Z\left(V;\frac{1}{q^r t}\right)=\pm q^{rk/2}t^k Z(V;t),\qquad k=\sum_{h=0}^{2r}(-1)^h B_h,$$

conjectured by Weil; see also Dwork [6], [7], Ireland [1], and Monsky [1]. The conjectures about the coefficients of the P_h and about the absolute values of the α_{hi} (the so-called *Riemann-Weil hypothesis*) remained unsettled. These results of Dwork for hypersurfaces were extended to the general case by Grothendieck [2], [3]. Alternative proofs were given by Lubkin [1], [2], and Lubkin [3] noted that the polynomials $P_h(q^{-h/2}t)$, $0 \leqslant h \leqslant 2r$, are quasi-self-reciprocal. The Riemann-Weil hypothesis was shown for certain cubic hypersurfaces by Bombieri and Swinnerton-Dyer [1] (see also Bombieri [2]) and Perel'muter [5], for certain quartic hypersurfaces by Deligne [1] and Dwork [10], for abelian varieties by Weil [3], [4], and for some other special types of varieties by Deligne [1], [2], Harder [1], Iskovskih [1], Manin [3], and Weil [9]. Finally, the general Riemann-Weil hypothesis was proved by Deligne [3] who also noted the implication for the number of points on the variety. Katz [3] gives an excellent account of the main ingredients in Deligne's proof, and a brief exposition can be found in Serre [2]. See Deligne [6] and Lubkin [4], [5] for extensions of the theory.

There are some equations in $n \geqslant 3$ indeterminates for which estimates for the number N of solutions that are of the same quality as those resulting from the Riemann-Weil hypothesis can be obtained by more elementary means. Apart from the obvious examples of linear, quadratic, and diagonal equations, there are the results of Davenport and Lewis [4] and Mordell [11], [17], [20], [21] which show that $N=p^2+O(p)$ for various cubic equations $f(x,y,z)=0$ over $\mathbb{F}_p$. Davenport and Lewis [4] have the same type of estimate for the equation $z^m=F(x,y)+b$ over $\mathbb{F}_p$, where F is a homogeneous polynomial, $b\in\mathbb{F}_p^*$, and certain trivial exceptions are excluded. Mordell [7] showed $N=p^{n-1}+O(p^{(n-1)/2})$ for equations in $n\leqslant 4$ indeterminates over $\mathbb{F}_p$ involving linear combinations of elementary symmetric polynomials. Mordell [14] proved $N=p^{n-1}+O(p^{n/2})$ for equations over $\mathbb{F}_p$ in any number n of indeterminates that are close to being diagonal equations. Perel'muter and Postnikov [1] considered $f_0(y)+f_1(y)x_1^{k_1}+\cdots+f_n(y)x_n^{k_n}=0$ over $\mathbb{F}_p$ and showed $N=p^n+O(p^{n/2})$ under certain conditions on the polynomials $f_0, f_1,\ldots,f_n$; see W. M. Schmidt [3, Ch. 4] for a generalization.

The estimates of Lang-Weil and Deligne for the number of points on varieties can be used to prove results on the distribution of the solutions of the equation $f(x_1,\ldots,x_n)=0$ or of a system of such equations. We refer to

Chalk and Williams [1], Myerson [4], R. A. Smith [2], and K. S. Williams [18]. For the general principle underlying the proofs of such results see Mordell [22] and Chalk [2].

As to further work on zeta-functions of varieties, we refer to Shimura and Taniyama [1, Ch. 4] and Waterhouse and Milne [1] for abelian varieties, to Manin [4, Ch. 4] and Swinnerton-Dyer [2] for the case of cubic surfaces, to Perel'muter [4] for the hypersurfaces $y^2 = f(x_1,\ldots,x_n)$ and $y^p - y = f(x_1,\ldots,x_n)$ over $\mathbb{F}_p$, to V. N. Kuznecov [1] for the hypersurfaces $y^p - y = f_0(x)+f_1(x)x_1^{k_1} + \cdots + f_n(x)x_n^{k_n}$ over $\mathbb{F}_q$, and also to Bayer and Neukirch [1], Dwork [4], [10], Schneider [1], Swinnerton-Dyer [1], and Taniyama [1]. Good accounts of the theory of zeta-functions can be found in B. Mazur [1] and Thomas [1]. Katz [2] and Koblitz [1] studied the behavior of the zeta-function over a family of varieties. In Curtis [1] one finds a connection between zeta-functions of varieties and the character theory of certain finite groups. Spackman [1] gives an elementary proof of the Weil conjectures for varieties defined by a pair of diagonal equations. Haris [1] notes a connection between Deligne's work on the Weil conjectures and abstract Poisson formulas. Lang [2] discusses relations between points on varieties over finite fields and points on varieties over algebraic number fields. For further information on absolutely irreducible polynomials see Fredman [2], W. M. Schmidt [3, Ch. 5], and K. S. Williams [3].

The theory of zeta-functions of varieties can be extended to that of L-functions. These functions play the same role for the estimation of character sums as the zeta-functions do for the estimation of the number of points on varieties. Let V be an affine or projective variety over $\mathbb{F}_q$, let $\chi^{(s)}$ be the canonical additive character of $\mathbb{F}_{q^s}$, and define the character sum

$$C_s = C_s(V, f) = \Sigma\chi^{(s)}(f(Q_s)),$$

where Q_s runs through all points on V with coordinates in $\mathbb{F}_{q^s}$ (resp. $\mathbb{F}_{q^s}$-rational points in the projective case) and f is a regular function on V so that $f(Q_s) \in \mathbb{F}_{q^s}$ for all points Q_s—for example, a polynomial over $\mathbb{F}_q$. The corresponding *L-function* is defined by

$$L(V, f; t) = \exp\left(\sum_{s=1}^{\infty} (C_s/s)t^s\right).$$

The fact that L-functions are again rational functions was first proved by Grothendieck [2], [3]; see also Bombieri [3], [4], Dwork [9], Hooley [6], and Perel'muter [6]. Under suitable restrictions on V and f appropriate Weil conjectures can be enunciated. Some of them—for example, the functional equation and the factorization of L-functions—were already shown by Grothendieck [2], [3]; for the case of L-functions attached to algebraic function fields over $\mathbb{F}_q$ see Weissinger [1]. The analog of the Riemann-Weil hypothesis was, however, resolved only through the work of Deligne [3], [4],

[6]. Surveys of Deligne's results are presented in Katz [4] and Serre [3]. Further work on such L-functions has been carried out by Adolphson and Sperber [1], Bombieri [7], Hooley [6], Laumon [1], and Sperber [1], [3]. A similar theory can be developed by using multiplicative rather than additive characters, and even for hybrid exponential sums (see Katz [4], Perel'muter [3], and Serre [3]). The implications of this work for the estimation of exponential sums are discussed in the notes to Chapter 5, Sections 4 and 5.

EXERCISES

6.1. Find the number of solutions of $x^5+2x^4+x^2+1=0$ in $\mathbb{F}_5$.

6.2. Find the number of solutions of $x^6-3x^5-x^4+x^3-x=0$ in $\mathbb{F}_7$.

6.3. Let $f\in\mathbb{F}_q[x_1,\ldots,x_n]$ with $\deg(f)<n$, and let $(c_{i1},\ldots,c_{in})$, $i=1,\ldots,N$, be the solutions of $f(x_1,\ldots,x_n)=0$ in $\mathbb{F}_q^n$. Prove that

$$\sum_{i=1}^{N} c_{ij}^k=0 \quad \text{for } j=1,\ldots,n \quad \text{and} \quad k=0,1,\ldots,q-2.$$

6.4. A polynomial $f\in\mathbb{F}_q[x_1,\ldots,x_n]$ of degree n is called a *norm polynomial* if $(0,\ldots,0)$ is the only solution in $\mathbb{F}_q^n$ of the equation $f(x_1,\ldots,x_n)=0$ (this definition is suggested by Example 6.7). For a norm polynomial f and an arbitrary polynomial $g\in\mathbb{F}_q[x_1,\ldots,x_n]$ with $\deg(g)<n$, prove that the equation $f(x_1,\ldots,x_n)=g(x_1,\ldots,x_n)$ has at least one solution in $\mathbb{F}_q^n$.

6.5. Construct explicitly a norm polynomial in $\mathbb{F}_2[x_1,x_2,x_3]$ of degree 3.

6.6. For a homogeneous $f\in\mathbb{F}_q[x_1,\ldots,x_n]$ of degree $d\geqslant 1$ that is divisible by some x_i, prove that the number of solutions of $f(x_1,\ldots,x_n)=0$ in $\mathbb{F}_q^n$ is at most $dq^{n-1}-(d-1)q^{n-2}$.

6.7. Prove that the number $\omega(d)$ defined prior to Theorem 6.16 is equal to the binomial coefficient $\binom{n+d}{d}$.

6.8. Find a diagonal quadratic form equivalent to $x_1x_2+2x_1x_3-x_2x_3\in\mathbb{F}_5[x_1,x_2,x_3]$.

6.9. Find a diagonal quadratic form equivalent to $2x_1^2-5x_2^2+3x_3^2+x_1x_2-4x_1x_3-3x_2x_3\in\mathbb{F}_{11}[x_1,x_2,x_3]$.

6.10. Determine the number of solutions of $3x_1^2+x_3^2-2x_1x_2+x_1x_3+3x_2x_3=2$ in $\mathbb{F}_7^3$.

6.11. Determine the number of solutions of $2x_1^2-x_2^2+x_3^2-x_1x_2-2x_1x_3=2$ in $\mathbb{F}_5^3$.

6.12. Determine the number of solutions of $x_1^2-x_2^2+2x_3^2-x_1x_4-x_2x_3+2x_2x_4=0$ in $\mathbb{F}_5^4$.

6.13. Determine the number of solutions of $x_1^2+x_3^2+x_5^2-x_1x_2-x_2x_3+x_2x_4+x_4x_5=-1$ in $\mathbb{F}_3^5$.

6.14. Determine the number of solutions of $x_1^2+2x_2^2-2x_3^2-2x_1x_2+x_1x_3+2x_2x_3=1$ in $\mathbb{F}_{25}^3$.

6.15. Determine the number of solutions of $x_1^2-x_2^2-x_3^2+x_1x_2+2x_2x_3-x_1+2x_3=0$ in $\mathbb{F}_5^3$.

6.16. Determine the number of solutions of $x_2^2-x_3^2+x_4^2-x_1x_2-x_1x_3+x_1x_4+x_3x_4+x_2+x_4=1$ in $\mathbb{F}_3^4$.

6.17. Determine the number of solutions of $x_1^2+x_2^2+x_1x_2+x_1x_3=1$ in $\mathbb{F}_4^3$.

6.18. Determine the number of solutions of $x_1^2+x_2^2+x_1x_3+x_1x_4+x_2x_4+x_3x_4=1$ in $\mathbb{F}_8^4$.

6.19. Determine the number of solutions of $\theta^2x_2^2+\theta x_4^2+x_1x_2+x_1x_3+x_1x_4+x_2x_3=\theta$ in $\mathbb{F}_4^4$, where $\theta\in\mathbb{F}_4$ satisfies $\theta^2=\theta+1$.

6.20. Determine the number of solutions of $(\theta^2+\theta)x_2^2+x_3^2+\theta x_4^2+\theta^2x_1x_2+\theta x_1x_4+\theta x_2x_3+(\theta+1)x_2x_4=0$ in $\mathbb{F}_8^4$, where $\theta\in\mathbb{F}_8$ satisfies $\theta^3=\theta^2+1$.

6.21. Prove Lemma 6.24 by the method in the proof of Lemma 6.31.

6.22. If f is a quadratic form in $n\geqslant 3$ indeterminates over $\mathbb{F}_q$, prove that the equation $f(x_1,\ldots,x_n)=0$ has a nontrivial solution in $\mathbb{F}_q^n$. Prove also that the condition $n\geqslant 3$ cannot be replaced by $n\geqslant 2$.

6.23. Let $f\in\mathbb{F}_q[x_1,\ldots,x_n]$ be a nonzero quadratic form and $b\in\mathbb{F}_q$. Prove that the equation $f(x_1,\ldots,x_n)=b$ has a solution in $\mathbb{F}_q^n$ whenever: (i) q is even; or (ii) q is odd and the rank of the coefficient matrix of f is $\geqslant 2$.

6.24. Prove that two nondegenerate quadratic forms $f, g\in\mathbb{F}_q[x_1,\ldots,x_n]$, q odd, are equivalent if and only if $\eta(\det(f))=\eta(\det(g))$, where η is the quadratic character of $\mathbb{F}_q$. (*Hint*: Use Lemma 6.20 and Exercise 6.23 in the proof of the nontrivial implication.)

6.25. Prove that all nondegenerate quadratic forms $f\in\mathbb{F}_q[x_1,x_2]$, q odd, for which the equation $f(x_1,x_2)=0$ has a nontrivial solution in $\mathbb{F}_q^2$ are equivalent to each other.

6.26. Prove that every nondegenerate quadratic form $f\in\mathbb{F}_q[x_1,\ldots,x_n]$, q odd, is equivalent to $x_1x_2+x_3x_4+\cdots+x_{n-3}x_{n-2}+x_{n-1}^2+ax_n^2$ for some $a\in\mathbb{F}_q^*$ if n is even and equivalent to $x_1x_2+x_3x_4+\cdots+x_{n-2}x_{n-1}+ax_n^2$ for some $a\in\mathbb{F}_q^*$ if n is odd.

6.27. Let f be a nondegenerate quadratic form over $\mathbb{F}_q$, q odd, in an even number n of indeterminates and with $\det(f)=\Delta$. Prove that for every nontrivial additive character χ of $\mathbb{F}_q$,

$$\sum_{c_1,\ldots,c_n\in\mathbb{F}_q}\chi(f(c_1,\ldots,c_n))=q^{n/2}\eta\left((-1)^{n/2}\Delta\right),$$

where η is the quadratic character of $\mathbb{F}_q$.

6.28. Use the notation of Exercise 6.27, but suppose that the number n of indeterminates is odd. Prove that

$$\sum_{c_1,\ldots,c_n\in\mathbb{F}_q}\chi(f(c_1,\ldots,c_n))=q^{(n-1)/2}\eta\left((-1)^{(n-1)/2}\Delta\right)G(\eta,\chi).$$

6.29. Let f be a nondegenerate quadratic form over $\mathbb{F}_q$, q even, in an odd number n of indeterminates. Prove that for every nontrivial additive character χ of $\mathbb{F}_q$,

$$\sum_{c_1,\ldots,c_n \in \mathbb{F}_q} \chi(f(c_1,\ldots,c_n)) = 0.$$

6.30. Let $f \in \mathbb{F}_q[x_1,\ldots,x_n]$, q even, n even, be a nondegenerate quadratic form and let χ be a nontrivial additive character of $\mathbb{F}_q$. Prove that

$$\sum_{c_1,\ldots,c_n \in \mathbb{F}_q} \chi(f(c_1,\ldots,c_n)) = \varepsilon q^{n/2},$$

where $\varepsilon = 1$ if f is equivalent to $x_1x_2 + x_3x_4 + \cdots + x_{n-1}x_n$ and $\varepsilon = -1$ if f is equivalent to $x_1x_2 + x_3x_4 + \cdots + x_{n-1}x_n + x_{n-1}^2 + ax_n^2$ with $a \in \mathbb{F}_q$ and $\mathrm{Tr}_{\mathbb{F}_q}(a) = 1$.

6.31. Let $a_0, b_0 \in \mathbb{F}_q$, q odd, and let $a_1,\ldots,a_n \in \mathbb{F}_q^*$, $b_1,\ldots,b_n \in \mathbb{F}_q$, where $b_i \neq 0$ for at least one i, $1 \leqslant i \leqslant n$. Denote by N the number of common solutions in $\mathbb{F}_q^n$ of the equations

$$\begin{cases} a_1x_1^2 + \cdots + a_nx_n^2 = a_0 \\ b_1x_1 + \cdots + b_nx_n = b_0 \end{cases}$$

Put $a = a_1 \cdots a_n$, $b = b_1^2a_1^{-1} + \cdots + b_n^2a_n^{-1}$, $c = b_0^2 - a_0b$. For $b \neq 0$, $c = 0$ prove that

$$N = \begin{cases} q^{n-2} & \text{if } n \text{ is even,} \\ q^{n-2} + q^{(n-3)/2}(q-1)\eta\left((-1)^{(n-1)/2}ab\right) & \text{if } n \text{ is odd,} \end{cases}$$

where η is the quadratic character of $\mathbb{F}_q$.

6.32. Let N be as in Exercise 6.31. For $b \neq 0$, $c \neq 0$ prove that

$$N = \begin{cases} q^{n-2} + q^{(n-2)/2}\eta\left((-1)^{n/2}ac\right) & \text{if } n \text{ is even,} \\ q^{n-2} - q^{(n-3)/2}\eta\left((-1)^{(n-1)/2}ab\right) & \text{if } n \text{ is odd.} \end{cases}$$

6.33. Let N be as in Exercise 6.31. For $b = c = 0$ prove that

$$N = \begin{cases} q^{n-2} + v(a_0)q^{(n-2)/2}\eta\left((-1)^{n/2}a\right) & \text{if } n \text{ is even,} \\ q^{n-2} + q^{(n-1)/2}\eta\left((-1)^{(n-1)/2}a_0a\right) & \text{if } n \text{ is odd,} \end{cases}$$

where v is as in Definition 6.22.

6.34. Let N be as in Exercise 6.31. For $b = 0$, $c \neq 0$ prove that $N = q^{n-2}$.

6.35. Let $a_0, b_0 \in \mathbb{F}_q$, q odd, and let $a_1,\ldots,a_n, b_1,\ldots,b_n \in \mathbb{F}_q^*$. Denote by N

the number of common solutions in $\mathbb{F}_q^{2n}$ of the equations

$$\begin{cases} a_1x_1^2 + \cdots + a_nx_n^2 = a_0 \\ b_1x_1y_1 + \cdots + b_nx_ny_n = b_0. \end{cases}$$

Put $a = a_1 \cdots a_n$. Prove that

$$N = \begin{cases} q^{2n-2} + (v(a_0)+1)v(b_0)q^{n-2} \\ \quad + v(a_0)q^{(3n-4)/2}\eta\big((-1)^{n/2}a\big) & \text{if } n \text{ is even,} \\ q^{2n-2} + (v(a_0)+1)v(b_0)q^{n-2} \\ \quad + q^{(3n-3)/2}\eta\big((-1)^{(n-1)/2}a_0a\big) & \text{if } n \text{ is odd,} \end{cases}$$

where η is the quadratic character of $\mathbb{F}_q$ and v is as in Definition 6.22.

6.36. Let $b_1,\ldots,b_n$ be distinct elements of $\mathbb{F}_q$ and let $a, b \in \mathbb{F}_q$. Prove that the number N of common solutions in $\mathbb{F}_q^{2n}$ of the equations

$$\begin{cases} x_1y_1 + \cdots + x_ny_n = a \\ b_1x_1y_1 + \cdots + b_nx_ny_n = b \end{cases}$$

is given by

$$N = q^{2n-2} + q^{n-2}(q-1)\sum_{i=1}^{n} v(b - ab_i)$$
$$+ q^{n-2}(v(a)v(b) + v(a) + v(b)),$$

where v is as in Definition 6.22.

6.37. Let $e \in \mathbb{F}_q$, q odd, and $a, b, c, d \in \mathbb{F}_q^*$. Prove that the number N of solutions of $ax_1^2 + bx_2^2 + cx_3^2 = 2dx_1x_2x_3 + e$ in $\mathbb{F}_q^3$ is given by

$$N = q^2 + 1 + q[\eta(a) + \eta(b) + \eta(c) + \eta(e)]\eta(d^2e - abc),$$

where η is the quadratic character of $\mathbb{F}_q$.

6.38. Let $k, m \in \mathbb{N}$, $n = km$, and let $a_1,\ldots,a_m \in \mathbb{F}_q^*$, $b \in \mathbb{F}_q$. Prove that the number N of solutions of the k-linear equation

$$a_1x_1 \cdots x_k + a_2x_{k+1} \cdots x_{2k} + \cdots + a_mx_{n-k+1} \cdots x_n = b$$

in $\mathbb{F}_q^n$ is given by

$$N = q^{n-1} + v(b)q^{m-1}\left[q^{k-1} - (q-1)^{k-1}\right]^m,$$

where v is as in Definition 6.22.

6.39. Prove that the equation $x_1^3 + x_2^4 + x_3^9 = 0$ has q^2 solutions in $\mathbb{F}_q^3$.

6.40. Prove that the equation $x_1^{15} + x_2^{12} + x_3^6 + x_4^{18} = 0$ has 1,331 solutions in $\mathbb{F}_{11}^4$.

6.41. Prove directly that if $\gcd(k_i, q-1) = 1$ for some i, then (6.10) has q^{n-1} solutions in $\mathbb{F}_q^n$.

6.42. For a prime $p \equiv 1 \bmod 4$, prove that the number N of solutions of

$x_1^2 + x_2^4 = 1$ in $\mathbb{F}_p^2$ is given by

$$N = \begin{cases} p-1+2A \text{ if } p \equiv 1 \bmod 8, \\ p-1-2A \text{ if } p \equiv 5 \bmod 8, \end{cases}$$

where A is the uniquely determined integer $\equiv -1 \bmod 4$ in the representation $p = A^2 + B^2$ with $B \in \mathbb{Z}$.

6.43. Prove that for $q \equiv 1 \bmod 6$ and $b \in \mathbb{F}_q^*$ the number N of solutions of $x_1^2 + x_2^3 = b$ in $\mathbb{F}_q^2$ is given by

$$N = q + 2\eta(b)\mathrm{Re}[\lambda(b)J(\eta, \lambda)],$$

where η is the quadratic character of $\mathbb{F}_q$ and λ is a multiplicative character of $\mathbb{F}_q$ of order 3.

6.44. Prove that for $q \equiv 1 \bmod 4$ and $b \in \mathbb{F}_q^*$ the number N of solutions of $x_1^4 + x_2^4 = b$ in $\mathbb{F}_q^2$ is given by

$$N = \begin{cases} q-3+2\mathrm{Re}[(2\bar{\psi}(b)+\eta(b))J(\eta, \psi)] & \text{if } q \equiv 1 \bmod 8, \\ q+1+2\mathrm{Re}[(2\bar{\psi}(b)-\eta(b))J(\eta, \psi)] & \text{if } q \equiv 5 \bmod 8, \end{cases}$$

where η is the quadratic character of $\mathbb{F}_q$ and ψ is a multiplicative character of $\mathbb{F}_q$ of order 4.

6.45. Let $\mathbb{F}_q$ be a finite field with $q = p^e$, p prime, $e \in \mathbb{N}$, and let $k \in \mathbb{N}$. Prove that every element of $\mathbb{F}_q$ can be written as a sum of kth powers of elements of $\mathbb{F}_q$ if and only if k is not divisible by $(p^e - 1)/(p^d - 1)$ for any divisor d of e with $1 \leqslant d < e$.

6.46. Deduce from Exercise 6.45 that for $q > (k-1)^2$ every element of $\mathbb{F}_q$ can be written as a sum of kth powers of elements of $\mathbb{F}_q$.

6.47. For an integer $s > 1$, let $\beta \in E = \mathbb{F}_{q^s}$, $\beta \notin \mathbb{F}_q$. Prove that the equation

$$a_1 x_1^{(q^s-1)/(q-1)} + \cdots + a_n x_n^{(q^s-1)/(q-1)} = \beta$$

with $a_1, \ldots, a_n \in \mathbb{F}_q$ has no solution in E^n.

6.48. If $d_1 = \cdots = d_n = d$, prove that

$$M(d_1, \ldots, d_n) = \frac{d-1}{d}\left[(d-1)^{n-1} - (-1)^{n-1}\right].$$

6.49. If N_s is as in (6.15) and z is a complex variable, prove that

$$\zeta(z) = \exp\left(\sum_{s=1}^{\infty} (N_s/s) z^s\right)$$

is a rational function. Here

$$\exp(z) = \sum_{s=0}^{\infty} (1/s!) z^s$$

is the exponential function.

6.50. Prove that if N_s, $s=1,2,\ldots$, are complex numbers such that the function $\zeta(z)$ in Exercise 6.49 is rational, then the N_s are given by a formula of the type (6.15).

6.51. If $f \in \mathbb{F}_q[x_1,\ldots,x_n]$, q odd, is a nondegenerate quadratic form and $b \in \mathbb{F}_q$, determine the number N_s of solutions of $f(x_1,\ldots,x_n)=b$ in $\mathbb{F}_{q^s}^n$ and verify that it is of the form (6.15).

6.52. If N_s is defined as in Exercise 6.51, find an explicit representation of the function $\zeta(z)$ defined in Exercise 6.49 as a rational function.

6.53. For $a_1,\ldots,a_n$, $b \in \mathbb{F}_q$ let $N(a_1,\ldots,a_n,b)$ denote the number of solutions of $a_1x_1^{k_1}+\cdots+a_nx_n^{k_n}=b$ in $\mathbb{F}_q^n$, where $k_1,\ldots,k_n \in \mathbb{N}$ are fixed. Prove that the average value of $N(a_1,\ldots,a_n,b)$ is q^{n-1}, in the sense that

$$\frac{1}{q^{n+1}} \sum_{a_1,\ldots,a_n,\, b \in \mathbb{F}_q} N(a_1,\ldots,a_n,b) = q^{n-1}.$$

6.54. With the notation of Exercise 6.53, prove that

$$\sum_{a_1,\ldots,a_n,\, b \in \mathbb{F}_q} \left(N(a_1,\ldots,a_n,b)-q^{n-1}\right)^2 \leqslant q^{2n-1}(q-1)d_1\cdots d_n,$$

where $d_i=\gcd(k_i,q-1)$ for $1 \leqslant i \leqslant n$.

6.55. Prove that the number of solutions of $x_1^2+x_2^3+x_3^5=-2$ in $\mathbb{F}_q^3$ is a positive multiple of the characteristic of $\mathbb{F}_q$.

6.56. If p is the characteristic of $\mathbb{F}_q$ and $d=\gcd(k,q-1)$ divides $p-1$, prove that the diagonal equation $a_1x_1^k+\cdots+a_dx_d^k=b$ has a solution in $\mathbb{F}_q^d$ for any $a_1,\ldots,a_d \in \mathbb{F}_q^*$ and $b \in \mathbb{F}_q$.

6.57. Let p be the characteristic of $\mathbb{F}_q$, let k be a positive divisor of $p-1$, and let $f \in \mathbb{F}_q[x_1,\ldots,x_k]$ be homogeneous of degree k. Suppose that the number of solutions of $f(x_1,\ldots,x_k)=0$ in $\mathbb{F}_q^k$ is not divisible by p. Prove that for any $g \in \mathbb{F}_q[x_1,\ldots,x_k]$ with $\deg(g)<k$ the equation $f(x_1,\ldots,x_k)=g(x_1,\ldots,x_k)$ has at least one solution in $\mathbb{F}_q^k$.

6.58. Let p be the characteristic of $\mathbb{F}_q$, let k be a positive divisor of $p-1$, and let $g \in \mathbb{F}_q[x_1,\ldots,x_k]$ with $\deg(g)<k$. Prove that the equation $x_1\cdots x_k=g(x_1,\ldots,x_k)$ has at least one solution in $\mathbb{F}_q^k$.

6.59. Under the hypotheses of Exercise 6.58, prove that the equation $a_1x_1^k+\cdots+a_kx_k^k+ax_1\cdots x_k=g(x_1,\ldots,x_k)$ has at least one solution in $\mathbb{F}_q^k$ provided that $a \in \mathbb{F}_q^*$ and $a_1,\ldots,a_k \in \mathbb{F}_q$ are such that at least one $a_i=0$.

6.60. For the binomial coefficients $\binom{n}{j}$, $0 \leqslant j \leqslant n$, prove that

$$E_p\left(\binom{n}{j}\right)=\frac{1}{p-1}(s_j+s_{n-j}-s_n),$$

where s_m denotes the sum of digits in the representation of m to the base p.

6.61. Deduce from the result of Exercise 6.60 that if $n = p^k$, $k \in \mathbb{N}$, and $1 \leqslant j \leqslant n$, then $E_p\left(\binom{n}{j}\right) = k - E_p(j)$.

6.62. If p is a fixed prime, prove that $E_p\left(\binom{n}{j}\right) = 0$ for all j with $0 \leqslant j \leqslant n$ if and only if n is of the form $mp^t - 1$ with $1 \leqslant m \leqslant p$ and $t \geqslant 0$.

6.63. Prove that $j \in \mathbb{N}$ satisfies $\gcd\left(\binom{n}{j}, j\right) = 1$ for all n with $j \leqslant n < 2j$ if and only if $j = p^t$, where p is a prime and $t \geqslant 0$.

6.64. Prove that for odd q the equation $y^2 = x^{2k}$ with $k \in \mathbb{N}$ has $2q - 1$ solutions in $\mathbb{F}_q^2$. (Note that $y^2 - x^{2k}$ is not absolutely irreducible.)

6.65. Prove that for $q \equiv 3 \bmod 8$ the equation $y^2 = 2x^4 + 4x^2 + 2$ has no solution in $\mathbb{F}_q^2$. (Note that $y^2 - 2x^4 - 4x^2 - 2$ is irreducible, but not absolutely irreducible over $\mathbb{F}_q$.)

6.66. If $m \in \mathbb{N}$ and $f \in \mathbb{F}_q[x]$ with $\deg(f) = k \geqslant 1$ and $\gcd(m, k) = 1$, prove that $y^m - f(x)$ is absolutely irreducible.

6.67. Generalize Theorem 6.57 as follows. Let $m \in \mathbb{N}$, let $f \in \mathbb{F}_q[x]$ with $\deg(f) \geqslant 1$, and let

$$f(x) = a(x - \alpha_1)^{e_1} \cdots (x - \alpha_d)^{e_d}$$

be the factorization of f in its splitting field over $\mathbb{F}_q$, where $\alpha_1, \ldots, \alpha_d$ are the distinct roots of f. Put $t = \gcd(m, q - 1)$ and $D = \gcd(t, e_1, \ldots, e_d)$, and let r be the number of α_i in $\mathbb{F}_q$. Then the number N of solutions of the equation $y^m = f(x)$ in $\mathbb{F}_q^2$ satisfies

$$|N - Dq + (D - 1)r| \leqslant (t - D)(d - 1)q^{1/2}$$

if a is the Dth power of an element of $\mathbb{F}_q^*$ and $N = r$ otherwise.

6.68. Dirichlet's theorem on simultaneous approximation says that if $t_1, \ldots, t_n$ are real numbers, then there exist $(n+1)$-tuples $(s, m_1, \ldots, m_n) \in \mathbb{Z}^{n+1}$ with arbitrarily large $s > 0$ such that $|t_i - (m_i/s)| < s^{-(n+1)/n}$ for $1 \leqslant i \leqslant n$. Use this theorem to prove that if $\omega_1, \ldots, \omega_n$ are complex numbers, then for any $\varepsilon > 0$ there exist infinitely many $s \in \mathbb{N}$ with

$$\operatorname{Re}(\omega_1^s + \cdots + \omega_n^s) \geqslant (1 - \varepsilon)(|\omega_1|^s + \cdots + |\omega_n|^s).$$

6.69. Give an alternative proof of Lemma 6.55 by using the result of Exercise 6.68.

6.70. Under the hypotheses of Theorem 5.36 and using Theorem 5.37, prove that for every $\varepsilon > 0$ there exist infinitely many $s \in \mathbb{N}$ with

$$\left| \sum_{\gamma \in \mathbb{F}_{q^s}} \chi^{(s)}(f(\gamma)) \right| \geqslant (1 - \varepsilon)(n - 1)q^{s/2}.$$

6.71. Under the hypotheses of Theorem 5.39 and using Theorem 5.40, prove that for every $\varepsilon > 0$ there exist infinitely many $s \in \mathbb{N}$ with

$$\left| \sum_{\gamma \in \mathbb{F}_{q^s}} \psi^{(s)}(f(\gamma)) \right| \geqslant (1-\varepsilon)(d-1)q^{s/2}.$$

6.72. Let $a_1, a_2, b_1, b_2 \in \mathbb{F}_q^*$ with $a_1 b_2 \neq a_2 b_1$ and let $m, m_1, m_2 \in \mathbb{N}$. Prove that the number N of common solutions in $\mathbb{F}_q^3$ of the equations

$$\begin{cases} x_1^{m_1} = a_1 + b_1 x_3^m \\ x_2^{m_2} = a_2 + b_2 x_3^m \end{cases}$$

satisfies $|N - q| \leqslant Cq^{1/2}$ for some constant C independent of q.

6.73. If χ_1 is the canonical additive character of $\mathbb{F}_q$, $q = 2^s$, prove that the Kloosterman sum $K(\chi_1;\ 1,1)$ has the value

$$K(\chi_1;\ 1,1) = -\frac{1}{q}\left((-1+i\sqrt{7})^s + (-1-i\sqrt{7})^s\right).$$

Thus show $|K(\chi_1;\ 1,1)| \leqslant 2q^{1/2}$.

6.74. If E is a finite extension of $\mathbb{F}_q$, prove that the number N of $\gamma \in E^*$ with $\mathrm{Tr}_{E/\mathbb{F}_q}(\gamma) = \mathrm{Tr}_{E/\mathbb{F}_q}(\gamma^{-1}) = 0$ is given by

$$N = \frac{1}{q^2} \sum_{a, b \in \mathbb{F}_q} K(\mu_1;\ a, b),$$

where μ_1 is the canonical additive character of E.

6.75. For $E = \mathbb{F}_q$, $q = 2^s$, prove that the number N of $\gamma \in E^*$ with $\mathrm{Tr}_E(\gamma) = \mathrm{Tr}_E(\gamma^{-1}) = 0$ is given by

$$N = \frac{1}{4}\left[q - 3 - \frac{1}{q}(-1+i\sqrt{7})^s - \frac{1}{q}(-1-i\sqrt{7})^s\right].$$

6.76. Under the hypotheses of Theorem 5.43 and using Theorem 5.44, prove that for every $\varepsilon > 0$ there exist infinitely many $s \in \mathbb{N}$ with

$$|K(\chi^{(s)};\ a, b)| \geqslant (2-\varepsilon)q^{s/2}.$$

Chapter 7

Permutation Polynomials

The aim of this chapter is to present a survey of results on polynomials for which the associated polynomial functions are permutations of a given finite field $\mathbb{F}_q$. Polynomials of this type are called permutation polynomials and exist for any $\mathbb{F}_q$ since, more generally, every mapping of $\mathbb{F}_q$ into itself can be expressed by a polynomial.

A number of natural questions arise in connection with permutation polynomials. First, the determination of permutation polynomials is a nontrivial problem, and the criteria in Section 1 can facilitate this task. The conditions for arbitrary polynomials to be permutation polynomials are, however, rather complicated. Therefore, the various results for special types of polynomials in Section 2 are of greater interest.

Permutation polynomials induce permutations of $\mathbb{F}_q$ and thus correspond to elements of the symmetric group S_q on q letters. Thus, given a class of permutation polynomials of $\mathbb{F}_q$ that is closed under composition (or closed under composition modulo $x^q - x$), one may ask which subgroup of S_q is represented by this class. Section 3 is devoted to questions of this kind.

The remarkable relationship between permutation polynomials and exceptional polynomials is explored in Section 4. Applications of the theory of equations over finite fields appear in this context.

The concept of permutation polynomial is generalized in Section 5 by considering polynomials in several indeterminates. As single polynomials in $n \geqslant 2$ indeterminates cannot induce mappings of $\mathbb{F}_q^n$ into itself, the connection with permutations is lost in this case. In order to recover this connection, one has to consider systems of polynomials. This leads to the notion of an orthogonal system of polynomials. The basic properties of permutation polynomials in several indeterminates and orthogonal systems are developed.

1. CRITERIA FOR PERMUTATION POLYNOMIALS

A polynomial $f \in \mathbb{F}_q[x]$ is called a *permutation polynomial* of $\mathbb{F}_q$ if the associated polynomial function $f: c \mapsto f(c)$ from $\mathbb{F}_q$ into $\mathbb{F}_q$ is a permutation of $\mathbb{F}_q$. Obviously, if f is a permutation polynomial of $\mathbb{F}_q$, then the equation $f(x)=a$ has exactly one solution in $\mathbb{F}_q$ for each $a \in \mathbb{F}_q$. Because of the finiteness of $\mathbb{F}_q$, the definition of a permutation polynomial can be expressed in various other ways.

7.1. Lemma. *The polynomial $f \in \mathbb{F}_q[x]$ is a permutation polynomial of $\mathbb{F}_q$ if and only if one of the following conditions holds:*

(i) *the function $f: c \mapsto f(c)$ is onto;*
(ii) *the function $f: c \mapsto f(c)$ is one-to-one;*
(iii) *$f(x)=a$ has a solution in $\mathbb{F}_q$ for each $a \in \mathbb{F}_q$;*
(iv) *$f(x)=a$ has a unique solution in $\mathbb{F}_q$ for each $a \in \mathbb{F}_q$.*

If $\phi: \mathbb{F}_q \to \mathbb{F}_q$ is an arbitrary function from $\mathbb{F}_q$ into $\mathbb{F}_q$, then there exists a unique polynomial $g \in \mathbb{F}_q[x]$ with $\deg(g) < q$ representing ϕ, in the sense that $g(c)=\phi(c)$ for all $c \in \mathbb{F}_q$. The polynomial g can be found by computing the Lagrange interpolation polynomial for the given function ϕ (see Theorem 1.71), or by the formula

$$g(x) = \sum_{c \in \mathbb{F}_q} \phi(c)\left(1-(x-c)^{q-1}\right). \tag{7.1}$$

If ϕ is already given as a polynomial function, say $\phi: c \mapsto f(c)$ with $f \in \mathbb{F}_q[x]$, then g can be obtained from f by reduction modulo $x^q - x$, according to the following result.

7.2. Lemma. *For $f, g \in \mathbb{F}_q[x]$ we have $f(c)=g(c)$ for all $c \in \mathbb{F}_q$ if and only if $f(x) \equiv g(x) \operatorname{mod}(x^q - x)$.*

Proof. By the division algorithm we can write $f(x)-g(x) = h(x)(x^q - x)+r(x)$ with $h, r \in \mathbb{F}_q[x]$ and $\deg(r) < q$. Then $f(c)=g(c)$ for all $c \in \mathbb{F}_q$ if and only if $r(c)=0$ for all $c \in \mathbb{F}_q$, and the latter condition is equivalent to $r=0$. □

We shall now establish a useful criterion for permutation polynomials. The following lemma will be needed.

7.3. Lemma. *Let $a_0, a_1, \ldots, a_{q-1}$ be elements of $\mathbb{F}_q$. Then the following two conditions are equivalent:*

(i) $a_0, a_1, \ldots, a_{q-1}$ *are distinct;*

(ii) $\displaystyle\sum_{i=0}^{q-1} a_i^t = \begin{cases} 0 & \text{for } t = 0, 1, \ldots, q-2, \\ -1 & \text{for } t = q-1. \end{cases}$

Proof. For fixed i with $0 \leqslant i \leqslant q-1$, consider the polynomial

$$g_i(x) = 1 - \sum_{j=0}^{q-1} a_i^{q-1-j} x^j.$$

We have $g_i(a_i) = 1$ and $g_i(b) = 0$ for $b \in \mathbb{F}_q$ with $b \neq a_i$. Therefore the polynomial

$$g(x) = \sum_{i=0}^{q-1} g_i(x) = -\sum_{j=0}^{q-1} \left(\sum_{i=0}^{q-1} a_i^{q-1-j} \right) x^j$$

maps each element of $\mathbb{F}_q$ into 1 if and only if $\{a_0, \ldots, a_{q-1}\} = \mathbb{F}_q$. Since $\deg(g) < q$, Lemma 7.2 shows that the polynomial g maps each element of $\mathbb{F}_q$ into 1 if and only if $g(x) = 1$, which is equivalent to condition (ii). □

7.4. Theorem (Hermite's Criterion). *Let $\mathbb{F}_q$ be of characteristic p. Then $f \in \mathbb{F}_q[x]$ is a permutation polynomial of $\mathbb{F}_q$ if and only if the following two conditions hold:*

(i) *f has exactly one root in $\mathbb{F}_q$;*

(ii) *for each integer t with $1 \leqslant t \leqslant q-2$ and $t \not\equiv 0 \bmod p$, the reduction of $f(x)^t \bmod (x^q - x)$ has degree $\leqslant q-2$.*

Proof. Let f be a permutation polynomial of $\mathbb{F}_q$. Then (i) is trivial. The reduction of $f(x)^t \bmod (x^q - x)$ is some polynomial $\sum_{j=0}^{q-1} b_j^{(t)} x^j$, where $b_{q-1}^{(t)} = -\sum_{c \in \mathbb{F}_q} f(c)^t$ by (7.1). According to Lemma 7.3, $b_{q-1}^{(t)} = 0$ for $t = 1, 2, \ldots, q-2$, hence (ii) follows.

Conversely, let (i) and (ii) be satisfied. Then (i) implies $\sum_{c \in \mathbb{F}_q} f(c)^{q-1} = -1$, while (ii) implies $\sum_{c \in \mathbb{F}_q} f(c)^t = 0$ for $1 \leqslant t \leqslant q-2$, $t \not\equiv 0 \bmod p$. Using

$$\sum_{c \in \mathbb{F}_q} f(c)^{tp^j} = \left(\sum_{c \in \mathbb{F}_q} f(c)^t \right)^{p^j},$$

we get $\sum_{c \in \mathbb{F}_q} f(c)^t = 0$ for $1 \leqslant t \leqslant q-2$, and this identity holds trivially for $t = 0$. Lemma 7.3 implies that f is a permutation polynomial of $\mathbb{F}_q$. □

7.5. Corollary. *If $d > 1$ is a divisor of $q-1$, then there is no permutation polynomial of $\mathbb{F}_q$ of degree d.*

Proof. If $f \in \mathbb{F}_q[x]$ with $\deg(f) = d$, then $\deg(f^{(q-1)/d}) = q-1$, and so condition (ii) of Theorem 7.4 is not satisfied for $t = (q-1)/d$. □

It is clear from the proof of Theorem 7.4 that if $f \in \mathbb{F}_q[x]$ is a permutation polynomial of $\mathbb{F}_q$, then condition (ii) of that theorem holds also without the restriction $t \not\equiv 0 \bmod p$. Condition (i) may be replaced by other conditions, as for instance in the following result.

7.6. Theorem. *Let $\mathbb{F}_q$ be of characteristic p. Then $f \in \mathbb{F}_q[x]$ is a permutation polynomial of $\mathbb{F}_q$ if and only if the following two conditions hold:*

(i) *the reduction of $f(x)^{q-1} \bmod (x^q - x)$ has degree $q-1$;*
(ii) *for each integer t with $1 \leqslant t \leqslant q-2$ and $t \not\equiv 0 \bmod p$, the reduction of $f(x)^t \bmod (x^q - x)$ has degree $\leqslant q-2$.*

Proof. The necessity of (ii) follows from Theorem 7.4. In the notation of the proof of that theorem, we have

$$b_{q-1}^{(q-1)} = -\sum_{c \in \mathbb{F}_q} f(c)^{q-1},$$

thus if f is a permutation polynomial of $\mathbb{F}_q$, then $b_{q-1}^{(q-1)} = 1$, and (i) holds.

Conversely, let (i) and (ii) be satisfied. Then as in the proof of Theorem 7.4, (ii) implies $\sum_{c \in \mathbb{F}_q} f(c)^t = 0$ for $0 \leqslant t \leqslant q-2$, while (i) implies $\sum_{c \in \mathbb{F}_q} f(c)^{q-1} \neq 0$. Thus the polynomial

$$g(x) = -\sum_{j=0}^{q-1} \left(\sum_{c \in \mathbb{F}_q} f(c)^{q-1-j} \right) x^j$$

is a nonzero constant. If f were not a permutation polynomial of $\mathbb{F}_q$, then the argument in the proof of Lemma 7.3 would show that $g(b) = 0$ for some $b \in \mathbb{F}_q$, which is a contradiction. □

A criterion for permutation polynomials can also be given by using additive characters of the underlying finite field (see Chapter 5, Section 1).

7.7. Theorem. *The polynomial $f \in \mathbb{F}_q[x]$ is a permutation polynomial of $\mathbb{F}_q$ if and only if*

$$\sum_{c \in \mathbb{F}_q} \chi(f(c)) = 0 \tag{7.2}$$

for all nontrivial additive characters χ of $\mathbb{F}_q$.

Proof. If f is a permutation polynomial of $\mathbb{F}_q$ and χ is a nontrivial additive character of $\mathbb{F}_q$, then

$$\sum_{c \in \mathbb{F}_q} \chi(f(c)) = \sum_{c \in \mathbb{F}_q} \chi(c) = 0$$

by (5.9). Conversely, if χ_0 denotes the trivial additive character of $\mathbb{F}_q$ and

(7.2) holds for all $\chi \neq \chi_0$, then for any $a \in \mathbb{F}_q$ the number N of solutions of $f(x) = a$ in $\mathbb{F}_q$ is, according to (5.10), given by

$$N = \frac{1}{q} \sum_{c \in \mathbb{F}_q} \sum_{\chi} \chi(f(c)) \overline{\chi(a)} = 1 + \frac{1}{q} \sum_{\chi \neq \chi_0} \overline{\chi(a)} \sum_{c \in \mathbb{F}_q} \chi(f(c)) = 1,$$

hence f is a permutation polynomial of $\mathbb{F}_q$. □

2. SPECIAL TYPES OF PERMUTATION POLYNOMIALS

Some simple examples of permutation polynomials can be obtained from the following elementary results.

7.8. Theorem. (i) *Every linear polynomial over* $\mathbb{F}_q$ *is a permutation polynomial of* $\mathbb{F}_q$.

(ii) *The monomial* x^n *is a permutation polynomial of* $\mathbb{F}_q$ *if and only if* $\gcd(n, q-1) = 1$.

Proof. (i) Trivial. (ii) x^n is a permutation polynomial of $\mathbb{F}_q$ if and only if the function $c \in \mathbb{F}_q \mapsto c^n$ is onto $\mathbb{F}_q$, which happens if and only if $\gcd(n, q-1) = 1$ (use Theorem 1.15(ii)). □

7.9. Theorem. *Let* $\mathbb{F}_q$ *be of characteristic* p. *Then the* p-*polynomial*

$$L(x) = \sum_{i=0}^{m} a_i x^{p^i} \in \mathbb{F}_q[x]$$

is a permutation polynomial of $\mathbb{F}_q$ *if and only if* $L(x)$ *only has the root* 0 *in* $\mathbb{F}_q$.

Proof. From the discussion following Definition 3.49 we know that the function $L: c \in \mathbb{F}_q \mapsto L(c)$ is a linear operator on the vector space $\mathbb{F}_q$ over $\mathbb{F}_p$. Thus L is one-to-one if and only if the polynomial $L(x)$ only has the root 0 in $\mathbb{F}_q$. □

Further examples can be generated from the above by observing that the set of permutation polynomials is closed under composition—that is, if $f(x)$ and $g(x)$ are permutation polynomials of $\mathbb{F}_q$, then $f(g(x))$ is a permutation polynomial of $\mathbb{F}_q$. The next theorem yields another class of permutation polynomials.

7.10. Theorem. *Let* $r \in \mathbb{N}$ *with* $\gcd(r, q-1) = 1$ *and let* s *be a positive divisor of* $q-1$. *Let* $g \in \mathbb{F}_q[x]$ *be such that* $g(x^s)$ *has no nonzero root in* $\mathbb{F}_q$. *Then* $f(x) = x^r(g(x^s))^{(q-1)/s}$ *is a permutation polynomial of* $\mathbb{F}_q$.

Proof. We show that f satisfies the conditions of Theorem 7.4. Condition (i) is obvious. To prove (ii), take $t \in \mathbb{Z}$ with $1 \leqslant t \leqslant q-2$ and suppose first that t is not divisible by s. Note that $f(x)^t$ is a sum of terms whose exponents are of the form $rt + ms$, where $m \in \mathbb{Z}$, $m \geqslant 0$. Since

$\gcd(r, s) = 1$, these exponents are not divisible by s, and hence not by $q - 1$. Consequently, the reduction of $f(x)^t \bmod (x^q - x)$ has degree $\leqslant q - 2$. If t is divisible by s, say $t = ks$ with $k \in \mathbb{N}$, then

$$f(x)^t = x^{rt}(g(x^s))^{(q-1)k}.$$

With $h(x) = x^{rt}$ we have $f(c)^t = h(c)$ for $c \in \mathbb{F}_q^*$ since $g(c^s) \neq 0$, and also $f(0)^t = h(0)$. Then $f(x)^t \equiv x^{rt} \bmod (x^q - x)$ by Lemma 7.2, and since rt is not divisible by $q - 1$, the reduction of $f(x)^t \bmod (x^q - x)$ has degree $\leqslant q - 2$. □

From what we have noted after Theorem 7.9, it follows in particular that if $f \in \mathbb{F}_q[x]$ is a permutation polynomial of $\mathbb{F}_q$ and $b, c, d \in \mathbb{F}_q$ with $c \neq 0$, then $f_1(x) = cf(x + b) + d$ is again a permutation polynomial of $\mathbb{F}_q$. By choosing b, c, d suitably, we can obtain f_1 in *normalized form*—that is, f_1 is monic, $f_1(0) = 0$, and when the degree n of f_1 is not divisible by the characteristic of $\mathbb{F}_q$, the coefficient of x^{n-1} is 0. It suffices, therefore, to study normalized permutation polynomials. On the basis of Hermite's criterion, one obtains the list shown in Table 7.1 of all normalized permutation polynomials of degree $\leqslant 5$.

For odd q we can characterize the permutation polynomials of $\mathbb{F}_q$ of the form $x^{(q+1)/2} + ax$. Let η be the quadratic character of $\mathbb{F}_q$, with the standard convention $\eta(0) = 0$.

7.11. Theorem. *For odd q, the polynomial $x^{(q+1)/2} + ax \in \mathbb{F}_q[x]$ is a permutation polynomial of $\mathbb{F}_q$ if and only if $\eta(a^2 - 1) = 1$.*

TABLE 7.1

Normalized permutation polynomial of $\mathbb{F}_q$	q
x	any q
x^2	$q \equiv 0 \bmod 2$
x^3	$q \not\equiv 1 \bmod 3$
$x^3 - ax$ (a not a square)	$q \equiv 0 \bmod 3$
$x^4 \pm 3x$	$q = 7$
$x^4 + a_1x^2 + a_2x$ (if its only root in $\mathbb{F}_q$ is 0)	$q \equiv 0 \bmod 2$
x^5	$q \not\equiv 1 \bmod 5$
$x^5 - ax$ (a not a fourth power)	$q \equiv 0 \bmod 5$
$x^5 + ax$ ($a^2 = 2$)	$q = 9$
$x^5 \pm 2x^2$	$q = 7$
$x^5 + ax^3 \pm x^2 + 3a^2x$ (a not a square)	$q = 7$
$x^5 + ax^3 + 5^{-1}a^2x$ (a arbitrary)	$q \equiv \pm 2 \bmod 5$
$x^5 + ax^3 + 3a^2x$ (a not a square)	$q = 13$
$x^5 - 2ax^3 + a^2x$ (a not a square)	$q \equiv 0 \bmod 5$

Proof. Put $f(x)=x^{(q+1)/2}+ax$. We show that f is not one-to-one if and only if $\eta(a^2-1)\neq 1$. If $f(c)=f(0)=0$ for $c\in\mathbb{F}_q^*$, then $a=-c^{(q-1)/2}$, hence $\eta(a^2-1)=0$. If $f(b)=f(c)\neq 0$ for $b,c\in\mathbb{F}_q^*$, $b\neq c$, then

$$bc^{-1}=\left(a+c^{(q-1)/2}\right)\left(a+b^{(q-1)/2}\right)^{-1}.$$

If we had $\eta(b)=\eta(c)$, then $b^{(q-1)/2}=c^{(q-1)/2}$, hence $b=c$, a contradiction. Thus $\eta(b)\neq\eta(c)$, say without loss of generality $\eta(b)=-1$, $\eta(c)=1$. Then $b^{(q-1)/2}=-1$, $c^{(q-1)/2}=1$, so

$$-1=\eta(bc^{-1})=\eta\left((a+1)(a-1)^{-1}\right)=\eta((a+1)(a-1))=\eta(a^2-1).$$

Conversely, suppose $\eta(a^2-1)\neq 1$, then either $a^2-1=0$ or $\eta(a^2-1)=-1$. In the first case, we have $a=\pm 1$, and so there is a $c\in\mathbb{F}_q^*$ with $c^{(q-1)/2}=-a$, whence $f(c)=f(0)$. If $\eta(a^2-1)=-1$, set $b=(a+1)(a-1)^{-1}$. Then $\eta(b)=-1$, thus $b^{(q-1)/2}=-1$, and so

$$f(b)=\left(a+b^{(q-1)/2}\right)b=(a-1)b=a+1=f(1)$$

with $b\neq 1$. In both cases, f is not one-to-one. □

7.12. Remark. We have $\eta(a^2-1)=1$ if and only if

$$a=(c^2+1)(c^2-1)^{-1}$$

for some $c\in\mathbb{F}_q^*$ with $c^2\neq 1$. For if $\eta(a^2-1)=1$, then $a^2-1=b^2$ for some $b\in\mathbb{F}_q^*$, and so with $c=(a+1)b^{-1}$ we have $c\neq 0$, $c^2\neq 1$, and

$$(c^2+1)(c^2-1)^{-1}=\left[(a+1)^2+b^2\right]\left[(a+1)^2-b^2\right]^{-1}=a.$$

Conversely, if $a=(c^2+1)(c^2-1)^{-1}$, $c\in\mathbb{F}_q^*$, $c^2\neq 1$, then $a^2-1=4c^2(c^2-1)^{-2}$, hence $\eta(a^2-1)=1$. □

*7.13. **Theorem.** If $a\in\mathbb{F}_q^*$, q odd, then $x^{(q+1)/2}+ax$ is not a permutation polynomial of any $\mathbb{F}_{q^r}$ with $r>1$.*

Proof. If r is even, the result follows from Corollary 7.5. If r is odd, then with $m=(q-1)/2$ we have $q^r\equiv -1 \bmod (m+1)$, so that $q^r=k(m+1)+m$ for some $k\in\mathbb{N}$. We note that $k(m+1)\equiv m+1 \bmod q$ and $\gcd(m+1,q)=1$ imply $k\equiv 1 \bmod q$. Because of Theorem 7.4, it will suffice to show that the reduction of

$$\left(x^{m+1}+ax\right)^{k+m-1}\bmod\left(x^{q^r}-x\right)$$

has degree q^r-1. Now

$$\begin{aligned}\left(x^{m+1}+ax\right)^{k+m-1}&=\sum_{j=0}^{k+m-1}\binom{k+m-1}{j}a^j x^{(m+1)(k+m-j-1)+j}\\&=\sum_{j=0}^{k+m-1}\binom{k+m-1}{j}a^j x^{q^r+m^2-m-1-jm}.\end{aligned}$$

For $j \geqslant m$, the corresponding exponents of x are $\leqslant q^r - 2$. For $j \leqslant m-2$, the corresponding exponents of x are easily seen to be $\geqslant q^r$ and $\leqslant 2q^r - 3$, so that after reduction of these terms $\operatorname{mod}(x^{q^r} - x)$ we get monomials of degree $\leqslant q^r - 2$. The only remaining term is the one for $j = m-1$—namely,

$$\binom{k+m-1}{m-1} a^{m-1} x^{q^r-1}.$$

It suffices then to prove that the binomial coefficient above is not divisible by the characteristic p of $\mathbb{F}_q$. If s_n denotes the sum of digits in the representation of n to the base p, then $k \equiv 1 \bmod q$, $m < q$, and $m \not\equiv 0 \bmod p$ imply that $s_{k+m-1} = s_{m-1} + s_k$, hence Lemma 6.39 yields

$$E_p\left(\binom{k+m-1}{m-1}\right) = \frac{1}{p-1}\left(s_{m-1} + s_k - s_{k+m-1}\right) = 0,$$

which is the desired fact. □

Theorem 7.13 suggests that polynomials over $\mathbb{F}_q$ that are permutation polynomials of *all* finite extensions of $\mathbb{F}_q$ are likely to be rare. In fact, the polynomials with this property can be classified completely and are indeed of a very special form.

7.14. Theorem. *A polynomial $f \in \mathbb{F}_q[x]$ is a permutation polynomial of all finite extensions of $\mathbb{F}_q$ if and only if it is of the form $f(x) = ax^{p^h} + b$, where $a \neq 0$, p is the characteristic of $\mathbb{F}_q$, and h is a nonnegative integer.*

Proof. The sufficiency is an immediate consequence of Theorem 7.8 and the remark following Theorem 7.9. To prove the necessity, we note first that if f is a permutation polynomial of $\mathbb{F}_q$, then for every $c \in \mathbb{F}_q$ the equation $f(x) = c$ has a unique solution $d \in \mathbb{F}_q$. Thus

$$f(x) - c = (x-d)^k g(x),$$

where $k \in \mathbb{N}$, $g \in \mathbb{F}_q[x]$, and either $\deg(g) = 0$ or g is a product of irreducible polynomials g_i in $\mathbb{F}_q[x]$ with $\deg(g_i) \geqslant 2$. If r is a multiple of some $\deg(g_i)$, then g_i has a root in $\mathbb{F}_{q^r}$, and so f is not a permutation polynomial of $\mathbb{F}_{q^r}$. Hence we must have

$$f(x) - c = a(x-d)^k \text{ with } a \neq 0, \tag{7.3}$$

that is, for each $c \in \mathbb{F}_q$ there is a $d \in \mathbb{F}_q$ depending on c such that this identity holds. Using this for $c = 0$ and $c = 1$, we obtain

$$a(x-d_0)^k - a(x-d_1)^k = 1,$$

and replacing x by $x + d_1$, this becomes

$$a(x + d_1 - d_0)^k - ax^k = 1.$$

Expanding by the binomial theorem, we get

$$\binom{k}{j} \equiv 0 \bmod p \quad \text{for } 0 < j < k. \tag{7.4}$$

We have $p^h \leqslant k < p^{h+1}$ for some $h \in \mathbb{Z}$, $h \geqslant 0$. If $k \neq p^h$, then Lemma 6.39 yields with $j = p^h$,

$$E_p\left(\binom{k}{j}\right) = \frac{1}{p-1}(s_j + s_{k-j} - s_k) = 0,$$

where s_n denotes the sum of digits in the representation of n to the base p. As this contradicts (7.4), we must have $k = p^h$, and the rest follows from (7.3). □

7.15. Corollary. *If $f \in \mathbb{F}_q[x]$ is not of the form $ax^{p^h} + b$, then there are infinitely many extension fields $\mathbb{F}_{q^r}$ of $\mathbb{F}_q$ such that f is not a permutation polynomial of $\mathbb{F}_{q^r}$.*

Proof. If f is not a permutation polynomial of $\mathbb{F}_q$, it cannot be a permutation polynomial of any $\mathbb{F}_{q^r}$. If f is a permutation polynomial of $\mathbb{F}_q$, the result follows from an inspection of the proof of Theorem 7.14. □

We introduce now a special class of polynomials called Dickson polynomials that have some interesting properties and also yield new examples of permutation polynomials. Let x_1, x_2 be indeterminates and $k \in \mathbb{N}$. Then we have seen in the proof of Theorem 5.46 that Waring's formula yields

$$x_1^k + x_2^k = \sum_{j=0}^{\lfloor k/2 \rfloor} \frac{k}{k-j}\binom{k-j}{j}(-x_1x_2)^j(x_1+x_2)^{k-2j}. \tag{7.5}$$

This holds over any commutative ring R with identity. For $a \in R$ we define the *Dickson polynomial* $g_k(x, a)$ over R by

$$g_k(x, a) = \sum_{j=0}^{\lfloor k/2 \rfloor} \frac{k}{k-j}\binom{k-j}{j}(-a)^j x^{k-2j}. \tag{7.6}$$

If we work over the complex numbers, then these polynomials are closely related to the well-known Chebyshev polynomials of the first kind $T_k(x) = \cos(k \arccos x)$. For if we substitute $x_1 = e^{i\theta}$, $x_2 = e^{-i\theta}$ in (7.5), then $2\cos k\theta = g_k(2\cos\theta, 1)$ by (7.6), hence

$$g_k(2x, 1) = 2T_k(x). \tag{7.7}$$

Because of this connection, Dickson polynomials are sometimes also called *Chebyshev polynomials*. The identity (7.7) can be used to define Chebyshev polynomials of the first kind $T_k(x)$ over any field of characteristic $\neq 2$.

If we consider a Dickson polynomial $g_k(x, a)$ over a field F, then in the field of rational functions over F in the indeterminate y we have the identity

$$g_k\left(y+\frac{a}{y},a\right)=y^k+\frac{a^k}{y^k}, \tag{7.8}$$

which follows from (7.5) by substituting $x_1=y$, $x_2=a/y$. The definition of Dickson polynomials yields also the formula

$$g_k(x,ab^2)=\sum_{j=0}^{\lfloor k/2\rfloor}\frac{k}{k-j}\binom{k-j}{j}(-a)^j b^k b^{-(k-2j)}x^{k-2j}=b^k g_k(b^{-1}x,a) \tag{7.9}$$

for any $a,b\in F$ with $b\neq 0$. Hence, if $F=\mathbb{F}_q$, q even, then every Dickson polynomial $g_k(x,a)$, $a\in\mathbb{F}_q^*$, can be expressed in terms of $g_k(x,1)$. If $F=\mathbb{F}_q$, q odd, then every Dickson polynomial $g_k(x,a)$, $a\in\mathbb{F}_q^*$, can be expressed in terms of either $g_k(x,1)$ or $g_k(x,c)$, c being a fixed nonsquare in $\mathbb{F}_q$. For odd q, the Dickson polynomials $g_k(x,a)$, $a\in\mathbb{F}_q^*$, can also be expressed in terms of the Chebyshev polynomials of the first kind $T_k(x)$ defined by (7.7). In fact, if $\beta\in\mathbb{F}_{q^2}$ is such that $\beta^2=a$, then (7.7) and (7.9) imply

$$g_k(x,a)=\beta^k g_k(\beta^{-1}x,1)=2\beta^k T_k\left((2\beta)^{-1}x\right).$$

In general, the case $a=0$ is not of great interest since $g_k(x,0)=x^k$.

7.16. Theorem. *The Dickson polynomial $g_k(x,a)$, $a\in\mathbb{F}_q^*$, is a permutation polynomial of $\mathbb{F}_q$ if and only if* $\gcd(k,q^2-1)=1$.

Proof. Suppose $g_k(b,a)=g_k(c,a)$ for some $b,c\in\mathbb{F}_q$. We can find $\beta,\gamma\in\mathbb{F}_{q^2}^*$ such that $\beta+a\beta^{-1}=b$, $\gamma+a\gamma^{-1}=c$. Then (7.8) yields $\beta^k+a^k\beta^{-k}=\gamma^k+a^k\gamma^{-k}$, hence $(\beta^k-\gamma^k)(\beta^k\gamma^k-a^k)=0$, and so $\beta^k=\gamma^k$ or $\beta^k=(a\gamma^{-1})^k$. Now if $\gcd(k,q^2-1)=1$, then x^k is a permutation polynomial of $\mathbb{F}_{q^2}$ by Theorem 7.8(ii), which implies $\beta=\gamma$ or $\beta=a\gamma^{-1}$. In either case, it follows that $b=c$, and so $g_k(x,a)$ is a permutation polynomial of $\mathbb{F}_q$.

Now suppose that $\gcd(k,q^2-1)=d>1$. If d is even, then q is odd and k is even. Since (7.6) shows that $g_k(x,a)$ contains only even powers of x, we have $g_k(c,a)=g_k(-c,a)$ for $c\in\mathbb{F}_q^*$, but $c\neq -c$, hence $g_k(x,a)$ is not a permutation polynomial of $\mathbb{F}_q$. If d is odd, then there exists an odd prime r dividing d. Then r divides k, and either $q-1$ or $q+1$ is divisible by r, so that we distinguish two cases accordingly. In the first case, the equation $x^r=1$ has r solutions in $\mathbb{F}_q$, thus there exists $b\in\mathbb{F}_q$, $b\neq 1,a$, with $b^r=1$. Then also $b^k=1$, and so (7.8) yields

$$g_k(b+ab^{-1},a)=1+a^k=g_k(1+a,a).$$

Since $b+ab^{-1}=1+a$ would imply $b=1$ or $b=a$, we have $b+ab^{-1}\neq 1+a$, hence $g_k(x,a)$ is not a permutation polynomial of $\mathbb{F}_q$. In the second case, let $\gamma\in\mathbb{F}_{q^2}$ be a solution of $x^{q+1}=a$. Since $x^r=1$ has r solutions in $\mathbb{F}_{q^2}$, there exists $\beta \in \mathbb{F}_{q^2}$, $\beta \neq 1, a\gamma^{-2}$, with $\beta^r = 1$. Then also $\beta^{q+1} = 1$ and $\beta^k = 1$, hence

$$g_k(\gamma + a\gamma^{-1}, a) = g_k(\beta\gamma + a(\beta\gamma)^{-1}, a)$$

by (7.8). Moreover, $\gamma + a\gamma^{-1} = \gamma + \gamma^q \in \mathbb{F}_q$ and $\beta\gamma + a(\beta\gamma)^{-1} = \beta\gamma + (\beta\gamma)^q \in \mathbb{F}_q$, as well as $\beta\gamma + a(\beta\gamma)^{-1} \neq \gamma + a\gamma^{-1}$, for otherwise $\beta = 1$ or $\beta = a\gamma^{-2}$. Thus $g_k(x, a)$ is not a permutation polynomial of $\mathbb{F}_q$. □

7.17. **Corollary.** *If $a \in \mathbb{F}_q^*$ and $\gcd(k, q^2 - 1) = 1$, then*

$$\sum_{c \in \mathbb{F}_q} \chi(g_k(c, a)) = 0$$

for every nontrivial additive or multiplicative character χ of $\mathbb{F}_q$.

Proof. Since $g_k(x, a)$ is a permutation polynomial of $\mathbb{F}_q$ by Theorem 7.16, we have

$$\sum_{c \in \mathbb{F}_q} \chi(g_k(c, a)) = \sum_{c \in \mathbb{F}_q} \chi(c),$$

and so the result follows from either (5.9) or (5.37). □

The character sums occurring in Corollary 7.17, with χ being the quadratic character of $\mathbb{F}_q$, q odd, and $k \in \mathbb{N}$ arbitrary, have been studied extensively and are called *Brewer sums* (compare with the notes to Chapter 5, Section 5).

As a generalization of the definition in (7.6), Dickson polynomials in several indeterminates will be introduced in the last section of this chapter.

3. GROUPS OF PERMUTATION POLYNOMIALS

Permutation polynomials of $\mathbb{F}_q$ of degree $< q$ can be combined by the operation of composition and subsequent reduction modulo $x^q - x$. It is convenient to write

$$\langle g(x)\rangle\langle f(x)\rangle = \langle h(x)\rangle$$

whenever $f(g(x)) \equiv h(x) \bmod (x^q - x)$. Under this operation, the set of permutation polynomials of $\mathbb{F}_q$ of degree $< q$ forms a group, which is isomorphic to S_q, the symmetric group on q letters. Thus, the symmetric group S_q and its subgroups can be represented as groups of permutation polynomials.

7.18. **Theorem.** *For $q > 2$, S_q is generated by x^{q-2} and all linear polynomials over $\mathbb{F}_q$.*

Proof. We note first that all these polynomials are permutation polynomials of $\mathbb{F}_q$ by Theorem 7.8. Now every permutation of $\mathbb{F}_q$ can be expressed as a product of transpositions. It is, in fact, sufficient to consider transpositions of the form $(0a)$, $a \in \mathbb{F}_q^*$, because $(bc) = (0b)(0c)(0b)$ for any transposition (bc) of S_q. The polynomial

$$f_a(x) = -a^2\Big[\big((x-a)^{q-2} + a^{-1}\big)^{q-2} - a\Big]^{q-2}$$

represents the transposition $(0a)$ and is a composition of linear polynomials and x^{q-2}. □

The value of these generators found for S_q lies in the fact that they are simple as polynomials; it is evident from the expression for $f_a(x)$ that simplicity as polynomials and simplicity as permutations are not equivalent.

7.19. Theorem. *If $q > 2$ and c is a fixed primitive element of $\mathbb{F}_q$, then S_q is generated by cx, $x+1$, and x^{q-2}.*

Proof. Let $a, b \in \mathbb{F}_q^*$; then $a = c^s$, $b = c^t$ with integers $s > t \geqslant 1$. The assertion of the theorem follows from Theorem 7.18 and the identities $\langle ax \rangle = \langle c^s x \rangle = \langle cx \rangle^s$, $\langle ax + b \rangle = \langle cx \rangle^{s-t} \langle x+1 \rangle \langle cx \rangle^t$. □

By an elaboration of these methods we may find generators for the alternating group A_q, the subgroup of S_q consisting of the even permutations. We call a permutation polynomial of $\mathbb{F}_q$ *even* if the induced permutation of $\mathbb{F}_q$ is even.

7.20. Lemma. *Let $q > 2$ and $a \in \mathbb{F}_q$. Then $x + a$ and $(x^{q-2} + a)^{q-2}$ are even permutation polynomials, and ax is an even permutation polynomial if and only if a is the square of an element of $\mathbb{F}_q^*$. Furthermore, x^{q-2} is an even permutation polynomial if and only if $q \equiv 3 \bmod 4$.*

Proof. The permutation induced by $x + a$ is composed of p^{e-1} cycles of length p, where $q = p^e$ and p is the characteristic of $\mathbb{F}_q$. Thus, if p is odd, or if $q = 2^e$ and $e > 1$, then $x + a$ is even. Since

$$\langle (x^{q-2} + a)^{q-2} \rangle = \langle x^{q-2} \rangle \langle x + a \rangle \langle x^{q-2} \rangle,$$

$(x^{q-2} + a)^{q-2}$ is even. Furthermore, ax induces a permutation if and only if $a \neq 0$, and in this case we have $\langle ax \rangle = \langle cx \rangle^s$, where c is a primitive element of $\mathbb{F}_q$ and $a = c^s$. The permutation induced by cx is a cycle of length $q - 1$. The criterion for the evenness of ax follows from this and the fact that every element of $\mathbb{F}_{2^e}$ is a square.

As a permutation, x^{q-2} is composed of disjoint transpositions containing all elements of $\mathbb{F}_q$ except $0, 1, -1$. It therefore contains $\frac{1}{2}(q-3)$ transpositions when q is odd and $\frac{1}{2}(q-2)$ when q is even. □

We now define the following sets of permutation polynomials of $\mathbb{F}_q$ for $q > 2$:

$$L_q = \{ax + b: \quad a \in \mathbb{F}_q^*, b \in \mathbb{F}_q\},$$

$$AL_q = \{a^2x + b: \quad a \in \mathbb{F}_q^*, b \in \mathbb{F}_q\},$$

$$Q_q = \{(x^{q-2} + a)^{q-2}: \quad a \in \mathbb{F}_q\}.$$

These sets form groups under the operation of composition modulo $x^q - x$ with orders $|L_q| = q(q-1)$, $|AL_q| = \frac{1}{2}q(q-1)$ for q odd and $q(q-1)$ for q even, and $|Q_q| = q$. The group Q_q is isomorphic to the additive group of $\mathbb{F}_q$. The following is easy to prove.

7.21. Theorem. *Let $q > 2$ and let c be a fixed primitive element of $\mathbb{F}_q$. Then:*

(i) *L_q is generated by cx and $x + 1$;*
(ii) *AL_q is generated by c^2x and $x + 1$;*
(iii) *A_q is generated by its subgroups AL_q and Q_q;*
(iv) *A_q is generated by c^2x, $x + 1$, and $(x^{q-2} + 1)^{q-2}$.*

Given a class of permutation polynomials closed under composition, we can ask which subgroup of S_q is represented by this class. For fixed $a \in \mathbb{F}_q$, we study first the set $P(a)$ of all Dickson polynomials $g_k(x, a)$ that are permutation polynomials of $\mathbb{F}_q$; that is,

$$P(0) = \{g_k(x,0): \quad k \in \mathbb{N}, \gcd(k, q-1) = 1\},$$

$$P(a) = \{g_k(x, a): \quad k \in \mathbb{N}, \gcd(k, q^2 - 1) = 1\} \text{ for } a \neq 0.$$

7.22. Theorem. *$P(a)$ is closed under composition of polynomials if and only if $a = 0$, 1, or -1.*

Proof. For $a \in \mathbb{F}_q$ and $k, m \in \mathbb{N}$ we have

$$g_k\left(g_m\left(y + \frac{a}{y}, a\right), a^m\right) = g_k\left(y^m + \frac{a^m}{y^m}, a^m\right)$$

$$= y^{km} + \frac{a^{km}}{y^{km}} = g_{km}\left(y + \frac{a}{y}, a\right)$$

by (7.8), hence

$$g_{km}(x, a) = g_k(g_m(x, a), a^m). \tag{7.10}$$

If $a \neq 0$ and $P(a)$ is closed under composition, then $g_k(g_m(x, a), a) \in P(a)$ for $\gcd(k, q^2 - 1) = \gcd(m, q^2 - 1) = 1$, thus $g_k(g_m(x, a), a) = g_{km}(x, a)$ and by (7.10),

$$g_k(g_m(x, a), a^m) = g_k(g_m(x, a), a).$$

Since $g_m(x, a)$ is nonconstant, it follows that

$$g_k(x, a^m) = g_k(x, a).$$

Comparing coefficients of x^{k-2} for $k > 1$, we get $a^m = a$ for all m with $\gcd(m, q^2 - 1) = 1$, hence $a^{-1} = a$, which implies $a = \pm 1$.

Conversely, if $a = 0$, 1, or -1, then $P(a)$ is closed under composition by (7.10). □

Therefore, in the three cases $a = 1$, $a = -1$, and $a = 0$, the set $G(a)$ of all permutations of $\mathbb{F}_q$ represented by the polynomials of $P(a)$ is an abelian subgroup of S_q. In the following we study the structure of this group $G(a)$.

Let $a = \pm 1$. For $c \in \mathbb{F}_q$ we can find $\gamma \in \mathbb{F}_{q^2}^*$ with $c = \gamma + a\gamma^{-1}$. Thus for $k \equiv m \bmod (q^2 - 1)$ by (7.8),

$$\begin{aligned} g_k(c, a) &= g_k(\gamma + a\gamma^{-1}, a) = \gamma^k + a^k\gamma^{-k} = \gamma^m + a^m\gamma^{-m} \\ &= g_m(\gamma + a\gamma^{-1}, a) = g_m(c, a). \end{aligned}$$

Hence, if $\gcd(k, q^2 - 1) = 1$, then $g_k(x, a)$ and $g_m(x, a)$ induce the same permutation of $\mathbb{F}_q$. Thus, if we assign to the residue class of $k \bmod (q^2 - 1)$ the permutation of $\mathbb{F}_q$ induced by $g_k(x, a)$, then we get an epimorphism of $R(q^2 - 1)$ onto $G(a)$, where $R(q^2 - 1)$ is the reduced residue class group $\bmod (q^2 - 1)$—that is, the group of units of the residue class ring $\mathbb{Z}/(q^2 - 1)$. By Theorem 1.23, it suffices now to determine the kernel $K(a)$ of this epimorphism.

If $k \in K(a)$, then $g_k(c, a) = c$ for all $c \in \mathbb{F}_q$, thus with γ as above, $\gamma^k + a^k\gamma^{-k} = \gamma + a\gamma^{-1}$. Because of $a^k = a$, we get $\gamma^k + a\gamma^{-k} = \gamma + a\gamma^{-1}$, hence $\gamma^k = \gamma$ or $\gamma^k = a\gamma^{-1}$, and so

$$\gamma^{k-1} = 1 \quad \text{or} \quad \gamma^{k+1} = a \quad \text{for all } \gamma \in \mathbb{F}_{q^2}^* \text{ with } \gamma + a\gamma^{-1} \in \mathbb{F}_q. \tag{7.11}$$

This is also sufficient for k to be in $K(a)$. Now $\gamma + a\gamma^{-1} \in \mathbb{F}_q$ if and only if $(\gamma + a\gamma^{-1})^q = \gamma + a\gamma^{-1}$, which is equivalent to $\gamma^{q-1} = 1$ or $\gamma^{q+1} = a$. Let $a = 1$, and let ζ be a primitive element of $\mathbb{F}_{q^2}$; then $\gamma = \zeta^{m(q+1)}$ or $\gamma = \zeta^{n(q-1)}$, $m, n \in \mathbb{Z}$. Thus from (7.11), $k \in K(1)$ if and only if k is a solution of one of the following four systems of congruences:

$$\begin{cases} k \equiv 1 & \bmod (q-1) \\ k \equiv 1 & \bmod (q+1) \end{cases} \qquad \begin{cases} k \equiv 1 & \bmod (q-1) \\ k \equiv -1 & \bmod (q+1) \end{cases}$$

$$\begin{cases} k \equiv -1 & \bmod (q-1) \\ k \equiv 1 & \bmod (q+1) \end{cases} \qquad \begin{cases} k \equiv -1 & \bmod (q-1) \\ k \equiv -1 & \bmod (q+1) \end{cases}$$

By solving these systems $\bmod (q^2 - 1)$, we obtain

$$K(1) = \{1, q, -q, -1\} \quad \text{for } q \text{ even}$$

and

$$K(1) = \{1, q, -q, -1, 1 + (q^2-1)/2, q + (q^2-1)/2,$$
$$-q + (q^2-1)/2, -1 + (q^2-1)/2\} \quad \text{for } q \text{ odd}.$$

The case $a = -1$ is treated similarly. We obtain

$$K(-1) = \{1, q\} \quad \text{for } q \equiv 3 \bmod 4$$

and

$$K(-1) = \{1, q, 1 + (q^2-1)/2, q + (q^2-1)/2\} \quad \text{for } q \equiv 1 \bmod 4.$$

The results may be summarized as follows.

7.23. Theorem. *For $a = \pm 1$, the group $G(a)$ is isomorphic to $R(q^2-1)/K(a)$, where $K(a)$ is as above. We have $|K(1)| = 2$ if $q = 2$, $|K(1)| = 4$ if $q > 2$, q even, $|K(1)| = 4$ if $q = 3$, $|K(1)| = 8$ if $q > 3$, q odd, $|K(-1)| = 2$ if $q \equiv 3 \bmod 4$, and $|K(-1)| = 4$ if $q \equiv 1 \bmod 4$. The group $G(0)$ is isomorphic to $R(q-1)$, the reduced residue class group* mod$(q-1)$.

Another interesting class of permutation polynomials is the following. Let $\mathbb{F}_{q^r}$ be an extension of $\mathbb{F}_q$ and consider linearized polynomials $L(x)$ of the form

$$L(x) = \sum_{s=0}^{r-1} \alpha_s x^{q^s} \in \mathbb{F}_{q^r}[x]. \tag{7.12}$$

By Theorem 7.9, $L(x)$ is a permutation polynomial of $\mathbb{F}_{q^r}$ if and only if $L(x)$ only has the root 0 in $\mathbb{F}_{q^r}$—that is, if and only if the linear operator on the vector space $\mathbb{F}_{q^r}$ over $\mathbb{F}_q$ induced by $L(x)$ is nonsingular. This linear operator is nonsingular precisely if $\gamma_0 = L(\beta_0)$, $\gamma_1 = L(\beta_1), \ldots, \gamma_{r-1} = L(\beta_{r-1})$ are linearly independent over $\mathbb{F}_q$ whenever $\beta_0, \beta_1, \ldots, \beta_{r-1} \in \mathbb{F}_{q^r}$ are linearly independent over $\mathbb{F}_q$. Now for $0 \leqslant i, j \leqslant r-1$ we get from (7.12),

$$\gamma_i^{q^j} = \sum_{s=0}^{r-1} \alpha_s^{q^j} \beta_i^{q^{s+j}}.$$

Using $\beta_i^{q^r} = \beta_i$ and setting $\alpha_t = \alpha_s$ if $t \equiv s \bmod r$, we obtain

$$\gamma_i^{q^j} = \sum_{s=0}^{r-1} \beta_i^{q^s} \alpha_{s-j}^{q^j}.$$

If Δ_1 and Δ_2 are the determinants in (3.13) formed with $\beta_0, \beta_1, \ldots, \beta_{r-1}$ and $\gamma_0, \gamma_1, \ldots, \gamma_{r-1}$, respectively, then it follows that

$$\Delta_2 = \Delta_1 \det(A),$$

where the $r \times r$ matrix A is given by

$$A = \begin{pmatrix} \alpha_0 & \alpha_{r-1}^q & \alpha_{r-2}^{q^2} & \cdots & \alpha_1^{q^{r-1}} \\ \alpha_1 & \alpha_0^q & \alpha_{r-1}^{q^2} & \cdots & \alpha_2^{q^{r-1}} \\ \alpha_2 & \alpha_1^q & \alpha_0^{q^2} & \cdots & \alpha_3^{q^{r-1}} \\ \vdots & \vdots & \vdots & & \vdots \\ \alpha_{r-1} & \alpha_{r-2}^q & \alpha_{r-3}^{q^2} & \cdots & \alpha_0^{q^{r-1}} \end{pmatrix}.$$

In view of Lemma 3.51, $L(x)$ is thus a permutation polynomial of $\mathbb{F}_{q^r}$ if and only if $\det(A) \neq 0$.

The set of $L(x)$ in (7.12) that are permutation polynomials of $\mathbb{F}_{q^r}$ constitutes a group under the operation of composition modulo $x^{q^r} - x$. This group is known as the *Betti-Mathieu group*. We state the following result without proof.

7.24. Theorem. *The Betti-Mathieu group is isomorphic to the general linear group $GL(r, \mathbb{F}_q)$ of nonsingular $r \times r$ matrices over $\mathbb{F}_q$ under matrix multiplication.*

4. EXCEPTIONAL POLYNOMIALS

One can introduce some geometric ideas into the study of permutation polynomials. The principal advantage in looking at the problem from this point of view is that one is able to make use of the powerful theorem of Lang and Weil (see Chapter 6, Notes), which estimates the number of rational points on an absolutely irreducible curve defined over a finite field.

Given a polynomial $f \in \mathbb{F}_q[x]$ of degree $d \geqslant 1$, we form the polynomial in two indeterminates

$$\Phi(x, y) = \frac{f(x) - f(y)}{x - y},$$

which has degree $d-1$. We define an *algebraic curve* C_φ over $\mathbb{F}_q$ to be a subset of $E \times E$ (the cartesian product of two copies of an algebraic extension E of $\mathbb{F}_q$) of the form

$$C_\varphi = \{(a, b) \in E \times E : \varphi(a, b) = 0\},$$

where $\varphi \in \mathbb{F}_q[x, y]$ is a nonzero polynomial in two indeterminates over $\mathbb{F}_q$. A point (a, b) on the curve C_φ is called a *rational point* if both a and b belong to $\mathbb{F}_q$. Of course, the number of rational points on C_φ will be finite as $\mathbb{F}_q \times \mathbb{F}_q$ is itself finite. With our notation above it follows that *a polynomial f is a permutation polynomial of $\mathbb{F}_q$ if and only if C_Φ contains no rational points off the line $y = x$.*

We recall that for a field K we have unique factorization in $K[x, y]$ into irreducibles, and that a polynomial in $K[x, y]$ of positive degree is called *absolutely irreducible* if it is irreducible over any algebraic extension of K.

7.25. Definition. A polynomial $f \in \mathbb{F}_q[x]$ of degree $\geqslant 2$ is said to be *exceptional* over $\mathbb{F}_q$ if no irreducible factor of

$$\Phi(x, y) = \frac{f(x) - f(y)}{x - y}$$

in $\mathbb{F}_q[x, y]$ is absolutely irreducible.

In other words, f is exceptional over $\mathbb{F}_q$ if every irreducible factor of $\Phi(x, y)$ in $\mathbb{F}_q[x, y]$ allows a proper factorization over some algebraic extension of $\mathbb{F}_q$.

The next theorem provides a connection between permutation polynomials and exceptional polynomials. We note first without proof that an exceptional polynomial is, in a sense, almost a permutation polynomial.

7.26. Lemma. *Let $f \in \mathbb{F}_q[x]$ be exceptional over $\mathbb{F}_q$, and let $V(f)$ denote the number of elements of the value set $\{f(c): c \in \mathbb{F}_q\}$ of f. Then $V(f) \geqslant q - A(d)$, where $A(d)$ is a constant only depending on the degree d of f.*

7.27. Theorem. *If $\mathbb{F}_q$ is of characteristic p and f is an exceptional polynomial over $\mathbb{F}_q$, where $p \geqslant B(d)$, a constant depending only on the degree d of f, then f is a permutation polynomial of $\mathbb{F}_q$.*

Proof. Using the notation of Lemma 7.26, we can write $V(f) = q - w$, where $0 \leqslant w \leqslant A(d)$. It suffices to prove that $w = 0$. We assume $w \geqslant 1$ and obtain a contradiction.

Let the distinct elements of the value set of f be $b_1, b_2, \ldots, b_{q-w}$ and let the remaining elements of $\mathbb{F}_q$ be $c_1, c_2, \ldots, c_w$. For $i = 1, 2, \ldots, q - w$, let m_i be the number of solutions of the equation $f(x) = b_i$ in $\mathbb{F}_q$, so that $\sum_{i=1}^{q-w} m_i = q$. Now each $m_i \geqslant 1$, hence

$$m_i \leqslant w + 1 \quad \text{for } i = 1, 2, \ldots, q - w. \tag{7.13}$$

For $t = 1, 2, \ldots, w$ we have

$$\sum_{c \in \mathbb{F}_q} f(c)^t = \sum_{i=1}^{q-w} m_i b_i^t.$$

If $p \geqslant B(d)$, where $B(d) = dA(d) + 2$, we have $q - 2 \geqslant p - 2 \geqslant dA(d) \geqslant dw$, so that for $t = 1, 2, \ldots, w$ we can write

$$f(x)^t = a_0^{(t)} + a_1^{(t)} x + \cdots + a_{q-2}^{(t)} x^{q-2}.$$

Then

$$\sum_{c \in \mathbb{F}_q} f(c)^t = \sum_{j=0}^{q-2} a_j^{(t)} \sum_{c \in \mathbb{F}_q} c^j = 0$$

by Lemma 7.3. Thus we have

$$\sum_{i=1}^{q-w} m_i b_i^t = 0 \quad \text{for } t = 1, 2, \ldots, w. \tag{7.14}$$

Set $m = \max(m_1, \ldots, m_{q-w})$, then $1 \leqslant m \leqslant w+1$ from (7.13). If s_j, $1 \leqslant j \leqslant m$, denotes the number of m_i with $m_i = j$, then $s_1 + \cdots + s_m = q - w$ and

$$\sum_{j=1}^{m} (j-1) s_j = \sum_{i=1}^{q-w} m_i - \sum_{j=1}^{m} s_j = q - (q-w) = w. \tag{7.15}$$

If we arrange $b_1, \ldots, b_{q-w}$ in such a way that $m_1 = \cdots = m_{s_1} = 1$, $m_{s_1+1} = \cdots = m_{s_1+s_2} = 2, \ldots, m_{s_1+\cdots+s_{m-1}+1} = \cdots = m_{s_1+\cdots+s_m} = m$, then (7.14) becomes

$$A_t = \sum_{j=1}^{m} j \sum_{i=s_1+\cdots+s_{j-1}+1}^{s_1+\cdots+s_j} b_i^t = 0 \quad \text{for } t = 1, 2, \ldots, w.$$

Thus

$$\sum_{k=1}^{w} c_k^t = \sum_{k=1}^{w} c_k^t + A_t = \sum_{c \in \mathbb{F}_q} c^t + \sum_{j=1}^{m} (j-1) \sum_{i=s_1+\cdots+s_{j-1}+1}^{s_1+\cdots+s_j} b_i^t.$$

Now $1 \leqslant t \leqslant w \leqslant dA(d) \leqslant q-2$, so $\sum_{c \in \mathbb{F}_q} c^t = 0$ and therefore

$$\sum_{k=1}^{w} c_k^t = \sum_{j=2}^{m} (j-1) \sum_{i=s_1+\cdots+s_{j-1}+1}^{s_1+\cdots+s_j} b_i^t. \tag{7.16}$$

We next consider the two polynomials

$$g(x) = \prod_{k=1}^{w} (x - c_k) \quad \text{and} \quad h(x) = \prod_{j=2}^{m} \prod_{i=1}^{s_j} \left(x - b_{s_1+\cdots+s_{j-1}+i}\right)^{j-1},$$

with $\deg(g) = w$ and $\deg(h) = w$ by (7.15). Let g_r, h_r, $0 \leqslant r \leqslant w$, denote the coefficient of x^{w-r} in $g(x)$ and $h(x)$, respectively, and let G_t, H_t, $1 \leqslant t \leqslant w$, denote the sum of the tth powers of all the roots of $g(x)$ and $h(x)$, respectively. Thus by (7.16) we have $G_t = H_t$ for $1 \leqslant t \leqslant w$. Newton's formula (see Theorem 1.75) yields

$$\sum_{i=0}^{t-1} G_{t-i} g_i + t g_t = 0 \quad \text{for } 1 \leqslant t \leqslant w. \tag{7.17}$$

Now $p \geqslant dA(d)+2 > A(d) \geqslant w$, so the coefficient of g_t in (7.17) does not vanish in $\mathbb{F}_q$. Hence the w equations (7.17) can be solved successively and uniquely for $g_1,\ldots,g_w$ in terms of $G_1,\ldots,G_w$—namely, $g_1 = -G_1$, $g_2 = \frac{1}{2}(G_1^2 - G_2)$, and so on. Similarly we obtain $h_1 = -H_1$, $h_2 = \frac{1}{2}(H_1^2 - H_2)$, and so on, and since $G_t = H_t$ for $1 \leqslant t \leqslant w$, we have $g_r = h_r$ for $0 \leqslant r \leqslant w$. Hence $g(x) = h(x)$ and $\{c_1,\ldots,c_w\}$ must be a rearrangement of $\{b_{s_1+1},\ldots,b_{q-w}\}$. This is clearly impossible as the b_i are distinct from the c_k by definition. This contradiction completes the proof. □

The preceding theorem shows that for finite fields of sufficiently large characteristic the property of being an exceptional polynomial is *sufficient* for being a permutation polynomial. We consider now the question under which assumptions the condition of being exceptional is also *necessary* for permutation polynomials.

The converse of Theorem 7.27 is true under some additional conditions that can be derived from the theorem of Lang and Weil, which we reformulate in the following way. Let N denote the number of rational points on the curve C_φ, where $\varphi \in \mathbb{F}_q[x, y]$ is absolutely irreducible and $\deg(\varphi) = d$. Then according to the theorem of Lang and Weil (see the notes to Chapter 6, Section 4),

$$|N - q| \leqslant (d-1)(d-2)q^{1/2} + C(d), \tag{7.18}$$

where $C(d)$ is a constant depending only on d. For our purposes we require only the following weak consequence of (7.18).

7.28. Lemma. *There exists a sequence $k_1, k_2,\ldots$ of positive integers having the following property: if $\varphi \in \mathbb{F}_q[x, y]$ is absolutely irreducible and $q \geqslant k_d$, where $d = \deg(\varphi)$, then either C_φ has a rational point (a, b) with $a \neq b$ or φ is of the form $c(y - x)$ for some $c \in \mathbb{F}_q$.*

Proof. For every $d \in \mathbb{N}$ choose k_d so that

$$q - (d-1)(d-2)q^{1/2} - C(d) > d$$

for all $q \geqslant k_d$. Then if φ is absolutely irreducible of degree d over $\mathbb{F}_q$ with $q \geqslant k_d$, it follows from (7.18) that C_φ contains at least $d+1$ rational points. If φ is not of the form $c(y - x)$, then the irreducibility of φ implies that φ is not divisible by $y - x$, and so $\varphi(x, x)$ is not the zero polynomial. Thus C_φ intersects the line $y = x$ in at most d rational points. Consequently, C_φ has a rational point (a, b) with $a \neq b$. □

To prove the following theorem, we use the characterization of permutation polynomials by algebraic curves and their rational points, given at the beginning of this section.

7.29. Theorem. *There exists a sequence $k_1, k_2,\ldots$ of positive integers such that for any finite field $\mathbb{F}_q$ of order $q \geqslant k_n$ with $\gcd(n, q) = 1$ the following* *statement is true: if $f \in \mathbb{F}_q[x]$ is a permutation polynomial of $\mathbb{F}_q$ with deg$(f) = n \geqslant 2$, then f is exceptional over $\mathbb{F}_q$.*

Proof. It is clear that $k_1, k_2, \ldots$ in Lemma 7.28 can be chosen so as to obtain a nondecreasing sequence $k_1 \leqslant k_2 \leqslant \cdots$. With k_n defined in this way, let $f \in \mathbb{F}_q[x]$ be a permutation polynomial of $\mathbb{F}_q$ such that the hypothesis of the theorem is satisfied. If

$$\Phi(x, y) = (f(x) - f(y))/(x - y) \in \mathbb{F}_q[x, y],$$

then C_Φ has no rational points off the line $y = x$. Suppose, by way of contradiction, that f is not exceptional; then $\Phi(x, y)$ has an absolutely irreducible factor $g(x, y)$ in $\mathbb{F}_q[x, y]$. If $g(x, y) = c(y - x)$, $c \in \mathbb{F}_q$, then $f(y) - f(x) = (y - x)^2 h(x, y)$ for some $h \in \mathbb{F}_q[x, y]$, hence

$$f'(y) = 2(y - x)h(x, y) + (y - x)^2(\partial h(x, y)/\partial y),$$

thus $f'(x) = 0$, a contradiction since $\gcd(n, q) = 1$. Therefore $g(x, y)$ is not of the form $c(y - x)$. With $d = \deg(g)$ we have $q \geqslant k_n \geqslant k_d$, hence Lemma 7.28 implies that $g(a, b) = 0$ for some $(a, b) \in \mathbb{F}_q^2$ with $a \neq b$. It follows that $\Phi(a, b) = 0$, which is a contradiction. □

If $\gcd(n, q) > 1$—that is, if the characteristic p of $\mathbb{F}_q$ divides n—then the result of Theorem 7.29 need not hold. For example, x^p is a permutation polynomial of $\mathbb{F}_q$, but $(x^p - y^p)/(x - y) = (x - y)^{p-1}$ shows that x^p is not exceptional over $\mathbb{F}_q$.

If we combine Theorems 7.27 and 7.29, we get the following characterization of permutation polynomials for finite fields of sufficiently large characteristic.

7.30. Corollary. *For every integer $n \geqslant 2$ there is a constant K_n such that for any finite field $\mathbb{F}_q$ of characteristic $\geqslant K_n$ the following holds: a polynomial $f \in \mathbb{F}_q[x]$ of degree n is a permutation polynomial of $\mathbb{F}_q$ if and only if f is exceptional over $\mathbb{F}_q$.*

The following results, which depend ultimately on the theorem of Lang and Weil, yield information about the nonexistence of permutation polynomials of a given degree n in certain finite fields.

7.31. Theorem. *There exists a sequence $k_1, k_2, \ldots$ of positive integers with the following property: given a positive integer n, if $\mathbb{F}_q$ is a finite field of order $q \geqslant k_n$ with $\gcd(n, q) = 1$, and if $\mathbb{F}_q$ contains an nth root of unity $\zeta \neq 1$, then there are no permutation polynomials of $\mathbb{F}_q$ of degree n.*

Proof. Let $f \in \mathbb{F}_q[x]$ be an arbitrary polynomial of degree n and put

$$\Phi(x, y) = (f(x) - f(y))/(x - y).$$

Starting from the factorization of Φ into irreducibles in $\mathbb{F}_q[x, y]$ and forming successive algebraic extensions, we obtain an algebraic extension E

of $\mathbb{F}_q$ and a factorization

$$\Phi = a_n g_1 \cdots g_r, \tag{7.19}$$

where each $g_i \in E[x, y]$ is monic in x and absolutely irreducible, and where a_n is the leading coefficient of f. For $1 \leqslant i \leqslant r$ let h_i be the homogeneous part of highest degree of g_i. Then

$$\frac{x^n - y^n}{x - y} = h_1 \cdots h_r,$$

as the left-hand side is the homogeneous part of highest degree of $a_n^{-1}\Phi$. We also have

$$\frac{x^n - y^n}{x - y} = (x - \zeta_1 y) \cdots (x - \zeta_{n-1} y),$$

where $\zeta_1, \ldots, \zeta_{n-1}$ are the nth roots of unity $\neq 1$ over $\mathbb{F}_q$, which are all distinct by Theorem 2.42(i). It follows that $x - \zeta y \in \mathbb{F}_q[x, y]$ divides exactly one of the h_i, say h_1.

Let σ be the ring automorphism from $E[x, y]$ onto itself defined by

$$\sigma\left(\sum_{j,k} \alpha_{jk} x^j y^k\right) = \sum_{j,k} \alpha_{jk}^q x^j y^k.$$

We apply σ to (7.19) and note that $\sigma(\Phi) = \Phi$ and $\sigma(a_n) = a_n$ since $\Phi \in \mathbb{F}_q[x, y]$ and $a_n \in \mathbb{F}_q$. Therefore, by unique factorization, σ permutes the polynomials g_i, so that $\sigma(g_1) = g_m$ for some m, $1 \leqslant m \leqslant r$, and hence $\sigma(h_1) = h_m$. As $x - \zeta y$ divides h_1, it also divides $h_m = \sigma(h_1)$ since $\sigma(x - \zeta y) = x - \zeta y$. This implies $m = 1$, thus $\sigma(g_1) = g_1$. Hence the coefficients of g_1 all belong to $\mathbb{F}_q$, and so g_1 is absolutely irreducible over $\mathbb{F}_q$.

We choose again $k_1, k_2, \ldots$ in Lemma 7.28 so as to get a nondecreasing sequence $k_1 \leqslant k_2 \leqslant \cdots$. Then let $d = \deg(g_1)$ and $q \geqslant k_n \geqslant k_d$. Since h_1 is divisible by $x - \zeta y$ and $\zeta \neq 1$, g_1 cannot be of the form $c(y - x)$, $c \in \mathbb{F}_q$. Hence Lemma 7.28 shows that $g_1(a, b) = 0$ for some $(a, b) \in \mathbb{F}_q^2$ with $a \neq b$, thus $\Phi(a, b) = 0$ by (7.19). Consequently, f cannot be a permutation polynomial of $\mathbb{F}_q$. □

7.32. Corollary. *Let n be a positive even integer. If $\mathbb{F}_q$ is of order $q \geqslant k_n$ with* $\gcd(n, q) = 1$, *then there are no permutation polynomials of $\mathbb{F}_q$ of degree n.*

Proof. Take $\zeta = -1$ in Theorem 7.31. □

Since the multiplicative group of $\mathbb{F}_q$ is cyclic of order $q - 1$, $\mathbb{F}_q$ contains an nth root of unity $\zeta \neq 1$ if and only if $\gcd(n, q-1) > 1$. Therefore, Theorem 7.31 leads to the following criterion.

7.33. Corollary. *Let $n \in \mathbb{N}$. If $q \geqslant k_n$ and* $\gcd(n, q) = 1$, *then there exist permutation polynomials of $\mathbb{F}_q$ of degree n if and only if* $\gcd(n, q-1) = 1$.

Proof. The necessity follows from Theorem 7.31 and the considerations above. Conversely, if $\gcd(n, q-1)=1$, then Theorem 7.8(ii) shows that x^n is a permutation polynomial of $\mathbb{F}_q$ of degree n. □

5. PERMUTATION POLYNOMIALS IN SEVERAL INDETERMINATES

Let $n \geqslant 1$ and let $\mathbb{F}_q[x_1,\ldots,x_n]$ be the ring of polynomials in n indeterminates over $\mathbb{F}_q$. Let $\mathbb{F}_q^n$ denote the cartesian product of n copies of $\mathbb{F}_q$. It is natural to define a permutation polynomial in n indeterminates over $\mathbb{F}_q$ to be a polynomial $f \in \mathbb{F}_q[x_1,\ldots,x_n]$ for which the number of solutions of the equation $f(x_1,\ldots,x_n)=a$ in $\mathbb{F}_q^n$ is the same for all $a \in \mathbb{F}_q$. This common number N of solutions must be $N=q^{n-1}$, because of the fact that $q^n = \sum_{a \in \mathbb{F}_q} N = qN$. Thus we are led to the following definition.

7.34. Definition. A polynomial $f \in \mathbb{F}_q[x_1,\ldots,x_n]$ is called a *permutation polynomial* in n indeterminates over $\mathbb{F}_q$ if the equation $f(x_1,\ldots,x_n)=a$ has q^{n-1} solutions in $\mathbb{F}_q^n$ for each $a \in \mathbb{F}_q$.

In the case $n > 1$ we cannot use the interpretation that a permutation polynomial $f(x_1,\ldots,x_n)$ over $\mathbb{F}_q$ induces a permutation of $\mathbb{F}_q^n$, because the associated mapping is not a mapping from $\mathbb{F}_q^n$ into itself. The next definition, however, enables us to consider functions from $\mathbb{F}_q^n$ into $\mathbb{F}_q^m$ induced by systems of polynomials in several indeterminates.

7.35. Definition. A system of polynomials

$$f_1,\ldots,f_m \in \mathbb{F}_q[x_1,\ldots,x_n], \quad 1 \leqslant m \leqslant n,$$

is said to be *orthogonal* in $\mathbb{F}_q$ if the system of equations

$$f_1(x_1,\ldots,x_n)=a_1,\ldots,f_m(x_1,\ldots,x_n)=a_m$$

has q^{n-m} solutions in $\mathbb{F}_q^n$ for each $(a_1,\ldots,a_m) \in \mathbb{F}_q^m$.

In the special case $m=n$ this means that the orthogonal system $f_1,\ldots,f_n$ induces a permutation of $\mathbb{F}_q^n$. Using the terminology established in Definition 7.35, we could as well say that f is a permutation polynomial if f alone forms an orthogonal system. It follows immediately from Definition 7.35 that every nonempty subsystem of an orthogonal system of polynomials is again orthogonal. In particular, *every polynomial occurring in an orthogonal system is a permutation polynomial.* On the other hand, the following theorem shows that every orthogonal system of m polynomials in n indeterminates with $m < n$ can be extended to an orthogonal system containing more polynomials. We note first that every mapping $\tau: \mathbb{F}_q^n \to \mathbb{F}_q$ can be represented by a polynomial $g(x_1,\ldots,x_n)$ over $\mathbb{F}_q$ of degree $< q$ in each indeterminate, being given by the formula

$$g(x_1,\ldots,x_n) = \sum_{(c_1,\ldots,c_n)\in\mathbb{F}_q^n} \tau(c_1,\ldots,c_n)\left(1-(x_1-c_1)^{q-1}\right)\cdots\left(1-(x_n-c_n)^{q-1}\right). \tag{7.20}$$

It is easily checked that $g(c_1,\ldots,c_n)=\tau(c_1,\ldots,c_n)$ for all $(c_1,\ldots,c_n)\in\mathbb{F}_q^n$.

7.36. Theorem. *For every orthogonal system $f_1,\ldots,f_m \in \mathbb{F}_q[x_1,\ldots,x_n]$, $1 \leqslant m < n$, in $\mathbb{F}_q$ and every r, $1 \leqslant r \leqslant n-m$, there exist polynomials $f_{m+1},\ldots,$ $f_{m+r} \in \mathbb{F}_q[x_1,\ldots,x_n]$ such that $f_1,\ldots,f_{m+r}$ form an orthogonal system in $\mathbb{F}_q$.*

Proof. It suffices to show the theorem for $r=1$. For $(a_1,\ldots,a_m)\in\mathbb{F}_q^m$ put

$$S(a_1,\ldots,a_m)=\{(c_1,\ldots,c_n)\in\mathbb{F}_q^n : f_i(c_1,\ldots,c_n)=a_i \quad \text{for } 1\leqslant i\leqslant m\}.$$

By hypothesis, each $S(a_1,\ldots,a_m)$ has q^{n-m} elements. Decompose each $S(a_1,\ldots,a_m)$ in an arbitrary way into q pairwise disjoint subsets $S(a_1,\ldots,a_m,a)$, $a\in\mathbb{F}_q$, each of them having q^{n-m-1} elements. We construct a mapping $\tau:\mathbb{F}_q^n\to\mathbb{F}_q$ in the following way. A given $(c_1,\ldots,c_n)\in\mathbb{F}_q^n$ lies in a uniquely determined $S(a_1,\ldots,a_m,a)$; define $\tau(c_1,\ldots,c_n)=a$. By (7.20) the mapping τ can be represented by a polynomial $f_{m+1}(x_1,\ldots,x_n)$ over $\mathbb{F}_q$, and this polynomial meets all requirements. □

A necessary and sufficient condition for a system of polynomials to be orthogonal can be given in terms of characters. We use the notation for additive characters introduced in Theorem 5.7.

7.37. Theorem. *The system $f_1,\ldots,f_m\in\mathbb{F}_q[x_1,\ldots,x_n]$, $1\leqslant m\leqslant n$, is orthogonal in $\mathbb{F}_q$ if and only if*

$$\sum_{(c_1,\ldots,c_n)\in\mathbb{F}_q^n} \chi_{b_1}(f_1(c_1,\ldots,c_n))\cdots\chi_{b_m}(f_m(c_1,\ldots,c_n))=0$$

for all additive characters $\chi_{b_1},\ldots,\chi_{b_m}$ of $\mathbb{F}_q$ with $(b_1,\ldots,b_m)\neq(0,\ldots,0)$.

Proof. For $(a_1,\ldots,a_m)\in\mathbb{F}_q^m$ let $N(a_1,\ldots,a_m)$ denote the number of solutions in $\mathbb{F}_q^n$ of the system

$$f_1(x_1,\ldots,x_n)=a_1,\ldots,f_m(x_1,\ldots,x_n)=a_m.$$

If the system $f_1,\ldots,f_m$ is orthogonal in $\mathbb{F}_q$, then

$$\sum_{(c_1,\ldots,c_n)\in\mathbb{F}_q^n} \chi_{b_1}(f_1(c_1,\ldots,c_n))\cdots\chi_{b_m}(f_m(c_1,\ldots,c_n))$$

$$= \sum_{(a_1,\ldots,a_m)\in\mathbb{F}_q^m} N(a_1,\ldots,a_m)\chi_{b_1}(a_1)\cdots\chi_{b_m}(a_m)$$

$$= q^{n-m}\left(\sum_{a_1\in\mathbb{F}_q}\chi_{b_1}(a_1)\right)\cdots\left(\sum_{a_m\in\mathbb{F}_q}\chi_{b_m}(a_m)\right)=0$$

provided that at least one $b_i \neq 0$, as follows from (5.9).

Conversely, if the condition of the theorem is satisfied, then for every $(a_1,\ldots,a_m)\in\mathbb{F}_q^m$ we get from (5.10),

$$N(a_1,\ldots,a_m)=\frac{1}{q^m}\sum_{(c_1,\ldots,c_n)\in\mathbb{F}_q^n}\left(\sum_{b_1\in\mathbb{F}_q}\chi_{b_1}(f_1(c_1,\ldots,c_n))\overline{\chi_{b_1}(a_1)}\right)\cdots$$

$$\left(\sum_{b_m\in\mathbb{F}_q}\chi_{b_m}(f_m(c_1,\ldots,c_n))\overline{\chi_{b_m}(a_m)}\right)$$

$$=\frac{1}{q^m}\sum_{(b_1,\ldots,b_m)\in\mathbb{F}_q^m}\bar{\chi}_{b_1}(a_1)\cdots\bar{\chi}_{b_m}(a_m)$$

$$\sum_{(c_1,\ldots,c_n)\in\mathbb{F}_q^n}\chi_{b_1}(f_1(c_1,\ldots,c_n))\cdots\chi_{b_m}(f_m(c_1,\ldots,c_n))$$

$$=\frac{1}{q^m}\cdot q^n = q^{n-m}.$$ □

7.38. Corollary. *The polynomial $f\in\mathbb{F}_q[x_1,\ldots,x_n]$ is a permutation polynomial over $\mathbb{F}_q$ if and only if*

$$\sum_{(c_1,\ldots,c_n)\in\mathbb{F}_q^n}\chi(f(c_1,\ldots,c_n))=0$$

for all nontrivial additive characters χ of $\mathbb{F}_q$.

7.39. Corollary. *The system $f_1,\ldots,f_m\in\mathbb{F}_q[x_1,\ldots,x_n]$, $1\leqslant m\leqslant n$, is orthogonal in $\mathbb{F}_q$ if and only if for all $(b_1,\ldots,b_m)\in\mathbb{F}_q^m$ with $(b_1,\ldots,b_m)\neq(0,\ldots,0)$ the polynomial $b_1f_1+\cdots+b_mf_m$ is a permutation polynomial over $\mathbb{F}_q$.*

Proof. This follows from Theorem 7.37, Corollary 7.38, and the fact that $\chi_b(c)=\chi_1(bc)$ for all $b,c\in\mathbb{F}_q$. □

Let $(x_1^q-x_1,\ldots,x_n^q-x_n)$ be the ideal of $\mathbb{F}_q[x_1,\ldots,x_n]$ consisting of all polynomials of the form

$$g_1(x_1,\ldots,x_n)(x_1^q-x_1)+\cdots+g_n(x_1,\ldots,x_n)(x_n^q-x_n)$$

with $g_1,\ldots,g_n \in \mathbb{F}_q[x_1,\ldots,x_n]$. Then Lemma 7.2 can be generalized as follows.

7.40. Lemma. (i) *For every $f \in \mathbb{F}_q[x_1,\ldots,x_n]$ there exists a unique $g \in \mathbb{F}_q[x_1,\ldots,x_n]$ of degree $< q$ in each indeterminate with $f(c_1,\ldots,c_n) = g(c_1,\ldots,c_n)$ for all $(c_1,\ldots,c_n) \in \mathbb{F}_q^n$.*

(ii) *For $f, g \in \mathbb{F}_q[x_1,\ldots,x_n]$ we have $f(c_1,\ldots,c_n) = g(c_1,\ldots,c_n)$ for all $(c_1,\ldots,c_n) \in \mathbb{F}_q^n$ if and only if $f \equiv g \bmod (x_1^q - x_1,\ldots,x_n^q - x_n)$.*

(iii) *For every $f \in \mathbb{F}_q[x_1,\ldots,x_n]$ there exists a unique $g \in \mathbb{F}_q[x_1,\ldots,x_n]$ of degree $< q$ in each indeterminate with $f \equiv g \bmod (x_1^q - x_1,\ldots,x_n^q - x_n)$.*

Proof. (i) The existence of g follows from (7.20). To prove the uniqueness, it suffices to show that if $g \in \mathbb{F}_q[x_1,\ldots,x_n]$ is of degree $< q$ in each indeterminate and satisfies $g(c_1,\ldots,c_n) = 0$ for all $(c_1,\ldots,c_n) \in \mathbb{F}_q^n$, then g is the zero polynomial. We proceed by induction on n and note that the case $n = 1$ follows from Lemma 7.2. Let $n \geqslant 2$ and suppose the statement is shown for polynomials in $n-1$ indeterminates. If $g \in \mathbb{F}_q[x_1,\ldots,x_n]$ is of the indicated type, we can write

$$g(x_1,\ldots,x_n) = h_0(x_2,\ldots,x_n) + h_1(x_2,\ldots,x_n)x_1 + \cdots + h_{q-1}(x_2,\ldots,x_n)x_1^{q-1},$$

where each h_j is of degree $< q$ in each of the indeterminates $x_2,\ldots,x_n$. Let $(c_2,\ldots,c_n) \in \mathbb{F}_q^{n-1}$ be fixed; from $g(c, c_2,\ldots,c_n) = 0$ for all $c \in \mathbb{F}_q$ we get a system of q homogeneous linear equations for the elements $h_j(c_2,\ldots,c_n)$, $0 \leqslant j \leqslant q-1$, with a nonzero Vandermonde determinant. It follows that $h_j(c_2,\ldots,c_n) = 0$ for $0 \leqslant j \leqslant q-1$, and since $(c_2,\ldots,c_n) \in \mathbb{F}_q^{n-1}$ is arbitrary, the induction hypothesis implies that each $h_j = 0$, hence $g = 0$.

(ii) Put $J = (x_1^q - x_1,\ldots,x_n^q - x_n)$. If $f \equiv g \bmod J$, it is clear that $f(c_1,\ldots,c_n) = g(c_1,\ldots,c_n)$ for all $(c_1,\ldots,c_n) \in \mathbb{F}_q^n$. Conversely, if $f(c_1,\ldots,c_n) = g(c_1,\ldots,c_n)$ for all $(c_1,\ldots,c_n) \in \mathbb{F}_q^n$, then by using $x_i^k \equiv x_i^m \bmod J$, $1 \leqslant i \leqslant n$, whenever $k > m \geqslant 1$ and $k \equiv m \bmod (q-1)$, we obtain polynomials f_1, g_1 of degree $< q$ in each indeterminate and satisfying $f \equiv f_1 \bmod J$, $g \equiv g_1 \bmod J$. Then

$$f_1(c_1,\ldots,c_n) = f(c_1,\ldots,c_n) = g(c_1,\ldots,c_n) = g_1(c_1,\ldots,c_n)$$

for all $(c_1,\ldots,c_n) \in \mathbb{F}_q^n$, hence (i) implies $f_1 = g_1$, and so $f \equiv g \bmod J$.

(iii) This follows from (i) and (ii). □

The unique polynomial g in Lemma 7.40(iii) is called the *reduction of* $f \bmod (x_1^q - x_1,\ldots,x_n^q - x_n)$. We can now generalize Theorem 7.6 as follows.

7.41. Theorem. *Let $\mathbb{F}_q$ be of characteristic p. Then the system $f_1,\ldots,f_n \in \mathbb{F}_q[x_1,\ldots,x_n]$ is orthogonal in $\mathbb{F}_q$ if and only if the following two* *conditions are satisfied*:

(i) *in the reduction of*

$$f_1^{q-1}\cdots f_n^{q-1}\operatorname{mod}(x_1^q - x_1,\ldots,x_n^q - x_n)$$

the coefficient of $x_1^{q-1}\cdots x_n^{q-1}$ *is* $\neq 0$;

(ii) *in the reduction of*

$$f_1^{t_1}\cdots f_n^{t_n}\operatorname{mod}(x_1^q - x_1,\ldots,x_n^q - x_n)$$

the coefficient of $x_1^{q-1}\cdots x_n^{q-1}$ *is* 0 *whenever* $t_1,\ldots,t_n$ *are integers with* $0 \leqslant t_i \leqslant q-1$ *for* $1 \leqslant i \leqslant n$, *not all* $t_i = q-1$, *and at least one* $t_i \not\equiv 0 \operatorname{mod} p$.

Proof. Let $f_1,\ldots,f_n$ be orthogonal in $\mathbb{F}_q$ and $t_1,\ldots,t_n \in \mathbb{Z}$ with $0 \leqslant t_i \leqslant q-1$ for $1 \leqslant i \leqslant n$. By Lemma 7.40 and (7.20), the reduction g of

$$f_1^{t_1}\cdots f_n^{t_n}\operatorname{mod}(x_1^q - x_1,\ldots,x_n^q - x_n)$$

is given by

$$g(x_1,\ldots,x_n) = \sum_{(c_1,\ldots,c_n)\in\mathbb{F}_q^n} (f_1^{t_1}\cdots f_n^{t_n})(c_1,\ldots,c_n) \cdot\left(1-(x_1-c_1)^{q-1}\right)\cdots\left(1-(x_n-c_n)^{q-1}\right).$$

Thus the coefficient of $x_1^{q-1}\cdots x_n^{q-1}$ in g is given by

$$\begin{aligned}(-1)^n & \sum_{(c_1,\ldots,c_n)\in\mathbb{F}_q^n} (f_1^{t_1}\cdots f_n^{t_n})(c_1,\ldots,c_n)\\ &= (-1)^n \sum_{(c_1,\ldots,c_n)\in\mathbb{F}_q^n} f_1(c_1,\ldots,c_n)^{t_1}\cdots f_n(c_1,\ldots,c_n)^{t_n}\\ &= (-1)^n \sum_{(a_1,\ldots,a_n)\in\mathbb{F}_q^n} a_1^{t_1}\cdots a_n^{t_n} = (-1)^n\left(\sum_{a_1\in\mathbb{F}_q} a_1^{t_1}\right)\cdots\left(\sum_{a_n\in\mathbb{F}_q} a_n^{t_n}\right),\end{aligned}$$

and (i) and (ii) follow from Lemma 7.3.

Conversely, let (i) and (ii) be satisfied. Then, according to the calculation above, (i) implies

$$\sum_{(c_1,\ldots,c_n)\in\mathbb{F}_q^n} (f_1^{q-1}\cdots f_n^{q-1})(c_1,\ldots,c_n) \neq 0, \tag{7.21}$$

while (ii) implies

$$\sum_{(c_1,\ldots,c_n)\in\mathbb{F}_q^n} (f_1^{t_1}\cdots f_n^{t_n})(c_1,\ldots,c_n) = 0$$

for $t_1,\ldots,t_n$ as in (ii). Using

$$\sum_{(c_1,\ldots,c_n)\in\mathbb{F}_q^n} f(c_1,\ldots,c_n)^{tp^j} = \left(\sum_{(c_1,\ldots,c_n)\in\mathbb{F}_q^n} f(c_1,\ldots,c_n)^t\right)^{p^j},$$

we get

$$\sum_{(c_1,\ldots,c_n)\in\mathbb{F}_q^n} \left(f_1^{t_1}\cdots f_n^{t_n}\right)(c_1,\ldots,c_n)=0 \tag{7.22}$$

for $t_1,\ldots,t_n\in\mathbb{Z}$ with $0\leqslant t_i\leqslant q-1$ for $1\leqslant i\leqslant n$, not all $t_i=q-1$, and not all $t_i=0$. The identity (7.22) holds trivially for $t_1=\cdots=t_n=0$. To prove that $f_1,\ldots,f_n$ is orthogonal in $\mathbb{F}_q$, it suffices to show that for each $(a_1,\ldots,a_n)\in\mathbb{F}_q^n$ the number $N(a_1,\ldots,a_n)$ of solutions in $\mathbb{F}_q^n$ of the system

$$f_1(x_1,\ldots,x_n)=a_1,\ldots,f_n(x_1,\ldots,x_n)=a_n$$

is $\neq 0$. We shall show that $N(a_1,\ldots,a_n)$, regarded as an element of $\mathbb{F}_q$, is $\neq 0$. Indeed, by (7.21) and (7.22) we have

$$\begin{aligned}
N(a_1,\ldots,a_n) &= (-1)^n \sum_{(c_1,\ldots,c_n)\in\mathbb{F}_q^n} \prod_{i=1}^{n}\left[\left(f_i(c_1,\ldots,c_n)-a_i\right)^{q-1}-1\right]\\
&= (-1)^n \sum_{(c_1,\ldots,c_n)\in\mathbb{F}_q^n} \left[f_1^{q-1}\cdots f_n^{q-1} + \sum_{\substack{t_1,\ldots,t_n=0\\ \text{not all } t_i=q-1}}^{q-1} b_{t_1\cdots t_n} f_1^{t_1}\cdots f_n^{t_n}\right](c_1,\ldots,c_n)\\
&= (-1)^n \sum_{(c_1,\ldots,c_n)\in\mathbb{F}_q^n} \left(f_1^{q-1}\cdots f_n^{q-1}\right)(c_1,\ldots,c_n)\neq 0. \qquad \square
\end{aligned}$$

The principle stated in the following theorem can be used to construct new permutation polynomials from given ones.

7.42. Theorem. *Suppose $f\in\mathbb{F}_q[x_1,\ldots,x_n]$ is of the form*

$$f(x_1,\ldots,x_n)=g(x_1,\ldots,x_m)+h(x_{m+1},\ldots,x_n), \quad 1\leqslant m<n.$$

If at least one of g and h is a permutation polynomial over $\mathbb{F}_q$, then f is a permutation polynomial over $\mathbb{F}_q$. If q is prime, then the converse holds as well.

Proof. For $a\in\mathbb{F}_q$ let $N(a)$ be the number of solutions of $f(x_1,\ldots,x_n)=a$ in $\mathbb{F}_q^n$, and define $L(a)$, $M(a)$ similarly with reference to g, h, respectively. Then

$$N(a)=\sum_{a_1+a_2=a} L(a_1)M(a_2). \tag{7.23}$$

Suppose, for the sake of definiteness, that g is a permutation polynomial over $\mathbb{F}_q$. Then $L(a)=q^{m-1}$ for all $a\in\mathbb{F}_q$, hence (7.23) yields

$$N(a)=q^{m-1}\sum_{a_2\in\mathbb{F}_q}M(a_2)=q^{m-1}q^{n-m}=q^{n-1},$$

so that f is a permutation polynomial over $\mathbb{F}_q$.

For a prime p, suppose f is a permutation polynomial over $\mathbb{F}_p$. We want to show that either g or h is a permutation polynomial over $\mathbb{F}_p$. From (7.23) we get

$$\sum_{a_1+a_2=a}L(a_1)M(a_2)=p^{n-1}\quad\text{for all } a\in\mathbb{F}_p.$$

Writing down these identities for $a=-1,0,1,\ldots,p-2$, we arrive at a system of linear equations for $M(p-1), M(p-2),\ldots,M(0)$ with a determinant D whose (i,j) entry is $L(i+j-2)$, $1\leqslant i,j\leqslant p$, where $i+j-2$ is taken modulo p. If $D\neq 0$, then the system has a unique solution, namely $M(p-1)=M(p-2)=\cdots=M(0)=p^{n-m-1}$, and so h is a permutation polynomial over $\mathbb{F}_p$. Now assume $D=0$. We use the fact that $D=\pm R$, where R is the resultant of the two polynomials $G(x)=x^p-1$, $H(x)=L(0)x^{p-1}+L(1)x^{p-2}+\cdots+L(p-1)$ over the rationals. Thus $G(x)$ and $H(x)$ have a common root in some extension field of the rationals. But $G(x)=(x-1)Q_p(x)$, where $Q_p(x)$ is the irreducible pth cyclotomic polynomial (see Theorem 2.47(i)), and $H(1)=p^m\neq 0$. Therefore $Q_p(x)$ divides $H(x)$, and so $H(x)=L(0)Q_p(x)$. Equating coefficients yields $L(a)=L(0)=p^{m-1}$ for all $a\in\mathbb{F}_p$, and so g is a permutation polynomial over $\mathbb{F}_p$. □

7.43. Theorem. *If q is not prime, then for $1\leqslant m<n$ there exist polynomials $g(x_1,\ldots,x_m)$ and $h(x_{m+1},\ldots,x_n)$ over $\mathbb{F}_q$ such that $g(x_1,\ldots,x_m)+h(x_{m+1},\ldots,x_n)$ is a permutation polynomial over $\mathbb{F}_q$, but neither $g(x_1,\ldots,x_m)$ nor $h(x_{m+1},\ldots,x_n)$ is a permutation polynomial over $\mathbb{F}_q$.*

Proof. We have $q=p^e$ with p prime and $e>1$. The additive factor group $\mathbb{F}_q/\mathbb{F}_p$ has order $r=p^{e-1}$. We construct a system $a_1,\ldots,a_r$ of elements of $\mathbb{F}_q$ by choosing a representative from each coset. Let L and M have the same meaning as in the proof of Theorem 7.42. By (7.20) there exists a polynomial $g(x_1,\ldots,x_m)$ over $\mathbb{F}_q$ such that $L(a_j)=(1/r)q^m$ for $1\leqslant j\leqslant r$ and $L(c)=0$ for all other elements $c\in\mathbb{F}_q$, and a polynomial $h(x_{m+1},\ldots,x_n)$ over $\mathbb{F}_q$ such that

$$M(0)=M(1)=\cdots=M(p-1)=\frac{1}{p}q^{n-m}$$

and $M(d)=0$ for all other elements $d\in\mathbb{F}_q$. Neither g nor h is a permutation polynomial over $\mathbb{F}_q$. But $g+h$ is a permutation polynomial over $\mathbb{F}_q$, since every $a\in\mathbb{F}_q$ has a unique representation of the form $a=a_j+b$ with

$1 \leqslant j \leqslant r$ and $b \in \mathbb{F}_p$, and so the total number of solutions of the equation

$$g(x_1,\ldots,x_m)+h(x_{m+1},\ldots,x_n)=a=a_j+b$$

in $\mathbb{F}_q^n$ is equal to

$$\left(\frac{1}{r}q^m\right)\left(\frac{1}{p}q^{n-m}\right)=q^{n-1}. \qquad \square$$

Sometimes it is helpful to use the following one-to-one correspondence between orthogonal systems of m polynomials in $n=mk$ indeterminates over $\mathbb{F}_q$ and permutation polynomials in k indeterminates over $\mathbb{F}_{q^m}$, established by the next theorem.

7.44. Theorem. *If $n=mk$ with positive integers n, m, k, then there is a one-to-one correspondence between orthogonal systems in $\mathbb{F}_q$ consisting of m polynomials over $\mathbb{F}_q$ of degree $<q$ in each of their n indeterminates and permutation polynomials over $\mathbb{F}_{q^m}$ of degree $<q^m$ in each of their k indeterminates.*

Proof. Let $\{\omega_1,\ldots,\omega_m\}$ be a basis of $E=\mathbb{F}_{q^m}$ over $\mathbb{F}_q$. Each $(\gamma_1,\ldots,\gamma_k)\in E^k$ determines a unique $(c_1,\ldots,c_n)\in\mathbb{F}_q^n$ via

$$\gamma_i=c_{(i-1)m+1}\omega_1+c_{(i-1)m+2}\omega_2+\cdots+c_{im}\omega_m \quad \text{for } 1\leqslant i\leqslant k.$$

Suppose $f_1,\ldots,f_m$ are of degree $<q$ in each of their n indeterminates and form an orthogonal system in $\mathbb{F}_q$. By (7.20) and Lemma 7.40 there exists a unique polynomial g over E of degree $<q^m$ in each of its k indeterminates such that

$$g(\gamma_1,\ldots,\gamma_k)=f_1(c_1,\ldots,c_n)\omega_1+\cdots+f_m(c_1,\ldots,c_n)\omega_m \tag{7.24}$$

for all $(\gamma_1,\ldots,\gamma_k)\in E^k$. Then g is a permutation polynomial over E, since for $\alpha=a_1\omega_1+\cdots+a_m\omega_m\in E$, $a_1,\ldots,a_m\in\mathbb{F}_q$, we have $g(\gamma_1,\ldots,\gamma_k)=\alpha$ if and only if $f_j(c_1,\ldots,c_n)=a_j$ for $1\leqslant j\leqslant m$, and so there are $q^{n-m}=(q^m)^{k-1}$ solutions $(\gamma_1,\ldots,\gamma_k)$. On the other hand, if g is a given permutation polynomial over E of degree $<q^m$ in each of its k indeterminates, then an orthogonal system $f_1,\ldots,f_m$ in $\mathbb{F}_q$ of the type considered can be recovered uniquely, by using the polynomials over $\mathbb{F}_q$ of degree $<q$ in each indeterminate representing the coordinate functions with respect to $\{\omega_1,\ldots,\omega_m\}$ determined by (7.24). $\square$

In the special case $m=n$ we get the following simple consequence of this theorem.

7.45. Corollary. *There is a one-to-one correspondence between orthogonal systems in $\mathbb{F}_q$ consisting of n polynomials over $\mathbb{F}_q$ of degree $<q$ in each of their n indeterminates and permutation polynomials in one indeterminate over $\mathbb{F}_{q^n}$ of degree $<q^n$.*

We obtain nontrivial examples of orthogonal systems by generalizing the construction of Dickson polynomials in Section 2. For $n \in \mathbb{N}$, $a \in \mathbb{F}_q$, and $(c_1,\ldots,c_n) \in \mathbb{F}_q^n$, consider the polynomial

$$r(c_1,\ldots,c_n,z) = z^{n+1} - c_1 z^n + c_2 z^{n-1} + \cdots + (-1)^n c_n z + (-1)^{n+1} a \tag{7.25}$$

in the indeterminate z over $\mathbb{F}_q$. This polynomial has $n+1$ not necessarily distinct roots $\beta_1,\ldots,\beta_{n+1}$ in a suitable extension of $\mathbb{F}_q$. Now let $k \in \mathbb{N}$ and set

$$r_k(c_1,\ldots,c_n,z) = (z-\beta_1^k)\cdots(z-\beta_{n+1}^k).$$

Then

$$r_k(c_1,\ldots,c_n,z)$$
$$= z^{n+1} - \sigma_1(\beta_1^k,\ldots,\beta_{n+1}^k)z^n + \cdots + (-1)^{n+1}\sigma_{n+1}(\beta_1^k,\ldots,\beta_{n+1}^k),$$

where σ_i is the ith elementary symmetric polynomial in $n+1$ indeterminates (compare with Example 1.74). Since $\sigma_i(u_1^k,\ldots,u_{n+1}^k)$ is symmetric in the indeterminates $u_1,\ldots,u_{n+1}$, there exist integral polynomials $g_k^{(1)},\ldots,g_k^{(n+1)}$ in $n+1$ indeterminates such that

$$\sigma_i(u_1^k,\ldots,u_{n+1}^k) = g_k^{(i)}(\sigma_1(u_1,\ldots,u_{n+1}),\ldots,\sigma_{n+1}(u_1,\ldots,u_{n+1})) \quad \text{for } 1 \leqslant i \leqslant n+1.$$

As $\beta_1,\ldots,\beta_{n+1}$ are the roots of (7.25), we have

$$\sigma_i(\beta_1,\ldots,\beta_{n+1}) = c_i \quad \text{for } 1 \leqslant i \leqslant n,$$
$$\sigma_{n+1}(\beta_1,\ldots,\beta_{n+1}) = \beta_1 \cdots \beta_{n+1} = a,$$

and so

$$\sigma_i(\beta_1^k,\ldots,\beta_{n+1}^k) = g_k^{(i)}(c_1,\ldots,c_n,a) \quad \text{for } 1 \leqslant i \leqslant n+1.$$

Substituting this in $r_k(c_1,\ldots,c_n,z)$, we get

$$r_k(c_1,\ldots,c_n,z) = z^{n+1} - g_k^{(1)}(c_1,\ldots,c_n,a)z^n + \cdots$$
$$+ (-1)^n g_k^{(n)}(c_1,\ldots,c_n,a)z + (-1)^{n+1}a^k.$$

The polynomials

$$g_k^{(i)}(x_1,\ldots,x_n,a), \quad 1 \leqslant i \leqslant n,$$

are integral polynomials in $x_1,\ldots,x_n,a$ and are polynomials over $\mathbb{F}_q$ in $x_1,\ldots,x_n$. The latter polynomials are the *Dickson polynomials* in n indeterminates over $\mathbb{F}_q$. This definition, though not its motivation, can also be made meaningful for any commutative ring R with identity, by choosing $a \in R$. For $n = 1$ and $a \in R$, $g_k^{(1)}(x_1, a)$ is the Dickson polynomial in one indeterminate over R defined in (7.6).

An explicit expression for the polynomials $g_k^{(1)}(x_1,\ldots,x_n,a)$ can be obtained from Waring's formula (see Theorem 1.76). For example, in the case $n = 2$ we get

$$g_k^{(1)}(x, y, a) = \sum_{i=0}^{\lfloor k/2 \rfloor} \sum_{\substack{j=0 \\ 2i+3j \leqslant k}}^{\lfloor k/3 \rfloor} \frac{k(-1)^i}{k-i-2j} \binom{k-i-2j}{i+j} \binom{i+j}{i} a^j x^{k-2i-3j} y^i.$$

Let $g_k(a)$ denote the system consisting of the polynomials $g_k^{(1)}(x_1, \ldots, x_n, a), \ldots, g_k^{(n)}(x_1, \ldots, x_n, a)$. We have then the following generalization of Theorems 7.16 and 7.8(ii).

7.46. Theorem. *For $a \in \mathbb{F}_q^*$ the system $g_k(a)$ is orthogonal in $\mathbb{F}_q$ if and only if* $\gcd(k, q^s - 1) = 1$ *for $s = 1, 2, \ldots, n+1$. The system $g_k(0)$ is orthogonal in $\mathbb{F}_q$ if and only if* $\gcd(k, q^s - 1) = 1$ *for $s = 1, 2, \ldots, n$.*

According to a principle stated earlier in this section, each polynomial occurring in an orthogonal system $g_k(a)$ is a permutation polynomial in n indeterminates over $\mathbb{F}_q$. Another class of permutation polynomials in several indeterminates is obtained by considering linear and quadratic polynomials. We note first that the property of being a permutation polynomial over $\mathbb{F}_q$ is invariant under transformations of indeterminates of the form

$$x_i = \sum_{j=1}^{n} a_{ij} y_j + b_i \quad \text{for } 1 \leqslant i \leqslant n, \tag{7.26}$$

where $a_{ij}, b_i \in \mathbb{F}_q$ for $1 \leqslant i, j \leqslant n$ and the matrix with (i, j) entry a_{ij} is nonsingular. We call two polynomials in n indeterminates over $\mathbb{F}_q$ *equivalent* if one can be transformed into the other by a transformation of the type (7.26).

7.47. Theorem. *Let $f \in \mathbb{F}_q[x_1, \ldots, x_n]$ with* $\deg(f) \leqslant 2$ *and $n \geqslant 2$. For q odd, f is a permutation polynomial over $\mathbb{F}_q$ if and only if f is equivalent to a polynomial of the form $g(x_1, \ldots, x_{n-1}) + x_n$ for some $g \in \mathbb{F}_q[x_1, \ldots, x_{n-1}]$. For q even, f is a permutation polynomial over $\mathbb{F}_q$ if and only if f is equivalent to $g(x_1, \ldots, x_{n-1}) + x_n$ or $g(x_1, \ldots, x_{n-1}) + x_n^2$ for some $g \in \mathbb{F}_q[x_1, \ldots, x_{n-1}]$.*

For q odd and $f \in \mathbb{F}_q[x_1, \ldots, x_n]$ with $\deg(f) \leqslant 2$, let A be the coefficient matrix of the quadratic form occurring in f (see Chapter 6, Section 2) and let A' be the augmented matrix consisting of A and one more column containing the coefficients of the linear terms. It follows then easily from Theorem 7.47 that f is a permutation polynomial over $\mathbb{F}_q$ if and only if $\operatorname{rank}(A') > \operatorname{rank}(A)$.

NOTES

1. The general study of permutation polynomials started with Hermite [2] who considered the case of finite prime fields. Some early

results for this case can also be found in C. Jordan [2] and Serret [2]. Permutation polynomials of arbitrary finite fields were first studied by Dickson [2]. An account of this work can also be found in Dickson [7, Part I, Ch. 5]. The history of the subject until 1922 has been traced in Dickson [42, Ch. 18]. A modern treatment of permutation polynomials is presented in Lausch and Nöbauer [1, Ch. 4].

The fact that any function from $\mathbb{F}_q$ into $\mathbb{F}_q$ can be represented by a polynomial was first noted by Hermite [2] for q prime (see also Weber [4, Sec. 180] and Zsigmondy [3]) and by Dickson [2] for general q. Dickson [2] showed also that the condition $\deg(g) < q$ uniquely determines the polynomial g representing the function. Various methods of calculating g are discussed in Bernstein [2], Gill and Jacob [1], Székely and Mureşan [1], and Wesselkamper [1]. Zsigmondy [3], Dickson [2], and Carlitz [86] noted that permutation polynomials of $\mathbb{F}_q$ can be obtained by using an interpolation formula with a permutation function of $\mathbb{F}_q$. Polynomial representations of functions from $\mathbb{F}_p$ into itself taking only the values 0 and 1 are considered in Carlitz [123] and Cazacu [1]. Wesselkamper [2], [3] discusses the representation of functions defined on subsets of $\mathbb{F}_q$.

Finite fields are polynomially complete in the sense of the following definition: a ring R is *polynomially complete* if any function from R into itself can be represented by a polynomial over R. Kempner [1] showed that the only residue class rings $\mathbb{Z}/(m)$ that are polynomially complete are the finite prime fields; see also Bernstein [2]. Rédei and Szele [1] proved more generally that the only nonzero commutative rings that are polynomially complete are the finite fields, and Heisler [1] showed this without requiring commutativity. For a general discussion of polynomially complete algebraic structures we refer to Lausch and Nöbauer [1, Ch. 1]. Using a more general notion of polynomial over a ring R, Brawley and Carlitz [2] proved that every function from R into itself can be represented by such a polynomial if and only if R is the trivial ring of order 1 or 2 (i.e., $ab = 0$ for all $a, b \in R$) or the ring of $n \times n$ matrices over some $\mathbb{F}_q$ for some $n \in \mathbb{N}$. Polynomial functions over the latter type of ring were also considered in Brawley [5].

Some attention has been devoted to functions from a residue class ring $R_m = \mathbb{Z}/(m)$ into itself. If m is composite, then according to the result of Kempner [1] quoted above not every such function can be represented by a polynomial over R_m. Criteria for polynomial representability of a given function have been established by Kempner [1], Rédei and Szele [1], [2], and Carlitz [97]. The set P_m of all functions from R_m into itself that can be represented by a polynomial over R_m forms a ring under the usual addition and multiplication of functions. An easy application of the ring homomorphism theorem shows that P_m is isomorphic to $R_m[x]/I_m$, where $I_m = \{f \in R_m[x] : f(a) = 0 \text{ for all } a \in R_m\}$. The polynomials in the ideal I_m are called *residue polynomials* modulo m. Various characterizations of residue polynomials modulo m have been given in Aĭzenberg, Semion, and Citkin [1], Kempner [1], Litzinger [1], Niven and Warren [1], Rédei and Szele [1], and Singmaster [1]. Further results on polynomial functions over R_m can be found in Keller and Olson [1], Nöbauer [1], and Rédei and Szele [2]. For a discussion of residue polynomials over general rings see Lausch and Nöbauer [1, Ch. 3].

One part of Lemma 7.3—namely that (i) implies (ii)—is already covered by Lemma 6.3. For a stronger version of the converse see Carlitz and Lutz [1]. The criterion in Theorem 7.4 was shown in an essentially equivalent form by Hermite [2] for finite prime fields; the general case was established by Dickson [2]. L. J. Rogers [2] noted that if $q = p$, then one only needs to check condition (ii) for $1 \leqslant t \leqslant (q-1)/2$; this is not true for $q \neq p$ (see Dickson [7, Sec. 96]). An explicit form of Hermite's criterion for $\mathbb{F}_p$ in terms of the coefficients of f was given by London and Ziegler [1]. Corollary 7.5 was shown for $q = p$ in Dickson [1] and for the general case in Dickson [2]. The sufficiency part of Theorem 7.6 is due to Carlitz and Lutz [1]. Other criteria for permutation polynomials can be found in de Polignac [1], Raussnitz [1], and T. P. Vaughan [1].

For applications of permutation polynomials of finite fields to finite projective geometries we refer to Chapter 9, Section 3, and the notes to it. Levine and Brawley [2] show how permutation polynomials of finite fields can be used to construct cryptographic systems.

Permutation polynomials of residue class rings $\mathbb{Z}/(m)$ were considered by Nöbauer [1], [2], [4], [8]; see also Cavior [5], Keller and Olson [1], Niven [2], and Zane [1]. For the theory of permutation polynomials of more general residue class rings see Lausch and Nöbauer [1, Ch. 4]. Brawley, Carlitz, and Levine [2] (see also R. Matthews [1]) and Brawley [4] studied polynomials over $\mathbb{F}_q$ that induce permutations of the ring of $n \times n$ matrices over $\mathbb{F}_q$, and Brawley [3] considered the more general case where $\mathbb{F}_q$ is replaced by any finite commutative ring with identity. P. Chowla [1] and Corzatt [1] discussed polynomials that permute sets of integers. Rational functions that yield permutations of $\mathbb{F}_q$ were considered by Rédei [4] and later by Carlitz [86], S. D. Cohen [5], [6], [9], Gwehenberger [1], and Nöbauer [8], [11], with the latter author also treating $\mathbb{Z}/(m)$.

Permutation polynomials of $\mathbb{F}_q$ are described by the property $V(f) = q$, where $V(f)$ is the cardinality of the value set $\{f(c): c \in \mathbb{F}_q\}$ of a given $f \in \mathbb{F}_q[x]$. The quantity $V(f)$ has also been studied for arbitrary polynomials over $\mathbb{F}_q$. Exact formulas for $V(f)$ can be established for polynomials of small degree; the linear and quadratic case are easy, and for cubic polynomials and certain quartic polynomials see von Sterneck [1] and Kantor [1]. S. Chowla [7] posed the problem of estimating $V(f)$ and Birch and Swinnerton-Dyer [1] succeeded in establishing the following significant result: if $f \in \mathbb{F}_q[x]$ with $\deg(f) = n \geqslant 1$ is a "general" polynomial (in the sense that the Galois group of the equation $f(x)=y$ over $\overline{\mathbb{F}}_q(y)$, $\overline{\mathbb{F}}_q$ being the algebraic closure of $\mathbb{F}_q$, is the symmetric group S_n), then

$$V(f)=q\sum_{j=1}^{n}\frac{(-1)^{j-1}}{j!}+O(q^{1/2}),$$

where the implied constant depends only on n. For $n=4$ a result of this type was mentioned in S. Chowla [7] and shown elementarily in McCann and Williams [2]. Prior to Birch and Swinnerton-Dyer [1], Uchiyama [2] had proved the following weaker result: if $n \geqslant 4$ and $[f(x)-f(y)]/(x-y)$ is absolutely irreducible, then $V(f)>\frac{1}{2}q$ provided $\mathbb{F}_q$ is of sufficiently large characteristic. K. S. Williams [4] estimated the number of "general" polynomials over $\mathbb{F}_q$ of fixed degree n and gave exact formulas for small n. Results on the average value of $V(f)$ with f ranging over all monic polynomials in $\mathbb{F}_q[x]$ with $f(0)=0$ and of fixed degree n were shown by Carlitz [61] and Uchiyama [3], [6]; see also Carlitz and Uchiyama [1] and K. S. Williams [4]. The case where further coefficients of f are fixed was treated in Uchiyama [5] and S. D. Cohen [8]. The result of Birch and Swinnerton-Dyer [1] was extended to rational functions by S. D. Cohen [5]. If $f \in \mathbb{F}_q[x]$ with $\deg(f)=n \geqslant 1$, then from the fact that an equation $f(x)=d \in \mathbb{F}_q$ has at most n solutions in $\mathbb{F}_q$ it follows easily that $V(f) \geqslant \lfloor (q-1)/n \rfloor + 1$. Carlitz, Lewis, Mills, and Straus [1] showed that if $V(f)=\lfloor (q-1)/n \rfloor + 1 \geqslant 3$ and n is less than the characteristic of $\mathbb{F}_q$, then $q \equiv 1 \bmod n$ and f is of the form $f(x)=a(x-b)^n+c$ with $a, b, c \in \mathbb{F}_q$; see Mills [1], Mordell [16], and K. S. Williams [5] for related results.

The number of values of $f \in \mathbb{F}_p[x]$ occurring in sets $\{1, 2, \ldots, h\}$, $1 \leqslant h < p$, was investigated by McCann and Williams [1] for cubic f and by K. S. Williams [5] for the case where $[f(x)-f(y)]/(x-y)$ has no nonlinear absolutely irreducible factors. S. D. Cohen [7] established a result on the average number of values of $f \in \mathbb{F}_q[x]$ occurring in subsets of $\mathbb{F}_q$; this improved and generalized an earlier result of K. S. Williams [15]. For $f \in \mathbb{F}_p[x]$ that is not a permutation polynomial of $\mathbb{F}_p$, Mordell [19] showed an estimate for the least nonnegative residue k modulo p not appearing in the value set of f in the cubic case, Bombieri and Davenport [1] extended it to the general case, and Tietäväinen [5] improved the general estimate somewhat to obtain $k < C(n)p^{1/2}$ with a constant $C(n)$ only depending on $n=\deg(f)$. In the case $n=4$ the result of Bombieri and Davenport [1] was also proved by M. Hudson [1] and K. S. Williams [2]. For $f \in \mathbb{F}_p[x]$ of arbitrary degree $n \geqslant 1$ Mordell [19] showed that the least nonnegative residue l modulo p occurring in the value set of f satisfies $l \leqslant np^{1/2}\log p$; an analogous result for general $\mathbb{F}_q$ was obtained by Cavior [4] and then improved by Tietäväinen [4]. For further results on the distribution of elements in the value set of f see L.E. Mazur [1], McCann and Williams [2], Perel'muter [8], Tietäväinen [7], and K. S. Williams [6], [8].

The relationship between two polynomials over $\mathbb{F}_p$ having the same degree and the same value set was worked out by K. S. Williams [10] in the quadratic case and by McCann and Williams [3] in the cubic case. Relationships between polynomials and rational functions for which containment relations between their value sets are given were considered by S. D. Cohen [6], [9] and Fried [1], [5]. Dickson [23] initiated the study of polynomials f over $\mathbb{F}_q$, q odd, for which the value set consists only of nonzero squares, and Carlitz [29] showed for such f that if $\deg(f) = n$ and $q > (n-1)^2$, then $f = g^2$ with $g \in \mathbb{F}_q[x]$. See also Carlitz [77], [89] and Rédei [2] for related investigations and Birch and Lewis [2] for an application. An analog of the result of Carlitz [29] for polynomials over $\mathbb{F}_q$ whose values are nonzero dth powers, $q \equiv 1 \bmod d$, was established in Carlitz [38]. Ribenboim [1] considered such polynomials over algebraic function fields with finite fields of constants. Rédei [11, Ch. 1] characterized polynomials over $\mathbb{F}_q$ that have all their values in a given subfield of $\mathbb{F}_q$. Tanner [2] considered polynomials f over $\mathbb{F}_p$, p odd, with $f(c) = \pm 1$ for all $c \in \mathbb{F}_p^*$.

2. For Theorem 7.8 see, for example, Dickson [7, Part I, Ch. 5]. The cycle structure of monomial mappings was studied by Ahmad [1]. Theorem 7.9 was already noted by Mathieu [1]. Further criteria for linearized polynomials to be permutation polynomials are given in Section 3 and Exercise 7.13 (see also Carlitz [93]). Payne [1] posed the problem of determining all 2-polynomials $L(x)$ over $\mathbb{F}_q$, q even, such that both $L(x)$ and $L(x)/x$ are permutation polynomials of $\mathbb{F}_q$. Theorem 7.10 is due to L. J. Rogers [1] for finite prime fields and to Dickson [2] for general $\mathbb{F}_q$. For further work on permutation polynomials of this type and related ones we refer to Ahmad [2], Dickson [7, Part I, Ch. 5], Fillmore [1], Nöbauer [8], and Wells [1], [3]. The table of normalized permutation polynomials of $\mathbb{F}_q$ of degree $\leqslant 5$ is taken from Dickson [7, Part I, Ch. 5]. The classification of these permutation polynomials and of those of degree 6 for odd q can be found in Dickson [2]; for q prime it was already carried out in Dickson [1]. An alternative approach to the case in the 12th line of Table 7.1 for q prime is described in S. Chowla [23]. Some types of permutation polynomials of degree 7 were classified in Dickson [2] and some of degree 8 in Cavior [1]. Permutation polynomials of $\mathbb{F}_q$ were determined for $q = 5$ by Betti [1] and for $q = 7$ by Hermite [2], Brioschi [1], [3], L. J. Rogers [1], and Dickson [2]; see also Brioschi [1] and Dickson [1], [7, Part I, Ch. 5] for other small values of q.

Theorem 7.11 is due to Niederreiter and Robinson [2]. The sufficiency part of Theorem 7.11, with the condition $\eta(a^2 - 1) = 1$ in the equivalent form of Remark 7.12, was shown earlier by Carlitz [83]. A simpler form of this condition is given in Exercise 7.9. Further work on permutation polynomials of $\mathbb{F}_q$ of the form $x^{m+1} + ax$ with m a divisor of $q-1$ and of closely related forms can be found in Carlitz [83], [93], Carlitz and Wells [1], Lausch and Nöbauer [1, Ch. 4], and Niederreiter and Robinson [2]; see also Exercise 7.11. Theorems 7.13 and 7.14 are due to Carlitz [93]. Brioschi [1] and Grandi [2] treated permutation polynomials of $\mathbb{F}_p$ of the form $x^{p-1-s} + ax^{(p-1-2s)/2}$. The question when a polynomial of the form $x^{p-s} + ax^{(p-s+1)/2} + bx$ can be a permutation polynomial of $\mathbb{F}_p$ was considered in Brioschi [2] and Grandi [1]. See Carlitz [88] and Grandi [2] for other special results.

Dickson polynomials were introduced in Dickson [2]; see also Dickson [7, Part I, Ch. 5]. Theorem 7.16 is due to Nöbauer [10]. A somewhat weaker form of this criterion was shown earlier by Dickson [2]. An alternative proof of the sufficiency part is presented in K. S. Williams [25]. Corollary 7.17 was noted by P. Chowla [2]. Information on Brewer sums can be found in the notes to Chapter 5, Section 5. Dickson polynomials may be calculated by a simple recursion (see Exercise 7.15). For further work on Dickson polynomials over finite fields we refer to Dickson [2], [7, Part I, Ch. 5], Lausch and Nöbauer [1, Ch. 4], and K. S. Williams [25]. Dickson polynomials as permutation polynomials of $\mathbb{Z}/(m)$ are discussed in Lausch, Müller, and Nöbauer [1], Müller [1], and Nöbauer [8], [12]. A relationship between the rational permutation functions of Rédei [4] and Dickson polynomials was discovered by Carlitz [86]. See Carlitz [79] and Rosenberger [1] for some applications of Chebyshev polynomials over finite fields. For general information on Chebyshev polynomials we refer to the book of Rivlin [1].

Dickson polynomials are important in connection with a celebrated conjecture of Schur [4] to the effect that any $f \in \mathbb{Z}[x]$ that is a permutation polynomial of $\mathbb{F}_p$ (when considered modulo p) for infinitely many primes p must be a composition of binomials $ax^n + b$ and Dickson polynomials. Schur [4] settled the case where $\deg(f)$ is prime, thus refining a result of Dickson [2]. Wegner [1] settled the case where $\deg(f)$ is either a product of two odd primes or an odd prime power, and Kurbatov [3] settled the case where $\deg(f)$ is either a product of at most four distinct odd primes or a product of two odd prime powers. Kurbatov [1] showed that if $n = p_1 \cdots p_k$ with distinct odd primes p_i and if none of the p_i can be written as a linear combination of the other primes with nonnegative integer coefficients, then Schur's conjecture holds for $\deg(f) = n$. Finally, Schur's conjecture was proved completely by Fried [2] who even established an appropriate version for polynomials over an algebraic number field. Furthermore, it is shown in this paper that if $f \in \mathbb{Q}[x]$ is a composition of binomials $ax^n + b$ and Dickson polynomials and if $\deg(f)$ is coprime to 6, then f is a permutation polynomial of $\mathbb{F}_p$ for infinitely many primes p; compare also with Exercise 7.34. For related work see Schur [4], Wegner [3], Kurbatov [2], Fried [4], [5], and Niederreiter and Lo [1]. If R is a commutative ring with identity and $f \in R[x]$, then the set of all (prime) ideals J of R such that f is a permutation polynomial of R/J when considered modulo J is called the (*prime*) *permutation spectrum* of f. These notions were first considered for $R = \mathbb{Z}$ by Nöbauer [8], [9] and in the general case by Lausch and Nöbauer [1, Ch. 4]. Further remarks on the case $R = \mathbb{Z}$ can be found in Narkiewicz [2]. A detailed investigation of the case where R is the ring of integers in an algebraic number field was carried out by Niederreiter and Lo [1].

Carlitz [82] considered permutation polynomials of $\mathbb{F}_q$, q odd, with the additional property

$$(f(a)-f(b))^{(q-1)/2} = (a-b)^{(q-1)/2} \quad \text{for all } a, b \in \mathbb{F}_q$$

and showed that if $\deg(f) < q$, then $f(x) = cx^{p^j} + d$ with c a nonzero square in $\mathbb{F}_q$, $d \in \mathbb{F}_q$ arbitrary, and p the characteristic of $\mathbb{F}_q$; see also Goldberg [1]. An extension to polynomials in several indeterminates was given in Carlitz [84]. McConnel [1] generalized the work of Carlitz [82] in another direction and proved a result which can be phrased as follows: if G is a proper subgroup of $\mathbb{F}_q^*$, then $f \in \mathbb{F}_q[x]$ with $\deg(f) < q$ satisfies

$$(a-b)^{-1}(f(a)-f(b)) \in G \quad \text{for all } a \neq b \text{ in } \mathbb{F}_q$$

if and only if $f(x) = cx^{p^j} + d$ with $c \in G$, $d \in \mathbb{F}_q$, and $p^j \equiv 1 \bmod m$, where m is the index of G in $\mathbb{F}_q^*$. Alternative proofs of this theorem were presented by Bruen [1] for q prime and by Bruen and Levinger [1] in the general case. McConnel [1], [2], [3] obtained extensions to polynomials in several indeterminates. Grundhöfer [1] characterized all $f \in \mathbb{F}_q[x]$ satisfying

$$(a-b)(f(a)-f(b)) \in G \quad \text{for all } a \neq b \text{ in } \mathbb{F}_q.$$

Głazek [1] considered permutation polynomials of $\mathbb{F}_q$ which commute with automorphisms of $\mathbb{F}_q$.

If both $f(x)$ and $f(x)+x$ are permutation polynomials of $\mathbb{F}_q$, then f is called a *complete mapping polynomial* of $\mathbb{F}_q$. This notion was first used in Niederreiter and Robinson [1] and studied in detail in Niederreiter and Robinson [2]. Chowla and Zassenhaus [1] conjectured that if $f \in \mathbb{Z}[x]$ with $\deg(f) \geqslant 2$, p is a sufficiently large prime, and f is a permutation polynomial of $\mathbb{F}_p$ when considered modulo p, then $f(x)+ax$, $a \in \mathbb{F}_p^*$, is not a permutation polynomial of $\mathbb{F}_p$. In the same paper it is conjectured that if $f \in \mathbb{Z}[x]$ with $\deg(f) \geqslant 2$, p is a sufficiently large prime, and f is not a permutation polynomial of $\mathbb{F}_p$ when considered modulo p, then there exists $c \in \mathbb{F}_p$ such that $f(x)+c$ is irreducible over $\mathbb{F}_p$.

3. Theorem 7.18 is a result of Carlitz [49]. The special cases $q = 5$ and $q = 7$ were shown by Betti [1] and Dickson [2], respectively. A version of Theorem 7.18 relating to transpositions in arbitrary fields is given in Carlitz [90]. The notion of a crude permutation polynomial (see Carlitz [93] and Exercises 7.22–7.24) stems from Theorem 7.18. Theorems 7.19 and 7.21

are due to Wells [4]. For the result of Exercise 7.19 see Fryer [1]. Generators for S_{q+1} and A_{q+1} using rational functions over $\mathbb{F}_q$ were given by Wells [4]. Subgroups of S_p, p prime, generated by certain permutation polynomials of $\mathbb{F}_p$ are discussed in Fryer [2].

Theorems 7.22 and 7.23 are due to Nöbauer [10]. A discussion of the group $G(1)$ can also be found in Lausch and Nöbauer [1, Ch. 4]. Hule and Müller [1] characterized those groups $G(a)$ that are cyclic. Groups analogous to $G(a)$ pertaining to residue class rings $\mathbb{Z}/(m)$ have been studied by Nöbauer [2] for $a=0$ and by Lausch, Müller, and Nöbauer [1], Müller [1], and Nöbauer [12] for $a=\pm 1$. For generalizations to several indeterminates see the notes to Section 5.

The Betti-Mathieu group occurred first in the work of Betti [2], [3] and Mathieu [1]. The study of this group was taken up again by Dickson [2], [5], [7, Part I, Ch. 5]. The criterion that $L(x)$ is a permutation polynomial of $\mathbb{F}_{q^r}$ if and only if $\det(A) \neq 0$ is due to Dickson [2]. In this paper a one-to-one correspondence between the elements of the Betti-Mathieu group and those of $GL(r, \mathbb{F}_q)$ is set up. The fact that these two groups are isomorphic (see Theorem 7.24) was first shown by Bottema [1]; see also Carlitz [91]. Brawley, Carlitz, and Vaughan [1] and T. P. Vaughan [1] established isomorphisms between the algebra of linearized polynomials of the form (7.12) and the algebra of $r \times r$ matrices over $\mathbb{F}_q$; restriction to the group of units yields again Theorem 7.24. Brawley, Carlitz, and Vaughan [1] also studied the group of permutation polynomials obtained by restricting the coefficients α_s in (7.12) to a given subfield of $\mathbb{F}_{q^r}$.

Groups of permutation polynomials obtained from the polynomials in Theorem 7.10 and related polynomials were considered in Ahmad [2], Fillmore [1], Lausch and Nöbauer [1, Ch. 4], and Wells [1], [3]. Carlitz and Hayes [1] studied the group of all permutation polynomials of $\mathbb{F}_{q^r}$ with coefficients in $\mathbb{F}_q$; see R. Matthews [3] for an extension to several indeterminates. Niederreiter and Robinson [2] showed that the permutation polynomials of $\mathbb{F}_q$, q odd, of the form $ax^{(q+1)/2}+bx$ constitute a group under composition modulo $x^q - x$. For the results of Exercises 7.20 and 7.21 and similar results which show that most permutations of $\mathbb{F}_q$ that move only very few elements are represented by polynomials of degree $q-2$ we refer to Wells [5]. Further groups of permutation polynomials of $\mathbb{Z}/(m)$, apart from those already mentioned, have been studied in Nöbauer [1], [4].

Property (7.10) implies that Dickson polynomials with $a=1$ commute under composition, as do those with $a=0$. This has given rise to an extensive literature on characterizing the polynomials f, g over a field F with $f(g(x)) = g(f(x))$. The classical papers of Fatou [1], Julia [1], and Ritt [2] deal with the case where F is the field of complex numbers. An important concept is that of a *V-chain*, meaning a sequence of nonconstant polynomials over F that all commute among each other and that contains polynomials of all positive degrees. Block and Thielman [1] characterized all V-chains for $F = \mathbb{R}$, and Jacobsthal [3] showed that up to a natural type of equivalence all V-chains over a field F of characteristic 0 are given by the V-chains of Dickson polynomials with $a = 0$ and $a = 1$, respectively. Kautschitsch [1] proved an analogous result for arbitrary F; see also Lausch and Nöbauer [1, Ch. 4] and Lidl [7]. Polynomials over $\mathbb{F}_q$ that commute with a given linear polynomial were characterized by Mullen [13], the case of monic linear polynomials having been treated earlier by Wells [6]. Further results on polynomials commuting with a given polynomial can be found in Bertram [1], Boyce [1], and Kautschitsch [2]. A class of rational functions over $\mathbb{F}_q$ commuting under composition appears in Rédei [4].

The decomposition of a polynomial into indecomposable polynomials (with respect to the operation of composition) and the extent to which such decompositions can differ was investigated by Ritt [1] for polynomials over $\mathbb{C}$. The generalization to fields of characteristic 0 was carried out by Engstrom [3] and Levi [1]; see also Fried and MacRae [1], Dorey and Whaples [1], and Lausch and Nöbauer [1, Ch. 4]. A partial result for fields of characteristic $\neq 0$ is contained in Fried and MacRae [1], and the case of an algebraically closed field is treated in Fried [3] and Kljačko [1]. Dickson polynomials play again a role in this theory. For related work see Bremner and Morton [1], Crampton and Whaples [1], Dorey and Whaples [1], Lausch and Nöbauer [1, Ch. 3], and Nöbauer [7]. The operation of composition modulo $x^q - x$ was used by Carlitz [47], Cavior [2], and Mullen [1], [3], [5] to define equivalence relations among polynomials over $\mathbb{F}_q$ considered modulo $x^q - x$.

4. Exceptional polynomials were introduced by Davenport and Lewis [2] who also conjectured the relationship between these polynomials and permutation polynomials. MacCluer [1] proved that if $f \in \mathbb{F}_q[x]$ is exceptional and $\deg(f) < 2p$, where p is the characteristic of $\mathbb{F}_q$, then f is a permutation polynomial of $\mathbb{F}_q$. S. D. Cohen [5] showed that this result holds without any restriction on $\deg(f)$ and even established an appropriate version for rational functions over $\mathbb{F}_q$. The weaker result in Theorem 7.27 can be shown more elementarily and its proof follows K. S. Williams [9]. For Lemma 7.26 we refer to K. S. Williams [5]. Theorem 7.29 was proved for fields $\mathbb{F}_p$, p prime, by Davenport and Lewis [2] and quantitative versions for this case were established by Bombieri and Davenport [1] and Tietäväinen [5]. K. S. Williams [5] weakened the hypothesis that f be a permutation polynomial of $\mathbb{F}_p$ to $V(f) = p + O(1)$. The general form of Theorem 7.29 can be found in Hayes [5], where Theorem 7.31 is also shown. A stronger version of Theorem 7.29 that is even valid for rational functions over $\mathbb{F}_q$ is due to S. D. Cohen [5]. The question of whether Corollary 7.32 remains true if the condition $\gcd(n, q) = 1$ is replaced by $\gcd(2, q) = 1$ is a well-known problem in this area. The answer is obviously affirmative if n is a power of 2. The only other cases in which solutions have been published are $n = 6$

(Dickson [2]) and $n=10$ (Hayes [5]); see Lidl [7] for remarks on this problem.

Exceptional polynomials, and even rational functions, over finite fields were classified by Fried [5]. K. S. Williams [24] gave an interpretation of the number of absolutely irreducible factors of $[f(x)-f(y)]/(x-y)$ for sufficiently large q in terms of the number of $(a,b)\in\mathbb{F}_q^2$ with $f(a)=f(b)$ and $a\neq b$. For the case where f is a Dickson polynomial, K. S. Williams [25] obtained the factorization of $[f(x)-f(y)]/(x-y)$ over the algebraic closure of $\mathbb{F}_q$ and thus recovered the condition for a Dickson polynomial to be a permutation polynomial of $\mathbb{F}_q$. Further remarks on exceptional polynomials can be found in Davenport and Lewis [2]. K. S. Williams [5] called f *extremal of index* k if $[f(x)-f(y)]/(x-y)$ has no absolutely irreducible factors apart from precisely k linear factors and showed for such $f\in\mathbb{F}_p[x]$ that $V(f)=p/(k+1)+O(1)$ provided p is sufficiently large. A partial converse of this result was proved earlier by Mordell [16]. These results were improved and generalized considerably by S. D. Cohen [5].

5. The fact that every mapping from $\mathbb{F}_q^n$ into $\mathbb{F}_q$ can be represented by a polynomial in n indeterminates over $\mathbb{F}_q$ of degree $<q$ in each indeterminate was shown by Weber [5, Sec. 77] for the case where q is prime. The uniqueness of this representation was proved by Hurwitz [1] for this case. For general q both the representability and the uniqueness (see (7.20) and Lemma 7.40) were established by Dickson [24]. A result related to the uniqueness theorem is contained in Mather [1]. An analysis of the relationship between mappings and polynomials can also be found in Joly [5]. Convenient methods for calculating the polynomial representing a given mapping were discussed by Bernstein and Debely [1] and later by Benjauthrit and Reed [1], [2], Pradhan [1], Takahashi [1], Thayse [1], and Yin [1]. Varnum [1] and Lehti [1] introduced matrix methods for this purpose. The computational complexity of such interpolation procedures and of polynomial evaluation was studied by Strassen [1], [2]; see Mihaĭljuk [1], [2] for the special case of elementary symmetric polynomials. Expressions for characteristic functions of subsets of $\mathbb{F}_p^n$ are given in Cazacu [2], [3]; for related characteristic functions see Rosenberg [3]. Pizzarello [1] has a criterion for a polynomial in several indeterminates over $\mathbb{F}_q$ to vanish identically on a finite extension of $\mathbb{F}_q$. Mappings defined on subsets of $\mathbb{F}_p^n$ are considered in Bernstein [2] and Bernstein and Debely [1]. Mappings from $(\mathbb{Z}/(m))^n$ into $\mathbb{Z}/(m)$ and their representability by polynomials are studied in Bernstein [1], Bernstein and Debely [1], Carlitz [97], Kempner [2], and Rosenberg [2]. Polynomial mappings of this type which vanish identically are treated in Kempner [2], Lausch and Nöbauer [1, Ch. 3], Litzinger [1], and Nöbauer [3]. Nöbauer [5] showed that the only commutative rings R with identity for which any mapping from R^n into R can be represented by a polynomial over R are the finite fields, thus generalizing a result of Rédei

and Szele [1] for $n = 1$; see also Ceccherini [1]. Brawley and Carlitz [2] used a more general notion of polynomial and proved that if R is a nonzero ring and $n \geqslant 2$, then every mapping from R^n into R can be represented by such a "polynomial" over R in n indeterminates if and only if R is a matrix ring over a finite field. The representability of symmetric functions in countably many variables by polynomials over $\mathbb{F}_q$ in countably many indeterminates was discussed by Metropolis, Nicoletti, and Rota [1].

Permutation polynomials in several indeterminates and orthogonal systems appear implicitly in the work of Carlitz [47], [59]. The study of these concepts was taken up again by Nöbauer [6]. Orthogonal systems with $n = 2$ and q prime were considered in Kurbatov and Starkov [1]. Orthogonal systems with $m = n$ are also called *permutation polynomial vectors* since they induce permutations of $\mathbb{F}_q^n$. Theorem 7.36 is implicit in Carlitz [59]; our proof follows Niederreiter [2]. Theorem 7.37 is due to Carlitz [47] and Corollary 7.39 appears in Niederreiter [2]. For further criteria for orthogonal systems see Exercise 7.47 and Niederreiter [2], and for permutation polynomials see Mullen [2]. For finite prime fields a special criterion for permutation polynomials can be given (see Exercise 7.32 and Niederreiter [3]). Theorem 7.41 is a refinement of a result of Lidl and Niederreiter [1]. The first part of Theorem 7.42 is due to Nöbauer [6], the second part was shown by Lidl [1] in the special case $m = 1$, $n = 2$, and by Lidl and Niederreiter [1] in the general case. The latter paper contains also a proof of Theorem 7.43. Niederreiter [2] proved Theorem 7.44, whereas Corollary 7.45 is already implicit in Carlitz [47] and was also shown in the special case $n = 2$ and q prime by Kurbatov and Starkov [1]. These results allow the enumeration of orthogonal systems consisting of polynomials over $\mathbb{F}_q$ of degree $< q$ in each indeterminate (see Carlitz [59] and Niederreiter [2]). Fried [4] extended the theorem of MacCluer [1] to permutation polynomial vectors by showing that one obtains a permutation polynomial vector whenever the system of polynomials is "exceptional" in a certain sense. Formulas for the representation of permutations of $\mathbb{F}_q^n$ by permutation polynomial vectors are given in Lidl [2]. For permutation polynomials and permutation polynomial vectors over $\mathbb{Z}/(m)$ see Lidl [3] and Nöbauer [3], [6] and over more general residue class rings see Lausch and Nöbauer [1, Ch. 4]. Analogous concepts for rational functions in several indeterminates are considered in Lidl [3].

Dickson polynomials (or Chebyshev polynomials) in several indeterminates were introduced by Lidl and Wells [1] who also showed Theorem 7.46. Explicit formulas, generating functions, and recursion formulas for such Dickson polynomials can be found in Eier and Lidl [1], Lidl [8], and Lidl and Wells [1]. The orthogonal systems $g_k(a)$ in Theorem 7.46 are closed under composition if and only if $a = 0$, 1, or -1, according to a result of Lidl and Wells [1], which generalizes Theorem 7.22. Group-theoretic investigations analogous to those in Theorem 7.23 were carried out in

Lidl [4], [6], Lidl and Müller [1], and R. Matthews [1]. In the case $n = 2$ the systems $g_k(a)$ for which the Jacobian is nonzero at all points of $\mathbb{F}_q^2$ were characterized by Lidl [5]; see R. Matthews [2] for general n. A conjecture of Lidl and Wells [1] attributing to Dickson polynomials in several indeterminates the same role as that played by Dickson polynomials in one indeterminate in the conjecture of Schur [4] (see the notes to Section 2) was disproved by Fried [4]. Systems of Dickson polynomials in several indeterminates were studied further by R. Matthews [2].

Theorem 7.47 is due to Niederreiter [1]. The case of odd q was settled independently by Lidl [1]. For the criterion in terms of the ranks of the coefficient matrix and the augmented matrix see Niederreiter [3]. Mullen [7], [8], [10] studied *local permutation polynomials*—that is, polynomials over $\mathbb{F}_q$ in $n \geqslant 2$ indeterminates with the property that if arbitrary values in $\mathbb{F}_q$ are assigned to any $n-1$ indeterminates, then the resulting polynomial in the remaining indeterminate is always a permutation polynomial of $\mathbb{F}_q$. Generating systems for groups of permutation polynomial vectors under composition have been obtained by Lidl [2] and Lidl and Niederreiter [1]. The group of permutation polynomial vectors in $\mathbb{F}_{q^r}$ with coefficients in $\mathbb{F}_q$ is considered by R. Matthews [3]. An application of orthogonal systems in $\mathbb{F}_p$ to the study of p-Sylow subgroups of symmetric groups S_{p^n} appears in Kaloujnine [1]. Goodstein [1] shows how to generate all polynomials over $\mathbb{F}_q$ in several indeterminates by composition.

The theory of correspondences and of admissible polynomials described in Exercises 7.26–7.31 was developed by Carlitz [115], [117], [121]. The notion of a coset of a system of polynomials (see Exercise 7.49) occurs in Niederreiter [2]. Equivalence classes of polynomials and systems of polynomials in $\mathbb{F}_q[x_1, \ldots, x_n]$ considered modulo the ideal $(x_1^q - x_1, \ldots, x_n^q - x_n)$ were studied by Carlitz [47], [59] and later by Carlitz [110], Cavior [6], and Mullen [1], [2], [3], [9]. Analogous concepts for matrices over finite fields are treated in Brawley and Mullen [1], Chao [1], and Mullen [4], [6], [11], [12].

Value sets of polynomials in several indeterminates have received some attention. Kantor [1] gave a formula for the number of values attained by a quadratic form over $\mathbb{F}_p$, p an odd prime. K. S. Williams [3] obtained a sufficient condition for the value set of a polynomial over $\mathbb{F}_q$ to be $\mathbb{F}_q$. Asymptotic results on the distribution of values of polynomials over finite prime fields were shown by Tietäväinen [10] and K. S. Williams [7]. For the special case of elementary symmetric polynomials, detailed investigations have been carried out by Aberth [1], Akhtar [1], and Fine [1]; see also Birch [1] for a related result on elementary symmetric polynomials. Lower bounds for the number of values attained by diagonal forms were established by Chowla, Mann, and Straus [1] (see also Mann [3, Ch. 2]) and Diderrich and Mann [1]. Dickson [23], [28] considered homogeneous polynomials whose value sets contain only squares or only cubes. Some special results on value sets of systems of polynomials can be found in Perel'muter [7], Rédei [1], and Rédei and Weinert [1]. Questions about value sets are of course also connected with problems about the solvability of equations (see Chapter 6 for the latter topic).

EXERCISES

7.1. Let $b \in \mathbb{F}_q$ be a fixed element. Define
$$f_b(x) = 1 - \sum_{i=0}^{q-1} b^{q-1-i} x^i.$$
Prove in detail that $f_b(a) = 0$ if $a \in \mathbb{F}_q$, $a \neq b$, and $f_b(b) = 1$. Using (7.1), deduce that $\binom{q-1}{i} \equiv (-1)^i \bmod p$ for $0 \leqslant i \leqslant q-1$, where p is the characteristic of $\mathbb{F}_q$. (*Note*: This congruence for binomial coefficients can also be obtained from the identity $(x-1)^{q-1} = (x^q - 1)/(x-1)$.)

7.2. Prove that if q is a prime, then in condition (ii) of Theorem 7.4 it suffices to consider the integers t with $1 \leqslant t \leqslant (q-1)/2$. Give an example which shows that this is not true for $q = p^e$, $e > 1$.

7.3. Let $q = km + 1$, $k, m \in \mathbb{N}$. Prove that x^{m+1} is a permutation polynomial of $\mathbb{F}_q$ if and only if $\gcd(m+1, k) = 1$.

7.4. Prove that the polynomial $x^{p^i} - ax^{p^k}$ over the finite field $\mathbb{F}_q$ of characteristic p is a permutation polynomial of $\mathbb{F}_q$ if and only if a is not a $(p^i - p^k)$th power of an element of $\mathbb{F}_q^*$.

7.5. Let $\mathbb{F}_q$ be of characteristic p, let $r \in \mathbb{N}$, d a positive divisor of $p^r - 1$, and $a \in \mathbb{F}_q$. Prove that $x(x^d - a)^{(p^r-1)/d}$ is a permutation polynomial of $\mathbb{F}_q$ if and only if a is not a dth power of an element of $\mathbb{F}_q^*$.

7.6. Let $a \in \mathbb{F}_q$, q odd, and let $r \in \mathbb{N}$ with $\gcd(r, q-1) = 1$. Prove that $x^r(x^{(q-1)/2} - a)^2$ is a permutation polynomial of $\mathbb{F}_q$ if and only if $a \neq \pm 1$.

7.7. Determine all permutation polynomials of $\mathbb{F}_7$ of the form $x^r(x^3 - a)^2$, where $r \in \mathbb{N}$ and $a \in \mathbb{F}_7$.

7.8. If $q \equiv \pm 2 \bmod 5$, prove that $5x^5 + 5ax^3 + a^2 x$ is a permutation polynomial of $\mathbb{F}_q$ for all $a \in \mathbb{F}_q$.

7.9. For odd q, prove that $x^{(q+1)/2} + ax \in \mathbb{F}_q[x]$ is a permutation polynomial of $\mathbb{F}_q$ if and only if $a = 2^{-1}(c + c^{-1})$ for some $c \in \mathbb{F}_q^*$ with $c^2 \neq 1$.

7.10. Determine the least integer M such that for every finite field $\mathbb{F}_q$ with q odd and $q \geqslant M$ there exists an element $a \in \mathbb{F}_q^*$ for which $x^{(q+1)/2} + ax$ is a permutation polynomial of $\mathbb{F}_q$.

7.11. Let $m > 1$ be a divisor of $q - 1$. Prove that
$$x^{(q+m-1)/m} + ax \in \mathbb{F}_q[x]$$

is a permutation polynomial of $\mathbb{F}_q$ if and only if $(-a)^m \neq 1$ and

$$\left[(a+c^i)(a+c^j)^{-1}\right]^{(q-1)/m} \neq c^{j-i} \quad \text{for all } 0 \leqslant i < j < m,$$

where c is a fixed primitive mth root of unity in $\mathbb{F}_q$.

7.12. Let $q = p^e$, p an odd prime, and set $m = (q-1)/2$. Prove that $\gcd\left(\binom{m}{t}, p\right) = 1$ if and only if $t = b_0 + b_1 p + \cdots + b_{e-1} p^{e-1}$ with $0 \leqslant b_i \leqslant (p-1)/2$ for $0 \leqslant i \leqslant e-1$.

7.13. Let

$$f(x) = \sum_{i=0}^{n-1} c_i x^{q^i} \in \mathbb{F}_q[x].$$

Prove that f is a permutation polynomial of $\mathbb{F}_{q^n}$ if and only if

$$\gcd\left(\sum_{i=0}^{n-1} c_i x^i, x^n - 1\right) = 1.$$

7.14. Prove that over a field of characteristic $\neq 2$ the Dickson polynomial $g_k(x, a)$ is formally represented by

$$g_k(x, a) = \left(\frac{x + \sqrt{x^2 - 4a}}{2}\right)^k + \left(\frac{x - \sqrt{x^2 - 4a}}{2}\right)^k.$$

7.15. Prove that Dickson polynomials satisfy $g_1(x, a) = x$, $g_2(x, a) = x^2 - 2a$, and

$$g_{k+1}(x, a) = x g_k(x, a) - a g_{k-1}(x, a) \quad \text{for } k \geqslant 2.$$

7.16. Prove that Dickson polynomials satisfy $g_k(ax, a^2) = a^k g_k(x, 1)$.

7.17. In the notation of Theorem 5.46, prove that Kloosterman sums satisfy

$$K(\chi^{(s)}; a, b) = -g_s(-K, q),$$

where $g_s(x, q)$ is a Dickson polynomial over the real numbers.

7.18. Prove that the alternating group A_q is generated by its subgroups AL_q and Q_q, which are defined in the paragraph preceding Theorem 7.21.

7.19. Let p be an odd prime. Prove that the alternating group A_p is generated by $x + 1$ and mx^{p-2}, where m is any nonzero square in $\mathbb{F}_p$ if $p \equiv 3 \bmod 4$ and any nonsquare in $\mathbb{F}_p$ if $p \equiv 1 \bmod 4$. Otherwise $x + 1$ and mx^{p-2} generate S_p.

7.20. Prove that every transposition on $\mathbb{F}_q$, $q > 2$, is represented by a unique polynomial of degree $q - 2$.

7.21. Prove: (i) if $q \equiv 2 \bmod 3$, $q > 2$, then every 3-cycle on $\mathbb{F}_q$ is represented by a unique polynomial of degree $q - 2$; (ii) if $q \equiv 1 \bmod 3$, then all but $\frac{2}{3}q(q-1)$ 3-cycles on $\mathbb{F}_q$ are represented by polynomials of degree $q - 2$.

7.22. For $q > 2$, define a *crude permutation polynomial* of $\mathbb{F}_q$ to be a composition of finitely many polynomials over $\mathbb{F}_q$ that are either linear or equal to x^{q-2}. Prove that a crude permutation polynomial of $\mathbb{F}_q$ is a permutation polynomial of the extension $\mathbb{F}_{q^r}$ if and only if $\gcd(2^r - 1, q-2) = 1$.

7.23. Let f be a crude permutation polynomial of $\mathbb{F}_q$, $q > 4$. Prove that there are infinitely many extensions $\mathbb{F}_{q^r}$ for which f is a permutation polynomial and also infinitely many $\mathbb{F}_{q^r}$ for which f is not a permutation polynomial.

7.24. Show that the reduction modulo $x^q - x$ of a crude permutation polynomial of $\mathbb{F}_q$ is not necessarily a crude permutation polynomial and that different crude permutation polynomials can have the same reduction modulo $x^q - x$.

7.25. Prove that $G(1)$ is a homomorphic image of $G(-1)$, where $G(a)$ is as in Theorem 7.23.

7.26. By a *correspondence* Γ in $\mathbb{F}_q$ we mean a pair of partitions $A_0, A_1, \ldots, A_k$ and $B_0, B_1, \ldots, B_k$ of $\mathbb{F}_q$ with $A_i \neq \varnothing$, $B_i \neq \varnothing$ for $1 \leqslant i \leqslant k$. The integer k is called the *rank* of Γ. A polynomial $h \in \mathbb{F}_q[x, y]$ is said to be *admissible for* Γ if $h(a, b) = 0$ whenever $(a, b) \in A_i \times B_i$ for some $1 \leqslant i \leqslant k$ and $h(a, b) \neq 0$ otherwise. Prove that an admissible polynomial for Γ of the form $h(x, y) = f(x) - g(y)$ exists if and only if $k \neq q-1$ or $k = q-1$ and A_0 or $B_0 = \varnothing$. In the case $k = q-1$, $A_0 \neq \varnothing$, $B_0 \neq \varnothing$, prove that an admissible polynomial for Γ is given by

$$h(x, y) = \left(1 - f(x)^{q-1}\right)\left(1 - g(y)^{q-1}\right) + \left(f(x) - g(y)\right)^{q-1},$$

where f and g are suitable permutation polynomials of $\mathbb{F}_q$.

7.27. A polynomial $h \in \mathbb{F}_q[x, y]$ is called *admissible* if it is admissible for some correspondence in $\mathbb{F}_q$, and two admissible polynomials are called *equivalent* if they are admissible for the same correspondence. Prove: (i) two admissible polynomials $h_1(x, y)$ and $h_2(x, y)$ are equivalent if and only if

$$h_1(x, y)^{q-1} \equiv h_2(x, y)^{q-1} \operatorname{mod}(x^q - x, y^q - y);$$

(ii) the number of equivalence classes of admissible polynomials is equal to the number of correspondences.

7.28. Prove that if $h_1(x, y)$ and $h_2(x, y)$ are admissible for some correspondence Γ, then $h(x, y) = h_1(x, y) h_2(x, y)$ is also admissible for Γ.

7.29. Prove that if $h(x, y) = g(x - y)$ is admissible for a correspondence Γ in $\mathbb{F}_q$ and the number of distinct roots of g in $\mathbb{F}_q$ is $m > 0$, then m divides q and the rank of Γ is q/m.

7.30. Let $f(x)$ and $g(y)$ be polynomials over $\mathbb{F}_q$. Prove that $h(x, y) = f(x)g(y)$ is admissible in $\mathbb{F}_{q^r}$ for $r = 1, 2, \ldots$ if and only if at least one of the polynomials f and g is a constant.

7.31. Let $f_1(x)$, $f_2(x)$, $g_1(y)$, $g_2(y)$ be polynomials over $\mathbb{F}_q$. Prove that $h(x, y) = f_1(x)g_1(y) + f_2(x)g_2(y)$ is admissible in $\mathbb{F}_{q^r}$ for $r = 1, 2, \ldots$ if and only if $\gcd(f_1, f_2) = \gcd(g_1, g_2) = 1$.

7.32. For $f \in \mathbb{Z}[x_1, \ldots, x_n]$ and a prime p, we call f a *permutation polynomial* mod p if f, considered as a polynomial over $\mathbb{F}_p$, is a permutation polynomial over $\mathbb{F}_p$. Prove that f is a permutation polynomial mod p if and only if each congruence

$$f(x_1, \ldots, x_n) \equiv a \bmod p, \quad a = 0, 1, \ldots, p-1,$$

has at least one solution and

$$\sum_{a_1, \ldots, a_n = 0}^{p-1} f(a_1, \ldots, a_n)^{tp^{n-2}} \equiv 0 \bmod p^{n-1} \quad \text{for } t = 1, 2, \ldots, p-1.$$

7.33. Prove that $ax^n + b \in \mathbb{Z}[x]$, $a \neq 0$, is a permutation polynomial mod p for infinitely many primes p if and only if n is odd.

7.34. Let $g_k(x, a)$ be a Dickson polynomial over $\mathbb{Z}$ with $a \neq 0$. Prove that $g_k(x, a)$ is a permutation polynomial mod p for infinitely many primes p if and only if $\gcd(k, 6) = 1$.

7.35. Prove that $f \in \mathbb{Z}[x]$ is a permutation polynomial mod p for all primes p if and only if f is a linear polynomial with leading coefficient ± 1.

7.36. For $1 \leqslant m < n$, prove that $f \in \mathbb{F}_q[x_1, \ldots, x_m]$ is a permutation polynomial over $\mathbb{F}_q$ if and only if f is a permutation polynomial over $\mathbb{F}_q$ when regarded as an element of $\mathbb{F}_q[x_1, \ldots, x_n]$.

7.37. Prove the first part of Theorem 7.42 by using characters.

7.38. Let $f \in \mathbb{F}_q[x_1, \ldots, x_m]$ be a permutation polynomial over $\mathbb{F}_q$ and let $g \in \mathbb{F}_q[x_{m+1}, \ldots, x_n]$, where $1 \leqslant m < n$. Prove that

$$h(x_1, \ldots, x_n) = f(x_1, \ldots, x_m)g(x_{m+1}, \ldots, x_n)$$

is a permutation polynomial over $\mathbb{F}_q$ if and only if the equation $g(x_{m+1}, \ldots, x_n) = 0$ has no solution in $\mathbb{F}_q^{n-m}$.

7.39. Prove that

$$a_1 x_1^{k_1} + \cdots + a_n x_n^{k_n} \in \mathbb{F}_q[x_1, \ldots, x_n]$$

is a permutation polynomial over $\mathbb{F}_q$ if for some i, $1 \leqslant i \leqslant n$, we have $a_i \neq 0$ and $\gcd(k_i, q-1) = 1$.

7.40. Prove that if $f \in \mathbb{F}_q[x_1, \ldots, x_n]$ is a permutation polynomial over $\mathbb{F}_q$, then so is $bf + c$ for all $b \in \mathbb{F}_q^*$, $c \in \mathbb{F}_q$.

7.41. Prove that if $f \in \mathbb{F}_q[x_1, \ldots, x_n]$ is a permutation polynomial over $\mathbb{F}_q$, then so is f^k whenever $k \in \mathbb{N}$ with $\gcd(k, q-1) = 1$.

7.42. With the notation as in the paragraph following Theorem 7.47, prove that $f \in \mathbb{F}_q[x_1, \ldots, x_n]$ is a permutation polynomial over $\mathbb{F}_q$ if and only if $\operatorname{rank}(A') > \operatorname{rank}(A)$.

7.43. Let $f, g \in \mathbb{F}_q[x_1, \ldots, x_{n-1}]$ and suppose the number of solutions of $f(x_1, \ldots, x_{n-1}) = 0$ in $\mathbb{F}_q^{n-1}$ is not divisible by q. Let $h \in \mathbb{F}_q[x_1, \ldots, x_n]$ be such that $h(c_1, \ldots, c_{n-1}, x_n)$ is a permutation polynomial of $\mathbb{F}_q$ in x_n for all $c_1, \ldots, c_{n-1} \in \mathbb{F}_q$. Prove that

$$s(x_1, \ldots, x_n) = h(x_1, \ldots, x_n) f(x_1, \ldots, x_{n-1}) + g(x_1, \ldots, x_{n-1})$$

is not a permutation polynomial over $\mathbb{F}_q$.

7.44. Let $f \in \mathbb{F}_q[x_1, \ldots, x_{n-1}]$ be such that the number of solutions of $f(x_1, \ldots, x_{n-1}) = 0$ in $\mathbb{F}_q^{n-1}$ is divisible by q, and let h be as in Exercise 7.43. Prove that there exists $g \in \mathbb{F}_q[x_1, \ldots, x_{n-1}]$ such that

$$s(x_1, \ldots, x_n) = h(x_1, \ldots, x_n) f(x_1, \ldots, x_{n-1}) + g(x_1, \ldots, x_{n-1})$$

is a permutation polynomial over $\mathbb{F}_q$.

7.45. Prove that the Dickson polynomial $g_k^{(2)}(x, y, a)$ is given by

$$g_k^{(2)}(x, y, a) = \sum_{i=0}^{\lfloor k/2 \rfloor} \sum_{\substack{j=0 \\ 2i+3j \leqslant k}}^{\lfloor k/3 \rfloor} \frac{k(-1)^i}{k-i-2j} \binom{k-i-2j}{i+j} \binom{i+j}{i} a^{i+2j} x^i y^{k-2i-3j}.$$

7.46. Prove the generalization of Theorem 7.23 for Dickson polynomials in two indeterminates.

7.47. Prove that the system $f_1, \ldots, f_m \in \mathbb{F}_q[x_1, \ldots, x_n]$, $1 \leqslant m \leqslant n$, is orthogonal in $\mathbb{F}_q$ if and only if $g(f_1(x_1, \ldots, x_n), \ldots, f_m(x_1, \ldots, x_n))$ is a permutation polynomial in n indeterminates over $\mathbb{F}_q$ for all permutation polynomials $g(y_1, \ldots, y_m)$ in m indeterminates over $\mathbb{F}_q$.

7.48. Prove that for every system $f_1, \ldots, f_{n+1} \in \mathbb{F}_q[x_1, \ldots, x_n]$ there exist coefficients $b_1, \ldots, b_{n+1} \in \mathbb{F}_q$ not all zero such that $b_1 f_1 + \cdots + b_{n+1} f_{n+1}$ is not a permutation polynomial over $\mathbb{F}_q$.

7.49. A *coset* of a system of polynomials $f_1, \ldots, f_m \in \mathbb{F}_q[x_1, \ldots, x_n]$, $1 \leqslant m \leqslant n$, is a nonempty subset of $\mathbb{F}_q^n$ that is mapped by the system into a single element of $\mathbb{F}_q^m$. Let $f_1, \ldots, f_m$ be orthogonal in $\mathbb{F}_q$. Prove that the following two conditions for a polynomial $g \in \mathbb{F}_q[x_1, \ldots, x_n]$ are equivalent: (i) g is a permutation polynomial over $\mathbb{F}_q$ with all cosets of the system $f_1, \ldots, f_m$ being cosets of g as well; (ii) $g \equiv h(f_1, \ldots, f_m) \bmod (x_1^q - x_1, \ldots, x_n^q - x_n)$ for some permutation polynomial h in m indeterminates over $\mathbb{F}_q$.

Chapter 8

Linear Recurring Sequences

Sequences in finite fields whose terms depend in a simple manner on their predecessors are of importance for a variety of applications. Such sequences are easy to generate by recursive procedures, which is certainly an advantageous feature from the computational viewpoint, and they also tend to have useful structural properties. Of particular interest is the case where the terms depend linearly on a fixed number of predecessors, resulting in a so-called linear recurring sequence. These sequences are employed in coding theory (see Chapter 9) and other branches of electrical engineering. In these applications, the underlying field is often taken to be $\mathbb{F}_2$, but the theory can be developed quite generally for any finite field.

In Section 1 we show how to implement the generation of linear recurring sequences on special switching circuits called feedback shift registers. We discuss also some basic periodicity properties of such sequences. Section 2 introduces the concept of an impulse response sequence, which is of both practical and theoretical interest. Further relations to periodicity properties are found in this way, and also through the use of the so-called characteristic polynomial of a linear recurring sequence. Another application of the characteristic polynomial yields explicit formulas for the terms of a linear recurring sequence. Maximal period sequences are also defined in this section.

The theory of linear recurring sequences can be approached via linear algebra, ideal theory, or formal power series. An approach based on the latter is presented in Section 3. This leads to a computation-oriented way of introducing the minimal polynomial of a linear recurring sequence in the next section. The minimal polynomial is of crucial importance for the linear recurring sequence, since the order of the minimal polynomial gives the least period of the sequence.

In Section 5 we study the collection of all sequences satisfying a given linear recurrence relation. This information is useful in the discussion of operations with linear recurring sequences, such as termwise addition and multiplication for sequences in general finite fields and binary complementation for sequences in $\mathbb{F}_2$. We consider also the problem of determining the various least periods of the sequences generated by a fixed linear recurrence relation. Section 6 presents some determinantal criteria characterizing linear recurring sequences as well as the Berlekamp-Massey algorithm for the calculation of minimal polynomials.

Section 7 is devoted to distribution properties of linear recurring sequences. Exponential sums with linear recurring sequences are the main tools for studying such properties.

1. FEEDBACK SHIFT REGISTERS, PERIODICITY PROPERTIES

Let k be a positive integer, and let $a, a_0, \ldots, a_{k-1}$ be given elements of a finite field $\mathbb{F}_q$. A sequence $s_0, s_1, \ldots$ of elements of $\mathbb{F}_q$ satisfying the relation

$$s_{n+k} = a_{k-1}s_{n+k-1} + a_{k-2}s_{n+k-2} + \cdots + a_0 s_n + a \quad \text{for } n = 0, 1, \ldots \tag{8.1}$$

is called a (*kth-order*) *linear recurring sequence* in $\mathbb{F}_q$. The terms s_0, $s_1, \ldots, s_{k-1}$, which determine the rest of the sequence uniquely, are referred to as the *initial values*. A relation of the form (8.1) is called a (*kth-order*) *linear recurrence relation*. In the older literature one may also find the term "difference equation." We speak of a *homogeneous* linear recurrence relation if $a = 0$; otherwise the linear recurrence relation is *inhomogeneous*. The sequence $s_0, s_1, \ldots$ itself is called a *homogeneous*, or *inhomogeneous*, *linear recurring sequence* in $\mathbb{F}_q$, respectively.

The generation of linear recurring sequences can be implemented on a *feedback shift register*. This is a special kind of electronic switching circuit handling information in the form of elements of $\mathbb{F}_q$, which are represented suitably. Four types of devices are used. The first is an *adder*, which has two inputs and one output, the output being the sum in $\mathbb{F}_q$ of the two inputs. The second is a *constant multiplier*, which has one input and yields as the output the product of the input with a constant element of $\mathbb{F}_q$. The third is a *constant adder*, which is analogous to a constant multiplier, but adds a constant element of $\mathbb{F}_q$ to the input. The fourth type of device is a *delay element* ("flip-flop"), which has one input and one output and is regulated by an external synchronous clock so that its input at a particular time appears as its output one unit of time later. We shall not be concerned here with the physical realization of these devices. The representation of the components in circuit diagrams is shown in Figure 8.1.

A feedback shift register is built by interconnecting a finite number of adders, constant multipliers, constant adders, and delay elements along a closed loop in such a way that two outputs are never connected together. Actually, for the purpose of generating linear recurring sequences, it suffices to connect the components in a rather special manner. A feedback shift register that generates a linear recurring sequence satisfying (8.1) is shown in Figure 8.2.

At the outset, each delay element D_j, $j = 0, 1, \ldots, k-1$, contains the initial value s_j. If we think of the arithmetic operations and the transfer along the wires to be performed instantaneously, then after one time unit each D_j will contain s_{j+1}. Continuing in this manner, we see that the output of the feedback shift register is the string of element $s_0, s_1, s_2, \ldots,$ received in intervals of one time unit. In most of the applications the desired linear recurring sequence is homogeneous, in which case the constant adder is not needed.

8.1. Example. In order to generate a linear recurring sequence in $\mathbb{F}_5$ satisfying the homogeneous linear recurrence relation

$$s_{n+6} = s_{n+5} + 2s_{n+4} + s_{n+1} + 3s_n \quad \text{for } n = 0, 1, \ldots,$$

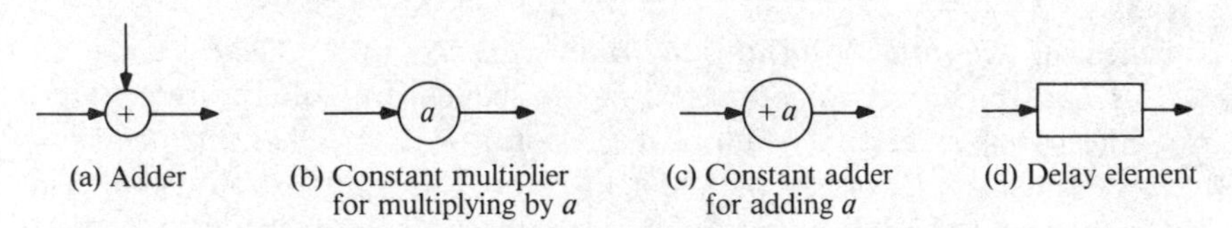

FIGURE 8.1 **The building blocks of feedback shift registers. (a) Adder. (b) Constant multiplier for multiplying by a. (c) Constant adder for adding a. (d) Delay element.**

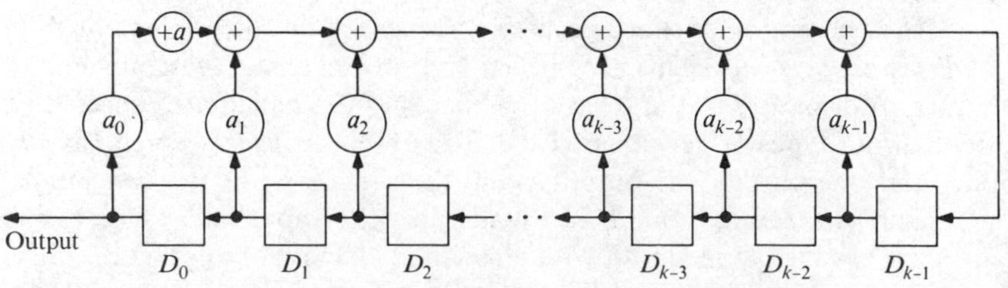

FIGURE 8.2

one may use the feedback shift register shown in Figure 8.3. Since $a_2 = a_3 = 0$, no connections are necessary at these points. □

8.2. Example. Consider the homogeneous linear recurrence relation

$$s_{n+7} = s_{n+4} + s_{n+3} + s_{n+2} + s_n, \quad n = 0, 1, \ldots, \text{ in } \mathbb{F}_2.$$

A feedback shift register corresponding to this linear recurrence relation is shown in Figure 8.4. Since multiplication by a constant in $\mathbb{F}_2$ either preserves or annihilates elements, the effect of a constant multiplier can be simulated by a wire connection or a disconnection. Therefore, a feedback shift register for the generation of binary homogeneous linear recurring sequences requires only delay elements, adders, and wire connections. □

Let $s_0, s_1, \ldots$ be a kth-order linear recurring sequence in $\mathbb{F}_q$ satisfying (8.1). As we have noted, this sequence can be generated by the feedback shift register in Figure 8.2. If n is a nonnegative integer, then after n time units the delay element D_j, $j = 0, 1, \ldots, k-1$, will contain s_{n+j}. It is therefore natural to call the row vector $\mathbf{s}_n = (s_n, s_{n+1}, \ldots, s_{n+k-1})$ the *nth state vector* of the linear recurring sequence (or of the feedback shift register). The state vector $\mathbf{s}_0 = (s_0, s_1, \ldots, s_{k-1})$ is also referred to as the *initial state vector*.

It is a characteristic feature of linear recurring sequences in finite fields that, after a possibly irregular behavior in the beginning, such sequences are eventually of a periodic nature (or ultimately periodic in the sense of Definition 8.3 below). Before studying this property in detail, we introduce some terminology and mention a few general facts about ultimately periodic sequences.

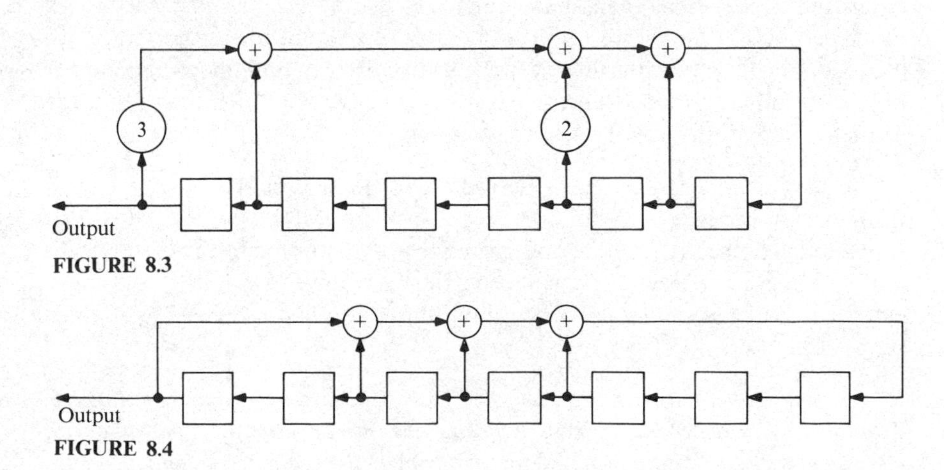

FIGURE 8.3

FIGURE 8.4

8.3. Definition. Let S be an arbitrary nonempty set, and let $s_0, s_1, \ldots$ be a sequence of elements of S. If there exist integers $r > 0$ and $n_0 \geqslant 0$ such that $s_{n+r} = s_n$ for all $n \geqslant n_0$, then the sequence is called *ultimately periodic* and r is called a *period* of the sequence. The smallest number among all the possible periods of an ultimately periodic sequence is called the *least period* of the sequence.

8.4. **Lemma.** *Every period of an ultimately periodic sequence is divisible by the least period.*

Proof. Let r be an arbitrary period of the ultimately periodic sequence $s_0, s_1, \ldots$ and let r_1 be its least period, so that we have $s_{n+r} = s_n$ for all $n \geqslant n_0$ and $s_{n+r_1} = s_n$ for all $n \geqslant n_1$ with suitable nonnegative integers n_0 and n_1. If r were not divisible by r_1, we could use the division algorithm for integers to write $r = mr_1 + t$ with integers $m \geqslant 1$ and $0 < t < r_1$. Then, for all $n \geqslant \max(n_0, n_1)$ we get

$$s_n = s_{n+r} = s_{n+mr_1+t} = s_{n+(m-1)r_1+t} = \cdots = s_{n+t},$$

and so t is a period of the sequence, which contradicts the definition of the least period. □

8.5. Definition. An ultimately periodic sequence $s_0, s_1, \ldots$ with least period r is called *periodic* if $s_{n+r} = s_n$ holds for all $n = 0, 1, \ldots$.

The following condition, which is sometimes found in the literature, is equivalent to the definition of a periodic sequence.

8.6. **Lemma.** *The sequence $s_0, s_1, \ldots$ is periodic if and only if there exists an integer $r > 0$ such that $s_{n+r} = s_n$ for all $n = 0, 1, \ldots$.*

Proof. The necessity of the condition is obvious. Conversely, if the condition is satisfied, then the sequence is ultimately periodic and has a least period r_1. Therefore, with a suitable n_0 we have $s_{n+r_1} = s_n$ for all $n \geqslant n_0$. Now let n be an arbitrary nonnegative integer, and choose an integer $m \geqslant n_0$ with $m \equiv n \bmod r$. Then $s_{n+r_1} = s_{m+r_1} = s_m = s_n$, which shows that the sequence is periodic in the sense of Definition 8.5. □

If $s_0, s_1, \ldots$ is ultimately periodic with least period r, then the least nonnegative integer n_0 such that $s_{n+r} = s_n$ for all $n \geqslant n_0$ is called the *preperiod*. The sequence is periodic precisely if the preperiod is 0.

We return now to linear recurring sequences in finite fields and establish the basic results concerning the periodicity behavior of such sequences.

8.7. **Theorem.** *Let $\mathbb{F}_q$ be any finite field and k any positive integer. Then every kth-order linear recurring sequence in $\mathbb{F}_q$ is ultimately periodic with least period r satisfying $r \leqslant q^k$, and $r \leqslant q^k - 1$ if the sequence is homogeneous.*

Proof. We note that there are exactly q^k distinct k-tuples of elements of $\mathbb{F}_q$. Therefore, by considering the state vectors $\mathbf{s}_m$, $0 \leqslant m \leqslant q^k$, of a given kth-order linear recurring sequence in $\mathbb{F}_q$, it follows that $\mathbf{s}_j = \mathbf{s}_i$ for some i and j with $0 \leqslant i < j \leqslant q^k$. Using the linear recurrence relation and induction, we arrive at $\mathbf{s}_{n+j-i} = \mathbf{s}_n$ for all $n \geqslant i$, which shows that the linear recurring sequence itself is ultimately periodic with least period $r \leqslant j - i \leqslant q^k$. In case the linear recurring sequence is homogeneous and no state vector is the zero vector, one can go through the same argument, but with q^k replaced by $q^k - 1$, to obtain $r \leqslant q^k - 1$. If, however, one of the state vectors of a homogeneous linear recurring sequence is the zero vector, then all subsequent state vectors are zero vectors, and so the sequence has least period $r = 1 \leqslant q^k - 1$. □

8.8. Example. The first-order linear recurring sequence $s_0, s_1, \ldots$ in $\mathbb{F}_p$, p prime, with $s_{n+1} = s_n + 1$ for $n = 0, 1, \ldots$ and arbitrary $s_0 \in \mathbb{F}_p$ shows that the upper bound for r in Theorem 8.7 may be attained. If $\mathbb{F}_q$ is any finite field and g is a primitive element of $\mathbb{F}_q$ (see Definition 2.9), then the first-order homogeneous linear recurring sequence $s_0, s_1, \ldots$ in $\mathbb{F}_q$ with $s_{n+1} = g s_n$ for $n = 0, 1, \ldots$ and $s_0 \neq 0$ has least period $r = q - 1$. Therefore, the upper bound for r in the homogeneous case may also be attained. Later on, we shall show that in any $\mathbb{F}_q$ and for any $k \geqslant 1$ there exist kth-order homogeneous linear recurring sequences with least period $r = q^k - 1$ (see Theorem 8.33). □

8.9. Example. For a first-order homogeneous linear recurring sequence in $\mathbb{F}_q$, it is easily seen that the least period divides $q - 1$. However, if $k \geqslant 2$, then the least period of a kth-order homogeneous linear recurring sequence need not divide $q^k - 1$. Consider, for instance, the sequence $s_0, s_1, \ldots$ in $\mathbb{F}_5$ with $s_0 = 0$, $s_1 = 1$, and $s_{n+2} = s_{n+1} + s_n$ for $n = 0, 1, \ldots$, which has least period 20, as is shown by inspection. □

8.10. Example. A linear recurring sequence in a finite field is ultimately periodic, but it need not be periodic, as is illustrated by a second-order linear recurring sequence $s_0, s_1, \ldots$ in $\mathbb{F}_q$ with $s_0 \neq s_1$ and $s_{n+2} = s_{n+1}$ for $n = 0, 1, \ldots$. □

An important sufficient condition for the periodicity of a linear recurring sequence is provided by the following result.

8.11. Theorem. *If $s_0, s_1, \ldots$ is a linear recurring sequence in a finite field satisfying the linear recurrence relation* (8.1), *and if the coefficient a_0 in* (8.1) *is nonzero, then the sequence $s_0, s_1, \ldots$ is periodic.*

Proof. According to Theorem 8.7, the given linear recurring sequence is ultimately periodic. If r is its least period and n_0 its preperiod, then $s_{n+r} = s_n$ for all $n \geqslant n_0$. Suppose we had $n_0 \geqslant 1$. From (8.1) with

$n = n_0 + r - 1$ and the fact that $a_0 \neq 0$, we obtain

$$\begin{aligned} s_{n_0 - 1 + r} &= a_0^{-1}\left(s_{n_0 + k - 1 + r} - a_{k-1}s_{n_0 + k - 2 + r} - \cdots - a_1 s_{n_0 + r} - a\right) \\ &= a_0^{-1}\left(s_{n_0 + k - 1} - a_{k-1}s_{n_0 + k - 2} - \cdots - a_1 s_{n_0} - a\right). \end{aligned}$$

Using (8.1) with $n = n_0 - 1$, we find the same expression for $s_{n_0 - 1}$, and so $s_{n_0 - 1 + r} = s_{n_0 - 1}$. This is a contradiction to the definition of the preperiod. □

Let $s_0, s_1, \ldots$ be a kth-order homogeneous linear recurring sequence in $\mathbb{F}_q$ satisfying the linear recurrence relation

$$s_{n+k} = a_{k-1}s_{n+k-1} + a_{k-2}s_{n+k-2} + \cdots + a_0 s_n \quad \text{for } n = 0, 1, \ldots, \tag{8.2}$$

where $a_j \in \mathbb{F}_q$ for $0 \leqslant j \leqslant k-1$. With this linear recurring sequence we associate the $k \times k$ matrix A over $\mathbb{F}_q$ defined by

$$A = \begin{pmatrix} 0 & 0 & 0 & \cdots & 0 & a_0 \\ 1 & 0 & 0 & \cdots & 0 & a_1 \\ 0 & 1 & 0 & \cdots & 0 & a_2 \\ \vdots & \vdots & \vdots & & \vdots & \vdots \\ 0 & 0 & 0 & \cdots & 1 & a_{k-1} \end{pmatrix}. \tag{8.3}$$

If $k = 1$, then A is understood to be the 1×1 matrix (a_0). We note that the matrix A depends only on the linear recurrence relation satisfied by the given sequence.

8.12. Lemma. *If $s_0, s_1, \ldots$ is a homogeneous linear recurring sequence in $\mathbb{F}_q$ satisfying* (8.2) *and A is the matrix in* (8.3) *associated with it, then for the state vectors of the sequence we have*

$$\mathbf{s}_n = \mathbf{s}_0 A^n \quad \textit{for } n = 0, 1, \ldots. \tag{8.4}$$

Proof. Since $\mathbf{s}_n = (s_n, s_{n+1}, \ldots, s_{n+k-1})$, one checks easily that $\mathbf{s}_{n+1} = \mathbf{s}_n A$ for all $n \geqslant 0$, so that (8.4) follows by induction. □

We note that the set of all nonsingular $k \times k$ matrices over $\mathbb{F}_q$ forms a finite group under matrix multiplication, called the *general linear group* $GL(k, \mathbb{F}_q)$.

8.13. Theorem. *If $s_0, s_1, \ldots$ is a kth-order homogeneous linear recurring sequence in $\mathbb{F}_q$ satisfying* (8.2) *with $a_0 \neq 0$, then the least period of the sequence divides the order of the associated matrix A from* (8.3) *in the general linear group $GL(k, \mathbb{F}_q)$.*

Proof. We have $\det A = (-1)^{k-1}a_0 \neq 0$, so that A is indeed an element of $GL(k,\mathbb{F}_q)$. If m is the order of A in $GL(k,\mathbb{F}_q)$, then from Lemma 8.12 we obtain $\mathbf{s}_{n+m} = \mathbf{s}_0 A^{n+m} = \mathbf{s}_0 A^n = \mathbf{s}_n$ for all $n \geqslant 0$, and so m is a period of the linear recurring sequence. The rest follows from Lemma 8.4. □

We remark that the above argument, together with Lemma 8.6, yields an alternative proof for Theorem 8.11 in the homogeneous case. From Theorem 8.13 it follows, in particular, that the least period of the sequence $s_0, s_1, \ldots$ divides the order of $GL(k,\mathbb{F}_q)$, which is known to be $q^{(k^2-k)/2}(q-1)(q^2-1)\cdots(q^k-1)$.

Let now $s_0, s_1, \ldots$ be a kth-order *inhomogeneous* linear recurring sequence in $\mathbb{F}_q$ satisfying (8.1). By using (8.1) with n replaced by $n+1$ and subtracting from the resulting identity the original form of (8.1) we obtain

$$s_{n+k+1} = b_k s_{n+k} + b_{k-1} s_{n+k-1} + \cdots + b_0 s_n \quad \text{for } n = 0, 1, \ldots, \tag{8.5}$$

where $b_0 = -a_0$, $b_j = a_{j-1} - a_j$ for $j = 1, 2, \ldots, k-1$, and $b_k = a_{k-1} + 1$. Therefore, the sequence $s_0, s_1, \ldots$ can be interpreted as a $(k+1)$st-order homogeneous linear recurring sequence in $\mathbb{F}_q$. Consequently, results on homogeneous linear recurring sequences yield information for the inhomogeneous case as well.

An alternative approach to the *inhomogeneous case* proceeds as follows. Let $s_0, s_1, \ldots$ be a kth-order inhomogeneous linear recurring sequence in $\mathbb{F}_q$ satisfying (8.1), and consider the $(k+1)\times(k+1)$ matrix C over $\mathbb{F}_q$ defined by

$$C = \begin{pmatrix} 1 & 0 & 0 & \cdots & 0 & a \\ 0 & 0 & 0 & \cdots & 0 & a_0 \\ 0 & 1 & 0 & \cdots & 0 & a_1 \\ 0 & 0 & 1 & \cdots & 0 & a_2 \\ \vdots & \vdots & \vdots & & \vdots & \vdots \\ 0 & 0 & 0 & \cdots & 1 & a_{k-1} \end{pmatrix}.$$

If $k = 1$, take

$$C = \begin{pmatrix} 1 & a \\ 0 & a_0 \end{pmatrix}.$$

We introduce modified state vectors by setting

$$\mathbf{s}'_n = (1, s_n, s_{n+1}, \ldots, s_{n+k-1}) \quad \text{for } n = 0, 1, \ldots.$$

Then it is easily seen that $\mathbf{s}'_{n+1} = \mathbf{s}'_n C$ for all $n \geqslant 0$, and so $\mathbf{s}'_n = \mathbf{s}'_0 C^n$ for all $n \geqslant 0$ by induction. If $a_0 \neq 0$ in (8.1), then $\det C = (-1)^{k-1}a_0 \neq 0$, so that the matrix C is an element of $GL(k+1,\mathbb{F}_q)$. One shows then as in the proof of Theorem 8.13 that the least period of $s_0, s_1, \ldots$ divides the order of C in $GL(k+1,\mathbb{F}_q)$.

2. IMPULSE RESPONSE SEQUENCES, CHARACTERISTIC POLYNOMIAL

Among all the homogeneous linear recurring sequences in $\mathbb{F}_q$ satisfying a given kth-order linear recurrence relation such as (8.2), we can single out one that yields the maximal value for the least period in this class of sequences. This is the *impulse response sequence* $d_0, d_1, \ldots$ determined uniquely by its initial values $d_0 = \cdots = d_{k-2} = 0$, $d_{k-1} = 1$ ($d_0 = 1$ if $k = 1$) and the linear recurrence relation

$$d_{n+k} = a_{k-1}d_{n+k-1} + a_{k-2}d_{n+k-2} + \cdots + a_0 d_n \quad \text{for } n = 0, 1, \ldots. \tag{8.6}$$

8.14. Example. Consider the linear recurrence relation

$$s_{n+5} = s_{n+1} + s_n, \quad n = 0, 1, \ldots, \text{ in } \mathbb{F}_2.$$

The impulse response sequence $d_0, d_1, \ldots$ corresponding to it is given by the string of binary digits

$$0\,0\,0\,0\,1\,0\,0\,0\,1\,1\,0\,0\,1\,0\,1\,0\,1\,1\,1\,1\,1\,0\,0\,0\,0\,1 \cdots$$

of least period 21. A feedback shift register generating this sequence is shown in Figure 8.5. We can think of this sequence as being obtained by starting with the state in which each delay element is "empty" (i.e., contains 0) and then sending the "impulse" 1 into the rightmost delay element. This explains the term "impulse response sequence." □

8.15. ***Lemma.*** *Let $d_0, d_1, \ldots$ be the impulse response sequence in $\mathbb{F}_q$ satisfying* (8.6), *and let A be the matrix in* (8.3). *Then two state vectors $\mathbf{d}_m$ and $\mathbf{d}_n$ are identical if and only if $A^m = A^n$.*

Proof. The sufficiency follows from Lemma 8.12. Conversely, suppose that $\mathbf{d}_m = \mathbf{d}_n$. From the linear recurrence relation (8.6) we obtain then $\mathbf{d}_{m+t} = \mathbf{d}_{n+t}$ for all $t \geqslant 0$. By Lemma 8.12 we get $\mathbf{d}_t A^m = \mathbf{d}_t A^n$ for all $t \geqslant 0$. But since the vectors $\mathbf{d}_0, \mathbf{d}_1, \ldots, \mathbf{d}_{k-1}$ obviously form a basis for the k-dimensional vector space $\mathbb{F}_q^k$ over $\mathbb{F}_q$, we conclude that $A^m = A^n$. □

8.16. ***Theorem.*** *The least period of a homogeneous linear recurring sequence in $\mathbb{F}_q$ divides the least period of the corresponding impulse response sequence.*

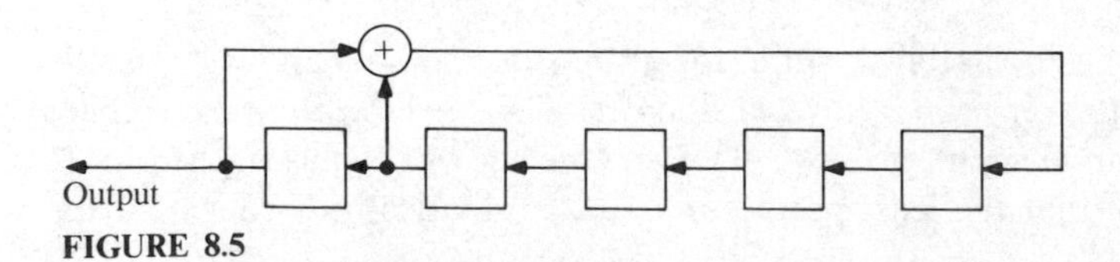

FIGURE 8.5

Proof. Let $s_0, s_1, \ldots$ be a homogeneous linear recurring sequence in $\mathbb{F}_q$ satisfying (8.2), let $d_0, d_1, \ldots$ be the corresponding impulse response sequence, and let A be the matrix in (8.3). If r is the least period of $d_0, d_1, \ldots$ and n_0 the preperiod, then $\mathbf{d}_{n+r} = \mathbf{d}_n$ for all $n \geqslant n_0$. It follows from Lemma 8.15 that $A^{n+r} = A^n$ for all $n \geqslant n_0$, and so $\mathbf{s}_{n+r} = \mathbf{s}_n$ for all $n \geqslant n_0$ by Lemma 8.12. Therefore, r is a period of $s_0, s_1, \ldots$, and an application of Lemma 8.4 completes the proof. □

8.17. Theorem. *If $d_0, d_1, \ldots$ is a kth-order impulse response sequence in $\mathbb{F}_q$ satisfying* (8.6) *with $a_0 \neq 0$ and A is the matrix in* (8.3) *associated with it, then the least period of the sequence is equal to the order of A in the general linear group $GL(k, \mathbb{F}_q)$.*

Proof. If r is the least period of $d_0, d_1, \ldots$, then r divides the order of A according to the Theorem 8.13. On the other hand, we have $\mathbf{d}_r = \mathbf{d}_0$ by Theorem 8.11, and so Lemma 8.15 yields $A^r = A^0$, which implies already the desired result. □

8.18. Example. For the linear recurrence relation $s_{n+5} = s_{n+1} + s_n$, $n = 0, 1, \ldots$, in $\mathbb{F}_2$ considered in Example 8.14 we have seen that the least period of the corresponding impulse response sequence is equal to 21, which is the same as the order of the matrix

$$A = \begin{pmatrix} 0 & 0 & 0 & 0 & 1 \\ 1 & 0 & 0 & 0 & 1 \\ 0 & 1 & 0 & 0 & 0 \\ 0 & 0 & 1 & 0 & 0 \\ 0 & 0 & 0 & 1 & 0 \end{pmatrix}$$

in $GL(5, \mathbb{F}_2)$. If the initial state vector of a linear recurring sequence in $\mathbb{F}_2$ satisfying the given linear recurrence relation is equal to one of the 21 different state vectors appearing in the impulse response sequence, then the least period is again 21 (since such a sequence is just a shifted impulse response sequence). If we choose the initial state vector $(1, 1, 1, 0, 1)$, we get the string of binary digits $1\ 1\ 1\ 0\ 1\ 0\ 0\ 1\ 1\ 1\ 0\ 1 \cdots$ of least period 7, and the same least period results from any one of the 7 different state vectors of this sequence in the role of the initial state vector. If the initial state vector is $(1, 1, 0, 1, 1)$, then we obtain the string of binary digits $1\ 1\ 0\ 1\ 1\ 0\ 1\ 1 \cdots$ of least period 3, and the same least period results if any one of the 3 different state vectors of this sequence is taken as the initial state vector. The initial state vector $(0, 0, 0, 0, 0)$ produces a sequence of least period 1. We have now exhausted all 32 possibilities for initial state vectors. □

8.19. Theorem. *Let $s_0, s_1, \ldots$ be a kth-order homogeneous linear recurring sequence in $\mathbb{F}_q$ with preperiod n_0. If there exist k state vectors $\mathbf{s}_{m_1}, \mathbf{s}_{m_2}, \ldots, \mathbf{s}_{m_k}$ with $m_j \geqslant n_0$ $(1 \leqslant j \leqslant k)$ that are linearly independent over $\mathbb{F}_q$,* *then both* $s_0, s_1, \ldots$ *and its corresponding impulse response sequence are periodic and they have the same least period.*

Proof. Let r be the least period of $s_0, s_1, \ldots$. For $1 \leqslant j \leqslant k$ we have $\mathbf{s}_{m_j} A^r = \mathbf{s}_{m_j + r} = \mathbf{s}_{m_j}$ by using Lemma 8.12, and so A^r is the $k \times k$ identity matrix over $\mathbb{F}_q$. Thus we get $\mathbf{s}_r = \mathbf{s}_0 A^r = \mathbf{s}_0$, which shows that $s_0, s_1, \ldots$ is periodic. Similarly, if $\mathbf{d}_n$ denotes the nth state vector of the impulse response sequence, then $\mathbf{d}_r = \mathbf{d}_0 A^r = \mathbf{d}_0$, and an application of Theorem 8.16 completes the proof. □

8.20. Example. The condition $m_j \geqslant n_0$ in Theorem 8.19 is needed since there are kth-order homogeneous linear recurring sequences that are not periodic but contain k linearly independent state vectors. Let $d_0, d_1, \ldots$ be the second-order impulse response sequence in $\mathbb{F}_q$ with $d_{n+2} = d_{n+1}$ for $n = 0, 1, \ldots$. The terms of this sequence are $0, 1, 1, 1, \ldots$. Clearly, the state vectors $\mathbf{d}_0$ and $\mathbf{d}_1$ are linearly independent over $\mathbb{F}_q$, but the sequence is not periodic (note that $n_0 = 1$ in this case). The converse of Theorem 8.19 is not true. Consider the third-order linear recurring sequence $s_0, s_1, \ldots$ in $\mathbb{F}_2$ with $s_{n+3} = s_n$ for $n = 0, 1, \ldots$ and $\mathbf{s}_0 = (1, 1, 0)$. Then both $s_0, s_1, \ldots$ and its corresponding impulse response sequence are periodic with least period 3, but any three state vectors of $s_0, s_1, \ldots$ are linearly dependent over $\mathbb{F}_2$. □

Let $s_0, s_1, \ldots$ be a kth-order homogeneous linear recurring sequence in $\mathbb{F}_q$ satisfying the linear recurrence relation

$$s_{n+k} = a_{k-1} s_{n+k-1} + a_{k-2} s_{n+k-2} + \cdots + a_0 s_n \quad \text{for } n = 0, 1, \ldots, \tag{8.7}$$

where $a_j \in \mathbb{F}_q$ for $0 \leqslant j \leqslant k-1$. The polynomial

$$f(x) = x^k - a_{k-1} x^{k-1} - a_{k-2} x^{k-2} - \cdots - a_0 \in \mathbb{F}_q[x]$$

is called the *characteristic polynomial* of the linear recurring sequence. It depends, of course, only on the linear recurrence relation (8.7). If A is the matrix in (8.3), then it is easily seen that $f(x)$ is identical with the characteristic polynomial of A in the sense of linear algebra—that is, $f(x) = \det(xI - A)$ with I being the $k \times k$ identity matrix over $\mathbb{F}_q$. On the other hand, the matrix A may be thought of as the companion matrix of the monic polynomial $f(x)$.

As a first application of the characteristic polynomial, we show how the terms of a linear recurring sequence may be represented explicitly in an important special case.

8.21. Theorem. *Let* $s_0, s_1, \ldots$ *be a kth-order homogeneous linear recurring sequence in* $\mathbb{F}_q$ *with characteristic polynomial* $f(x)$. *If the roots* $\alpha_1, \ldots, \alpha_k$ *of* $f(x)$ *are all distinct, then*

$$s_n = \sum_{j=1}^{k} \beta_j \alpha_j^n \quad \text{for } n = 0, 1, \ldots, \tag{8.8}$$

where $\beta_1,\ldots,\beta_k$ are elements that are uniquely determined by the initial values of the sequence and belong to the splitting field of $f(x)$ over $\mathbb{F}_q$.

Proof. The constants $\beta_1,\ldots,\beta_k$ can be determined from the system of linear equations

$$\sum_{j=1}^{k} \alpha_j^n \beta_j = s_n, \quad n = 0,1,\ldots,k-1.$$

Since the determinant of this system is a Vandermonde determinant, which is nonzero by the condition on $\alpha_1,\ldots,\alpha_k$, the elements $\beta_1,\ldots,\beta_k$ are uniquely determined and belong to the splitting field $\mathbb{F}_q(\alpha_1,\ldots,\alpha_k)$ of $f(x)$ over $\mathbb{F}_q$, as is seen from Cramer's rule. To prove the identity (8.8) for all $n \geqslant 0$, it suffices now to check whether the elements on the right-hand side of (8.8), with these specific values for $\beta_1,\ldots,\beta_k$, satisfy the linear recurrence relation (8.7). But

$$\sum_{j=1}^{k} \beta_j \alpha_j^{n+k} - a_{k-1} \sum_{j=1}^{k} \beta_j \alpha_j^{n+k-1} - a_{k-2} \sum_{j=1}^{k} \beta_j \alpha_j^{n+k-2} - \cdots - a_0 \sum_{j=1}^{k} \beta_j \alpha_j^n$$

$$= \sum_{j=1}^{k} \beta_j f(\alpha_j) \alpha_j^n = 0$$

for all $n \geqslant 0$, and the proof is complete. □

8.22. Example. Consider the linear recurring sequence $s_0, s_1, \ldots$ in $\mathbb{F}_2$ with $s_0 = s_1 = 1$ and $s_{n+2} = s_{n+1} + s_n$ for $n = 0,1,\ldots$. The characteristic polynomial is $f(x) = x^2 - x - 1 \in \mathbb{F}_2[x]$. If $\mathbb{F}_4 = \mathbb{F}_2(\alpha)$, then the roots of $f(x)$ are $\alpha_1 = \alpha$ and $\alpha_2 = 1 + \alpha$. Using the given initial values, we obtain $\beta_1 + \beta_2 = 1$ and $\beta_1 \alpha + \beta_2(1+\alpha) = 1$, hence $\beta_1 = \alpha$ and $\beta_2 = 1 + \alpha$. By Theorem 8.21 it follows that $s_n = \alpha^{n+1} + (1+\alpha)^{n+1}$ for all $n \geqslant 0$. Since $\beta^3 = 1$ for every nonzero $\beta \in \mathbb{F}_4$, we deduce that $s_{n+3} = s_n$ for all $n \geqslant 0$, which is in accordance with the fact that the least period of the sequence is 3. □

8.23. Remark. A formula similar to (8.8) is valid if the multiplicity of each root of $f(x)$ is at most the characteristic p of $\mathbb{F}_q$. In detail, let $\alpha_1,\ldots,\alpha_m$ be the distinct roots of $f(x)$, and suppose that each α_i, $i = 1,2,\ldots,m$, has multiplicity $e_i \leqslant p$ and that $e_i = 1$ if $\alpha_i = 0$. Then we have

$$s_n = \sum_{i=1}^{m} P_i(n) \alpha_i^n \quad \text{for } n = 0,1,\ldots,$$

where each P_i, $i = 1,2,\ldots,m$, is a polynomial of degree less than e_i whose coefficients are uniquely determined by the initial values of the sequence and belong to the splitting field of $f(x)$ over $\mathbb{F}_q$. The integer n is of course identified in the usual way with an element of $\mathbb{F}_q$. The reader familiar with differential equations will observe a certain analogy with the general solution of a homogeneous linear differential equation with constant coefficients. □

In case the characteristic polynomial is irreducible, the elements of the linear recurring sequence can be represented in terms of a suitable trace function (see Definition 2.22 and Theorem 2.23 for the definition and basic properties of trace functions).

8.24. Theorem. *Let $s_0, s_1, \ldots$ be a kth-order homogeneous linear recurring sequence in $K = \mathbb{F}_q$ whose characteristic polynomial $f(x)$ is irreducible over K. Let α be a root of $f(x)$ in the extension field $F = \mathbb{F}_{q^k}$. Then there exists a uniquely determined $\theta \in F$ such that*

$$s_n = \mathrm{Tr}_{F/K}(\theta\alpha^n) \quad \text{for } n = 0, 1, \ldots.$$

Proof. Since $\{1, \alpha, \ldots, \alpha^{k-1}\}$ constitutes a basis of F over K, we can define a uniquely determined linear mapping L from F into K by setting $L(\alpha^n) = s_n$ for $n = 0, 1, \ldots, k-1$. By Theorem 2.24 there exists a uniquely determined $\theta \in F$ such that $L(\gamma) = \mathrm{Tr}_{F/K}(\theta\gamma)$ for all $\gamma \in F$. In particular, we have

$$s_n = \mathrm{Tr}_{F/K}(\theta\alpha^n) \quad \text{for } n = 0, 1, \ldots, k-1.$$

It remains to show that the elements $\mathrm{Tr}_{F/K}(\theta\alpha^n)$, $n = 0, 1, \ldots$, form a homogeneous linear recurring sequence with characteristic polynomial $f(x)$. But if $f(x) = x^k - a_{k-1}x^{k-1} - \cdots - a_0 \in K[x]$, then using properties of the trace function we get

$$\begin{aligned}
&\mathrm{Tr}_{F/K}(\theta\alpha^{n+k}) - a_{k-1}\mathrm{Tr}_{F/K}(\theta\alpha^{n+k-1}) - \cdots - a_0\mathrm{Tr}_{F/K}(\theta\alpha^n) \\
&\quad = \mathrm{Tr}_{F/K}\left(\theta\alpha^{n+k} - a_{k-1}\theta\alpha^{n+k-1} - \cdots - a_0\theta\alpha^n\right) \\
&\quad = \mathrm{Tr}_{F/K}(\theta\alpha^n f(\alpha)) = 0
\end{aligned}$$

for all $n \geqslant 0$. □

Further relations between linear recurring sequences and their characteristic polynomials can be found on the basis of the following polynomial identity.

8.25. Theorem. *Let $s_0, s_1, \ldots$ be a kth-order homogeneous linear recurring sequence in $\mathbb{F}_q$ that satisfies the linear recurrence relation* (8.7) *and is periodic with period r. Let $f(x)$ be the characteristic polynomial of the sequence. Then the identity*

$$f(x)s(x) = (1 - x^r)h(x) \tag{8.9}$$

holds with

$$s(x) = s_0x^{r-1} + s_1x^{r-2} + \cdots + s_{r-2}x + s_{r-1} \in \mathbb{F}_q[x]$$

and

$$h(x)=\sum_{j=0}^{k-1}\sum_{i=0}^{k-1-j}a_{i+j+1}s_i x^j\in\mathbb{F}_q[x], \tag{8.10}$$

where we set $a_k=-1$.

Proof. We compare the coefficients on both sides of (8.9). For $0\leqslant t\leqslant k+r-1$, let c_t (resp. d_t) be the coefficient of x^t on the left-hand side (resp. right-hand side) of (8.9). Since $f(x)=-\sum_{i=0}^{k}a_i x^i$, we have

$$c_t=-\sum_{\substack{0\leqslant i\leqslant k,\,0\leqslant j\leqslant r-1\\ i+j=t}}a_i s_{r-1-j}\quad\text{for } 0\leqslant t\leqslant k+r-1. \tag{8.11}$$

We note also that the linear recurrence relation (8.7) may be written in the form

$$\sum_{i=0}^{k}a_i s_{n+i}=0\quad\text{for all } n\geqslant 0. \tag{8.12}$$

We distinguish now four cases. If $k\leqslant t\leqslant r-1$, then by (8.11) and (8.12),

$$c_t=-\sum_{i=0}^{k}a_i s_{r-1-t+i}=0=d_t.$$

If $t\leqslant r-1$ and $t<k$, then by (8.11), (8.12), and the periodicity of the given sequence,

$$\begin{aligned}c_t&=-\sum_{i=0}^{t}a_i s_{r-1-t+i}=\sum_{i=t+1}^{k}a_i s_{r-1-t+i}\\&=\sum_{i=t+1}^{k}a_i s_{i-t-1}=\sum_{i=0}^{k-1-t}a_{i+t+1}s_i=d_t.\end{aligned}$$

If $t\geqslant r$ and $t\geqslant k$, then by (8.11),

$$c_t=-\sum_{i=t-r+1}^{k}a_i s_{r-1-t+i}=-\sum_{i=0}^{k-1-t+r}a_{i+t-r+1}s_i=d_t.$$

If $r\leqslant t<k$, then by (8.11) and the periodicity of the given sequence,

$$\begin{aligned}c_t&=-\sum_{i=t-r+1}^{t}a_i s_{r-1-t+i}=-\sum_{i=0}^{r-1}a_{i+t-r+1}s_i\\&=\sum_{i=r}^{k-1-t+r}a_{i+t-r+1}s_i-\sum_{i=0}^{k-1-t+r}a_{i+t-r+1}s_i\\&=\sum_{i=0}^{k-1-t}a_{i+t+1}s_{i+r}-\sum_{i=0}^{k-1-t+r}a_{i+t-r+1}s_i\\&=\sum_{i=0}^{k-1-t}a_{i+t+1}s_i-\sum_{i=0}^{k-1-t+r}a_{i+t-r+1}s_i=d_t.\end{aligned}$$

□

In Lemma 3.1 we have seen that for any polynomial $f(x) \in \mathbb{F}_q[x]$ with $f(0) \neq 0$ there exists a positive integer e such that $f(x)$ divides $x^e - 1$. This gave rise to the definition of the order of f (see Definition 3.2). We give the following interpretation of $\operatorname{ord}(f)$.

8.26. Lemma. *Let*

$$f(x) = x^k - a_{k-1}x^{k-1} - a_{k-2}x^{k-2} - \cdots - a_0 \in \mathbb{F}_q[x]$$

with $k \geqslant 1$ *and* $a_0 \neq 0$. *Then* $\operatorname{ord}(f(x))$ *is equal to the order of the matrix* A *from* (8.3) *in the general linear group* $GL(k, \mathbb{F}_q)$.

Proof. Since A is the companion matrix of $f(x)$, the polynomial $f(x)$ is, in turn, the minimal polynomial of A. Consequently, if I is the $k \times k$ identity matrix over $\mathbb{F}_q$, then we have $A^e = I$ for some positive integer e if and only if $f(x)$ divides $x^e - 1$. The result follows now from the definitions of the order of $f(x)$ and the order of A. □

8.27. Theorem. *Let* $s_0, s_1, \ldots$ *be a homogeneous linear recurring sequence in* $\mathbb{F}_q$ *with characteristic polynomial* $f(x) \in \mathbb{F}_q[x]$. *Then the least period of the sequence divides* $\operatorname{ord}(f(x))$, *and the least period of the corresponding impulse response sequence is equal to* $\operatorname{ord}(f(x))$. *If* $f(0) \neq 0$, *then both sequences are periodic.*

Proof. If $f(0) \neq 0$, then in the light of Lemma 8.26 the result is essentially a restatement of Theorems 8.13 and 8.17. In this case, the periodicity property follows from Theorem 8.11. If $f(0) = 0$, then we write $f(x) = x^h g(x)$ as in Definition 3.2 and set $t_n = s_{n+h}$ for $n = 0, 1, \ldots$. Then $t_0, t_1, \ldots$ is a homogeneous linear recurring sequence with characteristic polynomial $g(x)$, provided that $\deg(g(x)) > 0$. Its least period is the same as that of the sequence $s_0, s_1, \ldots$. Therefore, by what we have already shown, the least period of $s_0, s_1, \ldots$ divides $\operatorname{ord}(g(x)) = \operatorname{ord}(f(x))$. The desired result concerning the impulse response sequence follows in a similar way. If $g(x)$ is constant, the theorem is trivial. □

We remark that for $f(0) \neq 0$ the least period of the impulse response sequence may also be obtained from the identity (8.9) in the following way. For the impulse response sequence with characteristic polynomial $f(x)$, the polynomial $h(x)$ in (8.10) is given by $h(x) = -1$. Therefore, if r is the least period of the impulse response sequence, then $f(x)$ divides $x^r - 1$ by (8.9), and so $r \geqslant \operatorname{ord}(f(x))$. On the other hand, r must divide $\operatorname{ord}(f(x))$ by the first part of Theorem 8.27, and so $r = \operatorname{ord}(f(x))$.

8.28. Theorem. *Let* $s_0, s_1, \ldots$ *be a homogeneous linear recurring sequence in* $\mathbb{F}_q$ *with nonzero initial state vector, and suppose the characteristic polynomial* $f(x) \in \mathbb{F}_q[x]$ *is irreducible over* $\mathbb{F}_q$ *and satisfies* $f(0) \neq 0$. *Then the sequence is periodic with least period equal to* $\operatorname{ord}(f(x))$.

Proof. The sequence is periodic and its least period r divides $\text{ord}(f(x))$ by Theorem 8.27. On the other hand, it follows from (8.9) that $f(x)$ divides $(x^r - 1)h(x)$. Since $s(x)$, and therefore $h(x)$, is a nonzero polynomial and since $\deg(h(x)) < \deg(f(x))$, the irreducibility of $f(x)$ implies that $f(x)$ divides $x^r - 1$, and so $r \geqslant \text{ord}(f(x))$. □

Now we present a different proof of Corollary 3.4, which we restate for convenience.

8.29. Theorem. *Let $f(x) \in \mathbb{F}_q[x]$ be irreducible over $\mathbb{F}_q$ with $\deg(f(x)) = k$. Then $\text{ord}(f(x))$ divides $q^k - 1$.*

Proof. We may assume without loss of generality that $f(0) \neq 0$ and that $f(x)$ is monic. We take a homogeneous linear recurring sequence in $\mathbb{F}_q$ that has $f(x)$ as its characteristic polynomial and has a nonzero initial state vector. According to Theorem 8.28, this sequence is periodic with least period $\text{ord}(f(x))$, so that altogether $\text{ord}(f(x))$ different state vectors appear in it. If $\text{ord}(f(x))$ is less than $q^k - 1$, the total number of nonzero k-tuples of elements of $\mathbb{F}_q$, we can choose such a k-tuple that does not appear as a state vector in the sequence above and use it as an initial state vector for another homogeneous linear recurring sequence in $\mathbb{F}_q$ with characteristic polynomial $f(x)$. None of the $\text{ord}(f(x))$ different state vectors of the second sequence is equal to a state vector of the first sequence, for otherwise the two sequences would be identical from some points onwards and the initial state vector of the second sequence would eventually appear as a state vector in the first sequence—a contradiction. By continuing to generate linear recurring sequences of the type above, we arrive at a partition of the set of $q^k - 1$ nonzero k-tuples of elements of $\mathbb{F}_q$ into subsets of cardinality $\text{ord}(f(x))$, and the conclusion of the theorem follows. □

8.30. Example. Consider the linear recurrence relation $s_{n+6} = s_{n+4} + s_{n+2} + s_{n+1} + s_n$, $n = 0, 1, \ldots$, in $\mathbb{F}_2$. The corresponding characteristic polynomial is $f(x) = x^6 - x^4 - x^2 - x - 1 \in \mathbb{F}_2[x]$. The polynomial $f(x)$ is irreducible over $\mathbb{F}_2$. Furthermore, $f(x)$ divides $x^{21} - 1$ and no polynomial $x^e - 1$ with $0 < e < 21$, so that $\text{ord}(f(x)) = 21$. The impulse response sequence corresponding to the linear recurrence relation is given by the string of binary digits

$$000001010010011001011000001 \cdots$$

of least period 21, as it should be. If $(0, 0, 0, 0, 1, 1)$ is taken as the initial state vector, we arrive at the string of binary digits

$$000011110110101011101000011 \cdots$$

of least period 21, and if $(0, 0, 0, 1, 0, 0)$ is taken as the initial state vector, we obtain the string of binary digits

$$000100011011111100111000100\cdots$$

of least period 21. Each one of the nonzero sextuples of elements of $\mathbb{F}_2$ appears as a state vector in exactly one of the three sequences. Any other nonzero initial state vector will produce a shifted version of one of the three sequences, which is again a sequence of least period 21. □

8.31. Example. If $f(x) \in \mathbb{F}_q[x]$ with $\deg(f(x)) = k$ is reducible, then $\operatorname{ord}(f(x))$ need not divide $q^k - 1$. Consider $f(x) = x^5 + x + 1 \in \mathbb{F}_2[x]$. Then $f(x)$ is reducible since

$$x^5 + x + 1 = (x^3 + x^2 + 1)(x^2 + x + 1).$$

It follows, for instance, from Theorem 8.27 and Example 8.14 that $\operatorname{ord}(f(x)) = 21$, and this is not a divisor of $2^5 - 1 = 31$. □

Linear recurring sequences whose least periods are very large are of particular importance in applications. We know from Theorem 8.7 that for a kth-order homogeneous linear recurring sequence in $\mathbb{F}_q$ the least period can be at most $q^k - 1$. In order to generate such sequences for which the least period is actually equal to $q^k - 1$, we have to use the notion of a primitive polynomial (see Definition 3.15).

8.32. Definition. A homogeneous linear recurring sequence in $\mathbb{F}_q$ whose characteristic polynomial is a primitive polynomial over $\mathbb{F}_q$ and which has a nonzero initial state vector is called a *maximal period sequence* in $\mathbb{F}_q$.

8.33. ***Theorem.*** *Every kth-order maximal period sequence in* $\mathbb{F}_q$ *is periodic and its least period is equal to the largest possible value for the least period of any kth-order homogeneous linear recurring sequence in* $\mathbb{F}_q$ *— namely,* $q^k - 1$.

Proof. The fact that the sequence is periodic and that the least period is $q^k - 1$ is a consequence of Theorem 8.28 and Theorem 3.16. The remaining assertion follows from Theorem 8.7. □

8.34. Example. The linear recurrence relation $s_{n+7} = s_{n+4} + s_{n+3} + s_{n+2} + s_n$, $n = 0, 1, \ldots$, in $\mathbb{F}_2$ considered in Example 8.2 has the polynomial $f(x) = x^7 - x^4 - x^3 - x^2 - 1 \in \mathbb{F}_2[x]$ as its characteristic polynomial. Since $f(x)$ is a primitive polynomial over $\mathbb{F}_2$, any sequence with nonzero initial state vector arising from this linear recurrence relation is a maximal period sequence in $\mathbb{F}_2$. If we choose one particular nonzero initial state vector, then the resulting sequence $s_0, s_1, \ldots$ has least period $2^7 - 1 = 127$ according to Theorem 8.33. Therefore, all possible nonzero vectors of $\mathbb{F}_2^7$ appear as state vectors in this sequence. Any other maximal period sequence arising from the given linear recurrence relation is just a shifted version of the sequence $s_0, s_1, \ldots$. □

3. GENERATING FUNCTIONS

So far, our approach to linear recurring sequences has employed only linear algebra, polynomial algebra, and the theory of finite fields. By using the algebraic apparatus of formal power series, other remarkable facts about linear recurring sequences can be established.

Given an arbitrary sequence $s_0, s_1, \ldots$ of elements of $\mathbb{F}_q$, we associate with it its *generating function*, which is a purely formal expression of the type

$$G(x) = s_0 + s_1 x + s_2 x^2 + \cdots + s_n x^n + \cdots = \sum_{n=0}^{\infty} s_n x^n \qquad (8.13)$$

with an indeterminate x. The underlying idea is that in $G(x)$ we have "stored" all the terms of the sequence in the correct order, so that $G(x)$ should somehow reflect the properties of the sequence. The name "generating function" is, strictly speaking, a misnomer since we do not consider $G(x)$ in any way as a function, but just as a formal object (in an obvious analogy, polynomials are essentially formal objects not to be confused with functions). The term is carried over from the case of real or complex sequences, where it may often turn out that the series analogous to the one in (8.13) is convergent after substitution of a real or complex number x_0 for x, thus enabling us to attach a meaning to $G(x_0)$. In our present situation, the question of the convergence or divergence of the expression in (8.13) is moot, since we think of $G(x)$ as being nothing but a hieroglyph for the sequence $s_0, s_1, \ldots$.

In general, an object of the type

$$B(x) = b_0 + b_1 x + b_2 x^2 + \cdots + b_n x^n + \cdots = \sum_{n=0}^{\infty} b_n x^n,$$

with $b_0, b_1, \ldots$ being a sequence of elements of $\mathbb{F}_q$, is called a *formal power series* (over $\mathbb{F}_q$). In this context, the terms $b_0, b_1, \ldots$ of the sequence are also called the *coefficients* of the formal power series. The adjective "formal" refers again to the idea that the convergence or divergence (whatever that may mean) of these expressions is irrelevant for their study. Two such formal power series

$$B(x) = \sum_{n=0}^{\infty} b_n x^n \quad \text{and} \quad C(x) = \sum_{n=0}^{\infty} c_n x^n$$

over $\mathbb{F}_q$ are considered identical if $b_n = c_n$ for all $n = 0, 1, \ldots$. The set of all formal power series over $\mathbb{F}_q$ is then in an obvious one-to-one correspondence with the set of all sequences of elements of $\mathbb{F}_q$. Thus, it seems as if we have not gained anything from the transition to formal power series (save a conceptual complication). The *raison d'être* of these objects is the fact that we can endow the set of all formal power series over $\mathbb{F}_q$ with a rich and interesting algebraic structure in a fairly natural way. This will be discussed in the sequel.

We note first that we may think of a polynomial

$$p(x) = p_0 + p_1 x + \cdots + p_k x^k \in \mathbb{F}_q[x]$$

as a formal power series over $\mathbb{F}_q$ by identifying it with

$$P(x) = p_0 + p_1 x + \cdots + p_k x^k + 0 \cdot x^{k+1} + 0 \cdot x^{k+2} + \cdots.$$

We introduce now the algebraic operations of addition and multiplication for formal power series in such a way that they extend the corresponding operations for polynomials. In detail, if

$$B(x) = \sum_{n=0}^{\infty} b_n x^n \quad \text{and} \quad C(x) = \sum_{n=0}^{\infty} c_n x^n$$

are two formal power series over $\mathbb{F}_q$, we define their *sum* to be the formal power series

$$B(x) + C(x) = \sum_{n=0}^{\infty} (b_n + c_n) x^n$$

and their *product* to be the formal power series

$$B(x)C(x) = \sum_{n=0}^{\infty} d_n x^n, \quad \text{where } d_n = \sum_{k=0}^{n} b_k c_{n-k} \quad \text{for } n = 0, 1, \ldots.$$

If $B(x)$ and $C(x)$ are both polynomials over $\mathbb{F}_q$, then the operations above obviously coincide with polynomial addition and multiplication, respectively. It should be observed at this point that the substitution principle, which is so useful in polynomial algebra, is not valid for formal power series, the simple reason being that the expression $B(a)$ with $a \in \mathbb{F}_q$ and $B(x)$ a formal power series over $\mathbb{F}_q$ may be meaningless. This is, of course, the price we have to pay for disregarding convergence questions.

8.35. Example. Let

$$B(x) = 2 + x^2$$

and

$$C(x) = 1 + x + x^2 + \cdots + x^n + \cdots = \sum_{n=0}^{\infty} 1 \cdot x^n$$

be formal power series over $\mathbb{F}_3$. Then

$$B(x) + C(x) = x + 2x^2 + x^3 + \cdots + x^n + \cdots = \sum_{n=0}^{\infty} d_n x^n$$

with $d_0 = 0$, $d_1 = 1$, $d_2 = 2$, and $d_n = 1$ for $n \geqslant 3$, and

$$B(x)C(x) = 2 + 2x + 0 \cdot x^2 + 0 \cdot x^3 + \cdots = 2 + 2x. \qquad \square$$

Addition of formal power series over $\mathbb{F}_q$ is clearly associative and commutative. The formal power series $0 = \sum_{n=0}^{\infty} 0 \cdot x^n$ serves as an identity element for addition, and if $B(x) = \sum_{n=0}^{\infty} b_n x^n$ is an arbitrary formal power series over $\mathbb{F}_q$, then it has the additive inverse $\sum_{n=0}^{\infty}(-b_n)x^n$, denoted by $-B(x)$. As usual, we shall write $B(x) - C(x)$ instead of $B(x) + (-C(x))$.

Evidently, multiplication of formal power series over $\mathbb{F}_q$ is commutative, and the formal power series $1 = 1 + 0 \cdot x + 0 \cdot x^2 + \cdots + 0 \cdot x^n + \cdots$ acts as a multiplicative identity. Multiplication is associative, for if

$$B(x) = \sum_{n=0}^{\infty} b_n x^n, \quad C(x) = \sum_{n=0}^{\infty} c_n x^n, \quad \text{and} \quad D(x) = \sum_{n=0}^{\infty} d_n x^n,$$

then $(B(x)C(x))D(x)$ and $B(x)(C(x)D(x))$ are both identical with

$$\sum_{n=0}^{\infty} \left(\sum_{(i,j,k) \in L(n)} b_i c_j d_k \right) x^n,$$

where $L(n)$ is the set of all ordered triples (i, j, k) of nonnegative integers with $i + j + k = n$. Furthermore, the distributive law is satisfied since

$$\begin{aligned}
B(x)(C(x) + D(x)) &= \sum_{n=0}^{\infty} \left(\sum_{k=0}^{n} b_k (c_{n-k} + d_{n-k}) \right) x^n \\
&= \sum_{n=0}^{\infty} \left(\sum_{k=0}^{n} b_k c_{n-k} + \sum_{k=0}^{n} b_k d_{n-k} \right) x^n \\
&= \sum_{n=0}^{\infty} \left(\sum_{k=0}^{n} b_k c_{n-k} \right) x^n + \sum_{n=0}^{\infty} \left(\sum_{k=0}^{n} b_k d_{n-k} \right) x^n \\
&= B(x)C(x) + B(x)D(x).
\end{aligned}$$

Altogether, we have shown that the set of all formal power series over $\mathbb{F}_q$, furnished with this addition and multiplication, is a commutative ring with identity, called the *ring of formal power series* over $\mathbb{F}_q$ and denoted by $\mathbb{F}_q[[x]]$. The polynomial ring $\mathbb{F}_q[x]$ is contained as a subring in $\mathbb{F}_q[[x]]$. We collect and extend the information on $\mathbb{F}_q[[x]]$ in the following theorem.

8.36. Theorem. *The ring $\mathbb{F}_q[[x]]$ of formal power series over $\mathbb{F}_q$ is an integral domain containing $\mathbb{F}_q[x]$ as a subring.*

Proof. It remains to verify that $\mathbb{F}_q[[x]]$ has no zero divisors—that is, that a product in $\mathbb{F}_q[[x]]$ can only be zero if one of the factors is zero. Suppose, on the contrary, that we have $B(x)C(x) = 0$ with

$$B(x) = \sum_{n=0}^{\infty} b_n x^n \neq 0 \quad \text{and} \quad C(x) = \sum_{n=0}^{\infty} c_n x^n \neq 0 \quad \text{in } \mathbb{F}_q[[x]].$$

Let k be the least nonnegative integer for which $b_k \neq 0$, and let m be the least nonnegative integer for which $c_m \neq 0$. Then the coefficient of x^{k+m} in $B(x)C(x)$ is $b_k c_m \neq 0$, which contradicts $B(x)C(x)=0$. □

It will be important for the applications to linear recurring sequences to find those $B(x) \in \mathbb{F}_q[[x]]$ that possess a multiplicative inverse—that is, for which there exists a $C(x) \in \mathbb{F}_q[[x]]$ with $B(x)C(x)=1$. These formal power series can, in fact, be characterized easily.

8.37. ***Theorem.*** *The formal power series*

$$B(x) = \sum_{n=0}^{\infty} b_n x^n \in \mathbb{F}_q[[x]]$$

has a multiplicative inverse if and only if $b_0 \neq 0$.

Proof. If

$$C(x) = \sum_{n=0}^{\infty} c_n x^n \in \mathbb{F}_q[[x]]$$

is such that $B(x)C(x)=1$, then the following infinite system of equations must be satisfied:

$$\begin{aligned} b_0c_0 &= 1 \\ b_0c_1 + b_1c_0 &= 0 \\ b_0c_2 + b_1c_1 + b_2c_0 &= 0 \\ &\vdots \\ b_0c_n + b_1c_{n-1} + \cdots + b_nc_0 &= 0 \\ &\vdots \end{aligned}$$

From the first equation we conclude that necessarily $b_0 \neq 0$. However, if this condition is satisfied, then c_0 is uniquely determined by the first equation. Passing to the second equation, we see that c_1 is then uniquely determined. In general, the coefficients $c_0, c_1, \ldots$ can be computed recursively from the first equation and the recurrence relation

$$c_n = -b_0^{-1} \sum_{k=1}^{n} b_k c_{n-k} \quad \text{for } n = 1, 2, \ldots.$$

The resulting formal power series $C(x)$ is then a multiplicative inverse of $B(x)$. □

If a multiplicative inverse of $B(x) \in \mathbb{F}_q[[x]]$ exists, then it is, of course, uniquely determined. We use the notation $1/B(x)$ for it. A product $A(x)(1/B(x))$ with $A(x) \in \mathbb{F}_q[[x]]$ will usually be written in the form $A(x)/B(x)$. Since $\mathbb{F}_q[[x]]$ is an integral domain, the familiar rules for operating with fractions hold. The multiplicative inverse of $B(x)$ or an expression $A(x)/B(x)$ can be computed by the algorithm in the proof of

Theorem 8.37. Long division also provides an effective means for accomplishing such computations.

8.38. Example. Let $B(x)=3+x+x^2$, considered as a formal power series over $\mathbb{F}_5$. Then $B(x)$ has a multiplicative inverse by Theorem 8.37. We compute $1/B(x)$ by long division:

$$\begin{array}{r|l}
 & 2+x \quad +4x^2 \quad +2x^4 \quad + \cdots \\
3+x+x^2 & 1+0\cdot x + 0\cdot x^2 + 0\cdot x^3+0\cdot x^4+0\cdot x^5+0\cdot x^6+\cdots \\
 & -1-2x - 2x^2 \\ \hline
 & \quad 3x + 3x^2 + 0\cdot x^3 \\
 & \quad -3x - x^2 - x^3 \\ \hline
 & \qquad 2x^2 + 4x^3+0\cdot x^4 \\
 & \qquad -2x^2 - 4x^3 - 4x^4 \\ \hline
 & \qquad\qquad x^4+0\cdot x^5+0\cdot x^6 \\
 & \qquad\qquad \vdots
\end{array}$$

Thus we get

$$\frac{1}{3+x+x^2}=2+x+4x^2+2x^4+\cdots. \qquad \square$$

8.39. Example. We compute $A(x)/B(x)$ in $\mathbb{F}_2[[x]]$, where

$$A(x)=1+x+x^2+x^3+\cdots=\sum_{n=0}^{\infty} 1\cdot x^n$$

and $B(x)=1+x+x^3$. Using long division, dropping the terms with zero coefficients, and recalling that $1=-1$ in $\mathbb{F}_2$, we get:

$$\begin{array}{r|l}
 & 1+x^2+x^3+x^7+\cdots \\
1+x+x^3 & 1+x+x^2+x^3+x^4+x^5+x^6+x^7+x^8+x^9+x^{10}+\cdots \\
 & 1+x \quad +x^3 \\ \hline
 & \quad x^2 \quad +x^4 +x^5 \\
 & \quad x^2+x^3 \quad +x^5 \\ \hline
 & \qquad x^3+x^4 \quad +x^6 \\
 & \qquad x^3+x^4 \quad +x^6 \\ \hline
 & \qquad\qquad x^7+x^8+x^9+x^{10} \\
 & \qquad\qquad \vdots
\end{array}$$

Therefore,

$$\frac{1+x+x^2+x^3+\cdots}{1+x+x^3}=1+x^2+x^3+x^7+\cdots. \qquad \square$$

In order to apply the theory of formal power series, we consider now a kth-order homogeneous linear recurring sequence $s_0, s_1, \ldots$ in $\mathbb{F}_q$ satisfying the linear recurrence relation (8.7) and define its *reciprocal characteristic polynomial* to be

$$f^*(x) = 1 - a_{k-1}x - a_{k-2}x^2 - \cdots - a_0 x^k \in \mathbb{F}_q[x]. \quad (8.14)$$

The characteristic polynomial $f(x)$ and the reciprocal characteristic polynomial are related by $f^*(x) = x^k f(1/x)$. The following basic identity can then be shown for the generating function of the given sequence.

8.40. Theorem. *Let $s_0, s_1, \ldots$ be a kth-order homogeneous linear recurring sequence in $\mathbb{F}_q$ satisfying the linear recurrence relation* (8.7), *let $f^*(x) \in \mathbb{F}_q[x]$ be its reciprocal characteristic polynomial, and let $G(x) \in \mathbb{F}_q[[x]]$ be its generating function in* (8.13). *Then the identity*

$$G(x) = \frac{g(x)}{f^*(x)} \quad (8.15)$$

holds with

$$g(x) = -\sum_{j=0}^{k-1} \sum_{i=0}^{j} a_{i+k-j} s_i x^j \in \mathbb{F}_q[x], \quad (8.16)$$

where we set $a_k = -1$. Conversely, if $g(x)$ is any polynomial over $\mathbb{F}_q$ with $\deg(g(x)) < k$ *and if $f^*(x) \in \mathbb{F}_q[x]$ is given by* (8.14), *then the formal power series $G(x) \in \mathbb{F}_q[[x]]$ defined by* (8.15) *is the generating function of a kth-order homogeneous linear recurring sequence in $\mathbb{F}_q$ satisfying the linear recurrence relation* (8.7).

Proof. We have

$$f^*(x)G(x) = -\left(\sum_{n=0}^{k} a_{k-n}x^n\right)\left(\sum_{n=0}^{\infty} s_n x^n\right)$$

$$= -\sum_{j=0}^{k-1}\left(\sum_{i=0}^{j} a_{i+k-j}s_i\right)x^j - \sum_{j=k}^{\infty}\left(\sum_{i=j-k}^{j} a_{i+k-j}s_i\right)x^j$$

$$= g(x) - \sum_{j=k}^{\infty}\left(\sum_{i=0}^{k} a_i s_{j-k+i}\right)x^j. \quad (8.17)$$

Thus, if the sequence $s_0, s_1, \ldots$ satisfies (8.7), then $f^*(x)G(x) = g(x)$ because of (8.12). Since $f^*(x)$ has a multiplicative inverse in $\mathbb{F}_q[[x]]$ by Theorem 8.37, the identity (8.15) follows. Conversely, we infer from (8.17) that $f^*(x)G(x)$ is equal to a polynomial of degree less than k only if

$$\sum_{i=0}^{k} a_i s_{j-k+i} = 0 \quad \text{for all } j \geqslant k.$$

But these identities just express the fact that the sequence $s_0, s_1, \ldots$ of coefficients of $G(x)$ satisfies the linear recurrence relation (8.7). □

One may summarize the theorem above by saying that the kth-order homogeneous linear recurring sequences with reciprocal characteristic polynomial $f^*(x)$ are in one-to-one correspondence with the fractions $g(x)/f^*(x)$ with $\deg(g(x)) < k$. The identity (8.15) can be used to compute the terms of a linear recurring sequence by long division.

8.41. Example. Consider the linear recurrence relation

$$s_{n+4} = s_{n+3} + s_{n+1} + s_n, \quad n = 0, 1, \ldots, \text{ in } \mathbb{F}_2.$$

Its reciprocal characteristic polynomial is

$$f^*(x) = 1 - x - x^3 - x^4 = 1 + x + x^3 + x^4 \in \mathbb{F}_2[x].$$

If the initial state vector is $(1, 1, 0, 1)$, then the polynomial $g(x)$ in (8.16) turns out to be $g(x) = 1 + x^2$. Therefore, the generating function $G(x)$ of the sequence can be obtained from the following long division:

$$\begin{array}{r|l}
 & 1 + x + x^3 + x^4 + x^6 + \cdots \\
\hline
1 + x + x^3 + x^4 & 1 \quad + x^2 \\
 & 1 + x \quad + x^3 + x^4 \\
\cline{2-2}
 & x + x^2 + x^3 + x^4 \\
 & x + x^2 \quad + x^4 + x^5 \\
\cline{2-2}
 & x^3 \quad + x^5 \\
 & x^3 + x^4 \quad + x^6 + x^7 \\
\cline{2-2}
 & x^4 + x^5 + x^6 + x^7 \\
 & x^4 + x^5 \quad + x^7 + x^8 \\
\cline{2-2}
 & x^6 \quad + x^8 \\
 & \quad \ddots
\end{array}$$

The result is

$$G(x) = \frac{1 + x^2}{1 + x + x^3 + x^4} = 1 + x + x^3 + x^4 + x^6 + \cdots,$$

which corresponds to the string of binary digits $1101101\cdots$ of least period 3. The impulse response sequence associated with the given linear recurrence relation can be obtained by observing that $g(x) = x^3$ in this case, so that an appropriate long division yields

$$G(x) = \frac{x^3}{1 + x + x^3 + x^4} = x^3 + x^4 + x^5 + x^9 + x^{10} + x^{11} + \cdots,$$

which corresponds to the string of binary digits $000111000111\cdots$ of least period 6. □

On the basis of the identity (8.15), we present now an *alternative proof* of Theorem 8.25. Since the sequence $s_0, s_1, \ldots$ is periodic with period r, its generating function $G(x)$ can be written in the form

$$G(x) = \left(s_0 + s_1x + \cdots + s_{r-1}x^{r-1}\right)\left(1 + x^r + x^{2r} + \cdots\right) = \frac{s^*(x)}{1 - x^r}$$

with $s^*(x) = s_0 + s_1x + \cdots + s_{r-1}x^{r-1}$. On the other hand, using the notation of Theorem 8.40 we have $G(x) = g(x)/f^*(x)$ by (8.15). By equating these expressions for $G(x)$, we arrive at the polynomial identity $f^*(x)s^*(x) = (1 - x^r)g(x)$. If $f(x)$ and $s(x)$ are as in (8.9), then

$$f(x)s(x) = x^k f^*\left(\frac{1}{x}\right)x^{r-1}s^*\left(\frac{1}{x}\right) = (x^r - 1)x^{k-1}g\left(\frac{1}{x}\right),$$

and a comparison of (8.10) and (8.16) shows that

$$x^{k-1}g\left(\frac{1}{x}\right) = -h(x), \tag{8.18}$$

which implies already (8.9).

4. THE MINIMAL POLYNOMIAL

Although we have not yet pointed it out, it is evident that a linear recurring sequence satisfies many other linear recurrence relations apart from the one by which it is defined. For instance, if the sequence $s_0, s_1, \ldots$ is periodic with period r, it satisfies the linear recurrence relations $s_{n+r} = s_n$ $(n = 0, 1, \ldots)$, $s_{n+2r} = s_n$ $(n = 0, 1, \ldots)$, and so on. The most extreme case is represented by the sequence $0, 0, 0, \ldots$, which satisfies any homogeneous linear recurrence relation. The following theorem describes the relationship between the various linear recurrence relations valid for a given homogeneous linear recurring sequence.

8.42. Theorem. *Let $s_0, s_1, \ldots$ be a homogeneous linear recurring sequence in $\mathbb{F}_q$. Then there exists a uniquely determined monic polynomial $m(x) \in \mathbb{F}_q[x]$ having the following property: a monic polynomial $f(x) \in \mathbb{F}_q[x]$ of positive degree is a characteristic polynomial of $s_0, s_1, \ldots$ if and only if $m(x)$ divides $f(x)$.*

Proof. Let $f_0(x) \in \mathbb{F}_q[x]$ be the characteristic polynomial of a homogeneous linear recurrence relation satisfied by the sequence, and let $h_0(x) \in \mathbb{F}_q[x]$ be the polynomial in (8.10) determined by $f_0(x)$ and the sequence. If $d(x)$ is the (monic) greatest common divisor of $f_0(x)$ and $h_0(x)$, then we can write $f_0(x) = m(x)d(x)$ and $h_0(x) = b(x)d(x)$ with $m(x), b(x) \in \mathbb{F}_q[x]$. We shall prove that $m(x)$ is the desired polynomial. Clearly, $m(x)$ is monic. Now let $f(x) \in \mathbb{F}_q[x]$ be an arbitrary characteristic polynomial of the given sequence, and let $h(x) \in \mathbb{F}_q[x]$ be the polynomial in (8.10) determined by $f(x)$ and the sequence. By applying Theorem 8.40, we obtain that the generating function $G(x)$ of the sequence satisfies

$$G(x) = \frac{g_0(x)}{f_0^*(x)} = \frac{g(x)}{f^*(x)}$$

with $g_0(x)$ and $g(x)$ determined by (8.16). Therefore $g(x)f_0^*(x) = g_0(x)f^*(x)$, and using (8.18) we arrive at

$$\begin{aligned} h(x)f_0(x) &= -x^{\deg(f(x))-1}g\left(\frac{1}{x}\right)x^{\deg(f_0(x))}f_0^*\left(\frac{1}{x}\right) \\ &= -x^{\deg(f_0(x))-1}g_0\left(\frac{1}{x}\right)x^{\deg(f(x))}f^*\left(\frac{1}{x}\right) = h_0(x)f(x). \end{aligned}$$

After division by $d(x)$ we have $h(x)m(x) = b(x)f(x)$, and since $m(x)$ and $b(x)$ are relatively prime, it follows that $m(x)$ divides $f(x)$.

Now suppose that $f(x) \in \mathbb{F}_q[x]$ is a monic polynomial of positive degree that is divisible by $m(x)$, say $f(x) = m(x)c(x)$ with $c(x) \in \mathbb{F}_q[x]$. Passing to reciprocal polynomials, we get $f^*(x) = m^*(x)c^*(x)$ in an obvious notation. We also have $h_0(x)m(x) = b(x)f_0(x)$, so that, using the relation (8.18), we obtain

$$\begin{aligned} g_0(x)m^*(x) &= -x^{\deg(f_0(x))-1}h_0\left(\frac{1}{x}\right)x^{\deg(m(x))}m\left(\frac{1}{x}\right) \\ &= -x^{\deg(m(x))-1}b\left(\frac{1}{x}\right)x^{\deg(f_0(x))}f_0\left(\frac{1}{x}\right). \end{aligned}$$

Since $\deg(b(x)) < \deg(m(x))$, the product of the first two factors on the right-hand side (negative sign included) is a polynomial $a(x) \in \mathbb{F}_q[x]$. Therefore, we have $g_0(x)m^*(x) = a(x)f_0^*(x)$. It follows then from Theorem 8.40 that the generating function $G(x)$ of the sequence satisfies

$$G(x) = \frac{g_0(x)}{f_0^*(x)} = \frac{a(x)}{m^*(x)} = \frac{a(x)c^*(x)}{m^*(x)c^*(x)} = \frac{a(x)c^*(x)}{f^*(x)}.$$

Since

$$\begin{aligned} \deg(a(x)c^*(x)) &= \deg(a(x)) + \deg(c^*(x)) \\ &< \deg(m(x)) + \deg(c(x)) = \deg(f(x)), \end{aligned}$$

the second part of Theorem 8.40 shows that $f(x)$ is a characteristic polynomial of the sequence. It is clear that there can only be one polynomial $m(x)$ with the indicated properties. □

The uniquely determined polynomial $m(x)$ over $\mathbb{F}_q$ associated with the sequence $s_0, s_1, \ldots$ according to Theorem 8.42 is called the *minimal polynomial* of the sequence. If $s_n = 0$ for all $n \geqslant 0$, the minimal polynomial is equal to the constant polynomial 1. For all other homogeneous linear recurring sequences, $m(x)$ is a monic polynomial with $\deg(m(x)) > 0$ that is, in fact, the characteristic polynomial of the linear recurrence relation of least possible order satisfied by the sequence. Another method of calculating the minimal polynomial will be introduced in Section 6.

8.43. Example. Let $s_0, s_1, \ldots$ be the linear recurring sequence in $\mathbb{F}_2$ with

$$s_{n+4} = s_{n+3} + s_{n+1} + s_n, \quad n = 0, 1, \ldots,$$

and initial state vector $(1,1,0,1)$. To find the minimal polynomial, we proceed as in the proof of Theorem 8.42. We may take $f_0(x) = x^4 - x^3 - x - 1 = x^4 + x^3 + x + 1 \in \mathbb{F}_2[x]$. Then by (8.10) the polynomial $h_0(x)$ is given by $h_0(x) = x^3 + x$. The greatest common divisor of $f_0(x)$ and $h_0(x)$ is $d(x) = x^2 + 1$, and so the minimal polynomial of the sequence is $m(x) = f_0(x)/d(x) = x^2 + x + 1$. One checks easily that the sequence satisfies the linear recurrence relation

$$s_{n+2} = s_{n+1} + s_n, \quad n = 0, 1, \ldots,$$

as it should according to the general theory. We note that $\operatorname{ord}(m(x)) = 3$, which is identical with the least period of the sequence (compare with Example 8.41). We shall see in Theorem 8.44 below that this is true in general. □

The minimal polynomial plays a decisive role in the determination of the least period of a linear recurring sequence. This is shown by the following result.

8.44. Theorem. *Let $s_0, s_1, \ldots$ be a homogeneous linear recurring sequence in $\mathbb{F}_q$ with minimal polynomial $m(x) \in \mathbb{F}_q[x]$. Then the least period of the sequence is equal to* $\operatorname{ord}(m(x))$.

Proof. If r is the least period of the sequence and n_0 its preperiod, then we have $s_{n+r} = s_n$ for all $n \geqslant n_0$. Therefore, the sequence satisfies the homogeneous linear recurrence relation

$$s_{n+n_0+r} = s_{n+n_0} \quad \text{for } n = 0, 1, \ldots.$$

Then, according to Theorem 8.42, $m(x)$ divides $x^{n_0+r} - x^{n_0} = x^{n_0}(x^r - 1)$, so that $m(x)$ is of the form $m(x) = x^h g(x)$ with $h \leqslant n_0$ and $g(x) \in \mathbb{F}_q[x]$, where $g(0) \neq 0$ and $g(x)$ divides $x^r - 1$. It follows from the definition of the order of a polynomial that $\operatorname{ord}(m(x)) = \operatorname{ord}(g(x)) \leqslant r$. On the other hand, r divides $\operatorname{ord}(m(x))$ by Theorem 8.27, and so $r = \operatorname{ord}(m(x))$. □

8.45. Example. Let $s_0, s_1, \ldots$ be the linear recurring sequence in $\mathbb{F}_2$ with $s_{n+5} = s_{n+1} + s_n$, $n = 0, 1, \ldots$, and initial state vector $(1,1,1,0,1)$. Following the method in the proof of Theorem 8.42, we take $f_0(x) = x^5 - x - 1 = x^5 + x + 1 \in \mathbb{F}_2[x]$ and get $h_0(x) = x^4 + x^3 + x^2$ from (8.10). Then $d(x) = x^2 + x + 1$, and so the minimal polynomial $m(x)$ of the sequence is given by $m(x) = f_0(x)/d(x) = x^3 + x^2 + 1$. We have $\operatorname{ord}(m(x)) = 7$, and so Theorem

8.44 implies that the least period of the sequence is 7 (compare with Example 8.18). □

The argument in the example above shows how to find the least period of a linear recurring sequence without evaluating its terms. The method is particularly effective if a table of orders of polynomials is available. Since such tables usually incorporate only irreducible polynomials (see Chapter 10, Section 2), the results in Theorems 3.8 and 3.9 may have to be used to find the order of a given polynomial (compare with Example 3.10).

8.46. Example. The method in Example 8.45 can also be applied to inhomogeneous linear recurring sequences. Let $s_0, s_1, \ldots$ be such a sequence in $\mathbb{F}_2$ with

$$s_{n+4} = s_{n+3} + s_{n+1} + s_n + 1 \quad \text{for } n = 0, 1, \ldots$$

and initial state vector $(1,1,0,1)$. According to (8.5), the sequence is also given by the homogeneous linear recurrence relation $s_{n+5} = s_{n+3} + s_{n+2} + s_n$, $n = 0, 1, \ldots$, with initial state vector $(1,1,0,1,0)$. Proceeding as in Example 8.45, we find that the characteristic polynomial

$$f(x) = x^5 + x^3 + x^2 + 1 = (x+1)^3(x^2 + x + 1) \in \mathbb{F}_2[x]$$

is in the present case identical with the minimal polynomial $m(x)$ of the sequence. Since $\operatorname{ord}((x+1)^3) = 4$ by Theorem 3.8 and $\operatorname{ord}(x^2 + x + 1) = 3$, it follows from Theorem 3.9 that $\operatorname{ord}(m(x)) = 12$. Therefore, the sequence $s_0, s_1, \ldots$ is periodic with least period 12. □

8.47. Example. Consider the linear recurring sequence $s_0, s_1, \ldots$ in $\mathbb{F}_2$ with

$$s_{n+4} = s_{n+2} + s_{n+1} \quad \text{for } n = 0, 1, \ldots$$

and initial state vector $(1,0,1,0)$. Then

$$f(x) = x^4 + x^2 + x = x(x^3 + x + 1) \in \mathbb{F}_2[x]$$

is a characteristic polynomial of the sequence, and since neither x nor $x^3 + x + 1$ is a characteristic polynomial, we have $m(x) = x^4 + x^2 + x$. The sequence is not periodic, but ultimately periodic with least period $\operatorname{ord}(m(x)) = 7$. □

8.48. Theorem. *Let $s_0, s_1, \ldots$ be a homogeneous linear recurring sequence in $\mathbb{F}_q$ and let b be a positive integer. Then the minimal polynomial $m_1(x)$ of the shifted sequence $s_b, s_{b+1}, \ldots$ divides the minimal polynomial $m(x)$ of the original sequence. If $s_0, s_1, \ldots$ is periodic, then $m_1(x) = m(x)$.*

Proof. To prove the first assertion, it suffices to show because of Theorem 8.42 that every homogeneous linear recurrence relation satisfied by the original sequence is also satisfied by the shifted sequence. But this is immediately evident. For the second part, let

$$s_{n+b+k} = a_{k-1}s_{n+b+k-1} + \cdots + a_0 s_{n+b}, \quad n = 0, 1, \ldots,$$

be a homogeneous linear recurrence relation satisfied by the shifted sequence. Let r be a period of $s_0, s_1, \ldots,$ so that $s_{n+r} = s_n$ for all $n \geqslant 0$, and choose an integer c with $cr \geqslant b$. Then, by using the linear recurrence relation with n replaced by $n + cr - b$ and invoking the periodicity property, we find that

$$s_{n+k} = a_{k-1}s_{n+k-1} + \cdots + a_0 s_n \quad \text{for all } n \geqslant 0,$$

that is, that the sequence $s_0, s_1, \ldots$ satisfies the same linear recurrence relation as the shifted sequence. By applying again Theorem 8.42, we conclude that $m_1(x) = m(x)$. □

8.49. Example. Let $s_0, s_1, \ldots$ be the linear recurring sequence in $\mathbb{F}_2$ considered in Example 8.47. Its minimal polynomial is $x^4 + x^2 + x$, whereas the minimal polynomial of the shifted sequence $s_1, s_2, \ldots$ is $x^3 + x + 1$, which is a proper divisor of $x^4 + x^2 + x$. This example shows that the second assertion in Theorem 8.48 need not hold if $s_0, s_1, \ldots$ is only ultimately periodic, but not periodic. □

8.50. Theorem. *Let $f(x) \in \mathbb{F}_q[x]$ be monic and irreducible over $\mathbb{F}_q$, and let $s_0, s_1, \ldots$ be a homogeneous linear recurring sequence in $\mathbb{F}_q$ not all of whose terms are* 0. *If the sequence has $f(x)$ as a characteristic polynomial, then the minimal polynomial of the sequence is equal to $f(x)$.*

Proof. Since the minimal polynomial $m(x)$ of the sequence divides $f(x)$ according to Theorem 8.42, the irreducibility of $f(x)$ implies that either $m(x) = 1$ or $m(x) = f(x)$. But $m(x) = 1$ holds only for the sequence all of whose terms are 0, and so the result follows. □

There is a general criterion for deciding whether the characteristic polynomial of the linear recurrence relation defining a given linear recurring sequence is already the minimal polynomial of the sequence.

8.51. Theorem. *Let $s_0, s_1, \ldots$ be a sequence in $\mathbb{F}_q$ satisfying a kth-order homogeneous linear recurrence relation with characteristic polynomial $f(x) \in \mathbb{F}_q[x]$. Then $f(x)$ is the minimal polynomial of the sequence if and only if the state vectors $\mathbf{s}_0, \mathbf{s}_1, \ldots, \mathbf{s}_{k-1}$ are linearly independent over $\mathbb{F}_q$.*

Proof. Suppose $f(x)$ is the minimal polynomial of the sequence. If $\mathbf{s}_0, \mathbf{s}_1, \ldots, \mathbf{s}_{k-1}$ were linearly dependent over $\mathbb{F}_q$, we would have $b_0\mathbf{s}_0 + b_1\mathbf{s}_1 + \cdots + b_{k-1}\mathbf{s}_{k-1} = \mathbf{0}$ with coefficients $b_0, b_1, \ldots, b_{k-1} \in \mathbb{F}_q$ not all of which are zero. Multiplying from the right by powers of the matrix A in (8.3) associated with the given linear recurrence relation yields

$$b_0\mathbf{s}_n + b_1\mathbf{s}_{n+1} + \cdots + b_{k-1}\mathbf{s}_{n+k-1} = \mathbf{0} \quad \text{for } n = 0, 1, \ldots,$$

because of (8.4). In particular, we obtain

$$b_0 s_n + b_1 s_{n+1} + \cdots + b_{k-1} s_{n+k-1} = 0 \text{ for } n = 0, 1, \ldots .$$

If $b_j = 0$ for $1 \leqslant j \leqslant k-1$, it follows that $s_n = 0$ for all $n \geqslant 0$, a contradiction to the fact that the minimal polynomial $f(x)$ of the sequence has positive degree. In the remaining case, let $j \geqslant 1$ be the largest index with $b_j \neq 0$. Then it follows that the sequence $s_0, s_1, \ldots$ satisfies a jth-order homogeneous linear recurrence relation with $j < k$, which again contradicts the assumption that $f(x)$ is the minimal polynomial. Therefore we have shown that $\mathbf{s}_0, \mathbf{s}_1, \ldots, \mathbf{s}_{k-1}$ are linearly independent over $\mathbb{F}_q$.

Conversely, suppose that $\mathbf{s}_0, \mathbf{s}_1, \ldots, \mathbf{s}_{k-1}$ are linearly independent over $\mathbb{F}_q$. Since $\mathbf{s}_0 \neq \mathbf{0}$, the minimal polynomial has positive degree. If $f(x)$ were not the minimal polynomial, the sequence $s_0, s_1, \ldots$ would satisfy an mth-order homogeneous linear recurrence relation with $1 \leqslant m < k$, say

$$s_{n+m} = a_{m-1} s_{n+m-1} + \cdots + a_0 s_n \quad \text{for } n = 0, 1, \ldots$$

with coefficients from $\mathbb{F}_q$. But this would imply $\mathbf{s}_m = a_{m-1}\mathbf{s}_{m-1} + \cdots + a_0 \mathbf{s}_0$, a contradiction to the given linear independence property. □

8.52. Corollary. *If $s_0, s_1, \ldots$ is an impulse response sequence for some homogeneous linear recurrence relation in $\mathbb{F}_q$, then its minimal polynomial is equal to the characteristic polynomial of that linear recurrence relation.*

Proof. This follows from Theorem 8.51 since the required linear independence property is obviously satisfied for an impulse response sequence. □

5. FAMILIES OF LINEAR RECURRING SEQUENCES

Let $f(x) \in \mathbb{F}_q[x]$ be a monic polynomial of positive degree. We denote the set of all homogeneous linear recurring sequences in $\mathbb{F}_q$ with characteristic polynomial $f(x)$ by $S(f(x))$. In other words, $S(f(x))$ consists of all sequences in $\mathbb{F}_q$ satisfying the homogeneous linear recurrence relation determined by $f(x)$. If $\deg(f(x)) = k$, then $S(f(x))$ contains exactly q^k sequences, corresponding to the q^k different choices for initial state vectors.

The set $S(f(x))$ may be considered as a vector space over $\mathbb{F}_q$ if operations for sequences are defined termwise. In detail, if σ is the sequence $s_0, s_1, \ldots$ and τ the sequence $t_0, t_1, \ldots$ in $\mathbb{F}_q$, then the sum $\sigma + \tau$ is taken to be the sequence $s_0 + t_0, s_1 + t_1, \ldots$. Furthermore, if $c \in \mathbb{F}_q$, then $c\sigma$ is defined as the sequence $cs_0, cs_1, \ldots$. It is seen immediately from the recurrence relation that $S(f(x))$ is closed under this addition and scalar multiplication. The required axioms are easily checked, and so $S(f(x))$ is indeed a vector space over $\mathbb{F}_q$. The role of the zero vector is played by the *zero sequence*, all of whose terms are 0. Since $S(f(x))$ has q^k elements, the dimension of the vector space is k. We obtain k linearly independent elements of $S(f(x))$ by choosing k linearly independent k-tuples $\mathbf{y}_1,\ldots,\mathbf{y}_k$ of elements of $\mathbb{F}_q$ and considering the sequences $\sigma_1,\ldots,\sigma_k$ belonging to $S(f(x))$, where each σ_j, $1 \leqslant j \leqslant k$, has $\mathbf{y}_j$ as its initial state vector. A natural choice for $\mathbf{y}_1,\ldots,\mathbf{y}_k$ is to take the standard basis vectors

$$\mathbf{e}_1 = (1,0,\ldots,0),\, \mathbf{e}_2 = (0,1,\ldots,0),\ldots,\mathbf{e}_k = (0,\ldots,0,1).$$

Another possibility that is often advantageous is to consider the impulse response sequence $d_0, d_1,\ldots$ belonging to $S(f(x))$ and to choose for $\mathbf{y}_1,\ldots,\mathbf{y}_k$ the state vectors $\mathbf{d}_0,\ldots,\mathbf{d}_{k-1}$ of this impulse response sequence.

In the following discussion, we shall explore the relationship between the various sets $S(f(x))$.

8.53. Theorem. *Let $f(x)$ and $g(x)$ be two nonconstant monic polynomials over $\mathbb{F}_q$. Then $S(f(x))$ is a subset of $S(g(x))$ if and only if $f(x)$ divides $g(x)$.*

Proof. Suppose $S(f(x))$ is contained in $S(g(x))$. Consider the impulse response sequence belonging to $S(f(x))$. This sequence has $f(x)$ as its minimal polynomial because of Corollary 8.52. By hypothesis, the sequence belongs also to $S(g(x))$. Therefore, according to Theorem 8.42, its minimal polynomial $f(x)$ divides $g(x)$. Conversely, if $f(x)$ divides $g(x)$ and $s_0, s_1,\ldots$ is any sequence belonging to $S(f(x))$, then the minimal polynomial $m(x)$ of the sequence divides $f(x)$ by Theorem 8.42. Consequently, $m(x)$ divides $g(x)$, and so another application of Theorem 8.42 shows that the sequence $s_0, s_1,\ldots$ belongs to $S(g(x))$. Therefore, $S(f(x))$ is a subset of $S(g(x))$. □

8.54. Theorem. *Let $f_1(x),\ldots,f_h(x)$ be nonconstant monic polynomials over $\mathbb{F}_q$. If $f_1(x),\ldots,f_h(x)$ are relatively prime, then the intersection*

$$S(f_1(x)) \cap \cdots \cap S(f_h(x))$$

consists only of the zero sequence. If $f_1(x),\ldots,f_h(x)$ have a (monic) greatest common divisor $d(x)$ of positive degree, then

$$S(f_1(x)) \cap \cdots \cap S(f_h(x)) = S(d(x)).$$

Proof. The minimal polynomial $m(x)$ of a sequence in the intersection must divide $f_1(x),\ldots,f_h(x)$. In the case of relative primality, $m(x)$ is necessarily the constant polynomial 1; but only the zero sequence has this minimal polynomial. In the second case, we conclude that $m(x)$ divides $d(x)$, and then Theorem 8.42 implies that $S(f_1(x)) \cap \cdots \cap S(f_h(x))$ is contained in $S(d(x))$. The fact that $S(d(x))$ is a subset of $S(f_1(x)) \cap \cdots \cap S(f_h(x))$ follows immediately from Theorem 8.53. □

We define $S(f(x))+S(g(x))$ to be the set of all sequences $\sigma+\tau$ with $\sigma\in S(f(x))$ and $\tau\in S(g(x))$. This definition can, of course, be extended to any finite number of such sets.

8.55. Theorem. *Let $f_1(x),\dots,f_h(x)$ be nonconstant monic polynomials over $\mathbb{F}_q$. Then*

$$S(f_1(x))+\cdots+S(f_h(x))=S(c(x)),$$

where $c(x)$ is the (monic) least common multiple of $f_1(x),\dots,f_h(x)$.

Proof. It suffices to consider the case $h=2$ since the general case follows easily by induction. We note first that, according to Theorem 8.53, each sequence belonging to $S(f_1(x))$ or to $S(f_2(x))$ belongs to $S(c(x))$, and since the latter is a vector space, it follows that $S(f_1(x))+S(f_2(x))$ is contained in $S(c(x))$. We compare now the dimensions of these vector spaces over $\mathbb{F}_q$. Writing $V_1=S(f_1(x))$ and $V_2=S(f_2(x))$ and letting $d(x)$ be the (monic) greatest common divisor of $f_1(x)$ and $f_2(x)$, we get

$$\begin{aligned}\dim(V_1+V_2)&=\dim(V_1)+\dim(V_2)-\dim(V_1\cap V_2)\\&=\deg(f_1(x))+\deg(f_2(x))-\deg(d(x)),\end{aligned}$$

where we have applied Theorem 8.54. But $c(x)=f_1(x)f_2(x)/d(x)$, and so

$$\dim(V_1+V_2)=\deg(c(x))=\dim(S(c(x))).$$

Therefore, the linear subspace $S(f_1(x))+S(f_2(x))$ has the same dimension as the vector space $S(c(x))$, and so $S(f_1(x))+S(f_2(x))=S(c(x))$. □

In the special case where $f(x)$ and $g(x)$ are relatively prime nonconstant monic polynomials over $\mathbb{F}_q$, we will have

$$S(f(x)g(x))=S(f(x))+S(g(x)).$$

Since, in this case, Theorem 8.54 shows that $S(f(x))\cap S(g(x))$ consists only of the zero sequence, $S(f(x)g(x))$ is (in the language of linear algebra) the direct sum of the linear subspaces $S(f(x))$ and $S(g(x))$. In other words, every sequence $\sigma\in S(f(x)g(x))$ can be expressed uniquely in the form $\sigma=\sigma_1+\sigma_2$ with $\sigma_1\in S(f(x))$ and $\sigma_2\in S(g(x))$.

Let us recall that $S(f(x))$ is a vector space over $\mathbb{F}_q$ whose dimension is equal to the degree of $f(x)$. This vector space has an interesting additional property: if the sequence $s_0, s_1,\dots$ belongs to $S(f(x))$, then for every integer $b\geqslant 0$ the shifted sequence $s_b, s_{b+1},\dots$ again belongs to $S(f(x))$. This follows, of course, immediately from the linear recurrence relation. We express this property by saying that $S(f(x))$ is *closed under shifts of sequences*. Taken together, the properties listed here characterize the sets $S(f(x))$ completely.

8.56. Theorem. *Let E be a set of sequences in $\mathbb{F}_q$. Then $E=S(f(x))$ for some monic polynomial $f(x)\in\mathbb{F}_q[x]$ of positive degree if and only if E is a* *vector space over* $\mathbb{F}_q$ *of positive finite dimension* (*under the usual addition and scalar multiplication of sequences*) *which is closed under shifts of sequences.*

Proof. We have already noted above that these conditions are necessary. To establish the converse, consider an arbitrary sequence $\sigma \in E$ that is not the zero sequence. If $s_0, s_1, \ldots$ are the terms of σ and $b \geqslant 0$ is an integer, we denote by $\sigma^{(b)}$ the shifted sequence $s_b, s_{b+1}, \ldots$. By hypothesis, the sequences $\sigma^{(0)}, \sigma^{(1)}, \sigma^{(2)}, \ldots$ all belong to E. But E is a finite set, and so there exist nonnegative integers $i < j$ with $\sigma^{(i)} = \sigma^{(j)}$. It follows that the original sequence σ satisfies the homogeneous linear recurrence relation $s_{n+j} = s_{n+i}$, $n = 0, 1, \ldots$. According to Theorem 8.42, the sequence σ has then a minimal polynomial $m_\sigma(x) \in \mathbb{F}_q[x]$ of positive degree k, say. The state vectors $\mathbf{s}_0, \mathbf{s}_1, \ldots, \mathbf{s}_{k-1}$ of the sequence σ are thus linearly independent over $\mathbb{F}_q$ by virtue of Theorem 8.51. Consequently, the sequences $\sigma^{(0)}, \sigma^{(1)}, \ldots, \sigma^{(k-1)}$ are linearly independent elements of $S(m_\sigma(x))$ and hence form a basis for $S(m_\sigma(x))$. Since $\sigma^{(0)}, \sigma^{(1)}, \ldots, \sigma^{(k-1)}$ belong to the vector space E, it follows that $S(m_\sigma(x))$ is a linear subspace of E. Letting E^* denote the set E with the zero sequence deleted and carrying out the argument above for every $\sigma \in E^*$, we arrive at the statement that the finite sum $\sum_{\sigma \in E^*} S(m_\sigma(x))$ of vector spaces is a linear subspace of E. On the other hand, it is trivial that E is contained in $\sum_{\sigma \in E^*} S(m_\sigma(x))$, and so $E = \sum_{\sigma \in E^*} S(m_\sigma(x))$. By invoking Theorem 8.55, we get

$$E = \sum_{\sigma \in E^*} S(m_\sigma(x)) = S(f(x)),$$

where $f(x)$ is the least common multiple of all the polynomials $m_\sigma(x)$ with σ running through E^*. □

It follows from Theorem 8.55 that the sum of two or more homogeneous linear recurring sequences in $\mathbb{F}_q$ is again a homogeneous linear recurring sequence. A characteristic polynomial of the sum sequence is also obtained from this theorem. In important special cases, the minimal polynomial and the least period of the sum sequence can be determined directly on the basis of the corresponding information for the original sequences.

8.57. Theorem. *For each* $i = 1, 2, \ldots, h$, *let* σ_i *be a homogeneous linear recurring sequence in* $\mathbb{F}_q$ *with minimal polynomial* $m_i(x) \in \mathbb{F}_q[x]$. *If the polynomials* $m_1(x), \ldots, m_h(x)$ *are pairwise relatively prime, then the minimal polynomial of the sum* $\sigma_1 + \cdots + \sigma_h$ *is equal to the product* $m_1(x) \cdots m_h(x)$.

Proof. It suffices to consider the case $h = 2$ since the general case follows then by induction. If $m_1(x)$ or $m_2(x)$ is the constant polynomial 1, the result is trivial. Similarly, if the minimal polynomial $m(x) \in \mathbb{F}_q[x]$ of $\sigma_1 + \sigma_2$ is the constant polynomial 1, we obtain a trivial case. Therefore, we assume that the polynomials $m_1(x)$, $m_2(x)$, and $m(x)$ have positive degrees.

Since

$$\sigma_1+\sigma_2\in S(m_1(x))+S(m_2(x))=S(m_1(x)m_2(x))$$

on account of Theorem 8.55, it follows that $m(x)$ divides $m_1(x)m_2(x)$. Now suppose that the terms of σ_1 are $s_0, s_1,\ldots,$ that those of σ_2 are $t_0, t_1,\ldots,$ and that

$$m(x)=x^k-a_{k-1}x^{k-1}-\cdots-a_0.$$

Then

$$s_{n+k}+t_{n+k}=a_{k-1}(s_{n+k-1}+t_{n+k-1})+\cdots+a_0(s_n+t_n)\quad \text{for } n=0,1,\ldots.$$

If we set

$$\begin{aligned} u_n &= s_{n+k}-a_{k-1}s_{n+k-1}-\cdots-a_0s_n \\ &= -t_{n+k}+a_{k-1}t_{n+k-1}+\cdots+a_0t_n \quad \text{for } n=0,1,\ldots \end{aligned}$$

and recall that $S(m_1(x))$ and $S(m_2(x))$ are vector spaces over $\mathbb{F}_q$ closed under shifts of sequences (see Theorem 8.56), then we can conclude that the sequence $u_0, u_1,\ldots$ belongs to both $S(m_1(x))$ and $S(m_2(x))$ and is thus the zero sequence, according to Theorem 8.54. But this shows that both $m_1(x)$ and $m_2(x)$ divide $m(x)$, hence $m_1(x)m_2(x)$ divides $m(x)$, and so $m(x)=m_1(x)m_2(x)$. □

If the minimal polynomials $m_1(x),\ldots,m_h(x)$ of the individual sequences $\sigma_1,\ldots,\sigma_h$ are not pairwise relatively prime, then the special nature of the sequences $\sigma_1,\ldots,\sigma_h$ has to be taken into account in order to determine the minimal polynomial of the sum sequence $\sigma=\sigma_1+\cdots+\sigma_h$. The most feasible method is based on the use of generating functions. Suppose that for $i=1,2,\ldots,h$ the generating function of σ_i is $G_i(x)\in\mathbb{F}_q[[x]]$. Then the generating function of σ is given by $G(x)=G_1(x)+\cdots+G_h(x)$. By Theorem 8.40, each $G_i(x)$ can be written as a fraction with, for instance, the reciprocal polynomial of $m_i(x)$ as denominator. We add these fractions, reduce the resulting fraction to lowest terms, and combine the second part of Theorem 8.40 and the method in the proof of Theorem 8.42 to find the minimal polynomial of σ. This technique yields also an alternative proof for Theorem 8.57.

8.58. Example. Let σ_1 be the impulse response sequence in $\mathbb{F}_2$ belonging to $S(x^4+x^3+x+1)$ and σ_2 the impulse response sequence in $\mathbb{F}_2$ belonging to $S(x^5+x^4+1)$. Then, according to Corollary 8.52, the corresponding minimal polynomials are

$$m_1(x)=x^4+x^3+x+1=(x^2+x+1)(x+1)^2\in\mathbb{F}_2[x]$$

and

$$m_2(x)=x^5+x^4+1=(x^2+x+1)(x^3+x+1)\in\mathbb{F}_2[x].$$

Using Theorem 8.40, the generating function $G(x)$ of the sum sequence $\sigma = \sigma_1 + \sigma_2$ turns out to be

$$G(x) = \frac{x^3}{(x^2+x+1)(x+1)^2} + \frac{x^4}{(x^2+x+1)(x^3+x^2+1)}$$

$$= \frac{x^3}{(x^3+x^2+1)(x+1)^2}.$$

By the second part of Theorem 8.40, the reciprocal polynomial $f_0(x) = (x^3+x+1)(x+1)^2$ of the denominator is a characteristic polynomial of σ. According to (8.18), the associated polynomial $h_0(x)$ is given by $h_0(x) = -x^4(1/x)^3 = -x$. Since $f_0(x)$ and $h_0(x)$ are relatively prime, the method in the proof of Theorem 8.42 yields the minimal polynomial

$$m(x) = (x^3+x+1)(x+1)^2$$

for σ. We note that $m(x)$ is a proper divisor of the least common multiple of $m_1(x)$ and $m_2(x)$, which is

$$(x^2+x+1)(x+1)^2(x^3+x+1). \qquad \square$$

From the information about the minimal polynomial contained in Theorem 8.57, one can immediately deduce a useful result concerning the least period of a sum sequence.

8.59. Theorem. *For each $i = 1, 2, \ldots, h$, let σ_i be a homogeneous linear recurring sequence in $\mathbb{F}_q$ with minimal polynomial $m_i(x) \in \mathbb{F}_q[x]$ and least period r_i. If the polynomials $m_1(x), \ldots, m_h(x)$ are pairwise relatively prime, then the least period of the sum $\sigma_1 + \cdots + \sigma_h$ is equal to the least common multiple of $r_1, \ldots, r_h$.*

Proof. We consider only the case $h = 2$, the general result following by induction. If r is the least period of $\sigma_1 + \sigma_2$, then $r = \text{ord}(m_1(x)m_2(x))$ by Theorems 8.44 and 8.57. An application of Theorem 3.9 shows that r is the least common multiple of $\text{ord}(m_1(x))$ and $\text{ord}(m_2(x))$, and so of r_1 and r_2. $\square$

8.60. Example. Let the sequences σ_1 and σ_2 be as in Example 8.58. Then the least periods of σ_1 and σ_2 are $r_1 = \text{ord}(m_1(x)) = 6$ and $r_2 = \text{ord}(m_2(x)) = 21$, respectively. The least period r of $\sigma_1 + \sigma_2$ is $r = \text{ord}(m(x)) = 14$. In these computations of orders we use, of course, Theorem 3.9. The arguments above have been carried out without having evaluated the terms of the sequences involved. In this special case we may, of course, compare the results with explicit computations of the least periods:

$$\begin{array}{rll} \sigma_1: & 000111000111000111000111000\cdots & \text{least period } r_1 = 6 \\ \sigma_2: & 000011111010100110001000001\cdots & \text{least period } r_2 = 21 \\ \hline \sigma_1+\sigma_2: & 000100111101100001001111101\cdots & \text{least period } r = 14 \end{array}$$

Notice that r is a proper divisor of the least common multiple of r_1 and r_2. □

8.61. Theorem. *For each $i = 1, 2, \ldots, h$, let σ_i be an ultimately periodic sequence in $\mathbb{F}_q$ with least period r_i. If $r_1, \ldots, r_h$ are pairwise relatively prime, then the least period of the sum $\sigma_1 + \cdots + \sigma_h$ is equal to the product $r_1 \cdots r_h$.*

Proof. It suffices to consider the case $h = 2$ since the general case follows then by induction. It is obvious that $r_1 r_2$ is a period of $\sigma_1 + \sigma_2$, so that the least period r of $\sigma_1 + \sigma_2$ divides $r_1 r_2$. Therefore, r is of the form $r = d_1 d_2$ with d_1 and d_2 being positive divisors of r_1 and r_2, respectively. In particular, $d_1 r_2$ is a period of $\sigma_1 + \sigma_2$. Consequently, if the terms of σ_1 are $s_0, s_1, \ldots$ and those of σ_2 are $t_0, t_1, \ldots$, then we have

$$s_{n+d_1 r_2} + t_{n+d_1 r_2} = s_n + t_n$$

for all sufficiently large n. But $t_{n+d_1 r_2} = t_n$ for all sufficiently large n, and so $s_{n+d_1 r_2} = s_n$ for all sufficiently large n. Therefore, r_1 divides $d_1 r_2$, and since r_1 and r_2 are relatively prime, r_1 divides d_1, which implies $d_1 = r_1$. Similarly, one shows that $d_2 = r_2$. □

In the finite field $\mathbb{F}_2$, there is an interesting operation on sequences called binary complementation. If σ is a sequence in $\mathbb{F}_2$, then its *binary complement*, denoted by $\bar{\sigma}$, is obtained by replacing each digit 0 in σ by 1 and each digit 1 in σ by 0. Binary complementation is, in fact, a special case of addition of sequences since the binary complement $\bar{\sigma}$ of σ arises by adding to σ the sequence all of whose terms are 1. Therefore, if σ is a homogeneous linear recurring sequence, then $\bar{\sigma}$ is one as well. Clearly, the least period of $\bar{\sigma}$ is the same as that of σ. The minimal polynomial of $\bar{\sigma}$ can be obtained from that of σ in an easy manner.

8.62. Theorem. *Let σ be a homogeneous linear recurring sequence in $\mathbb{F}_2$ with binary complement $\bar{\sigma}$. Write the minimal polynomial $m(x) \in \mathbb{F}_2[x]$ of σ in the form $m(x) = (x+1)^h m_1(x)$ with an integer $h \geqslant 0$ and $m_1(x) \in \mathbb{F}_2[x]$ satisfying $m_1(1) = 1$. Then the minimal polynomial $\bar{m}(x)$ of $\bar{\sigma}$ is given by $\bar{m}(x) = (x+1)m(x)$ if $h = 0$, $\bar{m}(x) = m_1(x)$ if $h = 1$, and $\bar{m}(x) = m(x)$ if $h > 1$.*

Proof. Let ε be the sequence in $\mathbb{F}_2$ all of whose terms are 1. Since $\bar{\sigma} = \sigma + \varepsilon$ and the minimal polynomial of ε is $x + 1$, the case $h = 0$ is settled by invoking Theorem 8.57. If $h \geqslant 1$, then $\bar{\sigma} = \sigma + \varepsilon \in S(m(x))$ because of Theorem 8.55, and so $\bar{m}(x)$ divides $m(x)$. If $\bar{m}(x)$ is the constant polynomial 1, then $\bar{\sigma}$ is necessarily the zero sequence and $\sigma = \varepsilon$, and the theorem holds. Therefore, we assume from now on that $\bar{m}(x)$ is of positive degree. We get $\sigma = \bar{\sigma} + \varepsilon \in S(\bar{m}(x)(x+1))$ because of Theorems 8.53 and 8.55, thus $m(x)$ divides $\bar{m}(x)(x+1)$, and so for $h \geqslant 1$ we have either $\bar{m}(x) = m(x)$ or $\bar{m}(x) = (x+1)^{h-1}m_1(x)$. If $h > 1$, it follows that $\sigma = \bar{\sigma} + \varepsilon \in S(\bar{m}(x))$, which yields $\bar{m}(x) = m(x)$. If $h = 1$, let the terms of σ be $s_0, s_1, \ldots$ and let

$$m_1(x) = x^k + a_{k-1}x^{k-1} + \cdots + a_0$$

be of positive degree, the excluded case being trivial. We set

$$u_n = s_{n+k} + a_{k-1}s_{n+k-1} + \cdots + a_0 s_n \quad \text{for } n = 0, 1, \ldots.$$

Since the sequence $s_0, s_1, \ldots$ has $m(x) = (x+1)m_1(x)$ as a characteristic polynomial, it follows easily that $u_{n+1} = u_n$ for all $n \geqslant 0$. Therefore, $u_n = u_0$ for all $n \geqslant 0$, and we must have $u_0 = 1$, for otherwise $m_1(x)$ would be a characteristic polynomial of σ. Consequently,

$$s_{n+k} + 1 = a_{k-1}s_{n+k-1} + \cdots + a_0 s_n \quad \text{for all } n \geqslant 0.$$

Since $m_1(1) = 1 + a_{k-1} + \cdots + a_0 = 1$, we obtain

$$s_{n+k} + 1 = a_{k-1}(s_{n+k-1} + 1) + \cdots + a_0(s_n + 1) \quad \text{for all } n \geqslant 0,$$

and this means that $m_1(x)$ is a characteristic polynomial of $\bar{\sigma}$. Thus, $\bar{m}(x) = m_1(x)$ in the case where $h = 1$. □

We recall that $S(f(x))$ denotes the set of all homogeneous linear recurring sequences in $\mathbb{F}_q$ with characteristic polynomial $f(x)$, where $f(x) \in \mathbb{F}_q[x]$ is a monic polynomial of positive degree. We want to determine the *positive integers that appear as least periods* of sequences from $S(f(x))$, and also, *for how many sequences* from $S(f(x))$ such a positive integer is attained as a least period.

The polynomial $f(x)$ can be written in the form $f(x) = x^h g(x)$, where $h \geqslant 0$ is an integer and $g(x) \in \mathbb{F}_q[x]$ with $g(0) \neq 0$. The case in which $g(x)$ is a constant polynomial can be dealt with immediately, since then every sequence from $S(f(x))$ has least period 1. If $h \geqslant 1$ and $g(x)$ is of positive degree, then, by the discussion following Theorem 8.55, every sequence $\sigma \in S(f(x))$ can be expressed uniquely in the form $\sigma = \sigma_1 + \sigma_2$ with $\sigma_1 \in S(x^h)$ and $\sigma_2 \in S(g(x))$. Apart from finitely many initial terms, all terms of σ_1 are zero, so that the least period of σ is equal to the least period of σ_2. Furthermore, a given sequence $\sigma_2 \in S(g(x))$ leads to q^h different sequences from $S(f(x))$ by adding to it all the q^h sequences from $S(x^h)$. Consequently, if $r_1, \ldots, r_t$ are the least periods of sequences from $S(g(x))$ and $N_1, \ldots, N_t$ are the corresponding numbers of sequences from $S(g(x))$ having these least periods, then, for $1 \leqslant i \leqslant t$, there are exactly $q^h N_i$ sequences belonging to $S(f(x))$ with least period r_i, and no other least periods occur among the sequences from $S(f(x))$.

We may assume from now on that $h=0$—that is, that $f(0)\neq 0$. Suppose first that $f(x)$ is irreducible over $\mathbb{F}_q$. Then, according to Theorems 8.44 and 8.50, every sequence from $S(f(x))$ with nonzero initial state vector has least period $\operatorname{ord}(f(x))$. Therefore, one sequence from $S(f(x))$ has least period 1 and $q^{\deg(f(x))}-1$ sequences from $S(f(x))$ have least period $\operatorname{ord}(f(x))$.

Next, we consider the case that $f(x)$ is a power of an irreducible polynomial. Thus, let $f(x)=g(x)^b$ with $g(x)\in\mathbb{F}_q[x]$ monic and irreducible over $\mathbb{F}_q$ and $b\geqslant 2$ an integer. The minimal polynomial of any sequence from $S(f(x))$ with nonzero initial state vector is then of the form $g(x)^c$ with $1\leqslant c\leqslant b$. According to Theorem 8.53, we have

$$S(g(x))\subseteq S\left(g(x)^2\right)\subseteq\cdots\subseteq S(f(x)).$$

Therefore, if $\deg(g(x))=k$, then there are q^k-1 sequences from $S(f(x))$ with minimal polynomial $g(x)$, $q^{2k}-q^k$ sequences from $S(f(x))$ with minimal polynomial $g(x)^2$, and, in general, for $c=1,2,\ldots,b$ there are $q^{ck}-q^{(c-1)k}$ sequences from $S(f(x))$ with minimal polynomial $g(x)^c$. By combining this information with Theorems 3.8 and 8.44, we arrive at the following result.

8.63. Theorem. *Let $f(x)=g(x)^b$ with $g(x)\in\mathbb{F}_q[x]$ monic and irreducible over $\mathbb{F}_q$, $g(0)\neq 0$, $\deg(g(x))=k$, $\operatorname{ord}(g(x))=e$, and b a positive integer. Let t be the smallest integer with $p^t\geqslant b$, where p is the characteristic of $\mathbb{F}_q$. Then $S(f(x))$ contains the following numbers of sequences with the following least periods: one sequence with least period 1, q^k-1 sequences with least period e, and for $b\geqslant 2$, $q^{kp^j}-q^{kp^{j-1}}$ sequences with least period ep^j $(j=1,2,\ldots,t-1)$ and $q^{kb}-q^{kp^{t-1}}$ sequences with least period ep^t.*

In the case of an arbitrary monic polynomial $f(x)\in\mathbb{F}_q[x]$ of positive degree with $f(0)\neq 0$, we start from the canonical factorization

$$f(x)=\prod_{i=1}^{h}g_i(x)^{b_i},$$

where the $g_i(x)$ are distinct monic irreducible polynomials over $\mathbb{F}_q$ and the b_i are positive integers. It follows then from Theorem 8.55 that

$$S(f(x))=S\left(g_1(x)^{b_1}\right)+\cdots+S\left(g_h(x)^{b_h}\right).$$

In fact, every sequence from $S(f(x))$ is obtained exactly once by forming all possible sums $\sigma_1+\cdots+\sigma_h$ with $\sigma_i\in S(g_i(x)^{b_i})$ for $1\leqslant i\leqslant h$. Since the least periods attained by sequences from $S(g_i(x)^{b_i})$ are known from Theorem 8.63, the analogous information about $S(f(x))$ can thus be deduced from Theorem 8.59.

8.64. Example. Let

$$f(x) = (x^2 + x + 1)^2(x^4 + x^3 + 1) \in \mathbb{F}_2[x].$$

According to Theorem 8.63, $S((x^2 + x + 1)^2)$ contains one sequence with least period 1, 3 sequences with least period 3, and 12 sequences with least period 6, whereas $S(x^4 + x^3 + 1)$ contains one sequence with least period 1 and 15 sequences with least period 15. Therefore, by forming all possible sums of sequences from $S((x^2 + x + 1)^2)$ and $S(x^4 + x^3 + 1)$ and using Theorem 8.59, we conclude that $S(f(x))$ contains one sequence with least period 1, 3 sequences with least period 3, 12 sequences with least period 6, 60 sequences with least period 15, and 180 sequences with least period 30. □

We have already investigated the behavior of linear recurring sequences under termwise addition. A similar theory can be developed for the operation of termwise multiplication, although it presents greater difficulties. If σ is the sequence of elements $s_0, s_1, \ldots$ of $\mathbb{F}_q$ and τ is the sequence of elements $t_0, t_1, \ldots$ of $\mathbb{F}_q$, then the product sequence $\sigma\tau$ has terms $s_0t_0, s_1t_1, \ldots$. Analogously, one defines the product of any finite number of sequences. Let S be the vector space over $\mathbb{F}_q$ consisting of all sequences of elements of $\mathbb{F}_q$, under the usual addition and scalar multiplication of sequences. For nonconstant monic polynomials $f_1(x), \ldots, f_h(x)$ over $\mathbb{F}_q$, let $S(f_1(x)) \cdots S(f_h(x))$ be the subspace of S spanned by all products $\sigma_1 \cdots \sigma_h$ with $\sigma_i \in S(f_i(x))$, $1 \leqslant i \leqslant h$. The following result is basic.

8.65. Theorem. *If $f_1(x), \ldots, f_h(x)$ are nonconstant monic polynomials over $\mathbb{F}_q$, then there exists a nonconstant monic polynomial $g(x) \in \mathbb{F}_q[x]$ such that*

$$S(f_1(x)) \cdots S(f_h(x)) = S(g(x)).$$

Proof. Set $E = S(f_1(x)) \cdots S(f_h(x))$. Since each $S(f_i(x))$, $1 \leqslant i \leqslant h$, contains a sequence with initial term 1, the vector space E contains a nonzero sequence. Furthermore, E is spanned by finitely many sequences and thus finite-dimensional. From the fact that each $S(f_i(x))$, $1 \leqslant i \leqslant h$, is closed under shifts of sequences it follows that E has the same property, and then the argument is complete by Theorem 8.56. □

8.66. Corollary. *The product of finitely many linear recurring sequences in $\mathbb{F}_q$ is again a linear recurring sequence in $\mathbb{F}_q$.*

Proof. By the remarks following (8.5), the given linear recurring sequences can be taken to be homogeneous. The result is then implicit in Theorem 8.65. □

The explicit determination of the polynomial $g(x)$ in Theorem 8.65 is, in general, not easy. There is, however, a special case that allows a simpler treatment of the problem.

For nonconstant polynomials $f_1(x),\ldots,f_h(x)$ over $\mathbb{F}_q$, we define $f_1(x)\vee\cdots\vee f_h(x)$ to be the monic polynomial whose roots are the distinct elements of the form $\alpha_1\cdots\alpha_h$, where each α_i is a root of $f_i(x)$ in the splitting field of $f_1(x)\cdots f_h(x)$ over $\mathbb{F}_q$. Since the conjugates (over $\mathbb{F}_q$) of such a product $\alpha_1\cdots\alpha_h$ are again elements of this form, it follows that $f_1(x)\vee\cdots\vee f_h(x)$ is a polynomial over $\mathbb{F}_q$.

8.67. Theorem. *For each $i=1,2,\ldots,h$, let $f_i(x)$ be a nonconstant monic polynomial over $\mathbb{F}_q$ without multiple roots. Then we have*

$$S(f_1(x))\cdots S(f_h(x))=S(f_1(x)\vee\cdots\vee f_h(x)).$$

We need a preparatory lemma and some notation for the proof of this result. For a finite extension field F of $\mathbb{F}_q$, let S_F be the vector space over F consisting of all sequences of elements of F, under termwise addition and scalar multiplication of sequences. Thus, in particular, $S_{\mathbb{F}_q}=S$. By the product $V_1\cdots V_h$ of h subspaces $V_1,\ldots,V_h$ of S_F we mean the subspace of S_F spanned by all products $\sigma_1\cdots\sigma_h$ with $\sigma_i\in V_i$, $1\leqslant i\leqslant h$. For a nonconstant monic polynomial $f(x)\in F[x]$, let $S_F(f(x))$ be the vector space over F consisting of all homogeneous linear recurring sequences in F with characteristic polynomial $f(x)$.

8.68. Lemma. *Let F be a finite extension field of $\mathbb{F}_q$, and let $f_1(x),\ldots,f_h(x)$ be nonconstant monic polynomials over $\mathbb{F}_q$. Then,*

$$S(f_1(x))\cdots S(f_h(x))=S\cap(S_F(f_1(x))\cdots S_F(f_h(x))).$$

Proof. Clearly, the vector space on the left-hand side is contained in the vector space on the right-hand side. To show the converse, we note first that each $S(f_i(x))$, $1\leqslant i\leqslant h$, spans $S_F(f_i(x))$ over F. Therefore, $S(f_1(x))\cdots S(f_h(x))$ spans $S_F(f_1(x))\cdots S_F(f_h(x))$ over F. Let $\rho_1,\ldots,\rho_m$ be a basis of $S(f_1(x))\cdots S(f_h(x))$ over $\mathbb{F}_q$, and let $\omega_1,\ldots,\omega_k$ be a basis of F over $\mathbb{F}_q$ with $\omega_1\in\mathbb{F}_q$. Then any $\sigma\in S_F(f_1(x))\cdots S_F(f_h(x))$ can be written in the form

$$\sigma=\sum_{i=1}^{k}\sum_{j=1}^{m}c_{ij}\omega_i\rho_j,$$

where the coefficients c_{ij} are in $\mathbb{F}_q$. Let the terms of the sequence ρ_j, $1\leqslant j\leqslant m$, be the elements $r_{j0},r_{j1},\ldots$ of $\mathbb{F}_q$. If now $\sigma\in S$, then for the terms s_n, $n=0,1,\ldots,$ of σ we get

$$s_n=\sum_{i=1}^{k}\left(\sum_{j=1}^{m}c_{ij}r_{jn}\right)\omega_i\in\mathbb{F}_q\quad\text{for } n=0,1,\ldots.$$

Since the coefficient of each ω_i is in $\mathbb{F}_q$, it follows from the definition of

$\omega_1, \ldots, \omega_k$ that $\sum_{j=1}^{m} c_{ij} r_{jn} = 0$ for $2 \leqslant i \leqslant k$ and all n. Consequently,

$$\sigma = \sum_{j=1}^{m} c_{1j}\omega_1\rho_j \in S(f_1(x)) \cdots S(f_h(x))$$

and the proof is complete. □

Proof of Theorem 8.67. Let F be the splitting field of $f_1(x) \cdots f_h(x)$ over $\mathbb{F}_q$. For $1 \leqslant i \leqslant h$, let α_i run through the roots of $f_i(x)$. Then by Theorem 8.55,

$$S_F(f_i(x)) = \sum_{\alpha_i} S_F(x - \alpha_i) \quad \text{for } 1 \leqslant i \leqslant h.$$

We note that we have the distributive law $V_1(V_2 + V_3) = V_1V_2 + V_1V_3$ for subspaces V_1, V_2, V_3 of S_F, which is shown by observing that the left-hand vector space is contained in the right-hand vector space (by the distributive law for sequences) and that $V_1V_2 \subseteq V_1(V_2 + V_3)$ and $V_1V_3 \subseteq V_1(V_2 + V_3)$ imply $V_1V_2 + V_1V_3 \subseteq V_1(V_2 + V_3)$. On the basis of the distributive law, it follows that

$$S_F(f_1(x)) \cdots S_F(f_h(x)) = \sum_{\alpha_1, \ldots, \alpha_h} S_F(x - \alpha_1) \cdots S_F(x - \alpha_h).$$

It is easy to check directly that

$$S_F(x - \alpha_1) \cdots S_F(x - \alpha_h) = S_F(x - \alpha_1 \cdots \alpha_h),$$

and so

$$\begin{aligned} S_F(f_1(x)) \cdots S_F(f_h(x)) &= \sum_{\alpha_1, \ldots, \alpha_h} S_F(x - \alpha_1 \cdots \alpha_h) \\ &= S_F(f_1(x) \vee \cdots \vee f_h(x)) \end{aligned}$$

by Theorem 8.55. The result of Theorem 8.67 follows now from Lemma 8.68. □

Theorem 8.67 shows, in particular, how to find a characteristic polynomial for the product of homogeneous linear recurring sequences, at least in the special case considered there. For this purpose, an *alternative argument* may be based on Theorem 8.21. It suffices to carry out the details for the product of two homogeneous linear recurring sequences. Let the sequence $s_0, s_1, \ldots$ belong to $S(f(x))$ and let $t_0, t_1, \ldots$ belong to $S(g(x))$. If $f(x)$ has only the simple roots $\alpha_1, \ldots, \alpha_k$ and $g(x)$ has only the simple roots $\beta_1, \ldots, \beta_m$, then by (8.8),

$$s_n = \sum_{i=1}^{k} b_i \alpha_i^n \quad \text{and} \quad t_n = \sum_{j=1}^{m} c_j \beta_j^n \quad \text{for } n = 0, 1, \ldots,$$

where the coefficients b_i and c_j belong to a finite extension field of $\mathbb{F}_q$. If $\gamma_1, \ldots, \gamma_r$ are the distinct values of the products $\alpha_i\beta_j$, $1 \leqslant i \leqslant k$, $1 \leqslant j \leqslant m$,

then

$$u_n = s_n t_n = \sum_{i=1}^{k} \sum_{j=1}^{m} b_i c_j (\alpha_i \beta_j)^n = \sum_{i=1}^{r} d_i \gamma_i^n \quad \text{for } n = 0, 1, \ldots,$$

with suitable coefficients $d_1, \ldots, d_r$ in a finite extension field of $\mathbb{F}_q$. Now let

$$h(x) = f(x) \vee g(x) = x^r - a_{r-1}x^{r-1} - \cdots - a_0 \in \mathbb{F}_q[x].$$

Then for $n = 0, 1, \ldots$ we have

$$u_{n+r} - a_{r-1}u_{n+r-1} - \cdots - a_0 u_n = \sum_{i=1}^{r} d_i \gamma_i^n h(\gamma_i) = 0,$$

and so the product sequence $u_0, u_1, \ldots$ has $h(x)$ as a characteristic polynomial.

8.69. Example. Consider the sequence $0, 1, 0, 1, \ldots$ in $\mathbb{F}_2$ with the least period 2 and minimal polynomial $(x-1)^2$. If we multiply this sequence with itself, we get back the same sequence. On the other hand, $(x-1)^2 \vee (x-1)^2 = x - 1$, which is not a characteristic polynomial of the product sequence. Therefore, the identity in Theorem 8.67 may cease to hold if some of the polynomials $f_i(x)$ are allowed to have multiple roots. □

There is an analog of Theorem 8.61 for multiplication of sequences. For obvious reasons, sequences for which all but finitely many terms are zero have to be excluded from consideration.

8.70. Theorem. *For each $i = 1, 2, \ldots, h$, let σ_i be an ultimately periodic sequence in $\mathbb{F}_q$ with infinitely many nonzero terms and with least period r_i. If $r_1, \ldots, r_h$ are pairwise relatively prime, then the least period of the product $\sigma_1 \cdots \sigma_h$ is equal to $r_1 \cdots r_h$.*

Proof. We consider only the case $h = 2$ since the general case follows then by induction. As in the proof of Theorem 8.61 one shows that the least period r of $\sigma_1\sigma_2$ must be of the form $r = d_1 d_2$ with d_1 and d_2 being positive divisors of r_1 and r_2, respectively. In particular, $d_1 r_2$ is a period of $\sigma_1\sigma_2$. Thus, if the terms of σ_1 are $s_0, s_1, \ldots$ and those of σ_2 are $t_0, t_1, \ldots$, then we have

$$s_{n+d_1r_2} t_n = s_{n+d_1r_2} t_{n+d_1r_2} = s_n t_n$$

for all sufficiently large n. Since there exists an integer b with $t_n \neq 0$ for all sufficiently large $n \equiv b \bmod r_2$, it follows that $s_{n+d_1r_2} = s_n$ for all such n. Now fix a sufficiently large n; by the Chinese remainder theorem, we can choose an integer $m \geqslant n$ with $m \equiv n \bmod r_1$ and $m \equiv b \bmod r_2$. Then

$$s_n = s_m = s_{m+d_1r_2} = s_{n+d_1r_2},$$

and so $d_1 r_2$ is a period of σ_1. Therefore, r_1 divides $d_1 r_2$, and since r_1 and r_2

are relatively prime, r_1 divides d_1, which implies $d_1 = r_1$. Similarly, one shows that $d_2 = r_2$. □

Multiplication of sequences can be used to describe the relation between homogeneous linear recurring sequences belonging to characteristic polynomials that are powers of each other. The case in which one of the characteristic polynomials is linear has to be considered first.

8.71. Lemma. *If c is a nonzero element of $\mathbb{F}_q$ and k is a positive integer, then*

$$S\left((x-c)^k\right) = S(x-c)S\left((x-1)^k\right).$$

Proof. Let the sequence $s_0, s_1, \ldots$ belong to $S(x-c)$, and let $t_0, t_1, \ldots$ belong to $S((x-1)^k)$. Then $s_n = c^n s_0$ for $n = 0, 1, \ldots$ and

$$\sum_{i=0}^{k} \binom{k}{i} (-1)^{k-i} t_{n+i} = 0 \quad \text{for } n = 0, 1, \ldots.$$

It follows that

$$\sum_{i=0}^{k} \binom{k}{i} (-c)^{k-i} s_{n+i} t_{n+i} = c^{n+k} s_0 \sum_{i=0}^{k} \binom{k}{i} (-1)^{k-i} t_{n+i} = 0$$

for $n = 0, 1, \ldots$, and so

$$\sum_{i=0}^{k} \binom{k}{i} (-c)^{k-i} x^i = (x-c)^k$$

is a characteristic polynomial of the product sequence $s_0 t_0, s_1 t_1, \ldots$. Consequently, the vector space $S(x-c)S((x-1)^k)$ is a subspace of $S((x-c)^k)$. Since $c \neq 0$, the first vector space has dimension k over $\mathbb{F}_q$ and is thus equal to $S((x-c)^k)$, which has the same dimension over $\mathbb{F}_q$. □

8.72. Theorem. *Let $f(x) \in \mathbb{F}_q[x]$ be a nonconstant monic polynomial with $f(0) \neq 0$ and without multiple roots, and let k be a positive integer. Then,*

$$S\left(f(x)^k\right) = S(f(x))S\left((x-1)^k\right).$$

Proof. Let F be the splitting field of $f(x)$ over $\mathbb{F}_q$. Then, with α running through the roots of $f(x)$, we get

$$S_F\left(f(x)^k\right) = \sum_{\alpha} S_F\left((x-\alpha)^k\right)$$

by Theorem 8.55. Using Lemma 8.71 and the distributive law shown in the proof of Theorem 8.67, we obtain

$$S_F\left(f(x)^k\right)=\sum_{\alpha} S_F\left((x-1)^k\right)S_F(x-\alpha)=S_F\left((x-1)^k\right)\sum_{\alpha} S_F(x-\alpha)$$
$$=S_F\left((x-1)^k\right)S_F(f(x)),$$

where we applied Theorem 8.55 in the last step. The desired result follows now from Lemma 8.68. □

6. CHARACTERIZATION OF LINEAR RECURRING SEQUENCES

It is an important problem to decide whether a given sequence of elements of $\mathbb{F}_q$ is a linear recurring sequence or not. From the theoretical point of view, the question can be settled immediately since *the linear recurring sequences in $\mathbb{F}_q$ are precisely the ultimately periodic sequences*. However, the periods of a linear recurring sequence (even of one of moderately low order) can be extremely long, so that in practice it may not be feasible to determine the nature of the sequence on the basis of this criterion. Alternative ways of characterizing linear recurring sequences employ techniques from linear algebra.

Let $s_0, s_1, \ldots$ be an arbitrary sequence of elements of $\mathbb{F}_q$. For integers $n \geqslant 0$ and $r \geqslant 1$, we introduce the *Hankel determinants*

$$D_n^{(r)}=\begin{vmatrix} s_n & s_{n+1} & \cdots & s_{n+r-1} \\ s_{n+1} & s_{n+2} & \cdots & s_{n+r} \\ \vdots & \vdots & & \vdots \\ s_{n+r-1} & s_{n+r} & \cdots & s_{n+2r-2} \end{vmatrix}.$$

It will transpire that linear recurring sequences can be characterized in terms of the vanishing of sufficiently many of these Hankel determinants.

8.73. Lemma. *Let $s_0, s_1, \ldots$ be an arbitrary sequence in $\mathbb{F}_q$, and let $n \geqslant 0$ and $r \geqslant 1$ be integers. Then $D_n^{(r)}=D_n^{(r+1)}=0$ implies $D_{n+1}^{(r)}=0$.*

Proof. For $m \geqslant 0$ define the vector $\mathbf{s}_m=(s_m, s_{m+1}, \ldots, s_{m+r-1})$. From $D_n^{(r)}=0$ it follows that the vectors $\mathbf{s}_n, \mathbf{s}_{n+1}, \ldots, \mathbf{s}_{n+r-1}$ are linearly dependent over $\mathbb{F}_q$. If $\mathbf{s}_{n+1}, \ldots, \mathbf{s}_{n+r-1}$ are already linearly dependent over $\mathbb{F}_q$, we immediately get $D_{n+1}^{(r)}=0$. Otherwise, $\mathbf{s}_n$ is a linear combination of $\mathbf{s}_{n+1}, \ldots, \mathbf{s}_{n+r-1}$. Set $\mathbf{s}'_m=(s_m, s_{m+1}, \ldots, s_{m+r})$ for $m \geqslant 0$. Then the vectors $\mathbf{s}'_n, \mathbf{s}'_{n+1}, \ldots, \mathbf{s}'_{n+r}$, being the row vectors of the vanishing determinant $D_n^{(r+1)}$, are linearly dependent over $\mathbb{F}_q$. If $\mathbf{s}'_n, \mathbf{s}'_{n+1}, \ldots, \mathbf{s}'_{n+r-1}$ are already linearly dependent over $\mathbb{F}_q$, then an application of the linear transformation

$$L_1:(a_0, a_1, \ldots, a_r) \in \mathbb{F}_q^{r+1} \mapsto (a_1, \ldots, a_r) \in \mathbb{F}_q^r$$

shows that $\mathbf{s}_{n+1}, \mathbf{s}_{n+2}, \ldots, \mathbf{s}_{n+r}$ are linearly dependent over $\mathbb{F}_q$, and so $D_{n+1}^{(r)} = 0$. Otherwise, $\mathbf{s}'_{n+r}$ is a linear combination of $\mathbf{s}'_n, \mathbf{s}'_{n+1}, \ldots, \mathbf{s}'_{n+r-1}$, and by an application of the linear transformation

$$L_2: (a_0, \ldots, a_{r-1}, a_r) \in \mathbb{F}_q^{r+1} \mapsto (a_0, \ldots, a_{r-1}) \in \mathbb{F}_q^r$$

we obtain that $\mathbf{s}_{n+r}$ is a linear combination of $\mathbf{s}_n, \mathbf{s}_{n+1}, \ldots, \mathbf{s}_{n+r-1}$. But in the case under consideration $\mathbf{s}_n$ is a linear combination of $\mathbf{s}_{n+1}, \ldots, \mathbf{s}_{n+r-1}$, so that the row vectors $\mathbf{s}_{n+1}, \ldots, \mathbf{s}_{n+r-1}, \mathbf{s}_{n+r}$ of $D_{n+1}^{(r)}$ are linearly dependent over $\mathbb{F}_q$, which implies $D_{n+1}^{(r)} = 0$. □

8.74. Theorem. *The sequence $s_0, s_1, \ldots$ in $\mathbb{F}_q$ is a linear recurring sequence if and only if there exists a positive integer r such that $D_n^{(r)} = 0$ for all but finitely many $n \geqslant 0$.*

Proof. Suppose $s_0, s_1, \ldots$ satisfies a kth-order homogeneous linear recurrence relation. For any fixed $n \geqslant 0$, consider the determinant $D_n^{(k+1)}$. Because of the linear recurrence relation, the $(k+1)$st row of $D_n^{(k+1)}$ is a linear combination of the first k rows, and so $D_n^{(k+1)} = 0$. The inhomogeneous case reduces to the homogeneous case by (8.5).

To show sufficiency, let $k+1$ be the least positive integer such that $D_n^{(k+1)} = 0$ for all but finitely many $n \geqslant 0$. If $k+1 = 1$, then we are done, and so we may assume $k \geqslant 1$. There is an integer $m \geqslant 0$ with $D_n^{(k+1)} = 0$ for all $n \geqslant m$. If we had $D_{n_0}^{(k)} = 0$ for some $n_0 \geqslant m$, then $D_n^{(k)} = 0$ for all $n \geqslant n_0$ by Lemma 8.73, which contradicts the definition of $k+1$. Therefore, $D_n^{(k)} \neq 0$ for all $n \geqslant m$. Setting $\mathbf{s}_n = (s_n, s_{n+1}, \ldots, s_{n+k})$, we note that for $n \geqslant m$ the vectors $\mathbf{s}_n, \mathbf{s}_{n+1}, \ldots, \mathbf{s}_{n+k}$, being the row vectors of $D_n^{(k+1)}$, are linearly dependent over $\mathbb{F}_q$. Since $D_n^{(k)} \neq 0$, the vectors $\mathbf{s}_n, \mathbf{s}_{n+1}, \ldots, \mathbf{s}_{n+k-1}$ are linearly independent over $\mathbb{F}_q$, and so $\mathbf{s}_{n+k}$ is a linear combination of $\mathbf{s}_n, \mathbf{s}_{n+1}, \ldots, \mathbf{s}_{n+k-1}$. It follows then by induction that each $\mathbf{s}_n$ with $n \geqslant m$ is a linear combination of $\mathbf{s}_m, \mathbf{s}_{m+1}, \ldots, \mathbf{s}_{m+k-1}$. The latter are k vectors in $\mathbb{F}_q^{k+1}$, therefore there exists a nonzero vector $(a_0, a_1, \ldots, a_k) \in \mathbb{F}_q^{k+1}$ with

$$a_0 s_n + a_1 s_{n+1} + \cdots + a_k s_{n+k} = 0 \quad \text{for } m \leqslant n \leqslant m+k-1.$$

This implies

$$a_0 s_n + a_1 s_{n+1} + \cdots + a_k s_{n+k} = 0 \quad \text{for all } n \geqslant m,$$

or

$$a_0 s_{n+m} + a_1 s_{n+m+1} + \cdots + a_k s_{n+m+k} = 0 \text{ for all } n \geqslant 0.$$

Thus, the sequence $s_0, s_1, \ldots$ satisfies a homogeneous linear recurrence relation of order at most $m+k$. □

8.75. Theorem. *The sequence $s_0, s_1, \ldots$ in $\mathbb{F}_q$ is a homogeneous linear recurring sequence with minimal polynomial of degree k if and only if $D_0^{(r)} = 0$ for all $r \geqslant k+1$ and $k+1$ is the least positive integer for which this holds.*

Proof. If a given linear recurring sequence is the zero sequence, the necessity of the condition is clear. Otherwise, we have $k \geqslant 1$, and $D_0^{(r)} = 0$ for all $r \geqslant k+1$ follows since the $(k+1)$st row of $D_0^{(r)}$ is a linear combination of the first k rows. Moreover, we get $D_0^{(k)} \neq 0$ from Theorem 8.51, and so the necessity of the condition is shown in all cases.

Conversely, suppose the condition on the Hankel determinants is satisfied. By using Lemma 8.73 and induction on n, one establishes that $D_n^{(r)} = 0$ for all $r \geqslant k+1$ and all $n \geqslant 0$. In particular, $D_n^{(k+1)} = 0$ for all $n \geqslant 0$, and so $s_0, s_1, \ldots$ is a linear recurring sequence by Theorem 8.74. If its minimal polynomial has degree d, then, by what we have already shown in the first part, we know that $D_0^{(r)} = 0$ for all $r \geqslant d+1$ and that $d+1$ is the least positive integer for which this holds. It follows that $d = k$. □

We note that if a homogeneous linear recurring sequence is known to have a minimal polynomial of degree $k \geqslant 1$, then the minimal polynomial is determined by the first $2k$ terms of the sequence. To see this, write down the equations (8.2) for $n = 0, 1, \ldots, k-1$, thereby obtaining a system of k linear equations for the unknown coefficients $a_0, a_1, \ldots, a_{k-1}$ of the minimal polynomial. The determinant of this system is $D_0^{(k)}$, which is $\neq 0$ by Theorem 8.51. Therefore, the system can be solved uniquely.

An important question is that of the actual *computation of the minimal polynomial* of a given homogeneous linear recurring sequence. To be sure, a method of finding the minimal polynomial was already presented in the course of the proof of Theorem 8.42. This method depends on the prior knowledge of a characteristic polynomial of the sequence and on the determination of a greatest common divisor in $\mathbb{F}_q[x]$. We shall now discuss a recursive algorithm (called *Berlekamp-Massey algorithm*) which produces the minimal polynomial after finitely many steps, provided we know an upper bound for the degree of the minimal polynomial.

Let $s_0, s_1, \ldots$ be a sequence of elements of $\mathbb{F}_q$ with generating function $G(x) = \sum_{n=0}^{\infty} s_n x^n$. For $j = 0, 1, \ldots$ we define polynomials $g_j(x)$ and $h_j(x)$ over $\mathbb{F}_q$, integers m_j, and elements b_j of $\mathbb{F}_q$ as follows. Initially, we set

$$g_0(x) = 1, \quad h_0(x) = x, \quad \text{and} \quad m_0 = 0. \tag{8.19}$$

Then we proceed recursively by letting b_j be the coefficient of x^j in $g_j(x)G(x)$ and setting:

$$\begin{aligned} g_{j+1}(x) &= g_j(x) - b_j h_j(x), \\ h_{j+1}(x) &= \begin{cases} b_j^{-1} x g_j(x) & \text{if } b_j \neq 0 \text{ and } m_j \geqslant 0, \\ x h_j(x) & \text{otherwise,} \end{cases} \\ m_{j+1} &= \begin{cases} -m_j & \text{if } b_j \neq 0 \text{ and } m_j \geqslant 0, \\ m_j + 1 & \text{otherwise.} \end{cases} \end{aligned} \tag{8.20}$$

If $s_0, s_1, \ldots$ is a homogeneous linear recurring sequence with a minimal polynomial of degree k, then it turns out that $g_{2k}(x)$ is equal to the reciprocal minimal polynomial. Thus, the minimal polynomial $m(x)$ itself is given by $m(x) = x^k g_{2k}(1/x)$. If it is only known that the minimal polynomial is of degree $\leqslant k$, then set $r = \lfloor k + \frac{1}{2} - \frac{1}{2} m_{2k} \rfloor$, where $\lfloor y \rfloor$ denotes the greatest integer $\leqslant y$, and *the minimal polynomial* $m(x)$ *is given by* $m(x) = x^r g_{2k}(1/x)$. In both cases, it is seen immediately from the algorithm that $m(x)$ depends only on the $2k$ terms $s_0, s_1, \ldots, s_{2k-1}$ of the sequence. Therefore, one may replace the generating function $G(x)$ in the algorithm by the polynomial

$$G_{2k-1}(x) = \sum_{n=0}^{2k-1} s_n x^n.$$

8.76. Example. The first 8 terms of a homogeneous linear recurring sequence in $\mathbb{F}_3$ of order $\leqslant 4$ are given by $0,2,1,0,1,2,1,0$. To find the minimal polynomial, we use the Berlekamp-Massey algorithm with

$$G_7(x) = 2x + x^2 + x^4 + 2x^5 + x^6 \in \mathbb{F}_3[x]$$

in place of $G(x)$. The computation is summarized in the following table.

j	$g_j(x)$	$h_j(x)$	m_j	b_j
0	1	x	0	0
1	1	x^2	1	2
2	$1+x^2$	$2x$	-1	1
3	$1+x+x^2$	$2x^2$	0	0
4	$1+x+x^2$	$2x^3$	1	2
5	$1+x+x^2+2x^3$	$2x+2x^2+2x^3$	-1	2
6	$1+x^3$	$2x^2+2x^3+2x^4$	0	1
7	$1+x^2+2x^3+x^4$	$x+x^4$	0	1
8	$1+2x+x^2+2x^3$		0	

Then, $r = \lfloor 4 + \frac{1}{2} - \frac{1}{2} m_8 \rfloor = 4$, and so $m(x) = x^4 + 2x^3 + x^2 + 2x$. The homogeneous linear recurrence relation of least order satisfied by the sequence is therefore $s_{n+4} = s_{n+3} + 2s_{n+2} + s_{n+1}$ for $n = 0, 1, \ldots$. □

8.77. Example. Find the homogeneous linear recurring sequence in $\mathbb{F}_2$ of least order whose first 8 terms are $1,1,0,0,1,0,1,1$. We use the Berlekamp-Massey algorithm with $G_7(x) = 1 + x + x^4 + x^6 + x^7 \in \mathbb{F}_2[x]$ in place of $G(x)$. The computation is summarized in the following table.

j	$g_j(x)$	$h_j(x)$	m_j	b_j
0	1	x	0	1
1	$1+x$	x	0	0
2	$1+x$	x^2	1	1
3	$1+x+x^2$	$x+x^2$	-1	1
4	1	x^2+x^3	0	1
5	$1+x^2+x^3$	x	0	0
6	$1+x^2+x^3$	x^2	1	0
7	$1+x^2+x^3$	x^3	2	0
8	$1+x^2+x^3$		3	

Then, $r=\lfloor 4+\frac{1}{2}-\frac{1}{2}m_8\rfloor=3$, and so $m(x)=x^3+x+1$. Therefore, the given terms form the initial segment of a homogeneous linear recurring sequence $s_0, s_1,\ldots$ satisfying $s_{n+3}=s_{n+1}+s_n$ for $n=0,1,\ldots,$ and no such sequence of lower order with these initial terms exists. □

We shall now prove, in general, that *the Berlekamp-Massey algorithm yields the minimal polynomial* after the indicated number of steps. To this end, we define auxiliary polynomials $u_j(x)$ and $v_j(x)$ over $\mathbb{F}_q$ recursively by setting

$$u_0(x)=0 \quad \text{and} \quad v_0(x)=-1, \tag{8.21}$$

and then for $j=0,1,\ldots,$

$$u_{j+1}(x)=u_j(x)-b_j v_j(x),$$

$$v_{j+1}(x)=\begin{cases} b_j^{-1}xu_j(x) & \text{if } b_j\neq 0 \text{ and } m_j\geqslant 0,\\ xv_j(x) & \text{otherwise.}\end{cases} \tag{8.22}$$

We claim that for each $j\geqslant 0$ we have

$$\deg(g_j(x))\leqslant \tfrac{1}{2}(j+1-m_j) \quad \text{and} \quad \deg(h_j(x))\leqslant \tfrac{1}{2}(j+2+m_j). \tag{8.23}$$

This is obvious for $j=0$ because of the initial conditions in (8.19), and assuming the inequalities to be shown for some $j\geqslant 0$, we get from (8.20) in the case where $b_j\neq 0$ and $m_j\geqslant 0$,

$$\begin{aligned}\deg(g_{j+1}(x)) &\leqslant \max(\deg(g_j(x)),\deg(h_j(x)))\\ &\leqslant \tfrac{1}{2}(j+2+m_j)=\tfrac{1}{2}(j+2-m_{j+1}).\end{aligned}$$

Otherwise,

$$\deg(g_{j+1}(x))\leqslant \tfrac{1}{2}(j+1-m_j)=\tfrac{1}{2}(j+2-m_{j+1}).$$

The same distinction of cases proves the second inequality in (8.23). A similar inductive argument shows that for each $j \geqslant 0$ we have

$$\deg(u_j(x)) \leqslant \tfrac{1}{2}(j-1-m_j) \quad \text{and} \quad \deg(v_j(x)) \leqslant \tfrac{1}{2}(j+m_j). \tag{8.24}$$

The auxiliary polynomials $u_j(x)$ and $v_j(x)$ are related to the polynomials $g_j(x)$ and $h_j(x)$ occurring in the algorithm by means of the following congruences, valid for each $j \geqslant 0$:

$$g_j(x)G(x) \equiv u_j(x) + b_j x^j \bmod x^{j+1}, \tag{8.25}$$

$$h_j(x)G(x) \equiv v_j(x) + x^j \bmod x^{j+1}. \tag{8.26}$$

Both (8.25) and (8.26) are true for $j = 0$ because of (8.19), (8.21), and the definition of b_0. Assuming that both congruences have been shown for some $j \geqslant 0$, we get

$$\begin{aligned} g_{j+1}(x)G(x) &= g_j(x)G(x) - b_j h_j(x)G(x) \\ &\equiv u_j(x) + b_j x^j + c_{j+1}x^{j+1} - b_j\left(v_j(x) + x^j + d_{j+1}x^{j+1}\right) \\ &\equiv u_{j+1}(x) + e_{j+1}x^{j+1} \bmod x^{j+2} \end{aligned}$$

with suitable coefficients $c_{j+1}, d_{j+1}, e_{j+1} \in \mathbb{F}_q$. Since $|m_j| \leqslant j$, as is seen easily by induction, we have $\deg(u_{j+1}(x)) \leqslant j$ from (8.24). Therefore, e_{j+1} is the coefficient of x^{j+1} in $g_{j+1}(x)G(x)$, and so $e_{j+1} = b_{j+1}$. The induction step for (8.26) is carried out similarly.

Next, one establishes by a straightforward induction argument that

$$h_j(x)u_j(x) - g_j(x)v_j(x) = x^j \quad \text{for each } j \geqslant 0. \tag{8.27}$$

Now let $s(x)$ and $u(x)$ be polynomials over $\mathbb{F}_q$ with $s(x)G(x) = u(x)$ and $s(0) = 1$. Then by (8.26),

$$\begin{aligned} h_j(x)u(x) - s(x)v_j(x) &= s(x)\left(h_j(x)G(x) - v_j(x)\right) \\ &\equiv s(x)x^j \equiv x^j \bmod x^{j+1}, \end{aligned}$$

and so for some $U_j(x) \in \mathbb{F}_q[x]$ we have

$$h_j(x)u(x) - s(x)v_j(x) = x^j U_j(x) \quad \text{with } U_j(0) = 1. \tag{8.28}$$

Similarly, one uses (8.25) to show that there exists $V_j(x) \in \mathbb{F}_q[x]$ with

$$g_j(x)u(x) - s(x)u_j(x) = x^j V_j(x). \tag{8.29}$$

Now suppose the minimal polynomial $m(x)$ of the given homogeneous linear recurring sequence satisfies $\deg(m(x)) \leqslant k$, and let $s(x)$ be the reciprocal minimal polynomial. Then $s(0) = 1$ and $\deg(s(x)) \leqslant k$, and from (8.15) we know that there exists $u(x) \in \mathbb{F}_q[x]$ with $s(x)G(x) = u(x)$ and $\deg(u(x)) \leqslant \deg(m(x)) - 1 \leqslant k - 1$. Consider (8.28) with $j = 2k$. Using (8.23) and (8.24), we obtain

$$\deg(h_{2k}(x)u(x)) \leqslant \tfrac{1}{2}(2k+2+m_{2k}) + k - 1 = 2k + \tfrac{1}{2}m_{2k}$$

and

$$\deg(s(x)v_{2k}(x)) \leqslant k + \tfrac{1}{2}(2k + m_{2k}) = 2k + \tfrac{1}{2}m_{2k},$$

and so

$$\deg(h_{2k}(x)u(x) - s(x)v_{2k}(x)) \leqslant 2k + \tfrac{1}{2}m_{2k}.$$

On the other hand,

$$\deg(h_{2k}(x)u(x) - s(x)v_{2k}(x)) = \deg(x^{2k}U_{2k}(x)) \geqslant 2k,$$

and these inequalities are only compatible if $m_{2k} \geqslant 0$. Using again (8.23) and (8.24), one verifies that $\deg(g_{2k}(x)u(x))$ and $\deg(s(x)u_{2k}(x))$ are both $\leqslant 2k - \frac{1}{2} - \frac{1}{2}m_{2k}$, hence (8.29) shows that

$$\deg(x^{2k}V_{2k}(x)) = \deg(g_{2k}(x)u(x) - s(x)u_{2k}(x)) < 2k.$$

But this is only possible if $V_{2k}(x)$ is the zero polynomial. Consequently, (8.29) yields $g_{2k}(x)u(x) = s(x)u_{2k}(x)$, and multiplying (8.28) for $j = 2k$ by $g_{2k}(x)$ leads to

$$\begin{aligned} &h_{2k}(x)g_{2k}(x)u(x) - s(x)g_{2k}(x)v_{2k}(x) \\ &\quad = s(x)(h_{2k}(x)u_{2k}(x) - g_{2k}(x)v_{2k}(x)) = x^{2k}U_{2k}(x)g_{2k}(x). \end{aligned}$$

Together with (8.27), we get $s(x) = U_{2k}(x)g_{2k}(x)$, which implies $u(x) = U_{2k}(x)u_{2k}(x)$. Since $s(x)$ is the reciprocal minimal polynomial, it follows from the second part of Theorem 8.40 that $s(x)$ and $u(x)$ are relatively prime. Because of this fact, $U_{2k}(x)$ must be a constant polynomial, and since $U_{2k}(0) = 1$ by (8.28), we actually have $U_{2k}(x) = 1$. Therefore $s(x) = g_{2k}(x)$, and as a by-product we obtain $u(x) = u_{2k}(x)$. If $\deg(m(x)) = k$, then

$$m(x) = x^k s\left(\frac{1}{x}\right) = x^k g_{2k}\left(\frac{1}{x}\right),$$

as we claimed earlier. If $\deg(m(x)) = t \leqslant k$, then we have $s(x) = g_{2t}(x)$, $u(x) = u_{2t}(x)$, and $m_{2t} \geqslant 0$. Clearly, $\max(\deg(s(x)), 1 + \deg(u(x))) \leqslant t$, and the second part of Theorem 8.40 implies that

$$t = \max(\deg(s(x)), 1 + \deg(u(x))).$$

It follows then from (8.23) and (8.24) that

$$t = \max(\deg(g_{2t}(x)), 1 + \deg(u_{2t}(x))) \leqslant t + \tfrac{1}{2} - \tfrac{1}{2}m_{2t},$$

and so $m_{2t} = 0$ or 1. Furthermore, we note that $g_j(x) = s(x)$ and $b_j = 0$ for all $j \geqslant 2t$, so that $m_j = m_{2t} + j - 2t$ for all $j \geqslant 2t$ by the definition of m_j. Setting $j = 2k$, we obtain $t = k + \frac{1}{2}m_{2t} - \frac{1}{2}m_{2k}$, and since $m_{2t} = 0$ or 1, we conclude that

$$t = \lfloor k + \tfrac{1}{2} - \tfrac{1}{2}m_{2k} \rfloor = r.$$

Therefore,

$$m(x) = x^r s\left(\frac{1}{x}\right) = x^r g_{2k}\left(\frac{1}{x}\right),$$

in accordance with our claim.

7. DISTRIBUTION PROPERTIES OF LINEAR RECURRING SEQUENCES

We are interested in the number of occurrences of a given element of $\mathbb{F}_q$ in either the full period or parts of the period of a linear recurring sequence in $\mathbb{F}_q$. In order to provide general information on this question, we first carry out a detailed study of exponential sums that involve linear recurring sequences. It will then become apparent that in the case of linear recurring sequences for which the least period is large, the elements of the underlying finite field appear about equally often in the full period and also in large segments of the full period.

Let $s_0, s_1, \ldots$ be a kth-order linear recurring sequence in $\mathbb{F}_q$ satisfying (8.1), let r be its least period and n_0 its preperiod, so that $s_{n+r} = s_n$ for $n \geqslant n_0$. With this sequence we associate a positive integer R in the following way. Consider the impulse response sequence $d_0, d_1, \ldots$ satisfying (8.6), let r_1 be its least period and n_1 its preperiod; then we set $R = r_1 + n_1$. Of course, R depends only on the linear recurrence relation (8.1) and not on the specific form of the sequence. If $s_0, s_1, \ldots$ is a homogeneous linear recurring sequence with characteristic polynomial $f(x) \in \mathbb{F}_q[x]$, then $r_1 = \operatorname{ord}(f(x))$, and if in addition $f(0) \neq 0$, then $R = \operatorname{ord}(f(x))$, as implied by Theorem 8.27. By the same theorem, r divides r_1 and $r \leqslant R$ in the homogeneous case.

In the exponential sums to be considered, we use additive characters of $\mathbb{F}_q$ as discussed in Chapter 5 and weights defined in terms of the function $e(t) = e^{2\pi i t}$ for real t.

8.78. Theorem. *Let $s_0, s_1, \ldots$ be a kth-order linear recurring sequence in $\mathbb{F}_q$ with least period r and preperiod n_0, and let R be the positive integer introduced above. Let χ be a nontrivial additive character of $\mathbb{F}_q$. Then for every integer h we have*

$$\left|\sum_{n=u}^{u+r-1} \chi(s_n) e\left(\frac{hn}{r}\right)\right| \leqslant \left(\frac{r}{R}\right)^{1/2} q^{k/2} \quad \text{for all } u \geqslant n_0. \tag{8.30}$$

In particular, we have

$$\left|\sum_{n=u}^{u+r-1} \chi(s_n)\right| \leqslant \left(\frac{r}{R}\right)^{1/2} q^{k/2} \quad \text{for all } u \geqslant n_0. \tag{8.31}$$

Proof. By changing the initial state vector from $\mathbf{s}_0$ to $\mathbf{s}_u$, which does not affect the upper bound in (8.30), we may assume, without loss of generality, that the sequence $s_0, s_1, \ldots$ is periodic and that $u = 0$. For a column vector $\mathbf{b} = (b_0, b_1, \ldots, b_{k-1})^{\mathrm{T}}$ in $\mathbb{F}_q^k$ and an integer h, we set

$$\sigma(\mathbf{b}; h) = \sigma(b_0, b_1, \ldots, b_{k-1}; h)$$
$$= \sum_{n=0}^{r-1} \chi(b_0 s_n + b_1 s_{n+1} + \cdots + b_{k-1} s_{n+k-1}) e\left(\frac{hn}{r}\right).$$

Since the general term of this sum has period r as a function of n, we can write

$$\sigma(\mathbf{b}; h) = \sum_{n=0}^{r-1} \chi(b_0 s_{n+1} + b_1 s_{n+2} + \cdots + b_{k-1} s_{n+k}) e\left(\frac{h(n+1)}{r}\right).$$

Using the linear recurrence relation (8.1), we get

$$|\sigma(\mathbf{b}; h)| = \left| \sum_{n=0}^{r-1} \chi(b_0 s_{n+1} + b_1 s_{n+2} + \cdots + b_{k-2} s_{n+k-1} + b_{k-1} a_0 s_n \right.$$
$$\left. + b_{k-1} a_1 s_{n+1} + \cdots + b_{k-1} a_{k-1} s_{n+k-1} + b_{k-1} a) e\left(\frac{hn}{r}\right) \right|$$
$$= \left| \sum_{n=0}^{r-1} \chi(b_{k-1} a_0 s_n + (b_0 + b_{k-1} a_1) s_{n+1} + \cdots \right.$$
$$\left. + (b_{k-2} + b_{k-1} a_{k-1}) s_{n+k-1}) e\left(\frac{hn}{r}\right) \right|$$
$$= |\sigma(b_{k-1} a_0, b_0 + b_{k-1} a_1, \ldots, b_{k-2} + b_{k-1} a_{k-1}; h)|.$$

This identity can be written in the form

$$|\sigma(\mathbf{b}; h)| = |\sigma(A\mathbf{b}; h)|,$$

where A is the matrix in (8.3). It follows by induction that

$$|\sigma(\mathbf{b}; h)| = |\sigma(A^j \mathbf{b}; h)| \quad \text{for all } j \geqslant 0. \tag{8.32}$$

Let $\mathbf{d}$ be the column vector $\mathbf{d} = (1, 0, \ldots, 0)^{\mathrm{T}}$ in $\mathbb{F}_q^k$, and let $\mathbf{d}_0, \mathbf{d}_1, \ldots$ be the state vectors of the impulse response sequence $d_0, d_1, \ldots$ satisfying (8.6). Then we claim that two state vectors $\mathbf{d}_m$ and $\mathbf{d}_n$ are identical if and only if $A^m\mathbf{d} = A^n\mathbf{d}$. For if $\mathbf{d}_m = \mathbf{d}_n$, then $A^m\mathbf{d} = A^n\mathbf{d}$ follows from Lemma 8.15. On the other hand, if $A^m\mathbf{d} = A^n\mathbf{d}$, then $A^{m+j}\mathbf{d} = A^{n+j}\mathbf{d}$, and so $A^m(A^j\mathbf{d}) = A^n(A^j\mathbf{d})$, for all $j \geqslant 0$. But since the vectors $\mathbf{d}$, $A\mathbf{d}$, $A^2\mathbf{d}, \ldots, A^{k-1}\mathbf{d}$ form a basis for the vector space $\mathbb{F}_q^k$ over $\mathbb{F}_q$, we get $A^m = A^n$, which implies $\mathbf{d}_m = \mathbf{d}_n$ by Lemma 8.15.

The distinct vectors in the sequence $\mathbf{d}_0, \mathbf{d}_1, \ldots$ are exactly given by $\mathbf{d}_0, \mathbf{d}_1, \ldots, \mathbf{d}_{R-1}$. Therefore, by what we have just shown, the distinct vectors among $\mathbf{d}$, $A\mathbf{d}$, $A^2\mathbf{d}, \ldots$ are exactly given by $\mathbf{d}$, $A\mathbf{d}, \ldots, A^{R-1}\mathbf{d}$. Using (8.32), we get

$$R|\sigma(\mathbf{d}; h)|^2 = \sum_{j=0}^{R-1} \left|\sigma(A^j\mathbf{d}; h)\right|^2 \leqslant \sum_{\mathbf{b}} |\sigma(\mathbf{b}; h)|^2, \tag{8.33}$$

where the last sum is taken over all column vectors $\mathbf{b}$ in $\mathbb{F}_q^k$. Now

$$\begin{aligned}
\sum_{\mathbf{b}} |\sigma(\mathbf{b}; h)|^2 &= \sum_{\mathbf{b}} \sigma(\mathbf{b}; h)\,\overline{\sigma(\mathbf{b}; h)} \\
&= \sum_{b_0, b_1, \ldots, b_{k-1} \in \mathbb{F}_q} \sum_{m,n=0}^{r-1} \chi(b_0(s_m - s_n) + b_1(s_{m+1} - s_{n+1}) \\
&\quad + \cdots + b_{k-1}(s_{m+k-1} - s_{n+k-1}))\, e\left(\frac{h(m-n)}{r}\right) \\
&= \sum_{m,n=0}^{r-1} e\left(\frac{h(m-n)}{r}\right) \\
&\quad \cdot \sum_{b_0, b_1, \ldots, b_{k-1} \in \mathbb{F}_9} \chi(b_0(s_m - s_n))\chi(b_1(s_{m+1} - s_{n+1})) \cdots \\
&\quad \cdot \chi(b_{k-1}(s_{m+k-1} - s_{n+k-1})) \\
&= \sum_{m,n=0}^{r-1} e\left(\frac{h(m-n)}{r}\right)\left(\sum_{b_0 \in \mathbb{F}_q} \chi(b_0(s_m - s_n))\right) \cdots \\
&\quad \cdot \left(\sum_{b_{k-1} \in \mathbb{F}_q} \chi(b_{k-1}(s_{m+k-1} - s_{n+k-1}))\right).
\end{aligned} \tag{8.34}$$

We note that for $c \in \mathbb{F}_q$ we have

$$\sum_{b \in \mathbb{F}_q} \chi(bc) = \begin{cases} 0 & \text{if } c \neq 0, \\ q & \text{if } c = 0, \end{cases}$$

according to (5.9). Therefore, in the last expression in (8.34) one only gets a contribution from those ordered pairs (m, n) for which simultaneously $s_m = s_n, \ldots, s_{m+k-1} = s_{n+k-1}$. But since $0 \leqslant m, n \leqslant r-1$, this is only possible for $m = n$. It follows that

$$\sum_{\mathbf{b}} |\sigma(\mathbf{b}; h)|^2 = rq^k.$$

By combining this with (8.33), we arrive at

$$|\sigma(\mathbf{d}; h)| \leqslant \left(\frac{r}{R}\right)^{1/2} q^{k/2},$$

which proves (8.30). The inequality (8.31) results from (8.30) by setting $h = 0$. □

8.79. Remark. Let χ be a nontrivial additive character of $\mathbb{F}_q$ and let ψ be an arbitrary multiplicative character of $\mathbb{F}_q$. Then the Gaussian sum

$$G(\psi, \chi) = \sum_{c \in \mathbb{F}_q^*} \psi(c)\chi(c)$$

can be considered as a special case of the sum in (8.30). To see this, let g be a primitive element of $\mathbb{F}_q$ and introduce the first-order linear recurring sequence $s_0, s_1, \ldots$ in $\mathbb{F}_q$ with $s_0 = 1$ and $s_{n+1} = gs_n$ for $n = 0, 1, \ldots$. Then $r = R = q-1$ and $n_0 = 0$. We note that $\psi(g) = e(h/r)$ for some integer h. Thus we can write

$$G(\psi, \chi) = \sum_{n=0}^{r-1} \chi(g^n)\psi(g^n) = \sum_{n=0}^{r-1} \chi(s_n) e\left(\frac{hn}{r}\right).$$

If ψ is nontrivial, then in this special case both sides of (8.30) are identical according to (5.15). □

The sums in Theorem 8.78 are extended over a full period of the given linear recurring sequence. An estimate for character sums over segments of the period can be deduced from this result. We need the following auxiliary inequality.

8.80. Lemma. *For any positive integers r and N we have*

$$\sum_{h=0}^{r-1} \left| \sum_{j=0}^{N-1} e\left(\frac{hj}{r}\right) \right| < \frac{2}{\pi} r \log r + \frac{2}{5} r + N. \tag{8.35}$$

Proof. The inequality is trivial for $r = 1$. For $r \geqslant 2$ we have

$$\left| \sum_{j=0}^{N-1} e\left(\frac{hj}{r}\right) \right| = \frac{|e(hN/r) - 1|}{|e(h/r) - 1|} \leqslant \frac{1}{\sin \pi \|h/r\|}$$
$$= \csc \pi \left\| \frac{h}{r} \right\| \quad \text{for } 1 \leqslant h \leqslant r-1,$$

where $\|t\|$ denotes the absolute distance from the real number t to the nearest integer. It follows that

$$\sum_{h=0}^{r-1} \left| \sum_{j=0}^{N-1} e\left(\frac{hj}{r}\right) \right| \leqslant \sum_{h=1}^{r-1} \csc \pi \left\| \frac{h}{r} \right\| + N \leqslant 2 \sum_{h=1}^{\lfloor r/2 \rfloor} \csc \frac{\pi h}{r} + N. \tag{8.36}$$

By comparing sums with integrals, we obtain

$$\sum_{h=1}^{\lfloor r/2 \rfloor} \csc \frac{\pi h}{r} = \csc \frac{\pi}{r} + \sum_{h=2}^{\lfloor r/2 \rfloor} \csc \frac{\pi h}{r} \leqslant \csc \frac{\pi}{r} + \int_1^{\lfloor r/2 \rfloor} \csc \frac{\pi x}{r} \, dx$$
$$\leqslant \csc \frac{\pi}{r} + \frac{r}{\pi} \int_{\pi/r}^{\pi/2} \csc t \, dt$$
$$= \csc \frac{\pi}{r} + \frac{r}{\pi} \log \cot \frac{\pi}{2r} \leqslant \csc \frac{\pi}{r} + \frac{r}{\pi} \log \frac{2r}{\pi}.$$

For $r \geqslant 6$ we have $(\pi/r)^{-1} \sin(\pi/r) \geqslant (\pi/6)^{-1} \sin(\pi/6)$, hence $\sin(\pi/r) \geqslant$

$3/r$. This implies

$$\sum_{h=1}^{\lfloor r/2 \rfloor} \csc\frac{\pi h}{r} \leqslant \frac{1}{\pi} r\log r + \left(\frac{1}{3} - \frac{1}{\pi}\log\frac{\pi}{2}\right) r \quad \text{for } r \geqslant 6,$$

and so

$$\sum_{h=1}^{\lfloor r/2 \rfloor} \csc\frac{\pi h}{r} < \frac{1}{\pi} r\log r + \frac{1}{5} r \quad \text{for } r \geqslant 6.$$

This inequality is easily checked for $r = 3$, 4, and 5, so that (8.35) holds for $r \geqslant 3$ in view of (8.36). For $r = 2$ the inequality (8.35) is shown by inspection. □

8.81. Theorem. *Let $s_0, s_1, \ldots$ be a kth-order linear recurring sequence in $\mathbb{F}_q$, and let r, n_0, and R be as in Theorem 8.78. Then, for any nontrivial additive character χ of $\mathbb{F}_q$ we have*

$$\left|\sum_{n=u}^{u+N-1} \chi(s_n)\right| < \left(\frac{r}{R}\right)^{1/2} q^{k/2}\left(\frac{2}{\pi}\log r + \frac{2}{5} + \frac{N}{r}\right) \quad \textit{for } u \geqslant n_0 \textit{ and } 1 \leqslant N \leqslant r.$$

Proof. We start from the identity

$$\sum_{n=u}^{u+N-1} \chi(s_n) = \sum_{n=u}^{u+r-1} \chi(s_n) \sum_{j=0}^{N-1} \frac{1}{r} \sum_{h=0}^{r-1} e\left(\frac{h(n-u-j)}{r}\right) \quad \text{for } 1 \leqslant N \leqslant r,$$

which is valid since the sum over j is 1 for $u \leqslant n \leqslant u+N-1$ and 0 for $u+N \leqslant n \leqslant u+r-1$. Rearranging terms, we get

$$\sum_{n=u}^{u+N-1} \chi(s_n) = \frac{1}{r} \sum_{h=0}^{r-1} \left(\sum_{j=0}^{N-1} e\left(\frac{-h(u+j)}{r}\right)\right)\left(\sum_{n=u}^{u+r-1} \chi(s_n) e\left(\frac{hn}{r}\right)\right),$$

and so by (8.30),

$$\left|\sum_{n=u}^{u+N-1} \chi(s_n)\right| \leqslant \frac{1}{r} \sum_{h=0}^{r-1} \left|\sum_{j=0}^{N-1} e\left(\frac{-h(u+j)}{r}\right)\right| \left|\sum_{n=u}^{u+r-1} \chi(s_n) e\left(\frac{hn}{r}\right)\right|$$

$$\leqslant \frac{1}{r}\left(\frac{r}{R}\right)^{1/2} q^{k/2} \sum_{h=0}^{r-1} \left|\sum_{j=0}^{N-1} e\left(\frac{hj}{r}\right)\right|.$$

An application of Lemma 8.80 yields the desired inequality. □

It should be noted that the inequalities in Theorems 8.78 and 8.81 are only of interest if the least period r of $s_0, s_1, \ldots$ is sufficiently large. For small r, these results are actually weaker than the trivial estimate

$$\left|\sum_{n=u}^{u+N-1} \chi(s_n)\right| \leqslant N \quad \text{for } 1 \leqslant N \leqslant r.$$

In order to obtain nontrivial statements, r should be somewhat larger than $q^{k/2}$.

Let $s_0, s_1, \ldots$ be a linear recurring sequence in $\mathbb{F}_q$ with least period r and preperiod n_0. For $b \in \mathbb{F}_q$ we denote by $Z(b)$ the number of $n, n_0 \leqslant n \leqslant n_0 + r - 1$, with $s_n = b$. Therefore $Z(b)$ is the number of occurrences of b in a full period of the linear recurring sequence.

If $s_0, s_1, \ldots$ is a kth-order maximal period sequence, then $Z(b)$ can be determined explicitly. We have $r = q^k - 1$ and $n_0 = 0$ according to Theorem 8.33, and so the state vectors $\mathbf{s}_0, \mathbf{s}_1, \ldots, \mathbf{s}_{r-1}$ of the sequence run exactly through all nonzero vectors in $\mathbb{F}_q^k$. Consequently, $Z(b)$ is equal to the number of nonzero vectors in $\mathbb{F}_q^k$ that have b as a first coordinate. Elementary counting arguments show then that $Z(b) = q^{k-1}$ for $b \neq 0$ and $Z(0) = q^{k-1} - 1$. Therefore, up to a slight aberration for the zero element, the elements of $\mathbb{F}_q$ occur equally often in a full period of a maximal period sequence.

In the general case, one cannot expect such an equitable distribution of elements. One may, however, estimate the deviation between the actual number of occurrences and the ideal number r/q. If r is sufficiently large, then this deviation is comparatively small.

8.82. Theorem. *Let $s_0, s_1, \ldots$ be a kth-order linear recurring sequence in $\mathbb{F}_q$ with least period r, and let R be as in Theorem 8.78. Then, for any $b \in \mathbb{F}_q$ we have*

$$\left| Z(b) - \frac{r}{q} \right| \leqslant \left(1 - \frac{1}{q}\right)\left(\frac{r}{R}\right)^{1/2} q^{k/2}.$$

Proof. For given $b \in \mathbb{F}_q$, let the real-valued function δ_b on $\mathbb{F}_q$ be defined by $\delta_b(b) = 1$ and $\delta_b(c) = 0$ for $c \neq b$. Because of (5.10), the function δ_b can be represented in the form

$$\delta_b(c) = \frac{1}{q} \sum_{\chi} \chi(c - b) \quad \text{for all } c \in \mathbb{F}_q,$$

where the sum is extended over all additive characters χ of $\mathbb{F}_q$. It follows that

$$\begin{aligned} Z(b) = \sum_{n=n_0}^{n_0+r-1} \delta_b(s_n) &= \sum_{n=n_0}^{n_0+r-1} \frac{1}{q} \sum_{\chi} \chi(s_n - b) \\ &= \frac{1}{q} \sum_{\chi} \bar{\chi}(b) \sum_{n=n_0}^{n_0+r-1} \chi(s_n). \end{aligned}$$

By separating the contribution from the trivial additive character of $\mathbb{F}_q$ and using an asterisk to indicate the deletion of this character from the range of summation, we get

$$Z(b)-\frac{r}{q}=\frac{1}{q}\sum_{\chi}{}^{*}\bar{\chi}(b)\sum_{n=n_0}^{n_0+r-1}\chi(s_n).$$

Thus, by using (8.31), we obtain

$$\left|Z(b)-\frac{r}{q}\right|\leqslant\frac{1}{q}\sum_{\chi}{}^{*}\left|\sum_{n=n_0}^{n_0+r-1}\chi(s_n)\right|\leqslant\left(1-\frac{1}{q}\right)\left(\frac{r}{R}\right)^{1/2}q^{k/2},$$

since there are $q-1$ nontrivial additive characters of $\mathbb{F}_q$. □

8.83. Corollary. *Let $s_0, s_1,\ldots$ be a homogeneous linear recurring sequence in $\mathbb{F}_q$ with least period r whose minimal polynomial $m(x)\in\mathbb{F}_q[x]$ has degree $k\geqslant 1$ and satisfies $m(0)\neq 0$. Then, for every $b\in\mathbb{F}_q$ we have*

$$\left|Z(b)-\frac{r}{q}\right|\leqslant\left(1-\frac{1}{q}\right)q^{k/2}.$$

Proof. We have $r=\operatorname{ord}(m(x))$ according to Theorem 8.44. Furthermore, $R=\operatorname{ord}(m(x))$ by a remark preceding Theorem 8.78, and Theorem 8.82 yields the desired result. □

If the linear recurring sequence has an irreducible minimal polynomial, then an alternative method based on Gaussian sums leads to somewhat better estimates. In the subsequent proof, we shall use the formulas for Gaussian sums in Theorem 5.11.

8.84. Theorem. *Let $s_0, s_1,\ldots$ be a homogeneous linear recurring sequence in $\mathbb{F}_q$ with least period r. Suppose the minimal polynomial $m(x)$ of the sequence is irreducible over $\mathbb{F}_q$, has degree k, and satisfies $m(0)\neq 0$. Let h be the least common multiple of r and $q-1$. Then,*

$$\left|Z(0)-\frac{(q^{k-1}-1)r}{q^k-1}\right|\leqslant\left(1-\frac{1}{q}\right)\left(\frac{r}{h}-\frac{r}{q^k-1}\right)q^{k/2} \tag{8.37}$$

and

$$\left|Z(b)-\frac{q^{k-1}r}{q^k-1}\right|\leqslant\left(\frac{r}{h}-\frac{r}{q^k-1}+\frac{h-r}{h}q^{1/2}\right)q^{(k/2)-1}\quad\text{for } b\neq 0. \tag{8.38}$$

Proof. Set $K=\mathbb{F}_q$, and let F be the splitting field of $m(x)$ over K. Let α be a fixed root of $m(x)$ in F; then $\alpha\neq 0$ because of $m(0)\neq 0$. By Theorem 8.24, there exists $\theta\in F$ such that

$$s_n=\operatorname{Tr}_{F/K}(\theta\alpha^n)\text{ for } n=0,1,\ldots. \tag{8.39}$$

We clearly have $\theta\neq 0$. Let λ' be the canonical additive character of K. Then, for any given $b \in K$, the character relation (5.9) yields

$$\frac{1}{q} \sum_{c \in K} \lambda'(c(b - s_n)) = \begin{cases} 1 & \text{if } s_n = b, \\ 0 & \text{if } s_n \neq b, \end{cases}$$

and so, together with (8.39),

$$Z(b) = \frac{1}{q} \sum_{n=0}^{r-1} \sum_{c \in K} \lambda'(bc) \lambda'(\mathrm{Tr}_{F/K}(-c\theta\alpha^n)).$$

If λ denotes the canonical additive character of F, then λ' and λ are related by $\lambda'(\mathrm{Tr}_{F/K}(\beta)) = \lambda(\beta)$ for all $\beta \in F$ (see (5.7)). Therefore,

$$\begin{aligned} Z(b) &= \frac{1}{q} \sum_{c \in K} \lambda'(bc) \sum_{n=0}^{r-1} \bar{\lambda}(c\theta\alpha^n) \\ &= \frac{r}{q} + \frac{1}{q} \sum_{c \in K^*} \lambda'(bc) \sum_{n=0}^{r-1} \bar{\lambda}(c\theta\alpha^n). \end{aligned} \tag{8.40}$$

Now by (5.17),

$$\bar{\lambda}(\beta) = \frac{1}{q^k - 1} \sum_{\psi} G(\bar{\psi}, \bar{\lambda}) \psi(\beta) \quad \text{for } \beta \in F^*,$$

where the sum is extended over all multiplicative characters ψ of F. For $c \in K^*$ it follows that

$$\begin{aligned} \sum_{n=0}^{r-1} \bar{\lambda}(c\theta\alpha^n) &= \frac{1}{q^k - 1} \sum_{n=0}^{r-1} \sum_{\psi} G(\bar{\psi}, \bar{\lambda}) \psi(c\theta\alpha^n) \\ &= \frac{1}{q^k - 1} \sum_{\psi} \psi(c\theta) G(\bar{\psi}, \bar{\lambda}) \sum_{n=0}^{r-1} \psi(\alpha)^n. \end{aligned}$$

The inner sum in the last expression is a finite geometric series that vanishes if $\psi(\alpha) \neq 1$, because of $\psi(\alpha)^r = \psi(\alpha^r) = \psi(1) = 1$. Therefore, we only have to sum over the set J of those characters ψ for which $\psi(\alpha) = 1$, and so

$$\sum_{n=0}^{r-1} \bar{\lambda}(c\theta\alpha^n) = \frac{r}{q^k - 1} \sum_{\psi \in J} \psi(c\theta) G(\bar{\psi}, \bar{\lambda}).$$

Substituting this in (8.40), we get

$$\begin{aligned} Z(b) &= \frac{r}{q} + \frac{r}{q(q^k - 1)} \sum_{c \in K^*} \lambda'(bc) \sum_{\psi \in J} \psi(c\theta) G(\bar{\psi}, \bar{\lambda}) \\ &= \frac{r}{q} + \frac{r}{q(q^k - 1)} \sum_{\psi \in J} \psi(\theta) G(\bar{\psi}, \bar{\lambda}) \sum_{c \in K^*} \psi(c) \lambda'(bc). \end{aligned}$$

If we consider the restriction ψ' of ψ to K^*, then the inner sum may be viewed as a Gaussian sum in K with an additive character $\lambda'_b(c) = \lambda'(bc)$ for

$c \in K$. Thus,

$$Z(b) = \frac{r}{q} + \frac{r}{q(q^k - 1)} \sum_{\psi \in J} \psi(\theta) G(\bar{\psi}, \bar{\lambda}) G(\psi', \lambda'_b). \tag{8.41}$$

Now let $b = 0$. Then λ'_b is the trivial additive character of K, and so the Gaussian sum $G(\psi', \lambda'_b)$ vanishes unless ψ' is trivial, in which case $G(\psi', \lambda'_b) = q - 1$. Consequently, it suffices to extend the sum in (8.41) over the set A of characters ψ for which $\psi(\alpha) = 1$ and ψ' is trivial, so that

$$Z(0) = \frac{r}{q} + \frac{(q-1)r}{q(q^k - 1)} \sum_{\psi \in A} \psi(\theta) G(\bar{\psi}, \bar{\lambda}).$$

The trivial multiplicative character contributes -1 to the sum, hence we get

$$Z(0) - \frac{(q^{k-1} - 1)r}{q^k - 1} = \frac{(q-1)r}{q(q^k - 1)} \sum_{\psi \in A}{}^{*} \psi(\theta) G(\bar{\psi}, \bar{\lambda}),$$

where the asterisk indicates that the trivial multiplicative character is deleted from the range of summation. Since λ is nontrivial, we have $|G(\bar{\psi}, \bar{\lambda})| = q^{k/2}$ for every nontrivial ψ, and so

$$\left| Z(0) - \frac{(q^{k-1} - 1)r}{q^k - 1} \right| \leqslant \frac{(q-1)r}{q(q^k - 1)} (|A| - 1) q^{k/2}. \tag{8.42}$$

Let H be the smallest subgroup of F^* containing α and K^*. The element α has order r in the cyclic group F^*, therefore $|H| = h$, the least common multiple of r and $q - 1$. Furthermore, we have $\psi \in A$ if and only if $\psi(\beta) = 1$ for all $\beta \in H$. In other words, A is the annihilator of H in $(F^*)^\wedge$ (see p. 189), and so

$$|A| = \frac{|F^*|}{|H|} = \frac{q^k - 1}{h} \tag{8.43}$$

by Theorem 5.6. The inequality (8.37) follows now from (8.42) and (8.43).

For $b \neq 0$, we go back to (8.41) and note first that the additive character λ'_b is then nontrivial. Therefore, the trivial multiplicative character contributes 1 to the sum in (8.41), so that we can write

$$Z(b) - \frac{q^{k-1} r}{q^k - 1} = \frac{r}{q(q^k - 1)} \sum_{\psi \in J}{}^{*} \psi(\theta) G(\bar{\psi}, \bar{\lambda}) G(\psi', \lambda'_b).$$

Now $G(\psi', \lambda'_b) = -1$ if ψ' is trivial and $|G(\psi', \lambda'_b)| = q^{1/2}$ if ψ' is nontrivial, which implies

$$\left| Z(b) - \frac{q^{k-1} r}{q^k - 1} \right| \leqslant \frac{r}{q^k - 1} \left(|A| - 1 + (|J| - |A|) q^{1/2} \right) q^{(k/2) - 1}.$$

Since J is the annihilator in $(F^*)^\wedge$ of the subgroup of F^* generated by α, we have $|J| = (q^k - 1)/r$ by Theorem 5.6. This is combined with (8.43) to complete the proof of (8.38). □

One can also obtain results about the distribution of elements in parts of the period. Let $s_0, s_1, \ldots$ be an arbitrary linear recurring sequence in $\mathbb{F}_q$ with least period r and preperiod n_0. For $b \in \mathbb{F}_q$, for $N_0 \geqslant n_0$ and $1 \leqslant N \leqslant r$, let $Z(b; N_0, N)$ be the number of n, $N_0 \leqslant n \leqslant N_0 + N - 1$, with $s_n = b$.

8.85. Theorem. *Let $s_0, s_1, \ldots$ be a kth-order linear recurring sequence in $\mathbb{F}_q$ with least period r and preperiod n_0, and let R be as in Theorem 8.78. Then, for any $b \in \mathbb{F}_q$ we have*

$$\left| Z(b; N_0, N) - \frac{N}{q} \right| \leqslant \left(1 - \frac{1}{q}\right)\left(\frac{r}{R}\right)^{1/2} q^{k/2}\left(\frac{2}{\pi}\log r + \frac{2}{5} + \frac{N}{r}\right)$$

for $N_0 \geqslant n_0$ and $1 \leqslant N \leqslant r$.

Proof. Proceeding as in the proof of Theorem 8.82 and using the same notation as there, we arrive at the identity

$$Z(b; N_0, N) - \frac{N}{q} = \frac{1}{q} \sum_{\chi}{}^{*} \bar{\chi}(b) \sum_{n=N_0}^{N_0+N-1} \chi(s_n).$$

On the basis of Theorem 8.81 we obtain then

$$\left| Z(b; N_0, N) - \frac{N}{q} \right| \leqslant \frac{1}{q} \sum_{\chi}{}^{*} \left| \sum_{n=N_0}^{N_0+N-1} \chi(s_n) \right|$$
$$\leqslant \left(1 - \frac{1}{q}\right)\left(\frac{r}{R}\right)^{1/2} q^{k/2}\left(\frac{2}{\pi}\log r + \frac{2}{5} + \frac{N}{r}\right),$$

since there are $q - 1$ nontrivial additive characters of $\mathbb{F}_q$. □

The method in the proof of Theorem 8.84 can also be adapted to produce results on the distribution of elements in parts of the period (compare with Exercises 8.69, 8.70, and 8.71).

NOTES

1. The theory of linear recurring sequences has a very long history, which Dickson [40, Ch. 17] traced from the year 1202 to 1918. Initially, the attention was devoted to linear recurring sequences of integers, notably the celebrated *Fibonacci sequence* $F_0, F_1, F_2, \ldots$ defined by $F_0 = 0$, $F_1 = 1$, and $F_{n+2} = F_{n+1} + F_n$ for $n = 0, 1, \ldots$. Later, linear recurring sequences of real or complex numbers were considered, especially in connection with the calculus of finite differences. The interest in linear recurring sequences in finite

fields arose when linear recurring sequences in $\mathbb{Z}$ were considered modulo a prime modulus p, thus obtaining linear recurring sequences in $\mathbb{F}_p$. Since the 1950s linear recurring sequences in finite fields have become important in electrical engineering because of their connections with switching circuits and coding theory. For a brief survey of the history of the subject concentrating on the development after 1918, see Selmer [3, Ch. 2].

Important classical papers on linear recurring sequences are Lucas [1] and d'Ocagne [1], and expositions of the subject can already be found in the books of Lucas [2, Chs. 17, 18] and Bachmann [5, Ch. 2]. The first notable contributions to the theory of linear recurring sequences in finite fields are due to Mantel [1] for $\mathbb{F}_p$ and to Scarpis [2] for general $\mathbb{F}_q$. Further work until the middle of the 20th century concentrated on linear recurring sequences in $\mathbb{Z}$ and $\mathbb{Z}/(m)$ (see Bell [1], Carmichael [1], [2], [3], Engstrom [1], [2], Hall [1], [2], [3], [4], Ward [2], [3], [4], [7], [8], [9], [11], [12], [14], [16], and particularly the fundamental paper of Ward [5]), but linear recurring sequences in arbitrary fields were considered in Ward [1] and linear recurring sequences in arbitrary commutative rings were discussed in Ward [13], [15]. The basic paper for the modern theory of linear recurring sequences in finite fields is Zierler [4]. Expository accounts of this theory can be found in the books of Birkhoff and Bartee [1, Ch. 13], Dornhoff and Hohn [1, Ch. 8], Gill [2], Golomb [4], Lüneburg [2], and Peterson and Weldon [1], in the lecture notes of Selmer [3], and in the survey article of Fillmore and Marx [1]. For detailed information on the Fibonacci sequence see Bachmann [5, Ch. 2], Jarden [1], Knuth [2, Ch. 1], Vorob'ev [1], and the journal *Fibonacci Quarterly*. Linear recurring sequences of real or complex numbers are treated for instance in the books of Gel'fond [1, Ch. 5], Ch. Jordan [1, Ch. 11], Markuševič [1], Milne-Thomson [1, Ch. 13], Montel [1], and Nörlund [1, Ch. 10].

The physical implementation of feedback shift registers and their design elements is described in McCluskey [1]. Roth [1] discusses the efficient design of feedback shift registers with $\mathbb{F}_2$ as the underlying field. The interplay between feedback shift registers and linear recurring sequences is stressed in Golomb [4], Peterson and Weldon [1], and Selmer [3]. Discussions of feedback shift registers in the general context of switching circuits and finite-state machines can be found in the books of Booth [1, Ch. 8], Gill [2], Golomb [4, Ch. 2], and Zadeh and Polak [1, Ch. 2]; see also Chapter 9, Section 5, for a more general viewpoint.

Theorem 8.7 is essentially due to Mantel [1]. Theorem 8.11 is a special case of a result of Ward [15]. In the context of linear recurring sequences in finite fields, the matrix A in (8.3), which is the companion matrix of the characteristic polynomial of the sequence, was introduced in Brenner [1] where one can also find Theorem 8.13. Subsequently, matrix methods were used extensively in this area; see Golomb [1], Birdsall and Ristenblatt [1], Elspas [1], Friedland [1], Stern and Friedland [1], and

Mendelsohn [1]. These methods have the advantage that they can also be applied to linear recurring sequences in much more general algebraic structures (see, e.g., Niederreiter [6]). The computational aspect of matrix methods was discussed by Kamal, Singh, Puri, and Nanda [1] and Latawiec [1].

For proofs of the formula for the order of $GL(k, \mathbb{F}_q)$ see, for example, Artin [7, Ch. 4], Carmichael [4, Ch. 10], Dickson [7, Part II, Ch. 1], and Newman [1, Ch. 7]. In these books one also finds formulas for the order of other matrix groups over $\mathbb{F}_q$, such as special linear groups, orthogonal groups, and symplectic groups. The group-theoretic aspect of such matrix groups is discussed for instance in Artin [5], [6], Carmichael [4, Ch. 10], Chevalley [2], Dickson [7, Part II], Dieudonné [2], and Dixon [1]. For a modern treatment of the representation theory of such groups see Srinivasan [1]. The formula for the order of $GL(k, \mathbb{F}_q)$ is a special case of the result that the number of $m \times n$ matrices over $\mathbb{F}_q$ of rank r is equal to

$$q^{(r^2-r)/2} \prod_{i=0}^{r-1} (q^{m-i}-1)(q^{n-i}-1)(q^{i+1}-1)^{-1} \quad \text{for } 1 \leqslant r \leqslant \min(m, n).$$

This was established by Landsberg [1] for q prime; see also Arghiriade and Peterfi [1], Boroş [1], and Fisher and Alexander [1] for proofs of this formula, as well as Porter and Riveland [1] for the case where $s \leqslant r$ linearly independent rows are fixed in the matrix. Klein [3] considers the number of $m \times n$ matrices over $\mathbb{F}_p$ for which all minors of size $\min(m, n)$, or all minors of size $\leqslant \min(m, n)$, are nonzero. A. Lee [1] showed that there exists no $(q-1) \times q$ matrix over $\mathbb{F}_q$ whose minors of size $q-1$ and $q-2$ are all nonzero. Carlitz and Hodges [4] enumerated rectangular matrices of given rank with prescribed ranks of submatrices and Brawley and Carlitz [1] and Fisher and Alexander [1] those with prescribed row and column sums. Other enumeration problems for rectangular matrices over $\mathbb{F}_q$ have been treated by Carlitz and Hodges [2], Daykin [2], J. D. Fulton [8], [10], Hodges [7], and Kim [1]. For square matrices over $\mathbb{F}_q$ of given size, further enumeration problems have been considered. Buckhiester [1] determined the number of such matrices with prescribed rank and trace (see also Johnson, Porter, and Varineau [1] for the special case of full rank), Reiner [1] and Gerstenhaber [1] found the number of matrices with given characteristic polynomial, and Carlitz and Hodges [3] gave a formula for the number of nonderogatory matrices. Fine and Herstein [1] and Gerstenhaber [1] prove that there are exactly q^{n^2-n} nilpotent $n \times n$ matrices over $\mathbb{F}_q$, and Bollman and Ramírez [1] enumerated nilpotent matrices over $\mathbb{Z}/(m)$ of given size and rank. Berlekamp [2] counted circulant matrices by rank and Carlitz [51], [54] and Carlitz and Hodges [1] enumerated respectively skew-symmetric, symmetric, and Hermitian matrices by rank; see also MacWilliams [3] for related formulas and alternative proofs, Brawley and Carlitz [1] for the case where prescribed row and column sums are used as additional constraints, and MacDougall [1] for an application. Feit and Fine [1] determined the number of ordered pairs of commuting $n \times n$ matrices over $\mathbb{F}_q$ and Carlitz [92] treated a related problem. Kung [1] obtained the number of nonsingular matrices commuting with a given block diagonal matrix. Equivalence and similarity classes of matrices were counted by Brawley [1], Carlitz [104], Carlitz and Hodges [3], and Gow [1]. Brawley and Mullen [1] enumerated the diagonalizable matrices having a prescribed number of distinct eigenvalues. For a given square matrix A over $\mathbb{F}_q$, Daykin [1] determined the number of distinct matrices $f(A)$ of prescribed rank with f running through $\mathbb{F}_q[x]$. For results on the number of solutions of matrix equations we refer to the notes for Chapter 6, Section 2. Closely related to these enumeration problems for matrices is the problem of counting subspaces of vector spaces over $\mathbb{F}_q$; here Dickson [7, Part I, Ch. 4] and Moore [4] have shown that the number of r-dimensional subspaces of an n-dimensional vector space over $\mathbb{F}_q$ is given by

$$\prod_{i=0}^{r-1} (q^{n-i}-1)(q^{r-i}-1)^{-1} \quad \text{for } 1 \leqslant r \leqslant n.$$

The least common multiple of the orders of all elements of $GL(k, \mathbb{F}_q)$ was determined by Niven [1] to be $p^e M$, where p is the characteristic of $\mathbb{F}_q$, e is the least integer with $p^e \geqslant k$, and M is the least common multiple of $q-1, q^2-1, \ldots, q^k-1$; this completed and generalized the work of Marshall [1]. Thus, the statement that the order of A in $GL(k, \mathbb{F}_q)$ divides

$$q^{(k^2-k)/2}(q-1)(q^2-1)\cdots(q^k-1)$$

can be refined to the statement that this order divides $p^e M$. An analog of Niven's result for $GL(k, \mathbb{Z}/(m))$ was shown by Davis [1] and Maxfield [1]. Niven [1] also gives an algorithm for determining the order of an element of $GL(k, \mathbb{F}_q)$. For further work on orders of matrices see Bollman [1], Dai [1], Fillmore and Marx [1], Gaiu [1], and Lüneburg [2, Chs. 32, 33].

Apart from the application of linear recurring sequences in finite fields to the analysis and synthesis of feedback shift registers, the other main application of such sequences arises in coding theory, especially in the theory of cyclic codes (compare with Chapter 9, Section 2). The pioneering work on the interplay between linear recurring sequences, feedback shift registers, and coding theory was done by Abramson [1], Green and San Soucie [1], Huffman [1], [2], Kasami [1], Mattson and Solomon [1], Peterson [1], Prange [2], Stern and Friedland [1], Yale [1], Zetterberg [1], and Zierler [1], [3]. See also Gabidulin [1], Massey [3], Mykkeltveit [1], and Zierler [5], as well as the books of Ash [1, Ch. 5], Lin [2, Ch. 4], and Peterson and Weldon [1]. For applications to computations in $\mathbb{F}_q$ and $\mathbb{F}_q[x]$ see Bartee and Schneider [1], Berlekamp [4, Ch. 2], Bhanu Murthy and Sampath [1], Gill [2, Ch. 6], Tanaka, Kasahara, Tezuka, and Kasahara [1], and Willett [6]. The algorithm of Mignotte [1] for the determination of the degree of the splitting field of a polynomial over $\mathbb{F}_q$ is based on properties of linear recurring sequences; see also Willett [5] for another connection between linear recurring sequences and factorization of polynomials. Properties of second-order linear recurring sequences in finite prime fields are important in the work of Niederreiter and Robinson [1] on finite Bol loops. Applications of linear recurring sequences in $\mathbb{F}_2$ to cryptography are discussed in Beker and Piper [1]. Sloane [2] mentions connections between cryptography and feedback shift registers. Surveys of applications of linear recurring sequences are presented in Golomb [3, Ch. 1], [4, Ch. 1]. Special applications of maximal period sequences will be mentioned in the notes to Section 2.

Linear recurring sequences were also considered in more general algebraic structures. Ward [1] looked at such sequences in arbitrary fields and later Ward [13], [15] initiated the study of linear recurring sequences in commutative rings. For further work on this topic we refer to Dade, Robinson, Taussky, and Ward [1], de Carli [1], Duparc [1], D. W. Robinson [4], and Shiue and Sheu [1]. More generally, linear recurring sequences in modules were considered by Nathanson [5] and Niederreiter [5], [6]. For recurring sequences of vectors see Bollman [1], Daykin [4], Selmer [3, Ch. 7], and Vince [2]. Periodicity properties of sequences in $\mathbb{F}_q$ and $\mathbb{Z}/(m)$ satisfying recurrence relations of the form

$$s_{n+k}=a_{k-1}(n)s_{n+k-1}+a_{k-2}(n)s_{n+k-2}+\cdots+a_0(n)s_n$$

with the $a_i(n)$ being periodic in n were studied by Nečaev [1], [3] and Polosuev [1]. Periodicity properties obtained from other types of recurrence relations were established by Duparc [2]. Linear recurring sequences in $\mathbb{F}_q$ constitute the one-dimensional case in a theory of linear recurring arrays over $\mathbb{F}_q$ developed by MacWilliams and Sloane [1], Nomura and Fukuda [1], Nomura, Miyakawa, Imai, and Fukuda [1], [2], [3], and Sakata [1], [2].

2. The special role played by impulse response sequences was already noted in the classical literature on linear recurring sequences; see, for example, Lucas [2, Ch. 17] and d'Ocagne [1]. Theorems 8.16 and 8.19 were shown by Ward [5] and Theorem 8.17 was proved by Speiser [1] for q prime. For other results pertaining to impulse response sequences see Ajtai [1], Kiss and Bui Minh Phong [1], D. W. Robinson [2], and Selmer [3, Chs. 3, 4]. In connection with Theorem 8.19 we note that Groth [1] uses the number of linearly independent state vectors to set up a measure of complexity for sequences in $\mathbb{F}_2$.

The notion of characteristic polynomial and the principle of Theorem 8.21 go back to Lagrange [1], [5] who showed an analogous theorem for linear recurring sequences of real numbers. The result mentioned in Remark 8.23 is well known for linear recurring sequences of real or complex numbers (see, e.g., Ch. Jordan [1, Ch. 11], Markuševič [1], and Milne-Thomson [1, Ch. 13]), and the proof for this case can be transferred to finite fields if only the restriction on the multiplicity of the roots of the characteristic polynomial is observed. Without this restriction, there are still ways of representing the terms of the sequence explicitly (see Fillmore and Marx [1]). The sequence resulting from Theorem 8.21 if all $\beta_j = 1$ has received some attention; see Selmer [2], [3, Ch. 5], Ward [3], [4], and Wegner [2], [4]. Theorem 8.24 can be found in van Lint [1, Ch. 3]. A more complicated formula holds in the case where the characteristic polynomial has no multiple roots; see Niederreiter [8] and Exercise 8.41. Arakelov and Varshamov [1] note that the general term of a kth-order homogeneous linear recurring sequence in $\mathbb{F}_q$ can be represented in the form

$$s_n = g_0(n)s_0 + \cdots + g_{k-1}(n)s_{k-1} \quad \text{with } g_i(n) \in \mathbb{F}_q$$

and they investigate the expressions $g_i(n)$. Algorithms for calculating s_n for large n are discussed in Gries and Levin [1], Miller and Brown [1], Pettorossi [1], Pettorossi and Burstall [1], Selmer [3, Ch. 5], Urbanek [1], and Wilson and Shortt [1]. Theorem 8.25 is essentially due to Ward [5]. For the result from linear algebra used in the proof of Lemma 8.26—namely, that f is the minimal polynomial of its companion matrix—see, for example, Hoffman and Kunze [1, Ch. 7]. Lemma 8.26 and Theorem 8.27 immediately imply results about the order of companion matrices in $GL(k, \mathbb{F}_q)$, such as the upper bound $q^k - 1$ for such an order which was obtained by Gupta [1]. Linear recurring sequences whose characteristic polynomials are trinomials were studied by Arakelov and Tenengol'c [1], Goldstein and Zierler [1], Lunnon, Pleasants, and Stephens [1], and Young [1]. Kumari [1] considered another special class of linear recurring sequences.

The first detailed study of maximal period sequences (also called *m-sequences* or *pseudo-noise sequences* in electrical engineering) was carried out by Golomb [1], but it was restricted to the field $\mathbb{F}_2$; see also Golomb [2], [4, Chs. 3, 4, 6] and Golomb and Welch [1] for this case. An in-depth investigation of such sequences for general $\mathbb{F}_q$ can be found in Zierler [4]; see also Selmer [3]. Daykin, Dresel, and Hilton [1] discussed second-order maximal period sequences. A number of papers have been devoted to the efficient generation of maximal period sequences; see Ball, Spittle, and Liu [1], Eier and Malleck [1], Harvey [1], Lempel [1], Lempel and Eastman [1], Möhrmann [1], [2], Scholefield [1], and Surböck and Weinrichter [1]. Various generalizations of maximal period sequences occur in MacWilliams and Sloane [1], Nečaev [1], Nomura, Miyakawa, Imai, and Fukuda [1], [3], and Sakata [1].

The construction of de Bruijn sequences using maximal period sequences (see Exercise 8.19) is due to Mantel [1]; see also Rees [1]. The existence of an (m, k) de Bruijn sequence for any m and k was first proved by Martin [1]; the special case $m = 2$ had been settled earlier by Flye Sainte-Marie [1]. These sequences are named after the work of de Bruijn [1]. For further work on de Bruijn sequences see, for example, Arazi [2], Fredricksen [1], Fredricksen and Kessler [1], Golomb [4, Ch. 6], Golomb and Welch [1], and Good [1], as well as the survey article of Fredricksen [2]. The related concept of a "code ring" was studied by Radchenko and

Filippov [1], [2]. Another combinatorial application of maximal period sequences arises in the theory of difference sets (compare with Definition 9.75); see Butson [1], Golomb [3, Ch. 4], Laxton and Anderson [1], and Selmer [3, Ch. 6]. The paper of Butson [1] also contains an application to the construction of Hadamard matrices (compare with Definition 9.86); see also MacWilliams and Sloane [1]. Bartee and Schneider [1] use the state vectors of a kth-order maximal period sequence in $\mathbb{F}_q$, together with the zero vector, to describe the elements of $\mathbb{F}_{q^k}$; see also MacWilliams and Sloane [1] and Mönnig [1]. Golomb [1] initiated the use of maximal period sequences as pseudo-random number generators; see also Golomb [3, Ch. 1], [4, Ch. 3], Knuth [3, Ch. 3], Niederreiter [7], [10], [12], [13], Pavlov and Pokhodzei [1], and Tausworthe [1] for this application. Some applications of maximal period sequences to coding theory occur in Green and San Soucie [1], Grushko [1], MacWilliams and Sloane [1], Weng [1], Yale [1], and Zierler [3]. For other applications of maximal period sequences we refer to Bartee and Schneider [1], Golomb [3, Ch. 2], Laxton and Anderson [1], Mohanty [1], Nadler and Sengupta [1], and Sagalovič [1].

3. The use of generating functions in the theory of linear recurring sequences in finite fields is due to Golomb [1] and Huffman [1], [2], and this viewpoint was later explored more fully by Friedland [1], Nazarov [1], Richalet [1], Stern and Friedland [1], and Zierler [4]. See also the expository accounts in Lüneburg [2, Chs. 24, 25] and Selmer [3, Ch. 3]. Formal power series over $\mathbb{F}_2$ representing "almost periodic" sequences were studied by Baum, Herzberg, Lomonaco, and Sweet [1]. More general sequences having algebraic functions over $\mathbb{F}_q$ as generating functions appear in Furstenberg [1].

Another approach to linear recurring sequences in $\mathbb{F}_q$ can be based on ideal theory; see Ward [5], Hall [3], Peterson [1], and Laksov [1] and the expositions in Peterson and Weldon [1, Ch. 7] and Selmer [3, Ch. 3]. The papers of Hemmati and Costello [1] and Ikai, Kosako, and Kojima [1], [2] combine the use of generating functions with that of arithmetic modulo ideals in $\mathbb{F}_q[x]$.

4. All the basic results on minimal polynomials can be found in the paper of Zierler [4]. For Theorem 8.44 see also Friedland and Stern [1]. Our proof of Theorem 8.42 has the advantage that it is constructive (see also Willett [1]). A quicker, but nonconstructive proof is outlined in Exercise 8.25 (see also Zierler [4]). For other approaches to the minimal polynomial see Laksov [1] and Selmer [3, Ch. 4]. Theorem 8.44 establishes an important link with the theory of orders of polynomials (see Chapter 3, Section 1). Theorem 8.51 is obviously connected with the determinantal criteria in Section 6. In this section one also finds another method for the calculation of the minimal polynomial.

Fitzpatrick [1] discusses the problem of generating a linear recurring sequence in $\mathbb{F}_2$ of prescribed period by a recurrence relation of least possible order. Many papers have been written on the least period of the Fibonacci sequence in $\mathbb{F}_p$ and $\mathbb{Z}/(m)$ (see, e.g., Barner [1], Catlin [1], Fulton and Morris [1], Halton [1], Kluyver [1], Mamangakis [1], D. W. Robinson [1], Stanley [1], [2], Täcklind [1], Vince [1], Vinson [1], and Wall [1]) and of more general second-order linear recurring sequences in $\mathbb{F}_p$ and $\mathbb{Z}/(m)$ (see, e.g., Bundschuh and Shiue [2], Kiss and Bui Minh Phong [1], D. W. Robinson [3], Smith and Hoggatt [1], Somer [2], [3], [4], Wyler [1], Yalavigi [1], [2], and Yalavigi and Krishna [1]). For the least periods of higher-order linear recurring sequences in such residue class rings see Carmichael [2], [3], Engstrom [1], [2], Hall [3], and Ward [2], [5].

5. The fundamental paper on the structure of the vector spaces $S(f(x))$ is Zierler [4], where one can find the Theorems 8.53, 8.54, 8.55, and 8.56 as well as results on the least period of sum sequences. $S(f(x))$ is also discussed in Fillmore and Marx [1] and Selmer [3, Chs. 3, 4]. The operation of binary complementation is studied in Selmer [3, Ch. 6], and Kumar and Kumari [1] consider the effect of carrying out binary complementation only at one or two positions within the period. Theorem 8.63 was shown by Ward [5] for finite prime fields. For the extension to arbitrary $f(x)$ described in the paragraph following Theorem 8.63 (compare with Example 8.64) one may also use the symbolic method outlined in a more general context in Chapter 9, Section 5. The distribution of least periods in $S(f(x))$, sometimes called the *cycle structure* of $S(f(x))$, is also discussed in Fillmore and Marx [1], Selmer [3, Ch. 4], and Zierler [4]. A parity question arising in this context was settled by Duvall and Kibler [1]. Ward [9] considers the distribution of least periods for linear recurrence relations over $\mathbb{Z}/(m)$. The problem of which least periods can appear at all for kth-order linear recurring sequences in $\mathbb{F}_q$ with given k and q is studied in Lüneburg [2, Chs. 32, 33].

The fact that the termwise multiplication of linear recurring sequences yields again a linear recurring sequence was already observed by d'Ocagne [1] who worked with sequences of real numbers and proved a weak form of Theorem 8.67—namely,

$$S(f_1(x))\cdots S(f_h(x)) \subseteq S(f_1(x) \vee \cdots \vee f_h(x)),$$

for this case. In the context of finite fields the operation of termwise multiplication was first studied by Selmer [3, Ch. 4]. A more thorough investigation was carried out in Zierler and Mills [1], where one can find the Theorems 8.67 and 8.72. It is also shown in this paper how to use these results in order to determine the polynomial $g(x)$ in Theorem 8.65 in the general case. A study of the relationship between $S(f(x))$ and $S(f(x)^k)$ appears already in Fillmore and Marx [1]. Some elementary remarks on the operation of termwise multiplication are contained in Brousseau [1]. Furstenberg [1] has an analog of Corollary 8.66 for more general types of sequences in $\mathbb{F}_q$.

An operation on sequences called *decimation* was introduced by Golomb [1] and is defined as follows: if σ is the sequence of elements $s_0, s_1, s_2, \ldots$ of $\mathbb{F}_q$ and $d \in \mathbb{N}$, then the decimated sequence $\sigma^{(d)}$ has the terms $s_0, s_d, s_{2d}, \ldots$. Thus $\sigma^{(d)}$ is obtained by taking every dth term of σ, starting from s_0. Special cases of this operation appear already in Hall [3] and Ward [3]. Detailed investigations of this operation were carried out in Golomb [2] and Zierler [4]. The interest has mainly focussed on the decimation of maximal period sequences because of the fact that any kth-order maximal period sequence in $\mathbb{F}_q$ can be obtained (up to shifts) from a single sequence of this type by a suitable decimation (see Golomb [2] and Selmer [3, Ch. 5]). For further work on this operation we refer to Arazi [1], Duvall and Mortick [1], Golomb [4, Chs. 3, 4], Pavlov and Pokhodzei [1], Selmer [3, Ch. 5], Surböck and Weinrichter [1], and Willett [2]. If $f(x)$ is a monic nonconstant polynomial over $\mathbb{F}_q$ with $f(0) \neq 0$, then a sequence $\sigma \in S(f(x))$ is called a *characteristic sequence* for $f(x)$ if $\sigma^{(q)} = \sigma$. This notion was introduced and studied by Gold [1], and tables of characteristic sequences for primitive polynomials over $\mathbb{F}_2$ were set up by Willett [4]. Willett [5] proved that the set of characteristic sequences for $f(x)$ forms a subspace of $S(f(x))$ whose dimension is equal to the number of distinct monic irreducible factors of $f(x)$.

Goka [1] considered the operation which turns the sequence of elements $s_0, s_1, s_2, \ldots$ of $\mathbb{F}_2$ into the sequence $s_0 + s_1, s_1 + s_2, s_2 + s_3, \ldots$ of sums of adjacent terms, and this was studied under the name "derivative" by Nathanson [1]. Inverse operations for it were treated in Nathanson [1], [2], and considerable generalizations appear in Nathanson [3], [5]. Various ways of decomposing periodic sequences in $\mathbb{F}_2$ are discussed in Hwang, Sheng, and Hsieh [1] and Weng [1].

6. A detailed account of the relations between linear recurring sequences and Hankel determinants is given in Pólya and Szegö [1, Sec. VII, Problems 17–29]. Theorem 8.75 was first shown by Kronecker [4] for sequences of real numbers, but his proof works for any field. Other versions of Kronecker's theorem were given later by d'Ocagne [1], Maillet [1], and Perrin [1]. Discussions of these determinantal criteria can also be found in Lüneburg [2, Ch. 26], Selmer [3, Ch. 4], and Willett [3].

The Berlekamp-Massey algorithm was discovered by Berlekamp [4] and Massey [4] in connection with a problem of coding theory (compare with Chapter 9, Section 2, and the notes to it). Burton [1] obtained a simplification for $\mathbb{F}_q$ with q even. Berlekamp, Fredricksen, and Proto [1] noted that while any $2k$ consecutive terms from a homogeneous linear recurring sequence in $\mathbb{F}_q$ having minimal polynomial of degree $k \geqslant 1$ will suffice to determine the minimal polynomial, no fewer than $2k$ terms can determine it if $q \neq 2$; if $q = 2$, then $2k-1$ terms will sometimes suffice, but $2k-2$ terms never will. See also Dillon and Morris [1] for further remarks on this question pertaining to the case $q = 2$. Gustavson [1] estimates the average number of additions and multiplications that are required in the Berlekamp-Massey algorithm. A discussion of the Berlekamp-Massey algorithm can also be found in Dornhoff and Hohn [1, Ch. 9]. The notes to Chapter 9, Section 2, contain further references on matters related to this algorithm.

7. The first paper on distribution properties is Scarpis [2], which studies $Z(0)$ for second-order linear recurring sequences in $\mathbb{F}_q$, q odd. Later, Ward [3] considered the distribution of elements in third-order linear recurring sequences in $\mathbb{F}_p$, and higher-order cases were first treated by Hall [3], [4]. The powerful method of exponential sums was first applied in this context by Korobov [1]. Theorems 8.78 and 8.81 are special cases of results of Niederreiter [5], [6]. The estimate (8.31) can be improved somewhat in the case occurring most frequently (see Exercise 8.66). The estimate in Theorem 8.81 is essentially best possible (see Niederreiter [5]). For further work on these exponential sums see Nečaev [5], [6] and Niederreiter [7], [10], as well as Nečaev [2] and Niederreiter [11] for the case of more general recurrence relations.

The simple formula for $Z(b)$ in the case of a maximal period sequence was first noted by Golomb [1], at least for $\mathbb{F}_2$. Theorem 8.82 is due to Niederreiter [6]. An estimate of this type was shown earlier by combinatorial arguments in Hall [4] for the case of a kth-order homogeneous linear recurring sequence in $\mathbb{F}_p$ with irreducible characteristic polynomial. Hall [3] had proved before that if in this case the least period exceeds $p^{k/2}$, then the element 0 must occur in the sequence. Selmer [3, Ch. 5] established analogs of Hall's results for the case where $q = 2$ and the characteristic polynomial is a product of two distinct irreducible polynomials over $\mathbb{F}_2$. Theorem 8.84 is due to McEliece [5]. An extension of this method to the case where the minimal polynomial has no repeated factors is given in Niederreiter [8]. The result in Theorem 8.85 on the distribution of elements in parts of the period was shown in Niederreiter [6], where it is also noted that this estimate is essentially best possible. The distribution of elements from a given subset of $\mathbb{F}_q$, such as the set of primitive elements of $\mathbb{F}_q$, in linear recurring sequences in $\mathbb{F}_q$ is considered in Korobov [1], Nečaev [4], Nečaev and Stepanova [1], Niederreiter [6], and Šparlinskiĭ [1]. Analogous questions for sequences satisfying more general recurrence relations were studied by Nečaev and Polosuev [1]. Much of this work can also be carried out for linear recurring sequences in $\mathbb{Z}/(m)$; see Nečaev [4] and Niederreiter [6].

Golomb [9] considered sequences in $\mathbb{F}_2$ with (not necessarily least) period $2^k - 1$ and the same number of occurrences of 0's and 1's as in the least period of kth-order maximal period sequences. Hemmati and Costello [1] constructed linear recurring sequences in $\mathbb{F}_q$ with $Z(0) = 0$. McEliece [4] showed congruences for $Z(b)$ modulo powers of the characteristic of $\mathbb{F}_q$. Research on the distribution of elements in linear recurring sequences of low order was already carried out by Scarpis [2], Ward [3], and Hall [2], and more recently by Bloom [1], Bruckner [1], Burr [1], Shah [1], and Zeckendorf [1], with some of this work also pertaining to $\mathbb{Z}/(m)$. Results on distribution properties of linear recurring sequences can be applied to coding theory (see McEliece [5] and Niederreiter [8]) and to pseudo-random numbers (see, e.g., Golomb [1], [4, Ch. 3] and Niederreiter [7], [10]).

Linear recurring sequences in $\mathbb{F}_q$ for which $Z(b)$ has the same value for all $b \in \mathbb{F}_q$ have received considerable attention. A sequence with this property is called *uniformly distributed* (or *equidistributed*) in $\mathbb{F}_q$ according to a definition first given by Gotusso [1]; see also Kuipers and Niederreiter [1, Ch. 5]. The study of equidistributed linear recurring sequences was initiated by Kuipers and Shiue [1], [2], [3], [4] who considered the second-order case for finite prime fields, but also for residue class rings $\mathbb{Z}/(m)$. In particular, the Fibonacci sequence is uniformly distributed in $\mathbb{Z}/(m)$ if and only if m is a power of 5 (necessity in Kuipers and Shiue [4], sufficiency in Niederreiter [4]). Equidistributed second-order linear recurring sequences in $\mathbb{Z}/(m)$ were characterized by Nathanson [4] for primes m, by Bundschuh and Shiue [1] for prime powers m (see also Webb and Long [1]), and by Bumby [1] for arbitrary m. For related work see Bundschuh [1], Cavior [7], Shiue [1], and Shiue and Hu [1]. Equidistributed linear recurring sequences in $\mathbb{F}_q$ of the second and third order have been characterized in Niederreiter and Shiue [1], and those of the fourth order in Niederreiter and Shiue [1], [2]. Knight and Webb [1] studied equidistributed linear recurring sequences in $\mathbb{Z}/(m)$ of the third order. Niederreiter and Shiue [1] showed that if a linear recurring sequence of arbitrary order is equidistributed in $\mathbb{F}_q$, then its minimal polynomial must necessarily have at least one multiple root $\neq 0$, and they also treated sequences whose minimal polynomial has a special factorization pattern. Results on equidistributed linear recurring sequences in $\mathbb{Z}/(m)$ of arbitrary order can be found in Kuipers [3], Niederreiter [11], and Rieger [1], [2], [3].

The question of the frequency of occurrence of individual elements in a linear recurring sequence can be generalized to that of the frequency of occurrence of given blocks of elements in blocks of consecutive terms of the sequence. For kth-order maximal period sequences in $\mathbb{F}_q$, the number of occurrences of a given block of length $\leqslant k$ over the full period can be determined by a straightforward combinatorial argument; see Golomb [1] for $q = 2$ and Zierler [4] for the general case. Further results on the distribution of consecutive elements and blocks of elements in linear recurring sequences were obtained in Feng [1], Fredricsson [1], Jordan and Wood [1], Laksov [1], Lindholm [1], Selmer [3, Ch. 5], and Zierler [4], and in connection with pseudo-random numbers generated by linear recurrence relations in Niederreiter [9], [12], [13]. A related topic is that of correlation functions of sequences which is important in electrical engineering. If $s_0, s_1, \ldots$ and $t_0, t_1, \ldots$ are two sequences in $\mathbb{F}_q$ of period r and χ is a nontrivial additive character of $\mathbb{F}_q$, then the corresponding *cross-correlation function* is defined by

$$C(h)=\sum_{n=0}^{r-1}\chi(s_n)\bar{\chi}(t_{n+h})\quad \text{for } h=0,1,\ldots,r-1,$$

where $\bar{\chi}$ denotes the conjugate character (compare with Chapter 5, Section 1). If the two sequences are identical, we speak of the *auto-correlation function*. For maximal period sequences in $\mathbb{F}_2$ the auto-correlation function was already calculated by Golomb [1], and an extension to arbitrary $\mathbb{F}_q$ is given in Zierler [4]. The cross-correlation function for two maximal period sequences in $\mathbb{F}_2$ was considered in Golomb [2]. For further work on correlation functions, see, for example, Feng [1], Gold [2], [3], Golomb [4, Chs. 3, 4, 6], [5], Golomb and Welch [1], Helleseth [2], Ipatov [1], Lee and Smith [1], Lempel, Cohn, and Eastman [1], Maritsas [1], McEliece [7], Mohanty [1], Selmer [3, Ch. 6], and the survey in Helleseth [1]. MacWilliams and Odlyzko [1] also treat a sort of correlation property for sequences in $\mathbb{F}_2$.

EXERCISES

8.1. Design a feedback shift register implementing the linear recurrence relation $s_{n+5}=s_{n+4}-s_{n+3}-s_{n+1}+s_n$, $n=0,1,\ldots$, in $\mathbb{F}_3$.

8.2. Design a feedback shift register implementing the linear recurrence relation $s_{n+7}=3s_{n+5}-2s_{n+4}+s_{n+3}+2s_n+1$, $n=0,1,\ldots$, in $\mathbb{F}_7$.

8.3. Let r be a period of the ultimately periodic sequence $s_0, s_1, \ldots$ and let n_0 be the least nonnegative integer such that $s_{n+r}=s_n$ for all $n\geqslant n_0$. Prove that n_0 is equal to the preperiod of the sequence.

8.4. Determine the order of the matrix

$$A=\begin{pmatrix}0&0&0&-1\\1&0&0&1\\0&1&0&1\\0&0&1&-1\end{pmatrix}$$

in the general linear group $GL(4,\mathbb{F}_3)$.

8.5. Obtain the results of Example 8.18 by the methods of Section 5.

8.6. Use (8.8) to give an explicit formula for the terms of the linear recurring sequence in $\mathbb{F}_3$ with $s_0=s_1=1$, $s_2=0$, and $s_{n+3}=-s_{n+1}+s_n$ for $n=0,1,\ldots$.

8.7. Use the result in Remark 8.23 to give an explicit formula for the terms of the linear recurring sequence in $\mathbb{F}_4$ with $s_0=s_1=s_2=0$, $s_3=1$, and $s_{n+4}=\alpha s_{n+3}+s_{n+1}+\alpha s_n$ for $n=0,1,\ldots$, where α is a primitive element of $\mathbb{F}_4$.

8.8. Prove that the terms s_n given by the formula in Remark 8.23 satisfy the homogeneous linear recurrence relation with characteristic polynomial $f(x)$.

8.9. Prove the result in Remark 8.23 for the case where $e_i \leqslant 2$ for $i = 1, 2, \ldots, m$ and $e_i = 1$ if $\alpha_i = 0$.

8.10. Represent the elements of the linear recurring sequence in $\mathbb{F}_2$ with $s_0 = 0$, $s_1 = s_2 = 1$, and $s_{n+3} = s_{n+2} + s_n$ for $n = 0, 1, \ldots$ in terms of a suitable trace function.

8.11. Prove Lemma 8.26 by using linear recurring sequences.

8.12. Determine the least period of the impulse response sequence in $\mathbb{F}_2$ satisfying the linear recurrence relation $s_{n+7} = s_{n+6} + s_{n+5} + s_{n+1} + s_n$ for $n = 0, 1, \ldots$.

8.13. Calculate the least period of the impulse response sequence associated with the linear recurrence relation $s_{n+10} = s_{n+7} + s_{n+2} + s_{n+1} + s_n$ in $\mathbb{F}_2$.

8.14. Prove Theorem 8.27 by using generating functions.

8.15. Find a linear recurring sequence of least order in $\mathbb{F}_2$ whose least period is 21.

8.16. Find a linear recurring sequence of least order in $\mathbb{F}_2$ whose least period is 24.

8.17. Let r be the least period of the *Fibonacci sequence* in $\mathbb{F}_q$—that is, of the sequence with $s_0 = 0$, $s_1 = 1$, and $s_{n+2} = s_{n+1} + s_n$ for $n = 0, 1, \ldots$. Let p be the characteristic of $\mathbb{F}_q$. Prove that $r = 20$ if $p = 5$, that r divides $p - 1$ if $p \equiv \pm 1 \bmod 5$, and that r divides $p^2 - 1$ in all other cases.

8.18. Construct a maximal period sequence in $\mathbb{F}_3$ of least period 80.

8.19. An (m, k) *de Bruijn sequence* is a finite sequence $s_0, s_1, \ldots, s_{N-1}$ with $N = m^k$ terms from a set of m elements such that the k-tuples $(s_n, s_{n+1}, \ldots, s_{n+k-1})$, $n = 0, 1, \ldots, N-1$, with subscripts considered modulo N are all different. Prove that if $d_0, d_1, \ldots$ is a kth-order impulse response sequence and maximal period sequence in $\mathbb{F}_q$, then $s_0 = 0$, $s_n = d_{n-1}$ for $1 \leqslant n \leqslant q^k - 1$ yields a (q, k) de Bruijn sequence.

8.20. Construct a (2,5) de Bruijn sequence.

8.21. Let $B(x) = 2 - x + x^3 \in \mathbb{F}_7[x]$. Calculate the first six nonzero terms of the formal power series $1/B(x)$.

8.22. Let

$$A(x) = -1 - x + x^2, \qquad B(x) = \sum_{n=0}^{\infty} (-1)^n x^n \in \mathbb{F}_3[[x]].$$

Calculate the first five nonzero terms of the formal power series $A(x)/B(x)$.

8.23. Consider the linear recurring sequence in $\mathbb{F}_3$ with $s_0 = s_1 = s_2 = 1$, $s_3 = s_4 = -1$, and $s_{n+5} = s_{n+4} + s_{n+2} - s_{n+1} + s_n$ for $n = 0, 1, \ldots$. Represent the generating function of the sequence in the form (8.15).

8.24. Calculate the first eight terms of the impulse response sequence associated with the linear recurrence relation $s_{n+5} = s_{n+3} + s_{n+2} + s_n$ in $\mathbb{F}_2$ by long division.

8.25. Let $s_0, s_1, \ldots$ be a homogeneous linear recurring sequence in $\mathbb{F}_q$. Prove that the set of all polynomials $f(x) = a_k x^k + \cdots + a_1 x + a_0 \in \mathbb{F}_q[x]$ such that $a_k s_{n+k} + \cdots + a_1 s_{n+1} + a_0 s_n = 0$ for $n = 0, 1, \ldots$ forms an ideal of $\mathbb{F}_q[x]$. Thus show the existence of a uniquely determined minimal polynomial of the sequence.

8.26. Consider the linear recurring sequence in $\mathbb{F}_2$ with $s_0 = s_3 = s_4 = s_5 = s_6 = 0$, $s_1 = s_2 = s_7 = 1$, and $s_{n+8} = s_{n+7} + s_{n+6} + s_{n+5} + s_n$ for $n = 0, 1, \ldots$. Use the method in the proof of Theorem 8.42 to determine the minimal polynomial of the sequence.

8.27. Consider the linear recurring sequence in $\mathbb{F}_5$ with $s_0 = s_1 = s_2 = 1$, $s_3 = -1$, and $s_{n+4} = 3s_{n+2} - s_{n+1} + s_n$ for $n = 0, 1, \ldots$. Use the method in the proof of Theorem 8.42 to determine the minimal polynomial of the sequence.

8.28. Prove that a homogeneous linear recurring sequence in a finite field is periodic if and only if its minimal polynomial $m(x)$ satisfies $m(0) \neq 0$.

8.29. Given a homogeneous linear recurring sequence in a finite field with minimal polynomial $m(x)$, prove that the preperiod of the sequence is equal to the multiplicity of 0 as a root of $m(x)$.

8.30. Prove Corollary 8.52 by using the construction of the minimal polynomial in the proof of Theorem 8.42.

8.31. Use the criterion in Theorem 8.51 to determine the minimal polynomial of the linear recurring sequence in $\mathbb{F}_2$ with $s_{n+6} = s_{n+3} + s_{n+2} + s_{n+1} + s_n$ for $n = 0, 1, \ldots$ and initial state vector $(1, 1, 1, 0, 0, 1)$.

8.32. Find the least period of the linear recurring sequence in Exercise 8.26.

8.33. Find the least period of the linear recurring sequence in Exercise 8.27.

8.34. Find the least period of the linear recurring sequence in $\mathbb{F}_2$ with $s_0 = s_1 = s_2 = s_6 = s_7 = 0$, $s_3 = s_4 = s_5 = s_8 = 1$, and $s_{n+9} = s_{n+7} + s_{n+4} + s_{n+1} + s_n$ for $n = 0, 1, \ldots$.

8.35. Find the least period of the linear recurring sequence in $\mathbb{F}_3$ with $s_0 = s_1 = 1$, $s_2 = s_3 = 0$, $s_4 = -1$, and $s_{n+5} = s_{n+4} - s_{n+3} + s_{n+2} + s_n$ for $n = 0, 1, \ldots$.

8.36. Find the least period of the linear recurring sequence in $\mathbb{F}_3$ with $s_{n+4} = s_{n+3} + s_{n+2} - s_n - 1$ for $n = 0, 1, \ldots$ and initial state vector $(0, -1, 1, 0)$.

8.37. Prove that a kth-order linear recurring sequence $s_0, s_1, \ldots$ in $\mathbb{F}_q$ has least period q^k exactly in the following cases:

(a) $k = 1$, q prime, $s_{n+1} = s_n + a$ for $n = 0, 1, \ldots$ with $a \in \mathbb{F}_q^*$;

(b) $k = 2$, $q = 2$, $s_{n+2} = s_n + 1$ for $n = 0, 1, \ldots$.

8.38. Given a homogeneous linear recurring sequence in $\mathbb{F}_q$ with a nonconstant minimal polynomial $m(x) \in \mathbb{F}_q[x]$ whose roots are nonzero and simple, prove that the least period of the sequence is equal to the least positive integer r such that $\alpha^r = 1$ for all roots α of $m(x)$.

8.39. Prove: if the homogeneous linear recurring sequence σ in $\mathbb{F}_q$ has minimal polynomial $f(x) \in \mathbb{F}_q[x]$ with $\deg(f(x)) = n \geqslant 1$, then every sequence in $S(f(x))$ can be expressed uniquely as a linear combination of $\sigma = \sigma^{(0)}$ and the shifted sequences $\sigma^{(1)}, \sigma^{(2)}, \ldots, \sigma^{(n-1)}$ with coefficients in $\mathbb{F}_q$.

8.40. Let $f_1(x), \ldots, f_k(x)$ be nonconstant monic polynomials over $\mathbb{F}_q$ that are pairwise relatively prime. Prove that $S(f_1(x) \cdots f_k(x))$ is the direct sum of the linear subspaces $S(f_1(x)), \ldots, S(f_k(x))$.

8.41. Let $s_0, s_1, \ldots$ be a homogeneous linear recurring sequence in $K = \mathbb{F}_q$ with characteristic polynomial $f(x) = f_1(x) \cdots f_r(x)$, where the $f_i(x)$ are distinct monic irreducible polynomials over K. For $i = 1, \ldots, r$, let α_i be a fixed root of $f_i(x)$ in its splitting field F_i over K. Prove that there exist uniquely determined elements $\theta_1 \in F_1, \ldots, \theta_r \in F_r$ such that

$$s_n = \mathrm{Tr}_{F_1/K}(\theta_1 \alpha_1^n) + \cdots + \mathrm{Tr}_{F_r/K}(\theta_r \alpha_r^n) \quad \text{for } n = 0, 1, \ldots.$$

8.42. With the notation of Exercise 8.41, prove that the sequence $s_0, s_1, \ldots$ has $f(x)$ as its minimal polynomial if and only if $\theta_i \neq 0$ for $1 \leqslant i \leqslant r$. Thus show that the number of sequences in $S(f(x))$ that have $f(x)$ as minimal polynomial is given by $(q^{k_1} - 1) \cdots (q^{k_r} - 1)$, where $k_i = \deg(f_i(x))$ for $1 \leqslant i \leqslant r$.

8.43. Let σ_1 and σ_2 be the impulse response sequences in $\mathbb{F}_2$ associated with the linear recurrence relations $s_{n+6} = s_{n+3} + s_n (n = 0, 1, \ldots)$ and $s_{n+3} = s_{n+1} + s_n (n = 0, 1, \ldots)$, respectively. Find the least period of $\sigma_1 + \sigma_2$.

8.44. Let σ_1 be the linear recurring sequence in $\mathbb{F}_3$ with $s_{n+3} = s_{n+2} - s_{n+1} - s_n$ for $n = 0, 1, \ldots$ and initial state vector $(0, 1, 0)$, and let σ_2 be the linear recurring sequence in $\mathbb{F}_3$ with $s_{n+5} = -s_{n+3} - s_{n+2} + s_n$ for $n = 0, 1, \ldots$ and initial state vector $(1, 1, 1, 0, 1)$. Use the method of Example 8.58 to determine the minimal polynomial of the sum sequence $\sigma_1 + \sigma_2$.

8.45. Find the least period of the sum sequence in Exercise 8.44.

8.46. Given a homogeneous linear recurring sequence in $\mathbb{F}_2$ with minimal polynomial $x^6 + x^5 + x^4 + 1 \in \mathbb{F}_2[x]$, determine the minimal polynomial of its binary complement.

8.47. Let $f(x) = x^9 + x^7 + x^4 + x^3 + x^2 + x + 1 \in \mathbb{F}_2[x]$. Determine the least periods of sequences from $S(f(x))$ and the number of sequences attaining each possible least period.

8.48. Let $f(x) = (x+1)^3(x^3 - x + 1) \in \mathbb{F}_3[x]$. Determine the least periods of sequences from $S(f(x))$ and the number of sequences attaining each possible least period.

8.49. Let $f(x)=x^5-2x^4-x^2-1\in\mathbb{F}_5[x]$. Determine the least periods of sequences from $S(f(x))$ and the number of sequences attaining each possible least period.

8.50. Find a monic polynomial $g(x)\in\mathbb{F}_3[x]$ such that

$$S(x+1)S(x^2+x-1)S(x^2-x-1)=S(g(x)).$$

8.51. Find a monic polynomial $g(x)\in\mathbb{F}_2[x]$ such that

$$S(x^2+x+1)S(x^5+x^4+1)=S(g(x)).$$

8.52. For odd q determine a monic $g(x)\in\mathbb{F}_q[x]$ for which

$$S((x-1)^2)S((x-1)^2)=S(g(x)).$$

What is the situation for even q?

8.53. Prove that $f\vee(gh)=(f\vee g)(f\vee h)$ for nonconstant polynomials $f,g,h\in\mathbb{F}_q[x]$, provided the two factors on the right-hand side are relatively prime.

8.54. Consider the impulse response sequence in $\mathbb{F}_2$ associated with the linear recurrence relation $s_{n+4}=s_{n+2}+s_n$, $n=0,1,\ldots$, and the linear recurring sequence in $\mathbb{F}_2$ with $s_{n+4}=s_n$, $n=0,1,\ldots$, and initial state vector $(0,1,1,1)$. Use these sequences to show that there is no analog of Theorem 8.59 for multiplication of sequences.

8.55. For $r\in\mathbb{N}$ and $f\in\mathbb{F}_q[x]$ with $\deg(f)>0$, let $\sigma_r(f)$ be the sum of the rth powers of the distinct roots of f. Prove that $\sigma_r(f\vee g)=\sigma_r(f)\sigma_r(g)$ for nonconstant polynomials $f,g\in\mathbb{F}_q[x]$, provided that the number of distinct roots of $f\vee g$ is equal to the product of the numbers of distinct roots of f and g, respectively.

8.56. Let $s_0,s_1,\ldots$ be an arbitrary sequence in $\mathbb{F}_q$, and let $n\geqslant 0$ and $r\geqslant 1$ be integers. Prove that if both Hankel determinants $D_{n+2}^{(r)}$ and $D_n^{(r+1)}$ are 0, then also $D_{n+1}^{(r)}=0$.

8.57. Prove that the sequence $s_0,s_1,\ldots$ in $\mathbb{F}_q$ is a homogeneous linear recurring sequence with minimal polynomial of degree k if and only if $D_n^{(k+1)}=0$ for all $n\geqslant 0$ and $k+1$ is the least positive integer for which this holds.

8.58. Give a complete proof for the second inequality in (8.23).

8.59. Prove the inequalities in (8.24).

8.60. Give a complete proof for (8.26).

8.61. Prove (8.27).

8.62. The first 10 terms of a homogeneous linear recurring sequence in $\mathbb{F}_2$ of order $\leqslant 5$ are given by $0,1,1,0,0,0,0,1,1,1$. Determine its minimal polynomial by the Berlekamp-Massey algorithm.

8.63. The first 8 terms of a homogeneous linear recurring sequence in $\mathbb{F}_5$ of order $\leqslant 4$ are given by $2,1,0,1,-2,0,-2,-1$. Determine its minimal polynomial by the Berlekamp-Massey algorithm.

8.64. The first 10 terms of a homogeneous linear recurring sequence in $\mathbb{F}_3$

of order $\leqslant 5$ are given by $1, -1, 0, -1, 0, 0, 0, 0, 1, 0$. Determine its minimal polynomial by the Berlekamp-Massey algorithm.

8.65. Find the homogeneous linear recurring sequence in $\mathbb{F}_5$ of least order whose first 10 terms are $2, 0, -1, -2, 0, 0, -2, 2, -1, -2$.

8.66. Suppose the conditions of Theorem 8.78 hold and assume in addition that the characteristic polynomial $f(x)$ of the sequence $s_0, s_1, \ldots$ satisfies $f(0) \neq 0$. Establish the following improvement of (8.31):

$$\left| \sum_{n=u}^{u+r-1} \chi(s_n) \right| \leqslant \left(\frac{r}{R} \right)^{1/2} (q^k - r)^{1/2} \quad \text{for all } u \geqslant 0.$$

(*Hint*: Note that $\mathbf{b} = \mathbf{0}$ can be excluded in (8.33).)

8.67. Suppose the conditions of Theorem 8.84 hold, let r be a multiple of $(q^k - 1)/(q-1)$ and let $(q^k - 1)/r$ and k be relatively prime. Prove that $Z(0) = (q^{k-1} - 1)r/(q^k - 1)$.

8.68. Suppose the conditions of Theorem 8.84 hold, let q be odd and $h = (q^k - 1)/2$. Prove that equality holds in (8.37).

8.69. Let $Z(b; N_0, N)$ be as in Theorem 8.85. Under the conditions of Theorem 8.84 and using the notation in the proof of this theorem, show that

$$Z(b; N_0, N) = \frac{N}{r} Z(b) + \frac{1}{q(q^k - 1)} \sum_{\substack{\psi \\ \psi(\alpha) \neq 1}} \psi(\theta) G(\bar{\psi}, \bar{\lambda}) G(\psi', \lambda'_b) \frac{\psi(\alpha)^{N_0 + N} - \psi(\alpha)^{N_0}}{\psi(\alpha) - 1}.$$

8.70. Deduce from the result of Exercise 8.69 that

$$\left| Z(0; N_0, N) - \frac{(q^{k-1} - 1)N}{q^k - 1} \right| \leqslant \left(1 - \frac{1}{q}\right)\left(\frac{N}{h} - \frac{N}{q^k - 1}\right) q^{k/2} + q^{(k/2)-1}\left(\frac{2}{\pi} \log \frac{h}{q-1} + \varepsilon_h\right),$$

where $\varepsilon_h = 0$ for $h = q - 1$ and $\varepsilon_h = \frac{2}{5}$ for $h > q - 1$.

8.71. Deduce from the result of Exercise 8.69 that

$$\left| Z(b; N_0, N) - \frac{q^{k-1} N}{q^k - 1} \right| \leqslant \left(\frac{2}{\pi} \log r + \frac{2}{5} + \frac{N(h - r)}{hr} \right) q^{(k-1)/2} + \left(\frac{N}{h} - \frac{N}{q^k - 1} \right) q^{(k/2)-1}$$

for $b \neq 0$.

Chapter 9

Applications of Finite Fields

One of the major applications of finite fields is coding theory. This theory has its origin in a famous theorem of Shannon that guarantees the existence of codes that can transmit information at rates close to capacity with an arbitrarily small probability of error. One purpose of algebraic coding theory—the theory of error-correcting and error-detecting codes—is to devise methods for the construction of such codes.

During the last two decades more and more abstract algebraic tools such as the theory of finite fields and the theory of polynomials over finite fields have influenced coding. In particular, the description of redundant codes by polynomials over $\mathbb{F}_q$ is a milestone in this development. The fact that one can use shift registers for coding and decoding establishes a connection with linear recurring sequences. In Sections 1 and 2 on algebraic coding theory, we do not consider any of the problems of the implementation or technical realization of the codes. We restrict ourselves to the discussion of basic properties of block codes, and in Section 2 we describe some of those aspects of cyclic codes that are closely connected with finite fields.

Section 3 contains some results on the use of finite fields in geometry—namely, affine and projective planes, and in particular projective planes with a finite number of points and lines.

Section 4 on combinatorics shows the variety of further applications of finite fields, especially their usefulness in problems of designs of experiments.

In the final Section 5 we give the definition of a linear modular system and show how finite fields are involved in this theory. A system is regarded as a structure into which something (matter, energy, or information) may be put at certain times and that itself puts out something at certain times. For instance, we may visualize a system as an electrical circuit whose input is a voltage signal and whose output is a current reading. Or we may think of a system as a network of switching elements whose input is an on/off setting of a number of input switches and whose output is the on/off pattern of an array of lights.

We emphasize that the applications are only described to give examples for the use of various properties of finite fields. Therefore, the examples contain rather the algebraic and combinatorial aspects, without regard to their practical application or indeed other usefulness. For instance, we are not going to discuss the analysis of experimental design or the analysis or synthesis of linear modular systems, nor do we explain geometric properties that are not directly connected with finite fields.

1. LINEAR CODES

The problem of the communication of information—in particular the coding and decoding of information for the reliable transmission over a "noisy" channel—is of great importance today. Typically, one has to transmit a message which consists of a finite sequence of symbols that are elements of some finite alphabet. For instance, if this alphabet consists simply of 0 and 1, the message can be described as a binary number. Generally the alphabet is assumed to be a finite field. Now the transmission of finite sequences of elements of the alphabet over a communication channel need not be perfect in the sense that each bit of information is transmitted unaltered over this channel. As there is no ideal channel without "noise," the receiver of the transmitted message may obtain distorted information and may make errors in interpreting the transmitted signal.

One of the main problems of coding theory is to make the errors, which occur for instance because of noisy channels, extremely improbable. The methods to improve the reliability of transmission depend on properties of finite fields.

A basic idea in algebraic coding theory is to transmit *redundant* information together with the message one wants to communicate; that is, one extends the sequence of message symbols to a longer sequence in a systematic manner.

A simple model of a communication system is shown in Figure 9.1. We assume that the symbols of the message and of the coded message are elements of the same finite field $\mathbb{F}_q$. Coding means to encode a block of k message symbols $a_1a_2 \cdots a_k$, $a_i \in \mathbb{F}_q$, into a *code word* $c_1c_2 \cdots c_n$ of n

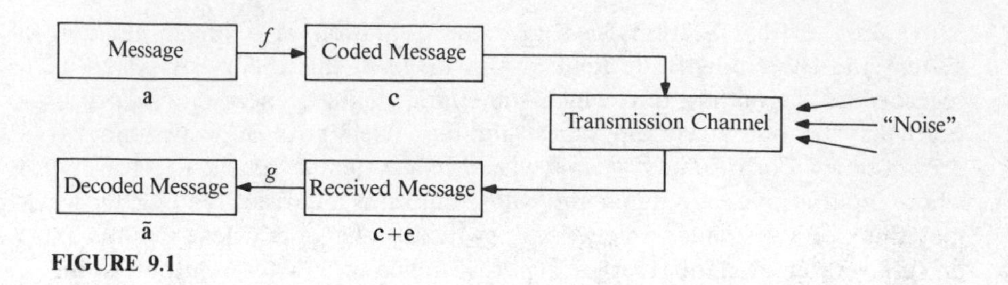

FIGURE 9.1

symbols $c_j \in \mathbb{F}_q$, where $n > k$. We regard the code word as an n-dimensional row vector $\mathbf{c}$ in $\mathbb{F}_q^n$. Thus f in Figure 9.1 is a function from $\mathbb{F}_q^k$ into $\mathbb{F}_q^n$, called a *coding scheme*, and g: $\mathbb{F}_q^n \to \mathbb{F}_q^k$ is a *decoding scheme*.

A simple type of coding scheme arises when each block $a_1 a_2 \cdots a_k$ of message symbols is encoded into a code word of the form

$$a_1 a_2 \cdots a_k c_{k+1} \cdots c_n,$$

where the first k symbols are the original *message symbols* and the additional $n-k$ symbols in $\mathbb{F}_q$ are *control symbols*. Such coding schemes are often presented in the following way. Let H be a given $(n-k)\times n$ matrix with entries in $\mathbb{F}_q$ that is of the special form

$$H = (A, I_{n-k}),$$

where A is an $(n-k)\times k$ matrix and I_{n-k} is the identity matrix of order $n-k$. The control symbols $c_{k+1}, \ldots, c_n$ can then be calculated from the system of equations

$$H\mathbf{c}^{\mathrm{T}} = \mathbf{0}$$

for code words $\mathbf{c}$. The equations of this system are called *parity-check equations*.

9.1. Example. Let H be the following 3×7 matrix over $\mathbb{F}_2$:

$$H = \begin{pmatrix} 1 & 0 & 1 & 1 & 1 & 0 & 0 \\ 1 & 1 & 0 & 1 & 0 & 1 & 0 \\ 1 & 1 & 1 & 0 & 0 & 0 & 1 \end{pmatrix}.$$

Then the control symbols can be calculated by solving $H\mathbf{c}^{\mathrm{T}} = \mathbf{0}$, given c_1, c_2, c_3, c_4:

$$\begin{aligned} c_1 \phantom{{}+c_2} + c_3 + c_4 + c_5 \phantom{{}+c_6+c_7} &= 0 \\ c_1 + c_2 \phantom{{}+c_3} + c_4 \phantom{{}+c_5} + c_6 \phantom{{}+c_7} &= 0 \\ c_1 + c_2 + c_3 \phantom{{}+c_4+c_5+c_6} + c_7 &= 0 \end{aligned}$$

The control symbols c_5, c_6, c_7 can be expressed as

$$\begin{aligned} c_5 &= c_1 + c_3 + c_4 \\ c_6 &= c_1 + c_2 + c_4 \\ c_7 &= c_1 + c_2 + c_3 \end{aligned}$$

Thus the coding scheme in this case is the linear map from $\mathbb{F}_2^4$ into $\mathbb{F}_2^7$ given by

$$(a_1, a_2, a_3, a_4) \mapsto (a_1, a_2, a_3, a_4, a_1 + a_3 + a_4, a_1 + a_2 + a_4, a_1 + a_2 + a_3).$$

□

In general, we use the following terminology in connection with coding schemes that are given by linear maps.

9.2. Definition. Let H be an $(n-k)\times n$ matrix of rank $n-k$ with entries in $\mathbb{F}_q$. The set C of all n-dimensional vectors $\mathbf{c} \in \mathbb{F}_q^n$ such that $H\mathbf{c}^{\mathrm{T}} = \mathbf{0}$ is called a *linear* (n, k) *code* over $\mathbb{F}_q$; n is called the *length* and k the *dimension* of the code. The elements of C are called *code words* (or *code vectors*), the matrix H is a *parity-check matrix* of C. If $q = 2$, C is called a *binary code*. If H is of the form (A, I_{n-k}), then C is called a *systematic code*.

We note that the set C of solutions of the system $H\mathbf{c}^{\mathrm{T}} = \mathbf{0}$ of linear equations is a subspace of dimension k of the vector space $\mathbb{F}_q^n$. Since the code words form an additive group, C is also called a *group code*. Moreover, C can be regarded as the null space of the matrix H.

9.3. Example (*Parity-Check Code*). Let $q = 2$ and let the given message be $a_1 \cdots a_k$, then the coding scheme f is defined by

$$f\colon a_1 \cdots a_k \mapsto b_1 \cdots b_{k+1},$$

where $b_i = a_i$ for $i = 1, \ldots, k$ and

$$b_{k+1} = \begin{cases} 0 & \text{if } \sum_{i=1}^{k} a_i = 0, \\ 1 & \text{if } \sum_{i=1}^{k} a_i = 1. \end{cases}$$

Hence it follows that the sum of digits of any code word $b_1 \cdots b_{k+1}$ is 0. If the sum of digits of the received word is 1, then the receiver knows that a transmission error must have occurred. Let $n = k+1$, then this code is a binary linear $(n, n-1)$ code with parity-check matrix $H = (11 \cdots 1)$. □

9.4. Example (*Repetition Code*). In a repetition code each code word consists of only one message symbol a_1 and $n-1$ control symbols $c_2 = \cdots = c_n$ all equal to a_1; that is, a_1 is repeated $n-1$ times. This is a linear $(n, 1)$ code with parity-check matrix $H = (-\mathbf{1}, I_{n-1})$. □

The parity-check equations $H\mathbf{c}^{\mathrm{T}} = \mathbf{0}$ with $H = (A, I_{n-k})$ imply

$$\mathbf{c}^{\mathrm{T}} = \begin{pmatrix} I_k \\ -A \end{pmatrix} \mathbf{a}^{\mathrm{T}} = \left[\mathbf{a}\left(I_k, -A^{\mathrm{T}}\right)\right]^{\mathrm{T}},$$

where $\mathbf{a} = a_1 \cdots a_k$ is the message and $\mathbf{c} = c_1 \cdots c_n$ is the code word. This leads to the following definition.

9.5. Definition. The $k \times n$ matrix $G = (I_k, -A^{\mathrm{T}})$ is called the *canonical generator matrix* of a linear (n, k) code with parity-check matrix $H = (A, I_{n-k})$.

From $H\mathbf{c}^{\mathrm{T}} = \mathbf{0}$ and $\mathbf{c} = \mathbf{a}G$ it follows that H and G are related by

$$GH^{\mathrm{T}} = 0. \tag{9.1}$$

The code C is equal to the row space of the canonical generator matrix G. More generally, any $k \times n$ matrix G whose row space is equal to C is called a *generator matrix* of C.

9.6. Example. The canonical generator matrix for the code defined by H in Example 9.1 is given by

$$G = \begin{pmatrix} 1 & 0 & 0 & 0 & 1 & 1 & 1 \\ 0 & 1 & 0 & 0 & 0 & 1 & 1 \\ 0 & 0 & 1 & 0 & 1 & 0 & 1 \\ 0 & 0 & 0 & 1 & 1 & 1 & 0 \end{pmatrix}. \qquad \square$$

9.7. Definition. If $\mathbf{c}$ is a code word and $\mathbf{y}$ is the received word after communication through a "noisy" channel, then $\mathbf{e} = \mathbf{y} - \mathbf{c} = e_1 \cdots e_n$ is called the *error word* or the *error vector*.

9.8. Definition. Let $\mathbf{x}, \mathbf{y}$ be two vectors in $\mathbb{F}_q^n$. Then:

(i) the *Hamming distance* $d(\mathbf{x}, \mathbf{y})$ between $\mathbf{x}$ and $\mathbf{y}$ is the number of coordinates in which $\mathbf{x}$ and $\mathbf{y}$ differ;

(ii) the (*Hamming*) *weight* $w(\mathbf{x})$ of $\mathbf{x}$ is the number of nonzero coordinates of $\mathbf{x}$.

Thus $d(\mathbf{x}, \mathbf{y})$ gives the number of errors if $\mathbf{x}$ is the transmitted code word and $\mathbf{y}$ is the received word. It follows immediately that $w(\mathbf{x}) = d(\mathbf{x}, \mathbf{0})$ and $d(\mathbf{x}, \mathbf{y}) = w(\mathbf{x} - \mathbf{y})$. The proof of the following lemma is left as an exercise.

9.9. Lemma. *The Hamming distance is a metric on* $\mathbb{F}_q^n$; *that is, for all* $\mathbf{x}, \mathbf{y}, \mathbf{z} \in \mathbb{F}_q^n$ *we have*:

(i) $d(\mathbf{x}, \mathbf{y}) = 0$ *if and only if* $\mathbf{x} = \mathbf{y}$;

(ii) $d(\mathbf{x}, \mathbf{y}) = d(\mathbf{y}, \mathbf{x})$;

(iii) $d(\mathbf{x}, \mathbf{z}) \leqslant d(\mathbf{x}, \mathbf{y}) + d(\mathbf{y}, \mathbf{z})$.

In decoding received words $\mathbf{y}$, one usually tries to find the code word $\mathbf{c}$ such that $w(\mathbf{y}-\mathbf{c})$ is as small as possible, that is, one assumes that it is more likely that few errors have occurred rather than many. Thus in decoding we are looking for a code word $\mathbf{c}$ that is closest to $\mathbf{y}$ according to the Hamming distance. This rule is called *nearest neighbor decoding*.

9.10. Definition. For $t \in \mathbb{N}$ a code $C \subseteq \mathbb{F}_q^n$ is called *t-error-correcting* if for any $\mathbf{y} \in \mathbb{F}_q^n$ there is at most one $\mathbf{c} \in C$ such that $d(\mathbf{y},\mathbf{c}) \leqslant t$.

If $\mathbf{c} \in C$ is transmitted and at most t errors occur, then we have $d(\mathbf{y},\mathbf{c}) \leqslant t$ for the received word $\mathbf{y}$. If C is t-error-correcting, then for all other code words $\mathbf{z} \neq \mathbf{c}$ we have $d(\mathbf{y},\mathbf{z}) > t$, which means that $\mathbf{c}$ is closest to $\mathbf{y}$ and nearest neighbor decoding gives the correct result. Therefore, one aim in coding theory is to construct codes with code words "far apart." On the other hand, one tries to transmit as much information as possible. To reconcile these two aims is one of the problems of coding.

9.11. Definition. The number

$$d_C = \min_{\substack{\mathbf{u},\mathbf{v} \in C \\ \mathbf{u} \neq \mathbf{v}}} d(\mathbf{u},\mathbf{v}) = \min_{\mathbf{0} \neq \mathbf{c} \in C} w(\mathbf{c})$$

is called the *minimum distance* of the linear code C.

9.12. Theorem. *A code C with minimum distance d_C can correct up to t errors if $d_C \geqslant 2t+1$.*

Proof. A *ball* $B_t(\mathbf{x})$ of radius t and center $\mathbf{x} \in \mathbb{F}_q^n$ consists of all vectors $\mathbf{y} \in \mathbb{F}_q^n$ such that $d(\mathbf{x},\mathbf{y}) \leqslant t$. The nearest neighbor decoding rule ensures that each received word with t or fewer errors must be in a ball of radius t and center the transmitted code word. To correct t errors, the balls with code words $\mathbf{x}$ as centers must not overlap. If $\mathbf{u} \in B_t(\mathbf{x})$ and $\mathbf{u} \in B_t(\mathbf{y})$, $\mathbf{x},\mathbf{y} \in C$, $\mathbf{x} \neq \mathbf{y}$, then

$$d(\mathbf{x},\mathbf{y}) \leqslant d(\mathbf{x},\mathbf{u}) + d(\mathbf{u},\mathbf{y}) \leqslant 2t,$$

a contradiction to $d_C \geqslant 2t+1$. □

9.13. Example. The code of Example 9.1 has minimum distance $d_C = 3$ and therefore can correct one error. □

The following lemma is often useful in determining the minimum distance of a code.

9.14. Lemma. *A linear code C with parity-check matrix H has minimum distance $d_C \geqslant s+1$ if and only if any s columns of H are linearly independent.*

Proof. Assume there are s linearly dependent columns of H, then $H\mathbf{c}^{\mathrm{T}} = \mathbf{0}$ and $w(\mathbf{c}) \leqslant s$ for suitable $\mathbf{c} \in C$, $\mathbf{c} \neq \mathbf{0}$, hence $d_C \leqslant s$. Similarly, if any s columns of H are linearly independent, then there is no $\mathbf{c} \in C$, $\mathbf{c} \neq \mathbf{0}$, of weight $\leqslant s$, hence $d_C \geqslant s+1$. □

Next we describe a simple decoding algorithm for linear codes. Let C be a linear (n, k) code over $\mathbb{F}_q$. The vector space $\mathbb{F}_q^n/C$ consists of all cosets $\mathbf{a}+C=\{\mathbf{a}+\mathbf{c}:\mathbf{c}\in C\}$ with $\mathbf{a}\in\mathbb{F}_q^n$. Each coset contains q^k vectors and $\mathbb{F}_q^n$ can be regarded as being partitioned into cosets of C—namely,

$$\mathbb{F}_q^n = (\mathbf{a}^{(0)}+C)\cup(\mathbf{a}^{(1)}+C)\cup\cdots\cup(\mathbf{a}^{(s)}+C),$$

where $\mathbf{a}^{(0)}=\mathbf{0}$ and $s=q^{n-k}-1$. A received vector $\mathbf{y}$ must be in one of the cosets, say in $\mathbf{a}^{(i)}+C$. If the code word $\mathbf{c}$ was transmitted, then the error is given by $\mathbf{e}=\mathbf{y}-\mathbf{c}=\mathbf{a}^{(i)}+\mathbf{z}\in\mathbf{a}^{(i)}+C$ for suitable $\mathbf{z}\in C$. This leads to the following decoding scheme.

9.15. Decoding of Linear Codes. All possible error vectors $\mathbf{e}$ of a received vector $\mathbf{y}$ are the vectors in the coset of $\mathbf{y}$. The most likely error vector is the vector $\mathbf{e}$ with minimum weight in the coset of $\mathbf{y}$. Thus we decode $\mathbf{y}$ as $\mathbf{x}=\mathbf{y}-\mathbf{e}$.

The implementation of this procedure can be facilitated by the *coset-leader algorithm* for error correction of linear codes.

9.16. Definition. Let $C\subseteq\mathbb{F}_q^n$ be a linear (n,k) code and let $\mathbb{F}_q^n/C$ be the factor space. An element of minimum weight in a coset $\mathbf{a}+C$ is called a *coset leader* of $\mathbf{a}+C$. If several vectors in $\mathbf{a}+C$ have minimum weight, we choose one of them as coset leader.

Let $\mathbf{a}^{(1)},\ldots,\mathbf{a}^{(s)}$ be the coset leaders of the cosets $\neq C$ and let $\mathbf{c}^{(1)}=\mathbf{0}$, $\mathbf{c}^{(2)},\ldots,\mathbf{c}^{(q^k)}$ be all code words in C. Consider the following array:

$$\begin{array}{cccc}
\mathbf{c}^{(1)} & \mathbf{c}^{(2)} & \cdots & \mathbf{c}^{(q^k)} \\
\mathbf{a}^{(1)}+\mathbf{c}^{(1)} & \mathbf{a}^{(1)}+\mathbf{c}^{(2)} & \cdots & \mathbf{a}^{(1)}+\mathbf{c}^{(q^k)} \\
\vdots & \vdots & & \vdots \\
\mathbf{a}^{(s)}+\mathbf{c}^{(1)} & \mathbf{a}^{(s)}+\mathbf{c}^{(2)} & \cdots & \mathbf{a}^{(s)}+\mathbf{c}^{(q^k)}
\end{array}
\begin{array}{l}
\}\ \text{row of code words} \\
\left.\begin{array}{l} \\ \\ \\ \end{array}\right\}\ \text{remaining cosets}
\end{array}$$

(first column, under brace: column of coset leaders)

If a word $\mathbf{y}=\mathbf{a}^{(i)}+\mathbf{c}^{(j)}$ is received, then the decoder decides that the error $\mathbf{e}$ is the corresponding coset leader $\mathbf{a}^{(i)}$ and decodes $\mathbf{y}$ as the code word $\mathbf{x}=\mathbf{y}-\mathbf{e}=\mathbf{c}^{(j)}$; that is, $\mathbf{y}$ is decoded as the code word in the column of $\mathbf{y}$. The coset of $\mathbf{y}$ can be determined by evaluating the so-called syndrome of $\mathbf{y}$.

9.17. Definition. Let H be the parity-check matrix of a linear (n,k) code C. Then the vector $S(\mathbf{y})=H\mathbf{y}^{\mathrm{T}}$ of length $n-k$ is called the *syndrome* of $\mathbf{y}$.

9.18. Theorem. *For* $\mathbf{y}, \mathbf{z} \in \mathbb{F}_q^n$ *we have:*

(i) $S(\mathbf{y}) = \mathbf{0}$ *if and only if* $\mathbf{y} \in C$;
(ii) $S(\mathbf{y}) = S(\mathbf{z})$ *if and only if* $\mathbf{y} + C = \mathbf{z} + C$.

Proof. (i) follows immediately from the definition of C in terms of H. For (ii) note that $S(\mathbf{y}) = S(\mathbf{z})$ if and only if $H\mathbf{y}^{\mathrm{T}} = H\mathbf{z}^{\mathrm{T}}$ if and only if $H(\mathbf{y}-\mathbf{z})^{\mathrm{T}} = \mathbf{0}$ if and only if $\mathbf{y} - \mathbf{z} \in C$ if and only if $\mathbf{y} + C = \mathbf{z} + C$. □

If $\mathbf{e} = \mathbf{y} - \mathbf{c}$, $\mathbf{c} \in C$, $\mathbf{y} \in \mathbb{F}_q^n$, then

$$S(\mathbf{y}) = S(\mathbf{c}+\mathbf{e}) = S(\mathbf{c}) + S(\mathbf{e}) = S(\mathbf{e}) \tag{9.2}$$

and $\mathbf{y}$ and $\mathbf{e}$ are in the same coset. The coset leader of that coset also has the same syndrome. We have the following decoding algorithm.

9.19. Coset-Leader Algorithm. Let $C \subseteq \mathbb{F}_q^n$ be a linear (n, k) code and let $\mathbf{y}$ be the received vector. To correct errors in $\mathbf{y}$, calculate $S(\mathbf{y})$ and find the coset leader, say $\mathbf{e}$, with syndrome equal to $S(\mathbf{y})$. Then decode $\mathbf{y}$ as $\mathbf{x} = \mathbf{y} - \mathbf{e}$. Here $\mathbf{x}$ is the code word with minimum distance to $\mathbf{y}$.

9.20. Example. Let C be a binary linear $(4,2)$ code with generator matrix G and parity-check matrix H:

$$G = \begin{pmatrix} 1 & 0 & 1 & 0 \\ 0 & 1 & 1 & 1 \end{pmatrix}, \qquad H = \begin{pmatrix} 1 & 1 & 1 & 0 \\ 0 & 1 & 0 & 1 \end{pmatrix}.$$

The corresponding array of cosets is:

message row	00	10	01	11	
code words	0000	1010	0111	1101	$\binom{0}{0}$
other cosets	1000	0010	1111	0101	$\binom{1}{0}$
	0100	1110	0011	1001	$\binom{1}{1}$
	0001	1011	0110	1100	$\binom{0}{1}$
	coset leaders				syndromes

If $\mathbf{y} = 1110$ is received, we could look where in the array $\mathbf{y}$ occurs. But for large arrays this is very time consuming. Therefore we find $S(\mathbf{y})$ first—namely, $S(\mathbf{y}) = H\mathbf{y}^{\mathrm{T}} = \binom{1}{1}$—and decide that the error is equal to the coset leader 0100 that also has syndrome $\binom{1}{1}$. The original code word was most likely the word 1010 and the original message was 10. □

In large linear codes it is practically impossible to find coset leaders with minimum weight; for example, a linear (50,20) code over $\mathbb{F}_2$ has some 10^9 cosets. Therefore it is necessary to construct special codes in order to overcome such difficulties. First we note the following.

9.21. *Theorem.* *In a binary linear* (n,k) *code with parity-check matrix* H *the syndrome is the sum of those columns of* H *that correspond to positions where errors have occurred.*

Proof. Let $\mathbf{y} \in \mathbb{F}_2^n$ be the received vector, $\mathbf{y} = \mathbf{x} + \mathbf{e}$, $\mathbf{x} \in C$; then from (9.2) we have $S(\mathbf{y}) = H\mathbf{e}^{\mathrm{T}}$. Let $i_1, i_2, \ldots$ be the error coordinates in $\mathbf{e}$, say $\mathbf{e} = 0 \cdots 01_{i_1}0 \cdots 01_{i_2}0 \cdots$, then $S(\mathbf{y}) = \mathbf{h}_{i_1} + \mathbf{h}_{i_2} + \cdots$, where $\mathbf{h}_i$ denotes the ith column of H. □

If all columns of H are different, then a single error in the ith position of the transmitted word yields $S(\mathbf{y}) = \mathbf{h}_i$, thus one error can be corrected. To simplify the process of error location, the following class of codes is useful.

9.22. Definition. A binary code C_m of length $n = 2^m - 1$, $m \geqslant 2$, with an $m \times (2^m - 1)$ parity-check matrix H is called a *binary Hamming code* if the columns of H are the binary representations of the integers $1, 2, \ldots, 2^m - 1$.

9.23. *Lemma.* C_m *is a 1-error-correcting code of dimension* $2^m - m - 1$.

Proof. By definition of the parity-check matrix H of C_m, the rank of H is m. Also, any two columns of H are linearly independent. Since H contains with any two of its columns also their sum, the minimum distance of C_m equals 3 by Lemma 9.14. Thus C_m is 1-error-correcting by Theorem 9.12. □

9.24. Example. Let C_3 be the (7,4) Hamming code with parity-check matrix

$$H = \begin{pmatrix} 0 & 0 & 0 & 1 & 1 & 1 & 1 \\ 0 & 1 & 1 & 0 & 0 & 1 & 1 \\ 1 & 0 & 1 & 0 & 1 & 0 & 1 \end{pmatrix}.$$

If the syndrome of a received word $\mathbf{y}$ is, say, $S(\mathbf{y}) = (1 \quad 0 \quad 1)^{\mathrm{T}}$, then we know that an error must have occurred in the fifth position, since 101 is the binary representation of 5. □

Hamming codes can also be defined in the nonbinary case—that is, over arbitrary finite fields $\mathbb{F}_q$. Here the parity-check matrix H is an $m \times (q^m - 1)/(q-1)$ matrix that has pairwise linearly independent columns. Such a matrix defines a linear $((q^m-1)/(q-1),\ (q^m-1)/(q-1) - m)$ code of minimum distance 3.

Next we describe some relationships between the length n of code words, the number k of information or message symbols, and the minimum distance d_C of a linear code over $\mathbb{F}_q$.

9.25. Theorem (Hamming Bound). *Let C be a t-error-correcting code over $\mathbb{F}_q$ of length n with M code words. Then*

$$M\left(1+\binom{n}{1}(q-1)+\cdots+\binom{n}{t}(q-1)^t\right) \leqslant q^n.$$

Proof. There are $\binom{n}{m}(q-1)^m$ vectors with n coordinates in $\mathbb{F}_q$ of weight m. The balls of radius t centered at the code words are all pairwise disjoint and each of the M balls contains

$$1+\binom{n}{1}(q-1)+\cdots+\binom{n}{t}(q-1)^t$$

vectors of all the q^n vectors in $\mathbb{F}_q^n$. □

9.26. Theorem (Plotkin Bound). *For a linear (n, k) code C over $\mathbb{F}_q$ of minimum distance d_C we have*

$$d_C \leqslant \frac{nq^{k-1}(q-1)}{q^k-1}.$$

Proof. Let $1 \leqslant i \leqslant n$ be such that C contains a code word with nonzero ith component. Let D be the subspace of C consisting of all code words with ith component zero. In C/D there are q elements which correspond to q choices for the ith component of a code word. Thus $|C|/|D| = |C/D|$ implies $|D| = q^{k-1}$. By counting along the components, the sum of the weights of the code words in C is then seen to be $\leqslant nq^{k-1}(q-1)$. The minimum distance d_C of the code is the minimum nonzero weight and therefore must satisfy the inequality given in the theorem since the total number of code words of nonzero weight is $q^k - 1$. □

9.27. Theorem (Gilbert-Varshamov Bound). *There exists a linear (n, k) code over $\mathbb{F}_q$ with minimum distance $\geqslant d$ whenever*

$$q^{n-k} > \sum_{i=0}^{d-2} \binom{n-1}{i}(q-1)^i.$$

Proof. We prove this theorem by constructing an $(n-k)\times n$ parity-check matrix H for such a code. We choose the first column of H as any nonzero $(n-k)$-tuple over $\mathbb{F}_q$. The second column is any $(n-k)$-tuple over $\mathbb{F}_q$ that is not a scalar multiple of the first column. In general, suppose $j-1$ columns have been chosen so that any $d-1$ of them are linearly independent. There are at most

$$\sum_{i=0}^{d-2} \binom{j-1}{i}(q-1)^i$$

vectors obtained by linear combinations of $d-2$ or fewer of these $j-1$

columns. If the inequality of the theorem holds, then it will be possible to choose a jth column that is linearly independent of any $d-2$ of the first $j-1$ columns. The construction can be carried out in such a way that H has rank $n-k$. The resulting code has minimum distance $\geqslant d$ by Lemma 9.14. □

We define the dual code of a given linear code C by means of the following concepts. Let $\mathbf{u}=(u_1,\ldots,u_n)$, $\mathbf{v}=(v_1,\ldots,v_n)\in\mathbb{F}_q^n$, then $\mathbf{u}\cdot\mathbf{v}=u_1v_1+\cdots+u_nv_n$ denotes the *dot product* of $\mathbf{u}$ and $\mathbf{v}$. If $\mathbf{u}\cdot\mathbf{v}=0$, then $\mathbf{u}$ and $\mathbf{v}$ are called *orthogonal*.

9.28. Definition. Let C be a linear (n,k) code over $\mathbb{F}_q$. Then its *dual* (or *orthogonal*) *code* $C^\perp$ is defined as

$$C^\perp=\{\mathbf{u}\in\mathbb{F}_q^n:\mathbf{u}\cdot\mathbf{v}=0\quad\text{for all }\mathbf{v}\in C\}.$$

The code C is a k-dimensional subspace of $\mathbb{F}_q^n$, the dimension of $C^\perp$ is $n-k$. $C^\perp$ is a linear $(n,n-k)$ code. It is easy to show that $C^\perp$ has generator matrix H if C has parity-check matrix H and that $C^\perp$ has parity-check matrix G if C has generator matrix G.

Considerable information on a code is obtained from the weight enumeration. For instance, to determine decoding error probabilities or in certain decoding algorithms it is important to know the distribution of the weights of code words. There is a fundamental connection between the weight distribution of a linear code and of its dual code. This will be derived in the following theorem.

9.29. Definition. Let A_i denote the number of code words $\mathbf{c}\in C$ of weight i, $0\leqslant i\leqslant n$. Then the polynomial

$$A(x,y)=\sum_{i=0}^{n}A_ix^iy^{n-i}$$

in the indeterminates x and y over the complex numbers is called the *weight enumerator* of C.

We shall need characters of finite fields, as discussed in Chapter 5.

9.30. Definition. Let χ be a nontrivial additive character of $\mathbb{F}_q$ and let $\mathbf{v}\cdot\mathbf{u}$ denote the dot product of $\mathbf{v},\mathbf{u}\in\mathbb{F}_q^n$. We define for fixed $\mathbf{v}\in\mathbb{F}_q^n$ the mapping $\chi_{\mathbf{v}}:\mathbb{F}_q^n\to\mathbb{C}$ by

$$\chi_{\mathbf{v}}(\mathbf{u})=\chi(\mathbf{v}\cdot\mathbf{u})\quad\text{for }\mathbf{u}\in\mathbb{F}_q^n.$$

If V is a vector space over $\mathbb{C}$ and f a mapping from $\mathbb{F}_q^n$ into V, then we define $g_f:\mathbb{F}_q^n\to V$ by

$$g_f(\mathbf{u})=\sum_{\mathbf{v}\in\mathbb{F}_q^n}\chi_{\mathbf{v}}(\mathbf{u})f(\mathbf{v})\quad\text{for }\mathbf{u}\in\mathbb{F}_q^n.$$

9.31. Lemma. *Let E be a subspace of $\mathbb{F}_q^n$, $E^\perp$ its orthogonal complement, $f:\mathbb{F}_q^n\to V$ a mapping from $\mathbb{F}_q^n$ into a vector space V over $\mathbb{C}$ and χ a* *nontrivial additive character of* $\mathbb{F}_q$. *Then*

$$\sum_{\mathbf{u}\in E} g_f(\mathbf{u}) = |E| \sum_{\mathbf{v}\in E^\perp} f(\mathbf{v}).$$

Proof.

$$\sum_{\mathbf{u}\in E} g_f(\mathbf{u}) = \sum_{\mathbf{u}\in E}\sum_{\mathbf{v}\in\mathbb{F}_q^n} \chi_{\mathbf{v}}(\mathbf{u})f(\mathbf{v}) = \sum_{\mathbf{v}\in\mathbb{F}_q^n}\sum_{\mathbf{u}\in E} \chi(\mathbf{v}\cdot\mathbf{u})f(\mathbf{v})$$

$$= |E| \sum_{\mathbf{v}\in E^\perp} f(\mathbf{v}) + \sum_{\mathbf{v}\notin E^\perp}\sum_{c\in\mathbb{F}_q}\sum_{\substack{\mathbf{u}\in E\\ \mathbf{v}\cdot\mathbf{u}=c}} \chi(c)f(\mathbf{v}).$$

For fixed $\mathbf{v}\notin E^\perp$, $\mathbf{u}\in E\mapsto \mathbf{v}\cdot\mathbf{u}$ is a nontrivial linear functional on E, thus

$$\sum_{\mathbf{u}\in E} g_f(\mathbf{u}) = |E|\sum_{\mathbf{v}\in E^\perp} f(\mathbf{v}) + \frac{|E|}{q}\sum_{\mathbf{v}\notin E^\perp} f(\mathbf{v})\sum_{c\in\mathbb{F}_q}\chi(c) = |E|\sum_{\mathbf{v}\in E^\perp} f(\mathbf{v}),$$

by using (5.9). □

We apply this lemma with V as the space of polynomials in two indeterminates x and y over $\mathbb{C}$ and the mapping f defined as $f(\mathbf{v}) = x^{w(\mathbf{v})}y^{n-w(\mathbf{v})}$, where $w(\mathbf{v})$ denotes the weight of $\mathbf{v}\in\mathbb{F}_q^n$.

9.32. Theorem (MacWilliams Identity). *Let C be a linear (n,k) code over $\mathbb{F}_q$ and $C^\perp$ its dual code. If $A(x,y)$ is the weight enumerator of C and $A^\perp(x,y)$ is the weight enumerator of $C^\perp$, then*

$$A^\perp(x,y) = q^{-k}A(y-x, y+(q-1)x).$$

Proof. Let $f\colon \mathbb{F}_q^n \to \mathbb{C}[x,y]$ be as given above, then the weight enumerator of $C^\perp$ is

$$A^\perp(x,y) = \sum_{\mathbf{v}\in C^\perp} f(\mathbf{v}).$$

Let g_f be as in Definition 9.30 and for $v\in\mathbb{F}_q$ define

$$|v| = \begin{cases} 1 & \text{if } v\neq 0,\\ 0 & \text{if } v = 0.\end{cases}$$

For $\mathbf{u} = (u_1,\ldots,u_n)\in\mathbb{F}_q^n$ we have

$$g_f(\mathbf{u}) = \sum_{\mathbf{v}\in\mathbb{F}_q^n} \chi(\mathbf{v}\cdot\mathbf{u})x^{w(\mathbf{v})}y^{n-w(\mathbf{v})}$$

$$= \sum_{v_1,\ldots,v_n\in\mathbb{F}_q} \chi(u_1v_1+\cdots+u_nv_n)x^{|v_1|+\cdots+|v_n|}y^{(1-|v_1|)+\cdots+(1-|v_n|)}$$

$$= \sum_{v_1,\ldots,v_n\in\mathbb{F}_q}\prod_{i=1}^{n}\left[\chi(u_iv_i)x^{|v_i|}y^{1-|v_i|}\right]$$

$$= \prod_{i=1}^{n}\sum_{v\in\mathbb{F}_q}\left[\chi(u_iv)x^{|v|}y^{1-|v|}\right].$$

For $u_i = 0$ we have $\chi(u_i v) = \chi(0) = 1$, hence the corresponding factor in the product is $(q-1)x + y$. For $u_i \neq 0$ the corresponding factor is

$$y + x \sum_{v \in \mathbb{F}_q^*} \chi(v) = y - x.$$

Therefore,

$$g_f(\mathbf{u}) = (y-x)^{w(\mathbf{u})}(y+(q-1)x)^{n-w(\mathbf{u})}.$$

Lemma 9.31 implies

$$|C|A^{\perp}(x,y) = |C| \sum_{\mathbf{v} \in C^{\perp}} f(\mathbf{v}) = \sum_{\mathbf{u} \in C} g_f(\mathbf{u}) = A(y-x, y+(q-1)x).$$

Finally, $|C| = q^k$ by hypothesis. □

9.33. Corollary. *Let $x = z$ and $y = 1$ in the weight enumerators $A(x, y)$ and $A^{\perp}(x, y)$ and denote the resulting polynomials by $A(z)$ and $A^{\perp}(z)$, respectively. Then the MacWilliams identity can be written in the form*

$$A^{\perp}(z) = q^{-k}(1+(q-1)z)^n A\left(\frac{1-z}{1+(q-1)z}\right).$$

9.34. Example. Let C_m be the binary Hamming code of length $n = 2^m - 1$ and dimension $n - m$ over $\mathbb{F}_2$. The dual code $C_m^{\perp}$ has as its generator matrix the parity-check matrix H of C_m, which consists of all nonzero column vectors of length m over $\mathbb{F}_2$. $C_m^{\perp}$ consists of the zero vector and $2^m - 1$ vectors of weight 2^{m-1}. Thus the weight enumerator of $C_m^{\perp}$ is

$$y^n + (2^m - 1)x^{2^{m-1}}y^{2^{m-1}-1}.$$

By Theorem 9.32 the weight enumerator for C_m is given by

$$A(x,y) = \frac{1}{n+1}\left[(y+x)^n + n(y-x)^{(n+1)/2}(y+x)^{(n-1)/2}\right].$$

Let $A(z) = A(z,1)$—that is, $A(z) = \sum_{i=0}^{n} A_i z^i$—then one can verify that $A(z)$ satisfies the differential equation

$$(1-z^2)\frac{dA(z)}{dz} + (1+nz)A(z) = (1+z)^n$$

with initial condition $A(0) = A_0 = 1$. This is equivalent to

$$iA_i = \binom{n}{i-1} - A_{i-1} - (n-i+2)A_{i-2} \quad \text{for } i = 2, 3, \ldots, n$$

with initial conditions $A_0 = 1$, $A_1 = 0$. □

2. CYCLIC CODES

Cyclic codes are a special class of linear codes that can be implemented fairly simply and whose mathematical structure is reasonably well known.

9.35. Definition. A linear (n, k) code C over $\mathbb{F}_q$ is called *cyclic* if $(a_0, a_1, \ldots, a_{n-1}) \in C$ implies $(a_{n-1}, a_0, \ldots, a_{n-2}) \in C$.

From now on we impose the restriction $\gcd(n, q) = 1$ and let $(x^n - 1)$ be the ideal generated by $x^n - 1 \in \mathbb{F}_q[x]$. Then all elements of $\mathbb{F}_q[x]/(x^n - 1)$ can be represented by polynomials of degree less than n and clearly this residue class ring is isomorphic to $\mathbb{F}_q^n$ as a vector space over $\mathbb{F}_q$. An isomorphism is given by

$$(a_0, a_1, \ldots, a_{n-1}) \leftrightarrow a_0 + a_1 x + \cdots + a_{n-1} x^{n-1}.$$

Because of this isomorphism, we denote the elements of $\mathbb{F}_q[x]/(x^n - 1)$ either as polynomials of degree $< n$ modulo $x^n - 1$ or as vectors or words over $\mathbb{F}_q$. We introduce multiplication of polynomials modulo $x^n - 1$ in the usual way; that is, if $f \in \mathbb{F}_q[x]/(x^n - 1)$, $g_1, g_2 \in \mathbb{F}_q[x]$, then $g_1 g_2 = f$ means that $g_1 g_2 \equiv f \operatorname{mod}(x^n - 1)$.

A cyclic (n, k) code C can be obtained by multiplying each message of k coordinates (identified with a polynomial of degree $< k$) by a fixed polynomial $g(x)$ of degree $n - k$ with $g(x)$ a divisor of $x^n - 1$. The polynomials $g(x), xg(x), \ldots, x^{k-1}g(x)$ correspond to code words of C. A generator matrix of C is given by

$$G = \begin{pmatrix} g_0 & g_1 & \cdots & g_{n-k} & & 0 & 0 & \cdots & 0 \\ 0 & g_0 & g_1 & \cdots & & g_{n-k} & 0 & \cdots & 0 \\ \vdots & \vdots & \vdots & & & & & & \vdots \\ 0 & 0 & 0 & \cdots & 0 & g_0 & g_1 & \cdots & g_{n-k} \end{pmatrix},$$

where $g(x) = g_0 + g_1 x + \cdots + g_{n-k} x^{n-k}$. The rows of G are obviously linearly independent and rank $(G) = k$, the dimension of C. If

$$h(x) = (x^n - 1)/g(x) = h_0 + h_1 x + \cdots + h_k x^k,$$

then we see that the matrix

$$H = \begin{pmatrix} 0 & 0 & \cdots & & 0 & h_k & h_{k-1} & \cdots & & h_0 \\ 0 & 0 & \cdots & 0 & h_k & h_{k-1} & & \cdots & h_0 & 0 \\ \vdots & \vdots & & & & & & & & \vdots \\ h_k & h_{k-1} & \cdots & & h_0 & 0 & & \cdots & & 0 \end{pmatrix}$$

is a parity-check matrix for C. The code with generator matrix H is the dual code of C, which is again cyclic.

Since we are using the terminologies of vectors $(a_0, a_1, \ldots, a_{n-1})$ and polynomials $a_0 + a_1 x + \cdots + a_{n-1} x^{n-1}$ over $\mathbb{F}_q$ synonymously, we can interpret C as a subset of the factor ring $\mathbb{F}_q[x]/(x^n - 1)$.

9.36. Theorem. *The linear code C is cyclic if and only if C is an ideal of $\mathbb{F}_q[x]/(x^n-1)$.*

Proof. If C is an ideal and $(a_0, a_1, \ldots, a_{n-1}) \in C$, then also

$$x\left(a_0 + a_1 x + \cdots + a_{n-1}x^{n-1}\right) = (a_{n-1}, a_0, \ldots, a_{n-2}) \in C.$$

Conversely, if $(a_0, a_1, \ldots, a_{n-1}) \in C$ implies $(a_{n-1}, a_0, \ldots, a_{n-2}) \in C$, then for every $a(x) \in C$ we have $xa(x) \in C$, hence also $x^2 a(x) \in C$, $x^3 a(x) \in C$, and so on. Therefore also $b(x)a(x) \in C$ for any polynomial $b(x)$; that is, C is an ideal. □

Every ideal of $\mathbb{F}_q[x]/(x^n-1)$ is principal; in particular, every nonzero ideal C is generated by the monic polynomial of lowest degree in the ideal, say $g(x)$, where $g(x)$ divides $x^n - 1$.

9.37. Definition. Let $C = (g(x))$ be a cyclic code. Then $g(x)$ is called the *generator polynomial* of C and $h(x) = (x^n-1)/g(x)$ is called the *parity-check polynomial* of C.

Let $x^n - 1 = f_1(x) f_2(x) \cdots f_m(x)$ be the decomposition of $x^n - 1$ into monic irreducible factors over $\mathbb{F}_q$. Since we assume $\gcd(n, q) = 1$, there are no multiple factors. If $f_i(x)$ is irreducible over $\mathbb{F}_q$, then $(f_i(x))$ is a maximal ideal and the cyclic code generated by $f_i(x)$ is called a *maximal cyclic code*. The code generated by $(x^n-1)/f_i(x)$ is called an *irreducible cyclic code*. We can find all cyclic codes of length n over $\mathbb{F}_q$ by factoring $x^n - 1$ as above and taking any of the $2^m - 2$ nontrivial monic factors of $x^n - 1$ as a generator polynomial.

If $h(x)$ is the parity-check polynomial of a cyclic code $C \subseteq \mathbb{F}_q[x]/(x^n-1)$ and $v(x) \in \mathbb{F}_q[x]/(x^n-1)$, then $v(x) \in C$ if and only if $v(x)h(x) \equiv 0 \bmod (x^n-1)$. A message polynomial $a(x) = a_0 + a_1 x + \cdots + a_{k-1}x^{k-1}$ is encoded by C into $w(x) = a(x)g(x)$, where $g(x)$ is the generator polynomial of C. If we divide the received polynomial $v(x)$ by $g(x)$, and if there is a nonzero remainder, we know that an error occurs. The canonical generator matrix of C can be obtained as follows. Let $\deg((g(x)) = n-k$. Then there are unique polynomials $a_j(x)$ and $r_j(x)$ with $\deg(r_j(x)) < n-k$ such that

$$x^j = a_j(x)g(x) + r_j(x).$$

Consequently, $x^j - r_j(x)$ is a code polynomial, and so is $g_j(x) = x^k(x^j - r_j(x))$ considered modulo $x^n - 1$. The polynomials $g_j(x)$, $j = n-k, \ldots, n-1$, are linearly independent and form the canonical generator matrix

$$(I_k, -R),$$

where I_k is the $k \times k$ identity matrix and R is the $k \times (n-k)$ matrix whose ith row is the vector of coefficients of $r_{n-k-1+i}(x)$.

9.38. Example. Let $n=7$, $q=2$. Then

$$x^7-1=(x+1)(x^3+x+1)(x^3+x^2+1).$$

Thus $g(x)=x^3+x^2+1$ generates a cyclic (7,4) code with parity-check polynomial $h(x)=x^4+x^3+x^2+1$. The corresponding canonical generator matrix and parity-check matrix is, respectively,

$$G=\begin{pmatrix}1&0&0&0&1&0&1\\0&1&0&0&1&1&1\\0&0&1&0&1&1&0\\0&0&0&1&0&1&1\end{pmatrix},$$

$$H=\begin{pmatrix}1&1&1&0&1&0&0\\0&1&1&1&0&1&0\\1&1&0&1&0&0&1\end{pmatrix}. \qquad \square$$

We recall from Chapter 8 that if $f\in\mathbb{F}_q[x]$ is a polynomial of the form

$$f(x)=f_0+f_1x+\cdots+f_kx^k,\quad f_0\neq 0, f_k=1,$$

then the solutions of the linear recurrence relation

$$\sum_{j=0}^{k} f_j a_{i+j}=0,\quad i=0,1,\ldots,$$

are periodic of period n. The set of the n-tuples of the first n terms of each possible solution, considered as polynomials modulo x^n-1, is the ideal generated by $g(x)$ in $\mathbb{F}_q[x]/(x^n-1)$, where $g(x)$ is the reciprocal polynomial of $(x^n-1)/f(x)$ of degree $n-k$. Thus *linear recurrence relations can be used to generate code words of cyclic codes*, and this generation process can be implemented on feedback shift registers.

9.39. Example. Let $f(x)=x^3+x+1$, a factor of x^7-1 over $\mathbb{F}_2$. The associated linear recurrence relation is $a_{i+3}+a_{i+1}+a_i=0$, which gives rise to a (7,3) cyclic code, which encodes 111, say, as 1110010. The generator polynomial is the reciprocal polynomial of $(x^7-1)/f(x)$; that is, $g(x)=x^4+x^3+x^2+1$. $\square$

Cyclic codes can also be described by prescribing certain roots of all code polynomials in a suitable extension field of $\mathbb{F}_q$. The requirement that all code polynomials are multiples of $g(x)$, a generator polynomial, simply means that they are all 0 at the roots of $g(x)$. Let $\alpha_1,\ldots,\alpha_s$ be elements of a finite extension field of $\mathbb{F}_q$ and $p_i(x)$ be the minimal polynomial of α_i over $\mathbb{F}_q$ for $i=1,2,\ldots,s$. Let $n\in\mathbb{N}$ be such that $\alpha_i^n=1$, $i=1,2,\ldots,s$, and define $g(x)=\operatorname{lcm}(p_1(x),\ldots,p_s(x))$. Thus $g(x)$ divides x^n-1. If $C\subseteq\mathbb{F}_q^n$ is the cyclic code with generator polynomial $g(x)$, then we have $v(x)\in C$ if and only if $v(\alpha_i)=0$, $i=1,2,\ldots,s$. As an example of the concurrence of the description of a cyclic code by a generator polynomial or by roots of code polynomials we prove the following result, which uses the concept of equivalence of codes in Exercise 9.10.

9.40. Theorem. *The binary cyclic code of length $n = 2^m - 1$ for which the generator polynomial is the minimal polynomial over $\mathbb{F}_2$ of a primitive element of $\mathbb{F}_{2^m}$ is equivalent to the binary $(n, n-m)$ Hamming code.*

Proof. Let α denote a primitive element of $\mathbb{F}_{2^m}$ and let

$$p(x) = (x-\alpha)(x-\alpha^2)\cdots(x-\alpha^{2^{m-1}})$$

be the minimal polynomial of α over $\mathbb{F}_2$. We now consider the cyclic code C generated by $p(x)$. We construct an $m \times (2^m - 1)$ matrix H for which the jth column is $(c_0, c_1, \ldots, c_{m-1})^{\mathrm{T}}$ if

$$\alpha^{j-1} = \sum_{i=0}^{m-1} c_i \alpha^i, \quad j = 1, 2, \ldots, 2^m - 1,$$

where $c_i \in \mathbb{F}_2$. If $\mathbf{a} = (a_0, a_1, \ldots, a_{n-1})$ and $a(x) = a_0 + a_1 x + \cdots + a_{n-1}x^{n-1} \in \mathbb{F}_2[x]$, then the vector $H\mathbf{a}^{\mathrm{T}}$ corresponds to the element $a(\alpha)$ expressed in the basis $\{1, \alpha, \ldots, \alpha^{m-1}\}$. Consequently, $H\mathbf{a}^{\mathrm{T}} = \mathbf{0}$ holds exactly when $p(x)$ divides $a(x)$, so H is a parity-check matrix of C. Since the columns of H are a permutation of the binary representations of the numbers $1, 2, \ldots, 2^m - 1$, the proof is complete. □

9.41. Example. The polynomial $x^4 + x + 1$ is primitive over $\mathbb{F}_2$ and thus has a primitive element α of $\mathbb{F}_{16}$ as a root. If we use vector notation for the 15 elements $\alpha^j \in \mathbb{F}_{16}^*$, $j = 0, 1, \ldots, 14$, expressed in the basis $\{1, \alpha, \alpha^2, \alpha^3\}$ and we form a 4×15 matrix with these vectors as columns, then we get the parity-check matrix of a code equivalent to the $(15, 11)$ Hamming code. A message $(a_0, a_1, \ldots, a_{10})$ is encoded into a code polynomial

$$w(x) = a(x)(x^4 + x + 1),$$

where $a(x) = a_0 + a_1 x + \cdots + a_{10}x^{10}$. Now suppose the received polynomial contains one error; that is, $w(x) + x^{e-1}$ is received when $w(x)$ is transmitted. Then the syndrome is $w(\alpha) + \alpha^{e-1} = \alpha^{e-1}$ and the decoder is led to the conclusion that there is an error in the eth position. □

9.42. Theorem. *Let $C \subseteq \mathbb{F}_q[x]/(x^n - 1)$ by a cyclic code with generator polynomial g and let $\alpha_1, \ldots, \alpha_{n-k}$ be the roots of g. Then $f \in \mathbb{F}_q[x]/(x^n - 1)$ is a code polynomial if and only if the coefficient vector $(f_0, \ldots, f_{n-1})$ of f is in the null space of the matrix*

$$H = \begin{pmatrix} 1 & \alpha_1 & \alpha_1^2 & \cdots & \alpha_1^{n-1} \\ \vdots & \vdots & \vdots & & \vdots \\ 1 & \alpha_{n-k} & \alpha_{n-k}^2 & \cdots & \alpha_{n-k}^{n-1} \end{pmatrix}. \tag{9.3}$$

Proof. Let $f(x)=f_0+f_1x+\cdots+f_{n-1}x^{n-1}$; then $f(\alpha_i)=f_0+f_1\alpha_i+\cdots+f_{n-1}\alpha_i^{n-1}=0$ for $1\leqslant i\leqslant n-k$, that is,

$$(1,\alpha_i,\ldots,\alpha_i^{n-1})(f_0,f_1,\ldots,f_{n-1})^{\mathrm{T}}=0 \quad \text{for } 1\leqslant i\leqslant n-k,$$

if and only if $H(f_0,f_1,\ldots,f_{n-1})^{\mathrm{T}}=\mathbf{0}$. □

We recall from Section 1 that for error correction we have to determine the syndrome of the received word $\mathbf{y}$. In the case of cyclic codes, the syndrome, which is a column vector of length $n-k$, can often be replaced by a simpler entity serving the same purpose. For instance, let α be a primitive nth root of unity in $\mathbb{F}_{q^m}$ and let the generator polynomial g be the minimal polynomial of α over $\mathbb{F}_q$. Since g divides $f\in\mathbb{F}_q[x]/(x^n-1)$ if and only if $f(\alpha)=0$, it suffices to replace the matrix H in (9.3) by

$$H=\begin{pmatrix}1 & \alpha & \alpha^2 & \cdots & \alpha^{n-1}\end{pmatrix}.$$

Then the role of the syndrome is played by $S(\mathbf{y})=H\mathbf{y}^{\mathrm{T}}$, and $S(\mathbf{y})=y(\alpha)$ since $\mathbf{y}=(y_0,y_1,\ldots,y_{n-1})$ can be regarded as a polynomial $y(x)$ with coefficients y_i. In the following we use the notation $\mathbf{w}$ for a transmitted word and $\mathbf{v}$ for a received word, and we write $w(x)$ and $v(x)$, respectively, for the corresponding polynomials. Suppose $e^{(j)}(x)=x^{j-1}$ with $1\leqslant j\leqslant n$ is an error polynomial with a single error, and let $\mathbf{v}=\mathbf{w}+\mathbf{e}^{(j)}$ be the received word. Then

$$v(\alpha)=w(\alpha)+e^{(j)}(\alpha)=e^{(j)}(\alpha)=\alpha^{j-1}.$$

$e^{(j)}(\alpha)$ is called the *error-location number*. $S(\mathbf{v})=\alpha^{j-1}$ indicates the error uniquely, since $e^{(i)}(\alpha)\neq e^{(j)}(\alpha)$ for $1\leqslant i\leqslant n$ with $i\neq j$.

Before describing a general class of cyclic codes and their decoding, we consider a special example to motivate the theory.

9.43. Example. Let $\alpha\in\mathbb{F}_{16}$ be a root of $x^4+x+1\in\mathbb{F}_2[x]$, then α and α^3 have the minimal polynomials $m^{(1)}(x)=x^4+x+1$ and $m^{(3)}(x)=x^4+x^3+x^2+x+1$ over $\mathbb{F}_2$, respectively. Both $m^{(1)}(x)$ and $m^{(3)}(x)$ are divisors of $x^{15}-1$. Hence we can define a binary cyclic code C with generator polynomial $g=m^{(1)}m^{(3)}$. Since g divides $f\in\mathbb{F}_2[x]/(x^{15}-1)$ if and only if $f(\alpha)=f(\alpha^3)=0$, it suffices to replace the matrix H in (9.3) by

$$H=\begin{pmatrix}1 & \alpha & \alpha^2 & \cdots & \alpha^{14}\\ 1 & \alpha^3 & \alpha^6 & \cdots & \alpha^{42}\end{pmatrix}.$$

We shall show (see Theorem 9.45 and Example 9.47) that the minimum distance of C is $\geqslant 5$, therefore C can correct up to 2 errors. C is a cyclic $(15,7)$ code. Let

$$S_1=\sum_{i=0}^{14} v_i\alpha^i \quad\text{and}\quad S_3=\sum_{i=0}^{14} v_i\alpha^{3i}$$

be the components of $S(\mathbf{v}) = H\mathbf{v}^{\mathrm{T}}$. Then $\mathbf{v} \in C$ if and only if $S(\mathbf{v}) = H\mathbf{v}^{\mathrm{T}} = \mathbf{0}$ if and only if $S_1 = S_3 = 0$. If we use binary notation to represent elements of $\mathbb{F}_{16}$, then H attains the form

$$H = \begin{pmatrix} 1 & 0 & 0 & 0 & 1 & 0 & 0 & 1 & 1 & 0 & 1 & 0 & 1 & 1 & 1 \\ 0 & 1 & 0 & 0 & 1 & 1 & 0 & 1 & 0 & 1 & 1 & 1 & 1 & 0 & 0 \\ 0 & 0 & 1 & 0 & 0 & 1 & 1 & 0 & 1 & 0 & 1 & 1 & 1 & 1 & 0 \\ 0 & 0 & 0 & 1 & 0 & 0 & 1 & 1 & 0 & 1 & 0 & 1 & 1 & 1 & 1 \\ 1 & 0 & 0 & 0 & 1 & 1 & 0 & 0 & 0 & 1 & 1 & 0 & 0 & 0 & 1 \\ 0 & 0 & 0 & 1 & 1 & 0 & 0 & 0 & 1 & 1 & 0 & 0 & 0 & 1 & 1 \\ 0 & 0 & 1 & 0 & 1 & 0 & 0 & 1 & 0 & 1 & 0 & 0 & 1 & 0 & 1 \\ 0 & 1 & 1 & 1 & 1 & 0 & 1 & 1 & 1 & 1 & 0 & 1 & 1 & 1 & 1 \end{pmatrix}.$$

The columns of H are calculated as follows: the first four entries of the first column are the coefficients in $1 = 1 \cdot \alpha^0 + 0 \cdot \alpha^1 + 0 \cdot \alpha^2 + 0 \cdot \alpha^3$, the first four entries of the second column are the coefficients in $\alpha = 0 \cdot \alpha^0 + 1 \cdot \alpha^1 + 0 \cdot \alpha^2 + 0 \cdot \alpha^3$, and so on; the last four entries of the first column are the coefficients in $1 = 1 \cdot \alpha^0 + 0 \cdot \alpha^1 + 0 \cdot \alpha^2 + 0 \cdot \alpha^3$, the last four entries of the second column are the coefficients in $\alpha^3 = 0 \cdot \alpha^0 + 0 \cdot \alpha^1 + 0 \cdot \alpha^2 + 1 \cdot \alpha^3$, and so on. We use $\alpha^4 + \alpha + 1 = 0$ in the calculations.

Suppose the received vector $\mathbf{v} = (v_0, \ldots, v_{14})$ has at most two errors; for example, $e(x) = x^{a_1} + x^{a_2}$ with $0 \leqslant a_1, a_2 \leqslant 14$, $a_1 \neq a_2$. Then we have

$$S_1 = \alpha^{a_1} + \alpha^{a_2}, \qquad S_3 = \alpha^{3a_1} + \alpha^{3a_2}.$$

Let $\eta_1 = \alpha^{a_1}$, $\eta_2 = \alpha^{a_2}$ be the error-location numbers, then

$$S_1 = \eta_1 + \eta_2, \qquad S_3 = \eta_1^3 + \eta_2^3,$$

therefore

$$S_3 = S_1^3 + S_1^2 \eta_1 + S_1 \eta_1^2,$$

hence

$$1 + S_1 \eta_1^{-1} + \left(S_1^2 + S_3 S_1^{-1}\right) \eta_1^{-2} = 0.$$

If two errors occurred, then η_1^{-1} and η_2^{-1} are roots of the polynomial

$$s(x) = 1 + S_1 x + \left(S_1^2 + S_3 S_1^{-1}\right) x^2. \tag{9.4}$$

If only one error occurred, then $S_1 = \eta_1$ and $S_3 = \eta_1^3$, hence $S_1^3 + S_3 = 0$; that is,

$$s(x) = 1 + S_1 x. \tag{9.5}$$

If no error occurred, then $S_1 = S_3 = 0$ and the correct code word $\mathbf{w}$ has been received.

To summarize, we first evaluate the syndrome $S(\mathbf{v}) = H\mathbf{v}^{\mathrm{T}}$ of the received vector $\mathbf{v}$, then determine $s(x)$ and find the errors via the roots of $s(x)$. The polynomial in (9.5) has a root in $\mathbb{F}_{16}$ whenever $S_1 \neq 0$. If $s(x)$ in (9.4) has no roots in $\mathbb{F}_{16}$, then we know that the error $e(x)$ has more than two error locations and therefore cannot be corrected by the given (15,7) code.

More specifically, suppose

$$\mathbf{v} = 100111000000000$$

is the received word. Then $S(\mathbf{v}) = \begin{pmatrix} S_1 \\ S_3 \end{pmatrix}$ is given by

$$S_1 = 1 + \alpha^3 + \alpha^4 + \alpha^5 = \alpha^2 + \alpha^3, \qquad S_3 = 1 + \alpha^9 + \alpha^{12} + \alpha^{15} = 1 + \alpha^2.$$

For the polynomial $s(x)$ in (9.4) we obtain

$$\begin{aligned} s(x) &= 1 + (\alpha^2 + \alpha^3)x + \left[1 + \alpha + \alpha^2 + \alpha^3 + (1 + \alpha^2)(\alpha^2 + \alpha^3)^{-1}\right]x^2 \\ &= 1 + (\alpha^2 + \alpha^3)x + (1 + \alpha + \alpha^3)x^2. \end{aligned}$$

We determine the roots of $s(x)$ by trial and error and find α and α^7 as roots. Hence we have $\eta_1^{-1} = \alpha$, $\eta_2^{-1} = \alpha^7$, thus $\eta_1 = \alpha^{14}$, $\eta_2 = \alpha^8$. Therefore, we know that errors must have occurred in the positions corresponding to x^8 and x^{14}, that is, in the 9th and 15th position of $\mathbf{v}$. The transmitted code word must have been

$$\mathbf{w} = 100111001000001.$$

The code word $\mathbf{w}$ is decoded by dividing the corresponding polynomial by the generator polynomial g. This gives $1 + x^3 + x^5 + x^6$ with remainder 0. Hence the original message was 1001011. □

9.44. Definition. Let b be a nonnegative integer and let $\alpha \in \mathbb{F}_{q^m}$ be a primitive nth root of unity, where m is the multiplicative order of q modulo n. A *BCH code* over $\mathbb{F}_q$ of length n and *designed distance* d, $2 \leqslant d \leqslant n$, is a cyclic code defined by the roots

$$\alpha^b, \alpha^{b+1}, \ldots, \alpha^{b+d-2}$$

of the generator polynomial.

If $m^{(i)}(x)$ denotes the minimal polynomial of α^i over $\mathbb{F}_q$, then the generator polynomial $g(x)$ of a BCH code is of the form

$$g(x) = \operatorname{lcm}\left(m^{(b)}(x), m^{(b+1)}(x), \ldots, m^{(b+d-2)}(x)\right).$$

Some special cases of the general Definition 9.44 are also important. If $b = 1$, the corresponding BCH codes are called *narrow-sense* BCH codes. If $n = q^m - 1$, the BCH codes are called *primitive*. If $n = q - 1$, a BCH code of length n over $\mathbb{F}_q$ is called a *Reed-Solomon code*.

9.45. Theorem. *The minimum distance of a BCH code of designed distance d is at least d.*

Proof. The BCH code is the null space of the matrix

$$H = \begin{pmatrix} 1 & \alpha^b & \alpha^{2b} & \cdots & \alpha^{(n-1)b} \\ 1 & \alpha^{b+1} & \alpha^{2(b+1)} & \cdots & \alpha^{(n-1)(b+1)} \\ \vdots & \vdots & \vdots & & \vdots \\ 1 & \alpha^{b+d-2} & \alpha^{2(b+d-2)} & \cdots & \alpha^{(n-1)(b+d-2)} \end{pmatrix}.$$

We show that any $d-1$ columns of this matrix are linearly independent. Take the determinant of any $d-1$ distinct columns of H, then we obtain

$$\begin{vmatrix} \alpha^{bi_1} & \alpha^{bi_2} & \cdots & \alpha^{bi_{d-1}} \\ \alpha^{(b+1)i_1} & \alpha^{(b+1)i_2} & \cdots & \alpha^{(b+1)i_{d-1}} \\ \vdots & \vdots & & \vdots \\ \alpha^{(b+d-2)i_1} & \alpha^{(b+d-2)i_2} & \cdots & \alpha^{(b+d-2)i_{d-1}} \end{vmatrix}$$

$$= \alpha^{b(i_1+i_2+\cdots+i_{d-1})} \begin{vmatrix} 1 & 1 & \cdots & 1 \\ \alpha^{i_1} & \alpha^{i_2} & \cdots & \alpha^{i_{d-1}} \\ \vdots & \vdots & & \vdots \\ \alpha^{i_1(d-2)} & \alpha^{i_2(d-2)} & \cdots & \alpha^{i_{d-1}(d-2)} \end{vmatrix}$$

$$= \alpha^{b(i_1+i_2+\cdots+i_{d-1})} \prod_{1 \leqslant k < j \leqslant d-1} (\alpha^{i_j} - \alpha^{i_k}) \neq 0.$$

Therefore the minimum distance of the code is at least d. □

9.46. Example. Let $m^{(1)}(x) = x^4 + x + 1$ be the minimal polynomial over $\mathbb{F}_2$ of a primitive element $\alpha \in \mathbb{F}_{16}$. We represent the powers α^i, $0 \leqslant i \leqslant 14$, as linear combinations of $1, \alpha, \alpha^2, \alpha^3$ and thus obtain a parity-check matrix H of a code equivalent to the $(15, 11)$ Hamming code:

$$H = \begin{pmatrix} 1 & 0 & 0 & 0 & 1 & 0 & 0 & 1 & 1 & 0 & 1 & 0 & 1 & 1 & 1 \\ 0 & 1 & 0 & 0 & 1 & 1 & 0 & 1 & 0 & 1 & 1 & 1 & 1 & 0 & 0 \\ 0 & 0 & 1 & 0 & 0 & 1 & 1 & 0 & 1 & 0 & 1 & 1 & 1 & 1 & 0 \\ 0 & 0 & 0 & 1 & 0 & 0 & 1 & 1 & 0 & 1 & 0 & 1 & 1 & 1 & 1 \end{pmatrix}$$

$$= (1 \quad \alpha \quad \alpha^2 \quad \alpha^3 \quad \alpha^4 \quad \alpha^5 \quad \alpha^6 \quad \alpha^7 \quad \alpha^8 \quad \alpha^9 \quad \alpha^{10} \quad \alpha^{11} \quad \alpha^{12} \quad \alpha^{13} \quad \alpha^{14}).$$

This code can also be regarded as a narrow-sense BCH code of designed distance $d = 3$ over $\mathbb{F}_2$ (note that α^2 is also a root of $m^{(1)}(x)$). Its minimum distance is also 3, and it can therefore correct one error. In order to decode a received vector $\mathbf{v} \in \mathbb{F}_2^{15}$, we have to find the syndrome $H\mathbf{v}^{\mathrm{T}}$. For this cyclic $(15, 11)$ code the syndrome is given as $v(\alpha)$ in the basis $\{1, \alpha, \alpha^2, \alpha^3\}$. It is obtained by dividing $v(x)$ by $m^{(1)}(x)$, say $v(x) = a(x)m^{(1)}(x) + r(x)$ with $\deg(r(x)) < 4$, for then $v(\alpha) = r(\alpha)$; that is, the components of the syndrome are equal to the coefficients of $r(x)$.

For instance, let

$$\mathbf{v} = 010110001011101,$$

then $r(x) = 1 + x$, hence

$$H\mathbf{v}^{\mathrm{T}} = (1100)^{\mathrm{T}} = 1 + \alpha.$$

Next we have to find the error $\mathbf{e}$ with weight $w(\mathbf{e}) \leqslant 1$ and having the same syndrome. Thus we must determine the exponent j, $0 \leqslant j \leqslant 14$, such that $\alpha^j = H\mathbf{v}^{\mathrm{T}}$. In our numerical example $j = 4$, thus in the received vector $\mathbf{v}$ the fifth position is in error and the transmitted word was

$$\mathbf{w} = 010100001011101. \qquad \square$$

9.47. Example. Let $q = 2$, $n = 15$, and $d = 4$. Then $x^4 + x + 1$ is irreducible over $\mathbb{F}_2$ and its roots are primitive elements of $\mathbb{F}_{16}$. If α is such a root, then α^2 is a root, and α^3 is then a root of $x^4 + x^3 + x^2 + x + 1$. Thus a narrow-sense BCH code with $d = 4$ is generated by

$$g(x) = (x^4 + x + 1)(x^4 + x^3 + x^2 + x + 1).$$

This is also a generator for a BCH code with $d = 5$, since α^4 is a root of $x^4 + x + 1$. The dimension of this code is $15 - \deg(g(x)) = 7$. This code was considered in greater detail in Example 9.43. $\square$

BCH codes are very powerful since for any positive integer d we can construct a BCH code of minimum distance $\geqslant d$. To find a BCH code for a larger minimum distance, we have to increase the length n and hence increase the number m—that is, the degree of $\mathbb{F}_{q^m}$ over $\mathbb{F}_q$. A BCH code of designed distance $d \geqslant 2t + 1$ will correct t or fewer errors, but at the same time, in order to achieve the desired minimum distance, we must use code words of great length.

We describe now a general *decoding algorithm for BCH codes*. Let us denote by $w(x)$, $v(x)$, and $e(x)$ the transmitted code polynomial, the received polynomial, and the error polynomial, respectively, so that $v(x) = w(x) + e(x)$. First we have to obtain the syndrome of $\mathbf{v}$,

$$S(\mathbf{v}) = H\mathbf{v}^{\mathrm{T}} = (S_b, S_{b+1}, \ldots, S_{b+d-2})^{\mathrm{T}},$$

where

$$S_j = v(\alpha^j) = w(\alpha^j) + e(\alpha^j) = e(\alpha^j) \quad \text{for } b \leqslant j \leqslant b + d - 2.$$

If $r \leqslant t$ errors occur, then

$$e(x) = \sum_{i=1}^{r} c_i x^{a_i},$$

where $a_1, \ldots, a_r$ are distinct elements of $\{0, 1, \ldots, n-1\}$. The elements $\eta_i = \alpha^{a_i} \in \mathbb{F}_{q^m}$ are called *error-location numbers*, the elements $c_i \in \mathbb{F}_q^*$ are called *error* *values*. Thus we obtain for the syndrome of $\mathbf{v}$,

$$S_j = e(\alpha^j) = \sum_{i=1}^{r} c_i \eta_i^j \quad \text{for } b \leqslant j \leqslant b+d-2.$$

Because of the computational rules in $\mathbb{F}_{q^m}$ we have

$$S_j^q = \left(\sum_{i=1}^{r} c_i \eta_i^j \right)^q = \sum_{i=1}^{r} c_i^q \eta_i^{jq} = \sum_{i=1}^{r} c_i \eta_i^{jq} = S_{jq}. \tag{9.6}$$

The unknown quantities are the pairs (η_i, c_i), $i = 1, \ldots, r$, the coordinates S_j of the syndrome $S(\mathbf{v})$ are known since they can be calculated from the received vector $\mathbf{v}$. In the binary case any error is completely characterized by the η_i alone, since in this case all c_i are 1.

In the next stage of the decoding algorithm we determine the coefficients σ_i defined by the polynomial identity

$$\prod_{i=1}^{r} (\eta_i - x) = \sum_{i=0}^{r} (-1)^i \sigma_{r-i} x^i$$

$$= \sigma_r - \sigma_{r-1} x + \cdots + (-1)^r \sigma_0 x^r.$$

Thus $\sigma_0 = 1$ and $\sigma_1, \ldots, \sigma_r$ are the elementary symmetric polynomials in $\eta_1, \ldots, \eta_r$. Substituting η_i for x gives

$$(-1)^r \sigma_r + (-1)^{r-1} \sigma_{r-1} \eta_i + \cdots + (-1) \sigma_1 \eta_i^{r-1} + \eta_i^r = 0 \quad \text{for } i = 1, \ldots, r.$$

Multiplying by $c_i \eta_i^j$ and summing these equations for $i = 1, \ldots, r$ yields

$$(-1)^r \sigma_r S_j + (-1)^{r-1} \sigma_{r-1} S_{j+1} + \cdots + (-1) \sigma_1 S_{j+r-1} + S_{j+r} = 0$$
$$\text{for } j = b, b+1, \ldots, b+r-1.$$

9.48. *Lemma.* *The system of equations*

$$\sum_{i=1}^{r} c_i \eta_i^j = S_j, \quad j = b, b+1, \ldots, b+r-1$$

in the unknowns c_i is solvable if the η_i are distinct elements of $\mathbb{F}_{q^m}^*$.

Proof. The determinant of the system is

$$\begin{vmatrix} \eta_1^b & \eta_2^b & \cdots & \eta_r^b \\ \eta_1^{b+1} & \eta_2^{b+1} & \cdots & \eta_r^{b+1} \\ \vdots & \vdots & & \vdots \\ \eta_1^{b+r-1} & \eta_2^{b+r-1} & \cdots & \eta_r^{b+r-1} \end{vmatrix} = \eta_1^b \eta_2^b \cdots \eta_r^b \prod_{1 \leqslant i < j \leqslant r} (\eta_j - \eta_i) \neq 0. \quad \square$$

9.49. Lemma. *The system of equations*

$$(-1)^r \sigma_r S_j + (-1)^{r-1} \sigma_{r-1} S_{j+1} + \cdots + (-1)\sigma_1 S_{j+r-1} + S_{j+r} = 0,$$
$$j = b, b+1, \ldots, b+r-1$$

in the unknowns $(-1)^i\sigma_i$, $i = 1, 2, \ldots, r$, *is solvable uniquely if and only if* r *errors occur.*

Proof. The matrix of the system can be decomposed as follows:

$$\begin{pmatrix} S_b & S_{b+1} & \cdots & S_{b+r-1} \\ S_{b+1} & S_{b+2} & \cdots & S_{b+r} \\ \vdots & \vdots & & \vdots \\ S_{b+r-1} & S_{b+r} & \cdots & S_{b+2r-2} \end{pmatrix} = VDV^{\mathrm{T}},$$

where

$$V = \begin{pmatrix} 1 & 1 & \cdots & 1 \\ \eta_1 & \eta_2 & \cdots & \eta_r \\ \vdots & \vdots & & \vdots \\ \eta_1^{r-1} & \eta_2^{r-1} & \cdots & \eta_r^{r-1} \end{pmatrix}$$

and

$$D = \begin{pmatrix} c_1\eta_1^b & 0 & \cdots & 0 \\ 0 & c_2\eta_2^b & \cdots & 0 \\ \vdots & \vdots & & \vdots \\ 0 & 0 & \cdots & c_r\eta_r^b \end{pmatrix}.$$

The matrix of the given system of equations is nonsingular if and only if V and D are nonsingular. V as a Vandermonde matrix is nonsingular if and only if the η_i, $i = 1, \ldots, r$, are distinct and D is nonsingular if and only if all the η_i and c_i are nonzero. Both conditions are satisfied if and only if r errors occur. □

We introduce the *error-locator polynomial* that is closely related to the considerations above:

$$s(x) = \prod_{i=1}^{r} (1 - \eta_i x) = \sum_{i=0}^{r} (-1)^i \sigma_i x^i,$$

where the σ_i are as above. The roots of $s(x)$ are $\eta_1^{-1}, \eta_2^{-1}, \ldots, \eta_r^{-1}$. In order to find these roots, we can use a search method due to Chien. First we want to know if α^{n-1} is an error-location number—that is, if $\alpha = \alpha^{-(n-1)}$ is a root of $s(x)$. To test this we form

$$-\sigma_1\alpha+\sigma_2\alpha^2+\cdots+(-1)^r\sigma_r\alpha^r.$$

If this is equal to -1, then α^{n-1} is an error-location number since then $s(\alpha)=0$. More generally, α^{n-m} is tested for $m=1,2,\ldots,n$ in the same way. In the binary case, the discovery of error locations is equivalent to correcting errors. We summarize the BCH decoding algorithm, writing now τ_i for $(-1)^i\sigma_i$.

9.50. BCH Decoding. Suppose at most t errors occur in transmitting a code word **w**, using a BCH code of designed distance $d \geqslant 2t+1$.

Step 1. Determine the syndrome of the received word **v**,

$$S(\mathbf{v})=(S_b, S_{b+1},\ldots,S_{b+d-2})^{\mathrm{T}}.$$

Let

$$S_j=\sum_{i=1}^{r} c_i\eta_i^j,\quad b\leqslant j\leqslant b+d-2.$$

Step 2. Determine the maximum number $r\leqslant t$ such that the system of equations

$$S_{j+r}+S_{j+r-1}\tau_1+\cdots+S_j\tau_r=0,\quad b\leqslant j\leqslant b+r-1,$$

in the τ_i has a nonsingular coefficient matrix, thus obtaining the number r of errors that have occurred. Then set up the error-locator polynomial

$$s(x)=\prod_{i=1}^{r}(1-\eta_i x)=\sum_{i=0}^{r}\tau_i x^i.$$

Find the coefficients τ_i from the S_j.

Step 3. Solve $s(x)=0$ by substituting the powers of α into $s(x)$. Thus find the error-location numbers η_i (Chien search).

Step 4. Introduce the η_i in the first r equations of Step 1 to determine the error values c_i. Then find the transmitted word **w** from $w(x)=v(x)-e(x)$.

9.51. Remark. We note that the difficult step in this algorithm is Step 2. There are various methods to perform this step, one possibility is to use the Berlekamp-Massey algorithm of Chapter 8 to determine the unknown coefficients τ_i in the linear recurrence relation for the S_j. □

9.52. Example. Consider a BCH code with designed distance $d=5$ that is able to correct any single or double error. In this case, let $b=1$, $n=15$, $q=2$. If $m^{(i)}(x)$ denotes the minimal polynomial of α^i over $\mathbb{F}_2$, where the primitive element $\alpha \in \mathbb{F}_{16}$ is a root of $x^4 + x + 1$, then

$$m^{(1)}(x) = m^{(2)}(x) = m^{(4)}(x) = m^{(8)}(x) = 1 + x + x^4,$$

$$m^{(3)}(x) = m^{(6)}(x) = m^{(12)}(x) = m^{(9)}(x) = 1 + x + x^2 + x^3 + x^4.$$

Therefore a generator polynomial of the BCH code will be

$$g(x) = m^{(1)}(x)m^{(3)}(x) = 1 + x^4 + x^6 + x^7 + x^8.$$

The code is a (15,7) code, with parity-check polynomial

$$h(x) = (x^{15} - 1)/g(x) = 1 + x^4 + x^6 + x^7.$$

We take the vectors corresponding to

$$g(x), xg(x), x^2g(x), x^3g(x), x^4g(x), x^5g(x), x^6g(x)$$

as the basis of the (15,7) BCH code and obtain the generator matrix

$$G = \begin{pmatrix} 1 & 0 & 0 & 0 & 1 & 0 & 1 & 1 & 1 & 0 & 0 & 0 & 0 & 0 & 0 \\ 0 & 1 & 0 & 0 & 0 & 1 & 0 & 1 & 1 & 1 & 0 & 0 & 0 & 0 & 0 \\ 0 & 0 & 1 & 0 & 0 & 0 & 1 & 0 & 1 & 1 & 1 & 0 & 0 & 0 & 0 \\ 0 & 0 & 0 & 1 & 0 & 0 & 0 & 1 & 0 & 1 & 1 & 1 & 0 & 0 & 0 \\ 0 & 0 & 0 & 0 & 1 & 0 & 0 & 0 & 1 & 0 & 1 & 1 & 1 & 0 & 0 \\ 0 & 0 & 0 & 0 & 0 & 1 & 0 & 0 & 0 & 1 & 0 & 1 & 1 & 1 & 0 \\ 0 & 0 & 0 & 0 & 0 & 0 & 1 & 0 & 0 & 0 & 1 & 0 & 1 & 1 & 1 \end{pmatrix}.$$

Suppose now that the received word $\mathbf{v}$ is

$$1\ 0\ 0\ 1\ 0\ 0\ 1\ 1\ 0\ 0\ 0\ 0\ 1\ 0\ 0,$$

or as a polynomial,

$$v(x) = 1 + x^3 + x^6 + x^7 + x^{12}.$$

We calculate the syndrome according to Step 1, using (9.6) to simplify the work:

$$S_1 = e(\alpha) = v(\alpha) = 1,$$

$$S_2 = e(\alpha^2) = v(\alpha^2) = 1,$$

$$S_3 = e(\alpha^3) = v(\alpha^3) = \alpha^4,$$

$$S_4 = e(\alpha^4) = v(\alpha^4) = 1.$$

The largest possible system of linear equations in the unknowns τ_i (Step 2) is then of the form

$$S_2\tau_1 + S_1\tau_2 = S_3,$$

$$S_3\tau_1 + S_2\tau_2 = S_4,$$

or

$$\tau_1 + \tau_2 = \alpha^4,$$
$$\alpha^4\tau_1 + \tau_2 = 1.$$

This system clearly has a nonsingular coefficient matrix. Therefore two errors must have occurred—that is, $r = 2$. We solve this system of equations and obtain $\tau_1 = 1$, $\tau_2 = \alpha$. Substituting these values into $s(x)$ and recalling $\tau_0 = 1$ gives

$$s(x) = 1 + x + \alpha x^2.$$

As roots in $\mathbb{F}_{16}$ we find $\eta_1^{-1} = \alpha^8$, $\eta_2^{-1} = \alpha^6$, hence $\eta_1 = \alpha^7$, $\eta_2 = \alpha^9$. Therefore, we know that errors must have occurred in positions 8 and 10 of the code word. We correct these errors in the received polynomial and obtain

$$\begin{aligned} w(x) &= v(x) - e(x) \\ &= (1 + x^3 + x^6 + x^7 + x^{12}) - (x^7 + x^9) \\ &= 1 + x^3 + x^6 + x^9 + x^{12}. \end{aligned}$$

The corresponding code word is

$$1\ 0\ 0\ 1\ 0\ 0\ 1\ 0\ 0\ 1\ 0\ 0\ 1\ 0\ 0.$$

The initial message can be recovered by dividing the corrected polynomial —that is, the transmitted code polynomial $w(x)$—by $g(x)$. This gives

$$w(x)/g(x) = 1 + x^3 + x^4,$$

which yields the corresponding message word 1001100. □

3. FINITE GEOMETRIES

In this section we describe the use of finite fields in geometric problems. In a sense, coding theory can also be regarded as part of geometry and combinatorics, since it studies the question of packing spheres in a metric space of finite cardinality, usually a finite-dimensional vector space over $\mathbb{F}_q$.

A projective plane consists of a set of points and a set of lines together with an incidence relation that allows us to state for every point and for every line either that the point is on the line or is not on the line. In order to have a proper definition, certain axioms have to be satisfied.

9.53. Definition. A *projective plane* is defined as a set of elements, called *points*, together with distinguished sets of points, called *lines*, as well as a relation I, called *incidence*, between points and lines subject to the following conditions:

(i) every pair of distinct lines is incident with a unique point (i.e., to every pair of distinct lines there is one point contained in both lines, called their *intersection*);

(ii) every pair of distinct points is incident with a unique line (i.e., to every pair of distinct points there is exactly one line which contains both points);
(iii) there exist four points such that no three of them are incident with a single line (i.e., there exist four points such that no three of them are on the same line).

It follows that each line contains at least three points and that through each point there must be at least three lines. If the set of points is finite, we speak of a *finite projective plane*. From the three axioms above one deduces that (iii) holds also with the concepts of "point" and "line" interchanged. This establishes a *principle of duality* between points and lines, from which one can derive the following result.

9.54. Theorem. *Let Π be a finite projective plane. Then:*

(i) *there is an integer $m \geqslant 2$ such that every point (line) of Π is incident with exactly $m+1$ lines (points) of Π;*
(ii) *Π contains exactly m^2+m+1 points (lines).*

9.55. Example. The simplest finite projective plane is that with $m=2$; there are precisely three lines through each point and three points on each line. Altogether there are 7 points and 7 lines in the plane. This projective plane is called the *Fano plane* and it may be illustrated as shown in Figure 9.2. The points are A, B, C, D, E, F, and G and the lines are ADC, AGE, AFB, CGF, CEB, DGB, and DEF. Since straightness is not a meaningful concept in a finite plane, the subset DEF is a line in the finite projective plane. □

The integer m in Theorem 9.54 is called the *order* of the finite projective plane. We will see that finite projective planes of order m exist for every integer m of the form $m=p^n$, where p is a prime. It is known that there is no plane for $m=6$, but it is not known whether a plane exists for

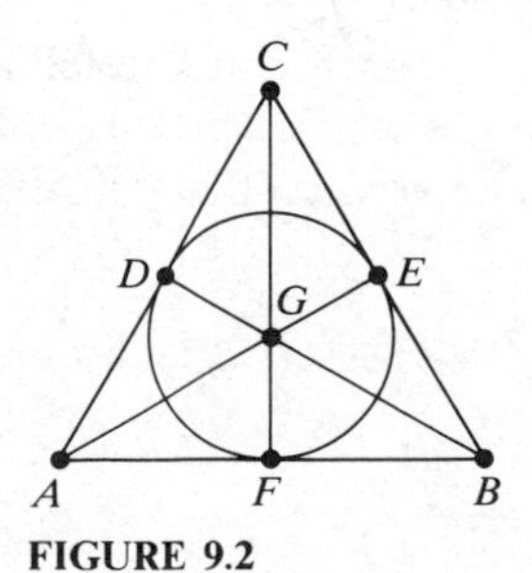

FIGURE 9.2

$m = 10$. Many planes have been found for $m = 9$, but no plane has yet been found for which m is not a power of a prime.

In ordinary analytic geometry we represent points of the plane as ordered pairs (x, y) of real numbers and lines are sets of points that satisfy real equations of the form $ax + by + c = 0$ with a and b not both 0. Now the field of real numbers can be replaced by any other field, in particular a finite field. This type of geometry is known as affine geometry (or euclidean geometry) and leads to the concept of an affine plane.

9.56. Definition. An *affine plane* is a triple $(\mathcal{P}, \mathcal{L}, I)$ consisting of a set $\mathcal{P}$ of points, a set $\mathcal{L}$ of lines, and an incidence relation I such that:

(i) every pair of distinct points is incident with a unique line;
(ii) every point $P \in \mathcal{P}$ not on a line $L \in \mathcal{L}$ lies on a unique line $M \in \mathcal{L}$ which does not intersect L;
(iii) there exist four points such that no three of them are incident with a single line.

The proof of the following theorem is straightforward.

9.57. Theorem. *Let K be any field. Let $\mathcal{P}$ denote the set of ordered pairs (x, y) with $x, y \in K$, and let $\mathcal{L}$ consist of those subsets L of $\mathcal{P}$ which satisfy linear equations, i.e., $L \in \mathcal{L}$ if for some $a, b, c \in K$ with $(a, b) \neq (0,0)$ we have $L = \{(x, y) : ax + by + c = 0\}$. A point $P \in \mathcal{P}$ is incident with a line $L \in \mathcal{L}$ if and only if $P \in L$. Then $(\mathcal{P}, \mathcal{L}, I)$ is an affine plane, denoted by $AG(2, K)$.*

It can be shown readily that if $|K| = m$, then each line of $AG(2, K)$ contains exactly m points. We can construct a projective plane from $AG(2, K)$ by adding a line to it (and, conversely, we can obtain an affine plane from any projective plane by deleting one line and all the points on it).

We change the notation in $AG(2, K)$ and rename all the points as $(x, y, 1)$, that is, (x, y, z) with $z = 1$, and use the equation $ax + by + cz = 0$ with $(a, b) \neq (0,0)$ as the equation of a line. Now add the set of points

$$L_\infty = \{(1,0,0)\} \cup \{(x,1,0) : x \in K\}$$

to $\mathcal{P}$ to form a new set $\mathcal{P}' = \mathcal{P} \cup L_\infty$. The points of L_∞ can be represented by the equation $z = 0$ and so can be interpreted as a line. Let this new line L_∞ be added to $\mathcal{L}$ to form the set $\mathcal{L}' = \mathcal{L} \cup \{L_\infty\}$. With the natural extended notion of incidence, it can be verified that $(\mathcal{P}', \mathcal{L}', I')$ satisfies all the axioms for a projective plane.

9.58. Theorem. *Let $AG(2, K) = (\mathcal{P}, \mathcal{L}, I)$ and let*

$$\mathcal{P}' = \mathcal{P} \cup \{(1,0,0)\} \cup \{(x,1,0) : x \in K\} = \mathcal{P} \cup L_\infty,$$

$$\mathcal{L}' = \mathcal{L} \cup \{L_\infty\},$$

and let the extended incidence relation be denoted by I'. Then $(\mathcal{P}', \mathcal{L}', I')$ is a projective plane, denoted by $PG(2, K)$.

9.59. Example. The plane $PG(2, \mathbb{F}_2)$—that is, the projective plane over the field $\mathbb{F}_2$—has seven points: $(0,0,1)$, $(1,0,1)$, $(0,1,1)$, and $(1,1,1)$ with $z \neq 0$ and the three distinct points on the line $z = 0$, namely, $(1,0,0)$, $(0,1,0)$, and $(1,1,0)$. It can be verified that $PG(2, \mathbb{F}_2)$ also contains seven lines and that this projective plane is the Fano plane of Example 9.55. □

In constructing $PG(2, K)$, every line of $AG(2, K)$ must meet the new line L_∞, so there will be an additional point on each line; also L_∞ contains $m+1$ points if K contains m elements. Since for every prime power $m = p^n = q$ there are finite fields $\mathbb{F}_q$, we have the following theorem.

9.60. Theorem. *For every prime power $q = p^n$, p prime, $n \in \mathbb{N}$, there exists a finite projective plane of order q—namely, $PG(2, \mathbb{F}_q)$.*

The additional line L_∞ added to an affine plane to obtain a projective plane is sometimes called the *line at infinity*. If two lines intersect on L_∞, they are called *parallel*.

Next we present without proof two interesting theorems, which hold in all projective planes that can be represented analytically in terms of fields. Two triangles $\triangle A_1B_1C_1$ and $\triangle A_2B_2C_2$ are said to be *in perspective* from a point O if the lines A_1A_2, B_1B_2, and C_1C_2 pass through O. Points on the same line are said to be *collinear*.

9.61. Theorem (Desargues's Theorem). *If $\triangle A_1B_1C_1$ and $\triangle A_2B_2C_2$ are in perspective from O, then the intersections of the lines A_1B_1 and A_2B_2, of A_1C_1 and A_2C_2, and of B_1C_1 and B_2C_2, are collinear.*

The theorem is illustrated in Figure 9.3; the intersections of corresponding lines are P, Q, and R and are collinear.

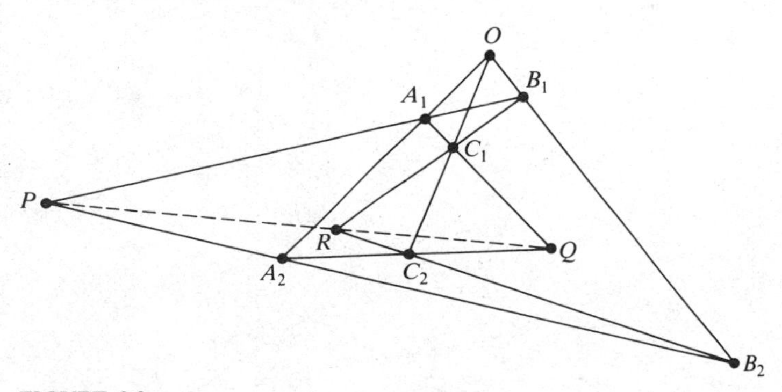

FIGURE 9.3

9.62. *Theorem* (Theorem of Pappus). *If A_1, B_1, C_1 are points of a line and A_2, B_2, C_2 are points of another line in the same plane, and if A_1B_2 and A_2B_1 intersect in P, A_1C_2 and A_2C_1 intersect in Q, and B_1C_2 and B_2C_1 intersect in R, then P, Q, and R are collinear.*

The theorem is illustrated in Figure 9.4. Both theorems play an important role in projective geometry. If Desargues's theorem holds in some projective plane, then coordinates can be defined in terms of elements from a division ring. Here we define a point as an ordered triple (x_0, x_1, x_2) of three *homogeneous coordinates*, where the x_i are elements of a division ring R, not all of them simultaneously 0. The triples (ax_0, ax_1, ax_2), $0 \neq a \in R$, shall denote the same point. Thus each point is represented in $m-1$ ways if $|R| = m$, and because there are $m^3 - 1$ possible triples of coordinates, the total number of different points is

$$(m^3 - 1)/(m-1) = m^2 + m + 1.$$

A line is defined as the set of all those points whose coordinates satisfy an equation of the form $x_0 + a_1x_1 + a_2x_2 = 0$, or of the form $x_1 + a_2x_2 = 0$, or of the form $x_2 = 0$, where $a_i \in R$. There are $m^2 + m + 1$ such lines in the plane and it is straightforward to show that the points and lines thus defined satisfy the axioms of a finite projective plane.

From Theorem 2.55—that is, Wedderburn's theorem—we know that any finite division ring is a field, a finite field $\mathbb{F}_q$. In that case the equation of any line can be written as $a_0x_0 + a_1x_1 + a_2x_2 = 0$, where the a_i are not simultaneously 0, and $(aa_0)x_0 + (aa_1)x_1 + (aa_2)x_2 = 0$ with $a \in \mathbb{F}_q^*$ is the same line. The line connecting the points (y_0, y_1, y_2) and (z_0, z_1, z_2) may then also be defined as the set of all points with coordinates

$$(ay_0 + bz_0, ay_1 + bz_1, ay_2 + bz_2),$$

where a and b are in $\mathbb{F}_q$, not both equal to 0. There are $q^2 - 1$ such triples,

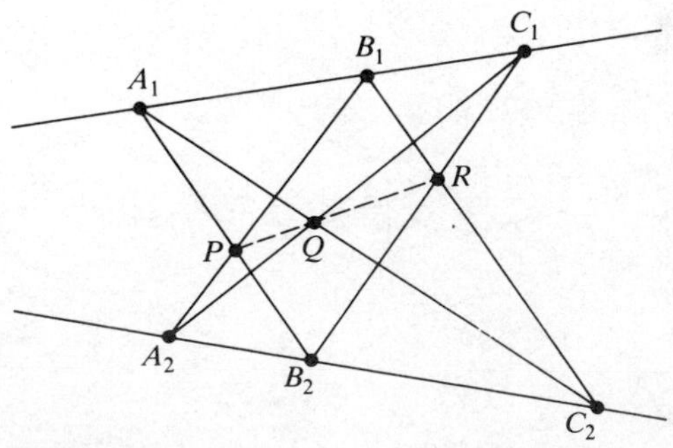

FIGURE 9.4

and since simultaneous multiplication of a and b by the same nonzero element produces the same point, they yield $q+1$ different points.

In $PG(2, \mathbb{F}_q)$ Desargues's theorem and its converse hold, and the proof relies on commutativity of multiplication in $\mathbb{F}_q$. In general, Desargues's theorem and its converse do not both apply if the coordinatizing ring does not have commutativity of multiplication. Thus Wedderburn's theorem plays an important role in this context.

A projective plane in which Desargues's theorem holds is called *Desarguesian*; otherwise it is called *non-Desarguesian*. Desarguesian planes of order m exist only if m is the power of a prime, and up to isomorphism there exists only one Desarguesian plane for any given prime power $m = p^n$. A finite Desarguesian plane can always be coordinatized by a finite field. Since such fields exist only when the order is a prime power, a projective plane with exactly $m+1$ points on each line, m not a prime power, will have to be non-Desarguesian. It is not known whether such planes for m not a prime power exist. If it can be proved that up to isomorphism there exists only one finite projective plane of order m, and if m is a prime power, then this plane must be Desarguesian. This is the case for $m = 2, 3, 4, 5, 7$, and 8. For m prime, only Desarguesian planes are known. But it has been shown that for all prime powers $m = p^n$, $n \geqslant 2$, except for 4 and 8, there exist non-Desarguesian planes of order m.

The theorem of Pappus implies the theorem of Desargues. If the theorem of Pappus holds in some projective plane, then the multiplication in the coordinatizing ring is necessarily commutative. The theorem of Pappus holds in $PG(2, \mathbb{F}_q)$ for any prime power q. A finite Desarguesian plane also satisfies the theorem of Pappus.

A remarkable distinction between the properties of a $PG(2, \mathbb{F}_q)$ with q even and a $PG(2, \mathbb{F}_q)$ with q odd is given in the following theorem.

9.63. Theorem. *The diagonal points of a complete quadrangle in $PG(2, \mathbb{F}_q)$ are collinear if and only if q is even.*

Proof. We assume, without loss of generality, that the vertices of the quadrangle are $(1,0,0)$, $(0,1,0)$, $(0,0,1)$, and $(1,1,1)$. Its six sides are $x_2 = 0$, $x_1 = 0$, $x_1 - x_2 = 0$, $x_0 = 0$, $x_0 - x_2 = 0$, and $x_0 - x_1 = 0$, while the three diagonal points are $(1,1,0)$, $(1,0,1)$, and $(0,1,1)$. The line through the first two points contains all points with coordinates $(a+b, a, b)$, where $(a, b) \neq (0,0)$, and the third point is one of these if and only if $a = b$ and $a + b = 0$. In a finite field $\mathbb{F}_q$ this is only possible if the characteristic is 2. □

The latter case is illustrated in Example 9.55. Let the vertices of the complete quadrangle be C, D, E, G. In this case, the diagonal points are A, F, B, and they are collinear.

We introduce now concepts analogous to those with which we are familiar in analytic geometry, and we restrict ourselves to Desarguesian planes, coordinatized by a finite field $\mathbb{F}_q$.

Let the equations of two distinct lines be

$$a_{01}x_0 + a_{11}x_1 + a_{21}x_2 = 0,$$
$$a_{02}x_0 + a_{12}x_1 + a_{22}x_2 = 0. \tag{9.7}$$

Let the point of intersection of these two lines be P. All lines through P form a pencil and each line in this pencil has an equation of the form

$$(ra_{01} + sa_{02})x_0 + (ra_{11} + sa_{12})x_1 + (ra_{21} + sa_{22})x_2 = 0,$$

where $r, s \in \mathbb{F}_q$ are not both 0. There are $q+1$ lines in the pencil: the two lines (9.7) given above corresponding to $s = 0$ and $r = 0$, respectively, and those corresponding to $q - 1$ different ratios rs^{-1} with $r \neq 0$ and $s \neq 0$. Let another pencil through a point $Q \neq P$ be given by

$$(rb_{01} + sb_{02})x_0 + (rb_{11} + sb_{12})x_1 + (rb_{21} + sb_{22})x_2 = 0.$$

A projective correspondence between the lines of the two pencils is defined by letting a line of the first, given by a pair (r, s), correspond to the line of the second pencil that belongs to the same pair. Two corresponding lines meet in a unique point, except when the line PQ corresponds to itself, and the coordinates of all the points satisfy the equation

$$(a_{01}x_0 + a_{11}x_1 + a_{21}x_2)(b_{02}x_0 + b_{12}x_1 + b_{22}x_2)$$
$$-(a_{02}x_0 + a_{12}x_1 + a_{22}x_2)(b_{01}x_0 + b_{11}x_1 + b_{21}x_2) = 0, \tag{9.8}$$

obtained by eliminating r and s from the equations of the two pencils.

9.64. Definition. The set of points whose coordinates satisfy equation (9.8) is called a *conic*. If the line PQ corresponds to itself under the correspondence above, then the conic is called *degenerate*. It consists then of the $2q+1$ points of two intersecting lines. A *nondegenerate* conic consists of the $q+1$ points of intersection of corresponding lines. A line that has precisely one point in common with a conic is called a *tangent* of it; a line that has two points in common is a *secant*.

The equation of a nondegenerate conic is quadratic, therefore it cannot have more than two points in common with any line. Take one point of a nondegenerate conic and connect it by lines to the other q points. Then the resulting lines are secants and the remaining one of the $q+1$ lines through that point must be a tangent.

The $q+1$ points of a nondegenerate conic thus have the property that no three of them are collinear. It can be shown that any set of $q+1$ points in a $PG(2, \mathbb{F}_q)$, q odd, such that no three of them are collinear is a nondegenerate conic.

The following theorem, which we prove only in part, exhibits a difference between conics in Desarguesian planes of odd and of even order.

9.65. Theorem. (i) *In a Desarguesian plane of odd order there pass two or no tangents of a nondegenerate conic through a point not on the conic.*

(ii) *In a Desarguesian plane of even order all the tangents of a nondegenerate conic meet in a single point.*

Proof. We prove (ii) as an example of how properties of finite fields are used in the theory of finite projective planes. Assume without loss of generality that three points on a nondegenerate conic in a plane of even order are $A(1,0,0)$, $B(0,1,0)$, $C(0,0,1)$ and that the tangents through these three points are, respectively, $x_1 - k_0 x_2 = 0$, $x_2 - k_1 x_0 = 0$, $x_0 - k_2 x_1 = 0$. Let $P(t_0, t_1, t_2)$ be another point of the conic. None of the t_i can be 0, because then P would be on a line through two of the points A, B, and C, contradicting the fact that no three points of the conic are collinear. Therefore we can write $x_1 - t_1 t_2^{-1} x_2 = 0$ for PA, $x_2 - t_2 t_0^{-1} x_0 = 0$ for PB, and $x_0 - t_0 t_1^{-1} x_1 = 0$ for PC.

Consider the equation for the line PA. As we choose for P the various points of the conic, leaving out A, B, and C, the ratio $t_1 t_2^{-1}$ runs through the elements of $\mathbb{F}_q$ apart from 0 and k_0. Since

$$\prod_{c \in \mathbb{F}_q^*} (x - c) = x^{q-1} - 1,$$

the product of all nonzero elements of $\mathbb{F}_q$ is $(-1)^q$. Thus, multiplying the product of the $q-2$ values $t_1 t_2^{-1}$ assumes by k_0, we obtain $(-1)^q = 1$, since q is even. We have

$$k_0 \prod t_1 t_2^{-1} = 1, \quad k_1 \prod t_2 t_0^{-1} = 1, \quad \text{and} \quad k_2 \prod t_0 t_1^{-1} = 1,$$

where the product extends over all points of the conic except A, B, and C. Multiplying the three products above we get $k_0 k_1 k_2 = 1$. Therefore the points $(1, k_0 k_1, k_1)$, $(k_2, 1, k_1 k_2)$, and $(k_0 k_2, k_0, 1)$ are identical. The three tangents at A, B, and C pass through this point; and because these points were arbitrary, any three tangents meet in the same point. □

There are interesting connections between permutation polynomials of finite fields (see Chapter 7) and finite projective planes. We describc one of these connections.

9.66. Definition. An *oval* in $PG(2, \mathbb{F}_q)$, q even, is a set of $q+2$ points of $PG(2, \mathbb{F}_q)$ no three of which are collinear.

An example of an oval is obtained by taking the $q+1$ points of a nondegenerate conic in $PG(2, \mathbb{F}_q)$, q even, and adding the point at which all its tangents meet (compare with Theorem 9.65(ii)). The following theorem gives a canonical form for ovals.

9.67. ***Theorem.*** *Any oval in $PG(2, \mathbb{F}_q)$, q even and $q > 2$, can be written in the form*

$$A(f) = \{(f(c), c, 1) : c \in \mathbb{F}_q\} \cup \{(1,0,0), (0,1,0)\},$$

where $f \in \mathbb{F}_q[x]$ *is such that*:

(i) f *is a permutation polynomial of* $\mathbb{F}_q$ *with* $\deg(f) < q$ *and*

$$f(0) = 0, f(1) = 1;$$

(ii) *for each* $a \in \mathbb{F}_q$, $g_a(x) = [f(x+a) + f(a)]/x$ *is a permutation polynomial of* $\mathbb{F}_q$ *with* $g_a(0) = 0$.

Conversely, every such set $A(f)$ *is an oval.*

Proof. Let D be an oval in $PG(2, \mathbb{F}_q)$. We can arrange the coordinatization in such a way that $P_0(1,0,0)$, $P_1(0,1,0)$, $P_2(0,0,1)$, and $P_3(1,1,1)$ are points of D. Then D has no other points on the line P_0P_1, hence the q points of D different from P_0, P_1 are of the form $(d_i, c_i, 1)$, $1 \leqslant i \leqslant q$, with $d_i, c_i \in \mathbb{F}_q$. Since each line through P_0 contains only one other point of D, we have $c_i \neq c_j$ for $i \neq j$, and since each line through P_1 contains only one other point of D, we have $d_i \neq d_j$ for $i \neq j$. Thus

$$\{c_1, \ldots, c_q\} = \{d_1, \ldots, d_q\} = \mathbb{F}_q,$$

and there exists a permutation polynomial f of $\mathbb{F}_q$ with $f(c_i) = d_i$ for $1 \leqslant i \leqslant q$ and $\deg(f) < q$ by (7.1). Since $P_2, P_3 \in D$, we have $f(0) = 0$, $f(1) = 1$. Thus $D = A(f)$ with f satisfying condition (i).

It remains to show that (ii) is equivalent to no three points of $A(f) \backslash \{P_0, P_1\}$ being collinear. The latter property holds if and only if

$$\begin{vmatrix} f(b) & b & 1 \\ f(c) & c & 1 \\ f(d) & d & 1 \end{vmatrix} \neq 0$$

for all distinct $b, c, d \in \mathbb{F}_q$. This means

$$[f(b) + f(c)](b+c)^{-1} \neq [f(b) + f(d)](b+d)^{-1}.$$

Equivalently, for each $a \in \mathbb{F}_q$, $[f(t) + f(a)](t+a)^{-1}$ takes a different value in $\mathbb{F}_q^*$ for each $t \in \mathbb{F}_q$ with $t \neq a$. Substituting $x + a$ for t yields that the polynomial

$$g_a(x) = [f(x+a) + f(a)]/x$$

defines a permutation of $\mathbb{F}_q^*$. Since $\deg(g_a) \leqslant q - 2$, we get from (7.1),

$$g_a(x) = \sum_{c \in \mathbb{F}_q} g_a(c)\left(1 - (x-c)^{q-1}\right).$$

Thus, comparing the coefficients of x^{q-1},

$$\begin{aligned} 0 = -\sum_{c \in \mathbb{F}_q} g_a(c) &= g_a(0) + \sum_{c \in \mathbb{F}_q^*} g_a(c) = g_a(0) + \sum_{c \in \mathbb{F}_q^*} c \\ &= g_a(0) + \sum_{c \in \mathbb{F}_q} c = g_a(0), \end{aligned}$$

where we used Lemma 7.3 in the last step. It follows that g_a is a permutation polynomial of $\mathbb{F}_q$. □

9.68. Corollary. *If $A(f)$ with $f(x)=\sum_{i=1}^{q-1} b_i x^i$ is an oval in $PG(2,\mathbb{F}_q)$, q even and $q>2$, then f is necessarily of the form*

$$f(x)=\sum_{j=1}^{(q-2)/2} b_{2j}x^{2j}.$$

Proof. For all $a\in\mathbb{F}_q$, condition (ii) of Theorem 9.67 yields

$$0=g_a(0)=b_1+b_3a^2+b_5a^4+\cdots+b_{q-1}a^{q-2},$$

which implies $b_1=b_3=b_5=\cdots=b_{q-1}=0$. □

9.69. Corollary. *The set $A(x^k)$ with $1\leqslant k<q$ is an oval in $PG(2,\mathbb{F}_q)$, q even and $q>2$, if and only if the following conditions hold*: (i) $\gcd(k,q-1)=1$; (ii) $\gcd(k-1,q-1)=1$; (iii) $[(x+1)^k+1]/x$ *is a permutation polynomial of* $\mathbb{F}_q$.

Proof. (i) is equivalent to condition (i) of Theorem 9.67, because of Theorem 7.8(ii). Similarly, (ii) is equivalent to condition (ii) of Theorem 9.67 for $a=0$. For $a\in\mathbb{F}_q^*$ we have

$$\begin{aligned} g_a(x) &= \left[(x+a)^k+a^k\right]/x = a^{k-1}\left[(a^{-1}x+1)^k+1\right]/(a^{-1}x)\\ &= a^{k-1}g_1(a^{-1}x), \end{aligned}$$

hence g_a is a permutation polynomial of $\mathbb{F}_q$ if and only if g_1 is one. Furthermore, if g_1 is a permutation polynomial of $\mathbb{F}_q$, then from $g_1(0)\in\mathbb{F}_2$, $g_1(1)=1$, it follows that $g_1(0)=0$, hence $g_a(0)=0$. □

Analogs of the concept of a projective plane can be defined for dimensions higher than 2.

9.70. Definition. A *projective space*, or a *projective geometry*, or an *m-space* is a set of points, together with distinguished sets of points, called lines, subject to the following conditions:

(i) There is a unique line through any pair of distinct points.
(ii) A line that intersects two lines of a triangle intersects the third line as well.
(iii) Every line contains at least three points.
(iv) Define a k-space as follows. A 0-space is a point. If $A_0,\ldots,A_k$ are points not all in the same $(k-1)$-space, then all points collinear with A_0 and any point in the $(k-1)$-space defined by $A_1,\ldots,A_k$ form a k-space. Thus a line is a 1-space, and all the

other spaces are defined recursively. Axiom (iv) demands: If $k < m$, then not all points considered are in the same k-space.

(v) There exists no $(m+1)$-space in the set of points considered.

We say that an m-space has m dimensions, and if we refer to a k-space as a subspace of a projective space of higher dimension, we call it a *k-flat*. An $(m-1)$-flat in a projective space of m dimensions is called a *hyperplane*. A 2-space is a projective plane in the sense of Definition 9.53. It can be proved that in any 2-flat in a projective space of at least three dimensions the theorem of Desargues (Theorem 9.61) is always valid. Desargues's theorem can only fail to be true in projective planes that cannot be embedded in a projective space of at least three dimensions.

A projective space containing only finitely many points is called a *finite projective space* (or *finite projective geometry*, or *finite m-space*). In analogy with $PG(2,\mathbb{F}_q)$, we can construct the finite m-space $PG(m,\mathbb{F}_q)$. Define a point as an ordered $(m+1)$-tuple $(x_0, x_1, \ldots, x_m)$, where the coordinates $x_i \in \mathbb{F}_q$ are not simultaneously 0. The $(m+1)$-tuples $(ax_0, ax_1, \ldots, ax_m)$ with $a \in \mathbb{F}_q^*$ define the same point. There are therefore $(q^{m+1}-1)/(q-1)$ points in $PG(m,\mathbb{F}_q)$.

A k-flat in $PG(m,\mathbb{F}_q)$ is the set of all those points whose coordinates satisfy $m-k$ linearly independent homogeneous linear equations

$$\begin{array}{lcl} a_{10}x_0 & + \cdots + a_{1m}x_m & = 0 \\ \vdots & \vdots & \vdots \\ a_{m-k,0}x_0 & + \cdots + a_{m-k,m}x_m & = 0 \end{array}$$

with coefficients $a_{ij} \in \mathbb{F}_q$. Alternatively, a k-flat consists of all those points with coordinates

$$(a_0x_{00} + \cdots + a_kx_{k0}, \ldots, a_0x_{0m} + \cdots + a_kx_{km})$$

with the $a_i \in \mathbb{F}_q$ not simultaneously 0 and the $k+1$ given points

$$(x_{00}, \ldots, x_{0m}), \ldots, (x_{k0}, \ldots, x_{km})$$

being linearly independent; that is, the matrix

$$\begin{pmatrix} x_{00} & \cdots & x_{0m} \\ x_{10} & \cdots & x_{1m} \\ \vdots & & \vdots \\ x_{k0} & \cdots & x_{km} \end{pmatrix}$$

has rank $k+1$. The number of points in a k-flat is $(q^{k+1}-1)/(q-1)$; there are $q+1$ points on a line and q^2+q+1 on a plane. That $PG(m,\mathbb{F}_q)$ satisfies the five axioms for an m-space is easily verified.

We know that in $\mathbb{F}_{q^{m+1}}$ all powers of a primitive element α can be represented as polynomials in α of degree at most m with coefficients in $\mathbb{F}_q$. If

$$\alpha^i = a_m\alpha^m + \cdots + a_0,$$

we may consider α^i as representing a point in $PG(m, \mathbb{F}_q)$ with coordinates $(a_0, \ldots, a_m)$. Two powers α^i, α^j represent the same point if and only if $\alpha^i = a\alpha^j$ for some $a \in \mathbb{F}_q^*$—that is, if and only if

$$i \equiv j \bmod (q^{m+1} - 1)/(q-1).$$

A k-flat S through $k+1$ linearly independent points represented by $\alpha^{i_0}, \ldots, \alpha^{i_k}$ will contain all points represented by $\sum_{r=0}^{k} a_r\alpha^{i_r}$, $a_r \in \mathbb{F}_q$ not simultaneously 0. For each $h = 0, 1, \ldots, v-1$ with $v = (q^{m+1}-1)/(q-1)$, the points $\sum_{r=0}^{k} a_r\alpha^{i_r+h}$, $a_r \in \mathbb{F}_q$ not simultaneously 0, form k-flats, and we denote the k-flat with given h by S_h. We have $S_v = S_0 = S$ because $\alpha^v \in \mathbb{F}_q$. Let j be the least positive integer for which $S_j = S$. Then from $S_{nj} = S$ for all $n \in \mathbb{N}$ it follows that j divides v, say $v = tj$. We call j the *cycle* of S.

If α^{d_0} is a point of the k-flat S, then so are the points with exponents

$$d_0, d_0 + j, \ldots, d_0 + (t-1)j,$$

because $S_{nj} = S$ for $n = 0, 1, \ldots, t-1$. Further points on S can be written with the following exponents of α:

$$\begin{matrix} d_1, & d_1 + j & , \ldots, d_1 + (t-1)j \\ \vdots & \vdots & \vdots \\ d_{u-1}, & d_{u-1} + j & , \ldots, d_{u-1} + (t-1)j, \end{matrix}$$

where $d_{r_1} - d_{r_2}$ is not divisible by j for $r_1 \neq r_2$. The number of all these distinct points is $tu = (q^{k+1}-1)/(q-1)$.

If $tj = (q^{m+1}-1)/(q-1)$ and $tu = (q^{k+1}-1)/(q-1)$ are relatively prime, then $t = 1$, $j = v$, and all k-flats have cycle v. This is the case for $k = m-1$, and for $k = 1$ when m is even.

9.71. Example. Consider $PG(3, \mathbb{F}_2)$ with 15 points, 35 lines, 15 planes, and $q^{m+1} = 16$. Using a root $\alpha \in \mathbb{F}_{16}$ of the primitive polynomial $x^4 + x + 1$ over $\mathbb{F}_2$, we can establish a correspondence between the powers of α and the points of $PG(3, \mathbb{F}_2)$. We obtain:

$A(0,0,0,1) \cdots \alpha^3$	$F(0,1,1,0) \cdots \alpha^5$	$K(1,0,1,1) \cdots \alpha^{13}$
$B(0,0,1,0) \cdots \alpha^2$	$G(0,1,1,1) \cdots \alpha^{11}$	$L(1,1,0,0) \cdots \alpha^4$
$C(0,0,1,1) \cdots \alpha^6$	$H(1,0,0,0) \cdots \alpha^0$	$M(1,1,0,1) \cdots \alpha^7$
$D(0,1,0,0) \cdots \alpha^1$	$I(1,0,0,1) \cdots \alpha^{14}$	$N(1,1,1,0) \cdots \alpha^{10}$
$E(0,1,0,1) \cdots \alpha^9$	$J(1,0,1,0) \cdots \alpha^8$	$O(1,1,1,1) \cdots \alpha^{12}$

The plane

$$S = S_0 = \{a_0\alpha^0 + a_1\alpha^1 + a_2\alpha^2 : \quad a_0, a_1, a_2 \in \mathbb{F}_2 \text{ not all } 0\}$$

is the same as the plane $x_3 = 0$. It contains the points B, D, F, H, J, L, and N. It has cycle 15, as has any other hyperplane. The plane

$$S_1 = \{a_0\alpha^1 + a_1\alpha^2 + a_2\alpha^3 : \quad a_0, a_1, a_2 \in \mathbb{F}_2 \text{ not all } 0\}$$

is the same as the plane $x_0 = 0$ and contains the points A, B, C, D, E, F, and G; and so on. The line

$$\{a_0\alpha^3 + a_1\alpha^8 : \quad a_0, a_1 \in \mathbb{F}_2 \text{ not both } 0\},$$

that is, the line AJK, has cycle 5, the lines ABC and ADE both have cycle 15, and this accounts for all the $5 + 15 + 15 = 35$ lines. □

A *finite affine* (or *euclidean*) *geometry*, denoted by $AG(m, \mathbb{F}_q)$, is the set of flats that remain when a hyperplane with all its flats is removed from $PG(m, \mathbb{F}_q)$. Those flats that were removed are called *flats at infinity*. Those remaining flats that intersect in a flat at infinity are called *parallel*. It is convenient to consider the excluded hyperplane as the one whose equation is $x_m = 0$. Then we may fix x_m for all points in $AG(m, \mathbb{F}_q)$ at 1, and consider only the remaining coordinates as those of a point in $AG(m, \mathbb{F}_q)$. Since there are $q^m + \cdots + q + 1$ points in $PG(m, \mathbb{F}_q)$, and the $q^{m-1} + \cdots + q + 1$ points of a hyperplane were removed, there remain q^m points in $AG(m, \mathbb{F}_q)$.

A k-flat within $AG(m, \mathbb{F}_q)$ contains all those q^k points that satisfy a system of equations of the form

$$a_{i0}x_0 + \cdots + a_{i,m-1}x_{m-1} + a_{im} = 0, \quad i = 1, \ldots, m-k,$$

where the coefficient matrix has rank $m - k$. In particular, a hyperplane is defined by

$$a_0x_0 + \cdots + a_{m-1}x_{m-1} + a_m = 0,$$

where $a_0, \ldots, a_{m-1}$ are not all 0. If $a_0, \ldots, a_{m-1}$ are kept constant and a_m runs through all elements of $\mathbb{F}_q$, then we obtain a pencil of parallel hyperplanes.

4. COMBINATORICS

In this section we describe some of the useful aspects of finite fields in combinatorics.

There is a close connection between finite geometries and *designs*. The designs we wish to consider consist of two nonempty sets of objects, with an incidence relation between objects of different sets. For instance, the objects may be points and lines, with a given point lying or not lying on a given line. The terminology that is normally used in this area has its origin in the applications in statistics, in connection with the design of experiments. The two types of objects are called *varieties* (in early applications these were plants or fertilizers) and *blocks*. The number of varieties will, as a rule, be denoted by v, and the number of blocks by b.

A design for which every block is incident with the same number k of varieties and every variety is incident with the same number r of blocks is called a *tactical configuration*. Clearly

$$vr = bk. \tag{9.9}$$

If $v = b$, and hence $r = k$, the tactical configuration is called *symmetric*. For instance, the points and lines of a $PG(2, \mathbb{F}_q)$ form a symmetric tactical configuration with $v = b = q^2 + q + 1$ and $r = k = q + 1$. The property of a finite projective plane that every pair of distinct points is incident with a unique line may serve to motivate the following definition.

9.72. Definition. A tactical configuration is called a *balanced incomplete block design* (*BIBD*), or (v, k, λ) *block design*, if $v \geqslant k \geqslant 2$ and every pair of distinct varieties is incident with the same number λ of blocks.

If for a fixed variety a_1 we count in two ways all the ordered pairs (a_2, B) with a variety $a_2 \neq a_1$ and a block B incident with a_1, a_2, we obtain the identity

$$r(k-1) = \lambda(v-1) \tag{9.10}$$

for any (v, k, λ) block design. Thus, the parameters b and r of a BIBD are determined by v, k, and λ because of (9.9) and (9.10).

9.73. Example. Let the set of varieties be $\{0,1,2,3,4,5,6\}$ and let the blocks be the subsets $\{0,1,3\}$, $\{1,2,4\}$, $\{2,3,5\}$, $\{3,4,6\}$, $\{4,5,0\}$, $\{5,6,1\}$, and $\{6,0,2\}$, with the obvious incidence relation between varieties and blocks. This is a symmetric BIBD with $v = b = 7$, $r = k = 3$, and $\lambda = 1$. It is equivalent to the Fano plane in Example 9.55. A BIBD with $k = 3$ and $\lambda = 1$ is called a *Steiner triple system*. □

9.74. Example. More generally, a BIBD is obtained by taking the points of a projective geometry $PG(m, \mathbb{F}_q)$ or of an affine geometry $AG(m, \mathbb{F}_q)$ as varieties and its t-flats for some fixed t, $1 \leqslant t < m$, as blocks. In the projective case, the parameters of the resulting BIBD are as follows:

$$v = \frac{q^{m+1}-1}{q-1}, \quad b = \prod_{i=1}^{t+1} \frac{q^{m-t+i}-1}{q^i-1}, \quad r = \prod_{i=1}^{t} \frac{q^{m-t+i}-1}{q^i-1},$$

$$k = \frac{q^{t+1}-1}{q-1}, \quad \lambda = \prod_{i=1}^{t-1} \frac{q^{m-t+i}-1}{q^i-1},$$

where the last product is interpreted to be 1 if $t = 1$. The BIBD is symmetric in case $t = m - 1$—that is, if the blocks are the hyperplanes of $PG(m, \mathbb{F}_q)$.

In the affine case, the parameters of the resulting BIBD are as follows:

$$v = q^m, \quad b = q^{m-t} \prod_{i=1}^{t} \frac{q^{m-t+i}-1}{q^i-1}, \quad r = \prod_{i=1}^{t} \frac{q^{m-t+i}-1}{q^i-1},$$

$$k = q^t, \quad \lambda = \prod_{i=1}^{t-1} \frac{q^{m-t+i}-1}{q^i-1},$$

with the same convention for $t = 1$ as above. Such a BIBD is never symmetric. □

A tactical configuration can be described by its *incidence matrix*. This is a matrix A of v rows and b columns, where the rows correspond to the varieties and the columns to the blocks. We number the varieties and blocks, and if the ith variety is incident with the jth block, we define the (i, j) entry of A to be the integer 1, otherwise 0. The sum of entries in any row is r and that in any column is k.

If A is the incidence matrix of a (v, k, λ) block design, then the inner product of two different rows of A is λ. Thus, if A^{T} denotes the transpose of A, then

$$AA^{\mathrm{T}} = \begin{pmatrix} r & \lambda & \cdots & \lambda \\ \lambda & r & \cdots & \lambda \\ \vdots & \vdots & & \vdots \\ \lambda & \lambda & \cdots & r \end{pmatrix} = (r-\lambda)I + \lambda J,$$

where I is the $v \times v$ identity matrix and J is the $v \times v$ matrix with all entries equal to 1. We compute the determinant of AA^{T} by subtracting the first column from the others and then adding to the first row the sum of the others. The result is

$$\det(AA^{\mathrm{T}}) = \begin{vmatrix} rk & 0 & 0 & \cdots & 0 \\ \lambda & r-\lambda & 0 & \cdots & 0 \\ \lambda & 0 & r-\lambda & \cdots & 0 \\ \vdots & \vdots & \vdots & & \vdots \\ \lambda & 0 & 0 & \cdots & r-\lambda \end{vmatrix} = rk(r-\lambda)^{v-1},$$

where we have used (9.10). If $v = k$, the design is trivial, since each block is incident with all v varieties. If $v > k$, then $r > \lambda$ by (9.10), and so AA^{T} is of rank v. The matrix A cannot have smaller rank, hence we obtain

$$b \geqslant v. \tag{9.11}$$

By (9.9), we must also have $r \geqslant k$.

For a *symmetric* (v, k, λ) block design we have $r = k$, hence $AJ = JA$, and so A commutes with $(r-\lambda)I + \lambda J = AA^{\mathrm{T}}$. Since A is nonsingular if

$v > k$, we get $A^{T}A = AA^{T} = (r-\lambda)I + \lambda J$. It follows that *any two distinct blocks have exactly* λ *varieties in common*. This holds trivially if $v = k$.

We have seen that the conditions (9.9) and (9.10), and furthermore (9.11) in the nontrivial case, are necessary for the existence of a BIBD with parameters v, b, r, k, λ. These conditions are, however, not sufficient for the existence of such a design. For instance, a BIBD with $v = b = 43$, $r = k = 7$, and $\lambda = 1$ is known to be impossible.

The varieties and blocks of a symmetric (v, k, λ) block design with $k \geqslant 3$ and $\lambda = 1$ satisfy the conditions for points and lines of a finite projective plane. The converse is also true. Thus, *the concepts of a symmetric* $(v, k, 1)$ *block design with* $k \geqslant 3$ *and of a finite projective plane are equivalent*.

Consider the BIBD in Example 9.73 and interpret the varieties $0,1,2,3,4,5,6$ as integers modulo 7. Each block of this design has the property that the differences between its distinct elements yield all nonzero residues modulo 7. This suggests the following definition.

9.75. Definition. A set $D = \{d_1, \ldots, d_k\}$ of $k \geqslant 2$ distinct residues modulo v is called a (v, k, λ) *difference set* if for every $d \not\equiv 0 \bmod v$ there are exactly λ ordered pairs (d_i, d_j) with $d_i, d_j \in D$ such that $d_i - d_j \equiv d \bmod v$.

The following results provide a connection between difference sets, designs, and finite projective planes.

9.76. Theorem. *Let* $\{d_1, \ldots, d_k\}$ *be a* (v, k, λ) *difference set. Then with all residues modulo* v *as varieties, the blocks*

$$B_t = \{d_1 + t, \ldots, d_k + t\}, \quad t = 0, 1, \ldots, v-1,$$

form a symmetric (v, k, λ) *block design under the obvious incidence relation.*

Proof. A residue a modulo v occurs exactly in the blocks with subscripts $a - d_1, \ldots, a - d_k$ modulo v, thus every variety is incident with the same number k of blocks. For a pair of distinct residues a, c modulo v, we have $a, c \in B_t$ if and only if $a \equiv d_i + t \bmod v$ and $c \equiv d_j + t \bmod v$ for some d_i, d_j. Consequently, $a - c \equiv d_i - d_j \bmod v$, and conversely, for every solution (d_i, d_j) of the last congruence, both a and c occur in the block with subscript $a - d_i$ modulo v. By hypothesis, there are exactly λ solutions (d_i, d_j) of this congruence, and so all the conditions for a symmetric (v, k, λ) block design are satisfied. □

9.77. Corollary. *Let* $\{d_1, \ldots, d_k\}$ *be a* $(v, k, 1)$ *difference set with* $k \geqslant 3$. *Then the residues modulo* v *and the blocks* B_t, $t = 0, 1, \ldots, v-1$, *from Theorem* 9.76 *satisfy the conditions for points and lines of a finite projective plane of order* $k - 1$.

Proof. This follows from Theorem 9.76 and the observation above that symmetric $(v, k, 1)$ block designs with $k \geqslant 3$ are finite projective planes.

□

It follows from Theorem 9.76 and (9.10) that the parameters v, k, λ of a difference set are linked by the identity $k(k-1)=\lambda(v-1)$. This can also be seen directly from the definition of a difference set.

9.78. Example. The set $\{0,1,2,4,5,8,10\}$ of residues modulo 15 is a $(15,7,3)$ difference set. The blocks

$$B_t=\{t,t+1,t+2,t+4,t+5,t+8,t+10\},\quad t=0,1,\ldots,14,$$

form a symmetric (15, 7, 3) block design according to Theorem 9.76. The blocks of this design can be interpreted as the 15 planes of the projective geometry $PG(3,\mathbb{F}_2)$, with the 15 residues representing the points. Each plane is a Fano plane $PG(2,\mathbb{F}_2)$. The lines of the block B_t can be obtained by cyclically permuting the points of the line

$$L_t=B_t\cap B_{t-4}=\{t,t+1,t+4\}$$

in the plane B_t according to the permutation

$$t\to t+1\to t+2\to t+4\to t+5\to t+10\to t+8\to t.$$

For instance, the lines in the plane $B_0=\{0,1,2,4,5,10,8\}$ are

$$\{0,1,4\},\{1,2,5\},\{2,4,10\},\{4,5,8\},\{5,10,0\},\{10,8,1\},\{8,0,2\}.\qquad\square$$

Examples of difference sets can be obtained from finite projective geometries. As in the discussion preceding Example 9.71, we identify points of $PG(m,\mathbb{F}_q)$ with powers of α, where α is a primitive element of $\mathbb{F}_{q^{m+1}}$ and the exponents of α are considered modulo $v=(q^{m+1}-1)/(q-1)$. Let S be any hyperplane of $PG(m,\mathbb{F}_q)$. Then S has cycle v, and so the hyperplanes $S_h=\alpha^h S$, $h=0,1,\ldots,v-1$, are distinct. These are already all hyperplanes of $PG(m,\mathbb{F}_q)$, since v is also the total number of hyperplanes. Thus, the following is the complete list of hyperplanes of $PG(m,\mathbb{F}_q)$, with the points contained in them indicated by the corresponding exponents of α:

$$\begin{array}{llll}
S_0 & : d_1 & d_2 & \cdots & d_k\\
S_1 & : d_1+1 & d_2+1 & \cdots & d_k+1\\
\vdots & \quad\vdots & \vdots & & \vdots\\
S_{v-1} & : d_1+v-1 & d_2+v-1 & \cdots & d_k+v-1
\end{array}$$

Here $k=(q^m-1)/(q-1)$, the number of points in a hyperplane. If we look for those rows that contain a particular value, say 0, then we obtain the k hyperplanes through α^0. These k rows are given by:

$$\begin{array}{llll}
d_1-d_1 & d_2-d_1 & \cdots & d_k-d_1\\
d_1-d_2 & d_2-d_2 & \cdots & d_k-d_2\\
\vdots & \vdots & & \vdots\\
d_1-d_k & d_2-d_k & \cdots & d_k-d_k
\end{array}$$

Any point $\neq \alpha^0$ appears in as many of those k hyperplanes as there are hyperplanes through two distinct points—that is, $\lambda = (q^{m-1}-1)/(q-1)$ of them—so that the off-diagonal entries repeat each nonzero residue modulo v precisely λ times. Hence $\{d_1,\ldots,d_k\}$ is a (v,k,λ) difference set. We summarize this result as follows.

9.79. Theorem. *The points in any hyperplane of $PG(m,\mathbb{F}_q)$ determine a (v,k,λ) difference set with parameters*

$$v = \frac{q^{m+1}-1}{q-1}, \quad k = \frac{q^m-1}{q-1}, \quad \lambda = \frac{q^{m-1}-1}{q-1}.$$

9.80. Example. Consider the hyperplane $x_1 = 0$ of $PG(3,\mathbb{F}_2)$ in Example 9.71. It contains the points A, B, C, H, I, J, K, and so the corresponding exponents of α yield the $(15, 7, 3)$ difference set $\{0,2,3,6,8,13,14\}$. □

Another branch of combinatorics in which finite fields are useful is the theory of orthogonal latin squares.

9.81. Definition. An array

$$L = (a_{ij}) = \begin{pmatrix} a_{11} & a_{12} & \cdots & a_{1n} \\ a_{21} & a_{22} & \cdots & a_{2n} \\ \vdots & \vdots & & \vdots \\ a_{n1} & a_{n2} & \cdots & a_{nn} \end{pmatrix}$$

is called a *latin square* of order n if each row and each column contains every element of a set of n elements exactly once. Two latin squares (a_{ij}) and (b_{ij}) of order n are said to be *orthogonal* if the n^2 ordered pairs (a_{ij}, b_{ij}) are all different.

9.82. Theorem. *A latin square of order n exists for every positive integer n.*

Proof. Consider (a_{ij}) with $a_{ij} \equiv i + j \bmod n$, $1 \leqslant a_{ij} \leqslant n$. Then $a_{ij} = a_{ik}$ implies $i + j \equiv i + k \bmod n$, and so $j \equiv k \bmod n$, which means $j = k$ since $1 \leqslant i, j, k \leqslant n$. Similarly, $a_{ij} = a_{kj}$ implies $i = k$. Thus the elements of each row and each column are distinct. □

Orthogonal latin squares were first studied by Euler. He conjectured that there did not exist pairs of orthogonal latin squares of order n if n is twice an odd integer. This was disproved in 1959 by the construction of a pair of orthogonal latin squares of order 22.

For some values of n, more than two latin squares of order n exist that are mutually orthogonal (i.e., orthogonal in pairs). We shall show that if $n = q$, a prime power, then there exist $q-1$ mutually orthogonal latin squares of order q, by using the existence of finite fields of order q.

9.83. ***Theorem.*** *Let $a_0 = 0, a_1, a_2, \ldots, a_{q-1}$ be the elements of $\mathbb{F}_q$. Then the arrays*

$$L_k = \begin{pmatrix} a_0 & a_1 & \cdots & a_{q-1} \\ a_k a_1 & a_k a_1 + a_1 & \cdots & a_k a_1 + a_{q-1} \\ a_k a_2 & a_k a_2 + a_1 & \cdots & a_k a_2 + a_{q-1} \\ \vdots & \vdots & & \vdots \\ a_k a_{q-1} & a_k a_{q-1} + a_1 & \cdots & a_k a_{q-1} + a_{q-1} \end{pmatrix}, \quad k = 1, \ldots, q-1,$$

form a set of $q-1$ mutually orthogonal latin squares of order q.

Proof. Each L_k is clearly a latin square. Let $a_{ij}^{(k)} = a_k a_{i-1} + a_{j-1}$ be the (i, j) entry of L_k. For $k \neq m$, suppose

$$\left(a_{ij}^{(k)}, a_{ij}^{(m)}\right) = \left(a_{gh}^{(k)}, a_{gh}^{(m)}\right) \text{ for some } 1 \leqslant i, j, g, h \leqslant q.$$

Then

$$(a_k a_{i-1} + a_{j-1}, a_m a_{i-1} + a_{j-1}) = (a_k a_{g-1} + a_{h-1}, a_m a_{g-1} + a_{h-1}),$$

and so

$$a_k(a_{i-1} - a_{g-1}) = a_{h-1} - a_{j-1}, \quad a_m(a_{i-1} - a_{g-1}) = a_{h-1} - a_{j-1}.$$

Since $a_k \neq a_m$, it follows that $a_{i-1} = a_{g-1}, a_{h-1} = a_{j-1}$, hence $i = g, j = h$. Thus the ordered pairs of corresponding entries from L_k and L_m are all different, and so L_k and L_m are orthogonal. □

9.84. Example. A set of four mutually orthogonal latin squares of order 5 is given below, using the construction in Theorem 9.83:

$$L_1 = \begin{pmatrix} 0 & 1 & 2 & 3 & 4 \\ 1 & 2 & 3 & 4 & 0 \\ 2 & 3 & 4 & 0 & 1 \\ 3 & 4 & 0 & 1 & 2 \\ 4 & 0 & 1 & 2 & 3 \end{pmatrix} \quad L_2 = \begin{pmatrix} 0 & 1 & 2 & 3 & 4 \\ 2 & 3 & 4 & 0 & 1 \\ 4 & 0 & 1 & 2 & 3 \\ 1 & 2 & 3 & 4 & 0 \\ 3 & 4 & 0 & 1 & 2 \end{pmatrix}$$

$$L_3 = \begin{pmatrix} 0 & 1 & 2 & 3 & 4 \\ 3 & 4 & 0 & 1 & 2 \\ 1 & 2 & 3 & 4 & 0 \\ 4 & 0 & 1 & 2 & 3 \\ 2 & 3 & 4 & 0 & 1 \end{pmatrix} \quad L_4 = \begin{pmatrix} 0 & 1 & 2 & 3 & 4 \\ 4 & 0 & 1 & 2 & 3 \\ 3 & 4 & 0 & 1 & 2 \\ 2 & 3 & 4 & 0 & 1 \\ 1 & 2 & 3 & 4 & 0 \end{pmatrix}.$$ □

The following result, which also yields information for the case where the order n of the latin squares is not a prime power, is proved in the same way as Theorem 9.83.

9.85. Theorem. *Let $q_1,\ldots,q_s$ be prime powers and let*

$$a_0^{(i)}=0,\, a_1^{(i)},\, a_2^{(i)},\ldots,a_{q_i-1}^{(i)}$$

be the elements of $\mathbb{F}_{q_i}$. Define the s-tuples

$$b_k=\left(a_k^{(1)},\ldots,a_k^{(s)}\right)\quad \text{for } 0\leqslant k\leqslant r=\min_{1\leqslant i\leqslant s}(q_i-1),$$

and let $b_{r+1},\ldots,b_{n-1}$ with $n=q_1\cdots q_s$ be the remaining s-tuples that can be formed by taking in the ith coordinate an element of $\mathbb{F}_{q_i}$. These s-tuples are added and multiplied by adding and multiplying their coordinates. Then the arrays

$$L_k=\begin{pmatrix} b_0 & b_1 & \cdots & b_{n-1}\\ b_kb_1 & b_kb_1+b_1 & \cdots & b_kb_1+b_{n-1}\\ b_kb_2 & b_kb_2+b_1 & \cdots & b_kb_2+b_{n-1}\\ \vdots & \vdots & & \vdots\\ b_kb_{n-1} & b_kb_{n-1}+b_1 & \cdots & b_kb_{n-1}+b_{n-1}\end{pmatrix},\quad k=1,\ldots,r,$$

form a set of r mutually orthogonal latin squares of order n.

Tactical configurations and latin squares are of use in the *design of statistical experiments*. For example, suppose that n varieties of wheat are to be compared as to their mean yield on a certain type of soil. At our disposal is a rectangular field subdivided into n^2 plots. However, even if we are careful in the selection of our field, differences in soil fertility will occur on it. Thus, if all the plots of the first row are occupied by the first variety, it may very well be that the first row is of high fertility and we might obtain a high yield for the first variety although it is not superior to the other varieties. We shall be less likely to vitiate our comparisons if we set every variety once in every row and once in every column. In other words, the varieties should be planted on the n^2 plots in such a way that a latin square of order n is formed.

It is often desirable to test at the same time other factors influencing the yield. For instance, we might want to apply n different fertilizers and evaluate their effectiveness. We will then arrange fertilizers and varieties on the n^2 plots in such a way that both the arrangement of fertilizers and the arrangement of varieties form a latin square of order n, and such that every fertilizer is applied exactly once to every variety. Thus, in the language of combinatorics, the latin squares of fertilizer and variety arrangements should be orthogonal. Similar applications exist for balanced incomplete block designs.

As another example for a combinatorial concept allowing applications of finite fields, we introduce so-called Hadamard matrices. These matrices are useful in coding theory, in communication theory, and physics because of Hadamard transforms, and also in problems of determination of weights, resistances, voltages, and so on.

9.86. Definition. A *Hadamard matrix* H_n is an $n \times n$ matrix with integer entries ± 1 that satisfies

$$H_n H_n^{\mathrm{T}} = nI.$$

Since $H_n^{-1} = (1/n)H_n^{\mathrm{T}}$, we also have $H_n^{\mathrm{T}} H_n = nI$. Thus, any two distinct rows and any two distinct columns of H_n are orthogonal. The determinant of a Hadamard matrix attains a bound due to Hadamard. We have $\det(H_n H_n^{\mathrm{T}}) = n^n$, and so $|\det(H_n)| = n^{n/2}$, while Hadamard's result states that $|\det(M)| \leqslant n^{n/2}$ for any real $n \times n$ matrix M with entries of absolute value $\leqslant 1$.

Changing the signs of rows or columns leaves the defining property unaltered, so we may assume that H_n is *normalized*—that is, that all entries in the first row and first column are $+1$. It is easily seen that the order n of a Hadamard matrix (a_{ij}) can only be 1, 2, or a multiple of 4. For we have

$$\sum_{j=1}^{n} (a_{1j} + a_{2j})(a_{1j} + a_{3j}) = \sum_{j=1}^{n} a_{1j}^2 = n$$

for $n \geqslant 3$ and every term in the first sum is either 0 or 4, hence the result follows. It is conjectured that a Hadamard matrix H_n exists for all those n.

9.87. Example. Hadamard matrices of the lowest orders are:

$$H_1 = (1), \quad H_2 = \begin{pmatrix} 1 & 1 \\ 1 & -1 \end{pmatrix}, \quad H_4 = \begin{pmatrix} 1 & 1 & 1 & 1 \\ 1 & -1 & 1 & -1 \\ 1 & 1 & -1 & -1 \\ 1 & -1 & -1 & 1 \end{pmatrix}. \qquad \square$$

We describe now a construction method for Hadamard matrices using finite fields.

9.88. Theorem. *Let $a_1, \ldots, a_q$ be the elements of $\mathbb{F}_q$, $q \equiv 3 \bmod 4$, and let η be the quadratic character of $\mathbb{F}_q$. Then the matrix*

$$H = \begin{pmatrix} 1 & 1 & 1 & 1 & \cdots & 1 \\ 1 & -1 & b_{12} & b_{13} & \cdots & b_{1q} \\ 1 & b_{21} & -1 & b_{23} & \cdots & b_{2q} \\ 1 & b_{31} & b_{32} & -1 & \cdots & b_{3q} \\ \vdots & \vdots & \vdots & \vdots & & \vdots \\ 1 & b_{q1} & b_{q2} & b_{q3} & \cdots & -1 \end{pmatrix}$$

with $b_{ij} = \eta(a_j - a_i)$ for $1 \leqslant i, j \leqslant q, i \neq j$, is a Hadamard matrix of order $q+1$.

Proof. Since all entries are ± 1, it suffices to show that the inner product of any two distinct rows is 0. The inner product of the first row with the $(i+1)$st row, $1 \leqslant i \leqslant q$, is

$$1+(-1)+\sum_{j \neq i} b_{ij} = \sum_{j \neq i} \eta(a_j - a_i) = \sum_{c \in \mathbb{F}_q^*} \eta(c) = 0$$

by (5.12). The inner product of the $(i+1)$st row with the $(k+1)$st row, $1 \leqslant i < k \leqslant q$, is

$$\begin{aligned} 1 - b_{ki} - b_{ik} &+ \sum_{j \neq i,k} b_{ij} b_{kj} \\ &= 1 - \eta(a_i - a_k) - \eta(a_k - a_i) + \sum_{j \neq i,k} \eta(a_j - a_i)\eta(a_j - a_k) \\ &= 1 - [1 + \eta(-1)]\eta(a_i - a_k) + \sum_{c \in \mathbb{F}_q} \eta((c - a_i)(c - a_k)) = 0, \end{aligned}$$

since $\eta(-1) = -1$ for $q \equiv 3 \bmod 4$ by Remark 5.13 and the last sum is -1 by Theorem 5.48. □

If H_n is a Hadamard matrix of order n, then

$$\begin{pmatrix} H_n & H_n \\ H_n & -H_n \end{pmatrix}$$

is one of order $2n$. Therefore, Hadamard matrices of orders $2^h(q+1)$ with $h \geqslant 0$ and prime powers $q \equiv 3 \bmod 4$ can be obtained in this manner. By starting from the Hadamard matrix H_1 in Example 9.87, one can also obtain Hadamard matrices of orders 2^h, $h \geqslant 0$.

5. LINEAR MODULAR SYSTEMS

System theory is a discipline that aims at providing a common abstract basis and unified conceptual framework for studying the behavior of various types and forms of systems. It is a collection of methods as well as special techniques and algorithms for dealing with problems in system analysis, synthesis, identification, optimization, and other areas. It is mainly the mathematical structure of a system that is of interest to a system theorist, and not its physical form or area of applications, or whether a system is electrical, mechanical, economic, biological, chemical, and so on. What matters to the theorist is whether it is linear or nonlinear, discrete-time or continuous-time, deterministic or stochastic, discrete-state or continuous-state, and so on.

In the introduction to this chapter we gave an informal description of systems. We present now a rigorous definition of finite-state systems, which provide an idealized model for a large number of physical devices and phenomena. Ideas and techniques developed for finite-state systems have also been found useful in such diverse problems as the investigation of human nervous activity, the analysis of English syntax, and the design of digital computers.

9.89. Definition. A (complete, deterministic) *finite-state system* $\mathfrak{M}$ is defined by the following:

(1) A finite, nonempty set $U=\{\alpha_1,\alpha_2,\ldots,\alpha_h\}$, called the *input alphabet* of $\mathfrak{M}$. An element of U is called an *input symbol.*
(2) A finite, nonempty set $Y=\{\beta_1,\beta_2,\ldots,\beta_s\}$, called the *output alphabet* of $\mathfrak{M}$. An element of Y is called an *output symbol.*
(3) A finite, nonempty set $S=\{\sigma_1,\sigma_2,\ldots,\sigma_r\}$, called the *state set* of $\mathfrak{M}$. An element of S is called a *state.*
(4) A *next-state function* f that maps the set of all ordered pairs (σ_i,α_j) into S.
(5) An *output function* g that maps the set of all ordered pairs (σ_i,α_j) into Y.

A finite-state system $\mathfrak{M}$ can be interpreted as a device whose input, output, and state at time t are denoted by $u(t)$, $y(t)$, and $s(t)$, respectively, where these variables are defined for integers t only and assume values taken from U, Y, and S, respectively. Given the state and input of $\mathfrak{M}$ at time t, f specifies the state at time $t+1$ and g the output at time t:

$$s(t+1)=f(s(t),u(t)),$$

$$y(t)=g(s(t),u(t)).$$

Linear modular systems constitute a special class of finite-state systems, where the input and output alphabets and the state set carry the structure of a vector space over a finite field $\mathbb{F}_q$ and the next-state and output functions are linear. Linear modular systems have found wide applications in computer control circuitry, implementation of error-correcting codes, random number generation, and other digital tasks.

9.90. Definition. A *linear modular system* (*LMS*) $\mathfrak{M}$ of order n over $\mathbb{F}_q$ is defined by the following:

(1) A k-dimensional vector space U over $\mathbb{F}_q$, called *input space* of $\mathfrak{M}$, the elements of which are called *inputs* and are written as column vectors.
(2) An m-dimensional vector space Y over $\mathbb{F}_q$, called *output space* of $\mathfrak{M}$, the elements of which are called *outputs* and are written as column vectors.

(3) An n-dimensional vector space S over $\mathbb{F}_q$, called *state space* of $\mathfrak{M}$, the elements of which are called *states* and are written as column vectors.

(4) Four *characterizing matrices* over $\mathbb{F}_q$:

$$A = (a_{ij})_{n \times n}, \quad B = (b_{ij})_{n \times k},$$

$$C = (c_{ij})_{m \times n}, \quad D = (d_{ij})_{m \times k}.$$

The matrix A is called the *characteristic matrix* of $\mathfrak{M}$.

(5) A rule relating the state at time $t+1$ and output at time t to the state and input at time t:

$$\mathbf{s}(t+1) = A\mathbf{s}(t) + B\mathbf{u}(t),$$

$$\mathbf{y}(t) = C\mathbf{s}(t) + D\mathbf{u}(t).$$

An LMS over $\mathbb{F}_q$ can be simulated by a switching circuit incorporating adders, constant multipliers, and delay elements (compare with Chapter 8, Section 1). It is convenient here to use adders summing also more than two field elements. Thus, an *adder* has two or more inputs

$$u_1(t), u_2(t), \ldots, u_r(t) \in \mathbb{F}_q$$

and a single output

$$y_1(t) = u_1(t) + u_2(t) + \cdots + u_r(t).$$

A *constant multiplier* with a constant $a \in \mathbb{F}_q$ has a single input $u_1(t) \in \mathbb{F}_q$ and a single output $y_1(t) = au_1(t)$. A *delay element* has a single input $u_1(t) \in \mathbb{F}_q$ and a single output $y_1(t) = u_1(t-1)$. Symbolically, these components are represented as shown in Figure 9.5.

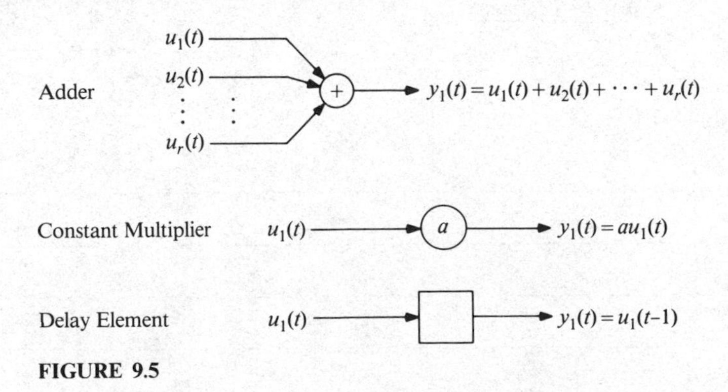

FIGURE 9.5

We describe now how we can obtain a realization of an LMS $\mathfrak{M}$ as a circuit simulating the operations of $\mathfrak{M}$:

1. Draw k input terminals labelled $u_1,\ldots,u_k$, m output terminals labelled $y_1,\ldots,y_m$, and n delay elements, where the output of the ith delay element is $s_i = s_i(t)$ and its input is $s_i' = s_i(t+1)$.
2. Insert an adder in front of each output terminal y_i and each delay element.
3. The inputs to the adder associated with the ith delay element are the s_j, each applied via a constant multiplier with constant a_{ij}, $1 \leqslant i, j \leqslant n$, and the u_j, each applied via a constant multiplier with constant b_{ij}, $1 \leqslant j \leqslant k$.
4. The inputs to the adder associated with the output terminal y_i, $1 \leqslant i \leqslant m$, are the s_j, each applied via a constant multiplier with constant c_{ij}, $1 \leqslant j \leqslant n$, and the u_j, each applied via a constant multiplier with constant d_{ij}, $1 \leqslant j \leqslant k$.

If we define

$$\mathbf{u}(t) = \begin{pmatrix} u_1 \\ \vdots \\ u_k \end{pmatrix}, \quad \mathbf{y}(t) = \begin{pmatrix} y_1 \\ \vdots \\ y_m \end{pmatrix}, \quad \mathbf{s}(t) = \begin{pmatrix} s_1 \\ \vdots \\ s_n \end{pmatrix}, \quad \mathbf{s}(t+1) = \begin{pmatrix} s_1' \\ \vdots \\ s_n' \end{pmatrix},$$

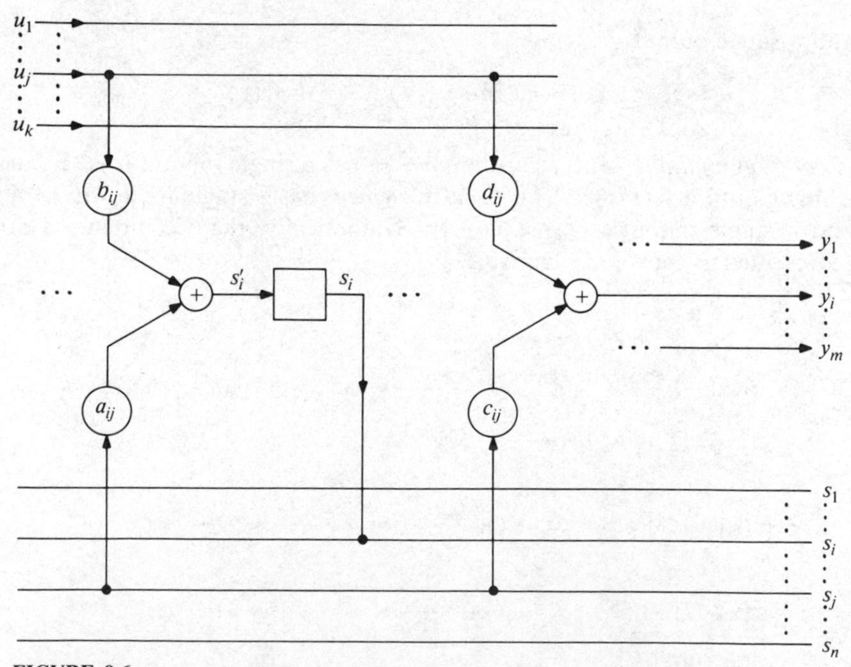

FIGURE 9.6

then the operation of the circuit represented in Figure 9.6 is precisely that described in Definition 9.90(5).

9.91. Example. Let the characterizing matrices of a fourth-order LMS over $\mathbb{F}_3$ be:

$$A = \begin{pmatrix} 0 & 2 & 0 & 0 \\ 1 & 0 & 2 & 1 \\ 0 & 1 & 1 & 0 \\ 2 & 0 & 1 & 1 \end{pmatrix}, \quad B = \begin{pmatrix} 1 \\ 0 \\ 0 \\ 0 \end{pmatrix}, \quad C = \begin{pmatrix} 0 & 0 & 2 & 1 \\ 0 & 2 & 0 & 0 \end{pmatrix}, \quad D = \begin{pmatrix} 0 \\ 1 \end{pmatrix}.$$

Then its realization as a circuit is shown in Figure 9.7. □

Conversely, we can describe an arbitrary switching circuit with a finite number of adders, constant multipliers, and delay elements over $\mathbb{F}_q$ as an LMS over $\mathbb{F}_q$ as follows (provided every closed loop contains at least one delay element):

1. Locate in the given circuit all delay elements and all external input and output terminals, and label them as in Figure 9.6.
2. Trace the paths from s_j to s_i' and compute the product of the multiplier constants encountered along each path and add the products. Let a_{ij} denote this sum.
3. Let b_{ij} denote the corresponding sum for the paths from u_j to s_i', c_{ij} for the paths from s_j to y_i, d_{ij} for the paths from u_j to y_i.

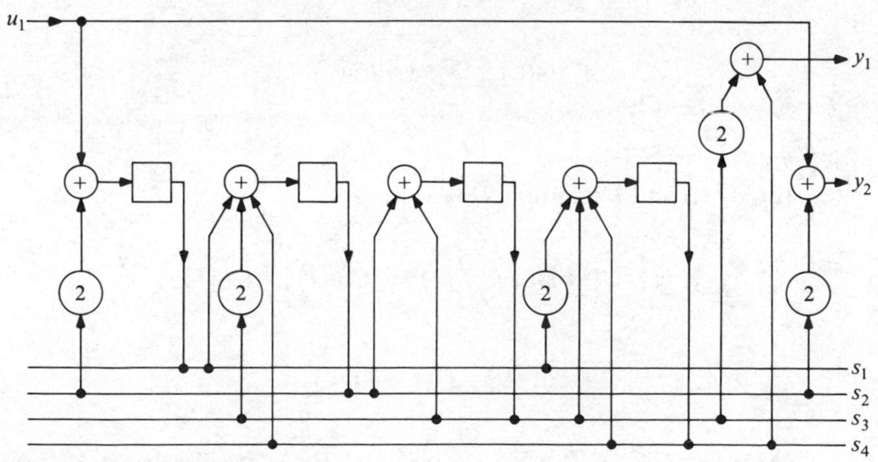

FIGURE 9.7

Then the circuit is the realization of an LMS over $\mathbb{F}_q$ with characterizing matrices A, B, C, D.

The states and the outputs of an LMS depend on the initial state $\mathbf{s}(0)$ and the sequence of inputs $\mathbf{u}(t)$, $t = 0, 1, \ldots$. The dependence on these data can be expressed explicitly.

9.92. Theorem (General Response Formula). *For an LMS with characterizing matrices A, B, C, D we have*:

(i) $\mathbf{s}(t) = A^t\mathbf{s}(0) + \sum_{i=0}^{t-1} A^{t-i-1}B\mathbf{u}(i)$ *for* $t = 1, 2, \ldots,$

(ii) $\mathbf{y}(t) = CA^t\mathbf{s}(0) + \sum_{i=0}^{t} H(t-i)\mathbf{u}(i)$ *for* $t = 0, 1, \ldots,$

where

$$H(t) = \begin{cases} D & \text{if } t = 0, \\ CA^{t-1}B & \text{if } t \geqslant 1. \end{cases}$$

Proof. (i) Let $t = 0$ in Definition 9.90(5), then

$$\mathbf{s}(1) = A\mathbf{s}(0) + B\mathbf{u}(0),$$

which proves (i) for $t = 1$. Assume (i) is true for some $t \geqslant 1$, then

$$\begin{aligned} \mathbf{s}(t+1) &= A\left(A^t\mathbf{s}(0) + \sum_{i=0}^{t-1} A^{t-i-1}B\mathbf{u}(i)\right) + B\mathbf{u}(t) \\ &= A^{t+1}\mathbf{s}(0) + \sum_{i=0}^{t} A^{t-i}B\mathbf{u}(i) \end{aligned}$$

proves (i) for $t+1$.

(ii) By (i) and Definition 9.90(5) we have

$$\begin{aligned} \mathbf{y}(t) &= C\left(A^t\mathbf{s}(0) + \sum_{i=0}^{t-1} A^{t-i-1}B\mathbf{u}(i)\right) + D\mathbf{u}(t) \\ &= CA^t\mathbf{s}(0) + \sum_{i=0}^{t} H(t-i)\mathbf{u}(i), \end{aligned}$$

where $H(t-i) = CA^{t-i-1}B$ when $t - i \geqslant 1$ and $H(t-i) = D$ when $t - i = 0$.

□

By Theorem 9.92(ii) we can decompose the output of an LMS into two components, the *free component*

$$\mathbf{y}(t)_{\text{free}} = CA^t\mathbf{s}(0)$$

obtained in case $\mathbf{u}(t) = \mathbf{0}$ for all $t \geqslant 0$, and the *forced component*

$$\mathbf{y}(t)_{\text{forced}} = \sum_{i=0}^{t} H(t-i)\mathbf{u}(i)$$

obtained by setting $\mathbf{s}(0) = \mathbf{0}$. Given any input sequence $\mathbf{u}(t)$, $t = 0, 1, \ldots$, and an initial state $\mathbf{s}(0)$, these two components can be found separately and then added up.

In the remainder of this section we study the states of an LMS in the *input-free case*—that is, when $\mathbf{u}(t) = \mathbf{0}$ for all $t \geqslant 0$. Some simple graph-theoretic language will be useful. Given an LMS $\mathfrak{M}$ of order n over $\mathbb{F}_q$ with characteristic matrix A, the *state graph* of $\mathfrak{M}$, or of A, is an oriented graph with q^n vertices, one for each possible state of $\mathfrak{M}$. An arrow points from state $\mathbf{s}_1$ to state $\mathbf{s}_2$ if and only if $\mathbf{s}_2 = A\mathbf{s}_1$. In this case we say that $\mathbf{s}_1$ *leads to* $\mathbf{s}_2$. A *path* of length r in a state graph is a sequence of r arrows $b_1, b_2, \ldots, b_r$ and $r+1$ vertices $v_1, v_2, \ldots, v_{r+1}$ such that b_i points from v_i to v_{i+1}, $i = 1, 2, \ldots, r$. If the v_i are distinct except $v_{r+1} = v_1$, the path is called a *cycle* of length r. If v_i is the only vertex leading to v_{i+1}, $i = 1, 2, \ldots, r-1$, and the only vertex leading to v_1 is v_r, then the cycle is called a *pure cycle*. For example, a pure cycle of length 8 is given as shown in Figure 9.8.

The *order* of a given state $\mathbf{s}$ is the least positive integer t such that $A^t\mathbf{s} = \mathbf{s}$. Thus, the order of $\mathbf{s}$ is the length of the cycle which inciudes $\mathbf{s}$. In the following, let A be nonsingular—that is, $\det(A) \neq 0$. It is clear that in this case the corresponding state graph consists of pure cycles only. The order of the characteristic matrix A is the least positive integer t such that $A^t = I$, the $n \times n$ identity matrix.

9.93. Lemma. *If $t_1, \ldots, t_K$ are the orders of the possible states of an LMS with nonsingular characteristic matrix A, then the order of A is* $\operatorname{lcm}(t_1, \ldots, t_K)$.

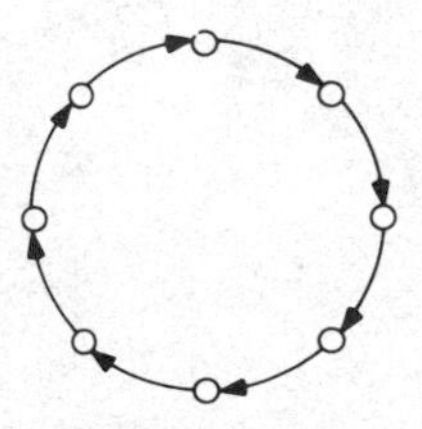

FIGURE 9.8

Proof. Let t be the order of A and $t' = \text{lcm}(t_1, \ldots, t_K)$. Since $A^t\mathbf{s} = \mathbf{s}$ for every $\mathbf{s}$, t must be a multiple of t'. Also, $(A^{t'} - I)\mathbf{s} = \mathbf{0}$ for all $\mathbf{s}$, hence $A^{t'} = I$. Thus $t' \geqslant t$, and therefore $t = t'$. □

9.94. Lemma. *If A has the form*

$$A = \begin{pmatrix} A_1 & 0 \\ 0 & A_2 \end{pmatrix}$$

with square matrices A_1 and A_2, and $\begin{pmatrix} \mathbf{s}_1 \\ \mathbf{0} \end{pmatrix}$ and $\begin{pmatrix} \mathbf{0} \\ \mathbf{s}_2 \end{pmatrix}$ are two states, partitioned according to the partition of A, with orders t_1 and t_2, respectively, then the order of $\mathbf{s} = \begin{pmatrix} \mathbf{s}_1 \\ \mathbf{s}_2 \end{pmatrix}$ is $\text{lcm}(t_1, t_2)$.

Proof. This follows immediately from the fact that $A^t\begin{pmatrix} \mathbf{s}_1 \\ \mathbf{s}_2 \end{pmatrix} = \begin{pmatrix} \mathbf{s}_1 \\ \mathbf{s}_2 \end{pmatrix}$ if and only if $A_1^t\mathbf{s}_1 = \mathbf{s}_1$ and $A_2^t\mathbf{s}_2 = \mathbf{s}_2$. □

Let $\mathfrak{M}$ be an LMS with nonsingular characteristic matrix A. Up to isomorphisms (i.e., one-to-one and onto mappings τ such that $\tau(\mathbf{s}_1)$ leads to $\tau(\mathbf{s}_2)$ whenever $\mathbf{s}_1$ leads to $\mathbf{s}_2$) the state graph of $\mathfrak{M}$ is characterized by the formal sum

$$\Sigma = (n_1, t_1) + (n_2, t_2) + \cdots + (n_R, t_R),$$

which indicates that n_i is the number of cycles of length t_i. Σ is called the *cycle sum* of $\mathfrak{M}$, or of A, and each ordered pair (n_i, t_i) is called a *cycle term*. Cycle terms are assumed to commute with respect to $+$, and we observe the convention $(n', t) + (n'', t) = (n' + n'', t)$.

Consider a matrix A of the form

$$A = \begin{pmatrix} A_1 & 0 \\ 0 & A_2 \end{pmatrix}$$

with square matrices A_1 and A_2, and suppose the state graph of A_i has n_i cycles of length t_i, $i = 1, 2$. Hence there are $n_1 t_1$ states of the form $\begin{pmatrix} \mathbf{s}_1 \\ \mathbf{0} \end{pmatrix}$ of order t_1, and $n_2 t_2$ states of the form $\begin{pmatrix} \mathbf{0} \\ \mathbf{s}_2 \end{pmatrix}$ of order t_2. By Lemma 9.94 the state graph of A must contain $n_1 n_2 t_1 t_2$ states of order $\text{lcm}(t_1, t_2)$ and hence

$$n_1 n_2 t_1 t_2 / \text{lcm}(t_1, t_2) = n_1 n_2 \gcd(t_1, t_2)$$

cycles of length $\text{lcm}(t_1, t_2)$.

The product of two cycle terms is the cycle term defined by

$$(n_1, t_1) \cdot (n_2, t_2) = (n_1 n_2 \gcd(t_1, t_2), \text{lcm}(t_1, t_2)).$$

The product of two cycle sums is defined as the formal sum of all possible products of cycle terms from the two given cycle sums. In other words, the product is calculated by the distributive law.

9.95. Theorem. *If*

$$A = \begin{pmatrix} A_1 & 0 \\ 0 & A_2 \end{pmatrix}$$

and the cycle sums of A_1 and A_2 are Σ_1 and Σ_2, respectively, then the cycle sum of A is $\Sigma_1 \Sigma_2$.

Our aim is to give a procedure for computing the cycle sum of an LMS over $\mathbb{F}_q$ with nonsingular characteristic matrix A. We need some basic facts about matrices. The *characteristic polynomial* of a square matrix M over $\mathbb{F}_q$ is defined by $\det(xI - M)$. The *minimal polynomial* $m(x)$ of M is the monic polynomial over $\mathbb{F}_q$ of least degree such that $m(M) = 0$, the zero matrix. For a monic polynomial

$$g(x) = x^k + a_{k-1}x^{k-1} + \cdots + a_1 x + a_0$$

over $\mathbb{F}_q$, its *companion matrix* is given by

$$M(g(x)) = \begin{pmatrix} 0 & 0 & 0 & \cdots & 0 & -a_0 \\ 1 & 0 & 0 & \cdots & 0 & -a_1 \\ 0 & 1 & 0 & \cdots & 0 & -a_2 \\ \vdots & \vdots & \vdots & & \vdots & \vdots \\ 0 & 0 & 0 & \cdots & 1 & -a_{k-1} \end{pmatrix}.$$

Then $g(x)$ is the characteristic polynomial and the minimal polynomial of $M(g(x))$.

Let M be a square matrix over $\mathbb{F}_q$ with the monic elementary divisors $g_1(x), \ldots, g_w(x)$. Then the product $g_1(x) \cdots g_w(x)$ is equal to the characteristic polynomial of M, and M is similar to

$$M^* = \begin{pmatrix} M(g_1(x)) & 0 & \cdots & 0 \\ 0 & M(g_2(x)) & \cdots & 0 \\ \vdots & \vdots & & \vdots \\ 0 & 0 & \cdots & M(g_w(x)) \end{pmatrix},$$

that is, $M = P^{-1} M^* P$ for some nonsingular matrix P over $\mathbb{F}_q$. The matrix M^* is called the *rational canonical form* of M and the submatrices $M(g_i(x))$ are called the *elementary blocks* of M^*.

Now let the nonsingular matrix A be the characteristic matrix of an LMS over $\mathbb{F}_q$. For the purpose of computing its cycle sum, A can be replaced by a similar matrix. Thus, we consider the rational canonical form A^* of A. Extending Theorem 9.95 by induction, we obtain the following. Let $g_1(x), \ldots, g_w(x)$ be the monic elementary divisors of A and let Σ_i be the cycle sum of the companion matrix $M(g_i(x))$; then the cycle sum Σ of A^*, and so of A, is given by

$$\Sigma = \Sigma_1 \Sigma_2 \cdots \Sigma_w.$$

Let the characteristic polynomial $f(x)$ of A have the canonical factorization

$$f(x) = \prod_{j=1}^{r} p_j(x)^{e_j},$$

where the $p_j(x)$ are distinct monic irreducible polynomials over $\mathbb{F}_q$. Then the elementary divisors of A are of the form

$$p_j(x)^{e_{j1}}, p_j(x)^{e_{j2}}, \ldots, p_j(x)^{e_{jk_j}}, \quad j = 1, 2, \ldots, r,$$

where

$$e_{j1} \geqslant e_{j2} \geqslant \cdots \geqslant e_{jk_j} > 0, \quad e_{j1} + e_{j2} + \cdots + e_{jk_j} = e_j.$$

The minimal polynomial of A is equal to

$$m(x) = \prod_{j=1}^{r} p_j(x)^{e_{j1}}.$$

It remains to consider the question of determining the cycle sum of a typical elementary block $M(g_i(x))$ of A^*, where $g_i(x)$ is of the form $p(x)^e$ for some monic irreducible factor $p(x)$ of $f(x)$. The following result provides the required information.

9.96. *Theorem.* *Let $p(x)$ be a monic irreducible polynomial over $\mathbb{F}_q$ of degree d and let $t_h = \operatorname{ord}(p(x)^h)$. Then the cycle sum of $M(p(x)^e)$ is given by*

$$(1,1) \dotplus \left(\frac{q^d - 1}{t_1}, t_1 \right) \dotplus \left(\frac{q^{2d} - q^d}{t_2}, t_2 \right) \dotplus \cdots \dotplus \left(\frac{q^{ed} - q^{(e-1)d}}{t_e}, t_e \right).$$

In summary, we obtain the following *procedure for determining the cycle sum* of an LMS $\mathfrak{M}$ over $\mathbb{F}_q$ with nonsingular characteristic matrix A:

C1. Find the elementary divisors of A, say $g_1(x), \ldots, g_w(x)$.

C2. Let $g_i(x) = f_i(x)^{m_i}$, where $f_i(x)$ is monic and irreducible over $\mathbb{F}_q$. Find the orders $t_1^{(i)} = \operatorname{ord}(f_i(x))$.

C3. Evaluate the orders $t_h^{(i)} = \operatorname{ord}(f_i(x)^h)$ for $i = 1, 2, \ldots, w$ and $h = 1, 2, \ldots, m_i$ by the formula $t_h^{(i)} = t_1^{(i)} p^{c_h}$, where p is the characteristic of $\mathbb{F}_q$ and c_h is the least integer such that $p^{c_h} \geqslant h$ (see Theorem 3.8).

C4. Determine the cycle sum Σ_i of $M(g_i(x))$ for $i = 1, 2, \ldots, w$ according to Theorem 9.96.

C5. The cycle sum Σ of $\mathfrak{M}$ is given by $\Sigma = \Sigma_1 \Sigma_2 \cdots \Sigma_w$.

9.97. Example. Let the characteristic matrix of an LMS $\mathfrak{M}$ over $\mathbb{F}_2$ be given as

$$A=\begin{pmatrix} 0 & 0 & 1 & 0 & 0 \\ 1 & 0 & 1 & 0 & 0 \\ 0 & 1 & 1 & 0 & 0 \\ 0 & 0 & 0 & 0 & 1 \\ 0 & 0 & 0 & 1 & 1 \end{pmatrix}.$$

Here

$$g_1(x)=x^3+x^2+x+1=(x+1)^3,\quad f_1(x)=x+1,\quad m_1=3,$$
$$g_2(x)=x^2+x+1,\quad f_2(x)=x^2+x+1,\quad m_2=1.$$

Steps C2 and C3 yield $t_1^{(1)}=1$, $t_2^{(1)}=2$, $t_3^{(1)}=4$, $t_1^{(2)}=3$. Hence by Theorem 9.96,

$$\sum\nolimits_1=(1,1)+(1,1)+(1,2)+(1,4)=(2,1)+(1,2)+(1,4),$$
$$\sum\nolimits_2=(1,1)+(1,3),$$

and so

$$\begin{aligned}\textstyle\sum=\sum_1\sum_2&=[(2,1)+(1,2)+(1,4)][(1,1)+(1,3)]\\&=(2,1)+(1,2)+(2,3)+(1,4)+(1,6)+(1,12).\end{aligned}$$

Thus the state graph of $\mathfrak{M}$ consists of two cycles of length 1, one cycle of length 2, two cycles of length 3, and one cycle each of length 4, 6, and 12. □

From C5 it follows that the state orders realizable by $\mathfrak{M}$ are given by

$$\operatorname{lcm}\left(t_{h_1}^{(1)},t_{h_2}^{(2)},\ldots,t_{h_w}^{(w)}\right)$$

for every combination of integers $h_1,\ldots,h_w$, $0\leqslant h_i\leqslant m_i$. If one wishes to compute all possible state orders realizable by $\mathfrak{M}$, without computing its cycle sum, one uses the following theorem.

9.98. *Theorem.* *Let $\mathfrak{M}$ be an LMS with nonsingular characteristic matrix A. Let the canonical factorization of the minimal polynomial of A be*

$$m(x)=p_1(x)^{b_1}\cdots p_r(x)^{b_r}$$

and let $t_h^{(j)}=\operatorname{ord}(p_j(x)^h)$. Then the state orders realizable by $\mathfrak{M}$ are given by all the integers of the form

$$\operatorname{lcm}\left(t_{h_1}^{(1)},t_{h_2}^{(2)},\ldots,t_{h_r}^{(r)}\right)\quad \textit{with } 0\leqslant h_j\leqslant b_j\quad \textit{for } 1\leqslant j\leqslant r.$$

NOTES

1. The theorem of Shannon mentioned in the introduction to this chapter was shown in Shannon [1] (see also Shannon and Weaver [1]). This work marks the beginning of information theory as a mathematical discipline. Proofs of Shannon's theorem and expository accounts of information theory can be found in Abramson [2], Ash [1], Guiaşu [1] (with detailed bibliography), McEliece [6], and Wolfowitz [1]. For coding in the general context of information theory we refer to Balakrishnan [1], Gallager [1], Ingels [1], Lucky, Salz, and Weldon [1], McEliece [6], and Slepian [4].

The first nontrivial example of an error-correcting code over a finite field appeared in the fundamental paper of Shannon [1]. This code would nowadays be called the (7,4) Hamming code, and its construction was credited to Hamming [1]. Codes of length 5 and minimum distance at least 2 over an alphabet of 26 letters were studied earlier by Friedman and Mendelsohn [1]. The important early contributions to the general theory of linear codes are contained in Golay [1], Hamming [1], Muller [1], Reed [1], and Slepian [1], [2], [3]. For a brief history of algebraic coding theory we refer to an excellent collection of important papers, put together and edited by Blake [1].

Detailed treatments of algebraic coding theory can be found in the books of Berlekamp [4], Blake and Mullin [1], Duske and Jürgensen [1], Lin [2], MacWilliams and Sloane [2] (with extensive bibliography), McEliece [6], Peterson and Weldon [1], Udalov and Suprun [1], van Lint [1], and von Ammon and Tröndle [1]. Some books on applied algebra also contain material on algebraic coding theory; see, for example, Birkhoff and Bartee [1], Dornhoff and Hohn [1], Lidl and Pilz [1], and Lidl and Wiesenbauer [1]. Survey articles on coding theory are Berlekamp [8], Dobrushin [1], Kautz and Levitt [1], and Sloane [1]. The books edited by Berlekamp [9] and Mann [5] represent interesting collections of papers on coding theory.

Hamming codes were introduced by Golay [1] and Hamming [1]. With regard to the various bounds on codes, see Hamming [1] for the Hamming bound, Plotkin [1] for the Plotkin bound, Gilbert [1] and Varshamov [1] for the Gilbert-Varshamov bound, and Singleton [1] for the Singleton bound (Exercise 9.5). Theorem 9.32 is due to MacWilliams [1]. Our proof follows van Lint [1]. See Berlekamp [4, Ch. 16], Chang and Wolf [1], and McEliece [6, Ch. 7] for other proofs. MacWilliams, Sloane, and Goethals [1] give an analog of the result for nonlinear codes. The identity in Exercise 9.19 is due to Pless [1].

Perfect codes (see Exercises 9.8 and 9.9) have been studied extensively. In addition to the perfect linear codes listed in these two exercises, there are two perfect linear codes discovered by Golay [1]—namely, a (23, 12) code over $\mathbb{F}_2$ and an (11, 6) code over $\mathbb{F}_3$. Tietäväinen [14] has shown that a (linear or nonlinear) perfect code $C \subseteq \mathbb{F}_q^n$ either contains just one code word, or is equal to $\mathbb{F}_q^n$, or is a binary repetition code of odd length, or has the same parameters (i.e., length, number of code words, and minimum distance) as one of the Hamming or Golay codes; see also Zinov'ev and Leont'ev [1] and Tietäväinen [15]. It is known that any code with the same parameters as one of the Golay codes is equivalent to that Golay code (Delsarte and Goethals [3], MacWilliams and Sloane [2, Ch. 20]). Lindström [1], Schönheim [2], and J. L. Vasil'ev [1] constructed nonlinear perfect codes with the same parameters as Hamming codes. See MacWilliams and Sloane [2, Ch. 6] and van Lint [3], [4] for excellent surveys of perfect codes.

The interrelation between coding and combinatorics is advantageous for both disciplines and there are numerous instances in which the techniques of one have proved results applicable to the other. For example, a result equivalent to the Hamming bound for codes was already established before the advent of coding theory by C. R. Rao [1] who worked in the context of combinatorial design theory. Many interesting results relating coding theory and combinatorics can be found in Assmus and Mattson [1], [2], Blake [2], Cameron and van Lint [1], [2], and MacWilliams and Sloane [2]. Finite geometries were used by Rudolph [1], Lin [1], P. Delsarte [1], and Sachar [1], among others, to construct and analyze codes. See Berlekamp [4, Ch. 15] and Peterson and Weldon [1, Ch. 10] for expository accounts of finite geometry codes.

2. Cyclic codes were introduced by Prange [1]. Other early work on cyclic codes was carried out by Abramson [1], Green and San Soucie [1], Peterson and Brown [1], Prange [2], and Yale [1]. Elspas and Short [1] studied the relationship between the performance of a cyclic code and the canonical factorization of its generator polynomial. Zetterberg [1] considered irreducible cyclic codes.

Relations between polynomial arithmetic modulo $x^n - 1$ and cyclic codes of length n were investigated by MacWilliams [2] and Peterson and Brown [1]. Polynomial arithmetic modulo $x^n - 1$ is also connected with the algebra of $n \times n$ circulant matrices (Karlin [1]). For the interplay between linear recurring sequences, shift registers, and cyclic codes see Abramson [1], Berlekamp [4, Ch. 5], Green and San Soucie [1], Peterson and Weldon [1, Ch. 8], Prange [2], Yale [1], Zetterberg [1], and Zierler [5].

Bose-Chaudhuri-Hocquenghem (BCH) codes were introduced by Hocquenghem [1] and Bose and Ray-Chaudhuri [1] in the binary case and Gorenstein and Zierler [1] in the nonbinary case. It was shown by Peterson [1] that the BCH codes are cyclic. Further basic work on BCH codes was carried out by Bose and Ray-Chaudhuri [2] and Mattson and Solomon [1]. For generalizations of the BCH bound (Theorem 9.45) see Hartmann and Tzeng [1]. Results on the true minimum distance and the weight distribution of BCH codes can be found in Berlekamp [5], Goldman, Kliman, and Smola [1], MacWilliams and Sloane [2, Ch. 9], Peterson [2], and Peterson and Weldon [1, Ch. 9].

The first decoding procedure for BCH codes was described by Peterson [1]. Other procedures were proposed by Berlekamp [1], Forney [1], Gorenstein and Zierler [1], and Massey [2], before Berlekamp [4] and Massey [4] developed their efficient algorithm (see also Chapter 8, Section 6). For small numbers of errors an improvement was obtained by C. L. Chen [2]. Connections between continued fractions, the Euclidean algorithm, and the Berlekamp-Massey algorithm were explored by Mills [4], Reed, Scholtz, Truong, and Welch [1], Reed and Truong [4], Reed, Truong, and Miller [3], and Welch and Scholtz [1]. The Euclidean algorithm and the Berlekamp-Massey algorithm can also be used for the decoding of other types of codes (Goppa [1], Helgert [1], Mandelbaum [2], [3], N. J. Patterson [1], Retter [1], Sarwate [1], Sugiyama, Kasahara, Hirasawa, and Namekawa [1], [2]). Michelson [1] discusses the computer implementation of decoding procedures for BCH codes. For the Chien search (Step 3 in 9.50) see Chien [1].

Reed-Solomon codes were first studied by Reed and Solomon [1]. For further information on Reed-Solomon codes and their decoding see Liu, Reed, and Truong [2], MacWilliams and Sloane [2, Ch. 10], Mandelbaum [1], Reed, Scholtz, Truong, and Welch [1], Reed, Truong, and Miller [3], and Reed, Truong, and Welch [1]. Blahut [1] gives a survey of the applications of discrete Fourier transforms to the decoding of Reed-Solomon and other codes. For reversible codes (see Exercise 9.33) we refer to MacWilliams and Sloane [2, Ch. 7] and Massey [1]. The class of polynomial codes, which includes BCH codes and finite geometry codes, was introduced by Kasami, Lin, and Peterson [1]; see also P. Delsarte [2], Gore and Cooper [1], and Peterson and Weldon [1, Ch. 10].

Information on the weight distribution in cyclic codes can be found in Baumert and McEliece [1], Berlekamp [4, Ch. 16], C. L. Chen [1], Delsarte and Goethals [1], Hartmann, Riek, and Longobardi [1], Hartmann, Tzeng, and Chien [1], Helleseth, Kløve, and Mykkeltveit [1], MacWilliams and Seery [1], MacWilliams and Sloane [2, Ch. 8], and Peterson and Weldon [1, Appendix D]. An approach to the problem of weight distribution based on Gaussian sums (Baumert and McEliece [1], McEliece [5], McEliece and Rumsey [1]) leads to a general inequality for the weights of code words in cyclic codes (Niederreiter [8]).

3. The most comprehensive account of projective geometries over finite fields is given in Hirschfeld [5]. Finite projective planes are discussed in most books on projective geometry; for example, Baer [1], Blumenthal [1], Horadam [1], Hughes and Piper [1], Pickert [1], Segre [6], and Veblen and Young [1]. For finite geometries see especially Albert and Sandler [1],

Berman and Fryer [1], Carmichael [4, Ch. 11], Dembowski [2], Hall [6], [8], Kárteszi [1], Segre [2], Vajda [1], and van Lint [2].

The Fano plane in Example 9.55 was discovered by Fano [1]. The nonexistence of a projective plane of order 6 follows from the work of Tarry [1]. More generally, Bruck and Ryser [1] have shown that if $m \equiv 1, 2 \bmod 4$, then a finite projective plane of order m can only exist if m can be written as a sum of two squares of integers (see also Hall [8, Ch. 12]). Theorem 9.60 is due to Veblen and Bussey [1]. For a detailed discussion of conics and ovals see Hirschfeld [5, Chs. 7, 8], where one can also find a proof of Theorem 9.65(i). Theorem 9.67 and its corollaries are due to Segre [1], [8]; see also Hirschfeld [3]. For a related connection with permutation polynomials see Hirschfeld [2].

For the introduction of coordinates in a finite Desarguesian plane a method used by Hilbert [3] is adapted. If one is concerned with the problem of introducing a coordinate system in any projective plane, one is led to the concept of a *ternary ring* (Albert and Sandler [1], Hall [6], [8]). *Veblen-Wedderburn systems* constitute an important special type of ternary rings. If the multiplication in a Veblen-Wedderburn system is associative, then the system is called a *near-field*. Every finite field is a near-field, and all finite near-fields were determined by Zassenhaus [1]. For detailed information on near-fields see Pilz [1]. A Veblen-Wedderburn system in which both distributive laws hold is called a *semifield* or a *nonassociative division ring* (Albert [2]). We refer to Albert and Sandler [1], Hall [8], Hughes [1], Knuth [1], H. Neumann [1], and Veblen and Wedderburn [1] for the construction of finite non-Desarguesian planes.

Finite fields were used by Crowe [1] in the construction of finite hyperbolic planes. For the applications of finite geometries to coding theory see Assmus and Mattson [2], Berlekamp [4, Ch. 15], Cameron and van Lint [1], [2], P. Delsarte [1], Lin [1], Peterson and Weldon [1, Ch. 10], Rudolph [1], and Sachar [1].

4. Most of the concepts described in this section can be found in textbooks on combinatorics; see, for example, Hall [8], Ryser [1], and Street and Wallis [1].

The definition of a balanced incomplete block design can be generalized to that of a t-design. A tactical configuration is called a *t-design* with parameters (v, k, λ) if $v \geqslant k \geqslant t \geqslant 1$ and every set of t distinct varieties is incident with the same number λ of blocks. A (v, k, λ) block design is the same as a 2-design with parameters (v, k, λ). An outstanding problem in this area is that of the existence of nontrivial t-designs with $t > 5$ (a trivial t-design would be one in which every set of k distinct varieties is a block).

An important necessary condition for the existence of symmetric BIBD's was established by Bruck, Ryser, and Chowla. If a symmetric (v, k, λ) block design exists, then: (i) if v is even, $k - \lambda$ is a square; (ii) if v

is odd, the equation $z^2=(k-\lambda)x^2+(-1)^{(v-1)/2}\lambda y^2$ has a solution in integers x, y, z not all zero. This result was proved in the case $\lambda=1$ by Bruck and Ryser [1] and by Chowla and Ryser [1] in the general case. See also Hall [8, Ch. 10], Ryser [1, Ch. 8], [3], and Shrikhande [1]. Further references for suggested reading on designs are R. C. Bose [2], Bridges and Ryser [1], Cameron [1], Cameron and van Lint [1], [2], Dembowski [1], [2], Hanani [1], Hughes [2], Lüneburg [1], Ryser [2], van Lint and Ryser [1], and Wilson [1], [2]. Relations between designs and coding theory are discussed in Assmus and Mattson [1], [2], Blake [2], Cameron and van Lint [1], [2], and MacWilliams and Sloane [2].

The difference sets in Theorem 9.79 were discovered by Singer [1] and are thus often referred to as *Singer difference sets*. For excellent surveys of difference sets see Baumert [1], Hall [5], [8], Mann [3], [4], and Storer [1]. Further results can be found in Bruck [1], Evans and Mann [1], Gordon, Mills, and Welch [1], Hall [7], E. Lehmer [3], MacWilliams and Mann [1], McEliece [1], Menon [2], Turyn [1], and Whiteman [12]. An application of certain Singer difference sets to coding theory is presented in Graham and MacWilliams [1]. There are interesting connections between difference sets on the one hand and Gaussian sums, Jacobi sums, and cyclotomy on the other; see Baumert [1, Ch. 5], Baumert and Fredricksen [1], Baumert, Mills, and Ward [1], Berndt and Chowla [1], Berndt and Evans [1], S. Chowla [4], Evans [4], [10], Hall [5], [7], E. Lehmer [3], Mann [3], Menon [2], Muskat and Whiteman [1], Storer [1], Whiteman [10], [11], and Yamamoto [3].

The standard reference for latin squares is Dénes and Keedwell [1]; see also Childs [1], Hall [8], Mann [2], Ryser [1], Street and Wallis [1], and Vajda [2]. Theorem 9.83 is due to MacNeish [1]; see also Mann [1], [2] and Ryser [1]. Orthogonal latin squares were first studied by Euler [1] who conjectured that there does not exist a pair of orthogonal latin squares of order n for $n\equiv 2 \bmod 4$. This was confirmed by Tarry [1] for the case $n=6$. Bose and Shrikhande [2] disproved Euler's conjecture by constructing a pair of orthogonal latin squares of order 22. Shortly afterwards, Parker [1] found a pair of orthogonal latin squares of order 10. Finally, Bose, Shrikhande, and Parker [1] proved that for any $n>6$ there exists a pair of orthogonal latin squares of order n. See Bose and Shrikhande [3] and Parker [2] for related work. R. C. Bose [1] and W. L. Stevens [1] showed that there exists a finite projective plane of order n if and only if there exists a set of $n-1$ mutually orthogonal latin squares of order n. It should be noted that a set of mutually orthogonal latin squares of order n cannot contain more than $n-1$ elements.

Applications of tactical configurations and orthogonal latin squares to the design of statistical experiments are discussed in Mann [2], Raghavarao [1], and Vajda [2]. For the original introduction of experimental design see Fisher [1].

Hadamard matrices are considered in many books on combinatorics (see, e.g., Hall [8], van Lint [2]) and there are several construction methods available (Baumert and Hall [1], Ehlich [1], Paley [3], Wallis, Street, and Wallis [1]). Applications of Hadamard matrices to coding theory are discussed in Bose and Shrikhande [1], Golomb and Baumert [1], and MacWilliams and Sloane [2, Ch. 2], and the latter reference also contains a survey of applications to other areas. For related types of matrices see Belevitch [1], Butson [1], Delsarte, Goethals, and Seidel [1], Goethals and Seidel [1], MacWilliams [4], and Wallis, Street, and Wallis [1].

5. The following are excellent sources of further information on finite-state systems (or finite-state machines) and linear modular systems: Arbib, Falb, and Kalman [1], Booth [1], Dornhoff and Hohn [1, Chs. 1, 8], Gill [1], [2], Harrison [1], Zadeh and Desoer [1], and Zadeh and Polak [1, Ch. 2]. The latter source contains also many references on linear modular systems. Some of the classical papers on the subject have been collected and edited by Kautz [1]; see also Crowell [1], Elspas [1], Friedland [1], and Huffman [1] for important contributions. Conditions under which a finite-state machine can be realized by a linear modular system were investigated by Eichner [1] and Hartfiel and Maxson [1], among others. Matluk and Gill [1] showed how to decompose linear modular systems over $\mathbb{Z}/(m)$ into linear modular systems over finite fields; see also Bollman [2] for linear modular systems over $\mathbb{Z}/(m)$. Kalman [1], [2] studied linear modular systems from the viewpoint of dynamical systems. For a detailed discussion of rational canonical forms of matrices we refer to Dornhoff and Hohn [1, Ch. 7] and Herstein [4, Ch. 6].

We mention briefly some other applications of finite fields. The theoretical analysis of switching circuits can be based on finite field arithmetic (Green and Taylor [1], [3], Moisil [1], [2], [3], [4], Moisil and Popovici [1], Murakami and Reed [1], Rudeanu [1], Vaida [1]). Finite fields are used in the calculation of switching functions (Benjauthrit and Reed [1], [2], Davio, Deschamps, and Thayse [1], Labunec and Sitnikov [1], Pradhan [1], Takahashi [1], Thayse [1], Yin [1]) and of general logic functions (Karpovsky [1]). Mendelsohn [2] uses finite fields to model quasigroup identities. There are various ways of employing properties of finite fields in cryptography (Beker and Piper [1], [2], Brawley and Levine [1], Cooper [1], Diffie and Hellman [1], Hartwig and Levine [1], Herlestam and Johannesson [1], Hershey [1], Konheim [1], Krishnamurthy and Ramachandran [1], Levine and Brawley [1], [2], Levine and Hartwig [1], Pohlig and Hellman [1], Sloane [2]). Redinbo [1] discusses applications of finite fields to array processors, Nicholson [1] has applications to the calculation of finite Fourier transforms, and English [1] presents applications to finite-state algorithms.

EXERCISES

9.1. Determine all code words, the minimum distance, and a parity-check matrix of the binary linear $(5,3)$ code that is defined by the generator matrix

$$G = \begin{pmatrix} 0 & 1 & 0 & 0 & 1 \\ 0 & 0 & 1 & 0 & 1 \\ 1 & 0 & 0 & 1 & 1 \end{pmatrix}.$$

9.2. Prove: a linear code can detect s or fewer errors if and only if its minimum distance is $\geqslant s+1$.

9.3. Prove that the Hamming distance is a metric on $\mathbb{F}_q^n$.

9.4. Let H be a parity-check matrix of a linear code. Prove that the code has minimum distance d if and only if any $d-1$ columns of H are linearly independent and there exist d linearly dependent columns.

9.5. If a linear (n, k) code has minimum distance d, prove that $n-k+1 \geqslant d$ (Singleton bound).

9.6. Let G_1 and G_2 be generator matrices for a linear (n_1, k) code and (n_2, k) code with minimum distance d_1 and d_2, respectively. Show that the linear codes with generator matrices

$$\begin{pmatrix} G_1 & 0 \\ 0 & G_2 \end{pmatrix} \quad \text{and} \quad (G_1, \quad G_2)$$

are $(n_1+n_2, 2k)$ codes and (n_1+n_2, k) codes, respectively, with minimum distances $\min(d_1, d_2)$ and $d \geqslant d_1 + d_2$, respectively.

9.7. Prove: given k and d, then for a binary linear (n, k) code to have minimum distance $d = d_0$ we must have

$$n \geqslant d_0 + d_1 + \cdots + d_{k-1},$$

where $d_{i+1} = \lfloor (d_i+1)/2 \rfloor$ for $i = 0, 1, \ldots, k-2$. Here $\lfloor x \rfloor$ denotes the largest integer $\leqslant x$.

9.8. A code $C \subseteq \mathbb{F}_q^n$ is called *perfect* if for some integer t the balls $B_t(\mathbf{c})$ of radius t centered at code words $\mathbf{c}$ are pairwise disjoint and "fill" the space $\mathbb{F}_q^n$—that is,

$$\bigcup_{\mathbf{c} \in C} B_t(\mathbf{c}) = \mathbb{F}_q^n.$$

Prove that in the binary case all Hamming codes and all repetition codes of odd length are perfect codes.

9.9. Using the definition of Exercise 9.8, prove that all Hamming codes over $\mathbb{F}_q$ are perfect.

9.10. Two linear (n, k) codes C_1 and C_2 over $\mathbb{F}_q$ are called *equivalent* if the code words of C_1 can be obtained from the code words of C_2 by applying a fixed permutation to the coordinate places of all words in C_2. Let G be a generator matrix for a linear code C. Show that any permutation of the rows of G or any permutation of the columns of G gives a generator matrix of a linear code which is equivalent to C.

9.11. Use the definition of equivalent codes in Exercise 9.10 to show that the binary linear codes with generator matrices

$$G_1 = \begin{pmatrix} 1 & 1 & 1 & 0 \\ 0 & 1 & 1 & 0 \\ 0 & 0 & 1 & 1 \end{pmatrix} \quad \text{and} \quad G_2 = \begin{pmatrix} 1 & 0 & 1 & 1 \\ 0 & 1 & 1 & 1 \\ 1 & 0 & 0 & 1 \end{pmatrix},$$

respectively, are equivalent.

9.12. Let C be a linear (n, k) code. Prove that the dimension of $C^{\perp}$ is $n-k$.

9.13. Prove that $(C^{\perp})^{\perp} = C$ for any linear code C.

9.14. Prove $(C_1 + C_2)^{\perp} = C_1^{\perp} \cap C_2^{\perp}$ for any linear codes C_1, C_2 over $\mathbb{F}_q$ of the same length.

9.15. If C is the binary $(n, 1)$ repetition code, prove that $C^{\perp}$ is the $(n, n-1)$ parity-check code.

9.16. Determine a generator matrix and all code words of the $(7, 3)$ code which is dual to the binary Hamming code C_3.

9.17. Determine the dual code $C^{\perp}$ to the code given in Exercise 9.1. Find the table of cosets of $\mathbb{F}_2^5$ modulo $C^{\perp}$, determine the coset leaders and syndromes. If $\mathbf{y} = 01001$ is a received word, which message was probably sent?

9.18. Apply Theorem 9.32 to the binary linear code $C = \{000, 011, 101, 110\}$; that is, find its dual code, determine the weight enumerators, and verify the MacWilliams identity.

9.19. Let C be a binary linear (n, k) code with weight enumerator

$$A(x, y) = \sum_{i=0}^{n} A_i x^i y^{n-i}$$

and let

$$A^{\perp}(x, y) = \sum_{i=0}^{n} A_i^{\perp} x^i y^{n-i}$$

be the weight enumerator of the dual code $C^{\perp}$. Show the following identity for $r = 0, 1, \ldots$:

$$\sum_{i=0}^{n} i^r A_i = \sum_{i=0}^{n} (-1)^i A_i^{\perp} \sum_{t=0}^{r} t!S(r, t) 2^{k-t} \binom{n-i}{n-t},$$

where

$$S(r, t) = \frac{1}{t!} \sum_{j=0}^{t} (-1)^{t-j} \binom{t}{j} j^r$$

is a Stirling number of the second kind and the binomial coefficient $\binom{m}{h}$ is defined to be 0 whenever $h > m$ or $h < 0$. Write down the identity for $r = 0$, 1, and 2.

9.20. Let $n = (q^m - 1)/(q-1)$ and β a primitive nth root of unity in $\mathbb{F}_{q^m}$, $m \geqslant 2$. Prove that the null space of the matrix $H = (1 \quad \beta \quad \beta^2 \quad \cdots \quad \beta^{n-1})$ is a code over $\mathbb{F}_q$ with minimum distance at least 3 if and only if $\gcd(m, q-1) = 1$.

9.21. Let α be a primitive element of $\mathbb{F}_9$ with minimal polynomial $x^2 - x - 1$ over $\mathbb{F}_3$. Find a generator polynomial for a BCH code of length 8 and dimension 4 over $\mathbb{F}_3$. Determine the minimum distance of this code.

9.22. Find a generator polynomial for a BCH code of dimension 12 and designed distance $d = 5$ over $\mathbb{F}_2$.

9.23. Determine the dimension of a 5-error-correcting BCH code over $\mathbb{F}_3$ of length 80.

9.24. Find the generator polynomial for a 3-error-correcting binary BCH code of length 15 by using the primitive element α of $\mathbb{F}_{16}$ with $\alpha^4 = \alpha^3 + 1$.

9.25. Determine a generator polynomial g for a $(31, 31-\deg(g))$ binary BCH code with designed distance $d = 9$.

9.26. Let m and t be any two positive integers. Show that there exists a binary BCH code of length $2^m - 1$ which corrects all combinations of t or fewer errors using not more than mt control symbols.

9.27. Describe a Reed-Solomon $(15, 13)$ code over $\mathbb{F}_{16}$ by determining its generator polynomial and the number of errors it will correct.

9.28. Prove that the minimum distance of a Reed-Solomon code with generator polynomial

$$g(x) = \prod_{i=1}^{d-1} (x - \alpha^i)$$

is equal to d.

9.29. Determine if the dual of an arbitrary BCH code is a BCH code. Is the dual of an arbitrary Reed-Solomon code a Reed-Solomon code?

9.30. Find the error locations in Example 9.43, given that the syndrome of a received vector is $(10010110)^{\mathrm{T}}$. Find a generator matrix for this code.

9.31. Let a binary 2-error-correcting BCH code of length 31 be defined by the root α of $x^5 + x^2 + 1$ in $\mathbb{F}_{32}$. Suppose the received word has the syndrome $(1110011101)^{\mathrm{T}}$. Find the error polynomial.

9.32. Let α be a primitive element of $\mathbb{F}_{16}$ with $\alpha^4 = \alpha + 1$, and let $g(x) = x^{10} + x^8 + x^5 + x^4 + x^2 + x + 1$ be the generator polynomial of a binary $(15, 5)$ BCH code. Suppose the word $\mathbf{v} = 000101100100011$ is received. Determine the corrected code word and the message word.

9.33. A code C is called *reversible* if $(a_0, a_1, \ldots, a_{n-1}) \in C$ implies $(a_{n-1}, \ldots, a_1, a_0) \in C$. (a) Prove that a cyclic code $C = (g(x))$ is reversible if and only if with each root of $g(x)$ also the reciprocal value of that root is a root of $g(x)$. (b) Prove that any cyclic code over $\mathbb{F}_q$ of length n is reversible if -1 is a power of q modulo n.

9.34. Given a cyclic (n,k) code, a linear $(n-m, k-m)$ code is obtained by omitting the last m rows and columns in the generator matrix of the cyclic code described prior to Theorem 9.36. Show that the resulting code is in general not cyclic, but that it has at least the same minimum distance as the original code. (*Note*: Such an $(n-m, k-m)$ code is called a *shortened cyclic code*.)

9.35. List the points and lines of $PG(2,\mathbb{F}_3)$. Draw a diagram showing all the intersections. Enumerate the points on L_∞ and the families of parallel lines in $AG(2,\mathbb{F}_3)$.

9.36. In $PG(2,\mathbb{F}_4)$ consider the quadrangle $A(1,1,1+\beta)$, $B(0,1,\beta)$, $C(1,1,\beta)$, $D(1,1+\beta,\beta)$, where β is a primitive element of $\mathbb{F}_4$. Find its diagonal points and verify that they are collinear.

9.37. There are six points in $PG(2,\mathbb{F}_4)$, no three of which are collinear. Four of them are the points A, B, C, D of Exercise 9.36. Find the other two points.

9.38. Find the equation of the conic consisting of the points A, B, C, D of Exercise 9.36 and $E(1,1+\beta,1+\beta)$, determine all its tangents and the point where they meet.

9.39. Show that for a nondegenerate conic in $PG(2,\mathbb{F}_5)$ the tangents do not all meet in the same point.

9.40. Prove: if L is a set of points of $PG(2,\mathbb{F}_q)$ such that every line of $PG(2,\mathbb{F}_q)$ contains a point of L, then $|L| \geqslant q+1$ with equality if and only if L is a line.

9.41. Prove that among any $m+3$ points of a finite projective plane of order m one can find three collinear ones. (*Note*: This shows that an oval in $PG(2,\mathbb{F}_q)$, q even, contains the maximum number of points with the property that no three of them are collinear.)

9.42. If two ovals in $PG(2,\mathbb{F}_q)$, q even, have more than half their points in common, show that they coincide.

9.43. For q even, a nondegenerate conic in $PG(2,\mathbb{F}_q)$ plus the point where its tangents meet is called a *regular oval*. Show that for $q=2$ and 4 every oval in $PG(2,\mathbb{F}_q)$ is regular.

9.44. Let $q=2^h$ and $1 \leqslant n < h$. Prove that the set $A(x^{2^n})$ defined as in Theorem 9.67 is an oval in $PG(2,\mathbb{F}_q)$ if and only if $\gcd(n,h)=1$.

9.45. In $PG(2,\mathbb{F}_q)$ with $q=2^h$, $h>1$, show that:
(a) if $\deg(f)=2$, then $A(f)$ is an oval if and only if $A(f)=A(x^2)$;
(b) if $\deg(f)=4$, then $A(f)$ is an oval if and only if h is odd and $A(f)=A(x^4)$.

9.46. Let $A(f)$ be as in Theorem 9.67. Then $A(f)$ is called a *translation oval* if it is an oval and if f induces an endomorphism of $\mathbb{F}_q$ as an additive group. Prove that $A(f)$ is a translation oval if and only if

the following conditions hold:
(a) $f(a+b)=f(a)+f(b)$ for all $a, b \in \mathbb{F}_q$;
(b) f is a permutation polynomial of $\mathbb{F}_q$ with $\deg(f)<q$ and $f(1)=1$;
(c) $f(x)/x$ is a permutation polynomial of $\mathbb{F}_q$ with constant term 0.
Prove also that if $\deg(f)<q$, then f satisfies (a) if and only if f is a p-polynomial, where p is the characteristic of $\mathbb{F}_q$.

9.47. Let $q=2^h$ and $1 \leqslant n < h$. Prove that $A(x^{2^n})$ is a translation oval in $PG(2,\mathbb{F}_q)$ if $\gcd(n,h)=1$.

9.48. Determine the number of points, lines, planes, and hyperplanes of $PG(4,\mathbb{F}_3)$. How many planes are there through a given line?

9.49. In $PG(4,\mathbb{F}_3)$ determine the 3-flats through the plane given by $(1,0,0,0,0)$, $(0,0,1,0,0)$, and $(0,0,0,0,1)$.

9.50. Prove that the number of k-flats of $PG(m,\mathbb{F}_q)$, $1 \leqslant k < m$, or also within a fixed m-flat of a projective geometry over $\mathbb{F}_q$ of higher dimension, is equal to

$$\frac{(q^{m+1}-1)(q^m-1)\cdots(q^{m-k+1}-1)}{(q^{k+1}-1)(q^k-1)\cdots(q-1)}.$$

9.51. Show that the following system of blocks forms a BIBD and evaluate the parameters v, b, r, k, and λ:

$$\begin{array}{cccc} \{1,2,3\} & \{1,4,7\} & \{1,5,9\} & \{1,6,8\} \\ \{4,5,6\} & \{2,5,8\} & \{2,6,7\} & \{2,4,9\} \\ \{7,8,9\} & \{3,6,9\} & \{3,4,8\} & \{3,5,7\} \end{array}$$

9.52. Solve the following special case of the Kirkman Schoolgirl Problem. A schoolmistress takes 9 girls for a daily walk, the girls arranged in rows of 3 girls. Plan the walk for 4 consecutive days so that no girl walks with any of her classmates in any triplet more than once.

9.53. In a school of b boys, t athletic teams of k boys each are formed in such a way that every boy plays on the same number of teams. Also, the arrangement is such that each pair of boys plays together the same number of times. On how many teams does a boy play and how often do two boys play on the same team?

9.54. Prove: if v is even for a symmetric (v,k,λ) block design, then $k-\lambda$ is a square.

9.55. Verify that $\{0,1,2,3,5,7,12,13,16\}$ is a difference set of residues modulo 19. Determine the parameters v, k, and λ.

9.56. Show that $\{0,4,5,7\}$ is a difference set of residues modulo 13 which yields $PG(2,\mathbb{F}_3)$.

9.57. Prove the following generalization of Theorem 9.76. Let

$$\{d_{i1},\ldots,d_{ik}\}, \quad i=1,\ldots,s,$$

be a system of (v, k, λ) difference sets. Then with all residues modulo v as varieties, the vs blocks

$$\{d_{i1}+t,\ldots,d_{ik}+t\}, \quad t=0,1,\ldots,v-1 \text{ and } i=1,\ldots,s,$$

form a $(v, k, \lambda s)$ block design.

9.58. Let $L^{(k)}=(a_{ij}^{(k)})$, where $a_{ij}^{(k)} \equiv i+jk \bmod 9$, $0 \leqslant a_{ij}^{(k)} < 9$ for $1 \leqslant i, j \leqslant 9$. Which of the arrays $L^{(k)}$, $k=1,2,\ldots,8$, are latin squares? Are $L^{(2)}$ and $L^{(5)}$ orthogonal?

9.59. A latin square of order n is said to be in *normalized* form if the first row and the first column are both the ordered set $\{1,2,\ldots,n\}$. How many normalized latin squares of each order $n \leqslant 4$ are there?

9.60. Let L be a latin square of order m with entries in $\{1,2,\ldots,m\}$ and M a latin square of order n with entries in $\{1,2,\ldots,n\}$. From L and M construct a latin square of order mn with entries in $\{1,2,\ldots,m\}\times\{1,2,\ldots,n\}$.

9.61. Construct three mutually orthogonal latin squares of order 4.

9.62. Prove that for $n \geqslant 2$ there can be at most $n-1$ mutually orthogonal latin squares of order n.

9.63. A *magic square* of order n consists of the integers 1 to n^2 arranged in an $n \times n$ array such that the sums of entries in rows, columns, and diagonals are all the same. Let $A=(a_{ij})$ and $B=(b_{ij})$ be two orthogonal latin squares of order n with entries in $\{0,1,\ldots,n-1\}$ such that the sum of entries in each of the diagonals of A and B is $n(n-1)/2$. Show that $M=(na_{ij}+b_{ij}+1)$ is a magic square of order n. Construct a magic square of order 4 from two orthogonal latin squares obtained in Exercise 9.61.

9.64. Determine Hadamard matrices of orders 8 and 12.

9.65. If H_m and H_n are Hadamard matrices, show that there exists a Hadamard matrix H_{mn}.

9.66. Show that from a normalized Hadamard matrix of order $4t$, $t \geqslant 2$, one can construct a symmetric $(4t-1, 2t-1, t-1)$ block design.

9.67. Prove that the state graph of an LMS over $\mathbb{F}_q$ with nonsingular characteristic matrix consists of pure cycles only.

9.68. Prove that the state graphs of similar characteristic matrices over $\mathbb{F}_q$ are isomorphic. (*Note*: Two matrices A, B over $\mathbb{F}_q$ are *similar* if there exists a nonsingular matrix P over $\mathbb{F}_q$ such that $B=PAP^{-1}$.)

9.69. Suppose the characteristic matrix A of an LMS $\mathfrak{M}$ over $\mathbb{F}_2$ has the minimal polynomial $(x+1)^5(x^3+x+1)^3$. What are the state orders realizable by $\mathfrak{M}$?

9.70. Determine the orders of all states in the LMS $\mathfrak{M}$ of Example 9.97.

9.71. Suppose the characteristic matrix A of an LMS $\mathfrak{M}$ over $\mathbb{F}_q$ is *nonderogatory*; that is, its minimal polynomial is equal to its characteristic polynomial. Let the minimal polynomial of A be of the form $p(x)^e$, where $p(x)$ is a monic irreducible polynomial over $\mathbb{F}_q$ of degree d. Without using Theorem 9.96, prove that the cycle sum of $\mathfrak{M}$ is given by the expression in that theorem.

9.72. Calculate the cycle sum of the LMS $\mathfrak{M}$ over $\mathbb{F}_3$ given in Example 9.91.

9.73. Prove Theorem 9.98.

Chapter 10

Tables

In this chapter we collect tables that facilitate the computation in finite fields and tables of irreducible and primitive polynomials. The description of these tables is given in Sections 1 and 2, respectively.

1. COMPUTATION IN FINITE FIELDS

Multiplication and division of nonzero elements of $\mathbb{F}_q$ can be performed using a notion analogous to logarithms. The term *index* is used rather than logarithm. If b is a primitive element of $\mathbb{F}_q$, then for any $a \in \mathbb{F}_q^*$ there exists a unique integer r with $0 \leqslant r < q-1$ such that $a = b^r$. We write $r = \operatorname{ind}_b(a)$, or simply $r = \operatorname{ind}(a)$ if b is kept fixed. The index function satisfies the following basic rules:

$$\operatorname{ind}(ac) \equiv \operatorname{ind}(a) + \operatorname{ind}(c) \bmod (q-1),$$

$$\operatorname{ind}(ac^{-1}) \equiv \operatorname{ind}(a) - \operatorname{ind}(c) \bmod (q-1).$$

The inverse function of the index function, corresponding to taking antilogarithms, is denoted by $\exp_b$ or simply exp, and we have:

$$\exp(r) = b^r, \quad \exp(\operatorname{ind}(a)) = a, \quad \operatorname{ind}(\exp(r)) = r.$$

Given a table of the ind and exp function, it is easy to carry out addition, subtraction, multiplication, and division in $\mathbb{F}_q$. Addition and subtraction are performed by using the vector space structure of $\mathbb{F}_q$ over its prime subfield $\mathbb{F}_p$, multiplication and division are performed by using the rules for the index function and the exp and ind table to convert from one notation to the other. Table A provides a complete list of the nonzero elements and their indices for the finite fields $\mathbb{F}_q$ with q composite and $q \leqslant 128$. In the exp column the parentheses and commas of the vector notation for the following element of $\mathbb{F}_q$ with $q = p^n$ have been dropped:

$$a = (a_1, \ldots, a_n) = a_1 b^{n-1} + a_2 b^{n-2} + \cdots + a_n, \quad 0 \leqslant a_i < p.$$

10.1. Example. As an example for the use of Table A we calculate

$$[(b+1)+(2b+2)b](b+2)^{-1}+b$$

in the field $\mathbb{F}_9$. Working with the portion of the table pertaining to this field, we get

$$\operatorname{ind}((2b+2)b) \equiv \operatorname{ind}(2b+2)+\operatorname{ind}(b) \equiv 3+1 \equiv 4 \bmod 8,$$
$$(2b+2)b = \exp(4) = 2.$$

Thus, $(b+1)+(2b+2)b = b$ and

$$\operatorname{ind}\left([(b+1)+(2b+2)b](b+2)^{-1}\right) \equiv \operatorname{ind}(b)-\operatorname{ind}(b+2) \equiv 1-6 \equiv 3 \bmod 8,$$
$$[(b+1)+(2b+2)b](b+2)^{-1} = \exp(3) = 2b+2.$$

The final result is $(2b+2)+b = 2$. □

Table B affords another possibility of doing arithmetic in finite fields. In the first two columns it provides a table of *Jacobi's logarithm* $L(n)$ for the fields $\mathbb{F}_{2^k}$ with $2 \leqslant k \leqslant 6$ (compare with Exercise 2.8). The symbol $n \to s$ means that $L(n) = s$ with respect to a fixed primitive element b. In characteristic 2 the value $L(0)$ is undefined. The elements b^n are multiplied in the obvious way and added according to the rule

$$b^m + b^n = b^{m+L(n-m)}$$

given in Exercise 2.8. The symbol "+" preceding the value of n indicates that b^n is a primitive element.

10.2. Example. We use Table B to calculate

$$(b^6 + b^{25} + b^{44})(1+b^{35})^{-1} + b^{28}$$

in the field $\mathbb{F}_{64}$. We have $b^6 + b^{25} = b^{6+L(19)} = b^{40}$ and $b^{40} + b^{44} = b^{40+L(4)} = b^{72}$. Since $1 + b^{35} = b^{L(35)} = b^{31}$, we get

$$(b^6 + b^{25} + b^{44})(1+b^{35})^{-1} = b^{72}b^{-31} = b^{41}.$$

Furthermore, since the argument of the function L and the exponent of b are considered modulo 63, we obtain $b^{41} + b^{28} = b^{41+L(-13)} = b^{41+L(50)} =$

$b^{101} = b^{38}$, which is the final result and happens to be a primitive element of $\mathbb{F}_{64}$. □

The remainder of Table B provides information about *minimal* and *characteristic polynomials* and about *dual bases*. We take the lines

$$+20 \to 26[100001] \quad 26 \quad 6 \quad 49 \quad 29 \quad 9 \quad 46:19$$
$$21 \to 42[101011] \qquad [11]$$

from the table for $\mathbb{F}_{64}$ over $\mathbb{F}_2$ as illustrations. The symbol $[a_1\ a_2 \cdots a_m]$ indicates that $x^m + a_1 x^{m-1} + a_2 x^{m-2} + \cdots + a_m$ is the characteristic polynomial of the element with respect to the given field extension. Thus, $x^6 + x^5 + 1$ is the characteristic polynomial of b^{20} over $\mathbb{F}_2$ and $x^6 + x^5 + x^3 + x + 1$ is that of b^{21} over $\mathbb{F}_2$. If b^n is a defining element of the extension, then the set of integers between the characteristic polynomial and the colon describes the dual basis of the polynomial basis determined by b^n. If b^n is not a defining element, then the minimal polynomial of b^n with respect to the given extension is listed in the bracket notation explained above. For instance, b^{20} is a defining element of $\mathbb{F}_{64}$ over $\mathbb{F}_2$ and the dual basis of the polynomial basis $\{1, b^{20}, b^{40}, b^{60}, b^{80}, b^{100}\}$ is $\{b^{26}, b^6, b^{49}, b^{29}, b^9, b^{46}\}$. On the other hand, b^{21} is not a defining element of $\mathbb{F}_{64}$ over $\mathbb{F}_2$ and the minimal polynomial of b^{21} over $\mathbb{F}_2$ is $x^2 + x + 1$, so that $b^{21} \in \mathbb{F}_4$. If b^n is not only a defining element, but also determines a *normal basis* of the given extension, then the integer after the colon describes the element determining the *dual normal basis*. For instance, b^{20} determines the normal basis

$$\{b^{20}, (b^{20})^2, (b^{20})^4, (b^{20})^8, (b^{20})^{16}, (b^{20})^{32}\}$$

of $\mathbb{F}_{64}$ over $\mathbb{F}_2$, and its dual basis is given by

$$\{b^{19}, (b^{19})^2, (b^{19})^4, (b^{19})^8, (b^{19})^{16}, (b^{19})^{32}\}.$$

Elements in subfields except $\mathbb{F}_2$ are denoted in the table by capital letters whose meaning becomes clear upon inspection of the data for minimal polynomials. For example, in the table for $\mathbb{F}_{64}$ the letter X stands for $b^{21} \in \mathbb{F}_4$ and D stands for $b^{27} \in \mathbb{F}_8$.

2. TABLES OF IRREDUCIBLE POLYNOMIALS

Table C lists all monic irreducible polynomials of degree n over prime fields $\mathbb{F}_p$ for small values of n and p. The extent of the table may be summarized as follows: $p = 2$ and $n \leqslant 11$, $p = 3$ and $n \leqslant 7$, $p = 5$ and $n \leqslant 5$, $p = 7$ and $n \leqslant 4$. The polynomial $a_0 x^n + a_1 x^{n-1} + \cdots + a_n$ is abbreviated in the form $a_0\ a_1 \cdots a_n$ with $a_0 = 1$. The left-hand column, headed by the value of n, lists all monic irreducible polynomials f for the degree n and the modulus p

concerned. The right-hand column, headed by e, contains the corresponding value of $\operatorname{ord}(f)$.

Table D lists one primitive polynomial over $\mathbb{F}_2$ for each degree $n \leqslant 100$. In this table only the degrees of the separate terms in the polynomial are given; thus 6 1 0 stands for $x^6 + x + 1$.

Table E lists all primitive polynomials $x^2 + a_1x + a_2$ of degree 2 over $\mathbb{F}_p$ for $11 \leqslant p \leqslant 31$. For smaller primes all quadratic primitive polynomials can be obtained from Table C by locating the polynomials f over $\mathbb{F}_p$ with $\operatorname{ord}(f) = p^2 - 1$.

Table F lists one primitive polynomial of degree n over $\mathbb{F}_p$ for all values of $n \geqslant 2$ and p with $p < 50$ and $p^n < 10^9$. The polynomial $x^n + a_1x^{n-1} + a_2x^{n-2} + \cdots + a_n$ is listed in the form $a_1\ a_2 \cdots a_n$.

NOTES

1. Table A is due to Alanen and Knuth [2]. The first extensive table of this type is that of Jacobi [3] who tabulated primitive roots modulo p and indices for all primes $p < 1000$. A much more limited table was published earlier by Crelle [1]. Desmarest [1] lists a primitive root modulo p for each $p < 10000$. Wertheim [1] tabulated the least positive primitive root modulo p for all $p < 6200$, Cunningham, Woodall, and Creak [1] extended this table to $p \leqslant 25409$ (see also Albert [3, Appendix I] for an extract from this table), and Western and Miller [1] extended it to $p \leqslant 50021$. Litver and Judina [1] calculated a primitive root modulo p for each $p \leqslant 1001321$. Osborn [1] tabulated all primitive roots modulo p for all primes $p < 1000$ and Hauptman, Vegh, and Fisher [1] extended this table to all $p < 5000$. Western and Miller [1] provide tables of indices with respect to the least positive primitive root modulo p for all $p \leqslant 50021$. See also Andręe [1] for another table of indices for finite prime fields. Bussey [1], [2] gave tables of indices for nonprime finite fields prior to Alanen and Knuth [2]; see also Albert [3, Appendix III] for extracts from Bussey's tables. Algorithms for the calculation of indices are discussed in Herlestam and Johannesson [1], Pohlig and Hellman [1], Pollard [3], and Zierler [9].

Table B just reflects a part of more extensive calculations carried out by Conway [1]. Jacobi's logarithm was introduced in Jacobi [2] where it is tabulated for finite prime fields $\mathbb{F}_p$ with $p \leqslant 103$. Further references on computation in finite fields can be found in the notes to Chapter 4, Section 1.

2. Table C is reprinted from Church [1]. Irreducible polynomials over $\mathbb{F}_p$, $p \leqslant 19$, of small degree were already listed by C. Jordan [3]. A slight extension of Table C was calculated by Garakov [3] who raised the bound on n by 1 for each p and also included the cases $p = 11$ and $n \leqslant 4$. Another

extension was obtained by Chang and Godwin [1] who treated the cases $11 \leqslant p \leqslant 37$ for $n = 2$ and $11 \leqslant p \leqslant 19$ for $n = 3$. Some irreducible trinomials over $\mathbb{F}_p$ were listed by Mortimer and Williams [1]. Irreducible polynomials over nonprime finite fields were tabulated by Green and Taylor [2] who dealt with the following cases: $n \leqslant 5$ for $\mathbb{F}_4$, $n \leqslant 3$ for $\mathbb{F}_8$ and $\mathbb{F}_9$, and $n = 2$ for $\mathbb{F}_{16}$. More extensive tabulations of irreducible polynomials have been carried out for $\mathbb{F}_2$ because of the importance of this case for applications. The tables of Garakov [1] and Golomb [1], [4, Ch. 3] for $\mathbb{F}_2$ do not extend beyond Table C. Marsh [1] lists all irreducible polynomials over $\mathbb{F}_2$ with $n \leqslant 19$ (see also Albert [3, Appendix IV] and Peterson and Weldon [1, Appendix C] for excerpts from this table) and Mossige [1] treats the case $10 \leqslant n \leqslant 20$. Peterson and Weldon [1, Appendix C] list one irreducible polynomial over $\mathbb{F}_2$ for each n with $17 \leqslant n \leqslant 34$ and each possible order. Lists of irreducible trinomials over $\mathbb{F}_2$ can be found in Fredricksen and Wisniewski [1], Golomb [4, Ch. 5], Golomb, Welch, and Hales [1], Zierler [7], and Zierler and Brillhart [1], [2].

Table D is reprinted from E. J. Watson [1] and Tables E and F are reprinted from Alanen and Knuth [2]. A short table of primitive polynomials over finite prime fields $\mathbb{F}_p$ with $p \leqslant 11$ can already be found in Dickson [7, Part I, Ch. 3]. Bussey [1], [2] lists one primitive polynomial over $\mathbb{F}_p$ of degree n for each $n \geqslant 2$ and $p^n < 1000$; see also Albert [3, Appendix III] and Heuzé [1]. Alanen and Knuth [1] list one primitive polynomial over $\mathbb{F}_p$ of degree n for each p and n with $11 \leqslant p \leqslant 17$ and $3 \leqslant n \leqslant 5$. The paper of Sugimoto [1] contains a table of primitive polynomials over $\mathbb{F}_p$ for $3 \leqslant p \leqslant 47$. Green and Taylor [2] consider nonprime finite fields and list one primitive polynomial for each of the following cases: $n \leqslant 11$ for $\mathbb{F}_4$, $n \leqslant 7$ for $\mathbb{F}_8$ and $\mathbb{F}_9$, and $n \leqslant 5$ for $\mathbb{F}_{16}$. Beard and West [1] tabulate primitive polynomials of a special type. Table D was extended by Stahnke [1] who lists one primitive polynomial over $\mathbb{F}_2$ for each degree $n \leqslant 168$. Primitive trinomials over $\mathbb{F}_2$ can be found in Rodemich and Rumsey [1], Zierler [6], and Zierler and Brillhart [1], [2].

Lloyd [1] tabulated the canonical factorization of all polynomials over $\mathbb{F}_2$ of degree $\leqslant 4$ and over $\mathbb{F}_3$ of degree $\leqslant 3$. This was extended in Lloyd [2] to cover all polynomials over $\mathbb{F}_p$ for $p = 2, 3, 5, 7$ of degree $\leqslant 11, 11, 8, 6$, respectively, and another extension appears in Lloyd and Remmers [1]. Factorization tables for the binomials $x^n - 1$ can be found in Beard and West [2] and McEliece [3]. Factorization tables for trinomials are given in Beard and West [3], Golomb [4, Ch. 5], Golomb, Welch, and Hales [1], Mortimer and Williams [1], and Zierler [7].

TABLE A

exp	ind
$GF(2^2)$	
01	0
10	1
11	2
$GF(2^3)$	
001	0
010	1
100	2
101	3
111	4
011	5
110	6
$GF(2^4)$	
0001	0
0010	1
0100	2
1000	3
1001	4
1011	5
1111	6
0111	7
1110	8
0101	9
1010	10
1101	11
0011	12
0110	13
1100	14
$GF(2^5)$	
00001	0
00010	1
00100	2
01000	3
10000	4
01001	5
10010	6
01101	7
11010	8
11101	9
10011	10
01111	11
11110	12
10101	13
00011	14
00110	15
01100	16
11000	17
11001	18
11011	19
11111	20
10111	21
00111	22
01110	23
11100	24
10001	25
01011	26
10110	27
00101	28
01010	29
10100	30
$GF(2^6)$	
000001	0
000010	1
000100	2
001000	3
010000	4
100000	5
100001	6
100011	7
100111	8
101111	9
111111	10
011111	11
111110	12
011101	13
111010	14
010101	15
101010	16
110101	17
001011	18
010110	19
101100	20
111001	21
010011	22
100110	23
101101	24
111011	25
010111	26
101110	27
111101	28
011011	29
110110	30
001101	31
011010	32
110100	33
001001	34
010010	35
100100	36
101001	37
110011	38
000111	39
001110	40
011100	41
111000	42
010001	43
100010	44
100101	45
101011	46
110111	47
001111	48
011110	49
111100	50
011001	51
110010	52
000101	53
001010	54
010100	55
101000	56
110001	57
000011	58
000110	59
001100	60
011000	61
110000	62
$GF(2^7)$	
0000001	0
0000010	1
0000100	2
0001000	3
0010000	4
0100000	5
1000000	6
0000011	7
0000110	8
0001100	9
0011000	10
0110000	11
1100000	12
1000011	13
0000101	14
0001010	15
0010100	16
0101000	17
1010000	18
0100011	19
1000110	20
0001111	21
0011110	22
0111100	23
1111000	24
1110011	25
1100101	26
1001001	27
0010001	28
0100010	29
1000100	30
0001011	31
0010110	32
0101100	33
1011000	34
0110011	35
1100110	36
1001111	37
0011101	38
0111010	39
1110100	40
1101011	41
1010101	42
0101001	43
1010010	44
0100111	45
1001110	46
0011111	47
0111110	48
1111100	49
1111011	50
1110101	51
1101001	52
1010001	53

exp	*ind*	*exp*	*ind*	*exp*	*ind*	*exp*	*ind*
$GF(2^7)$		$GF(2^7)$		$GF(3^3)$		$GF(3^4)$	
0100001	54	1110110	101	112	8	1102	27
1000010	55	1101111	102	222	9	0021	28
0000111	56	1011101	103	121	10	0210	29
0001110	57	0111001	104	012	11	2100	30
0011100	58	1110010	105	120	12	2002	31
0111000	59	1100111	106	002	13	1022	32
1110000	60	1001101	107	020	14	2221	33
1100011	61	0011001	108	200	15	0212	34
1000101	62	0110010	109	201	16	2120	35
0001001	63	1100100	110	211	17	2202	36
0010010	64	1001011	111	011	18	0022	37
0100100	65	0010101	112	110	19	0220	38
1001000	66	0101010	113	202	20	2200	39
0010011	67	1010100	114	221	21	0002	40
0100110	68	0101011	115	111	22	0020	41
1001100	69	1010110	116	212	23	0200	42
0011011	70	0101111	117	021	24	2000	43
0110110	71	1011110	118	210	25	1002	44
1101100	72	0111111	119			2021	45
1011011	73	1111110	120	$GF(3^4)$		1212	46
0110101	74	1111111	121			1121	47
1101010	75	1111101	122	0001	0	0211	48
1010111	76	1111001	123	0010	1	2110	49
0101101	77	1110001	124	0100	2	2102	50
1011010	78	1100001	125	1000	3	2022	51
0110111	79	1000001	126	2001	4	1222	52
1101110	80			1012	5	1221	53
1011111	81	$GF(3^2)$		2121	6	1211	54
0111101	82			2212	7	1111	55
1111010	83	01	0	0122	8	0111	56
		10	1	1220	9	1110	57
1110111	84	21	2	1201	10	0101	58
1101101	85	22	3	1011	11	1010	59
1011001	86	02	4	2111	12	2101	60
0110001	87	20	5	2112	13	2012	61
1100010	88	12	6	2122	14	1122	62
1000111	89	11	7	2222	15	0221	63
0001101	90			0222	16	2210	64
0011010	91	$GF(3^3)$		2220	17	0102	65
0110100	92			0202	18	1020	66
1101000	93	001	0	2020	19	2201	67
1010011	94	010	1	1202	20	0012	68
0100101	95	100	2	1021	21	0120	69
1001010	96	102	3	2211	22	1200	70
0010111	97	122	4	0112	23	1001	71
0101110	98	022	5	1120	24	2011	72
1011100	99	220	6	0201	25	1112	73
0111011	100	101	7	2010	26	0121	74

TABLE A (*Cont.*)

$GF(3^4)$

exp	*ind*
1210	75
1101	76
0011	77
0110	78
1100	79

$GF(5^2)$

exp	*ind*
01	0
10	1
43	2
42	3
32	4
44	5
02	6
20	7
31	8
34	9
14	10
33	11
04	12
40	13
12	14
13	15
23	16
11	17
03	18
30	19
24	20
21	21
41	22
22	23

$GF(5^3)$

exp	*ind*
001	0
010	1
100	2
403	3
132	4
223	5
031	6
310	7
304	8
244	9
241	10
211	11
411	12
212	13
421	14
312	15
324	16
444	17
042	18
420	19
302	20
224	21
041	22
410	23
202	24
321	25
414	26
242	27
221	28
011	29
110	30
003	31
030	32
300	33
204	34
341	35
114	36
043	37
430	38
402	39
122	40
123	41
133	42
233	43
131	44
213	45
431	46
412	47
222	48
021	49
210	50
401	51
112	52
023	53
230	54
101	55
413	56
232	57
121	58
113	59
033	60
330	61
004	62
040	63
400	64
102	65
423	66
332	67
024	68
240	69
201	70
311	71
314	72
344	73
144	74
343	75
134	76
243	77
231	78
111	79
013	80
130	81
203	82
331	83
014	84
140	85
303	86
234	87
141	88
313	89
334	90
044	91
440	92
002	93
020	94
200	95
301	96
214	97
441	98
012	99
120	100
103	101
433	102
432	103
422	104
322	105
424	106
342	107
124	108
143	109
333	110
034	111
340	112
104	113
443	114
032	115
320	116
404	117
142	118
323	119
434	120
442	121
022	122
220	123

$GF(7^2)$

exp	*ind*
01	0
10	1
64	2
53	3
56	4
16	5
54	6
66	7
03	8
30	9
45	10
12	11
14	12
34	13
15	14
44	15
02	16
20	17
51	18
36	19
35	20
25	21
31	22
55	23
06	24
60	25
13	26

exp	*ind*	*exp*	*ind*	*exp*	*ind*	*exp*	*ind*
$GF(7^2)$		$GF(11^2)$		$GF(11^2)$		$GF(11^2)$	
24	27	07	12	30	49	06	84
21	28	70	13			60	85
61	29	46	14			52	86
		25	15	81	50	89	87
23	30	38	16	4*	51	1*	88
11	31	51	17	65	52	94	89
04	32	79	18	*2	53		
40	33	26	19	37	54	63	90
32	34			41	55	82	91
65	35	48	20	85	56	5*	92
63	36	45	21	8*	57	59	93
43	37	15	22	2*	58	49	94
62	38	44	23	88	59	55	95
33	39	05	24			09	96
		50	25	0*	60	90	97
05	40	69	26	*0	61	23	98
50	41	32	27	17	62	18	99
26	42	*1	28	64	63		
41	43	27	29	92	64	74	100
42	44			43	65	86	101
52	45	58	30	*5	66	9*	102
46	46	39	31	67	67	13	103
22	47	61	32	12	68	24	104
		62	33	14	69	28	105
		72	34			68	106
$GF(11^2)$		66	35	34	70	22	107
		02	36	11	71	08	108
01	0	20	37	04	72	80	109
10	1	98	38	40	73		
*4	2	*3	39	75	74		
57	3			96	75	3*	110
29	4	47	40	83	76	71	111
78	5	35	41	6*	77	56	112
16	6	21	42	42	78	19	113
54	7	*8	43	95	79	84	114
*9	8	97	44			7*	115
*7	9	93	45	73	80	36	116
		53	46	76	81	31	117
87	10	99	47	*6	82	91	118
**	11	03	48	77	83	33	119

The symbol * denotes the element 10 in $\mathbb{F}_{11}$.

TABLE B

$\mathbb{F}_4$ over $\mathbb{F}_2$

```
 0 → * [01]  [1]
+1 → 2 [11]2 0:1
+2 → 1 [11]1 0:2
```

$\mathbb{F}_8$ over $\mathbb{F}_2$

```
 0 → * [111]   [1]
+1 → 5 [101]4 3 5:1
+2 → 3 [101]1 6 3:2
+3 → 2 [011]0 6 3:-
+4 → 6 [101]2 5 6:4
+5 → 1 [011]0 3 5:-
+6 → 4 [011]0 5 6:-
```

$\mathbb{F}_{16}$ over $\mathbb{F}_2$ / over $\mathbb{F}_4$

```
  0 →  * [0001]          [1]        [01]      [1]
 +1 →  4 [0011]14  2   1 0: -       [1X] 4  0: 1
 +2 →  8 [0011]13  4   2 0: -       [1Y] 8  0: 2
  3 → 14 [1111]14 10   1 2:11       [Y1] 2  5:13
 +4 →  1 [0011]11  8   4 0: -       [1X] 1  0: 4
  5 → 10 [0101]          [11]       [0Y]     [X]
  6 → 13 [1111]13  5   2 4: 7       [X1] 4 10:11
 +7 →  9 [1001] 9  2 10 1: 6        [XX] 8 10:12
 +8 →  2 [0011] 7  1   8 0: -       [1Y] 2  0: 8
  9 →  7 [1111] 7  5   8 1:13       [X1] 1 10:14

 10 →  5 [0101]          [11]       [0X]     [Y]
+11 → 12 [1001]12  1   5 8: 3       [YY] 4  5: 6
 12 → 11 [1111]11 10   4 8:14       [Y1] 8  5: 7
+13 →  6 [1001] 6  8 10 4: 9        [XX] 2 10: 3
+14 →  3 [1001] 3  4   5 2:12       [YY] 1  5: 9
```

$\mathbb{F}_{32}$ over $\mathbb{F}_2$

```
  0        * [10011]              [1]
 +1 → 19 [10111]16  3  6  5 17: 1
 +2 →  7 [10111] 1  6 12 10  3: 2
 +3 → 11 [01001] 0 28 25  6  3: -
 +4 → 14 [10111] 2 12 24 20  6: 4
 +5 → 29 [01111] 4 28  5 14  9: -
 +6 → 22 [01001] 0 25 19 12  6: -
 +7 →  2 [00101]27 20 17 10  3: -
 +8 → 28 [10111] 4 24 17  9 12: 8
 +9 → 15 [01111] 1  7  9 19 10: -

+10 → 27 [01111] 8 25 10 28 18: -
+11 →  3 [11101]23 12  7  6  3:15
+12 → 13 [01001] 0 19  7 24 12: -
+13 → 12 [11101]30 17 28 24 12:29
+14 →  4 [00101]23  9  3 20  6: -
+15 →  9 [11011]26 20  5 19 10:11
```

TABLE B

$\mathbb{F}_{32}$	over $\mathbb{F}_2$
+16 → 25	[10111] 8 17 3 18 24:16
+17 → 21	[01001] 0 14 28 3 17: -
+18 → 30	[01111] 2 14 18 7 20: -
+19 → 1	[00101]29 10 24 5 17: -
+20 → 23	[01111]16 19 20 25 5: -
+21 → 17	[11101]27 6 19 3 17:23
+22 → 6	[11101]15 24 14 12 6:30
+23 → 20	[11011]13 10 18 25 5:21
+24 → 26	[01001] 0 7 14 17 24: -
+25 → 16	[00101]30 5 12 18 24: -
+26 → 24	[11101]29 3 25 17 24:27
+27 → 10	[11011]22 5 9 28 18:26
+28 → 8	[00101]15 18 6 9 12: -
+29 → 5	[11011]11 18 20 14 9:13
+30 → 18	[11011]21 9 10 7 20:22

$\mathbb{F}_{64}$	over $\mathbb{F}_2$	over $\mathbb{F}_4$	over $\mathbb{F}_8$
0 → *	[010101] [1]	[111] [1]	[01] [1]
+ 1 → 8	[101101]44 43 58 54 53 45: -	[XXX]47 54 27: 8	[1A] 8 0: 1
+ 2 → 16	[101101]25 23 53 45 43 27: -	[YYY]31 45 54:16	[1B] 16 0: 2
3 → 53	[010111]60 47 46 43 3 0: -	[0Y1] 0 6 3: -	[FD]42 18:39
+ 4 → 32	[101101]50 46 43 27 23 54: -	[XXX]62 27 45:32	[1C] 32 0: 4
+ 5 → 38	[100001]38 33 28 23 18 43:52	[XYY]23 56 49: -	[CF] 4 27:59
6 → 43	[010111]57 31 29 23 6 0: -	[0X1] 0 12 6: -	[DE]21 36:15
7 → 62	[001001]42 35 28 0 56 49: -	[00X] 0 56 49: -	[E1] 2 9:25
+ 8 → 1	[101101]37 29 23 54 46 45: -	[YYY]61 54 27: 1	[1A] 1 0: 8
9 → 45	[010001] [101]	[101] 36 27 45: 9	[0B] [A]
+10 → 13	[100001]13 3 56 46 36 23:41	[YXX]46 49 35: -	[AD] 8 54:55
+11 → 51	[110011]37 14 3 55 36 48:11	[X1Y]55 48 24:25	[AC]16 54:56
12 → 23	[010111]51 62 58 46 12 0: -	[0Y1] 0 24 12: -	[EF]42 9:30
+13 → 10	[100111]10 7 59 46 33 23:17	[XYX]43 49 35: -	[AE]32 54:58
14 → 61	[001001]21 7 56 0 49 35: -	[00Y] 0 49 35: -	[F1] 4 18:50
15 → 44	[110101]51 36 46 31 47 3:15	[Y01] 18 3 33:57	[CA]21 27: 6
+16 → 2	[101101]11 58 46 45 29 27: -	[XXX]59 45 54: 2	[1B] 2 0:16
+17 → 41	[100001]41 24 7 53 36 58:13	[XYY]53 14 28: -	[AD] 1 54:62
18 → 27	[010001] [101]	[101] 9 54 27:18	[0C] [B]
+19 → 34	[100111]34 49 62 43 24 53:20	[XYX]58 28 56: -	[BF] 8 45:46
+20 → 26	[100001]26 6 49 29 9 46:19	[XYY]29 35 7: -	[BE]16 45:47
21 → 42	[101011] [11]	[XY1] [X]	[11] 42 0:21
+22 → 39	[110011]11 28 6 47 9 33:22	[Y1X]47 33 48:50	[BA]32 45:49
+23 → 12	[000011]40 29 6 46 23 0: -	[1XY]31 24 12:58	[EB] 4 9:41
24 → 46	[010111]39 61 53 29 24 0: -	[0X1] 0 48 24: -	[FD]21 18:60
+25 → 30	[110011]44 49 24 62 36 6:25	[Y1X]62 6 3:11	[AC] 2 54: 7

TABLE B (*Cont.*)

$\mathbb{F}_{64}$	over $\mathbb{F}_2$	over $\mathbb{F}_4$	over $\mathbb{F}_8$
+26 → 20	[100111]20 14 55 29 3 46:34	[YXY]23 35 7: -	[BF] 1 45:53
27 → 18	[000101] [011]	[011] 0 54 27: -	[OE] [D]
28 → 59	[001001]42 14 49 0 35 7: -	[00X] 0 35 7: -	[D1] 8 36:37
+29 → 48	[000011]34 53 24 58 29 0: -	[1XY]61 33 48:43	[DA]16 36:38
30 → 25	[110101]39 9 29 62 31 6:30	[X01] 36 6 3:51	[AB]42 54:12
+31 → 35	[011011]25 29 61 0 24 56: -	[11X] 46 28 56: -	[DD]32 36:40
+32 → 4	[101101]22 53 29 27 58 54: -	[YYY]55 27 45: 4	[1C] 4 0:32
33 → 58	[010111]30 55 23 53 33 0: -	[0X1] 0 3 33: -	[EF]21 9:51
+34 → 19	[100001]19 48 14 43 9 53:26	[YXX]43 28 56: -	[BE] 2 45:61
35 → 31	[001001]21 49 14 0 28 56: -	[00Y] 0 28 56: -	[D1] 1 36:44
36 → 54	[010001] [101]	[101] 18 45 54:36	[OA] [C]
+37 → 57	[110011]50 7 33 59 18 24:37	[Y1X]59 24 12:44	[CB] 8 27:28
+38 → 5	[100111] 5 35 61 23 48 43:40	[YXY]53 56 49: -	[CD]16 27:29
39 → 22	[110101]57 18 23 47 55 33:39	[X01] 9 33 48:60	[BC]42 45: 3
+40 → 52	[100001]52 12 35 58 18 29:38	[YXX]58 7 14: -	[CF]32 27:31
+41 → 17	[100111]17 56 31 53 12 58:10	[YXY]29 14 28: -	[AE] 4 54:23
42 → 21	[101011] [11]	[YX1] [Y]	[11] 21 0:42
+43 → 6	[000011]20 46 3 23 43 0: -	[1YX] 47 12 6:29	[DA] 2 36:52
+44 → 15	[110011]22 56 12 31 18 3:44	[X1Y]31 3 33:37	[CB] 1 27:35
45 → 9	[000101] [011]	[011] 0 27 45: -	[OD] [F]
+46 → 24	[000011]17 58 12 29 46 0: -	[1YX] 62 48 24:53	[FC] 8 18:19
+47 → 49	[011011]44 46 62 0 12 28: -	[11Y] 23 14 28: -	[FF]16 18:20
48 → 29	[010111]15 59 43 58 48 0: -	[0Y1] 0 33 48: -	[DE]42 36:57
49 → 47	[001001]42 56 7 0 14 28: -	[00X] 0 14 28: -	[F1] 32 18:22
+50 → 60	[110011]25 35 48 61 9 12:50	[X1Y]61 12 6:22	[BA] 4 45:14
51 → 11	[110101]60 9 43 55 59 48:51	[Y01] 36 48 24:30	[AB]21 54:33
+52 → 40	[100111]40 28 27 58 6 29: 5	[XYX]46 7 14: -	[CD] 2 27:43
+53 → 3	[000011]10 23 33 43 53 0: -	[1XY]55 6 3:46	[FC] 1 18:26
54 → 36	[000101] [011]	[011] 0 45 54: -	[OF] [E]
+55 → 56	[011011]22 23 31 0 6 14: -	[11X] 43 7 14: -	[EE] 8 9:10
56 → 55	[001001]21 28 35 0 7 14: -	[00Y] 0 7 14: -	[E1] 16 9:11
57 → 37	[110101]30 36 53 59 61 24:57	[X01] 18 24 12:15	[CA]42 27:48
+58 → 33	[000011] 5 43 48 53 58 0: -	[1YX] 59 3 33:23	[EB]32 9:13
+59 → 28	[011011]11 43 47 0 3 7: -	[11Y] 53 35 7: -	[DD] 4 36: 5
60 → 50	[110101]15 18 58 61 62 12:60	[Y01] 9 12 6:39	[BC]21 45:24
+61 → 14	[011011]37 53 55 0 33 35: -	[11X]58 49 35: -	[EF] 2 18:34
+62 → 7	[011011]50 58 59 0 48 49: -	[11Y]29 56 49: -	[EE] 1 9:17

TABLE C

Irreducible Polynomials for the Modulus 2

$n = 1$	e
10	1
11	1

$n = 2$	e
111	3

$n = 3$	e
1011	7
1101	7

$n = 4$	e
10011	15
11001	15
11111	5

$n = 5$	e
100101	31
101001	31
101111	31
110111	31
111011	31
111101	31

$n = 6$	e
1000011	63
1001001	9
1010111	21
1011011	63
1100001	63
1100111	63
1101101	63
1110011	63
1110101	21

$n = 7$	e
10000011	127
10001001	127
10001111	127
10010001	127
10011101	127
10100111	127
10101011	127
10111001	127
10111111	127
11000001	127
11001011	127
11010011	127
11010101	127
11100101	127
11101111	127
11110001	127
11110111	127
11111101	127

$n = 8$	e
100011011	51
100011101	255
100101011	255
100101101	255
100111001	17
100111111	85
101001101	255
101011111	255
101100011	255
101100101	255
101101001	255
101110001	255
101110111	85
101111011	85
110000111	255
110001011	85
110001101	255
110011111	51
110100011	85
110101001	255
110110001	51
110111101	85
111000011	255
111001111	255
111010111	17
111011101	85
111100111	255
111110011	51
111110101	255
111111001	85

$n = 9$	e
1000000011	73
1000010001	511
1000010111	73
1000011011	511
1000100001	511
1000101101	511
1000110011	511
1001001011	73
1001011001	511
1001011111	511
1001100101	73
1001101001	511
1001101111	511
1001110111	511
1001111101	511
1010000111	511
1010010101	511
1010011001	73
1010100011	511
1010100101	511
1010101111	511
1010110111	511
1010111101	511
1011001111	511
1011010001	511
1011011011	511
1011110101	511
1011111001	511
1100000001	73
1100010011	511
1100010101	511
1100011111	511
1100100011	511
1100110001	511
1100111011	511
1101001001	73
1101001111	511
1101011011	511
1101100001	511
1101101011	511
1101101101	511
1101110011	511
1101111111	511
1110000101	511
1110001111	511
1110100001	73
1110110101	511
1110111001	511
1111000111	511
1111001011	511
1111001101	511
1111010101	511
1111011001	511
1111100011	511
1111101001	511
1111111011	511

$n = 10$	e
10000001001	1023
10000001111	341
10000011011	1023
10000011101	341
10000100111	1023
10000101101	1023
10000110101	93
10001000111	341
10001010011	341
10001100011	341
10001100101	1023
10001101111	1023
10010000001	1023
10010001011	1023
10010011001	341
10010101001	33
10010101111	341
10011000101	1023
10011001001	341
10011010111	1023
10011100111	1023
10011101101	341
10011110011	1023
10011111111	1023
10100001011	93
10100001101	1023
10100011001	1023
10100011111	341
10100100011	1023
10100110001	1023
10100111101	1023
10101000011	1023
10101010111	1023
10101100001	93
10101100111	341
10101101011	1023
10110000101	1023
10110001111	1023
10110010111	1023
10110011011	341
10110100001	1023
10110101011	341
10110111001	341
10111000001	341

TABLE C **(*Cont.*)**

Irreducible Polynomials for the Modulus 2

10111000111	1023	11111011011	1023	100111100101	2047	101111101101	2047
10111100101	1023	11111101011	341	100111101111	89	110000001011	2047
10111110111	1023	11111110011	1023	100111110111	2047	110000001101	2047
10111111011	1023	11111111001	1023	101000000001	2047	110000011001	2047
11000010011	1023	11111111111	11	101000000111	2047	110000011111	2047
11000010101	1023			101000010011	2047	110000110001	89
11000100011	33	$n = 11$	e	101000010101	2047	110001010111	2047
11000100101	1023			101000101001	2047	110001100001	2047
11000110001	341	100000000101	2047	101001001001	2047	110001101011	2047
11000110111	1023	100000010111	2047	101001100001	2047	110001110011	2047
11001000011	1023	100000101011	2047	101001101101	2047	110001110101	23
11001001111	1023	100000101101	2047	101001111001	2047	110010000101	2047
11001010001	341	100001000111	2047	101001111111	2047	110010001001	2047
11001011011	1023	100001100011	2047	101010000101	2047	110010010111	2047
11001111001	1023	100001100101	2047	101010010001	2047	110010011011	2047
11001111111	1023	100001110001	2047	101010011101	2047	110010011101	2047
11010000101	93	100001111011	2047	101010100111	2047	110010110011	2047
11010001001	1023	100010001101	2047	101010101011	2047	110010111111	2047
11010100111	93	100010010101	2047	101010110011	2047	110011000111	2047
11010101101	341	100010011111	2047	101010110101	2047	110011001101	2047
11010110101	1023	100010101001	2047	101011010101	2047	110011010011	2047
11010111111	341	100010110001	2047	101011011111	2047	110011010101	2047
11011000001	1023	100011000011	89	101011100011	23	110011100011	2047
11011001101	341	100011001111	2047	101011101001	2047	110011101001	2047
11011010011	1023	100011010001	2047	101011101111	2047	110011110111	2047
11011011111	1023	100011100001	2047	101011110001	2047	110100000011	2047
11011110111	341	100011100111	2047	101011111011	2047	110100001111	2047
11011111101	1023	100011101011	2047	101100000011	2047	110100011101	2047
11100001111	341	100011110101	2047	101100001001	2047	110100100111	2047
11100010001	341	100100001101	2047	101100010001	2047	110100101101	2047
11100010111	1023	100100010011	2047	101100110011	2047	110101000001	2047
11100011101	1023	100100100101	2047	101100111111	2047	110101000111	2047
11100100001	1023	100100101001	2047	101101000001	2047	110101010101	2047
11100101011	93	100100110111	89	101101001011	2047	110101011001	2047
11100110101	341	100100111011	2047	101101011001	2047	110101100011	2047
11100111001	1023	100100111101	2047	101101011111	2047	110101101111	2047
11101000111	1023	100101000101	2047	101101100101	2047	110101110001	2047
11101001101	1023	100101001001	2047	101101101111	2047	110110010011	2047
11101010101	1023	100101010001	2047	101101111101	2047	110110011111	2047
11101011001	1023	100101011011	2047	101110000111	2047	110110101001	2047
11101100011	1023	100101110011	2047	101110001011	2047	110110111011	2047
11101111011	341	100101110101	2047	101110010011	2047	110110111101	2047
11101111101	1023	100101111111	2047	101110010101	2047	110111001001	2047
11110000001	341	100110000011	2047	101110101111	2047	110111010111	2047
11110000111	341	100110001111	2047	101110110111	2047	110111011011	2047
11110001101	1023	100110101011	2047	101110111101	2047	110111100001	2047
11110010011	1023	100110101101	2047	101111001001	2047	110111100111	2047
11110101001	341	100110111001	2047	101111011011	2047	110111110101	2047
11110110001	1023	100111000111	2047	101111011101	2047	110111111111	89
11111000101	341	100111011001	2047	101111100111	2047	111000000101	2047

Irreducible Polynomials for the Modulus 2

111000011101	2047	111001111011	2047	111011111001	2047	111110010001	2047
111000100001	2047	111001111101	2047	111100001011	2047	111110010111	2047
111000100111	2047	111010000001	2047	111100011001	2047	111110011011	2047
111000101011	2047	111010010011	2047	111100110001	2047	111110100111	2047
111000110011	2047	111010011111	2047	111100110111	2047	111110101101	2047
111000111001	2047	111010100011	2047	111101011101	2047	111110110101	2047
111001000111	2047	111010111011	2047	111101101011	2047	111111001101	2047
111001001011	2047	111011001001	89	111101101101	2047	111111010011	2047
111001010101	2047	111011001111	2047	111101110101	2047	111111100101	2047
111001011111	2047	111011011101	2047	111101111001	89	111111101001	2047
111001110001	2047	111011110011	2047	111110000011	2047	111111111011	89

Irreducible Polynomials for the Modulus 3

$n=1$	e
10	1
11	2
12	1

$n=2$	e
101	4
112	8
122	8

$n=3$	e
1021	26
1022	13
1102	13
1112	13
1121	26
1201	26
1211	26
1222	13

$n=4$	e
10012	80
10022	80
10102	16
10111	40
10121	40
10202	16
11002	80
11021	20
11101	40
11111	5
11122	80
11222	80
12002	80
12011	20
12101	40
12112	80
12121	10
12212	80

$n=5$	e
100021	242
100022	121
100112	121
100211	242
101011	242
101012	121
101102	121
101122	121
101201	242
101221	242
102101	242
102112	121
102122	11
102202	121
102211	242
102221	22
110002	121
110012	121
110021	242
110101	242
110111	242
110122	121
111011	242
111121	242
111211	242
111212	121
112001	242
112022	121
112102	11
112111	242
112201	242
112202	121
120001	242
120011	242
120022	121
120202	121
120212	121
120221	242
121012	121
121111	242
121112	121
121222	121
122002	121
122021	242
122101	242
122102	121
122201	22
122212	121

$n=6$	e
1000012	728
1000022	728
1000111	364
1000121	364
1000201	52
1001012	728
1001021	364
1001101	91
1001122	104
1001221	182
1002011	364
1002022	728
1002101	182
1002112	104
1002211	91
1010201	52
1010212	728
1010222	728
1011001	91
1011011	364
1011022	728
1011122	728
1012001	182
1012012	728
1012021	364
1012112	728
1020001	52
1020101	52
1020112	728
1020122	728
1021021	364
1021102	56
1021112	728
1021121	91
1022011	364
1022102	56
1022111	182
1022122	728
1100002	728
1100012	56
1100111	364
1101002	728
1101011	28
1101101	364
1101112	728
1101212	728
1102001	364
1102111	91
1102121	91
1102201	364
1102202	728
1110001	364
1110011	364
1110122	728
1110202	728
1110221	182
1111012	728
1111021	182
1111111	7
1111112	728
1111222	728
1112011	91
1112201	182
1112222	728
1120102	728
1120121	91
1120222	728
1121012	728
1121102	728
1121122	104
1121212	728
1121221	364
1122001	91
1122002	104
1122122	104
1122202	728
1122221	364
1200002	728
1200022	56
1200121	364
1201001	364
1201111	182
1201121	182
1201201	364
1201202	728
1202002	728
1202021	28
1202101	364
1202122	728
1202222	728
1210001	364
1210021	364
1210112	728
1210202	728
1210211	91
1211021	182
1211201	91
1211212	728

TABLE C (*Cont.*)

Irreducible Polynomials for the Modulus 3

1212011	91
1212022	728
1212121	14
1212122	728
1212212	728
1220102	728
1220111	182
1220212	728
1221001	182
1221002	104
1221112	104
1221202	728
1221211	364
1222022	728
1222102	728
1222112	104
1222211	364
1222222	728

$n = 7$	e
10000102	1093
10000121	2186
10000201	2186
10000222	1093
10001011	2186
10001012	1093
10001102	1093
10001111	2186
10001201	2186
10001212	1093
10002112	1093
10002122	1093
10002211	2186
10002221	2186
10010122	1093
10010222	1093
10011002	1093
10011101	2186
10011211	2186
10012001	2186
10012022	1093
10012111	2186
10012202	1093
10020121	2186
10020221	2186
10021001	2186
10021112	1093
10021202	1093
10022002	1093
10022021	2186
10022101	2186
10022212	1093
10100011	2186
10100012	1093
10100102	1093
10100122	1093
10100201	2186
10100221	2186
10101101	2186
10101112	1093
10101202	1093
10101211	2186
10102102	1093
10102201	2186
10110022	1093
10110101	2186
10110211	2186
10111001	2186
10111102	1093
10111121	2186
10111201	2186
10112002	1093
10112012	1093
10112021	2186
10112111	2186
10112122	1093
10120021	2186
10120112	1093
10120202	1093
10121002	1093
10121102	1093
10121201	2186
10121222	1093
10122001	2186
10122011	2186
10122022	1093
10122212	1093
10122221	2186
10200001	2186
10200002	1093
10200101	2186
10200112	1093
10200202	1093
10200211	2186
10201021	2186
10201022	1093
10201121	2186
10201222	1093
10202011	2186
10202012	1093
10210001	2186
10210121	2186
10210202	1093
10211101	2186
10211111	2186
10211122	1093
10211221	2186
10212011	2186
10212022	1093
10212101	2186
10212112	1093
10212212	1093
10220002	1093
10220101	2186
10220222	1093
10221122	1093
10221202	1093
10221212	1093
10221221	2186
10222012	1093
10222021	2186
10222111	2186
10222202	1093
10222211	2186
11000101	2186
11000222	1093
11001022	1093
11001112	1093
11001211	2186
11002012	1093
11002022	1093
11002121	2186
11002202	1093
11010001	2186
11010022	1093
11010121	2186
11010221	2186
11011111	2186
11011202	1093
11012002	1093
11012102	1093
11012212	1093
11020021	2186
11020022	1093
11020102	1093
11020112	1093
11020201	2186
11020222	1093
11021111	2186
11021122	1093
11021201	2186
11021212	1093
11022101	2186
11022122	1093
11022211	2186
11022221	2186
11100002	1093
11100022	1093
11100121	2186
11100212	1093
11101012	1093
11101022	1093
11101102	1093
11101111	2186
11101121	2186
11102002	1093
11102111	2186
11102222	1093
11110001	2186
11110012	1093
11110111	2186
11110112	1093
11110211	2186
11110222	1093
11111011	2186
11111021	2186
11111201	2186
11111222	1093
11112011	2186
11112221	2186
11120102	1093
11120111	2186
11120122	1093
11120212	1093
11120221	2186
11121001	2186
11121101	2186
11121202	1093
11122021	2186
11122112	1093
11122201	2186
11122222	1093
11200201	2186
11200202	1093
11201012	1093
11201021	2186
11201101	2186
11201111	2186
11201221	2186
11201222	1093
11202002	1093
11202121	2186
11202211	2186
11202212	1093
11210002	1093
11210011	2186
11210021	2186
11210101	2186
11211001	2186
11211022	1093
11211122	1093
11211212	1093
11211221	2186
11212012	1093
11212112	1093
11212202	1093
11220001	2186
11220112	1093
11220211	2186
11221022	1093
11221102	1093
11221112	1093
11221121	2186
11222011	2186
11222102	1093
11222122	1093
11222201	2186
11222221	2186
12000121	2186
12000202	1093
12001021	2186
12001112	1093
12001211	2186
12002011	2186
12002021	2186
12002101	2186
12002222	1093
12010021	2186
12010022	1093
12010102	1093
12010121	2186
12010201	2186
12010211	2186
12011102	1093
12011111	2186
12011212	1093
12011221	2186
12012112	1093
12012122	1093

Irreducible Polynomials for the Modulus 3

12012202	1093	12101201	2186	12112211	2186	12201121	2186	12212122	1093
12012221	2186	12101212	1093	12120002	1093	12201122	1093	12212201	2186
12020002	1093	12101222	1093	12120011	2186	12201202	1093	12212221	2186
12020021	2186	12102001	2186	12120112	1093	12201212	1093	12220001	2186
12020122	1093	12102121	2186	12120121	2186	12202001	2186	12220012	1093
12020222	1093	12102212	1093	12120211	2186	12202111	2186	12220022	1093
12021101	2186	12110111	2186	12120212	1093	12202112	1093	12220202	1093
12021212	1093	12110122	1093	12121012	1093	12202222	1093	12221002	1093
12022001	2186	12110201	2186	12121022	1093	12210002	1093	12221021	2186
12022111	2186	12110212	1093	12121102	1093	12210112	1093	12221111	2186
12022201	2186	12110221	2186	12121121	2186	12210211	2186	12221122	1093
12100001	2186	12111002	1093	12122012	1093	12211021	2186	12221221	2186
12100021	2186	12111101	2186	12122122	1093	12211201	2186	12222011	2186
12100111	2186	12111202	1093	12200101	2186	12211211	2186	12222101	2186
12100222	1093	12112022	1093	12200102	1093	12211222	1093	12222211	2186
12101011	2186	12112102	1093	12201011	2186	12212012	1093		
12101021	2186	12112121	2186	12201022	1093	12212102	1093		

Irreducible Polynomials for the Modulus 5

$n=1$	e
10	1
11	2
12	4
13	4
14	1

$n=2$	e
102	8
103	8
111	3
112	24
123	24
124	12
133	24
134	12
141	6
142	24

$n=3$	e		
1011	62	1113	124
1014	31	1114	31
1021	62	1131	62
1024	31	1134	31
1032	124	1141	62
1033	124	1143	124
1042	124	1201	62
1043	124	1203	124
1101	62	1213	124
1102	124	1214	31
		1222	124
		1223	124
		1242	124
		1244	31
		1302	124
		1304	31
		1311	62
		1312	124
		1322	124
		1323	124
		1341	62
		1343	124
		1403	124
		1404	31
		1411	62
		1412	124
		1431	62
		1434	31
		1442	124
		1444	31

$n=4$	e						
		11013	624	12022	624	13102	208
		11023	624	12033	624	13121	52
10002	16	11024	104	12042	624	13124	312
10003	16	11032	624	12102	208	13131	26
10014	312	11041	52	12121	13	13133	624
10024	312	11042	624	12123	624	13201	78
10034	312	11101	78	12131	52	13203	624
10044	312	11113	624	12134	312	13232	624
10102	48	11114	312	12201	39	13234	312
10111	78	11124	104	12203	624	13241	156
10122	624	11133	208	12211	156	13302	624
10123	624	11142	208	12222	624	13314	104
10132	624	11202	624	12224	312	13322	624
10133	624	11212	624	12302	624	13323	208
10141	39	11213	208	12311	39	13334	312
10203	48	11221	156	12312	208	13341	78
10221	39	11222	208	12324	312	13342	208
10223	208	11234	104	12332	624	13401	156
10231	78	11244	312	12333	208	13413	208
10233	208	11301	156	12344	104	13423	624
10303	48	11303	624	12401	156	13424	312
10311	156	11321	39	12414	104	13432	208
10313	208	11342	624	12422	208	13444	104
10341	156	11344	312	12433	624	14004	312
10343	208	11402	208	12434	312	14011	52
10402	48	11411	13	12443	208	14012	624
10412	624	11414	312	13004	312	14022	624
10413	624	11441	52	13012	624	14033	624
10421	156	11443	624	13023	624	14034	104
10431	156	12004	312	13031	13	14043	624
10442	624	12013	624	13032	624	14101	39
10443	624	12014	104	13043	624	14112	208
11004	312	12021	26	13044	104	14123	208

TABLE C (*Cont.*)

Irreducible Polynomials for the Modulus 5

14134	104	101033	284	103014	781	110123	3124	112034	781	114014	781
14143	624	101103	3124	103022	3124	110131	1562	112104	781	114024	781
14144	312	101104	781	103023	3124	110142	3124	112113	3124	114033	3124
14202	624	101141	1562	103101	1562	110144	781	112133	3124	114044	781
14214	312	101142	3124	103104	781	110202	284	112142	3124	114102	3124
14224	104	101203	3124	103111	1562	110213	3124	112143	284	114132	3124
14231	156	101204	781	103112	3124	110232	3124	112201	1562	114141	1562
14232	208	101212	3124	103143	3124	110243	3124	112212	3124	114201	1562
14242	624	101213	284	103144	71	110244	781	112214	781	114204	71
14243	208	101301	1562	103211	1562	110301	1562	112234	781	114233	3124
14301	156	101302	3124	103212	3124	110303	3124	112241	1562	114242	3124
14303	624	101312	284	103221	1562	110322	3124	112243	3124	114314	781
14312	624	101313	3124	103223	3124	110331	1562	112301	1562	114321	1562
14314	312	101401	1562	103232	3124	110333	3124	112311	1562	114322	3124
14331	78	101402	3124	103233	3124	110343	3124	112313	3124	114331	1562
14402	208	101443	3124	103313	3124	110403	3124	112314	781	114343	3124
14411	52	101444	781	103314	781	110411	1562	112323	3124	114401	1562
14413	624	102001	1562	103322	3124	110421	1562	112334	71	114403	3124
14441	26	102004	781	103324	781	110432	3124	112342	3124	114424	781
14444	312	102012	3124	103332	3124	110441	1562	112422	3124	114431	22
		102013	3124	103333	3124	110442	3124	112433	3124	114434	781
$n = 5$	e	102021	1562	103401	1562	110444	781	112441	1562	114442	3124
		102024	781	103404	781	111003	284	113002	3124	120003	3124
100041	1562	102112	3124	103413	3124	111013	3124	113004	781	120013	3124
100042	3124	102114	781	103414	781	111021	1562	113034	781	120042	3124
100043	3124	102121	1562	103441	142	111022	3124	113044	781	120104	71
100044	781	102122	3124	103442	3124	111024	781	113103	3124	120111	1562
100102	3124	102131	1562	104021	1562	111032	3124	113111	1562	120134	781
100114	781	102134	781	104024	781	111044	781	113134	781	120141	1562
100124	71	102202	3124	104031	142	111102	3124	113142	284	120143	3124
100132	3124	102203	3124	104034	71	111114	781	113143	3124	120201	1562
100143	3124	102211	1562	104101	1562	111123	3124	113211	1562	120212	3124
100201	1562	102213	3124	104103	3124	111212	44	113222	3124	120222	3124
100212	3124	102242	284	104111	142	111224	781	113224	71	120234	781
100222	284	102244	781	104114	781	111231	1562	113231	1562	120242	3124
100231	1562	102302	3124	104202	3124	111234	781	113241	1562	120243	3124
100244	781	102303	3124	104204	781	111301	1562	113243	284	120244	781
100304	781	102312	3124	104241	1562	111311	1562	113304	781	120321	1562
100313	3124	102314	781	104243	3124	111312	3124	113312	3124	120332	3124
100323	284	102341	1562	104301	1562	111324	71	113321	1562	120343	3124
100334	781	102343	284	104303	3124	111334	781	113323	3124	120344	781
100341	1562	102411	1562	104342	3124	111401	142	113324	781	120401	1562
100403	3124	102413	3124	104344	781	111404	781	113332	3124	120402	3124
100411	1562	102423	3124	104402	3124	111423	3124	113342	3124	120424	781
100421	142	102424	781	104404	781	111431	1562	113412	3124	120431	1562
100433	3124	102431	1562	104411	1562	111433	3124	113422	3124	120432	3124
100442	3124	102434	781	104414	71	111442	3124	113434	781	120441	1562
101022	3124	103002	3124	110004	781	112012	3124	114001	1562	121002	3124
101023	3124	103003	3124	110014	781	112023	3124	114011	1562	121012	3124
101032	284	103011	1562	110041	1562	112032	3124	114012	3124	121013	3124

Irreducible Polynomials for the Modulus 5

121014	781	123034	781	130103	3124	132042	3124	134022	3124	141023	3124
121023	3124	123102	3124	130104	781	132102	3124	134023	3124	141024	781
121031	1562	123113	3124	130121	1562	132111	1562	134031	1562	141033	3124
121043	3124	123114	781	130133	3124	132122	3124	134042	3124	141041	1562
121102	3124	123133	3124	130134	781	132123	3124	134103	3124	141101	1562
121103	284	123141	1562	130144	781	132124	781	134111	1562	141104	71
121131	1562	123142	3124	130224	781	132131	1562	134113	3124	141122	3124
121144	781	123224	781	130233	3124	132141	1562	134122	284	141132	3124
121201	1562	123231	1562	130241	1562	132204	781	134132	3124	141134	781
121202	3124	123242	3124	130242	3124	132213	3124	134201	1562	141143	3124
121223	3124	123303	3124	130304	781	132232	3124	134212	3124	141204	781
121232	44	123311	1562	130313	3124	132241	142	134224	781	141213	3124
121233	3124	123331	1562	130323	3124	132244	781	134302	3124	141214	781
121244	781	123341	142	130331	1562	132311	1562	134303	284	141221	142
121304	781	123344	781	130341	1562	132321	1562	134324	781	141231	1562
121334	781	123402	3124	130342	3124	132332	3124	134333	3124	141313	44
121342	3124	123411	1562	130343	3124	132413	3124	134334	781	141321	1562
121413	3124	123412	3124	130401	142	132421	1562	134341	1562	141331	1562
121422	3124	123413	3124	130414	781	132422	284	134411	22	141334	781
121424	781	123421	1562	130431	1562	132433	3124	134422	3124	141403	3124
121432	3124	123433	284	130442	3124	132443	3124	134432	3124	141411	1562
121441	1562	123444	781	130444	781	132444	71	134433	3124	141422	3124
122003	3124	124001	142	131003	3124	133011	1562	140001	1562	142013	3124
122004	781	124011	1562	131011	1562	133024	781	140011	1562	142022	3124
122033	3124	124022	3124	131012	3124	133031	1562	140044	781	142031	1562
122043	3124	124023	3124	131013	3124	133032	3124	140102	3124	142033	3124
122112	3124	124024	781	131022	3124	133103	3124	140114	781	142123	3124
122123	284	124034	781	131034	781	133112	3124	140124	781	142132	3124
122124	781	124043	3124	131042	3124	133113	3124	140133	3124	142144	781
122132	3124	124114	11	131112	3124	133114	781	140141	1562	142204	781
122141	142	124123	3124	131121	1562	133124	781	140143	3124	142211	1562
122142	3124	124132	3124	131123	3124	133132	284	140144	781	142212	3124
122214	781	124133	3124	131133	3124	133141	1562	140202	3124	142214	781
122224	781	124202	284	131144	781	133202	3124	140204	781	142222	3124
122233	3124	124203	3124	131201	1562	133214	781	140223	3124	142231	142
122301	1562	124221	1562	131231	1562	133234	781	140232	3124	142243	3124
122312	3124	124231	1562	131243	3124	133241	1562	140234	781	142304	781
122333	3124	124232	3124	131303	3124	133244	71	140242	3124	142311	1562
122341	1562	124244	781	131304	781	133321	1562	140303	284	142313	3124
122344	71	124304	781	131322	3124	133334	781	140312	3124	142331	1562
122403	3124	124313	3124	131332	3124	133343	3124	140333	3124	142342	3124
122414	781	124321	1562	131333	44	133403	3124	140341	1562	142344	781
122421	1562	124402	3124	131341	1562	133411	1562	140342	3124	142401	1562
122422	3124	124412	3124	131402	284	133412	3124	140422	3124	142412	3124
122423	3124	124414	781	131403	3124	133432	3124	140434	781	142432	3124
122434	781	124423	284	131434	781	133443	3124	140441	1562	142442	284
122444	781	124433	3124	131441	1562	133444	781	140443	3124	142443	3124
123014	781	130002	3124	132001	1562	134004	71	141002	284	143001	1562
123021	1562	130012	3124	132002	3124	134014	781	141012	3124	143003	3124
123033	3124	130043	3124	132032	3124	134021	1562	141021	1562	143031	1562

TABLE C (*Cont.*)

Irreducible Polynomials for the Modulus 5

143041	1562	143224	781	143344	781	144013	3124	144131	1562	144301	142
143113	3124	143233	3124	143402	3124	144014	781	144134	11	144304	781
143123	3124	143243	3124	143414	781	144021	1562	144143	3124	144332	3124
143131	1562	143314	781	143431	1562	144032	3124	144211	1562	144343	3124
143201	1562	143321	142	143442	3124	144041	1562	144223	3124	144403	3124
143213	3124	143323	3124	143443	284	144102	3124	144224	781	144433	3124
143221	1562	143334	781	144004	781	144104	781	144234	781	144444	781
143222	3124	143342	284	144011	1562	144121	1562	144242	3124		

Irreducible Polynomials for the Modulus 7

$n=1$	e										
		1021	38	1304	342	1552	342	10135	2400	10524	1200
		1026	19	1306	57	1556	57	10145	2400	10525	2400
10	1	1032	342	1311	38	1563	171	10151	200	10531	80
11	2	1035	171	1314	342	1564	342	10161	400	10533	2400
12	6	1041	114	1322	342	1565	171	10162	600	10536	800
13	3	1046	57	1325	171	1566	57	10203	96	10541	80
14	6	1052	342	1333	171	1604	342	10205	96	10543	2400
15	3	1055	171	1334	342	1606	57	10211	200	10546	800
16	1	1062	342	1335	171	1612	342	10214	1200	10554	1200
		1065	171	1336	19	1615	171	10224	600	10555	2400
$n=2$	e	1101	114	1341	38	1621	114	10236	800	10565	2400
		1103	171	1343	171	1623	171	10246	800	10603	96
101	4	1112	342	1352	342	1632	342	10254	600	10606	32
102	12	1115	171	1354	342	1636	19	10261	200	10613	2400
104	12	1124	342	1362	342	1641	114	10264	1200	10621	400
113	48	1126	57	1366	57	1644	342	10305	96	10623	2400
114	24	1131	38	1401	114	1653	171	10306	32	10632	240
116	16	1135	171	1403	171	1654	342	10316	800	10635	2400
122	24	1143	171	1413	171	1655	171	10322	1200	10636	800
123	48	1146	57	1416	19	1656	57	10326	800	10642	240
125	48	1151	114	1422	342	1662	342	10333	2400	10645	2400
131	8	1152	342	1425	171	1664	342	10334	240	10646	800
135	48	1153	171	1431	38			10335	2400	10651	400
136	16	1154	342	1432	342	$n=4$	e	10343	2400	10653	2400
141	8	1163	171	1433	171			10344	240	10663	2400
145	48	1165	171	1434	342	10011	400	10345	2400	11001	400
146	16	1201	38	1444	342	10012	1200	10352	1200	11003	480
152	24	1203	171	1446	19	10014	1200	10356	800	11013	2400
153	48	1214	342	1453	171	10023	480	10366	800	11026	800
155	48	1216	57	1455	171	10025	480	10405	96	11031	400
163	48	1223	171	1461	114	10026	160	10406	32	11042	75
164	24	1226	57	1465	171	10053	480	10412	1200	11054	300
166	16	1233	171	1504	342	10055	480	10414	600	11056	800
		1235	171	1506	19	10056	160	10422	600	11062	1200
$n=3$	e	1242	342	1511	114	10061	400	10433	2400	11063	2400
		1245	171	1513	171	10062	1200	10443	2400	11101	400
1002	18	1251	114	1521	114	10064	1200	10452	600	11103	2400
1003	9	1255	171	1524	342	10103	96	10462	1200	11105	2400
1004	18	1261	114	1532	342	10106	32	10464	600	11111	5
1005	9	1262	342	1534	342	10111	400	10503	96	11112	1200
1011	114	1263	171	1542	342	10112	600	10505	96	11124	75
1016	57	1264	342	1545	171	10121	200	10515	2400	11134	240

Irreducible Polynomials for the Modulus 7

11136	800	11556	800	12266	800	12665	480	13432	1200	14125	2400
11141	400	11562	1200	12303	2400	13004	1200	13434	1200	14132	1200
11152	1200	11566	160	12304	240	13005	480	13436	800	14145	2400
11153	2400	11602	240	12311	200	13011	400	13441	20	14156	800
11161	100	11605	2400	12323	2400	13015	2400	13443	2400	14165	2400
11163	480	11614	600	12325	2400	13022	300	13445	2400	14204	1200
11166	800	11625	2400	12332	120	13023	2400	13455	2400	14205	2400
11201	200	11626	800	12345	480	13031	50	13456	800	14206	800
11204	600	11631	40	12346	800	13044	1200	13465	2400	14211	400
11213	2400	11643	2400	12351	100	13053	2400	13501	80	14214	15
11223	2400	11646	160	12354	600	13065	2400	13506	800	14222	75
11225	2400	11652	600	12356	800	13103	2400	13512	600	14232	240
11232	60	11653	2400	12361	200	13106	800	13513	2400	14233	2400
11233	2400	11654	300	12363	2400	13115	2400	13516	800	14244	1200
11236	800	11664	600	12365	2400	13126	800	13521	200	14251	400
11241	400	11665	2400	12402	1200	13135	2400	13522	300	14255	2400
11244	1200	11666	800	12403	2400	13142	1200	13525	2400	14263	2400
11245	2400	12002	1200	12406	800	13151	400	13533	480	14264	300
11252	240	12006	160	12412	15	13155	2400	13535	2400	14265	480
11254	300	12016	800	12414	1200	13161	400	13544	120	14302	1200
11266	160	12025	2400	12421	25	13166	160	13553	2400	14314	150
11321	200	12032	1200	12431	80	13204	1200	13556	800	14325	2400
11323	2400	12044	75	12435	2400	13205	2400	13562	600	14335	2400
11324	150	12051	100	12442	1200	13206	800	13611	40	14341	25
11331	400	12055	2400	12454	1200	13213	2400	13612	1200	14346	800
11332	300	12064	1200	12456	800	13214	300	13616	800	14353	2400
11334	600	12066	800	12462	300	13215	480	13623	480	14354	1200
11351	200	12101	200	12465	2400	13221	400	13624	600	14361	400
11355	2400	12102	600	12466	160	13225	2400	13626	800	14363	480
11356	160	12116	800	12521	50	13234	1200	13641	100	14402	600
11362	120	12123	2400	12522	600	13242	240	13642	600	14404	600
11364	1200	12126	800	12526	800	13243	2400	13644	1200	14415	2400
11365	2400	12134	60	12531	200	13252	150	13652	75	14425	2400
11405	2400	12135	2400	12532	1200	13261	400	13654	600	14426	800
11406	800	12136	800	12534	300	13264	30	13655	2400	14431	20
11412	1200	12141	400	12552	600	13302	1200	14004	1200	14433	2400
11415	480	12142	1200	12553	2400	13311	400	14005	480	14435	2400
11422	1200	12143	2400	12555	480	13313	480	14015	2400	14442	1200
11423	2400	12151	100	12561	400	13323	2400	14023	2400	14444	1200
11434	1200	12154	240	12563	2400	13324	1200	14034	1200	14446	800
11443	2400	12165	480	12564	120	13331	50	14041	25	14451	80
11455	2400	12203	2400	12601	400	13336	800	14052	300	14452	300
11463	2400	12205	2400	12612	150	13345	2400	14053	2400	14463	480
11504	1200	12213	480	12626	800	13355	2400	14061	400	14501	80
11511	50	12214	1200	12636	800	13364	75	14065	2400	14506	800
11523	2400	12224	1200	12643	2400	13402	600	14103	2400	14512	600
11533	2400	12226	800	12644	75	13404	600	14106	800	14523	2400
11542	75	12231	400	12652	1200	13413	480	14111	400	14526	800
11545	2400	12246	800	12655	2400	13421	80	14116	160	14534	120
11551	400	12253	2400	12664	1200	13422	300	14121	400	14543	480

TABLE C (*Cont.*)

Irreducible Polynomials for the Modulus 7

14545	2400	15121	100	15353	2400	15622	1200	16204	600	16453	2400
14551	200	15124	240	15355	2400	15625	2400	16216	160	16462	1200
14552	300	15131	400	15361	200	15633	2400	16222	240	16465	480
14555	2400	15132	1200	15402	1200	15634	150	16224	300	16504	1200
14562	600	15133	2400	15403	2400	15646	800	16231	400	16512	1200
14563	2400	15144	60	15406	800	15656	800	16234	1200	16516	160
14566	800	15145	2400	15412	300	15662	75	16235	2400	16521	400
14622	150	15146	800	15415	2400	16001	400	16242	60	16526	800
14624	600	15153	2400	15416	160	16003	480	16243	2400	16532	150
14625	2400	15156	800	15424	1200	16012	1200	16246	800	16535	2400
14631	100	15166	800	15426	800	16013	2400	16253	2400	16543	2400
14632	600	15203	2400	15432	1200	16024	300	16255	2400	16553	2400
14634	1200	15205	2400	15441	80	16026	800	16263	2400	16561	25
14653	480	15216	800	15445	2400	16032	150	16312	120	16602	240
14654	600	15223	2400	15451	50	16041	400	16314	1200	16605	2400
14656	800	15236	800	15462	30	16056	800	16315	2400	16614	600
14661	40	15241	400	15464	1200	16063	2400	16321	200	16615	2400
14662	1200	15254	1200	15511	400	16101	400	16325	2400	16616	800
14666	800	15256	800	15513	2400	16103	2400	16326	160	16622	600
15002	1200	15263	480	15514	120	16105	2400	16341	400	16623	2400
15006	160	15264	1200	15522	600	16111	100	16342	300	16624	300
15014	1200	15303	2400	15523	2400	16113	480	16344	600	16633	2400
15016	800	15304	240	15525	480	16116	800	16351	200	16636	160
15021	100	15311	200	15541	200	16122	1200	16353	2400	16641	40
15025	2400	15313	2400	15542	1200	16123	2400	16354	75	16655	2400
15034	150	15315	2400	15544	300	16131	400	16405	2400	16656	800
15042	1200	15321	100	15551	25	16144	240	16406	800	16664	600
15055	2400	15324	600	15552	600	16146	800	16413	2400		
15066	800	15326	800	15556	800	16154	150	16425	2400		
15101	200	15335	480	15601	400	16161	10	16433	2400		
15102	600	15336	800	15614	1200	16162	1200	16444	1200		
15115	480	15342	120	15615	480	16201	200	16452	1200		

TABLE D

1	0						51	6	3	1	0		
2	1	0					52	3	0				
3	1	0					53	6	2	1	0		
4	1	0					54	6	5	4	3	2	0
5	2	0					55	6	2	1	0		
6	1	0					56	7	4	2	0		
7	1	0					57	5	3	2	0		
8	4	3	2	0			58	6	5	1	0		
9	4	0					59	6	5	4	3	1	0
10	3	0					60	1	0				
11	2	0					61	5	2	1	0		
12	6	4	1	0			62	6	5	3	0		
13	4	3	1	0			63	1	0				
14	5	3	1	0			64	4	3	1	0		
15	1	0					65	4	3	1	0		
16	5	3	2	0			66	8	6	5	3	2	0
17	3	0					67	5	2	1	0		
18	5	2	1	0			68	7	5	1	0		
19	5	2	1	0			69	6	5	2	0		
20	3	0					70	5	3	1	0		
21	2	0					71	5	3	1	0		
22	1	0					72	6	4	3	2	1	0
23	5	0					73	4	3	2	0		
24	4	3	1	0			74	7	4	3	0		
25	3	0					75	6	3	1	0		
26	6	2	1	0			76	5	4	2	0		
27	5	2	1	0			77	6	5	2	0		
28	3	0					78	7	2	1	0		
29	2	0					79	4	3	2	0		
30	6	4	1	0			80	7	5	3	2	1	0
31	3	0					81	4	0				
32	7	5	3	2	1	0	82	8	7	6	4	1	0
33	6	4	1	0			83	7	4	2	0		
34	7	6	5	2	1	0	84	8	7	5	3	1	0
35	2	0					85	8	2	1	0		
36	6	5	4	2	1	0	86	6	5	2	0		
37	5	4	3	2	1	0	87	7	5	1	0		
38	6	5	1	0			88	8	5	4	3	1	0
39	4	0					89	6	5	3	0		
40	5	4	3	0			90	5	3	2	0		
41	3	0					91	7	6	5	3	2	0
42	5	4	3	2	1	0	92	6	5	2	0		
43	6	4	3	0			93	2	0				
44	6	5	2	0			94	6	5	1	0		
45	4	3	1	0			95	6	5	4	2	1	0
46	8	5	3	2	1	0	96	7	6	4	3	2	0
47	5	0					97	6	0				
48	7	5	4	2	1	0	98	7	4	3	2	1	0
49	6	5	4	0			99	7	5	4	0		
50	4	3	2	0			100	8	7	2	0		

TABLE E

$p=11 \quad n=2 \quad q=121 \quad 120=2^3\cdot 3\cdot 5 \quad \phi(120)/2=16$

a_1	a_2	a_1	a_2	a_1	a_2	a_1	a_2
4	2	2	6	1	7	1	8
5	2	3	6	4	7	3	8
6	2	8	6	7	7	8	8
7	2	9	6	10	7	10	8

$p=13 \quad n=2 \quad q=169 \quad 168=2^3\cdot 3\cdot 7 \quad \phi(168)/2=24$

a_1	a_2	a_1	a_2	a_1	a_2	a_1	a_2
1	2	2	6	2	7	4	11
4	2	3	6	3	7	5	11
6	2	4	6	6	7	6	11
7	2	9	6	7	7	7	11
9	2	10	6	10	7	8	11
12	2	11	6	11	7	9	11

$p=17 \quad n=2 \quad q=289 \quad 288=2^5\cdot 3^2 \quad \phi(288)/2=48$

a_1	a_2	a_1	a_2	a_1	a_2	a_1	a_2
1	3	2	6	1	10	2	12
6	3	6	6	3	10	3	12
7	3	8	6	4	10	5	12
10	3	9	6	13	10	12	12
11	3	11	6	14	10	14	12
16	3	15	6	16	10	15	12
3	5	1	7	2	11	4	14
5	5	4	7	7	11	6	14
8	5	5	7	8	11	7	14
9	5	12	7	9	11	10	14
12	5	13	7	10	11	11	14
14	5	16	7	15	11	13	14

$p=19 \quad n=2 \quad q=361 \quad 360=2^3\cdot 3^2\cdot 5 \quad \phi(360)/2=48$

a_1	a_2	a_1	a_2	a_1	a_2	a_1	a_2
1	2	10	3	3	13	11	14
4	2	11	3	4	13	12	14
7	2	12	3	6	13	13	14
8	2	18	3	9	13	18	14
11	2	2	10	10	13	4	15
12	2	4	10	13	13	5	15
15	2	6	10	15	13	6	15
18	2	9	10	16	13	9	15
1	3	10	10	1	14	10	15
7	3	13	10	6	14	13	15
8	3	15	10	7	14	14	15
9	3	17	10	8	14	15	15

$p=23 \quad n=2 \quad q=529 \quad 528=2^4\cdot 3\cdot 11 \quad \phi(528)/2=80$

a_1	a_2	a_1	a_2	a_1	a_2	a_1	a_2	a_1	a_2
2	5	2	10	1	14	3	17	4	20
4	5	3	10	3	14	4	17	7	20
5	5	6	10	5	14	6	17	8	20
8	5	10	10	10	14	11	17	10	20
15	5	13	10	13	14	12	17	13	20
18	5	17	10	18	14	17	17	15	20

TABLE E (*Cont.*)

$p = 23$ $n = 2$ $q = 529$ $528 = 2^4 \cdot 3 \cdot 11$ $\phi(528)/2 = 80$

a_1	a_2	a_1	a_2	a_1	a_2	a_1	a_2	a_1	a_2
19	5	20	10	20	14	19	17	16	20
21	5	21	10	22	14	20	17	19	20
1	7	3	11	5	15	1	19	5	21
2	7	7	11	9	15	2	19	6	21
4	7	8	11	10	15	7	19	7	21
9	7	9	11	11	15	11	19	9	21
14	7	14	11	12	15	12	19	14	21
19	7	15	11	13	15	16	19	16	21
21	7	16	11	14	15	21	19	17	21
22	7	20	11	18	15	22	19	18	21

$p = 29$ $n = 2$ $q = 841$ $840 = 2^3 \cdot 3 \cdot 5 \cdot 7$ $\phi(840)/2 = 96$

a_1	a_2	a_1	a_2	a_1	a_2	a_1	a_2	a_1	a_2	a_1	a_2
5	2	1	8	6	11	7	15	2	19	5	26
7	2	7	8	9	11	9	15	4	19	6	26
11	2	10	8	10	11	11	15	7	19	8	26
14	2	14	8	11	11	12	15	8	19	12	26
15	2	15	8	18	11	17	15	21	19	17	26
18	2	19	8	19	11	18	15	22	19	21	26
22	2	22	8	20	11	20	15	25	19	23	26
24	2	28	8	23	11	22	15	27	19	24	26
1	3	3	10	1	14	4	18	3	21	2	27
2	3	5	10	3	14	8	18	4	21	3	27
9	3	9	10	8	14	13	18	6	21	6	27
14	3	10	10	13	14	14	18	12	21	13	27
15	3	19	10	16	14	15	18	17	21	16	27
20	3	20	10	21	14	16	18	23	21	23	27
27	3	24	10	26	14	21	18	25	21	26	27
28	3	26	10	28	14	25	18	26	21	27	27

$p = 31$ $n = 2$ $q = 961$ $960 = 2^6 \cdot 3 \cdot 5$ $\phi(960)/2 = 128$

a_1	a_2	a_1	a_2	a_1	a_2	a_1	a_2	a_1	a_2	a_1	a_2	a_1	a_2	a_1	a_2
2	3	2	11	1	12	1	13	1	17	2	21	1	22	1	24
5	3	3	11	3	12	4	13	2	17	5	21	4	22	3	24
6	3	4	11	4	12	6	13	3	17	7	21	5	22	4	24
7	3	5	11	10	12	8	13	6	17	8	21	7	22	5	24
8	3	6	11	11	12	9	13	7	17	11	21	9	22	7	24
10	3	9	11	12	12	10	13	8	17	12	21	10	22	8	24
14	3	11	11	14	12	12	13	9	17	13	21	14	22	12	24
15	3	15	11	15	12	13	13	11	17	15	21	15	22	13	24
16	3	16	11	16	12	18	13	20	17	16	21	16	22	18	24
17	3	20	11	17	12	19	13	22	17	18	21	17	22	19	24
21	3	22	11	19	12	21	13	23	17	19	21	21	22	23	24
23	3	25	11	20	12	22	13	24	17	20	21	22	22	24	24
24	3	26	11	21	12	23	13	25	17	23	21	24	22	26	24
25	3	27	11	27	12	25	13	28	17	24	21	26	22	27	24
26	3	28	11	28	12	27	13	29	17	26	21	27	22	28	24
29	3	29	11	30	12	30	13	30	17	29	21	30	22	30	24

TABLE F

p^n	$a_1a_2a_3\cdots a_n$	p^n	$a_1a_2a_3\cdots a_n$	p^n	$a_1a_2\cdots a_{n-1}a_n$
2^2	11	5^2	12	19^2	1 2
2^3	101	5^3	102	19^3	10 16
2^4	1001	5^4	1013	19^4	100 2
2^5	01001	5^5	00102	19^5	0001 16
2^6	100001	5^6	100002	19^6	00001 3
2^7	0000011	5^7	1000002	19^7	010000 9
2^8	11000011	5^8	00101003		
2^9	000100001	5^9	011000003	23^2	1 7
2^{10}	0010000001	5^{10}	1010000003	23^3	10 16
2^{11}	01000000001	5^{11}	10000000002	23^4	001 11
2^{12}	110000010001	5^{12}	000010010003	23^5	1000 18
2^{13}	1100100000001			23^6	10000 7
2^{14}	11000000000101				
2^{15}	100000000000001	7^2	13	29^2	1 3
2^{16}	1010000000010001	7^3	112	29^3	01 18
2^{17}	00100000000000001	7^4	1103	29^4	100 2
2^{18}	000000100000000001	7^5	10004	29^5	0100 26
2^{19}	1100100000000000001	7^6	110003	29^6	00001 3
2^{20}	00100000000000000001	7^7	0100004		
2^{21}	010000000000000000001	7^8	10000003		
2^{22}	1000000000000000000001	7^9	100001002	31^2	1 12
2^{23}	00001000000000000000001	7^{10}	1100000003	31^3	01 28
2^{24}	110000100000000000000001			31^4	100 13
2^{25}	0010000000000000000000001			31^5	0100 20
2^{26}	11000100000000000000000001	11^2	17	31^6	10000 12
2^{27}	110010000000000000000000001	11^3	105		
2^{28}	0010000000000000000000000001	11^4	0012		
2^{29}	01000000000000000000000000001	11^5	01109	37^2	1 5
		11^6	100017	37^3	10 24
		11^7	1000005	37^4	001 2
		11^8	00010012	37^5	0001 32
3^2	12				
3^3	201				
3^4	1002	13^2	1 2	41^2	1 12
3^5	10101	13^3	10 7	41^3	01 35
3^6	100002	13^4	101 2	41^4	001 17
3^7	1010001	13^5	0101 11	41^5	1000 35
3^8	00100002	13^6	10100 6		
3^9	010100001	13^7	001000 6	43^2	1 3
3^{10}	1010000002	13^8	0110000 2	43^3	01 40
3^{11}	10000010001			43^4	001 20
3^{12}	100010000002			43^5	1000 40
3^{13}	1000001000001	17^2	1 3		
3^{14}	10000000000002	17^3	01 14		
3^{15}	100000000010001	17^4	100 5	47^2	1 13
3^{16}	0000001000000002	17^5	1000 14	47^3	10 42
3^{17}	10000000100000001	17^6	10000 3	47^4	100 5
3^{18}	100000000000100002	17^7	000100 14	47^5	0001 42

Bibliography

Note. In the case of authors publishing in the 18th and 19th century, secondary sources such as collected works have also been listed whenever available, since the original sources may not be easily accessible.

ABDULLAEV, I.:

[1] Elliptic curves and the representation of numbers by quaternary quadratic forms (Russian), *Dokl. Akad. Nauk UzSSR* **1973**, no. 1, 3–4.

[2] Elliptic curves, and the representation of numbers by certain quadratic forms in four variables (Russian), *Izv. Akad. Nauk UzSSR Ser. Fiz.-Mat. Nauk* **18**, no. 1, 59–60 (1974).

ABDULLAEV, I., and KOGAN, L. A.:

[1] Elliptic curves and the representation of numbers by positive quadratic forms (Russian), *Dokl. Akad. Nauk UzSSR* **1971**, no. 6, 3–4.

ABERTH, O.:

[1] The elementary symmetric functions in a finite field of prime order, *Illinois J. Math.* **8**, 132–138 (1964).

ABRAMSON, N. M.:

[1] Error-correcting codes from linear sequential circuits, *Proc. Fourth London Symp. on Information Theory* (C. Cherry, ed.), pp. 26–40, Butterworths, London, 1961.

[2] *Information Theory and Coding*, McGraw-Hill, New York, 1963.

ADLEMAN, L. M.:

[1] On distinguishing prime numbers from composite numbers, *Proc. 21st Annual Symp. on Foundations of Computer Science* (Syracuse, N.Y., 1980), pp. 387–406, IEEE Computer Society, Long Beach, Cal., 1980.

ADLEMAN, L. M., MANDERS, K., and MILLER, G.:

[1] On taking roots in finite fields, *Proc. 18th Annual Symp. on Foundations of Computer Science* (Providence, R.I., 1977), pp. 175–178, IEEE Computer Society, Long Beach, Cal., 1977.

ADLEMAN, L. M., POMERANCE, C., and RUMELY, R. S.:

[1] On distinguishing prime numbers from composite numbers, *Ann. of Math.* (to appear).

ADOLPHSON, A., and SPERBER, S.:

[1] Exponential sums on the complement of a hypersurface, *Amer. J. Math.* **102**, 461–487 (1980).

AGARWAL, R. C., and BURRUS, C. S.:

[1] Number theoretic transforms to implement fast digital convolution, *Proc. IEEE* **63**, 550–560 (1975).

AGOU, S.:

[1] Sur la décomposition de certains idéaux premiers, *Publ. Dép. Math. Lyon* **7**, no. 1, 41–46 (1969).

[2] Formules explicites intervenant dans la division euclidienne des polynômes à coefficients dans un anneau unitaire et applications diverses, *Publ. Dép. Math. Lyon* **8**, no. 1, 107–121 (1971).

[3] Polynômes sur un corps fini, *Bull. Sci. Math.* (2)**95**, 327–330 (1971).

[4] Sur l'irréductibilité des polynômes à coefficients dans un corps fini, *C. R. Acad. Sci. Paris Sér. A* **272**, 576–577 (1971).

[5] Sur des formules explicites intervenant dans la division euclidienne des polynômes et leurs conséquences, *C. R. Acad. Sci. Paris Sér. A* **273**, 209–211 (1971).

[6] Une démonstration de la loi de réciprocité quadratique, *Publ. Dép. Math. Lyon* **9**, no. 3, 55–57 (1972).

[7] Sur l'irréductibilité de certains polynômes à plusieurs indéterminées et à coefficients dans un corps fini, *Publ. Dép. Math. Lyon* **12**, no. 1, 5–12 (1975).

[8] Factorisation des polynômes à coefficients dans un corps fini, *Publ. Dép. Math. Lyon* **13**, no. 1, 63–71 (1976).

[9] Critères d'irréductibilité des polynômes composés à coefficients dans un corps fini, *Acta Arith.* **30**, 213–223 (1976).

[10] Factorisation sur un corps fini F_{p^n} des polynômes composés $f(X^s)$ lorsque $f(X)$ est un polynôme irréductible de $F_{p^n}[X]$, *L'Enseignement Math.* (2)**22**, 305–312 (1976).

[11] Factorisation sur un corps fini K des polynômes composés $f(X^s)$ lorsque $f(X)$ est un polynôme irréductible de $K[X]$, *C. R. Acad. Sci. Paris Sér. A* **282**, 1067–1068 (1976).

[12] Polynômes irréductibles primitifs à coefficients dans un corps fini, *Publ. Dép. Math. Lyon* **14**, no. 4, 17–20 (1977).

[13] Irréductibilité des polynômes $f(X^{p^r} - aX)$ sur un corps fini F_{p^s}, *J. reine angew. Math.* **292**, 191–195 (1977).

[14] Factorisation sur un corps fini F_{p^n} des polynômes composés $f(X^{p^r} - aX)$ lorsque $f(X)$ est un polynôme irréductible de $F_{p^n}[X]$, *J. Number Theory* **9**, 229–239 (1977).

[15] Irréductibilité des polynômes $f(X^{p^{2r}} - aX^{p^r} - bX)$ sur un corps fini F_{p^s}, *J. Number Theory* **10**, 64–69 (1978).

[16] Irréductibilité des polynômes $f(X^{p^{2r}} - aX^{p^r} - bX)$ sur un corps fini $\mathbb{F}_{p^s}$, *J. Number Theory* **11**, 20 (1979).

[17] Sur le degré minimum de certains polynômes hyponormaux sur un corps fini, *Proc. Queen's Number Theory Conf.* (Kingston, Ont., 1979), Queen's Papers in Pure and Appl. Math., no. 54, pp. 115–118, Queen's Univ., Kingston, Ont., 1980.

[18] Irréductibilité des polynômes $f(\sum_{i=0}^{m} a_i X^{p^{ri}})$ sur un corps fini F_{p^s}, *Canad. Math. Bull.* **23**, 207–212 (1980).

[19] Sur la factorisation des polynômes $f(X^{p^{2r}} - aX^{p^r} - bX)$ sur un corps fini F_{p^s}, *J. Number Theory* **12**, 447–459 (1980).

[20] Sur une classe de polynômes hyponormaux sur un corps fini, *Acta Arith.* **39**, 105–111 (1981).

AHMAD, S.:

[1] Cycle structure of automorphisms of finite cyclic groups, *J. Combinatorial Theory* **6**, 370–374 (1969).

[2] Split dilations of finite cyclic groups with applications to finite fields, *Duke Math. J.* **37**, 547–554 (1970).

AHO, A. V., HOPCROFT, J. E., and ULLMAN, J. D.:

[1] *The Design and Analysis of Computer Algorithms*, Addison-Wesley, Reading, Mass., 1975.

AĬZENBERG, N. N., SEMION, I. V., and CITKIN, A. I.:

[1] Polynomial representations of logical functions (Russian), *Avtomat. i Vyčisl. Tehn.* **1971**, no. 2, 6–13; *Automat. Control* **5**, no. 2, 5–11 (1971).

AJTAI, M.:

[1] Divisibility properties of recurring sequences, *Compositio Math.* **21**, 43–51 (1969).

AKHTAR, S.:

[1] Values of symmetric functions, *Panjab Univ. J. Math. (Lahore)* **2**, 43–61 (1969).

AKULINIČEV, N. M.:

[1] Bounds for rational trigonometric sums of a special type (Russian), *Dokl. Akad. Nauk SSSR* **161**, 743–745 (1965); *Soviet Math. Dokl.* **6**, 480–482 (1965).

ALADOV, N. S.:

[1] On the distribution of quadratic residues and quadratic nonresidues of a prime number p in the series $1, 2, \ldots, p-1$ (Russian), *Mat. Sb.* **18**, 61–75 (1896).

ALANEN, J. D., and KNUTH, D. E.:

[1] A table of minimum functions for generating Galois fields of $GF(p^n)$, *Sankhyā Ser. A* **23**, 128 (1961).

[2] Tables of finite fields, *Sankhyā Ser. A* **26**, 305–328 (1964).

ALBERT, A. A.:

[1] Symmetric and alternate matrices in an arbitrary field, I, *Trans. Amer. Math. Soc.* **43**, 386–436 (1938).

[2] On nonassociative division algebras, *Trans. Amer. Math. Soc.* **72**, 296–309 (1952).

[3] *Fundamental Concepts of Higher Algebra*, Univ. of Chicago Press, Chicago, 1956.

[4] On certain trinomial equations in finite fields, *Ann. of Math.* (2)**66**, 170–178 (1957).

[5] On certain polynomial systems, *Scripta Math.* **28**, 15–19 (1967).

ALBERT, A. A., and SANDLER, R.:

[1] *An Introduction to Finite Projective Planes*, Holt, Rinehart and Winston, New York, 1968.

ALMKVIST, G.:

[1] Invariants, mostly old ones, *Pacific J. Math.* **86**, 1–13 (1980).

ALTHAUS, H. L., and LEAKE, R. J.:

[1] Inverse of a finite-field Vandermonde matrix, *IEEE Trans. Information Theory* **IT-15**, 173 (1969).

ANANIASHVILI, G. G., VARSHAMOV, R. R., GOROVOĬ, V. P., and PARHOMENKO, P. P.:

[1] On the question of the decomposition of polynomials over the field $GF(2)$ (Russian), *Soobšč. Akad. Nauk Gruzin. SSR* **41**, 129–134 (1966).

ANDERSON, D. D.:

[1] Problem 6201, *Amer. Math. Monthly* **85**, 203 (1978); Solution, *ibid.* **86**, 869–870 (1979).

ANDERSON, D. R., and STIFFLER, J. J.:

[1] Lower bounds for the maximum moduli of certain classes of trigonometric sums, *Duke Math. J.* **30**, 171–176 (1963).

ANDREE, R. V.:

[1] *A Table of Indices and Power Residues for all Primes and Prime Powers below 2000*, W. W. Norton, New York, 1962.

ANDREWS, G. E.:

[1] A note on the Bombieri-Selberg formula for algebraic curves, *Portugal. Math.* **27**, 75–81 (1968).

ANDRIANOV, A. N.:

[1] The representation of numbers by certain quadratic forms in connection with the theory of elliptic curves (Russian), *Izv. Akad. Nauk SSSR Ser. Mat.* **29**, 227–238 (1965).

ANDRUHAEV, H. M.:

[1] A sum of Kloosterman type (Russian), *Certain Problems in the Theory of Fields* (Russian), pp. 60–66, Izdat. Saratov. Univ., Saratov, 1964.

[2] Generalized Kloosterman sums on the Gaussian field and their evaluation (Russian), *Naučn. Trudy Krasnod. Gos. Ped. Inst.* **118**, 29–40 (1969).

ANKENY, N. C.:

[1] The least quadratic non-residue, *Ann. of Math.* (2)**55**, 65–72 (1952).

[2] Equations in finite fields, *Proc. Nat. Acad. Sci. U.S.A.* **40**, 1072–1073 (1954).

[3] Criterion for rth power residuacity, *Pacific J. Math.* **10**, 1115–1124 (1960).

ANKENY, N. C., and ERDÖS, P.:

[1] The insolubility of classes of diophantine equations, *Amer. J. Math.* **76**, 488–496 (1954).

APOSTOL, T. M.:

[1] Dirichlet L-functions and character power sums, *J. Number Theory* **2**, 223–234 (1970).

[2] *Introduction to Analytic Number Theory*, Springer-Verlag, New York-Heidelberg-Berlin, 1976.

ARAKELOV, V. A., and TENENGOL'C, G. M.:

[1] Certain properties of recurrent periodic sequences (Russian), *Trudy Vyčisl. Centra Akad. Nauk Armjan. SSR i Erevan. Gos. Univ. Mat. Voprosy Kibernet. i Vyčisl. Tehn.* **6**, 18–28 (1970).

ARAKELOV, V. A., and VARSHAMOV, R. R.:

[1] On the investigation of the algebraic structure of periodic recurrent sequences (Russian), *Izv. Akad. Nauk Armjan. SSR Ser. Mat.* **6**, 379–385 (1971).

ARAZI, B.:

[1] Decimation of m-sequences leading to any desired phase shift, *Electron. Lett.* **13**, 213–215 (1977).

[2] On the synthesis of de Bruijn sequences, *Information and Control* **49**, 81–90 (1981).

ARBIB, M. A., FALB, P. L., and KALMAN, R. E.:

[1] *Topics in Mathematical System Theory*, McGraw-Hill, New York, 1968.

ARF, C.:

[1] Untersuchungen über quadratische Formen in Körpern der Charakteristik 2 (Teil I), *J. reine angew. Math.* **183**, 148–167 (1941).

ARGHIRIADE, E., and PETERFI, I.:

[1] On matrices with elements from a finite field (Romanian), *Lucrăr. Şti. Inst. Ped. Timişoara Mat.-Fiz.* **1960**, 19–23.

ARHIPOV, G. I., KARACUBA, A. A., and ČUBARIKOV, V. N.:

[1] Multiple trigonometric sums (Russian), *Trudy Mat. Inst. Steklov.* **151** (1980); *Proc. Steklov Inst. Math. 1982*, no. 2, American Math. Society, Providence, R.I., 1982.

ARMITAGE, J. V.:

[1] On the genus of curves over finite fields, *Mathematika* **9**, 115–117 (1962).

[2] The product of N linear forms in a field of series and the Riemann hypothesis for curves, *Bull. Soc. Math. France Mém.* **25**, pp. 17–27, Soc. Math. France, Paris, 1971.

ARNDT, F.:

[1] Einfacher Beweis für die Irreduzibilität einer Gleichung in der Kreisteilung, *J. reine angew. Math.* **56**, 178–181 (1858).

ARNOUX, G.:

[1] *Arithmétique graphique. Introduction à l'étude des fonctions arithmétiques*, Gauthier-Villars, Paris, 1906.

ARTIN, E.:

[1] Quadratische Körper im Gebiet der höheren Kongruenzen. I, II, *Math. Z.* **19**, 153–206, 207–246 (1924).

[2] Über einen Satz von Herrn J. H. Maclagan Wedderburn, *Abh. Math. Sem. Univ. Hamburg* **5**, 245–250 (1927).

[3] Linear mappings and the existence of a normal basis, *Studies and Essays Presented to R. Courant on His 60th Birthday*, pp. 1–5, Interscience, New York, 1948.

[4] The influence of J. H. M. Wedderburn on the development of modern algebra, *Bull. Amer. Math. Soc.* **56**, 65–72 (1950).

[5] The orders of the linear groups, *Comm. Pure Appl. Math.* **8**, 355–365 (1955).

[6] The orders of the classical simple groups, *Comm. Pure Appl. Math.* **8**, 455–472 (1955).

[7] *Geometric Algebra*, Interscience, New York, 1957.

[8] *Galois Theory*, 2nd ed., Univ. of Notre Dame Press, South Bend, Ind., 1966.

[9] *Algebraic Numbers and Algebraic Functions*, Gordon and Breach, New York, 1967.

ARWIN, A.:

[1] Über das Auflösen der Kongruenzen von dem dritten und vierten Grade nach einem Primzahlmodulus, *Lunds Univ. Årsskrift N. F.* (*Avd. 2*) **12**, no. 3 (1915), 29 pp.

[2] Über Kongruenzen von dem fünften und höheren Graden nach einem Primzahlmodulus, *Ark. Mat. Astron. Fys.* **14**, no. 7 (1919), 46 pp.

ASH, R.:

[1] *Information Theory*, Wiley-Interscience, New York, 1965.

ASSMUS, E. F., Jr., and MATTSON, H. F., Jr.:

[1] On tactical configurations and error-correcting codes, *J. Combinatorial Theory* **2**, 243–257 (1967).

[2] Coding and combinatorics, *SIAM Rev.* **16**, 349–388 (1974).

AUSLANDER, L., and TOLIMIERI, R.:

[1] Is computing with the finite Fourier transform pure or applied mathematics?, *Bull. Amer. Math. Soc.* (*N.S.*) **1**, 847–897 (1979).

AVANESOV, E. T.:

[1] On a problem of W. Mnich (Russian), *Mat.-Fyz. Časopis Sloven. Akad. Vied* **15**, 280–284 (1965).

AX, J.:

[1] Zeroes of polynomials over finite fields, *Amer. J. Math.* **86**, 255–261 (1964).

[2] Solving diophantine problems modulo every prime, *Ann. of Math.* (2)**85**, 161–183 (1967).

[3] The elementary theory of finite fields, *Ann. of Math.* (2)**88**, 239–271 (1968).

AX, J., and KOCHEN, S.:

[1] Diophantine problems over local fields. I, *Amer. J. Math.* **87**, 605–630 (1965).

AYOUB, R.:

[1] *An Introduction to the Analytic Theory of Numbers*, American Math. Society, Providence, R.I., 1963.

BABAEV, G., and ISMOILOV, D.:

[1] The number of solutions of a pair of congruences (Russian), *Dokl. Akad. Nauk Tadžik. SSR* **22**, 404–407 (1979).

BACHMANN, P.:

[1] *Die Lehre von der Kreistheilung*, Teubner, Leipzig, 1872.

[2] *Zahlentheorie*, Band 2: *Die analytische Zahlentheorie*, Teubner, Leipzig, 1894.

[3] *Die Arithmetik der quadratischen Formen*, Teubner, Leipzig, 1898.

[4] *Niedere Zahlentheorie*, *Erster Teil*, Teubner, Leipzig, 1902.

[5] *Niedere Zahlentheorie*, *Zweiter Teil*, Teubner, Leipzig, 1910.

[6] *Das Fermatproblem in seiner bisherigen Entwicklung*, de Gruyter, Berlin, 1919; Springer-Verlag, Berlin-Heidelberg-New York, 1976.

BAER, R.:

[1] *Linear Algebra and Projective Geometry*, Academic Press, New York, 1952.

BAJOGA, B. G.:

[1] Generation of irreducible polynomials from trinomials over $GF(2)$. II, *Information and Control* **37**, 5–18 (1978).

BAJOGA, B. G., and WALBESSER, W. J.:
[1] Generation of irreducible polynomials from trinomials over $GF(2)$. I, *Information and Control* **30**, 396–407 (1976).

BALAKRISHNAN, A. V.:
[1] *Communication Theory*, McGraw-Hill, New York, 1968.

BALDISSERRI, N.:
[1] Sul numero dei punti di cubiche ellittiche, a moltiplicazione complessa, ridotte modulo p, *Boll. Un. Mat. Ital.* (5)**16A**, 367–373 (1979).

BALL, J. R., SPITTLE, A. H., and LIU, H. T.:
[1] High-speed m-sequence generation: a further note, *Electron. Lett.* **11**, 107–108 (1975).

BALLIEU, R.:
[1] Factorisation des polynômes cyclotomiques modulo un nombre premier, *Ann. Soc. Sci. Bruxelles Sér. I* **68**, 140–144 (1954).

BAMBAH, R. P., and CHOWLA, S.:
[1] On the sign of the Gaussian sum, *Proc. Nat. Inst. Sci. India Part A* **13**, 175–176 (1947).

BARNER, K.:
[1] Zur Fibonacci-Folge modulo p, *Monatsh. Math.* **69**, 97–104 (1965).
[2] Zur Reziprozität quadratischer Charaktersummen in algebraischen Zahlkörpern, *Monatsh. Math.* **71**, 369–384 (1967).

BARNETT, S.:
[1] Greatest common divisor of several polynomials, *Proc. Cambridge Philos. Soc.* **70**, 263–268 (1971).

BARRUCAND, P.:
[1] Sommes de Gauss et séries singulières de Hardy pour les cubes, *C. R. Acad. Sci. Paris* **250**, 4249–4251 (1960).

BARTEE, T. C., and SCHNEIDER, D. I.:
[1] Computation with finite fields, *Information and Control* **6**, 79–98 (1963).

BASSALYGO, L. A.:
[1] Note on fast multiplication of polynomials over Galois fields (Russian), *Problemy Peredači Informacii* **14**, no. 1, 101–102 (1978); *Problems of Information Transmission* **14**, 71–72 (1978).

BATEMAN, P. T., CHOWLA, S., and ERDÖS, P.:
[1] Remarks on the size of $L(1, \chi)$, *Publ. Math. Debrecen* **1**, 165–182 (1950).

BATEMAN, P. T., and DUQUETTE, A. L.:
[1] The analogue of the Pisot-Vijayaraghavan numbers in fields of formal power series, *Illinois J. Math.* **6**, 594–606 (1962).

BAUM, L. E., HERZBERG, N. P., LOMONACO, S. J., Jr., and SWEET, M. M.:
[1] Fields of almost periodic sequences, *J. Combinatorial Theory Ser. A* **22**, 169–180 (1977).

BAUM, L. E., and NEUWIRTH, L. P.:

[1] Decomposition of vector spaces over $GF(2)$ into disjoint equidimensional affine spaces, *J. Combinatorial Theory Ser. A* **18**, 88–100 (1975).

BAUM, L. E., and SWEET, M. M.:

[1] Continued fractions of algebraic power series in characteristic 2, *Ann. of Math.* (2)**103**, 593–610 (1976).

[2] Badly approximable power series in characteristic 2, *Ann. of Math.* (2)**105**, 573–580 (1977).

BAUMERT, L. D.:

[1] *Cyclic Difference Sets*, Lecture Notes in Math., vol. 182, Springer-Verlag, Berlin-Heidelberg-New York, 1971.

BAUMERT, L. D., and FREDRICKSEN, H.:

[1] The cyclotomic numbers of order eighteen with applications to difference sets, *Math. Comp.* **21**, 204–219 (1967).

BAUMERT, L. D., and HALL, M., Jr.:

[1] A new construction for Hadamard matrices, *Bull. Amer. Math. Soc.* **71**, 169–170 (1965).

BAUMERT, L. D., and MCELIECE, R. J.:

[1] Weights of irreducible cyclic codes, *Information and Control* **20**, 158–175 (1972).

BAUMERT, L. D., MILLS, W. H., and WARD, R. L.:

[1] Uniform cyclotomy, *J. Number Theory* **14**, 67–82 (1982).

BAYER, P., and NEUKIRCH, J.:

[1] On values of zeta functions and l-adic Euler characteristics, *Invent. Math.* **50**, 35–64 (1978).

BEARD, J. T. B., Jr.:

[1] Matrix fields over prime fields, *Duke Math. J.* **39**, 313–321 (1972).

[2] Matrix fields over finite extensions of prime fields, *Duke Math. J.* **39**, 475–484 (1972).

[3] The number of matrix fields over $GF(q)$, *Acta Arith.* **25**, 315–329 (1974).

[4] A rational canonical form for matrix fields, *Acta Arith.* **25**, 331–335 (1974).

[5] Computing in $GF(q)$, *Math. Comp.* **28**, 1159–1166 (1974).

[6] Unitary perfect polynomials over $GF(q)$, *Atti Accad. Naz. Lincei Rend. Cl. Sci. Fis. Mat. Natur.* (8)**62**, 417–422 (1977).

BEARD, J. T. B., Jr., BULLOCK, A. T., and HARBIN, M. S.:

[1] Infinitely many perfect and unitary perfect polynomials, *Atti Accad. Naz. Lincei Rend. Cl. Sci. Fis. Mat. Natur.* (8)**63**, 294–303 (1977).

BEARD, J. T. B., Jr., DOYLE, J. K., and MANDELBERG, K. I.:

[1] Square-separable primes and unitary perfect polynomials, *Atti Accad. Naz. Lincei Rend. Cl. Sci. Fis. Mat. Natur.* (8)**68**, 397–401 (1980).

BEARD, J. T. B., Jr., and HARBIN, M. S.:
[1] Nonsplitting unitary perfect polynomials over $GF(q)$, *Atti Accad. Naz. Lincei Rend. Cl. Sci. Fis. Mat. Natur.* (8)**66**, 179–185 (1979).

BEARD, J. T. B., Jr., and MCCONNEL, R.:
[1] Matrix fields over the integers modulo m, *Linear Algebra Appl.* **14**, 95–105 (1976).

BEARD, J. T. B., Jr., O'CONNELL, J. R., Jr., and WEST, K. I.:
[1] Perfect polynomials over $GF(q)$, *Atti Accad. Naz. Lincei Rend. Cl. Sci. Fis. Mat. Natur.* (8)**62**, 283–291 (1977).

BEARD, J. T. B., Jr., and WEST, K. I.:
[1] Some primitive polynomials of the third kind, *Math. Comp.* **28**, 1166–1167 (1974).
[2] Factorization tables for $x^n - 1$ over $GF(q)$, *Math. Comp.* **28**, 1167–1168 (1974).
[3] Factorization tables for trinomials over $GF(q)$, *Math. Comp.* **30**, 179–183 (1976).

BEDOCCHI, E.:
[1] Cubiche ellittiche su F_p, *Boll. Un. Mat. Ital.* (5)**17B**, 269–277 (1980).
[2] Classi di isomorfismo delle cubiche di F_q, *Rend. Circ. Mat. Palermo* (2) **30**, 397–415 (1981).

BEEGER, N. G. W. H.:
[1] Sur l'identité de M. G. Rados, *Rend. Circ. Mat. Palermo* **51**, 312–314 (1927).

BEKER, H., and PIPER, F. C.:
[1] Shift register sequences, *Combinatorics* (Swansea, 1981), London Math. Soc. Lecture Note Series, no. 52, pp. 56–79, Cambridge Univ. Press, Cambridge, 1981.
[2] *Cipher Systems. The Protection of Communications*, Northwood Books, London, 1982.

BELEVITCH, V.:
[1] Conference networks and Hadamard matrices, *Ann. Soc. Sci. Bruxelles Sér. I* **82**, 13–32 (1968).

BELL, E. T.:
[1] Notes on recurring series of the third order, *Tôhoku Math. J.* **24**, 168–184 (1925).

BELLMAN, R.:
[1] A note on the solution of polynomial congruences, *Boll. Un. Mat. Ital.* (3)**19**, 60–63 (1964).

BENJAUTHRIT, B., and REED, I. S.:
[1] Galois switching functions and their applications, *IEEE Trans. Computers* **C-25**, 78–86 (1976).
[2] On the fundamental structure of Galois switching functions, *IEEE Trans. Computers* **C-27**, 757–762 (1978).

BERGER, T. R., and REINER, I.:

[1] A proof of the normal basis theorem, *Amer. Math. Monthly* **82**, 915–918 (1975).

BERGSTRÖM, H.:

[1] Die Klassenzahlformel für reelle quadratische Zahlkörper mit zusammengesetzter Diskriminante als Produkt verallgemeinerter Gaußscher Summen, *J. reine angew. Math.* **186**, 91–115 (1944/45).

BERGUM, G. E., and JORDAN, J. H.:

[1] The distribution of quadratic residues in fields of order p^2, *Math. Mag.* **45**, 194–200 (1972).

BERLEKAMP, E. R.:

[1] On decoding binary Bose-Chaudhuri-Hocquenghem codes, *IEEE Trans. Information Theory* **IT-11**, 577–579 (1965).

[2] Distribution of cyclic matrices in a finite field, *Duke Math. J.* **33**, 45–48 (1966).

[3] Factoring polynomials over finite fields, *Bell System Tech. J.* **46**, 1853–1859 (1967).

[4] *Algebraic Coding Theory*, McGraw-Hill, New York, 1968.

[5] Weight enumeration theorems, *Proc. Sixth Allerton Conf. on Circuit and Systems Theory*, pp. 161–170, Univ. of Illinois Press, Urbana, Ill., 1968.

[6] Factoring polynomials over large finite fields, *Math. Comp.* **24**, 713–735 (1970).

[7] Factoring polynomials, *Proc. Third Southeastern Conf. on Combinatorics, Graph Theory, and Computing* (Boca Raton, Fla., 1972), pp. 1-7, Utilitas Math., Winnipeg, Man., 1972.

[8] A survey of coding theory, *J. Royal Statist. Soc. Ser. A* **135**, 44–73 (1972).

[9] *Key Papers in the Development of Coding Theory*, IEEE Press, New York, 1974.

[10] An analog to the discriminant over fields of characteristic two, *J. Algebra* **38**, 315–317 (1976).

BERLEKAMP, E. R., FREDRICKSEN, H., and PROTO, R. C.:

[1] Minimum conditions for uniquely determining the generator of a linear sequence, *Utilitas Math.* **5**, 305–315 (1974).

BERLEKAMP, E. R., RUMSEY, H., and SOLOMON, G.:

[1] On the solution of algebraic equations over finite fields, *Information and Control* **10**, 553–564 (1967).

BERMAN, G., and FRYER, K. D.:

[1] *Introduction to Combinatorics*, Academic Press, New York, 1972.

BERNDT, B. C.:

[1] On Gaussian sums and other exponential sums with periodic coefficients, *Duke Math. J.* **40**, 145–156 (1973).

BERNDT, B. C., and CHOWLA, S.:

[1] The reckoning of certain quartic and octic Gauss sums, *Glasgow Math. J.* **18**, 153–155 (1977).

BERNDT, B. C., and EVANS, R. J.:

[1] Sums of Gauss, Jacobi, and Jacobsthal, *J. Number Theory* **11**, 349–398 (1979).

[2] Sums of Gauss, Eisenstein, Jacobi, Jacobsthal, and Brewer, *Illinois J. Math.* **23**, 374–437 (1979).

[3] Half Gauss sums, *Math. Ann.* **249**, 115–125 (1980).

[4] The determination of Gauss sums, *Bull. Amer. Math. Soc.* (*N.S.*) **5**, 107–129 (1981); Corrigendum, *ibid.* **7**, 441 (1982).

BERNSTEIN, B. A.:

[1] A general theory of representation of finite operations and relations, *Bull. Amer. Math. Soc.* **32**, 533–536 (1926).

[2] Modular representations of finite algebras, *Proc. International Math. Congress* (Toronto, 1924), vol. 1, pp. 207–216, Univ. of Toronto Press, Toronto, 1928.

BERNSTEIN, B. A., and DEBELY, N.:

[1] A practical method for the modular representation of finite operations and relations, *Bull. Amer. Math. Soc.* **38**, 110–114 (1932).

BERTRAM, E. A.:

[1] Polynomials which commute with a Tchebycheff polynomial, *Amer. Math. Monthly* **78**, 650–653 (1971).

BETTI, E.:

[1] Sopra la risolubilità per radicali delle equazioni algebriche irriduttibili di grado primo, *Ann. Sci. Mat. Fis.* **2**, 5–19 (1851).

[2] Sulla risoluzione delle equazioni algebriche, *Ann. Sci. Mat. Fis.* **3**, 49–115 (1852).

[3] Sopra la teorica delle sostituzioni, *Ann. Sci. Mat. Fis.* **6**, 5–34 (1855).

BEYER, G.:

[1] Über eine Klasseneinteilung aller kubischen Restcharaktere, *Abh. Math. Sem. Univ. Hamburg* **19**, 115–116 (1954).

BHANU MURTHY, B. S., and SAMPATH, S.:

[1] An application of linear feedback shift registers in the computation of polynomial arithmetic, *Internat. J. Electron.* **45**, 177–185 (1978).

BHASKARAN, M.:

[1] Sums of mth powers in algebraic and abelian number fields, *Arch. Math.* **17**, 497–504 (1966); Correction, *ibid.* **22**, 370–371 (1971).

BIERSTEDT, R. G., and MILLS, W. H.:

[1] On the bound for a pair of consecutive quartic residues of a prime, *Proc. Amer. Math. Soc.* **14**, 628–632 (1963).

BILHARZ, H.:

[1] Primdivisor mit vorgegebener Primitivwurzel, *Math. Ann.* **114**, 476–492 (1937).

BINI, D., and CAPOVANI, M.:

[1] Lower bounds of the complexity of linear algebras, *Inform. Process. Lett.* **9**, no. 1, 46–47 (1979).

BIRCH, B. J.:

[1] Waring's problem for $\mathfrak{p}$-adic number fields, *Acta Arith.* **9**, 169–176 (1964).

[2] How the number of points of an elliptic curve over a fixed prime field varies, *J. London Math. Soc.* **43**, 57–60 (1968).

BIRCH, B. J., and LEWIS, D. J.:

[1] $\mathfrak{p}$-adic forms, *J. Indian Math. Soc.* **23**, 11–32 (1959).

[2] Systems of three quadratic forms, *Acta Arith.* **10**, 423–442 (1965).

BIRCH, B. J., LEWIS, D. J., and MURPHY, T. G.:

[1] Simultaneous quadratic forms, *Amer. J. Math.* **84**, 110–115 (1962).

BIRCH, B. J., and SWINNERTON-DYER, H. P. F.:

[1] Note on a problem of Chowla, *Acta Arith.* **5**, 417–423 (1959).

[2] Notes on elliptic curves. II, *J. reine angew. Math.* **218**, 79–108 (1965).

BIRDSALL, T. G., and RISTENBLATT, M. P.:

[1] Introduction to linear shift-register generated sequences, EDG Tech. Report No. 90, Univ. of Michigan Research Institute, Ann Arbor, Mich., 1958.

BIRKHOFF, G., and BARTEE, T. C.:

[1] *Modern Applied Algebra*, McGraw-Hill, New York, 1970.

BIRKHOFF, G., and MACLANE, S.:

[1] *A Survey of Modern Algebra*, 4th ed., Macmillan, New York, 1977.

BIRKHOFF, G. D., and VANDIVER, H. S.:

[1] On the integral divisors of $a^n - b^n$, *Ann. of Math.* (2)**5**, 173–180 (1904).

BLAHUT, R. E.:

[1] Algebraic codes in the frequency domain, *Algebraic Coding Theory and Applications* (G. Longo, ed.), CISM Courses and Lectures, vol. 258, pp. 447–494, Springer-Verlag, Vienna, 1979.

BLAKE, I. F.:

[1] *Algebraic Coding Theory: History and Development*, Dowden-Hutchinson-Ross, Stroudsburg, Penn., 1973.

[2] Codes and designs, *Math. Mag.* **52**, 81–95 (1979).

BLAKE, I. F., and MULLIN, R. C.:

[1] *The Mathematical Theory of Coding*, Academic Press, New York, 1975.

BLANCHARD, A.:

[1] *Les corps non commutatifs*, Presses Univ. de France, Paris, 1972.

BLANKINSHIP, W. A.:

[1] A new version of the Euclidean algorithm, *Amer. Math. Monthly* **70**, 742–745 (1963).

BLOCK, H. D., and THIELMAN, H.P.:

[1] Commutative polynomials, *Quart. J. Math.* (2)**2**, 241–243 (1951).

BLOKH, E. L.:

[1] A method of decoding Bose-Chaudhuri triple-error-correcting codes (Russian), *Izv. Akad. Nauk SSSR Tekh. Kibern.* **1964**, no. 3, 30–37; *Engineering Cybernetics* **1964**, no. 3, 23–32.

BLOOM, D. M.:

[1] On periodicity in generalized Fibonacci sequences, *Amer. Math. Monthly* **72**, 856–861 (1965).

BLUMENTHAL, L. M.:

[1] *A Modern View of Geometry*, W. H. Freeman, San Francisco, 1961.

BOCHNER, S.:

[1] Remarks on Gaussian sums and Tauberian theorems, *J. Indian Math. Soc.* **15**, 97–104 (1951).

BOLLMAN, D.:

[1] Some periodicity properties of transformations on vector spaces over residue class rings, *J. Soc. Indust. Appl. Math.* **13**, 902–912 (1965).

[2] Some periodicity properties of modules over the ring of polynomials with coefficients in a residue class ring, *SIAM J. Appl. Math.* **14**, 237–241 (1966).

BOLLMAN, D., and RAMÍREZ, H.:

[1] On the number of nilpotent matrices over Z_m, *J. reine angew. Math.* **238**, 85–88 (1969).

BOMBIERI, E.:

[1] Sull'analogo della formula di Selberg nei corpi di funzioni, *Atti Accad. Naz. Lincei Rend. Cl. Sci. Fis. Mat. Natur.* (8)**35**, 252–257 (1963).

[2] Nuovi risultati sulla geometria di una ipersuperficie cubica a tre dimensioni, *Rend. Mat. e Appl.* (5)**25**, 22–28 (1966).

[3] On exponential sums in finite fields, *Les Tendances Géométriques en Algèbre et Théorie des Nombres*, pp. 37–41, Edition du Centre National de la Recherche Scientifique, Paris, 1966.

[4] On exponential sums in finite fields, *Amer. J. Math.* **88**, 71–105 (1966).

[5] Counting points on curves over finite fields (d'après S. A. Stepanov), *Séminaire Bourbaki 1972/73*, Exp. 430, Lecture Notes in Math., vol. 383, pp. 234–241, Springer-Verlag, Berlin-Heidelberg-New York, 1974.

[6] Hilbert's 8th problem: an analogue, *Proc. Symp. Pure Math.*, vol. 28, pp. 269–274, American Math. Society, Providence, R.I., 1976.

[7] On exponential sums in finite fields, II, *Invent. Math.* **47**, 29–39 (1978).

BOMBIERI, E., and DAVENPORT, H.:

[1] On two problems of Mordell, *Amer. J. Math.* **88**, 61–70 (1966).

BOMBIERI, E.., and SWINNERTON-DYER, H. P. F.:

[1] On the local zeta function of a cubic threefold, *Ann. Scuola Norm. Sup. Pisa* (3) **21**, 1–29 (1967).

BOOTH, T. L.:
[1] *Sequential Machines and Automata Theory*, Wiley, New York, 1967.

BOREL, E., and DRACH, J.:
[1] *Introduction à l'étude de la théorie des nombres et de l'algèbre supérieure*, Nony, Paris, 1895.

BOREVICH, Z. I., and SHAFAREVICH, I. R.:
[1] *Number Theory*, Academic Press, New York, 1966.

BORHO, W.:
[1] Kettenbrüche im Galoisfeld, *Abh. Math. Sem. Univ. Hamburg* **39**, 76–82 (1973).

BORODIN, A., and MUNRO, I.:
[1] *The Computational Complexity of Algebraic and Numeric Problems*, American Elsevier, New York, 1975.

BOROŞ, E.:
[1] On matrices with elements of a finite field (Romanian), *Lucrăr. Şti. Inst. Ped. Timişoara Mat.-Fiz.* **1961**, 41–47.

BOROSH, I., MORENO, C. J., and PORTA, H.:
[1] Elliptic curves over finite fields. I, *Proc. Number Theory Conference* (Boulder, Colo., 1972), pp. 147–155, Univ. of Colorado, Boulder, Colo., 1972.
[2] Elliptic curves over finite fields. II, *Math. Comp.* **29**, 951–964 (1975).

BOSE, N. K.:
[1] A criterion to determine if two multivariable polynomials are relatively prime, *Proc. IEEE* **60**, 134–135 (1972).

BOSE, R. C.:
[1] On the application of the properties of Galois fields to the problem of construction of hyper-Graeco-Latin squares, *Sankhyā* **3**, 323–338 (1938).
[2] On the construction of balanced incomplete block designs, *Ann. of Eugenics* **9**, 353–399 (1939).

BOSE, R. C., CHOWLA, S., and RAO, C. R.:
[1] On the integral order (mod p) of quadratics $x^2 + ax + b$, with applications to the construction of minimum functions for $GF(p^2)$, and to some number theory results, *Bull. Calcutta Math. Soc.* **36**, 153–174 (1944).
[2] Minimum functions in Galois fields, *Proc. Nat. Acad. Sci. India Sect. A* **15**, 191–192 (1945).
[3] On the roots of a well-known congruence, *Proc. Nat. Acad. Sci. India Sect. A* **15**, 193 (1945).

BOSE, R. C., and RAY-CHAUDHURI, D. K.:
[1] On a class of error correcting binary group codes, *Information and Control* **3**, 68–79 (1960).
[2] Further results on error correcting binary group codes, *Information and Control* **3**, 279–290 (1960).

BOSE, R. C., and SHRIKHANDE, S. S.:
[1] A note on a result in the theory of code construction, *Information and Control* **2**, 183–194 (1959).
[2] On the falsity of Euler's conjecture about the non-existence of two orthogonal Latin squares of order $4t+2$, *Proc. Nat. Acad. Sci. U.S.A.* **45**, 734–737 (1959).
[3] On the construction of sets of mutually orthogonal Latin squares and the falsity of a conjecture of Euler, *Trans. Amer. Math. Soc.* **95**, 191–209 (1960).

BOSE, R. C., SHRIKHANDE, S. S., and PARKER, E. T.:
[1] Further results on the construction of mutually orthogonal Latin squares and the falsity of Euler's conjecture, *Canad. J. Math.* **12**, 189–203 (1960).

BOTTEMA, O.:
[1] On the Betti-Mathieu group (Dutch), *Nieuw Arch. Wisk.* (2)**16**, no. 4, 46–50 (1930).

BOURBAKI, N.:
[1] *Algèbre*, Ch. V, Actualités Sci. Ind., no. 1102, Hermann, Paris, 1950.
[2] *Algèbre*, Ch. VIII, Actualités Sci. Ind., no. 1261, Hermann, Paris, 1958.

BOVEY, J. D.:
[1] On the congruence $a_1x_1^k + \cdots + a_sx_s^k \equiv N \pmod{p^n}$, *Acta Arith.* **23**, 257–269 (1973).
[2] $\Gamma^*(8)$, *Acta Arith.* **25**, 145–150 (1974).
[3] A note on Waring's problem in p-adic fields, *Acta Arith.* **29**, 343–351 (1976).
[4] A new upper bound for Waring's problem $(\bmod\, p)$, *Acta Arith.* **32**, 157–162 (1977).

BOYARSKY, M.:
[1] p-adic gamma functions and Dwork cohomology, *Trans. Amer. Math. Soc.* **257**, 359–369 (1980).

BOYCE, W. M.:
[1] On polynomials which commute with a given polynomial, *Proc. Amer. Math. Soc.* **33**, 229–234 (1972).

BRAHANA, H. R.:
[1] On cubic congruences, *Bull. Amer. Math. Soc.* **39**, 962–969 (1933).
[2] Note on irreducible quartic congruences, *Trans. Amer. Math. Soc.* **38**, 395–400 (1935).

BRANDIS, A.:
[1] Ein gruppentheoretischer Beweis für die Kommutativität endlicher Divisionsringe, *Abh. Math. Sem. Univ. Hamburg* **26**, 234–236 (1963).

BRAUER, A.:
[1] Über Sequenzen von Potenzresten, *Sitzungsber. Preuß. Akad. Wiss. Phys.-Math. Kl.* **1928**, 9–16.

[2] Über den kleinsten quadratischen Nichtrest, *Math. Z.* **33**, 161–176 (1931).
[3] Über Sequenzen von Potenzresten. II, *Sitzungsber. Preuß. Akad. Wiss. Phys.-Math. Kl.* **1931**, 329–341.
[4] Über die Verteilung der Potenzreste, *Math. Z.* **35**, 39–50 (1932).
[5] Combinatorial methods in the distribution of kth power residues, *Proc. Conf. Combinatorial Math. and Its Appl.* (Chapel Hill, N.C., 1967), pp. 14–37, Univ. of North Carolina Press, Chapel Hill, N.C., 1969.

BRAUER, R.:
[1] A note on systems of homogeneous algebraic equations, *Bull. Amer. Math. Soc.* **51**, 749–755 (1945).

BRAUN, H.:
[1] Geschlechter quadratischer Formen, *J. reine angew. Math.* **182**, 32–49 (1940).

BRAWLEY, J. V.:
[1] Enumeration of canonical sets by rank, *Amer. Math. Monthly* **74**, 175–177 (1967).
[2] Certain sets of involutory matrices and their groups, *Duke Math. J.* **36**, 473–478 (1969).
[3] Polynomials over a ring that permute the matrices over that ring, *J. Algebra* **38**, 93–99 (1976).
[4] The number of polynomial functions which permute the matrices over a finite field, *J. Combinatorial Theory Ser. A* **21**, 147–154 (1976).
[5] A note on polynomial matrix functions over a finite field, *Linear Algebra Appl.* **28**, 35–38 (1979).

BRAWLEY, J. V., and CARLITZ, L.:
[1] Enumeration of matrices with prescribed row and column sums, *Linear Algebra Appl.* **6**, 165–174 (1973).
[2] A characterization of the $n \times n$ matrices over a finite field, *Amer. Math. Monthly* **80**, 670–672 (1973); Addendum, *ibid.* **80**, 1041–1043 (1973).

BRAWLEY, J. V., CARLITZ, L., and LEVINE, J.:
[1] Power sums of matrices over a finite field, *Duke Math. J.* **41**, 9–24 (1974).
[2] Scalar polynomial functions on the $n \times n$ matrices over a finite field, *Linear Algebra Appl.* **10**, 199–217 (1975).

BRAWLEY, J. V., CARLITZ, L., and VAUGHAN, T. P.:
[1] Linear permutation polynomials with coefficients in a subfield, *Acta Arith.* **24**, 193–199 (1973).

BRAWLEY, J. V., and GAMBLE, R. O.:
[1] Involutory matrices over finite commutative rings, *Linear Algebra Appl.* **21**, 175–188 (1978).

BRAWLEY, J. V., and HANKINS, M.:
[1] On the distribution by rank of bases for vector spaces of matrices over a finite field, *Linear Algebra Appl.* **39**, 91–101 (1981).

BRAWLEY, J. V., and LEVINE, J.:

[1] Equivalence classes of linear mappings with applications to algebraic cryptography. I, II, *Duke Math. J.* **39**, 121–132, 133–142 (1972).

[2] Equivalence classes of involutory mappings, *Duke Math. J.* **39**, 211–217 (1972).

BRAWLEY, J. V., and MULLEN, G. L.:

[1] A note of equivalence classes of matrices over a finite field, *Internat. J. Math. and Math. Sci.* **4**, 279–287 (1981).

BREMNER, A., and MORTON, P.:

[1] Polynomial relations in characteristic p, *Quart. J. Math.* (2)**29**, 335–347 (1978).

BRENNER, J. L.:

[1] Linear recurrence relations, *Amer. Math. Monthly* **61**, 171–173 (1954).

BRENNER, J. L., and CARLITZ, L.:

[1] Covering theorems for finite nonabelian simple groups. III. Solutions of the equation $\alpha x^2 + \beta t^2 + \gamma t^{-2} = a$ in a finite field, *Rend. Sem. Mat. Univ. Padova* **55**, 81–90 (1976).

BRESSOUD, D. M.:

[1] On the value of Gaussian sums, *J. Number Theory* **13**, 88–94 (1981).

BREWER, B. W.:

[1] On the quadratic reciprocity law, *Amer. Math. Monthly* **58**, 177–179 (1951).

[2] On certain character sums, *Trans. Amer. Math. Soc.* **99**, 241–245 (1961).

[3] On primes of the form $u^2 + 5v^2$, *Proc. Amer. Math. Soc.* **17**, 502–509 (1966).

BRIDGES, W. G., and RYSER, H. J.:

[1] Combinatorial designs and related systems, *J. Algebra* **13**, 432–446 (1969).

BRILLHART, J.:

[1] Some modular results on the Euler and Bernoulli polynomials, *Acta Arith.* **21**, 173–181 (1972).

BRILLHART, J., LEHMER, D. H., and LEHMER, E.:

[1] Bounds for pairs of consecutive seventh and higher power residues, *Math. Comp.* **18**, 397–407 (1964).

BRILLHART, J., LEHMER, D. H., and SELFRIDGE, J. L.:

[1] New primality criteria and factorizations of $2^m \pm 1$, *Math. Comp.* **29**, 620–647 (1975).

BRILLHART, J., and SELFRIDGE, J. L.:

[1] Some factorizations of $2^n \pm 1$ and related results, *Math. Comp.* **21**, 87–96 (1967); Corrigendum, *ibid.* **21**, 751 (1967).

BRIOSCHI, F.:

[1] Des substitutions de la forme $\theta(r) \equiv \varepsilon(r^{n-2} + ar^{(n-3)/2})$ pour un nombre n premier de lettres, *Math. Ann.* **2**, 467–470 (1870).

[2] Un teorema nella teorica delle sostituzioni, *Rend. Reale Ist. Lombardo Sci. Lett.* (2)**12**, 483–485 (1879).

[3] Sur les fonctions de sept lettres, *C. R. Acad. Sci. Paris* **95**, 665–669, 814–817, 1254–1257 (1882).

BROUSSEAU, A.:

[1] Recursion relations of products of linear recursion sequences, *Fibonacci Quart.* **14**, 159–166 (1976).

BROWKIN, J.:

[1] On zeros of forms, *Bull. Acad. Polon. Sci. Sér. Sci. Math. Astronom. Phys.* **17**, 611–616 (1969).

[2] *Theory of Fields* (Polish), Biblioteka Matematyczna, vol. 49, PWN, Warsaw, 1977.

BROWN, E.:

[1] The first proof of the quadratic reciprocity law, revisited, *Amer. Math. Monthly* **88**, 257–264 (1981).

BROWN, H., and ZASSENHAUS, H.:

[1] Some empirical observations on primitive roots, *J. Number Theory* **3**, 306–309 (1971).

BROWN, W. S.:

[1] On Euclid's algorithm and the computation of polynomial greatest common divisors, *J. Assoc. Comput. Mach.* **18**, 478–504 (1971).

BRUCK, R. H.:

[1] Difference sets in a finite group, *Trans. Amer. Math. Soc.* **78**, 464–481 (1955).

[2] Computational aspects of certain combinatorial problems, *Proc. Symp. Applied Math.*, vol. 6, pp. 31–43, McGraw-Hill, New York, 1956.

BRUCK, R. H., and RYSER, H.J.:

[1] The nonexistence of certain finite projective planes, *Canad. J. Math.* **1**, 88–93 (1949).

BRUCKNER, G.:

[1] Fibonacci sequence modulo a prime $p \equiv 3 \pmod 4$, *Fibonacci Quart.* **8**, 217–220 (1970).

BRUEN, A.:

[1] Permutation functions on a finite field, *Canad. Math. Bull.* **15**, 595–597 (1972).

BRUEN, A., and LEVINGER, B.:

[1] A theorem on permutations of a finite field, *Canad. J. Math.* **25**, 1060–1065 (1973).

BRUGGEMAN, R. W.:

[1] *Fourier Coefficients of Automorphic Forms*, Lecture Notes in Math., vol. 865, Springer-Verlag, Berlin-Heidelberg-New York, 1981.

BRUHAT, F.:

[1] *Lectures on Some Aspects of p-adic Analysis*, Tata Institute of Fundamental Research, Bombay, 1963.

BU, T.:

[1] Partitions of a vector space, *Discrete Math.* **31**, 79–83 (1980).

BUCKHIESTER, P. G.:

[1] The number of $n \times n$ matrices of rank r and trace α over a finite field, *Duke Math. J.* **39**, 695–699 (1972).

[2] Gauss sums and the number of solutions to the matrix equation $XAX^{\mathrm{T}} = 0$ over $GF(2^y)$, *Acta Arith.* **23**, 271–278 (1973).

[3] The number of solutions to the matrix equation $XAX^{\mathrm{T}} = C$, A and C nonalternate and of full rank, over $GF(2^y)$, *Math. Nachr.* **63**, 37–41 (1974).

[4] Rank r solutions to the matrix equation $XAX^{\mathrm{T}} = C$, A nonalternate, C alternate, over $GF(2^y)$, *Canad. J. Math.* **26**, 78–90 (1974).

[5] Rank r solutions to the matrix equation $XAX^{\mathrm{T}} = C$, A alternate, over $GF(2^y)$, *Trans. Amer. Math. Soc.* **189**, 201–209 (1974).

[6] Rank r solutions to the matrix equation $XAX^{\mathrm{T}} = C$, A and C nonalternate, over $GF(2^y)$, *Math. Nachr.* **63**, 413–422 (1974).

BUHŠTAB, A. A.:

[1] On those numbers in an arithmetic progression all prime factors of which are small in order of magnitude (Russian), *Dokl. Akad. Nauk SSSR* **67**, 5–8 (1949).

BUMBY, R. T.:

[1] A distribution property for linear recurrence of the second order, *Proc. Amer. Math. Soc.* **50**, 101–106 (1975).

BUNDSCHUH, P.:

[1] On the distribution of Fibonacci numbers, *Tamkang J. Math.* **5**, 75–79 (1974).

[2] Transzendenzmasse in Körpern formaler Laurentreihen, *J. reine angew. Math.* **299/300**, 411–432 (1978).

BUNDSCHUH, P., and SHIUE, J.-S.:

[1] Solution of a problem on the uniform distribution of integers, *Atti Accad. Naz. Lincei Rend. Cl. Sci. Fis. Mat. Natur.* (8)**55**, 172–177 (1973).

[2] A generalization of a paper by D. D. Wall, *Atti Accad. Naz. Lincei Rend. Cl. Sci. Fis. Mat. Natur.* (8)**56**, 135–144 (1974).

BURDE, K.:

[1] Verteilungseigenschaften von Potenzresten, *J. reine angew. Math.* **249**, 133–172 (1971).

[2] p-dimensionale Vektoren modulo p. I, *J. reine angew. Math.* **268/269**, 302–314 (1974).

[3] Sequenzen der Länge 2 von Restklassencharakteren, *J. reine angew. Math.* **272**, 194–202 (1975).

[4] Über allgemeine Sequenzen der Länge 3 von Legendresymbolen, *J. reine angew. Math.* **272**, 203–216 (1975).

[5] p-dimensionale Vektoren modulo p. II, *J. reine angew. Math.* **278/279**, 353–364 (1975).
[6] Zur Herleitung von Reziprozitätsgesetzen unter Benutzung von endlichen Körpern, *J. reine angew. Math.* **293/294**, 418–427 (1977).
[7] Potenzen von Galoisfeldern, *J. reine angew. Math.* **307/308**, 194–220 (1979).
[8] Pythagoräische Tripel und Reziprozität in Galoisfeldern, *J. Number Theory* **12**, 278–282 (1980).
[9] Ein Reziprozitätsgesetz in Galoisfeldern, *J. Number Theory* **13**, 66–87 (1981).

BURGESS, D. A.:
[1] The distribution of quadratic residues and non-residues, *Mathematika* **4**, 106–112 (1957).
[2] On character sums and primitive roots, *Proc. London Math. Soc.* (3)**12**, 179–192 (1962).
[3] On character sums and L-series, *Proc. London Math. Soc.* (3)**12**, 193–206 (1962).
[4] A note on the distribution of residues and non-residues, *J. London Math. Soc.* **38**, 253–256 (1963).
[5] On character sums and L-series. II, *Proc. London Math. Soc.* (3)**13**, 524–536 (1963).
[6] On Dirichlet characters of polynomials, *Proc. London Math. Soc.* (3)**13**, 537–548 (1963).
[7] Character sums and primitive roots in finite fields, *Proc. London Math. Soc.* (3)**17**, 11–25 (1967).
[8] On the quadratic character of a polynomial, *J. London Math. Soc.* **42**, 73–80 (1967).
[9] A note on character sums of binary quadratic forms, *J. London Math. Soc.* **43**, 271–274 (1968).
[10] A note on character sums over finite fields, *J. reine angew. Math.* **255**, 80–82 (1972).
[11] Dirichlet characters and polynomials, *Proc. Internat. Conf. on Number Theory* (Moscow, 1971), *Trudy Mat. Inst. Steklov.* **132**, 203–205 (1973); *Proc. Steklov. Inst. Math.* **132**, pp. 234–236, American Math. Society, Providence, R.I., 1975.

BURR, S. A.:
[1] On moduli for which the Fibonacci sequence contains a complete system of residues, *Fibonacci Quart.* **9**, 497–504 (1971).

BURTON, H. O.:
[1] Inversionless decoding of binary BCH codes, *IEEE Trans. Information Theory* **IT-17**, 464–466 (1971).

BUSSEY, W. H.:
[1] Galois field tables for $p^n \leqslant 169$, *Bull. Amer. Math. Soc.* **12**, 22–38 (1905).
[2] Tables of Galois fields of order less than 1000, *Bull. Amer. Math. Soc.* **16**, 188–206 (1909).

BUTLER, M. C. R.:

[1] On the reducibility of polynomials over finite fields, *Quart. J. Math.* (2)**5**, 102–107 (1954).

[2] The irreducible factors of $f(x^m)$ over a finite field, *J. London Math. Soc.* **30**, 480–482 (1955).

BUTSON, A. T.:

[1] Relations among generalized Hadamard matrices, relative difference sets, and maximal length linear recurring sequences, *Canad. J. Math.* **15**, 42–48 (1963).

BYERS, G. C.:

[1] Class number relations for quadratic forms over $GF[q, x]$, *Duke Math. J.* **21**, 445–461 (1954).

CAILLER, C.:

[1] Sur les congruences du troisième degré, *L'Enseignement Math.* **10**, 474–487 (1908).

CALABI, E., and WILF, H. S.:

[1] On the sequential and random selection of subspaces over a finite field, *J. Combinatorial Theory Ser. A* **22**, 107–109 (1977).

CALLAHAN, T., and SMITH, R. A.:

[1] *L*-functions of a quadratic form, *Trans. Amer. Math. Soc.* **217**, 297–309 (1976).

CALMET, J., and LOOS, R.:

[1] An improvement of Rabin's probabilistic algorithm for generating irreducible polynomials over $GF(p)$, *Inform. Process. Lett.* **11**, 94–95 (1980).

[2] A SAC-2 implementation of arithmetic and root finding over large finite fields, Tech. Report, Univ. of Karlsruhe, 1983.

CAMERON, P. J.:

[1] Extending symmetric designs, *J. Combinatorial Theory Ser. A* **14**, 214–220 (1973).

CAMERON, P. J., HALL, J. I., VAN LINT, J. H., SPRINGER, T. A., and VAN TILBORG, H. C. A.:

[1] Translates of subgroups of the multiplicative group of a finite field, *Indag. Math.* **37**, 285–289 (1975).

CAMERON, P. J., and SEIDEL, J. J.:

[1] Quadratic forms over $GF(2)$, *Indag. Math.* **35**, 1–8 (1973).

CAMERON, P. J., and VAN LINT, J. H.:

[1] *Graph Theory, Coding Theory and Block Designs*, London Math. Soc. Lecture Note Series, no. 19, Cambridge Univ. Press, London, 1975.

[2] *Graphs, Codes, and Designs*, London Math. Soc. Lecture Note Series, no. 43, Cambridge Univ. Press, Cambridge, 1980.

CAMION, P.:

[1] A proof of some properties of Reed-Muller codes by means of the normal basis theorem, *Proc. Conf. Combinatorial Math. and Its Appl.* (Chapel Hill,

N.C., 1967), pp. 371–376, Univ. of North Carolina Press, Chapel Hill, N.C., 1969.

[2] Un algorithme de construction des idempotents primitifs d'idéaux d'algèbres sur F_q, *C. R. Acad. Sci. Paris Sér. A* **291**, 479–482 (1980).

[3] Factorisation des polynômes de $\mathbb{F}_q[X]$, *Rev. CETHEDEC* **1981**, no. 2, 5–21.

CAMPBELL, A. D.:

[1] Plane cubic curves in the Galois fields of order 2^n, *Ann. of Math.* (2)**27**, 395–406 (1926).

[2] Pencils of conics in the Galois fields of order 2^n, *Amer. J. Math.* **45**, 401–406 (1927).

[3] Plane cubic curves in the Galois fields of order ρ^n, $\rho > 3$, *Messenger of Math.* **58**, 33–48 (1928).

[4] The discriminant of the *m*-ary quadratic in the Galois fields of order 2^n, *Ann. of Math.* (2)**29**, 395–398 (1928).

[5] Nets of conics in the Galois fields of order 2^n, *Bull. Amer. Math. Soc.* **34**, 481–489 (1928).

[6] Pencils of quadrics in the Galois fields of order 2^n, *Tôhoku Math. J.* **34**, 236–248 (1931).

[7] Apolarity in the Galois fields of order 2^n, *Bull. Amer. Math. Soc.* **38**, 52–56 (1932).

[8] Plane quartic curves in the Galois fields of order 2^n, *Tôhoku Math. J.* **37**, 88–93 (1933).

[9] Pseudo-covariants of an *n*-ic in *m* variables in a Galois field that consists of terms of this *n*-ic, *Bull. Amer. Math. Soc.* **39**, 252–256 (1933).

[10] Note on cubic surfaces in the Galois fields of order 2^n, *Bull. Amer. Math. Soc.* **39**, 406–410 (1933).

[11] Pseudo-covariants of *n*-ics in a Galois field, *Tôhoku Math. J.* **43**, 17–29 (1937).

CANADAY, E. F.:

[1] The sum of the divisors of a polynomial, *Duke Math. J.* **8**, 721–737 (1941).

CANTOR, D. G., and ZASSENHAUS, H.:

[1] A new algorithm for factoring polynomials over finite fields, *Math. Comp.* **36**, 587–592 (1981).

CAPELLI, A.:

[1] Sulla riduttibilità delle equazioni algebriche I, *Rend. Accad. Sci. Fis. Mat. Napoli* (3)**3**, 243–252 (1897).

[2] Sulla riduttibilità delle equazioni algebriche II, *Rend. Accad. Sci. Fis. Mat. Napoli* (3)**4**, 84–90 (1898).

[3] Sulla riduttibilità della funzione $x^n - A$ in un campo qualunque di razionalità, *Math. Ann.* **54**, 602–603 (1901).

CAR, M.:

[1] Le problème de Waring pour l'anneau des polynômes sur un corps fini, *C. R. Acad. Sci. Paris Sér. A* **273**, 141–144 (1971).

[2] Le problème de Goldbach pour l'anneau des polynômes sur un corps fini, *C. R. Acad. Sci. Paris Sér. A* **273**, 201–204 (1971).

[3] Le problème de Waring pour l'anneau des polynômes sur un corps fini, *Sém. Théorie des Nombres 1972–1973*, Exp. 6, 13 pp., Univ. Bordeaux I, Talence, 1973.

[4] La méthode des sommes trigonométriques pour $\mathbb{F}_q[X]$, *Journées de Théorie Additive des Nombres* (Bordeaux, 1977), pp. 19–33, Univ. Bordeaux I, Talence, 1978.

[5] Normes dans $\mathbb{F}_q[X]$ de polynômes de $\mathbb{F}_{q^h}[X]$, *C. R. Acad. Sci. Paris Sér. A* **288**, 669–672 (1979); Correction, *ibid.* **288**, 1049 (1979).

[6] Sommes de carrés et d'irréductibles dans $\mathbb{F}_q[X]$, *Ann. Fac. Sci. Toulouse* **3**, 129–166 (1981).

[7] Factorisation dans $\mathbb{F}_q[X]$, *C. R. Acad. Sci. Paris Sér. I* **294**, 147–150 (1982).

CARCANAGUE, J.:

[1] Propriétés des q-polynômes, *C. R. Acad. Sci. Paris Sér. A* **265**, 415–418 (1967).

[2] q-polynômes abéliens sur un corps K, *C. R. Acad. Sci. Paris Sér. A* **265**, 496–499 (1967).

CAREY, F. S.:

[1] Notes on the division of the circle, *Quart. J. Pure Appl. Math.* **26**, 322–371 (1893).

CARLITZ, L.:

[1] The arithmetic of polynomials in a Galois field, *Proc. Nat. Acad. Sci. U.S.A.* **17**, 120–122 (1931).

[2] The arithmetic of polynomials in a Galois field, *Amer. J. Math.* **54**, 39–50 (1932).

[3] On polynomials in a Galois field, *Bull. Amer. Math. Soc.* **38**, 736–744 (1932).

[4] On a theorem of higher reciprocity, *Bull. Amer. Math. Soc.* **39**, 155–160 (1933).

[5] On the representation of a polynomial in a Galois field as the sum of an even number of squares, *Trans. Amer. Math. Soc.* **35**, 397–410 (1933).

[6] On polynomials in a Galois field: Some formulae involving divisor functions, *Proc. London Math. Soc.* (2)**38**, 116–124 (1935).

[7] On certain functions connected with polynomials in a Galois field, *Duke Math. J.* **1**, 137–168 (1935).

[8] On the representation of a polynomial in a Galois field as the sum of an odd number of squares, *Duke Math. J.* **1**, 298–315 (1935).

[9] A theorem on higher congruences, *Bull. Amer. Math. Soc.* **41**, 844–846 (1935).

[10] On certain higher congruences, *Bull. Amer. Math. Soc.* **41**, 907–914 (1935).

[11] On certain equations in relative-cyclic fields, *Duke Math. J.* **2**, 650–659 (1936).

[12] On factorable polynomials in several indeterminates, *Duke Math. J.* **2**, 660–670 (1936).

[13] Sums of squares of polynomials, *Duke Math. J.* **3**, 1–7 (1937).

[14] An arithmetic function, *Bull. Amer. Math. Soc.* **43**, 271–276 (1937).

[15] Some formulae for factorable polynomials in several indeterminates, *Bull. Amer. Math. Soc.* **43**, 299–304 (1937).
[16] An analogue of the von Staudt-Clausen theorem, *Duke Math. J.* **3**, 503–517 (1937).
[17] Criteria for certain higher congruences, *Amer. J. Math.* **59**, 618–628 (1937).
[18] A class of polynomials, *Trans. Amer. Math. Soc.* **43**, 167–182 (1938).
[19] Some sums involving polynomials in a Galois field, *Duke Math. J.* **5**, 941–947 (1939).
[20] A set of polynomials, *Duke Math. J.* **6**, 486–504 (1940).
[21] Linear forms and polynomials in a Galois field, *Duke Math. J.* **6**, 735–749 (1940).
[22] An analogue of the Staudt-Clausen theorem, *Duke Math. J.* **7**, 62–67 (1940).
[23] An analogue of the Bernoulli polynomials, *Duke Math. J.* **8**, 405–412 (1941).
[24] The reciprocal of certain series, *Duke Math. J.* **9**, 234–243 (1942).
[25] The reciprocal of certain types of Hurwitz series, *Duke Math. J.* **9**, 629–642 (1942).
[26] Some topics in the arithmetic of polynomials, *Bull. Amer. Math. Soc.* **48**, 679–691 (1942).
[27] The singular series for sums of squares of polynomials, *Duke Math. J.* **14**, 1105–1120 (1947).
[28] Representations of arithmetic functions in $GF[p^n, x]$, *Duke Math. J.* **14**, 1121–1137 (1947).
[29] A problem of Dickson's, *Duke Math. J.* **14**, 1139–1140 (1947).
[30] Representations of arithmetic functions in $GF[p^n, x]$. II, *Duke Math. J.* **15**, 795–801 (1948).
[31] Finite sums and interpolation formulas over $GF[p^n, x]$, *Duke Math. J.* **15**, 1001–1012 (1948).
[32] Some applications of a theorem of Chevalley, *Duke Math. J.* **18**, 811–819 (1951).
[33] Diophantine approximation in fields of characteristic p, *Trans. Amer. Math. Soc.* **72**, 187–208 (1952).
[34] Some problems involving primitive roots in a finite field, *Proc. Nat. Acad. Sci. U.S.A.* **38**, 314–318 (1952); Errata, *ibid.* **38**, 618 (1952).
[35] Primitive roots in a finite field, *Trans. Amer. Math. Soc.* **73**, 373–382 (1952).
[36] Note on an arithmetic function, *Amer. Math. Monthly* **59**, 386–387 (1952).
[37] Sums of primitive roots in a finite field, *Duke Math. J.* **19**, 459–469 (1952).
[38] A problem of Dickson, *Duke Math. J.* **19**, 471–474 (1952).
[39] The number of solutions of certain equations in a finite field, *Proc. Nat. Acad. Sci. U.S.A.* **38**, 515–519 (1952); Errata, *ibid.* **38**, 618 (1952).
[40] A theorem of Dickson on irreducible polynomials, *Proc. Amer. Math. Soc.* **3**, 693–700 (1952).
[41] Distribution of primitive roots in a finite field, *Quart. J. Math.* (2)**4**, 4–10 (1953).
[42] Note on a conjecture of André Weil, *Proc. Amer. Math. Soc.* **4**, 5–9 (1953).

[43] Some special equations in a finite field, *Pacific J. Math.* **3**, 13–24 (1953).
[44] A theorem of Stickelberger, *Math. Scand.* **1**, 82–84 (1953).
[45] A reciprocity formula for weighted quadratic partitions, *Math. Scand.* **1**, 286–288 (1953).
[46] Weighted quadratic partitions over a finite field, *Canad. J. Math.* **5**, 317–323 (1953).
[47] Invariantive theory of equations in a finite field, *Trans. Amer. Math. Soc.* **75**, 405–427 (1953).
[48] A note on partitions in $GF[q, x]$, *Proc. Amer. Math. Soc.* **4**, 464–469 (1953).
[49] Permutations in a finite field, *Proc. Amer. Math. Soc.* **4**, 538 (1953).
[50] Certain special equations in a finite field, *Monatsh. Math.* **58**, 5–12 (1954).
[51] Representations by skew forms in a finite field, *Arch. Math.* **5**, 19–31 (1954).
[52] The number of solutions of some equations in a finite field, *Portugal. Math.* **13**, 25–31 (1954).
[53] A note on modular invariants, *Nieuw Arch. Wisk.* (3)**2**, 28–31 (1954).
[54] Representations by quadratic forms in a finite field, *Duke Math. J.* **21**, 123–137 (1954).
[55] A problem involving quadratic forms in a finite field, *Math. Nachr.* **11**, 135–142 (1954).
[56] Pairs of quadratic equations in a finite field, *Amer. J. Math.* **76**, 137–154 (1954).
[57] Sums of primitive roots of the first and second kind in a finite field, *Math. Nachr.* **12**, 155–172 (1954).
[58] The number of solutions of some special equations in a finite field, *Pacific J. Math.* **4**, 207–217 (1954).
[59] Invariant theory of systems of equations in a finite field, *J. Analyse Math.* **3**, 382–413 (1954).
[60] The number of solutions of a special quadratic congruence, *Portugal. Math.* **14**, 9–14 (1955).
[61] On the number of distinct values of a polynomial with coefficients in a finite field, *Proc. Japan Acad.* **31**, 119–120 (1955).
[62] The number of solutions of certain types of equations in a finite field, *Pacific J. Math.* **5**, 177–181 (1955).
[63] The number of solutions of some equations in a finite field, *J. Math. Soc. Japan* **7**, 209–223 (1955).
[64] A special symmetric equation in a finite field, *Acta Math. Acad. Sci. Hungar.* **6**, 445–450 (1955).
[65] Solvability of certain equations in a finite field, *Quart. J. Math.* (2)**7**, 3–4 (1956).
[66] A note on nonsingular forms in a finite field, *Proc. Amer. Math. Soc.* **7**, 27–29 (1956).
[67] An application of a theorem of Stickelberger, *Simon Stevin* **31**, 27–30 (1956).
[68] Sets of primitive roots, *Compositio Math.* **13**, 65–70 (1956).
[69] Class number formulas for quadratic forms over $GF[q, x]$, *Duke Math. J.* **23**, 225–235 (1956).

[70] A special quartic congruence, *Math. Scand.* **4**, 243–246 (1956).

[71] The number of solutions of a particular equation in a finite field, *Publ. Math. Debrecen* **4**, 379–383 (1956).

[72] Weighted quadratic partitions over $GF[q, x]$, *Duke Math. J.* **23**, 493–505 (1956).

[73] Note on a quartic congruence, *Amer. Math. Monthly* **63**, 569–571 (1956).

[74] A note on Gauss' sum, *Proc. Amer. Math. Soc.* **7**, 910–911 (1956).

[75] The number of points on certain cubic surfaces over a finite field, *Boll. Un. Mat. Ital.* (3)**12**, 19–21 (1957).

[76] Some cyclotomic determinants, *Bull. Calcutta Math. Soc.* **49**, 49–51 (1957).

[77] Some theorems on polynomials, *Ark. Mat.* **3**, 351–353 (1957).

[78] A theorem of Dickson on nonvanishing cubic forms in a finite field, *Proc. Amer. Math. Soc.* **8**, 975–977 (1957).

[79] Quadratic residues and Tchebycheff polynomials, *Portugal. Math.* **18**, 193–198 (1959).

[80] Some cyclotomic matrices, *Acta Arith.* **5**, 293–308 (1959).

[81] A note on exponential sums, *Acta Sci. Math. Szeged* **21**, 135–143 (1960).

[82] A theorem on permutations in a finite field, *Proc. Amer. Math. Soc.* **11**, 456–459 (1960).

[83] Some theorems on permutation polynomials, *Bull. Amer. Math. Soc.* **68**, 120–122 (1962).

[84] A theorem on "ordered" polynomials in a finite field, *Acta Arith.* **7**, 167–172 (1962).

[85] Some identities over a finite field, *Quart. J. Math.* (2)**13**, 299–303 (1962).

[86] A note on permutation functions over a finite field, *Duke Math. J.* **29**, 325–332 (1962).

[87] Solvability of certain equations in a finite field, *Acta Arith.* **7**, 389–397 (1962).

[88] A note on finite fields, *Proc. Amer. Math. Soc.* **13**, 546–549 (1962).

[89] Note on a problem of Dickson, *Proc. Amer. Math. Soc.* **14**, 98–100 (1963).

[90] A note on permutations in an arbitrary field, *Proc. Amer. Math. Soc.* **14**, 101 (1963).

[91] A note on the Betti-Mathieu group, *Portugal. Math.* **22**, 121–125 (1963).

[92] Classes of pairs of commuting matrices over a finite field, *Amer. Math. Monthly* **70**, 192–195 (1963).

[93] Permutations in finite fields, *Acta Sci. Math. Szeged* **24**, 196–203 (1963).

[94] Simultaneous representations in quadratic and linear forms over $GF[q, x]$, *Duke Math. J.* **30**, 259–270 (1963).

[95] The distribution of irreducible polynomials in several indeterminates, *Illinois J. Math.* **7**, 371–375 (1963).

[96] A property of irreducible polynomials related to Mersenne primes, *Univ. Nac. Tucumán Rev. Ser. A* **15**, 43–46 (1964).

[97] Functions and polynomials (mod p^n), *Acta Arith.* **9**, 67–78 (1964).

[98] A note on multiple Kloosterman sums, *J. Indian Math. Soc.* **29**, 197–200 (1965).

[99] The distribution of irreducible polynomials in several indeterminates II, *Canad. J. Math.* **17**, 261–266 (1965).

[100] A note on multiple exponential sums, *Pacific J. Math.* **15**, 757–765 (1965).

[101] A conjecture concerning a certain system of equations in a finite field, *Rev. Roum. Math. Pures Appl.* **11**, 277–282 (1966).

[102] A note on quadrics over a finite field, *Duke Math. J.* **33**, 453–458 (1966).

[103] A note on irreducible cubics mod p, *Norske Vid. Selsk. Forh.* (*Trondheim*) **40**, 25–30 (1967).

[104] Restricted product of the characteristic polynomials of matrices over a finite field, *Illinois J. Math.* **11**, 128–133 (1967).

[105] Some theorems on irreducible reciprocal polynomials over a finite field, *J. reine angew. Math.* **227**, 212–220 (1967).

[106] A note on Gauss's sum, *Le Matematiche* (*Catania*) **23**, 147–150 (1968).

[107] Some formulas related to Gauss's sum, *Rend. Sem. Mat. Univ. Padova* **41**, 222–226 (1968).

[108] A note on exponential sums, *Pacific J. Math.* **30**, 35–37 (1969).

[109] Gauss sums over finite fields of order 2^n, *Acta Arith.* **15**, 247–265 (1969).

[110] A theorem on sets of polynomials over a finite field, *Acta Arith.* **15**, 267–268 (1969).

[111] Kloosterman sums and finite field extensions, *Acta Arith.* **16**, 179–193 (1969).

[112] Factorization of a special polynomial over a finite field, *Pacific J. Math.* **32**, 603–614 (1970).

[113] Reduction formulas for certain multiple exponential sums, *Czechoslovak Math. J.* **20**, 616–627 (1970).

[114] The number of solutions of certain matrix equations over a finite field, *Math. Nachr.* **56**, 105–109 (1973).

[115] Correspondences in a finite field. I, *Acta Arith.* **27**, 101–123 (1975).

[116] A note on sums of three squares in $GF[q, x]$, *Math. Mag.* **48**, 109–110 (1975).

[117] Correspondences in a finite field. II, *Indiana Univ. Math. J.* **24**, 785–811 (1975).

[118] A theorem on lacunary polynomials in a finite field, *Amer. Math. Monthly* **83**, 37–38 (1976).

[119] Some theorems on polynomials over a finite field, *Amer. Math. Monthly* **84**, 29–32 (1977).

[120] A theorem on linear exponential sums, *Univ. Beograd. Publ. Elektrotehn. Fak. Ser. Mat. Fiz.* **577-598**, 55–56 (1977).

[121] Functions and correspondences in a finite field, *Bull. Amer. Math. Soc.* **83**, 139–165 (1977).

[122] A note on exponential sums, *Math. Scand.* **42**, 39–48 (1978).

[123] Polynomial characteristic functions for $GF(p)$ and irregular primes, *Rocky Mountain J. Math.* **8**, 583–587 (1978).

[124] Explicit evaluation of certain exponential sums, *Math. Scand.* **44**, 5–16 (1979).

[125] Evaluation of some exponential sums over a finite field, *Math. Nachr.* **96**, 319–339 (1980).

CARLITZ, L., and COHEN, E.:

[1] Divisor functions of polynomials in a Galois field, *Duke Math. J.* **14**, 13–20 (1947).

[2] Cauchy products of divisor functions in $GF[p^n, x]$, *Duke Math. J.* **14**, 707–722 (1947).

[3] The number of representations of a polynomial in certain special quadratic forms, *Duke Math. J.* **15**, 219–228 (1948).

CARLITZ, L., and CORSON, H. H.:

[1] Some special equations in a finite field, *Proc. Nat. Acad. Sci. U.S.A.* **41**, 752–754 (1955).

[2] Some special equations in a finite field, *Monatsh. Math.* **60**, 114–122 (1956).

CARLITZ, L., and HAYES, D. R.:

[1] Permutations with coefficients in a subfield, *Acta Arith.* **21**, 131–135 (1972).

CARLITZ, L., and HODGES, J. H.:

[1] Representations by Hermitian forms in a finite field, *Duke Math. J.* **22**, 393–405 (1955).

[2] Distribution of bordered symmetric, skew and hermitian matrices in a finite field, *J. reine angew. Math.* **195**, 192–201 (1956).

[3] Distribution of matrices in a finite field, *Pacific J. Math.* **6**, 225–230 (1956).

[4] Enumeration of matrices of given rank with submatrices of given rank, *Linear Algebra Appl.* **16**, 285–291 (1977).

CARLITZ, L., LEWIS, D. J., MILLS, W. H., and STRAUS, E. G.:

[1] Polynomials over finite fields with minimum value sets, *Mathematika* **8**, 121–130 (1961).

CARLITZ, L., and LONG, A. F., Jr.:

[1] The factorization of $Q(L(x_1),\ldots,L(x_k))$ over a finite field where $Q(x_1,\ldots,x_k)$ is of first degree and $L(x)$ is linear, *Acta Arith.* **32**, 407–420 (1977).

CARLITZ, L., and LUTZ, J. A.:

[1] A characterization of permutation polynomials over a finite field, *Amer. Math. Monthly* **85**, 746–748 (1978).

CARLITZ, L., and UCHIYAMA, S.:

[1] Bounds for exponential sums, *Duke Math. J.* **24**, 37–41 (1957).

CARLITZ, L., and WELLS, C.:

[1] The number of solutions of a special system of equations in a finite field, *Acta Arith.* **12**, 77–84 (1966).

CARMICHAEL, R. D.:

[1] On the numerical factors of the arithmetic forms $\alpha^n \pm \beta^n$, *Ann. of Math.* (2)**15**, 30–70 (1913).

[2] On sequences of integers defined by recurrence relations, *Quart. J. Pure Appl. Math.* **48**, 343–372 (1920).

[3] A simple principle of unification in the elementary theory of numbers, *Amer. Math. Monthly* **36**, 132–143 (1929).

[4] *Introduction to the Theory of Groups of Finite Order*, Ginn & Co., Boston, 1937; Dover, New York, 1956.

CARTIER, P.:
[1] Sur une généralisation des symboles de Legendre-Jacobi, *L'Enseignement Math.* (2)**16**, 31–48 (1970).

CASSELS, J. W. S.:
[1] Diophantine equations with special reference to elliptic curves, *J. London Math. Soc.* **41**, 193–291 (1966).
[2] On the determination of generalized Gauss sums, *Arch. Math. (Brno)* **5**, 79–84 (1969).
[3] On Kummer sums, *Proc. London Math. Soc.* (3)**21**, 19–27 (1970).
[4] On cubic trigonometric sums, *Actes du Congrès International des Mathématiciens* (Nice, 1970), vol. 1, pp. 377–379, Gauthier-Villars, Paris, 1971.
[5] Trigonometric sums and elliptic functions, *Algebraic Number Theory* (S. Iyanaga, ed.), pp. 1–7, Japan Soc. for the Promotion of Science, Tokyo, 1977.

CATLIN, P. A.:
[1] A lower bound for the period of the Fibonacci series modulo *m*, *Fibonacci Quart.* **12**, 349–350 (1974).

CAUCHY, A.-L.:
[1] Recherches sur les nombres, *J. de l'Ecole Polytechnique* **9**, 99–116 (1813); *Oeuvres (II)*, vol. 1, pp. 39–63, Gauthier-Villars, Paris, 1905.
[2] Mémoire sur la théorie des nombres, *Bull. Sci. Math. de M. Férussac* **12**, 205–221 (1829); *Oeuvres (II)*, vol. 2, pp. 88–107, Gauthier-Villars, Paris, 1958.
[3] Sur la résolution des équivalences dont les modules se réduisent à des nombres premiers, *Exercices de Math.* **4** (1829); *Oeuvres (II)*, vol. 9, pp. 298–341, Gauthier-Villars, Paris, 1891.
[4] Mémoire sur la théorie des nombres, *Mém. Acad. Sci. Inst. de France* **17** (1840); *Oeuvres (I)*, vol. 3, Gauthier-Villars, Paris, 1911.
[5] Méthode simple et nouvelle pour la détermination complète des sommes alternées, formées avec les racines primitives des équations binômes, *C. R. Acad. Sci. Paris* **10**, 560–572 (1840); *J. Math. Pures Appl.* **5**, 154–168 (1840); *Oeuvres (I)*, vol. 5, pp. 152–166, Gauthier-Villars, Paris, 1885.

CAVIOR, S. R.:
[1] A note on octic permutation polynomials, *Math. Comp.* **17**, 450–452 (1963).
[2] Equivalence classes of functions over a finite field, *Acta Arith.* **10**, 119–136 (1964).
[3] Exponential sums related to polynomials over the $GF(p)$, *Proc. Amer. Math. Soc.* **15**, 175–178 (1964).
[4] On the least non-negative trace of a polynomial over a finite field, *Boll. Un. Mat. Ital.* (3)**20**, 120–121 (1965).
[5] Uniform distribution of polynomials modulo *m*, *Amer. Math. Monthly* **73**, 171–172 (1966).

[6] Equivalence classes of sets of polynomials over a finite field, *J. reine angew. Math.* **225**, 191–202 (1967).

[7] Uniform distribution (mod m) of recurrent sequences, *Fibonacci Quart.* **15**, 265–267 (1977).

CAZACU, C.:

[1] Application of two-valued logic in the theory of numbers (Romanian), *An. Şti. Univ. "Al. I. Cuza" Iaşi Secţ. I (N. S.)* **6**, 481–492 (1960).

[2] Predicates in finite fields (Russian), *An. Şti. Univ. "Al. I. Cuza" Iaşi Secţ. I (N. S.)* **11**, 221–238 (1965).

[3] Predicates with quantifiers in finite fields (Russian), *An. Şti. Univ. "Al. I. Cuza" Iaşi Secţ. I (N. S.)* **13**, 241–247 (1967).

CAZACU, C., and SIMOVICI, D.:

[1] A new approach of some problems concerning polynomials over finite fields, *Information and Control* **22**, 503–511 (1973).

CECCHERINI, P. V.:

[1] Some new results on certain finite structures, *Atti Accad. Naz. Lincei Rend. Cl. Sci. Fis. Mat. Natur.* (8)**56**, 840–855 (1974).

CECCHERINI, P. V., and HIRSCHFELD, J. W. P.:

[1] On the number of zeros over a finite field of certain symmetric polynomials, *Canad. Math. Bull.* **23**, 327–332 (1980).

CHALK, J. H. H.:

[1] The number of solutions of congruences in incomplete residue systems, *Canad. J. Math.* **15**, 291–296 (1963).

[2] The Vinogradov-Mordell-Tietäväinen inequalities, *Indag. Math.* **42**, 367–374 (1980).

CHALK, J. H. H., and SMITH, R. A.:

[1] On Bombieri's estimate for exponential sums, *Acta Arith.* **18**, 191–212 (1971).

CHALK, J. H. H., and WILLIAMS, K. S.:

[1] The distribution of solutions of congruences, *Mathematika* **12**, 176–192 (1965); Corrigendum and Addendum, *ibid.* **16**, 98–100 (1969).

CHANG, J. A., and GODWIN, H. J.:

[1] A table of irreducible polynomials and their exponents, *Proc. Cambridge Philos. Soc.* **65**, 513–522 (1969).

CHANG, S. C., and WOLF, J. K.:

[1] A simple derivation of the MacWilliams identity for linear codes, *IEEE Trans. Information Theory* **IT-26**, 476–477 (1980).

CHANG, T.-H.:

[1] Lösung der Kongruenz $x^2 \equiv a (\mathrm{mod}\, p)$ nach einem Primzahlmodul $p = 4n+1$, *Math. Nachr.* **22**, 136–142 (1960).

CHAO, C. Y.:

[1] On equivalence classes of matrices, *Bull. Malaysian Math. Soc.* (2)**4**, 29–36 (1981).

CHÂTELET, F.:

[1] Classification des courbes de genre un, dans le corps des restes, module p, *C. R. Acad. Sci. Paris* **208**, 487–489 (1939).

[2] Les courbes de genre 1 dans un champ de Galois, *C. R. Acad. Sci. Paris* **224**, 1616–1618 (1947).

CHEBYSHEV, P. L.:

[1] *Theory of Congruences* (Russian), St. Petersburg, 1849.

CHEN, C. L.:

[1] Computer results on the minimum distance of some binary cyclic codes, *IEEE Trans. Information Theory* **IT-16**, 359–360 (1970).

[2] High-speed decoding of BCH codes, *IEEE Trans. Information Theory* **IT-27**, 254–256 (1981).

CHEN, J. M., and LI, X. M.:

[1] The structure of the polynomials over the finite field defined by $Q[f]$-matrix (Chinese), *Acta Math. Sinica* **20**, 294–297 (1977).

CHEN, J. R.:

[1] On the representation of a natural number as a sum of terms of the form $x(x+1)\cdots(x+k-1)/k!$ (Chinese), *Acta Math. Sinica* **9**, 264–270 (1959).

[2] On Professor Hua's estimate of exponential sums, *Sci. Sinica* **20**, 711–719 (1977).

CHERLY, J.:

[1] Addition theorems in $F_q[x]$, *J. reine angew. Math.* **293/294**, 223–227 (1977).

[2] A lower bound theorem in $F_q[x]$, *J. reine angew. Math.* **303/304**, 253–264 (1978).

[3] On complementary sets of group elements, *Arch. Math.* **35**, 313–318 (1980).

CHEVALLEY, C.:

[1] Démonstration d'une hypothèse de M. Artin, *Abh. Math. Sem. Univ. Hamburg* **11**, 73–75 (1936).

[2] Sur certains groupes simples, *Tôhoku Math. J.* (2) **7**, 14–66 (1955).

CHIEN, R. T.:

[1] Cyclic decoding procedures for Bose-Chaudhuri-Hocquenghem codes, *IEEE Trans. Information Theory* **IT-10**, 357–363 (1964).

CHIEN, R. T., and CUNNINGHAM, B. D.:

[1] Hybrid methods for finding roots of a polynomial with application to BCH decoding, *IEEE Trans. Information Theory* **IT-15**, 329–335 (1969).

CHILDS, L.:

[1] *A Concrete Introduction to Higher Algebra*, Springer-Verlag, New York-Heidelberg-Berlin, 1979.

CHILDS, L., and ORZECH, M.:

[1] On modular group rings, normal bases, and fixed points, *Amer. Math. Monthly* **88**, 142–145 (1981).

CHINBURG, T.:

[1] 'Easier' Waring problem for commutative rings, *Acta Arith.* **35**, 303–331 (1979).

CHOR, B.-Z.:

[1] Arithmetic of finite fields, *Inform. Process. Lett.* **14**, 4–6 (1982).

CHOWLA, I.:

[1] The number of solutions of a congruence in two variables, *Proc. Nat. Acad. Sci. India Sect. A* **4**, 654–655 (1936).

[2] On the number of solutions of some congruences in two variables, *Proc. Nat. Acad. Sci. India Sect. A* **5**, 40–44 (1937).

[3] Generalization of a theorem of Dickson, *Proc. Nat. Acad. Sci. India Sect. A* **8**, 223–226 (1938).

[4] On Waring's problem (mod p), *Proc. Nat. Acad. Sci. India Sect. A* **13**, 195–220 (1943).

CHOWLA, P.:

[1] On some polynomials which represent every natural number exactly once, *Norske Vid. Selsk. Forh. (Trondheim)* **34**, 8–9 (1961).

[2] A new proof and generalization of some theorems of Brewer, *Norske Vid. Selsk. Forh. (Trondheim)* **41**, 1–3 (1968).

CHOWLA, P., and CHOWLA, S.:

[1] On the integer points on some special hyper-elliptic curves over a finite field, *J. Number Theory* **8**, 280–281 (1976).

[2] On kth power residues, *J. Number Theory* **10**, 351–353 (1978).

CHOWLA, S.:

[1] Some formulae of the Gauss sum type, *Tôhoku Math. J.* **30**, 226–234 (1929); Corrigenda, *ibid.* **32**, 109–110 (1930).

[2] Some formulae of the Gauss sum type (II), *Tôhoku Math. J.* **32**, 352–353 (1930).

[3] A theorem on characters. II, *J. Indian Math. Soc.* **19**, 279–284 (1932).

[4] A property of biquadratic residues, *Proc. Nat. Acad. Sci. India Sect. A* **14**, 45–46 (1944).

[5] A formula similar to Jacobsthal's for the explicit value of x in $p = x^2 + y^2$ where p is a prime of the form $4k + 1$, *Proc. Lahore Philos. Soc.* **7** (1945), 2 pp.

[6] The last entry in Gauss' diary, *Proc. Nat. Acad. Sci. U.S.A.* **35**, 244–246 (1949).

[7] The Riemann zeta and allied functions, *Bull. Amer. Math. Soc.* **58**, 287–305 (1952).

[8] Some results in number-theory, *Norske Vid. Selsk. Forh. (Trondheim)* **33**, 43–44 (1960).

[9] A generalization of Meyer's theorem on indefinite quadratic forms in five or more variables, *J. Indian Math. Soc.* **25**, 41 (1961).

[10] On the congruence $\sum_{i=1}^{s} a_i x_i^k \equiv 0 (\bmod p)$, *J. Indian Math. Soc.* **25**, 47–48 (1961).

[11] On a formula of Jacobsthal, *Norske Vid. Selsk. Forh. (Trondheim)* **34**, 105–106 (1961).

[12] Some conjectures in elementary number theory, *Norske Vid. Selsk. Forh.* (*Trondheim*) **35**, 13 (1962).

[13] On Gaussian sums, *Norske Vid. Selsk. Forh.* (*Trondheim*) **35**, 66–67 (1962).

[14] On Gaussian sums, *Proc. Nat. Acad. Sci. U.S.A.* **48**, 1127–1128 (1962).

[15] On a conjecture of Artin. I, II, *Norske Vid. Selsk. Forh.* (*Trondheim*) **36**, 135–138, 139–141 (1963).

[16] *The Riemann Hypothesis and Hilbert's Tenth Problem*, Gordon and Breach, New York, 1965.

[17] A note on the construction of finite Galois fields $GF(p^n)$, *J. Math. Anal. Appl.* **15**, 53–54 (1966).

[18] An algebraic proof of the law of quadratic reciprocity, *Norske Vid. Selsk. Forh.* (*Trondheim*) **39**, 59 (1966).

[19] On the class-number of the function field $y^2 = f(x)$ over $GF(p)$. I, II, *Norske Vid. Selsk. Forh.* (*Trondheim*) **39**, 86–88 (1966); *ibid.* **40**, 7–10 (1967).

[20] Observation on a theorem of Stark, *Norske Vid. Selsk. Forh.* (*Trondheim*) **40**, 34–36 (1967).

[21] On some character sums, *Norske Vid. Selsk. Forh.* (*Trondheim*) **40**, 62–66 (1967).

[22] On Kloosterman's sum, *Norske Vid. Selsk. Forh.* (*Trondheim*) **40**, 70–72 (1967).

[23] On substitution polynomials (mod p), *Norske Vid. Selsk. Forh.* (*Trondheim*) **41**, 4–6 (1968).

CHOWLA, S., COWLES, J., and COWLES, M.:

[1] On the number of zeros of diagonal cubic forms, *J. Number Theory* **9**, 502–506 (1977).

[2] Congruence properties of the number of solutions of some equations, *J. reine angew. Math.* **298**, 101–103 (1978).

[3] The number of zeroes of $x^3 + y^3 + cz^3$ in certain finite fields, *J. reine angew. Math.* **299/300**, 406–410 (1978).

[4] On the difference of cubes (mod p), *Acta Arith.* **37**, 61–65 (1980).

CHOWLA, S., and HASSE, H.:

[1] On a paper of Bombieri, *Norske Vid. Selsk. Forh.* (*Trondheim*) **41**, 30–33 (1968).

CHOWLA, S., MANN, H. B., and STRAUS, E. G.:

[1] Some applications of the Cauchy-Davenport theorem, *Norske Vid. Selsk. Forh.* (*Trondheim*) **32**, 74–80 (1959).

CHOWLA, S., and RYSER, H. J.:

[1] Combinatorial problems, *Canad. J. Math.* **2**, 93–99 (1950).

CHOWLA, S., and SHIMURA, G.:

[1] On the representation of zero by a linear combination of k-th powers, *Norske Vid. Selsk. Forh.* (*Trondheim*) **36**, 169–176 (1963).

CHOWLA, S., and SMITH, R. A.:

[1] On certain functional equations, *Norske Vid. Selsk. Forh.* (*Trondheim*) **40**, 43–47 (1967).

CHOWLA, S., and VIJAYARAGHAVAN, T.:

[1] The complete factorization (mod p) of the cyclotomic polynomial of order p^2-1, *Proc. Nat. Acad. Sci. India Sect. A* **14**, 101–105 (1944).

CHOWLA, S., and ZASSENHAUS, H.:

[1] Some conjectures concerning finite fields, *Norske Vid. Selsk. Forh.* (*Trondheim*) **41**, 34–35 (1968).

CHURCH, R.:

[1] Tables of irreducible polynomials for the first four prime moduli, *Ann. of Math.* (2)**36**, 198–209 (1935).

CICCHESE, M.:

[1] Sulle cubiche di un piano di Galois, *Atti Accad. Naz. Lincei Rend. Cl. Sci. Fis. Mat. Natur.* (8)**32**, 38–42 (1962).

[2] Sulle cubiche di un piano di Galois, *Rend. Mat. e Appl.* (5)**24**, 291–330 (1965).

[3] Sulle cubiche di un piano lineare $S_{2,q}$, con $q \equiv 1 (\mathrm{mod}\, 3)$, *Atti Accad. Naz. Lincei Rend. Cl. Sci. Fis. Mat. Natur.* (8)**48**, 584–588 (1970).

[4] Sulle cubiche di un piano lineare $S_{2,q}$, con $q \equiv 1 (\mathrm{mod}\, 3)$, *Rend. Mat.* (6)**4**, 349–383 (1971).

CIPOLLA, M.:

[1] Un metodo per la risoluzione della congruenza di secondo grado, *Rend. Accad. Sci. Fis. Mat. Napoli* (3)**9**, 153–163 (1903).

[2] Formule di risoluzione della congruenza binomia quadratica e biquadratica, *Rend. Accad. Sci. Fis. Mat. Napoli* (3)**11**, 13–17 (1905).

[3] Sulle funzioni simmetriche delle soluzioni comuni a più congruenze secondo un modulo primo, *Periodico di Mat.* **22**, 36–41 (1907).

[4] Sulla risoluzione apiristica delle congruenze binomie secondo un modulo primo, *Math. Ann.* **63**, 54–61 (1907).

[5] Formule di risoluzione apiristica delle equazioni di grado qualunque in un corpo finito, *Rend. Circ. Mat. Palermo* **54**, 199–206 (1930).

CLAASEN, H. L.:

[1] The group of units in $GF(q)[x]/(a(x))$, *Indag. Math.* **39**, 245–255 (1977).

[2] The multiplications in $GF(q)[x]/(a(x))$ considered as linear transformations, *Linear Algebra Appl.* **22**, 105–123 (1978).

CLAY, J. R., and MALONE, J. J., Jr.:

[1] The near-rings with identities on certain finite groups, *Math. Scand.* **19**, 146–150 (1966).

COHEN, E.:

[1] Sums of an even number of squares in $GF[p^n, x]$, I, II, *Duke Math. J.* **14**, 251–267, 543–557 (1947).

[2] Sums of an odd number of squares in $GF[p^n, x]$, *Duke Math. J.* **15**, 501–511 (1948).
[3] An extension of Ramanujan's sums, *Duke Math. J.* **16**, 85–90 (1949).
[4] Sums of products of polynomials in a Galois field, *Duke Math. J.* **18**, 425–430 (1951).
[5] Rings of arithmetic functions, *Duke Math. J.* **19**, 115–129 (1952).
[6] Arithmetic functions of polynomials, *Proc. Amer. Math. Soc.* **3**, 352–358 (1952).
[7] Representations by cubic congruences, *Proc. Nat. Acad. Sci. U.S.A.* **39**, 119–121 (1953).
[8] Congruence representations in algebraic number fields, *Trans. Amer. Math. Soc.* **75**, 444–470 (1953).
[9] The number of solutions of certain cubic congruences, *Pacific J. Math.* **5**, 877–886 (1955).
[10] Simultaneous pairs of linear and quadratic equations in a Galois field, *Canad. J. Math.* **9**, 74–78 (1957).
[11] The number of simultaneous solutions of a quadratic equation and a pair of linear equations over a Galois field, *Rev. Roum. Math. Pures Appl.* **8**, 297–303 (1963).
[12] Linear and quadratic equations in a Galois field with applications to geometry, *Duke Math. J.* **32**, 633–641 (1965).
[13] Quadratic congruences with an odd number of summands, *Amer. Math. Monthly* **73**, 138–143 (1966).

COHEN, H., and LENSTRA, H. W., Jr.:
[1] Primality testing and Jacobi sums, Report 82-18, Dept. of Math., Univ. of Amsterdam, 1982.

COHEN, S. D.:
[1] The distribution of irreducible polynomials in several indeterminates over a finite field, *Proc. Edinburgh Math. Soc.* (2)**16**, 1–17 (1968).
[2] On irreducible polynomials of certain types in finite fields, *Proc. Cambridge Philos. Soc.* **66**, 335–344 (1969).
[3] Further arithmetical functions in finite fields, *Proc. Edinburgh Math. Soc.* (2)**16**, 349–363 (1969).
[4] Some arithmetical functions in finite fields, *Glasgow Math. J.* **11**, 21–36 (1970).
[5] The distribution of polynomials over finite fields, *Acta Arith.* **17**, 255–271 (1970).
[6] The distribution of polynomials over finite fields, II, *Acta Arith.* **20**, 53–62 (1972).
[7] Uniform distribution of polynomials over finite fields, *J. London Math. Soc.* (2)**6**, 93–102 (1972).
[8] The values of a polynomial over a finite field, *Glasgow Math. J.* **14**, 205–208 (1973).
[9] Value sets of functions over finite fields, *Acta Arith.* **39**, 339–359 (1981).
[10] The irreducibility of compositions of linear polynomials over a finite field, *Compositio Math.* **47**, 149–152 (1982).

COLLINS, G. E.:

[1] Computing multiplicative inverses in $GF(p)$, *Math. Comp.* **23**, 197–200 (1969).

[2] The calculation of multivariate polynomial resultants, *J. Assoc. Comput. Mach.* **18**, 515–532 (1971).

[3] Computer algebra of polynomials and rational functions, *Amer. Math. Monthly* **80**, 725–755 (1973).

CONSTANTIN, J., and COURTEAU, B.:

[1] Partitions linéaires arguésiennes d'un espace vectoriel, *Discrete Math.* **33**, 139–147 (1981).

CONWAY, J. H.:

[1] A tabulation of some information concerning finite fields, *Computers in Mathematical Research* (R. F. Churchhouse and J.-C. Herz, eds.), pp. 37–50, North-Holland, Amsterdam, 1968.

COOPER, R. H.:

[1] Linear transformations in Galois fields and their application to cryptography, *Cryptologia* **4**, 184–188 (1980).

CORDES, C. M.:

[1] A note on Pall partitions over finite fields, *Linear Algebra Appl.* **12**, 81–85 (1975).

[2] Some results on totally isotropic subspaces and five-dimensional quadratic forms over $GF(q)$, *Canad. J. Math.* **27**, 271–275 (1975).

CORDONE, G.:

[1] Sulla congruenza generale di 4^0 grado secondo un modulo primo, *Rend. Circ. Mat. Palermo* **9**, 209–243 (1895).

CORNACCHIA, G.:

[1] Sulla congruenza $x^n + y^n \equiv z^n (\text{mod. } p)$, *Giorn. Mat. Battaglini* **47**, 219–268 (1909).

CORSON, H. H.:

[1] On some special systems of equations, *Pacific J. Math.* **6**, 449–452 (1956).

CORZATT, C. E.:

[1] Permutation polynomials over the rational numbers, *Pacific J. Math.* **61**, 361–382 (1975).

CRAMPTON, T. H. M., and WHAPLES, G.:

[1] Additive polynomials. II, *Trans. Amer. Math. Soc.* **78**, 239–252 (1955).

CRAVEN, T., and CSORDAS, G.:

[1] Multiplier sequences for fields, *Illinois J. Math.* **21**, 801–817 (1977).

CRELLE, A. L.:

[1] Table des racines primitives etc. pour les nombres premiers depuis 3 jusqu'à 101, precédée d'une note sur le calcul de cette table, *J. reine angew. Math.* **9**, 27–53 (1832).

CROWE, D. W.:
[1] The trigonometry of $GF(2^{2n})$ and finite hyperbolic planes, *Mathematika* **11**, 83–88 (1964).

CROWELL, R. H.:
[1] Graphs of linear transformations over finite fields, *J. Soc. Indust. Appl. Math.* **10**, 103–112 (1962).

ČUBARIKOV, V. N.:
[1] Multiple rational trigonometric sums and multiple integrals (Russian), *Mat. Zametki* **20**, 61–68 (1976); *Math. Notes* **20**, 589–593 (1976).

CUNNINGHAM, A. J. C.:
[1] Factorisation of $(y^n \mp 1)$, $y > 12$, *Messenger of Math.* **57**, 72–80 (1927).

CUNNINGHAM, A. J. C., and WOODALL, H. J.:
[1] *Factorization of* $y^n \pm 1$, $y = 2,3,5,6,7,10,11,12$ *up to High Powers* (n), Hodgson, London, 1925.

CUNNINGHAM, A. J. C., WOODALL, H. J., and CREAK, T. G.:
[1] On least primitive roots, *Proc. London Math. Soc.* **21**, 343–358 (1923).

ČUPONA, G.:
[1] On periodic fields (Macedonian), *Bull. Soc. Math. Phys. Macédoine* **11**, 5–8 (1960).

CURTIS, C. W.:
[1] Representations of finite groups of Lie type, *Bull. Amer. Math. Soc. (N. S.)* **1**, 721–757 (1979).

CZARNOTA, A.:
[1] Congruences satisfied by a sum of powers of primitive roots with respect to a prime modulus (Polish), *Prace Mat.* **8**, 131–142 (1963/64).

DADE, E. C., ROBINSON, D. W., TAUSSKY, O., and WARD, M.:
[1] Divisors of recurrent sequences, *J. reine angew. Math.* **214/215**, 180–183 (1964).

DAI, Z. D.:
[1] The period of a circulant over a finite field (Chinese), *Acta Math. Sinica* **23**, 70–77 (1980).

DAI, Z. D., and FENG, X.:
[1] Notes on finite geometries and the construction of PBIB designs. IV. Some "Anzahl" theorems in orthogonal geometry over finite fields of characteristic not 2, *Sci. Sinica* **13**, 2001–2004 (1964).
[2] Studies in finite geometries and the construction of incomplete block designs. IV. Some "Anzahl" theorems in orthogonal geometry over finite fields of characteristic $\neq 2$ (Chinese), *Acta Math. Sinica* **15**, 545–558 (1965); *Chinese Math. Acta* **7**, 265–280 (1965).

DALEN, K.:
[1] On a theorem of Stickelberger, *Math. Scand.* **3**, 124–126 (1955).

DALLA, R. H., and PORTER, A. D.:

[1] A consideration by rank of the matrix equation $AX_1 \cdots X_n = B$, *Atti Accad. Naz. Lincei Rend. Cl. Sci. Fis. Mat. Natur.* (8)**52**, 301–311 (1972).

[2] The matrix equation $U_1 \cdots U_n A V_1 \cdots V_m = B$ over a finite field, *Math. Nachr.* **57**, 321–335 (1973).

DARBI, G.:

[1] Sulla riducibilità delle equazioni algebriche, *Ann. Mat. Pura Appl.* (4)**4**, 185–208 (1927).

DAVENPORT, H.:

[1] On the distribution of quadratic residues (mod p), *J. London Math. Soc.* **6**, 49–54 (1931).

[2] On the distribution of l-th power residues (mod p), *J. London Math. Soc.* **7**, 117–121 (1932).

[3] On the distribution of quadratic residues (mod p). II, *J. London Math. Soc.* **8**, 46–52 (1933).

[4] On certain exponential sums, *J. reine angew. Math.* **169**, 158–176 (1933).

[5] On the addition of residue classes, *J. London Math. Soc.* **10**, 30–32 (1935).

[6] On primitive roots in finite fields, *Quart. J. Math.* **8**, 308–312 (1937).

[7] On character sums in finite fields, *Acta Math.* **71**, 99–121 (1939).

[8] *Multiplicative Number Theory*, Markham, Chicago, 1967.

[9] Bases for finite fields, *J. London Math. Soc.* **43**, 21–39 (1968); Addendum, *ibid.* **44**, 378 (1969).

[10] A property of polynomials over a finite field, *Mathematika* **22**, 151–153 (1975).

DAVENPORT, H., and ERDÖS, P.:

[1] The distribution of quadratic and higher residues, *Publ. Math. Debrecen* **2**, 252–265 (1952).

DAVENPORT, H., and HASSE, H.:

[1] Die Nullstellen der Kongruenzzetafunktionen in gewissen zyklischen Fällen, *J. reine angew. Math.* **172**, 151–182 (1935).

DAVENPORT, H., and HEILBRONN, H.:

[1] On an exponential sum, *Proc. London Math. Soc.* (2)**41**, 449–453 (1936).

DAVENPORT, H., and LEWIS, D. J.:

[1] Exponential sums in many variables, *Amer. J. Math.* **84**, 649–665 (1962).

[2] Notes on congruences (I), *Quart. J. Math.* (2)**14**, 51–60 (1963).

[3] Character sums and primitive roots in finite fields, *Rend. Circ. Mat. Palermo* (2)**12**, 129–136 (1963).

[4] Notes on congruences (II), *Quart. J. Math.* (2)**14**, 153–159 (1963).

[5] Homogeneous additive equations, *Proc. Royal Soc. London Ser. A* **274**, 443–460 (1963).

[6] Cubic equations of additive type, *Phil. Trans. Royal Soc. London Ser. A* **261**, 97–136 (1966).

[7] Notes on congruences (III), *Quart. J. Math.* (2)**17**, 339–344 (1966).

[8] Simultaneous equations of additive type, *Phil. Trans. Royal Soc. London Ser. A* **264**, 557–595 (1969).

DAVIDA, G. I.:

[1] Inverse of elements of a Galois field, *Electron. Lett.* **8**, 518–520 (1972).

DAVIO, M., DESCHAMPS, J.-P., and THAYSE, A.:

[1] *Discrete and Switching Functions*, McGraw-Hill, New York, 1978.

DAVIS, A. S.:

[1] The Euler-Fermat theorem for matrices, *Duke Math. J.* **18**, 613–617 (1951).

DAYKIN, D. E.:

[1] On the rank of the matrix $f(A)$ and the enumeration of certain matrices over a finite field, *J. London Math. Soc.* **35**, 36–42 (1960).

[2] Distribution of bordered persymmetric matrices in a finite field, *J. reine angew. Math.* **203**, 47–54 (1960).

[3] The irreducible factors of $(cx+d)x^{q^m}-(ax+b)$ over $GF(q)$, *Quart. J. Math.* (2)**14**, 61–64 (1963).

[4] On linear sequences over a finite field, *Amer. Math. Monthly* **70**, 637–642 (1963).

[5] Polynomials over a finite field, *J. London Math. Soc.* **40**, 326–331 (1965).

[6] Generation of irreducible polynomials over a finite field, *Amer. Math. Monthly* **72**, 646–648 (1965).

DAYKIN, D. E., DRESEL, L. A. G., and HILTON, A. J. W.:

[1] The structure of second order sequences in a finite field, *J. reine angew. Math.* **270**, 77–96 (1974).

DE BRUIJN, N. G.:

[1] A combinatorial problem, *Indag. Math.* **8**, 461–467 (1946).

DE CARLI, D. J.:

[1] A generalized Fibonacci sequence over an arbitrary ring, *Fibonacci Quart.* **8**, 182–184, 198 (1970).

DEDEKIND, R.:

[1] Abriss einer Theorie der höhern Congruenzen in Bezug auf einen reellen Primzahl-Modulus, *J. reine angew. Math.* **54**, 1–26 (1857); *Gesammelte Math. Werke*, vol. 1, pp. 40–66, Vieweg, Braunschweig, 1930.

[2] Beweis für die Irreductibilität der Kreistheilungsgleichungen, *J. reine angew. Math.* **54**, 27–30 (1857); *Gesammelte Math. Werke*, vol. 1, pp. 68–71, Vieweg, Braunschweig, 1930.

[3] Über den Zusammenhang zwischen der Theorie der Ideale und der Theorie der höheren Kongruenzen, *Abh. Kgl. Ges. Wiss. Göttingen* **23**, 1–23 (1878); *Gesammelte Math. Werke*, vol. 1, pp. 202–230, Vieweg, Braunschweig, 1930.

DE GROOTE, R.:

[1] Les cubiques dans un plan projectif sur un corps de caractéristique trois, *Acad. Roy. Belg. Bull. Cl. Sci.* (5)**59**, 1140–1155 (1973).

[2] Les cubiques dans un plan projectif sur un corps fini de caractéristique 3, *Acad. Roy. Belg. Bull. Cl. Sci.* (5)**60**, 43–57 (1974).

DE GROOTE, R., and HIRSCHFELD, J. W. P.:
[1] The number of points on an elliptic cubic curve over a finite field, *European J. Combin.* **1**, 327–333 (1980).

DELIGNE, P.:
[1] La conjecture de Weil pour les surfaces $K3$, *Invent. Math.* **15**, 206–226 (1972).
[2] Les intersections complètes de niveau de Hodge un, *Invent. Math.* **15**, 237–250 (1972).
[3] La conjecture de Weil. I, *Inst. Hautes Etudes Sci. Publ. Math.* **43**, 273–307 (1974).
[4] Applications de la formule des traces aux sommes trigonométriques, *Cohomologie Etale* (*Séminaire de Géométrie Algébrique du Bois-Marie SGA* $4\frac{1}{2}$), Lecture Notes in Math., vol. 569, pp. 168–232, Springer-Verlag, Berlin-Heidelberg-New York, 1977.
[5] Sommes de Gauss cubiques et revêtements de $SL(2)$ (d'après S. J. Patterson), *Séminaire Bourbaki 1978/79*, Exp. 539, Lecture Notes in Math., vol. 770, pp. 244–277, Springer-Verlag, Berlin-Heidelberg-New York, 1980.
[6] La conjecture de Weil. II, *Inst. Hautes Etudes Sci. Publ. Math.* **52**, 137–252 (1980).

DELSARTE, J.:
[1] Nombre de solutions des équations polynômiales sur un corps fini (d'après A. Weil), *Séminaire Bourbaki 1950/1951*, Exp. 39, Benjamin, New York, 1966.

DELSARTE, P.:
[1] A geometric approach to a class of cyclic codes, *J. Combinatorial Theory* **6**, 340–359 (1969).
[2] On cyclic codes that are invariant under the general linear group, *IEEE Trans. Information Theory* **IT-16**, 760–769 (1970).
[3] Bilinear forms over a finite field, with applications to coding theory, *J. Combinatorial Theory Ser. A* **25**, 226–241 (1978).

DELSARTE, P., and GOETHALS, J.-M.:
[1] Irreducible binary cyclic codes of even dimension, *Proc. Second Chapel Hill Conf. on Combinatorial Math. and Its Appl.* (Chapel Hill, N.C., 1970), pp. 100–113, Univ. of North Carolina Press, Chapel Hill, N.C., 1970.
[2] Alternating bilinear forms over $GF(q)$, *J. Combinatorial Theory Ser. A* **19**, 26–50 (1975).
[3] Unrestricted codes with the Golay parameters are unique, *Discrete Math.* **12**, 211–224 (1975).

DELSARTE, P., GOETHALS, J.-M., and SEIDEL, J. J.:
[1] Orthogonal matrices with zero diagonal II, *Canad. J. Math.* **23**, 816–832 (1971).

DELSARTE, P., and MCELIECE, R. J.:

[1] Zeros of functions in finite abelian group algebras, *Amer. J. Math.* **98**, 197–224 (1976).

DE MATHAN, B.:

[1] Sur un théorème métrique d'équirépartition mod 1 dans un corps de séries formelles sur un corps fini, *C. R. Acad. Sci. Paris Sér. A* **265**, 289–291 (1967).

[2] Théorème de Koksma dans un corps de séries formelles sur un corps fini, *Sém. Delange-Pisot-Poitou 1967/68, Théorie des Nombres*, Exp. 4, Secrétariat Math., Paris, 1969.

[3] Approximations diophantiennes dans un corps local, *Bull. Soc. Math. France Suppl. Mém.* **21** (1970).

DEMBOWSKI, P.:

[1] Möbiusebenen gerader Ordnung, *Math. Ann.* **157**, 179–205 (1964).

[2] *Finite Geometries*, Springer-Verlag, Berlin-Heidelberg-New York, 1968.

DEM'JANOV, V. B.:

[1] On the representation of zero by forms of the type $\sum_{i=1}^{m} a_i X_i^n$ (Russian), *Dokl. Akad. Nauk SSSR* **105**, 203–205 (1955).

[2] Pairs of quadratic forms over a complete field with discrete norm with a finite field of residue classes (Russian), *Izv. Akad. Nauk SSSR Ser. Mat.* **20**, 307–324 (1956).

DÉNES, J., and KEEDWELL, A. D.:

[1] *Latin Squares and Their Applications*, Academic Press, New York, 1974.

DE POLIGNAC, C.:

[1] Sur la représentation analytique des substitutions, *Bull. Soc. Math. France* **9**, 59–67 (1881).

DESHOUILLERS, J.-M.:

[1] Sur la répartition modulo 1 des puissances d'un élément de $\mathbb{F}_q((X))$, *Proc. Queen's Number Theory Conf.* (Kingston, Ont., 1979), Queen's Papers in Pure and Appl. Math., no. 54, pp. 437–439, Queen's Univ., Kingston, Ont., 1980.

[2] La répartition modulo 1 des puissances de rationnels dans l'anneau des séries formelles sur un corps fini, *Sém. Théorie des Nombres 1979–1980*, Exp. 5, Univ. Bordeaux I, Talence, 1980.

[3] La répartition modulo 1 des puissances d'un élément dans $F_q((X))$, *Recent Progress in Analytic Number Theory* (H. Halberstam and C. Hooley, eds.), vol. 2, pp. 69–72, Academic Press, London, 1981.

DESHOUILLERS, J.-M., and IWANIEC, H.:

[1] Kloosterman sums and Fourier coefficients of cusp forms, *Invent. Math.* **70**, 219–288 (1982).

DESMAREST, E.:

[1] *Théorie des nombres*, Paris, 1852.

DEURING, M.:

[1] Galoissche Theorie und Darstellungstheorie, *Math. Ann.* **107**, 140–144 (1933).

[2] Die Typen der Multiplikatorenringe elliptischer Funktionenkörper, *Abh. Math. Sem. Univ. Hamburg* **14**, 197–272 (1941).

[3] The zeta-functions of algebraic curves and varieties, *J. Indian Math. Soc.* **20**, 89–101 (1956).

[4] *Lectures on the Theory of Algebraic Functions of one Variable*, Lecture Notes in Math., vol. 314, Springer-Verlag, Berlin-Heidelberg-New York, 1973.

DICKEY, L. J., KAIRIES, H.-H., and SHANK, H. S.:

[1] Analogs of Bernoulli polynomials in fields Z_p, *Aequationes Math.* **14**, 401–404 (1976).

DICKSON, L. E.:

[1] Analytic functions suitable to represent substitutions, *Amer. J. Math.* **18**, 210–218 (1896).

[2] The analytic representation of substitutions on a power of a prime number of letters with a discussion of the linear group, *Ann. of Math.* **11**, 65–120, 161–183 (1897).

[3] Higher irreducible congruences, *Bull. Amer. Math. Soc.* **3**, 381–389 (1897).

[4] Determination of the structure of all linear homogeneous groups in a Galois field which are defined by a quadratic invariant, *Amer. J. Math.* **21**, 193–256 (1899).

[5] Certain subgroups of the Betti-Mathieu group, *Amer. J. Math.* **22**, 49–54 (1900).

[6] Proof of the existence of the Galois field of order p^r for every integer r and prime number p, *Bull. Amer. Math. Soc.* **6**, 203–204 (1900).

[7] *Linear Groups with an Exposition of the Galois Field Theory*, Teubner, Leipzig, 1901; Dover, New York, 1958.

[8] On finite algebras, *Göttinger Nachr.* **1905**, 358–393.

[9] Criteria for the irreducibility of functions in a finite field, *Bull. Amer. Math. Soc.* **13**, 1–8 (1906).

[10] On the theory of equations in a modular field, *Bull. Amer. Math. Soc.* **13**, 8–10 (1906).

[11] On quadratic, hermitian and bilinear forms, *Trans. Amer. Math. Soc.* **7**, 275–292 (1906).

[12] Invariants of binary forms under modular transformations, *Trans. Amer. Math. Soc.* **8**, 205–232 (1907).

[13] Invariants of the general quadratic form modulo 2, *Proc. London Math. Soc.* (2)**5**, 301–324 (1907).

[14] On the last theorem of Fermat, *Messenger of Math.* (2)**38**, 14–32 (1908).

[15] On the canonical forms and automorphs of ternary cubic forms, *Amer. J. Math.* **30**, 117–128 (1908).

[16] On triple algebras and ternary cubic forms, *Bull. Amer. Math. Soc.* **14**, 160–169 (1908).

[17] On the congruence $x^n + y^n + z^n \equiv 0 (\mathrm{mod}\ p)$, *Amer. Math. Monthly* **15**, 217–222 (1908).
[18] Invariantive reduction of quadratic forms in the $GF[2^n]$, *Amer. J. Math.* **30**, 263–281 (1908).
[19] On higher congruences and modular invariants, *Bull. Amer. Math. Soc.* **14**, 313–318 (1908).
[20] On families of quadratic forms in a general field, *Quart. J. Pure Appl. Math.* **39**, 316–333 (1908).
[21] On the last theorem of Fermat. II, *Quart. J. Pure Appl. Math.* **40**, 27–45 (1908).
[22] Rational reduction of a pair of binary quadratic forms; their modular invariants, *Amer. J. Math.* **31**, 103–146 (1909).
[23] Definite forms in a finite field, *Trans. Amer. Math. Soc.* **10**, 109–122 (1909).
[24] General theory of modular invariants, *Trans. Amer. Math. Soc.* **10**, 123–158 (1909).
[25] On the congruence $x^n + y^n + z^n \equiv 0 (\mathrm{mod}\ p)$, *J. reine angew. Math.* **135**, 134–141 (1909).
[26] Lower limit for the number of sets of solutions of $x^e + y^e + z^e \equiv 0 (\mathrm{mod}\ p)$, *J. reine angew. Math.* **135**, 181–188 (1909).
[27] A theory of invariants, *Amer. J. Math.* **31**, 337–354 (1909).
[28] On the representation of numbers by modular forms, *Bull. Amer. Math. Soc.* **15**, 338–347 (1909).
[29] An invariantive investigation of irreducible binary modular forms, *Trans. Amer. Math. Soc.* **12**, 1–18 (1911).
[30] A fundamental system of invariants of the general modular linear group with a solution of the form problem, *Trans. Amer. Math. Soc.* **12**, 75–98 (1911).
[31] Note on cubic equations and congruences, *Ann. of Math.* (2)**12**, 149–152 (1911).
[32] On non-vanishing forms, *Quart. J. Pure Appl. Math.* **42**, 162–171 (1911).
[33] Congruencial theory of functions of several variables, *Bull. Amer. Math. Soc.* **17**, 293–294 (1911).
[34] Proof of the finiteness of modular covariants, *Trans. Amer. Math. Soc.* **14**, 299–310 (1913).
[35] *On Invariants and the Theory of Numbers*, American Math. Society, New York, 1914.
[36] The invariants, seminvariants and linear covariants of the binary quartic form modulo 2, *Ann. of Math.* (2)**15**, 114–117 (1914).
[37] Modular invariants of the system of a binary cubic, quadratic and linear form, *Quart. J. Pure Appl. Math.* **45**, 373–384 (1914).
[38] Recent progress in the theories of modular and formal invariants and in modular geometry, *Proc. Nat. Acad. Sci. U.S.A.* **1**, 1–4 (1915).
[39] Projective classification of cubic surfaces modulo 2, *Ann. of Math.* (2)**16**, 139–157 (1915).
[40] *History of the Theory of Numbers*, vol. 1, Carnegie Institute, Washington, D.C., 1919.

[41] *History of the Theory of Numbers*, vol. 2, Carnegie Institute, Washington, D.C., 1920.
[42] *History of the Theory of Numbers*, vol. 3, Carnegie Institute, Washington, D.C., 1923.
[43] Ternary quadratic forms and congruences, *Ann. of Math.* (2)**28**, 333–341 (1927).
[44] Cyclotomy and trinomial congruences, *Trans. Amer. Math. Soc.* **37**, 363–380 (1935).
[45] Cyclotomy when e is composite, *Trans. Amer. Math. Soc.* **38**, 187–200 (1935).
[46] Cyclotomy, higher congruences, and Waring's problem, *Amer. J. Math.* **57**, 391–424 (1935).
[47] Cyclotomy, higher congruences, and Waring's problem II, *Amer. J. Math.* **57**, 463–474 (1935).
[48] Congruences involving only e-th powers, *Acta Arith.* **1**, 161–167 (1936).

DICKSON, L. E., MITCHELL, H. H., VANDIVER, H. S., and WAHLIN, G. E.:
[1] *Algebraic Numbers*, National Research Council, Washington, D.C., 1923.

DIDERRICH, G. T., and MANN, H. B.:
[1] Representations by k-th powers in $GF(q)$, *J. Number Theory* **4**, 269–273 (1972).

DIEUDONNÉ, J.:
[1] Pseudo-discriminant and Dickson invariant, *Pacific J. Math.* **5**, 907–910 (1955).
[2] *La géométrie des groupes classiques*, 3rd ed., Springer-Verlag, Berlin-Heidelberg-New York, 1971.

DIFFIE, W., and HELLMAN, M. E.:
[1] New directions in cryptography, *IEEE Trans. Information Theory* **IT-22**, 644–654 (1976).

DIJKSMA, A.:
[1] The measure theoretic approach to uniform distribution of sequences in $GF[q,x]$, *Mathematica (Cluj)* **11**, 221–240 (1969).
[2] Uniform distribution of polynomials over $GF\{q,x\}$ in $GF[q,x]$, Part I, *Indag. Math.* **31**, 376–383 (1969).
[3] Uniform distribution of polynomials over $GF\{q,x\}$ in $GF[q,x]$, Part II, *Indag. Math.* **32**, 187–195 (1970).
[4] Metrical theorems concerning uniform distribution in $GF[q,x]$ and $GF\{q,x\}$, *Nieuw Arch. Wisk.* (3)**18**, 279–293 (1970).

DILLON, J. F., and MORRIS, R. A.:
[1] On a paper of E. R. Berlekamp, H. M. Fredricksen and R. C. Proto: "Minimum conditions for uniquely determining the generator of a linear sequence" (Utilitas Math. 5 (1974), 305–315), *Utilitas Math.* **5**, 317–322 (1974).

DIRICHLET, G. L.:
[1] Über eine neue Anwendung bestimmter Integrale auf die Summation endlicher oder unendlicher Reihen, *Abh. Königl. Preuss. Akad. Wiss.* **1835**, 391–407; *Werke*, vol. 1, pp. 237–256, Reimer, Berlin, 1889.

DIXON, J. D.:
[1] *The Structure of Linear Groups*, Van Nostrand Reinhold, London, 1971.

DOBRUSHIN, R. L.:
[1] Survey of Soviet research in information theory, *IEEE Trans. Information Theory* **IT-18**, 703–724 (1972).

D'OCAGNE, M.:
[1] Mémoire sur les suites récurrentes, *J. de l'Ecole Polytechnique* **64**, 151–224 (1894).

DOČEV, K., and DIMITROV, D.:
[1] Certain properties of homogeneous equations in finite fields (Bulgarian), *Ann. Univ. Sofia Fac. Math. Méc.* **64**, 269–276 (1969/70).

DODSON, M. M.:
[1] Homogeneous additive congruences, *Phil. Trans. Royal Soc. London Ser. A* **261**, 163–210 (1967).
[2] On Waring's problem in $GF[p]$, *Acta Arith.* **19**, 147–173 (1971).
[3] On a function due to S. Chowla, *J. Number Theory* **5**, 287–292 (1973).
[4] On Waring's Problem in p-adic fields, *Acta Arith.* **22**, 315–327 (1973).

DODSON, M. M., and TIETÄVÄINEN, A.:
[1] A note on Waring's problem in $GF(p)$, *Acta Arith.* **30**, 159–167 (1976).

DODUNEKOV, S. M.:
[1] Essentially distinct irreducible polynomials over finite fields (Russian), *Ann. Univ. Sofia Fac. Math. Méc.* **66**, 169–175 (1971/72).
[2] Goppa codes (Russian), *Ann. Univ. Sofia Fac. Math. Méc.* **68**, 317–322 (1973/74).

DOREY, F., and WHAPLES, G.:
[1] Prime and composite polynomials, *J. Algebra* **28**, 88–101 (1974).

DÖRGE, K.:
[1] Zur Verteilung der quadratischen Reste, *Jber. Deutsch. Math.-Verein.* **38**, 41–49 (1929).

DORNHOFF, L. L., and HOHN, F. E.:
[1] *Applied Modern Algebra*, Macmillan, New York, 1978.

DRESS, F.:
[1] Fonctions arithmétiques sur l'anneau des polynômes à coefficients dans un corps fini, *Sém. Delange-Pisot 1962/63*, Exp. 13, Secrétariat Math., Paris, 1967.

DUBOIS, E., and PAYSAN-LE ROUX, R.:
[1] Approximations simultanées dans un corps de séries formelles, *C. R. Acad. Sci. Paris Sér. A* **274**, 437–440 (1972).

DUNN, K. B., and LIDL, R.:

[1] Iterative roots of functions over finite fields, *Math. Nachr.* (to appear).

DUNTON, M.:

[1] Nontrivial solutions of $ax^3 + by^3 \equiv c(\text{mod}\, p)$, *Norske Vid. Selsk. Forh. (Trondheim)* **33**, 45–46 (1960).

[2] Bounds for pairs of cubic residues, *Proc. Amer. Math. Soc.* **16**, 330–332 (1965).

DUPARC, H. J. A.:

[1] Periodicity properties of recurring sequences. I, II, *Indag. Math.* **16**, 331–342, 473–485 (1954).

[2] Periodicity properties of certain sets of integers, *Indag. Math.* **17**, 449–458 (1955).

DUSKE, J., and JÜRGENSEN, H.:

[1] *Codierungstheorie*, Reihe Informatik, vol. 13, Bibliographisches Institut, Mannheim, 1977.

DUVALL, P. F., and KIBLER, R. E.:

[1] On the parity of the frequency of cycle lengths of shift register sequences, *J. Combinatorial Theory Ser. A* **18**, 357–361 (1975).

DUVALL, P. F., and MORTICK, J. C.:

[1] Decimation of periodic sequences, *SIAM J. Appl. Math.* **21**, 367–372 (1971).

DWORK, B.:

[1] On the congruence properties of the zeta function of algebraic varieties, *J. reine angew. Math.* **203**, 130–142 (1960).

[2] On the rationality of the zeta function of an algebraic variety, *Amer. J. Math.* **82**, 631–648 (1960).

[3] On the zeta function of a hypersurface I, *Inst. Hautes Etudes Sci. Publ. Math.* **12**, 5–68 (1962).

[4] A deformation theory for the zeta function of a hypersurface, *Proc. International Congress of Math.* (Stockholm, 1962), pp. 247–259, Institut Mittag-Leffler, Djursholm, 1963.

[5] On the zeta function of a hypersurface. II, *Ann. of Math.* (2)**80**, 227–299 (1964).

[6] Analytic theory of the zeta function of algebraic varieties, *Arithmetical Algebraic Geometry* (Proc. Conf. Purdue Univ., 1963), pp. 18–32, Harper and Row, New York, 1965.

[7] On zeta functions of hypersurfaces, *Les Tendances Géométriques en Algèbre et Théorie des Nombres*, pp. 77–82, Edition du Centre National de la Recherche Scientifique, Paris, 1966.

[8] On the zeta function of a hypersurface. III, *Ann. of Math.* (2)**83**, 457–519 (1966).

[9] On the rationality of zeta functions and *L*-series, *Proc. Conf. Local Fields* (Driebergen, 1966), pp. 40–55, Springer-Verlag, Berlin-Heidelberg-New York, 1967.

[10] p-adic cycles, *Inst. Hautes Etudes Sci. Publ. Math.* **37**, 27–115 (1969).
[11] Bessel functions as p-adic functions of the argument, *Duke Math. J.* **41**, 711–738 (1974).

DYE, R. H.:
[1] On the Arf invariant, *J. Algebra* **53**, 36–39 (1978).

DYN'KIN, V. N., and AGARONOV, D. A.:
[1] A method of decomposition of polynomials in a finite field (Russian), *Problemy Peredači Informacii* **6**, no. 3, 82–86 (1970); *Problems of Information Transmission* **6**, 257–261 (1970).

EHLICH, H.:
[1] Neue Hadamard-Matrizen, *Arch. Math.* **16**, 34–36 (1965).

EICHLER, M.:
[1] *Einführung in die Theorie der algebraischen Zahlen und Funktionen*, Birkhäuser Verlag, Basel, 1963.

EICHNER, L.:
[1] Lineare Realisierbarkeit endlicher Automaten über endlichen Körpern, *Acta Inform.* **3**, 75–100 (1973).

EIER, R., and LIDL, R.:
[1] Tschebyscheffpolynome in einer und zwei Variablen, *Abh. Math. Sem. Univ. Hamburg* **41**, 17–27 (1974).

EIER, R., and MALLECK, H.:
[1] Anwendung von Multiplextechniken bei der Erzeugung von schnellen Pseudozufallsfolgen, *Nachrichtentechn. Z.* **28**, 227–231 (1975).

EISENSTEIN, G.:
[1] Beiträge zur Kreistheilung, *J. reine angew. Math.* **27**, 269–278 (1844); *Math. Werke*, vol. 1, pp. 45–54, Chelsea, New York, 1975.
[2] Beweis des Reciprocitätssatzes für die cubischen Reste in der Theorie der aus dritten Wurzeln der Einheit zusammengesetzten complexen Zahlen, *J. reine angew. Math.* **27**, 289–310 (1844); *Math. Werke*, vol. 1, pp. 59–80, Chelsea, New York, 1975.
[3] La loi de réciprocité tirée des formules de Mr. Gauss, sans avoir déterminé préalablement le signe du radical, *J. reine angew. Math.* **28**, 41–43 (1844); *Math. Werke*, vol. 1, pp. 114–116, Chelsea, New York, 1975.
[4] Lois de réciprocité, *J. reine angew. Math.* **28**, 53–67 (1844); *Math. Werke*, vol. 1, pp. 126–140, Chelsea, New York, 1975.
[5] Zur Theorie der quadratischen Zerfällung der Primzahlen $8n+3$, $7n+2$ und $7n+4$, *J. reine angew. Math.* **37**, 97–126 (1848); *Math. Werke*, vol. 2, pp. 506–535, Chelsea, New York, 1975.
[6] Lehrsätze, *J. reine angew. Math.* **39**, 180–182 (1850); *Math. Werke*, vol. 2, pp. 620–622, Chelsea, New York, 1975.

ELISTRATOV, I. V.:
[1] On the number of solutions of certain equations in finite fields (Russian), *Works of Young Scientists* (Russian), pp. 27–30, Izdat. Saratov. Univ., Saratov, 1964.

[2] On the number of solutions of certain equations in finite fields (Russian), *Certain Problems in the Theory of Fields* (Russian), pp. 48–59, Izdat. Saratov. Univ., Saratov, 1964.

[3] Elementary proof of Hasse's theorem (Russian), *Studies in Number Theory* (Russian), no. 1, pp. 21–26, Izdat. Saratov. Univ., Saratov, 1966.

[4] The number of classes and the location of zeros of the $Z(u)$-function (Russian), *Volž. Mat. Sb. Vyp.* **4**, 58–65 (1966).

[5] Some questions of the theory of algebraic functions (Russian), *Studies in Number Theory* (Russian), no. 4, pp. 17–34, Izdat. Saratov. Univ., Saratov, 1972.

[6] An estimate of the number of solutions of a certain equation in a finite field (Russian), *Dokl. Akad. Nauk Tadžik. SSR* **17**, no. 8, 3–6 (1974).

[7] An elementary approach to the estimation of rational trigonometric sums (Russian), *Litovsk. Mat. Sb.* **17**, 91–110 (1977).

ELLIOTT, P. D. T. A.:

[1] Some notes on k-th power residues, *Acta Arith.* **14**, 153–162 (1968).

[2] A restricted mean value theorem, *J. London Math. Soc.* (2)**1**, 447–460 (1969).

[3] On the mean value of $f(p)$, *Proc. London Math. Soc.* (3)**21**, 28–96 (1970).

[4] The distribution of power residues and certain related results, *Acta Arith.* **17**, 141–159 (1970).

[5] A remark on the Dirichlet values of a completely reducible polynomial (mod p), *J. Number Theory* **13**, 12–17 (1981).

ELLISON, W. J.:

[1] Waring's problem, *Amer. Math. Monthly* **78**, 10–36 (1971).

ELSPAS, B.:

[1] The theory of autonomous linear sequential networks, *IRE Trans. Circuit Theory* **CT-6**, 45–60 (1959).

ELSPAS, B., and SHORT, R. A.:

[1] A note on optimum burst-error-correcting codes, *IRE Trans. Information Theory* **IT-8**, 39–42 (1962).

EMRE, E., and HÜSEYIN, Ö.:

[1] Relative primeness of multivariable polynomials, *IEEE Trans. Circuits and Systems* **CAS-22**, 56–57 (1975).

ENGLISH, W. R.:

[1] Synthesis of finite state algorithms in a Galois field $GF[p^n]$, *IEEE Trans. Computers* **C-30**, 225–229 (1981).

ENGSTROM, H. T.:

[1] Periodicity in sequences defined by linear recurrence relations, *Proc. Nat. Acad. Sci. U.S.A.* **16**, 663–665 (1930).

[2] On sequences defined by linear recurrence relations, *Trans. Amer. Math. Soc.* **33**, 210–218 (1931).

[3] Polynomial substitutions, *Amer. J. Math.* **63**, 249–255 (1941).

ENNOLA, V.:
[1] Note on an equation in a finite field, *Ann. Acad. Sci. Fenn. Ser. AI* **314** (1962), 6 pp.

ERDÖS, P.:
[1] Some recent advances and current problems in number theory, *Lectures on Modern Mathematics*, vol. 3, pp. 196–244, Wiley, New York, 1965.

ESCOTT, E. B.:
[1] Cubic congruences with three real roots, *Ann. of Math.* (2)**11**, 86–92 (1910).

ESTERMANN, T.:
[1] Vereinfachter Beweis eines Satzes von Kloosterman, *Abh. Math. Sem. Univ. Hamburg* **7**, 82–98 (1930).
[2] On the sign of the Gaussian sum, *J. London Math. Soc.* **20**, 66–67 (1945).
[3] On Kloosterman's sum, *Mathematika* **8**, 83–86 (1961).
[4] A new application of the Hardy-Littlewood-Kloosterman method, *Proc. London Math. Soc.* (3)**12**, 425–444 (1962).

EULER, L.:
[1] Recherches sur une nouvelle espèce de quarrés magiques, *Verh. Zeeuwsch Genootsch. der Wetensch. Vlissingen* **9**, 85–239 (1782).

EVANS, R. J.:
[1] Generalizations of a theorem of Chowla on Gaussian sums, *Houston J. Math.* **3**, 343–349 (1977).
[2] Resolution of sign ambiguities in Jacobi and Jacobsthal sums, *Pacific J. Math.* **81**, 71–80 (1979).
[3] Unambiguous evaluations of bidecic Jacobi and Jacobsthal sums, *J. Australian Math. Soc. Ser. A* **28**, 235–240 (1979).
[4] Bioctic Gauss sums and sixteenth power residue difference sets, *Acta Arith.* **38**, 37–46 (1980).
[5] Note on intersections of translates of powers in finite fields, *Hokkaido Math. J.* **9**, 135–137 (1980).
[6] The 2^r-th power character of 2, *J. reine angew. Math.* **315**, 174–189 (1980).
[7] Identities for products of Gauss sums over finite fields, *L'Enseignement Math.* (2)**27**, 197–209 (1981).
[8] Pure Gauss sums over finite fields, *Mathematika* **28**, 239–248 (1981).
[9] Rational reciprocity laws, *Acta Arith.* **39**, 281–294 (1981).
[10] Twenty-fourth power residue difference sets, *Math. Comp.* (to appear).

EVANS, R. J., and HILL, J. R.:
[1] The cyclotomic numbers of order sixteen, *Math. Comp.* **33**, 827–835 (1979).

EVANS, T. A., and MANN, H. B.:
[1] On simple difference sets, *Sankhyā* **11**, 357–364 (1951).

FADINI, A.:
[1] Un'interpretazione mediante algebre dei campi finiti di Galois di ordine p^n, *Rend. Accad. Sci. Fis. Mat. Napoli* (4)**19**, 42–44 (1952).

FAIRCLOTH, O. B.:

[1] Summary of new results concerning the solutions of equations in finite fields, *Proc. Nat. Acad. Sci. U.S.A.* **37**, 619–622 (1951).

[2] On the number of solutions of some general types of equations in a finite field, *Canad. J. Math.* **4**, 343–351 (1952).

FAIRCLOTH, O. B., and VANDIVER, H. S.:

[1] On multiplicative properties of a generalized Jacobi-Cauchy cyclotomic sum, *Proc. Nat. Acad. Sci. U.S.A.* **36**, 260–267 (1950).

[2] On certain diophantine equations in rings and fields, *Proc. Nat. Acad. Sci. U.S.A.* **38**, 52–57 (1952).

FANO, G.:

[1] Sui postulati fondamentali della geometria proiettiva in uno spazio a un numero qualunque di dimensioni, *Giorn. Mat. Battaglini* **30**, 106–132 (1892).

FATEMAN, R. J.:

[1] Polynomial multiplication, powers and asymptotic analysis: Some comments, *SIAM J. Computing* **3**, 196–213 (1974).

FATOU, P.:

[1] Sur les fonctions qui admettent plusieurs théorèmes de multiplication, *C. R. Acad. Sci. Paris* **173**, 571–573 (1921).

FEIT, W., and FINE, N. J.:

[1] Pairs of commuting matrices over a finite field, *Duke Math. J.* **27**, 91–94 (1960).

FEIT, W., and REES, E.:

[1] A criterion for a polynomial to factor completely over the integers, *Bull. London Math. Soc.* **10**, 191–192 (1978).

FENG, K. Q.:

[1] Pseudo-random properties of linear shift register sequences (Chinese), *Acta Math. Sinica* **19**, 192–202 (1976).

FENG, X., and DAI, Z. D.:

[1] Notes on finite geometries and the construction of PBIB designs. V. Some "Anzahl" theorems in orthogonal geometry over finite fields of characteristic 2, *Sci. Sinica* **13**, 2005–2008 (1964).

[2] Studies in finite geometries and the construction of incomplete block designs. V. Some "Anzahl" theorems in orthogonal geometry over finite fields of characteristic 2 (Chinese), *Acta Math. Sinica* **15**, 664–682 (1965); *Chinese Math. Acta* **7**, 392–410 (1965).

FIDUCCIA, C. M., and ZALCSTEIN, Y.:

[1] Algebras having linear multiplicative complexities, *J. Assoc. Comput. Mach.* **24**, 311–331 (1977).

FILLMORE, J. P.:

[1] A note on split dilations defined by higher residues, *Proc. Amer. Math. Soc.* **18**, 171–174 (1967).

FILLMORE, J. P., and MARX, M. L.:

[1] Linear recursive sequences, *SIAM Rev.* **10**, 342–353 (1968).

FINE, N. J.:

[1] On the asymptotic distribution of the elementary symmetric functions (mod p), *Trans. Amer. Math. Soc.* **69**, 109–129 (1950).

FINE, N. J., and HERSTEIN, I. N.:

[1] The probability that a matrix be nilpotent, *Illinois J. Math.* **2**, 499–504 (1958).

FINE, N. J., and NIVEN, I.:

[1] The probability that a determinant be congruent to a (mod m), *Bull. Amer. Math. Soc.* **50**, 89–93 (1944).

FISHER, R. A.:

[1] *The Design of Experiments*, Oliver and Boyd, Edinburgh, 1942.

FISHER, S. D., and ALEXANDER, M. N.:

[1] Matrices over a finite field, *Amer. Math. Monthly* **73**, 639–641 (1966).

FITZPATRICK, G. B.:

[1] Synthesis of binary ring counters of given periods, *J. Assoc. Comput. Mach.* **7**, 287–297 (1960).

FLYE SAINTE-MARIE, C.:

[1] Réponse à la question 48, *L'Interméd. Math.* **1**, 107–110 (1894).

FOMENKO, O. M.:

[1] An application of Eichler's reduction formula to the representation of numbers by certain quaternary quadratic forms (Russian), *Mat. Zametki* **9**, 71–76 (1971); *Math. Notes* **9**, 41–44 (1971).

FORNEY, G. D., Jr.:

[1] On decoding BCH codes, *IEEE Trans. Information Theory* **IT-11**, 549–557 (1965).

FORSYTH, A. R.:

[1] Primitive roots of prime numbers and their residues, *Messenger of Math.* (2)**13**, 169–192 (1884).

FRALEIGH, J. B.:

[1] *A First Course in Abstract Algebra*, Addison-Wesley, Reading, Mass., 1967.

FRAME, J. S.:

[1] A short proof of quadratic reciprocity, *Amer. Math. Monthly* **85**, 818–819 (1978).

FRATTINI, G.:

[1] Intorno ad un teorema di Lagrange, *Atti Reale Accad. Lincei Rend.* (4)**1**, 136–142 (1885).

[2] Sulle congruenze omogenee e simmetriche con un numero primo di variabili, *Periodico di Mat.* **29**, 49–53 (1913).

FRAY, R., and GILMER, R.:

[1] On solvability by radicals of finite fields, *Math. Ann.* **199**, 279–291 (1972).

FREDMAN, M. L.:

[1] Congruence formulas obtained by counting irreducibles, *Pacific J. Math.* **35**, 613–624 (1970).

[2] The distribution of absolutely irreducible polynomials in several indeterminates, *Proc. Amer. Math. Soc.* **31**, 387–390 (1972).

FREDRICKSEN, H.:

[1] A class of nonlinear de Bruijn cycles, *J. Combinatorial Theory Ser. A* **19**, 192–199 (1975).

[2] A survey of full length nonlinear shift register cycle algorithms, *SIAM Rev.* **24**, 195–221 (1982).

FREDRICKSEN, H., and KESSLER, I.:

[1] Lexicographic compositions and de Bruijn sequences, *J. Combinatorial Theory Ser. A* **22**, 17–30 (1977).

FREDRICKSEN, H., and WISNIEWSKI, R.:

[1] On trinomials $x^n + x^2 + 1$ and $x^{8l \pm 3} + x^k + 1$ irreducible over $GF(2)$, *Information and Control* **50**, 58–63 (1981).

FREDRICSSON, S. A.:

[1] Pseudo-randomness properties of binary shift register sequences, *IEEE Trans. Information Theory* **IT-21**, 115–120 (1975).

FRIED, M.:

[1] Arithmetical properties of value sets of polynomials, *Acta Arith.* **15**, 91–115 (1969).

[2] On a conjecture of Schur, *Michigan Math. J.* **17**, 41–55 (1970).

[3] On a theorem of Ritt and related Diophantine problems, *J. reine angew. Math.* **264**, 40–55 (1973).

[4] On a theorem of MacCluer, *Acta Arith.* **25**, 121–126 (1974).

[5] Arithmetical properties of function fields (II). The generalized Schur problem, *Acta Arith.* **25**, 225–258 (1974).

FRIED, M., and MACRAE, R. E.:

[1] On the invariance of chains of fields, *Illinois J. Math.* **13**, 165–171 (1969).

FRIED, M., and SACERDOTE, G.:

[1] Solving diophantine problems over all residue class fields of a number field and all finite fields, *Ann. of Math.* (2)**104**, 203–233 (1976).

FRIEDLAND, B.:

[1] Linear modular sequential circuits, *IRE Trans. Circuit Theory* **CT-6**, 61–68 (1959).

FRIEDLAND, B., and STERN, T. E.:

[1] On periodicity of states in linear modular sequential circuits, *IRE Trans. Information Theory* **IT-5**, 136–137 (1959).

FRIEDLANDER, J. B.:

[1] A note on primitive roots in finite fields, *Mathematika* **19**, 112–114 (1972).

[2] On the least kth power non-residue in an algebraic number field, *Proc. London Math. Soc.* (3)**26**, 19–34 (1973).

[3] Character sums in quadratic fields, *Proc. London Math. Soc.* (3)**28**, 99–111 (1974).

FRIEDMAN, W. F., and MENDELSOHN, C. J.:

[1] Notes on code words, *Amer. Math. Monthly* **39**, 394–409 (1932).

FRÖBERG, C.-E.:

[1] New results on the Kummer conjecture, *BIT* **14**, 117–119 (1974).

FRÖHLICH, A.:

[1] Non-abelian Jacobi sums, *Number Theory and Algebra* (H. Zassenhaus, ed.), pp. 71–75, Academic Press, New York, 1977.

[2] Stickelberger without Gauss sums, *Algebraic Number Fields* (A. Fröhlich, ed.), pp. 589–607, Academic Press, London, 1977.

FRÖHLICH, A., and TAYLOR, M. J.:

[1] The arithmetic theory of local Galois Gauss sums for tame characters, *Phil. Trans. Royal Soc. London Ser. A* **298**, 141–181 (1980).

FROLOV, M.:

[1] Sur les racines primitives, *Bull. Soc. Math. France* **21**, 113–128 (1893); *ibid.* **22**, 241–245 (1894).

FRYER, K. D.:

[1] Note on permutations in a finite field, *Proc. Amer. Math. Soc.* **6**, 1–2 (1955).

[2] A class of permutation groups of prime degree, *Canad. J. Math.* **7**, 24–34 (1955).

FULTON, J. D.:

[1] Symmetric involutory matrices over finite fields and modular rings of integers, *Duke Math. J.* **36**, 401–407 (1969).

[2] Stochastic involutions over a finite field, *Duke Math. J.* **39**, 391–399 (1972).

[3] Characterization and enumeration of linear classes of involutions over a finite field, *Linear Algebra Appl.* **6**, 119–127 (1973).

[4] Linear classes of involutions over fields of characteristic two, *Linear Algebra Appl.* **6**, 129–142 (1973).

[5] Representations by quadratic forms of arbitrary rank in a finite field of characteristic two, *Linear and Multilinear Algebra* **4**, 89–101 (1976).

[6] Representations by Hermitian forms in a finite field of characteristic two, *Canad. J. Math.* **29**, 169–179 (1977).

[7] Representations by quadratic forms in a finite field of characteristic two, *Math. Nachr.* **77**, 237–243 (1977).

[8] Generalized inverses of matrices over a finite field, *Discrete Math.* **21**, 23–29 (1978).

[9] Gauss sums and solutions to simultaneous equations over $GF(2^y)$, *Acta Arith.* **35**, 17–24 (1979).

[10] Generalized inverses of matrices over fields of characteristic two, *Linear Algebra Appl.* **28**, 69–76 (1979).

FULTON, J. D., and MORRIS, W. L.:
[1] On arithmetical functions related to the Fibonacci numbers, *Acta Arith.* **16**, 105–110 (1969).

FULTON, W.:
[1] A fixed point formula for varieties over finite fields, *Math. Scand.* **42**, 189–196 (1978).

FURQUIM DE ALMEIDA, F.:
[1] On a formula of Cipolla (Portuguese), *Summa Brasil. Math.* **1**, no. 10, 207–219 (1946).
[2] The law of quadratic reciprocity (Portuguese), *Bol. Soc. Mat. São Paulo* **3**, no. 1–2, 3–8 (1948).

FURSTENBERG, H.:
[1] Algebraic functions over finite fields, *J. Algebra* **7**, 271–277 (1967).

FURTADO GOMIDE, E.:
[1] On the theorem of Artin-Weil (Portuguese), *Bol. Soc. Mat. São Paulo* **4**, 1–18 (1949).

GAAL, L.:
[1] *Classical Galois Theory with Examples*, Markham, Chicago, 1971.

GABIDULIN, E. M.:
[1] Codes invariant with respect to linear transformations (Russian), *Radiotehn. i Elektron.* **11**, 433–438 (1966); *Radio Engrg. Electron. Phys.* **11**, 365–369 (1966).

GAIU, E.:
[1] Congruences of matrices having integer elements (Romanian), *Gaz. Mat. Fiz. Ser. A* **11**, 334–337 (1959).

GALLAGER, R. G.:
[1] *Information Theory and Reliable Communication*, Wiley, New York, 1968.

GALOIS, E.:
[1] Sur la théorie des nombres, *Bull. Sci. Math. de M. Férussac* **13**, 428–435 (1830); *J. Math. Pures Appl.* **11**, 398–407 (1846); *Oeuvres math.*, pp. 15–23, Gauthier-Villars, Paris, 1897.

GANTMACHER, F. R.:
[1] *The Theory of Matrices*, vol. 1, Chelsea, New York, 1959.

GARAKOV, G. A.:
[1] An algorithm for determining irreducible binary polynomials and their exponents (Russian), *Izv. Akad. Nauk Armjan. SSR Ser. Fiz.-Mat. Nauk* **17**, no. 5, 7–16 (1964).
[2] A certain property of the primitive elements of the field Z_p^* (Russian), *Dokl. Akad. Nauk Armjan. SSR* **46**, 213–216 (1968).

[3] Tables of irreducible polynomials over the field $GF(p)$ ($p \leqslant 11$) (Russian), *Trudy Vyčisl. Centra Akad. Nauk Armjan. SSR i Erevan. Gos. Univ. Mat. Voprosy Kibernet. i Vyčisl. Tehn.* **6**, 112–142 (1970).

GAUSS, C. F.:

[1] *Disquisitiones Arithmeticae*, Fleischer, Leipzig, 1801; *Werke*, vol. 1, Königl. Gesellschaft der Wissenschaften, Göttingen, 1863; *Untersuchungen über Höhere Arithmetik* (H. Maser, ed.), pp. 1–453, Springer, Berlin, 1889; Yale Univ. Press, New Haven, Conn., 1966.

[2] Summatio quarumdam serierum singularium, *Comment. Soc. Reg. Sci. Gottingensis* **1** (1811); *Werke*, vol. 2, pp. 11–45, Königl. Gesellschaft der Wissenschaften, Göttingen, 1876; *Untersuchungen über Höhere Arithmetik* (H. Maser, ed.), pp. 463–495, Springer, Berlin, 1889.

[3] Theoria residuorum biquadraticorum. Commentatio prima, *Comment. Soc. Reg. Sci. Gottingensis* **6** (1828); *Werke*, vol. 2, pp. 65–92, Königl. Gesellschaft der Wissenschaften, Göttingen, 1876; *Untersuchungen über Höhere Arithmetik* (H. Maser, ed.), pp. 511–533, Springer, Berlin, 1889.

[4] Analysis residuorum: Caput octavum. Disquisitiones generales de congruentiis, *Werke*, vol. 2, pp. 212–240, Königl. Gesellschaft der Wissenschaften, Göttingen, 1876; *Untersuchungen über Höhere Arithmetik* (H. Maser, ed.), pp. 602–629, Springer, Berlin, 1889.

[5] Disquisitionum circa aequationes puras ulterior evolutio, *Werke*, vol. 2, pp. 243–265, Königl. Gesellschaft der Wissenschaften, Göttingen, 1876; *Untersuchungen über Höhere Arithmetik* (H. Maser, ed.), pp. 630–652, Springer, Berlin, 1889.

GAY, D., and VÉLEZ, W. Y.:

[1] On the degree of the splitting field of an irreducible binomial, *Pacific J. Math.* **78**, 117–120 (1978).

GEGENBAUER, L.:

[1] Die Bedingungen für die Existenz einer bestimmten Anzahl von Wurzeln einer Congruenz, *Sitzungsber. Wien Abt. II* **95**, 165–169 (1887).

[2] Über Congruenzen, *Sitzungsber. Wien Abt. II* **95**, 610–617 (1887).

[3] Über ein Theorem des Herrn Pepin, *Sitzungsber. Wien Abt. II* **95**, 838–842 (1887).

[4] Zur Theorie der Congruenzen, *Sitzungsber. Wien Abt. II* **98**, 652–672 (1889).

[5] Zur Theorie der Congruenzen mit mehreren Unbekannten, *Sitzungsber. Wien Abt. II* **99**, 790–813 (1890).

[6] Einige mathematische Theoreme, *Sitzungsber. Wien Abt. II* **102**, 549–564 (1893).

[7] Ueber Congruenzen in Bezug auf einen Primzahlmodul, *Monatsh. Math. Phys.* **5**, 230–232 (1894).

GEIJSEL, J. M.:

[1] *Transcendence in Fields of Positive Characteristic*, Mathematical Centre Tracts, vol. 91, Mathematisch Centrum, Amsterdam, 1979.

GEL'FAND, S. I.:

[1] Irreducible polynomials over a finite field (Russian), *Uspehi Mat. Nauk* **24**, no. 4, 193–194 (1969).

[2] Representations of the full linear group over a finite field (Russian), *Mat. Sb.* **83**, 15–41 (1970); *Math. USSR Sbornik* **12**, 13–39 (1970).

GEL'FOND, A. O.:

[1] *Differenzenrechnung*, VEB Deutscher Verlag der Wissenschaften, Berlin, 1958.

GEL'FOND, A. O., and LINNIK, YU. V.:

[1] *Elementary Methods in the Analytic Theory of Numbers*, Pergamon Press, Oxford, 1966.

GERJETS, M. S., and BERGUM, G. E.:

[1] The distribution of primitive roots in fields of order p^2, *Bull. Calcutta Math. Soc.* **68**, 53–62 (1976).

GERST, I., and BRILLHART, J.:

[1] On the prime divisors of polynomials, *Amer. Math. Monthly* **78**, 250–266 (1971).

GERSTENHABER, M.:

[1] On the number of nilpotent matrices with coefficients in a finite field, *Illinois J. Math.* **5**, 330–333 (1961).

GHENT, K. S.:

[1] Sums of values of a polynomial multiplied by constants, *Duke Math. J.* **3**, 518–528 (1937).

GILBERT, E. N.:

[1] A comparison of signalling alphabets, *Bell System Tech. J.* **31**, 504–522 (1952).

GILL, A.:

[1] *Introduction to the Theory of Finite-State Machines*, McGraw-Hill, New York, 1962.

[2] *Linear Sequential Circuits: Analysis, Synthesis, and Applications*, McGraw-Hill, New York, 1966.

GILL, A., and JACOB, J. P.:

[1] On a mapping polynomial for Galois fields, *Quart. Appl. Math.* **24**, 57–62 (1966).

GILLETT, J. R.:

[1] Character sums of polynomials to a prime modulus, *Proc. London Math. Soc.* (3)**27**, 205–221 (1973).

GILMER, R.:

[1] Finite rings having a cyclic multiplicative group of units, *Amer. J. Math.* **85**, 447–452 (1963).

GILMER, R., and MOTT, J. L.:

[1] An algebraic proof of a theorem of A. Robinson, *Proc. Amer. Math. Soc.* **29**, 461–466 (1971).

GIUDICI, R. E.:

[1] Quadratic residues in $GF(p^2)$, *Math. Mag.* **44**, 153–157 (1971).

[2] Residui quadratici in un campo di Galois, *Atti Accad. Naz. Lincei Rend. Cl. Sci. Fis. Mat. Natur.* (8)**52**, 461–466 (1972).

GIUDICI, R. E., and MARGAGLIO, C.:

[1] On the factorization of polynomials of the fourth degree (Spanish), *Sci. Valparaiso* **147**, 70–76 (1976).

[2] A geometric characterization of the generators in a quadratic extension of a finite field, *Rend. Sem. Mat. Univ. Padova* **62**, 103–114 (1980).

GIUDICI, R. E., MUSKAT, J. B., and ROBINSON, S. F.:

[1] On the evaluation of Brewer's character sums, *Trans. Amer. Math. Soc.* **171**, 317–347 (1972).

GŁAZEK, K.:

[1] On weak automorphisms of finite fields, *Finite Algebra and Multiple-Valued Logic* (Szeged, 1979), Colloq. Math. Soc. János Bolyai, vol. 28, pp. 275–300, North-Holland, Amsterdam, 1981.

GLENN, O. E.:

[1] Theorems of finiteness in formal concomitant theory, modulo p, *Proc. International Math. Congress* (Toronto, 1924), vol. 1, pp. 331–345, Univ. of Toronto Press, Toronto, 1928.

GOETHALS, J.-M.:

[1] Nonlinear codes defined by quadratic forms over $GF(2)$, *Information and Control* **31**, 43–74 (1976).

GOETHALS, J.-M., and SEIDEL, J. J.:

[1] Orthogonal matrices with zero diagonal, *Canad. J. Math.* **19**, 1001–1010 (1967).

GOGIA, S. K., and LUTHAR, I. S.:

[1] Norms from certain extensions of $F_q(T)$, *Acta Arith.* **38**, 325–340 (1981).

GOKA, T.:

[1] An operator on binary sequences, *SIAM Rev.* **12**, 264–266 (1970).

GOLAY, M. J. E.:

[1] Notes on digital coding, *Proc. IRE* **37**, 657 (1949).

GOLD, R.:

[1] Characteristic linear sequences and their coset functions, *SIAM J. Appl. Math.* **14**, 980–985 (1966).

[2] Optimal binary sequences for spread spectrum multiplexing, *IEEE Trans. Information Theory* **IT-13**, 619–621 (1967).

[3] Maximal recursive sequences with 3-valued recursive cross-correlation functions, *IEEE Trans. Information Theory* **IT-14**, 154–156 (1968).

GOLDBERG, M.:

[1] The group of the quadratic residue tournament, *Canad. Math. Bull.* **13**, 51–54 (1970).

GOLDMAN, H. D., KLIMAN, M., and SMOLA, H.:

[1] The weight structure of some Bose-Chaudhuri codes, *IEEE Trans. Information Theory* **IT-14**, 167–169 (1968).

GOLDSTEIN, R. M., and ZIERLER, N.:

[1] On trinomial recurrences, *IEEE Trans. Information Theory* **IT-14**, 150–151 (1968).

GOLOMB, S. W.:

[1] Sequences with randomness properties, Glenn L. Martin Co. Final Report, Baltimore, Md., 1955; reprinted in Golomb [4].

[2] Structural properties of *PN* sequences, Technical Report, Jet Propulsion Lab., California Institute of Technology, Pasadena, Cal., 1958; reprinted in Golomb [4].

[3] *Digital Communications with Space Applications*, Prentice-Hall, Englewood Cliffs, N.J., 1964.

[4] *Shift Register Sequences*, Holden-Day, San Francisco, 1967.

[5] Theory of transformation groups of polynomials over $GF(2)$ with applications to linear shift register sequences, *Information Sciences* **1**, 87–109 (1968).

[6] Irreducible polynomials, synchronization codes, primitive necklaces, and the cyclotomic algebra, *Proc. Conf. Combinatorial Math. and Its Appl.* (Chapel Hill, N.C., 1967), pp. 358–370, Univ. of North Carolina Press, Chapel Hill, N.C., 1969.

[7] Cyclotomic polynomials and factorization theorems, *Amer. Math. Monthly* **85**, 734–737 (1978); Corrections, *ibid.* **88**, 338–339 (1981).

[8] Obtaining specified irreducible polynomials over finite fields, *SIAM J. Algebraic Discrete Methods* **1**, 411–418 (1980).

[9] On the classification of balanced binary sequences of period $2^n - 1$, *IEEE Trans. Information Theory* **IT-26**, 730–732 (1980).

GOLOMB, S. W., and BAUMERT, L. D.:

[1] The search for Hadamard matrices, *Amer. Math. Monthly* **70**, 12–17 (1963).

GOLOMB, S. W., and LEMPEL, A.:

[1] Second order polynomial recursions, *SIAM J. Appl. Math.* **33**, 587–592 (1977).

GOLOMB, S. W., REED, I. S., and TRUONG, T. K.:

[1] Integer convolutions over the finite field $GF(3 \cdot 2^n + 1)$, *SIAM J. Appl. Math.* **32**, 356–365 (1977).

GOLOMB, S. W., and WELCH, L. R.:

[1] Nonlinear shift register sequences, Memo 20-149, Jet Propulsion Lab., California Institute of Technology, Pasadena, Cal., 1957; reprinted in Golomb [4].

GOLOMB, S. W., WELCH, L. R., and HALES, A.:

[1] On the factorization of trinomials over *GF*(2), Memo 20-189, Jet Propulsion Lab., California Institute of Technology, Pasadena, Cal., 1959; reprinted in Golomb [4].

GOOD, I. J.:

[1] Normal recurring decimals, *J. London Math. Soc.* **21**, 167–169 (1946).

GOODSTEIN, R. L.:

[1] Polynomial generators over Galois fields, *J. London Math. Soc.* **36**, 29–32 (1961).

GOPPA, V. D.:

[1] Decoding and Diophantine approximations (Russian), *Problemy Uprav. i Teor. Informacii* **5**, 195–206 (1976); *Problems of Control and Information Theory* **5**, no. 3, 1–12 (1976).

[2] Codes on algebraic curves (Russian), *Dokl. Akad. Nauk SSSR* **259**, 1289–1290 (1981); *Soviet Math. Dokl.* **24**, 170–172 (1981).

GORBOV, A. N., and ŠMIDT, R. A.:

[1] The Klein resolvent for an equation of degree five over a field of characteristic that divides 5! (Russian), *Zap. Naučn. Sem. Leningrad. Otdel. Mat. Inst. Steklov.* **46**, 36–40 (1974).

GORDON, B., MILLS, W. H., and WELCH, L. R.:

[1] Some new difference sets, *Canad. J. Math.* **14**, 614–625 (1962).

GORDON, J. A.:

[1] Very simple method to find the minimum polynomial of an arbitrary nonzero element of a finite field, *Electron. Lett.* **12**, 663–664 (1976).

GORE, W. C., and COOPER, A. B.:

[1] Comments on polynomial codes, *IEEE Trans. Information Theory* **IT-16**, 635–638 (1970).

GORENSTEIN, D. C., and ZIERLER, N.:

[1] A class of error-correcting codes in p^m symbols, *J. Soc. Indust. Appl. Math.* **9**, 207–214 (1961).

GOSS, D.:

[1] Von Staudt for $F_q[T]$, *Duke Math. J.* **45**, 885–910 (1978).

[2] Modular forms for $F_r[T]$, *J. reine angew. Math.* **317**, 16–39 (1980).

[3] The algebraist's upper half-plane, *Bull. Amer. Math. Soc.* (*N.S.*) **2**, 391–415 (1980).

GOTUSSO, L.:

[1] Successioni uniformemente distribuite in corpi finiti, *Atti Sem. Mat. Fis. Univ. Modena* **12**, 215–232 (1962/63).

GOW, R.:

[1] The number of equivalence classes of nondegenerate bilinear and sesquilinear forms over a finite field, *Linear Algebra Appl.* **41**, 175–181 (1981).

GRAHAM, R. L.:

[1] On quadruples of consecutive kth power residues, *Proc. Amer. Math. Soc.* **15**, 196–197 (1964).

GRAHAM, R. L., and MACWILLIAMS, F. J.:

[1] On the number of information symbols in difference-set cyclic codes, *Bell System Tech. J.* **45**, 1057–1070 (1966).

GRANDET-HUGOT, M.:

[1] Une propriété des "nombres de Pisot" dans un corps de séries formelles, *C. R. Acad. Sci. Paris Sér. A* **265**, 39–41 (1967); Errata, *ibid.* **265**, 551 (1967).

[2] Éléments algébriques remarquables dans un corps de séries formelles, *Acta Arith.* **14**, 177–184 (1968).

GRANDI, A.:

[1] Un teorema sulla rappresentazione analitica delle sostituzioni sopra un numero primo di elementi, *Giorn. Mat. Battaglini* **19**, 238–244 (1881).

[2] Generalizzazione di un teorema sulla rappresentazione analitica delle sostituzioni, *Rend. Reale Ist. Lombardo Sci. Lett.* (2)**16**, 101–111 (1883).

GRANT, H. S.:

[1] A generalization of a cyclotomic formula, *Bull. Amer. Math. Soc.* **42**, 550–556 (1936).

GRAS, G.:

[1] Sommes de Gauss sur les corps finis, *Publ. Math. Fac. Sci. Besançon* **1977–1978**, no. 1, 71 pp.

GRAY, J. F.:

[1] Diagonal forms of odd degree over a finite field, *Michigan Math. J.* **7**, 297–301 (1960).

GREBENJUK, D. G.:

[1] On algebraic integers which depend on an irreducible equation of fourth degree (Russian), *Bull. Sredne-Aziatskogo Gos. Univ. Tashkent* **11**, 19–43 (1925).

[2] Application of complex Voronoĭ numbers to the solution to congruences of fourth degree (Russian), *Akad. Nauk Uzbek. SSR Trudy Inst. Mat.* **1962**, no. 26, 57–80.

GREEN, D. H., and TAYLOR, I. S.:

[1] Modular representation of multiple-valued logic systems, *Proc. IEE* **121**, 409–418 (1974).

[2] Irreducible polynomials over composite Galois fields and their applications in coding techniques, *Proc. IEE* **121**, 935–939 (1974).

[3] Multiple-valued switching circuit design by means of generalised Reed-Muller expansions, *Digital Process.* **2**, 63–81 (1976).

GREEN, J. H., Jr., and SAN SOUCIE, R. L.:

[1] An error-correcting encoder and decoder of high efficiency, *Proc. IRE* **46**, 1741–1744 (1958).

GREENBERG, M. J.:

[1] *Lectures on Forms in Many Variables*, Benjamin, New York, 1969.

GREENWOOD, R. E., and GLEASON, A. M.:

[1] Combinatorial relations and chromatic graphs, *Canad. J. Math.* **7**, 1–7 (1955).

GRIES, D., and LEVIN, G.:

[1] Computing Fibonacci numbers (and similarly defined functions) in log time, *Inform. Process. Lett.* **11**, no. 2, 68–69 (1980).

GROSS, B. H., and KOBLITZ, N.:

[1] Gauss sums and the p-adic Γ-function, *Ann. of Math.* (2)**109**, 569–581 (1979).

GROTH, E. J.:

[1] Generation of binary sequences with controllable complexity, *IEEE Trans. Information Theory* **IT-17**, 288–296 (1971).

GROTHENDIECK, A.:

[1] Sur une note de Mattuck-Tate, *J. reine angew. Math.* **200**, 208–215 (1958).

[2] Formule de Lefschetz et rationalité des fonctions L, *Séminaire Bourbaki 1964/65*, Exp. 279, Benjamin, New York, 1966; *Dix exposés sur la cohomologie des schémas*, Advanced Studies in Pure Math., vol. 3, pp. 31–45, North-Holland, Amsterdam, 1968.

[3] Cohomologie l-adique et fonctions L, *Séminaire de Géométrie Algébrique du Bois-Marie 1965–66* (*SGA 5*), Lecture Notes in Math., vol. 589, Springer-Verlag, Berlin-Heidelberg-New York, 1977.

GRUNDHÖFER, T.:

[1] Über Abbildungen mit eingeschränktem Differenzenprodukt auf einem endlichen Körper, *Arch. Math.* **37**, 59–62 (1981).

GRUSHKO, I. I.:

[1] An approach to the problem of the error-correcting possibilities of group codes (Russian), *Radiotehn. i Elektron.* **9**, 1749–1756 (1964); *Radio Engrg. Electron. Phys.* **9**, 1448–1454 (1964).

GUERRIER, W. J.:

[1] The factorization of the cyclotomic polynomials mod p, *Amer. Math. Monthly* **75**, 46 (1968).

GUIAŞU, S.:

[1] *Information Theory with Applications*, McGraw-Hill, New York, 1977.

GUINAND, A. P.:

[1] Gauss sums and primitive characters, *Quart. J. Math.* **16**, 59–63 (1945).

GUNJI, H., and ARNON, D.:

[1] On polynomial factorization over finite fields, *Math. Comp.* **36**, 281–287 (1981).

GUPTA, H.:

[1] On a problem in matrices, *Proc. Nat. Inst. Sci. India A* **30**, 556–560 (1964).

GUSTAVSON, F. G.:

[1] Analysis of the Berlekamp-Massey linear feedback shift-register synthesis algorithm, *IBM J. Res. Develop.* **20**, 204–212 (1976).

GWEHENBERGER, G.:

[1] Über den Grad von rationalen Funktionen, die Permutationen darstellen, *Monatsh. Math.* **75**, 215–222 (1971).

HALL, M., Jr.:

[1] Divisibility sequences of third order, *Amer. J. Math.* **58**, 577–584 (1936).

[2] Divisors of second-order sequences, *Bull. Amer. Math. Soc.* **43**, 78–80 (1937).

[3] An isomorphism between linear recurring sequences and algebraic rings, *Trans. Amer. Math. Soc.* **44**, 196–218 (1938).

[4] Equidistribution of residues in sequences, *Duke Math. J.* **4**, 691–695 (1938).

[5] A survey of difference sets, *Proc. Amer. Math. Soc.* **7**, 975–986 (1956).

[6] *The Theory of Groups*, Macmillan, New York, 1959.

[7] Characters and cyclotomy, *Proc. Symp. Pure Math.*, vol. 8, pp. 31–43, American Math. Society, Providence, R.I., 1965.

[8] *Combinatorial Theory*, Blaisdell, Waltham, Mass., 1967.

HALTON, J. H.:

[1] On the divisibility properties of Fibonacci numbers, *Fibonacci Quart.* **4**, 217–240 (1966).

HAMMING, R. W.:

[1] Error detecting and error correcting codes, *Bell System Tech. J.* **29**, 147–160 (1950).

HANANI, H.:

[1] Balanced incomplete block designs and related designs, *Discrete Math.* **11**, 255–369 (1975).

HANNEKEN, C. B.:

[1] Irreducible quintic congruences, *Duke Math. J.* **22**, 107–118 (1955).

[2] Irreducible congruences over $GF(p)$, *Proc. Amer. Math. Soc.* **10**, 18–26 (1959).

[3] Irreducible sextic congruences, *Duke Math. J.* **26**, 81–93 (1959).

[4] Irreducible congruences of prime power degree, *Trans. Amer. Math. Soc.* **153**, 167–179 (1971).

[5] Irreducible congruences over $GF(2)$, *Trans. Amer. Math. Soc.* **193**, 291–301 (1974).

HARDER, G.:

[1] Eine Bemerkung zu einer Arbeit von P. E. Newstead, *J. reine angew. Math.* **242**, 16–25 (1970).

HARDMAN, N. R., and JORDAN, J. H.:

[1] The distribution of quadratic residues in fields of order p^2, *Math. Mag.* **42**, 12–17 (1969).

HARDY, G. H., and LITTLEWOOD, J. E.:

[1] Some problems of "Partitio Numerorum"; I: A new solution of Waring's problem, *Göttinger Nachr.* **1920**, 33–54.

[2] A new solution of Waring's problem, *Quart. J. Math.* **48**, 272–293 (1920).

[3] Some problems of 'Partitio Numerorum': IV. The singular series in Waring's problem and the value of the number $G(k)$, *Math. Z.* **12**, 161–188 (1922).

[4] Some problems of "partitio numerorum" (VIII): The number $\Gamma(k)$ in Waring's problem, *Proc. London Math. Soc.* (2)**28**, 518–542 (1927).

HARIS, S. J.:

[1] Number theoretical developments arising from the Siegel formula, *Bull. Amer. Math. Soc. (N.S.)* **2**, 417–433 (1980).

HARRISON, M. A.:

[1] *Lectures on Linear Sequential Machines*, Academic Press, New York, 1969.

HARTFIEL, D. J., and MAXSON, C. J.:

[1] A semigroup characterization of a linearly realizable automaton over $GF(p)$, *J. Comput. System Sci.* **14**, 150–155 (1977).

HARTMANN, C. R. P., RIEK, J. R., Jr., and LONGOBARDI, R. J.:

[1] Weight distributions of some classes of binary cyclic codes, *IEEE Trans. Information Theory* **IT-21**, 345–350 (1975).

HARTMANN, C. R. P., and TZENG, K. K.:

[1] Generalizations of the BCH bound, *Information and Control* **20**, 489–498 (1972).

HARTMANN, C. R. P., TZENG, K. K., and CHIEN, R. T.:

[1] Some results on the minimum distance structure of cyclic codes, *IEEE Trans. Information Theory* **IT-18**, 402–409 (1972).

HARTWIG, R. E., and LEVINE, J.:

[1] Applications of the Drazin inverse to the Hill cryptographic system, III, IV, *Cryptologia* **5**, 67–77, 213–228 (1981).

HARVEY, J. T.:

[1] High-speed m-sequence generation, *Electron. Lett.* **10**, 480–481 (1974).

HASSE, H.:

[1] Zwei Bemerkungen zu der Arbeit "Zur Arithmetik der Polynome" von U. Wegner in den Mathematischen Annalen, Bd. 105, S. 628–631, *Math. Ann.* **106**, 455–456 (1932).

[2] Beweis des Analogons der Riemannschen Vermutung für die Artinschen und F. K. Schmidtschen Kongruenzzetafunktionen in gewissen elliptischen Fällen, *Göttinger Nachr.* **1933**, 253–262.

[3] Über die Kongruenzzetafunktionen, *Sitzungsber. Preuss. Akad. Wiss. Berlin Math.-Phys. Kl.* **17**, 250–263 (1934).

[4] Abstrakte Begründung der komplexen Multiplikation und Riemannsche Vermutung in Funktionenkörpern, *Abh. Math. Sem. Univ. Hamburg* **10**, 325–348 (1934).

[5] Theorie der relativ-zyklischen algebraischen Funktionenkörper, insbesondere bei endlichem Konstantenkörper, *J. reine angew. Math.* **172**, 37–54 (1935).

[6] Zur Theorie der abstrakten elliptischen Funktionenkörper, *Göttinger Nachr.* **1935**, 119–129.

[7] Theorie der höheren Differentiale in einem algebraischen Funktionenkörper mit vollkommenem Konstantenkörper bei beliebiger Charakteristik, *J. reine angew. Math.* **175**, 50–54 (1936).

[8] Zur Theorie der abstrakten elliptischen Funktionenkörper I, II, III, *J. reine angew. Math.* **175**, 55–62, 69–88, 193–208 (1936).

[9] Produktformeln für verallgemeinerte Gaußsche Summen und ihre Anwendung auf die Klassenzahlformel für reelle quadratische Zahlkörper, *Math. Z.* **46**, 303–314 (1940).

[10] Allgemeine Theorie der Gaußschen Summen in algebraischen Zahlkörpern, *Abh. Deutsch. Akad. Wiss. Berlin Math.-Naturw. Kl.* **1951**, no. 1, 4–23.

[11] *Über die Klassenzahl abelscher Zahlkörper*, Akademie-Verlag, Berlin, 1952.

[12] Gaußsche Summen zu Normalkörpern über endlich-algebraischen Zahlkörpern, *Abh. Deutsch. Akad. Wiss. Berlin Math.-Naturw. Kl.* **1952**, no. 1, 1–19.

[13] Artinsche Führer, Artinsche L-Funktion und Gaußsche Summen über endlich-algebraischen Zahlkörpern, *Acta Salmanticensia Ciencias Sec. Mat.* **4**, 1–113 (1954).

[14] Der 2^n-te Potenzcharakter von 2 im Körper der 2^n-ten Einheitswurzeln, *Rend. Circ. Mat. Palermo* (2)**7**, 185–243 (1958).

[15] *Vorlesungen über Zahlentheorie*, 2nd ed., Springer-Verlag, Berlin-Göttingen-Heidelberg-New York, 1964.

[16] *Bericht über neuere Untersuchungen und Probleme aus der Theorie der algebraischen Zahlkörper. Teil II: Reziprozitätsgesetz*, 2nd ed., Physica-Verlag, Würzburg, 1965.

[17] Modular functions and elliptic curves over finite fields, *Rend. Mat. e Appl.* (5)**25**, 248–266 (1966).

[18] *The Riemann Hypothesis in Function Fields*, Univ. of Pennsylvania Press, Philadelphia, 1969.

HAUPTMAN, H., VEGH, E., and FISHER, J.:

[1] *Table of all Primitive Roots for Primes Less than 5000*, Naval Research Laboratory, Washington, D.C., 1970.

HAUSNER, A.:

[1] On the quadratic reciprocity theorem, *Arch. Math.* **12**, 182–183 (1961).

HAYASHI, H. S.:

[1] The number of solutions of certain quintic congruences, *Duke Math. J.* **33**, 747–756 (1966).

[2] On a criterion for power residuacity, *Mem. Fac. Sci. Kyushu Univ. Ser. A* **27**, 211–220 (1973).

HAYES, D. R.:

[1] A polynomial analog of the Goldbach conjecture, *Bull. Amer. Math. Soc.* **69**, 115–116 (1963); Correction, *ibid.* **69**, 493 (1963).

[2] The distribution of irreducibles in $GF[q, x]$, *Trans. Amer. Math. Soc.* **117**, 101–127 (1965).
[3] A polynomial generalized Gauss sum, *J. reine angew. Math.* **222**, 113–119 (1966).
[4] The expression of a polynomial as a sum of three irreducibles, *Acta Arith.* **11**, 461–488 (1966).
[5] A geometric approach to permutation polynomials over a finite field, *Duke Math. J.* **34**, 293–305 (1967).
[6] The Galois group of $x^n + x - t$, *Duke Math. J.* **40**, 459–461 (1973).

HAZLETT, O. C.:
[1] A symbolic theory of formal modular covariants, *Trans. Amer. Math. Soc.* **24**, 286–311 (1922).
[2] Annihilators of modular invariants and covariants, *Ann. of Math.* (2)**23**, 198–211 (1923).
[3] Notes on formal modular protomorphs, *Amer. J. Math.* **49**, 181–188 (1927).
[4] On formal modular invariants, *J. Math. Pures Appl.* (9)**9**, 327–332 (1930).

HEATH-BROWN, D. R., and PATTERSON, S. J.:
[1] The distribution of Kummer sums at prime arguments, *J. reine angew. Math.* **310**, 111–130 (1979).

HECKE, E.:
[1] Über die *L*-Funktionen und den Dirichletschen Primzahlsatz für einen beliebigen Zahlkörper, *Göttinger Nachr.* **1917**, 299–318.
[2] Reziprozitätsgesetz und Gaußsche Summen in quadratischen Zahlkörpern, *Göttinger Nachr.* **1919**, 265–278.
[3] Eine neue Art von Zetafunktionen und ihre Beziehungen zur Verteilung der Primzahlen. II, *Math. Z.* **6**, 11–51 (1920).
[4] *Vorlesungen über die Theorie der algebraischen Zahlen*, Akademische Verlagsgesellschaft, Leipzig, 1923.

HEISLER, J.:
[1] A characterization of finite fields, *Amer. Math. Monthly* **74**, 537–538 (1967); Correction, *ibid.* **74**, 1211 (1967).
[2] Diagonal forms over finite fields, *J. Number Theory* **6**, 50–51 (1974).

HELGERT, H. J.:
[1] Decoding of alternant codes, *IEEE Trans. Information Theory* **IT-23**, 513–514 (1977).

HELLESETH, T.:
[1] Some results about the cross-correlation function between two maximal linear sequences, *Discrete Math.* **16**, 209–232 (1976).
[2] A note on the cross-correlation function between two binary maximal length linear sequences, *Discrete Math.* **23**, 301–307 (1978).

HELLESETH, T., KLØVE, T., and MYKKELTVEIT, J.:
[1] The weight distribution of irreducible cyclic codes with block length $n_1((q^l - 1)/N)$, *Discrete Math.* **18**, 179–211 (1977).

HELVERSEN-PASOTTO, A.:
[1] Série discrète de $GL(3, F_q)$ et sommes de Gauss, *C. R. Acad. Sci. Paris Sér. A* **275**, 263–266 (1972).
[2] L'identité de Barnes pour les corps finis, *Sém. Delange-Pisot-Poitou 1977/78, Théorie des Nombres*, Exp. 22, 12 pp., Secrétariat Math., Paris, 1978.
[3] L'identité de Barnes pour les corps finis, *C. R. Acad. Sci. Paris Sér. A* **286**, 297–300 (1978).
[4] Darstellungen von $GL(3, F_q)$ und Gaußsche Summen, *Math. Ann.* **260**, 1–21 (1982).

HEMMATI, F., and COSTELLO, D. J., Jr.:
[1] An algebraic construction for q-ary shift register sequences, *IEEE Trans. Computers* **C-27**, 1192–1195 (1978).

HENSEL, K.:
[1] Über die Darstellung der Zahlen eines Gattungsbereiches für einen beliebigen Primdivisor, *J. reine angew. Math.* **103**, 230–237 (1888).
[2] Über die zu einem algebraischen Körper gehörigen Invarianten, *J. reine angew. Math.* **129**, 68–85 (1905).

HERGET, W.:
[1] Über die Funktionalgleichung $f(x) = d^{m-1}\sum_{i=0}^{d-1} f(x + i/d)$ in den Körpern Z_p, *Manuscripta Math.* **23**, 131–141 (1978).
[2] Bernoulli-Polynome in den Restklassenringen Z_n, *Glasnik Mat.* (3)**14**, 27–33 (1979).

HERGLOTZ, G.:
[1] Zur letzten Eintragung im Gaußschen Tagebuch, *Ber. Math.-Phys. Kl. Sächs. Akad. Wiss. Leipzig* **73**, 271–276 (1921).

HERLESTAM, T., and JOHANNESSON, R.:
[1] On computing logarithms over $GF(2^p)$, *BIT* **21**, 326–334 (1981).

HERMITE, C.:
[1] Sur la théorie des formes quadratiques. II, *J. reine angew. Math.* **47**, 343–368 (1854); *Oeuvres*, vol. 1, pp. 234–263, Gauthier-Villars, Paris, 1905.
[2] Sur les fonctions de sept lettres, *C. R. Acad. Sci. Paris* **57**, 750–757 (1863); *Oeuvres*, vol. 2, pp. 280–288, Gauthier-Villars, Paris, 1908.

HERSHEY, J. E.:
[1] Implementation of Mitre public key cryptographic system, *Electron. Lett.* **16**, 930–931 (1980).

HERSTEIN, I. N.:
[1] An elementary proof of a theorem of Jacobson, *Duke Math. J.* **21**, 45–48 (1954).
[2] Wedderburn's theorem and a theorem of Jacobson, *Amer. Math. Monthly* **68**, 249–251 (1961).
[3] *Noncommutative Rings*, Carus Math. Monographs, no. 15, Math. Assoc. of America, Washington, D.C., 1968.

[4] *Topics in Algebra*, 2nd ed., Xerox College Publ., Lexington, Mass., 1975.

HEUZÉ, G.:

[1] Sur les corps finis, *Math. Sci. Humaines* **47**, 57–59 (1974).

HILBERT, D.:

[1] Über diophantische Gleichungen, *Göttinger Nachr.* **1897**, 48–54.

[2] Die Theorie der algebraischen Zahlkörper, *Jber. Deutsch. Math.-Verein.* **4**, 175–546 (1897).

[3] *Grundlagen der Geometrie*, Teubner, Leipzig, 1899; Open Court, Chicago, 1971.

HINZ, J. G.:

[1] Einige Bemerkungen zum Beweis eines Satzes von J. H. Maclagan-Wedderburn, *J. reine angew. Math.* **290**, 109–112 (1977).

HIRSCHFELD, J. W. P.:

[1] A curve over a finite field, the number of whose points is not increased by a quadratic extension of the field, and sub-Hermitian forms, *Atti Accad. Naz. Lincei Rend. Cl. Sci. Fis. Mat. Natur.* (8)**42**, 365–367 (1967).

[2] Rational curves on quadrics over finite fields of characteristic two, *Rend. Mat.* (6)**4**, 773–795 (1971).

[3] Ovals in Desarguesian planes of even order, *Ann. Mat. Pura Appl.* (4)**102**, 79–89 (1975).

[4] Cyclic projectivities in $PG(n, q)$, *Teorie Combinatorie* (Rome, 1973), vol. 1, pp. 201–211, Accad. Naz. dei Lincei, Rome, 1976.

[5] *Projective Geometries over Finite Fields*, Clarendon Press, Oxford, 1979.

HOCQUENGHEM, A.:

[1] Codes correcteurs d'erreurs, *Chiffres* **2**, 147–156 (1959).

HODGES, J. H.:

[1] Exponential sums for symmetric matrices in a finite field, *Math. Nachr.* **14**, 331–339 (1955).

[2] Representations by bilinear forms in a finite field, *Duke Math. J.* **22**, 497–509 (1955).

[3] Weighted partitions for symmetric matrices in a finite field, *Math. Z.* **66**, 13–24 (1956).

[4] Exponential sums for skew matrices in a finite field, *Arch. Math.* **7**, 116–121 (1956).

[5] The matric equation $AX = B$ in a finite field, *Amer. Math. Monthly* **63**, 243–244 (1956).

[6] Weighted partitions for general matrices over a finite field, *Duke Math. J.* **23**, 545–552 (1956).

[7] Distribution of bordered matrices in a finite field, *J. reine angew. Math.* **198**, 10–13 (1957).

[8] Weighted partitions for skew matrices over a finite field, *Arch. Math.* **8**, 16–22 (1957).

[9] Some matrix equations over a finite field, *Ann. Mat. Pura Appl.* (4)**44**, 245–250 (1957).

[10] Weighted partitions for Hermitian matrices over a finite field, *Math. Nachr.* **17**, 93–100 (1958).
[11] Scalar polynomial equations for matrices over a finite field, *Duke Math. J.* **25**, 291–296 (1958).
[12] The matrix equation $X^2 - I = 0$ over a finite field, *Amer. Math. Monthly* **65**, 518–520 (1958).
[13] Some determinantal equations over a finite field, *Math. Z.* **72**, 355–361 (1960).
[14] A note on systems of matrix equations over a finite field, *Portugal. Math.* **21**, 99–106 (1962).
[15] Some polynomial equations for determinants over a finite field, *Monatsh. Math.* **66**, 322–330 (1962).
[16] Generalized weighted m-th power partitions over a finite field, *Duke Math. J.* **29**, 405–412 (1962).
[17] Simultaneous pairs of linear and quadratic matrix equations over a finite field, *Math. Z.* **84**, 38–44 (1964).
[18] A bilinear matrix equation over a finite field, *Duke Math. J.* **31**, 661–666 (1964).
[19] The matrix equation $AXC = B$ over a finite field, *Riv. Mat. Univ. Parma* (2)**6**, 79–81 (1965).
[20] Determinantal equations related to Hermitian forms over a finite field, *Monatsh. Math.* **69**, 215–224 (1965).
[21] A symmetric matrix equation over a finite field, *Math. Nachr.* **30**, 221–228 (1965).
[22] A skew matrix equation over a finite field, *Arch. Math.* **17**, 49–55 (1966).
[23] Uniform distribution of sequences in $GF[q, x]$, *Acta Arith.* **12**, 55–75 (1966).
[24] An Hermitian matrix equation over a finite field, *Duke Math. J.* **33**, 123–129 (1966).
[25] Some pairs of matrix equations over a finite field, *Scripta Math.* **27**, 289–301 (1966).
[26] Uniform distribution of polynomial-generated sequences in $GF[q, x]$, *Ann. Mat. Pura Appl.* (4)**82**, 135–142 (1969).
[27] On uniform distribution of sequences in $GF\{q, x\}$ and $GF[q, x]$, *Ann. Mat. Pura Appl.* (4)**85**, 287–294 (1970).
[28] Note on some partitions of a rectangular matrix, *Atti Accad. Naz. Lincei Rend. Cl. Sci. Fis. Mat. Natur.* (8)**59**, 662–666 (1975).
[29] Ranked partitions of rectangular matrices over finite fields, *Atti Accad. Naz. Lincei Rend. Cl. Sci. Fis. Mat. Natur.* (8)**60**, 6–12 (1976).
[30] Note on a linear matrix equation over a finite field, *Atti Accad. Naz. Lincei Rend. Cl. Sci. Fis. Mat. Natur.* (8)**63**, 304–309 (1977).

HOFFMAN, K., and KUNZE, R.:
[1] *Linear Algebra*, 2nd ed., Prentice-Hall, Englewood Cliffs, N.J., 1971.

HOHLER, P.:
[1] Eine zahlentheoretische Konstruktion der Galois-Felder $GF(p^2)$, *Elemente der Math.* **31**, 64–66 (1976).

HOLZER, L.:
[1] *Zahlentheorie*, vol. 1, Teubner, Leipzig, 1958.

HONG, S. J., and BOSSEN, D. C.:
[1] On some properties of self-reciprocal polynomials, *IEEE Trans. Information Theory* **IT-21**, 462–464 (1975).

HOOLEY, C.:
[1] An asymptotic formula in the theory of numbers, *Proc. London Math. Soc.* (3)**7**, 396–413 (1957).
[2] On the distribution of the roots of polynomial congruences, *Mathematika* **11**, 39–49 (1964).
[3] *Applications of Sieve Methods to the Theory of Numbers*, Cambridge Univ. Press, Cambridge, 1976.
[4] On another sieve method and the numbers that are a sum of two hth powers, *Proc. London Math. Soc.* (3) **43**, 73–109 (1981).
[5] On Waring's problem for two squares and three cubes, *J. reine angew. Math.* **328**, 161–207 (1981).
[6] On exponential sums and certain of their applications, *Journées Arithmétiques 1980* (J. V. Armitage, ed.), London Math. Soc. Lecture Note Series, no. 56, pp. 92–122, Cambridge Univ. Press, Cambridge, 1982.

HOPF, H.:
[1] Über die Verteilung quadratischer Reste, *Math. Z.* **32**, 222–231 (1930).

HORADAM, A. F.:
[1] *A Guide to Undergraduate Projective Geometry*, Pergamon Press Australia, Rushcutters Bay, N.S.W., 1970.

HORÁKOVÁ, K., and SCHWARZ, Š.:
[1] Cyclic matrices and algebraic equations over a finite field (Russian), *Mat.-Fyz. Časopis Sloven. Akad. Vied* **12**, 36–46 (1962).

HOUNDONOUGBO, V.:
[1] Développement en fraction continue sur $K(X)$. Fonction profondeur, *C. R. Acad. Sci. Paris Sér. A* **286**, 1037–1039 (1978).
[2] Mesure de répartition d'une suite $(\theta^n)_{n \in \mathbb{N}^*}$ dans un corps de séries formelles sur un corps fini, *C. R. Acad. Sci. Paris Sér. A* **288**, 997–999 (1979).

HUA, L.-K.:
[1] On Waring's problem with polynomial summands, *Amer. J. Math.* **58**, 553–562 (1936).
[2] On a generalized Waring's problem, *Proc. London Math. Soc.* (2)**43**, 161–182 (1937).
[3] On an exponential sum, *J. London Math. Soc.* **13**, 54–61 (1938).
[4] On Waring's problem with cubic polynomial summands, *Science Reports National Tsing Hua Univ.* **4**, 55–83 (1940).
[5] On Waring's problem with cubic polynomial summands, *J. Indian Math. Soc.* **4**, 127–135 (1940).
[6] On an exponential sum, *J. Chinese Math. Soc.* **2**, 301–312 (1940).

[7] Sur une somme exponentielle, *C. R. Acad. Sci. Paris* **210**, 520–523 (1940).

[8] Sur le problème de Waring relatif à un polynome du troisième degré, *C. R. Acad. Sci. Paris* **210**, 650–652 (1940).

[9] *Additive Theory of Prime Numbers* (Russian), *Trudy Mat. Inst. Steklov.* **22**(1947); American Math. Society, Providence, R.I., 1965.

[10] On the number of solutions of Tarry's problem, *Acta Sci. Sinica* **1**, 1–76 (1952).

[11] On exponential sums, *Sci. Record* (*N.S.*) **1**, 1–4 (1957).

[12] *Die Abschätzung von Exponentialsummen und ihre Anwendung in der Zahlentheorie*, Enzyklopädie der Math. Wissenschaften, Band I2, Heft 13, Teil I, Teubner, Leipzig, 1959.

HUA, L.-K., and MIN, S. H.:

[1] On the number of solutions of certain congruences, *Science Reports National Tsing Hua Univ.* **4**, 113–133 (1940).

[2] On a double exponential sum, *Acad. Sinica Science Record* **1**, 23–25 (1942).

[3] On a double exponential sum, *Science Reports National Tsing Hua Univ.* **4**, 484–518 (1947).

HUA, L.-K., and VANDIVER, H. S.:

[1] On the existence of solutions of certain equations in a finite field, *Proc. Nat. Acad. Sci. U.S.A.* **34**, 258–263 (1948).

[2] Characters over certain types of rings with application to the theory of equations in a finite field, *Proc. Nat. Acad. Sci. U.S.A.* **35**, 94–99 (1949).

[3] On the number of solutions of some trinomial equations in a finite field, *Proc. Nat. Acad. Sci. U.S.A.* **35**, 477–481 (1949).

[4] On the nature of the solutions of certain equations in a finite field, *Proc. Nat. Acad. Sci. U.S.A.* **35**, 481–487 (1949).

HUDSON, M.:

[1] On the least non-residue of a polynomial, *J. London Math. Soc.* **41**, 745–749 (1966).

HUDSON, R. H.:

[1] On the distribution of k-th power nonresidues, *Duke Math. J.* **39**, 85–88 (1972).

[2] A bound for the first occurrence of three consecutive integers with equal quadratic character, *Duke Math. J.* **40**, 33–39 (1973).

[3] A note on Dirichlet characters, *Math. Comp.* **27**, 973–975 (1973).

[4] On the least kth power non-residue, *Ark. Mat.* **12**, 217–220 (1974).

[5] Power residues and nonresidues in arithmetic progressions, *Trans. Amer. Math. Soc.* **194**, 277–289 (1974).

[6] A sharper bound for the least pair of consecutive k-th power non-residues of non-principal characters (mod p) of order $k > 3$, *Acta Arith.* **30**, 133–135 (1976).

HUDSON, R. H., and WILLIAMS, K. S.:

[1] Resolution of ambiguities in the evaluation of cubic and quartic Jacobsthal sums, *Pacific J. Math.* **99**, 379–386 (1982).

HUFFMAN, D. A.:

[1] The synthesis of linear sequential coding networks, *Proc. Third London Symp. on Information Theory* (C. Cherry, ed.), pp. 71–95, Butterworths, London, 1956.

[2] A linear circuit viewpoint on error-correcting codes, *IRE Trans. Information Theory* **IT-2**, no. 3, 20–28 (1956).

HUGHES, D. R.:

[1] A class of non-Desarguesian projective planes, *Canad. J. Math.* **9**, 378–388 (1957).

[2] On t-designs and groups, *Amer. J. Math.* **87**, 761–778 (1965).

HUGHES, D. R., and PIPER, F. C.:

[1] *Projective Planes*, Springer-Verlag, New York-Heidelberg-Berlin, 1973.

HULE, H., and MÜLLER, W. B.:

[1] Cyclic groups of permutations induced by polynomials over Galois fields (Spanish), *An. Acad. Brasil. Ci.* **45**, 63–67 (1973).

HULL, R.:

[1] The numbers of solutions of congruences involving only kth powers, *Trans. Amer. Math. Soc.* **34**, 908–937 (1932).

HURWITZ, A.:

[1] Über höhere Kongruenzen, *Archiv Math. Phys.* (3)**5**, 17–27 (1903).

[2] Über die Kongruenz $ax^e + by^e + cz^e \equiv 0 (\text{mod. } p)$, *J. reine angew. Math.* **136**, 272–292 (1909).

HUSTON, R. E.:

[1] Asymptotic generalizations of Waring's theorem, *Proc. London Math. Soc.* (2)**39**, 82–115 (1935).

HWANG, J. C., SHENG, C. L., and HSIEH, C. C.:

[1] On the modulo-two-sum decomposition of binary sequences of finite periods, *Internat. J. Electron.* **39**, 97–104 (1975).

IGUSA, J.:

[1] On the theory of algebraic correspondences and its application to the Riemann hypothesis in function fields, *J. Math. Soc. Japan* **1**, 147–197 (1949).

IHARA, Y.:

[1] Some remarks on the number of rational points of algebraic curves over finite fields, *J. Fac. Sci. Univ. Tokyo Sect. IA* **28**, 721–724 (1981).

IKAI, T., KOSAKO, H., and KOJIMA, Y.:

[1] Subsequences in linear recurring sequences, *Electron. Commun. Japan* **53**, no. 12, 159–166 (1970).

[2] Nonperiod-length subsequences including a cyclic subspace. Subsequences in linear recurring sequences, *Systems-Computers-Controls* **2**, no. 4, 34–41 (1971).

INGELS, F. M.:

[1] *Information and Coding Theory*, Intext Educ. Publ., San Francisco-Toronto-London, 1971.

IPATOV, V. P.:

[1] Contribution to the theory of sequences with perfect periodic autocorrelation properties (Russian), *Radiotehn. i Elektron.* **25**, 723–727 (1980); *Radio Engrg. Electron. Phys.* **25**, no. 4, 31–34 (1980).

IRELAND, K.:

[1] On the zeta function of an algebraic variety, *Amer. J. Math.* **89**, 643–660 (1967).

IRELAND, K., and ROSEN, M. I.:

[1] *Elements of Number Theory*, Bogden & Quigley, Tarrytown-on-Hudson, N.Y., 1972.

ISHIMURA, S.:

[1] On Gaussian sums associated with a character of order 5 and a rational prime number $p \equiv 1 (\mathrm{mod}\, 5)$, *J. Tsuda College* **8**, 27–35 (1976).

ISKOVSKIH, V. A.:

[1] Verification of the Riemann hypothesis for certain local zeta-functions (Russian), *Uspehi Mat. Nauk* **28**, no. 3, 181–182 (1973).

IVANOV, I. I.:

[1] On two congruences (Russian), *Ž. Leningrad. Fiz.-Mat. Obšč.* **1**, 37–38 (1926).

IWANIEC, H.:

[1] Mean values for Fourier coefficients of cusp forms and sums of Kloosterman sums, *Journées Arithmétiques 1980* (J. V. Armitage, ed.), London Math. Soc. Lecture Note Series, no. 56, pp. 306–321, Cambridge Univ. Press, Cambridge, 1982.

IWASAWA, K.:

[1] A note on Jacobi sums, *Symposia Math.*, vol. 15, pp. 447–459, Academic Press, London, 1975.

JACOBI, C. G. J.:

[1] Brief an Gauss vom 8. Februar 1827, *Gesammelte Werke*, vol. 7, pp. 393–400, Reimer, Berlin, 1891.

[2] Über die Kreistheilung und ihre Anwendung auf die Zahlentheorie, *Monatsber. Königl. Akad. Wiss. Berlin* **1837**, 127–136; *J. reine angew. Math.* **30**, 166–182 (1846); *Gesammelte Werke*, vol. 6, pp. 254–274, Reimer, Berlin, 1891.

[3] *Canon Arithmeticus*, Typis Academicis, Berlin, 1839; expanded edition, Akademie-Verlag, Berlin, 1956.

JACOBSON, N.:

[1] Structure theory for algebraic algebras of bounded degree, *Ann. of Math.* (2)**46**, 695–707 (1945).

[2] *Lectures in Abstract Algebra*, vol. 3: *Theory of Fields and Galois Theory*, Van Nostrand, New York, 1964.

JACOBSTHAL, E.:
[1] Anwendungen einer Formel aus der Theorie der quadratischen Reste, Dissertation, Berlin, 1906.
[2] Über die Darstellung der Primzahlen der Form $4n+1$ als Summe zweier Quadrate, *J. reine angew. Math.* **132**, 238–245 (1907).
[3] Über vertauschbare Polynome, *Math. Z.* **63**, 243–276 (1955).

JAMISON, R. E.:
[1] Covering finite fields with cosets of subspaces, *J. Combinatorial Theory Ser. A* **22**, 253–266 (1977).

JÄNICHEN, W.:
[1] Über einen zahlentheoretischen Satz von Hurwitz, *Math. Z.* **17**, 277–292 (1923).

JARDEN, D.:
[1] *Recurring Sequences*, 2nd ed., Riveon Lematematika, Jerusalem, 1966.

JEGER, M.:
[1] Irreduzible Polynome als kombinatorische Figuren, *Elemente der Math.* **28**, 86–92 (1973).

JOHNSEN, J.:
[1] On the distribution of powers in finite fields, *J. reine angew. Math.* **251**, 10–19 (1971).
[2] On the large sieve method in $GF[q,x]$, *Mathematika* **18**, 172–184 (1971).

JOHNSON, L. S., PORTER, A. D., and VARINEAU, V. J.:
[1] Commutators over finite fields, *Publ. Math. Debrecen* **25**, 259–264 (1978).

JOLY, J.-R.:
[1] Sommes de puissances d-ièmes dans un anneau commutatif, *Acta Arith.* **17**, 37–114 (1970).
[2] Sommes de carrés dans certains anneaux principaux, *Bull. Sci. Math.* (2)**94**, 85–95 (1970).
[3] Nombre de solutions de certaines équations diagonales sur un corps fini, *C. R. Acad. Sci. Paris Sér. A* **272**, 1549–1552 (1971).
[4] Démonstration cyclotomique de la loi de réciprocité cubique, *Bull. Sci. Math.* (2)**96**, 273–278 (1972).
[5] Equations et variétés algébriques sur un corps fini, *L'Enseignement Math.* (2)**19**, 1–117 (1973).

JORDAN, C.:
[1] Sur les congruences du second degré, *C. R. Acad. Sci. Paris* **62**, 687–690 (1866); *Oeuvres*, vol. 3, pp. 363–365, Gauthier-Villars, Paris, 1962.
[2] *Traité des substitutions et des équations algébriques*, Gauthier-Villars, Paris, 1870.
[3] Sur la résolution des équations les unes par les autres, *C. R. Acad. Sci. Paris* **72**, 283–290 (1871); *Oeuvres*, vol. 1, pp. 277–284, Gauthier-Villars,

Paris, 1961.

[4] Sur les sommes de Gauss à plusieurs variables, *C. R. Acad. Sci. Paris* **73**, 1316–1319 (1871); *Oeuvres*, vol. 3, pp. 367–369, Gauthier-Villars, Paris, 1962.

[5] Sur les formes réduites des congruences du second degré, *C. R. Acad. Sci. Paris* **74**, 1093–1095 (1872); *Oeuvres*, vol. 3, pp. 371–373, Gauthier-Villars, Paris, 1962.

[6] Sur la forme canonique des congruences du second degré et le nombre de leurs solutions, *J. Math. Pures Appl.* (2)**17**, 368–402 (1872); *Oeuvres*, vol. 3, pp. 375–409, Gauthier-Villars, Paris, 1962.

[7] Sur le nombre des solutions de la congruence $|a_{ik}| \equiv A \bmod M$, *J. Math. Pures Appl.* (6)**7**, 409–416 (1911); *Oeuvres*, vol. 3, pp. 543–550, Gauthier-Villars, Paris, 1962.

JORDAN, CH.:

[1] *Calculus of Finite Differences*, Chelsea, New York, 1950.

JORDAN, H. F., and WOOD, D. C. M.:

[1] On the distribution of sums of successive bits of shift-register sequences, *IEEE Trans. Computers* **C-22**, 400–408 (1973).

JORDAN, J. H.:

[1] Pairs of consecutive power residues or non-residues, *Canad. J. Math.* **16**, 310–314 (1964).

[2] The distribution of cubic and quintic non-residues, *Pacific J. Math.* **16**, 77–85 (1966).

[3] Character sums in $Z(i)/(p)$, *Proc. London Math. Soc.* (3)**17**, 1–10 (1967).

[4] The distribution of kth power residues and nonresidues, *Proc. Amer. Math. Soc.* **19**, 678–680 (1968).

[5] The distribution of k-th power non-residues, *Duke Math. J.* **37**, 333–340 (1970).

JULIA, G.:

[1] Mémoire sur la permutabilité des fractions rationnelles, *Ann. Sci. Ecole Norm. Sup.* (3)**39**, 131–215 (1922).

JUNG, F. R.:

[1] Solutions of some systems of equations over a finite field with applications to geometry, *Duke Math. J.* **39**, 189–202 (1972).

[2] On conics over a finite field, *Canad. J. Math.* **26**, 1281–1288 (1974).

KACZYNSKI, T. J.:

[1] Another proof of Wedderburn's theorem, *Amer. Math. Monthly* **71**, 652–653 (1964).

KALMAN, R. E.:

[1] Mathematical description of linear dynamical systems, *SIAM J. Control* **1**, 152–192 (1963).

[2] Algebraic aspects of the theory of dynamical systems, *Differential Equations and Dynamical Systems* (J. K. Hale and J. P. LaSalle, eds.), pp. 133–146, Academic Press, New York, 1967.

KALOUJNINE, L.:

[1] La structure des p-groupes de Sylow des groupes symétriques finis, *Ann. Sci. Ecole Norm. Sup.* (3)**65**, 239–276 (1948).

KAMAL, A. K., SINGH, H., PURI, S., and NANDA, N. K.:

[1] On the evaluation of transition matrices in finite fields, *Internat. J. Systems Sci.* **6**, 561–564 (1975).

KAMKE, E.:

[1] Zur Arithmetik der Polynome, *Math. Z.* **19**, 247–264 (1924).

KANTOR, R.:

[1] Über die Anzahl inkongruenter Werte ganzer, rationaler Funktionen, *Monatsh. Math. Phys.* **26**, 24–39 (1915).

KAPLAN, P.:

[1] Démonstration des lois de réciprocité quadratique et biquadratique, *J. Fac. Sci. Univ. Tokyo Sect. I* **16**, 115–145 (1969).

KARACUBA, A. A.:

[1] Tarry's problem for a system of congruences (Russian), *Mat. Sb.* (*N.S.*) **55**, 209–220 (1961).

[2] On systems of congruences (Russian), *Izv. Akad. Nauk SSSR Ser. Mat.* **29**, 935–944 (1965).

[3] Asymptotic formulae for a certain class of trigonometric sums (Russian), *Dokl. Akad. Nauk SSSR* **169**, 9–11 (1966); *Soviet Math. Dokl.* **7**, 845–848 (1966).

[4] Theorems on the mean and complete trigonometric sums (Russian), *Izv. Akad. Nauk SSSR Ser. Mat.* **30**, 183–206 (1966).

[5] Estimates of complete trigonometric sums (Russian), *Mat. Zametki* **1**, 199–208 (1967); *Math. Notes* **1**, 133–139 (1967).

[6] Character sums and primitive roots in finite fields (Russian), *Dokl. Akad. Nauk SSSR* **180**, 1287–1289 (1968); *Soviet Math. Dokl.* **9**, 755–757 (1968).

[7] Trigonometric sums (Russian), *Dokl. Akad. Nauk SSSR* **189**, 31–34 (1969); *Soviet Math. Dokl.* **10**, 1334–1337 (1969).

[8] Estimates of character sums (Russian), *Izv. Akad. Nauk SSSR Ser. Mat.* **34**, 20–30 (1970); *Math. USSR Izv.* **4**, 19–29 (1970).

[9] Lower bounds for sums of the characters of polynomials (Russian), *Mat. Zametki* **14**, 67–72 (1973); *Math. Notes* **14**, 593–596 (1973).

KARAMATA, J., and TOMIĆ, M.:

[1] Sur une inégalité de Kusmin-Landau relative aux sommes trigonométriques et son application à la somme de Gauss, *Acad. Serbe Sci. Publ. Inst. Math.* **3**, 207–218 (1950).

KARLIN, M.:

[1] New binary coding results by circulants, *IEEE Trans. Information Theory* **IT-15**, 81–92 (1969).

KARPOVSKY, M. G.:

[1] *Finite Orthogonal Series in the Design of Digital Devices*, Wiley, New York, 1976.

KÁRTESZI, F.:
[1] *Introduction to Finite Geometries*, North-Holland, Amsterdam, 1976.

KASAMI, T.:
[1] Systematic codes using binary shift register sequences, *J. Info. Processing Soc. Japan* **1**, 198–200 (1960).

KASAMI, T., LIN, S., and PETERSON, W. W.:
[1] Polynomial codes, *IEEE Trans. Information Theory* **IT-14**, 807–814 (1968).

KATZ, N. M.:
[1] On a theorem of Ax, *Amer. J. Math.* **93**, 485–499 (1971).
[2] Travaux de Dwork, *Séminaire Bourbaki 1971/72*, Exp. 409, Lecture Notes in Math., vol. 317, pp. 167–200, Springer-Verlag, Berlin-Heidelberg-New York, 1973.
[3] An overview of Deligne's proof of the Riemann hypothesis for varieties over finite fields, *Proc. Symp. Pure Math.*, vol. 28, pp. 275–305, American Math. Society, Providence, R.I., 1976.
[4] *Sommes exponentielles*, Astérisque, no. 79, Soc. Math. France, Paris, 1980.
[5] Crystalline cohomology, Dieudonné modules, and Jacobi sums, *Automorphic Forms, Representation Theory and Arithmetic* (Bombay, 1979), Tata Inst. Fund. Res. Studies in Math., vol. 10, pp. 165–246, Tata Institute of Fundamental Research, Bombay, 1981.

KAUTSCHITSCH, H.:
[1] Kommutative Teilhalbgruppen der Kompositionshalbgruppe von Polynomen und formalen Potenzreihen, *Monatsh. Math.* **74**, 421–436 (1970).
[2] Über vertauschbare Polynome mit vorgegebenen Gradzahlen, *Arch. Math.* **27**, 611–619 (1976).

KAUTZ, W. H.:
[1] *Linear Sequential Switching Circuits – Selected Technical Papers*, Holden-Day, San Francisco, 1965.

KAUTZ, W. H., and LEVITT, K. N.:
[1] A survey of progress in coding theory in the Soviet Union, *IEEE Trans. Information Theory* **IT-15**, 197–245 (1969).

KELLER, G., and OLSON, F. R.:
[1] Counting polynomial functions (mod p^n), *Duke Math. J.* **35**, 835–838 (1968).

KEMPFERT, H.:
[1] On the factorization of polynomials, *J. Number Theory* **1**, 116–120 (1969).

KEMPNER, A. J.:
[1] Polynomials and their residue systems, *Trans. Amer. Math. Soc.* **22**, 240–266, 267–288 (1921).
[2] Polynomials of several variables and their residue systems, *Trans. Amer. Math. Soc.* **27**, 287–298 (1925).

KHADZHIIVANOV, N. G., and NENOV, N. D.:

[1] The number of nontrivial solutions of the Fermat equation $x^n + y^n = z^n$ in a Galois field (Russian), *C. R. Acad. Bulgare Sci.* **32**, 557–560 (1979).

KIEFE, C.:

[1] Sets definable over finite fields: their zeta-functions, *Trans. Amer. Math. Soc.* **223**, 45–59 (1976).

KIM, J. B.:

[1] The number of generalized inverses of a matrix, *Algebraic Theory of Semigroups* (Szeged, 1976), Colloq. Math. Soc. János Bolyai, vol. 20, pp. 277–280, North-Holland, Amsterdam, 1979.

KISS, P., and BUI MINH PHONG:

[1] On a function concerning second-order recurrences, *Ann. Univ. Sci. Budapest. Eötvös Sect. Math.* **21**, 119–122 (1978).

KLEIMAN, S. L.:

[1] Algebraic cycles and the Weil conjectures, *Dix exposés sur la cohomologie des schémas*, Advanced Studies in Pure Math., vol. 3, pp. 359–386, North-Holland, Amsterdam, 1968.

KLEIN, F.:

[1] Zur Theorie der linearen Kongruenzensysteme, *J. reine angew. Math.* **159**, 238–245 (1928).

[2] Zur Theorie der Systeme von Potenzproduktkongruenzen, *J. reine angew. Math.* **164**, 141–150 (1931).

[3] Über rechteckige Matrizen, bei denen die Determinanten maximaler Reihenanzahl teilerfremd zu einem Modul sind, *Jber. Deutsch. Math.-Verein.* **40**, 233–238 (1931).

KLINGENBERG, W., and WITT, E.:

[1] Über die Arfsche Invariante quadratischer Formen mod 2, *J. reine angew. Math.* **193**, 121–122 (1954).

KLJAČKO, A. A.:

[1] Monodromy groups of polynomial mappings (Russian), *Studies in Number Theory* (Russian), no. 6, pp. 82–91, Izdat. Saratov. Univ., Saratov, 1975.

KLOBE, W.:

[1] Über eine untere Abschätzung der n-ten Kreisteilungspolynome $g_n(z) = \prod_{d|n}(z^d - 1)^{\mu(n/d)}$, *J. reine angew. Math.* **187**, 68–69 (1949).

KLOOSTERMAN, H. D.:

[1] On the representation of numbers in the form $ax^2 + by^2 + cz^2 + dt^2$, *Acta Math.* **49**, 407–464 (1926).

[2] Asymptotische Formeln für die Fourierkoeffizienten ganzer Modulformen, *Abh. Math. Sem. Univ. Hamburg* **5**, 337–352 (1927).

[3] Thetareihen in total-reellen algebraischen Zahlkörpern, *Math. Ann.* **103**, 279–299 (1930).

[4] On the singular series in Waring's problem and in the problem of the

representation of integers as a sum of powers of primes, *Indag. Math.* **1**, 51–56 (1939).

[5] The behaviour of general theta functions under the modular group and the characters of binary modular congruence groups. II, *Ann. of Math.* (2)**47**, 376–447 (1946).

[6] The law of quadratic reciprocity, *Indag. Math.* **27**, 163–164 (1965).

KLÖSGEN, W.:

[1] *Untersuchungen über Fermatsche Kongruenzen*, Gesellschaft für Mathematik und Datenverarbeitung, no. 36, Bonn, 1970.

KLUYVER, J. C.:

[1] Problem 139 (Dutch), *Wiskundige Opgaven* **14**, 278–280 (1928).

KNEE, D., and GOLDMAN, H. D.:

[1] Quasi-self-reciprocal polynomials and potentially large minimum distance BCH codes, *IEEE Trans. Information Theory* **IT-15**, 118–121 (1969).

KNESER, A.:

[1] Arithmetische Begründung einiger algebraischer Fundamentalsätze, *J. reine angew. Math.* **102**, 20–55 (1888).

KNIGHT, M. J., and WEBB, W. A.:

[1] Uniform distribution of third order linear recurrence sequences, *Acta Arith.* **36**, 7–20 (1980).

KNIŽNERMAN, L. A., and SOKOLINSKIĬ, V. Z.:

[1] Some estimates for rational trigonometric sums and sums of Legendre symbols (Russian), *Uspehi Mat. Nauk* **34**, no. 3, 199–200 (1979).

KNOPFMACHER, J.:

[1] *Abstract Analytic Number Theory*, North-Holland, Amsterdam, 1975.

[2] *Analytic Arithmetic of Algebraic Function Fields*, Lecture Notes in Pure and Appl. Math., vol. 50, Dekker, New York, 1979.

KNOPP, M. I.:

[1] Automorphic forms of nonnegative dimension and exponential sums, *Michigan Math. J.* **7**, 257–287 (1960).

KNUTH, D. E.:

[1] Finite semi-fields and projective planes, *J. Algebra* **2**, 182–217 (1965).

[2] *The Art of Computer Programming*, vol. 1: *Fundamental Algorithms*, Addison-Wesley, Reading, Mass., 1968.

[3] *The Art of Computer Programming*, vol. 2: *Seminumerical Algorithms*, Addison-Wesley, Reading, Mass., 1969; 2nd ed., Addison-Wesley, Reading, Mass., 1981.

KOBLITZ, N.:

[1] *p*-adic variation of the zeta-function over families of varieties defined over finite fields, *Compositio Math.* **31**, 119–218 (1975).

[2] *p-adic Numbers, p-adic Analysis, and Zeta-Functions*, Springer-Verlag, New York-Heidelberg-Berlin, 1977.

[3] *p-adic Analysis: A Short Course on Recent Work*, London Math. Soc. Lecture Note Series, no. 46, Cambridge Univ. Press, Cambridge, 1980.

[4] The p-adic approach to solutions of equations over finite fields, *Amer. Math. Monthly* **87**, 115–118 (1980).

KOCHENDÖRFFER, R.:

[1] *Introduction to Algebra*, Wolters-Noordhoff, Groningen, 1972.

KONDO, T.:

[1] On Gaussian sums attached to the general linear groups over finite fields, *J. Math. Soc. Japan* **15**, 244–255 (1963).

KONHEIM, A. G.:

[1] *Cryptography. A Primer*, Wiley, New York, 1981.

KORNBLUM, H.:

[1] Über die Primfunktionen in einer arithmetischen Progression, *Math. Z.* **5**, 100–111 (1919).

KÖRNER, O., and STÄHLE, H.:

[1] Remarks on Hua's estimate of complete trigonometrical sums, *Acta Arith.* **35**, 353–359 (1979).

KOROBOV, N. M.:

[1] The distribution of non-residues and of primitive roots in recurrence series (Russian), *Dokl. Akad. Nauk SSSR* **88**, 603–606 (1953).

[2] Estimates of trigonometric sums and their applications (Russian), *Uspehi Mat. Nauk* **13**, no. 4, 185–192 (1958).

[3] Estimation of rational trigonometric sums (Russian), *Dokl. Akad. Nauk SSSR* **118**, 231–232 (1958).

[4] On zeros of the $\zeta(s)$ function (Russian), *Dokl. Akad. Nauk SSSR* **118**, 431–432 (1958).

[5] Double trigonometric sums and their applications to the estimation of rational sums (Russian), *Mat. Zametki* **6**, 25–34 (1969); *Math. Notes* **6**, 472–478 (1969).

[6] Estimate of a sum of Legendre symbols (Russian), *Dokl. Akad. Nauk SSSR* **196**, 764–767 (1971); *Soviet Math. Dokl.* **12**, 241–245 (1971).

[7] Complete systems of congruences (Russian), *Acta Arith.* **21**, 357–366 (1972).

KOROBOV, N. M., and MIT'KIN, D. A.:

[1] Lower bounds of complete trigonometric sums (Russian), *Vestnik Moskov. Univ. Ser. I Mat. Meh.* **1977**, no. 5, 54–57; *Moscow Univ. Math. Bull.* **32**, no. 5, 43–45 (1977).

KOUTSKÝ, K.:

[1] On the quadratic character of numbers and on the generalization of a theorem of Lagrange on the distribution of quadratic residues (Czech), *Rozpravy Ceské Akad. Ved* **39** (1930), no. 43, 21 pp.

[2] On the distribution of power residues for a prime modulus (Czech), *Čas. Pěst. Mat. Fys.* **59**, 65–82 (1930).

KOZEL, P. T., and ŠAKLEINA, T. A.:

[1] The number of isotropic subspaces in a space with an orthogonal metric (Russian), *Vestnik Beloruss. Gos. Univ. Ser. I* **1975**, no. 1, 11–15.

KRAÏTCHIK, M.:

[1] *Recherches sur la théorie des nombres. II: Factorisation*, Gauthier-Villars, Paris, 1929.

[2] On the factorization of $2^n \pm 1$, *Scripta Math.* **18**, 39–52 (1952).

KRASNER, M.:

[1] Sur la primitivité des corps $\mathfrak{p}$-adiques, *Mathematica (Cluj)* **13**, 72–191 (1937).

[2] Sur la représentation exponentielle dans les corps relativement galoisiens de nombres $\mathfrak{p}$-adiques, *Acta Arith.* **3**, 133–173 (1939).

KRÄTZEL, E.:

[1] Kubische und biquadratische Gaußsche Summen, *J. reine angew. Math.* **228**, 159–165 (1967).

KRISHNAMURTHY, E. V.:

[1] Exact inversion of a rational polynomial matrix using finite field transforms, *SIAM J. Appl. Math.* **35**, 453–464 (1978).

KRISHNAMURTHY, E. V., and RAMACHANDRAN, V.:

[1] A cryptographic system based on finite field transforms, *Proc. Indian Acad. Sci. Sect. A* **89**, no. 2, 75–93 (1980).

KRONECKER, L.:

[1] Mémoire sur les facteurs irréductibles de l'expression $x^n - 1$, *J. Math. Pures Appl.* **19**, 177–192 (1854); *Werke*, vol. 1, pp. 75–92, Teubner, Leipzig, 1895.

[2] Sur une formule de Gauss, *J. Math. Pures Appl.* (2)**1**, 392–395 (1856); *Werke*, vol. 4, pp. 171–175, Teubner, Leipzig, 1929.

[3] Über den vierten Gauss'schen Beweis des Reciprocitätsgesetzes für die quadratischen Reste, *Monatsber. Preuss. Akad. Wiss. Berlin* **1880**, 686–698, 854–860; *Werke*, vol. 4, pp. 275–294, Teubner, Leipzig, 1929.

[4] Zur Theorie der Elimination einer Variabeln aus zwei algebraischen Gleichungen, *Monatsber. Preuss. Akad. Wiss. Berlin* **1881**, 535–600; *Werke*, vol. 2, pp. 113–192, Teubner, Leipzig, 1897.

[5] Grundzüge einer arithmetischen Theorie der algebraischen Grössen, *J. reine angew. Math.* **92**, 1–122 (1882); *Werke*, vol. 2, pp. 237–387, Teubner, Leipzig, 1897.

[6] Zur Theorie der Abelschen Gleichungen, *J. reine angew. Math.* **93**, 338–364 (1882); *Werke*, vol. 4, pp. 131–162, Teubner, Leipzig, 1929.

[7] Über einige Anwendungen der Modulsysteme auf elementare algebraische Fragen, *J. reine angew. Math.* **99**, 329–371 (1886); *Werke*, vol. 3, part 1, pp. 145–208, Teubner, Leipzig, 1899.

[8] Ein Fundamentalsatz der allgemeinen Arithmetik, *J. reine angew. Math.* **100**, 490–510 (1887); *Werke*, vol. 3, part 1, pp. 209–240, Teubner, Leipzig, 1899.

[9] Summirung der Gauss'schen Reihen $\sum_{h=0}^{h=n-1} e^{2h^2\pi i/n}$, *J. reine angew. Math.* **105**, 267–268 (1889); *Werke*, vol. 4, pp. 295–300, Teubner, Leipzig, 1929.

KUBOTA, R. M.:

[1] Waring's problem for $F_q[x]$, *Dissertationes Math.* **117**, 1–60 (1974).

KUBOTA, T.:
[1] Über quadratische Charaktersummen, *Nagoya Math. J.* **19**, 15–25 (1961).
[2] Local relation of Gauss sums, *Acta Arith.* **6**, 285–294 (1961).
[3] Über eine Verallgemeinerung der Reziprozität der Gaußschen Summen, *Math. Z.* **82**, 91–100 (1963).
[4] Some arithmetical applications of an elliptic function, *J. reine angew. Math.* **214/215**, 141–145 (1964).
[5] On a special kind of Dirichlet series, *J. Math. Soc. Japan* **20**, 193–207 (1968).
[6] Some results concerning reciprocity law and real analytic automorphic functions, *Proc. Symp. Pure Math.*, vol. 20, pp. 382–395, American Math. Society, Providence, R.I., 1971.

KÜHNE, H.:
[1] Eine Wechselbeziehung zwischen Functionen mehrerer Unbestimmten, die zu Reciprocitätsgesetzen führt, *J. reine angew. Math.* **124**, 121–133 (1902).
[2] Angenäherte Auflösung von Congruenzen nach Primmodulsystemen in Zusammenhang mit den Einheiten gewisser Körper, *J. reine angew. Math.* **126**, 102–115 (1903).
[3] Bemerkungen zu der Abhandlung des Herrn Hurwitz: Über höhere Kongruenzen, *Archiv Math. Phys.* (3)**6**, 174–176 (1904).

KUIPERS, L.:
[1] A remark on a theorem of L. Carlitz, *Mat. Vesnik* **9**, 113–116 (1972).
[2] A remark on asymptotic distribution in $GF[p^r, x]$, *Rev. Roum. Math. Pures Appl.* **18**, 1217–1221 (1973).
[3] Einige Bemerkungen zu einer Arbeit von G. J. Rieger, *Elemente der Math.* **34**, 32–34 (1979).

KUIPERS, L., and NIEDERREITER, H.:
[1] *Uniform Distribution of Sequences*, Wiley-Interscience, New York, 1974.

KUIPERS, L., and SCHEELBEEK, P. A. J.:
[1] Uniform distribution of sequences from direct products of groups, *Ann. Sc. Norm. Sup. Pisa* **22**, 599–606 (1968).

KUIPERS, L., and SHIUE, J.-S.:
[1] On the distribution modulo m of sequences of generalized Fibonacci numbers, *Tamkang J. Math.* **2**, 181–186 (1971).
[2] A distribution property of a linear recurrence of the second order, *Atti Accad. Naz. Lincei Rend. Cl. Sci. Fis. Mat. Natur.* (8)**52**, 6–10 (1972).
[3] A distribution property of the sequence of Lucas numbers, *Elemente der Math.* **27**, 10–11 (1972).
[4] A distribution property of the sequence of Fibonacci numbers, *Fibonacci Quart.* **10**, 375–376 (1972).

KUMAR, I. J., and KUMARI, M.:
[1] Local complementation of periodic sequences over $GF(2)$, *J. Combin. Inform. System Sci.* **6**, 178–186 (1981).

KUMARI, M.:
[1] Concatenation properties of δ-sequences over $GF(2)$, *J. Inform. Optim. Sci.* **2**, 147–160 (1981).

KUMMER, E. E.:

[1] Eine Aufgabe, betreffend die Theorie der cubischen Reste, *J. reine angew. Math.* **23**, 285–286 (1842); *Collected Papers*, vol. 1, pp. 143–144, Springer-Verlag, Berlin-Heidelberg-New York, 1975.

[2] De residuis cubicis disquisitiones nonnullae analyticae, *J. reine angew. Math.* **32**, 341–359 (1846); *Collected Papers*, vol. 1, pp. 145–163, Springer-Verlag, Berlin-Heidelberg-New York, 1975.

[3] Über die Zerlegung der aus Wurzeln der Einheit gebildeten complexen Zahlen in ihre Primfactoren, *J. reine angew. Math.* **35**, 327–367 (1847); *Collected Papers*, vol. 1, pp. 211–251, Springer-Verlag, Berlin-Heidelberg-New York, 1975.

[4] Allgemeine Reciprocitätsgesetze für beliebig hohe Potenzreste, *Monatsber. Königl. Akad. Wiss. Berlin* **1850**, 154–165; *Collected Papers*, vol. 1, pp. 345–357, Springer-Verlag, Berlin-Heidelberg-New York, 1975.

[5] Mémoire sur la théorie des nombres complexes composés de racines de l'unité et de nombres entiers, *J. Math. Pures Appl.* **16**, 377–498 (1851); *Collected Papers*, vol. 1, pp. 363–484, Springer-Verlag, Berlin-Heidelberg-New York, 1975.

[6] Über die Ergänzungssätze zu den allgemeinen Reciprocitätsgesetzen, *J. reine angew. Math.* **44**, 93–146 (1852); *Collected Papers*, vol. 1, pp. 485–538, Springer-Verlag, Berlin-Heidelberg-New York, 1975.

[7] Theorie der idealen Primfaktoren der complexen Zahlen, welche aus den Wurzeln der Gleichung $\omega^n = 1$ gebildet sind, wenn n eine zusammengesetzte Zahl ist, *Math. Abh. Königl. Akad. Wiss. Berlin* **1856**, 1–47; *Collected Papers*, vol. 1, pp. 583–629, Springer-Verlag, Berlin-Heidelberg-New York, 1975.

KUNERT, D.:

[1] Ein neuer Beweis für die Reziprozitätsformel der Gaußschen Summen in beliebigen algebraischen Zahlkörpern, *Math. Z.* **40**, 326–347 (1936).

KUNG, J. P. S.:

[1] The cycle structure of a linear transformation over a finite field, *Linear Algebra Appl.* **36**, 141–155 (1981).

KURBATOV, V. A.:

[1] On polynomials which produce substitutions for infinitely many primes (Russian), *Sverdlovsk. Gos. Ped. Inst. Učen. Zap.* **4**, 79–121 (1947).

[2] Generalizations of Schur's theorem concerning a class of algebraic functions (Russian), *Mat. Sb.* **21**, 133–141 (1947); *Amer. Math. Soc. Transl.* (2)**37**, 1–11 (1964).

[3] On the monodromy group of an algebraic function (Russian), *Mat. Sb.* **25**, 51–94 (1949); *Amer. Math. Soc. Transl.* (2)**36**, 17–62 (1964).

KURBATOV, V. A., and STARKOV, N. G.:

[1] The analytic representation of permutations (Russian), *Sverdlovsk. Gos. Ped. Inst. Učen. Zap.* **31**, 151–158 (1965).

KUSTAANHEIMO, P., and QVIST, B.:

[1] On differentiation in Galois fields, *Suom. Tiedeak. Toimituk. Helsinki Ser. A I* **137** (1952), 12 pp.

KUTZKO, P. C.:

[1] The cyclotomy of finite commutative P.I.R.'s, *Illinois J. Math.* **19**, 1–17 (1975).

KUZNECOV, N. V.:

[1] The Petersson conjecture for cusp forms of weight zero and the Linnik conjecture. Sums of Kloosterman sums (Russian), *Mat. Sb.* (*N.S.*) **111**, 334–383 (1980); *Math. USSR-Sbornik* **39**, 299–342 (1981).

KUZNECOV, V. N.:

[1] Z-functions of a certain class of Artin-Schreier coverings (Russian), *Ural. Gos. Univ. Mat. Zap.* **10**, no. 1, 24–36 (1976).

LABUNEC, V. G., and SITNIKOV, O. P.:

[1] Harmonic analysis of Boolean functions and k-ary logic functions on finite fields (Russian), *Izv. Akad. Nauk SSSR Tehn. Kibernet.* **1975**, no. 1, 141–148; *Engrg. Cybernetics* **13**, 112–119 (1975).

LAFFEY, T. J.:

[1] Infinite rings with all proper subrings finite, *Amer. Math. Monthly* **81**, 270–272 (1974).

LAGRANGE, J.-L.:

[1] Sur l'intégration d'une équation différentielle à différences finies, qui contient la théorie des suites récurrentes, *Misc. Taurinensia* **1** (1759); *Oeuvres*, vol. 1, pp. 23–36, Gauthier-Villars, Paris, 1867.

[2] Nouvelle méthode pour résoudre les problèmes indéterminés en nombres entiers, *Mémoires Acad. Roy. Berlin* **24** (1770); *Oeuvres*, vol. 2, pp. 655–726, Gauthier-Villars, Paris, 1868.

[3] Démonstration d'un théorème d'arithmétique, *Nouv. Mémoires Acad. Roy. Berlin* **1770**, 123–133; *Oeuvres*, vol. 3, pp. 189–201, Gauthier-Villars, Paris, 1869.

[4] Réflexions sur la résolution algébrique des équations, *Nouv. Mémoires Acad. Roy. Berlin* **1770**, 134–215; *ibid.* **1771**, 138–254; *Oeuvres*, vol. 3, pp. 205–421, Gauthier-Villars, Paris, 1869.

[5] Recherches sur les suites récurrentes dont les termes varient de plusieurs manières différentes, ou sur l'intégration des équations linéaires aux différences finies et partielles; et sur l'usage de ces équations dans la théorie des hasards, *Nouv. Mémoires Acad. Roy. Berlin* **1775**, 183–272; *Oeuvres*, vol. 4, pp. 151–251, Gauthier-Villars, Paris, 1869.

LAKKIS, K.:

[1] Die galoisschen Gauss'schen Summen von Hasse, *Bull. Soc. Math. Grèce* **7**, 183–371 (1966).

[2] Die verallgemeinerten Gaußschen Summen, *Arch. Math.* **17**, 505–509 (1966).

[3] Die lokalen verallgemeinerten Gauss'schen Summen, *Bull. Soc. Math. Grèce* **8**, 143–150 (1967).

LAKSOV, D.:

[1] Linear recurring sequences over finite fields, *Math. Scand.* **16**, 181–196 (1965).

LAL, M.:

[1] On the separability of multivariable polynomials, *Proc. IEEE* **63**, 718–719 (1975).

LAMPRECHT, E.:

[1] Allgemeine Theorie der Gaußschen Summen in endlichen kommutativen Ringen, *Math. Nachr.* **9**, 149–196 (1953).

[2] Gaußsche Summen in endlichen Ringen und ihre Anwendungen, *Bericht Math.-Tagung* (Berlin, 1953), pp. 179–185, Deutscher Verlag der Wissenschaften, Berlin, 1953.

[3] Struktur und Relationen allgemeiner Gaußscher Summen in endlichen Ringen, *J. reine angew. Math.* **197**, 1–26, 27–48 (1957).

LANDAU, E.:

[1] Abschätzungen von Charaktersummen, Einheiten und Klassenzahlen, *Göttinger Nachr.* **1918**, 79–97.

[2] *Vorlesungen über Zahlentheorie*, vol. 1, part 2, Hirzel, Leipzig, 1927.

[3] Über das Vorzeichen der Gaußschen Summe, *Göttinger Nachr.* **1928**, 19–20.

[4] Zum Waringschen Problem. III, *Math. Z.* **32**, 699–702 (1930).

[5] *Elementary Number Theory*, 2nd ed., Chelsea, New York, 1958.

LANDSBERG, G.:

[1] Ueber eine Anzahlbestimmung und eine damit zusammenhängende Reihe, *J. reine angew. Math.* **111**, 87–88 (1893).

[2] Zur Theorie der Gaussschen Summen und der linearen Transformation der Thetafunctionen, *J. reine angew. Math.* **111**, 234–253 (1893).

LANG, S.:

[1] *Abelian Varieties*, Interscience, New York, 1959.

[2] Some theorems and conjectures in diophantine equations, *Bull. Amer. Math. Soc.* **66**, 240–249 (1960).

[3] *Algebraic Number Theory*, Addison-Wesley, Reading, Mass., 1970.

[4] *Algebra*, Addison-Wesley, Reading, Mass., 1971.

[5] *Cyclotomic Fields*, Springer-Verlag, New York-Heidelberg-Berlin, 1978.

[6] *Cyclotomic Fields*, vol. 2, Springer-Verlag, New York-Heidelberg-Berlin, 1980.

LANG, S., and WEIL, A.:

[1] Number of points of varieties in finite fields, *Amer. J. Math.* **76**, 819–827 (1954).

LATAWIEC, K. J.:

[1] On different time-domain solutions of the problem of generating shifted linear binary sequences, *Problemy Uprav. i Teor. Informacii* **6**, 223–230 (1977).

LAUMON, G.:

[1] Majorations de sommes trigonométriques (d'après P. Deligne et N. Katz), *Astérisque*, no. 82–83, pp. 221–258, Soc. Math. France, Paris, 1981.

LAUSCH, H., MÜLLER, W. B., and NÖBAUER, W.:

[1] Über die Struktur einer durch Dicksonpolynome dargestellten Permutationsgruppe des Restklassenringes modulo n, *J. reine angew. Math.* **261**, 88–99 (1973).

LAUSCH, H., and NÖBAUER, W.:

[1] *Algebra of Polynomials*, North-Holland, Amsterdam, 1973.

LAXTON, R. R., and ANDERSON, J. A.:

[1] Linear recurrences and maximal length sequences, *Math. Gaz.* **56**, 299–309 (1972).

LEAHEY, W.:

[1] Sums of squares of polynomials with coefficients in a finite field, *Amer. Math. Monthly* **74**, 816–819 (1967).

LEBEDEV, S. S.:

[1] Estimation of a trigonometric sum (Russian), *Vestnik Moskov. Univ. Ser. I Mat. Meh.* **1961**, no. 3, 22–28.

LEBESGUE, V. A.:

[1] Recherches sur les nombres. I, II, III, *J. Math. Pures Appl.* **2**, 253–292 (1837); *ibid.* **3**, 113–131, 132–144 (1838).

[2] Démonstration de quelques formules d'un mémoire de M. Jacobi, *J. Math. Pures Appl.* **19**, 289–300 (1854).

[3] Démonstration de l'irréductibilité de l'équation aux racines primitives de l'unité, *J. Math. Pures Appl.* (2)**4**, 105–110 (1859).

[4] Note sur les congruences, *C. R. Acad. Sci. Paris* **51**, 9–13 (1860).

[5] Sur une congruence du deuxième degré à plusieurs inconnues, *C. R. Acad. Sci. Paris* **62**, 868–872 (1866).

LEE, A.:

[1] Über einige Extremalaufgaben bezüglich endlicher Körper, *Acta Math. Acad. Sci. Hungar.* **13**, 235–243 (1962).

LEE, J. J., and SMITH, D. R.:

[1] Families of shift-register sequences with impulsive correlation properties, *IEEE Trans. Information Theory* **IT-20**, 255–261 (1974).

LEE, M. A.:

[1] Some irreducible polynomials which are reducible mod p for all p, *Amer. Math. Monthly* **76**, 1125 (1969).

LEHMER, D. H.:

[1] Tests for primality by the converse of Fermat's theorem, *Bull. Amer. Math. Soc.* **33**, 327–340 (1927).

[2] Ar extended theory of Lucas' functions, *Ann. of Math.* (2)**31**, 419–448

[3] v factorizations of $2^n \pm 1$, *Bull. Amer. Math. Soc.* **39**, 105–108

number sieve, *Amer. Math. Monthly* **40**, 401–406 (1933).

[5] A machine for combining sets of linear congruences, *Math. Ann.* **109**, 661–667 (1934).
[6] On the series for the partition function, *Trans. Amer. Math. Soc.* **43**, 271–295 (1938).
[7] A factorization theorem applied to a test for primality, *Bull. Amer. Math. Soc.* **45**, 132–137 (1939).
[8] On certain character matrices, *Pacific J. Math.* **6**, 491–499 (1956).
[9] Power character matrices, *Pacific J. Math.* **10**, 895–907 (1960).
[10] Computer technology applied to the theory of numbers, *Studies in Number Theory* (W. J. LeVeque, ed.), pp. 117–151, Prentice-Hall, Englewood Cliffs, N.J., 1969.
[11] Incomplete Gauss sums, *Mathematika* **23**, 125–135 (1976).

LEHMER, D. H., and LEHMER, E.:
[1] On the cubes of Kloosterman sums, *Acta Arith.* **6**, 15–22 (1960).
[2] On runs of residues, *Proc. Amer. Math. Soc.* **13**, 102–106 (1962).
[3] The cyclotomy of Kloosterman sums, *Acta Arith.* **12**, 385–407 (1967).
[4] The cyclotomy of hyper-Kloosterman sums, *Acta Arith.* **14**, 89–111 (1968).

LEHMER, D. H., LEHMER, E., and MILLS, W. H.:
[1] Pairs of consecutive power residues, *Canad. J. Math.* **15**, 172–177 (1963).

LEHMER, D. H., LEHMER, E., MILLS, W. H., and SELFRIDGE, J. L.:
[1] Machine proof of a theorem on cubic residues, *Math. Comp.* **16**, 407–415 (1962).

LEHMER, E.:
[1] On the quintic character of 2, *Bull. Amer. Math. Soc.* **55**, 62–63 (1949).
[2] The quintic character of 2 and 3, *Duke Math. J.* **18**, 11–18 (1951).
[3] On residue difference sets, *Canad. J. Math.* **5**, 425–432 (1953).
[4] On the number of solutions of $u^k + D \equiv w^2 \pmod{p}$, *Pacific J. Math.* **5**, 103–118 (1955).
[5] On the location of Gauss sums, *Math. Tables Aids Comput.* **10**, 194–202 (1956).
[6] On Euler's criterion, *J. Austral. Math. Soc.* **1**, 64–70 (1959).
[7] On Jacobi functions, *Pacific J. Math.* **10**, 887–893 (1960).
[8] Artiads characterized, *J. Math. Anal. Appl.* **15**, 118–131 (1966).

LEHMER, E., and VANDIVER, H. S.:
[1] On the computation of the number of solutions of certain trinomial congruences, *J. Assoc. Comput. Mach.* **4**, 505–510 (1957).

LEHTI, R.:
[1] Evaluation matrices for polynomials in Galois fields, *Soc. Sci. Fenn. Comment. Phys.-Math.* **22**, no. 3, 1959.

LEMPEL, A.:
[1] Analysis and synthesis of polynomials and sequences over $GF(2)$, *IEEE Trans. Information Theory* **IT-17**, 297–303 (1971).
[2] Matrix factorization over $GF(2)$ and trace-orthogonal bases of $GF(2^n)$, *SIAM J. Computing* **4**, 175–186 (1975).

LEMPEL, A., COHN, M., and EASTMAN, W. L.:

[1] A class of balanced binary sequences with optimal autocorrelation properties, *IEEE Trans. Information Theory* **IT-23**, 38–42 (1977).

LEMPEL, A., and EASTMAN, W. L.:

[1] High speed generation of maximal length sequences, *IEEE Trans. Computers* **C-20**, 227–229 (1971).

LEMPEL, A., and WINOGRAD, S.:

[1] A new approach to error-correcting codes, *IEEE Trans. Information Theory* **IT-23**, 503–508 (1977).

LENSKOĬ, D. N.:

[1] On the arithmetic of polynomials over a finite field (Russian), *Volž. Mat. Sb. Vyp.* **4**, 155–159 (1966).

[2] On the arithmetic of polynomials over a finite field. II (Russian), *Studies in Number Theory* (Russian), no. 1, pp. 27–34, Izdat. Saratov. Univ., Saratov, 1966.

LENSTRA, A. K.:

[1] Lattices and factorization of polynomials, *SIGSAM Bull.* **15**, no. 3, 15–16 (1981).

LENSTRA, A. K., LENSTRA, H. W., Jr., and LOVÁSZ, L.:

[1] Factoring polynomials with rational coefficients, *Math. Ann.* **261**, 515–534 (1982).

LENSTRA, H. W., Jr.:

[1] Primitive normal bases for finite fields, unpublished manuscript, 1977.

[2] Primality testing algorithms (after Adleman, Rumely and Williams), *Séminaire Bourbaki 1980/81*, Exp. 576, Lecture Notes in Math., vol. 901, pp. 243–257, Springer-Verlag, Berlin-Heidelberg-New York, 1981.

LEONARD, P. A.:

[1] On constructing quartic extensions of $GF(p)$, *Norske Vid. Selsk. Forh. (Trondheim)* **40**, 96–97 (1967).

[2] On factorizations of certain trinomials, *Norske Vid. Selsk. Forh. (Trondheim)* **42**, 56–62 (1969).

[3] On factoring quartics (mod p), *J. Number Theory* **1**, 113–115 (1969).

[4] A note on cubics over $GF(2^n)$, *Norske Vid. Selsk. Skr. (Trondheim)* **1974**, no. 1, 2 pp.

[5] Factorization of general polynomials, *J. Number Theory* **6**, 335–338 (1974).

LEONARD, P. A., MORTIMER, B. C., and WILLIAMS, K. S.:

[1] The eleventh power character of 2, *J. reine angew. Math.* **286**, 213–222 (1976).

LEONARD, P. A., and WILLIAMS, K. S.:

[1] Quartics over $GF(2^n)$, *Proc. Amer. Math. Soc.* **36**, 347–350 (1972).

[2] A diophantine system of Dickson, *Atti Accad. Naz. Lincei Rend. Cl. Sci. Fis. Mat. Natur.* (8)**56**, 145–150 (1974).

[3] The cyclotomic numbers of order seven, *Proc. Amer. Math. Soc.* **51**, 295–300 (1975).
[4] Jacobi sums and a theorem of Brewer, *Rocky Mountain J. Math.* **5**, 301–308 (1975); Erratum, *ibid.* **6**, 509 (1976).
[5] The cyclotomic numbers of order eleven, *Acta Arith.* **26**, 365–383 (1975).
[6] A rational sixteenth power reciprocity law, *Acta Arith.* **33**, 365–377 (1977).
[7] Evaluation of certain Jacobsthal sums, *Boll. Un. Mat. Ital.* (5)**15**, 717–723 (1978).

LEOPOLDT, H.-W.:
[1] Zur Arithmetik in abelschen Zahlkörpern, *J. reine angew. Math.* **209**, 54–71 (1962).

LERCH, M.:
[1] Zur Theorie der Gaußschen Summen, *Math. Ann.* **57**, 554–567 (1903).

LEVI, H.:
[1] Composite polynomials with coefficients in an arbitrary field of characteristic zero, *Amer. J. Math.* **64**, 389–400 (1942).

LEVINE, J., and BRAWLEY, J. V.:
[1] Involutory commutants with some applications to algebraic cryptography. I, II, *J. reine angew. Math.* **224**, 20–43 (1966); *ibid.* **227**, 1–24 (1967).
[2] Some cryptographic applications of permutation polynomials, *Cryptologia* **1**, 76–92 (1977).

LEVINE, J., and HARTWIG, R. E.:
[1] Applications of the Drazin inverse to the Hill cryptographic system, I, II, *Cryptologia* **4**, 71–85, 150–168 (1980).

LEWIS, D. J.:
[1] Singular quartic forms, *Duke Math. J.* **21**, 39–44 (1954).
[2] Cubic congruences, *Michigan Math. J.* **4**, 85–95 (1957).
[3] Diagonal forms over finite fields, *Norske Vid. Selsk. Forh.* (*Trondheim*) **33**, 61–65 (1960).

LEWIS, D. J., and SCHUUR, S. E.:
[1] Varieties of small degree over finite fields, *J. reine angew. Math.* **262/263**, 293–306 (1973).

LIANG, J. J.:
[1] On the solutions of trinomial equations over finite fields, *Bull. Calcutta Math. Soc.* **70**, 379–382 (1978).

LIBRI, G.:
[1] Mémoire sur la théorie des nombres, *J. reine angew. Math.* **9**, 169–188 (1832).
[2] Mémoire sur la théorie des nombres, *J. reine angew. Math.* **9**, 261–276 (1832).

LIDL, R.:

[1] Über Permutationspolynome in mehreren Unbestimmten, *Monatsh. Math.* **75**, 432–440 (1971).

[2] Über die Darstellung von Permutationen durch Polynome, *Abh. Math. Sem. Univ. Hamburg* **37**, 108–111 (1972).

[3] Über Permutationsfunktionen in mehreren Unbestimmten, *Acta Arith.* **20**, 291–296 (1972).

[4] Tschebyscheffpolynome und die dadurch dargestellten Gruppen, *Monatsh. Math.* **77**, 132–147 (1973).

[5] Reguläre Polynome über endlichen Körpern, *Beiträge zur Algebra und Geometrie* **2**, 55–59 (1974).

[6] Über die Struktur einer durch Tschebyscheffpolynome in zwei Variablen dargestellten Permutationsgruppe, *Beiträge zur Algebra und Geometrie* **3**, 41–48 (1974).

[7] Einige ungelöste Probleme bei endlichen Körpern, *Math. Balkanica* **4**, 409–414 (1974).

[8] Tschebyscheffpolynome in mehreren Variablen, *J. reine angew. Math.* **273**, 178–198 (1975).

LIDL, R., and MÜLLER, W. B.:

[1] Über Permutationsgruppen, die durch Tschebyscheff-Polynome erzeugt werden, *Acta Arith.* **30**, 19–25 (1976).

LIDL, R., and NIEDERREITER, H.:

[1] On orthogonal systems and permutation polynomials in several variables, *Acta Arith.* **22**, 257–265 (1973).

LIDL, R., and PILZ, G.:

[1] *Angewandte abstrakte Algebra I, II*, Bibliographisches Institut, Mannheim, 1982.

LIDL, R., and WELLS, C.:

[1] Chebyshev polynomials in several variables, *J. reine angew. Math.* **255**, 104–111 (1972).

LIDL, R., and WIESENBAUER, J.:

[1] *Ringtheorie und Anwendungen*, Akademische Verlagsgesellschaft, Wiesbaden, 1980.

LIN, S.:

[1] On a class of cyclic codes, *Error Correcting Codes* (H. B. Mann, ed.), pp. 131–148, Wiley, New York, 1968.

[2] *An Introduction to Error-Correcting Codes*, Prentice-Hall, Englewood Cliffs, N.J., 1970.

LINDGREN, H.:

[1] Polynomial solutions of binomial congruences, *J. Austral. Math. Soc.* **1**, 257–280 (1960).

LINDHOLM, J. H.:

[1] An analysis of the pseudo-randomness properties of subsequences of long m-sequences, *IEEE Trans. Information Theory* **IT-14**, 569–576 (1968).

LINDSTRÖM, B.:

[1] On group and nongroup perfect codes in q symbols, *Math. Scand.* **25**, 149–158 (1969).

LINNIK, YU. V.:

[1] Some remarks on estimates of trigonometric sums (Russian), *Uspehi Mat. Nauk* **14**, no. 3, 153–160 (1959).

[2] Additive problems and eigenvalues of the modular operators, *Proc. International Congress of Math.* (Stockholm, 1962), pp. 270–284, Institut Mittag-Leffler, Djursholm, 1963.

LITVER, E. L., and JUDINA, G. E.:

[1] Primitive roots for the first million primes and their powers (Russian), *Mathematical Analysis and Its Applications* (Russian), vol. 3, pp. 106–109, Izdat. Rostov. Univ., Rostov-on-Don, 1971.

LITZINGER, M.:

[1] A basis for residual polynomials in n variables, *Trans. Amer. Math. Soc.* **37**, 216–225 (1935).

LIU, K. Y., REED, I. S., and TRUONG, T. K.:

[1] Fast number-theoretic transforms for digital filtering, *Electron. Lett.* **12**, 644–646 (1976).

[2] High-radix transforms for Reed-Solomon codes over Fermat primes, *IEEE Trans. Information Theory* **IT-23**, 776–778 (1977).

LLOYD, D. B.:

[1] Factorization of the general polynomial by means of its homomorphic congruential functions, *Amer. Math. Monthly* **71**, 863–870 (1964).

[2] The use of finite polynomial rings in the factorization of the general polynomial, *J. Res. Nat. Bur. Standards Sect. B* **69**, 189–212 (1965).

LLOYD, D. B., and REMMERS, H.:

[1] Polynomial factor tables over finite fields, *Math. Algorithms* **2**, 85–99 (1967).

LONDON, D., and ZIEGLER, Z.:

[1] Functions over the residue field modulo a prime, *J. Austral. Math. Soc.* **7**, 410–416 (1967).

LONG, A. F., Jr.:

[1] Some theorems on factorable irreducible polynomials, *Duke Math. J.* **34**, 281–291 (1967).

[2] Classification of irreducible factorable polynomials over a finite field, *Acta Arith.* **12**, 301–313 (1967).

[3] Factorization of irreducible polynomials over a finite field with the substitution $x^{p^r} - x$ for x, *Duke Math. J.* **40**, 63–76 (1973).

[4] Factorization of irreducible polynomials over a finite field with the substitution $x^{q^r} - x$ for x, *Acta Arith.* **25**, 65–80 (1973).
[5] A theorem on factorable irreducible polynomials in several variables over a finite field with the substitution $x_i^{q^r} - x_i$ for x_i, *Math. Nachr.* **63**, 123–130 (1974).

LONG, A. F., Jr., and VAUGHAN, T. P.:
[1] Factorization of $Q(h(T)(x))$ over a finite field, where $Q(x)$ is irreducible and $h(T)(x)$ is linear. II, *Linear Algebra Appl.* **11**, 53–72 (1975).
[2] Factorization of $Q(h(T)(x))$ over a finite field, where $Q(x)$ is irreducible and $h(T)(x)$ is linear. I, *Linear Algebra Appl.* **13**, 207–221 (1976).

LONG, C. T., and WEBB, W. A.:
[1] Normality in $GF\{q, x\}$, *Atti Accad. Naz. Lincei Rend. Cl. Sci. Fis. Mat. Natur.* (8)**54**, 848–853 (1973).

LOWE, R. D., and ZELINSKY, D.:
[1] Which Galois fields are pure extensions?, *Math. Student* **21**, 37–41 (1953).

LOXTON, J. H.:
[1] Products related to Gauss sums, *J. reine angew. Math.* **268/269**, 53–67 (1974).
[2] On the determination of Gauss sums, *Sém. Delange-Pisot-Poitou 1976/77, Théorie des Nombres*, Exp. 27, 12 pp., Secrétariat Math., Paris, 1977.
[3] Some conjectures concerning Gauss sums, *J. reine angew. Math.* **297**, 153–158 (1978).

LOXTON, J. H., and SMITH, R. A.:
[1] On Hua's estimate for exponential sums, *J. London Math. Soc.* (2) **26**, 15–20 (1982).
[2] Estimates for multiple exponential sums, *J. Austral. Math. Soc. Ser. A* **33**, 125–134 (1982).

LUBELSKI, S.:
[1] Zur Theorie der höheren Kongruenzen, *J. reine angew. Math.* **162**, 63–68 (1930).
[2] Zur Reduzibilität von Polynomen in der Kongruenztheorie, *Acta Arith.* **1**, 169–183 (1936).
[3] Über zwei Wegnersche Sätze, *Izv. Akad. Nauk SSSR Ser. Mat.* **5**, 395–398 (1941).

LUBKIN, S.:
[1] On a conjecture of André Weil, *Amer. J. Math.* **89**, 443–548 (1967).
[2] A p-adic proof of Weil's conjectures, *Ann. of Math.* (2)**87**, 105–194, 195–255 (1968).
[3] A result on the Weil zeta function, *Trans. Amer. Math. Soc.* **139**, 297–300 (1969).
[4] Finite generations of lifted p-adic homology with compact supports. Generalization of the Weil conjectures to singular, noncomplete algebraic varieties, *Algebraic Geometry* (Copenhagen, 1978), pp. 317–373, Lecture

Notes in Math., vol. 732, Springer-Verlag, Berlin-Heidelberg-New York, 1979.

[5] Finite generation of lifted p-adic homology with compact supports. Generalization of the Weil conjectures to singular, noncomplete algebraic varieties, *J. Number Theory* **11**, 412–464 (1979).

LUCAS, E.:

[1] Théorie des fonctions numériques simplement périodiques, *Amer. J. Math.* **1**, 184–240, 289–321 (1878).

[2] *Théorie des nombres*, Gauthier-Villars, Paris, 1891.

LUCKY, R. W., SALZ, J., and WELDON, E. J., Jr.:

[1] *Principles of Data Communication*, McGraw-Hill, New York, 1968.

LUH, J.:

[1] On the representation of vector spaces as a finite union of subspaces, *Acta Math. Acad. Sci. Hungar.* **23**, 341–342 (1972).

LÜNEBURG, H.:

[1] *Transitive Erweiterungen endlicher Permutationsgruppen*, Lecture Notes in Math., vol. 84, Springer-Verlag, Berlin-Heidelberg-New York, 1969.

[2] *Galoisfelder, Kreisteilungskörper und Schieberegisterfolgen*, Bibliographisches Institut, Mannheim, 1979.

LÜNEBURG, H., and PLAUMANN, P.:

[1] Die Funktionalgleichung von Gołab und Schinzel in Galoisfeldern, *Arch. Math.* **28**, 55–59 (1977).

LUNNON, W. F., PLEASANTS, P. A. B., and STEPHENS, N. M.:

[1] Arithmetic properties of Bell numbers to a composite modulus, *Acta Arith.* **35**, 1–16 (1979).

L'VOV, I. V.:

[1] An application of the Chevalley-Warning theorem in the theory of rings (Russian), *Ural. Gos. Univ. Mat. Zap.* **11**, no. 1, 110–124 (1978).

MACCLUER, C. R.:

[1] On a conjecture of Davenport and Lewis concerning exceptional polynomials, *Acta Arith.* **12**, 289–299 (1967).

MACDOUGALL, J. A.:

[1] Bivectors over a finite field, *Canad. Math. Bull.* **24**, 489–490 (1981).

MACNEISH, H. F.:

[1] Euler squares, *Ann. of Math.* (2)**23**, 221–227 (1922).

MACWILLIAMS, F. J.:

[1] A theorem on the distribution of weights in a systematic code, *Bell System Tech. J.* **42**, 79–94 (1963).

[2] The structure and properties of binary cyclic alphabets, *Bell System Tech. J.* **44**, 303–332 (1965).

[3] Orthogonal matrices over finite fields, *Amer. Math. Monthly* **76**, 152–164 (1969).

[4] Orthogonal circulant matrices over finite fields, and how to find them, *J. Combinatorial Theory* **10**, 1–17 (1971).

MACWILLIAMS, F. J., and MANN, H. B.:
[1] On the p-rank of the design matrix of a difference set, *Information and Control* **12**, 474–488 (1968).

MACWILLIAMS, F. J., and ODLYZKO, A. M.:
[1] Pelikán's conjecture and cyclotomic cosets, *J. Combinatorial Theory Ser. A* **22**, 110–114 (1977).

MACWILLIAMS, F. J., and SEERY, J.:
[1] The weight distributions of some minimal cyclic codes, *IEEE Trans. Information Theory* **IT-27**, 796–806 (1981).

MACWILLIAMS, F. J., and SLOANE, N. J. A.:
[1] Pseudo-random sequences and arrays, *Proc. IEEE* **64**, 1715–1729 (1976).
[2] *The Theory of Error-Correcting Codes*, North-Holland, Amsterdam, 1977.

MACWILLIAMS, F. J., SLOANE, N. J. A., and GOETHALS, J.-M.:
[1] The MacWilliams identities for nonlinear codes, *Bell System Tech. J.* **51**, 803–819 (1972).

MADDEN, D. J.:
[1] Polynomials and primitive roots in finite fields, *J. Number Theory* **13**, 499–514 (1981).

MAILLET, E.:
[1] Des conditions pour que l'échelle d'une suite récurrente soit irréductible, *Nouv. Ann. Math.* (3)**14**, 152–157, 197–206 (1895).

MALYŠEV, A. V.:
[1] A generalization of Kloosterman sums and their estimates (Russian), *Vestnik Leningrad. Univ.* **15**, no. 13, 59–75 (1960).
[2] Gauss and Kloosterman sums (Russian), *Dokl. Akad. Nauk SSSR* **133**, 1017–1020 (1960); *Soviet Math. Dokl.* **1**, 928–932 (1961).
[3] On the representation of integers by positive quadratic forms (Russian), *Trudy Mat. Inst. Steklov.* **65** (1962).
[4] The Fourier coefficients of modular forms (Russian), *Zap. Naučn. Sem. Leningrad. Otdel. Mat. Inst. Steklov.* **1**, 140–163 (1966).

MAMANGAKIS, S. E.:
[1] Remarks on the Fibonacci series modulo m, *Amer. Math. Monthly* **68**, 648–649 (1961).

MANDELBAUM, D.:
[1] On decoding of Reed-Solomon codes, *IEEE Trans. Information Theory* **IT-17**, 707–712 (1971).
[2] Some results in decoding of certain maximal-distance and BCH codes, *Information and Control* **20**, 232–243 (1972).
[3] A method for decoding of generalized Goppa codes, *IEEE Trans. Information Theory* **IT-23**, 137–140 (1977); Addition, *ibid.* **IT-24**, 268 (1978).

MANIN, YU. I.:

[1] On congruences of the third degree modulo a prime (Russian), *Izv. Akad. Nauk SSSR Ser. Mat.* **20**, 673–678 (1956); *Amer. Math. Soc. Transl.* (2)**13**, 1–7 (1960).

[2] On the arithmetic of rational surfaces (Russian), *Dokl. Akad. Nauk SSSR* **152**, 46–49 (1963); *Soviet Math. Dokl.* **4**, 1243–1247 (1963).

[3] Correspondences, motifs and monoidal transformations (Russian), *Mat. Sb. (N.S.)* **77**, 475–507 (1968).

[4] *Cubic Forms*, North-Holland, Amsterdam, 1974.

[5] What is the maximum number of points on a curve over F_2?, *J. Fac. Sci. Univ. Tokyo Sect. IA* **28**, 715–720 (1981).

MANN, H. B.:

[1] The construction of orthogonal Latin squares, *Ann. Math. Statist.* **13**, 418–423 (1942).

[2] *Analysis and Design of Experiments*, Dover, New York, 1949.

[3] *Addition Theorems*, Wiley-Interscience, New York, 1965.

[4] Recent advances in difference sets, *Amer. Math. Monthly* **74**, 229–235 (1967).

[5] *Error Correcting Codes*, Proc. Symp. Math. Research Center (Univ. of Wisconsin, Madison, Wis., 1968), Wiley, New York, 1968.

[6] The solution of equations by radicals, *J. Algebra* **29**, 551–554 (1974).

MANTEL, W.:

[1] Residues of recurring series (Dutch), *Nieuw Arch. Wisk.* (2)**1**, 172–184 (1894).

[2] Problem 91 (Dutch), *Wiskundige Opgaven* **12**, 213–214 (1918).

MARITSAS, D. G.:

[1] On the statistical properties of a class of linear product feedback shift-register sequences, *IEEE Trans. Computers* **C-22**, 961–962 (1973).

MARKOVIČ, O. F.:

[1] Investigation of a singular sum in systems of Waring type by the elementary smoothing method (Russian), *Kuĭbyšev. Gos. Ped. Inst. Naučn. Trudy* **215**, 30–37 (1978).

MARKUŠEVIČ, A. I.:

[1] *Recurrence Sequences* (Russian), 2nd ed., Izdat. "Nauka," Moscow, 1975; 1st ed., VEB Deutscher Verlag der Wissenschaften, Berlin, 1955.

MAROULAS, J., and BARNETT, S.:

[1] Greatest common divisor of generalized polynomials and polynomial matrices, *Linear Algebra Appl.* **22**, 195–210 (1978).

MARSH, R. W.:

[1] *Table of Irreducible Polynomials over GF*(2) *through Degree 19*, Office of Techn. Serv., U.S. Dept. of Commerce, Washington, D.C., 1957.

MARSH, R. W., and GLEASON, A. M.:

[1] Problem 4709, *Amer. Math. Monthly* **63**, 669 (1956); Solution, *ibid.* **64**, 747–748 (1957).

MARSH, R. W., MILLS, W. H., WARD, R. L., RUMSEY, H., and WELCH, L. R.:

[1] Round trinomials, *Pacific J. Math.* **96**, 175–192 (1981).

MARSHALL, J. B.:

[1] On the extension of Fermat's theorem to matrices of order n, *Proc. Edinburgh Math. Soc.* (2)**6**, 85–91 (1939).

MARTIN, M. H.:

[1] A problem in arrangements, *Bull. Amer. Math. Soc.* **40**, 859–864 (1934).

MARTINET, J.:

[1] Character theory and Artin L-functions, *Algebraic Number Fields* (A. Fröhlich, ed.), pp. 1–87, Academic Press, London, 1977.

MASSEY, J. L.:

[1] Reversible codes, *Information and Control* **7**, 369–380 (1964).

[2] Step-by-step decoding of the Bose-Chaudhuri-Hocquenghem codes, *IEEE Trans. Information Theory* **IT-11**, 580–585 (1965).

[3] Some algebraic and distance properties of convolutional codes, *Error Correcting Codes* (H. B. Mann, ed.), pp. 89–109, Wiley, New York, 1968.

[4] Shift-register synthesis and BCH decoding, *IEEE Trans. Information Theory* **IT-15**, 122–127 (1969).

MATEOS MATEOS, F.:

[1] Classification of congruent symmetric matrices defined over a finite field (Spanish), *Gac. Mat. Madrid* **30**, 74–85 (1978).

MATHER, M.:

[1] The number of non-homogeneous lattice points in plane subsets, *Math. Proc. Cambridge Philos. Soc.* **83**, 25–29 (1978).

MATHIEU, E.:

[1] Mémoire sur l'étude des fonctions de plusieurs quantités, sur la manière de les former et sur les substitutions qui les laissent invariables, *J. Math. Pures Appl.* (2)**6**, 241–323 (1861).

MATLUK, M. M., and GILL, A.:

[1] Decomposition of linear sequential circuits over residue class rings, *J. Franklin Inst.* **294**, 167–180 (1972).

MATTHEWS, C. R.:

[1] Gauss sums and elliptic functions. I. The Kummer sum, *Invent. Math.* **52**, 163–185 (1979).

[2] Gauss sums and elliptic functions. II. The quartic sum, *Invent. Math.* **54**, 23–52 (1979).

MATTHEWS, K. R.:

[1] Waring's theorem for polynomials over a finite field, Dissertation, Univ. of Queensland, 1966.

MATTHEWS, R.:
[1] Some generalisations of Chebyshev polynomials and their induced group structure over a finite field, *Acta Arith.* **41**, 323–335 (1982).
[2] The structure of the group of permutations induced by Chebyshev polynomial vectors over the ring of integers mod m, *J. Austral. Math. Soc. Ser. A* **32**, 88–103 (1982).
[3] Orthogonal systems of polynomials over a finite field with coefficients in a subfield, *Papers in Algebra, Analysis and Statistics* (R. Lidl, ed.), Contemporary Math., vol. 9, pp. 295–302, American Math. Society, Providence, R.I., 1982.

MATTSON, H. F., Jr., and SOLOMON, G.:
[1] A new treatment of Bose-Chaudhuri codes, *J. Soc. Indust. Appl. Math.* **9**, 654–669 (1961).

MATTUCK, A., and TATE, J.:
[1] On the inequality of Castelnuovo-Severi, *Abh. Math. Sem. Univ. Hamburg* **22**, 295–299 (1958).

MATVEEVA, M. V.:
[1] Solution of equations of the third degree in a field of characteristic 3 (Russian), *Problemy Peredači Informacii* **4**, no. 4, 76–78 (1968); *Problems of Information Transmission* **4**, no. 4, 64–66 (1968).

MAXFIELD, M. W.:
[1] The order of a matrix under multiplication (modulo m), *Duke Math. J.* **18**, 619–621 (1951).

MAXSON, C. J.:
[1] A new characterization of finite prime fields, *Canad. Math. Bull.* **11**, 381–382 (1968).

MAZUR, B.:
[1] Eigenvalues of Frobenius acting on algebraic varieties over finite fields, *Proc. Symp. Pure Math.*, vol. 29, pp. 231–261, American Math. Society, Providence, R.I., 1975.

MAZUR, L. E.:
[1] Consecutive residues and nonresidues of polynomials (Russian), *Mat. Zametki* **7**, 97–107 (1970); *Math. Notes* **7**, 59–65 (1970).

MCCANN, K., and WILLIAMS, K. S.:
[1] On the residues of a cubic polynomial (mod p), *Canad. Math. Bull.* **10**, 29–38 (1967).
[2] The distribution of the residues of a quartic polynomial, *Glasgow Math. J.* **8**, 67–88 (1967).
[3] Cubic polynomials with the same residues (mod p), *Proc. Cambridge Philos. Soc.* **64**, 655–658 (1968).

MCCLELLAN, J. H., and PARKS, T. W.:
[1] Eigenvalue and eigenvector decomposition of the discrete Fourier transform, *IEEE Trans. Audio Electroacoust.* **AU-20**, 66–74 (1972).

MCCLELLAN, J. H., and RADER, C. M.:

[1] *Number Theory in Digital Signal Processing*, Prentice-Hall, Englewood Cliffs, N.J., 1979.

MCCLUSKEY, E. J.:

[1] *Introduction to the Theory of Switching Circuits*, McGraw-Hill, New York, 1965.

MCCONNEL, R.:

[1] Pseudo-ordered polynomials over a finite field, *Acta Arith.* **8**, 127–151 (1963).

[2] Functions over finite fields preserving m-th powers, *Duke Math. J.* **36**, 465–472 (1969).

[3] Functions over finite fields satisfying coordinate ψ-conditions, *Duke Math. J.* **39**, 297–312 (1972).

MCCRIMMON, K.:

[1] A note on finite division rings, *Proc. Amer. Math. Soc.* **23**, 598–600 (1969).

MCDONALD, B. R.:

[1] *Finite Rings with Identity*, Dekker, New York, 1974.

MCELIECE, R. J.:

[1] A generalization of difference sets, *Canad. J. Math.* **19**, 206–211 (1967).

[2] Factorization of polynomials over finite fields, *Math. Comp.* **23**, 861–867 (1969).

[3] Table of polynomials of period e over $GF(p)$, *Math. Comp.* **23**, microfiche suppl. C1–C6 (1969).

[4] On periodic sequences from $GF(q)$, *J. Combinatorial Theory Ser. A* **10**, 80–91 (1971).

[5] Irreducible cyclic codes and Gauss sums, *Combinatorics* (M. Hall, Jr., and J. H. van Lint, eds.), pp. 185–202, Reidel, Dordrecht-Boston, 1975.

[6] *The Theory of Information and Coding*, Encyclopedia of Math. and Its Appl., vol. 3, Addison-Wesley, Reading, Mass., 1977.

[7] Correlation properties of sets of sequences derived from irreducible cyclic codes, *Information and Control* **45**, 18–25 (1980).

MCELIECE, R. J., and RUMSEY, H.:

[1] Euler products, cyclotomy, and coding, *J. Number Theory* **4**, 302–311 (1972).

MCELIECE, R. J., and SHEARER, J. B.:

[1] A property of Euclid's algorithm and an application to Padé approximation, *SIAM J. Appl. Math.* **34**, 611–615 (1978).

MCGETTRICK, A. D.:

[1] A result in the theory of Weierstrass elliptic functions, *Proc. London Math. Soc.* (3)**25**, 41–54 (1972).

[2] On the biquadratic Gauss sum, *Proc. Cambridge Philos. Soc.* **71**, 79–83 (1972).

MCLAIN, K. E., and EDGAR, H. M.:

[1] A note on Golomb's "Cyclotomic polynomials and factorization theorems," *Amer. Math. Monthly* **88**, 753 (1981).

MEIJER, H. G., and DIJKSMA, A.:

[1] On uniform distribution of sequences in $GF[q, x]$ and $GF\{q, x\rangle$, *Duke Math. J.* **37**, 507–514 (1970).

MENDELSOHN, N. S.:

[1] Congruence relationships for integral recurrences, *Canad. Math. Bull.* **5**, 281–284 (1962).

[2] Algebraic construction of combinatorial designs, *Congr. Numer.* **13**, 157–168 (1975).

MENON, P. K.:

[1] On Gauss's sum, *J. Indian Math. Soc.* **16**, 31–36 (1952).

[2] On certain sums connected with Galois fields and their applications to difference sets, *Math. Ann.* **154**, 341–364 (1964).

METROPOLIS, N., NICOLETTI, G., and ROTA, G.-C.:

[1] A new class of symmetric functions, *Mathematical Analysis and Applications* (L. Nachbin, ed.), Advances in Math. Suppl. Studies, vol. 7B, pp. 563–575, Academic Press, New York, 1981.

METSÄNKYLÄ, R.:

[1] On kth power coset representatives mod p, *Ann. Acad. Sci. Fenn. Ser. A I* **557** (1973), 6 pp.

MEYER, K.:

[1] Äquivalenz von quadratischen Formen über endlichen Körpern, *Abh. Math. Sem. Univ. Hamburg* **37**, 79–85 (1972).

MICHELSON, A. M.:

[1] Computer implementation of decoders for several BCH codes, *Proc. Symp. on Computer Processing in Communications* (Polytechnic Inst. of Brooklyn, New York, 1969), pp. 401–413, Polytechnic Press, Brooklyn, N.Y., 1969.

MIGNOSI, G.:

[1] Risoluzione apiristica della equazione generale cubica in un corpo numerico finito, *Rend. Circ. Mat. Palermo* **53**, 411–427 (1929).

[2] Sulla risoluzione apiristica delle equazioni algebriche in un corpo numerico finito, *Rend. Accad. Sci. Fis. Mat. Napoli* (3)**35**, 218–233 (1930).

[3] Eliminazione nei sistemi di equazioni algebriche in un corpo finito, *Scritti matematici offerti a Luigi Berzolari*, pp. 249–260, Istituto matematico della Università, Pavia, 1936.

[4] Risoluzione apiristica dei sistemi di equazioni algebriche nei corpi finiti, *Atti Accad. Naz. Lincei Rend. Cl. Sci. Fis. Mat. Natur.* (8)**2**, 250–257 (1947).

[5] Estensione ai corpi finiti di una formula di Rados, *Atti Áccad. Naz. Lincei Rend. Cl. Sci. Fis. Mat. Natur.* (8)**7**, 216–219 (1949).

[6] Ancora sopra una estensione ai corpi finiti di una formula di Rados, *Atti*

Accad. Naz. Lincei Rend. Cl. Sci. Fis. Mat. Natur. (8)**7**, 284–289 (1949).

[7] Sulla enumerazione delle radici della più generale equazione algebrica in un corpo finito, *Convegno Reticoli e Geometrie Proiettive* (Palermo-Messina, 1957), pp. 99–108, Edizioni Cremonese, Rome, 1958.

MIGNOTTE, M.:

[1] Suites récurrentes linéaires, *Sém. Delange-Pisot-Poitou 1973/74, Théorie des Nombres*, Exp. G14, Secrétariat Math., Paris, 1975.

[2] Un algorithme sur la décomposition des polynômes dans un corps fini, *C. R. Acad. Sci. Paris Sér. A* **280**, 137–139 (1975).

[3] Factorisation des polynômes sur un corps fini, *Astérisque*, no. 38–39, pp. 149–157, Soc. Math. France, Paris, 1976.

[4] Algorithmes relatifs à la décomposition des polynômes, *Theoret. Comput. Sci.* **1**, 227–235 (1976).

[5] Calcul des racines d-ièmes dans un corps fini, *C. R. Acad. Sci. Paris Sér. A* **290**, 205–206 (1980).

MIHAĬLJUK, M. V.:

[1] On the complexity of calculating the elementary symmetric functions in finite fields (Russian), *Dokl. Akad. Nauk SSSR* **244**, 1072–1076 (1979); *Soviet Math. Dokl.* **20**, 170–174 (1979).

[2] Calculation of a basis of symmetric functions in finite fields (Russian), *Mat. Zametki* **30**, 291–304 (1981); *Math. Notes* **29/30**, 634–641 (1981).

MILLER, J. C. P.:

[1] On factorisation, with a suggested new approach, *Math. Comp.* **29**, 155–172 (1975).

MILLER, J. C. P., and BROWN, D. J. S.:

[1] An algorithm for evaluation of remote terms in a linear recurrence sequence, *Comput. J.* **9**, 188–190 (1966).

MILLER, R. L.:

[1] Necklaces, symmetries and self-reciprocal polynomials, *Discrete Math.* **22**, 25–33 (1978).

MILLER, R. L., REED, I. S., and TRUONG, T. K.:

[1] A theorem for computing primitive elements in the field of complex integers of a characteristic Mersenne prime, *IEEE Trans. Acoust. Speech Signal Process.* **29**, 119–120 (1981).

MILLS, W. H.:

[1] Polynomials with minimal value sets, *Pacific J. Math.* **14**, 225–241 (1964).

[2] Bounded consecutive residues and related problems, *Proc. Symp. Pure Math.*, vol. 8, pp. 170–174, American Math. Society, Providence, R.I., 1965.

[3] The degree of factors of certain polynomials over finite fields, *Proc. Amer. Math. Soc.* **25**, 860–863 (1970).

[4] Continued fractions and linear recurrences, *Math. Comp.* **29**, 173–180 (1975).

MILLS, W. H., and ZIERLER, N.:
[1] On a conjecture of Golomb, *Pacific J. Math.* **28**, 635–640 (1969).

MILNE, J. S.:
[1] *Étale Cohomology*, Princeton Univ. Press, Princeton, N.J., 1980.
[2] Some estimates from étale cohomology, *J. reine angew. Math.* **328**, 208–220 (1981).

MILNE-THOMSON, L. M.:
[1] *The Calculus of Finite Differences*, Macmillan, London, 1933.

MIN, S. H.:
[1] On a system of congruences, *J. London Math. Soc.* **22**, 47–53 (1947).
[2] On systems of algebraic equations and certain multiple exponential sums, *Quart. J. Math.* **18**, 133–142 (1947).

MIRIMANOFF, D.:
[1] Sur les congruences du troisième degré, *L'Enseignement Math.* **9**, 381–384 (1907).

MIRIMANOFF, D., and HENSEL, K.:
[1] Sur la relation $\left(\frac{D}{p}\right)=(-1)^{n-h}$ et la loi de réciprocité, *J. reine angew. Math.* **129**, 86–87 (1905).

MIRONCHIKOV, E. T.:
[1] A class of double-error-correcting codes and their realization (Russian), *Avt. Telem. i Pribor.* **1963**, no. 3, 251–255.

MITCHELL, H. H.:
[1] On the generalized Jacobi-Kummer cyclotomic function, *Trans. Amer. Math. Soc.* **17**, 165–177 (1916).
[2] On the congruence $cx^\lambda + 1 \equiv dy^\lambda$ in a Galois field, *Ann. of Math.* (2)**18**, 120–131 (1917).

MITCHELL, O. H.:
[1] Some theorems in numbers, *Amer. J. Math.* **4**, 25–38 (1881).

MIT'KIN, D. A.:
[1] On the estimation of a rational trigonometric sum with a prime denominator (Russian), *Vestnik Moskov. Univ. Ser. I Mat. Meh.* **27**, no. 5, 50–58 (1972); *Moscow Univ. Math. Bull.* **27**, no. 5, 40–47 (1972).
[2] Estimation of the sum of Legendre symbols of polynomials of even degree (Russian), *Mat. Zametki* **14**, 73–81 (1973); *Math. Notes* **14**, 597–602 (1973).
[3] Lower bounds of sums of Legendre symbols and lower bounds of trigonometric sums (Russian), *Uspehi Mat. Nauk* **30**, no. 5, 214 (1975).
[4] Existence of rational points on a hyperelliptic curve over a finite prime field (Russian), *Vestnik Moskov. Univ. Ser. I Mat. Meh.* **30**, no. 6, 86–90 (1975); *Moscow Univ. Math. Bull.* **30**, no. 5–6, 124–127 (1975).

[5] On estimates for rational trigonometric sums of a special form (Russian), *Dokl. Akad. Nauk SSSR* **224**, 760–763 (1975); *Soviet Math. Dokl.* **16**, 1296–1300 (1975).

MOENCK, R. T.:
[1] On the efficiency of algorithms for polynomial factoring, *Math. Comp.* **31**, 235–250 (1977).

MOHANTY, N. C.:
[1] Binary and ternary signals with small cross correlations, *Inform. Sci.* **13**, 35–50 (1977).

MÖHRMANN, K. H.:
[1] Erzeugung von binären Quasi-Zufallsfolgen hoher Taktfrequenz durch Multiplexen, *Siemens Research and Development Reports* **3**, no. 4, 218–224 (1974).
[2] Realisierung von Scramblern für PCM-Signale hoher Taktfrequenz, *Siemens Research and Development Reports* **6**, no. 1, 1–5 (1977).

MOISIL, G. C.:
[1] L'emploi des imaginaires de Galois dans la théorie des mécanismes automatiques. I–V, VII–IX (Romanian. French summary), *Com. Acad. R. P. Romîne* **4**, 581–585, 587–589 (1954); *ibid.* **5**, 959–963 (1955); *ibid.* **6**, 505–508, 509–513, 621–623, 625–626, 1055–1058 (1956).
[2] Synthèse des schémas à relais idéaux, à l'aide des corps d'imaginaires de Galois (Romanian. French summary), *Acad. R. P. Romîne. Bul. Şti. Secţ. Şti. Mat. Fiz.* **8**, 429–453 (1956).
[3] Sur la théorie algébrique de certains circuits électriques, *J. Math. Pures Appl.* (9)**36**, 313–324 (1957).
[4] *The Algebraic Theory of Switching Circuits*, Pergamon Press, Oxford, 1969.

MOISIL, G. C., and POPOVICI, C. P.:
[1] Analyse et synthèse des schémas à commande directe, à l'aide des imaginaires de Galois (Romanian. French summary), *Acad. R. P. Romîne. Bul. Şti. Secţ. Şti. Mat. Fiz.* **8**, 455–467 (1956).

MÖNNIG, P.:
[1] Ein Beitrag zur Darstellung endlicher Körper, *Math.-Phys. Semesterber.* **17**, 46–56 (1970).

MONSKY, P.:
[1] *p-adic Analysis and Zeta Functions*, Lectures in Math., Dept. of Math., Kyoto Univ., 1970.

MONTEL, P.:
[1] *Leçons sur les récurrences et leurs applications*, Gauthier-Villars, Paris, 1957.

MONTGOMERY, H. L.:
[1] *Topics in Multiplicative Number Theory*, Lecture Notes in Math., vol. 227, Springer-Verlag, Berlin-Heidelberg-New York, 1971.

[2] Distribution questions concerning a character sum, *Topics in Number Theory* (Debrecen, 1974), Colloq. Math. Soc. János Bolyai, vol. 13, pp. 195–203, North-Holland, Amsterdam, 1976.

MONTGOMERY, H. L., and VAUGHAN, R. C.:

[1] Exponential sums with multiplicative coefficients, *Invent. Math.* **43**, 69–82 (1977).

MOORE, E. H.:

[1] A doubly-infinite system of simple groups, *Bull. New York Math. Soc.* **3**, 73–78 (1893).

[2] A doubly-infinite system of simple groups, *Math. Papers read at the Congress of Mathematics* (Chicago, 1893), pp. 208–242, Chicago, 1896.

[3] A two-fold generalization of Fermat's theorem, *Bull. Amer. Math. Soc.* **2**, 189–199 (1896).

[4] *The Subgroups of the Generalized Finite Modular Group*, Decennial Publications, Chicago, 1903.

MORDELL, L. J.:

[1] On a simple summation of the series $\sum_{s=0}^{n-1} e^{2s^2\pi i/n}$, *Messenger of Math.* **48**, 54–56 (1918).

[2] *Three Lectures on Fermat's Last Theorem*, Cambridge Univ. Press, Cambridge, 1921.

[3] On the reciprocity formula for the Gauss's sums in the quadratic field, *Proc. London Math. Soc.* (2)**20**, 289–296 (1922).

[4] On a sum analogous to a Gauss's sum, *Quart. J. Math.* **3**, 161–167 (1932).

[5] The number of solutions of some congruences in two variables, *Math. Z.* **37**, 193–209 (1933).

[6] Thoughts on number theory, *J. London Math. Soc.* **21**, 58–74 (1946).

[7] Note on the linear symmetric congruence in n variables, *Canad. J. Math.* **5**, 433–438 (1953).

[8] Note on simultaneous quadratic congruences, *Math. Scand.* **5**, 21–26 (1957).

[9] On the number of solutions in incomplete residue sets of quadratic congruences, *Arch. Math.* **8**, 153–157 (1957).

[10] On Lehmer's congruence associated with cubes of Kloosterman's sums, *J. London Math. Soc.* **36**, 335–339 (1961).

[11] On a cubic congruence in three variables, *Acta Arith.* **8**, 1–9 (1962).

[12] The sign of the Gaussian sum, *Illinois J. Math.* **6**, 177–180 (1962).

[13] On a cyclotomic resolvent, *Arch. Math.* **13**, 486–487 (1962).

[14] On a special polynomial congruence and exponential sums, *Calcutta Math. Soc. Golden Jubilee Commemoration Volume*, Part I, pp. 29–32, Calcutta Math. Soc., Calcutta, 1963.

[15] On a cubic exponential sum in three variables, *Amer. J. Math.* **85**, 49–52 (1963).

[16] A congruence problem of E. G. Straus, *J. London Math. Soc.* **38**, 108–110 (1963).

[17] On a cubic congruence in three variables (III), *J. London Math. Soc.* **38**, 351–355 (1963).

[18] On a cubic exponential sum in two variables, *J. London Math. Soc.* **38**, 356–358 (1963).
[19] On the least residue and non-residue of a polynomial, *J. London Math. Soc.* **38**, 451–453 (1963).
[20] On a cubic congruence in three variables. II, *Proc. Amer. Math. Soc.* **14**, 609–614 (1963).
[21] On the congruence $ax^3 + by^3 + cz^3 + dxyz \equiv n (\mathrm{mod}\ p)$, *Duke Math. J.* **31**, 123–126 (1964).
[22] Incomplete exponential sums and incomplete residue systems for congruences, *Czechoslovak Math. J.* **14**, 235–242 (1964).
[23] *Diophantine Equations*, Academic Press, London, 1969.
[24] Some exponential sums in several variables, *Monatsh. Math.* **73**, 348–353 (1969).
[25] Cubic polynomials with the same residues mod p, *Proc. London Math. Soc.* (3)**21**, 129–144 (1970).
[26] On some exponential sums related to Kloosterman sums, *Acta Arith.* **21**, 65–69 (1972).
[27] A finite evaluation of a special exponential sum, *Proc. Cambridge Philos. Soc.* **71**, 75–78 (1972).
[28] On rational functions representing all residues mod p, *J. London Math. Soc.* (2)**5**, 166–168 (1972).
[29] Rational functions representing all residues mod p. II, *Proc. Amer. Math. Soc.* **35**, 411–412 (1972).
[30] On Salié's sum, *Glasgow Math. J.* **14**, 25–26 (1973).
[31] Some exponential sums, *Proc. Internat. Conf. on Number Theory* (Moscow, 1971), *Trudy Mat. Inst. Steklov.* **132**, 30–34 (1973); *Proc. Steklov Inst. Math.* **132**, pp. 29–34, American Math. Society, Providence, R.I., 1975.

MORENO, C. J.:
[1] Sur le problème de Kummer, *L'Enseignement Math.* (2)**20**, 45–51 (1974).

MORENO, O.:
[1] Counting traces of powers over $GF(2^m)$, *Congr. Numer.* **29**, 673–679 (1980).
[2] On primitive elements of trace equal to 1 in $GF(2^m)$, *Discrete Math.* **41**, 53–56 (1982).

MORLAYE, B.:
[1] Equations diagonales non homogènes sur un corps fini, *C. R. Acad. Sci. Paris Sér. A* **272**, 1545–1548 (1971).
[2] Démonstration élémentaire d'un théorème de Davenport et Hasse, *L'Enseignement Math.* (2)**18**, 269–276 (1972).

MOROZ, B. Z.:
[1] The distribution of power residues and non-residues (Russian), *Vestnik Leningrad. Univ.* **16**, no. 19, 164–169 (1961).

MORTIMER, B. C., and WILLIAMS, K. S.:
[1] Note on a paper of S. Uchiyama, *Canad. Math. Bull.* **17**, 289–293 (1974).

MORTON, P.:

[1] On the eigenvectors of Schur's matrix, *J. Number Theory* **12**, 122–127 (1980).

MOSES, J.:

[1] Algebraic structures and their algorithms, *Algorithms and Complexity* (J. F. Traub, ed.), pp. 301–319, Academic Press, New York, 1976.

MOSSIGE, S.:

[1] Table of irreducible polynomials over $GF(2)$ of degrees 10 through 20, *Math. Comp.* **26**, 1007–1009 (1972).

MOUSOURIS, N., and PORTER, A. D.:

[1] The symmetric matrix equation $X_n' \cdots X_1' A X_1 \cdots X_n = B$, *Atti Accad. Naz. Lincei Rend. Cl. Sci. Fis. Mat. Natur.* (8)**62**, 126–130 (1977).

[2] The Hermitian matric equation $U_n^* \cdots U_1^* A U_1 \cdots U_n = B$, *Rend. Mat.* (6)**11**, 387–392 (1978).

MULLEN, G. L.:

[1] Equivalence classes of functions over a finite field, *Acta Arith.* **29**, 353–358 (1976).

[2] Permutation polynomials in several variables over finite fields, *Acta Arith.* **31**, 107–111 (1976).

[3] Equivalence classes of polynomials over finite fields, *Acta Arith.* **31**, 113–123 (1976).

[4] Equivalence classes of matrices over finite fields, *Linear Algebra Appl.* **27**, 61–68 (1979).

[5] Weak equivalence of functions over a finite field, *Acta Arith.* **35**, 259–272 (1979).

[6] Equivalence classes of matrices over a finite field, *Internat. J. Math. and Math. Sci.* **2**, 487–491 (1979).

[7] Local permutation polynomials over Z_p, *Fibonacci Quart.* **18**, 104–108 (1980).

[8] Local permutation polynomials in three variables over Z_p, *Fibonacci Quart.* **18**, 208–214 (1980).

[9] Equivalence classes of sets of functions over a finite field, *Acta Arith.* **36**, 323–329 (1980).

[10] Local permutation polynomials over a finite field, *Norske Vid. Selsk. Skrifter* **1981**, no. 1, 1–4.

[11] Permutation matrices and matrix equivalence over a finite field, *Internat. J. Math. and Math. Sci.* **4**, 503–512 (1981).

[12] Matrix equivalence over finite fields, *Acta Arith.* **41**, 133–139 (1982).

[13] Polynomials over finite fields which commute with linear permutations, *Proc. Amer. Math. Soc.* **84**, 315–317 (1982).

MULLER, D. E.:

[1] Application of Boolean algebra to switching circuit design and error detection, *IRE Trans. Electron. Comp.* **EC-3**, 6–12 (1954).

MÜLLER, W. B.:

[1] Über eine Klasse von durch Dickson-Polynome dargestellten Gruppen, *Rings, Modules and Radicals* (Keszthely, 1971), Colloq. Math. Soc. János Bolyai, vol. 6, pp. 361–376, North-Holland, Amsterdam, 1973.

MURAKAMI, H., and REED, I. S.:

[1] Recursive realization of finite impulse filters using finite field arithmetic, *IEEE Trans. Information Theory* **IT-23**, 232–242 (1977).

MURZAEV, E. A.:

[1] The selection of multiple factors of polynomials over commutative finite fields (Russian), *Volž. Mat. Sb. Vyp.* **5**, 255–259 (1966).

MUSKAT, J. B.:

[1] On certain prime power congruences, *Abh. Math. Sem. Univ. Hamburg* **26**, 102–110 (1963).

[2] On the solvability of $x^e \equiv e (\mathrm{mod}\, p)$, *Pacific J. Math.* **14**, 257–260 (1964).

[3] Criteria for solvability of certain congruences, *Canad. J. Math.* **16**, 343–352 (1964).

[4] The cyclotomic numbers of order fourteen, *Acta Arith.* **11**, 263–279 (1966).

[5] Reciprocity and Jacobi sums, *Pacific J. Math.* **20**, 275–280 (1967).

[6] On Jacobi sums of certain composite orders, *Trans. Amer. Math. Soc.* **134**, 483–502 (1968).

[7] Use of computers in cyclotomy, *Computers in Number Theory* (A. O. L. Atkin and B. J. Birch, eds.), pp. 141–147, Academic Press, London, 1971.

MUSKAT, J. B., and STREET, A. P.:

[1] Sum-free cyclotomic classes in finite fields, *Proc. Third Manitoba Conf. Numerical Math.* (Winnipeg, Man., 1973), pp. 399–406, Utilitas Math., Winnipeg, Man., 1974.

MUSKAT, J. B., and WHITEMAN, A. L.:

[1] The cyclotomic numbers of order twenty, *Acta Arith.* **17**, 185–216 (1970).

MUSKAT, J. B., and ZEE, Y.-C.:

[1] Sign ambiguities of Jacobi sums, *Duke Math. J.* **40**, 313–334 (1973).

MUSSER, D. R.:

[1] Multivariate polynomial factorization, *J. Assoc. Comput. Mach.* **22**, 291–308 (1975).

MYERSON, G.:

[1] On the number of zeros of diagonal cubic forms, *J. Number Theory* **11**, 95–99 (1979).

[2] A combinatorial problem in finite fields. I, *Pacific J. Math.* **82**, 179–187 (1979).

[3] A combinatorial problem in finite fields. II, *Quart. J. Math.* (2)**31**, 219–231 (1980).

[4] The distribution of rational points on varieties defined over a finite field, *Mathematika* **28**, 153–159 (1981).

[5] Period polynomials and Gauss sums for finite fields, *Acta Arith.* **39**, 251–264 (1981).

MYKKELTVEIT, J.:

[1] Nonlinear recurrences and arithmetic codes, *Information and Control* **33**, 193–209 (1977).

NADLER, M., and SENGUPTA, A.:

[1] Shift-register code for indexing applications, *Comm. Assoc. Comput. Mach.* **2**, no. 10, 40–43 (1959).

NAGAHARA, T., and TOMINAGA, H.:

[1] Elementary proofs of a theorem of Wedderburn and a theorem of Jacobson, *Abh. Math. Sem. Univ. Hamburg* **41**, 72–74 (1974).

NAGATA, M.:

[1] On the number of solutions of $x_1^2 + x_2^2 + \cdots + x_n^2 = a$ in a finite field (Japanese), *Sûgaku* **14**, 98–99 (1962/63).

[2] *Field Theory*, Dekker, New York, 1977.

NAGELL, T.:

[1] On the solvability of some congruences, *Norske Vid. Selsk. Forh.* (*Trondheim*) **27**, 1–5 (1954).

NARKIEWICZ, W.:

[1] *Elementary and Analytic Theory of Algebraic Numbers*, Monografie Mat., vol. 57, PWN, Warsaw, 1974.

[2] Uniform distribution of sequences of integers, *Journées Arithmétiques 1980* (J. V. Armitage, ed.), London Math. Soc. Lecture Note Series, no. 56, pp. 202–210, Cambridge Univ. Press, Cambridge, 1982.

NASHIER, B. S., and RAJWADE, A. R.:

[1] Determination of a unique solution of the quadratic partition for primes $p \equiv 1 (\bmod 7)$, *Pacific J. Math.* **72**, 513–521 (1977).

NATHANSON, M. B.:

[1] Derivatives of binary sequences, *SIAM J. Appl. Math.* **21**, 407–412 (1971).

[2] Integrals of binary sequences, *SIAM J. Appl. Math.* **23**, 84–86 (1972).

[3] Shift dynamical systems over finite fields, *Proc. Amer. Math. Soc.* **34**, 591–594 (1972).

[4] Linear recurrences and uniform distribution, *Proc. Amer. Math. Soc.* **48**, 289–291 (1975).

[5] Difference operators and periodic sequences over finite modules, *Acta Math. Acad. Sci. Hungar.* **28**, 219–224 (1976).

NAZAROV, I. A.:

[1] A mathematical apparatus for the analysis and synthesis of linear-sequential coding networks (Russian), *Izv. Leningrad. Elektrotekh. Inst.* **39**, 153–162 (1959).

NEČAEV, V. I.:

[1] The group of nonsingular matrices over a finite field, and recursive sequences (Russian), *Dokl. Akad. Nauk SSSR* **152**, 275–277 (1963); *Soviet Math. Dokl.* **4**, 1296–1298 (1963).

[2] A best possible estimate of trigonometric sums for recursive functions with nonconstant coefficients (Russian), *Dokl. Akad. Nauk SSSR* **154**, 520–522 (1964); *Soviet Math. Dokl.* **5**, 124–126 (1964).

[3] Linear recurrent congruences with periodic coefficients (Russian), *Mat. Zametki* **3**, 625–632 (1968); *Math. Notes* **3**, 397–402 (1968).

[4] Recurrent sequences (Russian), *Moskov. Gos. Ped. Inst. Učen. Zap.* **375**, 103–123 (1971).

[5] Trigonometric sums for recurrent sequences of elements of a finite field (Russian), *Mat. Zametki* **11**, 597–607 (1972); *Math. Notes* **11**, 362–367 (1972).

[6] Trigonometric sums for recurrent sequences (Russian), *Dokl. Akad. Nauk SSSR* **206**, 811–814 (1972); *Soviet Math. Dokl.* **13**, 1320–1324 (1972).

[7] An estimate of the complete rational trigonometric sum (Russian), *Mat. Zametki* **17**, 839–849 (1975); *Math. Notes* **17**, 504–511 (1975).

NEČAEV, V. I., and POLOSUEV, A. M.:

[1] The distribution of non-residues and primitive roots in a sequence which satisfies a finite-difference equation with polynomial coefficients (Russian), *Vestnik Moskov. Univ. Ser. I Mat. Meh.* **1964**, no. 6, 75–84.

NEČAEV, V. I., and STEPANOVA, L. L.:

[1] The distribution of nonresidues and primitive roots in recurrence sequences over a field of algebraic numbers (Russian), *Uspehi Mat. Nauk* **20**, no. 3, 197–203 (1965).

NEČAEV, V. I., and TOPUNOV, V. L.:

[1] Estimation of the modulus of complete rational trigonometric sums of degree three and four (Russian), *Trudy Mat. Inst. Steklov.* **158**, 125–129 (1981).

NEIKIRK, L. I.:

[1] A geometric representation of the Galois field, *Bull. Amer. Math. Soc.* **14**, 323–325 (1908).

NEUMANN, H.:

[1] On some finite non-desarguesian planes, *Arch. Math.* **6**, 36–40 (1954).

NEUMANN, O.:

[1] Über die Kongruenz $ax^4 + 1 \equiv cz^2 (\text{mod. } p)$, *Monatsber. Deutsch. Akad. Wiss. Berlin* **11**, 699–703 (1969).

NEWMAN, M.:

[1] *Integral Matrices*, Academic Press, New York, 1972.

NICHOLSON, P. J.:

[1] Algebraic theory of finite Fourier transforms, *J. Computer and Syst. Sci.* **5**, 524–547 (1971).

NIEDERREITER, H.:

[1] Permutation polynomials in several variables over finite fields, *Proc. Japan Acad.* **46**, 1001–1005 (1970).

[2] Orthogonal systems of polynomials in finite fields, *Proc. Amer. Math. Soc.* **28**, 415–422 (1971).

[3] Permutation polynomials in several variables, *Acta Sci. Math. Szeged* **33**, 53–58 (1972).

[4] Distribution of Fibonacci numbers mod 5^k, *Fibonacci Quart.* **10**, 373–374 (1972).

[5] Some new exponential sums with applications to pseudo-random numbers, *Topics in Number Theory* (Debrecen, 1974), Colloquia Math. Soc. János Bolyai, vol. 13, pp. 209–232, North-Holland, Amsterdam, 1976.

[6] On the cycle structure of linear recurring sequences, *Math. Scand.* **38**, 53–77 (1976).

[7] On the distribution of pseudo-random numbers generated by the linear congruential method. III, *Math. Comp.* **30**, 571–597 (1976).

[8] Weights of cyclic codes, *Information and Control* **34**, 130–140 (1977).

[9] Statistical tests for linear congruential pseudo-random numbers, *COMPSTAT 1978: Proceedings in Computational Statistics* (Leiden, 1978), pp. 398–404, Physica-Verlag, Vienna, 1978.

[10] Quasi-Monte Carlo methods and pseudo-random numbers, *Bull. Amer. Math. Soc.* **84**, 957–1041 (1978).

[11] Verteilung von Resten rekursiver Folgen, *Arch. Math.* **34**, 526–533 (1980).

[12] Statistical independence properties of Tausworthe pseudo-random numbers, *Proc. Third Caribbean Conf. on Combinatorics and Computing* (Cave Hill, Barbados, 1981), pp. 163–168, Univ. of the West Indies, Cave Hill, Barbados, 1981.

[13] Statistical tests for Tausworthe pseudo-random numbers, *Probability and Statistical Inference* (W. Grossmann, G. C. Pflug, and W. Wertz, eds.), pp. 265–274, Reidel, Dordrecht, 1982.

[14] Richard Dedekind and the development of the theory of finite fields, *Abh. Braunschweig. Wissenschaftl. Gesellschaft* **33**, 183–187 (1982).

NIEDERREITER, H., and LO, S. K.:

[1] Permutation polynomials over rings of algebraic integers, *Abh. Math. Sem. Univ. Hamburg* **49**, 126–139 (1979).

NIEDERREITER, H., and ROBINSON, K. H.:

[1] Bol loops of order *pq*, *Math. Proc. Cambridge Philos. Soc.* **89**, 241–256 (1981).

[2] Complete mappings of finite fields, *J. Austral. Math. Soc. Ser. A* **33**, 197–212 (1982).

NIEDERREITER, H., and SHIUE, J.-S.:

[1] Equidistribution of linear recurring sequences in finite fields, *Indag. Math.* **80**, 397–405 (1977).

[2] Equidistribution of linear recurring sequences in finite fields, II, *Acta Arith.* **38**, 197–207 (1980).

NISNEVICH, L. B.:

[1] On the number of points of an algebraic variety in a finite prime field (Russian), *Dokl. Akad. Nauk SSSR* **99**, 17–20 (1954).

NIVEN, I.:

[1] Fermat's theorem for matrices, *Duke Math. J.* **15**, 823–826 (1948).

[2] Uniform distribution of sequences of integers, *Trans. Amer. Math. Soc.* **98**, 52–61 (1961).

NIVEN, I., and WARREN, L. J.:

[1] A generalization of Fermat's theorem, *Proc. Amer. Math. Soc.* **8**, 306–313 (1957).

NÖBAUER, W.:

[1] Über Gruppen von Restklassen nach Restpolynomidealen, *Sitzungsber. Österr. Akad. Wiss. Abt. II* **162**, 207–233 (1953).

[2] Über eine Gruppe der Zahlentheorie, *Monatsh. Math.* **58**, 181–192 (1954).

[3] Gruppen von Restklassen nach Restpolynomidealen in mehreren Unbestimmten, *Monatsh. Math.* **59**, 118–145 (1955).

[4] Gruppen von Restpolynomidealrestklassen nach Primzahlpotenzen, *Monatsh. Math.* **59**, 194–202 (1955).

[5] Bemerkungen über die Darstellung von Abbildungen durch Polynome und rationale Funktionen, *Monatsh. Math.* **68**, 138–142 (1964).

[6] Zur Theorie der Polynomtransformationen und Permutationspolynome, *Math. Ann.* **157**, 332–342 (1964).

[7] Über die Vollideale und Permutationspolynome eines Galoisfeldes, *Acta Math. Acad. Sci. Hungar.* **16**, 37–42 (1965).

[8] Über Permutationspolynome und Permutationsfunktionen für Primzahlpotenzen, *Monatsh. Math.* **69**, 230–238 (1965).

[9] Polynome, welche für gegebene Zahlen Permutationspolynome sind, *Acta Arith.* **11**, 437–442 (1966).

[10] Über eine Klasse von Permutationspolynomen und die dadurch dargestellten Gruppen, *J. reine angew. Math.* **231**, 215–219 (1968).

[11] Darstellungen von Permutationen durch Polynome und rationale Funktionen, *Ber. Math. Forschungsinst. Oberwolfach*, vol. 5, pp. 89–100, Bibliographisches Institut, Mannheim, 1971.

[12] Über Gruppen von Dickson-Polynomfunktionen und einige damit zusammenhängende zahlentheoretische Fragen, *Monatsh. Math.* **77**, 330–344 (1973).

NOMURA, T., and FUKUDA, A.:

[1] Linear recurring planes and two-dimensional cyclic codes, *Electron. Commun. Japan* **54**, no. 3, 23–30 (1971).

NOMURA, T., MIYAKAWA, H., IMAI, H., and FUKUDA, A.:

[1] A method of construction and some properties of planes having maximum area matrix, *Electron. Commun. Japan* **54**, no. 5, 18–25 (1971).

[2] Some properties of the $\gamma\beta$-plane and its extension to three-dimensional space, *Electron. Commun. Japan* **54**, no. 8, 27–34 (1971).

[3] A theory of two-dimensional linear recurring arrays, *IEEE Trans. Information Theory* **IT-18**, 775–785 (1972).

NORDON, D.:

[1] Zéros non singuliers des formes quadratiques, *C. R. Acad. Sci. Paris Sér. A* **277**, 295–297 (1973).

[2] Zéros communs non singuliers de deux formes quadratiques, *Sém. Théorie des Nombres 1972–1973*, Exp. 9, 45 pp., Univ. Bordeaux I, Talence, 1973.

[3] Zéros communs non singuliers de deux formes quadratiques, *Acta Arith.* **30**, 109–119 (1976).

NÖRLUND, N. E.:

[1] *Vorlesungen über Differenzenrechnung*, Springer, Berlin, 1924.

NORTON, K. K.:

[1] Upper bounds for kth power coset representatives modulo n, *Acta Arith.* **15**, 161–179 (1969).

[2] On the distribution of kth power residues and non-residues modulo n, *J. Number Theory* **1**, 398–418 (1969).

[3] Numbers with small prime factors, and the least kth power non-residue, *Memoirs Amer. Math. Soc.*, no. 106, American Math. Society, Providence, R.I., 1971.

[4] On the distribution of power residues and non-residues, *J. reine angew. Math.* **254**, 188–203 (1972).

[5] On character sums and power residues, *Trans. Amer. Math. Soc.* **167**, 203–226 (1972); Erratum, *ibid.* **174**, 507 (1972).

[6] Bounds for sequences of consecutive power residues. I, *Proc. Symp. Pure Math.*, vol. 24, pp. 213–220, American Math. Society, Providence, R.I., 1973.

NUSSBAUMER, H. J.:

[1] *Fast Fourier Transform and Convolution Algorithms*, Springer-Verlag, Berlin-Heidelberg-New York, 1981.

NYMANN, J. E.:

[1] Groups and fields in Z_n, *Elemente der Math.* **30**, 82–84 (1975).

O'CONNOR, R. E.:

[1] Quadratic and linear congruence, *Bull. Amer. Math. Soc.* **45**, 792–798 (1939).

O'CONNOR, R. E., and PALL, G.:

[1] The quaternion congruence $\bar{t}at \equiv b \pmod g$, *Amer. J. Math.* **61**, 487–508 (1939).

ODLYZKO, A. M., and STANLEY, R. P.:

[1] Enumeration of power sums modulo a prime, *J. Number Theory* **10**, 263–272 (1978).

ODONI, R. W. K.:

[1] The statistics of Weil's trigonometric sums, *Proc. Cambridge Philos. Soc.* **74**, 467–471 (1973).

OLSON, L. D.:

[1] Hasse invariants and anomalous primes for elliptic curves with complex multiplication, *J. Number Theory* **8**, 397–414 (1976).

OLTRAMARE, G.:

[1] Considérations générales sur les racines des nombres premiers, *J. reine angew. Math.* **45**, 303–344 (1853).

ONO, T.:

[1] Gauss transforms and zeta-functions, *Ann. of Math.* (2)**91**, 332–361 (1970).

[2] A remark on Gaussian sums and algebraic groups, *J. Math. Kyoto Univ.* **13**, 139–142 (1973).

[3] On certain numerical invariants of mappings over finite fields. I, *Proc. Japan Acad. Ser. A* **56**, 342–347 (1980).

[4] On certain numerical invariants of mappings over finite fields. II, *Proc. Japan Acad. Ser. A* **56**, 397–400 (1980).

[5] On certain numerical invariants of mappings over finite fields. III, *Proc. Japan Acad. Ser. A* **56**, 441–444 (1980).

[6] On certain numerical invariants of mappings over finite fields. IV, *Proc. Japan Acad. Ser. A* **57**, 66–71 (1981).

[7] On certain numerical invariants of mappings over finite fields. V, *Proc. Japan Acad. Ser. A* **57**, 121–125 (1981).

[8] On a generalization of Jacobi sums, *J. Fac. Sci. Univ. Tokyo Sect. IA* **28**, 823–828 (1981).

ORE, Ö.:

[1] Über höhere Kongruenzen, *Norsk Mat. Forenings Skrifter Ser. I* **1922**, no. 7, 15 pp.

[2] Über die Reduzibilität von algebraischen Gleichungen, *Skrifter Norske Vid. Akad. Oslo* **1923**, no. 1.

[3] Note sur une identité dans la théorie des congruences supérieures, *Rend. Circ. Mat. Palermo* **48**, 37–40 (1924).

[4] Theory of non-commutative polynomials, *Ann. of Math.* (2)**34**, 480–508 (1933).

[5] On a special class of polynomials, *Trans. Amer. Math. Soc.* **35**, 559–584 (1933); Errata, *ibid.* **36**, 275 (1934).

[6] Contributions to the theory of finite fields, *Trans. Amer. Math. Soc.* **36**, 243–274 (1934).

[7] Some studies on cyclic determinants, *Duke Math. J.* **18**, 343–354 (1951).

OSBORN, R.:

[1] *Tables of Primitive Roots of Odd Primes Less than 1000*, Univ. of Texas Press, Austin, Tex., 1961.

PALEY, R. E. A. C.:

[1] A theorem on characters, *J. London Math. Soc.* **7**, 28–32 (1932).

[2] Theorems on polynomials in a Galois field, *Quart. J. Math.* **4**, 52–63 (1933).

[3] On orthogonal matrices, *J. Math. Phys.* **12**, 311–320 (1933).

PARKER, E. T.:

[1] Orthogonal Latin squares, *Proc. Nat. Acad. Sci. U.S.A.* **45**, 859–862 (1959).

[2] Construction of some sets of mutually orthogonal Latin squares, *Proc. Amer. Math. Soc.* **10**, 946–949 (1959).

PARNAMI, J. C., AGRAWAL, M. K., and RAJWADE, A. R.:

[1] On the 4-power stufe of a field, *Rend. Circ. Mat. Palermo* (2) **30**, 245–254 (1981).

[2] A congruence relation between the coefficients of the Jacobi sum, *Indian J. Pure Appl. Math.* **12**, 804–806 (1981).

[3] Jacobi sums and cyclotomic numbers for a finite field, *Acta Arith.* **41**, 1–13 (1982).

PARSON, L. A.:

[1] Generalized Kloosterman sums and the Fourier coefficients of cusp forms, *Trans. Amer. Math. Soc.* **217**, 329–350 (1976).

PATTERSON, N. J.:

[1] Thc algcbraic decoding of Goppa codes, *IEEE Trans. Information Theory* **IT-21**, 203–207 (1975).

PATTERSON, S. J.:

[1] A cubic analogue of the theta series. I, II, *J. reine angew. Math.* **296**, 125–161, 217–220 (1977).

[2] On Dirichlet series associated with cubic Gauss sums, *J. reine angew. Math.* **303/304**, 102–125 (1978).

[3] On the distribution of Kummer sums, *J. reine angew. Math.* **303/304**, 126–143 (1978).

[4] The distribution of general Gauss sums at prime arguments, *Recent Progress in Analytic Number Theory* (H. Halberstam and C. Hooley, eds.), vol. 2, pp. 171–182, Academic Press, London, 1981.

PAVLOV, A. I., and POKHODZEI, B. B.:

[1] Pseudo-random numbers generated by linear recurrence relations over a finite field (Russian), *Ž. Vyčisl. Mat. i Mat. Fiz.* **19**, 836–842 (1979); *U.S.S.R. Comp. Math. and Math. Phys.* **19**, no. 4, 38–44 (1979).

PAYNE, S. E.:

[1] Linear transformations of a finite field, *Amer. Math. Monthly* **78**, 659–660 (1971).

PEARSON, E. H., and VANDIVER, H. S.:

[1] On a new problem concerning trinomial congruences involving rational integers, *Proc. Nat. Acad. Sci. U.S.A.* **39**, 1278–1285 (1953).

PELE, R. L.:

[1] Some remarks on the vector subspaces of cyclic Galois extensions, *Acta Math. Acad. Sci. Hungar.* **20**, 237–240 (1969).

PELLEGRINO, G.:

[1] Sui campi di Galois, di ordine dispari, che ammettono terne di elementi quadrati (non quadrati) consecutivi, *Boll. Un. Mat. Ital.* (5)**17B**, 1482–1495 (1980).

[2] Sulle sostituzioni lineari, sui campi finiti di ordine dispari, che conservano oppure scambiano il carattere quadratico degli elementi trasformati, *Boll. Un. Mat. Ital.* (6) **1B**, 211–223 (1982).

PELLET, A.-E.:

[1] Sur les fonctions irréductibles suivant un module premier et une fonction modulaire, *C. R. Acad. Sci. Paris* **70**, 328–330 (1870).

[2] Sur la décomposition d'une fonction entière en facteurs irréductibles suivant un module premier, *C. R. Acad. Sci. Paris* **86**, 1071–1072 (1878).

[3] Résolution d'une classe de congruences, *C. R. Acad. Sci. Paris* **88**, 417–418 (1879).

[4] Sur une classe d'équations dont toutes les racines peuvent s'exprimer linéairement en fonction de l'une d'elles, *Bull. Sci. Math.* (2)**4**, 262–265 (1880).

[5] Sur les fonctions irréductibles suivant un module premier, *C. R. Acad. Sci. Paris* **90**, 1339–1341 (1880).

[6] Méthode nouvelle pour diviser le cercle en parties égales, *C. R. Acad. Sci. Paris* **93**, 838–840 (1881).

[7] Sur les fonctions irréductibles suivant un module premier, *C. R. Acad. Sci. Paris* **93**, 1065–1066 (1881).

[8] Mémoire sur la théorie algébrique des équations, *Bull. Soc. Math. France* **15**, 61–103 (1887).

[9] Sur les fonctions réduites suivant un module premier, *Bull. Soc. Math. France* **17**, 156–167 (1889).

PEPIN, T.:

[1] Sur diverses tentatives de démonstration du théorème de Fermat, *C. R. Acad. Sci. Paris* **91**, 366–368 (1880).

PEREL'MUTER, G. I.:

[1] Evaluation of a sum containing primes (Russian), *Dokl. Akad. Nauk SSSR* **144**, 48–51 (1962); *Soviet Math. Dokl.* **3**, 663–667 (1962).

[2] On certain character sums (Russian), *Uspehi Mat. Nauk* **18**, no. 2, 145–149 (1963).

[3] The problem of estimation of certain arithmetic sums (Russian), *Certain Problems in the Theory of Fields* (Russian), pp. 6–15, Izdat. Saratov. Univ., Saratov, 1964.

[4] On some sums and manifolds connected with them (Russian), *Works of Young Scientists* (Russian), pp. 69–72, Izdat. Saratov. Univ., Saratov, 1964.

[5] The Z-function of a class of cubic surfaces (Russian), *Studies in Number Theory* (Russian), no. 1, pp. 49–58, Izdat. Saratov. Univ., Saratov, 1966.

[6] Rationality of L-functions of a class of algebraic varieties (Russian), *Studies in Number Theory* (Russian), no. 1, pp. 59–62, Izdat. Saratov. Univ., Saratov, 1966.

[7] The smallest non-residue of a polynomial along an algebraic curve (Russian), *Studies in Number Theory* (Russian), no. 3, pp. 64–68, Izdat. Saratov. Univ., Saratov, 1969.

[8] On a certain conjecture of K. Williams (Russian), *Dokl. Akad. Nauk SSSR* **184**, 282–284 (1969); *Soviet Math. Dokl.* **10**, 67–69 (1969).

[9] Estimate of a sum along an algebraic curve (Russian), *Mat. Zametki* **5**, 373–380 (1969); *Math. Notes* **5**, 223–227 (1969).

[10] Estimation of a multiple sum with Legendre symbol for a cubic polynomial (Russian), *Studies in Number Theory* (Russian), no. 6, pp. 129–131, Izdat. Saratov. Univ., Saratov, 1975.

[11] Estimation of a multiple sum with the Legendre symbol (Russian), *Mat. Zametki* **18**, 421–427 (1975); *Math. Notes* **18**, 840–844 (1975).

[12] Estimation of a multiple sum with the Legendre symbol for a polynomial of odd degree (Russian), *Mat. Zametki* **20**, 815–824 (1976); *Math. Notes* **20**, 1015–1020 (1976).

PEREL'MUTER, G. I., and POSTNIKOV, A. G.:

[1] The number of solutions of a certain congruence (Russian), *Acta Arith.* **21**, 103–110 (1972).

PERKINS, J. C.:

[1] Rank r solutions to the matrix equation $XX^{\mathrm{T}} = 0$ over a field of characteristic two, *Math. Nachr.* **48**, 69–76 (1971).

[2] Gauss sums and the matrix equation $XX^{\mathrm{T}} = 0$ over fields of characteristic two, *Acta Arith.* **19**, 205–214 (1971).

PERKINS, J. C., and FULTON, J. D.:

[1] Symmetric involutions over fields of characteristic 2, *Duke Math. J.* **38**, 697–702 (1971).

PERLIS, S.:

[1] Normal bases of cyclic fields of prime-power degree, *Duke Math. J.* **9**, 507–517 (1942).

PERRIN, R.:

[1] Sur la résolution des équations numériques au moyen des suites récurrentes, *C. R. Acad. Sci. Paris* **119**, 990–993 (1894).

PERRON, O.:

[1] Bemerkungen über die Verteilung der quadratischen Reste, *Math. Z.* **56**, 122–130 (1952).

PETERSON, W. W.:

[1] Encoding and error-correction procedures for the Bose-Chaudhuri codes, *IRE Trans. Information Theory* **IT-6**, 459–470 (1960).

[2] Some new results on finite fields and their application to the theory of BCH codes, *Proc. Conf. Combinatorial Math. and Its Appl.* (Chapel Hill, N.C., 1967), pp. 329–334, Univ. of North Carolina Press, Chapel Hill, N.C., 1969.

PETERSON, W. W., and BROWN, D. T.:

[1] Cyclic codes for error detection, *Proc. IRE* **49**, 228–235 (1961).

PETERSON, W. W., and WELDON, E. J., Jr.:

[1] *Error-Correcting Codes*, 2nd ed., M.I.T. Press, Cambridge, Mass., 1972.

PETERSSON, H.:

[1] Über die Entwicklungskoeffizienten der automorphen Formen, *Acta Math.* **58**, 169–215 (1932).

PETR, K.:

[1] Über die Reduzibilität eines Polynoms mit ganzzahligen Koeffizienten nach einem Primzahlmodul, *Časopis Pěst. Mat. Fys.* **66**, 85–94 (1937).

PETTERSON, E. L.:

[1] Eine Bedingung für die irreduziblen Faktoren von gewissen Polynomen modulo eines Primzahlprodukts, *Jber. Deutsch. Math.-Verein.* **45**, 169–172 (1935).

[2] Über einen Satz von Ö. Ore, *J. reine angew. Math.* **172**, 217–218 (1935).

[3] Über die Irreduzibilität ganzzahliger Polynome nach einem Primzahlmodul, *J. reine angew. Math.* **175**, 209–220 (1936).

PETTOROSSI, A.:

[1] Derivation of an $O(k^2 \log n)$ algorithm for computing order-k Fibonacci numbers from the $O(k^3 \log n)$ matrix multiplication method, *Inform. Process. Lett.* **11**, no. 4–5, 172–179 (1980).

PETTOROSSI, A., and BURSTALL, R. M.:

[1] Deriving very efficient algorithms for evaluating linear recurrence relations using the program transformation technique, *Acta Informatica* **18**, 181–206 (1982).

PICKERT, G.:

[1] *Projektive Ebenen*, 2nd ed., Springer-Verlag, Berlin-Heidelberg-New York, 1975.

PIEPER, H.:

[1] *Variationen über ein zahlentheoretisches Thema von Carl Friedrich Gauss*, Birkhäuser-Verlag, Basel, 1978.

PILZ, G.:

[1] *Near Rings: The Theory and Its Applications*, North-Holland, Amsterdam, 1977.

PIMENOV, N. V.:

[1] The oscillation of the sign of the remainder in the formula for the number of points of an algebraic curve (Russian), *Ukrain. Mat. Ž.* **28**, 546–551 (1976).

PIUMA, C. M.:

[1] Intorno ad una congruenza di modulo primo, *Ann. Mat. Pura Appl.* (2)**11**, 237–245 (1883).

PIZZARELLO, G.:

[1] Sui polinomi in n indeterminate sopra un campo finito, *Ricerca (Napoli)* (4)**28**, no. 3, 3–7 (1977).

PLESKEN, W.:

[1] Counting with groups and rings, *J. reine angew. Math.* **334**, 40–68 (1982).

PLESS, V.:

[1] Power moment identities on weight distributions in error correcting codes, *Information and Control* **6**, 147–152 (1963).

[2] On the invariants of a vector subspace of a vector space over a field of characteristic two, *Proc. Amer. Math. Soc.* **16**, 1062–1067 (1965).

PLOTKIN, M.:

[1] Binary codes with specified minimum distances, *IRE Trans. Information Theory* **IT-6**, 445–450 (1960).

POCKLINGTON, H. C.:

[1] The direct solution of the quadratic and cubic binomial congruences with prime moduli, *Proc. Cambridge Philos. Soc.* **19**, 57–59 (1917).

[2] Quadratic and higher reciprocity of modular polynomials, *Proc. Cambridge Philos. Soc.* **40**, 212–214 (1944).

POHLIG, S. C., and HELLMAN, M. E.:

[1] An improved algorithm for computing logarithms over $GF(p)$ and its cryptographic significance, *IEEE Trans. Information Theory* **IT-24**, 106–110 (1978).

POLKINGHORN, F., Jr.:

[1] Decoding of double and triple error correcting Bose-Chaudhuri codes, *IEEE Trans. Information Theory* **IT-12**, 480–481 (1966).

POLLARD, J. M.:

[1] The fast Fourier transform in a finite field, *Math. Comp.* **25**, 365–374 (1971).

[2] Implementation of number-theoretic transforms, *Electron. Lett.* **12**, 378–379 (1976).

[3] Monte Carlo methods for index computation (mod p), *Math. Comp.* **32**, 918–924 (1978).

POLOSUEV, A. M.:

[1] Certain arithmetic properties of recurrent functions with variable coefficients (Russian), *Mat. Zametki* **1**, 45–52 (1967); *Math. Notes* **1**, 29–33 (1967).

PÓLYA, G.:

[1] Über die Verteilung der quadratischen Reste und Nichtreste, *Göttinger Nachr.* **1918**, 21–29.

PÓLYA, G., and SZEGÖ, G.:

[1] *Aufgaben und Lehrsätze aus der Analysis II*, Springer, Berlin, 1925.

POPOVICI, C. P.:

[1] Integral polynomials irreducible modulo p (Russian), *Rev. Math. Pures Appl.* **4**, 369–379 (1959).

[2] Irreducible polynomials modulo p (Romanian), *Acad. R. P. Romîne Fil. Iaşi Stud. Cerc. Şti. Mat.* **11**, 13–23 (1960).

PORTER, A. D.:

[1] Systems of bilinear and quadratic equations in a finite field, *Ann. Mat. Pura Appl.* (4)**68**, 21–29 (1965).

[2] Systems of one quadratic and two bilinear equations in a finite field, *Publ. Math. Debrecen* **13**, 117–121 (1966).

[3] Pairs of bilinear and quadratic equations in a finite field, *Monatsh. Math.* **70**, 155–160 (1966).

[4] Special equations in a finite field, *Math. Nachr.* **32**, 277–279 (1966).

[5] Trilinear equations in a finite field, *Atti Accad. Naz. Lincei Rend. Cl. Sci. Fis. Mat. Natur.* (8)**40**, 361–365 (1966).

[6] Pairs of bilinear equations in a finite field, *Canad. J. Math.* **18**, 561–565 (1966).

[7] Some systems of equations in a finite field, *Math. Z.* **100**, 141–145 (1967).

[8] Orthogonal similarity for skew matrices in $GF(q)$, *Atti. Accad. Naz. Lincei Rend. Cl. Sci. Fis. Mat. Natur.* (8)**42**, 757–762 (1967).

[9] Generalized quadratic forms in $GF(q)$, *Arch. Math.* **19**, 615–620 (1968).

[10] The matric equation $AX_1 \cdots X_a = B$, *Atti Accad. Naz. Lincei Rend. Cl. Sci. Fis. Mat. Natur.* (8)**44**, 727–732 (1968).

[11] Simultaneous equations in a finite field, *Publ. Math. Debrecen* **16**, 99–110 (1969).

[12] Some partitions of a skew matrix, *Ann. Mat. Pura Appl.* (4)**82**, 115–120 (1969).

[13] Orthogonal similarity in a finite field, *Math. Nachr.* **40**, 327–331 (1969).

[14] Generalized bilinear forms in a finite field, *Duke Math. J.* **37**, 55–60 (1970).

[15] Systems of four equations with a matric application in a finite field, *Portugal. Math.* **31**, 121–131 (1972).

[16] A matrix form of an exponential sum, *Rend. Mat.* (6)**5**, 803–818 (1972).

[17] An exponential sum in a finite field, *Publ. Math. Debrecen* **20**, 53–62 (1973).

[18] The matric equation $A_1X_1 + \cdots + A_mX_m = B$ in $GF(q)$, *J. Natur. Sci. and Math.* **13**, 115–124 (1973).

[19] Some partitions of a rectangular matrix, *Atti Accad. Naz. Lincei Rend. Cl. Sci. Fis. Mat. Natur.* (8)**56**, 667–671 (1974).

[20] Some partitions of a Hermitian matrix, *Linear Algebra Appl.* **12**, 231–239 (1975).

[21] Solvability of the matrix equation $AX = B$, *Linear Algebra Appl.* **13**, 177–184 (1976).

[22] Some partitions of a symmetric matrix, *Math. Nachr.* **84**, 179–183 (1978).

PORTER, A. D., and ADAMS, J.:

[1] Similarity and orthogonal similarity in a finite field, *Duke Math. J.* **35**, 519–524 (1968).

PORTER, A. D., and HANSON, L. A.:

[1] Unitary similarity of normal matrices in $GF(q)$, *Math. Nachr.* **49**, 351–357 (1971).

[2] Orthogonal similarity of normal matrices in $GF(q)$, *Duke Math. J.* **38**, 795–803 (1971).

PORTER, A. D., and MOUSOURIS, N.:

[1] Ranked solutions of some matric equations, *Linear and Multilinear Algebra* **6**, 145–151 (1978).

[2] Ranked solutions of $AXC = B$ and $AX = B$, *Linear Algebra Appl.* **24**, 217–224 (1979).

[3] Ranked solutions of the matric equation $A_1X_1 = A_2X_2$, *Internat. J. Math. and Math. Sci.* **3**, 293–304 (1980).

[4] Exponential sums and rectangular partitions, *Linear Algebra Appl.* **29**, 347–355 (1980).

[5] Partitions of a symmetric matrix over a finite field, *Linear and Multilinear Algebra* **10**, 329–341 (1981).

PORTER, A. D., and RIVELAND, A. A.:

[1] A generalized skew equation over a finite field, *Math. Nachr.* **69**, 291–296 (1975).

POSTNIKOV, A. G.:

[1] Ergodic problems in the theory of congruences and of diophantine approximations (Russian), *Trudy Mat. Inst. Steklov.* **82** (1966); *Proc. Steklov Inst. Math.* **82**, American Math. Society, Providence, R.I., 1967.

POSTNIKOV, A. G., and STEPANOV, S. A.:

[1] On the theory of Jacobsthal sums (Russian), *Trudy Mat. Inst. Steklov.* **142**, 208–214 (1976); *Proc. Steklov Inst. Math. 1979*, no. 3, pp. 225–231, American Math. Society, Providence, R.I., 1979.

PRABHU, K. A., and BOSE, N. K.:

[1] Number of irreducible q-ary polynomials in several variables with prescribed degrees, *IEEE Trans. Circuits and Systems* **CAS-26**, 973–975 (1979).

PRADHAN, D. K.:

[1] A theory of Galois switching functions, *IEEE Trans. Computers* **C-27**, 239–248 (1978).

PRANGE, E.:

[1] Cyclic error-correcting codes in two symbols, Tech. Note AFCRC-TN-57-103, Air Force Cambridge Research Center, Bedford, Mass., 1957.

[2] Some cyclic error-correcting codes with simple decoding algorithms, Tech. Note AFCRC-TN-58-156, Air Force Cambridge Research Center, Bedford, Mass., 1958.

PREPARATA, F. P., and SARWATE, D. V.:

[1] Computational complexity of Fourier transforms over finite fields, *Math. Comp.* **31**, 740–751 (1977).

PREŠIĆ, M. D.:

[1] A method for solving equations in finite fields, *Mat. Vesnik* **7**, 507–509 (1970).

PROSKURIN, N. V.:

[1] The summation formulas for general Kloosterman sums (Russian), *Zap. Naučn. Sem. Leningrad. Otdel. Mat. Inst. Steklov.* **82**, 103–135 (1979).

[2] On the conjecture of Yu. V. Linnik (Russian), *Zap. Naučn. Sem. Leningrad. Otdel. Mat. Inst. Steklov.* **91**, 94–118 (1979); *J. Soviet Math.* **17**, 2147–2162 (1981).

[3] General Kloosterman sums (Russian), Preprint LOMI R-3-80, Akad. Nauk SSSR, Mat. Inst. Leningrad. Otdel., Leningrad, 1980, 36 pp.

RABER, N. C.:

[1] A geometric approach to counting distribution of squares in a finite field, *Geom. Dedicata* **4**, 297–303 (1975).

RABIN, M. O.:

[1] Probabilistic algorithms in finite fields, *SIAM J. Computing* **9**, 273–280 (1980).

RABUNG, J. R., and JORDAN, J. H.:

[1] Consecutive power residues or nonresidues, *Math. Comp.* **24**, 737–740 (1970).

RADCHENKO, A. N., and FILIPPOV, V. I.:

[1] Shift registers with logical feedback and their use as counting and code devices (Russian), *Avt. i Telem.* **20**, 1507–1514 (1959); *Automation and Remote Control* **20**, 1467–1473 (1959).

[2] Logical feedback in shift registers (Russian), *Avt. Telem. i Pribor.* **1960**, no. 3, 257–267.

RADEMACHER, H.:

[1] The Fourier coefficients of the modular invariant $J(\tau)$, *Amer. J. Math.* **60**, 501–512 (1938).

[2] Fourier expansions of modular forms and problems of partition, *Bull. Amer. Math. Soc.* **46**, 59–73 (1940).

RADOS, G.:

[1] Zur Theorie der Congruenzen höheren Grades, *J. reine angew. Math.* **99**, 258–260 (1886).

[2] Sur une théorie des congruences à plusieurs variables, *Ann. Sci. Ecole Norm. Sup.* (3)**27**, 217–231 (1910).

[3] Sur la théorie des congruences de degré supérieur, *Ann. Sci. Ecole Norm. Sup.* (3)**30**, 395–412 (1913).

[4] Ein Satz über Kongruenzen höheren Grades, *Acta Lit. Scient. Univ. Hung.* **1**, 1–5 (1922).

[5] Sur une identité remarquable de la théorie des congruences binomes, *Rend. Circ. Mat. Palermo* **46**, 308–314 (1922).

RAGHAVARAO, D.:
[1] *Constructions and Combinatorial Problems in Design of Experiments*, Wiley, New York, 1971.

RAJWADE, A. R.:
[1] Arithmetic on curves with complex multiplication by $\sqrt{-2}$, *Proc. Cambridge Philos. Soc.* **64**, 659–672 (1968).
[2] Arithmetic on curves with complex multiplication by the Eisenstein integers, *Proc. Cambridge Philos. Soc.* **65**, 59–73 (1969).
[3] On rational primes p congruent to 1(mod 3 or 5), *Proc. Cambridge Philos. Soc.* **66**, 61–70 (1969).
[4] A note on the number of solutions N_p of the congruence $y^2 \equiv x^3 - Dx \pmod p$, *Proc. Cambridge Philos. Soc.* **67**, 603–606 (1970).
[5] The number of solutions of the congruence $y^2 \equiv x^6 - a \pmod p$, *Indian J. Pure Appl. Math.* **4**, 325–332 (1973).
[6] On the congruence $y^2 \equiv x^5 - a \pmod p$, *Proc. Cambridge Philos. Soc.* **74**, 473–475 (1973).
[7] Certain classical congruences via elliptic curves, *J. London Math. Soc.* (2)**8**, 60–62 (1974).
[8] Notes on the congruence $y^2 \equiv x^5 - a \pmod p$, *L'Enseignement Math.* (2)**21**, 49–56 (1975).
[9] The Diophantine equation $y^2 = x(x^2 + 21Dx + 112D^2)$ and the conjectures of Birch and Swinnerton-Dyer, *J. Austral. Math. Soc. Ser. A* **24**, 286–295 (1977).

RAJWADE, A. R., and PARNAMI, J. C.:
[1] A new cubic character sum, *Acta Arith.* **40**, 347–356 (1982).

RAKTOE, B. L.:
[1] Generalized combining of elements from finite fields, *Ann. Math. Statist.* **41**, 1763–1767 (1970).

RALSTON, T.:
[1] On the distribution of squares in a finite field, *Geom. Dedicata* **8**, 207–212 (1979).

RANKIN, R. A.:
[1] *Modular Forms and Functions*, Cambridge Univ. Press, Cambridge, 1977.

RAO, C. R.:
[1] Factorial experiments derivable from combinatorial arrangements of arrays, *J. Royal Statist. Soc. Suppl.* **9**, 128–139 (1947).

RAO, K. N.:
[1] A congruence equation in $GF[p^n, x]$ and some related arithmetical identities, *Duke Math. J.* **33**, 783–789 (1966).
[2] Some applications of Carlitz's η-sum, *Acta Arith.* **12**, 213–221 (1967).
[3] Algebras of quadratic residues (mod P) in $GF[p^n, x]$, *Boll. Un. Mat. Ital.* (4)**1**, 680–686 (1968).

RAUSSNITZ, G.:

[1] *Math. Naturw. Ber. Ungarn* **1**, 266–278 (1882/83).

RAUTER, H.:

[1] Studien zur Theorie des Galoisschen Körpers über dem Körper der rationalen Funktionen einer Unbestimmten t mit Koeffizienten aus einem beliebigen endlichen Körper von p^{m_0} Elementen, *J. reine angew. Math.* **159**, 117–132 (1928); Bemerkungen, *ibid.* **159**, 228 (1928).

[2] Höhere Kreiskörper, *J. reine angew. Math.* **159**, 220–227 (1928).

RÉDEI, L.:

[1] Über einige Mittelwertfragen im quadratischen Zahlenkörper, *J. reine angew. Math.* **174**, 15–55 (1936).

[2] Über einige merkwürdige Polynome in endlichen Körpern mit zahlentheoretischen Beziehungen, *Acta Sci. Math. Szeged* **11**, 39–54 (1946).

[3] Zur Theorie der Gleichungen in endlichen Körpern, *Acta Sci. Math. Szeged* **11**, 63–70 (1946).

[4] Über eindeutig umkehrbare Polynome in endlichen Körpern, *Acta Sci. Math. Szeged* **11**, 85–92 (1946).

[5] Über die Gleichungen dritten und vierten Grades in endlichen Körpern, *Acta Sci. Math. Szeged* **11**, 96–105 (1946).

[6] Bemerkung zu meiner Arbeit "Über die Gleichungen dritten und vierten Grades in endlichen Körpern", *Acta Sci. Math. Szeged* **11**, 184–190 (1947).

[7] Zwei Lückensätze über Polynome in endlichen Primkörpern mit Anwendung auf die endlichen Abelschen Gruppen und die Gaussischen Summen, *Acta Math.* **79**, 273–290 (1947).

[8] Kurzer Beweis eines Satzes von Vandiver über endliche Körper, *Publ. Math. Debrecen* **1**, 99–100 (1949).

[9] A short proof of a theorem of Š. Schwarz concerning finite fields, *Časopis Pěst. Mat. Fys.* **75**, 211–212 (1950).

[10] *Algebra*, Geest & Portig, Leipzig, 1959; Pergamon Press, London, 1967.

[11] *Lückenhafte Polynome über endlichen Körpern*, Birkhäuser Verlag, Basel-Stuttgart, 1970; Akadémiai Kiadó, Budapest, 1973.

RÉDEI, L., and SZELE, T.:

[1] Algebraisch-zahlentheoretische Betrachtungen über Ringe. I, *Acta Math.* **79**, 291–320 (1947).

[2] Algebraisch-zahlentheoretische Betrachtungen über Ringe. II, *Acta Math.* **82**, 209–241 (1950).

RÉDEI, L., and TURÁN, P.:

[1] Zur Theorie der algebraischen Gleichungen über endlichen Körpern, *Acta Arith.* **5**, 223–225 (1959).

RÉDEI, L., and WEINERT, H. J.:

[1] Ein Gleichverteilungssatz für Systeme homogener Linearformen modulo p, *Acta Sci. Math. Szeged* **27**, 41–43 (1966).

REDINBO, G. R.:

[1] Finite field arithmetic on an array processor, *IEEE Trans. Computers* **C-28**, 461–471 (1979).

REE, R.:

[1] Proof of a conjecture of S. Chowla, *J. Number Theory* **3**, 210–212 (1971); Erratum, *ibid.* **4**, 223 (1972).

REED, I. S.:

[1] A class of multiple-error-correcting codes and the decoding scheme, *IRE Trans. Information Theory* **PGIT-4**, 38–49 (1954).

REED, I. S., SCHOLTZ, R. A., TRUONG, T. K., and WELCH, L. R.:

[1] The fast decoding of Reed-Solomon codes using Fermat theoretic transforms and continued fractions, *IEEE Trans. Information Theory* **IT-24**, 100–106 (1978).

REED, I. S., and SOLOMON, G.:

[1] Polynomial codes over certain finite fields, *J. Soc. Indust. Appl. Math.* **8**, 300–304 (1960).

REED, I. S., and TRUONG, T. K.:

[1] The use of finite fields to compute convolutions, *IEEE Trans. Information Theory* **IT-21**, 208–213 (1975).

[2] Convolutions over residue classes of quadratic integers, *IEEE Trans. Information Theory* **IT-22**, 468–475 (1976); Correction, *ibid.* **IT-23**, 544 (1977).

[3] Fast Mersenne-prime transforms for digital filtering, *Proc. Inst. Electr. Engrs.* **125**, 433–440 (1978).

[4] Simple proof of the continued fraction algorithm for decoding Reed-Solomon codes, *Proc. Inst. Electr. Engrs.* **125**, 1318–1320 (1978).

REED, I. S., TRUONG, T. K., and MILLER, R. L.:

[1] Fast algorithm for computing a primitive $2^{p+1}p$th root of unity in $GF[(2^p-1)^2]$, *Electron. Lett.* **14**, 493–494 (1978).

[2] Simple method for computing elements of order $2^k n$, where $n \mid 2^{p-1}-1$ and $2 \leqslant k \leqslant p+1$, in $GF[(2^p-1)^2]$, *Electron. Lett.* **14**, 697–698 (1978).

[3] Decoding of B.C.H. and R.S. codes with errors and erasures using continued fractions, *Electron. Lett.* **15**, 542–544 (1979).

[4] A new algorithm for computing primitive elements in the field of Gaussian complex integers modulo a Mersenne prime, *IEEE Trans. Acoust. Speech Signal Process.* **27**, 561–563 (1979).

REED, I. S., TRUONG, T. K., and WELCH, L. R.:

[1] The fast decoding of Reed-Solomon codes using Fermat transforms, *IEEE Trans. Information Theory* **IT-24**, 497–499 (1978).

REES, D.:

[1] Note on a paper by I. J. Good, *J. London Math. Soc.* **21**, 169–172 (1946).

REICH, D.:

[1] A p-adic fixed point formula, *Amer. J. Math.* **91**, 835–850 (1969).

REINER, I.:

[1] On the number of matrices with given characteristic polynomial, *Illinois J. Math.* **5**, 324–329 (1961).

RELLA, T.:

[1] Lineare Operatoren in endlichen Kongruenzkörpern, *Monatsh. Math. Phys.* **32**, 139–150 (1922).

REŠETUHA, I. V.:

[1] A certain question in the theory of cubic remainders (Russian), *Mat. Zametki* **7**, 469–476 (1970); *Math. Notes* **7**, 284–288 (1970).

[2] Generalized sums for characters and their applications to reciprocity laws (Russian), *Ukrain. Mat. Ž.* **23**, 270–276 (1971); *Ukrainian Math. J.* **23**, 235–240 (1971).

RETTER, C. T.:

[1] Decoding Goppa codes with a BCH decoder, *IEEE Trans. Information Theory* **IT-21**, 112 (1975).

RHIN, G.:

[1] Quelques résultats métriques dans un corps de séries formelles sur un corps fini, *Sém. Delange-Pisot-Poitou 1967/68, Théorie des Nombres*, Exp. 21, Secrétariat Math., Paris, 1969.

[2] Généralisation d'un théorème de I. M. Vinogradov à un corps de séries formelles sur un corps fini, *C. R. Acad. Sci. Paris Sér. A* **272**, 567–569 (1971).

[3] Répartition modulo 1 dans un corps de séries formelles sur un corps fini, *Dissertationes Math.* **95** (1972).

RIBENBOIM, P.:

[1] Polynomials whose values are powers, *J. reine angew. Math.* **268/269**, 34–40 (1974).

[2] *13 Lectures on Fermat's Last Theorem*, Springer-Verlag, New York-Heidelberg-Berlin, 1979.

RICE, B.:

[1] Some good fields and rings for computing number-theoretic transforms, *IEEE Trans. Acoust. Speech Signal Process.* **27**, 432–433 (1979).

[2] Winograd convolution algorithms over finite fields, *Congr. Numer.* **29**, 827–857 (1980).

RICHALET, J.:

[1] Operational calculus for finite rings, *IEEE Trans. Circuit Theory* **CT-12**, 558–570 (1965).

RIEGER, G. J.:

[1] Sur les nombres de Cullen, *Sém. Théorie des Nombres 1976–1977*, Exp. 16, 9 pp., Univ. Bordeaux I, Talence, 1977.

[2] Bemerkungen über gewisse nichtlineare Kongruenzen, *Elemente der Math.* **32**, 113–115 (1977).

[3] Über Lipschitz-Folgen, *Math. Scand.* **45**, 168–176 (1979).

RITT, J. F.:

[1] Prime and composite polynomials, *Trans. Amer. Math. Soc.* **23**, 51–66 (1922).

[2] Permutable rational functions, *Trans. Amer. Math. Soc.* **25**, 399–448 (1923).

RIVELAND, A. A., and PORTER, A. D.:

[1] The skew matric equation $X_n' \cdots X_1' A X_1 \cdots X_n = B$, *Rend. Mat.* (6)**9**, 633–638 (1976).

[2] The skew matrix equation $X_n' \cdots X_1' A X_1 \cdots X_n = B$, *Atti Accad. Naz. Lincei Rend. Cl. Sci. Fis. Mat. Natur.* (8)**60**, 751–755 (1976).

RIVLIN, T. J.:

[1] *The Chebyshev Polynomials*, Wiley, New York, 1974.

ROBERT, A.:

[1] *Elliptic Curves*, Lecture Notes in Math., vol. 326, Springer-Verlag, Berlin-Heidelberg-New York, 1973.

ROBINSON, A.:

[1] *Introduction to Model Theory and to the Metamathematics of Algebra*, North-Holland, Amsterdam, 1963.

ROBINSON, D. W.:

[1] The Fibonacci matrix modulo m, *Fibonacci Quart.* **1**, no. 2, 29–36 (1963).

[2] A note on linear recurrent sequences modulo m, *Amer. Math. Monthly* **73**, 619–621 (1966).

[3] Iteration of the modular period of a second order linear recurrent sequence, *Acta Arith.* **22**, 249–256 (1972/73).

[4] The rank and period of a linear recurrent sequence over a ring, *Fibonacci Quart.* **14**, 210–214 (1976).

ROBINSON, S. F.:

[1] Theorems on Brewer sums, *Pacific J. Math.* **25**, 587–596 (1968).

ROCCI, E.:

[1] Sulla distribuzione dei residui quadratici di un numero primo nella serie naturale, *Giorn. Mat. Battaglini* **65**, 112–134 (1927).

RODEMICH, E. R., and RUMSEY, H.:

[1] Primitive trinomials of high degree, *Math. Comp.* **22**, 863–865 (1968).

ROGERS, K.:

[1] Cyclotomic polynomials and division rings, *Monatsh. Math.* **69**, 239–242 (1965).

[2] An elementary proof of a theorem of Jacobson, *Abh. Math. Sem. Univ. Hamburg* **35**, 223–229 (1971); Berichtigung, *ibid.* **37**, 268 (1972).

ROGERS, L. J.:

[1] On the analytic representation of heptagrams, *Proc. London Math. Soc.* **22**, 37–52 (1890).

[2] Note on functions proper to represent a substitution of a prime number of letters, *Messenger of Math.* (2)**21**, 44–47 (1891).

ROHRBACH, H.:
[1] Die Charaktere der binären Kongruenzgruppen mod p^2, *Schr. Math. Sem. Inst. Angew. Math. Univ. Berlin* **1**, 33–94 (1932).

ROQUETTE, P.:
[1] Riemannsche Vermutung in Funktionenkörpern, *Arch. Math.* **4**, 6–16 (1953).
[2] Arithmetischer Beweis der Riemannschen Vermutung in Kongruenzfunktionenkörpern beliebigen Geschlechts, *J. reine angew. Math.* **191**, 199–252 (1953).

ROSATI, L. A.:
[1] Sul numero dei punti di una superficie cubica in uno spazio lineare finito, *Boll. Un. Mat. Ital.* (3)**11**, 412–418 (1956).

ROSENBERG, I. G.:
[1] Sums of Legendre symbols. I, II (Czech), *Sb. Vysoké. Učení Tech. Brno* **1962**, no. 1–2, 183–190; no. 3–4, 311–314.
[2] Polynomial functions over finite rings, *Glasnik Mat.* (3)**10**, 25–33 (1975).
[3] Characteristic polynomials in *GF*(2) of zero-one inequalities and equations, *Utilitas Math.* **7**, 323–343 (1975).

ROSENBERGER, G.:
[1] Über Tschebyscheff-Polynome, Nicht-Kongruenzuntergruppen der Modulgruppe und Fibonacci-Zahlen, *Math. Ann.* **246**, 193–203 (1980).

ROTA, G.-C.:
[1] On the foundations of combinatorial theory, I. Theory of Moebius functions, *Z. Wahrscheinlichkeitstheorie* **2**, 340–368 (1964).

ROTH, H. H.:
[1] Linear binary shift register circuits utilizing a minimum number of mod-2 adders, *IEEE Trans. Information Theory* **IT-11**, 215–220 (1965).

RUDEANU, S.:
[1] L'emploi des imaginaires de Galois dans la théorie des mécanismes automatiques. X (Romanian. French summary), *Acad. R. P. Romîne. Stud. Cerc. Mat.* **9**, 217–287 (1958).

RUDOLPH, L. D.:
[1] A class of majority-logic decodable codes, *IEEE Trans. Information Theory* **IT-13**, 305–307 (1967).

RUTHERFORD, D. E.:
[1] *Modular Invariants*, Cambridge Tracts in Math. and Math. Physics, no. 27, Cambridge Univ. Press, London, 1932.

RYSER, H. J.:
[1] *Combinatorial Mathematics*, Carus Math. Monographs, no. 14, Math. Assoc. of America, New York, 1963.
[2] Symmetric designs and related configurations, *J. Combinatorial Theory Ser. A* **12**, 98–111 (1972).

[3] The existence of symmetric block designs, *J. Combinatorial Theory Ser. A* **32**, 103–105 (1982).

SACHAR, H.:
[1] The F_p span of the incidence matrix of a finite projective plane, *Geom. Dedicata* **8**, 407–415 (1979).

SAGALOVIČ, JU. L.:
[1] Sequences of maximal length as codes of automaton states (Russian), *Problemy Peredači Informacii* **12**, no. 4, 70–73 (1976); *Problems of Information Transmission* **12**, 296–299 (1976).

SAKATA, S.:
[1] General theory of doubly periodic arrays over an arbitrary finite field and its applications, *IEEE Trans. Information Theory* **IT-24**, 719–730 (1978).
[2] On determining the independent point set for doubly periodic arrays and encoding two-dimensional cyclic codes and their duals, *IEEE Trans. Information Theory* **IT-27**, 556–565 (1981).

SALIÉ, H.:
[1] Über die Kloostermanschen Summen $S(u, v; q)$, *Math. Z.* **34**, 91–109 (1932).
[2] Zur Abschätzung der Fourierkoeffizienten ganzer Modulformen, *Math. Z.* **36**, 263–278 (1933).
[3] Über die Verteilung der quadratischen Reste, *Math. Z.* **37**, 594–602 (1933).

SAMPSON, J. H., and WASHNITZER, G.:
[1] Numerical equivalence and the zeta-function of a variety, *Amer. J. Math.* **81**, 735–748 (1959).

SANSONE, G.:
[1] La risoluzione apiristica delle congruenze biquadratiche, *Atti Accad. Naz. Lincei Rend. Cl. Sci. Fis. Mat. Natur.* (6)**6**, 573–578 (1927).
[2] Nuove formule risolutive delle congruenze cubiche, *Rend. Accad. Sci. Fis. Mat. Napoli* (4)**35**, 54–81 (1929).
[3] La risoluzione apiristica delle congruenze cubiche, *Ann. Mat. Pura Appl.* (4)**6**, 127–160 (1929).
[4] Sul problema della risoluzione apiristica delle congruenze di grado qualunque rispetto ad un modulo primo, e la risoluzione apiristica delle congruenze di quarto grado, *Mem. Accad. Naz. Lincei* (6)**3**, 220–260 (1929).
[5] La risoluzione apiristica delle congruenze cubiche, *Ann. Mat. Pura Appl.* (4)**7**, 1–32 (1930).

SÁRKÖZY, A.:
[1] Some remarks concerning irregularities of distribution of sequences of integers in arithmetic progressions. IV, *Acta Math. Acad. Sci. Hungar.* **30**, 155–162 (1977).

SARWATE, D. V.:

[1] On the complexity of decoding Goppa codes, *IEEE Trans. Information Theory* **IT-23**, 515–516 (1977).

[2] Semi-fast Fourier transforms over $GF(2^m)$, *IEEE Trans. Computers* **C-27**, 283–285 (1978).

SATO, M., and YORINAGA, M.:

[1] Numerical experiments on a conjecture of B. C. Mortimer and K. S. Williams, *Proc. Japan Acad.* **49**, 791–794 (1973).

ŠATUNOVSKIĬ, S. O.:

[1] Conditions for the existence of n distinct roots of a congruence of nth degree for a prime modulus (Russian), *Izv. Fiz.-Mat. Obšč. Kazan* (2)**12**, no. 3, 33–49 (1902).

SCARPIS, U.:

[1] Intorno alla risoluzione per radicali di un'equazione algebrica in un campo di Galois, *Periodico di Mat.* (3)**9**, 73–79 (1912).

[2] Successioni ricorrenti in un campo di Galois, *Ann. Mat. Pura Appl.* (3)**18**, 245–286 (1912).

[3] Intorno all'interpretazione della Teoria di Galois in un campo di razionalità finito, *Ann. Mat. Pura Appl.* (3)**23**, 41–60 (1914).

SCHAAR, M.:

[1] Mémoire sur la théorie des résidus quadratiques, *Acad. Roy. Sci. Lettres Beaux Arts Belgique* **24** (1850), 14 pp.

SCHANUEL, S. H.:

[1] An extension of Chevalley's theorem to congruences modulo prime powers, *J. Number Theory* **6**, 284–290 (1974).

SCHMID, H. L.:

[1] Relationen zwischen verallgemeinerten Gaußschen Summen, *J. reine angew. Math.* **176**, 189–191 (1937).

[2] Kongruenzzetafunktionen in zyklischen Körpern, *Abh. Preuss. Akad. Wiss. Math.-Naturw. Kl.* **1941**, no. 14, 1–30.

SCHMID, H. L., and TEICHMÜLLER, O.:

[1] Ein neuer Beweis für die Funktionalgleichung der L-Reihen, *Abh. Math. Sem. Univ. Hamburg* **15**, 85–96 (1943).

SCHMIDT, F. K.:

[1] Allgemeine Körper im Gebiet der höheren Kongruenzen, Dissertation, Freiburg i. Br., 1925.

[2] Zur Zahlentheorie in Körpern von der Charakteristik p, *Sitzungsber. Phys.-Med. Soz. Erlangen* **58/59**, 159–172 (1926/27).

[3] Analytische Zahlentheorie in Körpern der Charakteristik p, *Math. Z.* **33**, 1–32 (1931).

SCHMIDT, W. M.:

[1] Zur Methode von Stepanov, *Acta Arith.* **24**, 347–367 (1973).

[2] A lower bound for the number of solutions of equations over finite fields, *J. Number Theory* **6**, 448–480 (1974).

[3] *Equations over Finite Fields: An Elementary Approach*, Lecture Notes in Math., vol. 536, Springer-Verlag, Berlin-Heidelberg-New York, 1976.

SCHNEIDER, P.:

[1] On the values of the zeta function of a variety over a finite field, *Compositio Math.* **46**, 133–143 (1982).

SCHOLEFIELD, P. H. R.:

[1] Shift registers generating maximum-length sequences, *Electronic Technology* **37**, 389–394 (1960).

SCHÖNEMANN, T.:

[1] Ueber die Congruenz $x^2 + y^2 \equiv 1 (\text{mod } p)$, *J. reine angew. Math.* **19**, 93–112 (1839).

[2] Theorie der symmetrischen Functionen der Wurzeln einer Gleichung. Allgemeine Sätze über Congruenzen nebst einigen Anwendungen derselben, *J. reine angew. Math.* **19**, 289–308 (1839).

[3] Grundzüge einer allgemeinen Theorie der höhern Congruenzen, deren Modul eine reelle Primzahl ist, *J. reine angew. Math.* **31**, 269–325 (1846).

[4] Über einige von Herrn Dr. Eisenstein aufgestellte Lehrsätze, *J. reine angew. Math.* **40**, 185–187 (1850).

SCHÖNHAGE, A.:

[1] Schnelle Berechnung von Kettenbruchentwicklungen, *Acta Informatica* **1**, 139–144 (1971).

[2] Schnelle Multiplikation von Polynomen über Körpern der Charakteristik 2, *Acta Informatica* **7**, 395–398 (1977).

SCHÖNHEIM, J.:

[1] Formules pour résoudre la congruence $x^2 \equiv a (\text{mod } P)$ dans des cas encore inconnus et leur application pour déterminer directement des racines primitives de certains nombres premiers (Romanian. French summary), *Acad. R. P. Romîne Fil. Cluj. Stud. Cerc. Mat. Fiz.* **7**, no. 1–4, 51–58 (1956).

[2] On linear and nonlinear single-error-correcting q-nary perfect codes, *Information and Control* **12**, 23–26 (1968).

SCHUPFER, F.:

[1] Su due proposizioni di teoria dei numeri, *Rend. Mat. e Appl.* (5)**5**, 246–251 (1946).

SCHUR, I.:

[1] Über die Kongruenz $x^m + y^m \equiv z^m (\text{mod. } p)$, *Jber. Deutsch. Math.-Verein.* **25**, 114–117 (1916).

[2] Einige Bemerkungen zu der vorstehenden Arbeit des Herrn G. Pólya: Über die Verteilung der quadratischen Reste und Nichtreste, *Göttinger Nachr.* **1918**, 30–36.

[3] Über die Gaußschen Summen, *Göttinger Nachr.* **1921**, 147–153.

[4] Über den Zusammenhang zwischen einem Problem der Zahlentheorie und einem Satz über algebraische Funktionen, *Sitzungsber. Preuß. Akad. Wiss. Berlin Math.-Naturwiss. Kl.* **1923**, 123–134.

SCHWARZ, Š.:

[1] Contribution à la réductibilité des polynômes dans la théorie des congruences, *Věstník Královské České Spol. Nauk. Třída Matemat.-Přírodověd.* **1939**, no. 7, 1–7.

[2] A contribution to the arithmetic of finite fields (Slovak), *Prírodoved. Príloha Techn. Obzoru Sloven.* **1**, no. 8, 75–81 (1940).

[3] Sur le nombre des racines et des facteurs irréductibles d'une congruence donnée, *Časopis Pěst. Mat. Fys.* **69**, 128–145 (1940).

[4] A contribution to the reducibility of binomial congruences (Slovak), *Časopis Pěst. Mat. Fys.* **71**, 21–31 (1946).

[5] On Waring's problem for finite fields, *Quart. J. Math.* **19**, 123–128 (1948).

[6] On the equation $a_1x_1^k + a_2x_2^k + \cdots + a_kx_k^k + b = 0$ in finite fields, *Quart. J. Math.* **19**, 160–163 (1948).

[7] On the reducibility of binomial congruences and on the bound of the least integer belonging to a given exponent mod p, *Časopis Pěst. Mat. Fys.* **74**, 1–16 (1949).

[8] On equations of the form $c_1x_1^k + \cdots + c_sx_s^k = c$ in finite fields (Slovak), *Časopis Pěst. Mat. Fys.* **74**, 175–176 (1949).

[9] On universal forms in finite fields, *Časopis Pěst. Mat. Fys.* **75**, 45–50 (1950).

[10] On a type of universal forms in discretely normed fields, *Acta Sci. Math. Szeged* **17**, 5–19 (1956).

[11] On the reducibility of polynomials over a finite field, *Quart. J. Math.* (2)**7**, 110–124 (1956).

[12] On a class of polynomials over a finite field (Russian), *Mat.-Fyz. Časopis Sloven. Akad. Vied* **10**, 68–80 (1960).

[13] On the number of irreducible factors of a polynomial over a finite field (Russian), *Czechoslovak Math. J.* **11**, 213–225 (1961).

[14] A remark on algebraic equations over a finite field (Russian), *Mat.-Fyz. Časopis Sloven. Akad. Vied* **12**, 224–229 (1962).

[15] On a system of congruences. A remark on the preceding paper of J. Sedláček (Slovak), *Mat.-Fyz. Časopis Sloven. Akad. Vied* **13**, 103–104 (1963).

SCHWERING, K.:

[1] Zur Theorie der arithmetischen Functionen, welche von Jacobi $\psi(\alpha)$ genannt werden, *J. reine angew. Math.* **93**, 334–337 (1882).

SCOGNAMIGLIO, G.:

[1] Algebre di matrici atte a rappresentare campi di Galois, *Giorn. Mat. Battaglini* (6)**2**, 37–48 (1964).

SCORZA, G.:

[1] La risoluzione apiristica delle congruenze binomie e la formula di interpolazione di Lagrange, *Atti Accad. Naz. Lincei Rend. Cl. Sci. Fis. Mat. Natur.* (6)**3**, 390–394 (1926).

SCOTT, W. R.:

[1] *Group Theory*, Prentice-Hall, Englewood Cliffs, N.J., 1964.

SEDLÁČEK, J.:

[1] Some remarks on the problem of W. Mnich (Czech), *Mat.-Fyz. Časopis Sloven. Akad. Vied* **13**, 97–102 (1963).

SEGAL, B.:

[1] Character sums and their application (Russian), *Izv. Akad. Nauk SSSR Ser. Mat.* **5**, 401–410 (1941).

SEGRE, B.:

[1] Sui k-archi nei piani finiti di caratteristica due, *Rev. Math. Pures Appl.* **2**, 289–300 (1957).

[2] Le geometrie di Galois, *Ann. Mat. Pura Appl.* (4)**48**, 1–97 (1959).

[3] Sulla teoria delle equazioni e delle congruenze algebriche. I, II, *Atti Accad. Naz. Lincei Rend. Cl. Sci. Fis. Mat. Natur.* (8)**27**, 155–161, 303–311 (1959).

[4] Sistemi di equazioni nei campi di Galois, *Convegno Teoria dei Gruppi Finiti e Applicazioni* (Florence, 1960), pp. 66–80, Edizioni Cremonese, Rome, 1960.

[5] Sul numero delle soluzioni di un qualsiasi sistema di equazioni algebriche sopra un campo finito, *Atti Accad. Naz. Lincei Rend. Cl. Sci. Fis. Mat. Natur.* (8)**28**, 271–277 (1960).

[6] *Lectures on Modern Geometry*, Edizioni Cremonese, Rome, 1961.

[7] Geometry and algebra in Galois spaces, *Abh. Math. Sem. Univ. Hamburg* **25**, 129–139 (1962).

[8] Ovali e curve σ nei piani di Galois di caratteristica due, *Atti Accad. Naz. Lincei Rend. Cl. Sci. Fis. Mat. Natur.* (8)**32**, 785–790 (1962).

[9] Intorno ad una congettura di Lang e Weil, *Atti Accad. Naz. Lincei Rend. Cl. Sci. Fis. Mat. Natur.* (8)**34**, 337–339 (1963).

[10] Arithmetische Eigenschaften von Galois-Räumen. I, *Math. Ann.* **154**, 195–256 (1964).

[11] Forme e geometrie hermitiane, con particolare riguardo al caso finito, *Ann. Mat. Pura Appl.* (4)**70**, 1–201 (1965).

SEGRE, B., and BARTOCCI, U.:

[1] Ovali ed altre curve nei piani di Galois di caratteristica due, *Acta Arith.* **18**, 423–449 (1971).

SELBERG, A.:

[1] Über die Fourierkoeffizienten elliptischer Modulformen negativer Dimension, *Neuvième Congrès des Mathématiciens Scandinaves* (Helsinki, 1938), pp. 320–322, Merc. Kirjapaino, Helsinki, 1939.

[2] On the estimation of Fourier coefficients of modular forms, *Proc. Symp. Pure Math.*, vol. 8, pp. 1–15, American Math. Society, Providence, R.I., 1965.

SELFRIDGE, J. L., NICOL, C. A., and VANDIVER, H. S.:

[1] On diophantine equations which have no solutions, *Proc. Nat. Acad. Sci. U.S.A.* **42**, 264–266 (1956).

SELMER, E. S.:

[1] The diophantine equation $ax^3 + by^3 + cz^3 = 0$, *Acta Math.* **85**, 203–362 (1951).

[2] On Newton's equations for the power sums, *Nordisk Tidskr. for Informationsbehandling (BIT)* **6**, 158–160 (1966).

[3] *Linear Recurrence Relations over Finite Fields*, Univ. of Bergen, 1966.

SENGENHORST, P.:

[1] Über Körper der Charakteristik p, *Math. Z.* **24**, 1–39 (1926); Bemerkungen, *ibid.* **26**, 495 (1927).

SERGEEV, È. A.:

[1] The splitting of the polynomials $f_n(x)$ over finite fields (Russian), *Kuban. Gos. Univ. Naučn. Trudy Vyp.* **166**, 20–33 (1973).

SEROUSSI, G., and LEMPEL, A.:

[1] Factorization of symmetric matrices and trace-orthogonal bases in finite fields, *SIAM J. Computing* **9**, 758–767 (1980).

SERRE, J.-P.:

[1] *Cours d'arithmétique*, Presses Univ. de France, Paris, 1970.

[2] Valeurs propres des endomorphismes de Frobenius (d'après P. Deligne), *Séminaire Bourbaki 1973/74*, Exp. 446, Lecture Notes in Math., vol. 431, pp. 190–204, Springer-Verlag, Berlin-Heidelberg-New York, 1975.

[3] Majorations des sommes exponentielles, *Astérisque*, no. 41–42, pp. 111–126, Soc. Math. France, Paris, 1977.

SERRET, J.-A.:

[1] *Cours d'algèbre supérieure*, 2nd ed., Mallet-Bachelier, Paris, 1854.

[2] *Cours d'algèbre supérieure*, 3rd ed., Gauthier-Villars, Paris, 1866.

[3] Mémoire sur la théorie des congruences suivant un module premier et suivant une fonction modulaire irréductible, *Mém. Acad. Sci. Inst. de France* **35**, 617–688 (1866).

[4] Détermination des fonctions entières irréductibles, suivant un module premier, dans le cas où le degré est égal au module, *J. Math. Pures Appl.* (2)**18**, 301–304 (1873).

[5] Sur les fonctions entières irréductibles suivant un module premier, dans le cas où le degré est une puissance du module, *J. Math. Pures Appl.* (2)**18**, 437–451 (1873).

SHADER, L. E.:

[1] Arithmetical functions associated with unitary divisors in $GF[q, x]$. I, II, *Ann. Mat. Pura Appl.* (4)**86**, 79–85, 87–97 (1970).

[2] On the number of solutions of congruences and equations in $GF[q, x]$, *Portugal. Math.* **30**, 181–190 (1971).

[3] On the number of solutions of a congruence in $GF[q, x]$, *Portugal. Math.* **32**, 9–16 (1973).

[4] Closed form expressions for several Ramanujan sums, *Portugal. Math.* **32**, 147–153 (1973).

SHAH, A. P.:

[1] Fibonacci sequence modulo *m*, *Fibonacci Quart.* **6**, 139–141 (1968).

SHANKS, D.:

[1] Two theorems of Gauss, *Pacific J. Math.* **8**, 609–612 (1958).

[2] Five number-theoretic algorithms, *Proc. Second Manitoba Conf. on Numerical Math.* (Winnipeg, Man., 1972), pp. 51–70, Utilitas Math., Winnipeg, Man., 1973.

SHANNON, C. E.:

[1] A mathematical theory of communication, *Bell System Tech. J.* **27**, 379–423, 623–656 (1948).

SHANNON, C. E., and WEAVER, W.:

[1] *A Mathematical Theory of Communication*, Univ. of Illinois Press, Urbana, Ill., 1949.

SHEHADEH, N. M.:

[1] On the distribution of the coefficients of some polynomials, *SIAM J. Appl. Math.* **16**, 958–963 (1968).

SHIMURA, G.:

[1] *Introduction to the Arithmetic Theory of Automorphic Functions*, Princeton Univ. Press, Princeton, N.J., 1971.

SHIMURA, G., and TANIYAMA, Y.:

[1] *Complex Multiplication of Abelian Varieties and Its Applications to Number Theory*, Math. Society of Japan, Tokyo, 1961.

SHIRATANI, K.:

[1] On the Gauss-Hecke sums, *J. Math. Soc. Japan* **16**, 32–38 (1964).

SHIUE, J.-S.:

[1] A remark of a paper by Bundschuh, *Tamkang J. Math.* **4**, 129–130 (1973).

SHIUE, J.-S., and HU, M. H.:

[1] Some remarks on the uniform distribution of a linear recurrence of the second order, *Tamkang J. Math.* **4**, 101–103 (1973).

SHIUE, J.-S., and SHEU, T. L.:

[1] On the periodicity of linear recurrence of second order in commutative rings, *Tamkang J. Math.* **4**, 105–107 (1973).

SHIVA, S. G. S., and ALLARD, P. E.:

[1] A few useful details about a known technique for factoring $1+X^{2^q-1}$, *IEEE Trans. Information Theory* **IT-16**, 234–235 (1970).

SHRIKHANDE, S. S.:

[1] The impossibility of certain symmetrical balanced incomplete block designs, *Ann. Math. Statist.* **21**, 106–111 (1950).

SIEGEL, C. L.:

[1] Über die analytische Theorie der quadratischen Formen, *Ann. of Math.* (2)**36**, 527–606 (1935).

[2] Generalization of Waring's problem to algebraic number fields, *Amer. J. Math.* **66**, 122–136 (1944).

[3] Über das quadratische Reziprozitätsgesetz in algebraischen Zahlkörpern, *Göttinger Nachr.* **1960**, no. 1, 1–16.

SILVA, J. A.:

[1] A theorem on cyclic matrices, *Duke Math. J.* **18**, 821–825 (1951).

[2] Representation of arithmetic functions in $GF[p^n, x]$ with values in an arbitrary field, *Duke Math. J.* **19**, 31–44 (1952).

SIMMONS, G.:

[1] On the number of irreducible polynomials of degree d over $GF(p)$, *Amer. Math. Monthly* **77**, 743–745 (1970).

SIMS, C. C.:

[1] The role of algorithms in the teaching of algebra, *Topics in Algebra* (M. F. Newman, ed.), Lecture Notes in Math., vol. 697, pp. 95–107, Springer-Verlag, Berlin-Heidelberg-New York, 1978.

SINGER, J.:

[1] A theorem in finite projective geometry and some applications to number theory, *Trans. Amer. Math. Soc.* **43**, 377–385 (1938).

SINGH, S.:

[1] Bounds of cubic residues in A. P., *Indian J. Pure Appl. Math.* **1**, 265–268 (1970).

[2] The number of decompositions of an integer as a sum of two squares in $GF(p^n)$, *Indian J. Pure Appl. Math.* **4**, 606–611 (1973).

[3] Stufe of a finite field, *Fibonacci Quart.* **12**, 81–82 (1974).

[4] Analysis of each integer as sum of two cubes in a finite integral domain, *Indian J. Pure Appl. Math.* **6**, 29–35 (1975).

[5] Bound for the solutions of a Diophantine equation in prime Galois fields, *Indian J. Pure Appl. Math.* **8**, 1428–1430 (1977).

[6] Integer points on special hyper-elliptic curves in $GF(p)$, *Indian J. Pure Appl. Math.* **10**, 1213–1215 (1979).

SINGH, S., and RAJWADE, A. R.:

[1] The number of solutions of the congruence $y^2 \equiv x^4 - a \pmod p$, *L'Enseignement Math.* (2)**20**, 265–273 (1974).

SINGLETON, R. C.:

[1] Maximum distance q-nary codes, *IEEE Trans. Information Theory* **IT-10**, 116–118 (1964).

SINGMASTER, D.:

[1] On polynomial functions $(\bmod m)$, *J. Number Theory* **6**, 345–352 (1974).

SKOLEM, T.:

[1] Zwei Sätze über kubische Kongruenzen, *Norske Vid. Selsk. Forh.* **10**, 89–92 (1937).

[2] Die Anzahl der Wurzeln der Kongruenz $x^3 + ax + b \equiv 0(\text{mod } p)$ für die verschiedenen Paare a, b, *Norske Vid. Selsk. Forh.* **14**, 161–164 (1942).
[3] Unlösbarkeit von Gleichungen, deren entsprechende Kongruenz für jeden Modul lösbar ist, *Avh. Norske Vid. Akad. Oslo I* **1942**, no. 4, 28 pp.
[4] The general congruence of 4th degree modulo p, p prime, *Norsk Mat. Tidsskr.* **34**, 73–80 (1952).
[5] On a certain connection between the discriminant of a polynomial and the number of its irreducible factors (mod p), *Norsk Mat. Tidsskr.* **34**, 81–85 (1952).
[6] Remarks on proofs by cyclotomic formulas of reciprocity laws for power residues, *Math. Scand.* **9**, 229–242 (1961).

SLEPIAN, D.:
[1] A note on two binary signaling alphabets, *IRE Trans. Information Theory* **IT-2**, 84–86 (1956).
[2] A class of binary signaling alphabets, *Bell System Tech. J.* **35**, 203–234 (1956).
[3] Some further theory of group codes, *Bell System Tech. J.* **39**, 1219–1252 (1960).
[4] *Key Papers in the Development of Information Theory*, IEEE Press, New York, 1974.

SLOANE, N. J. A.:
[1] A survey of constructive coding theory, and a table of binary codes of highest known rate, *Discrete Math.* **3**, 265–294 (1972).
[2] Error-correcting codes and cryptography, *The Mathematical Gardner* (D. A. Klarner, ed.), pp. 346–382, Prindle, Weber & Schmidt, Boston, 1981.

SMALL, C.:
[1] Waring's problem mod n, *Amer. Math. Monthly* **84**, 12–25 (1977).
[2] Sums of powers in large finite fields, *Proc. Amer. Math. Soc.* **65**, 35–36 (1977).
[3] Solution of Waring's problem mod n, *Amer. Math. Monthly* **84**, 356–359 (1977).

SMITH, C., and HOGGATT, V. E., Jr.:
[1] Primitive periods of generalized Fibonacci sequences, *Fibonacci Quart.* **14**, 343–347 (1976).

SMITH, H. J. S.:
[1] Report on the theory of numbers. Part II, *Report of the British Association for 1860*, pp. 120–169; *Collected Math. Papers*, vol. 1, pp. 93–162, Chelsea, New York, 1965.

SMITH, R. A.:
[1] The circle problem in an arithmetic progression, *Canad. Math. Bull.* **11**, 175–184 (1968).
[2] The distribution of rational points on hypersurfaces defined over a finite field, *Mathematika* **17**, 328–332 (1970).

[3] On n-dimensional Kloosterman sums, *C. R. Math. Rep. Acad. Sci. Canada* **1**, 173–176 (1979).
[4] On n-dimensional Kloosterman sums, *J. Number Theory* **11**, 324–343 (1979).
[5] A generalization of Kuznietsov's identity for Kloosterman sums, *C. R. Math. Rep. Acad. Sci. Canada* **2**, 315–320 (1980).
[6] Estimates for exponential sums, *Proc. Amer. Math. Soc.* **79**, 365–368 (1980).

SMITS, T. H. M.:
[1] On the group of units of $GF(q)[x]/(a(x))$, *Indag. Math.* **44**, 355–358 (1982).

SNAPPER, E.:
[1] Quadratic spaces over finite fields and codes, *J. Combinatorial Theory Ser. A* **27**, 263–268 (1979).

SOKOLOVSKIĬ, A. V.:
[1] Lower bounds in the "large sieve" (Russian), *Zap. Naučn. Sem. Leningrad. Otdel. Mat. Inst. Steklov.* **91**, 125–133 (1979); *J. Soviet Math.* **17**, 2166–2173 (1981).

SOMER, L.:
[1] Problem E 2377, *Amer. Math. Monthly* **79**, 906 (1972); Solution, *ibid.* **81**, 282–283 (1974).
[2] Fibonacci-like groups and periods of Fibonacci-like sequences, *Fibonacci Quart.* **15**, 35–41 (1977).
[3] The divisibility properties of primary Lucas recurrences with respect to primes, *Fibonacci Quart.* **18**, 316–334 (1980).
[4] Possible periods of primary Fibonacci-like sequences with respect to a fixed odd prime, *Fibonacci Quart.* **20**, 311–333 (1982).

SPACKMAN, K. W.:
[1] Simultaneous solutions to diagonal equations over finite fields, *J. Number Theory* **11**, 100–115 (1979).
[2] On the number and distribution of simultaneous solutions to diagonal congruences, *Canad. J. Math.* **33**, 421–436 (1981).

ŠPARLINSKIĬ, I. E.:
[1] Distribution of nonresidues and primitive roots in recurrent sequences (Russian), *Mat. Zametki* **24**, 603–613 (1978); *Math. Notes* **24**, 823–828 (1978).

SPEISER, A.:
[1] Die Zerlegung von Primzahlen in algebraischen Zahlkörpern, *Trans. Amer. Math. Soc.* **23**, 173–178 (1922).

SPERBER, S.:
[1] p-adic hypergeometric functions and their cohomology, *Duke Math. J.* **44**, 535–589 (1977).

[2] Congruence properties of the hyperkloosterman sum, *Compositio Math*. **40**, 3–33 (1980).

[3] On the *L*-functions associated with certain exponential sums, *J. Number Theory* **12**, 141–153 (1980).

SPRINGER, T. A.:

[1] Note on quadratic forms in characteristic 2, *Nieuw Arch. Wisk*. (3)**10**, 1–10 (1962).

[2] The zeta function of a cuspidal representation of a finite group $GL_n(k)$, *Lie Groups and Their Representations* (I. M. Gelfand, ed.), pp. 645–648, Akadémiai Kiadó, Budapest, 1975.

[3] Caractères quadratiques de groupes abéliens finis et sommes de Gauss, *Colloque sur les Formes Quadratiques* (Montpellier, 1975), *Bull. Soc. Math. France Suppl. Mém*. **48**, 103–115 (1976).

[4] Trigonometric sums, Green functions of finite groups and representations of Weyl groups, *Invent. Math*. **36**, 173–207 (1976).

SRINIVASAN, B.:

[1] *Representations of Finite Chevalley Groups*, Lecture Notes in Math., vol. 764, Springer-Verlag, Berlin-Heidelberg-New York, 1979.

STAHNKE, W.:

[1] Primitive binary polynomials, *Math. Comp*. **27**, 977–980 (1973).

STANLEY, T. E.:

[1] A note on the sequence of Fibonacci numbers, *Math. Mag*. **44**, 19–22 (1971).

[2] Some remarks on the periodicity of the sequence of Fibonacci numbers, *Fibonacci Quart*. **14**, 52–54 (1976).

STARK, H. M.:

[1] On the Riemann hypothesis in hyperelliptic function fields, *Proc. Symp. Pure Math*., vol. 24, pp. 285–302, American Math. Society, Providence, R.I., 1973.

STEČKIN, S. B.:

[1] An estimate of a complete rational trigonometric sum (Russian), *Trudy Mat. Inst. Steklov*. **143**, 188–207 (1977).

STEINITZ, E.:

[1] Algebraische Theorie der Körper, *J. reine angew. Math*. **137**, 167–309 (1910).

STEMMLER, R. M.:

[1] The easier Waring problem in algebraic number fields, *Acta Arith*. **6**, 447–468 (1961).

STEPANOV, S. A.:

[1] The number of points of a hyperelliptic curve over a finite prime field (Russian), *Izv. Akad. Nauk SSSR Ser. Mat*. **33**, 1171–1181 (1969); *Math. USSR Izv*. **3**, 1103–1114 (1969).

[2] Elementary method in the theory of congruences for a prime modulus, *Acta Arith*. **17**, 231–247 (1970).

[3] Estimation of Weil sums with prime denominator (Russian), *Izv. Akad. Nauk SSSR Ser. Mat.* **34**, 1015–1037 (1970); Correction, *ibid*. **35**, 965–966 (1971); *Math. USSR Izv.* **4**, 1017–1041 (1970).

[4] An estimation of Kloosterman sums (Russian), *Izv. Akad. Nauk SSSR Ser. Mat.* **35**, 308–323 (1971); *Math. USSR Izv.* **5**, 319–336 (1971).

[5] On estimating rational trigonometric sums with prime denominator (Russian), *Trudy Mat. Inst. Steklov.* **112**, 346–371 (1971); *Proc. Steklov Inst. Math.* **112**, pp. 358–385, American Math. Society, Providence, R.I., 1973.

[6] An elementary proof of the Hasse-Weil theorem for hyperelliptic curves, *J. Number Theory* **4**, 118–143 (1972).

[7] Congruences in two unknowns (Russian), *Izv. Akad. Nauk SSSR Ser. Mat.* **36**, 683–711 (1972); *Math. USSR Izv.* **6**, 677–704 (1972).

[8] Constructive method in the theory of equations over finite fields, *Proc. Internat. Conf. on Number Theory* (Moscow, 1971), *Trudy Mat. Inst. Steklov.* **132**, 237–246 (1973); *Proc. Steklov Inst. Math.* **132**, pp. 271–281, American Math. Society, Providence, R.I., 1975.

[9] Rational points of algebraic curves over finite fields (Russian), *Current Problems of Analytic Number Theory* (Proc. Summer School Analytic Number Theory, Minsk, 1972) (Russian), pp. 223–243, Izdat. "Nauka i Tehnika," Minsk, 1974.

[10] An elementary method in the theory of equations over finite fields (Russian), *Proc. International Congress of Math.* (Vancouver, B.C., 1974), vol. 1, pp. 383–391, Canad. Math. Congress, Montreal, Que., 1975; *Amer. Math. Soc. Transl.* (2)**109**, 13–20 (1977).

[11] On lower estimates of incomplete character sums of polynomials (Russian), *Trudy Mat. Inst. Steklov.* **143**, 175–177 (1977); *Proc. Steklov Inst. Math. 1980*, no. 1, pp. 187–189, American Math. Society, Providence, R.I., 1980.

[12] Equations over finite fields (Russian), *Mat. Zametki* **21**, 271–279 (1977); *Math. Notes* **21**, 147–152 (1977).

[13] An elementary method in algebraic number theory (Russian), *Mat. Zametki* **24**, 425–431 (1978); *Math. Notes* **24**, 728–731 (1978).

[14] Proof of the Davenport-Hasse relations (Russian), *Mat. Zametki* **27**, 3–6 (1980); *Math. Notes* **27**, 3–4 (1980).

STEPHENS, N. M.:

[1] On a conjecture of Chowla and Chowla, *J. Number Theory* **9**, 276–277 (1977).

[2] Dirichlet characters and polynomials, *Bull. London Math. Soc.* **11**, 52–54 (1979).

STERN, M. A.:

[1] Bemerkungen über höhere Arithmetik, *J. reine angew. Math.* **6**, 147–158 (1830).

STERN, T. E., and FRIEDLAND, B.:

[1] Application of modular sequential circuits to single error-correcting p-nary codes, *IRE Trans. Information Theory* **IT-5**, 114–123 (1959).

STEVENS, H.:

[1] Linear homogeneous equations over finite rings, *Canad. J. Math.* **16**, 532–538 (1964).

STEVENS, H., and KUTY, L.:

[1] Applications of an elementary theorem to number theory, *Arch. Math.* **19**, 37–42 (1968).

STEVENS, W. L.:

[1] The completely orthogonalized Latin square, *Ann. of Eugenics* **9**, 82–93 (1939).

STICKELBERGER, L.:

[1] Ueber eine Verallgemeinerung der Kreistheilung, *Math. Ann.* **37**, 321–367 (1890).

[2] Über eine neue Eigenschaft der Diskriminanten algebraischer Zahlkörper, *Verhandl. Ersten Intern. Math.-Kongr.* (Zürich, 1897), pp. 182–193, Teubner, Leipzig, 1898.

STORER, T.:

[1] *Cyclotomy and Difference Sets*, Markham, Chicago, 1967.

[2] On the unique determination of the cyclotomic numbers for Galois fields and Galois domains, *J. Combinatorial Theory* **2**, 296–300 (1967).

[3] Mixed cyclotomy, prime-power circulants, and cyclotomy modulo $p = ef + 1$ for composite e, *J. Number Theory* **1**, 280–290 (1969).

[4] Extensions of cyclotomic theory, *Proc. Symp. Pure Math.*, vol. 20, pp. 123–134, American Math. Society, Providence, R.I., 1971.

STRASSEN, V.:

[1] Berechnungen in partiellen Algebren endlichen Typs, *Computing* **11**, 181–196 (1973).

[2] Computational complexity over finite fields, *SIAM J. Computing* **5**, 324–331 (1976).

STREET, A. P., and WALLIS, W. D.:

[1] *Combinatorial Theory: An Introduction*, Charles Babbage Research Centre, Winnipeg, Man., 1977.

STREET, A. P., and WHITEHEAD, E. G., Jr.:

[1] Sum-free sets, difference sets and cyclotomy, *Combinatorial Mathematics* (D. A. Holton, ed.), Lecture Notes in Math., vol. 403, pp. 109–124, Springer-Verlag, Berlin-Heidelberg-New York, 1974.

SUGIMOTO, E.:

[1] A short note on new indexing polynomials of finite fields, *Information and Control* **41**, 243–246 (1979).

SUGIYAMA, Y., KASAHARA, M., HIRASAWA, S., and NAMEKAWA, T.:

[1] A method for solving key equation for decoding Goppa codes, *Information and Control* **27**, 87–99 (1975).

[2] Further results on Goppa codes and their applications for constructing efficient binary codes, *IEEE Trans. Information Theory* **IT-22**, 518–526 (1976).

SURBÖCK, F., and WEINRICHTER, H.:

[1] Interlacing properties of shift-register sequences with generator polynomials irreducible over $GF(p)$, *IEEE Trans. Information Theory* **IT-24**, 386–389 (1978).

SWAN, R. G.:

[1] Factorization of polynomials over finite fields, *Pacific J. Math.* **12**, 1099–1106 (1962).

SWIFT, J. D.:

[1] Construction of Galois fields of characteristic two and irreducible polynomials, *Math. Comp.* **14**, 99–103 (1960).

SWINNERTON-DYER, H. P. F.:

[1] The conjectures of Birch and Swinnerton-Dyer, and of Tate, *Proc. Conf. Local Fields* (Driebergen, 1966), pp. 132–157, Springer-Verlag, Berlin-Heidelberg-New York, 1967.

[2] The zeta function of a cubic surface over a finite field, *Proc. Cambridge Philos. Soc.* **63**, 55–71 (1967).

[3] Applications of algebraic geometry to number theory, *Proc. Symp. Pure Math.*, vol. 20, pp. 1–52, American Math. Society, Providence, R.I., 1971.

SZALAY, M.:

[1] On the distribution of primitive roots mod p (Hungarian), *Mat. Lapok* **21**, 357–362 (1970).

[2] On the distribution of the primitive roots of a prime, *J. Number Theory* **7**, 184–188 (1975).

SZÉKELY, I., and MUREŞAN, T.:

[1] Interpolation with respect to a prime modulus (Romanian), *Bul. Şti. Inst. Politehn. Cluj* **8**, 105–109 (1965).

SZELE, T.:

[1] An elementary proof of the fundamental theorem for finite fields (Hungarian), *Mat. Lapok* **7**, 249–254 (1956).

SZYMICZEK, K.:

[1] Sums of powers of generators of a finite field, *Colloq. Math.* **20**, 59–63 (1969).

TÄCKLIND, S.:

[1] Über die Periodizität der Lösungen von Differenzenkongruenzen, *Ark. Mat. Astr. Fys.* **30A**, no. 22 (1944), 9 pp.

TAKAHASHI, I.:

[1] Switching functions constructed by Galois extension fields, *Information and Control* **48**, 95–108 (1981).

TAMARKINE, J., and FRIEDMANN, A.:

[1] Sur les congruences du second degré et les nombres de Bernoulli, *Math. Ann.* **62**, 409–412 (1906).

TANAKA, H., KASAHARA, M., TEZUKA, Y., and KASAHARA, Y.:

[1] Computation over Galois fields using shiftregisters, *Information and Control* **13**, 75–84 (1968).

TANIYAMA, Y.:

[1] Distribution of positive 0-cycles in absolute classes of an algebraic variety with finite constant field, *Sci. Papers Coll. Gen. Ed. Univ. Tokyo* **8**, 123–137 (1958).

TANNER, H. W. L.:

[1] On the binomial equation $x^p - 1 = 0$: Quinquisection, *Proc. London Math. Soc.* **18**, 214–234 (1887).

[2] On some square roots of unity for a prime modulus, *Messenger of Math.* (2)**21**, 139–144 (1892).

[3] On complex primes formed with the fifth roots of unity, *Proc. London Math. Soc.* **24**, 223–272 (1893).

TARRY, G.:

[1] Le problème des 36 officiers, *C. R. Assoc. Française Avancement Sci. Nat.* **1**, 122–123 (1900); *ibid.* **2**, 170–203 (1901).

TATE, J.:

[1] Algebraic cycles and poles of zeta functions, *Arithmetical Algebraic Geometry* (Proc. Conf. Purdue Univ., 1963), pp. 93–110, Harper and Row, New York, 1965.

[2] The arithmetic of elliptic curves, *Invent. Math.* **23**, 179–206 (1974).

TAUSWORTHE, R. C.:

[1] Random numbers generated by linear recurrence modulo two, *Math. Comp.* **19**, 201–209 (1965).

TAYLOR, D. E.:

[1] Some classical theorems on division rings, *L'Enseignement Math.* (2)**20**, 293–298 (1974).

TAYLOR, M. J.:

[1] Local Gauss sums, *Sém. Théorie des Nombres 1978–79*, Exp. 8, 9 pp., Univ. Bordeaux I, Talence, 1979.

[2] Adams operations, local root numbers, and the Galois module structure of rings of integers, *Proc. London Math. Soc.* (3)**39**, 147–175 (1979).

TAZAWA, M.:

[1] A theorem on congruence, *Tôhoku Math. J.* **32**, 354–356 (1930).

TEICHMÜLLER, O.:

[1] Differentialrechnung bei Charakteristik p, *J. reine angew. Math.* **175**, 89–99 (1936).

TERJANIAN, G.:

[1] Sur les corps finis, *C. R. Acad. Sci. Paris Sér. A* **262**, 167–169 (1966).

THAYSE, A.:

[1] Differential calculus for functions from $(GF(p))^n$ into $GF(p)$, *Phillips Res. Rep.* **29**, 560–586 (1974).

THOMAS, A. D.:

[1] *Zeta-Functions: An Introduction to Algebraic Geometry*, Research Notes in Math., no. 12, Pitman, London, 1977.

THOUVENOT, S.:

[1] Propriétés arithmétiques déductibles d'une présentation simplifiée de la formule du binome, *C. R. Acad. Sci. Paris* **254**, 1550–1552 (1962).

THOUVENOT, S., and CHÂTELET, F.:

[1] Au sujet des congruences de degré supérieur à deux, *L'Enseignement Math.* (2)**13**, 89–98 (1967).

TIETÄVÄINEN, A.:

[1] On the non-trivial solvability of some systems of equations in finite fields, *Ann. Univ. Turku Ser. AI* **71** (1964), 5 pp.

[2] On the non-trivial solvability of some equations and systems of equations in finite fields, *Ann. Acad. Sci. Fenn. Ser. AI* **360** (1965), 38 pp.

[3] On systems of linear and quadratic equations in finite fields, *Ann. Acad. Sci. Fenn. Ser. AI* **382** (1965), 5 pp.

[4] On the trace of a polynomial over a finite field, *Ann. Univ. Turku Ser. AI* **87** (1966), 7 pp.

[5] On non-residues of a polynomial, *Ann. Univ. Turku Ser. AI* **94** (1966), 6 pp.

[6] On systems of equations in finite fields, *Ann. Acad. Sci. Fenn. Ser. AI* **386** (1966), 10 pp.

[7] On the solvability of equations in incomplete finite fields, *Ann. Univ. Turku Ser. AI* **102** (1967), 13 pp.

[8] On pairs of additive equations, *Ann. Univ. Turku Ser. AI* **112** (1967), 7 pp.

[9] On diagonal forms over finite fields, *Ann. Univ. Turku Ser. AI* **118**, no. 1 (1968), 10 pp.

[10] On the distribution of the residues of a polynomial, *Ann. Univ. Turku Ser. AI* **120** (1968), 4 pp.

[11] On a homogeneous congruence of odd degree, *Ann. Univ. Turku Ser. AI* **131** (1969), 6 pp.

[12] On a problem of Chowla and Shimura, *J. Number Theory* **3**, 247–252 (1971).

[13] Note on Waring's problem (mod p), *Ann. Acad. Sci. Fenn. Ser. AI* **554** (1973), 7 pp.

[14] On the nonexistence of perfect codes over finite fields, *SIAM J. Appl. Math.* **24**, 88–96 (1973).

[15] A short proof for the nonexistence of unknown perfect codes over $GF(q)$, $q > 2$, *Ann. Acad. Sci. Fenn. Ser. AI* **580** (1974), 6 pp.

[16] Proof of a conjecture of S. Chowla, *J. Number Theory* **7**, 353–356 (1975).

TIETZE, U. P.:

[1] Zur Theorie quadratischer Formen über Körpern der Charakteristik 2, *J. reine angew. Math.* **268/269**, 388–390 (1974).

TONELLI, A.:

[1] Bemerkung über die Auflösung quadratischer Congruenzen, *Göttinger Nachr.* **1891**, 344–346.

TORNHEIM, L.:

[1] Sums of n-th powers in fields of prime characteristic, *Duke Math. J.* **4**, 359–362 (1938).

T'U, K. C.:

[1] The structure of Q-matrices and the reducibility of polynomials over a Galois field (Chinese), *Acta Math. Sinica* **17**, 46–59 (1974).

[2] Canonical forms of a class of ternary forms over $GF(2)$ (Chinese), *Acta Math. Sinica* **23**, 1–10 (1980).

TURYN, R.:

[1] Sequences with small correlation, *Error Correcting Codes* (H. B. Mann, ed.), pp. 195–228, Wiley, New York, 1968.

TUŠKINA, T. A.:

[1] A numerical experiment on the calculation of the Hasse invariant for certain curves (Russian), *Izv. Akad. Nauk SSSR Ser. Mat.* **29**, 1203–1204 (1965); *Amer. Math. Soc. Transl.* (2)**66**, 204–205 (1968).

UCHIYAMA, S.:

[1] Sur les polynômes irréductibles dans un corps fini, I, *Proc. Japan Acad.* **30**, 523–527 (1954).

[2] Sur le nombre des valeurs distinctes d'un polynôme à coefficients dans un corps fini, *Proc. Japan Acad.* **30**, 930–933 (1954).

[3] Note on the mean value of $V(f)$, *Proc. Japan Acad.* **31**, 199–201 (1955).

[4] Sur les polynômes irréductibles dans un corps fini, II, *Proc. Japan Acad.* **31**, 267–269 (1955).

[5] Note on the mean value of $V(f)$. II, *Proc. Japan Acad.* **31**, 321–323 (1955).

[6] Note on the mean value of $V(f)$. III, *Proc. Japan Acad.* **32**, 97–98 (1956).

[7] On a multiple exponential sum, *Proc. Japan Acad.* **32**, 748–749 (1956).

[8] On a conjecture of K. S. Williams, *Proc. Japan Acad.* **46**, 755–757 (1970).

UDALOV, A. P., and SUPRUN, B. A.:

[1] *Redundant Coding for the Transmission of Information Using Binary Codes* (Russian), Izdat. "Svjaz'," Moscow, 1964.

ULBRICH, K.-H.:

[1] Über Endomorphismen, deren Minimalpolynom mit dem charakteristischen Polynom übereinstimmt, *J. reine angew. Math.* **299/300**, 385–387 (1978).

URBANEK, F. J.:

[1] An O(log n) algorithm for computing the nth element of the solution of a difference equation, *Inform. Process. Lett.* **11**, no. 2, 66–67 (1980).

USOL'CEV, L. P.:

[1] Estimates of large deviations in certain problems concerning an incomplete system of residues (Russian), *Dokl. Akad. Nauk SSSR* **143**, 539–542 (1962); *Soviet Math. Dokl.* **3**, 440–443 (1962).

USPENSKY, J. V., and HEASLET, M. A.:

[1] *Elementary Number Theory*, McGraw-Hill, New York, 1939.

VAIDA, D.:

[1] L'emploi des imaginaires de Galois dans la théorie des mécanismes automatiques. VI (Romanian. French summary), *Acad. R. P. Romîne. Bul. Şti. Secţ. Şti. Mat. Fiz.* **8**, 21–29 (1956).

VAIDYANATHASWAMY, R.:

[1] The quadratic reciprocity of polynomials modulo p, *J. Indian Math. Soc.* **17**, 185–196 (1928).

[2] The algebra of cubic residues, *J. Indian Math. Soc.* (*N.S.*) **21**, 57–66 (1957).

VAJDA, S.:

[1] *Patterns and Configurations in Finite Spaces*, Hafner, New York, 1967.

[2] *The Mathematics of Experimental Design*, Hafner, New York, 1967.

VAN DER CORPUT, J. G.:

[1] Sur un certain système de congruences. I, II, *Indag. Math.* **1**, 168–176, 254–259 (1939).

VAN DER WAERDEN, B. L.:

[1] Noch eine Bemerkung zu der Arbeit "Zur Arithmetik der Polynome" von U. Wegner in Math. Ann. 105, S. 628–631, *Math. Ann.* **109**, 679–680 (1934).

[2] *Algebra*, vol. 1, 7th ed., Springer-Verlag, Berlin-Heidelberg-New York, 1966.

[3] *Algebra*, vol. 2, 5th ed., Springer-Verlag, Berlin-Heidelberg-New York, 1967.

VAN DE VOOREN-VAN VEEN, J.:

[1] On the number of irreducible equations of degree n in $GF(p)$ and the decomposability of the cyclotomic polynomials in $GF(p)$ (Dutch), *Simon Stevin* **31**, 80–82 (1957).

VANDIVER, H. S.:

[1] Note on trinomial congruences and the first case of Fermat's last theorem, *Ann. of Math.* (2)**27**, 54–56 (1925).

[2] Algorithms for the solution of the quadratic congruence, *Amer. Math. Monthly* **36**, 83–86 (1929).

[3] Some theorems in finite field theory with applications to Fermat's last theorem, *Proc. Nat. Acad. Sci. U.S.A.* **30**, 362–367 (1944).

[4] On trinomial congruences and Fermat's last theorem, *Proc. Nat. Acad. Sci.*

U.S.A. **30**, 368–370 (1944).

[5] New types of relations in finite field theory I, II, *Proc. Nat. Acad. Sci. U.S.A.* **31**, 50–54, 189–194 (1945).

[6] On the number of solutions of certain non-homogeneous trinomial equations in a finite field, *Proc. Nat. Acad. Sci. U.S.A.* **31**, 170–175 (1945).

[7] On the number of solutions of some general types of equations in a finite field, *Proc. Nat. Acad. Sci. U.S.A.* **32**, 47–52 (1946).

[8] On classes of diophantine equations of higher degrees which have no solutions, *Proc. Nat. Acad. Sci. U.S.A.* **32**, 101–106 (1946).

[9] Cyclotomy and trinomial equations in a finite field, *Proc. Nat. Acad. Sci. U.S.A.* **32**, 317–319 (1946).

[10] On some special trinomial equations in a finite field, *Proc. Nat. Acad. Sci. U.S.A.* **32**, 320–326 (1946).

[11] Limits for the number of solutions of certain general types of equations in a finite field, *Proc. Nat. Acad. Sci. U.S.A.* **33**, 236–242 (1947).

[12] Applications of cyclotomy to the theory of nonhomogeneous equations in a finite field, *Proc. Nat. Acad. Sci. U.S.A.* **34**, 62–66 (1948).

[13] Congruence methods as applied to diophantine analysis, *Math. Mag.* **21**, 185–192 (1948).

[14] Cyclotomic power characters and trinomial equations in a finite field, *Proc. Nat. Acad. Sci. U.S.A.* **34**, 196–203 (1948).

[15] Quadratic relations involving the numbers of solutions of certain types of equations in a finite field, *Proc. Nat. Acad. Sci. U.S.A.* **35**, 681–685 (1949).

[16] On a generalization of a Jacobi exponential sum associated with cyclotomy, *Proc. Nat. Acad. Sci. U.S.A.* **36**, 144–151 (1950).

[17] On cyclotomy and extensions of Gaussian type quadratic rclations involving numbers of solutions of conditional equations in finite fields, *Proc. Nat. Acad. Sci. U.S.A.* **38**, 981–991 (1952).

[18] New types of trinomial congruence criteria applying to Fermat's last theorem, *Proc. Nat. Acad. Sci. U.S.A.* **40**, 248–252 (1954).

[19] On trinomial equations in a finite field, *Proc. Nat. Acad. Sci. U.S.A.* **40**, 1008–1010 (1954).

[20] On the properties of certain trinomial equations in a finite field, *Proc. Nat. Acad. Sci. U.S.A.* **41**, 651–653 (1955).

[21] Relation of the theory of certain trinomial equations in a finite field to Fermat's last theorem, *Proc. Nat. Acad. Sci. U.S.A.* **41**, 770–775 (1955).

[22] On cyclotomic relations and trinomial equations in a finite field, *Proc. Nat. Acad. Sci. U.S.A.* **41**, 775–780 (1955).

[23] Diophantine equations in certain rings, *Proc. Nat. Acad. Sci. U.S.A.* **42**, 656–665 (1956); Errata, *ibid.* **43**, 252–253 (1957).

[24] The rapid computing machine as an instrument in the discovery of new relations in the theory of numbers, *Proc. Nat. Acad. Sci. U.S.A.* **44**, 459–464 (1958).

[25] On distribution problems involving the number of solutions of certain trinomial congruences, *Proc. Nat. Acad. Sci. U.S.A.* **45**, 1635–1641 (1959).

VAN LINT, J. H.:

[1] *Coding Theory*, Lecture Notes in Math., vol. 201, Springer-Verlag, Berlin-Heidelberg-New York, 1971.

[2] *Combinatorial Theory Seminar Eindhoven University of Technology*, Lecture Notes in Math., vol. 382, Springer-Verlag, Berlin-Heidelberg-New York, 1974.

[3] Recent results on perfect codes and related topics, *Combinatorics*, Part 1: *Theory of Designs, Finite Geometry and Coding Theory*, Math. Centre Tracts, no. 55, pp. 158–178, Math. Centrum, Amsterdam, 1974.

[4] A survey of perfect codes, *Rocky Mountain J. Math.* **5**, 199–224 (1975).

VAN LINT, J. H., and RYSER, H. J.:

[1] Block designs with repeated blocks, *Discrete Math.* **3**, 381–396 (1972).

VAN METER, R. G.:

[1] The number of solutions of certain systems of equations in a finite field, *Duke Math. J.* **38**, 365–377 (1971).

[2] Generalized k-linear equations over a finite field, *Math. Nachr.* **53**, 63–67 (1972).

[3] The number of solutions of certain equations over a finite field, *Portugal. Math.* **32**, 119–124 (1973).

VARNUM, E. C.:

[1] Polynomial determination in a field of integers modulo p, *J. Computing Systems* **1**, 57–70 (1953).

VARSHAMOV, R. R.:

[1] Estimate of the number of signals in error correcting codes (Russian), *Dokl. Akad. Nauk SSSR* **117**, 739–741 (1957).

[2] A theorem on polynomial reducibility (Russian), *Dokl. Akad. Nauk SSSR* **156**, 1308–1311 (1964); *Soviet Phys. Dokl.* **9**, 426–428 (1964).

[3] A certain linear operator in a Galois field and its applications (Russian), *Studia Sci. Math. Hungar.* **8**, 5–19 (1973).

[4] Certain questions in the constructive theory of the reducibility of polynomials over finite fields (Russian), *Problemy Kibernet.* **27**, 127–134 (1973); Erratum, *ibid.* **28**, 280 (1974).

[5] Operator substitutions in a Galois field, and their application (Russian), *Dokl. Akad. Nauk SSSR* **211**, 768–771 (1973); *Soviet Math. Dokl.* **14**, 1095–1099 (1973).

VARSHAMOV, R. R., and ANANIASHVILI, G. G.:

[1] The theory of polynomial reducibility in finite fields (Russian), *Abstract and Structural Theory of the Construction of Switching Circuits* (Russian) (M. A. Gavrilov, ed.), pp. 134–138, Izdat. "Nauka," Moscow, 1966.

VARSHAMOV, R. R., and ANTONJAN, A. M.:

[1] A method for the synthesis of irreducible polynomials over finite fields (Russian), *Dokl. Akad. Nauk Armjan. SSR* **66**, 197–199 (1978).

VARSHAMOV, R. R., and GAMKRELIDZE, L. I.:

[1] On a method of construction of primitive polynomials over finite fields (Russian), *Soobšč. Akad. Nauk Gruzin. SSR* **99**, 61–64 (1980).

VARSHAMOV, R. R., and GARAKOV, G. A.:

[1] On the theory of selfdual polynomials over a Galois field (Russian), *Bull. Math. Soc. Sci. Math. R. S. Roumanie* (*N.S.*) **13**, 403–415 (1969).

VARSHAMOV, R. R., and OSTIANU, V. M.:

[1] The application of finite field theory to the theory of error-correcting codes and the synthesis of reliable switching structures (Russian), *Theory of Finite and Probabilistic Automata* (Russian), pp. 376–378, Izdat. "Nauka," Moscow, 1965.

VASIL'EV, J. L.:

[1] On nongroup close-packed codes (Russian), *Problemy Kibernet.* **8**, 337–339 (1962); *Probleme der Kybernetik* **8**, 375–378 (1965).

VASIL'EV, YU. P.:

[1] Computer description of finite fields (Russian), *Kibernetika* (*Kiev*) **1979**, no. 5, 133–135; *Cybernetics* **15**, 749–752 (1979).

VAUGHAN, R. C.:

[1] *The Hardy-Littlewood Method*, Cambridge Univ. Press, Cambridge, 1981.

VAUGHAN, T. P.:

[1] Polynomials and linear transformations over finite fields, *J. reine angew. Math.* **267**, 179–206 (1974).

[2] Linear transformations of a finite field, *Linear Algebra Appl.* **8**, 413–426 (1974).

VEBLEN, O., and BUSSEY, W. H.:

[1] Finite projective geometries, *Trans. Amer. Math. Soc.* **7**, 241–259 (1906).

VEBLEN, O., and WEDDERBURN, J. H. M.:

[1] Non-Desarguesian and non-Pascalian geometries, *Trans. Amer. Math. Soc.* **8**, 379–388 (1907).

VEBLEN, O., and YOUNG, J. W.:

[1] *Projective Geometry*, 2 vols., Ginn & Co., Boston, 1938.

VEGH, E.:

[1] Pairs of consecutive primitive roots modulo a prime, *Proc. Amer. Math. Soc.* **19**, 1169–1170 (1968).

[2] Primitive roots modulo a prime as consecutive terms of an arithmetic progression, *J. reine angew. Math.* **235**, 185–188 (1969).

[3] Arithmetic progressions of primitive roots of a prime. II, *J. reine angew. Math.* **244**, 108–111 (1970).

[4] A new condition for consecutive primitive roots of a prime, *Elemente der Math.* **25**, 113 (1970).

[5] A note on the distribution of the primitive roots of a prime, *J. Number Theory* **3**, 13–18 (1971).

[6] Arithmetic progressions of primitive roots of a prime. III, *J. reine angew. Math.* **256**, 130–137 (1972).

VENKATARAYUDU, T.:

[1] The algebra of the eth power residues, *J. Indian Math. Soc.* **3**, 73–81 (1938).

VERNER, L.:

[1] A singular series in characteristic p, *Bull. Acad. Polon. Sci. Sér. Sci. Math.* **26**, 957–961 (1978).

[2] A singular series in characteristic p, *Bull. Acad. Polon. Sci. Sér. Sci. Math.* **27**, 147–151 (1979).

VILANOVA, K.:

[1] Certain trinomial equations over finite fields (Russian), *Trudy Univ. Družby Narod.* **21**, no. 2, 17–31 (1967).

VINCE, A.:

[1] The Fibonacci sequence modulo N, *Fibonacci Quart.* **16**, 403–407 (1978).

[2] Period of a linear recurrence, *Acta Arith.* **39**, 303–311 (1981).

VINOGRADOV, A. I.:

[1] On cubic Gaussian sums (Russian), *Izv. Akad. Nauk SSSR Ser. Mat.* **31**, 123–148 (1967).

VINOGRADOV, I. M.:

[1] Sur la distribution des résidus et des non-résidus des puissances, *Ž. Fiz.-Mat. Obšč. Permsk. Gos. Univ.* **1918**, no. 1, 94–98.

[2] On a general theorem concerning the distribution of the residues and non-residues of powers, *Trans. Amer. Math. Soc.* **29**, 209–217 (1927).

[3] On the bound of the least non-residue of nth powers, *Trans. Amer. Math. Soc.* **29**, 218–226 (1927).

[4] On a certain trigonometrical expression and its applications in the theory of numbers (Russian), *C. R. Acad. Sci. URSS* **1933**, no. 5, 195–203.

[5] Some trigonometrical polynomials and their applications (Russian), *C. R. Acad. Sci. URSS* **1933**, no. 6, 249–254.

[6] New applications of trigonometrical polynomials (Russian), *C. R. Acad. Sci. URSS* **1934**, no. 1, 10–14.

[7] New asymptotical expressions (Russian), *C. R. Acad. Sci. URSS* **1934**, no. 1, 49–51.

[8] Some theorems on the distribution of indices and of primitive roots (Russian), *Trudy Mat. Inst. Steklov.* **5**, 87–93 (1934).

[9] Trigonometrical polynomials for complicated moduli (Russian), *C. R. Acad. Sci. URSS* **1934**, no. 1, 225–229.

[10] New theorems on the distribution of quadratic residues (Russian), *C. R. Acad. Sci. URSS* **1934**, no. 1, 289–290.

[11] On the distribution of primitive roots (Russian), *C. R. Acad. Sci. URSS* **1934**, no. 1, 366–369.

[12] A new improvement of the method of estimation of double sums (Russian), *Dokl. Akad. Nauk SSSR* **73**, 635–638 (1950).

VINSON, J.:

[1] The relation of the period modulo m to the rank of apparition of m in the Fibonacci sequence, *Fibonacci Quart.* **1**, no. 2, 37–45 (1963).

VIRY, G.:

[1] Factorisation des polynômes à plusieurs variables à coefficients entiers, *RAIRO Inform. Théor.* **12**, 305–318 (1978).

[2] Factorisation des polynômes à plusieurs variables, *RAIRO Inform. Théor.* **14**, 209–223 (1980).

VOGT, W. G., and BOSE, N. K.:

[1] A method to determine whether two polynomials are relatively prime, *IEEE Trans. Automatic Control* **AC-15**, 379–380 (1970).

VON AMMON, U., and TRÖNDLE, K.:

[1] *Mathematische Grundlagen der Codierung*, Oldenbourg, Munich, 1974.

VON GROSSCHMID, L.:

[1] Generalization of a theorem of Lagrange. Contribution to the theory of the distribution of quadratic residues (Hungarian), *Math. és Termész. Ért.* **36**, 165–191 (1918).

VON NEUMANN, J., and GOLDSTINE, H. H.:

[1] A numerical study of a conjecture of Kummer, *Math. Tables Aids Comput.* **7**, 133–134 (1953).

VON SCHRUTKA, L.:

[1] Ein Beweis für die Zerlegbarkeit der Primzahlen von der Form $6n+1$ in ein einfaches und ein dreifaches Quadrat, *J. reine angew. Math.* **140**, 252–265 (1911).

VON STERNECK, R. D.:

[1] Über die Anzahl inkongruenter Werte, die eine ganze Funktion dritten Grades annimmt, *Sitzungsber. Wien Abt. II* **116**, 895–904 (1907).

VOROB'EV, N. N.:

[1] *The Fibonacci Numbers* (Russian), 3rd ed., Izdat. "Nauka," Moscow, 1969; 1st ed., Heath, Boston, Mass., 1963.

VORONOÏ, G.:

[1] *On Integral Algebraic Numbers Depending on a Root of a Cubic Equation* (Russian), St. Petersburg, 1894.

[2] Sur une propriété du discriminant des fonctions entières, *Verhandl. Dritten Intern. Math.-Kongr.* (Heidelberg, 1904), pp. 186–189, Teubner, Leipzig, 1905.

WADE, L. I.:

[1] Certain quantities transcendental over $GF(p^n, x)$. I, II, *Duke Math. J.* **8**, 701–720 (1941); *ibid.* **10**, 587–594 (1943).

[2] Remarks on the Carlitz ψ-functions, *Duke Math. J.* **13**, 71–78 (1946).

[3] Transcendence properties of the Carlitz ψ-functions, *Duke Math. J.* **13**, 79–85 (1946).

WAGNER, C. G.:

[1] On the factorization of some polynomial analogues of binomial coefficients, *Arch. Math.* **24**, 50–52 (1973).

[2] Linear pseudo-polynomials over $GF[q, x]$, *Arch. Math.* **25**, 385–390 (1974).
[3] Polynomials over $GF(q, x)$ with integral-valued differences, *Arch. Math.* **27**, 495–501 (1976).

WALKER, G. L.:
[1] Fermat's theorem for algebras, *Pacific J. Math.* **4**, 317–320 (1954).

WALL, D. D.:
[1] Fibonacci series modulo m, *Amer. Math. Monthly* **67**, 525–532 (1960).

WALLIS, W. D., STREET, A. P., and WALLIS, J. S.:
[1] *Combinatorics: Room Squares, Sum-Free Sets, Hadamard Matrices*, Lecture Notes in Math., vol. 292, Springer-Verlag, Berlin-Heidelberg-New York, 1972.

WALUM, H.:
[1] Some averages of character sums, *Pacific J. Math.* **16**, 189–192 (1966).

WAMSLEY, J. W.:
[1] On a condition for commutativity of rings, *J. London Math. Soc.* (2)**4**, 331–332 (1971).

WAN, Z., and YANG, B.:
[1] Notes on finite geometries and the construction of PBIB designs. III. Some "Anzahl" theorems in unitary geometry over finite fields and their applications, *Sci. Sinica* **13**, 1006–1007 (1964).
[2] Studies in finite geometries and the construction of incomplete block designs. III. Some "Anzahl" theorems in unitary geometry over finite fields and their applications (Chinese), *Acta Math. Sinica* **15**, 533–544 (1965); *Chinese Math. Acta* **7**, 252–264 (1965).

WANG, P. S.:
[1] Factoring multivariate polynomials over algebraic number fields, *Math. Comp.* **30**, 324–336 (1976).
[2] An improved multivariate polynomial factoring algorithm, *Math. Comp.* **32**, 1215–1231 (1978).

WANG, P. S., and ROTHSCHILD, L. P.:
[1] Factoring multivariate polynomials over the integers, *Math. Comp.* **29**, 935–950 (1975).

WANG, Y.:
[1] A note on the least primitive root of a prime, *Sci. Record (N.S.)* **3**, 174–179 (1959).
[2] On the least primitive root of a prime (Chinese), *Acta Math. Sinica* **9**, 432–441 (1959); *Sci. Sinica* **10**, 1–14 (1961).
[3] Estimation and application of character sums (Chinese), *Shuxue Jinzhan* **7**, 78–83 (1964).

WARD, M.:
[1] The algebra of recurring series, *Ann. of Math.* (2)**32**, 1–9 (1931).
[2] The characteristic number of a sequence of integers satisfying a linear recursion relation, *Trans. Amer. Math. Soc.* **33**, 153–165 (1931).

[3] The distribution of residues in a sequence satisfying a linear recursion relation, *Trans. Amer. Math. Soc.* **33**, 166–190 (1931).
[4] Some arithmetical properties of sequences satisfying a linear recursion relation, *Ann. of Math.* (2)**32**, 734–738 (1931).
[5] The arithmetical theory of linear recurring series, *Trans. Amer. Math. Soc.* **35**, 600–628 (1933).
[6] Note on the period of a mark in a finite field, *Bull. Amer. Math. Soc.* **40**, 279–281 (1934).
[7] An arithmetical property of recurring series of the second order, *Bull. Amer. Math. Soc.* **40**, 825–828 (1934).
[8] Note on an arithmetical property of recurring series, *Math. Z.* **39**, 211–214 (1935).
[9] An enumerative problem in the arithmetic of linear recurring series, *Trans. Amer. Math. Soc.* **37**, 435–440 (1935).
[10] On the factorization of polynomials to a prime modulus, *Ann. of Math.* (2)**36**, 870–874 (1935).
[11] The null divisors of linear recurring series, *Duke Math. J.* **2**, 472–476 (1936).
[12] Linear divisibility sequences, *Trans. Amer. Math. Soc.* **41**, 276–286 (1937).
[13] Arithmetic functions on rings, *Ann. of Math.* (2)**38**, 725–732 (1937).
[14] The law of apparition of primes in a Lucasian sequence, *Trans. Amer. Math. Soc.* **44**, 68–86 (1938).
[15] Arithmetical properties of sequences in rings, *Ann. of Math.* (2)**39**, 210–219 (1938).
[16] Memoir on elliptic divisibility sequences, *Amer. J. Math.* **70**, 31–74 (1948).

WARNING, E.:
[1] Bemerkung zur vorstehenden Arbeit von Herrn Chevalley, *Abh. Math. Sem. Univ. Hamburg* **11**, 76–83 (1936).

WATERHOUSE, W. C.:
[1] Abelian varieties over finite fields, *Ann. Sci. Ecole Norm. Sup.* (4)**2**, 521–560 (1969).
[2] The sign of the Gaussian sum, *J. Number Theory* **2**, 363 (1970).
[3] The normal basis theorem, *Amer. Math. Monthly* **86**, 212 (1979).

WATERHOUSE, W. C., and MILNE, J. S.:
[1] Abelian varieties over finite fields, *Proc. Symp. Pure Math.*, vol. 20, pp. 53–64, American Math. Society, Providence, R.I., 1971.

WATSON, E. J.:
[1] Primitive polynomials (mod 2), *Math. Comp.* **16**, 368–369 (1962).

WATSON, G. L.:
[1] Cubic congruences, *Mathematika* **11**, 142–150 (1964).

WEBB, W. A.:
[1] On the representation of polynomials over finite fields as sums of powers and irreducibles, *Rocky Mountain J. Math.* **3**, 23–29 (1973).
[2] Numerical results for Waring's problem in $GF[q, x]$, *Math. Comp.* **27**, 193–196 (1973).

[3] Waring's problem in $GF[q,x]$, *Acta Arith.* **22**, 207–220 (1973).
[4] Uniformly distributed functions in $GF[q,x]$ and $GF\{q,x\}$, *Ann. Mat. Pura Appl.* (4)**95**, 285–291 (1973).

WEBB, W. A., and LONG, C. T.:
[1] Distribution modulo p^h of the general linear second order recurrence, *Atti Accad. Naz. Lincei Rend. Cl. Sci. Fis. Mat. Natur.* (8)**58**, 92–100 (1975).

WEBER, H.:
[1] Ueber die mehrfachen Gaussischen Summen, *J. reine angew. Math.* **74**, 14–56 (1872).
[2] Beweis des Satzes, dass jede eigentlich primitive quadratische Form unendlich viele Primzahlen darzustellen fähig ist, *Math. Ann.* **20**, 301–329 (1882).
[3] Die allgemeinen Grundlagen der Galois'schen Gleichungstheorie, *Math. Ann.* **43**, 521–549 (1893).
[4] *Lehrbuch der Algebra*, vol. 1, Vieweg, Braunschweig, 1895.
[5] *Lehrbuch der Algebra*, vol. 2, Vieweg, Braunschweig, 1896.
[6] Über Abel's Summation endlicher Differenzenreihen, *Acta Math.* **27**, 225–233 (1903).

WEDDERBURN, J. H. M.:
[1] A theorem on finite algebras, *Trans. Amer. Math. Soc.* **6**, 349–352 (1905).

WEGNER, U.:
[1] Über die ganzzahligen Polynome, die für unendlich viele Primzahlmoduln Permutationen liefern, Dissertation, Berlin, 1928.
[2] Über ein algebraisches Problem, *Math. Ann.* **105**, 779–785 (1931).
[3] Über einen Satz von Dickson, *Math. Ann.* **105**, 790–792 (1931).
[4] Über das Verhalten der Potenzsummen der Wurzeln einer algebraischen Gleichung hinsichtlich ihrer Gruppe, *J. reine angew. Math.* **173**, 185–190 (1935).

WEIL, A.:
[1] Sur les fonctions algébriques à corps de constantes fini, *C. R. Acad. Sci. Paris* **210**, 592–594 (1940).
[2] On the Riemann hypothesis in function fields, *Proc. Nat. Acad. Sci. U.S.A.* **27**, 345–347 (1941).
[3] *Sur les courbes algébriques et les variétés qui s'en déduisent*, Actualités Sci. Ind., no. 1041, Hermann, Paris, 1948.
[4] *Variétés abéliennes et courbes algébriques*, Actualités Sci. Ind., no. 1064, Hermann, Paris, 1948.
[5] On some exponential sums, *Proc. Nat. Acad. Sci. U.S.A.* **34**, 204–207 (1948).
[6] Numbers of solutions of equations in finite fields, *Bull. Amer. Math. Soc.* **55**, 497–508 (1949).
[7] Jacobi sums as "Grössencharaktere", *Trans. Amer. Math. Soc.* **73**, 487–495 (1952).
[8] Footnote to a recent paper, *Amer. J. Math.* **76**, 347–350 (1954).

[9] Abstract versus classical algebraic geometry, *Proc. International Congress of Math.* (Amsterdam, 1954), vol. 3, pp. 550–558, North-Holland, Amsterdam, 1956.
[10] Sommes de Jacobi et caractères de Hecke, *Göttinger Nachr.* **1974**, 1–14.
[11] La cyclotomie jadis et naguère, *L'Enseignement Math.* (2)**20**, 247–263 (1974).

WEINBERGER, P. J., and ROTHSCHILD, L. P.:
[1] Factoring polynomials over algebraic number fields, *ACM Trans. Math. Software* **2**, 335–350 (1976).

WEINSTEIN, L.:
[1] The hyper-Kloosterman sum, *L'Enseignement Math.* (2)**27**, 29–40 (1981).

WEISSINGER, J.:
[1] Theorie der Divisorenkongruenzen, *Abh. Math. Sem. Univ. Hamburg* **12**, 115–126 (1938).

WELCH, L. R., and SCHOLTZ, R. A.:
[1] Continued fractions and Berlekamp's algorithm, *IEEE Trans. Information Theory* **IT-25**, 19–27 (1979).

WELLS, C.:
[1] Groups of permutation polynomials, *Monatsh. Math.* **71**, 248–262 (1967).
[2] The number of solutions of a system of equations in a finite field, *Acta Arith.* **12**, 421–424 (1967).
[3] A generalization of the regular representation of finite abelian groups, *Monatsh. Math.* **72**, 152–156 (1968).
[4] Generators for groups of permutation polynomials over finite fields, *Acta Sci. Math. Szeged* **29**, 167–176 (1968).
[5] The degrees of permutation polynomials over finite fields, *J. Combinatorial Theory* **7**, 49–55 (1969).
[6] Polynomials over finite fields which commute with translations, *Proc. Amer. Math. Soc.* **46**, 347–350 (1974).

WELLS, J., and MUSKAT, J. B.:
[1] On the number of solutions of certain trinomial congruences, *Math. Comp.* **19**, 483–487 (1965).

WENDT, E.:
[1] Arithmetische Studien über den "letzten" Fermatschen Satz, welcher aussagt, dass die Gleichung $a^n = b^n + c^n$ für $n > 2$ in ganzen Zahlen nicht auflösbar ist, *J. reine angew. Math.* **113**, 335–347 (1894).

WENG, L.-J.:
[1] Decomposition of *m*-sequences and its applications, *IEEE Trans. Information Theory* **IT-17**, 457–463 (1971).

WERTHEIM, G.:
[1] *Anfangsgründe der Zahlenlehre*, Vieweg, Braunschweig, 1902.

WESSELKAMPER, T. C.:

[1] Divided difference methods for Galois switching functions, *IEEE Trans. Computers* **C-27**, 232–238 (1978).

[2] The algebraic representation of partial functions, *Discrete Appl. Math.* **1**, 137–142 (1979).

[3] The algebraic representation of partial functions, *Proc. Ninth Internat. Symp. on Multiple-Valued Logic* (Bath, 1979), pp. 290–293, IEEE, Long Beach, Cal., 1979.

WESTERN, A. E.:

[1] An extension of Eisenstein's law of reciprocity. I, II, *Proc. London Math. Soc.* (2)**6**, 16–28, 265–297 (1908).

[2] Some criteria for the residues of eighth and other powers, *Proc. London Math. Soc.* (2)**9**, 244–272 (1911).

WESTERN, A. E., and MILLER, J. C. P.:

[1] *Tables of Indices and Primitive Roots*, Royal Soc. Math. Tables, vol. 9, Cambridge Univ. Press, London, 1968.

WHAPLES, G.:

[1] Additive polynomials, *Duke Math. J.* **21**, 55–65 (1954).

WHITEMAN, A. L.:

[1] On a theorem of higher reciprocity, *Bull. Amer. Math. Soc.* **43**, 567–572 (1937).

[2] A note on Kloosterman sums, *Bull. Amer. Math. Soc.* **51**, 373–377 (1945).

[3] Theorems analogous to Jacobsthal's theorem, *Duke Math. J.* **16**, 619–626 (1949).

[4] Theorems on quadratic partitions, *Proc. Nat. Acad. Sci. U.S.A.* **36**, 60–66 (1950).

[5] Finite Fourier series and cyclotomy, *Proc. Nat. Acad. Sci. U.S.A.* **37**, 373–378 (1951).

[6] Cyclotomy and Jacobsthal sums, *Amer. J. Math.* **74**, 89–99 (1952).

[7] Finite Fourier series and equations in finite fields, *Trans. Amer. Math. Soc.* **74**, 78–98 (1953).

[8] The sixteenth power residue character of 2, *Canad. J. Math.* **6**, 364–373 (1954).

[9] The cyclotomic numbers of order sixteen, *Trans. Amer. Math. Soc.* **86**, 401–413 (1957).

[10] The cyclotomic numbers of order twelve, *Acta Arith.* **6**, 53–76 (1960).

[11] The cyclotomic numbers of order ten, *Proc. Symp. Appl. Math.*, vol. 10, pp. 95–111, American Math. Society, Providence, R.I., 1960.

[12] A family of difference sets, *Illinois J. Math.* **6**, 107–121 (1962).

[13] A theorem of Brewer on character sums, *Duke Math. J.* **30**, 545–552 (1963).

[14] Theorems on Brewer and Jacobsthal sums. I, *Proc. Symp. Pure Math.*, vol. 8, pp. 44–55, American Math. Society, Providence, R.I., 1965.

[15] Theorems on Brewer and Jacobsthal sums. II, *Michigan Math. J.* **12**, 65–80 (1965).

WHYBURN, C. T.:

[1] The distribution of r-th powers in finite fields, *J. reine angew. Math.* **245**, 183–187 (1970).

[2] An elementary note on character sums, *Duke Math. J.* **37**, 307–310 (1970).

WILEY, F. B.:

[1] Proof of the finiteness of the modular covariants of a system of binary forms and cogredient points, *Trans. Amer. Math. Soc.* **15**, 431–438 (1914).

WILLETT, M.:

[1] The minimum polynomial for a given solution of a linear recursion, *Duke Math. J.* **39**, 101–104 (1972).

[2] The index of an m-sequence, *SIAM J. Appl. Math.* **25**, 24–27 (1973).

[3] On a theorem of Kronecker, *Fibonacci Quart.* **14**, 27–29 (1976).

[4] Characteristic m-sequences, *Math. Comp.* **30**, 306–311 (1976).

[5] Factoring polynomials over a finite field, *SIAM J. Appl. Math.* **35**, 333–337 (1978).

[6] Arithmetic in a finite field, *Math. Comp.* **35**, 1353–1359 (1980).

WILLIAMS, H. C.:

[1] Some algorithms for solving $x^q \equiv N (\bmod p)$, *Proc. Third Southeastern Conf. on Combinatorics, Graph Theory, and Computing* (Boca Raton, Fla., 1972), pp. 451–462, Utilitas Math., Winnipeg, Man., 1972.

[2] Primality testing on a computer, *Ars Combinatoria* **5**, 127–185 (1978).

WILLIAMS, H. C., and ZARNKE, C. R.:

[1] Some algorithms for solving a cubic congruence modulo p, *Utilitas Math.* **6**, 285–306 (1974).

WILLIAMS, K. S.:

[1] On the number of solutions of a congruence, *Amer. Math. Monthly* **73**, 44–49 (1966).

[2] On the least non-residue of a quartic polynomial, *Proc. Cambridge Philos. Soc.* **62**, 429–431 (1966).

[3] Eisenstein's criteria for absolute irreducibility over a finite field, *Canad. Math. Bull.* **9**, 575–580 (1966).

[4] On general polynomials, *Canad. Math. Bull.* **10**, 579–583 (1967).

[5] On extremal polynomials, *Canad. Math. Bull.* **10**, 585–594 (1967).

[6] Pairs of consecutive residues of polynomials, *Canad. J. Math.* **19**, 655–666 (1967).

[7] A sum of fractional parts, *Amer. Math. Monthly* **74**, 978–980 (1967).

[8] Note on pairs of consecutive residues of polynomials, *Canad. Math. Bull.* **11**, 79–83 (1968).

[9] On exceptional polynomials, *Canad. Math. Bull.* **11**, 279–282 (1968).

[10] Quadratic polynomials with the same residues, *Amer. Math. Monthly* **75**, 969–973 (1968).

[11] Polynomials with irreducible factors of specified degree, *Canad. Math. Bull.* **12**, 221–223 (1969).

[12] Small solutions of the congruence $ax^2 + by^2 \equiv c (\bmod k)$, *Canad. Math. Bull.* **12**, 311–320 (1969).

[13] On two conjectures of Chowla, *Canad. Math. Bull.* **12**, 545–565 (1969).
[14] Note on factorable polynomials, *Canad. Math. Bull.* **12**, 589–595 (1969).
[15] Distinct values of a polynomial in subsets of a finite field, *Canad. J. Math.* **21**, 1483–1488 (1969).
[16] Finite transformation formulae involving the Legendre symbol, *Pacific J. Math.* **34**, 559–568 (1970).
[17] On a result of Libri and Lebesgue, *Amer. Math. Monthly* **77**, 610–613 (1970).
[18] A distribution property of the solutions of a congruence modulo a large prime, *J. Number Theory* **3**, 19–32 (1971).
[19] Note on the Kloosterman sum, *Proc. Amer. Math. Soc.* **30**, 61–62 (1971).
[20] A class of character sums, *J. London Math. Soc.* (2)**3**, 67–72 (1971).
[21] On Salié's sum, *J. Number Theory* **3**, 316–317 (1971).
[22] Note on Salié's sum, *Proc. Amer. Math. Soc.* **30**, 393–394 (1971).
[23] Small solutions of the congruence $a_1x_1^{l_1} + a_2x_2^{l_2} + a_0 \equiv 0(\bmod p)$, *Proc. Cambridge Philos. Soc.* **70**, 409–412 (1971).
[24] Note on the number of solutions of $f(x_1) = f(x_2) = \cdots = f(x_r)$ over a finite field, *Canad. Math. Bull.* **14**, 429–432 (1971).
[25] Note on Dickson's permutation polynomials, *Duke Math. J.* **38**, 659–665 (1971).
[26] Products of polynomials over a finite field, *Delta* (*Waukesha*) **3**, no. 2, 35–37 (1972).
[27] Exponential sums over $GF(2^n)$, *Pacific J. Math.* **40**, 511–519 (1972).
[28] The Kloosterman sum revisited, *Canad. Math. Bull.* **16**, 363–365 (1973).
[29] Elementary treatment of quadratic partition of primes $p \equiv 1(\bmod 7)$, *Illinois J. Math.* **18**, 608–621 (1974).
[30] Note on a cubic character sum, *Aequationes Math.* **12**, 229–231 (1975).
[31] On Euler's criterion for cubic nonresidues, *Proc. Amer. Math. Soc.* **49**, 277–283 (1975).
[32] Note on cubics over $GF(2^n)$ and $GF(3^n)$, *J. Number Theory* **7**, 361–365 (1975).
[33] Note on a result of Kaplan, *Proc. Amer. Math. Soc.* **56**, 34–36 (1976).
[34] A rational octic reciprocity law, *Pacific J. Math.* **63**, 563–570 (1976).
[35] Note on Brewer's character sum, *Proc. Amer. Math. Soc.* **71**, 153–154 (1978).
[36] Remark on an assertion of Chowla, *Norske Vid. Selsk. Skrifter* **1979**, no. 1, 3–4.
[37] Problem E 2760, *Amer. Math. Monthly* **86**, 128 (1979); Solution, *ibid.* **87**, 223–224 (1980).
[38] Evaluation of character sums connected with elliptic curves, *Proc. Amer. Math. Soc.* **73**, 291–299 (1979).

WILLIAMS, W. L. G.:
[1] Fundamental systems of formal modular seminvariants of the binary cubic, *Trans. Amer. Math. Soc.* **22**, 56–79 (1921).
[2] On the formal modular invariants of binary forms, *J. Math. Pures Appl.* (9)**4**, 169–192 (1925).
[3] Fundamental systems of formal modular protomorphs of binary forms, *Trans. Amer. Math. Soc.* **28**, 183–197 (1926).

[4] Formal modular invariants of forms in q variables, *Proc. International Math. Congress* (Toronto, 1924), vol. 1, pp. 347–359, Univ. of Toronto Press, Toronto, 1928.

[5] A summation theorem in the theory of numbers, *Trans. Roy. Soc. Canada* (3)**26**, 35–37 (1932).

WILSON, R. M.:

[1] An existence theory for pairwise balanced designs. I, II, *J. Combinatorial Theory Ser. A* **13**, 220–245, 246–273 (1972).

[2] An existence theory for pairwise balanced designs. III—Proof of the existence conjectures, *J. Combinatorial Theory Ser. A* **18**, 71–79 (1975).

WILSON, T. C., and SHORTT, J.:

[1] An O(log n) algorithm for computing general order-k Fibonacci numbers, *Inform. Process. Lett.* **10**, no. 2, 68–75 (1980).

WINOGRAD, S.:

[1] Some bilinear forms whose multiplicative complexity depends on the field of constants, *Math. Systems Theory* **10**, 169–180 (1977).

[2] On multiplication in algebraic extension fields, *Theoret. Comput. Sci.* **8**, 359–377 (1979).

WINTER, D. J.:

[1] *The Structure of Fields*, Springer-Verlag, New York-Heidelberg-Berlin, 1974.

WITT, E.:

[1] Über die Kommutativität endlicher Schiefkörper, *Abh. Math. Sem. Univ. Hamburg* **8**, 413 (1931).

[2] Über eine Invariante quadratischer Formen mod 2, *J. reine angew. Math.* **193**, 119–120 (1954).

WOLFMANN, J.:

[1] Un problème d'extrémum dans les espaces vectoriels binaires, *Ann. Discrete Math.* **9**, 261–264 (1980).

WOLFOWITZ, J.:

[1] *Coding Theorems of Information Theory*, Prentice-Hall, Englewood Cliffs, N.J., 1961.

WOLKE, D.:

[1] Eine Bemerkung über das Legendre-Symbol, *Monatsh. Math.* **77**, 267–275 (1973).

WYLER, O.:

[1] On second-order recurrences, *Amer. Math. Monthly* **72**, 500–506 (1965).

WYMAN, B. F.:

[1] What is a reciprocity law?, *Amer. Math. Monthly* **79**, 571–586 (1972).

YALAVIGI, C. C.:

[1] A conjecture of J. H. Halton, *Math. Education Ser. A* **4**, 125–126 (1970).

[2] Fibonacci series modulo m, *Math. Education Ser. A* **7**, 48–54 (1973).

YALAVIGI, C. C., and KRISHNA, H. V.:

[1] Periodic lengths of the generalized Fibonacci sequence modulo p, *Fibonacci Quart.* **15**, 150–152 (1977).

YALE, R. B.:

[1] Error correcting codes and linear recurring sequences, Report 34–77, M.I.T. Lincoln Laboratory, Lexington, Mass., 1958.

YAMADA, T.:

[1] On the Davenport-Hasse curves, *J. Math. Soc. Japan* **20**, 403–410 (1968).

YAMAMOTO, K.:

[1] On Gaussian sums with biquadratic residue characters, *J. reine angew. Math.* **219**, 200–213 (1965).

[2] On a conjecture of Hasse concerning multiplicative relations of Gaussian sums, *J. Combinatorial Theory* **1**, 476–489 (1966).

[3] On Jacobi sums and difference sets, *J. Combinatorial Theory* **3**, 146–181 (1967).

[4] The gap group of multiplicative relationships of Gaussian sums, *Symposia Math.*, vol. 15, pp. 427–440, Academic Press, London, 1975.

YAMAMOTO, Y., NAGANUMA, M., and DOI, K.:

[1] Experimental integer theory (Japanese), *Sûgaku* **18**, 95–103 (1966).

YAMAUCHI, M.:

[1] Some identities on the character sum containing $x(x-1)(x-\lambda)$, *Nagoya Math. J.* **42**, 109–113 (1971).

YIN, K. Z.:

[1] An inversion formula for switching functions (Chinese), *J. Math. Res. Exposition* **1981**, no. 1, 63–68.

YOKOYAMA, A.:

[1] On the Gaussian sum and the Jacobi sum with its application, *Tôhoku Math. J.* (2)**16**, 142–153 (1964).

YOSHIDA, H.:

[1] On an analogue of the Sato conjecture, *Invent. Math.* **19**, 261–277 (1973).

YOUNG, F. H.:

[1] Analysis of shift register counters, *J. Assoc. Comput. Mach.* **5**, 385–388 (1958).

ZADEH, L. A., and DESOER, C. A.:

[1] *Linear System Theory*, McGraw-Hill, New York, 1963.

ZADEH, L. A., and POLAK, E.:

[1] *System Theory*, McGraw-Hill, New York, 1969.

ZANE, B.:

[1] Uniform distribution modulo m of monomials, *Amer. Math. Monthly* **71**, 162–164 (1964).

ZASSENHAUS, H.:

[1] Über endliche Fastkörper, *Abh. Math. Sem. Univ. Hamburg* **11**, 187–220 (1936).

[2] A group-theoretic proof of a theorem of Maclagan-Wedderburn, *Proc. Glasgow Math. Assoc.* **1**, 53–63 (1952).

[3] The quadratic law of reciprocity and the theory of Galois fields, *Proc. Glasgow Math. Assoc.* **1**, 64–71 (1952).

[4] Über die Fundamentalkonstruktionen der endlichen Körpertheorie, *Jber. Deutsch. Math.-Verein.* **70**, 177–181 (1968).

[5] On Hensel factorization I, *J. Number Theory* **1**, 291–311 (1969).

[6] On Hensel factorization II, *Symposia Math.*, vol. 15, pp. 499–513, Academic Press, London, 1975.

[7] A remark on the Hensel factorization method, *Math. Comp.* **32**, 287–292 (1978).

ZECKENDORF, E.:

[1] Représentation graphique des suites récurrentes modulo p et premiers résultats, *Bull. Soc. Roy. Sci. Liège* **45**, 13–25 (1976).

ZEE, Y.-C.:

[1] The Jacobi sums of orders thirteen and sixty and related quadratic decompositions, *Math. Z.* **115**, 259–272 (1970).

[2] The Jacobi sums of order twenty-two, *Proc. Amer. Math. Soc.* **28**, 25–31 (1971).

ZETTERBERG, L.-H.:

[1] Cyclic codes from irreducible polynomials for correction of multiple errors, *IRE Trans. Information Theory* **IT-8**, 13–21 (1962).

ZIERLER, N.:

[1] Several binary-sequence generators, Tech. Report No. 95, M.I.T. Lincoln Laboratory, Lexington, Mass., 1955.

[2] On the theorem of Gleason and Marsh, *Proc. Amer. Math. Soc.* **9**, 236–237 (1958).

[3] On a variation of the first-order Reed-Muller codes, Report 34-80, M.I.T. Lincoln Laboratory, Lexington, Mass., 1958.

[4] Linear recurring sequences, *J. Soc. Indust. Appl. Math.* **7**, 31–48 (1959).

[5] Linear recurring sequences and error-correcting codes, *Error Correcting Codes* (H. B. Mann, ed.), pp. 47–59, Wiley, New York, 1968.

[6] Primitive trinomials whose degree is a Mersenne exponent, *Information and Control* **15**, 67–69 (1969).

[7] On $x^n + x + 1$ over $GF(2)$, *Information and Control* **16**, 502–505 (1970).

[8] Trinomials with non conjugate roots of the same prime order, *J. Combinatorial Theory Ser. A* **11**, 307–309 (1971).

[9] A conversion algorithm for logarithms on $GF(2^n)$, *J. Pure Appl. Algebra* **4**, 353–356 (1974).

ZIERLER, N., and BRILLHART, J.:

[1] On primitive trinomials (mod 2), *Information and Control* **13**, 541–554 (1968).

[2] On primitive trinomials (mod 2), II, *Information and Control* **14**, 566–569 (1969).

ZIERLER, N., and MILLS, W. H.:

[1] Products of linear recurring sequences, *J. Algebra* **27**, 147–157 (1973).

ZIMMER, H. G.:

[1] An elementary proof of the Riemann hypothesis for an elliptic curve over a finite field, *Pacific J. Math.* **36**, 267–278 (1971).

[2] *Computational Problems, Methods, and Results in Algebraic Number Theory*, Lecture Notes in Math., vol. 262, Springer-Verlag, Berlin-Heidelberg-New York, 1972.

ZINOV'EV, V. A., and LEONT'EV, V. K.:

[1] On non-existence of perfect codes over Galois fields (Russian), *Problemy Uprav. i Teor. Informacii* **2**, 123–132 (1973); *Problems of Control and Information Theory* **2**, no. 2, 16–24 (1973).

ŽMUD', È. M.:

[1] An invariant of quadratic forms over a Galois field of characteristic 2 (Russian), *Vestnik Har'kov. Gos. Univ.* **177**, no. 44, 77–86 (1979).

ZSIGMONDY, K.:

[1] Zur Theorie der Potenzreste, *Monatsh. Math. Phys.* **3**, 265–284 (1892).

[2] Ueber die Anzahl derjenigen ganzen ganzzahligen Functionen $n$ten Grades von x, welche in Bezug auf einen gegebenen Primzahlmodul eine vorgeschriebene Anzahl von Wurzeln besitzen, *Sitzungsber. Wien Abt. II* **103**, 135–144 (1894).

[3] Ueber wurzellose Congruenzen in Bezug auf einen Primzahlmodul, *Monatsh. Math. Phys.* **8**, 1–42 (1897).

List of Symbols

Note. Symbols that appear only in a restricted context are not listed. Wherever appropriate, a page reference is given.

$\mathbb{N}$	the set of natural numbers (= positive integers)
$\mathbb{Z}$	the set of integers
$\mathbb{Q}$	the set of rational numbers
$\mathbb{R}$	the set of real numbers
$\mathbb{C}$	the set of complex numbers
$S_1 \times \cdots \times S_n$	the set of all n-tuples $(s_1, \ldots, s_n)$ with $s_i \in S_i$ for $1 \leqslant i \leqslant n$
S^n	the set of all n-tuples $(s_1, \ldots, s_n)$ with $s_i \in S$ for $1 \leqslant i \leqslant n$
$\lvert S \rvert$	the cardinality (= number of elements) of the finite set S
$[s]$	the equivalence class of s, 4
$\operatorname{Re} z$	the real part of z
$\operatorname{Im} z$	the imaginary part of z
$\bar{z}$	the complex conjugate of z
$\lvert z \rvert$	the absolute value of z
$\log z$	the natural logarithm of z
$e(t)$	$e^{2\pi i t}$ for $t \in \mathbb{R}$

$\lfloor t \rfloor$	the greatest integer $\leqslant t \in \mathbb{R}$
$\max(k_1,\ldots,k_n)$	the maximum of $k_1,\ldots,k_n$
$\min(k_1,\ldots,k_n)$	the minimum of $k_1,\ldots,k_n$
$\gcd(k_1,\ldots,k_n)$	the greatest common divisor of $k_1,\ldots,k_n$
$\operatorname{lcm}(k_1,\ldots,k_n)$	the least common multiple of $k_1,\ldots,k_n$
$a \equiv b \bmod n$	a congruent to b modulo n, 4
$\phi(n)$	Euler's function of n, 7
$\mu(n)$	Moebius function of n, 92
$\binom{k}{i}$	binomial coefficient
$E_p(r)$	the largest j such that the prime power p^j divides $r \in \mathbb{N}$, 296
$\left(\frac{c}{p}\right)$	Legendre symbol, 191
$M(d_1,\ldots,d_n)$	the number of $(j_1,\ldots,j_n) \in \mathbb{Z}^n$ such that $1 \leqslant j_i \leqslant d_i - 1$ for $1 \leqslant i \leqslant n$ and $(j_1/d_1) + \cdots + (j_n/d_n) \in \mathbb{Z}$, 291
A^{T}	the transpose of the matrix A
$\det(A)$	the determinant of the matrix A
$\operatorname{Tr}(A)$	the trace of the matrix A
$\operatorname{rank}(A)$	the rank of the matrix A
$D_n^{(r)}$	Hankel determinant, 437
$\dim(V)$	the dimension of the vector space V
$\mathbb{Z}_n$	the group of integers modulo n, 5
S_n	the symmetric group on n letters, 357
A_n	the alternating group on n letters, 358
$GL(r,\mathbb{F}_q)$	the general linear group of nonsingular $r \times r$ matrices over $\mathbb{F}_q$, 362
$\lvert G \rvert$	the order of the finite group G, 5
$\langle a \rangle$	the cyclic group generated by a, 4, 6
aH	left coset modulo the subgroup H, 6
G/H	the factor group of the group G modulo the normal subgroup H, 9
$N(S)$	the normalizer of the nonempty subset S of a group, 10
$\ker f$	the kernel of the homomorphism f, 9, 14
(a)	the principal ideal generated by a, 13
$[a], a + J$	the residue class of the ring element a modulo the ideal J, 13
$a \equiv b \bmod J$	congruence of ring elements a, b modulo the ideal J, 13
R/J	the residue class ring of the ring R modulo the ideal J, 13
$\mathbb{Z}/(n)$	the ring of integers modulo n, 14

$R[x]$	the polynomial ring over the ring R, 19
$R[x_1,\ldots,x_n]$	the ring of polynomials over the ring R in n indeterminates, 28
$\deg(f)$	the degree of the polynomial f, 20, 29
$D(f)$	the discriminant of the polynomial f, 35
$\text{ord}(f)$	the order of the polynomial f, 84
$\det(f)$	the determinant of the quadratic form f, 280
f'	the derivative of the polynomial f, 27
$E^{(n)}(f)$	the nth hyperderivative of the polynomial f, 303
f^*	the reciprocal polynomial of f, 88
$R(f,g)$	the resultant of the polynomials f and g, 36
$\gcd(f_1,\ldots,f_n)$	the greatest common divisor of the polynomials $f_1,\ldots,f_n$, 22
$\text{lcm}(f_1,\ldots,f_n)$	the least common multiple of the polynomials $f_1,\ldots,f_n$, 23
$f_1(x)\vee \cdots \vee f_h(x)$	433
$L_1(x)\otimes L_2(x)$	symbolic multiplication of linearized polynomials $L_1(x)$ and $L_2(x)$, 114
$Q_n(x)$	the nth cyclotomic polynomial, 64
$g_k(x,a)$	Dickson polynomial, 355
$g_k^{(i)}(x_1,\ldots,x_n,a)$	Dickson polynomial in n indeterminates, 376
$\sigma_k(x_1,\ldots,x_n)$	the kth elementary symmetric polynomial in n indeterminates, 29

$\overline{K}$	the algebraic closure of the field K, 40
$K(M)$	the extension of K obtained by adjoining M, 30
$[L:K]$	the degree of the field L over K, 32
$K^{(n)}$	the nth cyclotomic field over K, 63
$E^{(n)}$	the set of nth roots of unity over K, 63
$\mathbb{F}_q, GF(q)$	the finite field of order q, 49, 73
$\mathbb{F}_q^*$	the multiplicative group of nonzero elements of $\mathbb{F}_q$, 50
$\text{Tr}_{F/K}(\alpha)$	the trace of $\alpha \in F$ over K, 54
$\text{Tr}_F(\alpha)$	the absolute trace of $\alpha \in F$, 54
$\text{N}_{F/K}(\alpha)$	the norm of $\alpha \in F$ over K, 57
$\Delta_{F/K}(\alpha_1,\ldots,\alpha_m)$	the discriminant of $\alpha_1,\ldots,\alpha_m \in F$ over K, 61
$N_q(d)$	the number of monic irreducible polynomials in $\mathbb{F}_q[x]$ of degree d, 91
$I(q,n;x)$	the product of all monic irreducible polynomials in $\mathbb{F}_q[x]$ of degree n, 94
$\Phi_q(f)$	the number of polynomials in $\mathbb{F}_q[x]$ whose degree is less than $\deg(f)$ and which are relatively prime to $f \in \mathbb{F}_q[x]$, 122

$V(f)$	the number of elements of $\{f(c): c \in \mathbb{F}_q\}$ for $f \in \mathbb{F}_q[x]$, 363
$[A_0, A_1, \ldots, A_i]$	continued fraction representing a rational function, 235
$v(b)$	$v(b) = -1$ for $b \in \mathbb{F}_q^*$, $v(0) = q-1$, 280
$N(f(x_1, \ldots, x_n) = b)$	the number of solutions of the equation $f(x_1, \ldots, x_n) = b$ in the underlying finite field, 277, 281
$Z(V; t)$	the zeta-function of the variety V, 336
$\mathbb{F}_q[[x]]$	the ring of formal power series over $\mathbb{F}_q$, 413
$S(f(x))$	the set of all homogeneous linear recurring sequences in $\mathbb{F}_q$ with characteristic polynomial $f(x)$, 423
$G^\wedge$	the set of characters of the finite abelian group G, 187
$\bar{\chi}$	the conjugate of the character χ, 187
χ_0	the trivial additive character of $\mathbb{F}_q$, 191
χ_1	the canonical additive character of $\mathbb{F}_q$, 190
ψ_0	the trivial multiplicative character of $\mathbb{F}_q$, 191
η	the quadratic character of $\mathbb{F}_q$ (q odd), 191
$G(\psi, \chi)$	Gaussian sum, 192
$J(\lambda_1, \ldots, \lambda_k)$	Jacobi sum, 205
$J_a(\lambda_1, \ldots, \lambda_k)$	205
$K(\chi; a, b)$	Kloosterman sum, 226
$K(\psi, \chi; a, b)$	generalized Kloosterman sum, 265
$H_n(a)$	Jacobsthal sum, 231
$I_n(a)$	231
$\Lambda_k(a)$	Brewer sum, 256
$E(\psi)$	Eisenstein sum, 264
$E_s(\psi; a)$	generalized Eisenstein sum, 264
$d(\mathbf{x}, \mathbf{y})$	the Hamming distance between $\mathbf{x}$ and $\mathbf{y}$, 474
$w(\mathbf{x})$	the Hamming weight of $\mathbf{x}$, 474
d_C	the minimum distance of the linear code C, 475
$C^\perp$	the dual code of C, 480
$S(\mathbf{y})$	the syndrome of $\mathbf{y}$, 476
$AG(2, K)$	the affine plane over the field K, 498
$PG(2, K)$	the projective plane over the field K, 499
$AG(m, \mathbb{F}_q)$	affine geometry over $\mathbb{F}_q$, 508
$PG(m, \mathbb{F}_q)$	projective geometry over $\mathbb{F}_q$, 506
$\square$	end of proof, end of example, end of remark

Author Index

Note. Numbers set in *italics* designate those page numbers on which the complete literature citations are given.

Subject Index